XXIII International Conference on Integrable Systems and Quantum Symmetries 2015 (ISQS-23)

Journal of Physics: Conference Series Volume 670

Prague, Czech Republic
23 – 27 June 2015

Editors:

Cestmir Burdik
Ondrej Navratil
Severin Posta

ISBN: 978-1-5108-1961-0
ISSN: 1742-6588

Printed from e-media with permission by:

Curran Associates, Inc.
57 Morehouse Lane
Red Hook, NY 12571

Some format issues inherent in the e-media version may also appear in this print version.

Printed by Curran Associates, Inc. (2016)

For permission requests, please contact the Institute of Physics at the address below.

Institute of Physics
Dirac House, Temple Back
Bristol BS1 6BE UK

Phone: 44 1 17 929 7481
Fax: 44 1 17 920 0979

techtracking@iop.org

Additional copies of this publication are available from:

Curran Associates, Inc.
57 Morehouse Lane
Red Hook, NY 12571 USA
Phone: 845-758-0400
Fax: 845-758-2634
Email: curran@proceedings.com
Web: www.proceedings.com

Table of contents

Volume 670

XXIII International Conference on Integrable Systems and Quantum Symmetries (ISQS-23)
23–27 June 2015, Prague, Czech Republic

Accepted papers received: 24 November 2015
Published online: 25 January 2016

Preface

Papers

The XXIII International Conference on Integrable Systems and Quantum Symmetries (ISQS-23), organized by the Department of Mathematics, Faculty of Nuclear Sciences and Physical Engineering, Czech Technical University Prague and the Bogoliubov Laboratory of Theoretical Physics of the Joint Institute for Nuclear Research, belongs to the successful series of conferences held at the Czech Technical University which began in 1992 and is devoted to problems of mathematical physics related to the theory of integrable systems, quantum groups and quantum symmetries. During the last 20 years, each of the conferences gathered around 60 scientists from all over the world. 56 papers of plenary lectures and contributions presented at ISQS-23 are published in the present issue of *Journal of Physics: Conference Series*.

Čestmír Burdík, Ondřej Navrátil, Severin Pošta
Editors

Peer review statement

All papers published in this volume of *Journal of Physics: Conference Series* have been peer reviewed through processes administered by the proceedings Editors. Reviews were conducted by expert referees to the professional and scientific standards expected of a proceedings journal published by IOP Publishing.

Journal of Physics: Conference Series **670** (2016) 012001 doi:10.1088/1742-6596/670/1/012001

Higher-Twist Mechanism and Inclusive Gluon Production in Pion-Proton Collisions

A. I. Ahmadov[1,2], C. Aydin[3], R. Myrzakulov[4], O. Uzun[3]

[1] Fachbereich C, Bergische Universität Wuppertal, 42097 Wuppertal, Germany,
[2] Department of Theoretical Physics, Baku State University, Z. Khalilov st. 23, AZ-1148, Baku, Azerbaijan,
[3] Department of Physics, Karadeniz Technical University, 61080, Trabzon, Turkey,
[4] Eurasion International Center for Theoretical Physics and Department of General Theoretical Physics, Eurasian National University, Astana 010008, Kazakhstan

E-mail: ahmadovazar@yahoo.com, coskun@ktu.edu.tr, rmyrzakulov@gmail.com, oguzhan_deu@hotmail.com

Abstract. We calculate the contribution of the higher-twist Feynman diagrams to the large-p_T inclusive gluon production cross section in πp collisions incase of the running coupling and frozen coupling approaches within perturbative and holographic QCD. The structure of infrared renormalon singularities of the higher-twist subprocess cross section are obtained and the resummed higher-twist cross sections (Borel sum) with the ones obtained in the framework of the frozen coupling approach and leading-twist cross section are compared and analyzed.

1. Introduction

It is well known that Quantum Chromodynamics (QCD) is the fundamental theory of the strong interactions.Therefore in order to describe the structure and dynamical properties of hadrons at the amplitude level many researchers have been studying QCD. The hadronic distribution amplitude in terms of internal structure degrees of freedoms plays a crucial role in QCD process predictions.

One of the basic problems in QCD is choosing the renormalization scale in running coupling constant $\alpha_s(Q^2)$. In principle, in perturbative QCD (pQCD) calculations, the argument of the running coupling constant in both the renormalization and factorization scale Q^2 should be taken as equal to the square of the momentum transfer of a hard gluon in a corresponding Feynman diagram [1]. In the perturbative QCD, the physical information of the inclusive gluon production is obtained efficiently; therefore, it can be directly compared to the experimental data.

It should be noted, that problem the existence of the higher-twist contribution is not yet settled. Also necessary to study the difference of the leading-twist results for the frozen and running coupling constant approaches and compare it with that of the higher-twist is important.

Take into account of this point the aim of this study is calculation and analysis of the inclusive gluon production in the pion-proton collisions using the frozen and running coupling constant approaches. Using this approaches the higher twist effects have been already calculated by many authors [2, 3, 4, 5, 6, 7, 8, 9, 10, 11, 12, 13, 14, 15, 16, 17, 18, 19, 20, 21, 22, 23].

Journal of Physics: Conference Series **670** (2016) 012001 doi:10.1088/1742-6596/670/1/012001

The contents of the paper is as follows. The related formulas for the calculation of the contributions of the higher-twist and leading-twist diagrams are provided in the next section. The formulas and analysis of the higher twist effects on the dependence of the pion distribution amplitudes by the running coupling constant approach are presented in Section 3, and the numerical results for the cross section and discussion of the dependence of the cross section on the pion distribution amplitudes are presented in Section 4. Finally, our conclusions and the highlights of the study are listed in Section 5.

2. HIGHER-TWIST AND LEADING-TWIST CONTRIBUTIONS TO INCLUSIVE GLUON PRODUCTION

The higher-twist Feynman diagrams for the inclusive gluon production in the pion-proton collision $\pi p \to gX$ are shown in Fig.1. For the process $\pi p \to gX$, we write invariant amplitude as in the form (as called by Brodsky-Lepage formula [24])

$$M(\hat{s},\hat{t}) = \int_0^1 dx_1 \int_0^1 dx_2 \delta(1-x_1-x_2)\Phi_M(x_1,x_2,Q^2)T_H(x_1,x_2;Q^2,\mu_R^2,\mu_F^2) \tag{1}$$

where T_H is the sum of the graphs contributing to the hard-scattering part of the subprocess. For the higher-twist, the subprocess is taken as $\pi q_p \to gq$, which contributes to $\pi p \to gX$, where q_p is a constituent of the initial proton target. As seen from Fig.1, the processes $\pi^+ p \to gX$ and $\pi^- p \to gX$ arise from subprocesses as $\pi^+ d_p \to gu$ and $\pi^- u_p \to gd$, respectively. The production of the hadronic gluon in the large transverse momentum is available at the high energy, especially at the Large Hadron Collider. Finally hadronic gluon is a product of the hard-scattering processes, before hadronization. In the final state, this hadronic gluon is fragmented to hadron. The main dynamical properties of the gluon, which carried one part of the four momentum, are close to the parent parton. In order to understand the parton kinematics one should consider the gluon production process [25]. The higher-twist cross section for $\pi p \to gX$ process has the form:

$$E\frac{d\sigma}{d^3p}(\pi p \to gX) = \int_0^1 dx \delta(\hat{s}+\hat{t}+\hat{u})\hat{s}G_{q/p}(x,Q^2)\frac{1}{\pi}\frac{d\sigma}{d\hat{t}}(\pi q_p \to gq), \tag{2}$$

where $G_{q/p}(x,Q^2)$ is the quark distribution function inside a proton.

For higher-twist subprocess $\pi q_p \to gq$ the Mandelstam invariant variables are writen in the form:

$$\hat{s} = (p_1+p_g)^2 = (p_2+p_\pi)^2, \quad \hat{t} = (p_g-p_\pi)^2, \quad \hat{u} = (p_1-p_\pi)^2. \tag{3}$$

Then the parton-level cross section within running coupling constant method becomes

$$\frac{d\sigma}{d\hat{t}}(\pi q_p \to gq) = \frac{256\pi^2}{81\hat{s}^2}[D(\hat{s},\hat{u})]^2\left(-\frac{\hat{t}}{\hat{s}^2}-\frac{\hat{t}}{\hat{u}^2}\right), \tag{4}$$

where

$$D(\hat{s},\hat{u}) = \int_0^1 dx \alpha_s^{3/2}(Q_1^2)\left[\frac{\Phi_\pi(x,Q_1^2)}{x(1-x)}\right] + \int_0^1 dx \alpha_s^{3/2}(Q_2^2)\left[\frac{\Phi_\pi(x,Q_2^2)}{x(1-x)}\right]. \tag{5}$$

By the way it must be denoted that as a special case we can directly get the result [26] from Eq(4) by the applying the frozen coupling constant approximation.

According BLM approach transfer momentum of the hard gluon in Fig.1 for s and u channels get the forms

$$Q_1^2 = (1-x)\hat{s} \quad \text{and} \quad Q_2^2 = -x\hat{u}, \tag{6}$$

respectively.

Journal of Physics: Conference Series **670** (2016) 012001 doi:10.1088/1742-6596/670/1/012001

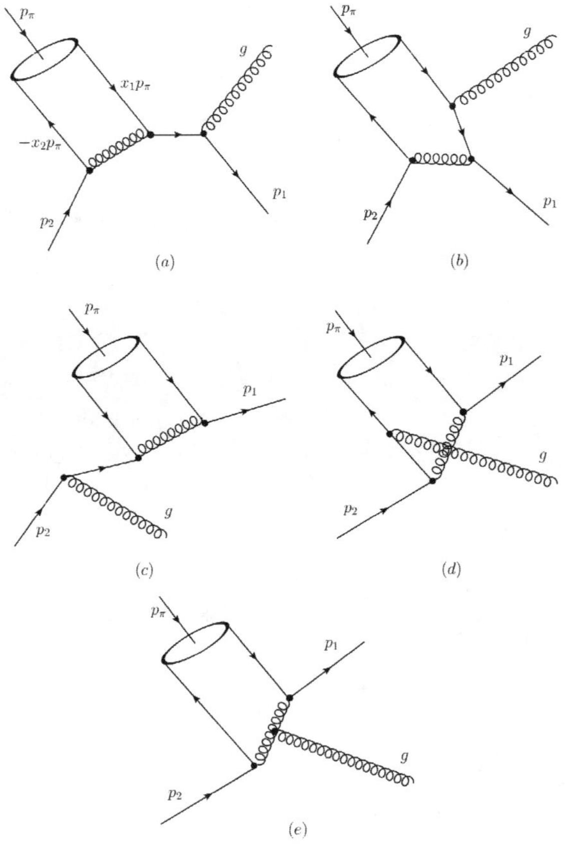

Figure 1. Full set of QCD Feynman diagrams for higher-twist subprocess $\pi q \to gq$.

In the soft regions $x \to 0$ and $x \to 1$ (for u and s channels) integrals (5) diverge, therefore in these regions for their calculations some regularization methods of $\alpha_s(Q^2)$ are needed. One of the simple method is called frozen coupling constant approximation for the regularization these singularity.

There are few forms of the pion distribution amplitude available in the literature. In the present numerical calculations, we use several choices, such as the asymptotic distribution amplitude derived in pQCD evalution [27], the Vega-Schmidt-Branz-Gutsche-Lyubovitskij (VSBGL) distribution amplitude [28], distribution amplitudes predicted by AdS/CFT [30, 29], the Chernyak-Zhitnitsky(CZ) [31], the Bakulev-Mikhailov-Stefanis (BMS) [32] and pion distribution amplitudes in which Gegenbauer coefficients C_2 and C_4 are extracted from BELLE experiment [33, 34]:

$$\Phi_{asy}(x, Q^2 \to \infty) = \sqrt{3} f_\pi x(1-x), \tag{7}$$

$$\Phi^{hol}_{VSBGL}(x, \mu_0^2) = \frac{A_1 k_1}{2\pi}\sqrt{x(1-x)}exp\left(-\frac{m^2}{2k_1^2 x(1-x)}\right), \tag{8}$$

$$\Phi^{hol}(x, \mu_0^2) = \frac{4}{\sqrt{3}\pi} f_\pi \sqrt{x(1-x)}, \tag{9}$$

Journal of Physics: Conference Series **670** (2016) 012001 doi:10.1088/1742-6596/670/1/012001

$$\Phi_{CZ}(x,\mu_0^2)=\Phi_{asy}(x)\left[C_0^{3/2}(2x-1)+\frac{2}{3}C_2^{3/2}(2x-1)\right], \quad (10)$$

$$\Phi_{BMS}(x,\mu_0^2)=\Phi_{asy}(x)\left[C_0^{3/2}(2x-1)+0.20C_2^{3/2}(2x-1)-0.14C_4^{3/2}(2x-1)\right], \quad (11)$$

$$\Phi_{BELLE}(x,\mu_0^2)=\Phi_{asy}(x)\left[C_0^{3/2}(2x-1)+0.12C_2^{3/2}(2x-1)+0.08C_4^{3/2}(2x-1)\right], \quad (12)$$

here $C_n^{\lambda}(2x-1)$ are Gegenbauer polynomials.

Substituting Eq.(4) into Eq.(2) then the differential cross section for the process $\pi p\to gX$ takes the form [26]

$$E\frac{d\sigma}{d^3p}(\pi p\to gX)=\frac{s}{s+u}xG_{q/p}(x,Q^2)\frac{256\pi}{81\hat{s}^2}\left[D(\hat{s},\hat{u})\right]^2\left(-\frac{\hat{t}}{\hat{s}^2}-\frac{\hat{t}}{\hat{u}^2}\right). \quad (13)$$

It should be noted that, as seen from Eq.(4) and Eq.(13), the higher-twist cross section is linear with respect to $\hat{t}$, so the cross section vanishes, if the scattering angle between the final gluon and incident pion is approximately equal to zero. From Eq.(13) we see that the higher-twist cross section proportional to $\hat{s}^{-3}$, which is equivalent to the higher-twist contributions to the $\pi p\to gX$ cross section have the form of $p_T^{-6}f(x_F,x_T)$.

In the expression (6) we fixed the variable x by taking it is mean value. So, average values for x we take $\overline{x}=1/2$. Thus, for the calculations higher-twist cross sections within frozen coupling constant approach we substitute $\overline{Q}^2=\hat{s}/2$ and $\overline{Q}^2=-\hat{u}/2$ in Eq.(13) for the transfer momentum of the hard gluon, respectively.

The extracting of higher-twist contribution from the inclusive gluon production cross section is also complicated. One can also consider the comparison of higher-twist corrections with leading-twist contributions. For the leading-twist subprocess in the inclusive gluon production, we take $q\bar{q}\to g\gamma$ as a subprocess of the quark-antiquark annihilation. The differential cross section for subprocess $q\bar{q}\to g\gamma$ is

$$\frac{d\sigma}{d\hat{t}}(q\bar{q}\to g\gamma)=\frac{8}{9}\pi\alpha_E\frac{e_q^2}{\hat{s}^2}\left(\alpha_s(-\hat{u})\frac{\hat{t}}{\hat{u}}+\alpha_s(-\hat{t})\frac{\hat{u}}{\hat{t}}\right). \quad (14)$$

As is seen from Eq.(14) leading-twist cross section strongly depend of the running coupling constant where the running coupling constant depends on the transfer momentum. However running coupling constant depends on the channels of the process. Here running coupling have been evaluated in the momentum subtraction scheme, for momentum scales u and t, which define the off-shell momenta carried by the quark propagators.

The leading-twist cross section for production of inclusive gluon is [35]

$$\Sigma_M^{LT}\equiv E\frac{d\sigma}{d^3p}(\pi p\to gX)=\int_0^1 dx_1\int_0^1 dx_2\delta(\hat{s}+\hat{t}+\hat{u})G_{\bar{q}/M}(x_1,Q_1^2)G_{q/p}(x_2,Q_2^2)\frac{\hat{s}}{\pi}\frac{d\sigma}{d\hat{t}}(q\bar{q}\to g\gamma), \quad (15)$$

where

$$\hat{s}=x_1x_2s,\ \hat{t}=x_1t,\ \hat{u}=x_2u.$$

3. HIGHER TWIST MECHANISM WITHIN PERTURBATIVE AND HOLOGRAPHIC QCD AND THE ROLE INFRARED RENORMALONS

Mainly object of this study is beside the calculations of the higher twist cross section with running coupling constant approach within holographic and perturbative QCD and renormalon effect's contribution to the cross section, and also comparisons between higher-twist cross sections which are calculated by the running coupling constant method and the principle maximum conformality approach. It should be noted that, in the exclusive processes, the coupling constant α_s runs not only due to the loop integration but also the integration in the process amplitude over the light-cone momentum fraction of hadron constituents. Therefore, it is worth noting that the renormalization scale according to Fig.1 should be chosen equal to $\mu^2_{R_1} = Q_1^2 = (1-x)\hat{s}$, and $\mu^2_{R_2} = Q_2^2 = -x\hat{u}$. The integral in Eq.(5) in the framework of the running coupling approach takes the form

$$D(\mu_R^2) = \int_0^1 \frac{\alpha_s^{3/2}((1-x)\hat{s})\Phi_M(x,\mu_F^2)dx}{x(1-x)} + \int_0^1 \frac{\alpha_s^{3/2}(-x\hat{u})\Phi_M(x,\mu_F^2)dx}{x(1-x)}. \tag{16}$$

At the leading order of perturbative QCD calculations the hard scattering amplitude $T_H(x_1, x_2; Q^2, \mu_R^2, \mu_F^2)$ does not depend on the factorization scale μ_F^2, but strongly depends on μ_R^2. The one-loop QCD correction to the hard scattering amplitude $T_H(x_1, x_2; Q^2, \mu_R^2, \mu_F^2)$ generates its explicit dependence on the scales μ_F^2 and μ_R^2.

As we noted above in the regions $x \to 0$ and $x \to 1$ the integral (16) diverges, because in this regions running coupling constants $\alpha_s((1-x)\hat{s})$ and $\alpha_s(-x\hat{u})$ have the infrared singularity. The other words the singularity of the integrand of x=0 and x=1 is due to only by $\alpha_s((1-x)\hat{s})$ and $\alpha_s(-x\hat{u})$. Thus for the regularization of the integral, by expressing the running coupling at scaling variable $\alpha_s(\mu_R^2)$, we use renormalization group equation with the fixed $\alpha_s(\hat{s})$ and $\alpha_s(-\hat{u})$ for s and u channels, respectively. The solution of renormalization group equation for the running coupling $\alpha \equiv \alpha_s/\pi$ is in the form [22]

$$\frac{\alpha(\lambda)}{\alpha} = \left[1 + \alpha\frac{\beta_0}{4}\ln\lambda\right]^{-1}. \tag{17}$$

Then, for $\alpha_s((1-x)\hat{s})$, we get

$$\alpha((1-x)s) = \frac{\alpha_s}{1 + ln(1-x)/t} \tag{18}$$

where $t = 4\pi/\alpha_s(Q^2)\beta_0 = 4/\alpha\beta_0$.

If we insert Eq.(18) into Eq.(16), we obtain

$$D(\hat{s},\hat{u}) = \int_0^1 dx \frac{\alpha_s^{3/2}((1-x)\hat{s})\Phi_\pi(x,Q_1^2)}{x(1-x)} + \int_0^1 dx \frac{\alpha_s^{3/2}(-x\hat{u})\Phi_\pi(x,Q_2^2)}{x(1-x)}$$

$$= \alpha_s^{3/2}(\hat{s})t_1^{3/2}\int_0^1 dx \frac{\Phi_\pi(x,Q_1^2)}{x(1-x)(t_1+\ln\lambda_1)^{3/2}} + \alpha_s^{3/2}(-\hat{u})t_2^{3/2}\int_0^1 dx \frac{\Phi_\pi(x,Q_2^2)}{x(1-x)(t_2+\ln\lambda_2)^{3/2}} \tag{19}$$

where $t_1 = 4\pi/\alpha_s(\hat{s})\beta_0$ and $t_2 = 4\pi/\alpha_s(-\hat{u})\beta_0$.

Although the integral (19) is still has singularity, this expression can be transformed to more convenient form by the change of variable as, $z = \ln\lambda$ and after applying the integral representation of $1/(t+z)^\nu$ [36, 37],

$$\frac{1}{(t+z)^\nu} = \frac{1}{\Gamma(\nu)}\int_0^\infty e^{-(t+z)u}u^{\nu-1}du, Re\nu > 0 \tag{20}$$

then the singularity in (19) disappears and we obtain

$$D(\hat{s},\hat{u}) = \frac{\alpha_s^{3/2}(-\hat{s})t_1^{3/2}}{\Gamma(\frac{3}{2})}\int_0^1\int_0^\infty \frac{\Phi_\pi(x,Q_1^2)e^{-(t_1+z_1)u}u^{1/2}dudx}{x(1-x)} +$$
$$+\frac{\alpha_s^{3/2}(-\hat{u})t_2^{3/2}}{\Gamma(\frac{3}{2})}\int_0^1\int_0^\infty \frac{\Phi_\pi(x,Q_2^2)e^{-(t_2+z_2)u}u^{1/2}dudx}{x(1-x)}, \tag{21}$$

Journal of Physics: Conference Series **670** (2016) 012001 doi:10.1088/1742-6596/670/1/012001

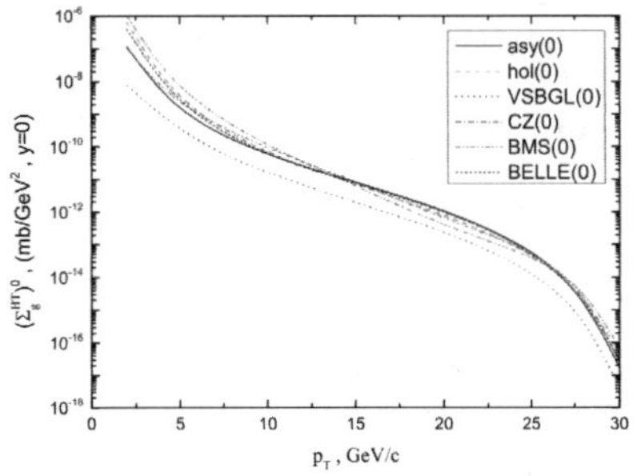

Figure 2. Higher-twist $\pi^+p \to gX$ inclusive gluon production cross section $(\Sigma_g^{HT})^0$ as a function of the transverse momentum of the gluon p_T at the c.m. energy $\sqrt{s} = 62.4\ GeV$.

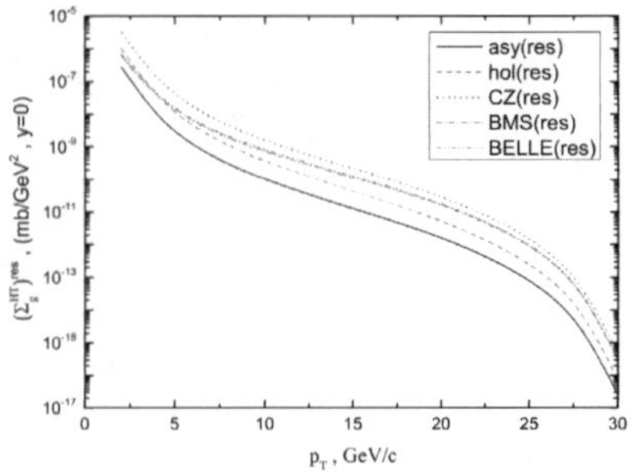

Figure 3. Higher-twist $\pi^+p \to gX$ inclusive gluon production cross section $(\Sigma_g^{HT})^{res}$ as a function of the transverse momentum of the gluon p_T at the c.m.energy $\sqrt{s} = 62.4\ GeV$.

Then the Eq.(21) can be written, for $\Phi^{hol}(x, Q^2)$, as

$$D(\hat{s}, \hat{u}) = \frac{32\sqrt{\pi} f_\pi}{\beta_0\sqrt{3\beta_0}\Gamma(\frac{3}{2})} \int_0^\infty du e^{-t_1 u} u^{1/2} B\left(\frac{1}{2}, \frac{1}{2} - u\right) + \frac{32\sqrt{\pi} f_\pi}{\beta_0\sqrt{3\beta_0}\Gamma(\frac{3}{2})} \int_0^\infty du e^{-t_2 u} u^{1/2} B\left(\frac{1}{2}, \frac{1}{2} - u\right), \tag{22}$$

for $\Phi_{asy}(x, Q^2 \to \infty)$ distribution amplitude, as

$$D(\hat{s}, \hat{u}) = \frac{8\pi\sqrt{3\pi} f_\pi}{\beta_0\sqrt{\beta_0}\Gamma(\frac{3}{2})} \int_0^\infty du e^{-t_1 u} \left[\frac{u^{1/2}}{1-u}\right] + \frac{8\pi\sqrt{3\pi} f_\pi}{\beta_0\sqrt{\beta_0}\Gamma(\frac{3}{2})} \int_0^\infty du e^{-t_2 u} \left[\frac{u^{1/2}}{1-u}\right], \tag{23}$$

for $\Phi_{CZ}(x, Q^2)$ distribution amplitude, as

$$\begin{aligned} D(\hat{s}, \hat{u}) \quad = \quad & \frac{8\pi\sqrt{3\pi} f_\pi}{\beta_0\sqrt{\beta_0}\Gamma(\frac{3}{2})} \int_0^\infty du e^{-t_1 u} u^{1/2} \left[\frac{1}{1-u} + 0.84\left[\frac{4}{1-u} - \frac{20}{2-u} + \frac{20}{3-u}\right]\left(\frac{\alpha_s(Q_1^2)}{\alpha_s(\mu_0^2)}\right)^{\frac{50}{81}}\right] + \\ & \frac{8\pi\sqrt{3\pi} f_\pi}{\beta_0\sqrt{\beta_0}\Gamma(\frac{3}{2})} \int_0^\infty du e^{-t_2 u} u^{1/2} \left[\frac{1}{1-u} + 0.84\left[\frac{4}{1-u} - \frac{20}{2-u} + \frac{20}{3-u}\right]\left(\frac{\alpha_s(Q_2^2)}{\alpha_s(\mu_0^2)}\right)^{\frac{50}{81}}\right]. \end{aligned} \tag{24}$$

4. NUMERICAL RESULTS AND DISCUSSION

We discuss the numerical results for higher-twist and renormalon mechanism with higher-twist contributions calculated in the context of the running and frozen coupling approaches on the dependence of the chosen pion distributions amplitudes in the inclusive gluon production process. For the numerical calculations, we take supprocess $\pi^+ d_p \to gu$ and $\pi^- u_p \to gd$ for $\pi^+ p \to gX$ and $\pi^- p \to gX$ process, respectively.

Inclusive direct gluon production represents a significant test case in which higher-twist terms dominate those of leading-twist in certain kinematic domains. For the dominant leading-twist subprocess for the gluon production, we take the quark-antiquark annihilation $q\bar{q} \to \gamma g$. In the numerical calculations, for the quark distribution functions inside the pion and proton we used expressions as given in [38, 39], respectively. Results obtained in our calculations are visualized in Figs. 2-7. In all figures we represent the choice of pion distribution amplitudes

Journal of Physics: Conference Series **670** (2016) 012001 doi:10.1088/1742-6596/670/1/012001

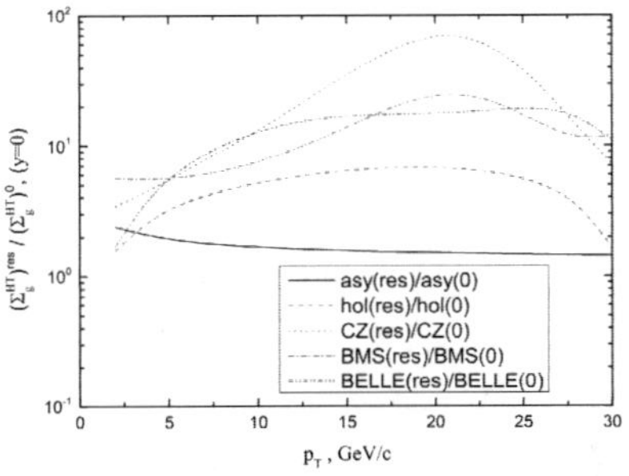

Figure 4. Ratios $(\Sigma_g^{HT})^{res}/(\Sigma_g^{HT})^0$, in the process $\pi^+p \to gX$, where higher-twist contribution are calculated for the gluon rapidity $y = 0$ at the c.m.energy $\sqrt{s} = 62.4\ GeV$ as function of the gluon transverse momentum, p_T.

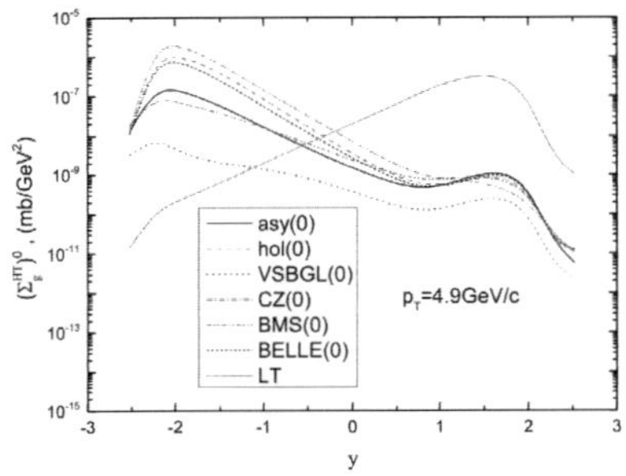

Figure 5. Higher-twist $\pi^+p \to gX$ inclusive gluon production cross section $(\Sigma_g^{HT})^0$, as function of the rapidity of the gluon y at the transverse momentum of the gluon $p_T = 4.9\ GeV/c$, at the c.m. energy $\sqrt{s} = 62.4\ GeV$.

Eqs.(7)-(12) by different line types: $\Phi_{asy}(x)$ as solid black line, $\Phi^{hol}(x)$ as dashed red line, $\Phi^{hol}_{VSBGL}(x)$ as dotted blue line, $\Phi_{CZ}(x,Q^2)$ as dash-dot magenta line, $\Phi_{BMS}(x,Q^2)$ as dash-double dot olive line, and $\Phi_{BELLE}(x,Q^2)$ as short dash navy line. Firstly, it is very interesting to compare the higher-twist cross sections obtained within holographic QCD with ones obtained within the perturbative QCD and also with the leading-twist cross section. In Fig.2 and Fig.3 we show higher-twist cross sections $(\Sigma_g^{HT})^0, (\Sigma_g^{HT})^{res}$ calculated in the context of the frozen (frozen cross section) and running coupling constant (resummed cross section) approaches as a function of the gluon transverse momentum p_T for the pion distribution amplitudes presented in Eqs.(2.6)-(2.11) at $y = 0$. It is seen from Fig.2 and Fig.3 that the higher-twist cross section is monotonically decreasing with an increase in the transverse momentum of the gluon. In the region $2\ GeV/c < p_T < 30\ GeV/c$ the resummed cross sections of the process $\pi^+p \to gX$ decreases in the range between $3,172 \cdot 10^{-6}\mu b/GeV^2$ to $4,912 \cdot 10^{-16}\mu b/GeV^2$.

In Fig.4 we show $(\Sigma_g^{HT})^{res}/(\Sigma_g^{HT})^0$, for the process $\pi^+p \to gX$ as a function of p_T for the pion distribution amplitudes presented in Eqs.(7)-(12) at $y = 0$. We see in Fig.4, that in the region $15\ GeV/c < p_T < 22\ GeV/c$, the ratio $(\Sigma_g^{HT})^{res}/(\Sigma_g^{HT})^0$ for $\Phi_{CZ}(x,Q^2)$ is enhanced by about two orders of magnitude relative to one for $\Phi_{asy}(x)$. However, the enhancement is one order of magnitude for $\Phi^{hol}(x)$ and half an order for $\Phi_{BMS}(x,Q^2)$ and $\Phi_{BELLE}(x,Q^2)$ pion distribution amplitudes. Through Fig.5 to Fig.7 the dependence of higher-twist cross sections $(\Sigma_g^{HT})^0$, $(\Sigma_g^{HT})^{res}$, ratios $(\Sigma_g^{HT})^{res}/(\Sigma_g^{HT})^0$ are shown for the processes $\pi^+p \to \gamma X$ and $\pi^-p \to \gamma X$ as a function of the rapidity of the gluon y at the transverse momentum of the gluon $p_T = 4.9\ GeV/c$. It is seen from figures in Fig.5 and Fig.6, that frozen and resummed cross sections for all distribution amplitudes of pion have two maxima, where the first maximum is approximately at the point $y = -2$ and second maximum is approximately at the point $y = 2$. Notice that distribution amplitude of frozen and resummed cross sections for $\Phi_{CZ}(x,Q^2)$ are enhanced by about half and two orders of magnitude relative to all other distribution amplitudes. As is seen from the figures cross sections vary slowly and smoothly with the angle of the scattering. We think that this feature of infrared renormalons may help theoretical interpretations of the future experimental data for the direct inclusive gluon production cross section in the pion-proton collisions. Higher-twist cross section obtained in our study should be observable at hadron collider.

Journal of Physics: Conference Series **670** (2016) 012001 doi:10.1088/1742-6596/670/1/012001

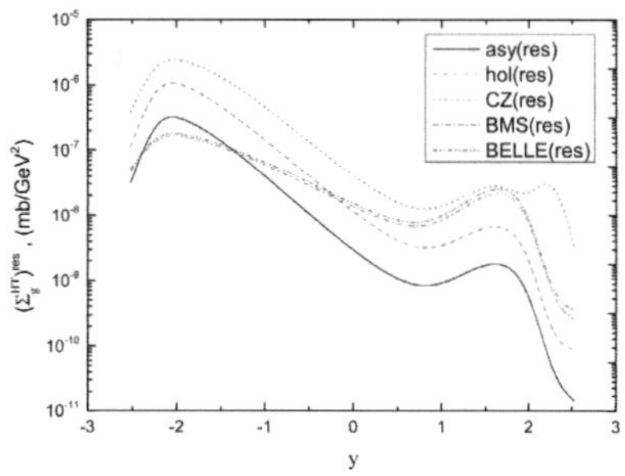

Figure 6. Higher-twist $\pi^+p \to gX$ inclusive gluon production cross section $(\Sigma_g^{HT})^{res}$, as a function of the rapidity of the gluon y at the transverse momentum of the gluon $p_T = 4.9\ GeV/c$, at the c.m. energy $\sqrt{s} = 62.4\ GeV$.

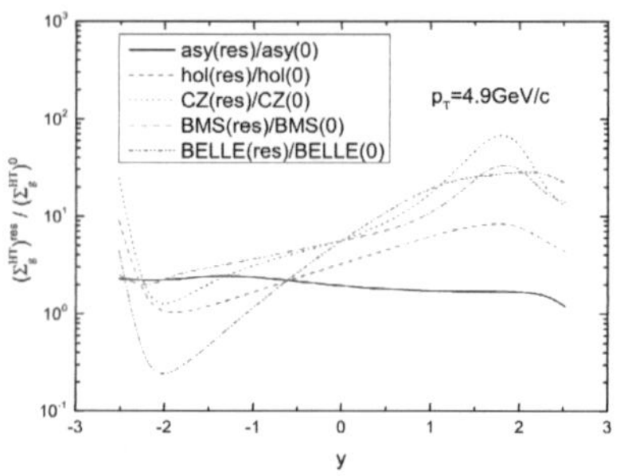

Figure 7. Ratio $(\Sigma_g^{HT})^{res}/(\Sigma_g^{HT})^0$, in the process $\pi^+p \to gX$, as function of the rapidity of the gluon y at the transverse momentum of the gluon $p_T = 4.9\ GeV/c$, at the c.m. energy $\sqrt{s} = 62.4\ GeV$.

5. CONCLUSIONS

In this study the inclusive single gluon production are calculated via higher twist mechanism within perturbative and holographic QCD. In the calculation of the cross sections the running and frozen coupling constant approaches are employed and infrared renormalon poles in the cross section expression are revealed. Infrared renormalon induced divergences are regularized by means of the principal value prescripton and the Borel sum for the higher twist cross section is found. It is observed that, the resummed higher-twist cross section differs from that found using the frozen coupling approximation, especially in some regions, considerably.

Concerning the study of the higher-twist contribution, it is primarily important to analyze its relative magnitude of contribution compared to the leading-twist contribution, since only leading-twist diagrams are commonly considered in usual studies of the hadron-hadron collision. However, in our studies the difference of the higher-twist results for the frozen and running coupling constant approaches have been studied with importance. The following results can be concluded from the experiments: the higher-twist contributions to single gluon production cross section in the pion-proton collisions have important phenomenological consequences. Therefore they will be helpful for detailed investigation dynamical properties of nucleon. Also the higher-twist gluon production cross section in the pion-proton collisions depends on the form of the pion distribution amplitudes and may be used for future study. Moreover the contributions of renormalon effects within holograpich QCD in these process are essential and may help to analyze experimental results. We compared frozen and resumed cross sections of the direct gluon production in the processes $\pi^-p \to gX$ and $\pi^+p \to gX$. Our calculations show in both cases, running and frozen coupling constant approaches that the inclusive gluon production cross section for the process $\pi^-p \to gX$ is suppress over the direct gluon production cross section of the process $\pi^+p \to gX$. Notice that, the direct gluon production spectrum can be measured with large precision, so results obtained in this study will helpful further tests of the hadron dynamics at large p_T. As is seen from Eqs.(4,13) higher- twist cross sections in both cases are proportional to the third power of $\alpha_s(Q^2)$, but the leading-twist is linearly proportional to $\alpha_s(Q^2)$. Therefore their ratios strongly depend on the $\alpha_s^2(Q^2)$.

Further investigations are needed in order to clarify the role of higher-twist effects in QCD. In hadron-hadron collisions, real gluons at high transverse momentum can serve as a short distance probe of the incident hadrons. Especially, the future experimental measurements will provide further tests of the dynamics of large-p_T hadron production beyond the leading twist.

Acknowledgments
One of the authors, A. I. A. thanks the organizers of the ISQS-23, in Prague especially to the Prof.Cestmir Burdik for the opportunity to speak at the conference and for their exceptional hospitality.

[1] S. J. Brodsky, G. L. Lepage, and P. B. Mackenize, Phys. Rev. D **28** (1983) 228.
[2] F. S. Sadykhov and A. I. Akhmedov, Russ. Phys. J. **38** (1995) 513.
[3] A. I. Ahmadov, I. Boztosun, R. Kh. Muradov, A. Soylu, and E. A. Dadashov, Int. J. Mod. Phys. E **15** (2006), 1209.
[4] A. I. Ahmadov, I. Boztosun, A. Soylu, and E. A. Dadashov, Int. J. Mod. Phys. E **17** (2008), 1041.
[5] A. I. Ahmadov, C. Aydin, Sh. M. Nagiyev, A. Hakan Yilmaz, and E. A. Dadashov, Phys. Rev. D **80** 016003 (2009) 016003.
[6] A. I. Ahmadov, C. Aydin, E. A. Dadashov, and Sh. M. Nagiyev, Phys. Rev. D **81** (2010) 054016.
[7] A. I. Ahmadov, R. M. Burjaliyev, Int. J. Mod. Phys. E **20** (2011) 1243.
[8] A. I. Ahmadov, Sh. M. Nagiyev, and E. A. Dadashov , Int. J. Mod. Phys. E **21** (2012) 1250014.
[9] A. I. Ahmadov, C. Aydin, and F. Keskin, Phys. Rev. D **85** (2012) 034009.
[10] A. I. Ahmadov, C. Aydin, and F. Keskin, Ann. Phys. **327** (2012) 1472.
[11] A. I. Ahmadov, C. Aydin, and O. Uzun, Phys. Rev. D **87** (2013) 014006.
[12] A. I. Ahmadov, C. Aydin, and O. Uzun, Phys. Rev. D **89** (2014) 014018.
[13] J. A. Bagger and J. F. Gunion, Phys. Rev. D **29** (1984) 40.
[14] A. Bagger and J. F. Gunion, Phys. Rev. D **25** (1982) 2287.
[15] V. N. Baier and A. Grozin, Phys. Lett.B **96** (1980) 181; S. Gupta, Phys. Rev. D**24** (1981) 1169.
[16] S. S. Agaev, Phys. Lett.B **360** (1995) 117; **369**, (1996) 379(E).
[17] G.'t. Hooft, in The Whys of Subnuclear Physics, Erice, 1977, edited by A. Zichichi (Plenum, New York, 1979) p.94
[18] A. H. Mueller, Nucl. Phys.B **250** (1985)327; Phys. Lett. **B 308** (1993) 355.
[19] V. I. Zakharov, Nucl. Phys.B **385** (1992) 452.
[20] M. Beneke, Phys. Rep. **317** (1999) 1.
[21] W. Greiner, S. Schramm and E. Stein, Quantum Chromodynamics, 3rd edn.(Berlin, Springer, 2007), pp.553.
[22] H. Contopanagos and G. Sterman, Nucl. Phys.B **419** (1994) 77.
[23] A. L. Kataev, JETP Lett. **81** (2005) 608; Mod. Phys. Lett. **A 20** (2005) 2007.
[24] G. L. Lepage and S. J. Brodsky, Phys. Rev. D **22** (1980) 2157.
[25] J.F. Owens, Rev. Modern Phys. **59** (1987) 465.
[26] E. L. Berger and S. J. Brodsky, Phys. Rev. D **24** (1981) 2428.
[27] G. P. Lepage and S. J. Brodsky, Phys. Lett.B **87** (1979) 359.
[28] A. Vega, I. Schmidt, T. Branz, T.Gutsche, V. Lyubovitskij, Phys. Rev. D **80** (2009) 055014.
[29] S. J. Brodsky and G. F. de Teramond, Phys. Rev. D **77** (2008) 056007.
[30] S. J. Brodsky, Proc. Sci., LHC07 (2007) 002.
[31] V. L. Chernyak and A. R. Zhitnitsky, Phys. Rep. **112** (1984) 173.
[32] A. P. Bakulev, S. V. Mikhailov and N. G. Stefanis, Phys. Lett, **B 578** (2004) 91; A. P. Bakulev, S.V. Mikhailov, A.V. Pimikov, and N. G. Stefanis, Phys. Rev. D **86** (2012) 031501.
[33] S. S. Agaev, V. M. Braun, N. Offer. F. A. Porkort, Phys. Rev. D**86** (2012) 077504.
[34] S. Uehara et al. [Belle Collaboration], Phys. Rev. D **86** 092007 (2012) 092007.
[35] E. L. Berger, Phys. Rev. D **26** (1982) 105.
[36] J. Zinn-Justin, Phys. Rept. **70** (1981) 109.
[37] A. Erdelyi, Higher Transcendental Functions (McGrow-Hill Book Company, New York, 1953), Vol.2.
[38] S.-i Nam, Phys. Rev. D **86** (2012) 074005.
[39] A.D. Martin, W. J. Stirling, R.S. Thorne, G. Watt, Eur. Phys. J. C **63** (2009) 189.

Journal of Physics: Conference Series **670** (2016) 012002 doi:10.1088/1742-6596/670/1/012002

Solutions in the $\mathbb{C}P^N$ Skyrme type model

Yuki Amari[1], Paweł Klimas[2], Nobuyuki Sawado[1], Yuta Tamaki[1]

[1] Department of Physics, Tokyo University of Science, Noda, Chiba 278-8510, Japan
[2] Universidade Federal de Santa Catarina, Trindade, 88040-900, Florianópolis, SC, Brazil

E-mail: amalyamary@gmail.com,sawado@ph.noda.tus.ac.jp

Abstract. The extended Skyrme-Faddeev model possesses vortex solutions in a (3+1) dimensional Minkowski space-time with target space CP^N. The right choice of the potential leads to solutions with finite energy per unit of length. Such solutions have a form of waves propagating along vortices with the speed of light. Some possible applications are considered, such as the quantization scheme and the vortex-fermion system.

1. Introduction

The Skyrme type model is a large class of soliton models which is based on the non-linear sigma model in 2 or 3 spatial dimensions. The standard Skyrme model was originated by T.H.R. Skyrme [1, 2, 3] and it has soliton solutions which describe the hadrons after suitable quantization scheme [4]. The Skyrme-Faddeev model is a modification of the Skyrme model and supports the knotted solitons [5] as well as the vortices [6]. The significance of this model has increased noticeably when it has been conjectured that the model can be seen as a low-energy effective classical model of the underlying Yang-Mills theory [7, 8]. Remarkably, the exact soliton (vortex) solution of the model has been found within the integrable sector [9]. It has been shown that also in the case of the complex projective target space CP^N the extended Skyrme-Faddeev model (in which it has been imposed some special constraints for the parameters of the model) possesses an exact soliton solutions in the integrable sector [10]. It has not been clear, however, if the presence of the solutions in the considered model is related to the particular choice of the coupling constants like in the case of the exact solution or it is rather a general property of the model. We investigate the existence of the solutions of the model inside/outside the integrable sector, especially in absence of some particular relations between coupling constants. The key role is playing by the potential which usually work as a stabilizer for the solution. Many CP^N - type models are usually considered in context of instantons, i.e., the scale invariant solutions. The potential breaks such invariance and then the stable solutions appear. In the present paper we construct the potentials for combinations of the positive/negative winding numbers $\{n_i\}$.

There are several interesting topics related with the present solutions; the collective coordinate quantizations and the construction of the vortex-fermion systems. As was pointed out, the solitons describe the particle state after suitable quantization scheme. For the present model we employ the rotational zero-mode and discuss how the spectrum is estimated. The problem of the fermion can be more phenomenological. The existence of transverse zero energy modes of the Dirac equation in the presence of a string can make the string superconducting [11].

Journal of Physics: Conference Series **670** (2016) 012002 doi:10.1088/1742-6596/670/1/012002

We propose two independent coupling scheme for the fermions and check how they work in two and three spatial dimensions, respectively.

The study of such models is promising and could be important for understanding some aspects of the strong coupling sector of the Yang-Mills theory.

2. The model

We start with a Skyrme type model

$$\mathcal{L} = -\frac{M^2}{2}\mathrm{Tr}(X^{-1}\partial_\mu X)^2 + \frac{1}{e^2}\mathrm{Tr}(X^{-1}\partial_\mu X X^{-1}\partial_\nu X)^2 \\ +\frac{\beta}{2}\left[\mathrm{Tr}(X^{-1}\partial_\mu X)^2\right]^2 + \gamma\left[\mathrm{Tr}(X^{-1}\partial_\mu X X^{-1}\partial_\nu X)\right]^2 - V \quad (1)$$

where M is a coupling constant with dimension of mass whereas the coupling constants e^{-2}, β, γ are dimensionless. The variable $X(t,\vec{x})$ is an $SU(N+1)$- valued scalar. We shall discuss its explicit form below. All terms containing derivatives in the Lagrangian (1) possess the global symmetry $SU(N+1)_L \times SU(N+1)_R$ corresponding to transformations $X \mapsto g_L X g_R^\dagger$, where g_L, g_R are constant elements of $SU(N+1)$. Such a symmetry originates in the way how the variable X enters to the Lagrangian. In fact it is independent on particular form of X. One can make whole Lagrangian (1) invariant under this transformation for appropriate choice of the potential V. The first term is quadratic in X and corresponds with the Lagrangian of the CP^N model. The quartic term proportional to e^{-2} is the Skyrme term, and together with the other quartic terms they support the stability of the solution. Similarly, the form of V is (by construction) such that it is consistent with the stability requirement. Quite naturally X can be expressed as $X \sim \xi_L\xi_R^\dagger$, where the chiral transformation is given via $\xi_L \to \xi_L' = g_L\xi_L$, and $\xi_R \to \xi_R' = g_R\xi_R$.

In this paper we explore solutions with the CP^N target space. The standard Skyrme-Faddeev model on the CP^1 target space is parametrized by a real triplet unit vector $\vec{n}$, which has only two independent degrees of freedom. The target space (coset space) in the case of some higher dimensional $SU(N+1)$ Lie groups, i.e. $N \geqq 2$, can be chosen in several nonequivalent ways. Recently it has been proposed some formulation of the extended Skyrme-Faddeev model on the CP^N target space [10]. The coset space $CP^N = SU(N+1)/SU(N) \otimes U(1)$ is an example of a symmetric space and it can be naturally parametrized in terms of so called *principal variable*

$$X(g) = g\sigma(g)^{-1}, \qquad g \in SU(N+1) \quad (2)$$

where σ being the order two automorphism under which the subgroup $SU(N)\otimes U(1)$ is invariant i.e. $\sigma(h) = h$ for $h \in SU(N)\otimes U(1)$. The coordinate $X(g)$ defined above satisfies $X(gh) = X(g)$. The Lagrangian has a global left symmetry such that $g \to \bar{g}g$ with $\bar{g}, g \in SU(N+1)$, which implies that $X \to \bar{g}X\sigma(\bar{g})^{-1}$ and then $X^{-1}\partial_\mu X \to \sigma(\bar{g})X^{-1}\partial_\mu X\sigma(\bar{g})^{-1}$. In addition it has a right local $SU(N)\times U(1)$ symmetry such that $g \to gk$ with $k \in SU(N)\times U(1)$ and $g \in SU(N+1)$, which implies that X is invariant.

There is an additional symmetry $X \to \mathcal{P}X$. The prefactor $\mathcal{P}$ is a constant, unitary matrix (in our considerations it is diagonal) **and seems to be essentially appeared by a choice of the gauge ξ_L, ξ_R.** The standard choice $\mathcal{P}_{\mathrm{I}} = (\mathbf{1}_{N+1\times N+1})$ coincides with definition used in [10]. On the other hand, if one choose

$$\mathcal{P}_{\mathrm{II}} = \begin{pmatrix} \mathbf{1}_{N\times N} & 0 \\ 0 & -1 \end{pmatrix}$$

the resulting matrix $X_{\rm H} \equiv \mathcal{P}_{\rm H} g\sigma(g)^{-1}$ becomes Hermitian. Notable thing is that both definitions of $\mathcal{P}$ have the same classical solutions because they are concerned with the symmetry of the model.

2.1. The parametrization

Let us shortly discuss the parametrization of the model. According to the previous paper [10] one can parametrize the model in terms of N complex fields u_i, where $i = 1, \ldots, N$. Assuming $(N+1)$-dimensional defining representation where the $SU(N+1)$ valued element g is of the form

$$g \equiv \frac{1}{\vartheta}\begin{pmatrix} \Delta & i\,u \\ i\,u^\dagger & 1 \end{pmatrix} \qquad \vartheta \equiv \sqrt{1 + u^\dagger \cdot u} \tag{3}$$

and where Δ is the Hermitian $N \times N$-matrix

$$\Delta_{ij} = \vartheta\,\delta_{ij} - \frac{u_i\,u_j^*}{1+\vartheta} \quad \text{which satisfies} \quad \Delta \cdot u = u \quad u^\dagger \cdot \Delta = u^\dagger .$$

For g given by (3), one can has $\sigma(g) = g^{-1}$ [10], and so the principal variable $X_{\rm I}(g) := \mathcal{P}_{\rm I} g\sigma(g)^{-1}$ is given by

$$X_{\rm I}(g) = \mathcal{P}_{\rm I} g^2 = \begin{pmatrix} \mathbf{1}_{N\times N} & 0 \\ 0 & -1 \end{pmatrix} + \frac{2}{\vartheta^2}\begin{pmatrix} -u\otimes u^\dagger & iu \\ iu^\dagger & 1 \end{pmatrix}$$

Also for $X_{\rm H}(g) := \mathcal{P}_{\rm H} g\sigma(g)^{-1}$ we have

$$X_{\rm H}(g) = \mathcal{P}_{\rm H} g^2 = \begin{pmatrix} \mathbf{1}_{N\times N} & 0 \\ 0 & 1 \end{pmatrix} + \frac{2}{\vartheta^2}\begin{pmatrix} -u\otimes u^\dagger & iu \\ -iu^\dagger & -1 \end{pmatrix}$$

Apparently the $X_{\rm H}$ is Hermitian while the $X_{\rm I}$ is not.

The Lagrangian (1) reads

$$\mathcal{L} = -\frac{1}{2}\Big[M^2\eta_{\mu\nu} + C_{\mu\nu}\Big]\tau^{\nu\mu} - V \tag{4}$$

where the symbols $C_{\mu\nu}$ and $\tau_{\mu\nu}$ read

$$C_{\mu\nu} := M^2\eta_{\mu\nu} - \frac{4}{e^2}\Big[(\beta e^2 - 1)\tau^\rho_\rho\eta_{\mu\nu} + (\gamma e^2 - 1)\tau_{\mu\nu} + (\gamma e^2 + 2)\tau_{\nu\mu}\Big], \tag{5}$$

$$\tau_{\mu\nu} := -\frac{4}{\vartheta^4}\Big[\vartheta^2\partial_\nu u^\dagger \cdot \partial_\mu u - (\partial_\nu u^\dagger \cdot u)(u^\dagger \cdot \partial_\mu u)\Big]. \tag{6}$$

We shall discuss the specific form of the potential in the further part of the paper. It is enough to assume that $V = V(u^\dagger, u)$. The variation with respect to u_i^* leads to the equations

$$(1 + u^\dagger \cdot u)\partial^\mu(C_{\mu\nu}\partial^\nu u_i) - C_{\mu\nu}\Big[(u^\dagger \cdot \partial^\mu u)\partial^\nu u_i + (u^\dagger \cdot \partial^\nu u)\partial^\mu u_i\Big]$$
$$+\frac{u_i}{4}(1 + u^\dagger \cdot u)^2\left[\frac{\delta V}{\delta |u_i|^2} + \sum_{k=1}^N |u_k|^2 \frac{\delta V}{\delta |u_k|^2}\right] = 0. \tag{7}$$

In order to get the result (7) one needs to multiply the equation obtained directly from the variation with respect to u_i^* by the inverse matrix of Δ^2_{ki} i.e. $\Delta^{-2}_{ij} = \frac{1}{1+u^\dagger\cdot u}(\delta_{ij} + u_i u_j^\dagger)$. It leads to the term being a combination of partial derivatives of the potential. We introduce the dimensionless coordinates (t, ρ, φ, z) defined as

$$x^0 = r_0 t, \quad x^1 = r_0 \rho \cos\varphi, \quad x^2 = r_0 \rho \sin\varphi \quad x^3 = r_0 z \tag{8}$$

where the length scale r_0 is defined in terms of coupling constants M^2 and e^2 i.e.

$$r_0^2 = -\frac{4}{M^2 e^2}$$

and the light speed is $c = 1$ in the chosen unit system. The linear element ds^2 reads

$$ds^2 = r_0^2 (dt^2 - dz^2 - d\rho^2 - \rho^2 d\varphi^2).$$

The family of exact vortex solutions has been found for the model without potential $V = 0$ where in addition the coupling constants satisfy the condition $\beta e^2 + \gamma e^2 = 2$. The exact solutions have the form of vortices which depend on some specific combination of the coordinates i.e. one light-cone coordinate $x^3 + x^0$ and one complex coordinate $x^1 + ix^2$. The functions $u_i(x^3 + x^0, x^1 + ix^2)$ satisfy the zero curvature condition $\partial_\mu u_i \partial^\mu u_j = 0$ for all $i, j = 1, \ldots, N$ and therefore one can construct the infinite set of conserved currents.

The solutions can generally be written such as the form

$$u_j = f_j(\rho) e^{i(n_j \varphi + k_j \psi(w))} \tag{9}$$

where $\psi(w)$ is a real function of the light-cone coordinate and $f_i(\rho)$ is N-th element set of real functions. The constants n_i form the set of integer numbers and k_i are some real constants. The solutions in the integrable sector is thus holomorphic one

$$f_i(\rho) = c_i \rho^{n_i} \tag{10}$$

where c_i being complex constants.

We shall use the matrix notation for the convenience therefore we define two diagonal matrices

$$\lambda \equiv \mathrm{diag}(n_1, \ldots, n_N), \qquad \sigma \equiv \mathrm{diag}(k_1, \ldots, k_N). \tag{11}$$

In the matrix form the ansatz reads $u = f(\rho) \exp[i(\lambda\varphi + \sigma\psi(w))]$ where w is either $z + t$ or $z - t$. The components of $\tau_{\mu\nu}$ have the following form

$$\begin{aligned}
&\tau_{\rho\rho} \equiv \theta(\rho) = -\frac{4}{\vartheta^4}\left[\vartheta^2 f'^T.f' - (f'^T.f)(f^T.f')\right] \\
&\tau_{\varphi\varphi} \equiv \omega(\rho) = -\frac{4}{\vartheta^4}\left[\vartheta^2 f^T.\lambda^2.f - (f^T.\lambda.f)^2\right] \\
&\tau_{\varphi\rho} = -\tau_{\rho\varphi} \equiv i\zeta(\rho), \quad \zeta(\rho) = -\frac{4}{\vartheta^4}\left[\vartheta^2 f'^T.\lambda.f - (f^T.\lambda.f)(f'^T.f)\right] \\
&\tau_{t\rho} = -\tau_{\rho t} \equiv (i\partial_t\psi)\xi(\rho), \quad \tau_{z\rho} = -\tau_{\rho z} \equiv (i\partial_z\psi)\xi(\rho), \quad \xi(\rho) = -\frac{4}{\vartheta^4}\left[\vartheta^2 f'^T.\sigma.f - (f^T.\sigma.f)(f'^T.f)\right] \\
&\tau_{t\varphi} = \tau_{\varphi t} \equiv (\partial_t\psi)\eta(\rho), \quad \tau_{z\varphi} = \tau_{\varphi z} \equiv (\partial_z\psi)\eta(\rho), \quad \eta(\rho) = -\frac{4}{\vartheta^4}\left[\vartheta^2 f^T.\lambda.\sigma.f - (f^T.\lambda.f)(f^T.\sigma.f)\right] \\
&\tau_{tt} \equiv (\partial_t\psi)^2\chi(\rho), \quad \tau_{zz} \equiv (\partial_z\psi)^2\chi(\rho) \\
&\tau_{tz} = \tau_{zt} \equiv (\partial_t\psi)(\partial_z\psi)\chi(\rho), \quad \chi(\rho) = -\frac{4}{\vartheta^4}\left[\vartheta^2 f^T.\sigma^2.f - (f^T.\sigma.f)(f^T.\sigma.f)\right]
\end{aligned}$$

where the derivative with respect to ρ has been denoted by $\frac{d}{d\rho} =' $. The equations of motion written in dimensionless coordinates take the form

$$
\begin{aligned}
&(1+f^T.f)\left[\frac{1}{\rho}\left(\rho\tilde{C}_{\rho\rho}f_k'\right)' + \frac{i}{\rho}\left(\frac{\tilde{C}_{\rho\varphi}}{\rho}\right)'(\lambda.f)_k - \frac{1}{\rho^4}\tilde{C}_{\varphi\varphi}(\lambda^2.f)_k\right] \\
&-2\left[\tilde{C}_{\rho\rho}(f^T.f')f_k' - \frac{1}{\rho^4}\tilde{C}_{\varphi\varphi}(f^T.\lambda.f)(\lambda.f)_k\right] \\
&+\frac{f_k}{4}(1+f^T.f)^2\left[\frac{\delta\tilde{V}}{\delta f_k^2} + \sum_{i=1}^{N} f_i^2\frac{\delta\tilde{V}}{\delta f_i^2}\right] = 0
\end{aligned}
\tag{12}
$$

for each $k = 1, \ldots, N$, where we have introduced the symbols $\tilde{C}_{\mu\nu} := \frac{1}{r_0^2 M^2}C_{\mu\nu}$, and also $\tilde{V} := \frac{r_0^2}{M^2}V$. These of components $\tilde{C}_{\mu\nu}$ which appear in the equations of motion read

$$
\begin{aligned}
\tilde{C}_{\rho\rho} &= -1 + (\beta e^2 - 1)\left(\theta + \frac{\omega}{\rho^2}\right) + (2\gamma e^2 + 1)\theta \\
\tilde{C}_{\varphi\varphi} &= -\rho^2 + \rho^2(\beta e^2 - 1)\left(\theta + \frac{\omega}{\rho^2}\right) + (2\gamma e^2 + 1)\omega \\
\tilde{C}_{\varphi\rho} &= -\tilde{C}_{\rho\varphi} = -3i\zeta\,.
\end{aligned}
\tag{13}
$$

2.2. The topological property
According to discussions in [12] and also in [10], we can define the topological charge in the present model. The field u_i provide a mapping from x^1x^2 plane into CP^N. However, for the finiteness of the energy, the field goes to a constant. Then the plane should be compactified into S^2 and the solutions define the mapping $S^2 \to CP^N$ which is classified into the homotopy classes of $\pi_2(CP^N)$. There exists a theorem describing in [12], $\pi_2(G/H) = \pi_1(H)_G$ where $\pi_1(H)_G$ is the subset of $\pi_1(H)$ formed by closed paths in H which can be contracted to a point in G. Thus, in the present case, the topological charges are given by

$$\pi_1(SU(N)\otimes U(1))_{SU(N+1)}. \tag{14}$$

The topological charges are equal to the number of poles of u_i, including those at infinity. And then, it can be obtained as

$$Q_{\text{top}} = n_{\max} + |n_{\min}| \tag{15}$$

where the highest positive integer in the set $n_i, i = 1, 2, \cdots, N$ and $n_{\min}$ is the lowest negative integer in the same set.

3. The integrable sector and the potential

In [6] the authors argued that exact holomorphic solutions exist in the CP^1 model with special choice of potential V. We have tried to consider generalization to the CP^N case. Until now, however, we could solve the problem only for a quite limited case i.e., the first $K-$th integers n_i are equal and the remaining are all zero. The problem we shall solve is essentially an inverse problem. Therefore, first we assume holomorphic solutions and then explore appropriate potential which satisfies the equations of motion. We assume the solutions in the form

$$
\begin{aligned}
u_i &= c_i\rho^n e^{in\varphi} \quad &\text{for} \quad i = 1, \ldots K \\
u_i &= c_i \quad &\text{for} \quad i = K+1, \ldots N
\end{aligned}
\tag{16}
$$

where c_i are complex constants. The solutions satisfy the zero-curvature condition $\partial_\mu u_i \partial^\mu u_j = 0$. At least in this special case, the problem can be solved by the method of the separation of variables. For this reason we introduce the ansatz for potential of the form

$$V = \frac{\left(\sum_{k=1}^K |u_k|^2\right)^a Q(|u_{K+1}|^2, \dots, |u_N|^2)}{(1 + \sum_{j=1}^N |u_j|^2)^4}$$

where a is a free constant. Plugging the solution (16) into the Eq.(7), one obtain the set of equations

$$\begin{cases} \frac{1}{\tilde{\mu}^2}\frac{R(\rho)}{\rho^4} - \left\{ a\left(\sum_{j=1}^K |u_j|^2\right)^{a-1} + (a - 4 + \sum_{k=K+1}^N \frac{|u_k|^2}{Q}\frac{\delta Q}{\delta |u_k|^2})\left(\sum_{j=1}^K |u_j|^2\right)^a \right\} Q = 0 \\ \frac{\delta Q}{\delta |u_i|^2} + \sum_{k=K+1}^N |u_k|^2 \frac{\delta Q}{\delta |u_k|^2} + (a-4)Q = 0 \end{cases} \quad (17)$$

where the symbol $R(\rho)$ reads

$$R(\rho) = 32(\beta e^2 + \gamma e^2 - 2)n^4 \left(1 + \sum_{k=K+1}^K c_k^2\right)^2 \left[\alpha \sum_{k=1}^K |u_k|^2 - \epsilon\left(\sum_{k=1}^N |u_k|^2\right)^2\right],$$

and the coefficients α and ϵ have the form

$$\alpha := 2\left(1 - \frac{1}{n}\right) \qquad \epsilon := \frac{2}{1 + \sum_{k=K+1}^N c_k^2}\left(1 + \frac{1}{n}\right). \quad (18)$$

The parameter a is fixed as $a = \alpha$ by comparing a power of ρ in the first set of Eq.(17). In order to satisfy Eq.(17), the function Q has to satisfy the condition

$$\left[\frac{1}{Q}\frac{\delta Q}{\delta |u_i|^2}\right]_{|u_k|^2 = c_k^2} = \epsilon \quad \text{for} \quad i = K+1, \cdots, N \quad (19)$$

and the value of $\tilde{\mu}^2$ is determined in terms of other constants :

$$\tilde{\mu}^2 = 32(\beta e^2 + \gamma e^2 - 2)n^4 \frac{\left(\sum_{i=1}^K c_i^2\right)^{2/n}\left(1 + \sum_{i=K+1}^N c_i^2\right)^2}{Q(c_{K+1}^2, \cdots, c_N^2)}. \quad (20)$$

Any function Q which satisfies the condition (19) provide a set of analytical solutions. The simplest example of the integrable potential is

$$V = \frac{\left(\sum_{i=1}^K |u_i|^2\right)^\alpha \left(1 + \sum_{i=K+1}^N |u_i|^2\right)^{4-\alpha}}{\left(1 + \sum_{i=1}^N |u_i|^2\right)^4} \quad (21)$$

with the coefficient

$$\tilde{\mu}^2 = 32(\beta e^2 + \gamma e^2 - 2)n^4 \left(\sum_{i=1}^K c_i^2\right)^{2/n} \left(1 + \sum_{i=K+1}^N c_i^2\right)^{-2/n}. \quad (22)$$

The question whether there exist exact solutions for arbitrary combination of winding numbers is still an open problem.

Journal of Physics: Conference Series **670** (2016) 012002 doi:10.1088/1742-6596/670/1/012002

4. The typical classical solutions

Eq.(12) is highly nonlinear and therefore we solve it numerically. The standard successive overrelaxation method is used with a suitable rescaling of the coordinate $\rho := \sqrt{(1-y)/y}$ $(0 \leqq y \leqq 1)$ and the field $f_i := \sqrt{(1-g_i)/g_i}$ $(0 \leqq g_i \leqq 1)$.

4.1. The potential

The key point for the construction of the potential is careful analysis of the vacuum structure of the field. As a simple example, we start with the CP^1 ($O(3)$) case. The $O(3)$ σ model is usually defined as a vectorial triplet $\vec{n} = (n_1, n_2, n_3)$ with the constraint $\vec{n}\cdot\vec{n} = 1$. The well-known potential named "old-baby" type, i.e.potential with one vacuum, is of the form [13]

$$V(\vec{n}) = (1 - \vec{n}_\infty \cdot \vec{n}) \tag{23}$$

where $\vec{n}_\infty$ is a vacuum value of the field $\vec{n}$ at spatial infinity. For the choice $\vec{n}_\infty = (0,0,1)$ the potential becomes $V = (1 - n_3)$, which is most well-known and is used.

Generally speaking, a potential can be deduced from the asymptotic structure of solutions of the model. Moreover, the potential has to have the form such that the model has solutions for all qualitatively different combinations of the integers (n_1, n_2). In the following subsections, we give an explicit form of the potentials for basic combinations of (n_1, n_2). The crucial point is that we explore only such potentials that numerical solutions and its analytical counterparts, both characterised by the same set of winding numbers (n_1, n_2), have the same asymptotics i.e. both behave as $\sim (\rho^{n_1}, \rho^{n_2})$ at spatial infinity.

4.1.1. The case: $n_1 > n_2 > 0$ An assumption that the solution and its holomorphic counterpart have the same asymptotic behavior at spatial infinity causes that an inverse of the principal variable X goes to ${X_\infty}^{-1} := \mathrm{diag}(-1, 1, -1)$ for $\rho \to \infty$, which results in the following expression for the potential

$$V(u_i) = \mathrm{Tr}(1 - {X_\infty}^{-1}X) = 4\frac{1 + |u_2|^2}{1 + |u_1|^2 + |u_2|^2}. \tag{24}$$

Note that for $\rho \to 0$ an inverse of the principal variable goes to $X_0^{-1} := \mathrm{diag}(1, 1, 1)$, then the expression $\mathrm{Tr}(1 - X_0^{-1}X)$ can be included as the "new-baby" potential which has two vacua [14]. Finally, the following expression can be considered as a general form of the potential

$$\begin{aligned} V &= [\mathrm{Tr}(1 - X_0^{-1}X)]^a[\mathrm{Tr}(1 - X_\infty^{-1}X)]^b \\ &= \frac{(|u_1|^2 + |u_2|^2)^a(1 + |u_2|^2)^b}{(1 + |u_1|^2 + |u_2|^2)^{a+b}} = \frac{(f_1^2 + f_2^2)^a(1 + f_2^2)^b}{(1 + f_1^2 + f_2^2)^{a+b}} \end{aligned} \tag{25}$$

where the integers a, b satisfy $a \geqq 0, b > 0$.

4.1.2. The case: $n_1 > 0 > n_2$ Assuming that for $n_2 < 0$ the field u_2 behaves at zero as its holomorphic counterpart i.e. $\sim \rho^{n_2}$ one gets that it tends to diverge as $\rho \to 0$. Then inverse of the principal variable X goes to $X_0^{-1} := \mathrm{diag}(1, -1, -1)$ for $\rho \to 0$. The general form of the potential takes the form

$$V = \frac{(1 + f_1^2)^a(1 + f_2)^b}{(1 + f_1^2 + f_2^2)^{a+b}}, \tag{26}$$

where the integers satisfy $a \geqq 0, b > 0$.

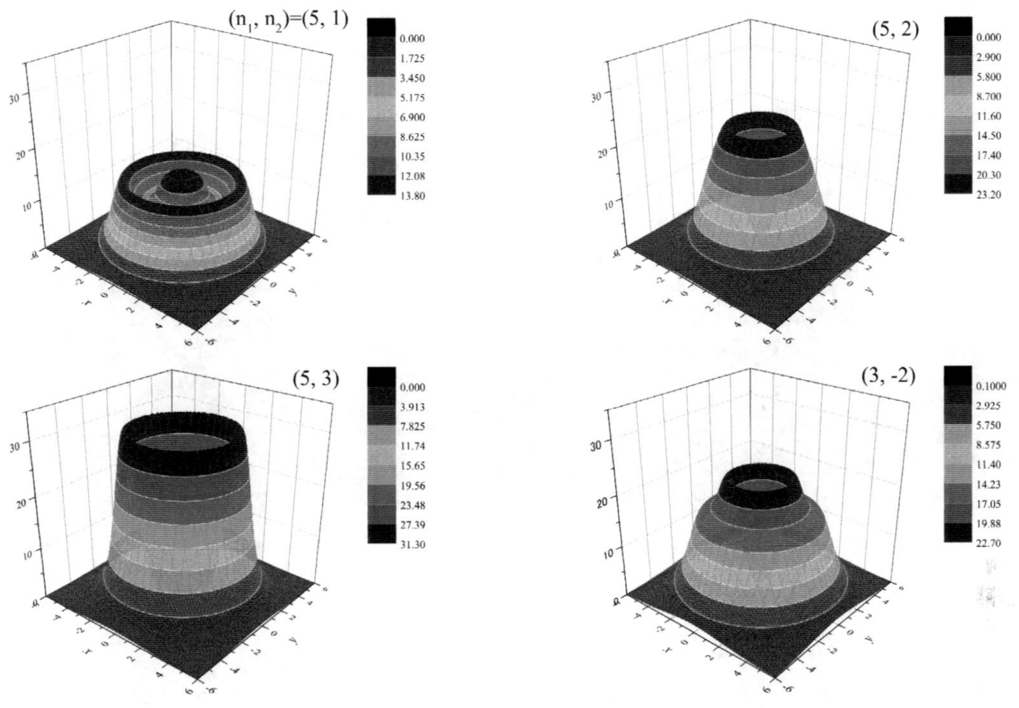

Figure 1. The plot of the CP^2 density of total energies $\mathcal{H}$ (in units of $4M^2$) for $Q_{\text{top}} = 5$ for different pairs (n_1, n_2) and $(k_1, k_2) = (1.0, 1.0)$. The other parameters are $(\beta e^2, \gamma e^2, \tilde{\mu}^2) = (2.0, 2.0, 1.0)$.

4.1.3. The case: $n_1, n_2 < 0, \quad |n_1| > |n_2|$ The asymptotic values of inverse of the principal variable are given by constant matrices $X_\infty^{-1} = \text{diag}(1,1,1)$ and $X_0^{-1} = \text{diag}(-1,1,-1)$. Then the potential is

$$V = \frac{(1+f_2^2)^a (f_1^2+f_2^2)^b}{(1+f_1^2+f_2^2)^{a+b}} \tag{27}$$

where the integers satisfy $a \geqq 0, b > 0$.

In Fig.1 we plot some cases of the energy density per unit of length $\mathcal{H}$ for $Q_{\text{top}} = 5$. For $N = 2$, it has two core structure where the constituent with the larger winding number n_i is located outward and the lower winding number is inward. It is thus easily see that if $n_1 \gg n_2$ the shape is a single torus, while $n_1 \sim n_2$ it looks like double torus.

In the following sections, we will give some possible applications of the vortex solutions.

5. The quantum state via a rigid body rotation

In this section, we explain a basic idea for quantization of vortices in the CP^2 model. A standard approach to the quantization of soliton solutions is based on the zero-mode quantization of the solution as a rigid body [4]. We start with the static case of (9), i.e., $u_j = f_j(\rho)e^{in_j\varphi}$ $j = 1, 2$. The constants n_j are some integers of both signs, however, we can study only the case $n_1 > n_2 > 0$ without loss of generality. We shall work with Hermitian variable X_{H} which is more convenient to study a problem of rotational zero-modes. In this section the subscript H is omitted for simplicity.

The Lagrangian (1) has a global $SU(3) \times SU(3)$ symmetry corresponding to the transformation $X \to \mathcal{A}X\mathcal{B}^\dagger$, where $\mathcal{A}, \mathcal{B} \in SU(3)$. Such symmetry is spontaneously broken to a residual one when Lagrangian is evaluated on solutions. There are two conditions for having residual symmetry. It must to preserve the topological charge and the asymptotic value of X i.e. X_∞. In terms of the Hermitian principal variable the topological charge is defined as

$$Q_{\text{top}} = \frac{1}{8\pi} \int d^2x \epsilon_{ij} \text{Tr}(X\partial_i X \partial_j X). \tag{28}$$

In order to preserve the topological charge, the transformation should be diagonal, i.e., $\mathcal{A} = \mathcal{B}$. The asymptotic form of the variable X reads $X_\infty = \text{diag}(-1, 1, 1)$ according to the leading behavior of u_j at spatial infinity. It follows that the unitary matrix $\mathcal{A}$ have to satisfy the condition

$$X_\infty \to \mathcal{A}X_\infty\mathcal{A}^\dagger = X_\infty. \tag{29}$$

If the transformation does not satisfy (29) then the resulting moments of inertia always diverge. The rotation matrices can be written in terms of the four Euler angles

$$\mathcal{A} = \mathcal{B} = e^{-i\lambda_- \alpha_1/2} e^{-i\lambda_7 \alpha_2/2} e^{-i\lambda_- \alpha_3/2} e^{-i\lambda_+ \alpha_4/2} \tag{30}$$

where the generators λ_- and λ_+ are defined as

$$\lambda_- := -(\lambda_3 - \sqrt{3}\lambda_8)/2, \quad \lambda_+ := -(\lambda_3 + \lambda_8/\sqrt{3}).$$

The generators $\{\lambda_6, \lambda_7, \lambda_-\}$ satisfies the $SU(2)$ algebra and the λ_+ commutes with them.

The standard procedure proceeds by promoting the parameter $\mathcal{A}$ to the status of dynamical variables $\mathcal{A}(t)$. Then the dynamical ansatz adopted in collective coordinate quantization is

$$X(\boldsymbol{r}; \mathcal{A}(t)) = \mathcal{A}(t)X(\boldsymbol{r})\mathcal{A}^\dagger(t). \tag{31}$$

Plugging (31) into the Lagrangian (1) and assuming that $\mathcal{A}^\dagger\dot{\mathcal{A}} = -i\lambda_a\Omega^a/2$, where the symbol Ω^a represent the angular velocities, one gets the effective Lagrangian in the form

$$L_{\text{eff}} = -M_{\text{cl}} + \frac{1}{2} I_{ab}\Omega^a\Omega^b \quad a, b = 6, 7, -, + \tag{32}$$

where M_{cl} is the mass of static solution . By virtue of the axial symmetry of solution, we get the relation $I_{66} = I_{77}$ and the off-diagonal components vanish except $I_{+-} = I_{-+}$. Through the Legendre transformation, one can derive the quantum Hamiltonian which coincide with the Hamiltonian of a rotating symmetric top. The following set of eigenfunctions

$$\psi = D^l_{m,-m'}(\alpha_1, \alpha_2, \alpha_3)e^{iY\alpha_4} \tag{33}$$

diagolalize the Hamiltonian, where $D^l_{m,-m'}(\alpha_1, \alpha_2, \alpha_3)$ is the $SU(2)$ Wigner D-function. The energy spectrum is labelled by quantum numbers l, m and Y being eigenvalues of the Hamiltonian.

6. Fermions coupled with the principal variable

The fermion-vortex system was initially studied by [15]. It was shown that the normalizable zero-mode of the fermion appears due to the Index theorem. The Lagrangian of such a system has the form

$$\mathcal{L} = \bar{\psi} i\gamma^\mu \partial_\mu \psi - \frac{1}{2} ig\phi\bar{\psi}\psi^c + \frac{1}{2} ig\phi\bar{\psi}^c\psi \tag{34}$$

where $\mu = 0, 1, 2, 3$ and a superscript c denotes charge conjugation. For the Dirac spinors and matrices, they keep four-components. However, due to the cylindrical symmetry of the vortex, there occurs an effective reduction to a two-space, one-time theory. The authors succeeded to find the analytical zero mode solutions of the spinors. A convenient redefinition of the Lagrangian in terms of Majorana spinors is presented in [16]. The Majorana fermions are defined as

$$\chi = \frac{1}{\sqrt{2}}[\psi_L + (\psi_L)^c], \quad \omega = \frac{1}{\sqrt{2}}[\psi_R + (\psi_R)^c] \tag{35}$$

where ψ_L, ψ_R are the left-, and right-handed spinors. One can see that the fermion Lagrangian can be rewritten in the form

$$\mathcal{L} = \bar{\chi} i\gamma^\mu \partial_\mu \chi - gf\phi\bar{\chi}\exp[i(n\theta + \frac{1}{2}\pi)\gamma_5]\chi + \bar{\omega} i\gamma^\mu \partial_\mu \omega - gf\phi\bar{\omega}\exp[-i(n\theta + \frac{1}{2}\pi), \gamma_5]\omega \tag{36}$$

where $n-$vortices have the form $\phi(\rho, \theta) = f(\rho)\exp(in\theta)$. If the fermion is a right-(or left-) handed field, χ(or ω) is identically zero. We can introduce the Lagrangian in a similar way for the CP^N vortices

$$\mathcal{L}_3 = \bar{\psi} i\gamma^\mu \partial_\mu \psi - m\bar{\psi} X^{\gamma_5} \psi \tag{37}$$

where $X^{\gamma_5} \equiv P_R X + P_L X^\dagger$, and $P_{L/R} = \frac{1}{2}(1 \mp \gamma_5)$. The corresponding Dirac equation has exactly $Q_{\rm top}$ normalizable modes.

The coupling of vortices to fermionic field can be studied in another approach following [17, 18, 19] where the fermionic sigma model is considered. The Lagrangian of such a model has the form

$$\mathcal{L} = \bar{\psi} i\gamma^\alpha \partial_\alpha \psi - m\bar{\psi}\vec{\tau}\cdot\vec{n}\psi \tag{38}$$

where $\alpha = 0, 1, 2$, $\vec{\tau}$ is a standard Pauli matrices and $\vec{n}$ is the real triplet scalar field. There are the normalizable modes on the plane and, what calls attention, a derivative expansion of the determinant corresponding to (38) indicates that the model has a proper topological charge of the bosonic part. Clearly, it is worth to study the fermion-CP^N vortex in this context. We point out that the field $\vec{\tau}\cdot\vec{n}$ is Hermitian, and therefore it is natural to employ the CP^N Hermitian field $X_{\rm H}$. The Lagrangian is thus

$$\mathcal{L}_2 = \bar{\psi} i\gamma^\alpha \partial_\alpha \psi - m\bar{\psi} X_{\rm H} \psi, \tag{39}$$

The corresponding Dirac equation can be solved numerically. The quantum number of the spinor consists of a funny combination of the angular momentum, the isospin and the Gell-Mann matrices

$$\mathcal{K} = l_3 + \frac{\tau_3}{2} - \frac{n_1}{2}\left(\lambda_3 + \frac{1}{\sqrt{3}}\lambda_8\right) + \frac{n_2}{2}\left(\lambda_3 - \frac{1}{\sqrt{3}}\lambda_8\right) \tag{40}$$

where l_3 is a third component of the orbital angular momentum and the λ_i stand for the Gell-Mann matrices. All the spectra are thus labelled by $\mathcal{K}$. For instance they are of the form

$$\begin{aligned} &\mathcal{K} = -\frac{7}{6}, \; -\frac{1}{6}, \; \frac{5}{6}, \; \frac{11}{6}, \cdots \quad (n_1, n_2) = (3, 1), (2, -1), \cdots \\ &\mathcal{K} = -\frac{11}{6}, \; -\frac{5}{6}, \; \frac{1}{6}, \; \frac{7}{6}, \cdots \quad (n_1, n_2) = (4, 1), (3, -1), \cdots \end{aligned} \tag{41}$$

In Fig.2 we present the angular averaged densities $\sim \psi^\dagger\psi$ coupled with the vortex $(n_1, n_2) = (-3, -2), (3, -1)$. The number of states equals to the topological charge $Q_{\rm top} = 3, 4$.

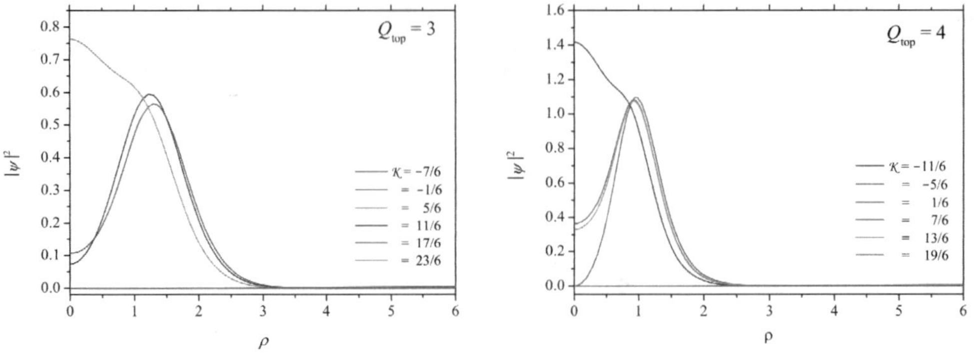

Figure 2. The angular averaged fermion densities coupled with the vortex with $(n_1, n_2) = (-3, -2)$ (left) and $(3, -1)$ (right).

7. Summary

We found solutions of the CP^N Skyrme-Faddeev type model. The model is written in terms of the principal variable which naturally describes the CP^N manifold. We observe that the model can be defined in several inequivalent ways which differ by the choice of X. For any such a choice the model has the same classical vortex solutions. However, the quantization and coupling to fermions fermions clearly depend on the choice of X (form of the prefactor $\mathcal{P}$). The existence of classical vortex solutions is very sensitive on the form of the potential. We proposed a systematic formulation of the potential and we manage to find solutions with many winding numbers. In the last sections, we gave some basic discussions of the collective quantization for the vortices and the coupling of fermions, i.e., fermion-CP^N vortex system.

Acknowledgments We express our gratitude to the conference organizers of ISQS-23 for kind accommodation and hospitality.

References

[1] T. H. R. Skyrme, Proc. Roy. Soc. Lond. A **247**, 260 (1958).
[2] T. H. R. Skyrme, Proc. Roy. Soc. Lond. A **260**, 127 (1961).
[3] T. H. R. Skyrme, Nucl. Phys. **31**, 556 (1962).
[4] G. S. Adkins, C. R. Nappi and E. Witten, Nucl. Phys. B **228**, 552 (1983).
[5] L. D. Faddeev and A. J. Niemi, Nature **387**, 58 (1997) [hep-th/9610193].
[6] L. A. Ferreira, J. Jaykka, N. Sawado and K. Toda, Phys. Rev. D **85**, 105006 (2012) [arXiv:1112.1085 [hep-th]].
[7] L. D. Faddeev and A. J. Niemi, Phys. Rev. Lett. **82**, 1624 (1999) [hep-th/9807069].
[8] H. Gies, Phys. Rev. D **63**, 125023 (2001) [hep-th/0102026].
[9] L. A. Ferreira, JHEP **0905**, 001 (2009) [arXiv:0809.4303 [hep-th]].
[10] L. A. Ferreira and P. Klimas, JHEP **1010**, 008 (2010) [arXiv:1007.1667 [hep-th]].
[11] E. Witten, Nucl. Phys. B **249**, 557 (1985).
[12] A. D'Adda, M. Luscher and P. Di Vecchia, Nucl. Phys. B **146**, 63 (1978).
[13] B. M. A. G. Piette, B. J. Schroers and W. J. Zakrzewski, Z. Phys. C **65**, 165 (1995) [hep-th/9406160].
[14] A. E. Kudryavtsev, B. M. A. Piette and W. J. Zakrzewski, Nonlinearity **11**, 783 (1998) [arXiv:hep-th/9709187].
[15] R. Jackiw and P. Rossi, Nucl. Phys. B **190**, 681 (1981).
[16] R. MacKenzie and A. M. Matheson, Phys. Lett. B **259**, 63 (1991).
[17] T.Jaroszewicz, Phys. Lett. B**146**, 337 (1984).
[18] M. Carena, S.Chaudhuri, C.E.M.Wagner, Phys. Rev. D**42**, 2120 (1990).

[19] Alexander G. Abanov, Paul B. Wiegmann, JHEP**10**, 030 (2001).

Symmetries of the Pais-Uhlenbeck oscillator on the Hamiltonian level

K. Andrzejewski

Department of Computer Science,
University of Łódź,
Pomorska 149/153, 90-236, Łódź, Poland

E-mail: k-andrzejewski@uni.lodz.pl

Abstract. In this survey we report on the symmetry properties of the Pais-Uhlenbeck oscillator. Especially, basing on the paper [Nucl. Phys. B 889 (2014) 333] we describe the Hamiltonian formalisms and its symmetry realizations on the Hamiltonian level.

1. Introduction

We present here a review of the results obtained in the paper [1] concerning the Hamiltonian approaches to the Pais-Uhlenbeck (PU) oscillator. We describe the three main Hamiltonian formalisms and relation between them as well as the realization of symmetries of the PU model in these frameworks.

We start with a very brief discussion of the ordinary mechanics: the free motion and the harmonic oscillator and their symmetries (the Schrödinger group). Next, we move, on the Lagrangian level, to their generalizations to the case of higher derivatives theory, especially to the PU model. The main part is devoted to the relations between the Hamiltonian formalisms for the PU model: the original due to Pais and Uhlenbeck, the Ostrogradski one and the third one based on the free higher derivatives theory. We also discuss the realization of the symmetries of the PU oscillator in those formalisms.

It is well known that the maximal symmetry group of the free Lagrangian

$$L = \frac{m}{2}\dot{\vec{x}}^2, \tag{1}$$

is the Schrödinger group

$$(SO(3,\mathbb{R}) \times SL(2,\mathbb{R})) \uplus A_6. \tag{2}$$

Its Lie algebra is generated by: $J^\alpha(SO(3,\mathbb{R}))$, $H, D, K(SL(2,\mathbb{R}))$ and $P^\beta, B^\gamma(A_6)$ $\alpha,\beta,\gamma = 1,2,3$; together with the following commutation rules

$$\begin{gathered}[H,D] = H, \quad [H,K] = 2D, \quad [D,K] = K, \\ [D,\vec{B}] = \frac{1}{2}\vec{B}, \quad [D,\vec{P}] = -\frac{1}{2}\vec{P}, \quad [K,\vec{P}] = \vec{B}, \quad [H,\vec{B}] = \vec{P}, \\ [J^\alpha, J^\beta] = \epsilon_{\alpha\beta\gamma}J^\gamma, \quad [J^\alpha, P^\beta] = \epsilon_{\alpha\beta\gamma}P^\gamma, \quad [J^\alpha, B^\beta] = \epsilon_{\alpha\beta\gamma}B^\gamma.\end{gathered} \tag{3}$$

Journal of Physics: Conference Series **670** (2016) 012003 doi:10.1088/1742-6596/670/1/012003

Moreover, it turns out that the maximal symmetry group of the harmonic oscillator Lagrangian

$$L = \frac{m}{2}\dot{\vec{x}}^2 - \frac{m\omega^2}{2}\vec{x}^2, \tag{4}$$

is isomorphic to the Schrödinger group. This isomorphism is related to the different choice of the generator of the time translation (Hamiltonian)

$$H \rightarrow H + \omega^2 K. \tag{5}$$

Consequently, for the harmonic oscillator the symmetry algebra enjoys the following commutation relations

$$\begin{aligned}
&[H, D] = H - 2\omega^2 K, \quad [H, K] = 2D, \quad [D, K] = K, \\
&[D, \vec{B}] = \frac{1}{2}\vec{B}, \quad [D, \vec{P}] = -\frac{1}{2}\vec{P}, \quad [K, \vec{P}] = \vec{B}, \\
&[H, \vec{P}] = -\omega^2 \vec{B}, \quad [H, \vec{B}] = \vec{P}, \\
&[J^\alpha, J^\beta] = \epsilon_{\alpha\beta\gamma} J^\gamma, \quad [J^\alpha, P^\beta] = \epsilon_{\alpha\beta\gamma} P^\gamma, \quad [J^\alpha, B^\beta] = \epsilon_{\alpha\beta\gamma} B^\gamma.
\end{aligned} \tag{6}$$

Although both algebra are isomorphic the change of the Hamiltonian alters the dynamics. Moreover, on the Hamiltonian level (and quantum level), the Schrödinger algebra is enlarged by the central charge M

$$[P^\alpha, B^\beta] = M\delta_{\alpha\beta}. \tag{7}$$

On the other hand, we observe the increasing interest in theories containing higher order time derivatives. Originally, these theories were proposed as a method for dealing with ultraviolet divergences [2] then they appeared in some physical models (see, e.g., [3]-[6]). The most basic theory involving higher time derivatives is the one defined by the following Lagrangian (generalization of the free motion)

$$\tilde{L} = \frac{1}{2}\left(\frac{d^n \vec{x}}{dt^n}\right)^2, \quad n = 1, 2, \ldots \tag{8}$$

The dynamical equation of motion, following from the action principle, takes the form

$$\frac{d^{2n}\vec{x}}{dt^{2n}} = 0. \tag{9}$$

It was shown (see [7]) that the maximal symmetry group of the Lagrangian (8) is a generalization of the Schrödinger group which is called the l-conformal Galilei group (see, e.g., [8]-[10]) with $l = n - \frac{1}{2}$; (for $n = 1$ we obtain the ordinary free dynamics and the Schrödinger symmetry). The structure of the Lie algebra of this group is as follows

$$\begin{aligned}
&[H, D] = H, \quad [H, K] = 2D, \quad [D, K] = K, \\
&[D, \vec{C}_p] = (p - \frac{2n-1}{2})\vec{C}_p, \quad [K, \vec{C}_p] = (p - 2n + 1)\vec{C}_{p+1}, \\
&[H, \vec{C}_p] = p\vec{C}_{p-1}, \\
&[J^{\alpha\beta}, J^{\gamma\delta}] = \delta^{\alpha\delta} J^{\gamma\beta} + \delta^{\alpha\gamma} J^{\beta\delta} + \delta^{\beta\gamma} J^{\delta\alpha} + \delta^{\beta\alpha} J^{\alpha\gamma}, \\
&[J^{\alpha\beta}, C_p^\gamma] = \delta^{\alpha\gamma} C_p^\beta - \delta^{\beta\gamma} C_p^\alpha.
\end{aligned} \tag{10}$$

There exists also a generalization of the harmonic oscillator to the case of higher derivatives. Such model was proposed by Pais and Uhlenbeck in their classical paper [12] (and then quite extensively studied – see among others, [13]-[25]). This model is described by the following Lagrangian,

$$L = -\frac{1}{2}\vec{x}\prod_{k=1}^{n}\left(\frac{d^2}{dt^2}+\omega_k^2\right)\vec{x}, \tag{11}$$

$n = 1, 2, \ldots$. which implies the following equation of motion

$$\prod_{k=1}^{n}\left(\frac{d^2}{dt^2}+\omega_k^2\right)\vec{x} = 0. \tag{12}$$

The general solution is of the form

$$\vec{x}(t) = \sum_{k=1}^{n}(\vec{\alpha}_k \cos\omega_k t + \vec{\beta}_k \sin\omega_k t), \tag{13}$$

where $\vec{\alpha}'s$ and $\vec{\beta}'s$ are some arbitrary constants.

In contrast to the ordinary harmonic oscillator the structure of the symmetry group of Lagrangian (11) depends on the values of $\omega's$ (see, [26]).

1. If the frequencies of oscillation are proportional to odd integers,

$$\omega_k = (2k-1)\omega, \quad \omega \neq 0, \quad k = 1, \ldots, n \tag{14}$$

then the maximal group of Noether symmetries of the system (11) is the l-conformal Newton-Hooke group with $l = n - \frac{1}{2}$. The Lie algebra of this group is spanned by $H, D, K, J^{\alpha\beta}$ and C_p^α, $\alpha, \beta = 1, 2, 3$, $p = 0, 1, \ldots, 2n-1$, satisfying the following commutation rules

$$\begin{aligned}
&[H, D] = H - 2\omega^2 K, \quad [H, K] = 2D, \quad [D, K] = K,\\
&[D, \vec{C}_p] = (p - \frac{2n-1}{2})\vec{C}_p, \quad [K, \vec{C}_p] = (p - 2n + 1)\vec{C}_{p+1},\\
&[H, \vec{C}_p] = p\vec{C}_{p-1} + (p - 2n + 1)\omega^2\vec{C}_{p+1},\\
&[J^{\alpha\beta}, J^{\gamma\delta}] = \delta^{\alpha\delta}J^{\gamma\beta} + \delta^{\alpha\gamma}J^{\beta\delta} + \delta^{\beta\gamma}J^{\delta\alpha} + \delta^{\beta\alpha}J^{\alpha\gamma},\\
&[J^{\alpha\beta}, C_p^\gamma] = \delta^{\alpha\gamma}C_p^\beta - \delta^{\beta\gamma}C_p^\alpha.
\end{aligned} \tag{15}$$

So, similarly to the ordinary case it is isomorphic to the symmetry group of the free higher derivative theory i.e., l-conformal Galilei group.

2. In the *generic* case the maximal symmetry group is simpler. Its Lie algebra consists of $H, J^{\alpha\beta}$ and $\vec{C}_k^\pm$, $k = 1, \ldots, n$ and the action of $J^{\alpha\beta}$ remains unchanged, only the commutation rules between H and $\vec{C}'s$ are modified

$$\begin{aligned}
&[H, \vec{C}_k^+] = -\omega_k\vec{C}_k^-,\\
&[H, \vec{C}_k^-] = \omega_k\vec{C}_k^+.
\end{aligned} \tag{16}$$

Moreover, both symmetry algebras posses the central extension:

– in the case of odd frequencies

$$[C_p^\alpha, C_q^\beta] = (-1)^p p! q! \delta_{\alpha\beta}\delta_{2n-1,p+q}, \tag{17}$$

– and in the generic case

$$[C_k^{+\alpha} C_j^{-\beta}] = \frac{\omega_k}{\rho_k}\delta_{kj}\delta^{\alpha\beta}, \tag{18}$$

which turns out to be relevant on the Hamiltonian level.

2. Hamiltonian approaches to PU oscillator

Much attention has been paid to the Hamiltonian formulations of the PU oscillator. There exists a few approaches.

1 Decomposition into the sum of the alternating harmonic oscillators proposed by Pais and Uhlenbeck in their original paper,

2 Ostrogradski Hamiltonian for theories with higher time derivatives,

3 An alternative one, applicable in the case of odd frequencies, which exhibits the group structure of the model.

Consequently, the question arises concerning the relations between them as well as the realization of the symmetry on the Hamiltonian level.

2.1. The Pais-Uhlenbeck approach

The first Hamiltonian approach was proposed by Pais and Uhlenbeck in Ref. [12]. The Hamiltonian turns into the sum of the Hamiltonians of decoupled harmonic oscillators with alternating sign. To show this we introduce new variables

$$\vec{x}_k = \Pi_k \vec{x}, \quad k = 1, \dots, n; \tag{19}$$

where Π_k is the projector operator:

$$\Pi_k = \sqrt{|\rho_k|} \prod_{\substack{i=1 \\ i \neq k}}^{n} \left(\frac{d^2}{dt^2} + \omega_i^2 \right), \tag{20}$$

and

$$\rho_k = \frac{1}{\prod\limits_{\substack{i=1 \\ i \neq k}}^{n} (\omega_i^2 - \omega_k^2)}, \quad k = 1, 2, \dots, n. \tag{21}$$

Then one finds

$$\vec{x} = \sum_{k=1}^{n} (-1)^{k-1} \sqrt{|\rho_k|} \vec{x}_k, \tag{22}$$

and

$$L = -\frac{1}{2} \sum_{k=1}^{n} (-1)^{k-1} \vec{x}_k \left(\frac{d^2}{dt^2} + \omega_k^2 \right) \vec{x}_k = \frac{1}{2} \sum_{k=1}^{n} (-1)^{k-1} (\dot{\vec{x}}_k^2 - \omega_k^2 \vec{x}_k^2) + t.d. \tag{23}$$

The corresponding Hamiltonian reads

$$H = \frac{1}{2} \sum_{k=1}^{n} (-1)^{k-1} (\vec{p}_k^2 + \omega_k^2 \vec{x}_k^2), \tag{24}$$

while the canonical equations of motion are of the form

$$\dot{\vec{x}}_k = (-1)^{k-1} \vec{p}_k, \quad \dot{\vec{p}}_k = (-1)^k \omega_k^2 \vec{x}_k. \tag{25}$$

Taking into account the form of the general solution (13) we see that

$$\begin{aligned} \vec{x}_k &= \frac{(-1)^{k-1}}{\sqrt{|\rho_k|}} (\vec{\alpha}_k \cos(\omega_k t) + \vec{\beta}_k \sin(\omega_k t)), \\ \vec{p}_k &= \frac{\omega_k}{\sqrt{|\rho_k|}} (\vec{\beta}_k \cos(\omega_k t) - \vec{\alpha}_k \sin(\omega_k t)). \end{aligned} \tag{26}$$

Consequently we have the correspondence between the set of solutions of the Lagrange equation (12) and the set of solutions of the canonical equations (25). Thus, we can transform the action of the group symmetry from the Lagrangian level to the Hamiltonian one As we will see soon the generators, obtained in this way, form the algebra which is the central extension of the symmetry algebra on the Lagrangian level.

In the generic case it is easy to find the form of the symmetry generators. First, we find that the infinitesimal action of $\vec{\mu}_k \vec{C}_k^+$ and $\vec{\nu}_k \vec{C}_k^-$, $k = 1, \ldots, n$, on the Lagrangian level, takes the form

$$\vec{x}'(t) = \vec{x}(t) + \sum_{k=1}^{n} (\vec{\mu}_k \cos \omega_k t + \vec{\nu}_k \sin \omega_k t). \tag{27}$$

Acting with Π_k and applying eq. (25) we find the infinitesimal action of $\vec{C}_k^{\pm}$ on the phase space; by virtue of

$$\delta(\cdot) = \epsilon\{\cdot, \text{Generator}\}, \tag{28}$$

we obtain the following generators:

$$\begin{aligned} \vec{C}_k^+ &= \frac{(-1)^{k-1}}{\sqrt{|\rho_k|}} \cos(\omega_k t)\vec{p}_k + \frac{\omega_k}{\sqrt{|\rho_k|}} \sin(\omega_k t)\vec{x}_k, \\ \vec{C}_k^- &= \frac{(-1)^{k-1}}{\sqrt{|\rho_k|}} \sin(\omega_k t)\vec{p}_k - \frac{\omega_k}{\sqrt{|\rho_k|}} \cos(\omega_k t)\vec{x}_k, \end{aligned} \tag{29}$$

which commute to the central charge according to eq. (18). Similarly, the angular momentum generators read

$$J^{\alpha\beta} = \sum_{k=1}^{n} (x_k^\alpha p_k^\beta - p_k^\alpha x_k^\beta). \tag{30}$$

Consequently, we obtain the desired centrally extended algebra (18).

In the odd case the symmetry group is reacher and, therefore, this case is much more interesting. Let us assume that the frequencies are odd , i.e., $\omega_k = (2k-1)\omega$, $k = 1, \ldots, n$. In this case the main point is that the numbers ρ_k can be explicitly computed; the final result reads

$$\rho_k = \frac{(-1)^{k-1}(2k-1)}{(4\omega^2)^{n-1}(n-k)!(n+k-1)!}, \quad k = 1, \ldots, n. \tag{31}$$

Next, the following Fourier expansion

$$\sin^p \omega t \cos^{2n-1-p} \omega t = \begin{cases} \sum\limits_{k=1}^{m} \gamma_{kp}^+ \cos(\omega_k t), & p \text{ - even}; \\ \sum\limits_{k=1}^{m} \gamma_{kp}^- \sin(\omega_k t), & p \text{ - odd}; \end{cases} \tag{32}$$

enables us to rewrite the infinitesimal action (27) in the equivalent form

$$\vec{x}'(t) = \vec{x}(t) + \frac{1}{\omega^p} \vec{\epsilon}_p \sin^p \omega t \cos^{2n-1-p} \omega t, \tag{33}$$

which gives suitable family of the generators $\vec{C}_p$, $p = 0, 1, 2, \ldots, 2n-1$ on the Lagrangian level, i.e., satisfying commutation rules of the l-conformal Newton-Hooke algebra (8). The Fourier coefficients (32) enjoy some useful identities (see [1] for more details) which are crucial for further consideration, Here, we write out two of them

$$\gamma_{kp}^+ = (-1)^{k-1} \gamma_{k,2n-1-p}^-, \qquad \beta_{pk}^+ = (-1)^{k-1} \beta_{2n-1-p,k}^-, \tag{34}$$

$$2p!(2n-1-p)!\beta^{\pm}_{pk} = 2^{2n-1}(n-k)!(n+k-1)!\gamma^{\pm}_{kp}, \tag{35}$$

where β^+, β^- are the inverse matrices to γ^+, γ^-.

In order to find the action of $\vec{C}_p$ in this form of Hamiltonian formalism, we use eqs. (32) together with (19) and (25), which yields

$$\vec{x}'_k = \vec{x}_k + \frac{(-1)^{k-1}\vec{\epsilon}_p}{\omega^p\sqrt{|\rho_k|}} \begin{cases} \gamma^+_{kp}\cos(2k-1)\omega t, & p \text{ - even;} \\ \gamma^-_{kp}\sin(2k-1)\omega t, & p \text{ - odd;} \end{cases} \tag{36}$$

$$\vec{p}'_k = \vec{p}_k + \frac{(2k-1)\omega\vec{\epsilon}_p}{\omega^p\sqrt{|\rho_k|}} \begin{cases} -\gamma^+_{kp}\sin(2k-1)\omega t, & p \text{ - even;} \\ \gamma^-_{kp}\cos(2k-1)\omega t, & p \text{ - odd.} \end{cases} \tag{37}$$

Using eq. (28) we derive the explicit expression for the generators $\vec{C}_p$

$$\vec{C}_p = \sum_{k=1}^{n} \frac{\omega^{-p}\gamma^+_{kp}}{\sqrt{|\rho_k|}} \Big((2k-1)\omega\sin((2k-1)\omega t)\vec{x}_k - (-1)^k\cos((2k-1)\omega t)\vec{p}_k\Big) \tag{38}$$

for p even, and

$$\vec{C}_p = \sum_{k=1}^{n} \frac{\omega^{-p}\gamma^-_{kp}}{\sqrt{|\rho_k|}} \Big((2k-1)\omega\cos((2k-1)\omega t)\vec{x}_k - (-1)^k\sin((2k-1)\omega t)\vec{p}_k\Big), \tag{39}$$

for p odd. Eqs. (38) and (39) can be inverted to yield $\vec{x}_k$ and $\vec{p}_k$ in terms of the generators $\vec{C}_p$

$$\begin{aligned} \vec{p}_k = {} & (-1)^{k-1}\sqrt{|\rho_k|}\cos((2k-1)\omega t)\sum_{p=0}^{2n-1}{}'' \beta^+_{pk}\omega^p\vec{C}_p \\ & + (-1)^{k-1}\sin((2k-1)\omega t)\sum_{p=0}^{2n-1}{}' \beta^-_{pk}\omega^p\vec{C}_p, \end{aligned} \tag{40}$$

$$\begin{aligned} \vec{x}_k = {} & \frac{\sqrt{|\rho_k|}}{(2k-1)\omega}\sin((2k-1)\omega t)\sum_{p=0}^{2n-1}{}'' \beta^+_{pk}\omega^p\vec{C}_p \\ & - \frac{\sqrt{|\rho_k|}}{(2k-1)\omega}\cos((2k-1)\omega t)\sum_{p=0}^{2n-1}{}' \beta^-_{pk}\omega^p\vec{C}_p, \end{aligned} \tag{41}$$

one and two primes $','' $ denote the sum over odd and even indices, respectively. Let us consider the dilatation generator. We find that the infinitesimal action of dilatation on coordinates is of the form ([26])

$$\vec{x}'(t) = \vec{x}(t) - \frac{\epsilon}{2\omega}\Big((2n-1)\omega\cos(2\omega t)\vec{x}(t) - \sin(2\omega t)\dot{\vec{x}}(t)\Big). \tag{42}$$

Substituting (22) and acting with Π_k we obtain the infinitesimal dilatation transformation on the phase space. According to eq. (28) infinitesimal canonical transformation is generated by

$$D = \frac{-1}{2\omega}\left(\omega A\cos(2\omega t) + B\sin(2\omega t)\right), \tag{43}$$

where

$$\begin{aligned}A &= -\sum_{k=1}^{n}\left(\sqrt{\left|\frac{\rho_{k-1}}{\rho_k}\right|}(n-k+1)\vec{x}_{k-1}+\sqrt{\left|\frac{\rho_{k+1}}{\rho_k}\right|}(n+k)\vec{x}_{k+1}\right)\vec{p}_k + n\vec{x}_1\vec{p}_1,\\ B &= -\sum_{k=1}^{n}(-1)^k\frac{n-k+1}{2k-3}\sqrt{\left|\frac{\rho_{k-1}}{\rho_k}\right|}(\vec{p}_k\vec{p}_{k-1}-(2k-1)(2k-3)\omega^2\vec{x}_k\vec{x}_{k-1})\\ &+\frac{1}{2}n(\omega^2\vec{x}_1^2-\vec{p}_1^2),\end{aligned} \tag{44}$$

and, by definition, $\vec{x}_0 = \vec{p}_0 = 0$. The meaning of the components A and B will become more clear later (see (63)). Similar calculations can be done for the conformal generator K. Namely, the infinitesimal conformal transformation, on the Lagrangian level, reads [26]

$$\vec{x}'(t) = \vec{x}(t) - \frac{\epsilon}{2\omega^2}\Big((2n-1)\omega\sin(2\omega t)\vec{x}(t) + (\cos(2\omega t)-1)\dot{\vec{x}}(t)\Big). \tag{45}$$

and the induced generator K on phase space takes form

$$K = \frac{1}{2\omega^2}\left(B\cos(2\omega t) - \omega A\sin(2\omega t) + H\right). \tag{46}$$

Finally, the angular momentum takes the same form as in the generic case

$$J^{\alpha\beta} = \sum_{k=1}^{n}(x_k^\alpha p_k^\beta - p_k^\alpha x_k^\beta). \tag{47}$$

It remains to verify that the obtained generators, do yield integrals of motion and define the centrally extended l-conformal Newton-Hooke algebra. First, we compute the commutators of $\vec{C}$'s. The only nontrivial case is $[C_p^\alpha, C_q^\beta]$ with p even and q odd. By virtue of eqs. (31), (34) and (35) we have

$$\begin{aligned}[C_p^\alpha, C_q^\beta] &= \frac{\omega(2\omega)^{2(n-1)}\delta^{\alpha\beta}}{\omega^{p+q}}\sum_{k=1}^{n}(-1)^{k-1}(n-k)!(n+k-1)!\gamma_{kp}^+\gamma_{kq}^-\\ &= \frac{p!(2n-1-p)!\omega^{2n-1}\delta^{\alpha\beta}}{\omega^{p+q}}\sum_{k=1}^{n}\beta_{2n-1-p,k}^-\gamma_{kq}^- =\\ &= p!q!\delta_{\alpha\beta}\delta_{2n-1,p+q}.\end{aligned} \tag{48}$$

In order to find the remaining commutators we observe that

$$\begin{aligned}[A,B] &= -2H,\\ [B,H] &= 2\omega^2 A,\\ [A,H] &= -2B.\end{aligned} \tag{49}$$

Finally, it is not hard to find the adjoint action of $H, D, K, J^{\alpha\beta}$ on $\vec{C}_p$ as well as the commutators involving angular momentum.

Having all the commutation rules one can verify that the obtained generators are constants of motion. This concludes the proof that for odd frequencies the symmetry generators form, on the Hamiltonian level, the centrally extended l-conformal Newton-Hooke algebra.

2.2. Ostrogradski approach

Now we use the Hamiltonian formalism proposed by Ostrogradski for higher derivatives theories [27]. To this end let us expand Lagrangian (11) in the sum of higher derivatives terms (here, $\vec{Q} = \vec{x}$)

$$L = -\frac{1}{2}\vec{Q}\prod_{k=1}^{n}\left(\frac{d^2}{dt^2} + \omega_k^2\right)\vec{Q} = \frac{1}{2}\sum_{k=0}^{n}(-1)^{k-1}\sigma_k(\vec{Q}^{(k)})^2 + t.d., \tag{50}$$

where

$$\sigma_k = \sum_{i_1<\ldots<i_{n-k}} \omega_{i_1}^2\cdots\omega_{i_{n-k}}^2, \quad k = 0,\ldots,n; \quad \sigma_n = 1. \tag{51}$$

Now, we introduce the Ostrogradski variables

$$\begin{aligned}\vec{Q}_k &= \vec{Q}^{(k-1)},\\ \vec{P}_k &= \sum_{j=0}^{n-k}\left(-\frac{d}{dt}\right)^j \frac{\partial L}{\partial\vec{Q}^{(k+j)}} = (-1)^{k-1}\sum_{j=k}^{n}\sigma_j\vec{Q}^{(2j-k)},\end{aligned} \tag{52}$$

for $k = 1,\ldots,n$. Then the Ostrogradski Hamiltonian takes the form

$$H = \frac{(-1)^{n-1}}{2}\vec{P}_n^2 + \sum_{k=2}^{n}\vec{P}_{k-1}\vec{Q}_k - \frac{1}{2}\sum_{k=1}^{n}(-1)^k\sigma_{k-1}\vec{Q}_k^2. \tag{53}$$

By virtue of eqs. (25) and (52), for $k = 1,\ldots,n$, we find that

$$\begin{aligned}\vec{Q}_k &= (-1)^{\frac{k-1}{2}}\sum_{j=1}^{n}\sqrt{|\rho_j|}(-1)^{j-1}\omega_j^{k-1}\vec{x}_j, \quad k-\text{odd},\\ \vec{Q}_k &= (-1)^{\frac{k}{2}-1}\sum_{j=1}^{n}\sqrt{|\rho_j|}\omega_j^{k-2}\vec{p}_j, \quad k-\text{even},\end{aligned} \tag{54}$$

and

$$\begin{aligned}\vec{P}_k &= (-1)^{\frac{k}{2}-1}\sum_{i=1}^{n}(-1)^{i-1}\sqrt{|\rho_i|}\left(\sum_{j=k}^{n}\sigma_j(-1)^j\omega_i^{2j-k}\right)\vec{x}_i, \quad k-\text{ even};\\ \vec{P}_k &= (-1)^{\frac{k-3}{2}}\sum_{i=1}^{n}\sqrt{|\rho_i|}\left(\sum_{j=k}^{n}\sigma_j(-1)^j\omega_i^{2j-k-1}\right)\vec{p}_i, \quad k-\text{ odd}.\end{aligned} \tag{55}$$

define a canonical transformation. The inverse transformation is of the form

$$\begin{aligned}\vec{x}_i &= {\sum_{k=1}^{n}}'(-1)^{\frac{k-3}{2}}\sum_{j=k}^{n}\sigma_j(-1)^j\omega_i^{2j-k-1}\sqrt{|\rho_i|}\vec{Q}_k + {\sum_{k=1}^{n}}''(-1)^{\frac{k}{2}}\sqrt{|\rho_i|}\omega_i^{k-2}\vec{P}_k,\\ \vec{p}_i &= {\sum_{k=1}^{n}}''(-1)^{\frac{k}{2}+i-1}\sum_{j=k}^{n}\sigma_j(-1)^j\omega_i^{2j-k}\sqrt{|\rho_i|}\vec{Q}_k + {\sum_{k=1}^{n}}'(-1)^{\frac{k+1}{2}+i}\sqrt{|\rho_i|}\omega_i^{k-1}\vec{P}_k.\end{aligned} \tag{56}$$

Now, we can find the symmetry generators in terms of Ostrogradski variables. First, we can check that the Hamiltonian (24) is transformed into the Ostrogradski one. Similarly, we check that the angular momentum transforms under (56) into Ostrogradski angular momentum

$$J^{\alpha\beta} = \sum_{k=1}^{n}(Q_k^\alpha P_k^\beta - Q_k^\beta P_k^\alpha). \tag{57}$$

Journal of Physics: Conference Series **670** (2016) 012003 doi:10.1088/1742-6596/670/1/012003

As far as the generators $\vec{C}'s$ are concerned we obtain the following expressions:
– in the case of generic frequencies

$$\begin{aligned}\vec{C}_k^+ &= \sum_{k=1}^{n}(\cos\omega_i t)^{(k-1)}\vec{P}_k - \sum_{k=1}^{n}\left((-1)^{k-1}\sum_{j=k}^{n}\sigma_j(\cos\omega_i t)^{2j-k}\right)\vec{Q}_k,\\ \vec{C}_k^- &= \sum_{k=1}^{n}(\sin\omega_i t)^{(k-1)}\vec{P}_k - \sum_{k=1}^{n}\left((-1)^{k-1}\sum_{j=k}^{n}\sigma_j(\sin\omega_i t)^{2j-k}\right)\vec{Q}_k,\end{aligned} \tag{58}$$

– and in the case of odd frequencies

$$\begin{aligned}\vec{C}_p = \frac{1}{\omega^p}\sum_{k=1}^{n}&\Big(\vec{P}_k(\sin^p\omega t\cos^{2n-1-p}\omega t)^{(k-1)}\\ &+(-1)^k\vec{Q}_k\sum_{j=k}^{n}\sigma_j(\sin^p\omega t\cos^{2n-1-p}\omega t)^{(2j-k)}\Big),\end{aligned} \tag{59}$$

which agrees with the definitions of the Ostrogradski canonical variables (52) and the action of $\vec{C}'s$ on Q (eqs. (27) and (33)). Similar reasoning can be done for the remaining two generators D and K in the odd case. Then, they become bilinear forms in the Ostrogradski variables; however the explicit form of the coefficients is less transparent.

2.3. Algebraic approach to the odd case

Since the free higher derivatives theory possesses also the $SL(2,R)$ symmetry it is instructive to consider, as in the case of the ordinary oscillator, the new Hamiltonian by adding the conformal generator to the Hamiltonian of the free higher derivatives theory. We show that the new Hamiltonian, obtained in this way, corresponds to the PU model with odd frequencies.

Denoting by $\vec{q}_m, \vec{\pi}_m$, $m=0,\ldots,n-1$ the phase space coordinates of the free higher derivatives theory, the generators $\tilde{H}, \tilde{K}$ (at $t=0$ in the free case) take the form (see, [28, 29])

$$\begin{aligned}\tilde{H} &= \frac{(-1)^{n+1}}{2}\vec{\pi}_{n-1}^2 - \sum_{m=1}^{n-1}\vec{q}_m\vec{\pi}_{m-1},\\ \tilde{K} &= (-1)^{n+1}\frac{n^2}{2}\vec{q}_{n-1}^2 + \sum_{m=0}^{n-2}(2n-1-m)(m+1)\vec{q}_m\vec{\pi}_{m+1}.\end{aligned} \tag{60}$$

Thus

$$\begin{aligned}H = \tilde{H}+\omega^2\tilde{K} &= \frac{(-1)^{n+1}}{2}\vec{\pi}_{n-1}^2 - \sum_{m=1}^{n-1}\vec{q}_m\vec{\pi}_{m-1}\\ &+(-1)^{n+1}\frac{n^2\omega^2}{2}\vec{q}_{n-1}^2 + \sum_{m=0}^{n-2}(2n-1-m)(m+1)\omega^2\vec{q}_m\vec{\pi}_{m+1}.\end{aligned} \tag{61}$$

Now, we show that eq. (61) gives the PU Hamiltonian and we find the remaining generators in terms of q_m, π_m. To this end let us define

$$
\begin{aligned}
\vec{x}_k &= (-1)^k \left(\sum_{m=0}^{n-1}{}'' \frac{\omega^{-m}}{m!\sqrt{|\rho_k|}} \gamma^+_{km} \vec{q}_m + \sum_{m=0}^{n-1}{}' \frac{m!\omega^m \sqrt{|\rho_k|}}{(2k-1)\omega} \beta^+_{2n-1-m,k} \vec{\pi}_m \right), \\
\vec{p}_k &= (-1)^k \left(\sum_{m=0}^{n-1}{}' \frac{\omega^{-m}(2k-1)\omega}{m!\sqrt{|\rho_k|}} \gamma^+_{k,2n-1-m} \vec{q}_m + \sum_{m=0}^{n-1}{}'' m!\omega^m \sqrt{|\rho_k|} \beta^+_{mk} \vec{\pi}_m \right),
\end{aligned}
\tag{62}
$$

for $k = 1, \ldots, n$. By direct calculations one can check that eqs. (62) define a canonical transformation and that the PU Hamiltonian (24) (with odd frequencies) is transformed into the one given by eq. (61). The remaining generators can be also transformed

$$
\begin{aligned}
A &= -2\tilde{D}, \\
B &= -\tilde{H} + \omega^2 \tilde{K},
\end{aligned}
\tag{63}
$$

and, consequently, we obtain an interpretation of A and B. Using eqs. (63) one checks that H, D, K take the form

$$
\begin{aligned}
H &= \tilde{H} + \omega^2 \tilde{K}, \\
D &= \tilde{D} \cos 2\omega t + \frac{1}{2\omega} (\tilde{H} - \omega^2 \tilde{K}) \sin 2\omega t, \\
K &= \frac{1}{2} (1 + \cos 2\omega t) \tilde{K} + \frac{1}{2\omega^2} (1 - \cos 2\omega t) \tilde{H} + \frac{\sin 2\omega t}{\omega} \tilde{D}.
\end{aligned}
\tag{64}
$$

The angular momentum reads

$$
J^{\alpha\beta} = \sum_{m=0}^{n-1} (q^\alpha_m \pi^\beta_m - q^\beta_m \pi^\alpha_m),
\tag{65}
$$

i.e., takes the same form as the one for the free theory. Finally,

$$
\vec{C}_p = \sum_{r=0}^{n-1} \left((-1)^{r-1} r! a_{pr}(t) \vec{\pi}_r + (2n-1-r)! a_{p,2n-1-r}(t) \vec{q}_r \right),
\tag{66}
$$

where

$$
a_{pr}(t) = \omega^{r-p} \sum_{k=1}^{n} \beta^\pm_{rk} \gamma^\pm_{kp} \cos{(2k-1)\omega t},
\tag{67}
$$

upper (lower) sign corresponds to p, r even (odd); and

$$
a_{pr}(t) = \mp \omega^{r-p} \sum_{k=1}^{n} \beta^\mp_{rk} \gamma^\pm_{kp} \sin{(2k-1)\omega t},
\tag{68}
$$

upper (lower) sign corresponds to p even and r odd (p odd and r even).
So, we expressed the PU symmetry generators in terms of the ones for free theory.

Summarizing, we focused on the Hamiltonian approaches to the PU model and its symmetries. We described the form of the symmetry generators in the original Pais and Uhlenbeck approach. We have shown that the resulting algebra is the central extension of the one obtained on the

Lagrangian level: in the case of odd frequencies it is the centrally extended l-conformal Newton-Hooke algebra. Next, we wrote out the canonical transformation leading from the decoupled oscillators approach to the Ostrogradski Hamiltonian formalism, which enables us to find the symmetry generators in the Ostrogradski approach. Let us note that both the above approaches, do not distinguish the odd frequencies. A deeper insight is attained by noting that for odd frequencies an alternative Hamiltonian formalism can be constructed: to the Hamiltonian of the free higher derivatives theory we added the conformal generator. As a result, we obtain the new Hamiltonian, which turns out to be related, by a canonical transformation, to the PU one with odd frequencies. Then the symmetry generators can be directly expressed in terms of the variables (integrals of motion) of the free higher derivatives theory.

Acknowledgments

The author would like to thank Joanna Gonera, Piotr Kosiński and Paweł Maślanka for helpful comments and remarks. This work is supported from the earmarked subsidies of MNiSzW for Young Scientists.

[1] Andrzejewski K 2014 *Nucl. Phys.* B **889** 333
[2] Thiring W 1950 *Phys. Rev.* **77** 570
[3] Stelle K 1977 *Phys. Rev.* D **16** 953
[4] Mironov A, Morozov A 2008 *Int. J. Mod. Phys.* A **23** 4677
[5] Galakhov D 2008 *JETP Lett.* **87** 452
[6] Douglas M, Nekrasov N 2001 *Rev. Mod. Phys.* **73** 977
[7] Andrzejewski K, Gonera J 2013 *Phys. Lett.* B **721** 319
[8] Negro J, del Olmo M, Rodriguez-Marco A 1997 *J. Math. Phys.* **38** 3810
[9] Galajinsky A, Masterov I 2011 *Phys. Lett. B* **702** 265
[10] Duval C, Horvathy P 2011 *J. Phys. A* **44** 335203
[11] Fedoruk S, Ivanov E, Lukierski J 2011 *Phys. Rev.* D **83** 085013
[12] Pais A, Uhlenbeck G 1950 *Phys. Rev.* **79** 145
[13] Hawking S, Hertog T 2002 *Phys. Rev.* D **65** 103515
[14] Andrzejewski K, Gonera J, Maślanka P 2011 *Prog. Theor. Phys.* **125** 247
[15] Sarioglu O, Tekin B 2006 *Class. Quant. Grav.* **23** 7541
[16] Nesterenko V 2007 *Phys. Rev.* D **75** 087703
[17] Smilga A 2009 *SIGMA* **5** 017
[18] Andrzejewski K, Gonera J, Machalski P, Bolonek-Lasoń K 2012 *Phys. Lett. B* **706** 427
[19] Bagchi B, Choudhury A, Guha P 2013 *Mod. Phys. Lett.* A **28** 1375001
[20] Jiménez J, Di Dio E, Durrer R 2013 *JHEP* **1304** 030
[21] Pavšič M 2013 *Phys. Rev.* D **87** 107502
[22] Andrzejewski K 2014 *Phys. Lett.* B **738** 405
[23] Masterov I 2014 *J. Math. Phys.* **55** 102901
[24] Aizawa N, Kuznetsova Z, Toppan F 2015 *J. Math. Phys.* **56** 031701
[25] Galajinsky A, Masterov I 2015 *Nucl. Phys.* B **896** 244
[26] Andrzejewski K, Galajinsky A, Gonera J, Masterov I 2014 *Nucl. Phys.* B **885** 150
[27] Ostrogradski M 1850 *Mem. Acad. St. Petersburg* **4** 385
[28] Andrzejewski K, Gonera J, Maślanka P 2012 *Phys. Rev.* D **86** 065009
[29] Gomis J, Kamimura K 2012 *Phys. Rev.* D **85** 045023

Journal of Physics: Conference Series **670** (2016) 012004 doi:10.1088/1742-6596/670/1/012004

Functions of positive type on phase space, between classical and quantum, and beyond

Paolo Aniello

Dipartimento di Fisica dell'Università di Napoli "Federico II" and INFN – Sezione di Napoli, Complesso Universitario di Monte S. Angelo, via Cintia, I-80126 Napoli, Italy

E-mail: paolo.aniello@na.infn.it

Abstract. Functions of positive type on locally compact abelian groups can be defined as positive functionals on group algebras, and play a remarkable role in probability theory and in classical statistical mechanics. By Bochner's celebrated theorem, indeed, they are Fourier-Stieltjes transforms of finite positive measures. Hence, a properly normalized nonzero function of positive type on (the group of translations on) phase space provides a realization of a classical state, so it may be called a function of *classical* positive type. A similar result holds in the quantum setting as well, where a generalized kind of functions of positive type on phase space — the so-called functions of *quantum* positive type — are related, via the Fourier-Plancherel transform, to the Wigner quasi-probability distributions. In this paper, we will argue that, as in the classical setting, the notion of function of quantum positive type is of a group-theoretical nature. Exploiting an interesting interplay between functions of classical and quantum positive type, we will then provide an interesting characterization of a class of semi-groups of operators that describe the evolution of certain open quantum systems which are of interest in quantum information science. Finally, a suitable extension of this framework to generalized phase spaces that are relevant for current applications will be proposed.

1. Introduction

In the standard Hilbert space formulation of quantum mechanics, the physical states are realized as normalized, positive trace class operators (the so-called density or statistical operators) [1]. In the — more abstract — algebraic formulation, on the other hand, states can be defined as positive linear functionals on a C*-algebra of observables which, by the celebrated Gelfand-Naimark theorem, will be isomorphic to a C*-algebra of bounded operators in a Hilbert space [2,3]. This more abstract approach to physical states allows one, among other things, to encompass in a unique framework both the quantum and the *classical* case (i.e., classical statistical mechanics), where the complex Radon phase-space measures provide a realization of the bounded functionals on the commutative algebra of continuous complex functions on phase space that vanish at infinity [4].

More precisely, in this context the classical states are the probability measures on phase space, while the classical observables are the real functions of the relevant algebra (the selfadjoint part of the $*$-algebra of observables). Probability measures, being set functions, are often rather awkward objects to deal with, so it may be convenient to replace them with their Fourier — i.e., Fourier-Stieltjes — transforms, which are 'ordinary' bounded continuous functions. Actually, this is a standard technique in probability theory, where the Fourier-Stieltjes transforms of

Journal of Physics: Conference Series **670** (2016) 012004 doi:10.1088/1742-6596/670/1/012004

probability measures are usually called *characteristic functions* [5].

The Fourier transform is in a natural way a group-theoretical object [6, 7] — clearly, in the case at hand, the group of phase-space translations is involved — hence it is not surprising that characteristic functions on phase space admit an *intrinsic* group-theoretical characterization. Indeed, by Bochner's theorem, they coincide with the properly normalized, nonzero (i.e., not identically zero) continuous *functions of positive type* on phase space [6, 7]. More generally, functions of positive type on locally compact groups play a remarkable role in abstract harmonic analysis [7, 8]. They can be defined as positive functionals on a suitable group algebra.

The circle of ideas between classical and quantum mechanics outlined above becomes even more intriguing by switching to the phase-space formulation of quantum mechanics [9–20]. In this approach, the physical states are implemented by *quasi*-probability distributions — the Wigner functions — and the Fourier-Plancherel transform of these distributions — the so-called *quantum characteristic functions* — turn out to be suitably normalized positive functionals on a certain noncommutative $*$-algebra of square integrable functions [21]. We will call the positive functionals on this algebra *functions of quantum positive type*.

It is known that quantum mechanics on phase space has an underlying group-theoretical framework [15, 19–21], based on the theory of square integrable representations [22–24]. This theoretical framework is actually at the root of the notion of function of quantum positive type [21], and the link — via the Fourier-Plancherel transform — between functions of this kind and quantum quasi-distributions can be regarded as a quantum version of Bochner's theorem [21, 25–29].

However, beside these striking analogies between the classical and the quantum setting, it is worth noting a few significant differences. Indeed, on the one hand, in the classical case there is a one-to-one correspondence between the physical states and the normalized nonzero functions of positive type, and it turns out that these functions are continuous (modulo unessential modifications on sets of zero measure). On the other hand, a nonzero function of *quantum* positive type may not be a quantum characteristic function, not even up to normalization, although the reverse implication is always true; otherwise stated, not every nonzero function of quantum positive type represents, up to normalization, a quantum state. One can show that *continuity* is precisely the condition that allows one to select, up to a positive constant factor, the quantum characteristic functions among all nonzero functions of quantum positive type. Moreover, whereas in the classical case functions of positive type are normalized in the sense of functionals, in the quantum case a different kind of normalization is associated with the physical states and normalization in the sense of functionals has the meaning of the square root of *purity* [30]. Therefore, the two normalization criteria coincide for *pure* states only.

The peculiar properties of functions of positive type allow one to note an interesting interplay between classical and quantum states. In fact, probability measures on phase space are a semigroup, with respect to the convolution product, and, equivalently, the corresponding functions of positive type form a semigroup with respect to the ordinary point-wise multiplication. One can prove, moreover, that taking the point-wise product of a function of positive type on phase space by a continuous function of *quantum* positive type one gets a function of the *latter* type. This result implies that one can obtain a semigroup of operators, acting on functions of quantum positive type (more precisely, on a Banach space generated by these functions) — a so-called *classical-quantum semigroup* [21, 31] — out of a one-parameter multiplication semigroup of positive definite functions.

A straightforward way to unveil the physical meaning of a classical-quantum semigroup is to 'quantize' it (a group-covariant quantization procedure strictly related to quantization *à la* Weyl is involved), i.e., to pass from phase-space functions to Hilbert space operators. This procedure yields a new semigroup of operators acting in the Banach space of trace class operators; more precisely, a *quantum dynamical semigroup*. As is well known, semigroups of operators of this kind describe the evolution of open quantum systems [32, 33]. The quantized version of a classical-quantum semigroup belongs, more specifically, to the class of *twirling semigroups* [34–37]. Because of the role that they play in quantum information, this type of twirling semigroups can be called *classical-noise semigroups* [21, 31].

The paper is organized as follows. In sect. 2, we will introduce the notions of function of classical and quantum positive type, and we will briefly comment about their physical meaning. In sect. 3, the interplay between these two kinds of functions of positive type will be analyzed. We will first show — see subsect. 3.1 — that with every one-parameter multiplication semigroup of functions of positive type is associated in a natural way a classical-quantum semigroup. Then, in subsect. 3.2, we will argue that group-covariant quantization/dequantization establishes a precise relation between functions of quantum positive type and Hilbert space operators. At that point, we will be able to show that a classical-quantum semigroup is mapped, via quantization, to a suitable quantum dynamical semigroup; see subsect. 3.3. Finally, in sect. 4, possible extensions of these results will be proposed.

2. Functions of positive type on phase space

The notion of function of positive type is group-theoretical; hence, we will consider, at first, the case of an abstract locally compact group G. Let $\mathrm{L}^1(G)$ be the Banach space of complex (i.e., $\mathbb{C}$-valued) functions on G, integrable with respect to a left-invariant Haar measure ν_G. $\mathrm{L}^1(G)$ becomes a Banach $*$-algebra if endowed with the convolution product $(\cdot) \circledcirc (\cdot)$,

$$(\varphi_1 \circledcirc \varphi_2)(g) := \int_G \varphi_1(h)\varphi_2(h^{-1}g)\, \mathrm{d}\nu_G(h), \tag{1}$$

and with the involution $\mathsf{I}\colon \varphi \mapsto \varphi^*$,

$$\varphi^*(g) := \Delta_G(g^{-1})\,\overline{\varphi(g^{-1})}, \tag{2}$$

where Δ_G denotes the modular function on G. Functions of positive type can be defined as (linear) functionals [7, 21].

Definition 1 *A* function of positive type *on G is a positive bounded functional on the Banach algebra $\big(\mathrm{L}^1(G), \circledcirc, \mathsf{I}\big)$, implemented by a function belonging to the Banach space of ν_G-essentially bounded functions $\mathrm{L}^\infty(G)$. Otherwise stated, a complex function χ on G is said to be of positive type if it belongs to $\mathrm{L}^\infty(G)$ and, for every $\varphi \in \mathrm{L}^1(G)$, satisfies the inequality*

$$\int_G \chi(g)\,(\varphi^* \circledcirc \varphi)(g)\, \mathrm{d}\nu_G(g) \ge 0. \tag{3}$$

It turns out that a function of positive type $\chi \in \mathrm{L}^\infty(G)$ agrees ν_G-almost everywhere with a continuous function (it goes without saying that functions agreeing ν_G-almost everywhere are identified) and

$$\|\chi\|_\infty = \chi(e), \tag{4}$$

where $\|\chi\|_\infty := \nu_G\text{-ess sup}_{g\in G}\,|\chi(g)|$ and $\chi(e)$ — with a slight abuse of notation — denotes the value at the identity $e \in G$ of the 'continuous version' of χ.

It can be shown, moreover, that for a bounded continuous function $\chi\colon G \to \mathbb{C}$ the following properties are equivalent:

P1) χ is of positive type;

P2) χ satisfies the inequality (3), for all $\varphi \in \mathsf{C_c}(G)$ (with $\mathsf{C_c}(G)$ denoting the linear space of continuous complex functions on G, with a compact support);

P3) χ satisfies the inequality

$$\int_G \int_G \chi(g^{-1}h)\,\overline{\varphi(g)}\,\varphi(h)\;\mathrm{d}\nu_G(g)\,\mathrm{d}\nu_G(h) \geq 0, \quad \forall \varphi \in \mathsf{C}_\mathsf{c}(G); \tag{5}$$

P4) χ is a *positive definite function*, i.e., it is such that

$$\sum_{j,k} \chi(g_j^{-1} g_k)\, \overline{c_j}\, c_k \geq 0, \tag{6}$$

for every finite subset $\{g_1, \ldots, g_m\}$ of G and arbitrary complex numbers $c_1, \ldots, c_m$.

We will now consider the case where the group G is *abelian* (e.g., a vector group). The Pontryagin dual of G [6, 7] — the group of all irreducible (necessarily one-dimensional) unitary representations of G, the unitary characters — will be denoted by $\widehat{G}$, and $\mathrm{CM}(\widehat{G})$ will indicate the Banach space of complex Radon measures on $\widehat{G}$. In this case, by Bochner's theorem, the previous list of equivalent properties will admit a further entry:

P5) χ is the Fourier-Stieltjes transform of a positive measure $\mu \in \mathrm{CM}(\widehat{G})$.

If the positive measure $\mu \in \mathrm{CM}(\widehat{G})$ is, in particular, a probability measure, the normalization condition $\mu(\widehat{G}) = 1$ translates into the constraint

$$\chi(0) = 1, \tag{7}$$

for the associated function of positive type (here 0 denotes the identity of the abelian group G). Clearly, by relation (4), the latter condition is precisely the normalization of χ as a functional on $\mathrm{L}^1(G)$. In probability theory, χ is often called the *characteristic function* of μ [5].

We will now focus on the case where G is the group of translations on the $(n+n)$-dimensional phase space — namely, the vector group $\mathbb{R}^n \times \mathbb{R}^n$. In this case, we will obviously identify the dual group $\widehat{G}$ with G itself, and it will be convenient to use the *symplectic* Fourier transform [13], instead of the ordinary one.

The physical relevance of functions of positive type is related to the fact that probability measures on phase space provide a realization of physical *states* in classical statistical mechanics [3]. In fact, a classical state can be defined as a normalized positive functional on the commutative C*-algebra of (classical) observables. By Gelfand theory — see, e.g., [38] — such an algebra is isomorphic to an algebra of continuous functions vanishing at infinity. Therefore, a natural choice is $\mathsf{C}_0(\mathbb{R}^n \times \mathbb{R}^n)$, the Banach space of continuous complex functions on $\mathbb{R}^n \times \mathbb{R}^n$ that vanish at infinity, endowed with the point-wise product and with the ordinary complex conjugation (involution). Clearly, the selfadjoint part $\mathsf{C}_0(\mathbb{R}^n \times \mathbb{R}^n; \mathbb{R})$ of the whole algebra — i.e., the real functions — contains the *true* observables of the theory. The dual space of $\mathsf{C}_0(\mathbb{R}^n \times \mathbb{R}^n)$ is the Banach space $\mathrm{CM}(\mathbb{R}^n \times \mathbb{R}^n)$ of complex Radon measures on $\mathbb{R}^n \times \mathbb{R}^n$ and, in particular, the convex set of physical states consists of the probability measures on $\mathbb{R}^n \times \mathbb{R}^n$. Moreover, the expectation value of an observable $f \in \mathsf{C}_0(\mathbb{R}^n \times \mathbb{R}^n; \mathbb{R})$ in the state $\mu \in \mathrm{CM}(\mathbb{R}^n \times \mathbb{R}^n)$ is given by the 'pairing'

$$\langle f \rangle_\mu = \int_{\mathbb{R}^n \times \mathbb{R}^n} f(q,p)\,\mathrm{d}\mu(q,p). \tag{8}$$

As already mentioned, it is often convenient to replace a probability measure $\mu \in \mathrm{CM}(\mathbb{R}^n \times \mathbb{R}^n)$ (a classical state) with its symplectic Fourier-Stieltjes transform $\widetilde{\mu}$,

$$\widetilde{\mu}(q,p) := \int_{\mathbb{R}^n \times \mathbb{R}^n} \mathrm{e}^{\mathrm{i}\omega((q,p),(q',p'))}\,\mathrm{d}\mu(q',p') \tag{9}$$

— with ω denoting the standard symplectic form on $\mathbb{R}^n \times \mathbb{R}^n$:

$$\omega((q,p),(q',p')) := (q,p)^{\mathsf{T}}\,\Omega\,(q',p') = q\cdot p' - p\cdot q', \quad \Omega = \begin{pmatrix} 0_n & I_n \\ -I_n & 0_n \end{pmatrix} \tag{10}$$

— which is a continuous function of positive type on phase space.

Remark 1 As is well known [7], given a unitary representation V of a locally compact group G and a vector Ψ in the Hilbert space $\mathcal{H}$ where V acts, the map

$$G \ni g \mapsto \langle \Psi, V(g)\Psi\rangle \tag{11}$$

is a function of positive type on G. Conversely, every function of positive type on G is of the form (11), for some unitary representation V and some vector Ψ. Keeping in mind this fact and considering the case where $G = \mathbb{R}^n \times \mathbb{R}^n$, it is not surprising that formula (9) admits the following interesting interpretation. Let $\mathsf{p}\colon \mathsf{B} \to \mathbb{R}^n \times \mathbb{R}^n$ be the constant Hilbert bundle over $\mathbb{R}^n \times \mathbb{R}^n$ such that $\mathsf{p}^{-1}(q,p) = \mathbb{C} \equiv \mathcal{H}_{q,p}$, for all $(q,p) \in \mathbb{R}^n \times \mathbb{R}^n$. Of course, the direct integral Hilbert space associated with this Hilbert bundle and with a probability measure μ on $\mathbb{R}^n \times \mathbb{R}^n$ is given by

$$\mathcal{H} := \int_{\mathbb{R}^n\times\mathbb{R}^n}^{\oplus} \mathcal{H}_{q,p}\,\mathrm{d}\mu(q,p) = \mathrm{L}^2(\mathbb{R}^n \times \mathbb{R}^n, \mu). \tag{12}$$

Let V be the unitary representation of $\mathbb{R}^n \times \mathbb{R}^n$ in $\mathcal{H}$ defined as a *direct integral* [39]

$$V := \int_{\mathbb{R}^n\times\mathbb{R}^n}^{\oplus} V_{q',p'}\,\mathrm{d}\mu(q',p') \tag{13}$$

of unitary characters

$$V_{q',p'}(q,p)\,z = \mathrm{e}^{\mathrm{i}(q\cdot p' - p\cdot q')}\,z, \quad z \in \mathbb{C} \equiv \mathcal{H}_{q',p'}. \tag{14}$$

We then have that $(V(q,p)\Phi)(q',p') = \exp(\mathrm{i}(q\cdot p' - p\cdot q'))\,\Phi(q',p')$ and

$$\widetilde{\mu}(q,p) = \langle \Psi, V(q,p)\Psi\rangle, \tag{15}$$

where the matrix element on the rhs of (15) is computed with respect to the constant function $\Psi \equiv 1$ in $\mathrm{L}^2(\mathbb{R}^n \times \mathbb{R}^n, \mu)$.

At this point, before going back to the role of functions of positive type in classical statistical mechanics, it is worth giving two simple examples. Consider first the *Dirac measure* δ_{q_0,p_0} concentrated at $(q_0,p_0) \in \mathbb{R}^n \times \mathbb{R}^n$. This is a pure — namely, an extremal — classical state. It is mapped by the symplectic Fourier-Stieltjes transform into the function of positive type $(q,p) \mapsto \mathrm{e}^{\mathrm{i}(q\cdot p_0 - p\cdot q_0)}$. Actually, *every* pure state on the C*-algebra $\mathsf{C}_0(\mathbb{R}^n \times \mathbb{R}^n)$ is a Dirac measure, and the associated function of positive type is characterized by the property that the corresponding unitary representation V — see the rhs of (15) — is irreducible; hence, a unitary character of the form (14).

Consider next a *Gaussian measure* γ on $\mathbb{R}^n \times \mathbb{R}^n$,

$$\mathrm{d}\gamma(q,p) = \pi^{-n}\det(\mathtt{M})^{-1/2}\exp\!\big(-\,(q-q_0,p-p_0)^{\mathsf{T}}\mathtt{M}^{-1}(q-q_0,p-p_0)\big)\,\mathrm{d}^n q\,\mathrm{d}^n p, \tag{16}$$

where $\mathtt{M}$ is a positive definite, symmetric, $2n \times 2n$ real matrix, the so-called *covariance matrix*. The symplectic Fourier-Stieltjes transform of this probability measure is the function of positive type

$$(q,p) \mapsto \widetilde{\gamma}(q,p) = \exp\!\big(-\,(q,p)^{\mathsf{T}}\,\widetilde{\mathtt{M}}\,(q,p)/4 + \mathrm{i}\,(q,p)^{\mathsf{T}}\,\Omega\,(q_0,p_0)\big), \tag{17}$$

where $\widetilde{\mathtt{M}} = -\Omega\,\mathtt{M}\,\Omega$.

Now, to complete our reasoning recall that the symplectic Fourier transform maps $\mathrm{L}^1(\mathbb{R}^n\times\mathbb{R}^n)$ into a dense linear subspace $\mathcal{F}\mathrm{L}^1(\mathbb{R}^n\times\mathbb{R}^n)$ of the Banach space $\mathsf{C}_0(\mathbb{R}^n\times\mathbb{R}^n)$:

$$\varphi \mapsto \breve{\varphi}, \quad \breve{\varphi}(q,p) := \int_{\mathbb{R}^n\times\mathbb{R}^n} \varphi(q',p')\, \mathrm{e}^{\mathrm{i}(q'\cdot p - p'\cdot q)}\, \mathrm{d}^n q'\, \mathrm{d}^n p'. \tag{18}$$

Thus, for every observable $f \in \mathcal{F}\mathrm{L}^1(\mathbb{R}^n\times\mathbb{R}^n)$ the expectation value $\langle f\rangle_\mu$ admits the following expression:

$$\langle f\rangle_\mu = \int_{\mathbb{R}^n\times\mathbb{R}^n} \varphi(q,p)\, \chi(q,p)\, \mathrm{d}^n q\, \mathrm{d}^n p =: \langle \varphi, \chi\rangle\,, \tag{19}$$

with $\breve{\varphi} = f$ and $\chi = \widetilde{\mu}$. Namely, the expectation value $\langle f\rangle_\mu$ can be expressed as a pairing $\langle\varphi,\chi\rangle$ between $\varphi \in \mathrm{L}^1(\mathbb{R}^n\times\mathbb{R}^n)$ and the function of positive type $\chi \in \mathrm{L}^\infty(\mathbb{R}^n\times\mathbb{R}^n)$, representing respectively an observable and a state.

Remark 2 The condition that $f \in \mathcal{F}\mathrm{L}^1(\mathbb{R}^n\times\mathbb{R}^n)$ be an observable — i.e., $f = \overline{f}$ — translates into the condition that φ be invariant with respect to the involution of the algebra $\mathrm{L}^1(\mathbb{R}^n\times\mathbb{R}^n)$: $\varphi = \mathsf{I}\,\varphi$; namely, that

$$\varphi(q,p) = \overline{\varphi(-q,-p)}. \tag{20}$$

The main ideas underlying the picture sketched above can be condensed as follows:

(i) A function of positive type can be defined, for every locally compact group G, as a positive functional on the group algebra $\mathrm{L}^1(G)$, and gives rise to various equivalent characterizations. In particular, in the case of an abelian group, a nonzero function of positive type can be characterized, up to normalization, as the Fourier-Stieltjes transform of a probability measure.

(ii) Therefore, there is a link between the notion of function of positive type and classical statistical mechanics, where the physical states are realized as probability measures on phase space, while the observables form the selfadjoint part of the C*-algebra $\mathsf{C}_0(\mathbb{R}^n\times\mathbb{R}^n)$.

(iii) Adopting a *characteristic function approach*, one can replace probability measures with their Fourier-Stieltjes transforms, i.e., with the corresponding normalized functions of positive type. The standard algebra of classical observables is then 'densely' replaced with the Banach $*$-algebra $\mathrm{L}^1(\mathbb{R}^n\times\mathbb{R}^n)$.

(iv) It is worth remarking once again that the *positive* functionals on the algebra $\mathrm{L}^1(\mathbb{R}^n\times\mathbb{R}^n)$ are precisely the Fourier-Stieltjes transform of the positive functionals on the C*-algebra $\mathsf{C}_0(\mathbb{R}^n\times\mathbb{R}^n)$. Hence, in the characteristic function approach $\mathrm{L}^1(\mathbb{R}^n\times\mathbb{R}^n)$ may be thought of, *directly*, as the algebra of physical observables and, in principle, there is no point in considering a larger algebra (say, the the universal C*-completion of $\mathrm{L}^1(\mathbb{R}^n\times\mathbb{R}^n)$).

The functions of positive type on phase space form a convex cone $\mathsf{P}_n \equiv \mathsf{P}(\mathbb{R}^n\times\mathbb{R}^n)$ in $\mathrm{L}^\infty(\mathbb{R}^n\times\mathbb{R}^n)$, and the characteristic functions — i.e., the normalized (nonzero) functions of positive type — form a convex subset $\breve{\mathsf{P}}_n$ of P_n. For reasons that will be clear soon, the elements of P_n will be also called in the following *functions of classical positive type*.

In fact, a quantum analogue of functions of classical positive type emerges by undertaking a phase-space approach to quantum mechanics. In particular, among the various phase-space formalisms proposed in the literature [19, 40–47], we will adopt here the classical approach developed by Weyl, Wigner, Groenewold and Moyal [9–12] that in the following will be referred to as the *WWGM formulation* of quantum mechanics.

A *pure* state $\hat{\rho}_\psi = |\psi\rangle\langle\psi|$, $\psi \in \mathrm{L}^2(\mathbb{R}^n)$, with $\|\psi\| = 1$ — a one-dimensional projection — is replaced in the WWGM formulation with a function $\varrho_\psi \colon \mathbb{R}^n \times \mathbb{R}^n \to \mathbb{C}$ defined by the celebrated classical formula ($\hbar = 1$) [10]:

$$\varrho_\psi(q,p) := (2\pi)^{-n} \int_{\mathbb{R}^n} \mathrm{e}^{-\mathrm{i}p\cdot x}\, \overline{\psi\Big(q - \frac{x}{2}\Big)}\, \psi\Big(q + \frac{x}{2}\Big)\ \mathrm{d}^n x . \tag{21}$$

This definition extends in a natural way to any density operator in $\mathrm{L}^2(\mathbb{R}^n)$ — the associated phase-space functions are usually called *Wigner functions* — and then to any trace class operator (recall that every trace class operator is a linear superposition of four density operators). This construction gives rise to a complex Banach space of functions which will be denoted by LW_n. The linear space LW_n contains a convex cone W_n consisting of those functions that are associated with *positive* trace class operators, and W_n contains a convex set $\breve{\mathsf{W}}_n$ consisting of the Wigner functions, characterized as those functions in W_n satisfying the normalization condition

$$\lim_{r\to+\infty} \int_{|q|^2+|p|^2<r} \varrho(q,p)\ \mathrm{d}^n q\, \mathrm{d}^n p = 1 = \mathrm{tr}(\hat{\rho}) \tag{22}$$

— see, e.g., [48] — where we have denoted by $\varrho \in \breve{\mathsf{W}}_n$ the Wigner function associated with the density operator $\hat{\rho}$.

We stress that the limit on the lhs of (22) is necessary for taking into account the fact that a Wigner function ϱ may *not* be integrable with respect to the Lebesgue measure [48]; accordingly, although always real, ϱ in general is *not* a genuine probability distribution, because it may take both positive and negative values. On the other hand, a Wigner function can be regarded as a *quasi-probability distribution*, since one can express the expectation value $\langle\hat{A}\rangle_{\hat{\rho}} = \mathrm{tr}(\hat{A}\,\hat{\rho})$ of an observable $\hat{A}$ in the state $\hat{\rho}$ as a 'classical' phase-space integral, i.e.,

$$\langle\hat{A}\rangle_{\hat{\rho}} = \int_{\mathbb{R}^n\times\mathbb{R}^n} \alpha(q,p)\, \varrho(q,p)\ \mathrm{d}^n q\, \mathrm{d}^n p, \tag{23}$$

where α is a real function — or generalized function [49] — suitably associated with the selfadjoint operator $\hat{A}$.

Remark 3 The characterization of those Wigner functions that are genuine probability distributions is an interesting problem [50–52]. By results due to Hudson [50] and Littlejohn [51], the Wigner functions of *pure* states assuming non-negative values only are those probability distributions associated with Gaussian measures, see (16), whose covariance matrix $\mathtt{M}$, besides being a positive definite symmetric matrix, is also symplectic: $\mathtt{M}^{\mathsf{T}}\Omega\,\mathtt{M} = \mathtt{M}\,\Omega\,\mathtt{M} = \Omega$.

Similarly to the classical setting, it is often useful to replace a quasi-probability distribution with its symplectic Fourier transform. In this regard, it is worth recalling that a Wigner function is always *square* integrable [19, 20]. Therefore, in this case the natural tool is the (symplectic) *Fourier-Plancherel transform* $\hat{\mathcal{F}}_{\mathrm{sp}}$, namely, the selfadjoint unitary operator in the Hilbert space $\mathrm{L}^2(\mathbb{R}^n \times \mathbb{R}^n)$ determined by

$$(\hat{\mathcal{F}}_{\mathrm{sp}} f)(q,p) = (2\pi)^{-n} \int_{\mathbb{R}^n\times\mathbb{R}^n} f(q',p')\, \mathrm{e}^{\mathrm{i}(q\cdot p' - p\cdot q')}\ \mathrm{d}^n q'\, \mathrm{d}^n p', \tag{24}$$

with $f \in \mathrm{L}^1(\mathbb{R}^n \times \mathbb{R}^n) \cap \mathrm{L}^2(\mathbb{R}^n \times \mathbb{R}^n)$. The linear space LW_n is mapped via $\hat{\mathcal{F}}_{\mathrm{sp}}$ onto a (dense) subspace LQ_n of $\mathrm{L}^2(\mathbb{R}^n \times \mathbb{R}^n)$ and the convex cone $\mathsf{W}_n \subset \mathsf{LW}_n$ is mapped onto $\mathsf{Q}_n \subset \mathsf{LQ}_n$. Since every function in LQ_n agrees almost everywhere with a continuous function, it will be convenient to regard LQ_n as a linear space of continuous functions.

As it will be clear soon, the convex cone Q_n can indeed be considered as a quantum analogue of P_n. By this analogy, we will call a function $\widetilde{\varrho}$ in Q_n, such that

$$\widetilde{\varrho} = (2\pi)^n \hat{\mathcal{F}}_{\mathrm{sp}}\, \varrho, \tag{25}$$

for some quasi-probability distribution $\varrho \in \breve{\mathsf{W}}_n$, the *quantum characteristic function* associated with ϱ. Hence, the quantum characteristic functions form a convex subset $\breve{\mathsf{Q}}_n = (2\pi)^n \hat{\mathcal{F}}_{\mathrm{sp}}\, \breve{\mathsf{W}}_n$ of the linear space LQ_n.

Remark 4 The factor $(2\pi)^n$ in (25) has been fixed in such a way that the quantum characteristic functions are those functions in Q_n satisfying the normalization condition

$$\widetilde{\varrho}(0) = 1, \tag{26}$$

with $0 \equiv (0,0)$ denoting the origin in $\mathbb{R}^n \times \mathbb{R}^n$; compare with (7).

A rather obvious problem arising in the WWGM formulation is to achieve an *intrinsic* characterization of the convex set $\breve{\mathsf{W}}_n$ of quasi-probability distributions or, equivalently, of the convex set $\breve{\mathsf{Q}}_n$ of quantum characteristic functions. Analyzing this problem one is lead to the notion of *function of quantum positive type*. Indeed, keeping in mind the classical case, it is natural to consider a suitable $*$-algebra of functions, and then to define the functions of positive type as positive functionals on this algebra. It is clear, however, that a *noncommutative* algebra should be involved in this case.

Recall that the Hilbert space $\mathrm{L}^2(\mathbb{R}^n \times \mathbb{R}^n)$ becomes a H*-algebra [20] once endowed with the *twisted convolution* $(\cdot) \circledast (\cdot)$, where

$$\big(\mathcal{A}_1 \circledast \mathcal{A}_2\big)(q,p) := (2\pi)^{-n} \int_{\mathbb{R}^n\times\mathbb{R}^n} \mathcal{A}_1(q',p')\, \mathcal{A}_2(q-q',p-p')\, \mathrm{e}^{\frac{\mathrm{i}}{2}(q\cdot p' - p\cdot q')}\, \mathrm{d}^n q'\, \mathrm{d}^n p', \tag{27}$$

$\mathcal{A}_1, \mathcal{A}_2 \in \mathrm{L}^2(\mathbb{R}^n \times \mathbb{R}^n)$, and with the involution $\mathsf{J}\colon \mathrm{L}^2(\mathbb{R}^n \times \mathbb{R}^n) \ni \mathcal{A} \mapsto \mathcal{A}^*$,

$$\mathcal{A}^*(q,p) := \overline{\mathcal{A}(-q,-p)}. \tag{28}$$

The involution J acts precisely as in the classical case, compare with (20), but of course in a different space. An element $\mathcal{A}$ of $\mathrm{L}^2(\mathbb{R}^n \times \mathbb{R}^n)$ such that $\mathcal{A} = \mathsf{J}\,\mathcal{A}$ is said to be *selfadjoint*.

To justify our choice of the twisted convolution as the relevant algebra operation, it is worth anticipating that it can be considered as an expression of the product of Hilbert space operators in terms of phase-space functions, see subsect. 3.2. A 'nonlocal' product of functions of this kind is often called a *star product* [19, 20].

We are now ready to give a formal definition of a function of positive type in the quantum setting. Clearly, the Hilbert space $\mathrm{L}^2(\mathbb{R}^n \times \mathbb{R}^n)$ is selfdual. Hence, in our approach the functions of quantum positive type, defined as functionals, are assumed to be square integrable [21]:

Definition 2 *A function of* quantum positive type *on* $\mathbb{R}^n \times \mathbb{R}^n$ *is a positive bounded functional on the* H*-*algebra* $\big(\mathrm{L}^2(\mathbb{R}^n \times \mathbb{R}^n), \circledast, \mathsf{J}\big)$, *implemented by a square integrable function. Otherwise stated, a complex function* $\mathcal{Q}$ *on* $\mathbb{R}^n \times \mathbb{R}^n$ *is said to be of quantum positive type if it belongs to* $\mathrm{L}^2(\mathbb{R}^n \times \mathbb{R}^n)$ *and, for every* $\mathcal{A} \in \mathrm{L}^2(\mathbb{R}^n \times \mathbb{R}^n)$, *satisfies the inequality*

$$\int_{\mathbb{R}^n\times\mathbb{R}^n} \mathcal{Q}(q,p)\, (\mathcal{A}^* \circledast \mathcal{A})(q,p)\, \mathrm{d}^n q\, \mathrm{d}^n p \geq 0. \tag{29}$$

As already mentioned, the twisted convolution of functions 'mimics' the product of operators, hence, it is not surprising that one should regard a function $\mathcal{Q}$ defined as above as a 'quantum' object. Nevertheless, the precise consequences of the previous definition are not immediately clear. We will now argue that, similarly to the classical setting, a *continuous* function of quantum positive type can be characterized by various equivalent properties, and one of these has a precise physical meaning.

It will be convenient to adopt the following concise notation: we set $z \equiv (q,p) \in \mathbb{R}^n \times \mathbb{R}^n$, $\mathrm{d}z \equiv \mathrm{d}^n q\, \mathrm{d}^n p$ and, for the symplectic form, $\omega(z,z') \equiv \omega((q,p),(q',p')) = q \cdot p' - p \cdot q'$.

One can prove the following facts; see [21] and references therein. If a *continuous* function $\mathcal{Q}$ is of quantum positive type, then it is bounded and

$$\|\mathcal{Q}\|_\infty = \mathcal{Q}(0). \tag{30}$$

This relation is the quantum analogue of (4). For a *continuous* function $\mathcal{Q}\colon \mathbb{R}^n \times \mathbb{R}^n \to \mathbb{C}$ the following properties are equivalent:

Q1) $\mathcal{Q}$ is of quantum positive type;

Q2) $\mathcal{Q}$ satisfies the inequality (29), for all $\mathcal{A} \in \mathsf{C_c}(\mathbb{R}^n \times \mathbb{R}^n)$;

Q3) $\mathcal{Q}$ satisfies the inequality

$$\int_{\mathbb{R}^n\times\mathbb{R}^n}\int_{\mathbb{R}^n\times\mathbb{R}^n} \mathcal{Q}(z-z')\,\overline{\mathcal{A}(z')}\,\mathcal{A}(z)\,\mathrm{e}^{\mathrm{i}\omega(z',z)/2}\,\mathrm{d}z\,\mathrm{d}z' \geq 0, \quad \forall \mathcal{A} \in \mathsf{C_c}(\mathbb{R}^n \times \mathbb{R}^n); \tag{31}$$

Q4) $\mathcal{Q}$ is a *quantum positive definite function*, i.e.,

$$\sum_{j,k} \mathcal{Q}(z_k - z_j)\,\mathrm{e}^{\mathrm{i}\omega(z_j,z_k)/2}\,\overline{c_j}\,c_k \geq 0, \tag{32}$$

for every finite subset $\{z_1,\ldots,z_m\}$ of $\mathbb{R}^n \times \mathbb{R}^n$ and arbitrary complex numbers $c_1,\ldots,c_m$;

Q5) $\mathcal{Q}$ is contained in the convex cone $\mathsf{Q}_n \subset \mathsf{LQ}_n$, i.e., it is a non-negative real multiple of the Fourier-Plancherel transform of a quasi-probability distribution.

The equivalence between the first and the last property is a quantum version of Bochner's theorem [21,25–29]. By this equivalence, Q_n can be regarded as the set of continuous functions of quantum positive type, and LQ_n as the complex vector space generated by linear superpositions of functions of this kind. By (26) and (30), the convex set $\breve{\mathsf{Q}}_n$ of quantum characteristic functions coincides with the set of those continuous functions of quantum positive type normalized in such a way that

$$\|\mathcal{Q}\|_\infty = \mathcal{Q}(0) = 1. \tag{33}$$

One can show moreover — see subsect. 3.2 — that the norm of every $\mathcal{Q} \in \breve{\mathsf{Q}}_n$, regarded as a functional (i.e., as an element of $\mathrm{L}^2(\mathbb{R}^n \times \mathbb{R}^n)$), verifies the inequality

$$\|\mathcal{Q}\|_2 \leq 1, \tag{34}$$

for a suitable normalization of the Haar measure (a multiple of the Lebesgue measure on $\mathbb{R}^n\times\mathbb{R}^n$). Indeed, in subsect. 3.2 it will be argued that the square of $\|\mathcal{Q}\|_2$ is nothing but the purity of the density operator associated with $\mathcal{Q}$. Therefore, the inequality (34) is saturated if and only if the state associated with $\mathcal{Q}$ is *pure*.

Remark 5 The analogy with the classical setting cannot be pushed too far. On the one hand, a function of positive type on a locally compact group is automatically continuous (modulo modifications on ν_G-null sets). On the other hand, a function of quantum positive type $\mathcal{Q}$ is *not*, in general, a continuous function, and continuity, together with the normalization condition (33), must be *imposed* in order to select the quantum characteristic functions. Moreover, the normalization of $\mathcal{Q}$ as a functional in general differs from (33) (its normalization as a state).

Remark 6 A comparison of relations (3), (5) and (6) with relations (29), (31) and (32), shows the central role played by the function $(\mathbb{R}^n \times \mathbb{R}^n) \times (\mathbb{R}^n \times \mathbb{R}^n) \ni (z, z') \mapsto \exp(\mathrm{i}\omega(z, z')/2)$. Of course, this is the *non-exact* $\mathbb{T}$-multiplier, for the abelian group $\mathbb{R}^n \times \mathbb{R}^n$, associated with the Weyl system and with the canonical commutation relations (expressed in the Weyl form) [53]. Accordingly, whereas an ordinary function of positive type on the vector group $\mathbb{R}^k$ is defined for every $k = 1, 2, \ldots$, the *quantum* notion of function of positive type involves the symplectic form ω — namely, a non-degenerate skew-symmetric bilinear form; hence, in this case the dimension k of the vector group must be even.

3. Interplay between classical and quantum

The notion of function of positive type on phase space — classical and quantum — allows one to note a remarkable interplay between classical and quantum states. To clarify this point, we first recall that the *convolution* $\mu_1 \circledcirc \mu_2$ of two positive measures $\mu_1, \mu_2 \in \mathrm{CM}(G)$ on a locally compact group G — defined by

$$(\mu_1 \circledcirc \mu_2)(\mathscr{E}) := \int_G \int_G \varkappa_{\mathscr{E}}(gh) \, \mathrm{d}\mu_1(g) \mathrm{d}\mu_2(h), \tag{35}$$

where $\varkappa_{\mathscr{E}}$ is the characteristic function of a Borel set $\mathscr{E} \subset G$ — is again a positive measure in $\mathrm{CM}(G)$ (a probability measure if μ_1, μ_2 are normalized) [7]. The convex subset of $\mathrm{CM}(G)$ formed by the probability measures, endowed with convolution, becomes a semigroup, whose identity is δ_e (the Dirac measure at the identity e of G). Assuming now that G is abelian, the convolution of two probability measures is mapped, via the Fourier-Stieltjes transform, to the point-wise multiplication of the corresponding characteristic functions [6]. Hence, by Bochner's theorem, the point-wise product $\chi_1 \chi_2$ of two (normalized) continuous functions of positive type χ_1 and χ_2 on G is a (normalized) continuous function of positive type too. In particular, in the case where $G = \mathbb{R}^n \times \mathbb{R}^n$, the set $\breve{\mathsf{P}}_n$ of normalized functions of classical positive type, endowed with the point-wise product, is a semigroup, the identity being the constant function $\chi \equiv 1$.

3.1. Classical-quantum semigroups

As we have seen, point-wise multiplication is a legitimate operation between functions of classical positive type. What happens if one multiplies a function of *classical* positive type by a continuous function of *quantum* positive type? The answer is contained in the following result [21]:

Theorem 1 *The point-wise product $\chi\,\mathcal{Q}$ of a function $\chi \in \mathsf{P}_n$ by a function $\mathcal{Q} \in \mathsf{Q}_n$ is contained in Q_n. In particular, if χ and $\mathcal{Q}$ are normalized, $\chi\,\mathcal{Q}$ belongs to $\breve{\mathsf{Q}}_n$.*

This result, which may be regarded at first sight as a mere mathematical curiosity, actually establishes a link between the notion of function of positive type and the theory of open quantum systems. Consider, indeed, a set $\{\chi_t\}_{t\in\mathbb{R}^+}$ of normalized continuous functions of positive type on $\mathbb{R}^n \times \mathbb{R}^n$ such that

$$\chi_t \, \chi_s = \chi_{t+s}, \ \forall t, s \geq 0, \quad \chi_0 \equiv 1, \tag{36}$$

where $\chi_t \, \chi_s$ is again a point-wise product; namely, a (one-parameter) *multiplication semigroup of functions of positive type.* One should also assume continuity, with respect to a suitable topology on $\breve{\mathsf{P}}_n$ [21], for the homomorphism $\mathbb{R}^+ \ni t \mapsto \chi_t \in \breve{\mathsf{P}}_n$. One-parameter semigroups of this kind can be classified because they are related, via the Fourier-Stieltjes transform, with the convolution semigroups of probability measures on $\mathbb{R}^n \times \mathbb{R}^n$ [34, 54].

Observe now that for every function of positive type $\chi \in \breve{\mathsf{P}}_n$ one can define a linear operator $\hat{\mathfrak{C}}_\chi$ in $\mathrm{L}^2(\mathbb{R}^n \times \mathbb{R}^n)$ by setting

$$(\hat{\mathfrak{C}}_\chi f)(q, p) := \chi(q, p) \, f(q, p). \tag{37}$$

It is clear that $\hat{\mathfrak{C}}_\chi$ is a (well defined) bounded operator because χ is a bounded continuous function. It will be shown — see Remark 8 below — that it maps the convex cone of functions of quantum positive type into itself. For every multiplication semigroup of functions of positive type $\{\chi_t\}_{t\in\mathbb{R}^+} \subset \check{\mathsf{P}}_n$ and for every $t \geq 0$, one can set

$$(\hat{\mathfrak{C}}_t f)(q,p) \equiv (\hat{\mathfrak{C}}_{\chi_t} f)(q,p) := \chi_t(q,p)\, f(q,p), \quad f \in \mathrm{L}^2(\mathbb{R}^n \times \mathbb{R}^n). \tag{38}$$

The set $\{\hat{\mathfrak{C}}_t\}_{t\in\mathbb{R}^+}$ is a (one-parameter) semigroup of operators [55]:

(i) $\hat{\mathfrak{C}}_t \hat{\mathfrak{C}}_s = \hat{\mathfrak{C}}_{t+s}$, $t, s \geq 0$;

(ii) $\hat{\mathfrak{C}}_0 = I$ (with I denoting the identity operator).

We will call the semigroups of operators $\{\hat{\mathfrak{C}}_t\}_{t\in\mathbb{R}^+}$ a *classical-quantum semigroup.*

Taking into account the contents Theorem 1, we will now consider a natural restriction of the semigroup of operators $\{\hat{\mathfrak{C}}_t\}_{t\in\mathbb{R}^+}$ to a linear subspace of $\mathrm{L}^2(\mathbb{R}^n \times \mathbb{R}^n)$. In fact, as mentioned in sect. 2, one can extend the convex cone Q_n of continuous functions of quantum positive type on $\mathbb{R}^n \times \mathbb{R}^n$, by taking (complex) linear superpositions, to a dense subspace LQ_n of $\mathrm{L}^2(\mathbb{R}^n \times \mathbb{R}^n)$. Since the point-wise product of a continuous function of classical positive type by a continuous function quantum positive type is again a function of the latter type, a semigroup of operators $\{\mathfrak{C}_t\}_{t\in\mathbb{R}^+}$ in LQ_n is defined by setting

$$(\mathfrak{C}_t\, \mathcal{Q})(q,p) := \chi_t(q,p)\, \mathcal{Q}(q,p), \quad \mathcal{Q} \in \mathsf{LQ}_n. \tag{39}$$

Note that here $\mathcal{Q}$, with a slight abuse with respect to our previous convention, denotes a *linear superposition* of four functions of quantum positive type, namely, $\mathcal{Q} = \mathcal{Q}_1 - \mathcal{Q}_2 + \mathrm{i}(\mathcal{Q}_3 - \mathcal{Q}_4)$, with $\mathcal{Q}_1, \ldots, \mathcal{Q}_4 \in \mathsf{Q}_n$. Obviously, by construction we have that $\mathfrak{C}_t\, \mathsf{Q}_n \subset \mathsf{Q}_n$ and $\mathfrak{C}_t\, \check{\mathsf{Q}}_n \subset \check{\mathsf{Q}}_n$.

Thus, it is now clear that the semigroup of operators $\{\mathfrak{C}_t\}_{t\in\mathbb{R}^+}$ — which will be called a *proper classical-quantum semigroup* — maps the convex set of quantum characteristic functions into itself. Nevertheless, the precise connection with the theory of open quantum systems needs to be further clarified. It will be unveiled by means of a suitable quantization procedure mapping phase-space functions to Hilbert space operators.

3.2. Group-theoretical quantization/dequantization

The procedure that allows one to associate with a Hilbert space operator a phase-space function (e.g., a Wigner function) involves a suitable *dequantization map.* Of course the reverse arrow is a *quantization map*, which transforms functions into operators. By these maps, one is able to set a (at least implicit) correspondence between the product of operators and a *star product* [19,20,56] of functions — the twisted convolution (27), in the case where quantum characteristic functions are involved. The star product is an essential ingredient for a self-consistent formulation of quantum mechanics in terms of phase-space functions, together with the related notion of function of quantum positive type. Our aim, now, is to briefly illustrate the meaning of this notion in terms of Hilbert space operators. As a byproduct, we will be able to highlight the precise connection of classical-quantum semigroups with quantum dynamical semigroups.

A fundamental tool for constructing a pair of 'group-covariant' quantization/dequantization maps is a *square integrable* (in general, projective) representation U of a locally compact group G in a Hilbert space $\mathcal{H}$; see [15,23,24]. One may think of $\mathcal{H}$ as the Hilbert space of a quantum-mechanical system and of G as a symmetry group. Let $\mathcal{B}_2(\mathcal{H})$ be the Hilbert space of Hilbert-Schmidt operators in $\mathcal{H}$. The representation U allows one to define the dequantization map as a linear isometry $\mathscr{D}\colon \mathcal{B}_2(\mathcal{H}) \to \mathrm{L}^2(G)$, see [15,19,20]. If the group G is unimodular and $\hat{\rho} \in \mathcal{B}_2(\mathcal{H})$ is a trace class operator, the function $\mathscr{D}\hat{\rho}$ associated with $\hat{\rho}$ is of the simple form

$$(\mathscr{D}\hat{\rho})(g) = d_U^{-1}\ \mathrm{tr}(U(g)^* \hat{\rho}), \tag{40}$$

where $d_U > 0$ is a constant which depends on U and on the normalization of the Haar measure on G. The quantization map $\mathscr{Q}$ associated with U is nothing but the adjoint of the map $\mathscr{D}$; i.e., it is the partial isometry $\mathscr{Q}$ defined by $\mathscr{Q} := \mathscr{D}^*\colon \mathrm{L}^2(G) \to \mathcal{B}_2(\mathcal{H})$. It is clear that $\mathrm{Ker}(\mathscr{Q}) = \mathrm{Ran}(\mathscr{D})^\perp$ (in general, the closed subspace $\mathrm{Ker}(\mathscr{Q})$ of $\mathrm{L}^2(G)$ is not trivial). The star product associated with U is then defined implicitly by

$$f_1 \star f_2 := \mathscr{D}((\mathscr{Q}f_1)(\mathscr{Q}f_2)), \quad f_1, f_2 \in \mathrm{L}^2(G), \tag{41}$$

were $(\mathscr{Q}f_1)(\mathscr{Q}f_2)$ is the ordinary composition of the operators $\mathscr{Q}f_1$ and $\mathscr{Q}f_2$. Note that for a pair of functions belonging to $\mathrm{Ran}(\mathscr{D})$ the star product can be thought of as the dequantized version of the product of operators. Explicit formulae for the star products are derived in [20].

Let now G be, in particular, the group $\mathbb{R}^n\times\mathbb{R}^n$ of phase-space translations (see [19,20] for the details). In this case, $\mathcal{H} = \mathrm{L}^2(\mathbb{R}^n)$, $\mathrm{L}^2(G) = \mathrm{L}^2(\mathbb{R}^n \times \mathbb{R}^n) \equiv \mathrm{L}^2(\mathbb{R}^n \times \mathbb{R}^n, (2\pi)^{-n}\mathrm{d}^n q\, \mathrm{d}^n p; \mathbb{C})$ (the Haar measure is normalized in such a way that $d_U = 1$ in (40)) and the representation U is a *Weyl system* [20, 53], namely, $U(q,p) := \exp(\mathrm{i}(p\cdot\hat{q} - q\cdot\hat{p}))$, where $\hat{q} = (\hat{q}_1, \dots, \hat{q}_n)$, $\hat{p} = (\hat{p}_1, \dots, \hat{p}_n)$, with $\hat{q}_j$, $\hat{p}_j$ denoting the standard j-th coordinate position and momentum operators in $\mathrm{L}^2(\mathbb{R}^n)$. The Weyl system U is a square integrable, genuinely projective representation,

$$U(q+\tilde{q}, p+\tilde{p}) = \mathsf{m}(q,p;\tilde{q},\tilde{p})\, U(q,p)\, U(\tilde{q},\tilde{p}), \quad \mathsf{m}(q,p;\tilde{q},\tilde{p}) \in \mathbb{T}, \tag{42}$$

where the (non-exact) multiplier m is of the form $\mathsf{m}(q,p;\tilde{q},\tilde{p}) := \exp(\mathrm{i}(q\cdot\tilde{p} - p\cdot\tilde{q})/2)$. It turns out that for every density operator $\hat{\rho}$ in $\mathrm{L}^2(\mathbb{R}^n)$ the function $\mathscr{D}\hat{\rho} = \mathrm{tr}(U(\cdot)^*\hat{\rho})$ coincides with the quantum characteristic function $\widetilde{\varrho}$ defined by (25), where ϱ is the Wigner distribution associated with $\hat{\rho}$; see [15, 19, 20]. Thus, as $\hat{\mathcal{F}}_{\mathrm{sp}} = \hat{\mathcal{F}}_{\mathrm{sp}}^*$, we have that $\varrho = (2\pi)^{-n}\hat{\mathcal{F}}_{\mathrm{sp}}\mathscr{D}\hat{\rho}$, and quantization *à la* Weyl is obtained composing $\mathscr{Q} = \mathscr{D}^*$ with the symplectic Fourier-Plancherel operator.

Moreover, the following facts hold [19, 20]:

- $\mathrm{Ran}(\mathscr{D}) = \mathrm{L}^2(\mathbb{R}^n \times \mathbb{R}^n)$, so that in this case the partial isometries $\mathscr{Q}$ and $\mathscr{D}$ are actually unitary operators. The unitary operator $\mathscr{Q}$ intertwines the involution $\mathsf{J}\colon \mathcal{A} \mapsto \mathcal{A}^*$ in $\mathrm{L}^2(\mathbb{R}^n \times \mathbb{R}^n)$, defined by (28), with the adjoining map $\hat{A} \mapsto \hat{A}^*$ in $\mathcal{B}_2(\mathrm{L}^2(\mathbb{R}^n))$, i.e.,

$$\mathscr{Q}\,\mathsf{J}\,\mathcal{A} = (\mathscr{Q}\,\mathcal{A})^*, \quad \forall \mathcal{A} \in \mathrm{L}^2(\mathbb{R}^n \times \mathbb{R}^n). \tag{43}$$

- The star product (41) is implemented by the twisted convolution, see (27).

Taking into account these points and the contents of sect. 2, one can also argue that

- Translated in terms of Hilbert space operators, the quantum positivity condition (29) amounts to requiring that $\hat{B} = \mathscr{Q}\,\mathcal{Q} \in \mathcal{B}_2(\mathrm{L}^2(\mathbb{R}^n))$ satisfies the inequality $\mathrm{tr}(\hat{B}\,\hat{A}^*\hat{A}) \geq 0$, for all $\hat{A} \in \mathcal{B}_2(\mathrm{L}^2(\mathbb{R}^n))$; hence, equivalently, that $\hat{B} \geq 0$.
- For the linear space LQ_n and for the convex cone $\mathsf{Q}_n \subset \mathsf{LQ}_n$, we have that

$$\mathsf{LQ}_n = \mathscr{D}\,\mathcal{B}_1(\mathrm{L}^2(\mathbb{R}^n)) \ \text{ and } \ \mathsf{Q}_n = \mathscr{D}\,\mathcal{B}_1(\mathrm{L}^2(\mathbb{R}^n))^+, \tag{44}$$

  where $\mathcal{B}_1(\mathrm{L}^2(\mathbb{R}^n))$ is the Banach space of trace class operators in $\mathrm{L}^2(\mathbb{R}^n)$ and $\mathcal{B}_1(\mathrm{L}^2(\mathbb{R}^n))^+$ the convex cone of *positive* trace class operators. It is clear, moreover, that for every $\hat{\rho} \in \mathcal{B}_1(\mathrm{L}^2(\mathbb{R}^n))^+$, $\|\mathscr{D}\hat{\rho}\|_\infty = (\mathscr{D}\hat{\rho})(0) = \mathrm{tr}(\hat{\rho})$.
- It follows that the unitary operator $\mathscr{Q} = \mathscr{D}^*$ maps a function of quantum positive type into a positive Hilbert-Schmidt operator and a *continuous* function of quantum positive type into a positive trace class operator (in particular, a *normalized continuous* function of quantum positive type into a density operator). By (43), a function of quantum positive type $\mathcal{Q}$ is selfadjoint: $\mathcal{Q} = \mathsf{J}\,\mathcal{Q}$.

- For every density operator $\hat{\rho}$ in $\mathrm{L}^2(\mathbb{R}^n)$, denoting by $\|\cdot\|_2$ the norm of the Hilbert space $\mathrm{L}^2(\mathbb{R}^n \times \mathbb{R}^n, (2\pi)^{-n}\mathrm{d}^n q\,\mathrm{d}^n p; \mathbb{C})$, we have that $\|\mathscr{D}\hat{\rho}\|_2 = \sqrt{\mathrm{tr}(\hat{\rho}^2)} \leq 1$. Hence, as mentioned in sect. 2 (see (34)), with a suitable choice of the Haar measure the norm, as a functional, of a normalized continuous function of quantum positive type coincides with the square root of the purity of the associated state.

Remark 7 Taking into account the previous facts, the contents of Theorem 1 can be slightly extended by including the following assertion: *for every* $\chi \in \mathsf{P}_n$ *and for every function of quantum positive type* $\mathcal{Q}$ *in* $\mathrm{L}^2(\mathbb{R}^n \times \mathbb{R}^n)$ (not necessarily contained in Q_n) *the point-wise product* $\chi\,\mathcal{Q}$ *is again function of the latter type.* We now give a sketch of the proof. Let us set $\hat{A} \equiv \mathscr{Q}\,\mathcal{Q} \in \mathcal{B}_2(\mathrm{L}^2(\mathbb{R}^n))$, $\hat{A} \geq 0$. Assuming that $\hat{A} \notin \mathcal{B}_1(\mathrm{L}^2(\mathbb{R}^n))^+$ (i.e., $\mathcal{Q} \notin \mathsf{Q}_n$), consider the (necessarily infinite) spectral decomposition $\hat{A} = \sum_j \alpha_j \hat{\Pi}_j$ — where $\alpha_1 > \alpha_2 > \cdots > 0$ and $\hat{\Pi}_j$ is a (finite rank) projection operator — which converges in the Hilbert-Schmidt norm. Then, the sequence $\{\mathcal{Q}_m\}_{m\in\mathbb{N}}$, with $\mathcal{Q}_m \equiv \mathscr{D}(\sum_{j=1}^m \alpha_j \hat{\Pi}_j)$, is contained in Q_n and $\mathcal{Q}_m \to \mathcal{Q}$ in $\mathrm{L}^2(\mathbb{R}^n \times \mathbb{R}^n)$. Therefore, we have that $\{\chi\,\mathcal{Q}_m\}_{m\in\mathbb{N}} \subset \mathsf{Q}_n$, $\chi\,\mathcal{Q}_m \to \chi\,\mathcal{Q}$ and $0 \leq \int (\chi\,\mathcal{Q}_m)(q,p)\,(\mathcal{A}^* \circledast \mathcal{A})(q,p)\,\mathrm{d}^n q\,\mathrm{d}^n p \to \int (\chi\,\mathcal{Q})(q,p)\,(\mathcal{A}^* \circledast \mathcal{A})(q,p)\,\mathrm{d}^n q\,\mathrm{d}^n p$.

Remark 8 By Remark 7, as anticipated in subsect. 3.1, for every function of positive type $\chi \in \breve{\mathsf{P}}_n$ the bounded operator $\hat{\mathfrak{C}}_\chi$ defined by (37) maps the convex cone of functions of quantum positive type into itself, and obviously the same property holds for each member of the semigroup of operators $\{\hat{\mathfrak{C}}_t\}_{t\in\mathbb{R}^+}$ defined by (38).

3.3. Unveiling the physical meaning of classical-quantum semigroups

We are now ready to unveil the physical meaning of a classical-quantum semigroup. Observe that with the projective representation U is associated an isometric representation $U \vee U$ of the group $\mathbb{R}^n \times \mathbb{R}^n$ acting in the Banach space $\mathcal{B}_1(\mathrm{L}^2(\mathbb{R}^n))$; i.e.,

$$U \vee U(q,p)\,\hat{\rho} := U(q,p)\,\hat{\rho}\,U(q,p)^*, \quad \forall \hat{\rho} \in \mathcal{B}_1(\mathrm{L}^2(\mathbb{R}^n)). \tag{45}$$

This is, of course, the canonical symmetry action of the group of phase-space translations on trace class operators, according to Wigner's theorem on symmetry transformations. Given a convolution semigroup $\{\mu_t\}_{t\in\mathbb{R}^+}$ of probability measures on $\mathbb{R}^n \times \mathbb{R}^n$, one can define a semigroup of operators $\{\mu_t[U \vee U]\}_{t\in\mathbb{R}^+}$ in $\mathcal{B}_1(\mathrm{L}^2(\mathbb{R}^n))$ by setting

$$\mu_t[U \vee U]\,\hat{\rho} := \int_{\mathbb{R}^n \times \mathbb{R}^n} U \vee U(q,p)\,\hat{\rho}\,\mathrm{d}\mu_t(q,p). \tag{46}$$

It can be shown that the *twirling semigroup* $\{\mu_t[U \vee U]\}_{t\in\mathbb{R}^+}$ is a quantum dynamical semigroup (namely, a completely positive, trace-preserving semigroup of operators in $\mathcal{B}_1(\mathrm{L}^2(\mathbb{R}^n))$); see [34–37]. More specifically, $\{\mu_t[U \vee U]\}_{t\in\mathbb{R}^+}$ belongs to the class of *classical-noise semigroups*; see [31] and references therein. The following result [21] establishes a precise connection between the semigroup of operators (46) and a classical-quantum semigroup:

Theorem 2 *Let* $\{\chi_t\}_{t\in\mathbb{R}^+}$ *be the one-parameter multiplication semigroup of functions of positive type related, via the symplectic Fourier-Stieltjes transform, to the convolution semigroup of probability measures* $\{\mu_t\}_{t\in\mathbb{R}^+}$ *— namely,* $\chi_t(q,p) = \int \exp(\mathrm{i}(q\cdot p' - p\cdot q'))\,\mathrm{d}\mu_t(q',p')$ *— and let* $\{\mathfrak{C}_t\}_{t\in\mathbb{R}^+}$ *be the proper classical-quantum semigroup associated with* $\{\chi_t\}_{t\in\mathbb{R}^+}$*, see (39). Then, the unitary operator* $\mathscr{Q}$ *intertwines the semigroup of operators* $\{\mathfrak{C}_t\}_{t\in\mathbb{R}^+}$ *with the quantum dynamical semigroup* $\{\mu_t[U \vee U]\}_{t\in\mathbb{R}^+}$*; i.e., for every* $\mathcal{Q} \in \mathsf{LQ}_n$*, we have:*

$$\mathscr{Q}\,\mathfrak{C}_t\,\mathcal{Q} = \mu_t[U \vee U]\,\mathscr{Q}\,\mathcal{Q}, \quad \forall t \geq 0. \tag{47}$$

Journal of Physics: Conference Series **670** (2016) 012004 doi:10.1088/1742-6596/670/1/012004

Remark 9 The isometric representation $U \vee U$ in $\mathcal{B}_1(\mathrm{L}^2(\mathbb{R}^n))$ admits a natural extension to a unitary representation in $\mathcal{B}_2(\mathrm{L}^2(\mathbb{R}^n))$. By this extension, one can define a new semigroup of operators in the Hilbert space $\mathcal{B}_2(\mathrm{L}^2(\mathbb{R}^n))$ acting in a way completely analogous to (46). The new semigroup of operators is unitarily equivalent, via the dequantization map $\mathscr{D}$, to the classical-quantum semigroup $\{\hat{\mathfrak{C}}_t\}_{t\in\mathbb{R}^+}$ in $\mathrm{L}^2(\mathbb{R}^n \times \mathbb{R}^n)$.

4. ...and beyond

Functions of positive type on phase space play a remarkable role in classical statistical mechanics and in the WWGM formulation of quantum mechanics since they are related to probability measures and to quasi-probability distributions. Both the 'classical' and the 'quantum' functions of positive type admit a simple and elegant group-theoretical characterization as positive functionals on suitable group algebras. In the classical case, on the one hand, the standard commutative convolution algebra $\big(\mathrm{L}^1(\mathbb{R}^n\times\mathbb{R}^n), \circledcirc, \mathsf{I}\big)$ is involved, and continuity of the associated positive type functions is a byproduct of their definition. On the other hand, the *continuous* functions of quantum positive type — equivalently, the non-negative real multiples of those functions representing physical states (the *quantum characteristic functions*) — are embedded in the convex cone of positive functionals on the twisted convolution algebra $\big(\mathrm{L}^2(\mathbb{R}^n\times\mathbb{R}^n), \circledast, \mathsf{J}\big)$. As we have seen, this embedding corresponds to the natural inclusion of the positive trace class operators in the positive Hilbert-Schmidt operators. It is also worth observing that the physically relevant normalization condition for the continuous functions of quantum positive type differs, in general, from their normalization as functionals; the coincidence of the two normalization criteria is indeed a distinguishing feature of the *pure* states.

It is clear that in the group-theoretical approach to quantization/dequantization outlined in subsect. 3.2 a natural problem is to achieve a generalized notion of function of positive type, suitable for an extension of the results valid for the group of translations on phase space to other groups. Consider, e.g., the case of a *unimodular* locally compact group G admitting a square integrable projective representation U, with multiplier m. It can be shown that, for every pair of functions f_1, f_2 *living in the range of the dequantization map* $\mathscr{D}$ (associated with U), and for a suitable normalization of the Haar measure ν_G, the star product (41) can be expressed as a m-twisted convolution product [20]:

$$(f_1 \star f_2)(g) = \int_G f_1(h)\, f_2(h^{-1}g)\, \overline{\mathsf{m}(h, h^{-1}g)}\; \mathrm{d}\nu_G(h). \tag{48}$$

Of course, this is a natural generalization of the standard twisted convolution (27). One can define a notion of function of positive type relatively to the algebra structure determined by this star product. Note, however, that in general the closed subspace $\mathrm{Ran}(\mathscr{D})^\perp = \mathrm{Ker}(\mathscr{Q})$ of $\mathrm{L}^2(G)$ is not trivial, and this aspect has to be taken into account when formulating the positivity condition. Another nontrivial aspect is related to the characterization of those functions of positive type (in the aforementioned generalized sense) that represent quantum states. The case of non-unimodular groups presents further delicate aspects, due to more complicated expressions of quantization/dequantization maps and of the associated star products [20], and the case of certain semidirect product groups that do not admit square integrable representations (e.g., the Poincaré group) entails other intricacies [64]. These natural developments of our work will be analyzed in detail elsewhere. Let us only mention, here, that an interesting and reasonably simple case is that of a group of the form $G = \mathbb{A} \times \widehat{\mathbb{A}}$, where $\mathbb{A}$ is a locally compact abelian group and $\widehat{\mathbb{A}}$ its Pontryagin dual. Like $\mathbb{R}^n \times \mathbb{R}^n$, also $\mathbb{A} \times \widehat{\mathbb{A}}$ may be thought of, simultaneously, as a (generalized) phase space and as the group of translations on this space. Actually, phase spaces of this sort — where $\mathbb{A}$ is, e.g., a cyclic groups or a finite field — and the associated representations of quantum states are of current interest in quantum information, quantum tomography etc.; see [57–63] and further references therein.

In this circle of ideas is also inscribed the notion of classical-quantum semigroup. We have first introduced this notion in [31], with the aim of highlighting connections with other classes of semigroups of operators and with quantum information science. Then, in [21] we have studied the link of classical-quantum semigroups with functions of positive type on phase space. As shown in subsect. 3.3, by quantizing a classical-quantum semigroup one obtains a *twirling semigroup*, generated in a natural way by a a Weyl system and by a convolution semigroup on the vector group $\mathbb{R}^n \times \mathbb{R}^n$. Actually, every pair formed by a projective representation of a locally compact group and by a convolution semigroup of probability measures on that group gives rise in a natural way to a twirling semigroup [34–36]. Thus, it would be interesting to study the construction — e.g., for groups of the form $G = \mathbb{A} \times \widehat{\mathbb{A}}$ — of suitable classical-quantum semigroups, a further development of our work that we plan to pursue.

References

[1] Prugovečki E 1981 *Quantum Mechanics in Hilbert Space* 2nd edn (New York: Academic)
[2] Emch G G 1984 *Mathematical and Conceptual Foundations of 20th-Century Physics* (Amsterdam: North-Holland)
[3] Strocchi F 2005 *An Introduction to the Mathematical Structure of Quantum Mechanics* (Singapore: World Scientific)
[4] Folland G B 1984 *Real Analysis* (New York: John Wiley & Sons)
[5] Klenke A 2014 *Probability Theory* 2nd edn (London: Springer)
[6] Rudin W 1962 *Fourier Analysis on Groups* (New York: Interscience)
[7] Folland G B 1995 *A Course in Abstract Harmonic Analysis* (Boca Raton, FL: CRC)
[8] Godement R 1948 Les fonctions de type positif et la theorie des groupes *Trans. Am. Math. Soc.* **63** 1
[9] Weyl H 1927 Quantenmechanik und Gruppentheorie *Z. Phys.* **46** 1
[10] Wigner E 1932 On the quantum correction for thermodinamic equilibrium *Phys. Rev.* **40** 749
[11] Groenewold H J 1946 On the principles of elementary quantum mechanics *Physica* **12** 405
[12] Moyal J E 1949 Quantum mechanics as a statistical theory *Proc. Camb. Phil. Soc.* **45** 99
[13] Folland G B 1989 *Harmonic Analysis in Phase Space* (Princeton, NJ: Princeton University Press)
[14] Hillery M, O'Connell R F, Scully M O and Wigner E P 1984 Distribution functions in physics: fundamentals *Phys. Rep.* **106** 121
[15] Ali S T, Antoine J P and Gazeau J P 2000 *Coherent States, Wavelets and Their Generalizations* (New York: Springer)
[16] Schleich W P 2001 *Quantum Optics in Phase Space* (Berlin: Wiley-VCH)
[17] Dodonov V V and Man'ko V I eds 2003 *Theory of Nonclassical States of Light* (London: Taylor & Francis)
[18] Aniello P, Man'ko V, Marmo G, Solimeno S and Zaccaria F 2000 On the coherent states, displacement operators and quasidistributions associated with deformed quantum oscillators *J. Opt. B: Quantum Semiclass. Opt.* **2** 718
[19] Aniello P, Man'ko V I and Marmo G 2008 Frame transforms, star products and quantum mechanics on phase space *J. Phys. A: Math. Theor.* **41** 285304
[20] Aniello P 2009 Star products: a group-theoretical point of view *J. Phys. A: Math. Theor.* **42** 475210
[21] Aniello P 2015 Playing with functions of positive type, classical and quantum *Phys. Scr.* **90** 074042
[22] Duflo M and Moore C C 1976 On the regular representation of a nonunimodular locally compact group *J. Funct. Anal.* **21** 209
[23] Aniello P, Cassinelli G, De Vito E and Levrero A 1998 Square-integrability of induced representations of semidirect products *Rev. Math. Phys.* **10** 301
[24] Aniello P 2006 Square integrable projective representations and square integrable representations modulo a relatively central subgroup *Int. J. Geom. Methods Mod. Phys.* **3** 233
[25] Kastler D 1965 The C*-algebras of free boson field *Commun. Math. Phys.* **1** 14
[26] Loupias G and Miracle-Sole S 1966 C*-algèbre des systèmes canoniques. I *Commun. Math. Phys.* **2** 31
[27] Loupias G and Miracle-Sole S 1967 C*-algèbre des systèmes canoniques. II *Ann. Inst. Henri Poincaré* A **6** 39
[28] Narcowich F J and O'Connell F R 1986 Necessary and sufficient conditions for a phase-space function to be a Wigner dstribution *Phys. Rev.* A **34** 1
[29] Narcowich F J 1989 Distributions of $\hbar$-positive type and applications *J. Math. Phys.* **30** 2565
[30] Lupo C, Aniello P and Scardicchio A 2008 Bipartite quantum systems: on the realignment criterion and beyond *J. Phys. A: Math. Theor.* **41** 415301
[31] Aniello P 2014 Classical-quantum semigroups, *J. Phys.: Conf. Ser.* **563** 012002

[32] Holevo A S 2001 *Statistical Structure of Quantum Theory* (Berlin: Springer)
[33] Breuer H-P and Petruccione F 2002 *The Theory of Open Quantum Systems* (Oxford: Oxford University Press)
[34] Aniello P, Kossakowski A, Marmo G and Ventriglia F 2010 Brownian motion on Lie groups and open quantum systems *J. Phys. A: Math. Theor.* **43** 265301
[35] Aniello P 2011 On a certain class of semigroups of operators *Open Syst. Inf. Dyn.* **18** 129
[36] Aniello P 2013 Quantum dynamical semigroups, group representations and convolution semigroups *Phys. Scr.* **T153** 014003
[37] Aniello P 2013 Operators versus functions: from quantum dynamical semigroups to tomographic semigroups *J. Phys.: Conf. Ser.* **474** 012005
[38] Murphy G J 1990 C^**-algebras and Operator Theory* (San Diego, CA: Academic)
[39] Dixmier J 1981 *Von Neumann Algebras* (Amsterdam: North-Holland)
[40] Husimi K 1940 Some formal properties of the density matrix *Proc. Phys. Math. Soc. Japan* **22** 264
[41] Kano Y 1965 A new phase-space distribution function in the statistical theory of the electromagnetic field *J. Math. Phys.* **6** 1913
[42] Sudarshan E C G 1963 Equivalence of semiclassical and quantum mechanical descriptions of statistical light beams *Phys. Rev. Lett.* **10** 277
[43] Mehta C L and Sudarshan E C G 1965 Relation between quantum and semiclassical description of optical coherence *Phys. Rev.* **138** B274
[44] Glauber R J 1963 Photon correlations *Phys. Rev. Lett.* **10** 84
[45] Cahill K E and Glauber R J 1969 Ordered expansions on boson amplitude operators *Phys. Rev.* **177** 1857
[46] Cahill K E and Glauber R J 1969 Density operators and quasi-probability distributions *Phys. Rev.* **177** 1882
[47] Mancini S, Man'ko V I and Tombesi P 1996 Symplectic tomography as classical approach to quantum systems *Phys. Lett.* A **213** 1
[48] Daubechies I 1983 Continuity statements and counterintuitive examples in connection with Weyl quantization *J. Math. Phys.* **24** 1453
[49] Daubechies I 1980 On the distributions corresponding to bounded operators in the Weyl quantization *Commun. Math. Phys.* **75** 229
[50] Hudson R L 1974 When is the Wigner quasi-probability density non-negative? *Rep. Math. Phys.* **6** 249
[51] Littlejohn R G 1986 The semiclassical evolution of wave packets *Phys. Rep.* **138** 193
[52] Bröcker T and Werner R F 1995 Mixed states with positive Wigner functions *J. Math. Phys.* **36** 62
[53] Aniello P 2010 On the notion of Weyl system *J. Russ. Laser Res.* **31** 102
[54] Heyer H 1977 *Probability Measures on Locally Compact Groups* (Berlin: Springer)
[55] Yosida K 1968 *Functional Analysis* (Berlin: Springer)
[56] Aniello P, Ibort A, Man'ko V I and Marmo G 2009 Remarks on the star product of functions on finite and compact groups *Phys. Lett.* A **373** 401
[57] Gibbons K S, Hoffman M J and Wootters W K 2004 Discrete phase space based on finite fields *Phys. Rev.* A **70** 062101
[58] Paz J P, Roncaglia A J and Saraceno M 2005 Qubits in phase space: Wigner-function approach to quantum-error correction and the mean-king problem *Phys. Rev.* A **72** 0122309
[59] Marchiolli M A, Ruzzi M and Galetti D 2005 Extended Cahill-Glauber formalism for finite-dimensional spaces. II. Applications in quantum tomography and quantum teleportation *Phys. Rev.* A **72** 042308
[60] Vourdas A 2008 Quantum mechanics on p-adic numbers *J. Phys. A: Math. Theor.* **41** 455303
[61] Vourdas A and Banderier C 2010 Symplectic transformations and quantum tomography in finite quantum systems *J. Phys. A: Math. Theor.* **43** 042001
[62] Patra M K and Braunstein S L 2011 Quantum Fourier transform, Heisenberg groups and quasi-probability distributions *New J. Phys.* **13** 063013
[63] Shalaby M and Vourdas A 2013 Mutually unbiased projectors and duality between lines and bases in finite quantum systems *Ann. Phys.* **337** 208
[64] Aniello P 2006 Extended wavelet transforms *Int. J. Geom. Methods Mod. Phys.* **3** 341

Journal of Physics: Conference Series **670** (2016) 012005 doi:10.1088/1742-6596/670/1/012005

Cotangent bundles over all the Hermitian symmetric spaces

Masato Arai[1] **and Kurando Baba**[2]

[1]Faculty of Science, Yamagata University, Kojirakawa-machi 1-4-12, Yamagata, Yamagata 990-8560, Japan,

[2]Department of General Education, Fukushima National College of Technology, Fukushima 970-8034, Japan

E-mail: arai@sci.kj.yamagata-u.ac.jp

Abstract. We construct the $\mathcal{N}=2$ supersymmetric nonlinear sigma models on the cotangent bundles over all the compact and non-compact Hermitian symmetric spaces. In order to construct them we use the projective superspace formalism which is an $\mathcal{N}=2$ off-shell superfield formulation in four-dimensional space-time. This formalism allows us to obtain the explicit expression of $\mathcal{N}=2$ supersymmetric nonlinear sigma models on the cotangent bundles over any Hermitian symmetric spaces in terms of the $\mathcal{N}=1$ superfields, once the Kähler potentials of the base manifolds are obtained. Starting with $\mathcal{N}=1$ supersymmetric Kähler nonlinear sigma models on the Hermitian symmetric spaces, we extend them into the $\mathcal{N}=2$ supersymmetric models by using the projective superspace formalism and derive the general formula for the cotangent bundles over all the compact and non-compact Hermitian symmetric spaces. We apply to the formula for the non-compact Hermitian symmetric space $E_7/E_6 \times U(1)$[1].

1. Introduction

Supersymmetry (SUSY) has an intimate relation to complex geometry in mathematics. Indeed, it is well known that target spaces of $\mathcal{N}=1$ and $\mathcal{N}=2$ SUSY nonlinear sigma models (NLSMs) must be Kähler [1] and hyperkähler manifolds [2], respectively. It is important to construct these manifolds because they frequently appear in field theories with/without supersymmetry, supergravity and superstring theories.

Recently there have been developments to construct $\mathcal{N}=2$ SUSY NLSMs in the projective superspace formalism [3, 4, 5, 6, 7], which is an $\mathcal{N}=2$ off-shell superfield formulation in four-dimensional space-time. In this formalism, $\mathcal{N}=2$ SUSY NLSMs on cotangent bundles over Kähler manifolds have been constructed [8, 9, 10, 11, 12, 13, 14]. A key observation in the developments is that once a certain $\mathcal{N}=1$ SUSY NLSM is obtained, this model can be extended into the $\mathcal{N}=2$ SUSY NLSM with use of the projective superspace formalism. In other words, if we have the $\mathcal{N}=1$ SUSY NLSM on the Kähler manifold, we can obtain $\mathcal{N}=2$ SUSY NLSM on the cotangent bundle over the Kähler manifold. The target space of the $\mathcal{N}=2$ SUSY NLSM is shown to be an open domain of the zero section of the cotangent bundle [8, 9]. Namely it is hyperkähler. Based on the observation in [8, 9], the $\mathcal{N}=2$ SUSY NLSMs on the cotangent bundles over the irreducible Hermitian symmetric spaces (HSSs) except the

[1] This talk is based on [16].

	classical type				exceptional type	
compact type	$\frac{U(n+m)}{U(n)\times U(m)}$	$\frac{SO(2n)}{U(n)}$	$\frac{Sp(n)}{U(n)}$	$\frac{SO(n+2)}{SO(n)\times U(1)}$	$\frac{E_6}{SO(10)\times U(1)}$	$\frac{E_7}{E_6\times U(1)}$
non-compact type	$\frac{U(n,m)}{U(n)\times U(m)}$	$\frac{SO^*(2n)}{U(n)}$	$\frac{Sp(n,\mathbf{R})}{U(n)}$	$\frac{SO_0(n,2)}{SO(n)\times U(1)}$	$\frac{E_{6(-14)}}{SO(10)\times U(1)}$	$\frac{E_{7(-25)}}{E_6\times U(1)}$

Table 1. Irreducible Hermitian symmetric spaces

non-compact exceptional types of them have been constructed [8, 9, 10, 11, 12, 13, 14]. The irreducible HSSs classified by Cartan [15] consist of compact type and non-compact type. They are summarized in Table 1.

In this talk we construct the $\mathcal{N}=2$ SUSY NLSM on the cotangent bundle over the non-compact HSS $E_{7(-25)}/E_6\times U(1)$. First we show how to derive general formula for the $\mathcal{N}=2$ SUSY NLSMs on cotangent bundles over all the compact and the non-compact exceptional HSSs in the framework of the projective superspace formalism [13, 14]. The method to construct the models in [8, 9, 10, 11, 12] is model-dependent and furthermore it is difficult to apply for the exceptional types of the HSS such as the compact type $E_7/E_6\times U(1)$ and the non-compact type $E_{7(-25)}/E_6\times U(1)$. Accordingly, the other method to construct the $\mathcal{N}=2$ SUSY NLSMs on cotangent bundles over the HSS has been developed in [13] and [14]. In this talk, we apply the method to the case for the $E_{7(-25)}/E_6\times U(1)$. Combined with the application to $\frac{E_{6(-14)}}{SO(10)\times U(1)}$ given in [16], we complete construction of the $\mathcal{N}=2$ SUSY NLSM on the cotangent bundle over all the HSSs.

2. $\mathcal{N}=2$ sigma models and the projective superspace

Projective superspace is described as $(x_\mu,\theta_{\alpha i},\bar\theta^i_{\dot\alpha},\zeta)$, where $\mu=0,1,2,3$ is a space-time index, $\alpha,\dot\alpha=1,2$ are spinor indices, $i=1,2$ is an $SU(2)_R$ index and ζ is the projective coordinate. Superfields are functions on its subspace, which are defined by the so-called projective condition [3] similar to the chiral condition in the four-dimensional $\mathcal{N}=1$ SUSY field theory. This condition makes a number of the Grassmann coordinates be half and integration measure for SUSY invariant action reduces to one on the full $\mathcal{N}=1$ superspace $z_M=(x_\mu,\theta_\alpha,\bar\theta_{\dot\alpha})$ with the projective coordinate ζ. A certain class of four-dimensional $\mathcal{N}=2$ NLSM is described in terms of $\mathcal{N}=1$ language as [17, 8, 9]

$$S[\Upsilon,\breve\Upsilon]=\frac{1}{2\pi i}\oint\frac{\mathrm{d}\zeta}{\zeta}\int \mathrm{d}^8z\,K(\Upsilon^I(z,\zeta),\breve\Upsilon^{\bar J}(z,\zeta)),\tag{1}$$

where I,J are indices of fields[2]. The contour encircles the origin of the ζ-plane in anti-clockwise direction. The action is written by the function of the superfields representing the polar multiplets Υ and $\breve\Upsilon$, which are called an arctic superfield and an antarctic superfield respectively. They are expanded with respect to ζ as

$$\Upsilon(z,\zeta)=\sum_{n=0}^{\infty}\Upsilon_n\zeta^n=\Phi+\Sigma\,\zeta+\mathcal{A}\,,\qquad \breve\Upsilon(\zeta)=\sum_{n=0}^{\infty}\bar\Upsilon_n(-\zeta)^{-n}\,,\tag{2}$$

where $\Upsilon_0\equiv\Phi$ is a chiral superfield ($\bar D_{\dot\alpha}\Phi=0$) and $\Upsilon_1=\Sigma$ is a complex linear superfield ($\bar D^2\Sigma=0$). An infinite set of unconstrained auxiliary fields is expressed as $\mathcal{A}$ which contains terms with an order higher than ζ. The antarctic superfield $\breve\Upsilon$ is a conjugate of Υ, which is the combination of the ordinary complex conjugate and the antipodal map $\zeta\mapsto -1/\zeta$ on the

[2] More general type of action has the form $K(\Upsilon,\breve\Upsilon,\zeta)$ [5, 6].

Journal of Physics: Conference Series **670** (2016) 012005 doi:10.1088/1742-6596/670/1/012005

Riemann sphere. Note that the action (1) is an $\mathcal{N}=2$ extension of the general $\mathcal{N}=1$ SUSY NLSM [1].

The action (2) is invariant under

$$K(\Upsilon^I, \breve{\Upsilon}^{\bar{J}}) \to K(\Upsilon^I, \breve{\Upsilon}^{\bar{J}}) + \Lambda(\Upsilon^I) + \breve{\Lambda}(\breve{\Upsilon}^{\bar{J}}), \tag{3}$$

and

$$\Upsilon^I \to f^I(\Upsilon^J), \tag{4}$$

respectively. The latter implies the reparametrization of the manifold, yielding the transformation law for Σ^I as

$$\Sigma^I = \left.\frac{d\Upsilon^I}{d\zeta}\right|_{\zeta=0} \to \left.\frac{df^I}{d\zeta}\right|_{\zeta=0} = \left.\frac{d\Upsilon^J}{d\zeta}\frac{df^I}{d\Upsilon^J}\right|_{\zeta=0} = \Sigma^J \left.\frac{df^I}{d\Upsilon^J}\right|_{\zeta=0}. \tag{5}$$

It is seen that Σ^I transforms as a tangent vector.

In order to represent the action (1) in terms of physical fields (Φ^I, Σ^J) only, we need to eliminate the auxiliary fields by using their equations of motion

$$\oint \frac{\mathrm{d}\zeta}{\zeta}\zeta^n \frac{\partial K(\Upsilon,\breve{\Upsilon})}{\partial \Upsilon^I} = \oint \frac{\mathrm{d}\zeta}{\zeta}\zeta^{-n}\frac{\partial K(\Upsilon,\breve{\Upsilon})}{\partial \breve{\Upsilon}^{\bar{I}}} = 0, \qquad n \geq 2. \tag{6}$$

Let $\Upsilon_*(\zeta) \equiv \Upsilon_*(\zeta;\Phi,\bar{\Phi},\Sigma,\bar{\Sigma})$ be a unique solution of the equation (6) with the initial conditions

$$\Upsilon_*(0) = \Phi, \qquad \left.\frac{d\Upsilon(\zeta)}{d\zeta}\right|_{\zeta=0} = \Sigma. \tag{7}$$

For a general Kähler manifold, it is possible to eliminate $\Upsilon_n (n \geq 2)$ and their conjugates by solving (6) perturbatively [18]. After eliminating all the auxiliary fields, the following form of the action is obtained

$$S_{\mathrm{tb}}[\Phi,\Sigma] = \int \mathrm{d}^8 z \left\{ K(\Phi,\bar{\Phi}) + \mathcal{L}(\Phi,\bar{\Phi},\Sigma,\bar{\Sigma}) \right\}, \tag{8}$$

where $\mathcal{L}(\Phi,\bar{\Phi},\Sigma,\bar{\Sigma})$ is the part describing the tangent space:

$$\begin{aligned}\mathcal{L}(\Phi,\bar{\Phi},\Sigma,\bar{\Sigma}) &= \sum_{n=1}^{\infty} \mathcal{L}^{(n)}(\Phi,\bar{\Phi},\Sigma,\bar{\Sigma}) \\ &= \sum_{n=1}^{\infty} \mathcal{L}_{I_1\cdots I_n \bar{J}_1\cdots\bar{J}_n}(\Phi,\bar{\Phi})\Sigma^{I_1}\ldots\Sigma^{I_n}\bar{\Sigma}^{\bar{J}_1}\ldots\bar{\Sigma}^{\bar{J}_n}.\end{aligned} \tag{9}$$

Here $\mathcal{L}_{I\bar{J}} = -g_{I\bar{J}}(\Phi,\bar{\Phi})$ while the tensors $\mathcal{L}_{I_1\cdots I_n\bar{J}_1\cdots\bar{J}_n}(n \geq 2)$ are functions of the metric $g_{I\bar{J}}(\Phi,\bar{\Phi})$, the Riemann tensor $R_{I\bar{J}K\bar{L}}(\Phi,\bar{\Phi})$ and its covariant derivative. The action (8) is written by the base manifold coordinate Φ and the tangent vector Σ. Therefore this action represents the $\mathcal{N}=2$ SUSY model on the tangent bundle over the Kähler manifold.

The rest of the work is to derive the Kähler potential of the cotangent bundle over the Kähler manifold. It is carried out by changing the tangent vectors Σ's in (8) into chiral one-forms, cotangent vectors Ψ's. It can be performed by the generalized Legendre transformation [5] as follows.

$$\begin{aligned}S_{\mathrm{tb}} &= \int d^8 z \left(K(\Phi,\bar{\Phi}) + \mathcal{L}(\Phi,\bar{\Phi},\Sigma,\bar{\Sigma})\right) \\ \rightsquigarrow S &= \int \mathrm{d}^8 z \left\{ K(\Phi,\bar{\Phi}) + \mathcal{L}(\Phi,\bar{\Phi},U,\bar{U}) + \Psi_I U^I + \bar{\Psi}_{\bar{I}}\bar{U}^{\bar{I}} \right\},\end{aligned} \tag{10}$$

where U is a complex unconstrained superfield and Ψ is a chiral superfield. This action goes back to the tangent bundle action (8) after eliminating the chiral superfields Ψ and $\bar{\Psi}$ by their equations of motion. On the other hand, eliminating U and $\bar{U}$ with the aid of their equations of motion, the action is written only in terms of Φ, Ψ and their conjugates:

$$S_{\text{ctb}}[\Phi,\Psi] = \int \mathrm{d}^8 z \left\{ K(\Phi,\bar{\Phi}) + \mathcal{H}(\Phi,\bar{\Phi},\Psi,\bar{\Psi}) \right\}, \tag{11}$$

where

$$\begin{aligned}\mathcal{H}(\Phi,\bar{\Phi},\Psi,\bar{\Psi}) &= \sum_{n=1}^{\infty} \mathcal{H}^{(n)}(\Phi,\bar{\Phi},\Psi,\bar{\Psi}) \\ &= \sum_{n=1}^{\infty} \mathcal{H}^{I_1\cdots I_n \bar{J}_1\cdots\bar{J}_n}(\Phi,\bar{\Phi}) \Psi_{I_1}\ldots\Psi_{I_n}\bar{\Psi}_{\bar{J}_1}\ldots\bar{\Psi}_{\bar{J}_n}, \end{aligned} \tag{12}$$

with $\mathcal{H}^{I\bar{J}}(\Phi,\bar{\Phi}) = g^{I\bar{J}}(\Phi,\bar{\Phi})$. Here $g^{I\bar{J}}$ is the inverse metric of $g_{I\bar{J}}$. The variables (Φ^I, Ψ_J) parameterize the cotangent bundle over the Kähler manifold and therefore the action gives the Kähler potential of the cotangent bundle over the Kähler manifold.

The explicit forms of $\mathcal{L}$ and $\mathcal{H}$ are obtained in [12] and [13, 14] for the case that the base manifold is the HSS respectively. Here we focus only on the derivation of $\mathcal{H}$, which we are interested in. The cotangent bundle action (11) has to be invariant under the following second SUSY transformations [12]

$$\delta\Phi^I = \frac{1}{2}\bar{D}^2\{\bar{\varepsilon}_{\dot{\alpha}}\bar{\theta}^{\dot{\alpha}}\,\Sigma^I(\Phi,\bar{\Phi},\Psi,\bar{\Psi})\}, \tag{13}$$

$$\delta\Psi_I = -\frac{1}{2}\bar{D}^2\left\{\bar{\varepsilon}_{\dot{\alpha}}\bar{\theta}^{\dot{\alpha}}\,K_I(\Phi,\bar{\Phi})\right\} + \frac{1}{2}\bar{D}^2\left\{\bar{\varepsilon}_{\dot{\alpha}}\bar{\theta}^{\dot{\alpha}}\,\Gamma^K{}_{IJ}(\Phi,\bar{\Phi})\,\Sigma^J(\Phi,\bar{\Phi},\Psi,\bar{\Psi})\right\}\Psi_K, \tag{14}$$

with

$$\Sigma^I(\Phi,\bar{\Phi},\Psi,\bar{\Psi}) = \frac{\partial}{\partial\Psi_I}\mathcal{H}(\Phi,\bar{\Phi},\Psi,\bar{\Psi}) := \mathcal{H}^I. \tag{15}$$

The requirement of invariance under such transformations can be shown to be equivalent to the following nonlinear equation [12]:

$$\mathcal{H}^I g_{I\bar{J}} - \frac{1}{2}\mathcal{H}^K\mathcal{H}^L R_{K\bar{J}L}{}^I\Psi_I = \bar{\Psi}_{\bar{J}}. \tag{16}$$

Eq. (16) implies that

$$\Psi_I\mathcal{H}^I - \mathcal{H}^K\mathcal{H}^L(R_\Psi)_{KL} = g^{I\bar{J}}\Psi_I\bar{\Psi}_{\bar{J}}, \quad (R_\Psi)_{KL} := \frac{1}{2}R_K{}^I{}_L{}^J\Psi_I\Psi_J. \tag{17}$$

By using the identities

$$\Psi_I\mathcal{H}^I = \bar{\Psi}_{\bar{I}}\mathcal{H}^{\bar{I}} = \sum_{n=1}^{\infty} n\mathcal{H}^{(n)}, \tag{18}$$

(17) is rewritten as

$$\mathcal{H}^{(1)} = g^{I\bar{J}}\Psi_I\bar{\Psi}_{\bar{J}}, \quad n\mathcal{H}^{(n)} - \sum_{p=1}^{n-1}\mathcal{H}^{(p)K}(R_\Psi)_{KL}\mathcal{H}^{(n-p)L} = 0, \quad n \geq 2. \tag{19}$$

In order to solve this, it is useful to define

$$\mathbf{R}_{\Psi,\bar{\Psi}} = \begin{pmatrix} 0 & (R_\Psi)_I{}^{\bar{J}} \\ (R_{\bar{\Psi}})_{\bar{I}}{}^{J} & 0 \end{pmatrix} = \begin{pmatrix} 0 & \frac{1}{2}R_I{}^{K\bar{J}L}\Psi_K\Psi_L \\ \frac{1}{2}R_{\bar{I}}{}^{\bar{K}J\bar{L}}\bar{\Psi}_{\bar{K}}\bar{\Psi}_{\bar{L}} & 0 \end{pmatrix} \tag{20}$$

and

$$G^{(2k+2)} = \Psi_I g^{I\bar{J}}\left((R_{\bar{\Psi}}R_\Psi)^k\right)_{\bar{J}}{}^{\bar{K}}(R_{\bar{\Psi}})_{\bar{K}}{}^{L}\Psi_L = \Psi^\dagger_{\bar{I}}g^{\bar{I}J}\left((R_\Psi R_{\bar{\Psi}})^k\right)_J{}^K(R_\Psi)_K{}^{\bar{L}}\bar{\Psi}_{\bar{L}}, \tag{21}$$

$$G^{(2k+1)} = \Psi_I g^{I\bar{J}}\left((R_{\bar{\Psi}}R_\Psi)^k\right)_{\bar{J}}{}^{\bar{K}}\bar{\Psi}_{\bar{K}} = \Psi^\dagger_{\bar{I}}g^{\bar{I}J}\left((R_\Psi R_{\bar{\Psi}})^k\right)_J{}^K\Psi_K, \tag{22}$$

with $k = 0, 1, 2\cdots$. They satisfy the identities

$$\Psi_I\frac{\partial G^{(n)}}{\partial \Psi_I} = \bar{\Psi}_{\bar{I}}\frac{\partial G^{(n)}}{\partial \bar{\Psi}_{\bar{I}}} = nG^{(n)}, \tag{23}$$

$$\begin{aligned} \frac{\partial G^{(2k+2)}}{\partial \Psi_I} &= (2k+2)g^{I\bar{J}}\left((R_{\bar{\Psi}}R_\Psi)^k\right)_{\bar{J}}{}^{\bar{K}}(R_{\bar{\Psi}})_{\bar{K}}{}^{L}\Psi_L \\ &= (2k+2)\Psi_J g^{J\bar{K}}\left((R_{\bar{\Psi}}R_\Psi)^k\right)_{\bar{K}}{}^{\bar{L}}(R_{\bar{\Psi}})_{\bar{L}}{}^{I}, \end{aligned} \tag{24}$$

$$\begin{aligned} \frac{\partial G^{(2k+1)}}{\partial \Psi_I} &= (2k+1)g^{I\bar{J}}\left((R_{\bar{\Psi}}R_\Psi)^k\right)_{\bar{J}}{}^{\bar{K}}\bar{\Psi}_{\bar{K}} \\ &= (2k+1)\Psi^\dagger_{\bar{J}}g^{\bar{J}K}\left((R_\Psi R_{\bar{\Psi}})^k\right)_K{}^I. \end{aligned} \tag{25}$$

Now if we introduce the ansatz

$$\mathcal{H} = \sum_{n=1}^{\infty} c_n G^{(n)}, \tag{26}$$

where c_n is a constant, we find that the differential equation (19) turns out to be the algebraic equation

$$nc_n - \sum_{p=1}^{n-1} p(n-p)c_p c_{n-p} = 0, \qquad c_1 = 1. \tag{27}$$

The equation (27) is universal and independent of the Hermitian symmetric space. Therefore, their solution can be deduced by considering any choice of the Hermitian symmetric space. For instance, the projective complex space $\mathbf{C}P^1$ can be used. These considerations lead to the solution [13]

$$\mathcal{H}(\Phi, \bar{\Phi}, \Psi, \bar{\Psi}) = \frac{1}{2}\mathbf{\Psi}^{\mathrm{T}}\mathbf{g}^{-1}\mathcal{F}(-\mathbf{R}_{\Psi,\bar{\Psi}})\mathbf{\Psi}, \tag{28}$$

where

$$\mathbf{\Psi} = \begin{pmatrix} \Psi_I \\ \bar{\Psi}_{\bar{I}} \end{pmatrix}, \quad \mathbf{g}^{-1} = \begin{pmatrix} 0 & g^{I\bar{J}} \\ g^{\bar{I}J} & 0 \end{pmatrix}, \tag{29}$$

$$\mathcal{F}(x) = \frac{1}{x}\left\{\sqrt{1+4x} - 1 - \ln\left(\frac{1+\sqrt{1+4x}}{2}\right)\right\}, \quad \mathcal{F}(0) = 1. \tag{30}$$

For our purpose to construct the $\mathcal{N}=2$ supersymmetric NLSMs on the cotangent bundles over exceptional Hermitian symmetric spaces, we shall rewrite (28) into a more convenient form [14]. First performing the Taylor expansion for (28) one can see that $\mathcal{H}$ has the same form as (26) with the coefficient c_n given by

$$c_n = \frac{(-1)^{n-1}\mathcal{F}^{(n-1)}(0)}{(n-1)!}. \tag{31}$$

Second we introduce the differential operators

$$\mathcal{R}_{\Psi,\bar{\Psi}} = -(R_\Psi)_I{}^{\bar{J}}\bar{\Psi}_{\bar{J}}\frac{\partial}{\partial\Psi_I}, \quad \bar{\mathcal{R}}_{\Psi,\bar{\Psi}} = -(R_{\bar{\Psi}})_{\bar{I}}{}^{J}\Psi_J\frac{\partial}{\partial\bar{\Psi}_{\bar{I}}}, \tag{32}$$

which satisfy the following identity

$$\mathcal{R}_{\Psi,\bar{\Psi}}\mathcal{H}^{(n)} = \bar{\mathcal{R}}_{\Psi,\bar{\Psi}}\mathcal{H}^{(n)}. \tag{33}$$

With the use of (32), we can prove that (21) and (22) are compactly written as

$$G^{(n+1)} = \frac{(-\mathcal{R}_{\Psi,\bar{\Psi}})^n}{n!}|\Psi|^2, \quad |\Psi|^2 := g^{I\bar{J}}\Psi_I\bar{\Psi}_{\bar{J}}, \quad n \geq 1. \tag{34}$$

Substituting (34) with (31) into (26), we find

$$\mathcal{H} = \sum_{n=0}^{\infty}\frac{\mathcal{F}^{(n)}(0)}{n!}\frac{(\mathcal{R}_{\Psi,\bar{\Psi}})^n}{n!}|\Psi|^2. \tag{35}$$

Making use of the following formula

$$\frac{x^n}{n!} = \oint_C \frac{d\xi}{2\pi i}\frac{e^{\xi x}}{\xi^{n+1}}, \tag{36}$$

where the contour C encircles the origin of the complex ξ-plane in the counterclockwise direction, we have

$$\mathcal{H} = \sum_{n=0}^{\infty}\frac{\mathcal{F}^{(n)}(0)}{n!}\oint_C\frac{d\xi}{2\pi i}\frac{e^{\xi\mathcal{R}_{\Psi,\bar{\Psi}}}}{\xi^{n+1}}|\Psi|^2. \tag{37}$$

Here the contour C must be chosen such that the function $e^{\xi\mathcal{R}_{\Psi,\bar{\Psi}}}|\Psi|^2$ is analytic. Keeping this in mind, one finds that (37) can be transformed into

$$\mathcal{H} = \oint_C\frac{d\xi}{2\pi i}\frac{\mathcal{F}(1/\xi)}{\xi}e^{\xi\mathcal{R}_{\Psi,\bar{\Psi}}}|\Psi|^2. \tag{38}$$

The function $\mathcal{F}(1/\xi)/\xi$ gives a branch cut between -4 and 0. Since the $\mathcal{H}$ is regular and analytic, the contour C has to be chosen so that it does not cross the branch cut. The resultant contour encircles $\xi=-4,0$ without crossing the cut and is bounded by poles of the factor $e^{\xi\mathcal{R}_{\Psi,\bar{\Psi}}}|\Psi|^2$. One can transform the contour C as $C'+\tilde{C}$ where C' encircles the poles that may arise from $e^{\xi\mathcal{R}_{\Psi,\bar{\Psi}}}|\Psi|^2$ in the counterclockwise direction and $\tilde{C}$ encircles those poles together with the branch cut in the clockwise direction. One can check that contribution from the contour $\tilde{C}$ is just a constant by substituting $\xi = Re^{i\theta}$ with $R\to\infty$. Therefore, this does not contribute to the Kähler metric and can be neglected. We finally have

$$\mathcal{H} = -\oint_{-C'}\frac{d\xi}{2\pi i}\frac{\mathcal{F}(1/\xi)}{\xi}e^{\xi\mathcal{R}_{\Psi,\bar{\Psi}}}|\Psi|^2, \tag{39}$$

where $-C'$ goes in the counterclockwise direction, yielding the minus sign in front of the integration.

3. Cotangent bundle

In this section, we derive the cotangent bundle action of $E_{7(-25)}/E_6 \times U(1)$ by using (39). To this end, we first derive the Kähler potential on $E_{7(-25)}/E_6 \times U(1)$, which was firstly obtained in [16]. The Kähler potential is written by the coordinates which are representation of the broken generator for the case when $E_{7(-25)}$ is broken down to $E_6 \times U(1)$. We shall write them as

$$\Phi^I \to \phi^i, \quad \bar{\Phi}^{\bar{I}} \to \bar{\phi}_i. \tag{40}$$

The transformation law for ϕ is obtained from the commutation relations:

$$[E^i, \phi^j] = \phi^i \phi^j - \frac{1}{2}\Gamma^{ijk}\Gamma_{klm}\phi^l\phi^m, \tag{41}$$

$$[E^i, \bar{\phi}_j] = \delta^i_j, \quad [\bar{E}_i, \phi^j] = \delta^j_i, \tag{42}$$

$$[\bar{E}_i, \bar{\phi}_j] = \bar{\phi}_i\bar{\phi}_j - \frac{1}{2}\Gamma_{ijk}\Gamma^{klm}\bar{\phi}_l\bar{\phi}_m, \tag{43}$$

$$[T_A, \phi^i] = -2i\rho(T_A)^i{}_j\phi^j, \quad [T_A, \bar{\phi}_i] = 2i\bar{\phi}_j\rho(T_A)^j{}_i, \tag{44}$$

$$[T, \phi^i] = -i\sqrt{\frac{2}{3}}\phi^i, \quad [T, \bar{\phi}_i] = i\sqrt{\frac{2}{3}}\bar{\phi}_i. \tag{45}$$

Here Γ^{ijk} is the invariant tensor of E_6. This is symmetric with respect to the indices (i, j, k) and the complex conjugate is defined as $(\Gamma^{ijk})^\dagger = \Gamma_{ijk}$. This satisfies the following identity

$$\Gamma_{ijk}\Gamma^{ljk} = 10\delta^l_i, \tag{46}$$

and the Springer relation [19]

$$\Gamma^{ijk}\Gamma_{jl(m}\Gamma_{pq)k} = \delta^i_{(l}\Gamma_{mpq)}. \tag{47}$$

One can see the closure of the algebra by checking the Jacobi identity.

The commutation relations (41)–(43) lead to the infinitesimal transformation law for ϕ^i:

$$\begin{aligned}\delta\phi^i &= \epsilon^j[\bar{E}_j, \phi^i] + \bar{\epsilon}_j[E^j, \phi^i] \\ &= \epsilon^i - (\bar{\epsilon}_j\phi^j)\phi^i + \frac{1}{2}\Gamma^{ijk}\bar{\epsilon}_j\Gamma_{klm}\phi^l\phi^m,\end{aligned} \tag{48}$$

where ϵ^i is the complex transformation parameter.

Now we look for the function being invariant under (48). To this end, we introduce the $E_6 \times U(1)$ invariants:

$$I_1 = \bar{\phi}_i\phi^i, \tag{49}$$

$$I_2 = (\Gamma_{ijk}\phi^j\phi^k)(\Gamma^{ilm}\bar{\phi}_l\bar{\phi}_m), \tag{50}$$

$$I_3 = \frac{1}{9}(\Gamma_{ijk}\phi^i\phi^j\phi^k)(\Gamma^{lmn}\bar{\phi}_l\bar{\phi}_m\bar{\phi}_n). \tag{51}$$

They transform under (48) as

$$\delta I_1 = (1 - I_1)(\bar{\epsilon}_i\phi^i) + \frac{1}{2}(\Gamma_{ijk}\phi^j\epsilon^k)(\Gamma^{ilm}\bar{\phi}_l\bar{\phi}_m) + \text{c.c.} \tag{52}$$

$$\delta I_2 = 2(\Gamma_{ijk}\phi^j\phi^k)(\Gamma^{ilm}\bar{\phi}_l\bar{\phi}_m) - (\bar{\epsilon}_i\phi^i)I_2 + \frac{1}{3}(\Gamma_{ijk}\phi^i\phi^j\phi^k)(\Gamma^{lmn}\bar{\phi}_l\bar{\phi}_m\bar{\phi}_n) + \text{c.c.} \tag{53}$$

$$\delta I_3 = -(\bar{\epsilon}_i\phi^i)I_3 + \frac{1}{3}(\Gamma_{ijk}\phi^i\phi^j\phi^k)(\Gamma^{lmn}\bar{\epsilon}_l\bar{\phi}_m\bar{\phi}_n) + \text{c.c.} \tag{54}$$

By using (52)–(54), one can check that the function

$$K = -\ln\left(1 - I_1 + \frac{1}{4}I_2 - \frac{1}{4}I_3\right) \tag{55}$$

transforms under the infinitesimal transformation (48) as

$$\delta K = (\bar{\epsilon}_i \phi^i) + \text{c.c.} \tag{56}$$

This shows that (55) is invariant under (48) up to the Kähler transformation (56). Thus we conclude that (55) is the Kähler potential of $E_{7(-25)}/E_6 \times U(1)$. The sign in front of the logarithm in (55) cannot be determined by the invariance under (48). It is just chosen so that positivity of the metric is ensured.

We are now ready to construct the cotangent bundle over $E_{7(-25)}/E_6 \times U(1)$. First we introduce the cotangent vectors:

$$\Psi \to \Psi_i, \quad \bar{\Psi}^{\bar{I}} \to \bar{\Psi}^i \tag{57}$$

Since we are considering the symmetric space, we set $\phi = \bar{\phi} = 0$ in calculations. The metric and the Riemann tensor at $\phi = \bar{\phi} = 0$ are derived from (55):

$$g_i{}^j\Big|_{\phi=\bar{\phi}=0} = \frac{\partial^2 K}{\partial\phi^i \partial\bar{\phi}_j}\Big|_{\phi=\bar{\phi}=0} = \delta_i{}^j, \tag{58}$$

$$R_i{}^j{}_k{}^l\Big|_{\phi=\bar{\phi}=0} = \delta_i{}^l\delta_k{}^j - \Gamma_{mik}\Gamma^{mjl} + \delta_k{}^l\delta_i{}^j. \tag{59}$$

Then the differential operator (32) at $\phi = \bar{\phi} = 0$ is obtained as

$$\mathcal{R}_{\Psi,\bar{\Psi}}\Big|_{\phi=\bar{\phi}=0} = -|\psi|^2\psi_i\frac{\partial}{\partial\psi_i} + \frac{1}{2}(\Gamma^{mik}\psi_i\psi_k)\Gamma_{mlj}\bar{\psi}^j\frac{\partial}{\partial\psi_l}, \quad |\psi|^2 := \psi_i\bar{\psi}^i, \tag{60}$$

where ψ and $\bar{\psi}$ are coordinates of the cotangent space at $\phi = \bar{\phi} = 0$. If we define the $E_6 \times U(1)$ invariants in terms of the cotangent vector

$$x := \psi_i\bar{\psi}^i, \tag{61}$$

$$y := (\Gamma^{ijk}\psi_j\psi_k)(\Gamma_{ilm}\bar{\psi}^l\bar{\psi}^m), \tag{62}$$

$$z := (\Gamma^{ijk}\psi_i\psi_j\psi_k)(\Gamma_{lmn}\bar{\psi}^l\bar{\psi}^m\bar{\psi}^n), \tag{63}$$

the differential operator (60) is rewritten as

$$\mathcal{R}_{\Psi,\bar{\Psi}}\Big|_{\phi=\bar{\phi}=0} = xD - \frac{1}{2}y\frac{\partial}{\partial x} - \frac{1}{3}z\frac{\partial}{\partial y}, \quad D := x\frac{\partial}{\partial x} + y\frac{\partial}{\partial y} + z\frac{\partial}{\partial z}. \tag{64}$$

The factor $e^{\xi\mathcal{R}_{\Psi,\bar{\Psi}}}x$ in (39) is calculated by using the Baker-Campbell-Hausdorff formula

$$\begin{aligned} e^{\xi\mathcal{R}_{\Psi,\bar{\Psi}}}x\Big|_{\phi=\bar{\phi}=0} &= e^{-\xi xD}e^{\frac{1}{2}\xi y\frac{\partial}{\partial x}}e^{\frac{1}{3}\xi z\frac{\partial}{\partial y}}e^{\frac{1}{12}\xi^2 z\frac{\partial}{\partial x}}e^{-\frac{1}{4}\xi^2 yD}e^{\frac{1}{18}\xi^3 zD}x \\ &= \frac{\partial}{\partial\xi}\ln\left(1 + \xi x + \frac{1}{4}\xi^2 y + \frac{1}{36}\xi^3 z\right). \end{aligned} \tag{65}$$

The poles arising from this factor contribute to the integral in (39), which are obtained from the equation

$$1 + \xi x + \frac{1}{4}\xi^2 y + \frac{1}{36}\xi^3 z = \frac{z}{36}(\xi - \xi_1)(\xi - \xi_2)(\xi - \xi_3) = 0. \tag{66}$$

The explicit expressions of the poles are given in [16]. The cotangent bundle part (39) is led to the form:

$$\begin{aligned} \mathcal{H} &= -\oint_{-C'} \frac{d\xi}{2\pi i} \frac{\mathcal{F}(1/\xi)}{\xi} \frac{\partial}{\partial \xi} \ln\left(1 + \xi x + \frac{1}{4}\xi^2 y + \frac{1}{36}\xi^3 z\right) \\ &= -\left(\frac{\mathcal{F}(1/\xi_1)}{\xi_1} + \frac{\mathcal{F}(1/\xi_2)}{\xi_2} + \frac{\mathcal{F}(1/\xi_3)}{\xi_3}\right). \end{aligned} \tag{67}$$

The result at an arbitrary point of Φ can be obtained by the following replacements

$$x \to (g^{-1})_i{}^j \Psi_j \bar{\Psi}^i, \tag{68}$$

$$\frac{1}{4}y \to \frac{1}{2}((g^{-1})_i{}^j \Psi_j \bar{\Psi}^i)^2 - \frac{1}{4}\tilde{R}_i{}^j{}_k{}^l \bar{\Psi}^i \Psi_j \bar{\Psi}^k \Psi_l, \tag{69}$$

$$\begin{aligned} -\frac{1}{36}z \to & -\frac{1}{6}((g^{-1})_i{}^j \Psi_j \bar{\Psi}^j)^3 + \frac{1}{4}((g^{-1})_i{}^j \Psi_j \bar{\Psi}^i)(\tilde{R}_k{}^l{}_m{}^n \bar{\Psi}^k \Psi_l \bar{\Psi}^m \Psi_n) \\ & - \frac{1}{12}|(g^{-1})_i{}^j \tilde{R}_j{}^k{}_l{}^m \Psi_k \bar{\Psi}^l \Psi_m|^2, \end{aligned} \tag{70}$$

where $\tilde{R}_i{}^j{}_k{}^l = (g^{-1})_i{}^m (g^{-1})_n{}^j (g^{-1})_k{}^p (g^{-1})_q{}^l R_m{}^n{}_p{}^q$.

4. Conclusion

We have constructed the $\mathcal{N} = 2$ SUSY NLSM on the cotangent bundle over the non-compact exceptional HSS $\mathcal{M} = E_{7(-25)}/E_6 \times U(1)$ by using the results elaborated in [13] and [14]. The point is to use the projective superspace formalism which is an $\mathcal{N} = 2$ off-shell superfield formulation. Once an $\mathcal{N} = 1$ SUSY NLSM on a certain Kähler manifold is obtained, it is possible to extend it to the $\mathcal{N} = 2$ SUSY model containing the corresponding $\mathcal{N} = 1$ SUSY NLSM. We first have derived the transformation law of the fields parameterizing $\mathcal{M}$ and have constructed the $\mathcal{N} = 1$ SUSY NLSMs on $\mathcal{M}$ invariant under the derived transformation laws. Second we have extended the $\mathcal{N} = 1$ SUSY NLSM to one with the $\mathcal{N} = 2$ SUSY model by using the explicit formula of the cotangent bundle over any HSS developed in [13, 14]. We have also constructed the $\mathcal{N} = 2$ SUSY NLSM on the cotangent bundle over $E_{6(-14)}/SO(10) \times U(1)$ in [16] (where more detailed derivation for $E_{7(-25)}/E_6 \times U(1)$ is given). By the series of works [8, 9, 11, 12, 13, 14, 16], we have completed constructing the Kähler potentials of the cotangent bundles over *all* the compact and non-compact HSSs listed in Table 1.

References

[1] Zumino B 1979 Supersymmetry and Kähler manifolds *Phys. Lett.* B **87** 203
[2] Alvarez-Gaumé L and Freedman D. Z. 1981 Geometrical structure and ultraviolet finiteness in the supersymmetric sigma model *Commun. Math. Phys.* **80** 443
[3] Karlhede A, Lindström U and Roček M 1984 Self-interacting tensor multiplets in N=2 superspace *Phys. Lett.* B **147** 297
[4] Hitchin N J, Karlhede A, Lindström U and Roček M 1987 Hyperkähler metrics and supersymmetry *Commun. Math. Phys.* **108** 535
[5] Lindström U and Roček M 1988 New hyperkähler metrics and new supermultiplets 1988 *Commun. Math. Phys.* **115** 21
[6] Lindström U and Roček M 1990 N=2 super Yang-Mills theory in projective superspace *Commun. Math. Phys.* **128** 191
[7] Gonzalez-Rey F, Lindström U, Roček M, Wiles S and von Unge R 1998 Feynman rules in N = 2 projective superspace. I: Massless hypermultiplets *Nucl. Phys.* B **516** 426
[8] Gates Jr S J and Kuzenko S M 1999 The CNM-hypermultiplet nexus *Nucl. Phys.* B **543** 122
[9] Gates Jr S J and Kuzenko S M 2000 4D N = 2 supersymmetric off-shell sigma models on the cotangent bundles of Kähler manifolds *Fortsch. Phys.* **48** 115

[10] Arai M and Nitta M 2006 Hyper-Kähler sigma models on (co)tangent bundles with SO(n) isometry *Nucl. Phys.* B **745** 208

[11] Arai M, Kuzenko S M and Lindström U 2007 Hyperkähler sigma models on cotangent bundles of Hermitian symmetric spaces using projective superspace J. High Energy Phys. JHEP02(2007)100

[12] Arai M, Kuzenko S M and Lindström U 2007 Polar supermultiplets, Hermitian symmetric spaces and hyperkahler metrics J. High Energy Phys. JHEP12(2007)008

[13] Kuzenko S M and Novak J 2008 Chiral formulation for hyperkahler sigma-models on cotangent bundles of symmetric spaces J. High Energy Phys. JHEP12(2008)072

[14] Arai M and Blaschke F 2013 Cotangent bundle over Hermitian symmetric space $E_7/E_6 \times U(1)$ from projective superspace J. High Energy Phys. JHEP02(2013)045

[15] Cartan É 1935 Sur les domaines bornés homogénes de l'espace de n variables complexes *Math. Sem. Univ. Hamburg* **11** 116

[16] Arai M and Baba K 2015 Supersymmetry and Cotangent Bundle over Non-compact Exceptional Hermitian Symmetric Space J. High Energy Phys. JHEP07(2015)169

[17] Kuzenko S M 1999 Projective superspace as a double punctured harmonic superspace *Int. J. Mod. Phys.* A **14** 1737

[18] Kuzenko S M and Linch W D 2006 On five-dimensional superspaces J. High Energy Phys. JHEP02(2006)038

[19] Springer T A 1962 *Proc. Kon. Ned. Akad. Wet.* A **65** 259

Journal of Physics: Conference Series **670** (2016) 012006 doi:10.1088/1742-6596/670/1/012006

Low energy dynamics of slender monopoles in non-Abelian superconductor

M Arai[1], F Blaschke[1], M Eto[1] and N Sakai[2]

[1] Department of Physics, Yamagata University, Yamagata 990-8560, Japan
[2] Department of Physics, and Research and Education Center for Natural Sciences, Keio University, Hiyoshi 4-1-1, Yokohama, Kanagawa 223-8521, Japan

E-mail: fblasch@sci.kj.yamagata-u.ac.jp

Abstract. Low energy dynamics of magnetic monopoles and anti-monopoles in the $U(2)_{\mathrm{C}}$ gauge theory is studied in the Higgs (non-Abelian superconducting) phase. The monopoles in this phase are slender ellipsoids, pierced by a vortex string. We investigate scattering of monopole with anti-monopole and find that they do not always decay into radiation, contrary to our naive intuition. They can repel, make bound states (magnetic mesons) or resonances. We point out that some part of solutions in $1+3$ dimensions can be mapped exactly onto the sine-Gordon system in $1+1$ dimensions in the first non-trivial order of rigid-body approximation and we provide analytic formulas for such solutions there.

1. Introduction

Among various topological solitons, magnetic monopoles are one of the most fascinating ones in high energy physics. If a monopole exists, electric charges in the universe are quantized by the Dirac's quantization condition [2]: product of electric charge $Q_{\mathrm{e}}e$ and magnetic charge $Q_{\mathrm{m}}g$ must be proportional to integers: $Q_{\mathrm{e}}eQ_{\mathrm{m}}g \propto n$. This implies that a weak electric coupling e corresponds to a strong magnetic coupling g. This strong-weak coupling duality is a powerful tool to understand strong coupling physics. Furthermore, monopoles are expected to play an important role to explain the confinement in QCD. It has been proposed that the vacuum is in dual color superconductor where magnetic monopoles condense [3, 4, 5]. It is, however, very difficult to verify this idea, since QCD is strongly coupled at low energies because of the asymptotic freedom.

In contrast, QCD becomes weakly coupled at high baryon density, and enters into the color-superconducting phase where di-quarks condense [6, 7]. Then magnetic monopoles should be confined and color magnetic fields form flux tubes. A distinctive feature of non-Abelian vortex compared to the usual Nielsen-Olesen Abelian vortex is that the non-Abelian vortex breaks a non-Abelian global symmetry of the vacuum state and non-Abelian orientational moduli [8, 9, 10] emerge. This orientational moduli has been found to give a confining state of monopole and anti-monopole, namely magnetic meson was predicted in high density QCD [11, 12, 13].

If we choose couplings to critical values which enable us to embed the theory into a supersymmetric one, we can have Bogomolnyi-Prasad-Sommerfield (BPS) solitons [14], which preserve a part of supersymmetry [15]. The dynamics of slowly moving BPS solitons are well-approximated by a geodesic motion on the moduli space [16]. This is called the moduli approximation. Although the moduli approximation is useful, one should note that it can be applied neither for scattering of BPS solitons with high momentum nor for non-BPS systems. Among non-BPS solutions, an interesting non-BPS "bound state" of a monopole and an anti-monopole in the Coulomb phase has been rigorously established [17]. Although these "bound states" eventually decay due to unstable modes, they can play a significant role in understanding the dynamics of monopole and anti-monopole system in the Coulomb phase.

If the non-Abelian gauge symmetry is completely broken, we are in the Higgs phase, namely the non-Abelian superconducting phase. In contrast to the Coulomb phase, monopoles in the Higgs phase have several distinctive features. Firstly, they are pierced by a vortex string. In fact, the static BPS monopole in the Higgs phase has been found as a static BPS kink solution in the $1+1$-dimensional low-energy effective field theory on the vortex [18]. Secondly, shape of a monopole is not spherical. There are two fundamental length scales: One is a transverse size $L_{\rm T}$ of the flux tube and the other is a length $L_{\rm L}$ of the monopole. Since a monopole resides on a vortex, its shape is generally not spherical depending on the ratio of the two scales. If we assume

$$L_{\rm T} \ll L_{\rm L}, \tag{1}$$

monopoles are of a slender ellipsoidal shape. This approximation has been used previously [19] to obtain an effective action of 1/4 BPS non-Abelian monopole-vortex complex. Another work to obtain effective action of monopole-vortex complex appeared recently [20].

In this work, we will consider a straight vortex string and we study head-on collisions of monopoles and anti-monopoles. In contrast to lots of studies on scattering of the BPS monopoles, there is very few studies about the dynamics of BPS monopoles and anti-monopoles, especially in the Higgs phase. This is mainly due to the inapplicability of the moduli approximation to the non-BPS monopoles and anti-monopoles system, which necessitates other methods such as numerical analysis. Instead of numerical methods, we consider a systematic expansion in powers of the ratio $L_{\rm T}/L_{\rm L}$ of length scales of the model, which allows us to obtain analytic solutions. At the first order of the expansion, we obtain the rigid-body approximation, where we can provide analytic solutions, which are very useful to understand the dynamics of monopole and anti-monopole system. The monopole and anti-monopole dynamics here is essentially $1+1$-dimensional. We will observe that a part of the dynamics can be mapped on to the integrable sine-Gordon model. We provide exact mapping between $1+3$-dimensional field configurations of gauge theory and $1+1$-dimensional field configurations of the sine-Gordon model.

2. Model & topological solitons

Let us consider a $U(2)_{\rm C}$ Yang-Mills-Higgs system

$$\mathcal{L} = {\rm Tr}\left[-\frac{1}{2g^2}F_{\mu\nu}F^{\mu\nu} + D_\mu H (D^\mu H)^\dagger + \frac{1}{g^2}D_\mu\Sigma D^\mu\Sigma\right] - V, \tag{2}$$

$$V = {\rm Tr}\left[\frac{g^2}{4}\left(HH^\dagger - v^2\mathbf{1}_2\right)^2 + (\Sigma H - HM)(\Sigma H - HM)^\dagger\right], \tag{3}$$

where the field strength and the covariant derivatives are defined by

$$F_{\mu\nu} = \partial_\mu A_\nu - \partial_\nu A_\mu + i\left[A_\mu, A_\nu\right], \tag{4}$$

$$D_\mu H = \partial_\mu H + iA_\mu H, \tag{5}$$

$$D_\mu\Sigma = \partial_\mu\Sigma + i\left[A_\mu, \Sigma\right]. \tag{6}$$

The N_F species of Higgs fields in the fundamental representation of the $U(2)_{\rm C}$ gauge group is denoted by a $2\times N_{\rm F}$ matrix H. We concentrate on $N_{\rm F}=2$ case in the following. Another Higgs field Σ in the adjoint representation of the $U(2)_{\rm C}$ gauge group is denoted by a real 2×2 matrix. The quartic scalar coupling is given in terms of the gauge coupling constant g, which allows our model to be embeded in a supersymmetric theory. The parameter v giving the vacuum expectation value of the Higgs field H comes from the so-called Fayet-Illiopoulos (FI) parameter in the supersymmetric context. We assume $v>0$ in what follows. We take the mass matrix M in the following form

$$M=\frac{m}{2}\sigma_3. \tag{7}$$

Global symmetry of the model depends on the mass parameter m. If $m=0$, the flavor symmetry is $SU(2)_{\rm F}$. If $m\neq 0$, the flavor symmetry reduces to $U(1)_{\rm F}\subset SU(2)_{\rm F}$ generated by the third component of $SU(2)_{\rm F}$. In this work, we consider $m\neq 0$ case unless stated otherwise.

The vacuum of the model is determined by the condition $V=0$:

$$HH^\dagger=v^2\mathbf{1}_2,\quad \Sigma=M. \tag{8}$$

The vacuum manifold is topologically a torus T^2 but we can make $U(2)_{\rm C}$ gauge transformations to bring the fields to the following representative value

$$H=v\mathbf{1}_2,\quad \Sigma=M. \tag{9}$$

Therefore all the points in the vacuum manifold are physically equivalent, and the vacuum moduli space consists of only one point. We call this vacuum the color-flavor locking (CFL) vacuum. The vacuum is in the Higgs phase, where the gauge symmetry is completely broken.

The model (2) admits rich topological excitations; vortex strings and magnetic monopoles. In the Higgs vacuum, magnetic field can only exist by having an unbroken normal vacuum in a small neighbourhood of the zero of the Higgs field. Hence magnetic field is squeezed into a vortex, which we call a vortex string. The vortex string is topologically stable due to a non-trivial fundamental homotopy group in the massive case

$$\pi_1(T^2)=\mathbb{Z}\times\mathbb{Z}. \tag{10}$$

There are two kinds of vortex quantum numbers, corresponding to two kinds of vortex strings, which we call the N-vortex and S-vortex. A magnetic monopole is a source of the conserved magnetic fluxes which are squeezed into vortex strings in this Higgs vacuum. Therefore a stable magnetic monopole is possible only as a composite soliton in the middle of a vortex string, but cannot exist as an isolated soliton, which can also be understood from the trivial homotopy

$$\pi_2(T^2)=0. \tag{11}$$

In the $U(2)_{\rm C}$ Yang-Mills Higgs theory, solutions for the magnetic monopole pierced by vortex strings have been found [18, 21], which preserve a quarter of supersymmetry charges when embedded into the supersymmetric theory.

The total energy E for 1/4 BPS configuration can be written as the sum of two topological charges representing the monopole energy $M_{\rm mono}$ and the vortex energy $M_{\rm vor}$

$$E \geq M_{\rm mono}+M_{\rm vor}, \tag{12}$$

$$M_{\rm mono} = \frac{1}{g^2}\int d^3x\ {\rm Tr}[\epsilon_{ijk}\partial_i\left(\Sigma F_{jk}\right)], \tag{13}$$

$$M_{\rm vor} = -v^2\int d^3x\ {\rm Tr}[F_{12}]. \tag{14}$$

This bound is saturated when the following BPS equations are satisfied

$$F_{12} - D_3\Sigma - \frac{g^2}{2}\left(HH^\dagger - v\mathbf{1}_2\right) = 0, \tag{15}$$
$$F_{23} - D_1\Sigma = 0, \tag{16}$$
$$F_{31} - D_2\Sigma = 0, \tag{17}$$
$$(D_1 + iD_2)H = 0, \tag{18}$$
$$D_3H + \Sigma H - HM = 0. \tag{19}$$

If we define a 2×2 matrix field S taking values in $GL(2, \mathbb{C})$ whose elements are functions of $x^{1,2,3}$ as

$$\bar{A} = -iS^{-1}\bar{\partial}S, \quad A_3 - i\Sigma = -iS^{-1}\partial_3 S, \tag{20}$$

we can solve the equations (16) – (19) in terms of a holomorphic matrix $H_0(z)$

$$H = vS^{-1}(x^1, x^2, x^3)H_0(z)e^{Mx^3}, \tag{21}$$

where we have defined

$$z = x^1 + ix^2, \quad \bar{z} = x^1 - ix^2, \quad \partial = \frac{\partial_1 - i\partial_2}{2}, \quad \bar{\partial} = \frac{\partial_1 + i\partial_2}{2}, \tag{22}$$
$$A = \frac{A_1 - iA_2}{2}, \quad \bar{A} = \frac{A_1 + iA_2}{2}, \quad D = \frac{D_1 - iD_2}{2}, \quad \bar{D} = \frac{D_1 + iD_2}{2}. \tag{23}$$

This method to solve the BPS equation is called the moduli matrix formalism [21, 1]. The following V-transformations leave the physical fields H in Eq. (21) and A_i and Σ in Eq. (20) unchanged.

$$S(x^1, x^2, x^3) \to V(z)S(x^1, x^2, x^3), \quad H_0(z) \to V(z)H_0(z), \qquad V(z) \in GL(2, \mathbb{C}), \tag{24}$$

where elements of the $GL(2, \mathbb{C})$ matrix $V(z)$ are holomorphic functions in z. Therefore the moduli space of the monopole vortex complex becomes the moduli matrices divided by the V-equivalence relation.

The $U(2)_{\mathrm{C}}$ gauge transformations act on S^{-1} from the left as

$$S^{-1} \to U_{\mathrm{C}} S^{-1}. \tag{25}$$

By defining $U(2)_{\mathrm{C}}$ gauge invariant matrices Ω and Ω_0,

$$\Omega = SS^\dagger, \quad \Omega_0 = H_0 e^{2Mx^3} H_0^\dagger, \tag{26}$$

we can cast the remaining BPS equation (15) into the following master equation

$$\frac{1}{g^2v^2}\left[4\bar{\partial}\left(\partial\Omega\Omega^{-1}\right) + \partial_3\left(\partial_3\Omega\Omega^{-1}\right)\right] = \mathbf{1}_2 - \Omega_0\Omega^{-1}. \tag{27}$$

This master equation should be solved with the boundary condition

$$\Omega \to \Omega_0 \qquad \text{as} \qquad |\vec{x}| \to \infty. \tag{28}$$

The $U(2)_{\mathrm{C}}$ gauge invariants Ω and Ω_0 are covariant under the V-transformations.

Journal of Physics: Conference Series **670** (2016) 012006 doi:10.1088/1742-6596/670/1/012006

3. Vortex strings

Before describing the monopole-vortex complex, let us first explain a simpler configuration of vortex strings without monopoles. We have two different types of Abelian vortices. If the complex zero is placed in the upper-left corner in the moduli matrix we obtain

$$H_0 = \begin{pmatrix} z & 0 \\ 0 & 1 \end{pmatrix}, \quad S = \begin{pmatrix} e^{\frac{\psi(x^1,x^2)}{2}} & 0 \\ 0 & 1 \end{pmatrix} e^{Mx^3}, \tag{29}$$

which is called an N-vortex, We find that the master equation for the N-vortex (29) reduces to the master equation for the Abelian vortex

$$\frac{4}{g^2v^2}\partial\bar{\partial}\psi = 1 - |z|^2 e^{-\psi}. \tag{30}$$

This equation has no known analytic solution, but can easily be solved numerically. The asymptotic behavior for ψ is given by

$$\psi \to \log|z|^2 + qK_0(gv|z|), \qquad \text{as} \quad |z| \to \infty, \tag{31}$$

where K_0 is the modified Bessel function of the second kind and a constant q can be obtained numerically.

For later convenience, let us decompose the magnetic field F_{12} in the $U(2)_{\rm C} = (U(1)_{\rm C0} \times SU(2)_{\rm C})/\mathbb{Z}_2$ gauge group into the field F^0_{12} for the overall $U(1)_{\rm C0}$ and the field F^{Σ}_{12} projected along the adjoint field Σ (this is identical to the third component $U(1)_{\rm C3}$ of $SU(2)_{\rm C}$ in the present case) as

$$F^0_{ij} = \mathrm{Tr}\left[F_{ij}\frac{\mathbf{1}_2}{2}\right], \quad F^{\Sigma}_{ij} = \mathrm{Tr}\left[F_{ij}\frac{\Sigma}{m}\right]. \tag{32}$$

We will call F^0_{12} as Abelian magnetic field and F^{Σ}_{12} as non-Abelain magnetic field. Note that these two magnetic fields are associated with $(U(1)_{\rm C0} \times U(1)_{\rm C3})/\mathbb{Z}_2 \subset U(2)_{\rm C}$ (asymptotically) which is not broken by the adjoint scalar field $\Sigma = (m/2)\sigma_3$. A linear combination of these $U(1)$ gauge symmetries is restored inside vortices. Therefore they are precisely the appropriate magnetic fields to measure the magnetic flux flowing to infinity through vortices. For the N-vortex, we obtain Abelian and non-Abelian magnetic fields as

$$F^0_{12} = -\partial\bar{\partial}\psi, \quad F^{\Sigma}_{12} = -\partial\bar{\partial}\psi. \tag{33}$$

We see that only the sum $F^0_{12} + F^{\Sigma}_{12}$ has nonvanishing magnetic field inside the N-vortex. This linear combination precisely corresponds to the restored $U(1)$ gauge symmetry inside the N-vortex.

Another possibility to place the zero of the Higgs field is at the lower-right corner of the moduli matrix as

$$H_0 = \begin{pmatrix} 1 & 0 \\ 0 & z \end{pmatrix}, \quad S = \begin{pmatrix} 1 & 0 \\ 0 & e^{\psi/2} \end{pmatrix} e^{Mx^3}, \tag{34}$$

where ψ is the same function as the N-vortex. We call this the S-vortex. The Abelian F^0_{12} and non-Abelian F^3_{12} magnetic fields of the S-vortex is given as

$$F^0_{12} = -\partial\bar{\partial}\psi, \quad F^{\Sigma}_{12} = +\partial\bar{\partial}\psi. \tag{35}$$

We see that only the difference $F^0_{12} - F^{\Sigma}_{12}$ has nonvanishing magnetic field inside the S-vortex. This linear combination is the restored $U(1)$ gauge symmetry inside the S-vortex.

4. Rigid-body approximation

In this section, we consider monopoles in the non-Abelian superconducting phase. We use a systematic expansion up to the next-to-leading order within an approximation, which we call the rigid-body approximation. The transverse size of the vortex string $L_{\rm T} = 1/(gv)$ is associated with a large mass scale gv, and the longitudinal monopole size $L_{\rm L} = 1/m$ is associated with a small mass scale. Therefore the condition $gv \gg m$ introduces hierarchal mass scales in the system: the thin vortex-string is generated at the high energy scale $\sim gv$, and the slender monopole is generated at the lower energy scale $\sim m$.

This picture allows us to understand the slender monopole as a kink in the 1+1 dimensional theory on the vortex world-sheet [18]. Assuming

$$\epsilon = \frac{m}{gv} \sim \frac{\partial_\alpha}{\partial_i} \ll 1, \quad (\alpha = 0, 3 \text{ and } i = 1, 2), \tag{36}$$

we expand the fields in power series of ϵ

$$H = H^{(0)} + H^{(2)} + \cdots, \tag{37}$$
$$A_i = A_i^{(0)} + A_i^{(2)} + \cdots, \quad (i = 1, 2), \tag{38}$$
$$A_\alpha = A_\alpha^{(1)} + A_\alpha^{(3)} + \cdots, \quad (\alpha = 0, 3), \tag{39}$$
$$\Sigma - M = \Sigma^{(1)} + \Sigma^{(3)} + \cdots, \tag{40}$$

where the superscript (n) indicates the n-th order in powers of ϵ. Note that H and A_i start from the zeroth order because they are nontrivial in the background vortex-string configuration. On the other hand, since A_α and $\Sigma - M$ vanish in the background vortex-string configuration, they start from the first order.

Zero-th order: background vortex string for $m = 0$ Retaining only the zero-th order fields in Eqs. (37) – (40), we find the following zero-th order reductions of full equations of motion

$$2(D_z D_{\bar z} H^{(0)} + D_{\bar z} D_z H^{(0)}) - \frac{g^2}{2}(H^{(0)} H^{(0)\dagger} - v^2 \mathbf{1}_2) H^{(0)} = 0, \tag{41}$$
$$-\frac{4}{g^2} D_{\bar z} F_{z\bar z}^{(0)} + i(H^{(0)} D_{\bar z} H^{(0)\dagger} - D_{\bar z} H^{(0)} H^{(0)\dagger}) = 0. \tag{42}$$

The zero-th order solutions can be compactly expressed in the moduli matrix formalism as

$$H^{(0)} = v S^{(0)-1} H_0(z), \quad \bar A^{(0)} = -i S^{(0)-1} \bar\partial S^{(0)}, \tag{43}$$

with the master equation for the vortex

$$\frac{4}{g^2 v^2} \bar\partial \left(\partial \Omega^{(0)} \Omega^{(0)-1} \right) = \mathbf{1}_2 - \Omega_0^{(0)} \Omega^{(0)-1}, \quad \Omega^{(0)} = S^{(0)} S^{(0)\dagger}, \quad \Omega_0^{(0)} = H_0^{(0)} H_0^{(0)\dagger}. \tag{44}$$

When $M = 0$, the flavor symmetry is enhanced from $U(1)_{\rm F}$ to $SU(2)_{\rm F}$ and the symmetry of the vacuum becomes $SU(2)_{\rm C+F}$. A single vortex spontaneously breaks this symmetry to $U(1)_{\rm C+F}$. Therefore, the Nambu-Goldstone zero modes ϕ appear as a moduli

$$\mathbb{C}P^1 = \frac{SU(2)_{\rm C+F}}{U(1)_{\rm C+F}} \simeq S^2. \tag{45}$$

By introducing the moduli parameter ϕ as an inhomogeneous coordinate of the moduli space $\mathbb{C}P^1 \simeq S^2$, we can express the generic moduli matrix H_0 with the moduli parameter $\phi \in \mathbb{C}$

as a color-flavor $SU(2)_{\rm C+F}$ rotation of the N-vortex solution together with an accompanying V-transformation as

$$H_0^{(0)} = \begin{pmatrix} z & 0 \\ -\phi & 1 \end{pmatrix} = V \begin{pmatrix} z & 0 \\ 0 & 1 \end{pmatrix} U, \tag{46}$$

$$S^{(0)} = \begin{pmatrix} \frac{e^{\frac{\psi}{2}}+z|\phi|^2}{1+|\phi|^2} & \frac{(e^{\frac{\psi}{2}}-z)\bar{\phi}}{1+|\phi|^2} \\ -\phi & 1 \end{pmatrix} = V \begin{pmatrix} e^{\frac{\psi}{2}} & 0 \\ 0 & 1 \end{pmatrix} U, \tag{47}$$

with

$$U = \frac{1}{\sqrt{1+|\phi|^2}} \begin{pmatrix} 1 & \bar{\phi} \\ -\phi & 1 \end{pmatrix} \in SU(2)_{\rm C+F}, \tag{48}$$

$$V = \frac{1}{\sqrt{1+|\phi|^2}} \begin{pmatrix} 1 & -\bar{\phi}z \\ 0 & 1+|\phi|^2 \end{pmatrix}. \tag{49}$$

Here we need the V-transformation $V(z)$, in order for $H_0^{(0)}$ to be a holomorphic function of the moduli parameter ϕ. The single vortex solution with the generic moduli ϕ can be obtained explicitly by inserting the N-vortex solution ψ into Eqs. (47) and (43).

At the zero-th order in $\epsilon \ll 1$, we obtained a moduli parameter ϕ in Eqs. (48) and (49) as a constant. However, our approximation allows the weak dependence of ϕ on x^0, x^3 from the beginning. Therefore, we should consider $\phi(x^0, x^3)$ to be a slowly varying function of x^0, x^3. Then, the vortex background configuration depends on x^0 and x^3 only through the moduli field.

$$H^{(0)}(x^1, x^2; \phi(x^0, x^3)), \quad A_i^{(0)}(x^1, x^2; \phi(x^0, x^3)) \qquad (i = 1, 2). \tag{50}$$

We can determine $A_\alpha^{(1)}$ and $\Sigma^{(1)}$ as

$$A_\alpha^{(1)} = i\left[(\delta_\alpha S^{(0)\dagger})S^{(0)\dagger-1} - S^{(0)-1}\delta_\alpha^\dagger S^{(0)}\right], \quad (\alpha = 0, 3), \tag{51}$$

$$\Sigma^{(1)} = M + i\left[(\delta_\phi S^{(0)\dagger})S^{(0)\dagger-1} - S^{(0)-1}\delta_\phi^\dagger S^{(0)}\right], \tag{52}$$

with

$$\delta_\alpha = \partial_\alpha \phi \frac{\delta}{\delta\phi}, \quad \delta_\alpha^\dagger = \partial_\alpha \bar{\phi} \frac{\delta}{\delta\bar{\phi}}, \quad \delta_\phi = -im\phi \frac{\delta}{\delta\phi}, \quad \delta_\phi^\dagger = im\bar{\phi}\frac{\delta}{\delta\bar{\phi}}. \tag{53}$$

The remaining task is to look for the appropriate configurations of $\phi(x^0, x^3)$ which minimize the energy of the solution. To this end, we plug $A_\alpha^{(1)}$ and $\Sigma^{(1)}$ into the original Lagrangian (2) and pick up terms up to the second order in ϵ. After a tedious calculation, one obtains the following expression, where the $x^{1,2}$ and $x^{0,3}$ dependence are factorized as

$$\mathcal{L} = -v^2 F_{12}(x^1, x^2) + \frac{F(x^1, x^2)}{g^2} \times \frac{|\partial_\alpha \phi(x^0, x^3)|^2 - m^2|\phi(x^0, x^3)|^2}{(1+|\phi(x^0, x^3)|^2)^2} + \mathcal{O}(\epsilon^4), \tag{54}$$

where we ignore unessential total derivative terms. The prefactor in the second term depends on only x^1 and x^2 and it is given by

$$F(x^1, x^2) = 4\partial\bar{\partial}\psi(x^1, x^2). \tag{55}$$

Hence, in order to minimize the action to the second order, we need to find a stationary point of

$$\mathcal{L}^{(2)} = \frac{F(x^1,x^2)}{g^2} \times \frac{|\partial_\alpha \phi(x^0,x^3)|^2 - m^2|\phi(x^0,x^3)|^2}{(1+|\phi(x^0,x^3)|^2)^2}. \tag{56}$$

Since the prefactor $F(x^1,x^2)$ is determined at the zero-th order, our task is basically to solve the massive non-linear sigma model in two dimensions with the target space $\mathbb{C}P^1$. Note that the process here is essentially the same as a well known derivation of a low energy effective action in the moduli approximation. To obtain the effective action, one just needs to integrate the Lagrangian over x^1 and x^2. The resulting overall coefficient is $4\pi = \int dx^1dx^2\ F(x^1,x^2)$ and thus

$$\mathcal{L}_{\text{eff}} = \frac{4\pi}{g^2}\frac{|\partial_\alpha \phi(x^0,x^3)|^2 - m^2|\phi(x^0,x^3)|^2}{(1+|\phi(x^0,x^3)|^2)^2}. \tag{57}$$

In summary, in order to solve the equations of motion to the first order, we just need to solve the equations of motion of the effective theory, and to plug the solution $\phi(x^0,x^3)$ into $H(x^1,x^2;\phi(x^0,x^3))$ and $A_{1,2}(x^1,x^2;\phi(x^0,x^3))$. The remaining fields $A^{(1)}_{0,3}(x^0,x^3)$ and $\Sigma^{(1)}(x^0,x^3)$ to the first order are obtained through Eqs. (51) and (52).

For later convenience, let us introduce another parametrization of $\mathbb{C}P^1$ in terms of polar angles $0 \le \Theta \le \pi$ and $0 \le \Phi \le 2\pi$ as

$$\phi = -e^{i\Phi}\tan\frac{\Theta}{2}. \tag{58}$$

The effective Lagrangian is rewritten as

$$\mathcal{L}_{\text{eff}} = \frac{\pi}{g^2}\left[\partial_\alpha\Theta\partial^\alpha\Theta + \sin^2\Theta\partial_\alpha\Phi\partial^\alpha\Phi - m^2\sin^2\Theta\right]. \tag{59}$$

The scalar potential $(\pi m^2/g^2)\sin^2\Theta$ is minimized at $\Theta = 0$ and $\Theta = \pi$. Clearly, these correspond to the N-vortex ($\phi = 0$) and the S-vortex ($\phi = \infty$).

4.1. Monopole

We are now ready to reconsider the slender magnetic monopole in the Higgs phase in our rigid-body approximation. Let us first look for an appropriate moduli configuration which minimizes the action in 1+3 dimensions by solving the equations of motion in the low energy effective theory:

$$\partial_\alpha\partial^\alpha\Theta + (m^2 - \partial_\alpha\Phi\partial^\alpha\Phi)\sin\Theta\cos\Theta = 0, \tag{60}$$

$$\partial^\alpha\left(\sin^2\Theta\partial_\alpha\Phi\right) = 0. \tag{61}$$

Eq. (61) admits a constant solution for Φ, say $\Phi = \eta$. In this study, we focus our attention to this class of solutions. Then the equation of motion reduces to the sine-Gordon equation

$$-\Theta'' + m^2\sin\Theta\cos\Theta = 0, \tag{62}$$

where the prime stands for the derivative in terms of x^3. The sine-Gordon model admits non-trivial topological excitations, kinks. The kinks interpolating $\Theta = 0$ and $\Theta = \pi$ are given by

$$\Theta = 2\arctan\exp\left(\pm m(x^3 - X_m)\right). \tag{63}$$

The solution with the plus sign is the kink connecting $\Theta = 0$ at $x^3 \to -\infty$ and $\Theta = \pi$ at $x^3 \to +\infty$, while that with minus sign is the anti-kink which connects $\Theta = \pi$ at $x^3 \to -\infty$ and 0 at $x^3 \to +\infty$.

From now on, we consider $X_m = \eta = 0$ case for simplicity. Combining this with Eqs. (43), (46), (47), and (48), we find

$$H^{(0)}(x^1, x^2, \phi(x^3)) = vU^\dagger(x^3)\begin{pmatrix} ze^{-\frac{\psi}{2}} & 0 \\ 0 & 1 \end{pmatrix}U(x^3), \tag{64}$$

$$\bar{A}^{(0)}(x^1, x^2, \phi(x^3)) = U^\dagger(x^3)\begin{pmatrix} -\frac{i}{2}\bar{\partial}\psi & 0 \\ 0 & 0 \end{pmatrix}U(x^3), \tag{65}$$

with

$$U(x^3) = \frac{1}{\sqrt{1+|\phi(x^3)|^2}}\begin{pmatrix} 1 & \bar{\phi}(x^3) \\ -\phi(x^3) & 1 \end{pmatrix}. \tag{66}$$

Furthermore, plugging $S^{(0)}$

$$S^{(0)} = \begin{pmatrix} \frac{e^{\frac{\psi}{2}}+z|\phi|^2}{1+|\phi|^2} & \frac{(e^{\frac{\psi}{2}}-z)\bar{\phi}}{1+|\phi|^2} \\ -\phi & 1 \end{pmatrix}, \qquad \phi = -\exp(\pm m x^3), \tag{67}$$

into the solutions $A_\alpha^{(1)}$ and $\Sigma^{(1)}$ given in Eqs. (51) and (52), we obtain the induced fields

$$A_3 \simeq \pm\frac{im}{2}\,\mathrm{sech}\, mx^3\, U^\dagger(x^3)\begin{pmatrix} 0 & 1 - ze^{-\frac{\psi}{2}} \\ \bar{z}e^{-\frac{\psi}{2}} - 1 & 0 \end{pmatrix}U(x^3), \tag{68}$$

$$\Sigma \simeq \frac{m}{2}\,\mathrm{sech}\, mx^3\, U^\dagger(x^3)\begin{pmatrix} \mp\sinh mx^3 & ze^{-\frac{\psi}{2}} \\ \bar{z}e^{-\frac{\psi}{2}} & \pm\sinh mx^3 \end{pmatrix}U(x^3). \tag{69}$$

The electric fields are given as:

$$F^0_{12} \simeq -\partial\bar{\partial}\psi, \quad F^0_{23} \simeq 0, \quad F^0_{31} \simeq 0, \tag{70}$$

and the magnetic fields are given as:

$$B_3^\Sigma = F_{12}^\Sigma \simeq \pm\partial\bar{\partial}\psi\tanh mx^3, \tag{71}$$

$$B_1^\Sigma = F_{23}^\Sigma \simeq \pm\frac{m}{4}\partial_1(r^2e^{-\psi})\,\mathrm{sech}^2\, mx^3, \tag{72}$$

$$B_2^\Sigma = F_{31}^\Sigma \simeq \pm\frac{m}{4}\partial_2(r^2e^{-\psi})\,\mathrm{sech}^2\, mx^3. \tag{73}$$

The solution with the lower sign connects the N-vortex as $x^3 \to +\infty$ and the S-vortex as $x^3 \to -\infty$, which is opposite to the configuration with upper sign. The corresponding monopole has the magnetic field F_{ij}^Σ pointing toward monopole, namely it is an anti-monopole in the Higgs phase.

Magnetic charges of the above solutions can be easily calculated

$$Q_\mathrm{m} = \frac{1}{g}\int d^3x\ \mathrm{div}\vec{B}^\Sigma = \frac{1}{g}\left[\int_{x^3\to\infty} dx^1dx^2\ (\pm\partial\bar{\partial}\psi) - \int_{x^3\to-\infty} dx^1dx^2\ (\mp\partial\bar{\partial}\psi)\right] = \pm\frac{4\pi}{g}, \tag{74}$$

where we used $\int dx^1dx^2\ \partial\bar{\partial}\psi = \pi$ and $r^2e^{-\psi} \to 1$ as $r \to \infty$. Here the factor $1/g$ is needed due to our notation that the gauge coupling is absorbed in the gauge field, see Eqs. (4) – (6). This magnetic charge precisely coincides with one of the 't Hooft-Polyakov monopole in the Coulomb phase [18].

Journal of Physics: Conference Series **670** (2016) 012006 doi:10.1088/1742-6596/670/1/012006

5. Dynamics of slender monopoles and anti-monopoles

In the following, we make full use of the similarity between our system and the sine-Gordon model. Let us denote another choice of the range of angles as

$$\tilde{\Theta} \in \mathbb{R} \ (\mathrm{mod}\ 2\pi), \qquad \tilde{\Phi} \in [0, \pi), \tag{75}$$

to parametrize the $\mathbb{C}P^1$ moduli ϕ

$$\phi(x^0, x^3) = -e^{i\tilde{\Phi}(x^0,x^3)} \tan \frac{\tilde{\Theta}(x^0, x^3)}{2}. \tag{76}$$

The equations of motion for $\tilde{\Theta}, \tilde{\Phi}$ are the same as those for Θ, Φ. Therefore, $\tilde{\Phi} = \text{const.}$ is a solution, to which we restrict ourselves in the following. Without loss of generality, the value of the constant Φ can be chosen as

$$\tilde{\Phi} = 0. \tag{77}$$

Then the equation of motion for $\tilde{\Theta}$ is reduced to

$$\partial_\alpha \partial^\alpha \tilde{\Theta} + m^2 \sin\tilde{\Theta} \cos\tilde{\Theta} = 0, \qquad \tilde{\Theta} \in \mathbb{R} \ (\mathrm{mod}\ 2\pi). \tag{78}$$

This is nothing but the sine-Gordon equation with a periodicity π in 1+1 dimensions.

Now we can compute all field configurations in $1 + 3$ dimensions with the help of the sine-Gordon field $\tilde{\Theta}$

$$F^0_{12} = -\partial\bar{\partial}\psi, \quad F^0_{23} = F^0_{31} = F^0_{01} = F^0_{02} = F^0_{03} = 0, \tag{79}$$

and

$$\begin{aligned}
F^\Sigma_{12} &= -\partial\bar{\partial}\psi \cos\tilde{\Theta}, &(80)\\
F^\Sigma_{23} &= \frac{1}{4}\partial_1(r^2 e^{-\psi})\partial_3\tilde{\Theta}\sin\tilde{\Theta}, &(81)\\
F^\Sigma_{31} &= \frac{1}{4}\partial_2(r^2 e^{-\psi})\partial_3\tilde{\Theta}\sin\tilde{\Theta}, &(82)\\
F^\Sigma_{01} &= \frac{1}{4}\partial_2(r^2 e^{-\psi})\partial_0\tilde{\Theta}\sin\tilde{\Theta}, &(83)\\
F^\Sigma_{02} &= -\frac{1}{4}\partial_1(r^2 e^{-\psi})\partial_0\tilde{\Theta}\sin\tilde{\Theta}, &(84)\\
F^\Sigma_{03} &= 0. &(85)
\end{aligned}$$

Here we define Abelian and non-Abelian electric fields in the same spirit as in Eq. (32)

$$F^0_{0i} = \mathrm{Tr}\left[F_{0i}\frac{\mathbf{1}_2}{2}\right], \quad F^\Sigma_{0i} = \mathrm{Tr}\left[F_{0i}\frac{\Sigma}{m}\right]. \tag{86}$$

Note that the electric field and magnetic field are orthogonal

$$\epsilon^{ijk} F^\Sigma_{ij} F^\Sigma_{0k} = 0. \tag{87}$$

Therefore, there is no energy dissipation.

The Hamiltonian density is decomposed into two parts: the energy density of the rigid vortex-string $\mathcal{H}^{(0)}_{\text{vortex}}$ and that of the dressed monopole $\mathcal{H}^{(2)}_{\text{dress}}$

$$\mathcal{H} = \mathcal{H}^{(0)}_{\text{vortex}} + \mathcal{H}^{(2)}_{\text{dress}} + \mathcal{O}(\epsilon^4). \tag{88}$$

The rigid vortex-string Hamiltonian density does not depend on x^0 and x^3

$$\begin{aligned}\mathcal{H}^{(0)}_{\text{vortex}} &= \text{Tr}\left[\frac{1}{g^2}(F^{(0)}_{12})^2 + (D_iH)^{(0)}(D_iH)^{(0)\dagger} + \frac{g^2}{4}\left(H^{(0)}H^{(0)\dagger} - v^2\mathbf{1}_2\right)^2\right] \\ &= \text{Tr}\left[\frac{1}{g^2}\left(F^{(0)}_{12} - \frac{g^2}{2}\left(H^{(0)}H^{(0)\dagger} - v^2\mathbf{1}_2\right)\right)^2 + 4(\bar{D}H)^{(0)}(\bar{D}H)^{(0)\dagger}\right. \\ &\quad \left. -v^2F^{(0)}_{12} + i\left\{\partial_1(H^{(0)}(D_2H)^{(0)\dagger}) - \partial_2(H^{(0)}(D_1H)^{(0)\dagger})\right\}\right] \\ &= 2v^2\mathcal{V}. \end{aligned} \tag{89}$$

where

$$\mathcal{V}(x^1, x^2) = \partial\bar{\partial}\psi - \frac{4}{g^2v^2}(\partial\bar{\partial})^2\psi \tag{90}$$

The dressed Hamiltonian density which depends on x^0 and x^3 is given by

$$\begin{aligned}\mathcal{H}^{(2)}_{\text{dress}} &= \text{Tr}\left[\frac{1}{g^2}\left\{(F^{(1)}_{23})^2 + (F^{(1)}_{31})^2 + (F^{(1)}_{01})^2 + (F^{(1)}_{02})^2 + (D_1\Sigma^{(1)})^2 + (D_2\Sigma^{(1)})^2\right\}\right. \\ &\quad \left. + D_0H^{(0)}(D_0H^{(0)})^\dagger + D_3H^{(0)}(D_3H^{(0)})^\dagger + (\Sigma^{(1)}H^{(0)} - H^{(0)}M)(\Sigma^{(1)}H^{(0)} - H^{(0)}M)^\dagger\right] \\ &= \frac{\mathcal{V}}{g^2}\left((\partial_0\tilde{\Theta})^2 + (\partial_3\tilde{\Theta})^2 + m^2\sin^2\tilde{\Theta}\right), \end{aligned} \tag{91}$$

where we used the master equation (30).

The topological charge density is given as

$$\mathcal{Q}_{\text{m}} = \frac{2\mathcal{V}}{g}\partial_3\tilde{\Theta}\sin\tilde{\Theta}. \tag{92}$$

From this expression, one can easily compute the magnetic charge as

$$Q_{\text{m}} = \int d^3x\ \mathcal{Q}_{\text{m}} = \frac{2}{g}\int dx^1dx^2\ \mathcal{V}\ \int dx^3\ \partial_3\tilde{\Theta}\sin\tilde{\Theta} = \frac{2\pi}{g}\left[-\cos\tilde{\Theta}\right]^{x_3=+\infty}_{x^3=-\infty}. \tag{93}$$

Here we used $\int dx^1dx^2\partial\bar{\partial}\psi = \pi$. As a check, one can compute the energy of the magnetic monopoles from the solutions given in Eq. (63)

$$Q_{\text{m}} = \pm\frac{4\pi}{g}. \tag{94}$$

Similarly, one may introduce a electric charge density by

$$\mathcal{Q}_{\text{e}} = \frac{1}{g}\partial^iF^{\Sigma}_{0i}. \tag{95}$$

But this is identically zero for any $\tilde{\Theta}(x^0, x^3)$. This matches with a naive intuition that the fixed azimuthal angle $\tilde{\Phi}$ does not generate any electric charges. Note, however, that this does not mean the electric fields themselves are zeros. One can easily find that rotation of $\vec{E}^{\Sigma} = (F^{\Sigma}_{10}, F^{\Sigma}_{20}, F^{\Sigma}_{30})$ are non-zero.

$$(\vec{\nabla}\times\vec{E}^{\Sigma})_3 = -\frac{4}{g^2v^2}(\partial\bar{\partial})^2\psi\ \partial_0\tilde{\Theta}\sin\tilde{\Theta}. \tag{96}$$

The other components are of higher order, so we ignore them.

Journal of Physics: Conference Series **670** (2016) 012006 doi:10.1088/1742-6596/670/1/012006

5.1. Two different species of slender monopoles

The zenith angle $\tilde{\Theta}$ takes values between 0 and 2π ($= \mathbb{R}$ mod 2π), and the sine-Gordon equation (78) is periodic with a period π. Therefore, there exist two sine-Gordon kinks: the one interpolates from 0 to π as $x^3 = -\infty \to +\infty$, and the other interpolates from π to 2π as $x^3 = -\infty \to +\infty$. Here we need to pay some attention to our terminology in translating the sine-Gordon kinks into monopoles in $1+3$ dimensions. Although these two configurations are both to be called kinks in the sense of the sine-Gordon model, the former connects the N-vortex and S-vortex from left to right, while the latter connects them from right to left. Namely, the former kink is the monopole (denoted as $\mathcal{M}_0$) and the latter kink is the anti-monopole ($\bar{\mathcal{M}}_\pi$). Similarly, the anti-kink interpolating from π to 0 as $x^3 = -\infty \to +\infty$ is the anti-monopole ($\bar{\mathcal{M}}_0$), while the other anti-kink interpolating from 2π to π as $x^3 = -\infty \to +\infty$ is the monopole ($\mathcal{M}_\pi$). Correspondence between the sine-Gordon (anti-)kinks and the slender (anti-)monopoles are depicted in Fig. 1. The configurations are given by

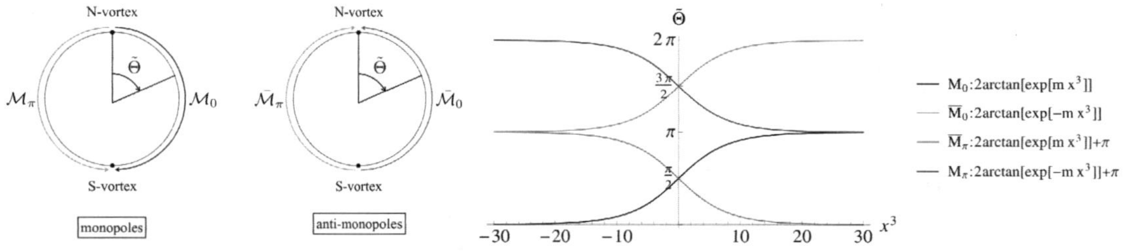

Figure 1. Correspondence between the sine-Gordon (anti-)kinks and the slender (anti-)monopoles.

$$\mathcal{M}_0 \quad : \quad \tilde{\Theta} = 2\arctan\exp(mx^3) + 2n\pi, \tag{97}$$

$$\bar{\mathcal{M}}_0 \quad : \quad \tilde{\Theta} = 2\arctan\exp(-mx^3) + 2n\pi, \tag{98}$$

$$\mathcal{M}_\pi \quad : \quad \tilde{\Theta} = 2\arctan\exp(-mx^3) + (2n+1)\pi, \tag{99}$$

$$\bar{\mathcal{M}}_\pi \quad : \quad \tilde{\Theta} = 2\arctan\exp(mx^3) + (2n+1)\pi, \tag{100}$$

with n being an integer.

5.2. Magnetic meson

It is well-known that the sine-Gordon model admits a bound state of kink and anti-kink, the so-called breather solution. In our case, it is nothing but a bound state of the slender monopole and anti-monopole, which we call the magnetic meson. The configuration is given by

$$\tilde{\Theta}(x^0, x^3) = 2\arctan\left(\frac{\eta\sin\omega x^0}{\cosh\eta\omega x^3}\right), \quad \eta = \sqrt{\frac{m^2}{\omega^2} - 1}, \quad \omega < m, \tag{101}$$

where ω is the frequency and $(\eta\omega)^{-1} = 1/\sqrt{m^2 - \omega^2}$ is the typical size of the magnetic meson. The mass of the meson depends on ω as

$$M_{\text{meson}} = 2M_{\text{mono}} \times \sqrt{1 - \frac{\omega^2}{m^2}} < 2M_{\text{mono}}. \tag{102}$$

The mass of the mesonic bound state is smaller than the sum of the masses of isolated monopole and anti-monopole.

We show how the magnetic meson varies in one period $T = \frac{2\pi}{\omega}$ in Fig. 2. The sources of outgoing magnetic field are identified as slender monopoles and those of incoming magnetic field as slender anti-monopoles. It is interesting to observe that the meson is made of $\mathcal{M}_0$ and $\bar{\mathcal{M}}_0$ at an instance (for example $t = T/4$), and that it transforms into a different meson made of $\mathcal{M}_\pi$ and $\bar{\mathcal{M}}_\pi$ at another instance (for example $t = 3T/4$). In Fig. 2, we also show the topological charge density $\mathcal{Q}_{\rm m}$ given in Eq. (92) together with the energy density of the electric field

$$\mathcal{E} = \frac{1}{g^2}\,\mathrm{Tr}\left[(F_{01})^2 + (F_{02})^2\right] = \frac{1}{g^2}\,|1 - z\partial\psi|^2\, e^{-\psi}(\partial_t\tilde{\Theta})^2. \tag{103}$$

As the monopole and anti-monopole approach each other, the magnetic energy density $\mathcal{M}$ decreases and the electric field energy density $\mathcal{E}$ grows. At the very instance of collision, the magnetic energy disappears and is transferred into the electric energy completely. The electric field is generated by the time variation (decrease) of the magnetic field as monopole and anti-monopole merge.

5.3. *Scattering of the slender monopole and anti-monopole*

Let us next study the head-on collision of the slender monopole and anti-monopole. There are two types of collisions: one type is the collision of $\mathcal{M}_0$ and $\bar{\mathcal{M}}_0$, ($\mathcal{M}_\pi$ and $\bar{\mathcal{M}}_\pi$) and the other type is that of $\mathcal{M}_0$ and $\bar{\mathcal{M}}_\pi$ ($\mathcal{M}_\pi$ and $\bar{\mathcal{M}}_0$).

Scattering of $\mathcal{M}_0$ and $\bar{\mathcal{M}}_0$ ($\mathcal{M}_\pi$ and $\bar{\mathcal{M}}_\pi$) The exact solution for the moduli field for the collision of a monopole and anti-monopole of the same species ($\mathcal{M}_0$ or $\mathcal{M}_\pi$) is given by

$$\tilde{\Theta} = 2\arctan\left(\frac{\sinh mu\gamma x^0}{u\cosh m\gamma x^3}\right), \quad \gamma = \frac{1}{\sqrt{1-u^2}}. \tag{104}$$

The parameter u corresponds to the velocity of the monopole. However, we should keep in mind that our approximation holds only for small velocities, that is

$$u \ll 1 \qquad (\gamma \simeq 1). \tag{105}$$

Since we are using the rigid-body approximation we cannot faithfully describe Lorentz boosted monopoles. Thus even though we can solve the 1+1-dimensional effective dynamics for arbitrary velocities, the full 1+3-dimensional dynamics is correctly represented only within the restriction of Eq. (105). A typical configuration is shown in Fig. 3. The slender magnetic monopole $\mathcal{M}_\pi$ comes from the left infinity and the anti-monopole $\bar{\mathcal{M}}_\pi$ comes from the right infinity. As they approach to the collision point, the magnetic energy decreases while the electric energy grows. After the collision, the magnetic energy grows as the monopole $\mathcal{M}_0$ (anti-monopole $\bar{\mathcal{M}}_0$) goes toward the left (right) infinity. Thus we find that the species of the monopole and the anti-monopole changes after the collision.

Scattering of $\mathcal{M}_0$ and $\bar{\mathcal{M}}_\pi$ (or $\mathcal{M}_\pi$ and $\bar{\mathcal{M}}_0$) The solution for the scattering of a monopole and an anti-monopole of different species is given by

$$\tilde{\Theta} = 2\arctan\left(\frac{u\sinh m\gamma x^3}{\cosh mu\gamma x^0}\right). \tag{106}$$

A typical configuration is shown in Fig. 4. In contrast to the previous type of scattering the species of monopoles ($\mathcal{M}_0$ or $\mathcal{M}_\pi$) do not change into different species during the collision. As shown in Fig. 4, the anti-monopole $\bar{\mathcal{M}}_\pi$ comes from the left infinity and reflects back toward the left infinity, while the monopole $\mathcal{M}_0$ comes from the right infinity and reflects back toward the right infinity.

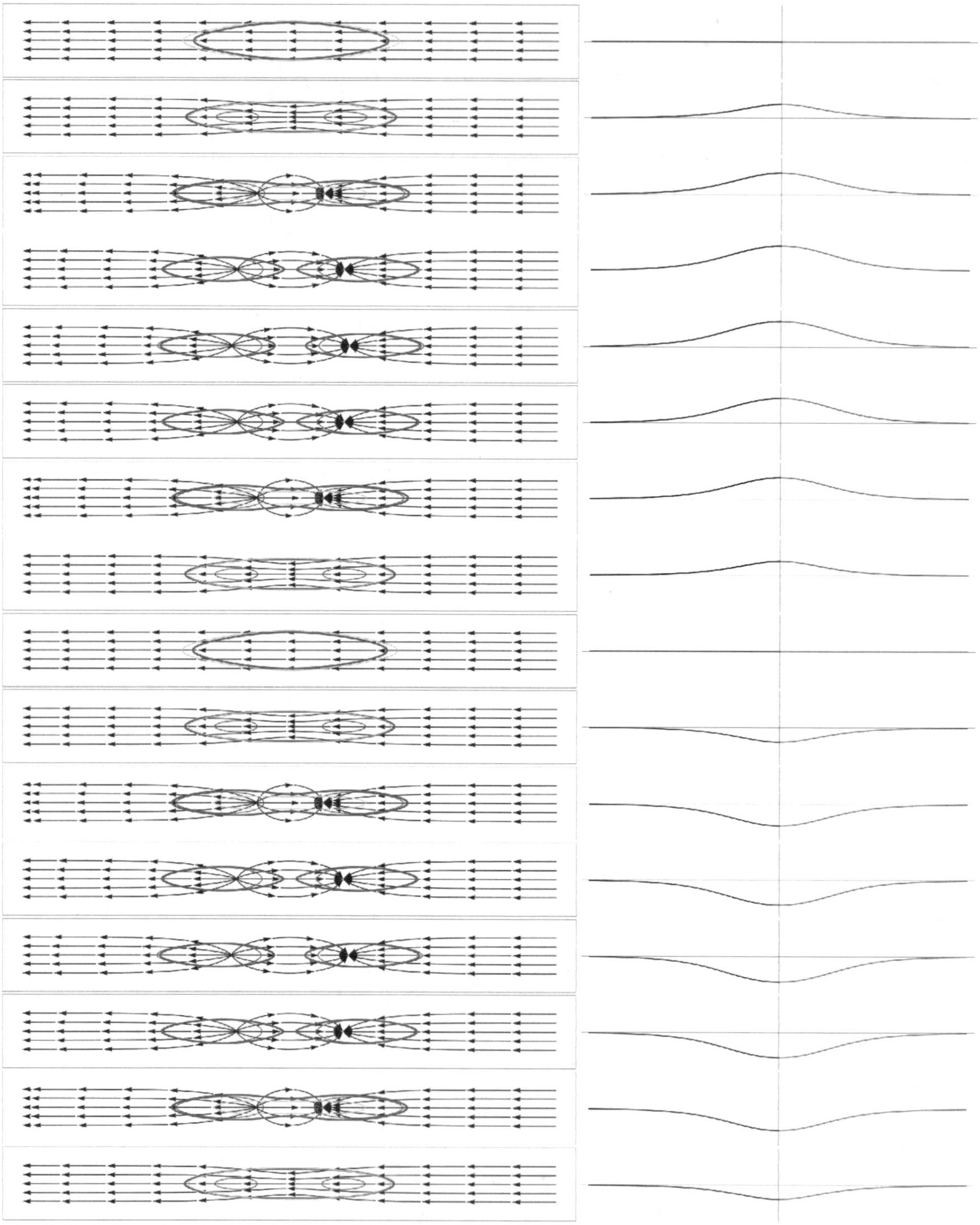

Figure 2. Snapshots of a single period of the magnetic meson. The top is at $t = 0$ and the bottom is at $t = T - \delta t$ with $\delta t = T/16$. The left panel shows the magnetic field $(F_{12}^{\Sigma}, F_{23}^{\Sigma})$ by blue streamlines and the topological charge densities, $\mathcal{M} = \pm 0.017$, electric energy density, $\mathcal{E} = 0.012$, and the dressed energy density $\mathcal{H}_{\rm dress} = 0.02$ by red/green/grey contours. In the right figures, $\tilde{\Theta}(x^3, t)$ is plotted. The curves are piecewise colored by black, red, blue and green for $\mathcal{M}_0$, $\bar{\mathcal{M}}_0$, $\mathcal{M}_\pi$ and $\bar{\mathcal{M}}_\pi$, respectively. We set $gv = 1$, $m = 1/5$ and $\omega = 1/10$. $x^1 \in [-3, 3]$ and $x^3 \in [-30, 30]$.

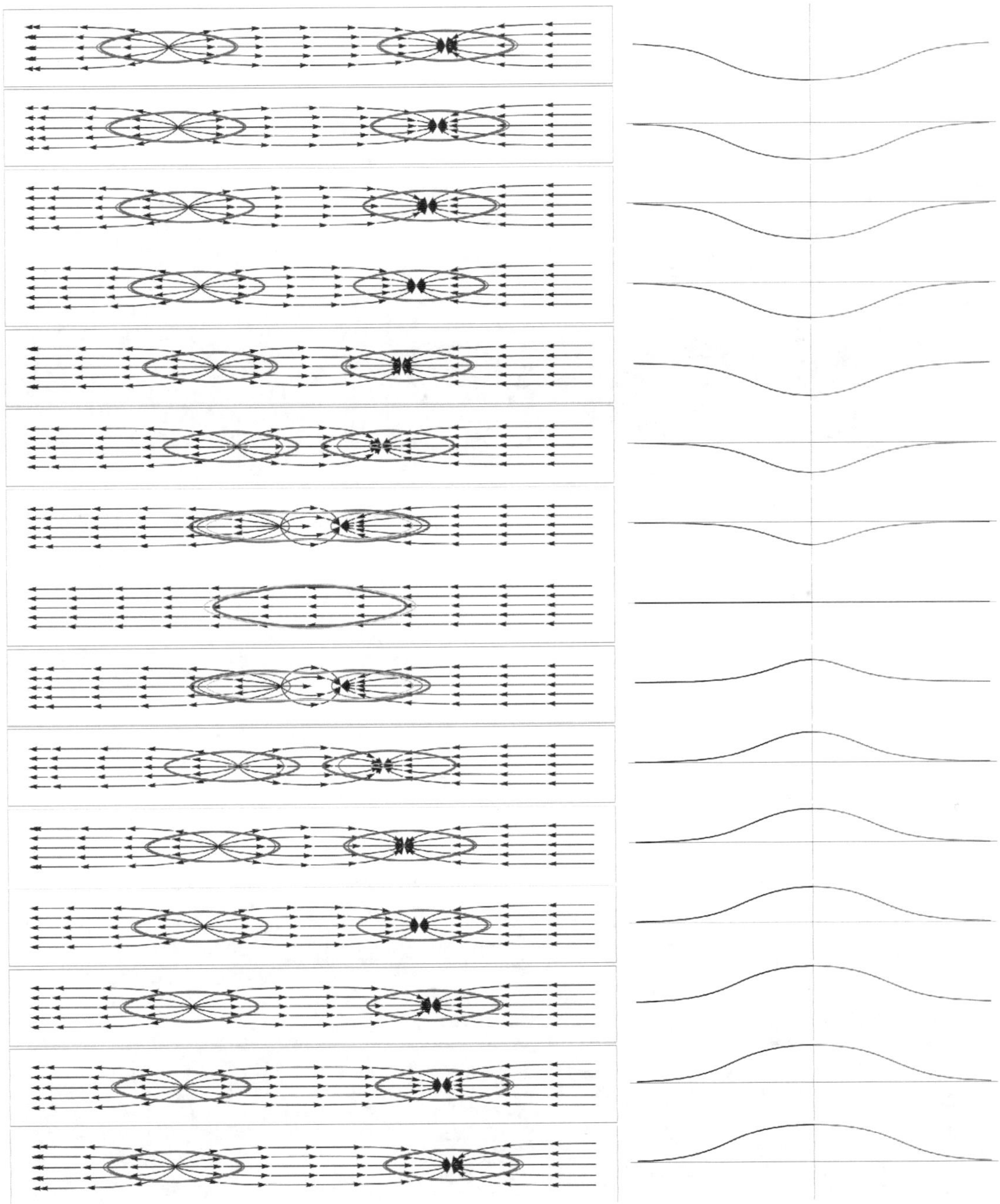

Figure 3. Snapshots (from $t = -42$ to $t = 42$ with $\delta t = 6$ interval) of scattering of the slender monopole and anti-monopole. The red/green/gray contours are $(\mathcal{M}, \mathcal{E}, \mathcal{H}_{\text{dress}}) = (\pm 0.017, 0.01, 0.02)$, see the caption of Fig. 2 for explanation. We set $gv = 1$, $m = 1/5$ and $u = 1/10$. $x^1 \in [-3, 3]$ and $x^3 \in [-30, 30]$.

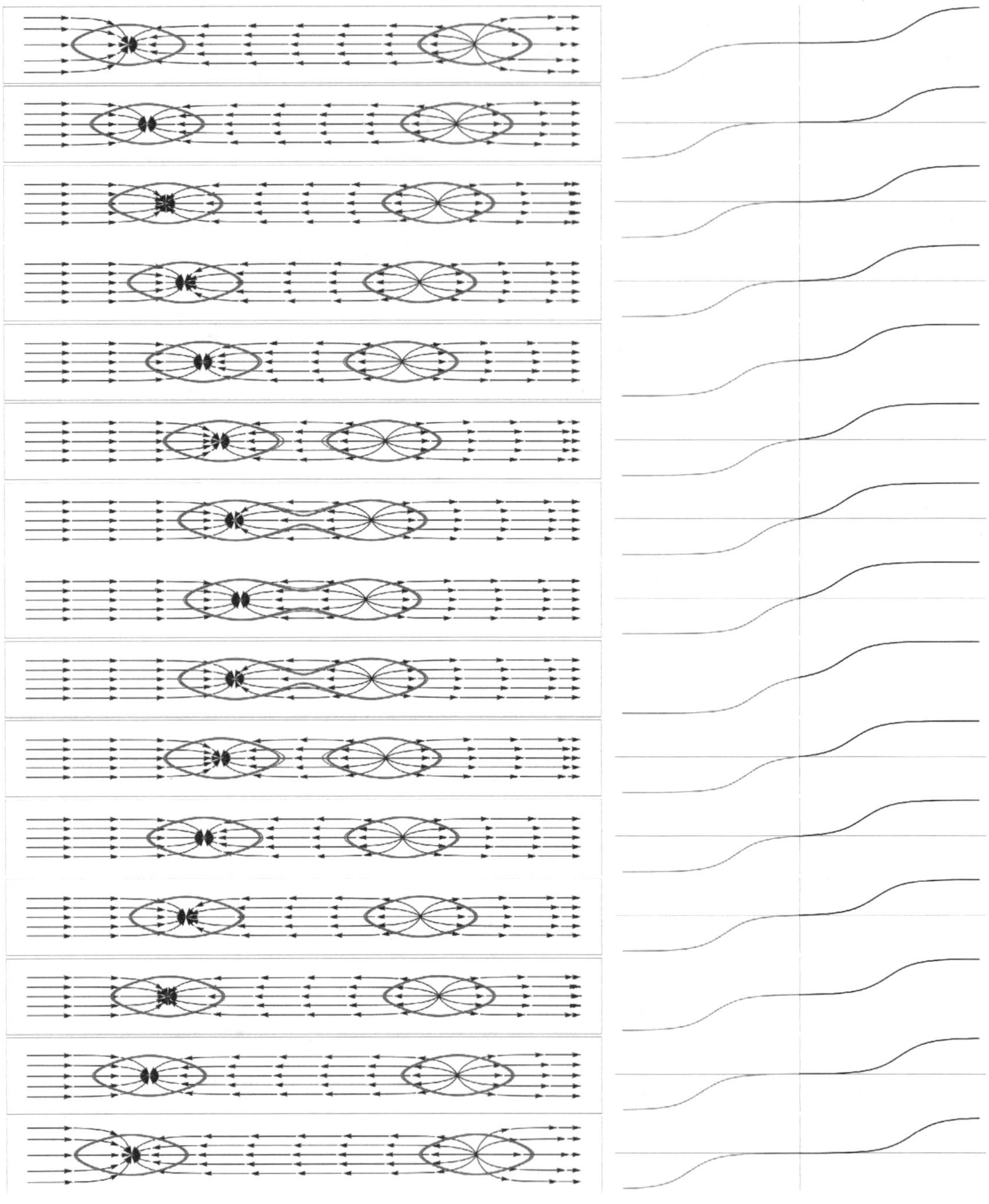

Figure 4. Snapshots (from $t = -70$ to $t = 70$ with $\delta t = 10$ interval) of scattering of the slender monopole and anti-monopole. The red/gray contours are $(\mathcal{M}, \mathcal{H}_{\rm dress}) = (\pm 0.017, 0.02)$. See the caption of Fig. 2 for explanation. We set $gv = 1$, $m = 1/3$ and $u = 1/3$. $x^1 \in [-3, 3]$ and $x^3 \in [-30, 30]$.

6. Conclusion

In this paper we have investigated the low energy dynamics of monopoles and anti-monopoles in the non-Abelian superconductor. We have restricted ourselves to the parameter region $m \ll gv$ where the monopoles are of slender ellipsoidal shape, confined on a vortex string, with the cross-section comparable to that of the monopole. For that reason, the scattering problem becomes essentially $1+1$ dimensional. Indeed, we have found that at least a part of the low energy dynamics is identical to the sine-Gordon system in $1+1$ dimensions up to the first order of the expansion in $\epsilon = m/(gv)$, when $\{m, \partial_0, \partial_3\} \ll \{gv, \partial_1, \partial_2\}$ holds. This observation is very useful because the sine-Gordon system is solvable. In the literature, only the static kink was identified with the monopole. In this paper, we have dealt with all the sine-Gordon solutions and have constructed the dictionary with which one can easily translate the dynamics of sine-Gordon kinks in 1+1 dimensions into the dynamics of monopoles in $1+3$ dimensions. A surprising fact is that the monopole and anti-monopole do not always decay into radiation when they make a head-on collision, although they are not protected by topology. We have studied three concrete examples: (1) the magnetic meson which is the bound state of the slender monopole and anti-monopole, (2) the scattering of the monopole and anti-monopole of the same species, and (3) the scattering of the monopole and anti-monopole of the different species. All these three examples show that the monopole and anti-monopole do not always annihilate. This observation may be counter-intuitive and remarkable.

Acknowledgments
F. B. is an international research fellow of the Japan Society for the Promotion of Science. This work was supported by Grant-in-Aid for JSPS Fellows, Grant Number 26004750.

References

[1] M. Eto, Y. Isozumi, M. Nitta, K. Ohashi and N. Sakai, "Solitons in the Higgs phase: The Moduli matrix approach," J. Phys. A **39** (2006) R315 [hep-th/0602170].
[2] P. A. M. Dirac, "Quantized Singularities in the Electromagnetic Field," Proc. Roy. Soc. Lond. A **133** (1931) 60.
[3] Y. Nambu, "Strings, Monopoles and Gauge Fields," Phys. Rev. D **10** (1974) 4262.
[4] S. Mandelstam, "Vortices And Quark Confinement In Nonabelian Gauge Theories," Phys. Lett. B **53** (1975) 476; "Vortices and Quark Confinement in Nonabelian Gauge Theories," Phys. Rept. **23** (1976) 245.
[5] G. 't Hooft, "Topology of the Gauge Condition and New Confinement Phases in Nonabelian Gauge Theories," Nucl. Phys. B **190** (1981) 455.
[6] M. G. Alford, K. Rajagopal and F. Wilczek, "QCD at finite baryon density: Nucleon droplets and color superconductivity," Phys. Lett. B **422** (1998) 247 [hep-ph/9711395]; "Color flavor locking and chiral symmetry breaking in high density QCD," Nucl. Phys. B **537** (1999) 443 [hep-ph/9804403].
[7] R. Rapp, T. Schäfer, E. V. Shuryak and M. Velkovsky, "Diquark Bose condensates in high density matter and instantons," Phys. Rev. Lett. **81** (1998) 53 [hep-ph/9711396].
[8] A. Hanany and D. Tong, "Vortices, instantons and branes," JHEP **0307** (2003) 037 [hep-th/0306150]; A. Hanany and D. Tong, "Vortex strings and four-dimensional gauge dynamics," JHEP **0404** (2004) 066 [hep-th/0403158].
[9] R. Auzzi, S. Bolognesi, J. Evslin, K. Konishi and A. Yung, "NonAbelian superconductors: Vortices and confinement in N=2 SQCD," Nucl. Phys. B **673** (2003) 187 [hep-th/0307287].
[10] M. Shifman and A. Yung, "NonAbelian string junctions as confined monopoles," Phys. Rev. D **70** (2004) 045004 [hep-th/0403149].
[11] A. Gorsky, M. Shifman and A. Yung, "Confined Magnetic Monopoles in Dense QCD," Phys. Rev. D **83** (2011) 085027 [arXiv:1101.1120 [hep-ph]].
[12] M. Eto, M. Nitta and N. Yamamoto, "Confined Monopoles Induced by Quantum Effects in Dense QCD," Phys. Rev. D **83** (2011) 085005 [arXiv:1101.2574 [hep-ph]].
[13] M. Eto, Y. Hirono, M. Nitta and S. Yasui, "Vortices and Other Topological Solitons in Dense Quark Matter," PTEP **2014** 1, 012D01 [arXiv:1308.1535 [hep-ph]].
[14] E. B. Bogomolny, "Stability Of Classical Solutions," Sov. J. Nucl. Phys. **24** (1976) 449 [Yad. Fiz. **24** (1976) 861]; M. K. Prasad and C. M. Sommerfield, "An Exact Classical Solution For The 'T Hooft Monopole And The Julia-Zee Dyon," Phys. Rev. Lett. **35** (1975) 760.
[15] E. Witten and D. I. Olive, "Supersymmetry Algebras That Include Topological Charges," Phys. Lett. B **78** (1978) 97.
[16] N. S. Manton, "A Remark On The Scattering Of Bps Monopoles," Phys. Lett. B **110** (1982) 54.

[17] C. H. Taubes, "The Existence of a Nonminimal Solution to the SU(2) Yang-Mills Higgs Equations on R**3," Commun. Math. Phys. **86** (1982) 257.
[18] D. Tong, "Monopoles in the higgs phase," Phys. Rev. D **69** (2004) 065003 [hep-th/0307302].
[19] M. Cipriani and T. Fujimori, "Effective Action of Non-Abelian Monopole-Vortex Complex," arXiv:1207.2070 [hep-th].
[20] J. Evslin, K. Konishi, M. Nitta, K. Ohashi and W. Vinci, "Non-Abelian Vortices with an Aharonov-Bohm Effect," JHEP **1401** (2014) 086 [arXiv:1310.1224 [hep-th]].
[21] Y. Isozumi, M. Nitta, K. Ohashi and N. Sakai, "All exact solutions of a 1/4 Bogomol'nyi-Prasad-Sommerfield equation," Phys. Rev. D **71** (2005) 065018 [arXiv:hep-th/0405129].

Journal of Physics: Conference Series **670** (2016) 012007 doi:10.1088/1742-6596/670/1/012007

Scattering in Twisted Yangians

Jean Avan[a], **Anastasia Doikou**[b1], **Nikos Karaiskos**[c]

[a] Laboratoire de Physique Théorique et Modélisation (CNRS UMR 8089), Université de Cergy-Pontoise, F-95302 Cergy-Pontoise, France

[b] Department of Mathematics, Heriot-Watt University, EH14 4AS, Edinburgh, United Kingdom

[c] Institute for Theoretical Physics, Leibniz University Hannover, Appelstraße 2, 30167 Hannover, Germany

E-mail: jean.avan@u-cery.fr, a.doikou@hw.ac.uk, nikolaos.karaiskos@itp.uni-hannover.de

Abstract.
We study the bulk and boundary scattering of the $\mathfrak{sl}(\mathcal{N})$ twisted Yangian spin chain via the solution of the Bethe ansatz equations in the thermodynamic limit. Explicit expressions for the scattering amplitudes are obtained and the factorization of the bulk scattering is shown. The issue of defects in twisted Yangians is also briefly discussed.

1. Introduction

We shall discuss here the bulk and boundary scattering in the context of twisted Yangians. Most of this material is described in more detail in [1, 2]. Here we are giving a brief review of the main results regarding basically the computation of the bulk and boundary scattering amplitudes in the case of the spin chain with twisted Yangian underlying algebra.

We shall deal henceforth with open spin chains, requiring introduction of boundary terms consistent with quantum integrability. These are related to generalized reflection algebras (quadratic algebras) à la Freidel-Maillet [3] extending the original construction of Cherednik [4] and Sklyanin [5] to a four matrix structure canonically expressed as:

$$A_{12}\, K_1\, B_{12}\, K_2 = K_2\, C_{12}\, K_1\, D_{12}\,, \tag{1}$$

with unitarity requirements

$$\begin{aligned} &A_{12}\, A_{21} = D_{12}\, D_{21} = \mathbb{I}_{12}\,, \\ &C_{12} = B_{21}. \end{aligned} \tag{2}$$

In the particular case when $A_{12} = D_{21} = R_{12}$ a given Yang-Baxter R matrix, and $B_{12} = C_{21} = \bar{R}_{21}$ (its soliton anti-soliton counter part), $\bar{R}_{12} \sim R_{12}^{t_1}$, (1) yields the so-called twisted Yangian structure if R is the simple Yangian solution of the Yang-Baxter equation [6].

[1] Based on a talk presented by AD, in *"Integrable systems and quantum symmetries"*, Prague, June 2015. This work is mainly based on [1, 2]

Spin chains based on such a twisted Yangian were first constructed and investigated in [7] for the first time whereas investigations were generalized in [8]. They were then considered in the thermodynamic limit in our previous paper [1]. They naturally exhibit soliton non-preserving boundary conditions due the choice of $B_{12} = C_{21}$ as a soliton–anti-soliton S-matrix and the subsequent conversion of a soliton into an anti-soliton by the building reflection matrix K.

2. Bethe ansatz e uations in twisted Yangian

Analytical Bethe ansatz techniques were applied in [7, 8] to obtain the spectrum and BAEs for the twisted Yangian. Throughout the text we consider the boundary matrices, c-number representations of the twisted Yangian (2) ($A_{12} = D_{21} = R_{12}$, and $B_{12} = C_{21} = \bar{R}_{21}$), to be proportional to unit. The spectrum of the $\mathfrak{sl}(\mathcal{N})$ twisted Yangian is then given by the following expression:

$$\Lambda(\lambda) = (a(\lambda)\,\bar{b}(\lambda))^L\, g_0(\lambda)\, A_0(\lambda) + (b(\lambda)\,\bar{b}(\lambda))^L \prod_{j=1}^{\mathcal{N}-2} g_j(\lambda)\, A_j(\lambda) + (\bar{a}(\lambda)\, b(\lambda))^L g_{\mathcal{N}-1}(\lambda)\, A_{\mathcal{N}-1}(\lambda) \tag{3}$$

where we define:

$$a(\lambda) = \lambda + i, \quad b(\lambda) = \lambda, \quad \bar{a}(\lambda) = \lambda + i\rho - i, \quad \bar{b}(\lambda) = \lambda + i\rho \tag{4}$$

g_l are terms due to boundary contributions

$$\begin{aligned} g_l(\lambda) &= \frac{\lambda + \frac{i\rho}{2} - \frac{i}{2}}{\lambda + \frac{i\rho}{2}} \\ g_{\frac{\mathcal{N}-1}{2}}(\lambda) &= 1, \quad \mathcal{N} \quad \text{odd} \\ g_{\mathcal{N}-l+1}(\lambda) &= g_l(-\lambda - i\rho) \\ \rho &= \frac{\mathcal{N}}{2}, \qquad l \in \{1, \dots \frac{\mathcal{N}}{2} - 1\} \end{aligned} \tag{5}$$

and A_l are the so called dressing functions defined as

$$\begin{aligned} A_0(\lambda) &= \prod_{j=1}^{M^{(1)}} \frac{\lambda + \lambda_j^{(1)} - \frac{i}{2}}{\lambda + \lambda_j^{(1)} + \frac{i}{2}} \frac{\lambda + \lambda_j^{(1)} - \frac{i}{2}}{\lambda + \lambda_j^{(1)} + \frac{i}{2}} \\ A_k(\lambda) &= \prod_{j=1}^{M^{(k)}} \frac{\lambda + \lambda_j^{(k)} - \frac{ik}{2} + i}{\lambda + \lambda_j^{(k)} + \frac{ik}{2}} \frac{\lambda + \lambda_j^{(k)} + \frac{ik}{2} + i}{\lambda + \lambda_j^{(k)} + \frac{ik}{2}} \\ &\times \prod_{j=1}^{M^{(k+1)}} \frac{\lambda + \lambda_j^{(k+1)} + \frac{ik}{2} - \frac{i}{2}}{\lambda + \lambda_j^{(k+1)} + \frac{i}{2}} \frac{\lambda - \lambda_j^{(k+1)} + \frac{ik}{2} - \frac{i}{2}}{\lambda - \lambda_j^{(k+1)} + \frac{i}{2}} \end{aligned} \tag{6}$$

$$A_k(\lambda) = A_{\mathcal{N}-k+1}(-\lambda - i\rho), \qquad k \in \{1, \dots, \frac{\mathcal{N}}{2} - 1\} \tag{7}$$

for $\mathcal{N} = 2n + 1$

$$\begin{aligned} A_n(\lambda) &= \prod_{j=1}^{M^{(n)}} \frac{\lambda + \lambda_j^{(n)} + \frac{in}{2} + i}{\lambda + \lambda_j^{(n)} + \frac{in}{2}} \frac{\lambda - \lambda_j^{(n)} + \frac{in}{2} + i}{\lambda - \lambda_j^{(n)} + \frac{in}{2}} \\ &\times \frac{\lambda + \lambda_j^{(n)} + \frac{in}{2} - \frac{i}{2}}{\lambda + \lambda_j^{(n)} + \frac{in}{2} + \frac{i}{2}} \frac{\lambda - \lambda_j^{(n)} + \frac{in}{2} - \frac{i}{2}}{\lambda - \lambda_j^{(n)} + \frac{in}{2} + \frac{i}{2}} \end{aligned} \tag{8}$$

Analyticity conditions imposed on the spectrum give rise to the associated Bethe ansatz equations presented below: defining

$$e_n(\lambda) = \frac{\lambda + \frac{in}{2}}{\lambda - \frac{in}{2}}, \tag{9}$$

the BAE read as follows:

- $\mathfrak{sl}(2n+1)$

$$\begin{aligned}
e_1^L(\lambda_i^{(1)}) &= -\prod_{j=1}^{M^{(1)}} e_2(\lambda_i^{(1)} - \lambda_j^{(1)})\, e_2(\lambda_i^{(1)} + \lambda_j^{(1)}) \prod_{j=1}^{M^{(2)}} e_{-1}(\lambda_i^{(1)} - \lambda_j^{(2)})\, e_{-1}(\lambda_i^{(1)} + \lambda_j^{(2)}),\\
1 &= -\prod_{j=1}^{M^{(\ell)}} e_2(\lambda_i^{(\ell)} - \lambda_j^{(\ell)})\, e_2(\lambda_i^{(\ell)} + \lambda_j^{(\ell)}) \prod_{\tau=\pm1}\prod_{j=1}^{M^{(\ell+\tau)}} e_{-1}(\lambda_i^{(\ell)} - \lambda_j^{(\ell+\tau)})\, e_{-1}(\lambda_i^{(\ell)} + \lambda_j^{(\ell+\tau)})\\
&\text{for} \quad \ell = 2,\dots,n-1,\\
e_{-\frac{1}{2}}(\lambda_i^{(n)}) &= -\prod_{j=1}^{M^{(n)}} e_{-1}(\lambda_i^{(n)} - \lambda_j^{(n)})\, e_{-1}(\lambda_i^{(n)} + \lambda_j^{(n)})\, e_2(\lambda_i^{(n)} - \lambda_j^{(n)})\, e_2(\lambda_i^{(n)} + \lambda_j^{(n)})\\
&\times \prod_{j=1}^{M^{(n-1)}} e_{-1}(\lambda_i^{(n)} - \lambda_j^{(n-1)})\, e_{-1}(\lambda_i^{(n)} + \lambda_j^{(n-1)}).
\end{aligned} \tag{10}$$

Note that in this case the Bethe ansatz equations are similar to the ones of the open $\mathfrak{osp}(1|2n)$ spin chain (see also [7], [8], [9]).

- $\mathfrak{sl}(2n)$

$$\begin{aligned}
e_1^L(\lambda_i^{(1)}) &= -\prod_{j=1}^{M^{(1)}} e_2(\lambda_i^{(1)} - \lambda_j^{(1)})\, e_2(\lambda_i^{(1)} + \lambda_j^{(1)}) \prod_{j=1}^{M^{(2)}} e_{-1}(\lambda_i^{(1)} - \lambda_j^{(2)})\, e_{-1}(\lambda_i^{(1)} + \lambda_j^{(2)}),\\
1 &= -\prod_{j=1}^{M^{(\ell)}} e_2(\lambda_i^{(\ell)} - \lambda_j^{(\ell)})\, e_2(\lambda_i^{(\ell)} + \lambda_j^{(\ell)}) \prod_{\tau=\pm1}\prod_{j=1}^{M^{(\ell+\tau)}} e_{-1}(\lambda_i^{(\ell)} - \lambda_j^{(\ell+\tau)})\, e_{-1}(\lambda_i^{(\ell)} + \lambda_j^{(\ell+\tau)})\\
&\ell = 2,\dots,n-1,\\
e_{-1}(\lambda_i^{(n)}) &= -\prod_{j=1}^{M^{(n)}} e_2(\lambda_i^{(n)} - \lambda_j^{(n)})\, e_2(\lambda_i^{(n)} + \lambda_j^{(n)})\\
&\times \prod_{j=1}^{M^{(n-1)}} e_{-1}^2(\lambda_i^{(n)} - \lambda_j^{(n-1)})\, e_{-1}^2(\lambda_i^{(n)} + \lambda_j^{(n-1)}).
\end{aligned} \tag{11}$$

As opposed to the $\mathfrak{sl}(2n+1)$ case the Bethe ansatz equations above do not reduce to any of the known forms of BAE, which makes the whole study even more intriguing.
Note that the numbers $M^{(l)}$ are associated to the eigenvalues of the diagonal generators S_l of the underlying algebra $\mathfrak{so}(n)$ (see [7, 8] for a detailed discussion on the underlying symmetry of the models), i.e.

$$S_1 = \frac{1}{2}M^{(0)} - M^{(1)}, \quad S_l = M^{(l-1)} - M^{(l)}, \quad S_l = \frac{1}{2}(E_{ll} - E_{\overline{ll}}), \quad l \le l \le \frac{\mathcal{N}-1}{2} \tag{12}$$

E_{ll} are the diagonal generators of $\mathfrak{sl}_{\mathcal{N}}$, and $\bar{l} = \mathcal{N} - l + 1$ the conjugate index.
It is also worth recalling that the corresponding numbers in the usual $\mathfrak{sl}(\mathcal{N})$ case are given by:

$$E_{ll} = M^{(l-1)} - M^{(l)}, \qquad M^{(0)} = 2L, \qquad M^{(\mathcal{N})} = 0, \qquad l \in \{1, 2, \dots, \mathcal{N}\} \tag{13}$$

By imposing $M^{(l)} = M^{(\mathcal{N}-1)}$ and considering the differences $E_{ll} - E_{\bar{l}\bar{l}}$ we end up with (12) in accordance to the folding of $\mathfrak{sl}(\mathcal{N})$ leading to the $\mathfrak{so}(n)$ algebra [7, 8].

3. Thermodynamics

The aim now is to consider the study of the BAEs at the thermodynamic limit. The ground state of the model consists of n filled Dirac seas, unlike the Yangian case, where the ground state consists of $2n+1$ or $2n$ filled seas respectively. As usual, an excitation corresponds to a hole in the Dirac sea. We perform our computations in the thermodynamic limit of the BAE, which is obtained according to the thermodynamic rule (for more details the interested reader is referred to e.g. [10, 11, 12] or [1] in a more relevant context)

$$\frac{1}{L}\sum_{j=1}^{M^{(\ell)}} f(\lambda_j^{(\ell)}) \to \int_0^\infty d\mu\, \sigma_\ell(\mu)\, f(\mu) - \frac{1}{L}\sum_{j=1}^{\nu^{(\ell)}} f(\lambda_j^{(\ell)}) - \frac{1}{2L} f(0)\,, \tag{14}$$

with $\nu^{(\ell)}$ holes of rapidities $\lambda_j^{(\ell)}$ in the ℓth Dirac sea σ_ℓ is the density in the ℓth sea. The last term is the halved contribution at 0^+ due to the boundaries. We shall focus here on the two-holes state, so that we can investigate both bulk and boundary scattering. In the thermodynamic limit the densities describing the state in the presence of holes (particle-like excitations) are given as:

$$\hat{\sigma}(\omega) = \hat{\varepsilon}(\omega) + \frac{1}{L}\,\hat{r}^{(1)}(\omega) \tag{15}$$

where $\hat{\sigma}$, $\hat{\varepsilon}^{(0)}$, $\hat{r}^{(i)}$ are n column-vectors. In fact the $r^{(1)}$ contribution is the one that will provide the bulk and boundary scattering amplitudes as will be transparent in the subsequent section.

It is also worth noting that from the BAEs in the thermodynamic limit we can compute the energy of the holes in each sea:

- $\mathfrak{sl}(2n+1)$

$$\hat{\varepsilon}^{(j)}(\omega) = \frac{\cosh(n + \frac{1}{2} - j)\frac{\omega}{2}}{\cosh(n + \frac{1}{2})\frac{\omega}{2}}, \qquad j \in \{1, 2 \dots, n\}$$

- $\mathfrak{sl}(2n)$

$$\begin{aligned} \hat{\varepsilon}^{(j)}(\omega) &= \frac{\cosh(n - j)\frac{\omega}{2}}{\cosh\frac{n\omega}{2}}, \qquad j \in \{1, 2 \dots, n-1\} \\ \hat{\varepsilon}^{(n)}(\omega) &= \frac{1}{2\cosh\frac{n\omega}{2}} \end{aligned} \tag{16}$$

The details here are omitted but we refer the interested reader for more details in [1, 2].

4. Scattering

The key element in this context is now the generalized quantization condition for the twisted Yangian introduced in [1]. We shall consider here the scattering of particle-like excitations in the first sea. This is in fact inspired by earliest studies on the formulation of the quantization for quantum integrable systems with different boundary conditions (see also [10], [11], [12])

In the twisted Yangian a modified isomonodromy condition is imposed on the two-holes state [1, 2]:

$$\Big(e^{i\mathcal{P}^{(\ell)}L}\,\mathbb{S}(\lambda_1,\lambda_2)-1\Big)|\lambda_1,\lambda_2\rangle=0\,, \tag{17}$$

$\mathcal{P}^{(\ell)}$ the momentum of the hole in the ℓth sea. The global scattering matrix $\mathbb{S}$ is given by:

$$\mathbb{S}(\lambda_1,\ \lambda_2)=\mathcal{K}^+(\lambda_1)\ \mathcal{S}(\lambda_1-\lambda_2)\ \mathcal{K}^-(\lambda_2)\ \mathcal{S}(\lambda_1+\lambda_2)\,, \tag{18}$$

The bulk scattering $\mathcal{S}$ is factorized as

$$\mathcal{S}(\lambda)=S(\lambda)\ \bar{S}(\lambda) \tag{19}$$

where S is the soliton-soliton scattering matrix, and $\bar{S}$ is the soliton–antisoliton scattering matrix in the $\mathfrak{sl}(\mathcal{N})$ spin chain.

The quantization condition is schematically depicted below:

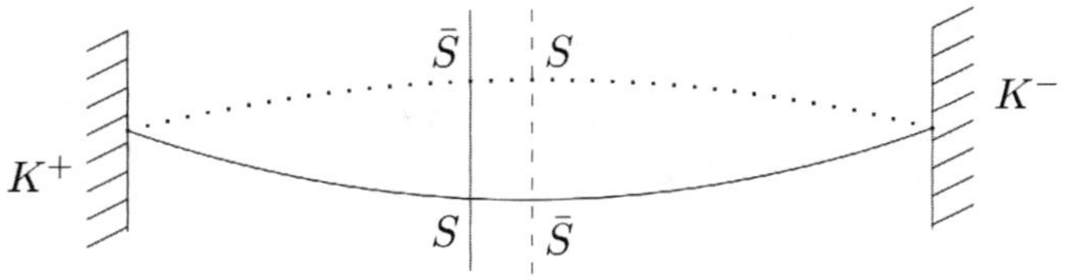

As already pointed out in the previous section we focus on the state with 2-holes in the first Dirac sea. If we now compare quantization condition with the density of the state (15), recall also:

$$\varepsilon^{(\ell)}(\lambda)=\frac{i}{2\pi}\frac{d\mathcal{P}^{(\ell)}(\lambda)}{d\lambda}$$

then we conclude:

$$\mathcal{S}_0(\lambda)=\exp\Big\{-\int_{-\infty}^{\infty}\frac{d\omega}{\omega}\,e^{-i\omega\lambda}\,\mathcal{B}_1(\omega)\Big\}$$
$$K_0^+(\lambda)\,K_0^-(\lambda)=\exp\Big\{-\int_{-\infty}^{\infty}\frac{d\omega}{\omega}\,\Big(e^{-i\omega\lambda}\,\mathcal{B}_2(\omega)+e^{-2i\omega\lambda}\,\mathcal{B}_1(\omega)\Big)\Big\},$$

where $\mathcal{S},\ K_0^{\pm}$ are eigenvalues of $\mathcal{S},\ \mathcal{K}^{\pm}$ respectively, and we define

$$\mathcal{B}_1(\omega)=\hat{a}_2(\omega)\,\hat{\mathcal{R}}_{11}(\omega)-\hat{a}_1(\omega)\,\hat{\mathcal{R}}_{12}(\omega)$$
$$\mathcal{B}_2(\omega)=\sum_{j=1}^{n}\Big(\hat{a}_2(\omega)-2\hat{a}_1(\omega)+\hat{a}_1(\omega)\delta_{i1}-\hat{a}_{\frac{1}{2}}(\omega)\delta_{in}\Big)\hat{\mathcal{R}}_{1i}(\omega)\,. \tag{20}$$

$$\hat{a}_n(\omega)=e^{-\frac{n|\omega|}{2}},\qquad \hat{\mathcal{R}}_{ij}(\omega)=e^{\frac{\omega}{2}}\frac{\sinh\Big(\min(\mathrm{i},\,\mathrm{j})\frac{\omega}{2}\Big)\ \cosh\Big(n+\frac{1}{2}-\max(i,j)\Big)\frac{\omega}{2}}{\cosh\Big(n+\frac{1}{2}\Big)\frac{\omega}{2}\ \sinh\frac{\omega}{2}}. \tag{21}$$

Explicit expressions for bulk and boundary scattering amplitudes are manifestly extracted above and the bulk scattering factorization: $\mathcal{S}(\lambda)=S(\lambda)\ \bar{S}(\lambda)$ may be then shown (we refer the interested reader to [1, 2] for more details).

5. Im lementing de ects

We shall briefly discuss here the case where a local defect is implemented.

$$\text{Define :} \quad X_k^+(\lambda) = \frac{\lambda + i\alpha_k - \frac{ik}{2}}{\lambda + i\alpha_{k+1} - \frac{ik}{2}}, \qquad X_k^-(\lambda) = \frac{\lambda + i\alpha_{\mathcal{N}-k+1} + \frac{i(\mathcal{N}-k)}{2}}{\lambda + i\alpha_{\mathcal{N}-k} + \frac{i(\mathcal{N}-k)}{2}}$$

then the BAEs via the analytical Bethe ansatz formulation read as

- $\mathfrak{sl}(2n+1)$

$$\begin{aligned}
& X_1^+(\lambda_i^{(1)} - \Theta)\, X_1^+(\lambda_i^{(1)} + \Theta)\, e_1^L(\lambda_i^{(1)}) = \\
& \quad - \prod_{j=1}^{M^{(1)}} e_2(\lambda_i^{(1)} - \lambda_j^{(1)})\, e_2(\lambda_i^{(1)} + \lambda_j^{(1)}) \prod_{j=1}^{M^{(2)}} e_{-1}(\lambda_i^{(1)} - \lambda_j^{(2)})\, e_{-1}(\lambda_i^{(1)} + \lambda_j^{(2)}), \\
& X_\ell^+(\lambda_i^{(l)} - \Theta)\, X_\ell^+(\lambda_i^{(l)} + \Theta) = \\
& \quad - \prod_{j=1}^{M^{(\ell)}} e_2(\lambda_i^{(\ell)} - \lambda_j^{(\ell)})\, e_2(\lambda_i^{(\ell)} + \lambda_j^{(\ell)}) \prod_{\tau=\pm 1} \prod_{j=1}^{M^{(\ell+\tau)}} e_{-1}(\lambda_i^{(\ell)} - \lambda_j^{(\ell+\tau)})\, e_{-1}(\lambda_i^{(\ell)} + \lambda_j^{(\ell+\tau)}) \\
& \quad \text{for} \qquad \ell = 2, \dots, n-1, \\
& X_n^+(\lambda_i^{(n)} - \Theta)\, X_n^+(\lambda_i^{(n)} + \Theta)\, e_{-\frac{1}{2}}(\lambda_i^{(n)}) = - \prod_{j=1}^{M^{(n-1)}} e_{-1}(\lambda_i^{(n)} - \lambda_j^{(n-1)})\, e_{-1}(\lambda_i^{(n)} + \lambda_j^{(n-1)}) \\
& \quad \times - \prod_{j=1}^{M^{(n)}} e_{-1}(\lambda_i^{(n)} - \lambda_j^{(n)})\, e_{-1}(\lambda_i^{(n)} + \lambda_j^{(n)})\, e_2(\lambda_i^{(n)} - \lambda_j^{(n)})\, e_2(\lambda_i^{(n)} + \lambda_j^{(n)})
\end{aligned} \tag{22}$$

Notice that the Bethe ansatz equations in the presence of defects are very similar to the ones presented in the previous section. The main difference is the existence of the extra contributions $X^\pm$ due to the presence of defects. In the thermodynamic limit these contributions will provide the transmission amplitudes.

We may now formulate a suitable quantization condition for the model in the presence of defects. In order to determine the relevant transmission matrix it suffices to consider a state with one hole in the first sea. Let us first introduce some notation and define the transmission amplitudes in $\mathfrak{sl}(\mathcal{N})$ [13] as

$$\begin{aligned}
T(\lambda - \Theta) &: \quad \text{soliton–defect scattering} \\
\bar{T}(\lambda - \Theta) &: \quad \text{soliton–anti-defect scattering} \\
T^*(\lambda + \Theta) &: \quad \text{anti-soliton–defect scattering} \\
\bar{T}^*(\lambda + \Theta) &: \quad \text{anti-soliton–anti-defect scattering}
\end{aligned} \tag{23}$$

The quantization condition for such a state reads as

$$\left(e^{i\mathcal{P}^{(l)}} \mathbb{S}(\lambda^{(l)}, \Theta) - 1\right) |\lambda^{(l)}, \Theta\rangle = 0, \tag{24}$$

where the global scattering amplitude is given by

$$\mathbb{S}(\lambda, \Theta) = \mathcal{K}^+(\lambda)\, T(\lambda - \Theta)\, \bar{T}(\lambda - \Theta)\, \mathcal{K}^-(\lambda)\, \bar{T}^*(\lambda + \Theta)\, T^*(\lambda + \Theta) \tag{25}$$

We shall not give further details on the derivation of transmission amplitudes in twisted Yangians. Explicit expressions of transmission amplitudes and their factorizations are provided in [2]. With this we conclude our presentation on the bulk and boundary scattering in the context of twisted Yangians.

Re erences

[1] J. Avan, A. Doikou and N. Karaiskos, J. Stat. Mech. (2015) P02007, ar iv:1410.5991 [hep-th].
[2] J. Avan, A. Doikou and N. Karaiskos, J. Stat. Mech. (2015) P05024, ar iv:1412.6480 [hep-th]
[3] L. Freidel, J. M. Maillet, Phys. Lett **B262** (1991) 268.
[4] I.V. Cherednik, Theor. Math. Phys. **61** (1984) 977.
[5] E.K. Sklyanin, J. Phys. **A21** (1988) 2375.
[6] G.I. Olshanski "Quantum Groups (1992)", Springer Lecture notes in Math. 1510
[7] A. Doikou, J. Phys. A **33** (2000) 8797, [hep-th 0006197].
[8] D. Arnaudon, J. Avan, N. Crampe, A. Doikou, L. Frappat and E. Ragoucy, J. Stat. Mech. **0408** (2004) P08005, [math-ph 0406021].
[9] D. Arnaudon, J. Avan, N. Crampe, A. Doikou, L. Frappat and E. Ragoucy, Nucl. Phys. B **687** (2004) 257, [math-ph 0310042].
[10] V. E. Korepin, Commun. Math. Phys. **76** (1980) 165.
[11] N. Andrei and C. Destri, Nucl. Phys. B **231** (1984) 445.
[12] M. T. Grisaru, L. Mezincescu and R. I. Nepomechie, J. Phys. A **28** (1995) 1027, [hep-th 9407089].
[13] A. Doikou, JHEP 08 (2013) 103, a iv:1304.5901 [hep-th].

Journal of Physics: Conference Series **670** (2016) 012008 doi:10.1088/1742-6596/670/1/012008

Ferrara–Porrati–Sagnotti approach and the one-dimensional supersymmetric model with PBGS

S. Bellucci[1], S. Krivonos[2], A. Sutulin[1,2]

[1] INFN-Laboratori Nazionali di Frascati, Via E. Fermi 40, 00044 Frascati, Italy
[2] Bogoliubov Laboratory of Theoretical Physics, JINR, 141980 Dubna, Russia

Abstract. We apply Ferrara–Porrati–Sagnotti approach to the case of one-dimensional supersymmetric model with $N = 2$ supersymmetry spontaneously broken to the $N = 1$ one. We explicitly demonstrate that only one superfield can be treated as the Goldstone one, while the second one has the meaning of the matter superfield. The general action for such a system is constructed and also two relevant particular cases are considered.

1. Introduction

In [1] *J Ba er and A Galperin* (*BG*) proposed the approach to construct the $N = 2, D = 4$ supersymmetric Born–Infeld theory. Combining two $N = 1$ supermultiplets, namely, $N = 1$ *vec r* and $N = 1$ *c iral* ones, they realized the transformations of additional $N = 1$ supersymmetry with parameters $\eta_\alpha, \bar{\eta}_{\dot{\alpha}}$ as

$$\delta\left(W\right)_\alpha = \left(1 - \frac{1}{4}\overline{D}^2\overline{X}\right)\eta_\alpha - \mathrm{i}\partial_{\alpha\dot{\alpha}}X\bar{\eta}^{\dot{\alpha}}. \quad \delta X = -2\left(W\right)^\alpha\eta_\alpha, \tag{1.1}$$

Due to the presence of the constant term in this transformation law, this additional supersymmetry is spontaneously broken. Keeping in mind, that from the vector superfield W_α Bianchi identity, for the $N = 1$ vector supermultiplet it follows that

$$D^2 W_\alpha \sim \partial_{\alpha\dot{\alpha}}\overline{W}^{\dot{\alpha}}, \tag{1.2}$$

then the trivial action

$$S = \int d^4x d^2\,\theta X \tag{1.3}$$

becomes invariant under transformations (1.1). However, the action (1.3) acquires a real meaning only after introducing the additional constraint (which is invariant under (1.1))

$$W \cdot W + X\left(1 - \frac{1}{4}\overline{D}^2\overline{X}\right) = 0. \tag{1.4}$$

Once again, due to the presence of the constant in (1.4), this constraint can be explicitly solved by expressing X in terms of the superfields $W_\alpha, \overline{W}_{\dot{\alpha}}$. The resulting action is just the $N = 2$ Born Infeld one.

In the recent paper [2] *S Ferrara, M P rra i and A Sa n i* (*FPS*) proposed the generalization of the *BG* construction to the cases of several $N = 1$ vector supermultiplets. The *FPS* approach includes w basic ingredients.

Firstly, the nonlinear constraint (1.4) is generalized to be

$$d_{abc}\left(W_b \cdot W_c + X_b\left(m_c - \frac{1}{4}\overline{D}^2\overline{X}_c\right)\right) = 0, \tag{1.5}$$

where $W_{a\alpha}$ and X_a are n-copies of $N = 1$ vector and chiral multiplets respectively, and the totally symmetric tensor d_{abc} and vector m_a are a set of constants.

The realization of hidden $N = 1$ supersymmetry is similar to the BG case and reads

$$\delta W_{a\alpha} = \left(m_a - \frac{1}{4}\overline{D}^2\overline{X}_a\right)\eta_\alpha - \mathrm{i}\partial_{\alpha\dot{\alpha}}X_a\bar{\eta}^{\dot{\alpha}}, \quad \delta X_a = -2W_a^\alpha\eta_\alpha. \tag{1.6}$$

The invariance of (1.5) with respect to (1.6) involves the new *addi i nal* constraint

$$d_{abc}W_{b\alpha}X_c = 0, \tag{1.7}$$

which was automatically satisfied in the BG case. The constraints (1.5), (1.7) can be solved to express the bosonic $N = 1$ chiral superfields X_a through the $N = 1$ vector supermultiplets $W_{a\alpha}$.

The next nontrivial step of the *FPS* approach yields the structure of the corresponding action, which proved to be more involved

$$S = \int d^4x d^2\theta\left[e^a X_a + C_{ab}\left(W_a \cdot W_b + X_a\left(m_b - \frac{1}{4}\overline{D}^2\overline{X}_b\right)\right) + c.c.\right], \tag{1.8}$$

where $e^a = const, C_{ab} = C_{ba} = const$. In fact, the additional term with C_{ab} is invariant with respect to S-transformations (1.6). Such a term does not exist in the case of one supermultiplet, but it proved to be *e en ial* in the cases of several supermultiplets. The action (1.8) with X_a being the solution of the constraints (1.5), (1.7) is treated as the many-field extension of $N = 2$ Born-Infeld theory.

In [2] the detailed analysis of the $n = 2$ case was presented, which can be divided into two subcases:

- $d_{111} = 1, d_{112} = -1 \quad I_4 = 0,$
- $d_{111} = 1, d_{122} = -1 \quad I_4 > 0,$

where I_4 is a quartic invariant. Then, the analysis of the next $n = 3$ case has been given in [3]. The above, quite short sketch of the *FPS* approach is enough to raise w interesting questions:

- whether the supermultiplets with *differen* constants d_{abc} are really *differen ?*
- whether the action (1.8) is *unique* as it happened in the $n = 1$ case?

The reasons for raising the first question are the following.

After resolving the constraints (1.5), (1.7), we will have in the theory n copies of $N = 1$ vector multiplets $W_{a\alpha}$ with highly nonlinear transformation properties under hidden $N = 1$ supersymmetry (1.6). The presence of the constants m_a in (1.6) at first sight means that all of them have Goldstone nature. But it cannot be so, because the spontaneous breaking of $N = 2$ to $N = 1$ supersymmetries is accompanied by only one Goldstone superfield [4, 5]. Therefore, it should exist a special basis in which only one superfield is a Goldstone one, while the remaining superfields have to be matter superfields with trivial transformation properties with respect to the broken S-supersymmetry. Due to close relations between linear and nonlinear realizations of partially broken supersymmetries [6], the same property has to be visible within the linear realizations method, the *FPS* approach is dealing with. Thus, the claim that different choices of the constants d_{abc} in the basic constraints (1.5), (1.7) must be equivalent, should be carefully analyzed.

The second question about the uniqueness of the action we are raising, is also related with the existence of the special basis in which we have one Goldstone superfield and an arbitrary number of matter superfields. In such a basis the general action will have a functional freedom. Thus, it has to be quite interesting to analyze this general action and understand the reasons that just select the *FPS* actions.

Unfortunately, the analysis of $N = 2$ extended Born-Infeld theory is quite involved, and the ideological part is hidden behind complicated calculations. Fortunately enough, the *FPS* approach, as we demonstrated in [7], is not limited by application to the Born-Infeld theory only, and the questions we raised above may be answered through the analysis of a much simpler system, which the *FPS* approach can be applied to.

In our paper [8] we explored the generalization of the *FPS* construction for the simplest model of super-particle which realized the partial breaking of $N = 2$ to $N = 1$ supersymmetry in one dimension. We explicitly established the transformations that relate two different $n = 2$ cases with the special basis discussed above, as well as the relations between them. We also constructed the general action for such a system containing two arbitrary functions and provided the explicit form of these functions reducing the general action to *FPS*-like ones.

In this contribution we review our results.

2. Partial breaking in one dimension and the su er article action

We begin with a simple example of a superparticle model describing $N = 2 \rightarrow N = 1$ partial breaking of global supersymmetry in $d = 1$, which was considered in paper [9] The anticommutation relation of $N = 1, d = 1$ Poincaré superalgebra reads

$$\{Q, Q\} = 2P\,, \tag{2.1}$$

and the coordinates (t, θ) of $N = 1$ superspace satisfy the following rules of complex conjugation: $t^\dagger = -t, \theta^\dagger = \theta$. Now we define the bosonic and fermionic superfields $v(t,\theta)$ and $\psi(t,\theta)$ related as

$$\psi = \frac{1}{2} Dv, \quad (v^\dagger = -v,\ \psi^\dagger = \psi), \tag{2.2}$$

where the spinor derivative D satisfies the relation

$$D = \frac{\partial}{\partial\theta} + \theta\partial_t, \quad \{D, D\} = 2\partial_t\,. \tag{2.3}$$

Linear reali a i n ne-par icle ca e

In full analogy with *BG* approach [1], we introduce two spinor superfields $\psi(t,\theta)$ and $\nu(t,\theta)$, assuming that an additional spontaneously broken $N = 1$ supersymmetry is realized on them as

$$\delta\psi = \varepsilon(1 - D\nu)\,, \quad \delta\nu = \varepsilon D\psi. \tag{2.4}$$

The presence of the constant shift in the first relation suggests the interpretation of $\psi(t,\theta)$ as the Goldstone fermion accompanying the $N = 2 \rightarrow N = 1$ breaking. The superfield $\nu(t,\theta)$, due to its transformation properties under the broken $N = 1$ supersymmetry, may be chosen as a *La ran ian den i* , since the integral over the superspace

$$S = \int dt d\theta\, \nu \tag{2.5}$$

is invariant with respect to both broken and unbroken supersymmetries.

To be meaningful, the action (2.5) should be accompanied by an additional constraint, which is invariant under transformations (2.4). One may easily check, that the corresponding invariant constraint looks quite similar to (1.4)

$$\psi D\psi - \nu\left(1 - D\nu\right) = 0 \tag{2.6}$$

with evident solution given by

$$\nu = \frac{2\psi D\psi}{1 + \sqrt{1 - 4(D\psi)^2}}\,. \tag{2.7}$$

It is easy to find that from (2.6) it follows that

$$\psi\nu = 0. \tag{2.8}$$

The bosonic limit of the action (2.5) with (2.7) corresponds to the action of a particle in $d = 1$

$$S_{bos} = \frac{1}{2}\int dt\left(1 - \sqrt{1 - (\dot{v})^2}\right). \tag{2.9}$$

Thus, this simplest system demonstrate a close analogy with the BG construction of the supersymmetric Born-Infeld action.

Linear reali a i n e ul ipar icle ca e

The above considered case can be easily extended to that of an arbitrary number of $N = 1$ superfields strictly following the FPS approach. Doing so, we firstly introduce a set of n superfields $\psi_a(t,\theta)$ and $\nu_a(t,\theta)$ transforming under implicit $N = 1$ supersymmetry as

$$\delta\psi_a = \varepsilon(m_a - D\nu_a)\,, \quad \delta\nu_a = \varepsilon D\psi_a, \tag{2.10}$$

where m_a are arbitrary constants. Now, again in full analogy with FPS approach, we impose the following generalization of the constraint (2.6)

$$d_{abc}\big(\psi_b D\psi_c - \nu_b\left(m_c - D\nu_c\right)\big) = 0. \tag{2.11}$$

The invariance of this constraint under the broken S-supersymmetry leads to an additional restriction

$$\partial_t\left(d_{abc}\psi_b\nu_c\right) = 0, \tag{2.12}$$

which can be reformulated as

$$d_{abc}\psi_b\nu_c = 0, \tag{2.13}$$

assuming that the constant of integration is equal to zero. Finally, the FPS-like generalization of the action reads

$$S = \int dt d\theta\left[e^a\nu_a + C_{ab}\big(\psi_a D\psi_b - \nu_a\left(m_b - D\nu_b\right)\big)\right], \tag{2.14}$$

where e^a and $C_{ab} = C_{ba}$ are arbitrary real constants.

After repeating the basic steps of the FPS approach, let us consider the system of two superparticles in the details.

3. Standard basis or $n = 2$ cases

In the standard approach of the Partial Breaking of Global Supersymmety, the two-particle model should contain only one Goldstone superfield (with the Goldstone fermion among their components). The rest of the fields must be matter ones, i.e. they should not transform with respect to the broken S-supersymmetry. Now we demonstrate how such a splitting works in a two-particle system for special choices of the components of tensor d_{abc}. We show that there is a special basis in which the systems in question, corresponding to a different choice of d_{abc}, can be described in a unified way.

e ca e wi $d_{111} = d_{222} = 1$

With such a choice of the symmetric tensor d_{abc}, the basic constraints (2.11), (2.13) have a splitting form

$$\psi_1 D\psi_1 - \nu_1 (m_1 - D\nu_1) = 0, \quad \psi_1\nu_1 = 0,$$
$$\psi_2 D\psi_2 - \nu_2 (m_2 - D\nu_2) = 0, \quad \psi_2\nu_2 = 0. \tag{3.1}$$

By rescaling of the variables ψ_a and ν_a in (3.1), one may always choose

$$m_1 = m_2 = 1, \tag{3.2}$$

then the transformations of fermions ψ_a and ν_a in (2.10) take the form

$$\delta\psi_1 = \epsilon (1 - D\nu_1), \quad \delta\nu_1 = \epsilon D\psi_1,$$
$$\delta\psi_2 = \epsilon (1 - D\nu_2), \quad \delta\nu_2 = \epsilon D\psi_2. \tag{3.3}$$

Similarly to the one particle case (2.7), the constraints (3.1) can be solved as

$$\nu_1 = \frac{2\psi_1 D\psi_1}{1 + \sqrt{1 - 4(D\psi_1)^2}}, \qquad \nu_2 = \frac{2\psi_2 D\psi_2}{1 + \sqrt{1 - 4(D\psi_2)^2}}. \tag{3.4}$$

The *FPS*-like action (1.8), invariant with respect to both S- and Q-supersymmetries, for this case reads

$$S = \int dt d\theta \left\{ e_1 \nu_1 + e_2\nu_2 + C_{12}\big(\psi_1 D\psi_2 - \nu_1(1 - D\nu_2) + \psi_2 D\psi_1 - \nu_2(1 - D\nu_1)\big)\right\}. \tag{3.5}$$

Now we introduce two Goldstone spinor superfields ξ_a that transform as

$$\delta\xi_1 = \epsilon + \epsilon\,\xi_1\,\partial_t\,\xi_1, \qquad \delta\xi_2 = \epsilon + \epsilon\,\xi_2\,\partial_t\,\xi_2. \tag{3.6}$$

It is easy to check that the *ilded field* ψ_a and ν_a

$$\psi_a = \psi_a - \xi_a (1 - D\nu_a), \quad \nu_a = \nu_a - \xi_a D\psi_a, \tag{3.7}$$

transform under (3.3), (3.6) as

$$\delta\psi_a = \epsilon\,\xi_a\,\partial_t\,\psi_a, \qquad \delta\nu_a = \epsilon\,\xi_a\,\partial_t\,\nu_a. \tag{3.8}$$

Thus, it is a covariant operation to put the superfields (3.7) equal to zero, which is equivalent to the relations

$$\psi_a - \xi_a (1 - D\nu_a) = 0, \qquad \nu_a - \xi_a D\psi_a = 0. \tag{3.9}$$

Journal of Physics: Conference Series **670** (2016) 012008 doi:10.1088/1742-6596/670/1/012008

Moreover, the constraints $\psi_a = \nu_a = 0$ contain some additional information: they express the superfields of the linear realization ψ_a and ν_a in terms of the Goldstone superfields ξ_a

$$\nu_a = \frac{\xi_a D\xi_a}{1 + D\xi_a D\xi_a}, \quad \psi_a = \frac{\xi_a}{1 + D\xi_a D\xi_a} \qquad \text{(no summation over } a). \tag{3.10}$$

Until now we have w Goldstone superfields $\xi_a(t,\theta)$, while we are expecting to have nl ne essential Goldstone fermionic superfield and ne a er superfield. This may be achieved by passing to the *new* superfields $\eta(t,\theta)$ and $\lambda(t,\theta)$

$$\eta = \frac{1}{2}(\xi_1 + \xi_2) + \frac{1}{4}\xi_1\xi_2\left(\xi_1 - \xi_2\right), \quad \lambda = \frac{1}{2}(\xi_1 - \xi_2) + \frac{1}{4}\xi_1\xi_2\left(\xi_1 + \xi_2\right), \tag{3.11}$$

which, in virtue of (3.6), transform as expected[1]

$$\delta\eta = \epsilon + \epsilon\eta\partial_t\eta, \quad \delta\lambda = \epsilon\eta\partial_t\lambda. \tag{3.12}$$

¿From (3.12) it follows that $\eta(t,\theta)$ is the fermionic Goldstone superfield accompanying the spontaneous breaking of $N = 2$ supersymmetry, while $\lambda(t,\theta)$ is a matter fermionic superfield. Finally, the superfields $\xi_a(t,\theta)$ as well as $\psi_a(t,\theta)$ and $\nu_a(t,\theta)$ can be expressed in terms of $\eta(t,\theta)$ and $\lambda(t,\theta)$:

$$\xi_1 = \eta + \lambda + \eta\lambda(\eta + \lambda), \quad \xi_2 = \eta - \lambda - \eta\lambda(\eta - \lambda), \tag{3.13}$$

$$\begin{aligned} \psi_1 &= \frac{\eta + \lambda}{1 + (D\eta + D\lambda)^2} + \eta\lambda(\eta + \lambda)\frac{1 - (D\eta + D\lambda)^2}{\left[1 + (D\eta + D\lambda)^2\right]^2}, \quad \psi_2 = \psi_1|_{\lambda \to -\lambda}, \\ \nu_1 &= (D\eta + D\lambda)\left(\frac{\eta + \lambda}{1 + (D\eta + D\lambda)^2} + \frac{2\eta\lambda(\eta + \lambda)}{\left[1 + (D\eta + D\lambda)^2\right]^2}\right), \quad \nu_2 = \nu_1|_{\lambda \to -\lambda}. \end{aligned} \tag{3.14}$$

Thus, the system of constraints (3.1) represents a n n- *andard* description of the essential Goldstone superfield $\eta(t,\theta)$ and a matter fermionic superfield $\lambda(t,\theta)$.

e ca e wi $d_{111} = 1, d_{122} = -1$

For these values of the constants the constraints (2.11) read

$$\begin{aligned} &\psi_1 D\psi_1 - \psi_2 D\psi_2 - \nu_1(m_1 - D\nu_1) + \nu_2(m_2 - D\nu_2) = 0, \\ &\psi_1 D\psi_2 + \psi_2 D\psi_1 - \nu_1(m_2 - D\nu_2) - \nu_2(m_1 - D\nu_1) = 0, \end{aligned} \tag{3.15}$$

while those in (2.13) take form

$$\psi_1\nu_1 - \psi_2\nu_2 = 0, \qquad \psi_1\nu_2 + \psi_2\nu_1 = 0. \tag{3.16}$$

Now we introduce the complex superfields ψ, ν and complex parameter m as

$$\begin{aligned} \psi &= \psi_1 + \mathrm{i}\psi_2, & \nu &= \nu_1 + \mathrm{i}\nu_2, & m &= m_1 + \mathrm{i}m_2, \\ \bar\psi &= \psi_1 - \mathrm{i}\psi_2, & \bar\nu &= \nu_1 - \mathrm{i}\nu_2, & \bar m &= m_1 - \mathrm{i}m_2, \end{aligned} \tag{3.17}$$

and then rewrite eqyations (3.15) and (3.16) in the splitting form

$$\begin{aligned} &\psi D\psi - \nu(m - D\nu) = 0, \quad \psi\nu = 0, \\ &\bar\psi D\bar\psi - \bar\nu(\bar m - D\bar\nu) = 0, \quad \bar\psi\bar\nu = 0. \end{aligned} \tag{3.18}$$

[1] This is the form-variation of the fields under implicit $N = 1$ supersymmetry: $\delta\mathcal{A} = \mathcal{A}'(t,\theta) - \mathcal{A}(t,\theta)$.

Journal of Physics: Conference Series **670** (2016) 012008 doi:10.1088/1742-6596/670/1/012008

The solution to the equations (3.18) read

$$\nu = \frac{2\psi D\psi}{m + \sqrt{m^2 - 4(D\psi)^2}}, \qquad \bar\nu = \frac{2\bar\psi D\bar\psi}{\bar m + \sqrt{\bar m^2 - 4(D\bar\psi)^2}}. \tag{3.19}$$

In terms of these variables the invariant *FPS* action (1.8) acquires a form

$$S = \int dt d\theta \left\{ e\,\nu + e^* \bar\nu + C_{12}\big(\psi D\bar\psi - \nu(\bar m - D\bar\nu) + \bar\psi D\psi - \bar\nu(m - D\nu)\big)\right\}, \tag{3.20}$$

where $e = \frac{1}{2}(e_1 - \mathrm{i}e_2)$. Let us note, that one may always choose the constants m_1, m_2 as [7]

$$m_1 = 1,\ m_2 = 0, \qquad \Rightarrow \qquad m = \bar m = 1. \tag{3.21}$$

With this choice of the parameter m, the complex superfields ψ and ν transform with respect to the broken S-supersymmetry as

$$\begin{aligned} \delta\psi &= \epsilon\,(1 - D\nu), & \delta\nu &= \epsilon D\psi, \\ \delta\bar\psi &= \epsilon\,(1 - D\bar\nu), & \delta\bar\nu &= \epsilon D\bar\psi, \end{aligned} \tag{3.22}$$

and obey the constraints

$$\begin{aligned} \psi D\psi - \nu\,(1 - D\nu) &= 0, & \psi\,\nu &= 0, \\ \bar\psi D\bar\psi - \bar\nu\,(1 - D\bar\nu) &= 0, & \bar\psi\,\bar\nu &= 0. \end{aligned} \tag{3.23}$$

So, we have a full analogy with the previously considered splitting case.
Let us introduce two Goldstone spinor superfields $\xi(t,\theta)$ and $\bar\xi(t,\theta)$ which transform as

$$\delta\xi = \epsilon + \epsilon\,\xi\,\partial_t\,\xi, \qquad \delta\bar\xi = \epsilon + \epsilon\,\bar\xi\,\partial_t\,\bar\xi. \tag{3.24}$$

Then the *ilded field* $\psi, \bar\psi$ and $\nu, \bar\nu$, which are defined by the following expressions

$$\psi = \psi - \xi\,(1 - D\nu), \quad \nu = \nu - \xi D\psi, \quad \bar\psi = \bar\psi - \bar\xi\,(1 - D\bar\nu), \quad \bar\nu = \bar\nu - \bar\xi D\bar\psi, \tag{3.25}$$

transform under (3.22), (3.24) as

$$\delta\psi = \epsilon\,\xi\,\partial_t\,\psi, \qquad \delta\nu = \epsilon\,\xi\,\partial_t\,\nu, \qquad \delta\bar\psi = \epsilon\,\bar\xi\,\partial_t\,\bar\psi, \qquad \delta\bar\nu = \epsilon\,\bar\xi\,\partial_t\,\bar\nu. \tag{3.26}$$

Thus, one may again to put these superfields equal to zero, that means

$$\psi - \xi\,(1 - D\nu) = 0, \ \ \bar\psi - \bar\xi\,(1 - D\bar\nu) = 0, \ \ \nu - \xi D\psi = 0, \ \ \bar\nu - \bar\xi D\bar\psi = 0. \tag{3.27}$$

It is obvious that the constraints (3.23) follow from (3.27). The constraints (3.27) can be solved as

$$\nu = \frac{\xi D\xi}{1 + D\xi D\xi}, \quad \psi = \frac{\xi}{1 + D\xi D\xi}, \quad \bar\nu = \frac{\bar\xi D\bar\xi}{1 + D\bar\xi D\bar\xi}, \quad \bar\psi = \frac{\bar\xi}{1 + D\bar\xi D\bar\xi}. \tag{3.28}$$

In full analogy with the previously considered case, we introduce the superfields $\eta(t,\theta)$ and $\lambda(t,\theta)$ by

$$\eta = \frac{1}{2}(\xi + \bar\xi) + \frac{1}{4}\xi\bar\xi(\xi - \bar\xi), \quad \lambda = \frac{\mathrm{i}}{2}(\xi - \bar\xi) + \frac{\mathrm{i}}{4}\xi\bar\xi(\xi + \bar\xi), \tag{3.29}$$

and find that they have the same transformation properties as in (3.12)

$$\delta\eta = \epsilon + \epsilon\eta\partial_t\eta, \qquad \delta\lambda = \epsilon\eta\partial_t\lambda. \tag{3.30}$$

The superfields $\xi(t,\theta), \bar{\xi}(t,\theta)$ as well as the genuine superfields $\psi(t,\theta), \nu(t,\theta)$ can be expressed in terms of the Goldstone superfield $\eta(t,\theta)$ and matter fermionic superfield $\lambda(t,\theta)$:

$$\xi = \eta - \mathrm{i}\lambda - \mathrm{i}\eta\lambda(\eta - \mathrm{i}\lambda), \qquad \bar{\xi} = \eta + \mathrm{i}\lambda + \mathrm{i}\eta\lambda(\eta + \mathrm{i}\lambda), \tag{3.31}$$

$$\psi = \frac{\eta - \mathrm{i}\lambda}{1 + (D\eta - \mathrm{i}D\lambda)^2} - \mathrm{i}\eta\lambda(\eta - \mathrm{i}\lambda)\frac{1 - (D\eta - \mathrm{i}D\lambda)^2}{\left[1 + (D\eta - \mathrm{i}D\lambda)^2\right]^2}, \quad \bar{\psi} = (\psi)^\dagger, \tag{3.32}$$
$$\nu = (D\eta - \mathrm{i}D\lambda)\left(\frac{\eta - \mathrm{i}\lambda}{1 + (D\eta - \mathrm{i}D\lambda)^2} - \mathrm{i}\frac{2\eta\lambda(\eta - \mathrm{i}\lambda)}{\left[1 + (D\eta - \mathrm{i}D\lambda)^2\right]^2}\right), \quad \bar{\nu} = (\nu)^\dagger.$$

Rela i n be ween e w ca e
Since the relations between the pairs of superfields (ξ_1, ξ_2) and (η, λ) as well as between $(\xi, \bar{\xi})$ and (η, λ) are invertible, one can express the spinor fields (ξ_1, ξ_2) through $(\xi, \bar{\xi})$ and vice versa

$$\xi = \frac{1}{2}(1 - \mathrm{i})\,\xi_1 + \frac{1}{2}(1 + \mathrm{i})\,\xi_2 + \frac{1}{2}\xi_1\xi_2\left(\xi_1 - \xi_2\right), \tag{3.33}$$
$$\bar{\xi} = \frac{1}{2}(1 + \mathrm{i})\,\xi_1 + \frac{1}{2}(1 - \mathrm{i})\,\xi_2 + \frac{1}{2}\xi_1\xi_2\left(\xi_1 - \xi_2\right),$$

$$\xi_1 = \frac{1}{2}(1 + \mathrm{i})\,\xi + \frac{1}{2}(1 - \mathrm{i})\,\bar{\xi} + \frac{1}{2}\xi\bar{\xi}\left(\xi - \bar{\xi}\right), \tag{3.34}$$
$$\xi_2 = \frac{1}{2}(1 - \mathrm{i})\,\xi + \frac{1}{2}(1 + \mathrm{i})\,\bar{\xi} + \frac{1}{2}\xi\bar{\xi}\left(\xi - \bar{\xi}\right).$$

Thus, at least in one dimension, the two cases are completely equivalent, because they are related by invertible fields redefinitions (3.33), (3.34). The only difference between these cases lies in the diverse definition of the invariant actions (3.5), (3.20). In fact, these two actions are not the most general ones. We will demonstrate this by constructing the most general action for the Goldstone fermion $\eta(t,\theta)$ and the matter fermionic superfield $\lambda(t,\theta)$.

4. Nonlinear realizations a roach

In this section we provide an alternative description of the system with a partial breaking of the global $N = 2, d = 1$ supersymmetry within the nonlinear realizations approach. It turns out that in the present case this approach is more suitable for the construction of the most general superfield action.

NLR appr ac in ne di en i n
We start with the $N = 2, d = 1$ Poincaré superalgebra with one central charge generator Z

$$\{Q, Q\} = 2P, \quad \{S, S\} = 2P, \quad \{Q, S\} = 2Z. \tag{4.1}$$

Introducing a coset element g as

$$g = e^{tP}e^{\theta Q}e^{qZ}e^{\eta S}, \tag{4.2}$$

and calculating the expression $g^{-1}dg = \omega_P P + \omega_Q Q + \omega_Z Z + \omega_S S$, one finds the explicit structure for the Cartan forms

$$\omega_P = dt - d\theta\theta - d\eta\eta, \quad \omega_Q = d\theta, \quad \omega_Z = dq - 2d\theta\eta, \quad \omega_S = d\eta. \tag{4.3}$$

The covariant derivatives can be found in a standard way, and now they read

$$\nabla_\theta = D + \eta D\eta \nabla_t, \quad \nabla_t = E^{-1}\partial_t, \tag{4.4}$$

where the explicit forms of the einbein E and its inverse E^{-1} are

$$E = 1 + \eta\partial_t\eta, \quad E^{-1} = 1 - \eta\nabla_t\eta. \tag{4.5}$$

These derivatives satisfy the following (anti)commutation relations

$$\begin{aligned} \{\nabla_\theta, \nabla_\theta\} &= 2(1 + \nabla_\theta\eta\nabla_\theta\eta)\nabla_t, \\ [\nabla_t, \nabla_\theta] &= 2\nabla_t\eta\nabla_\theta\eta\nabla_t. \end{aligned} \tag{4.6}$$

Acting on the coset element g (4.2) from the left by different elements g_0 of the $N=2, d=1$ Poincaré supergroup, one can find the transformation properties of the coordinates (t,θ) and the Goldstone superfield $\eta(t,\theta)$. In particular, under both unbroken Q- and broken S-supersymmetries one gets:

- Unbroken supersymmetry ($g_0 = e^{\epsilon Q}$):

$$\delta_Q t = -\epsilon\theta, \quad \delta_Q\theta = \epsilon, \tag{4.7}$$

- Broken supersymmetry ($g_0 = e^{\varepsilon S}$):

$$\delta_S t = -\varepsilon\eta, \quad \delta_S q = -2\varepsilon\theta, \quad \delta_S\eta = \varepsilon. \tag{4.8}$$

The final step is to express the Goldstone superfield $\eta(t,\theta)$ in terms of the bosonic superfield $q(t,\theta)$ by imposing the constraint [10]

$$\omega_Z|_{d\theta} = 0 \quad \Rightarrow \quad \eta = \frac{1}{2}\nabla_\theta q. \tag{4.9}$$

In order to describe the matter spinor superfield $\lambda(t,\theta)$ within the nonlinear realizations approach, it is necessary to postulate its transformation properties under the S-supersymmetry

$$\delta_S\lambda = 0. \tag{4.10}$$

It is obvious that, due to the transformation of the coordinate t under the S-supersymmetry in (4.8), the variation of $\lambda(t,\theta)$ reads

$$\delta\lambda = \varepsilon\eta\partial_t\lambda. \tag{4.11}$$

One may also define the bosonic superfield $\phi(t,\theta)$ with the transformation property

$$\delta_S\phi = 0, \tag{4.12}$$

which is related with $\lambda(t,\theta)$ in the same way as in (4.9)

$$\lambda = \frac{1}{2}\nabla_\theta\phi. \tag{4.13}$$

General ac i n
The most general *An a* for the $N=1$ superfield action describing the Goldstone fermionic superfield $\eta(t,\theta)$ and the matter fermionic superfield $\lambda(t,\theta)$ and having no dimensional constants reads

$$S = \int dtd\theta \big(1+\eta\eta\big)\Big[\eta\ F_1^{cov} + \lambda\ F_2^{cov} + \eta\eta\lambda\ F_3^{cov} + \eta\lambda\nabla_t\lambda\ F_4^{cov}\Big], \tag{4.14}$$

where $F_1^{cov}, F_2^{cov}, F_3^{cov}, F_4^{cov}$ are arbitrary functions depending on $\nabla_\theta\eta$ and $\nabla_\theta\lambda$ only. The reason for such a form of *An a* is quite understandable:

- with respect to S-supersymmetry transformation, $\delta_S t = -\varepsilon\eta$, $\delta_S\eta = \varepsilon$, $\delta_S\lambda = 0$, the improved measure $dtd\theta\big(1+\eta\eta\big)$ is invariant;
- the functions $F_1^{cov}, F_2^{cov}, F_3^{cov}, F_4^{cov}$ as well as $\nabla_t\lambda$ are also invariant with respect to the broken S-supersymmetry;
- the functions $F_1^{cov}, F_2^{cov}, F_3^{cov}, F_4^{cov}$ have to be further restricted by imposing the invariance of the action (4.14) with respect to the broken S-supersymmetry.

The variation of the action (4.14) with respect to S-supersymmetry reads

$$\delta_S S = \int dtd\theta\ \varepsilon\Big[\big(1+\eta\eta\big)F_1^{cov} + \eta\lambda F_3^{cov} + \lambda\lambda F_4^{cov}\Big]. \tag{4.15}$$

Thus, the function F_2^{cov} may be chosen to be an arbitrary function of its arguments.
To fix the functions $F_1^{cov}, F_3^{cov}, F_4^{cov}$ from the condition $\delta_S S = 0$ one has, firstly, to integrate over θ in (4.15) and then to replace the arguments of these functions, $\nabla_\theta\eta$ and $\nabla_\theta\lambda$, by

$$\nabla_\theta\eta = \big(1+\eta\eta\big)D\eta \equiv \big(1+\eta\eta\big)x, \quad \nabla_\theta\lambda = D\lambda + \eta\lambda D\eta \equiv y + \eta\lambda x. \tag{4.16}$$

Hence, the explicit expression for each of $F^{cov}[\nabla_\theta\eta, \nabla_\theta\lambda]$ takes the form

$$F^{cov}[\nabla_\theta\eta, \nabla_\theta\lambda] = F[x,y] + \eta\eta x(F[x,y])_x + \eta\lambda x(F[x,y])_y, \tag{4.17}$$

where, by definition,

$$x \equiv D\eta, \qquad y \equiv D\lambda. \tag{4.18}$$

Now

- performing the integration in (4.15) over θ,
- expanding the functions $F_1^{cov}, F_3^{cov}, F_4^{cov}$ as in (4.17),
- collecting the terms which are linear and cubic in the fermions,
- and assuming that the right hand side of variation is a total derivative with respect to t of an arbitrary function depending on x, y,

we will get the following three equations

$$\begin{aligned} \Big[\big(1+x^2\big)F_1\Big]_x - yF_3 &= a, && (a)\\ y\Big[\big(1+x^2\big)F_1\Big]_{yy} + \big(y^2F_4\big)_y &= 0, && (b)\\ \big(F_3\big)_y + \big(F_4\big)_x &= 0, && (c) \end{aligned} \tag{4.19}$$

where a is an arbitrary constant. The equation (4.19b) may be integrated once, giving

$$y\big(\widetilde{F}_1\big)_y - \widetilde{F}_1 + y^2 F_4 = G[x], \tag{4.20}$$

where

$$\widetilde{F}_1 \equiv \big(1+x^2\big) F_1, \tag{4.21}$$

and $G[x]$ is an arbitrary function depending on x. Then the equation (4.19a) reads

$$\big(\widetilde{F}_1\big)_x - yF_3 = a. \tag{4.22}$$

Differentiating the equation (4.20) over x and the equation (4.22) over y, and using the equation (4.19c) one may get

$$\big(G[x]\big)_x = -a \quad \Rightarrow \quad G[x] = -ax - b, \tag{4.23}$$

where b is a new constant. Finally, representing the function $\widetilde{F}_1$ as

$$\widetilde{F}_1 = ax + b + y\widehat{F}_1 \tag{4.24}$$

we will finish with the equations

$$\big(\widehat{F}_1\big)_y + F_4 = 0, \qquad \big(\widehat{F}_1\big)_x - F_3 = 0, \tag{4.25}$$

which define the functions F_3, F_4 in terms of an arbitrary function $\widehat{F}_1[x,y]$.
Thus, the general action (4.14) is invariant under the broken S-supersymmetry, if the functions F_1, F_3 and F_4 are expressed through arbitrary function F_1 as

$$F_1 = \frac{ax+b}{1+x^2} + \frac{y}{1+x^2}\widehat{F}_1, \quad F_3 = \big(\widehat{F}_1\big)_x, \quad F_4 = -\big(\widehat{F}_1\big)_y. \tag{4.26}$$

Thus, the most general action for the present system is defined up to two arbitrary functions $\widehat{F}_1$ and F_2.
Let us note that the constant b can be chosen equal to zero, because the action corresponding to the Lagrangian

$$L = \frac{b}{1+x^2} \tag{4.27}$$

is trivial. Thus, the final result for the covariant functions F^{cov} which enter the action (4.14) reads

$$F_1^{cov} = a\frac{\nabla_\theta\eta}{1+\big(\nabla_\theta\eta\big)^2} + \frac{\nabla_\theta\lambda}{1+\big(\nabla_\theta\eta\big)^2}\widehat{F}_1^{cov}, \quad F_3^{cov} = \big(\widehat{F}_1\big)_{\nabla_\theta\eta}, \quad F_4^{cov} = -\big(\widehat{F}_1\big)_{\nabla_\theta\lambda}, \tag{4.28}$$

with $\widehat{F}_1^{cov}$ and F_2^{cov} being arbitrary functions on $\nabla_\theta\eta$ and $\nabla_\theta\lambda$

$$\widehat{F}_1 = \widehat{F}_1[\nabla_\theta\eta, \nabla_\theta\lambda], \qquad F_2 = F_2[\nabla_\theta\eta, \nabla_\theta\lambda]. \tag{4.29}$$

5. Interesting cases

Having at hands the most general expression for the action (4.14), (4.28) invariant with respect to both Q- and S-supersymmetries, it is interesting to visualize the functions F_1, F_2 which reproduce the *FPS* actions (3.5), (3.20). Comparing (3.14) and (3.32) one may conclude, that these two *FPS* cases are related through the substitutions: $\lambda \to -\mathrm{i}\lambda, y \to -\mathrm{i}y$ (where $y = D\lambda$). Therefore, it is enough to consider the first case, with $d_{111} = d_{222} = 1$, only.

In order to simplify everything, it is useful to represent the integrand in (4.14) as follows

$$L = \eta F_1 + \lambda F_2 + \lambda\eta\eta\big(xF_2 + \widehat{F}_1\big)_x - \eta\lambda\lambda\big(xF_2 + \widehat{F}_1\big)_y, \tag{5.1}$$

with

$$F_1 \equiv a\frac{x}{1+x^2} + \frac{y}{1+x^2}\widehat{F}_1, \tag{5.2}$$

where the functions $\widehat{F}_1$ and F_2 depend on x, y variables (4.18). Comparing the integrands in the actions (3.5) and (5.1), we will get for the two special choices of the constant parameters e_a and C_{ab} of the *FPS* action the following result

$$\begin{aligned}
(a) \quad & if \quad e_2 = C_{ab} = 0, \qquad then \quad L_{FPS} = \nu_1 \quad \Rightarrow \\
& F_1 = F_2 = \frac{x+y}{1+(x+y)^2}, \quad a = 1 && (5.3)\\
(b) \quad & if \quad e_1 = e_2 = C_{ab} = 1, \qquad then \quad L_{FPS} = \psi_1 D\psi_2 + \nu_1 D\nu_2 \quad \Rightarrow \\
& F_1 = F_2 = \frac{(x-y)(1+x^2-y^2)}{(1+(x-y)^2)(1+(x+y)^2)}, \quad a = 1. && (5.4)
\end{aligned}$$

Unfortunately, the explicit form of the functions F_1, F_2 which corresponds to *FPS* actions (5.3), (5.4) is not informative enough to understand why these actions were selected.

Let us remind that the general action

$$S = \int dt d\theta (1+\eta\eta)\Big[\eta\, F_1^{cov} + \lambda\, F_2^{cov} + \eta\eta\lambda\, F_3^{cov} + \eta\lambda\nabla_t\lambda\, F_4^{cov}\Big], \tag{5.5}$$

with the restrictions

$$F_1^{cov} = a\frac{\nabla_\theta\eta}{1+(\nabla_\theta\eta)^2} + \frac{\nabla_\theta\lambda}{1+(\nabla_\theta\eta)^2}\widehat{F}_1^{cov}, \quad F_3^{cov} = \big(\widehat{F}_1\big)_{\nabla_\theta\eta}, \quad F_4^{cov} = -\big(\widehat{F}_1\big)_{\nabla_\theta\lambda} \tag{5.6}$$

contains two arbitrary functions F_1^{cov} and F_2^{cov}. So, as we expected, the invariance with respect to additional, spontaneously broken $N = 1$ supersymmetry does not fix the action in the many-particles case, in contrast with the one-particle case.

6. Conclusion

In this work we analysed in details the application of the *FPS* approach in $d = 1$.

We demonstrated that the resulting system describes the $N = 2, d = 1$ supersymmetric action for two particles in which one of $N = 1$ supersymmetries is spontaneously broken. The final actions possess the same features as the *FPS* ones.

Using the nonlinear realization approach we reconsider the system in the basis where only one superfield has the Goldstone nature while the second superfield can be treated as the matter one. Having at hands the transformations relating the two selected *FPS* cases with our more generic one, we established the field redefinitions which relate these cases. Namely, in this basis, the two *FPS* supermultiplets are related by the redefinition of the matter superfield $\lambda \to -\mathrm{i}\lambda$. This analysis leads to the conclusion that the only difference between two *FPS* cases lies in a various choice of actions, while the supermultiplets specified by the *FPS* constraints are really the same.

Going further on with the nonlinear realization approach, we constructed the most general action for the system of two $N = 1$ superfields possessing one additional hidden spontaneously broken $N = 1$ supersymmetry. This action contains two arbitrary functions and reduces to the *FPS* actions upon specification of these functions.

Unfortunately, the exact form of these functions corresponding to *FPS* actions is not very informative and it gives no hint about the reason why the *FPS* cases were selected. Of course, our consideration was strictly one-dimensional and therefore we cannot argue that all features we discussed will appear in the generalized supersymmetric Born-Infeld theory constructed in [2, 3]:

Nevertheless, we believe that our results and the generality of the nonlinear realization approach are yield reasonable tools to reconsider the questions:

- whether the supermultiplets with different constants d_{abc} are really different?
- which additional properties select the *FPS* actions?

in four dimensions.

Acknowledgments
The work of S.K. was supported by RSCF grant 14-11-00598. The work of A.S was partially supported by RFBR grants 15-02-06670 and 15-52-05022 Arm-a.

Re erences

[1] Bagger J and Galperin A 1997 *Phys. Rev. D* **55** 1091, (*Preprint* hep-th 9608177)
[2] Ferrara S, Porrati M and Sagnotti A 2014 *JHEP* **1412** 065 (*Preprint* hep-th 1411.4954)
[3] Ferrara S, Porrati M, Sagnotti A, Stora R and Yeranyan A 2014 Generalized Born–Infeld Actions and Projective Cubic Curves *Preprint* hep-th 1412.3337
[4] Coleman S R, Wess J and Zumino B 1969 *Phys.Rev.* **177** 2239
Callan C, Coleman S R, Wess J and Zumino B 1969 *Phys.Rev.* **177** 2247
[5] Volkov D V 1973 *Sov.J.Part.Nucl.* **4** 3
Ogievetsky V I 1974 *Proceedings of the Xth Winter School of Theoretical Physics in Karpacz, Vol.1* p 117
[6] Ivanov E A and Kapustnikov A.A 1978 *J. Phys. A* **11** 2375; 1982 *J. Phys. G* **8** 167
[7] Bellucci S, Krivonos S and Sutulin A 2014 Comments on $N = 2$ Born–Infeld Attractors *Preprint* hep-th 1411.5592
[8] Bellucci S, Krivonos S and Sutulin A 2015 *JHEP* **1504** 177 (*Preprint* hep-th 1502.06389)
[9] Delduc F, Ivanov E and Krivonos S 2000 *Nucl. Phys. B* **576** 196 (*Preprint* hep-th 9912222)
[10] Ivanov E A and Ogievetsky V I 1975 *Teor. Mat. Fiz.* **25** 164

Journal of Physics: Conference Series **670** (2016) 012009 doi:10.1088/1742-6596/670/1/012009

Supersymmetric versions and integrability of conformally parametrized surfaces

Sébastien Bertrand

Department of Mathematics and Statistics, Université de Montréal, Montréal CP 6128 (QC) H3C 3J7, Canada

Abstract. The main aim of this paper is twofold. One is the construction and analysis of supersymmetric (bosonic and fermionic) versions of the structural equations for conformally parametrized surfaces. The other is the investigation of integrability in the sense of soliton theory via a comparison of the symmetries of the system of differential equations and those of the associated linear problem. This paper consists of an overview of the results obtained in two previous works [1, 2].

1. Introduction

Over the last decades, applications of supersymmetry have expanded from particle physic to a large number of domains (see e.g. [3] and references therein). A number of supersymmetric (SUSY) extensions of classical and quantum physical models have been investigated, such as the Chaplygin gas model [4] (and references therein), the Born-Infled model [5], the Korteweg-de Vries equation (e.g. for waves in shallow water in (1+1)-dimensions) [6, 7, 8], the Kadomtsev-Petviashvili equation (e.g. for waves in shallow water in (2+1)-dimensions) [9] and the sine/sinh-Gordon equation (e.g. for crystal dislocation) [10, 11, 12, 13, 14, 15].

In differential geometry, parametrized surfaces are described in terms of a moving frame satisfying the Gauss-Weingarten (GW) equations and their compatibility conditions are the Gauss-Codazzi (GC) equations. The construction and analysis of such surfaces associated with integrable systems in several areas of mathematical physics provided new tools for the investigation of nonlinear phenomena described by these systems. The analysis of such SUSY versions of the structural equations for conformally parametrized surfaces are one of the main goals of this paper and have been studied in [1].

Methods to solve SUSY differential equations are not as well established as in the classical case. In order to establish new methods and obtain solutions, one can investigate special types of differential equations which are easier to solve in general, like integrable systems (in the sense of the inverse scattering method). In order to solve such equations, we can use the group theory approach, e.g. given in [16], and adapt it for Grassmann-valued coordinates and functions. This lead to the second objective of this paper, which is to formulate a conjecture that states that by comparing the symmetries of the (SUSY) differential equations with those of its associated linear problem, it is possible to know if the system is a candidate for integrability. This conjecture, first proposed for classical differential equations [17, 18, 19], can be extended to SUSY differential equations as stated in [2].

Journal of Physics: Conference Series **670** (2016) 012009 doi:10.1088/1742-6596/670/1/012009

The paper is organized as follow. At the end of this section, an outline of the classical Gauss-Codazzi (CGC) and classical Gauss-Weingarten (CGW) equations for conformally parametrized surfaces is given. In section 2, a conjecture states that the integrability can be determinated by comparing the symmetries of a system of equations and those of the associated linear problem. This conjecture can be extented to SUSY equations and is applied to the CGC equations and to the SUSY sinh-Gordon equation in this section as examples. In section 3, the bosonic conformal parametrization of a surface is studied and its integrability is investigated via the conjecture of section 2. In section 4, it is the fermionic conformal parametrization of a surface which is studied and its integrability is also investigated. More details on this work can be found in the papers [1, 2].

1.1. Classical conformally parametrized surfaces

Let us consider a smooth orientable surface $\mathcal{S}$ immersed in the 3-dimensional Euclidean space $\mathbb{R}^3$ which is conformally parametrized by the function $F = (F_1, F_2, F_3) : \mathcal{R} \rightarrow \mathbb{R}^3$. The normalization is given by

$$\langle \partial F, \partial F \rangle = \langle \bar{\partial} F, \bar{\partial} F \rangle = 0, \quad \langle \partial F, \bar{\partial} F \rangle = \frac{1}{2} e^u, \quad \langle \partial F, N \rangle = \langle \bar{\partial} F, N \rangle = 0, \quad \langle N, N \rangle = 1, \qquad (1)$$

where N is the normal unit vector and $\partial, \bar{\partial}$ are the partial derivatives with respect to the complex variables $z, \bar{z}$ $(z = x + iy)$. Here and subsequently, the form $\langle \cdot | \cdot \rangle$ is the traditionnal inner product in $\mathbb{R}^3$. Introducing the notion of the moving frame $\Omega = (\partial F, \bar{\partial} F, N)^T$ and assuming that the derivatives of each component of the moving frame can be written as linear combinations of ∂F, $\bar{\partial} F$ and N, we get the CGW equations

$$\begin{aligned}
\partial \begin{pmatrix} \partial F \\ \bar{\partial} F \\ N \end{pmatrix} &= \begin{pmatrix} \partial u & 0 & Q \\ 0 & 0 & \frac{1}{2} H e^u \\ -H & -2e^{-u} Q & 0 \end{pmatrix} \begin{pmatrix} \partial F \\ \bar{\partial} F \\ N \end{pmatrix}, \quad \partial \Omega = V_1 \Omega, \\
\bar{\partial} \begin{pmatrix} \partial F \\ \bar{\partial} F \\ N \end{pmatrix} &= \begin{pmatrix} 0 & 0 & \frac{1}{2} H e^u \\ 0 & \bar{\partial} u & \bar{Q} \\ -2e^{-u} \bar{Q} & -H & 0 \end{pmatrix} \begin{pmatrix} \partial F \\ \bar{\partial} F \\ N \end{pmatrix}, \quad \bar{\partial} \Omega = V_2 \Omega,
\end{aligned} \qquad (2)$$

where Q (which is associated with the Hopf differential $Q dz^2$) and the mean curvature H are defined by

$$Q = \langle \partial^2 F, N \rangle, \qquad H = 2e^{-u} \langle \partial \bar{\partial} F, N \rangle. \qquad (3)$$

From the compatibility condition of the CGW equations, we obtain the CGC equations,

$$\begin{aligned}
&\partial \bar{\partial} u + \tfrac{1}{2} H^2 e^u - 2 Q \bar{Q} e^{-u} = 0 \quad \text{(, the Gauss equation)}, \\
&\partial \bar{Q} = \tfrac{1}{2} e^u \bar{\partial} H, \quad \bar{\partial} Q = \tfrac{1}{2} e^u \partial H \quad \text{(, the Codazzi equations)}.
\end{aligned} \qquad (4)$$

2. Investigating integability via a comparison of symmetries

The conjecture stated further in this section was first proposed by Levy et al. [19] and then by Cieslinski [17, 18] for classical differential equations. The formulation proposed in this paper uses a new projection operator and is extended to SUSY differential equations. In order to formulate this conjecture, let us define some notation.

Let L_1 be the set of all vector fields associated with the Lie point symmetries of the original PDEs $\Delta = 0$. Let L_2 be the set of all vector fields associated with the Lie point symmetries of the linear problem (LP) $\Lambda = 0$. Let π_ρ be a projection operator acting on a set L of vector fields ω such that

$$\pi_\rho(L) = \{\omega' | \omega' = \omega \rho\}, \qquad (5)$$

where ρ is in the form of a dilation generator (e.g. $\rho = x_1\partial_{x_1} + x_2\partial_{x_2} + y\partial_y$).

Here and in what follows, in order to compare the sets L_1 and L_2, we will consider a ρ involving all independent and dependent variables of the PDEs $\Delta = 0$. This choice of ρ implies that π_ρ acts as an identity on the set L_1, i.e. $\pi_\rho(L_1) = L_1$. The common symmetries of the PDEs $\Delta = 0$ and LP $\Lambda = 0$ are the vector fields which span the set

$$L_3 = L_1 \cap \pi_\rho(L_2) \neq \emptyset. \tag{6}$$

One should note that the set L_3 is not necessarily an algebra. Also, let L_4 be the set of all vector fields that generate a symmetry of the PDEs $\Delta = 0$ but do not generate a symmetry of the LP $\Lambda = 0$. The set L_4 is define as

$$L_4 = L_1 \backslash L_3. \tag{7}$$

This set does not form necessarily an algebra.

Conjecture 1

(i) If $L_1 = \pi(L_2)$ then the PDEs $\Delta = 0$ are not integrable.

(ii) If the following conditions are satisfied

(a) $\pi(L_2)$ is a proper subset of L_1, that is

$$L_1 \supset \pi(L_2). \tag{8}$$

A free parameter can be introduced into the linear system using a symmetry transformation generated by one of the vector fields appearing in L_4.

(b) The transformation given in (a) acts in a nontivial way (i.e. cannot be eliminated through an L_1-valued gauge matrix function).

Then the system of PDEs $\Delta = 0$ is a candidate to be an integrable one.

In order to apply this conjecture to the CGC equations, we must compare the symmetry Lie algebra of these PDEs,

$$\begin{aligned} X(\eta) &= \eta(z)\partial_z + \eta'(z)(-2Q\partial_Q - U\partial_U), \\ Y(\zeta) &= \zeta(\bar{z})\partial_{\bar{z}} + \zeta'(\bar{z})(-2\bar{Q}\partial_{\bar{Q}} - U\partial_U), \\ e_0 &= -H\partial_H + Q\partial_Q + \bar{Q}\partial_{\bar{Q}} + 2U\partial_U, \end{aligned} \tag{9}$$

with the symmetry Lie algebra of the CGW equations,

$$\begin{aligned} &X(\eta) = \eta(z)\partial_z - \eta'(z)(U\partial_U + 2Q\partial_Q), \\ &Y(\zeta) = \zeta(\bar{z})\partial_{\bar{z}} - \zeta'(\bar{z})(U\partial_U + 2\bar{Q}\partial_{\bar{Q}}), \\ &\hat{e}_0 = -H\partial_H + Q\partial_Q + \bar{Q}\partial_{\bar{Q}} + 2U\partial_U + F_i\partial_{F_i}, \\ &T_i = \partial_{F_i}, \qquad \mathcal{D}_i = F_i\partial_{F_{(i)}} + N_i\partial_{N_{(i)}}, \qquad i = 1,2,3 \\ &R_{ij} = (F_i\partial_{F_j} - F_j\partial_{F_i}) + (N_i\partial_{N_j} - N_j\partial_{N_i}), \qquad i < j = 2,3 \\ &S_{ij} = (F_i\partial_{F_j} + F_j\partial_{F_i}) + (N_i\partial_{N_j} + N_j\partial_{N_i}). \end{aligned} \tag{10}$$

By comparing these algebras with ρ of the form

$$\rho = z\partial_z + \bar{z}\bar{\partial} + Q\partial_Q + \bar{Q}\partial_{\bar{Q}} + H\partial_H + u\partial_u, \tag{11}$$

we get that

$$L_1 = \pi_\rho(L_2). \tag{12}$$

Hence the classical CGC equations are not integrable.

As a second example, we consider the SUSY sinh-Gordon,

$$D_+D_-\Phi = i\sin\Phi, \tag{13}$$

where the derivatives $D_\pm$ are given by

$$D_\pm = \partial_{\theta^\pm} - i\theta^\pm\partial_{x_\pm}, \tag{14}$$

$\theta^\pm$ being the fermionic independent variables and $x_\pm$ being the bosonic independent variables. The associate LP is given by

$$D_\pm\Psi = B_\pm\Psi, \qquad \text{where} \quad \Psi = \begin{pmatrix} \psi_{11} & \psi_{12} & f_{13} \\ \psi_{21} & \psi_{22} & f_{23} \\ f_{31} & f_{32} & \psi_{33} \end{pmatrix},$$
$$B_+ = \frac{1}{2}\begin{pmatrix} 0 & 0 & ie^{i\Phi} \\ 0 & 0 & -ie^{-i\Phi} \\ -e^{-i\Phi} & e^{i\Phi} & 0 \end{pmatrix}, \qquad B_- = \begin{pmatrix} iD_-\Phi & 0 & -i \\ 0 & -iD_-\Phi & i \\ -1 & 1 & 0 \end{pmatrix}. \tag{15}$$

One should note that Ψ is an even-supermatrix and $B_\pm$ are odd-supermatrices.

The symmetries of the SUSY sinh-Gordon equation and its associated LP are given respectively by

$$P_\pm = \partial_{x_\pm}, \qquad J_\pm = \partial_{\theta^\pm} + i\theta^\pm\partial_{x_\pm}, \qquad \mathcal{K} = 2x_+\partial_{x_+} - 2x_-\partial_{x_-} + \theta^+\partial_{\theta^+} - \theta^-\partial_{\theta^-}, \tag{16}$$

and by

$$\begin{array}{lll} P_\pm = \partial_{x_\pm}, & J_\pm = \partial_{\theta^\pm} + i\theta^\pm\partial_{x_\pm}, & G_1 = \psi_{11}\partial_{\psi_{11}} + \psi_{21}\partial_{\psi_{21}} + f_{31}\partial_{f_{31}}, \\ G_2 = \psi_{12}\partial_{\psi_{12}} + \psi_{22}\partial_{\psi_{22}} + f_{32}\partial_{f_{32}}, & & G_3 = f_{13}\partial_{f_{13}} + f_{23}\partial_{f_{23}} + \psi_{33}\partial_{\psi_{33}}. \end{array} \tag{17}$$

Using conjecture 1 with ρ of the form

$$\rho = x_+\partial_{x_+} + x_-\partial_{x_-} + \theta^+\partial_{\theta^+} + \theta^-\partial_{\theta^-} + \Phi\partial_\Phi, \tag{18}$$

we get that $L_1 \supset \pi_\rho(L_2)$. So using the symmetry generator $\mathcal{K} \in L_4$, we can introduce a free parameter λ in such a way :

$$\begin{array}{ll} D_+\Psi = B_+\Psi, & D_-\Psi = B_-\Psi, \\ B_+ = \frac{1}{2\sqrt{\lambda}}\begin{pmatrix} 0 & 0 & ie^{i\Phi} \\ 0 & 0 & -ie^{-i\Phi} \\ -e^{-i\Phi} & e^{i\Phi} & 0 \end{pmatrix}, & B_- = \begin{pmatrix} iD_-\Phi & 0 & -i\sqrt{\lambda} \\ 0 & -iD_-\Phi & i\sqrt{\lambda} \\ -\sqrt{\lambda} & \sqrt{\lambda} & 0 \end{pmatrix}, \end{array} \tag{19}$$

where λ does not act trivially on the new linear spectral problem. Hence, the SUSY sinh-Gordon equation is a candidate to be integrable.

3. Bosonic conformal parametrization of a surface

Let the surface $\mathcal{S}$ be smooth, orientable and immersed in the superspace $\mathbb{R}^{(2,1|2)}$. We assume that the surface $\mathcal{S}$ is conformally parametrized by a bosonic superfield $F(x_+, x_-, \theta^+, \theta^-)$ with the normalization conditions

$$\langle D_\pm F, D_\pm F\rangle = 0, \qquad \langle D_+F, D_-F\rangle = \frac{1}{2}e^\phi f, \qquad \langle D_-F, D_+F\rangle = -\frac{1}{2}e^\phi f \tag{20}$$

where $D_\pm$ are defined as in equation (14), ϕ is a bosonic-valued function of $x_+, x_-, \theta^+, \theta^-$, and f is a bodiless bosonic-valued function of x_+, x_-. We can introduce a bosonic superfield N which acts as the classical normal unit vector

$$\langle D_\pm F, N\rangle = 0, \qquad \langle N, N\rangle = 1. \tag{21}$$

The tangent vectors $D_\pm F$ together with the normal superfield N form the SUSY moving frame $\Omega = (D_+F, D_-F, N)^T$. Assuming that the first derivatives of Ω can be written as linear combinations of the elements of Ω and with some calculation we get the bosonic case of the supersymmetric Gauss-Weingarten (bSUSYGW) equations

$$D_+\Omega = A_+\Omega, \qquad\qquad D_-\Omega = A_-\Omega,$$

$$A_+ = \begin{pmatrix} {\Gamma_{11}}^1 & {\Gamma_{11}}^2 & Q^+ f \\ -{\Gamma_{12}}^1 & -{\Gamma_{12}}^2 & -\frac{1}{2}e^\phi H f \\ H & 2e^{-\phi}Q^+ & 0 \end{pmatrix}, \quad A_- = \begin{pmatrix} {\Gamma_{12}}^1 & {\Gamma_{12}}^2 & \frac{1}{2}e^\phi H f \\ {\Gamma_{22}}^1 & {\Gamma_{22}}^2 & Q^- f \\ -2e^{-\phi}Q^- & H & 0 \end{pmatrix}. \tag{22}$$

where the fermionic functions ${\Gamma_{ij}}^k$ are the Christoffel symbols of second kind and where the $Q^\pm$ and the mean curvature H are defined by

$$\langle D_+D_+F, N\rangle = Q^+ f, \quad \langle D_-D_+F, N\rangle = -\langle D_+D_-F, N\rangle = \frac{1}{2}e^\phi H f, \quad \langle D_-D_-F, N\rangle = Q^-. \tag{23}$$

The zero curvature condition of the bSUSYGW equations

$$D_+A_- + D_-A_+ - \{EA_+, EA_-\} = 0, \tag{24}$$

where

$$E = \pm \begin{pmatrix} 1 & 0 & 0 \\ 0 & 1 & 0 \\ 0 & 0 & -1 \end{pmatrix}, \tag{25}$$

constitutes the bosonic case of the supersymmetric Gauss-Codazzi (bSUSYGC) equations and corresponds to the following six linearly independent equations

$$\begin{array}{ll}
(i) & D_-({\Gamma_{11}}^1) + D_+({\Gamma_{22}}^2) + D_+({\Gamma_{12}}^1) - D_-({\Gamma_{12}}^2) = 0, \\
(ii) & D_-({\Gamma_{11}}^1) - {\Gamma_{11}}^2{\Gamma_{22}}^1 + D_+({\Gamma_{12}}^1) + {\Gamma_{12}}^2{\Gamma_{12}}^1 + \frac{1}{2}H^2 e^\phi f - 2Q^+Q^-e^{-\phi}f = 0, \\
(iii) & Q^+{\Gamma_{22}}^2 - {\Gamma_{11}}^2 Q^- + D_-Q^+ - Q^+D_-\phi + \frac{1}{2}e^\phi D_+H = 0, \\
(iv) & Q^-{\Gamma_{11}}^1 - {\Gamma_{22}}^1 Q^+ + D_+Q^- - Q^-D_+\phi - \frac{1}{2}e^\phi D_-H = 0, \\
(v) & D_-({\Gamma_{11}}^2) - {\Gamma_{12}}^1{\Gamma_{11}}^2 - {\Gamma_{11}}^2{\Gamma_{22}}^2 - {\Gamma_{11}}^1{\Gamma_{12}}^2 + D_+({\Gamma_{12}}^2) + 2Q^+Hf = 0, \\
(vi) & D_+({\Gamma_{22}}^1) + {\Gamma_{12}}^2{\Gamma_{22}}^1 - {\Gamma_{22}}^1{\Gamma_{11}}^1 + {\Gamma_{22}}^2{\Gamma_{12}}^1 - D_-({\Gamma_{12}}^1) + 2Q^-Hf = 0.
\end{array} \tag{26}$$

We will now determine whether or not the bSUSYGC equations form an integrable system. In order to lighthen the Christoffel symbols's notation, we write them as

$$R^+ = {\Gamma_{11}}^1, \quad R^- = {\Gamma_{11}}^2, \quad S^+ = {\Gamma_{12}}^1, \quad S^- = {\Gamma_{12}}^2, \quad T^+ = {\Gamma_{22}}^1, \quad T^- = {\Gamma_{22}}^2. \tag{27}$$

The vector fields of the Lie point symmetries of the bSUSYGC equations are given by

$$\begin{array}{ll}
C_0 = H\partial_H + Q^+\partial_{Q^+} + Q^-\partial_{Q^-} - 2f\partial_f, & \\
K_0 = -H\partial_H + Q^+\partial_{Q^+} + Q^-\partial_{Q^-} + 2\partial_\phi, & \\
K_1^b = -2x_+\partial_{x_+} - \theta^+\partial_{\theta^+} + R^+\partial_{R^+} + 2R^-\partial_{R^-} + S^-\partial_{S^-} - T^+\partial_{T^+} + 2Q^+\partial_{Q^+} + \partial_\phi, & \\
K_2^b = -2x_-\partial_{x_-} - \theta^-\partial_{\theta^-} - R^-\partial_{R^-} + S^+\partial_{S^+} + 2T^+\partial_{T^+} + T^-\partial_{T^-} + 2Q^-\partial_{Q^-} + \partial_\phi, & \\
P_+ = \partial_{x_+}, & P_- = \partial_{x_-}, \\
J_+ = \partial_{\theta^+} + i\theta^+\partial_{x_+}, & J_- = \partial_{\theta^-} + i\theta^-\partial_{x_-}.
\end{array} \tag{28}$$

and the symmetries of the bSUSYGW equations are spanned by

$$\begin{aligned}
&P_\pm = \partial_{x_\pm}, \qquad J_\pm = \partial_{\theta^\pm} + i\theta^\pm \partial_{x_\pm},\\
&\hat{C}_0 = H\partial_H + Q^+\partial_{Q^+} + Q^-\partial_{Q^-} - 2f\partial_f + N_i\partial_{N_i},\\
&\hat{K}_0 = -H\partial_H + Q^+\partial_{Q^+} + Q^-\partial_{Q^-} + 2\partial_\phi - N_i\partial_{N_i},\\
&K_1^b = -2x_+\partial_{x_+} - \theta^+\partial_{\theta^+} + R^+\partial_{R^+} + 2R^-\partial_{R^-} + S^-\partial_{S^-} - T^+\partial_{T^+} + 2Q^+\partial_{Q^+} + \partial_\phi,\\
&K_2^b = -2x_-\partial_{x_-} - \theta^-\partial_{\theta^-} - R^-\partial_{R^-} + S^+\partial_{S^+} + 2T^+\partial_{T^+} + T^-\partial_{T^-} + 2Q^-\partial_{Q^-} + \partial_\phi,\\
&G_i = F_i\partial_{F_i} + N_i\partial_{N_i}, \qquad B_i = \partial_{F_i}, \qquad \text{for } i = 1,2,3\\
&R_{ij} = F_i\partial_{F_j} - F_j\partial_{F_i} + N_i\partial_{N_j} - N_j\partial_{N_i}, \qquad i < j = 2,3.
\end{aligned} \tag{29}$$

Using conjecture 1 with ρ involving all independent and dependent variables of the bSUSYGC equations, we see that $L_1 = \pi_\rho(L_2)$. Hence the bSUSYGC equations are not integrable.

4. Fermionic conformal parametrization of a surface

Let the surface $\mathcal{S}$ be smooth, orientable and immersed in the superspace $\mathbb{R}^{(1,1|3)}$. We assume that the surface $\mathcal{S}$ is conformally parametrized by a fermionic superfield $F(x_+, x_-, \theta^+, \theta^-)$ with the normalization conditions

$$\langle D_\pm F, D_\pm F\rangle = 0, \qquad \langle D_\pm F, D_\mp F\rangle = \frac{1}{2}e^\phi f, \tag{30}$$

where ϕ is a bosonic-valued function of $x_+, x_-, \theta^+, \theta^-$ and the bosonic-valued function $f(x_+, x_-)$ may be bodiless depending on the structure of F. The normalization on the bosonic superfield N is given by

$$\langle D_\pm F, N\rangle = 0, \qquad \langle N, N\rangle = 1. \tag{31}$$

The tangent vectors $D_\pm F$ together with the normal superfield N form the SUSY moving frame $\Omega = (D_+F, D_-F, N)^T$. Assuming that the first derivatives of the moving frame Ω can be written as linear combinations of the elements of Ω and with some calculation we get the fermionic case of the supersymemtric Gauss-Weingarten (fSUSYGW) equations

$$\begin{aligned}
&D_+\begin{pmatrix} D_+F \\ D_-F \\ N\end{pmatrix} = \begin{pmatrix} {\Gamma_{11}}^1 & 0 & Q^+f \\ 0 & 0 & -\frac{1}{2}e^\phi Hf \\ H & -2e^{-\phi}Q^+ & 0\end{pmatrix}\begin{pmatrix} D_+F \\ D_-F \\ N\end{pmatrix}, \quad D_+\Omega = A_+\Omega,\\
&D_-\begin{pmatrix} D_+F \\ D_-F \\ N\end{pmatrix} = \begin{pmatrix} 0 & 0 & \frac{1}{2}e^\phi Hf \\ 0 & {\Gamma_{22}}^2 & Q^-f \\ -2e^{-\phi}Q^- & -H & 0\end{pmatrix}\begin{pmatrix} D_+F \\ D_-F \\ N\end{pmatrix}, \quad D_-\Omega = A_-\Omega.
\end{aligned} \tag{32}$$

The fermionic case of the supersymmetric Gauss-Codazzi (fSUSYGC) equations, which are equivalent to the ZCC

$$D_+A_- + D_-A_+ - \{A_+, A_-\} = 0, \tag{33}$$

reduce to the following four linearly independent equations

$$\begin{aligned}
&(i) \quad D_+({\Gamma_{22}}^2) + D_-({\Gamma_{11}}^1) = 0,\\
&(ii) \quad D_-({\Gamma_{11}}^1) + 2e^{-\phi}Q^+Q^-f = 0,\\
&(iii) \quad D_+Q^- - \tfrac{1}{2}e^\phi D_-H + Q^-(D_+\phi - {\Gamma_{11}}^1) = 0,\\
&(iv) \quad D_-Q^+ + \tfrac{1}{2}e^\phi D_+H + Q^+(D_-\phi - {\Gamma_{22}}^2) = 0.
\end{aligned} \tag{34}$$

Using the abbreviated notation (27) and conjecture 1, we seek to know if the fSUSYGC equations are integrable. The symmetries of the fSUSYGC equations are spanned by

$$\begin{aligned}
&J_+ = \partial_{\theta^+} + i\theta^+\partial_{x_+}, \qquad J_- = \partial_{\theta^-} + i\theta^-\partial_{x_-}, \qquad W = \partial_H \\
&P_+ = \partial_{x_+}, \qquad P_- = \partial_{x_-}, \\
&C_0 = H\partial_H + Q^+\partial_{Q^+} + Q^-\partial_{Q^-} - 2f\partial_f, \\
&K_0 = -H\partial_H + Q^+\partial_{Q^+} + Q^-\partial_{Q^-} + 2\partial_\phi, \\
&K_1^f = -2x_+\partial_{x_+} - \theta^+\partial_{\theta^+} + 2Q^+\partial_{Q^+} + R^+\partial_{R^+} + \partial_\phi, \\
&K_2^f = -2x_-\partial_{x_-} - \theta^-\partial_{\theta^-} + 2Q^-\partial_{Q^-} + T^-\partial_{T^-} + \partial_\phi,
\end{aligned} \tag{35}$$

and the set of vector fields which generates symmetries of the fSUSYGW equations is given by

$$\begin{aligned}
&P_\pm = \partial_{x_\pm}, \qquad J_\pm = \partial_{\theta^\pm} + i\theta^\pm\partial_{x_\pm}, \\
&\hat{C}_0 = H\partial_H + Q^+\partial_{Q^+} + Q^-\partial_{Q^-} - 2f\partial_f + N_i\partial_{N_i}, \\
&\hat{K}_0 = -H\partial_H + Q^+\partial_{Q^+} + Q^-\partial_{Q^-} + 2\partial_\phi - N_i\partial_{N_i}, \\
&K_1^f = -2x_+\partial_{x_+} - \theta^+\partial_{\theta^+} + 2Q^+\partial_{Q^+} + R^+\partial_{R^+} + \partial_\phi, \\
&K_2^f = -2x_-\partial_{x_-} - \theta^-\partial_{\theta^-} + 2Q^-\partial_{Q^-} + T^-\partial_{T^-} + \partial_\phi, \\
&G_i = F_i\partial_{F_i} + N_i\partial_{N_i}, \qquad B_i = \partial_{F_i}, \qquad \text{for } i,j = 1,2,3 \\
&R_{ij} = F_i\partial_{F_j} - F_j\partial_{F_i} + N_i\partial_{N_j} - N_j\partial_{N_i}, \qquad i < j = 2,3.
\end{aligned} \tag{36}$$

Comparing these sets of generators, we see that the symmetry spanned by the vector field W is not common to both the fSUSYGC equations and the fSUSYGW equations. We can use this symmetry to incorporate a free fermionic parameter λ which cannot be eliminated through a gauge transformation. The potential matrices $A_\pm$ take the form

$$\begin{aligned}
A_+ &= \begin{pmatrix} {\Gamma_{11}}^1 & 0 & Q^+ f \\ 0 & 0 & -\frac{1}{2}e^\phi(H+\lambda)f \\ H+\lambda & -2e^{-\phi}Q^+ & 0 \end{pmatrix}, \\
A_- &= \begin{pmatrix} 0 & 0 & \frac{1}{2}e^\phi(H+\lambda)f \\ 0 & {\Gamma_{22}}^2 & Q^- f \\ -2e^{-\phi}Q^- & -H-\lambda & 0 \end{pmatrix}.
\end{aligned} \tag{37}$$

Therefore, the system of the fSUSYGC equation is a candidate to be integrable.

5. Conclusion

In this paper we have presented supersymmetric conformal parametrizations of surfaces, one for the bosonic case and another for the ferminonic case. Both the Gauss-Codazzi and Gauss-Weingarten equations have been given for the bosonic and fermionic cases. Moreover, a conjecture on the integrability of systems of differential equations in the sense of soliton theory has been presented. This conjecture allows us to know if a system of differential equations can be integrable by comparing the symmetries of the system with the symmetries of its linear problem. In this paper, examples of this conjecture have been applied to the classical Gauss-Codazzi equations, the supersymmetric sinh-Gordon equation, the bosonic case of the supersymmetric Gauss-Codazzi equations and the fermionic case of the supersymmetric Gauss-Codazzi equations.

6. Acknowledgments

The author wishes to thank Prof. A. M. Grundland (Université du Québec à Trois-Rivières & Université de Montréal) for his support and discussions on the subjet. This paper was supported by a research grant of AMG from NSERC. The author acknowledges a doctoral fellowship provided by the FQRNT of the Gouvernement du Québec.

References

[1] Bertrand S, Grundland A M and Hariton A J 2015 *J. Phys. A* **48** 175208, 37 (arXiv : 1504.08260) URL `http://stacks.iop.org/1751-8121/48/i=17/a=175208`

[2] Bertrand S, Grundland A M and Hariton A J 2015 *SIGMA. Symmetry, Integrability and Geometry: Methods and Applications* **11** 046 (arXiv : 1502.02948) URL `http://dx.doi.org/10.3842/SIGMA.2015.046`

[3] Varadarajan V S 2011 *Reflections on quanta, symmetries, and supersymmetries* (Springer, New York) ISBN 978-1-4419-0666-3 URL `http://dx.doi.org/10.1007/978-1-4419-0667-0`

[4] Jackiw R 2002 *Lectures on fluid dynamics : A particle theorist's view of supersymmetric, non-abelian, noncommutative fluid mechanics and d-branes* CRM Series in Mathematical Physics (Springer-Verlag, New York) ISBN 0-387-95422-8 URL `http://dx.doi.org/10.1007/978-1-4757-3665-6`

[5] Hariton A J 2006 *J. Phys. A* **39** 7105–7114 ISSN 0305-4470 URL `http://dx.doi.org/10.1088/0305-4470/39/22/021`

[6] Mathieu P 1988 *J. Math. Phys.* **29** 2499–2506 ISSN 0022-2488 URL `http://dx.doi.org/10.1063/1.528090`

[7] Labelle P and Mathieu P 1991 *J. Math. Phys.* **32** 923–927 ISSN 0022-2488 URL `http://dx.doi.org/10.1063/1.529351`

[8] Liu Q P and Mañas M 1998 *Phys. Lett. B* **436** 306–310 ISSN 0370-2693 URL `http://dx.doi.org/10.1016/S0370-2693(98)00852-1`

[9] Manin Y I and Radul A O 1985 *Comm. Math. Phys.* **98** 65–77 ISSN 0010-3616 URL `http://projecteuclid.org/euclid.cmp/1103942285`

[10] Coleman S 1975 *Phys. Rev. D* **11**(8) 2088–2097 URL `http://link.aps.org/doi/10.1103/PhysRevD.11.2088`

[11] Grammaticos B, Ramani A and Carstea A S 2001 *J. Phys. A* **34** 4881–4886 ISSN 0305-4470 URL `http://dx.doi.org/10.1088/0305-4470/34/23/307`

[12] Siddiq M and Hassan M 2005 *Europhys. Lett.* **70** 149–154 ISSN 0295-5075 URL `http://dx.doi.org/10.1209/epl/i2004-10475-y`

[13] Siddiq M, Hassan M and Saleem U 2006 *J. Phys. A* **39** 7313–7318 ISSN 0305-4470 URL `http://dx.doi.org/10.1088/0305-4470/39/23/010`

[14] Witten E 1984 *Comm. Math. Phys.* **92** 455–472 ISSN 0010-3616 URL `http://projecteuclid.org/euclid.cmp/1103940923`

[15] Grundland A M, Hariton A J and Šnobl L 2009 *J. Phys. A* **42** 335203, 23 ISSN 1751-8113 URL `http://dx.doi.org/10.1088/1751-8113/42/33/335203`

[16] Olver P J 1986 *Applications of Lie groups to differential equations* (*Graduate Texts in Mathematics* vol 107) (Springer-Verlag, New York) ISBN 0-387-96250-6 URL `http://dx.doi.org/10.1007/978-1-4684-0274-2`

[17] Cieśliński J, Goldstein P and Sym A 1994 *J. Phys. A* **27** 1645–1664 ISSN 0305-4470 URL `http://stacks.iop.org/0305-4470/27/1645`

[18] Cieśliński J 1992 *Nonlinear evolution equations and dynamical systems (Baia Verde, 1991)* (World Sci. Publ., River Edge, NJ) pp 260–268

[19] Levi D, Sym A and Tu G 1990 A working algorithm to isolate integrable surfaces in e^3 preprint DF-INFN, N761

Journal of Physics: Conference Series **670** (2016) 012010 doi:10.1088/1742-6596/670/1/012010

Solvable structures of a simple model of earthquakes

Mariusz Białecki

Institute of Geophysics, Polish Academy of Sciences, Warszawa, Poland

E-mail: bialecki@igf.edu.pl

Abstract.

This paper propose an extension of Random Domino Automaton by introducing extra parameter related to a very short-distance interactions between clusters which allows to isolate influence two mechanisms of clusters grow, namely enlarging and coalescence. In this setting statistically stationary state under assumption of independence of clusters is investigated and respective set of equations is derived.

Moreover, a special solvable case is studied in detail and it is shown how to derive Motzkin numbers out of the automaton for a family of parameters without considering a limit.

1. Introduction

Random Domino Automaton (RDA) [6, 7, 8] is a toy model of earthquakes — an extension of well known Drossel-Schwabl model of forest fires [10]. RDA is a slowly driven system, and their rules for accumulation and abrupt release of energy result in various types of frequency-size distributions of produced avalanches.

Such simple models, as for example [13, 14] are useful for analysis of specific properties of earthquakes. The Random Domino Automaton may give a possible mechanistic explanation of an universal earthquake recurrence time distribution [9]. So far, a finite version of RDA [4] was studied, and preliminary results [5] are promising. To explain the recurrence time distribution, it is necessary to solve an inverse problem for the model, which is already done for finite version of RDA [3]. The solution for more realistic cases will be published soon.

Moreover, the RDA model possess a interesting mathematical structure which is fully solvable for a specific choice of parameters and this choice leads to relation to Motzkin numbers [2]. Here we discuss the structure of the automaton and propose an extension of RDA in order to obtain Motzkin numbers for wider family of parameters, without considering a limit, as has been done originally.

2. The Random Domino Automaton

Random Domino Automaton comes from very simplified view of earthquakes. The space - one dimensional lattice - corresponds to boundaries of two tectonic plates moving with relative constant velocity. Due to irregularities of materials, relative motion can be locked at some places producing stress accumulation. Beyond some threshold of the stress, a relaxation takes place. A size of the relaxation depends on the nearby accumulated stress.

Energy in the automaton is represented by balls added to the randomly chosen cell (each one is equally possible) with a constant rate - one ball in one time step. If the chosen cell is empty, it becomes occupied with probability ν or the ball is scattered with probability $(1-\nu)$.

If the chosen place is already occupied, there are also two possibilities: the ball is scattered with probability $(1-\mu)$ or with probability μ the incoming ball triggers a relaxation, i.e. balls from the chosen cell and all adjacent occupied cells are removed. An example of relaxation of size **four** is presented in the diagram below. The up most line contains also probability of occupation of respective empty cells.

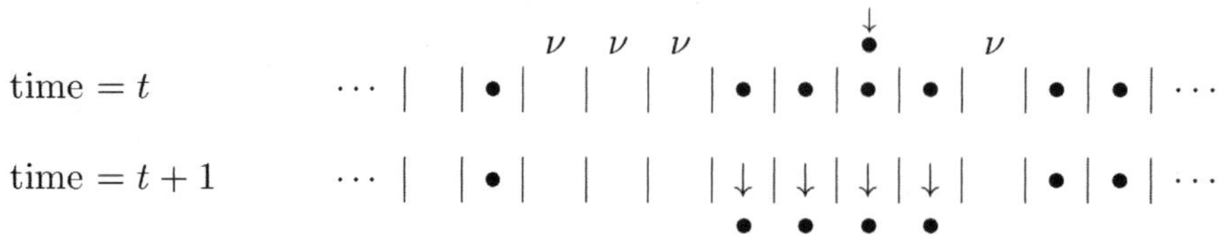

The stationary state of the system (it exists, see [4]) may be described by the distribution of clusters. The number of clusters of the length i, for $i = 1, 2, \ldots$, is denoted by n_i; the number of empty clusters of length 1 is denoted by n_1^0. Then the number of all clusters n and and the density ρ are

$$n = \sum_{i\geq 1} n_i, \qquad \rho = \frac{1}{N}\sum_{i\geq 1} i n_i. \tag{1}$$

The following set of equations for RDA is derived [7]

$$n_1 = \frac{1}{\frac{\mu_1}{\nu}+2}\left((1-\rho)N - 2n + n_1^0\right), \tag{2}$$

$$n_2 = \frac{2}{2\frac{\mu_2}{\nu}+2}\left(1-\frac{n_1^0}{n}\right) n_1, \tag{3}$$

$$n_i = \frac{1}{\frac{\mu_i}{\nu}i+2}\left(2n_{i-1}\left(1-\frac{n_1^0}{n}\right) + \frac{n_1^0}{n^2}\sum_{k=1}^{i-2} n_k n_{i-1-k}\right) \qquad for \quad i \geq 3, \tag{4}$$

and

$$n_1^0 = \frac{2n}{\left(3 + \frac{2}{\nu n}\sum_{i\geq 1}\mu_i i n_i\right)}. \tag{5}$$

The balance equation for the total number of clusters n and for the density ρ reads

$$(1-\rho)N - 2n = \sum_{i\geq 1}\frac{\mu_i}{\nu} n_i i, \tag{6}$$

$$(1-\rho)N = \sum_{i\geq 1}\frac{\mu_i}{\nu} n_i i^2. \tag{7}$$

The linear term on the righ hand side of equation (4), i.e. the term containing n_{i-1} stems from enlarging of a cluster of the size $i-1$ to the size i, like, for example, presented below.

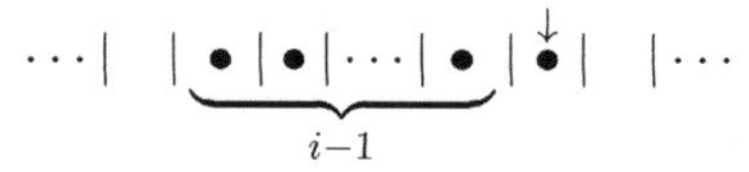

The nonlinear (quadratic in n_j) term of equation (4) results from merging of two clusters: one of size $k \in \{1, 2, \ldots, (i-2)\}$ and the second of the size $((i-1)-k)$, like presented below.

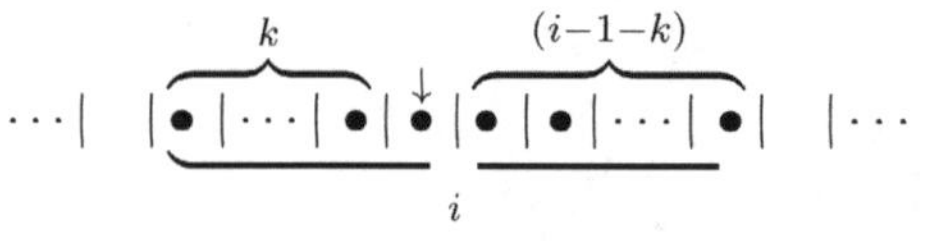

Journal of Physics: Conference Series **670** (2016) 012010 doi:10.1088/1742-6596/670/1/012010

The equations above are derived under the assumption of the independence for the order of clusters. In other words, we neglect correlations.

Now we consider a case $\mu = \mu_i = \delta/i$, which is related to equal probability of triggering relaxation for each cluster. For $\mu_i = \frac{\delta}{i}$, where $\delta = const$, the balance for ρ takes the following form

$$\rho = \frac{1}{(\theta + 1)}, \tag{8}$$

where $\theta = \frac{\delta}{\nu}$, which relates the density with the ratio of coefficients δ and ν only. Since $\theta \in (0, \infty)$, any density $\rho \in (0, 1)$ may be realized. The balance for n is reduced to $(1 - \rho)N = (2 + \theta)n$. Together with the balance equation for ρ it gives

$$n = N\frac{\theta}{(1 + \theta)(2 + \theta)}. \tag{9}$$

We define new variables c_i for $i = 0, 1, \ldots,$ by

$$c_i = \frac{\beta}{\alpha^{i+1}} n_{i+1}, \tag{10}$$

where

$$\alpha = \frac{2}{2 + \theta}\left(\frac{2\theta + 1}{2\theta + 3}\right), \qquad \beta = \frac{(\theta + 1)(\theta + 2)}{N\theta(2\theta + 1)}. \tag{11}$$

Then, the set of equations for n_i can be rewritten in the form

$$c_{m+2} = c_{m+1} + \sum_{k=0}^{m} c_k c_{m-k}, \tag{12}$$

which is valid for $m \geq 0$ $(m = i - 2)$ and initial data c_0 and c_1 are

$$c_0 = c_1 = \frac{1 + \frac{3}{2}\theta + \theta^2}{(1 + 2\theta)^2}. \tag{13}$$

The above recurrence generate Motzkin numbers [1], if it starts with $c_0 = c_1 = 1$. Thus, to obtain Motzkin numbers it is necessary to consider a limit $\theta \longrightarrow 0$, while θN is constant [2]. The generating function formalism gives explicit solution for any value of θ, namely for $m \geq 0$ given by

$$c_m = \frac{1}{2} \sum_{j=0}^{[\frac{m+2}{2}]} \frac{(2c_0 - \frac{1}{2})^j}{(m - j + 2)2^{m-j}} \binom{2(m-j)+1}{m-j+1} \binom{m-j+2}{j} \tag{14}$$

3. An extension of Random Domino Automaton

Now we propose generalisation of RDA to interacting RDA (iRDA), namely we introduce a parameter "c" controlling efficiency of some sort of short range interaction. In the evolution rule we change occupation rule for empty cell as follows.

- If the chosen cell is empty and two adjacent cells are occupied, it becomes occupied with probability c or the ball is scattered with probability $(1 - c)$ leaving the state of the automaton unchanged.
- If the chosen cell is empty and at least one adjacent cell is also empty, it becomes occupied with probability ν or the ball is scattered with probability $(1 - \nu)$.

All the other rules remain valid.

In the new setting an example of relaxation of size **four** looks as in the diagram below. Like previously, the upmost line present probability of occupation of respective empty cells.

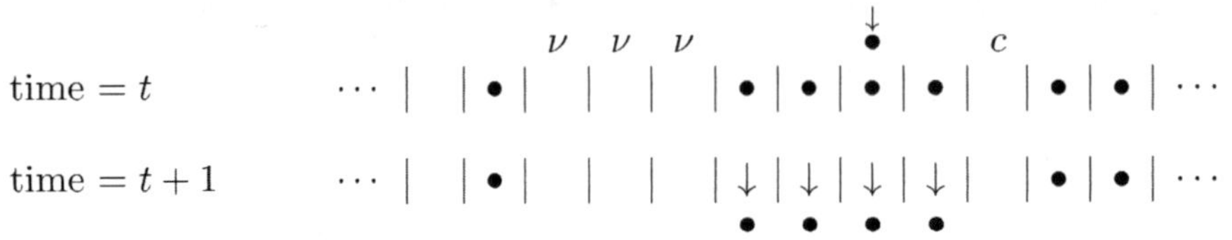

The set of equations for iRDA is in the form remarkably similar to the original set for RDA. It is of the form

$$n_1 = \frac{1}{\frac{\mu_1}{\nu} + 2 + 2(\xi - 1)\frac{n_1^0}{n}} \left((1-\rho)N - 2n + n_1^0\right), \tag{15}$$

$$n_2 = \frac{2}{2\frac{\mu_2}{\nu} + 2 + 2(\xi - 1)\frac{n_1^0}{n}} \left(1 - \frac{n_1^0}{n}\right) n_1, \tag{16}$$

$$n_i = \frac{1}{\frac{\mu_i}{\nu} i + 2 + 2(\xi - 1)\frac{n_1^0}{n}} \left(2n_{i-1}\left(1 - \frac{n_1^0}{n}\right) + \xi\frac{n_1^0}{n^2}\sum_{k=1}^{i-2} n_k n_{i-1-k}\right) \tag{17}$$

for $i \geq 3$, where $\xi = \frac{c}{\nu} \in [0, \infty)$ and

$$n_1^0 = \frac{2n}{\left(2 + \xi + \frac{2}{\nu n}\sum_{i\geq 1}\mu_i i n_i\right)}. \tag{18}$$

The balance equation for the total number of clusters n and for the density ρ reads

$$(1-\rho)N - 2n - (\xi - 1)n_1^0 = \sum_{i\geq 1}\frac{\mu_i}{\nu} n_i i, \tag{19}$$

$$(1-\rho)N + (\xi - 1)n_1^0 = \sum_{i\geq 1}\frac{\mu_i}{\nu} n_i i^2. \tag{20}$$

Naturally, the above equations for $\xi = 1$ reproduce former RDA equations.

The balance equations (20) and (19) (as well as equations for higher weighted moments of n_i defined below) can be also obtained from the above set (15)-(17). For n_i, $i = 1, 2, \ldots$, we define a moment of order γ and a weighted moment $\widehat{m}_\gamma$ of order γ by

$$m_\gamma = \frac{1}{N}\sum_{i\geq 1} n_i i^\gamma, \qquad \widehat{m}_\gamma = \frac{1}{N}\sum_{i\geq 1}\frac{\mu_i}{\nu} n_i i^\gamma. \tag{21}$$

Then, one obtains

$$\begin{aligned}\widehat{m}_{z+1} &= 1 - m_1 - 2(\xi - 1)\frac{n_1^0}{N m_0} m_z - 2m_0 - 2m_z + \\ &+ \frac{n_1^0}{N} + 2\left(1 - \frac{n_1^0}{N m_0}\right)\sum_{k=0}^{z}\binom{z}{k} m_k + \xi\frac{n_1^0}{N m_0}\sum_{l,p=1}^{l+p\leq z}\binom{z}{l+p}\binom{l+p}{l} m_l m_p.\end{aligned} \tag{22}$$

4. Special case: $\mu = \delta/i$, where $\delta =$ constant

In the case which refers to equal probability of triggering avalanche for each cluster, the parameters are fixed as follows: $\nu =$ const and $\mu = \frac{\delta}{i}$, where $\delta =$ const and $\theta = \frac{\delta}{\nu}$. Equations (20) and (19) give

$$\rho = \frac{1}{(\theta+1)}\left(1+\frac{n}{N}\frac{2(\xi-1)}{2(\theta+1)+\xi}\right), \tag{23}$$

$$n = N\frac{(1-\rho)}{2+\theta+\frac{2(\xi-1)}{2(\theta+1)+\xi}}, \tag{24}$$

where $\theta = \frac{\delta}{\nu} \in (0,\infty)$. Hence

$$\rho = 1-\frac{\theta}{\theta+1+K}, \tag{25}$$

$$n = \frac{N\theta}{(\theta+1+K)\left(\theta+2+\frac{2(\xi-1)}{2(\theta+1)+\xi}\right)}, \tag{26}$$

where

$$K = \frac{2(\xi-1)}{(2+\theta)\,(2(\theta+1)+\xi)+2(\xi-1)}. \tag{27}$$

Together with simple form of

$$n_1^0 = \frac{2n}{2(\theta+1)+\xi} \tag{28}$$

it allows to reduce the set of equations (15)-(17) to the following recurrence:

$$n_1 = \frac{1}{2+\theta+\frac{4(\xi-1)}{2(\theta+1)+\xi}}\left(\theta+\frac{2\xi}{2(\theta+1)+\xi}\right)n, \tag{29}$$

$$n_2 = \frac{2}{2+\theta+\frac{4(\xi-1)}{2(\theta+1)+\xi}}\left(\frac{2\theta+\xi}{2(\theta+1)+\xi}\right)n_1, \tag{30}$$

$$n_{i+1} = \frac{2}{2+\theta+\frac{4(\xi-1)}{2(\theta+1)+\xi}}\times \tag{31}$$
$$\times\left(\left(\frac{2\theta+\xi}{2(\theta+1)+\xi}\right)n_i+\xi\frac{\sum_{k=1}^{i-1}n_k n_{i-k}}{n(2(\theta+1)+\xi)}\right)$$

for $i \geq 2$ and where n is an explicit function of θ and ξ given by formula (26).

We define new variables c_i for $i = 0, 1, \ldots,$ by

$$c_i = \frac{\beta}{\alpha^{i+1}}n_{i+1}, \tag{32}$$

where

$$\alpha = \frac{2}{2+\theta+\frac{4(\xi-1)}{2(\theta+1)+\xi}}\left(\frac{2\theta+\xi}{2(\theta+1)+\xi}\right), \tag{33}$$

$$\beta = \xi\frac{1}{(2\theta+\xi)n}. \tag{34}$$

Then, the equation (31) can be rewritten in the form of Motzkin numbers recurrence (12). Initial data c_0 and c_1 are easily obtained from equations (29)-(30), when it is transformed according the rule of equation (32), namely

$$c_0 = c_1 = \xi \frac{\xi + (1 + \frac{\xi}{2})\theta + \theta^2}{(\xi + 2\theta)^2}. \tag{35}$$

Contrary to the previous case, equations (35) allow to obtain value 1 without considering a limit. Condition $c_0 = c_1 = 1$ is fulfiled for a family of parameters satisfying

$$\theta = \frac{\xi}{\xi - 4}(3 - \frac{\xi}{2}) \quad \text{or} \quad \theta = 0. \tag{36}$$

Thus for $\xi \in (4, 6)$ and respective value of θ iRDA corresponds exactly to the Motzkin numbers. For iRDA, a solution of the recurrence (12) is given by formula (14) but with $c_0 = c_1$ given by equations (35). Together with the formula (32), it gives explicit solution of equations (15)-(17) for the distribution n_is for any value of θ.

5. Conclusions

We have proposed an extension of Random Domino Automaton by introducing extra parameter related to very short distance interactions between clusters. Investigating statistically stationary state under assumption of independence of clusters, we have derived respective set of equations for clusters size distribution as well as balance equations for density and the total number of clusters.

We have also shown how to derive Motzkin Numbers out of the Automaton for a family of parameters without considering a limit.

By adjusting a value of the parameter related to very short distance interactions between clusters it is possible to separate of influence of two mechanisms of clusters grow. Thus it is straightforward to deduce other results presented in [7] accordingly and all the relations to other disciplines mentioned there are still applicable.

We point out that considering of hierarchical clustering in the spirit of [11] based on Random Domino Automaton, and in particular relation to branching numbers would be of interest and we intend to investigate this topic in future.

Acknowledgments

This work was partially supported by National Science Centre, project 2012/05/B/ST10/00598 and partially supported within statutory activities No. 3841/E-41/S/2015 of the Ministry of Science and Higher Education of Poland.

References

[1] Aigner M 2007 A Course in Enumeration (Berlin: Springer)
[2] Białecki M 2012 Motzkin numbers out of random domino automaton *Phys. Lett A* **376** 3098
[3] Białecki M 2013 From statistics of avalanches to microscopic dynamics parameters in a toy model of earthquakes *Acta Geophys.* **61** 1677
[4] Białecki M 2015 Properties of a finite stochastic cellular automaton toy model of earthquakes *Acta Geophys.* **63** 923
[5] Białecki M 2015 On mechanistic explanation of the shape of the universal curve of earthquake recurrence time distributions *Acta Geophys.* **63** in press DOI: 10.1515/acgeo-2015-0044.
[6] Białecki M and Czechowski Z 2010 Synchronization and triggering: from fracture to earthquake processes (GeoPlanet: Earth and Planetary Sciences) ed V De Rubeis et al (Berlin: Springer) chapter 5 pp 63–75 DOI: 10.1007/978-3-642-12300-9_5

[7] Białecki M and Czechowski Z 2013 On one-to-one dependence of rebound parameters on statistics of clusters: exponential and inverse-power distributions out of Random Domino Automaton *J. Phys. Soc. Jpn.* **82** 014003

[8] Białecki M and Czechowski Z 2014 Achievements, History and Challenges in Geophysics (GeoPlanet: Earth and Planetary Sciences) ed R Bialik et al (Berlin: Springer) chapter 13 pp 223–241 DOI: 10.1007/978-3-319-07599-0_13

[9] Corral A 2004 Unified Scaling Law for Earthquakes *Phys. Rev. Lett.* **92** 108501

[10] Drossel B and Schwabl F 1992 Self-Organized Critical Forest-Fire Model *Phys. Rev. Lett.* **69** 1629

[11] Gabrielov A, Newman W I and Turcotte D L 1999 Exactly soluble hierarchical clustering model: Inverse cascades. self-similarity, and scaling *Phys. Rev. E* **60** 5293

[12] Saichev A and D Sornette D 2006 Universal distribution of inter-earthquake times explained *Phys. Rev. Lett.* **97** 078501

[13] Tejedor A, Gomez J B and Pacheco A F 2009 Earthquake size-frequency statistics in a forest-fire model of individual faults *Phys. Rev. E* **79** 046102

[14] Tejedor A, Gomez J B and Pacheco A F 2010 Hierarchical model for disturbed seismicity *Phys. Rev. E* **82** 016118

Journal of Physics: Conference Series **670** (2016) 012011 doi:10.1088/1742-6596/670/1/012011

On the uniqueness of the octonionic instanton solution on conformally flat 8-manifolds

A H Bilge[1],

[1] Kadir Has University

E-mail: ayse.bilge@khas.edu.tr

Abstract. Let M be an 8-manifold and E be an $SO(8)$ bundle on M. In a previous paper [F. Ozdemir and A.H. Bilge, "Self-duality in dimensions $2n > 4$: equivalence of various definitions and the derivation of the octonionic instanton solution", ARI (1999) 51:247-253], we have shown that if the second Pontrjagin number p_2 of the bundle E is minimal, then the components of the curvature 2-form matrix F with respect to a local orthonormal frame are $F_{ij} = c_{ij}\omega_{ij}$, where c_{ij}'s are certain functions and the ω_{ij}'s are strong self-dual 2-forms such that for all distinct i, j, k, l, the products $\omega_{ij}\omega_{jk}$ are self dual and $\omega_{ij}\omega_{kl}$ are anti self-dual. We prove that if the c_{ij}'s are equal to each other and the manifold M is conformally flat, then the octonionic instanton solution given in [B.Grossman, T.W.Kephart, J.D.Stasheff, *Commun. Math. Phys.*, **96**, 431-437, (1984)] is unique in this class

1. Introduction

The set-up for gauge theory is based on vector bundles over differentiable manifolds [1]. Let M be a differentiable manifold with a Riemannian connection and E be a vector bundle on M with a structure group G. Let g be the Lie algebra of G. The connection on the vector bundle is defined locally by a g valued connection 1-form A. If F is the curvature of this connection, then the invariant polynomials of F are local representatives of the characteristic classes of the bundle E. Action integrals are given in terms of the inner products of the components of the curvature of the bundle; these integrals are bounded below by the integrals of the characteristic classes of vector bundles. Solutions for which the action integrals reach these topological lower bounds are minimizers of the action integrals. In our approach, we start with a topological lower bound and we use algebraic methods to characterize those solutions that saturate various inequalities among exterior products and inner products of forms. This procedure results in an action with the topological lower bound we started with.

We use the notion of Strong self-duality of a 2-form ω in even dimensions. These are building blocks for the solutions that saturate various inequalities. In this approach, we identify 2-forms and skew-symmetric matrices. A form is called strong self-dual , if the minimal polynomial of the corresponding skew-symmetric matrix A is $A^2 + \lambda^2 I = 0$ [2]. Equivalently, in $4n$ dimensions, ω^n is Hodge self-dual and in $2n$ dimensions, $\omega^n = k * \omega$ [3].

The octonionic instanton solution [4] is the minimizer of the the action $\int_M |F^2, F^2| dvol$ on S^8. In [3], we have derived this solution by the maximality of (minus) the second Pontrjagin class by the procedure described above. In the present work, we show that the octonionic instanton solution is unique in the class on solutions on conformally flat 8-manifolds that admit an $SO(8)$ vector bundle with an action that maximize $-p_2$.

Journal of Physics: Conference Series **670** (2016) 012011 doi:10.1088/1742-6596/670/1/012011

2. Preliminaries

The base manifold M is a Riemannian manifold, equipped with a torsion free metric connection B. Local sections of the cotangent bundle are denoted as $\{e_i\}$. The connection is given by

$$de_i = \sum B_{ij} \wedge e_j$$

where $B_{ij} = -B_{ji}$ are 1-forms. The curvature of the manifold is given by $R = dB - B \wedge B$.

The vector bundle E is an n-plane bundle on M. Local sections of E are denoted as s_i. The connection on E is given by

$$ds_i = \sum A_{ij} s_j$$

where A_{ij}'s are 1-forms; A takes vales in the Lie algebra of the structure group of E. The curvature of E is given by

$$F = dA - A \wedge A$$

and F satisfies the Bianchi identities

$$dF + F \wedge A - A \wedge F = 0.$$

Local representatives of the characteristic classes are computed as follows. If s_i, $i = 1, \ldots, N$ is a basis for local sections of E, then with respect to this basis, F is represented by an $N \times N$ matrix of 2-forms. As 2-forms belong to a commutative ring, we can compute the determinant

$$det(F + \lambda I) = \lambda^N + \sigma_1 \lambda^{N-1} + \sigma_2 \lambda^{N-2} + \cdots + \sigma_{N-1} \lambda + \sigma_N$$

The σ_i's are $2i$ forms that are proportional to the representatives of the Chern or Pontrjagin classes of E. If E is real, the odd ones, c_{2k+1} are trivial, only the even ones, $c_{2k} \sim p_k$ are nontrivial. The Euler Class is the square root of the determinant of F. Actions that involve the curvature F can be related to Pontrjagin numbers by relating inner products and exterior products.

In our approach, we start with a characteristic class say p_1 on a 4-dimensional manifold or a combination of p_2 and p_1^2 on an 8-dimensional manifold. We use conditions for the saturation of various inequalities to obtain an upper bound for the integrals of the characteristic classes as algebraic equations for the curvature. Then we try to solve for a connection that gives the curvature 2-form we determined by algebraic requirements. The solvability of this connection usually impose conditions on the base manifold hence it may determine the background metric. We will construct the action by maximizing $-p_2$ on conformally flat 8-manifolds by relating $\langle F^2, F^2 \rangle$ and traceF^4. Actually, it will turn out that trace$F^2 = 0$, hence traceF^4 is proportional to the second Pontrjagin class.

3. Strong Sel -Dual 2- orms

In 4-dimensions, 2-forms have a number of nice properties; they live in the middle dimension hence their Hodge duality is defined; self-dual 2-forms belong to a linear space; when $F = *F$, Yang-Mills equations are satisfied and finally the Yang-Mills equations form an elliptic system.

We noticed that the matrix of a self-dual 2-form has minimal polynomial $A^2 + \lambda^2 I = 0$, i.e, its eigenvalues are equal in absolute value. Strong self-duality of 2-forms in $2n$ dimensions is defined by the equality of the absolute values of the eigenvalues of the matrix of ω with respect to an orthonormal basis [2] We have also shown that it is equivalent to the self-duality in the Hodge sense of $\omega^{n/2}$ (used in [4] and to the equality $*\omega = k\omega^{n-1}$ (used by Trautman, in[5].

In 4-dimensions, the Yang-Mills action $\int \langle F, F \rangle = \int F \wedge *F$, where F is the curvature 2-form of an $SO(N)$ bundle, reaches the topological lower bound $\int F \wedge F$, provided that F is self-dual.

Journal of Physics: Conference Series **670** (2016) 012011 doi:10.1088/1742-6596/670/1/012011

In 4-dimensions self-dual and antiself-dual 2-forms are eigenspaces of the Hodge map and they form linear subspaces. In higher dimensions we look for linear subspaces of the set of strong self-dual 2 forms. In [6], we have shown that the dimension of maximal linear subspaces of strongly self-dual forms on a $2n$ manifold is equal to the number of linearly independent vector fields on S^{2n-1}. In eight dimensions, there are 7 dimensional maximal linear subspaces of strong self-dual 2-forms. We will use these these subspaces to construct the octonionic instanton solution.

4. Strong sel -duality and e uivalence o various ro erties.

Let ω_{ij} be the components of a 2-form in $2n$ dimensions with respect to some local orthonormal basis. We denote the 2-form ω and the skew-symmetric matrix consisting of its components with respect to some orthonormal basis by the same symbol. We recall the standard inequalities:

$$(\omega,\eta)^2 \leq (\omega,\omega)(\eta,\eta), \quad 2(\omega,\eta) \leq (\omega,\omega)+(\eta,\eta).$$

The invariant polynomials s_{2i} of ω can be expressed in terms of the elementary symmetric functions of the eigenvalues $\pm\lambda_k^2$'s. The inner products (ω^i,ω^i) and the s_{2i}'s are related as follows.

$$\begin{aligned}
s_2 &= (\omega,\omega) = \lambda_1^2+\lambda_2^2+\cdots+\lambda_n^2,\\
s_4 &= \tfrac{1}{(2!)^2}(\omega^2,\omega^2) = \lambda_1^2\lambda_2^2+\lambda_1^2\lambda_3^2+\cdots+\lambda_{n-1}^2\lambda_n^2,\\
s_6 &= \tfrac{1}{(3!)^2}(\omega^3,\omega^3) = \lambda_1^2\lambda_2^2\lambda_3^2+\lambda_1^2\lambda_2^2\lambda_4^2+\cdots+\lambda_{n-2}^2\lambda_{n-1}^2\lambda_n^2,\\
\cdots & \quad \cdots\\
s_{2n} &= \tfrac{1}{(n!)^2}(\omega^n,\omega^n) = \tfrac{1}{(n!)^2}\mid *\omega^n\mid^2 = \lambda_1^2\lambda_2^2\ldots\lambda_n^2
\end{aligned}$$

Defining the weighted elementary symmetric polynomials by $\binom{n}{i}q_i = s_{2i}$, one has the inequalities

$$q_1 \geq q_2^{1/2} \geq q_3^{1/3} \geq \cdots \geq q_n^{1/n}, \qquad q_{r-1}q_{r+1} \leq q_r^2, \quad 1 \leq r < n,$$

and the equalities hold iff all the λ_k's are equal [7]. This is a key result and our definition of strong self-duality,

Definition. Let ω be a 2-form in $2n$ dimensions, $\pm i\lambda_k$, $k=1,\ldots,n$ be its eigenvalues and η be the the square root of the determinant of ω, with a fixed choice of sign. Then ω is called *r n l elf-dual* (*r n l an i elf-dual*) if $\mid\lambda_1\mid=\mid\lambda_2\mid=\cdots=\mid\lambda_n\mid$, and $\eta>0$ ($\eta<0$).

The strong self-duality condition is equivalent to the matrix equation $\omega^2+\lambda^2 I=0$, where I is the identity matrix, and $\lambda^2=\frac{1}{2n}\mathrm{tr}\omega^2$. This definition gives quadratic equations for the ω_{ij}'s, hence the strong self-duality condition determines a nonlinear set.

In four dimensions, the matrices satisfying $\omega^2+\lambda I=0$ consist of the union of the usual self-dual and anti self-dual forms. In higher dimensions the set $\mathcal{S}_{2n}$ is an n^2-n+1 dimensional submanifold and the dimension of maximal linear submanifolds of $\mathcal{S}_{2n}$ is equal to the number of linearly independent vector fields on S^{2n-1}.

The equivalence of various definitions is given by the following Lemma [3].

Lemma. Let ω be a 2-form in $2n$ dimensions. Then

$$(n-1)(\omega,\omega)^2 - \frac{n}{2}(\omega^2,\omega^2) \geq 0, \quad (\omega^{n/2},\omega^{n/2}) \geq *\omega^n,$$

and equality holds if and only if all eigenvalues of ω are equal.

From this Lemma , we immediately have

Corollary. The 2-form ω is strongly self-dual iff $\omega^{n/2}$ is self-dual in the Hodge sense.

The strong self duality condition is also equivalent to the self-duality definition used by Trautman.

Pro osition. Let ω be a 2-form in $2n$ dimensions. Then $\omega^{n-1} = k * \omega$ where k is a constant, if and only if ω is strongly self-dual and $k = \frac{n!}{n^{n/2}}(\omega,\omega)^{\frac{n}{2}-1}$.

5. Linear subs aces o strongly sel -dual orms in eight dimensions.

In eight dimensions,

$$(\omega,\omega)^2 \geq \tfrac{2}{3}(\omega^2,\omega^2) \geq \tfrac{2}{3} * \omega^4$$

For strongly self-dual 2-forms, these inequalities are saturated and we also have

$$\omega^3 = \tfrac{3}{2}(\omega,\omega) * \omega.$$

By applying the equalities above to $\omega \pm \eta$, we obtain

$$\begin{aligned}&(\omega,\omega)^2 + (\eta,\eta)^2 + 2(\omega,\omega)(\eta,\eta)\\ &\geq \tfrac{3}{2}\left[(\omega^2,\omega^2) + (\eta^2,\eta^2)^2 + 2(\omega^2,\eta^2) \pm 4(\omega^2+\eta^2,\omega\eta) + 4(\omega\eta,\omega\eta)\right]\\ &\geq \tfrac{3}{2}\left[\omega^4 + \eta^4 + 6\omega^2\eta^2 \pm 4\omega^3\omega\eta \pm 4\omega\eta^3\right])\end{aligned}$$

Using these inequalities we obtain a series of results concerning the products involving strongly self-dual forms.

If ω is strongly self-dual and $(\omega,\eta) = 0$. Then $\omega^3\eta = 0$. When ω and η are both strongly self-dual $\omega^2 = *\omega^2$ and $\eta^2 = *\eta^2$, and

$$\begin{aligned}2(\omega,\omega)(\eta,\eta) &\geq \tfrac{3}{2}\left[2\omega^2\eta^2 4(\omega\eta,\omega\eta)\right]\\ &\geq \tfrac{3}{2}\left[+6\omega^2\eta^2\right]\end{aligned}$$

If ω, η, and $\omega \pm \eta$ are strongly self-dual and $(\omega,\eta) = 0$. Then

$$(\omega,\omega)(\eta,\eta) = 2(\omega^2,\eta^2) = 2\omega^2\eta^2.$$

Let ω and η be strongly self-dual and $(\omega,\eta) = 0$. Then

$$\omega\eta = *(\omega\eta)$$

if and only if $\omega \pm \eta$ is strongly self-dual.

Let ω and η, and $\omega \pm \eta$ be strongly self-dual and $(\omega,\eta) = 0$. Then

$$(\omega,\omega)(\eta,\eta) = 2(\omega^2,\eta^2) = 2\omega^2\eta^2.$$

Let ω, η and α be mutually orthogonal strongly self-dual 2-forms such that $\omega + \eta + \alpha$ is also strongly self-dual. Then

$$\omega\eta\alpha = 0.$$

Pro osition. Let $F = \sum \omega_a E_a$ where ω_a's belong to a linear subspace of strongly self-dual 2-forms and E_a's belong to a basis of the Lie algebra. Then (i) $F^2 = *F^2$ for any Lie algebra, (ii) $*F$ is proportional to kF^3 provided that $\mathrm{tr}E_a^2E_b$ is proportional to E_b.

We note that if k above is constant, then the Yang-Mills equations are automatically satisfied, however this condition means that each ω_a has constant norm.

Another important property of strongly self-dual 2-forms is that the multiplication by a strongly self-dual 2-form is nondegenerate. Let ω be strongly self-dual 2 form and η be any 2-form. Then $\omega\eta = 0$ implies that $\eta = 0$. As a result, the equation $\omega\eta = \alpha$ has a solution unique solution provided that α is in the image of the multiplication by ω, in other words if the equation has a solution, this solution is unique.

Using these results we know show that the solution constructed by Grossman is unique.

Journal of Physics: Conference Series **670** (2016) 012011 doi:10.1088/1742-6596/670/1/012011

6. The Grossman-Ke hart-Stasheff solution

In Grossman et.al, F in the form $F = f_0\sigma_{ij}$ where f_0 is a 2 form and the Σ_{ij}'s are a basis for Spin(8). As a result the action $\langle F^2, F^2\rangle$ is equal to the topological term F^4 as consequence of the properties of the Σ_{ij}'s.

We consider real $SO(N)$ bundles, hence locally F takes values in a skew-symmetric matrix. It follows that all principal minors of F are skew-symmetric matrices, their determinants are perfect squares and the σ_{2i}'s are sums of squares of i-fold products of the entries of F. Then, a local representative of the second Pontrjagin class is given by

$$-p_2 = \lambda\sigma_{i<j<k<l} - *(F_{ij}F_{kl} - F_{ik}F_{jl} + F_{il}F_{jk})^2,$$

where λ is a proportionality constant.

In [3] we proved the following theorem.

Theorem Let F be the local curvature 2-form of an 8-plane vector bundle. If the negative of the second Pontrjagin class, $-p_2$ is maximal, then

i. Each F_{ij} is strong self-dual,

ii. For distinct i, j, k, $F_{ij}F_{jk}$ is self-dual,

iii. For distinct i, j, k, l, $F_{ij}F_{kl}$ is anti self dual.

The theorem above implies that

$$F_{ij} = c_{ij}\omega_{ij},$$

where c_{ij}'s are functions and ω_{ij}'s are strong self-dual 2-forms. We use the structure of 7-dimensional linear subspaces to obtain the following set of 2-forms that form a local basis of Λ^2 [3].

$$\begin{aligned}
\omega_{12} &= e_{14} + e_{23} + e_{58} + e_{67} \\
\omega_{13} &= e_{13} - e_{24} - e_{57} + e_{68} \\
\omega_{14} &= e_{16} + e_{25} - e_{38} - e_{47} \\
\omega_{15} &= e_{15} - e_{26} + e_{37} - e_{48} \\
\omega_{16} &= e_{18} + e_{27} + e_{36} + e_{45} \\
\omega_{17} &= e_{17} - e_{28} - e_{35} + e_{46} \\
\omega_{18} &= e_{12} + e_{34} + e_{56} + e_{78} \\
\omega_{23} &= e_{12} + e_{34} - e_{56} - e_{78} \\
\omega_{24} &= -e_{17} - e_{28} - e_{35} - e_{46} \\
\omega_{25} &= -e_{18} + e_{27} + e_{36} - e_{45} \\
\omega_{26} &= e_{15} + e_{26} - e_{37} - e_{48} \\
\omega_{27} &= e_{16} - e_{25} + e_{38} - e_{47} \\
\omega_{28} &= -e_{13} + e_{24} - e_{57} + e_{68} \\
\omega_{34} &= -e_{18} + e_{27} - e_{36} + e_{45} \\
\omega_{35} &= e_{17} + e_{28} - e_{35} - e_{46} \\
\omega_{36} &= e_{16} - e_{25} - e_{38} + e_{47} \\
\omega_{37} &= -e_{15} - e_{26} - e_{37} - e_{48} \\
\omega_{38} &= e_{14} + e_{23} - e_{58} - e_{67} \\
\omega_{45} &= e_{12} - e_{34} + e_{56} - e_{78} \\
\omega_{46} &= -e_{13} - e_{24} - e_{57} - e_{68} \\
\omega_{47} &= -e_{14} + e_{23} + e_{58} - e_{67} \\
\omega_{48} &= -e_{15} + e_{26} + e_{37} - e_{48} \\
\omega_{56} &= -e_{14} + e_{23} - e_{58} + e_{67} \\
\omega_{57} &= e_{13} + e_{24} - e_{57} - e_{68}
\end{aligned}$$

Journal of Physics: Conference Series **670** (2016) 012011 doi:10.1088/1742-6596/670/1/012011

$$\begin{aligned}
\omega_{58} &= e_{16} + e_{25} + e_{38} + e_{47} \\
\omega_{67} &= e_{12} - e_{34} - e_{56} + e_{78} \\
\omega_{68} &= -e_{17} + e_{28} - e_{35} + e_{46} \\
\omega_{78} &= e_{18} + e_{27} - e_{36} - e_{45}
\end{aligned}$$

The way we number this set is as follows. The ω_{i8}s are the seven 2-forms obtained from Corrigan's equations (we could start with any such set)[8]. Then start with for example ω_{18} and find all the 2-forms that constitute a linear space together with ω_{18}, these will be placed to the first row and eight column. Actually given one row one can construct the matrix uniquely from the following requirement: given for example the first row, ω_{23} has to form a linear space together both with ω_{13} and ω_{12}, the structure of these spaces are discussed in [3], looking at intersections, we determine the matrix completely.

The properties of products of forms belonging to the same linear subspace, as discussed above, implies that F^3 is proportional to $*F$.

Note that the Bianchi identities are linear in the connection A. Actually, the connection of the base manifold also come into play in $d\omega_{ij}$, since

$$dF_{ij} + F_{ik} \wedge A_{kj} - A_{ik} \wedge F_{kj} = dc_{ij} \wedge \omega_{ij} + c_{ij} d\omega_{ij} + c_{ik}\omega_{ik} \wedge A_{kj} - c_{kj} A_{ik} \wedge \omega_{kj}$$

We start by the general case, with arbitrary c_{ij}. We solve the components of the dc_{ij} and the components of the connection of the manifold from the Bianchi identities. Then, the remaining components of the Bianchi identities are linear homogeneous equations for the components of the connection on the bundle. We have shown that, generically, this coefficient matrix of this homogeneous system is nonsingular, hence the connection of the bundle is trivial. On the other extreme, if we assume all c_{ij}'s are equal to each other, then the coefficient matrix is identically zero. It follows that the connection of the base manifold is determined by the connection on the bundle (or vice versa).

After this stage we assume that the base manifold is conformally flat. This amounts to assuming that $e_i = e^p dx_i$, hence $de_i = p_i e_i$, where p_i is the partial derivative of p, in the direction i. We can incorporate the common multiplicative function in F into the conformal factor. Finally we use the Coulomb gauge condition, $\nabla \cdot A = 0$ and we solve $F = dA - A \wedge A$ for A, in terms of the derivatives of p. The second derivatives of p satisfy $p_{ij} = 0$ for $i \neq j$ and

$$p_{ii} = -1 - \tfrac{1}{2}\left[p_1^2 + p_2^2 + p_3^2 + p_4^2 + p_5^2 + p_6^2 + p_7^2 + p_8^2\right],$$

fixing the base manifold completely.

[1] J.M.Milnor and J.D. Stasheff, *Characterisitic Classes*, Annals of Mathematics Studies, Princton University Press, Princton, New Jersey, (1974).
[2] A.H. Bilge, T. Dereli and S. Koçak, "Self-dual Yang-Mills fields in eight dimensions", *Lett. Math. Phys.*, **36**, 301, (1996).
[3] F. Ozdemir and A.H. Bilge, "Self-duality in dimensions $2n > 4$: equivalence of various definitions and the derivation of the octonionic instanton solution", ARI (1999) 51:247-253.
[4] B.Grossman, T.W.Kephart, J.D.Stasheff, *Commun. Math. Phys.*, **96**, 431-437, (1984).
[5] A.Trautman, *International Journal of Theoretical Physics*, **16**, 561-565, (1977).
[6] A.H. Bilge, T. Dereli and S. Koçak, "The Geometry of Self-dual 2-forms", *J. Math. Phys.*, 38:4804-4814..
[7] M. Marcus and H.Minc, *A Survey of Matrix Theory and Matrix Inequalities*, Allyn and Bacon Inc., Boston, (1964).
[8] E.Corrigan, C.Devchand, D.B.Fairlie and J.Nuyts, *Nuclear Physics*, **B214**, 452-464, (1983).

Quantizing the classical non-cooperative symmetric games with arbitrary number of strategies

Katarzyna Bolonek-Lasoń[1] **and Piotr Kosiński**[2]

[1] Department of Statistical Methods, Faculty of Economics and Sociology, University of Lodz, 41/43 Rewolucji 1905 St., 90-214 Lodz, Poland.

[2] Department of Computer Science, Faculty of Physics and Applied Informatics, University of Lodz, Pomorska 149/153, 90-236 Lodz, Poland.

E-mail: kbolonek@uni.lodz.pl

Abstract. Two-player N-strategy symmetric noncooperative games are considered. A general form of gate operators is found using group-theoretical methods. The stability group of initial state is introduced and calculated in the case of three strategies games. Its role in determining the Nash equilibria for maximally entangled games is stressed.

1. Introduction

We present here the review of results obtained in our papers [1] and [2]. The aim of the paper is to describe the generalization of the ELW game to the case of two-player and N-strategy games, in particular the construction of the gate operator using group-theoretical methods.

The first scheme of quantization of classical noncooperative symmetric games was proposed by Eisert, Wilkens and Lewenstein [3], [4]. Their method concerns two-player two-strategy (C-cooperate and D-defect) game. Classical game is defined in Table 1, where the letters r, p, s, t are the possible outcomes for the classical strategies C and D. The game is uniquely defined if the value of r, p, s, t are selected. For example, if the classical payoffs obey $t > r > p > s$, the Prisoner Dilemma emerges on the classical level. The Prisoner Dilemma is one of the most interesting game in economy and is set up in such a way that both players choose to protect themselves at the expense of the other player. As a result of a purely logical thought process to help oneself, both players find themselves in a worse state than if they had cooperated with each other in the decision-making process.

Table 1. The payoffs resulting from different ELW strategies.

Strategies		Payoffs	
player A	player B	player A	player B
C	C	r	r
C	D	s	t
D	C	t	s
D	D	p	p

Journal of Physics: Conference Series **670** (2016) 012012 doi:10.1088/1742-6596/670/1/012012

The procedure of quantization proceeds as follows. Firstly, we assign the classical strategies C and D to the basis vectors $|1\rangle$ and $|2\rangle$ of 2D complex Hilbert spaces, one ascribed to each player. The initial state of the game is given by

$$|\Psi_{in}\rangle = J\left(|1\rangle \otimes |1\rangle\right), \tag{1}$$

where J is an unitary operator known to both players. The J operator, called gate operator, is very important because it represents entanglement allowing for the genuinely quantum correlations. Strategic moves of both players are associated with unitary 2×2 operators U_A, U_B operating on their own qubits. The resulting final state of the game is given by

$$|\Psi_{out}\rangle = J^{+}\left(U_A \otimes U_B\right)|\Psi_{in}\rangle = J^{+}\left(U_A \otimes U_B\right) J\left(|1\rangle \otimes |1\rangle\right). \tag{2}$$

The expected payoffs are computed according to

$$\begin{aligned} \$_A &= rP_{11} + pP_{22} + tP_{21} + sP_{12} \\ \$_B &= rP_{11} + pP_{22} + sP_{21} + tP_{12}. \end{aligned} \tag{3}$$

where

$$P_{kk'} \equiv \left|\langle k \otimes k' | \Psi_{out}\rangle\right|^2, \qquad k, k' = 1, 2 \tag{4}$$

denotes the probability of obtaining one of the four outcomes.

The form of the gate operator J results from two assumptions: (a) to preserve the symmetry of the game J is symmetric with respect to the interchange of players; (b) the quantum game entails a faithful representation of its classical counterpart.
In the case of ELW the game J has form

$$J = \exp\left(-\frac{i\gamma}{2}\sigma_2 \otimes \sigma_2\right), \tag{5}$$

where $\gamma \in [0, \frac{\pi}{2}]$ is a real parameter and σ_2 is the second Pauli matrix. In fact, γ is a measure of the game's entanglement. Eisert et al. considered the particular case of the Prisoners' Dilemma [3]. They chose the following values of payoffs $r = 3$, $p = 1$, $t = 5$, $s = 0$ and restricted the strategic space to the 2-parameter set of unitary matrices

$$U(\theta, \phi) = \begin{pmatrix} e^{i\phi}\cos\frac{\theta}{2} & \sin\frac{\theta}{2} \\ -\sin\frac{\theta}{2} & e^{-i\phi}\cos\frac{\theta}{2} \end{pmatrix} \tag{6}$$

with $0 \le \theta \le \pi$, $0 \le \phi \le \frac{\pi}{2}$. This attitiude was criticized by Benjamin and Hayden [5]. They pointed out that there are no compelling reasons to impose such a restriction on the strategic space. In our considerations the set of allowed strategies is the whole $SU(N)$ group.

In the case of maximal entangled game $\gamma = \frac{\pi}{2}$, Eisert et al. constructed quantum strategy $Q \equiv U\left(0, \frac{\pi}{2}\right)$ which is optimal strategy for both players. What's more, $Q \times Q$ is not only Nash equilibrium but also Pareto optimal. So we could say that the players escape the dilemma. Since the papers [3] and [4] appeared quantum game theory has been a subject of intensive research [5]÷[54].

The paper is organized as follows. In Sect. 2, we introduce the generalization of the ELW game to the case of N strategies and we construct a wide class of entanglers for arbitrary N. The case of three strategies is discussed in Sect. 3, as an example. In Section 4 the stability group of initial state of the game is introduced and its role in determining the Nash equilibria is stressed. The last Section 5 is devoted to some conclusions.

2. Non-cooperative symmetric games with N strategies

The generalized ELW game can be described as follows. We have some classical non-cooperative two-player N-strategy game defined by a relevant payoff table. In order to quantize it, one ascribes to any player an N dimensional complex Hilbert space spanned by the vectors

$$|1\rangle = \begin{pmatrix} 1 \\ 0 \\ \vdots \\ 0 \end{pmatrix}, \quad \ldots, \quad |N\rangle = \begin{pmatrix} 0 \\ \vdots \\ 0 \\ 1 \end{pmatrix}. \tag{7}$$

The initial state is defined in the same way as in the $SU(2)$ case

$$|\Psi_{in}\rangle \equiv J\left(|1\rangle \otimes |1\rangle\right). \tag{8}$$

In order to preserve the symmetry of the game we assume that the gate operator J is symmetric with respect to the permutation of the factors entering the tensor product; further, we assume that the classical game is faithfully represented in its quantum counterpart. We demand that the set of allowed strategies is the whole $SU(N)$ group, unlike Eisert et. al who assumed that the set of strategies, in the case of 2-strategy game, belongs to the 2D submanifold of $SU(2)$, which itself is not a group. The strategic moves of the players are associated with unitary operators U_A and U_B operating on their own qubits and the final state of the game is given by

$$|\Psi_{out}\rangle = J^{+}\left(U_A \otimes U_B\right) J\left(|1\rangle \otimes |1\rangle\right). \tag{9}$$

The player's expected payoffs can be computed according to

$$\$^{A,B} = \sum_{k,k'=1}^{N} p_{k,k'}^{A,B} \left|\langle k, k' | \Psi_{out}\rangle\right|^2, \tag{10}$$

where $|k,k'\rangle \equiv |k\rangle \otimes |k'\rangle$, $k,k' = 1,...,N$ and $p_{k,k'}^{A,B}$ are classical payoffs of Alice and Bob, respectively. There exists one difference between original and generalized ELW game. The set of allowed gate operators J is parametrized by one real parameter γ (cf. Eq. (5)) in the case of SU(2) group. For general N we showed in [2] that J depends on a number of free parameters growing proportionally to N^2. As we have mentioned above, in order to find the form of gate operators we make two assumptions: the gate operator is symmetric under the exchange of the factors in the tensor product of Hilbert spaces ascribed to both players and all classical pure strategies are contained in the set of pure quantum ones. The second assumption is fulfilled if N matrices $U_k \in SU(N)$, $k = 1,...,N$ exist which satisfy

$$U_k |1\rangle = e^{i\phi_k} |k\rangle, \quad k = 1,...,N \tag{11}$$

$$[J, U_k \otimes U_l] = 0, \quad k,l = 1,...,N. \tag{12}$$

We want to leave as much freedom as possible in the choice of gate operator so we assume that $[U_k, U_l] = 0$, $k,l = 1,...,N$. We define the matrices U_k in form

$$U_k \equiv e^{\frac{i\pi(N-1)(k-1)}{N}} U^{k-1} \in SU(N), \qquad k = 1,...,N \tag{13}$$

where

$$U = \begin{pmatrix} 0 & 0 & \cdots\cdots & 0 & 1 \\ 1 & 0 & \cdots\cdots & 0 & 0 \\ 0 & 1 & \cdots\cdots & 0 & 0 \\ \cdots & \cdots & \cdots\cdots & \cdots & \cdots \\ 0 & 0 & \cdots\cdots & 1 & 0 \end{pmatrix} \tag{14}$$

Journal of Physics: Conference Series **670** (2016) 012012 doi:10.1088/1742-6596/670/1/012012

is the matrix representing the cyclic permutation of $12\ldots N$. All matrices U_k commute and eq. (11) is satisfied. In order to diagonalize the matrices U_k we must diagonalize the matrix U. The elements of matrix V diagonalizing U have the form

$$V_{ik} = \frac{1}{\sqrt{N}} \bar{\varepsilon}^{(i-1)(k-1)}, \qquad i,k = 1,...,N. \tag{15}$$

ε being the primitive N-th root of unity. Next we define

$$\widetilde{J} \equiv \left(V^+ \otimes V^+\right) J \left(V \otimes V\right). \tag{16}$$

The matrix V^+UV is diagonal matrix with different elements on the diagonal and operator $\widetilde{J}$ commute with V^+UV; therefore $\widetilde{J}$ must be diagonal and can be written as

$$\widetilde{J} = \exp\left(i \sum_{k=1}^{N-1} \lambda_k \left(\Lambda_k \otimes \Lambda_k\right) + i \sum_{k\neq l=1}^{N-1} \mu_{kl} \left(\Lambda_k \otimes \Lambda_l + \Lambda_l \otimes \Lambda_k\right) \right) \tag{17}$$

where Λ_i, $i = 1, ..., N-1$ is any basis in Cartan subalgebra of $SU(N)$; λ_k and $\mu_{kl} = \mu_{lk}$ are real parameters and determine the degree of entanlglement. The gate operator defined by eq. (17) depends on $N - 1 + \binom{N-1}{2} = \binom{N}{2}$ free parameters.

3. Two-player three-strategy game

In this section we present the example of quantization scheme presented above. In the special case $N = 3$ one can construct the gate operator J in full generality. Considering the game with three strategies; we assume that the matrices representing classical strategies satisfy the conditions

$$\begin{array}{ll} U_k \left|1\right\rangle = e^{i\varphi_k} \left|k\right\rangle, & k = 1,2,3 \\ \left[J, U_j \otimes U_k\right] = 0, & j,k = 1,2,3 \end{array} \tag{18}$$

$$\left[U_j, U_k\right] = 0. \tag{19}$$

Next, we assume that $U_1 = I$ and from eqs. (18) and (19) we find the final form of matrices U_2 and U_3

$$\begin{array}{l} U_2 = \begin{pmatrix} 0 & 0 & \varepsilon e^{-i\varphi_3} \\ e^{i\varphi_2} & 0 & 0 \\ 0 & \bar{\varepsilon} e^{i(\varphi_3 - \varphi_2)} & 0 \end{pmatrix} \\ U_3 = \begin{pmatrix} 0 & \varepsilon e^{-i\varphi_2} & 0 \\ 0 & 0 & \bar{\varepsilon} e^{i(\varphi_2 - \varphi_3)} \\ e^{i\varphi_3} & 0 & 0 \end{pmatrix}, \end{array} \tag{20}$$

where ε is any cubic root from unity, we assume $\varepsilon \neq 1$.
All matrices U_k can be diagonalized [2]

$$\begin{array}{l} \widetilde{U}_1 = V^+ U_1 V = I \\ \widetilde{U}_2 = V^+ U_2 V = diag\left(1, \varepsilon, \varepsilon^2\right) . \\ \widetilde{U}_3 = V^+ U_3 V = diag\left(\varepsilon, 1, \varepsilon^2\right). \end{array} \tag{21}$$

Journal of Physics: Conference Series **670** (2016) 012012 doi:10.1088/1742-6596/670/1/012012

using the matrix V defined by

$$V = \frac{1}{\sqrt{3}} \begin{pmatrix} 1 & 1 & 1 \\ e^{i\varphi_2} & \bar{\varepsilon} e^{i\varphi_2} & \varepsilon e^{i\varphi_2} \\ \bar{\varepsilon} e^{i\varphi_3} & e^{i\varphi_3} & \varepsilon e^{i\varphi_3} \end{pmatrix}, \qquad VV^+ = I. \tag{22}$$

The operator $\widetilde{J}$, defined by eq. (16), commutes with $\widetilde{U}_i$, $i = 1, 2, 3$ and has form

$$\widetilde{J} = \exp i \left(\tau \left(\Lambda \otimes \Lambda\right) + \rho \left(\Lambda \otimes \Delta + \Delta \otimes \Lambda\right) + \sigma \left(\Delta \otimes \Delta\right)\right), \tag{23}$$

where τ, ρ and σ are real parameters, while $\Lambda = \lambda_3$, $\Delta = \frac{1}{2}\left(\lambda_3 + \sqrt{3}\lambda_8\right)$ (λ_3 and λ_8 are Gell-Mann matrices).

4. The stability subgroup

It should be noted that different strategies may lead to the same outcome. For that reason we want to find the stability subgroup $G_s \in SU(N) \times SU(N)$ of the initial state $|\Psi_{in}\rangle$, i.e. the set of elements (pairs of strategies) $g \in SU(N) \times SU(N)$ such that

$$g \left|\Psi_{in}\right\rangle = \left|\Psi_{in}\right\rangle. \tag{24}$$

Then two games, (U_A, U_B) and (U'_A, U'_B), differing by an element $g \in G_s$,

$$\left(U'_A, U'_B\right) = \left(U_A, U_B\right) \cdot g \tag{25}$$

share the same final result. Consequently, the coset space $SU(N) \times SU(N)/G_s$ is the effective set of strategies.

The stability group depends on the degree of entanglement of the initial state. We consider the most interesting case of maximal entanglement. In order to find the form of elements of the stability group we write the initial state of the game as

$$\left|\Psi_{in}\right\rangle \equiv J\left(|1\rangle \otimes |1\rangle\right) \equiv F_{ij} \left|i\right\rangle \otimes \left|j\right\rangle, \tag{26}$$

where the summation over repeated indices is understood and $F_{ij} = F_{ji}$.
The corresponding density matrix reads

$$\rho_{in} = \left|\Psi_{in}\right\rangle \left\langle \Psi_{in}\right|. \tag{27}$$

The initial state is maximally entangled if the corresponding reduced density matrix is proportional to the unit matrix [2]

$$Tr_A \rho_{in} = \frac{1}{N} I, \qquad Tr_B \rho_{in} = \frac{1}{N} I. \tag{28}$$

Equations (28) imply that the matrix $\widetilde{F} \equiv \sqrt{N} F$ is unitary. After some calculations [2] we find that G_s consists of the elements of the form

$$\left(U, \widetilde{F} \overline{U} \widetilde{F}^+\right). \tag{29}$$

Therefore, the stability group G_s of $|\Psi_{in}\rangle$ is, up to a group automorphism, the diagonal subgroup of $SU(N) \times SU(N)$. Its Lie algebra induces the symmetric Cartan decomposition of $sU(N) \oplus sU(N)$.

Let (U_1, U_2) be a pair of strategies leading to the expected payoffs desired by one of the player, let's say Bob. Assume that Alice has chosen the strategy U_A. The pair of strategies $(U_1, U_2) \in SU(N) \times SU(N)$ can be written as [1],[2]

$$(U_1, U_2) = \left(U_A, U_2 \widetilde{F}\, \overline{U}_1^+ \,\overline{U}_A \widetilde{F}^+\right) \left(U_A^+ U_1, \widetilde{F}\, \overline{U}_A^+ \,\overline{U}_1 \widetilde{F}^+\right). \tag{30}$$

The second element on the right hand side of eq. (30) belongs to the stability group of initial state; therefore, the strategies (U_1, U_2) and $\left(U_A, U_2 \widetilde{F}\, \overline{U}_1^+ \,\overline{U}_A \widetilde{F}^+\right)$ lead to the same payoffs. We conclude that $U_2 \widetilde{F}\, \overline{U}_1^+ \,\overline{U}_A \widetilde{F}^+$ is the relevant counterstrategy to Alice's strategy U_A. As a result, there is no pure Nash equilibrium unless, in the table of payoff, there exists a pair of strategies leading to the optimal outcomes for both players. Then, we have only trivial pure Nash equilibria.

As an example we present the stability group in the case of three strategies. We have to find the values of parameters ρ, σ, τ for which the initial state is maximal entangled. Then, the reduced density matrix defined by the initial state $|\Psi_{in}\rangle$

$$\mathrm{Tr}_B \rho_{in} = \frac{1}{9} \left(\begin{array}{c|c|c} 3 & \begin{array}{c} e^{i(3\rho+\sigma+2\tau)}+ \\ +e^{-i(\rho+2\tau)}+ \\ +e^{-i(2\rho+\sigma)} \end{array} & \begin{array}{c} e^{i(3\rho+2\sigma+\tau)}+ \\ +e^{-i(2\rho+\tau)} \\ +e^{-i(\rho+2\sigma)} \end{array} \\ \hline \begin{array}{c} e^{-i(3\rho+\sigma+\tau)}+ \\ +e^{i(\rho+2\tau)}+ \\ +e^{2\rho+\sigma} \end{array} & 3 & \begin{array}{c} e^{i(\sigma-\tau)}+ \\ +e^{-i(\rho-\tau)}+ \\ +e^{i(\rho-\sigma)} \end{array} \\ \hline \begin{array}{c} e^{-i(3\rho+2\sigma+\tau)}+ \\ +e^{i(2\rho+\tau)}+ \\ +e^{i(\rho+2\sigma)} \end{array} & \begin{array}{c} e^{-i(\sigma-\tau)}+ \\ +e^{i(\rho-\tau)}+ \\ +e^{-i(\rho-\sigma)} \end{array} & 3 \end{array} \right). \tag{31}$$

must be proportional to the unit matrix, eq. (28). Therefore all off-diagonal elements of reduced density matrix defined by (31) vanish. From this condition, we find the following sets of parameters [2]

$$\begin{array}{l} \left\{ \begin{array}{c} \tau = \rho = \sigma - \frac{2\pi}{3} \\ \sigma = \frac{2\pi}{3}, \frac{8\pi}{9}, \frac{10\pi}{9}, \frac{4\pi}{3}, \frac{14\pi}{9}, \frac{16\pi}{9}, 2\pi \end{array} \right. \\ \left\{ \begin{array}{c} \tau = \rho = \sigma + \frac{2\pi}{3} \\ \sigma = 0, \frac{2\pi}{9}, \frac{4\pi}{9}, \frac{2\pi}{3}, \frac{8\pi}{9}, \frac{10\pi}{9}, \frac{4\pi}{3}, \frac{14\pi}{9}, \frac{16\pi}{9} \end{array} \right. \\ \left\{ \begin{array}{c} \tau = \sigma - \frac{2\pi}{3} \\ \rho = \sigma = \frac{2\pi}{3}, \frac{8\pi}{9}, \frac{10\pi}{9}, \frac{4\pi}{3}, \frac{14\pi}{9}, \frac{16\pi}{9}, 2\pi \end{array} \right. \\ \left\{ \begin{array}{c} \tau = \sigma + \frac{2\pi}{3} \\ \rho = \sigma = 0, \frac{2\pi}{9}, \frac{4\pi}{9}, \frac{2\pi}{3}, \frac{8\pi}{9}, \frac{10\pi}{9}, \frac{4\pi}{3}, \frac{14\pi}{9}, \frac{16\pi}{9} \end{array} \right. \\ \left\{ \begin{array}{c} \rho = \sigma - \frac{2\pi}{3} \\ \tau = \sigma = \frac{2\pi}{3}, \frac{8\pi}{9}, \frac{10\pi}{9}, \frac{4\pi}{3}, \frac{14\pi}{9}, \frac{16\pi}{9}, 2\pi \end{array} \right. \\ \left\{ \begin{array}{c} \rho = \sigma + \frac{2\pi}{3} \\ \tau = \sigma = 0, \frac{2\pi}{9}, \frac{4\pi}{9}, \frac{2\pi}{3}, \frac{8\pi}{9}, \frac{10\pi}{9}, \frac{4\pi}{3}, \frac{14\pi}{9}, \frac{16\pi}{9}. \end{array} \right. \end{array} \tag{32}$$

According to eq. (29), the generators of the stability subgroup have form

$$Y \otimes I - I \otimes \widetilde{F} \overline{Y} \widetilde{F}^+, \tag{33}$$

where Y runs over all generators of $SU(3)$. Using a fact that $\widetilde{F}$ is a symmetric matrix and substituting

$$Y \to Y \mp \widetilde{F} \overline{Y} \widetilde{F}^+ \equiv X \tag{34}$$

one easily finds that the generators can be put in the form

$$X \otimes I \pm I \otimes X. \tag{35}$$

As an example we write out the explicit form of generators of stability subgroup for one of the solutions (32), i.e. $\rho = \frac{2\pi}{3}$, $\sigma = \tau = 0$:

$$\begin{aligned}
G_1 &= \left(\lambda_1 - \sqrt{3}\lambda_2 + \frac{2}{\sqrt{3}}\lambda_8\right) \otimes I - I \otimes \left(\lambda_1 - \sqrt{3}\lambda_2 + \frac{2}{\sqrt{3}}\lambda_8\right) \\
G_2 &= \left(\sqrt{3}\lambda_2 + \lambda_3 + \lambda_4 - \frac{1}{\sqrt{3}}\lambda_8\right) \otimes I - I \otimes \left(\sqrt{3}\lambda_2 + \lambda_3 + \lambda_4 - \frac{1}{\sqrt{3}}\lambda_8\right) \\
G_3 &= \left(\lambda_3 + 2\lambda_6 + \frac{1}{\sqrt{3}}\lambda_8\right) \otimes I - I \otimes \left(\lambda_3 + 2\lambda_6 + \frac{1}{\sqrt{3}}\lambda_8\right) \\
G_4 &= (\lambda_2 + \lambda_5) \otimes I - I \otimes (\lambda_2 + \lambda_5) \\
G_5 &= \left(4\lambda_2 + \sqrt{3}\lambda_3 + 2\lambda_7 - 3\lambda_8\right) \otimes I - I \otimes \left(4\lambda_2 + \sqrt{3}\lambda_3 + 2\lambda_7 - 3\lambda_8\right) \\
G_6 &= \left(\lambda_1 - \frac{1}{2}\lambda_4 + \frac{1}{4}\lambda_6 - \frac{3\sqrt{3}}{4}\lambda_7 - \frac{\sqrt{3}}{2}\lambda_8\right) \otimes I + \\
&\quad + I \otimes \left(\lambda_1 - \frac{1}{2}\lambda_4 + \frac{1}{4}\lambda_6 - \frac{3\sqrt{3}}{4}\lambda_7 - \frac{\sqrt{3}}{2}\lambda_8\right) \\
G_7 &= \left(\lambda_2 - \frac{\sqrt{3}}{2}\lambda_4 - \lambda_5 - \frac{\sqrt{3}}{4}\lambda_6 + \frac{1}{4}\lambda_7 + \frac{3}{2}\lambda_8\right) \otimes I + \\
&\quad + I \otimes \left(\lambda_2 - \frac{\sqrt{3}}{2}\lambda_4 - \lambda_5 - \frac{\sqrt{3}}{4}\lambda_6 + \frac{1}{4}\lambda_7 + \frac{3}{2}\lambda_8\right) \\
G_8 &= \left(\lambda_3 - \lambda_4 - \frac{1}{2}\lambda_6 - \frac{\sqrt{3}}{2}\lambda_7\right) \otimes I + I \otimes \left(\lambda_3 - \lambda_4 - \frac{1}{2}\lambda_6 - \frac{\sqrt{3}}{2}\lambda_7\right).
\end{aligned} \tag{36}$$

5. Conclusions

We have generalized the ELW game to the case of N-strategies. The gate operator J is one of the most important elements of the quantum game because it introduces quantum correlations. In order to construct the operator J we assume that it preserves the symmetry of the classical game and the classical game is faithfully represented in its quantum counterpart.

The maximal entanglement game corresponds to the maximal stability subgroup of initial state. The elements differing by an element of the stability subgroup lead to the same outcome. The effective manifold of games is the coset space $SU(N) \times SU(N)/G_s$.

In the $N = 3$ case we found the most general form of the gate operator and we determined the values of parameters leading to the maximal entanglement of the game. As an example, the explicit form of generators of stability group G_s was given for one case.

References

[1] Bolonek-Lasoń K, Kosiński P 2013 *Prog. Theor. Exp. Phys.* **7**, 073A02
[2] Bolonek-Lasoń K 2015 *Prog. Theor. Exp. Phys.* **2**, 023A03
[3] Eisert J, Wilkens M, Lewenstein M 1999, *Phys. Rev. Lett.* **83**, 3077
[4] Eisert J, Wilkens M 2000, *J. Mod. Opt.* **47**, 2543
[5] Benjamin S, Hayden P 2001, *Phys. Rev. Lett.* **87(6)**, 069801
[6] Meyer D 1999, *Phys. Rev. Lett.* **82**, 1052

[7] Marinatto L, Weber T 2000, *Phys. Lett*, **A272**, 291
[8] Benjamin S 2000, *Phys. Lett.*, **A277**, 180
[9] Marinatto L, Weber T 2000, *Phys. Lett.* **A 277**, 183
[10] Iqbal A, Toor A 2001, *Phys. Lett* **A280**, 249
[11] Du J, Li H, Xu X, Zhou X, Han R 2001, *Phys. Lett* **A289**, 9
[12] Eisert J, Wilkens M, Lewenstein M 2001, *Phys. Rev.Lett.* **87**, 069802
[13] Flitney A, Abbott D 2000, *Fluct. Noise Lett.* **2**, R175
[14] Iqbal A, Toor A 2002, *Phys. Rev.* **A65**, 052328
[15] Du J, Li H, Xu X, Shi M, Wu J, Zhou X, Han R 2002, *Phys. Rev. Lett.* **88**, 137902
[16] van Enk S.J, Pike R 2002, *Phys. Rev* **A66**, 024306
[17] Du J, Xu X, Li H, Zhou X, Han R 2002, *Fluct. Noise Lett.* **2**, R189
[18] Guinea F, Martin-Delgado M.A 2003, *Journ. Phys.* **A36**, L197
[19] Flitney A, Abbott D 2003, *Proc. R. Soc. Lond.* **A459**, 2463
[20] Piotrowski E, Sladkowski J 2003, *Int. Journ. Theor. Phys.* **42**, 1089
[21] Du J, Li H, Xu X, Zhou X, Han R 2003, *Journ. Phys.* **A36**, 6551
[22] Zhou L, Kuang L 2003, *Phys. Lett* **A315**, 426
[23] Chen L, Ang H, Kiang D, Kwek L, Lo C 2003, *Phys. Lett.* **A316**, 317
[24] Lee C.F, Johnson N.F 2003, *Phys. Rev.* **A67**, 022311
[25] Shimamura J, Ozdemir S, Morikoshi F, Imoto N 2004, *Int. Journ. Quant. Inf.* **2**, 79
[26] Shimamura J, Ozdemir S, Morikoshi F, Imoto N 2004, *Phys. Lett. A* **328**, 20
[27] Ozdemir S, Shimamura J, Imoto N 2004, *Phys. Lett. A* **325**, 104
[28] Ozdemir S, Shimamura J, Morikoshi F, Imoto N 2004, *Phys. Lett. A* **333**, 218
[29] Ozdemir S, Shimamura J, Imoto N 2007, *New Journ. Phys.*, **9**, 43
[30] Landsburg S 2004, *Notices of the Am. Math. Soc.* **51**, 394
[31] Rosero A, "Classification of Quantum Symmetric Non-zero Sum 2×2 Games in the Eisert Scheme", quant-phys/0402117
[32] Nawaz A, Toor A 2004, *Journ. Phys.* **A37**, 11457
[33] Nawaz A, Toor A 2004, *Journ. Phys.* **A37**, 4437
[34] Iqbal A, "Studies in the theory of quantum games", quant-phys/0503176
[35] Flitney A, Abbott D 2005, *Journ. Phys.* **A38**, 449
[36] Ichikawa T, Tsutsui I 2007, *Ann. Phys.* **322**, 531
[37] Cheon T, Tsutsui I 2006, *Phys. Lett.* **A348**, 147
[38] Patel N 2007, *Nature* **445**, 144
[39] Ichikawa T, Tsutsui I, Cheon T 2008, *Journ. Phys. A: Math. and Theor.* **41**, 135303
[40] Flitney A, Hollenberg L 2007, *Phys. Lett.* **A363**, 381
[41] Nawaz A, "The generalized quantization schemes for games and its application to quantum information", arXiv:1012.1933
[42] Landsburg S 2011, *Proc. Am. Math. Soc.* **139**, 4413
[43] Landsburg S 2011, *Wiley Encyclopedia of operations Research and Management science*
[44] Schneider D 2011, *Journ. Phys.* **A44**, 095301
[45] Schneider D 2012, *Journ. Phys.* **A45**, 085303
[46] Avishai Y, "Some Topics in Quantum Games", arXiv:1306.0284
[47] Ramzan M 2013, *Quant. Inf. Process.* **12**, 577
[48] Ramzan M, Khan M.K 2013, *Fluctuation and Noise Letters* **12**, 1350025;
[49] Nawaz A 2013, *Chin. Phys. Lett.* **30(5)**, 050302
[50] Frackiewicz P 2013, *Acta Phys. Polonica B* **44**, 29
[51] Nawaz A, "Werner-like States and Strategies form of Quantum games", arXiv:1307.5508
[52] Nawaz A 2012, *J. Phys. A: Math. Theor. J. Phys.* **45**, 195304
[53] Avishai Y 2015, *Journal of Quantum Information Science* **5**, 16
[54] Brunner N, Linden N 2013, *Nature Communications* **4**, 2057

Journal of Physics: Conference Series **670** (2016) 012013 doi:10.1088/1742-6596/670/1/012013

Real and pseudoreal forms of D=4 complex Euclidean (super)algebras and super–Poincaré/ super–Euclidean r-matrices

A. Borowiec[1)], **J. Lukierski**[1) 1] **and V.N. Tolstoy**[2)]

[1)] Institute for Theoretical Physics, University of Wrocław, pl. Maxa Borna 9, 50–205 Wrocław, Poland

[2)] Lomonosov Moscow State University, Skobeltsyn Institute of Nuclear Physics, Moscow 119991, Russian Federation

Abstract. We provide the classification of real forms of complex D=4 Euclidean algebra $\varepsilon(4;\mathbb{C}) = \mathfrak{o}(4;\mathbb{C})) \ltimes \mathbf{T}^4_{\mathbb{C}}$ as well as (pseudo)real forms of complex D=4 Euclidean superalgebras $\varepsilon(4|N;\mathbb{C})$ for N=1,2. Further we present our results: N=1 and N=2 supersymmetric D=4 Poincaré and Euclidean r-matrices obtained by using D= 4 Poincaré r-matrices provided by Zakrzewski [1] For N=2 we shall consider the general superalgebras with two central charges.

1. Introduction

The classification of quantum deformations of Lorentz symmetries described by classical r-matrices was given firstly by Zakrzewski [2] (see also [3]), and has been further extended to the classification of classical r-matrices for Poincaré algebra in [1]. The classification of dual Hopf-algebraic quantum deformations of Poincaré group were presented in [4]. Subsequently, the infinitesimal r-matrix description of the deformations of Poincaré algebra presented in [1] has been extended in several papers to finite Hopf-algebraic deformations, with conclusive results obtained in [5]. Because majority of studied deformations were triangular, i.e. described by the twist deformations, they permitted (see e.g. [6, 7]) to derive explicit formulas for the non-commutative algebra of deformed space-time coordinates by the use of so-called star product realizations.

The study of deformations of spacetime supersymmetries and the corresponding deformed superspaces were less systematic, related mostly either with the supersymmetrization of simplest Abelian canonical twist deformation of Poincaré symmetries [8]–[10] or with the supersymmetric extension of κ-deformation [11, 12]. The supersymmetrization of such a canonical twist F for $N = 1$ Poincare superalgebra looks as follows

$$F = \exp\frac{1}{2}\theta^{\mu\nu}P_\mu \wedge P_\nu \quad \to \quad \mathcal{F} = \exp\frac{1}{2}\theta^{\mu\nu}P_\mu \wedge P_\nu \exp\xi^{\alpha\beta}Q_\alpha \wedge Q_\beta\,, \tag{1.1}$$

where $\theta_{\mu\nu} = -\theta_{\nu\mu}$, $\xi^{\alpha\beta} = \xi^{\beta\alpha}$ and $\hat{x}\wedge\hat{y} \equiv (\hat{x}\otimes\hat{y}) - (-1)^{|x||y|}(\hat{y}\otimes\hat{x})$ where $|x| = 0,1$ is the Z_2-grading of the superalgebra element $\hat{x}$ with definite grading. The classification of the supersymmetric triangular deformations for finite-dimensional simple Lie superalgebras was firstly given in a mathematical framework by one of the present authors [13]; one should add also that already long time ago the

[1] Presented by J. Lukierski

nontriangular Drinfeld-Jimbo deformation has been provided for all complex finite-dimensional simple Lie superalgebras in [14, 15, 16] but without discussion of its real forms. Recently however (see [17, 19]) the classification of Poincaré r-matrices in [2] has been extended to $D = 4, N = 1$ Poincare and Euclidean [19] supersymmetries, which provided physically important deformations of non-semisimple Lie superalgebras. New result in this paper is the presentation of the list of D=4 N=2 classical super-Poincaré and super-Euclidean r-matrices.

We add that there were also considered twisted Euclidean $N = 1$ and $N = 2$ supersymmetries (see e.g. [9]) but only under the assumption that the fermionic part of the twist factor was a supersymmetric enlargement of canonical Abelian twist (see (1.1)). In our recent paper by considering partial classification of $N = 1$ complex r−matrices and their pseudoreal (Euclidean) and real (Poincaré) forms [19] we have obtained large class of $N = 1$ supersymmetric twists which provide new D=4 superspaces with Lie-algebraic deformation of bosonic spacetime sector.

In quantum deformations approach the basic primary notion is the Hopf-algebraic deformation of (super)symmetries which subsequently implies the modification of (super)spacetime algebra. It follows that for classifying the possible deformations of (super)symmetries one should list the deformations of corresponding Hopf (super)algebras. In non-supersymmetric case the most complete discussion of such Hopf-algebraic approach to field theories deformed by canonical twists F (see (1.1)) was presented in [18]; for Poincaré and Euclidean supersymmetries such way of introducing deformations via supertwist, firstly advocated in [8, 9], was presented recently in [19, 20]. In present paper we supplement classical $N = 2$ super–Poincaré r−matrices which generate the corresponding N=2 supersymmetric twist (triangular) deformations. Subsequently using $\star$−product formulation [6, 7] one can provide effective formulae for the description of quantum-deformed N=2 superspaces. We add that the particular supersymmetric $N = 2$ twist deformations considered in earlier studies [21]–[25] appear as particular cases in the list of N=2 deformations presented in this paper.

In Sect. 2 and 3 we shall consider in some detail inhomogeneous $\mathfrak{o}(4;\mathbb{C})$ algebra with its real forms and the N=1,2 superextensions; in Sect. 4 and 5 we present our partial results describing N=1,2 supersymmetric Poincaré and Euclidean r-matrices. More detailed plan of our paper is the following: In Sect. 2 we shall consider complex $\mathfrak{o}(4)$ and inhomogeneous $i\mathfrak{o}(4)$ algebras and its $N = 1$ SUSY extension. We shall also provide their real and pseudoreal forms defining $D = 4$ Poincare, Euclidean and Kleinian algebras and corresponding $N = 1$ real/pseudoreal superalgebras. In Sect. 3 we shall discuss complex $N = 2$ superalgebras with two central charges and odd sector described by 8 independent complex supercharges $(Q_{\alpha}{}^{a}, Q_{\dot{\alpha}b},\ \alpha, \dot{\alpha} = 1,2\ a,b = 1,2)$. Further following [28]–[27] and by using suitable conjugations or pseudoconjugations (real or pseudoreal forms) we shall describe $N = 2$ real Poincaré and Kleinian superalgebras as well as complex selfconjugate Euclidean superalgebras. In order to compare with earlier results for N=0 (nonsupersymmetric case) and N=1 we shall provide in Sect. 4 the tables of classical Poincaré (super)-r-matrices given in [1] and [19][2]. In Sect. 5 we present new results: we use the set of real Poincaré classical r-matrices provided by Zakrzewski in [1], and describe the ones which do have N=2 supersymmetric extension; we present also N=2 D=4 Euclidean supersymmetric r-matrices. Finally in Sect. 6 we present final remarks.

In Appendix A we shall outline the general theory of conjugations and pseudoconjugations for Lie superalgebras.

The list of $N = 2$ complex supersymmetric r-matrices and their Kleinian $\mathfrak{o}(2,2)$ real counterparts satisfying suitable (pseudo)reality conditions will be presented in our next publications.

[2] We point out here that among 21 classical r−matrices obtained by Zakrzewski [1] there is one class which satisfies modified YB equation. We shall consider 20 classes of r−matrices which lead to triangular deformations.

2. Complex D=4 Euclidean algebra, its simple (N=1) supersymmetrization and their real or pseudoreal forms

2.1. Complex $\mathfrak{io}(4,\mathbb{C})$ *algebra and its real forms*

It is known that there are four real forms of complex orthogonal Lie algebras $\mathfrak{o}(4,\mathbb{C})$ corresponding to three different nondegenerate $D=4$ metric signatures and the fourth one which can be obtained by imposing quaternionic structure (see e.g. [29, 30]):

i) $\mathfrak{o}(4)\cong\mathfrak{o}(3)\oplus\mathfrak{o}(3)\cong\mathfrak{su}(2)\oplus\mathfrak{su}(2)$ – Euclidean case
ii) $\mathfrak{o}(3,1)\cong\mathfrak{sl}(2,\mathbb{C})_{\mathbb{R}}$ – Lorentzian case[3]
iii) $\mathfrak{o}(2,2)\cong\mathfrak{o}(2,1)\oplus\mathfrak{o}(2,1)\cong\mathfrak{sl}(2,\mathbb{R})\oplus\mathfrak{sl}(2,\mathbb{R})\cong$
$\mathfrak{su}(1,1)\oplus\mathfrak{su}(1,1)\cong\mathfrak{sp}(2,\mathbb{R})\oplus\mathfrak{sp}(2,\mathbb{R})$ – Kleinian case
iv) $\mathfrak{o}^*(4)\cong\mathfrak{o}(3)\oplus\mathfrak{o}(2,1)\cong\mathfrak{su}(2)\oplus\mathfrak{su}(1,1)$ – quaternionic case.

The quaternionic origin of the fourth real form follows from isomorphism with orthogonal quaternionic algebras [29]

$$\mathfrak{o}(2;\mathbb{H})=\mathfrak{o}^*(4),\qquad \mathfrak{o}(1;\mathbb{H})=\mathfrak{o}(2)\,. \tag{2.1}$$

In first three cases (Euclidean, Lorentzian, Kleinian) one can lift the corresponding real forms to inhomogeneous algebra $\varepsilon(4;\mathbb{C})=\mathfrak{o}(4;\mathbb{C}))\ltimes\mathbf{T}^4_{\mathbb{C}}$, where the complex Abelian generators describe complexified momentum fourvectors ($\mathcal{P}_\mu\in\mathbf{T}^4_{\mathbb{C}}$).

In fourth case one can introduce the momenta as complex $\mathfrak{o}^*(4)$ vectors described by second order fourcomponent $SU(2)\times SU(1,1)$ complex spinors. The corresponding inhomogeneous algebra $\mathfrak{o}^*(4)\ltimes\mathbf{T}^4_q$ can be endowed with quaternionic structure if we perform the following contraction of the quaternionic symmetric coset (see also (2.1))

$$\mathfrak{o}(3;\mathbb{H})=\mathfrak{o}(2;\mathbb{H})\oplus\mathfrak{o}(1;\mathbb{H})\oplus\frac{\mathfrak{o}(3;\mathbb{H})}{\mathfrak{o}(2;\mathbb{H})\oplus\mathfrak{o}(1;\mathbb{H})}\Rightarrow(\mathfrak{o}^*(4)\oplus\mathfrak{o}(2))\ltimes\mathbf{T}^4_q, \tag{2.2}$$

where $\mathbf{T}^4_q$ has eight real dimensions, i.e. the dimensionality of $\mathbf{T}^4_{\mathbb{C}}$ is not reduced. We add that the inhomogeneous $\mathfrak{o}^*(4)$ algebra does not play known significant role in the description of physically relevant $D=4$ geometries.

The most known real form of $\mathfrak{io}(4,\mathbb{C})$ is the $D=4$ Poincaré Lie algebra $p(3,1))=\mathfrak{o}(3,1)\ltimes\mathbf{T}^{1,3}$ generated by the Poincaré fourmomenta $P_\mu\in\mathbf{T}^{1,3}$ and six Lorentz rotations $L_{\mu\lambda}\in\mathfrak{o}(3,1)$ ($\mu,\nu=1,2,3,4$) and looks as follows:

$$\begin{aligned}
[L_{\mu\nu},L_{\lambda\rho}] &= i\big(g_{\nu\lambda}L_{\mu\rho}-g_{\nu\rho}L_{\mu\lambda}+g_{\mu\rho}L_{\nu\lambda}-g_{\mu\lambda}L_{\nu\rho}\big)\,, & L_{\mu\nu} &= -L_{\nu\mu}\,,\\
[L_{\mu\nu},P_\rho] &= i\big(g_{\nu\rho}P_\mu-g_{\mu\rho}P_\nu\big)\,, & [P_\mu,P_\nu] &= 0\,,
\end{aligned} \tag{2.3}$$

where $g_{\mu\nu}=\mathrm{diag}(-1,-1,-1,1)$ is the Minkowski (Lorentzian) metric. If we replace in (2.3) such a metric by the Euclidean one i.e. $g^E_{\mu\nu}=-\delta_{\mu\nu}=\mathrm{diag}(-1,-1,-1,-1)$, one gets the $D=4$ Euclidean algebra $\varepsilon(4)=\mathfrak{o}(4)\ltimes\mathbf{T}^4$, described by Euclidean generators $\mathcal{P}_\mu$, $\mathcal{L}_{\mu\nu}$:

$$\begin{aligned}
[\mathcal{L}_{\mu\nu},\mathcal{L}_{\lambda\rho}] &= -i\big(\delta_{\nu\lambda}\mathcal{L}_{\mu\rho}-\delta_{\nu\rho}\mathcal{L}_{\mu\lambda}+\delta_{\mu\rho}\mathcal{L}_{\nu\lambda}-\delta_{\mu\lambda}\mathcal{L}_{\nu\rho}\big)\,, & \mathcal{L}_{\mu\nu} &= -\mathcal{L}_{\nu\mu}\,,\\
[\mathcal{L}_{\mu\nu},\mathcal{P}_\rho] &= -i\big(\delta_{\nu\rho}\mathcal{P}_\mu-\delta_{\mu\rho}\mathcal{P}_\nu\big)\,, & [\mathcal{P}_\mu,\mathcal{P}_\nu] &= 0\,.
\end{aligned} \tag{2.4}$$

The Poincaré algebra can be obtained from Euclidean one by the following substitution ($r,s,t=1,2,3$)

$$L_{\mu\nu}=(\mathcal{L}_{rs},\ i\mathcal{L}_{0r})\quad,\qquad P_\mu=(\mathcal{P}_r,\ i\mathcal{P}_0) \tag{2.5}$$

[3] The real generators $\mathfrak{o}(3,1)$ can be described by complex–conjugated generators of $\mathfrak{o}(3;\mathbb{C})$ and $\overline{\mathfrak{o}(3;\mathbb{C})}$ (see (2.11)).

or equivalently $M_r = \mathcal{M}_r$, $N_r = -i\mathcal{N}_r$, where $M_r = \frac{1}{2}\varepsilon_{rst}L_{st}$, $\mathcal{M}_r = \frac{1}{2}\varepsilon_{rst}\mathcal{L}_{st}$ and $N_r = L_{4r}$, $\mathcal{N}_r = \mathcal{L}_{4r}$. The imaginary unit occurring in (2.5) effectively change in (2.3) the Euclidean metric $g^E_{\mu\nu}$ into the Lorentzian one $g_{\mu\nu}$.

If we choose the metric with Kleinian (neutral) signature $g^K_{\mu\nu} = \mathrm{diag}(1,-1,1,-1)$ one gets Lie algebra $i\mathfrak{o}(2,2)) = \mathfrak{o}(2,2) \ltimes \mathbf{T}^{2,2}$, which can be obtained by the following change of the Euclidean generators

$$M^K_i = i\mathcal{M}_i, \quad N^K_i = i\mathcal{N}_i \quad \text{(i=1,3)}, \quad M^K_2 = \mathcal{M}_2, \quad N^K_2 = \mathcal{N}_2, \quad P^K_\mu = (\mathcal{P}_1, i\mathcal{P}_2, \mathcal{P}_3, i\mathcal{P}_4). \tag{2.6}$$

For describing the $\mathfrak{o}^*(4)$ algebra with quaternionic structure one should split the real $\mathfrak{o}(4)$ generators as follows

$$\mathfrak{o}(4) = \mathfrak{u}(2) \oplus \frac{\mathfrak{o}(4)}{\mathfrak{u}(2)} \tag{2.7}$$

and multiply the coset generators by "i".

In order to embed three cases of inhomogeneous algebras into unified framework one can consider the complex $D = 4$ Euclidean algebra $\varepsilon(4;C)$ and introduce its three real forms (Euclidean, Poincaré and Kleinian). These real forms are introduced with the help of three non-isomorphic antilinear involutive conjugations $I_A \to I^\#_A$ ($I_A \in (\mathcal{L}_{\mu\nu}, \mathcal{P}_\mu)$), where $\# = \dagger,\ \ddagger,\ \oplus$. The following reality conditions imposed on the generators of $\mathcal{E}(4;C)$ $(r,s = 1,2,3; i = 1,3; k = 2,4)$

$$\mathcal{L}^\dagger_{rs} = \mathcal{L}_{rs}, \quad \mathcal{L}^\dagger_{4r} = -\mathcal{L}_{4r}, \qquad \mathcal{P}^\dagger_r = \mathcal{P}_r, \quad \mathcal{P}^\dagger_4 = -\mathcal{P}_4 \qquad \text{Poincaré case} \tag{2.8}$$

$$\mathcal{L}^\ddagger_{\mu\nu} = \mathcal{L}_{\mu\nu}, \qquad \mathcal{P}^\ddagger_\mu = \mathcal{P}_\mu \qquad \text{Euclidean case} \tag{2.9}$$

$$\mathcal{L}^\oplus_{13} = \mathcal{L}_{13}, \quad \mathcal{L}^\oplus_{24} = \mathcal{L}_{24}, \quad \mathcal{L}^\oplus_{ik} = -\mathcal{L}_{ik} \quad \mathcal{P}^\oplus_i = \mathcal{P}_i, \quad \mathcal{P}^\oplus_k = -\mathcal{P}_k \qquad \text{Kleinian case} \tag{2.10}$$

define respectively real $D = 4$ Poincaré, Euclidean and Kleinian algebras. We observe that first two real structures (conjugations) coincide on $\mathcal{E}(3)$ subalgebra.

Another convenient basis in complex $\mathcal{E}(4;\mathbb{C})$ algebra is obtained by introducing the pair of chiral (left-handed) and anti-chiral (right-handed) generators: $2M^\pm_r = \frac{1}{2}\varepsilon_{rst}\mathcal{L}_{st} \pm \mathcal{L}_{4r} \equiv \mathcal{M}_r \pm i\mathcal{N}_r$ describing two complex commuting $\mathfrak{o}_\pm(3,C) \equiv sl_\pm(2,C)$ subalgebras

$$[M^\pm_r, M^\pm_s] = i\varepsilon_{ijk}M^\pm_k, \qquad [M^\pm_r, M^\mp_s] = 0, \tag{2.11}$$

$$[M^\pm_r, \mathcal{P}_s] = \frac{i}{2}(\varepsilon_{rst}\mathcal{P}_t \mp \delta_{rs}\mathcal{P}_4), \qquad [M^\pm_r, \mathcal{P}_4] = \pm\frac{i}{2}\mathcal{P}_r. \tag{2.12}$$

The reality conditions (2.8–2.10) imposed on the complex generators $M^\pm$ look as follows (we provide also the fourth reality condition related with quaternionic structure; $r = 1,2,3; i = 1,3$)

$$(M^\pm_r)^\dagger = M^\mp_r \qquad \text{Poincaré case} \tag{2.13}$$

$$(M^\pm_r)^\ddagger = M^\pm_r \qquad \text{Euclidean case} \tag{2.14}$$

$$(M^\pm_i)^\oplus = -M^\pm_i, \qquad (M^\pm_2)^\oplus = M^\pm_2 \qquad \text{Kleinian case} \tag{2.15}$$

$$(M^+_r)^{\tilde\oplus} = M^+_r, \qquad (M^-_i)^{\tilde\oplus} = -M^-_i, \quad (M^-_2)^{\tilde\oplus} = M^-_2 \qquad \text{quaternionic case} \tag{2.16}$$

Further we shall consider the supersymmetric N=1,2 extensions of the Euclidean and Poincaré real forms (2.13–2.15). The complexifications of real Poincaré, Euclidean and Kleinian algebras are equvalent, and one can consider as well in place of reality constraints (2.8)–(2.10) the real forms of complexified Poincaré or Kleinian algebras in order to provide the real Poincaré, Euclidean and Kleinian algebras.

It is quite useful to work further with Lorentzian (Poincaré) canonical basis which is obtained after realification of the Cartan-Chevaley basis of $\mathfrak{sl}(2,\mathbb{C})$ [4]. In such a basis Lorentz generators are defined as follows (see [13, 19])

$$\begin{aligned} h &= -iN_3\,, & e_\pm &= -i(N_1 \mp M_2)\,, \\ h' &= iM_3\,, & e'_\pm &= i(M_1 \pm N_2)\,. \end{aligned} \tag{2.17}$$

One obtains the following description of $D=4$ Lorentz algebra

$$\begin{aligned} [h,e_\pm] &= \pm e_\pm, & [e_+,e_-] &= 2h, \\ [h,e'_\pm] &= \pm e'_\pm\,, & [h',e_\pm] &= \pm e'_\pm, & [e_\pm,e'_\mp] &= \pm 2h', \\ [h',e'_\pm] &= \mp e_\pm, & [e'_+,e'_-] &= -2h. \end{aligned} \tag{2.18}$$

The fourmomenta generators with the components

$$P_\mu = (P_1,\ P_2,\ P_\pm = P_4 \pm P_3) \tag{2.19}$$

extend (2.18) to the real Poincaré algebra as follows

$$\begin{aligned} [h,P_\pm] &= \pm P_\pm, & [h,P_i] &= 0 \quad (i=1,2), \\ [e_\pm,P_\pm] &= 0, & [e_\pm,P_\mp] &= 2P_1, & [e_\pm,P_1] &= P_\pm, & [e_\pm,P_2] &= 0, \end{aligned} \tag{2.20}$$

$$\begin{aligned} [h',P_\pm] &= 0, & [h',P_1] &= -P_2, & [h',P_2] &= P_1, \\ [e'_\pm,P_\pm] &= 0, & [e'_\pm,P_\mp] &= \mp 2P_2, & [e'_\pm,P_1] &= 0, & [e'_\pm,P_2] &= \mp P_\pm\,. \end{aligned} \tag{2.21}$$

Three reality conditions imposed on these canonical generators are now

$$x_A^\dagger = -x_A \quad \text{for } x_A \in (h,\, e_\pm,\, h',\, e'_\pm,\, P_\pm,\, P_1,\, P_2) \quad (\text{ Poincaré case }) \tag{2.22}$$

$$\begin{aligned} h^\ddagger &= h, \quad e_\pm^\ddagger = e_\mp, \quad h'^\ddagger = -h', \quad e'^\ddagger_\pm = -e'_\mp, \\ P_\pm^\ddagger &= -P_\mp, \quad P_i^\ddagger = P_i \quad (i=1,2) \qquad (\text{ Euclidean case }) \end{aligned} \tag{2.23}$$

$$\begin{aligned} h^\oplus &= -h, \quad e_\pm^\oplus = -e_\pm, \quad h'^\oplus = h', \quad e'^\oplus_\pm = e'_\pm, \\ P_\pm^\oplus &= -P_\pm, \quad \tilde{P}_\pm^\oplus = -\tilde{P}_\pm^\oplus, \quad \tilde{P}_\pm = P_1 \pm iP_2 \qquad (\text{ Kleinian case }) \end{aligned} \tag{2.24}$$

2.2. *Complex $D=4$ $N=1$ Euclidean superalgebra*

In this paper we shall consider the superalgebra generators in purely algebraic way, without reference to concrete realizations. In this subsection we shall recall the complex $D=4$ $N=1$ Euclidean superalgebra (see e.g. [31, 32, 18]) describing simple ($N=1$) supersymmetrization of $\mathfrak{e}(4;\mathbb{C})$ (inhomogeneous $\mathfrak{o}(4;\mathbb{C})$) complex algebra. Such superalgebra is obtained by adding to the generator of complex $D=4$ Euclidean algebra $\mathfrak{e}(4;C) = \mathfrak{o}(4;C) \ltimes \mathbf{T}_4^C$ four independent complex supercharges $Q_\alpha, \bar{Q}_{\dot\alpha}$ transforming

[4] Let $(h,e_\pm)$ be the Cartan-Weyl basis of $\mathfrak{sl}(2,\mathbb{C})$ with the commutation relations in the first line of (2.18). Setting $h' := ih, e'_\pm := ie_\pm$ we obtain all commutation relations (2.18). The real Lie algebra $\mathfrak{sl}(2,\mathbb{C})_\mathbb{R}$ generated by the elements $(h,e_\pm,h',e'_\pm)$ with the defining relations (2.17) is called realification of $\mathfrak{sl}(2,\mathbb{C})$.

as fundamental representations under "left" and "right" internal symmetry groups $SL_{+}(2;C)$ and $SL_{-}(2;C)$ [5]. The supercharges $Q_\alpha, Q_{\dot\alpha}$ extend the $D=4$ complex Euclidean algebra (2.4) by the following algebraic relations

$$\left\{Q_\alpha, \bar{Q}_{\dot\beta}\right\} = 2(\sigma^E_\mu)_{\alpha\dot\beta}\mathcal{P}^\mu \qquad \left\{Q_\alpha, Q_\beta\right\} = \left\{\bar{Q}_{\dot\alpha}, \bar{Q}_{\dot\beta}\right\} = 0 \tag{2.25}$$

$$\left[\mathcal{L}_{\mu\nu}, Q_\alpha\right] = -(\sigma^E_{\mu\nu})_\alpha{}^\beta Q_\beta \qquad \left[\mathcal{L}_{\mu\nu}, \bar{Q}_{\dot\alpha}\right] = \bar{Q}_{\dot\beta}(\tilde\sigma^E_{\mu\nu})^{\dot\beta}{}_{\dot\alpha} \tag{2.26}$$

$$\left[\mathcal{P}_\mu, Q_\alpha\right] = \left[\mathcal{P}_\mu, \bar{Q}_{\dot\alpha}\right] = 0, \tag{2.27}$$

where the Euclidean sigma matrices σ^E_μ are expressed by standard Pauli matrices $\sigma_r (r = 1,2,3)$ as follows

$$(\sigma^E_\mu)_{\alpha\dot\beta} = ((\sigma_r)_{\alpha\dot\beta}, i(I_2)_{\alpha\dot\beta}) \qquad r = 1,2,3 \tag{2.28}$$

and satisfy the known reality conditions under Hermitean matrix conjugation

$$(\sigma^E_r)^+ = \sigma^E_r \qquad (\sigma^E_0)^+ = -\sigma_0. \tag{2.29}$$

The matrices $\sigma^E_{\mu\nu}$ and $\tilde\sigma^E_{\mu\nu}$ describe the following realizations of the pair of commuting $sl_\pm(2;C)$ algebras

$$\begin{aligned} sl_+(2;C): \qquad (\sigma^E_{\mu\nu})_\alpha{}^\beta &= (\sigma_{rs}{}^E = \frac{1}{2}\epsilon_{rst}\sigma_t, \ \sigma_{0r}{}^E = \frac{1}{2}\sigma_r), \\ sl_-(2;C): \qquad (\tilde\sigma^E_{\mu\nu})_{\dot\alpha}{}^{\dot\beta} &= (\tilde\sigma_{rs}{}^E = \frac{1}{2}\epsilon_{rst}\sigma_t, \ \tilde\sigma^E_{0r} = -\frac{1}{2}\sigma_r). \end{aligned} \tag{2.30}$$

The complexified Euclidean fourmomenta $\mathcal{P}_{\alpha\dot\beta} \equiv (\sigma_\mu{}^E)_{\alpha\dot\beta}\mathcal{P}^\mu$ transform under the complex Euclidean rotations $A^\mu_\nu \in O(4;C)$ described by the product of two commuting Lorentz groups $SL_\pm(2;C)$ as follows ($S_\alpha{}^\beta \in SL_+(2;C)$, $\tilde{S}_{\dot\alpha}{}^{\dot\beta} \in S_-(2;C)$)

$$\mathcal{P}'_{\alpha\dot\beta} = S_\alpha{}^\gamma \tilde{S}_{\dot\beta}{}^{\dot\delta}\mathcal{P}_{\gamma\dot\delta} \quad \longleftrightarrow \quad \mathcal{P}'_\mu = A^\nu_\mu \mathcal{P}_\nu \tag{2.31}$$

where $\mathcal{P}_\mu = \frac{1}{2}\sigma^{\alpha\dot\beta}_\mu \mathcal{P}_{\alpha\dot\beta}$ and $A^\mu_\nu = \sigma^{\alpha\dot\beta}_\nu \sigma^\mu_{\gamma\dot\delta} S^\gamma_\alpha \tilde{S}^{\dot\delta}_{\dot\beta}$.

Finally one can check that the relations (2.25–2.27) are invariant under the complex rescaling transformation (c is a complex number)

$$Q'_\alpha = cQ_\alpha \qquad \bar{Q}'_{\dot\alpha} = c^{-1}\bar{Q}_{\dot\alpha}. \tag{2.32}$$

The rescaling (2.32) is described by $GL(1;C) = U(1) \times R$ Abelian group and represents the one-parameter complex internal symmetries of $D=4$ simple complex Euclidean superalgebra. The N=1 internal $gl(1;C)$ generator T satisfies the algebraic relation

$$[T, Q_\alpha] = Q_\alpha \qquad [T, \bar{Q}_{\dot\alpha}] = -\bar{Q}_{\dot\alpha}. \tag{2.33}$$

[5] We recall that $O(4;C) = O_+(3;C) \otimes O_-(3;C)$ has a spinorial covering $SL_+(2;C) \otimes SL_-(2;C)$. The $SL_+(2;C)$ spinors have undotted spinorial indices, and dotted indices characterize $SL_-(2;C)$ spinors. These two groups are also called chiral (left) and antichiral (right) projections of $o(4;C)$ group.

2.3. Real and pseudoreal forms of $D = 4$ $N = 1$ complex Euclidean superalgebra.

2.3.1. $D = 4$ $N = 1$ pseudoreal Euclidean superalgebra and its pseudoreal forms. The reality conditions (2.9) defining $D = 4$ Euclidean space-time algebra $\mathfrak{o}(4;\mathbb{R}) \ltimes \mathbf{T}_4$ can be extended to the sector of Euclidean spinorial supercharges if we use the corresponding spinorial covering group $\overline{O(4;\mathbb{R})} = SU_+(2) \oplus SU_-(2)$ which requires a pair of independent two-component complex $SU(2)$ spinors. This property of doubling of $D = 4$ Euclidean spinorial components in comparison with standard relativistic $D = 4$ case leads to the known conclusion that contrary to the Poincaré case the four-component real (Majorana) Euclidean spinors do not exist. In the algebraic framework one can however extend in odd supercharges sector the conjugation (2.9) as the pseudoconjugation (see Appendix A, (A.3b)) which should be consistent with $N = 1$ complex Euclidean algebra (2.25–2.27) under the assumption that the generators $\mathcal{P}_\mu$, $\mathcal{L}_{\mu\nu}$ are real Euclidean, i.e. satisfy the reality conditions (2.9). The $N = 1$ pseudoconjugation of Euclidean supercharges is an involution of fourth order (see also (A.3b)) satisfying the relation

$$(Q^\ddagger_\alpha)^\ddagger = -Q_\alpha, \qquad (\bar{Q}^\ddagger_{\dot\alpha})^\ddagger = -\bar{Q}_{\dot\alpha} \tag{2.34}$$

and it look as follows [24, 22, 25]

$$Q^\ddagger_\alpha = \varepsilon_{\alpha\beta} Q_\beta, \qquad \bar{Q}^\ddagger_{\dot\alpha} = \eta \varepsilon_{\dot\alpha\dot\beta} \bar{Q}_{\dot\beta}, \qquad \eta = \pm 1. \tag{2.35}$$

It can be shown that the map $Q_\alpha \to -\varepsilon_{\alpha\beta} Q^\ddagger_\beta, \bar{Q}_{\dot\alpha} \to -\eta\varepsilon_{\dot\alpha\dot\beta} Q^\ddagger_{\dot\beta}$ leaves the superalgebra (2.25–2.27) invariant provided that we choose the parameter q (see Appendix, A (A.1)) in accordance with the relation

$$\eta = (-1)^{q+1}. \tag{2.36}$$

We obtain the following two cases:

i) $\eta = -1, q = 0$ (standard choice)

In this case the product of odd (fermionic operators f_1, f_2 which are odd powers of supercharges are conjugated as follows

$$(f_1 f_2)^\ddagger = f_2^\ddagger f_1^\ddagger \tag{2.37}$$

and leads to standard supersymmetry scheme with pseudoconjugation $\ddagger$ which can be represented on complex Grassmannian variables $\theta_\alpha = \theta^1_\alpha + i\theta^2_\alpha$ as the complex conjugation $\theta_\alpha \to \theta^\star_\alpha = \theta^1_\alpha - i\theta^2_\alpha$. It should be added that all applications of pseudoconjugations to the description of real $N = 1$ Euclidean SUSY use the case *i*) (see e.g. [23]).

ii) $\eta = 1, q = 1$ (exotic choice)

In this case the product of odd (fermionic) operators f_1, f_2 which are odd powers of supercharges and Grassmann variables are conjugated as follows

$$(f_1 f_2)^\ddagger = -f_2^\ddagger f_1^\ddagger. \tag{2.38}$$

Such choice leads to nonstandard supersymmetry scheme, which can be realized in complex superspace described by odd Grassmann variables $\tilde{\theta}_A$, $\bar{\tilde{\theta}}_A$ with the products transformating under conjugation in the nonstandard way

$$(\tilde{\theta}_A \tilde{\theta}_B)^\ddagger = -\tilde{\theta}^\ddagger_B \tilde{\theta}^\ddagger_A, \qquad (\bar{\tilde{\theta}}_A \bar{\tilde{\theta}}_B)^\ddagger = -(\bar{\tilde{\theta}})^\ddagger_B (\bar{\tilde{\theta}})^\ddagger_A. \tag{2.39}$$

Such type of Grassmann variables was considered as mathematically consistent choice in [26, 27, 28] but it has not been applied in the literature to describe the physical supersymmetric systems, therefore exotic.

In the exotic case the conjugation $A \mapsto A^\ddagger$ (antilinear antiautomorphism of second order) in bosonic sector of the Euclidean superalgebra defining the reality condition (2.9) is lifted in fermionic sector of supercharges to the antilinear antiautomorphism of fourth order defining pseudoconjugation.

One can show that for both values $\eta = \pm 1$ in the scaling transformations (2.32) the parameter c should be real, what means that the invariance of (2.33) under the pseudoconjugation (2.35) is valid if $T^{\ddagger} = -T$, i.e. $N = 1$ internal symmetry $GL(1;\mathbb{C}) = U(1) \oplus O(1,1)$ is reduced to $GL(1;\mathbb{R}) = O(1,1)$ for pseudoreal $N = 1$ Euclidean superalgebra.

2.3.2. $D = 4$ $N = 1$ real Poincaré superalgebra $N = 1$ Poincaré superalgebra is obtained from the relations (2.25–2.27) after imposing the following extension of the Poincaré reality condition (2.8) to the supercharges sector

$$(Q_\alpha)^\dagger = \bar{Q}_{\dot{\alpha}}, \qquad (\bar{Q}_{\dot{\beta}})^\dagger = \eta Q_\beta \qquad \eta = \pm 1 . \tag{2.40}$$

If $\eta = 1$, $q = 0$ the formula (2.40) describe conjugation and lead to standard $N = 1$ real Poincaré supersymmetry. If $\eta = -1$ we get $q = 1$ and relations (2.40) describe pseudoconjugation (pseudoreality condition) defining exotic supersymmetry with Grassmann variables satisfying relations (2.38).

After introducing $P_i = \mathcal{P}_i$, $P_0 = i\mathcal{P}_0$ (see 2.5) one gets from (2.25–2.27) the well-known N=1 Poincaré superalgebra with the pair of supercharges represented by two-component complex Weyl spinors

$$\begin{aligned}
\{Q_\alpha, \bar{Q}_{\dot{\beta}}\} &= 2(\sigma_\mu)_{\alpha\dot{\beta}} P^\mu, & [P_\mu, Q^a_\alpha] &= [P_\mu, \bar{Q}_{\dot{\alpha}a}] = 0, \\
\{Q_\alpha, Q_\beta\} &= 0, & \{\bar{Q}_{\dot{\alpha}}, \bar{Q}_{\dot{\beta}}\} &= 0, \\
[L_{\mu\nu}, Q_\alpha] &= -(\sigma_{\mu\nu})_\alpha{}^\beta Q^a_\beta, & [L_{\mu\nu}, \bar{Q}_{\dot{\alpha}a}] &= \bar{Q}_{\dot{\beta}a}(\tilde{\sigma}_{\mu\nu})^{\dot{\beta}}_{\dot{\alpha}},
\end{aligned} \tag{2.41}$$

where the Lorentzian σ–matrices are defined as follows

$$\begin{aligned}
\sigma_\mu &= (\sigma_i = \sigma^E_i, \ \sigma_4 = i\sigma^E_4), \\
\sigma_{\mu\nu} &= (\sigma_{ij} = \sigma^E_{ij}, \ \sigma_{4i} = -i\sigma^E_{4i})], \\
\tilde{\sigma}_{\mu\nu} &= (\tilde{\sigma}_{ij} = \tilde{\sigma}^E_{ij}, \ \tilde{\sigma}_{4i} = -i\tilde{\sigma}^E_{4i}).
\end{aligned} \tag{2.42}$$

It is easy to see that for $\eta = \pm 1$ the pair of relations (2.33) is invariant under the conjugation (2.40) if $T^\dagger = T$, i.e. the internal symmetry of $N = 1$ Poincaré supersymmetry is described by the restriction of $GL(1;\mathbb{C})$ to $U(1)$.

2.3.3. $D = 4$ $N = 1$ real Kleinian superalgebra The $N = 1$ Kleinian $\mathfrak{o}(2,2)$ superalgebra is obtained after the following extension of the reality condition (2.10) to the supercharges sector

$$(Q_\alpha)^\oplus = Q_\alpha, \qquad (\bar{Q}_{\dot{\beta}})^\oplus = \eta \bar{Q}_{\dot{\beta}} \qquad \eta = \pm 1 , \tag{2.43}$$

i.e. the supercharges (2.43) form a pair of respectively $\mathfrak{o}_+(2,1) = \mathfrak{sl}_+(2;\mathbb{R})$ and $\mathfrak{o}_-(2,1) = \mathfrak{sl}_-(2;\mathbb{R})$ real spinors. Imposing on complex superalgebra (2.25–2.27) the reality condition (2.43) one gets the following real superalgebra

$$\{Q_\alpha, \bar{Q}_{\dot{\beta}}\} = 2(\tilde{\sigma}_\mu)_{\alpha\dot{\beta}} \tilde{P}^\mu , \tag{2.44}$$

where $\tilde{P}_\mu \tilde{P}^\mu = \tilde{P}_1^2 - \tilde{P}_2^2 + \tilde{P}_3^2 - \tilde{P}_4^2$ and

$$\tilde{P}_\mu = (\mathcal{P}_1, i\mathcal{P}_2, \mathcal{P}_3, i\mathcal{P}_4) \tag{2.45}$$

with $\tilde{\sigma}_\mu$ describing $\mathfrak{o}(2,2)$ real σ–matrices

$$\tilde{\sigma}_\mu = (\sigma_1, i\sigma_2, \sigma_3, -I_2) . \tag{2.46}$$

The inhomogeneous $\mathfrak{o}(2,2)$ algebra can be described more conveniently if we use the formulation of complex Euclidean algebra $\varepsilon(4;\mathbb{C})$ using formulae (2.11–2.12). In Kleinian case the "complex" generators $M_r^{\pm}$ become the set of real generators (i.e. generating real Lie superalgebra with real structure constants)

$$\begin{aligned} [M_r^{(+)}, Q^a_\alpha] &= -\tfrac{1}{2}(\tilde{\sigma}_r)_\alpha{}^\beta Q^a_\beta, & [M_r^{(-)}, Q^a_\alpha] &= 0, \\ [M_r^{(+)}, \bar{Q}_{\dot\alpha a}] &= 0, & [M_r^{(-)}, \bar{Q}_{\dot\alpha a}] &= \tfrac{1}{2}\bar{Q}_{\dot\beta a}(\tilde{\sigma}_r)^{\dot\beta}{}_{\dot\alpha}. \end{aligned} \tag{2.47}$$

We add that similarly as in Poincaré case, the choice $\eta = -1$, $q = 0$ leads to standard $N = 1$ Kleinian supersymmetry with (2.43) describing conjugation $\oplus$ (reality conditions), and if $\eta = 1$, $q = 1$ one gets (see also Sec. 2.3.1) exotic supersymmetry with pseudoconjugation implying the exotic antiautomorphism which leads to relations (2.38). Finally for both values of η the relations (2.33) are consistent with the reality condition (2.43) if $T^{\oplus} = -T$, i.e. real $N = 1$ Kleinian symmetry is endowed with $GL(1;\mathbb{R}) \equiv O(1,1)$ internal symmetry.

3. Complex $D = 4$ $N = 2$ Euclidean supersalgebra and its Euclidean, Poincaré and Kleinian (pseudo) real forms

3.1. Complex $D = 4$ $N = 2$ Euclidean superalgebra

In this subsection we shall describe $N = 2$ superization of complex inhomogeneous $\mathfrak{o}(4;\mathbb{C})$ algebra. Basic N=1 relations (2.25) are extended to $N = 2$ as follows $(a, b.. = 1, 2)$:

$$\left\{Q^a_\alpha, \bar{Q}_{\dot\beta b}\right\} = 2\delta^a_b(\sigma^E_\mu)_{\alpha\dot\beta}\mathcal{P}^\mu, \tag{3.1}$$

$$\left\{Q^a_\alpha, Q^b_\beta\right\} = \varepsilon_{\alpha\beta}\varepsilon^{ab}Z, \qquad \left\{\bar{Q}_{\dot\alpha a}, \bar{Q}_{\dot\beta b}\right\} = \varepsilon_{\dot\alpha\dot\beta}\varepsilon_{ab}\tilde{Z}, \tag{3.2}$$

where $Z, \tilde{Z}$ describe a pair of complex scalar central charges. The relations (3.1) are invariant under the internal symmetries $A \in GL(2,C)$, where

$$Q^a{}_\alpha \to (A)^a{}_b Q^b_\alpha \qquad \bar{Q}_{\dot\alpha a} \to \bar{Q}_{\dot\alpha b}(A^{-1})^b{}_a. \tag{3.3}$$

The presence of central charges breaks only internal symmetry $GL(2;\mathbb{C})$ to $SL(2;\mathbb{C})$, what follows from the relation $A\varepsilon A^T = \varepsilon$ valid for complex 2×2 matrices $A \in SL(2;\mathbb{C})$ due to the equivalence $SL(2;\mathbb{C}) \sim Sp(2;\mathbb{C})$. The $\mathfrak{gl}(2;\mathbb{C})$ generators $T_i^{\ j}$ $(i, j = 1, 2)$

$$[T_i^{\ j}, T_k^{\ l}] = i(\delta^j{}_k T_i^{\ l} - \delta^l{}_i T_c^{\ k}) \tag{3.4}$$

are restricted to $\mathfrak{sl}(2;\mathbb{C})$ by the condition $T_0 \equiv T_1{}^1 + T_2{}^2 = 0$.

One can describe $N = 2$ internal symmetries algebra $\mathfrak{gl}(2;\mathbb{C})$ in convenient way by replacing four generators $T_i^{\ j}$ by generators $T_A (A = 0, 1, 2, 3)$ adjusted to further description of $U(2)$ internal symmetry

$$T_A = \frac{i}{2}(\sigma_A)^i_j T_i^{\ j}, \tag{3.5}$$

where $\sigma_A = (\sigma_r, \sigma_0 = I_2)$; σ_r are three $2x2$ Hermitean Pauli matrices. The $\mathfrak{gl}(2;\mathbb{C})$ covariance relation of $N = 2$ supercharges looks as follows

$$[T_A, Q^a_\alpha] = -\frac{1}{2}(\sigma_A)^a{}_b Q^b_\alpha, \qquad [T_A, \bar{Q}_{\dot\alpha a}] = \frac{1}{2}\bar{Q}_{\dot\alpha b}(\sigma_A)^b{}_a, \tag{3.6}$$

where the $2x2$ matrices $(t_A)^b_a = \frac{1}{2}(\sigma_A)^b_a$ provide the fundamental realizations of $\mathfrak{gl}(2, C)$ generators (3.5)[6]. Using known relation $(r, s, t = 1, 2, 3)$

$$\sigma_r\sigma_s = \delta_{rs} + i\varepsilon_{rst}\sigma_t \tag{3.7}$$

[6] It corresponds to the realization $(T_i^{\ j})^b_a = i\delta_i^b\delta^j_a$ of the generators in the relations (3.4).

one gets from (3.4) and (3.5) the $N = 2$ internal symmetry algebra $\mathfrak{gl}(2)$

$$[T_r, T_s] = i\epsilon_{rst}T_t \qquad [T_0, T_r] = 0. \tag{3.8}$$

Three generators T_r span $\mathfrak{sl}(2;\mathbb{C})$ algebra which is preserved even in the case of nonvanishing central charges $Z, \tilde{Z}$; the Abelian generator T_0 describing coset $\frac{GL(2;\mathbb{C})}{SL(2;\mathbb{C})}$ is broken in the presence of central charges.

If we consider the chiral projections of Euclidean $\mathfrak{o}(4;\mathbb{C})$ generators (see (2.11)–(2.12)) one gets from (3.6) the commutators exposing the chiral (antichiral) nature of supercharges Q^i_α ($\bar{Q}_{\dot{\alpha}i}$)

$$\begin{aligned} [M^{(+)}_r, Q^a_\alpha] &= -\tfrac{1}{2}(\sigma_r)_\alpha{}^\beta Q^a_\beta, & [M^{(-)}_r, Q^a_\alpha] &= 0, \\ [M^{(+)}_r, \bar{Q}_{\dot{\alpha}a}] &= 0, & [M^{(-)}_r, \bar{Q}_{\dot{\alpha}a}] &= \tfrac{1}{2}\bar{Q}_{\dot{\beta}a}(\sigma_r)^{\dot{\beta}}{}_{\dot{\alpha}}. \end{aligned} \tag{3.9}$$

The vanishing commutators in (3.9) illustrate that the supercharges Q^a_α are left-handed (chiral) and the supercharges $\bar{Q}_{\dot{\alpha}a}$ are right-handed (antichiral).

3.2. $N = 2$ *real Poincaré superalgebras with central charges*

The reality conditions for supercharges can take the form (see e.g. [33, 34]):

$$(Q^a_\alpha)^\dagger = \bar{Q}_{\dot{\alpha}a}, \qquad (\bar{Q}_{\dot{\beta}b})^\dagger = \eta\, Q^b_\beta \qquad \eta = \pm 1. \tag{3.10}$$

The reality condition for well-known $N = 2$ real Poincaré superalgebra with central charges [33] is obtained if we put $q = 0$ (see (A.1)) and $\eta = 1$. In such a case we impose on the complex generators $\{\mathcal{L}_{\mu\nu}, \mathcal{P}_\mu, Q^i_\alpha, \bar{Q}_{\dot{\beta}j}, T_r = \frac{1}{2}(\sigma_r)^i{}_j T_i{}^j, Z_1, Z_2; i = 1,2; r = 1,2,3\}$ of centrally extended $N = 2$ complex Euclidean superalgebra the reality constraints which extend consistently the conjugation (2.4) in bosonic sector to odd superalgebra generators.

In particular in the representation which permits the Hermitean conjugation of supercharges the conjugation (3.10) can be seen as Hermitean conjugation. Further the reality constraints on the internal symmetry generators T_A (see (3.5) and (3.8)) and the central charges (Z_1, Z_2), which are consistent with the relations (3.1–3.2), (3.6) and (3.8) are the following

$$T_r^\dagger = T_r, \qquad T_0^\dagger = -T_0, \qquad Z_1^\dagger = Z_2, \qquad Z_2^\dagger = Z_1, \tag{3.11}$$

where the generator T_0 describes internal symmetry only in the case when central charges vanish. If the central charges are not vanishing from first set of the relations (3.11) one can see that the algebra $\mathfrak{sl}(2;\mathbb{C})$ is constrained to $\mathfrak{su}(2)$ algebra. If we use the formulae (2.42) for Minkowskian σ-matrices, it follows from (2.5) and (2.28) that $\sigma_\mu P^\mu = \sigma^E_\mu P^{E\mu}$ and we get the following real $N = 2$ Poincaré superalgebra with one complex central charge Z:

$$\begin{aligned} \{Q^a_\alpha, \bar{Q}_{\dot{\beta}b}\} &= 2\delta^a_b(\sigma_\mu)_{\alpha\dot{\beta}}P^\mu, \\ \{Q^a_\alpha, Q^b_\beta\} &= \epsilon^{ab}\epsilon_{\alpha\beta}Z, & \{\bar{Q}_{\dot{\alpha}a}, \bar{Q}_{\dot{\beta}b}\} &= \epsilon_{ab}\epsilon_{\dot{\alpha}\dot{\beta}}\bar{Z}, \\ [L_{\mu\nu}, Q^a_\alpha] &= -(\sigma_{\mu\nu})_\alpha{}^\beta Q^a_\beta, & [L_{\mu\nu}, \bar{Q}_{\dot{\alpha}a}] &= \bar{Q}_{\dot{\beta}a}(\tilde{\sigma}_{\mu\nu})^{\dot{\beta}}_{\dot{\alpha}}, \\ [T_r, Q^a_\alpha] &= -(t_r)^a{}_b Q^b_\alpha, & [T_r, \bar{Q}_{\dot{\alpha}a}] &= \bar{Q}_{\dot{\alpha}b}(t_r)^b{}_a, \\ [P_\mu, Q^a_\alpha] &= [P_\mu, \bar{Q}_{\dot{\alpha}a}] = 0, & [T_r, T_s] &= i\epsilon_{rsm}T_m. \end{aligned} \tag{3.12}$$

If $Z \neq 0$ the $N = 2$ internal symmetries (R-symmetries) are described by $\mathfrak{su}(2)$ algebra, with the fundamental 2×2 matrix realizations (see (2.25) and (3.12) (the fourth line)) described by Hermitean

Pauli matrices ($t_r \equiv \sigma_r$). If $Z = 0$ in relations (3.12) one can add fourth R-symmetry generator T_0 describing the extension of $\mathfrak{su}(2)$ to $\mathfrak{u}(2)$. In such a case the complex $GL(1;\mathbb{C})$ rescaling in complex N=2 Euclidean superalgebra generated by T_0 (see (3.5) and (3.8)); is restricted to $U(1)$ phase transformations

$$Q_\alpha^{l\,\prime} = (\exp^{i\xi})Q_\alpha^l, \qquad \bar{Q}'_{\dot\beta i} = (\exp^{-i\xi})\bar{Q}_{\dot\beta i}. \tag{3.13}$$

If $\eta = -1$ the map (3.10) describes a pseudoconjugation, but requires the exotic version ($q = 1$) of formula (A.1), i.e. we get the following relation between η in (3.10) and parameters q (see (A.1))

$$\eta = (-1)^q. \tag{3.14}$$

Further it can be shown that the antiautomorphism (3.10) leads to the following reality constraints on central supercharges $Z, \tilde{Z}$

$$Z^\dagger = -\eta\tilde{Z}, \qquad \tilde{Z}^\dagger = -\eta Z. \tag{3.15}$$

The antiautomorphism of complex $N = 2$ relations (see (3.1)–(3.8)) leads for both cases $\eta = \pm 1$ to the same restriction of internal $gl(2,C)$ to its subgroup: $SU(2)$ in the presence of central charge ($Z \neq 0$) and $U(2)$ if $Z = 0$. We see therefore that we obtain the same internal symmetry sectors for the standard conjugation ($\eta = 1$) and nonstandard pseudoconjugations ($\eta = -1$).

3.3. $N = 2$ pseudoconjugations and corresponding selfconjugate $N = 2$ Euclidean superalgebras

We shall consider now the $N = 2$ complex Euclidean superalgebra with the supercharges Q_α^a and $\bar{Q}_{\dot\alpha a}$ (see Sect. 3.1). The pseudoconjugation map (2.35) for $N = 1$ superalgebra can be extended to $N = 2$ as follows (see also [23, 24, 25]):

$$(Q_\alpha^a)^\ddagger = \varepsilon_{\alpha\beta}Q_\beta^a, \qquad (\bar{Q}_{\dot\alpha a})^\ddagger = \eta\varepsilon_{\dot\alpha\dot\beta}\bar{Q}_{\dot\beta a}, \qquad \eta = \pm 1. \tag{3.16}$$

In order to obtain the antiautomorphism of $N = 2$ complex superalgebra (see Sec. 3.1) we should postulate the relation (2.36) implying that $q = 0$ (see (A.1)) for $\eta = -1$ and $q = 1$ for $\eta = 1$. We get therefore two prolongations to fermionic sector of the Euclidean conjugation (2.8). The invariance of the relation (3.1–3.2) under the pseudoconjugation maps (3.16) implies the following reality conditions for central charges $Z, \tilde{Z}$

$$Z^\ddagger = (-1)^{q+1}Z, \qquad \tilde{Z}^\ddagger = (-1)^{q+1}\tilde{Z} \tag{3.17}$$

i.e. we obtain two real ($q = 1$) or imaginary ($q = 0$) central charges.

It is easy to show that the $N = 2$ supersymmetrization of complex $\mathfrak{io}(4;C)$ algebra becomes selfconjugate under pseudoconjugation (3.16) under the assumption that the Euclidean fourmomentum generators $\mathcal{P}_\mu$ are real (see (2.9)). Further, using the relations $\varepsilon\bar{\sigma}_i\varepsilon = \sigma_i$, we should choose for $\mathfrak{o}(4,C)$ generators the Euclidean reality conditions. The invariance of superalgebra under the pseudoconjugation (3.16) requires that (see (3.12))

$$T_i^j = (T_i^j)^\dagger \qquad (t_i^j)^d{}_c = -(\bar{t}_i^j)^d{}_c, \tag{3.18}$$

where $t \mapsto \bar{t}$ is the complex conjugation of 2×2 matrix elements describing fundamental realizations of $\mathfrak{gl}(2;\mathbb{C})$ generators T_i^j. If central charges are absent the reality constraints (3.18) restrict the complex internal $\mathfrak{gl}(2;\mathbb{C})$ algebra to its real form $\mathfrak{gl}(2;\mathbb{R})$. The presence of scalar central charges $Z, \tilde{Z}$ commuting with Euclidean $\mathfrak{o}(4)$ and internal symmetry generators reduces the internal symmetries $GL(2;R)$ to its subgroup $SL(2;R)$; i.e. the $\mathfrak{sl}(2;\mathbb{R})$ subalgebra of $\mathfrak{gl}(2;\mathbb{R})$ remains not broken for any value of central charges (3.17) and describes N=2 Euclidean R-symmetry.

3.4. $N=2$ conjugation and corresponding $N=2$ real Euclidean superalgebra

If $N=2$ it is possible to introduce as well the following conjugation

$$(Q^a_\alpha)^\dagger = \varepsilon_{\alpha\beta}\varepsilon^{ab}Q^b_\beta, \qquad (\bar{Q}_{\dot\alpha a})^\dagger = \eta\,\varepsilon_{\dot\alpha\dot\beta}\varepsilon_{ab}\bar{Q}_{\dot\beta b}, \qquad \eta=\pm1 \tag{3.19}$$

with involutive property

$$((Q^a_\alpha)^\dagger)^\dagger = Q^a_\alpha, \qquad ((\bar{Q}_{\dot\alpha a})^\dagger)^\dagger = \bar{Q}_{\dot\alpha a}, \tag{3.20}$$

which is an antilinear antiautomorphism describing the superextension of conjugation (2.9). It can be checked that (3.19) describes conjugation if $\eta=1$ and pseudoconjugation if $\eta=-1$. Due to the relation if $\eta=(-1)^q$ if $\eta=1$ (3.19) provides standard antiautomorphism of complex $N=2$ superalgebra with $q=0$; if $\eta=-1$ one should postulate nonstandard antiautomorphism (see (A.1)) with $q=1$. The restrictions on $N=2$ central charges $Z,\tilde{Z}$ which follow from the isomorphism of $N=2$ Euclidean superalgebra under the (pseudo)conjugation (3.19) are the following

$$Z^\dagger = \eta Z \qquad \tilde{Z}^\dagger = \eta\tilde{Z}, \qquad \eta=\pm1. \tag{3.21}$$

The covariance of relations (3.6) under the conjugation (3.19) implies that the generators T_A of internal symmetry $\mathfrak{gl}(2,C)$ satisfy the reality conditions ($A=r,0$)

$$(T_r)^\dagger = T_r \qquad T_0^\dagger = -T_0. \tag{3.22}$$

One can express in $N=2$ case the relations (3.6) consistently with (3.19) as follows

$$[T_0,Q^a_\alpha] = -\frac{1}{2}Q^a_\alpha \quad [T_r,Q^a_\alpha] = -\frac{1}{2}(\sigma_r)^a{}_b Q^b_\alpha. \tag{3.23}$$

The generators T_r and T_0 due to relations (3.22) and (3.5) describe the internal $N=2$ algebra $\mathfrak{su}(2)\oplus\mathfrak{u}(1)=\mathfrak{u}(2)$ (see also e.g. [25]).

In the realizations of superalgebra permitting to define the Hermitean conjugation of supercharges $Q^a_\alpha \to Q^\star_{\dot\alpha a}$, $\bar{Q}_{\dot\alpha a}\to\bar{Q}^{a\star}_\alpha$ the antilinearity of the automorphism (3.19) can be realized explicitly. If $\eta=1$ (i.e. $q=0$) one can introduce the following counterpart of the conjugation (3.19) which employs the Hermitean conjugation

$$(Q^a_\alpha)^\dagger = \varepsilon_{\alpha\beta}\varepsilon^{ab}Q^{b\star}_\beta \qquad (Q_{\dot\alpha a})^\dagger = \varepsilon_{\dot\alpha\dot\beta}\varepsilon_{ab}Q^\star_{\dot\beta b}. \tag{3.24}$$

In such a case one can define the following two complex–conjugated N=2 Euclidean superalgebras [32]:

– holomorphic N=2 Euclidean superalgebra $\mathcal{E}(4;2|\mathbb{C})$ generated by supercharges $Q^a_\alpha, \bar{Q}_{\dot\alpha a}$

– antiholomorphic N=2 Euclidean superalgebra $\overline{\mathcal{E}(4;2|\mathbb{C})}$ generated by Hermitean – conjugated supercharges $(Q^a_\alpha)^\star, (\bar{Q}_{\dot\alpha a})^\star$.

The reality conditions described by conjugation (3.24) maps $\mathcal{E}(4;2|\mathbb{C}) \longrightarrow \overline{\mathcal{E}(4;2|\mathbb{C})}$, i.e. they describe the inner automorphism of real N=2 Euclidean superalgebra $\mathcal{E}(4;2|\mathbb{C})\oplus\overline{\mathcal{E}(4;2|\mathbb{C})}$. Such form of N=2 Euclidean reality conditions has been employed in earlier applications, e.g. for the description of N=2 Euclidean supersymmetric field–theoretic models, formulated in complex N=2 superspace (see e.g. [36, 32, 37])

3.5. $N=2$ real Kleinian superalgebra with central charges

For $N=2$ Kleinian real supersymmetry we have the following possible reality conditions:

$$(Q^a_\alpha)^\oplus = Q^a_\alpha, \qquad (\bar{Q}_{\dot\alpha a})^\oplus = \eta\bar{Q}_{\dot\alpha a}, \quad \eta=\pm1 \tag{3.25}$$

which extend from $N=1$ to N=2 Kleinian reality condition (2.43). The $N=2$ Kleinian superalgebra for $q=0$ because of the condition $\eta=(-1)^q$ takes the following form (see also (2.45–2.46))

$$\left\{Q^a_\alpha, \bar{Q}_{\dot\beta b}\right\} = 2\delta^a_b\,(\tilde\sigma_\mu)_{\alpha\dot\beta}\tilde{\mathcal{P}}^\mu \tag{3.26}$$

$$\left\{Q^a_\alpha, Q^b_\beta\right\} = \varepsilon_{\alpha\beta}\varepsilon^{ab}Z, \qquad \left\{\bar{Q}_{\dot\alpha a}, \bar{Q}_{\dot\beta b}\right\} = \varepsilon_{\dot\alpha\dot\beta}\varepsilon_{ab}\tilde{Z}, \tag{3.27}$$

where the reality conditions for two central charges look as follows

$$Z^\oplus = \eta Z \qquad \tilde{Z}^\oplus = \eta\tilde{Z}. \tag{3.28}$$

The internal symmetry generator T_A $(A = 0, r)$ (see (3.5), (3.8)) satisfy for vanishing central charges the reality condition

$$(T_A)^\oplus = T_A\,.$$

If central charges do not vanish the R-symmetry $GL(2;\mathbb{R})$ is reduced to $SL(2;\mathbb{R})$.

4. Classical real Poincaré and Euclidean r-matrices and their N=1 superextensions

4.1. General remarks

We shall follow the method used in our previous paper [19] based on the following steps:

i) Consider Zakrzewski list of 21 real classical $r-$matrices satisfying classical Yang-Baxter (YB) [7], and use in their presentation the canonical Poincaré basis (Sec. 2.1);

ii) Remove the Poincaré reality conditions (see (2.4) or (2.17)) imposed in [1]. The generators $h, e_\pm, h', e'_\pm, P_1, P_2, P_\pm$ are becoming complex and we obtain corresponding class of classical $r-$matrices for complex inhomogeneous $\mathfrak{io}(4,C)$ algebra.

iii) Extend supersymmetrically the complex classical $r-$matrices obtained in ii) to N=1 and N=2 by adding suitable terms which depend on supercharges $Q^a_\alpha, \bar{Q}_{\dot\alpha a}$ $(a = 1\ldots N)$. For N=2 we consider as well terms in classical r-matrix which depend on complex $N = 2$ central charges $Z, \tilde{Z}$ (see (3.2)) in such a way that the supersymmetric N=2 complex $r-$matrices satisfy the classical super-YB equation.

In Sect. 4.2 we recall the Zakrzewski list of D=4 real Poincaré and D=4 self–conjugate Euclidean r-matrices from [1] as well as D=4 N=1 real super–Poincaré and pseudoreal super-Euclidean r-matrices obtained in [19]. In Sect. 5 we present new results for supersymmetric r-matrices with standard N=2 Poincaré reality conditions (see Sect. 3.2) and N=2 Euclidean (pseudo)reality conditons (see Sect. 3.3–4).

4.2. D=4 Poincaré real r-matrices

Let us present the real D=4 Poincaré r-matrices listed in [1] (see also [5]). Using the decomposition of $r \in \mathfrak{io}(3,1)\wedge\mathfrak{io}(3,1)$

$$r = a + b + c \tag{4.1}$$

where ($\mathbb{P}$ denotes the fourmomenta generators)

$$a \in \mathbb{P}\wedge\mathbb{P} \qquad b \in \mathbb{P}\wedge\mathfrak{o}(3,1) \qquad c \in \mathfrak{o}(3,1)\wedge\mathfrak{o}(3,1)\,. \tag{4.2}$$

Zakrzewski [1] obtained the following list

where we use Cartan–Chevaley basis for $\mathfrak{o}(3,1)$ (see (2.17–2.18)) and $P_\pm = P_0 \pm P_3$. Besides b_{P_+}, b_{P_2} are given by the expressions:

$$\begin{aligned} b_{P_+} &= P_1\wedge e_+ - P_2\wedge e'_+ + P_+\wedge h\,, \\ b_{P_2} &= 2P_1\wedge h' + P_-\wedge e'_+ - P_+\wedge e'_-\,, \end{aligned} \tag{4.3}$$

and provide r-matrices describing light-cone (b_{P_+}) and tachyonic (b_{P_2}) κ-deformation [38, 39].

[7] From the list of Zakrzewski's 21 cases of Poincaré $r-$matrices, given in [1] only one set, denoted with $\mathcal{N} = 6$, does not satisfy homogeneous classical YB equation, i.e. cannot be lifted to twisted Hopf algebra.

c	b	a	#	$\mathcal{N}$
$\gamma h' \wedge h$	0	$\alpha P_+ \wedge P_- + \tilde{\alpha} P_1 \wedge P_2$	2	1
$\gamma e'_+ \wedge e_+$	$\beta_1 b_{P_+} + \beta_2 P_+ \wedge h'$	0	1	2
	$\beta_1 b_{P_+}$	$\alpha P_+ \wedge P_1$	1	3
	$\gamma\beta_1 (P_1 \wedge e_+ + P_2 \wedge e'_+)$	$P_+ \wedge (\alpha_1 P_1 + \alpha_2 P_2) - \gamma\beta_1^2 P_1 \wedge P_2$	2	4
$\gamma(h \wedge e_+$ $-h' \wedge e'_+)$ $+\gamma_1 e'_+ \wedge e_+$	0	0	1	5
$\gamma h \wedge e_+$	$\beta_1 b_{P_2} + \beta_2 P_2 \wedge e_+$	0	1	6
0	$\beta_1 b_{P_+} + \beta_2 P_+ \wedge h'$	0	1	7
	$\beta_1 b_{P_+} + \beta_2 P_+ \wedge e_+$	0	1	8
	$P_1 \wedge (\beta_1 e_+ + \beta_2 e'_+) +$ $\beta_1 P_+ \wedge (h + \sigma e_+)$, $\sigma = 0, \pm 1$	$\alpha P_+ \wedge P_2$	2	9
	$\beta_1 (P_1 \wedge e'_+ + P_+ \wedge e_+)$	$\alpha_1 P_- \wedge P_1 + \alpha_2 P_+ \wedge P_2$	2	10
	$\beta_1 P_2 \wedge e_+$	$\alpha_1 P_+ \wedge P_1 + \alpha_2 P_- \wedge P_2$	1	11
	$\beta_1 P_+ \wedge e_+$	$P_- \wedge (\alpha P_+ + \alpha_1 P_1 + \alpha_2 P_2) + \tilde{\alpha} P_+ \wedge P_2$	3	12
	$\beta_1 P_0 \wedge h'$	$\alpha_1 P_0 \wedge P_3 + \alpha_2 P_1 \wedge P_2$	2	13
	$\beta_1 P_3 \wedge h'$	$\alpha_1 P_0 \wedge P_3 + \alpha_2 P_1 \wedge P_2$	2	14
	$\beta_1 P_+ \wedge h'$	$\alpha_1 P_0 \wedge P_3 + \alpha_2 P_1 \wedge P_2$	1	15
	$\beta_1 P_1 \wedge h$	$\alpha_1 P_0 \wedge P_3 + \alpha_2 P_1 \wedge P_2$	2	16
	$\beta_1 P_+ \wedge h$	$\alpha P_1 \wedge P_2 + \alpha_1 P_+ \wedge P_1$	1	17
	$P_+ \wedge (\beta_1 h + \beta_2 h')$	$\alpha_1 P_1 \wedge P_2$	1	18
	0	$\alpha_1 P_1 \wedge P_+$	0	19
		$\alpha_1 P_1 \wedge P_2$	0	20
		$\alpha_1 P_0 \wedge P_3 + \alpha_2 P_1 \wedge P_2$	1	21

Table 1. Real D=4 (pseudo)real Poincaré r-matrices (all satisfy homogeneous CYBE except $\mathcal{N}$=6).

4.3. N=1 D=4 (pseudo) real Poincaré and Euclidean supersymmetric r-matrices

In [19] we have presented the list of possible supersymmetric D=4 N=1 super–Poincaré r-matrices $r^{(1)}$. It appears that only 7 out of 21 classes of r-matrices present in Table 1 can be supersymmetrized. The N=1 super–Poincaré r-matrices decomposed as follows:

$$r^{(1)} = r + s = a + b + c + s\,, \tag{4.4}$$

where $s \in \mathbb{Q}^{(1)} \wedge \mathbb{Q}^{(1)}$ ($\mathbb{Q}^{(1)}$ denote N=1 Poincaré supercharges). The list of r-matrices (4.4) looks as follows:

c	b	a	s	$\mathcal{N}$
$\gamma e'_+ \wedge e_+$	$\beta_1 b_{P_+} + \beta_2 P_+ \wedge h'$	0	$\beta_1 \bar{Q}_1 \wedge Q_1$	2
	$\beta_1 b_{P_+}$	$\alpha P_+ \wedge P_1$	$\beta_1 \bar{Q}_1 \wedge Q_1$	3
$\gamma h \wedge e_+$	$\beta_1 b_{P_2} + \beta_2 P_2 \wedge e_+$	0	$i\beta_1 (Q_1 + \bar{Q}_1) \wedge (Q_2 - \bar{Q}_2)$	6
0	$\beta_1 b_{P_+} + \beta_2 P_+ \wedge h'$	0	$\beta_1 \bar{Q}_1 \wedge Q_1$	7
	$\beta_1 b_{P_+} + \beta_2 P_+ \wedge e_+$	0	$\beta_1 \bar{Q}_1 \wedge Q_1$	8
	$P_1 \wedge (\beta_1 e_+ + \beta_2 e'_+) +$	$\alpha P_+ \wedge P_2$	$\beta_1 \bar{Q}_1 \wedge Q_1$	9
	$\beta_1 P_+ \wedge h$	$\alpha_2 P_1 \wedge P_2 + \alpha_1 P_+ \wedge P_1$	$\beta_1 \bar{Q}_1 \wedge Q_1$	17

Table 2. Real D= 4 N=1 super–Poincaré r-matrices (all satisfy homogeneous CYBE except $\mathcal{N}$=6).

We see that the superextension is realized in all cases except $\mathcal{N}$=6 with the help of unique supersymmetric term $S = \beta_1 \bar{Q}_1 \wedge Q_1$, where β_1 is purely imaginary, which is invariant under the N=1 super–Poincaré conjugation.

The list of N=1 complex Euclidean supersymmetric r-matrices, which are self-conjugate under the pseudoconjugation (2.35) with $\eta = -1$ looks as follows:

Journal of Physics: Conference Series **670** (2016) 012013 doi:10.1088/1742-6596/670/1/012013

c	b	a	s	$\mathcal{N}$
$\gamma h' \wedge h$	0	$\alpha P_+ \wedge P_- + \tilde{\alpha} P_1 \wedge P_2$	$\beta Q_2 \wedge Q_1$	1
	$\beta_1 P_0 \wedge h'$	$\alpha_1 P_0 \wedge P_3 + \alpha_2 P_1 \wedge P_2$	$\beta Q_2 \wedge Q_1$	13
	$\beta_1 P_3 \wedge h'$	$\alpha_1 P_0 \wedge P_3 + \alpha_2 P_1 \wedge P_2$	$\beta Q_2 \wedge Q_1$	14
	$\beta_1 P_+ \wedge h'$	$\alpha_1 P_0 \wedge P_3 + \alpha_2 P_1 \wedge P_2$	$\beta Q_2 \wedge Q_1$	15
	$\beta_1 P_1 \wedge h$	$\alpha_1 P_0 \wedge P_3 + \alpha_2 P_1 \wedge P_2$	$\beta Q_2 \wedge Q_1$	16
	0	$\alpha_1 P_1 \wedge P_+$	$\eta^{\alpha\beta} Q_\alpha \wedge Q_\eta$	19
		$\alpha_2 P_1 \wedge P_2$	$\eta^{\alpha\beta} Q_\alpha \wedge Q_\beta$	20
		$\alpha_1 P_0 \wedge P_3 + \alpha_2 P_1 \wedge P_2$	$\eta^{\alpha\beta} Q_\alpha \wedge Q_\beta$	21

Table 3. Pseudoreal (selfconjugate under pseudoreality condition) D= 4 N=1 super–Euclidean r-matrices (all satisfy homogeneous CYBE).

where $s \in Q_c^{(1)} \wedge Q_c^{(1)}$ and $Q_c^{(1)}$ denote N=1 complex Euclidean supercharges $(Q_\alpha, \bar{Q}_{\dot{\alpha}})$ (see (2.25–2.27)). Due to inner automorphisms of N=1 Euclidean superalgebra the selfconjugate term can be chosen only as $\beta Q_1 \wedge \bar{Q}_{\dot{1}}$, with parameter β purely imaginary.

We observe that among supersymmetric r-matrices in Table 2 the r-matrices with $\mathcal{N}$=2,3 and 7,8 contain term b_{P_+} characterizing light-cone κ-deformation [38], and the super r-matrix $\mathcal{N}$=6 contains term b_{P_2} describing tachyonic κ–deformation [39]. It should be added that the standard "time–like" κ-deformation characterized by the term b_{P_0}, is not present in Tables 1–3. Further we comment that in Euclidean case without supersymmetry (Table 1) and with N=1 supersymmetry (Table 3), the r-matrices characterizing κ-deformations are not present.

5. N=2 extensions with central charges of N=1 super–Poincaré and super–Euclidean r-matrices

The N=2 superextensions of D=4 Poincaré and Euclidean r-matrices can be decomposed in the following way compare with formula (4.3)

$$r^{(2)} = \tilde{a} + \tilde{b} + \tilde{c} + \tilde{s}, \tag{5.1}$$

where $\tilde{a}, \tilde{b}, \tilde{c}$ contains contributions from central charges $\mathbb{Z} = (Z_1, Z_2)$ $(\mathbb{Z} \wedge \mathbb{Z} \equiv Z_1 \wedge Z_2)$

$$\begin{aligned} &\tilde{a} \in \mathbb{P} \wedge \mathbb{P} \\ &\tilde{b} \in \mathbb{P} \wedge (\mathfrak{o}(3,1) \oplus \mathbb{Z}) \\ &\tilde{c} \in \mathfrak{o}(3,1) \oplus \mathbb{Z} \wedge (\mathfrak{o}(3,1) \oplus \mathbb{Z}) \end{aligned} \tag{5.2}$$

and $\tilde{s} \in \mathbb{Q}^{(2)} \wedge \mathbb{Q}^{(2)}$, where $\mathbb{Q}^{(2)}$ denote N=2 supercharges.

We shall list in Sect. 5.1 the r-matrices (5.2) invariant under N=2 Poincaré reality condition (3.10)–(3.11) and in Sect. 5.2 the ones invariant under the pseudoconjugation (3.16) and conjugation (3.24).

5.1. N=2 real super–Poincaré r-matrices

Due to the reality condition (3.11) N=2 Poincaré superalgebra depends on one complex supercharge $\mathbb{Z} \equiv Z_1 = Z_2^\dagger$. In such a case to every Zakrzewski r-matrix (see Table 1) one can add unique term bilinear in central charges (see (5.2))

$$Z \wedge Z^\dagger \in \mathbb{Z} \wedge \mathbb{Z} \in \tilde{c}. \tag{5.3}$$

It follows from Table 2 that only Zakrzewski r-matrices with $\mathcal{N} = 2,3,6,7,8,9,17$ admit $N = 1$ supersymmetrization, realized by universal term $\bar{Q}_1 \wedge Q_1$. It appears that for N=2 D=4 super–Poincaré r-marices the term $\tilde{s}$ in formula (5.1) is also universal and described by (α real, χ, $\chi' = 0, \pm 1$)

$$\tilde{s} = \alpha(\chi Q_1^1 \wedge \bar{Q}_{\dot{1}}^1 + \chi' Q_1^2 \wedge \bar{Q}_{\dot{1}}^2). \tag{5.4}$$

Journal of Physics: Conference Series **670** (2016) 012013 doi:10.1088/1742-6596/670/1/012013

The term (5.4) provides N=2 supersymmetrization of r-matrices[8] described in Table 2.

If we denote $\tilde{b} = b + \Delta b$ $(\Delta b \in \mathbb{P} \wedge \mathbb{Z}$; see (5.2)) such a term is possible only for supersymmetric r-matrices if $N \geqslant 2$. These additional terms linearly dependent on central charges, can be added only for $\mathcal{N} = 2,6$ by the following universal expression ($p_+ = p_0 + p_3$; β complex)

$$\Delta b = p_+ \wedge (\beta Z + \bar{\beta} \bar{Z}) . \tag{5.5}$$

For $\mathcal{N} = 3,7,8,9,17$ one can add consistently with CYBE the following complex term belonging to $\mathfrak{o}(3,1) \wedge \mathbb{Z} \in \tilde{c}$ (β_1, β_2 – complex)

$$\Delta c = \beta_1 e_+ \wedge Z + \beta_2 e'_+ \wedge \bar{Z} . \tag{5.6}$$

Unfortunately, term (5.6) can not satisfy the Poincaré reality conditions.

Listed above supersymmetric r-matrices can be presented as a sum of subordinated r-matrices which are of super-Abelian and super-Jordanian types. The subordination enables us to construct a correct sequence of quantizations and to obtain the corresponding twists describing the quantum deformations. These twists are in general case the super-extensions of the twists obtained in [5].

5.2. *N=2 (pseudo) real super–Euclidean r-matrices*

Contrary to the N=2 super–Poincaré case, when there is only one conjugation providing reality condition (see Sect. 3.2[9]), in N=2 case we have two types of reality structure:

a) Defined by pseudoconjuation (3.16) (we consider $q = 0$ and $\eta = -1$), which is the straightforward extension of pseudoreality structure considered for N=1 in (2.35).

b) Defined by the conjugation (3.24). We further assume that the Hermitean conjugation $Q^a_\alpha \to (Q^a_\alpha)^\star$, $\bar{Q}^a_{\dot\alpha} \to (\bar{Q}^a_{\dot\alpha})^\star$ is well defined.[10] Then one can formulate N=2 Euclidean superalgebra in a Hermitean form if we impose the subsidiary condition which follows from (3.24) ($\alpha = 1,2$; $a,b = 1,2$)

$$Q^a_\alpha = (Q^a_\alpha)^\dagger \Rightarrow Q^a_\alpha = \varepsilon_{\alpha\beta}\varepsilon^{ab}(\bar{Q}^b_\beta)^\star \qquad \bar{Q}^a_{\dot\alpha} = (\bar{Q}^a_{\dot\alpha})^\dagger \Rightarrow \bar{Q}_{\dot\alpha a} = \varepsilon_{\dot\alpha\dot\beta}\varepsilon_{ab}Q_{\dot\beta b}, \tag{5.7}$$

or more explicitly

$$\begin{aligned} Q^1_\alpha &= \varepsilon_{\alpha\beta}(\bar{Q}^2_\beta)^\star & \qquad Q^2_\alpha &= -\varepsilon_{\alpha\beta}(\bar{Q}^1_\beta)^\star , \\ \bar{Q}^1_{\dot\alpha} &= \varepsilon_{\dot\alpha\dot\beta}(Q^2_\beta)^\star & \qquad \bar{Q}^2_{\dot\alpha} &= -\varepsilon_{\dot\alpha\dot\beta}(Q^1_\beta)^\star . \end{aligned} \tag{5.8}$$

We see that the supercharges $Q^2_\alpha, \bar{Q}^2_{\dot\alpha}$ can be expressed by $(\bar{Q}^1_{\dot\alpha})^\star$, $(Q^1_\alpha)^\star$ and N=2 superalgebra can be described by two pairs of complex Hermitean–conjugated supercharges which equivalently can be expressed as Hermitean–conjugated pair of four–component Dirac spinors (A=1,2,3,4)

$$\Psi_A = (Q^1_\alpha, \bar{Q}^1_{\dot\alpha}), \qquad \Psi^\star_A = ((Q^1_\alpha)^\star, (\bar{Q}^1_{\dot\alpha})^\star) . \tag{5.9}$$

We shall consider below separately the N=2 Euclidean r-matrices selfconjugate under pseudoconjugation (3.16) and conjugation (5.7).

[8] If χ or $\chi' = 0$ then we obtain only N=1 supersymmetrization.

[9] We considered only here the reality structure defined by standard antilinear antiinvolution ($q = 0$ in relation (A.1)) what corresponds to the choice $\eta = 1$ in (3.10).

[10] Such situation occurs in supersymmetric QFT, with supercharges realized as differential operators on superspace fields. In fact the conjugataion (3.24) can be defined for suitable class of realizations of supercharges $Q^a_\alpha, \bar{Q}^a_{\dot\alpha}$.

Journal of Physics: Conference Series **670** (2016) 012013 doi:10.1088/1742-6596/670/1/012013

5.2.1. N=2 super–Euclidean r-matrices selfconjugate under pseudoconjugation We consider the complex N=2 $\varepsilon(4;2|\mathbb{C})$ r-matrices which are invariant under the map (3.16); we choose the standard version of formula (A.1) with $q=0$ what implies $\eta=-1$. For N=2 we should take into consideration only the Poincaré r-matrices from Table 1 with $\mathcal{N}=1,13-16,19-21$ (see also Table 3) which allow N=1 supersymmetrization. If we consider the relations (3.17) with $q=0$, we get the pair of independent N=2 Euclidean central charges Z, $\tilde{Z}$ which are purely imaginary. They provide Euclidean counterpart of formula (5.3) describing universal contribution to N=2 super-r-matrices.

The fermionic part $\tilde{s}_E$ of N=2 super–Euclidean r-matrix (5.1) for $\mathcal{N}=1,13-16$ is described by the following pair of two forms bilinear in supercharges ($\alpha_1,\alpha_2,\tilde{\alpha}_1,\tilde{\alpha}_2$ are real)

$$\begin{aligned}\tilde{s}_{E;1} &= \alpha_1\, Q_1^1\wedge Q_2^1+\alpha_2\bar{Q}_1^2\wedge\bar{Q}_2^2\,,\\ \tilde{s}_{E;2} &= \tilde{\alpha}_1\, Q_1^2\wedge Q_2^2+\tilde{\alpha}_2\bar{Q}_1^1\wedge\bar{Q}_2^1\,.\end{aligned}\tag{5.10}$$

To either of two terms (5.10) one can add the unique term $\Delta c_E\in\mathfrak{o}(3,1)\wedge\mathbb{Z}$ which takes the form (α complex, β_1,β_2 real)

$$\Delta c_E=(\alpha h+\bar{\alpha}h')(\beta_1 Z+\beta_2\tilde{Z})\,.\tag{5.11}$$

For $\mathcal{N}=19-21$ we can again choose pair of the purely fermionic terms $\tilde{s}_E$, which are described by formulae (5.10); the terms Δc_E, linear in $Z,\tilde{Z}$, are however not universal, different for three cases $\mathcal{N}=19,20$ and 21.

5.2.2. N=2 real super-Euclidean r-matrices In this case the algebraic structure does not have a counterpart in the formulae obtained by Euclidean N=1 supersymmetryzation (see Table 3). The task consists in finding such N=2 complex $\varepsilon(4;2|\mathbb{C})$ r-matrices which are consistent with N=2 super–Euclidean reality conditions (5.8). We should mention that it is necessary to consider the N=2 supersymmetrization of Poincaré r-matrices for all $\mathcal{N}=1\ldots21$. We obtain the following list of fermionic and central charge dependent terms for varius choices of $\mathcal{N}$

α) $\mathcal{N}=1,13-16,18$

There are possible the following independent four fermionic two-forms $\tilde{s}_{E;k}$ ($k=1,2,3,4;\ \alpha_1,\ldots\alpha_6$ real)

$$\begin{aligned}\tilde{s}_{E;1} &= \alpha_1(Q_1^1\wedge Q_2^1+\bar{Q}_1^2\wedge\bar{Q}_2^2)\,,\\ \tilde{s}_{E;2} &= \alpha_2(Q_1^2\wedge Q_2^2+\bar{Q}_1^1\wedge Q_2^1)\,,\\ \tilde{s}_{E;3} &= i\alpha_3\, Q_1^1\wedge\bar{Q}_2^2+i\alpha_4 Q_2^1\wedge\bar{Q}_1^2\,,\\ \tilde{s}_{E;4} &= i\alpha_5\, Q_1^2\wedge\bar{Q}_2^1+i\alpha_6 Q_2^2\wedge\bar{Q}_1^1\,.\end{aligned}\tag{5.12}$$

The additional bosonic terms $\Delta c_E\in\mathfrak{o}(3,1)\wedge\mathbb{Z}$ which are linear in real central charges $Z,\tilde{Z}$ satisfying (for $\eta=1$) the reality conditions (3.21) is described again by formula (5.11).

β) $\mathcal{N}$=19–21.

For such values of $\mathcal{N}$ one can add any of four three–parameter term $\tilde{\tilde{s}}_{E;k}$ below, bilinear in supercharges ($\alpha_1,\ldots\alpha_{12}$ real):

$$
\begin{aligned}
\tilde{\tilde{s}}_{E;1} &= \alpha_1(Q^1_1 \wedge Q^1_1 + \bar{Q}^2_2 \wedge \bar{Q}^2_2) + \alpha_2(Q^1_1 \wedge Q^1_2 + \bar{Q}^2_1 \wedge Q^2_1) + \alpha_3(Q^1_2 \wedge Q^1_2 + \bar{Q}^2_1 \wedge \bar{Q}^2_1)\,,\\
\tilde{\tilde{s}}_{E;2} &= \alpha_4(Q^1_2 \wedge Q^1_2 + Q^2_1 \wedge \bar{Q}^2_1) + \alpha_5(Q^2_1 \wedge Q^2_2 + \bar{Q}^1_1 \wedge \bar{Q}^1_2) + \alpha_6(Q^2_2 \wedge Q^2_2 + \bar{Q}^1_1 \wedge \bar{Q}^1_1)\,,\\
\tilde{\tilde{s}}_{E;3} &= \alpha_7\, Q^1_1 \wedge \bar{Q}^2_2 + \alpha_8\, Q^1_2 \wedge \bar{Q}^2_1 + \alpha_9(Q^1_1 \wedge \bar{Q}^2_1 + Q^1_2 \wedge \bar{Q}^2_2)\,,\\
\tilde{\tilde{s}}_{E;4} &= \alpha_{10}\, Q^2_1 \wedge \bar{Q}^1_2 + \alpha_{11}\, Q^2_2 \wedge \bar{Q}^1_1 + \alpha_{12}(Q^2_1 \wedge \bar{Q}^1_1 + Q^2_2 \wedge \bar{Q}^1_2)\,.
\end{aligned}
\tag{5.13}
$$

The additional bosonic terms Δb_E linear in central charges and fourmomenta are different for considered three classes of r-matrices, namely ($\beta_1, \beta_2, \gamma_2, \gamma_3, \gamma_0$ real)

$$
\begin{aligned}
N = 19: &\quad \Delta b_E = p_1 \wedge (\beta_1 Z + \beta_2 \tilde{Z})\,,\\
N = 20: &\quad \Delta b_E = (p_1 + \gamma_2 p_2) \wedge (\beta_1 Z + \beta_2 \tilde{Z})\,,\\
N = 21: &\quad \Delta b_E = (p_1 + \gamma_2 p_2 + \gamma_3 p_3 + i\gamma_0 p_0) \wedge (\beta_1 Z + \beta_2 \tilde{Z})\,.
\end{aligned}
\tag{5.14}
$$

Finally we point out that among all 21 Zakrzewski classes of complex $\mathfrak{io}(4;\mathbb{C})$ r-matrices only for $\mathcal{N}$=5 we could not find the superextension of complex $\mathfrak{o}(4;\mathbb{C})$ r-matrices to $\varepsilon(4;N|\mathbb{C})$ ($N = 1,2$) supersymmetric r-matrices, i.e. we were not able for $\mathcal{N}$=5 to provide any consistent fermionic term $\tilde{s}$ in (5.1).

6. Final remarks

This paper provides firstly systematic discussion of real forms of $\mathfrak{o}(4;\mathbb{C})$, $\mathfrak{io}(4;\mathbb{C})$, $\varepsilon(4;1|\mathbb{C})$ and $\varepsilon(4;2|\mathbb{C})$, where $\varepsilon(4;N;\mathbb{C})$ describes complex D=4 N–extended Euclidean superalgebra. In Sect. 2 and 3 we consider the reality and pseudoreality conditions (reality constraints). To the Poincaré and Euclidean (pseudo)real forms we added also the (pseudo)real forms for the Kleinian signature $g_{\mu\nu} = \mathrm{diag}(1,-1,1,-1)$. In particular we also considered the (pseudo) reality conditions leading to exotic supersymmetry scheme, with odd (Grassmann) coordinates conjugated in nonstandard way (see e.g. (2.38)–(2.39), with $q = 1$).

Our second aim was to present the extension of Zakrzewski list of classical D=4 Poincaré r-matrices to Euclidean case and N=1,2 supersymmetrizations. The D=4 N=1 Poincaré and Euclidean supersymmetric classical r-matrices already considered in [19] were presented in Sect. 4. In Sect. 5 we describe partial results for the classification of D=4 N=2 supersymmetric r-matrices for various D=4 signatures. We consider in this paper N=2 Poincaré and Euclidean signatures and these results are new. It should be pointed out that we considered also the terms depending on a pair of N=2 central charges (for $\varepsilon(4;2|\mathbb{C})$); they are complex-conjugated for Poincaré and real for Euclidean cases.

We did not consider the deformations of N=2 Kleinian supersymmetry and corresponding classical r-matrices. More systematic approach, with more complete list of complex classical $\varepsilon(4;2|\mathbb{C})$ r-matrices and their various real forms will be considered in our subsequent publication.

Acknowledgments

The paper was presented at XXIIIth International Conference on Integrable Systems and Quantum Symmetries, Prague (June 2015). The presenting author (JL) would like to thank prof. Cestimir Burdik for warm hospitality. The paper has been supported by the Polish Science Center (NCN), project 2014/13/B/ST2/04043 (A.B. and J.L.), RFBR grant No 14-01-00474-a (V.N.T.).

Appendix A. Conjugations and pseudoconjugations of complex Lie superalgebras

In order to describe the supersymmetries of physical systems one should consider real and pseudoreal forms of complex Lie superalgebra L, which are defined in algebraic framework with help of the

Journal of Physics: Conference Series **670** (2016) 012013 doi:10.1088/1742-6596/670/1/012013

conjugations and pseudoconjugations. The conjugations and pseudoconjugations are usually defined as abstract antilinear antiautomorphisms $x \to x^\star$ of second and fourth order preserving the Z_2 grading of superalgebra and satisfying the properties ($x, y \in U_L$; $|x| = 0$ ($|x| = 1$) describes the parity of even (odd) element x), where U_L denotes enveloping algebra of L:

$$(xy)^\star = (-1)^{q|x||y|} y^\star x^\star \qquad q = 0,1 \tag{A.1}$$

$$(\alpha x + \beta y)^\star = \bar{\alpha} x^\star + \bar{\beta} y^\star \qquad \alpha, \beta \in C \tag{A.2}$$

where $q = 0,1$ defines two types of antilinear antiinvolution map in U_L and $\alpha \to \bar{\alpha}$, $\beta \to \bar{\beta}$ describe complex conjugation in C. Further

$$(x^\star)^\star = x \qquad \text{(conjugation)}, \tag{A.3a}$$

$$(x^\star)^\star = -x \qquad \text{(pseudoconjugation)}. \tag{A.3b}$$

For Lie superalgebras the property (A.3b) occurs only in odd parity sector, i.e. for any $x \in L$ both relations (A.3a) and (A.3b) can be written together as

$$(x^\star)^\star = (-1)^{p|x|} x, \tag{A.3c}$$

where $p = 0$ (resp. $p = 1$) denotes conjugations (resp. pseduconjugation) and we recall that $|x|$ describe the grading of superalgebra element x. The conjugations (A.3a) in matrix and Hilbert space realizations of superalgebra can be identified with the Hermitean conjugation, and pseudoconjugations in odd sector of the matrix superalgebras with $p = 1$ were introduced as graded Hermitean conjugation [26] (see also [27]). In the case of conjugations (A.3a) the Hermitean elements $L_R, \widetilde{L}_R$ of complex Lie algebra L are defined as follows

$$L_R \,:\, x_R = (x + x^\star) \qquad \widetilde{L}_R :\; \widetilde{x}_R = i(x - x^\star) \qquad x \in L, \tag{A.4}$$

where $x_R^+ = -x_R$ and $\tilde{x}_R^+ = -\tilde{x}_R$. In the case of conjugation the superalgebras $L_R, \widetilde{L}_R$ are the subsuperalgebras which can be defined as fixed points of the conjugation map, and $L = L_R \oplus \widetilde{L}_R$ provides the formula describing the realification of L. In the case of pseudoconjugations (A.3b) the elements (A.4) of complex Lie algebra satisfy the set of relations

$$x_R^\# = -(x - x^\#) = i\widetilde{x}_R \qquad \widetilde{x}_R^\# = -ix_R \tag{A.5}$$

with elements $x_R, \widetilde{x}_R$ satisfying involutive relations of fourth order

$$(x_R^\#)^\# = -x_R \qquad (x_L^\#)^\# = -x_L. \tag{A.6}$$

We comment that one uses for description of real supersymmetries in classical physics an alternative conjugations and pseudoconjugations, which are antilinear automorphism with the property

$$\overline{xy} = \bar{x}\bar{y} \quad ; \tag{A.7}$$

the remaining relations (A.2), (A.3a–b) are unchanged. One can add that authomorphisms (A.7) for classical physical systems are usually represented by complex conjugation, and antiauthomorphisms adjusted to quantum systems are realized as (graded) Hermitean conjugation in suitable Hilbert space framework.

References

[1] S. Zakrzewski, *Lett. Math. Phys.*, **32**, 11 (1994).
[2] S. Zakrzewski, *Commun. Math. Phys.*, **187**, 285 (1997); [q-alg/9602001v1].
[3] A. Mudrov, Yadernaya Fizyka **60**, 946 (1997).
[4] P. Podles, S.L. Woronowicz, *Commun. Math. Phys.*, **178**, 61 (1996); [hep-th/9412059].
[5] V.N. Tolstoy, *Bulg. J. Phys.* **35** (2008) 441–459; (Proceedings of VII International Workshop "Lie Theory and Its Applications in Physics". Ed. V.K. Dobrev et al, Heron Press, Sofia, 2008) [0712.3962 [math.QA]].
[6] C. Blohmann, *J. Math. Phys.* **44**, 4736 (2000); [0209180v1[math.QA]].
[7] P.P. Kulish, *Twists of quantum groups and noncommutative field theory*, [hep-th/0606056v1].
[8] Y. Kobayashi and S. Sasaki, *Int. J. Mod. Phys.*, **A20**, 7175 (2005); [hep-th/0410164v3].
[9] B.M. Zupnik, *Phys. Lett.*, **B627**, 208 (2005); [hep-th/0506043v3]; *Theor. Math. Phys.* **147**, 670 (2006); [hep-th/0602034v2].
[10] M. Ihl and C. Saemann, *JHEP*, **0601**, 065 (2006); hep-th/0506057.
[11] J. Lukierski, A. Nowicki, J. Sobczyk, *J. Phys.* **A26**, L1109 (1993).
[12] P. Kosinski, J. Lukierski, P. Maslanka and J. Sobczyk, *J. Phys.*, **A27**, 6827 (1994); [hep-th/9405076v1]; *ibid.* **A28**, 2255 (1995); [hep-th/9411115v1].
[13] V.N. Tolstoy, *Nankai Tracts in Mathematics "Differential Geometry and Physics". Proceedings of the 23-th International Conference of Differential Geometric Methods in Theoretical Physics (Tianjin, China, 20-26 August, 2005). Editors: Mo-Lin Ge and Weiping Zhang. Wold Scientific*, 2006, Vol. **10**, 443-452; [QA/0701079v1].
[14] V.N. Tolstoy, *Lecture Notes in Physics*, vol. 370, p.118 (1990)
[15] M. Chaichian, P.P. Kulish, *Phys. Lett.*, **B 234**, 72 (1990).
[16] S.M. Khoroshkin, V.N. Tolstoy, *Commun. Math. Phys.*, **141**, 599 (1991).
[17] A. Borowiec, J. Lukierski, V.N. Tolstoy, *New twisted quantum deformations of D=4 super-Poincaré algebra*, Proc. of Dubna Workshop "Supersymmetries and Quantum Symmetries (SQS'07)", 30.07–4.08.2007, ed. S. Fedoruk, E. ivanov; arXiv:0803.4167.
[18] G. Fiore, J. Wess *Phys. Rev.* **D75**, 105022 (2007); [hep-th/0701078].
[19] A. Borowiec, J. Lukierski, M. Mozrzymas, V. N. Tolstoy, JHEP 1206, 154 (2012); [hep-th/1112.1936].
[20] N. Seiberg, *JHEP*, **0306**, 010 (2003); [hep-th/0305248v1].
[21] S. Ferrara, M.A. Lledo, O. Macia, *JHEP* **0309**, 063 (2003); [hep-th/0307039].
[22] S. Ferrara, E. Sokhachev,*Phys. Lett.*, **B579**, 226 (2004); [hep-th/0308021].
[23] E. Ivanov, D. Lechtenfeld, B. Zupnik, [hep-th/0402062].
[24] A de Castro, E. Ivanov, O. Lechtenfeld, L. Quevedo, *Nucl. Phys.*, **B747**, 1 (2006); [hep-th/0510013].
[25] I.L. Buchbinder, E. Ivanov, O. Lechtenfeld, I.B. Samsonov, B.M. Zupnik, [hep-th/0709.3770].
[26] M. Scheunert, W. Nahm, V. Rittenberg, *Journ. Math. Phys.* **18**, 146 (1977).
[27] F.A. Berezin and V.N. Tolstoy, *Commun. Math. Phys.*, **78**, 409 (1981);
[28] Yu.T. Manin, *"Gauge Fields and Complex Geometry"* (in Russian), Ed. Nauka (Moscow), 1984.
[29] R. Gilmore, "Lie Groups, Lie Algebras and Some of Their Applications", ed. John Wiley and Sons, 1974.
[30] A. Barut, R. Rączka, "Theory of Group Representations and Applications", ed. PWN, Warszawa 1977.
[31] J. Lukierski, A. Nowicki, *Journ. Math. Phys.* **25**, 2545 (1984).
[32] J. Lukierski, *Czech. J. Phys.* **B37**, 359 (1987).
[33] R. Haag. J.T. Lopuszanski, M. Sohnius, *Nucl. Phys.* **B88**, 257 (1975)
[34] L. Frappat, A. Sciarrino, P. Sorba, *"Dictionary on Lie Superalgebras"*, [hep-th/9607161].
[35] S. Ferrara, C. Savoy, B. Zumino, *Phys. Lett.* **B100**, 393 (1981).
[36] B. Zumino, *Phys. Lett.* **B69**, 369 (1979).
[37] D. Butter, G. Inverso, I. Lodato, JHEP 1509(2015)088, arXiv:1505.03500 [hep-th].
[38] A. Ballesteros, F.J. Herranz, M.A. delOlmo, M. Santander, *Phys. Lett.* **B351**, 137 (1995).
[39] P. Kosiński, P. Maślanka, in *From Quantum Field Theory to Quantum Groups*, ed. B. Jancewicz and J. Sobczyk, World Scientific, 1996, p. 41.

Journal of Physics: Conference Series **670** (2016) 012014 doi:10.1088/1742-6596/670/1/012014

Deformed Heisenberg Algebra with a minimal length: Application to some molecular potentials

Djamil Bouaziz

Abstract. We review the essentials of the formalism of quantum mechanics based on a deformed Heisenbeg algebra, leading to the existence of a minimal length scale. We compute in this context, the energy spectra of the pseudoharmonic oscillator and Kratzer potentials by using a perturbative approach. We derive the molecular constants, which characterize the vibration–rotation energy levels of diatomic molecules, and investigate the effect of the minimal length on each of these parameters for both potentials. We confront our result to experimental data for the hydrogen molecule to estimate an order of magnitude of this fundamental scale in molecular physics.

Département de Physique, Université de Jijel, BP 98, Ouled Aissa, 18000 Jijel, Algeria

E-mail: djamilbouaziz@univ-jijel.dz

1. Introduction

This work, is a continuation of the recent studies [1, 2], where the effects of the minimal length on the vibration-rotation of diatomic molecules have been investigated through the pseudoharmonic oscillator (PHO) [1] and the Kratzer [2] interactions.

As it is well known, the minimal length is a prediction of quantum gravity [3, 4, 5] and string theory [6, 7, 8], where the Planck length is supposed to be a lower bound to all physical length scales [9, 10]. This fundamental scale is implemented by the modification of the Heisenberg uncertainty relation to the form: $(\Delta X)(\Delta P) \geq \frac{\hbar}{2}(1+\beta(\Delta P)^2+...)$, where β is a small positive parameter [4, 7, 11]. The minimal uncertainty in position (minimal length) implied by this generalized uncertainty principle (GUP) is then given by $(\Delta X)_{\text{min}} = \hbar\sqrt{\beta}$ [12].

The study of theoretical and physical implications of the GUP is still a hot topic. Many problems with great physical interest have been studied in connection with the GUP, see, for instance, Refs. [12, 13, 14, 15, 16, 17, 18, 19, 20, 21, 22, 23, 24, 25, 26, 27, 28, 29, 30, 31].

It has been in particular concluded that the presence of the minimal length in the formalism provides a natural ultraviolet regularization in quantum mechanics [26] and in quantum field theory [15]. Furthermore, upper bounds for this elementary length have been estimated; the values differ from one application to another and, mostly, belong to the range 10^{-6}– 10^{6} fm [27, 28]. The minimal length seems to depend on the energy scale of the problem and might therefore characterize the size of the system under study [13, 26]. The latter finding was behind the motivation of our recent investigations [1, 2] on the GUP effects in diatomic molecules, because the spatial extension of these systems is relatively large, and the effect of the minimal length may clearly manifest.

In Ref. [1], we studied the vibration-rotation energy levels of diatomic molecules in the presence of a minimal length by addressing the Schrödinger equation with the PHO potential.

Journal of Physics: Conference Series **670** (2016) 012014 doi:10.1088/1742-6596/670/1/012014

A more detailed investigation was carried out in Ref. [2], by taking the Kratzer interaction. It has been explicitly shown that the minimal length would have some physical importance in studying the spectra of diatomic molecules. Furthermore, an upper bound for the minimal length has been estimated of about 0.01 Å by confronting the correction of the GUP to an experimental result of the hydrogen molecule [2].

Here, we review the main results obtained in [1, 2]: We give the expression of the vibration-rotation energy spectrum of diatomic molecules in the presence of a minimal length by studying the deformed Schrödinger equation with the two potentials. In both cases, we apply the obtained formulas to compute the spectroscopic constants of diatomic molecules and investigate the effect of the GUP on these constants. We show how the importance that the deformation parameter can play in fitting the experimental results. Moreover, the expressions of the molecular constants derived from the energy spectrum of the PHO show the importance of the generalization of these studies to the case of a two-parameter deformed Heisenberg algebra, which is presented in Sec. II.

The rest of this paper is organized as follows. In Sec. II, we review the essentials of the formalism of quantum mechanics with a GUP. Sec. III is devoted to investigate the KP in this formalism: we derive the molecular constants, and investigate the effect of the minimal length on vibration-rotation energy levels of diatomic molecules. In Sec. VI, we perform the same study with the PHO interaction. The last section summarizes our main results and conclusions.

2. Deformed Heisenberg algebra with a minimal length: A review

As mentioned in Sec. I, to include a minimal length scale, the uncertainty relation can be modified to the form [8, 11] $(\Delta X)(\Delta P) \geq \frac{\hbar}{2}(1+\beta(\Delta P)^2+...)$, where β is a small positive parameter related to the minimal length by $(\Delta X)_{\min} = \hbar\sqrt{\beta}$. This generalized uncertainty principle (GUP) implies the modification of the commutation relations between the position and the momentum operators as [12]

$$[\widehat{X}, \widehat{P}] = i\hbar(1+\beta\widehat{P}^2). \tag{1}$$

The operators $\widehat{X}$ and $\widehat{P}$ can be represented in both coordinate and momentum spaces as follows:

in momentum space, the simplest realization is [12]

$$\widehat{X} = (1+\beta\widehat{p}^2)\widehat{x}, \qquad \widehat{P} = \widehat{p}, \tag{2}$$

where $\widehat{p} = p$ and $\widehat{x} = i\hbar\frac{\partial}{\partial p}$ represent the position and momentum operators of ordinary quantum mechanics.

In coordinate space one has [29]

$$\widehat{X} = \widehat{x}, \qquad \widehat{P} = (1+\tfrac{1}{3}\beta\widehat{p}^2)\widehat{p}, \tag{3}$$

where $\widehat{x} = x, \;\; \widehat{p} = -i\hbar\frac{\partial}{\partial x}$.

The formalism of of quantum mechanics with a minimal length has been extended to arbitrary dimensions (D) [12, 13, 14, 15]. The modified Heisenberg algebra reads

$$[\widehat{X}_i, \widehat{P}_j] = i\hbar[(1+\beta\widehat{P}^2)\delta_{ij} + \beta'\widehat{P}_i\widehat{P}_j], \quad (\beta,\beta') > 0,$$

$$[\widehat{P}_i, \widehat{P}_j] = 0, \tag{4}$$

$$[\widehat{X}_i, \widehat{X}_j] = i\hbar\frac{2\beta-\beta'+\beta(2\beta+\beta')\widehat{P}^2}{1+\beta\widehat{P}^2}(\widehat{P}_i\widehat{X}_j - \widehat{P}_j\widehat{X}_i).$$

The corresponding GUP is

$$(\Delta X_i)(\Delta P_i) \geq \tfrac{\hbar}{2}(1+\beta\sum_{j=1}^{N}[(\Delta P_j)^2 + \left\langle \widehat{P}_j \right\rangle^2] + \beta'[(\Delta P_i)^2 + \left\langle \widehat{P}_i \right\rangle^2]), \tag{5}$$

in which the minimal uncertainty in position (minimal length) is [13]

$$(\Delta X_i)_{\min} = \hbar\sqrt{D\beta + \beta'}, \quad \forall i. \tag{6}$$

The position and momentum operators are represented in momentum space as [13, 17]

$$\widehat{X}_i = i\hbar[(1+\beta p^2)\tfrac{\partial}{\partial p_i} + \beta' p_i p_j \tfrac{\partial}{\partial p_j} + \gamma p_i], \qquad \widehat{P}_i = p_i, \tag{7}$$

where γ is a small positive parameter related to β and β'.

In coordinate space, the simplest representation of the operators $\widehat{X}_i$ and $\widehat{P}_i$ is [16]

$$\widehat{X}_i = \widehat{x}_i, \qquad \widehat{P}_i = \widehat{p}_i\left(1+\beta\widehat{p}^2\right), \tag{8}$$

where $\widehat{x}_i$ and $\widehat{p}_i$ satisfy the standard commutation relations of ordinary quantum mechanics.

Representation (8) satisfies the deformed algebra in the case $\beta' = 2\beta$ up to the first order of β. It is especially more adequate in the treatment of the minimal length as a perturbation to the Schrödinger equation for a given interaction. In this special case the minimal length reads in 3-dimensions as

$$(\Delta X_i)_{\min} = \hbar\sqrt{5\beta}, \quad \forall i. \tag{9}$$

The Schrödinger equation with the representation (8) can be written as follows:

$$(\frac{\widehat{p}^2}{2\mu} + V(r) + \frac{\beta}{\mu}\widehat{P}^4)\psi(\mathbf{r}) = E\psi(\mathbf{r}), \tag{10}$$

where terms of order β^2 have been neglected.

In the following sections, we will use equation (10) to study the Kratzer and the PHO potentials. The main goal of this study is to derive the expressions of the spectroscopic constants of diatomic molecules in the presence of a minimal length by using the deformed energy spectrum of these two interactions.

3. Energy spectrum of Kratzer potential with a minimal length

We are interested to the deformed Schödinger equation (10) with the Kratzer's molecular potential (KP), which has the form [33]

$$V(r) = g_1/r^2 - g_2/r, \tag{11}$$

with $g_1 = D_e r_e^2$ and $g_2 = 2D_e r_e$, where D_e is the dissociation energy and r_e is the equilibrium internuclear distance of a given diatomic molecule.

It has been shown in detail in Ref. [2] that the minimal length corrections $\Delta E_{n\ell}$ can be analytically computed, and the energy spectrum of the KP in the presence of a minimal length is given by the following expression:

$$\begin{aligned} E_{n\ell} = &-\tfrac{\gamma^2 D_e}{(\lambda+n)^2} + \beta\mu D_e^2\left(\tfrac{2\gamma}{\lambda+n}\right)^4\left\{-\tfrac{3}{4} + \tfrac{\lambda+n}{\lambda-1/2}\left(1+\tfrac{\gamma^2}{2}\left(\tfrac{1}{(\lambda+n)^2} - \tfrac{2}{\lambda(\lambda-1)}\right)\right)\right. \\ &\left.+\tfrac{\gamma^4}{4}\tfrac{1}{(\lambda-\frac{1}{2})(\lambda-1)(\lambda-\frac{3}{2})(\lambda+n)}\left(1+\tfrac{3n(2\lambda+n)}{\lambda(2\lambda+1)}\right)\right\}, \end{aligned} \tag{12}$$

Journal of Physics: Conference Series **670** (2016) 012014 doi:10.1088/1742-6596/670/1/012014

with the notations

$$\gamma = \tfrac{r_e}{\hbar}\sqrt{2\mu D_e}, \quad \lambda = 1/2 + \sqrt{(\ell + 1/2)^2 + \gamma^2}, \tag{13}$$

where n and ℓ are, respectively, the radial (vibrational) and orbital (rotational) quantum numbers; and μ is the reduced mass of the molecule.

Formula (12) shows the effect of this deformed algebra on the energy levels of KP. As it has been outlined in Ref. [2], several features of diatomic molecules can be studied by using this formula. In the following, we apply Eq. (12) to investigate the effect of the minimal length on the spectroscopic constants, which characterize the rovibrational energy levels of diatomic molecules.

Spectroscopic constants of diatomic molecules

As we will see, the energy spectrum of the KP is similar to the well-known spectroscopic formula which defines the vibration-rotation energy spectrum of diatomic molecules [32, 33].

To this end, we observe that the dimensionless parameter γ, in Eq. (12), is so large for most molecules ($\gamma \gg 1$) [33], the levels $E_{n\ell}$ may then be expanded into powers of $1/\gamma$ as follow:

$$\begin{aligned} E_{n\ell} = D_e\Big(&-1 + 2(n+\tfrac{1}{2})\tfrac{1}{\gamma} + (\ell+\tfrac{1}{2})^2\tfrac{1}{\gamma^2} - 3(n+\tfrac{1}{2})^2\tfrac{1}{\gamma^2} + 4(n+\tfrac{1}{2})^3\tfrac{1}{\gamma^3} - 3(n+\tfrac{1}{2})(\ell+\tfrac{1}{2})^2\tfrac{1}{\gamma^3}\Big) \\ &+ \beta\mu D_e^2\Big(6\left\{(n+\tfrac{1}{2})^2+\tfrac{1}{4}\right\}\tfrac{1}{\gamma^2} + 2(n+\tfrac{1}{2})\left\{-\tfrac{1}{4}+4(\ell+\tfrac{1}{2})^2 - 15(n+\tfrac{1}{2})^2\right\}\tfrac{1}{\gamma^3}\Big) + \ldots \end{aligned} \tag{14}$$

This formula shows the different parts of the undeformed spectrum ($\beta = 0$), and the corrections caused by the minimal length. The effect of this GUP on each kind of the rovibrational energy was discussed in detail in Ref. [2]. Here, we will suggest some ideas on how to get information about the order of the deformation parameter β from formula (14), and then apply it to extract the spectroscopic constants of diatomic molecules, which is a way to give values of β for any molecule.

We now have basically two approaches to deal with the parameter β in formula (14): The first is to consider that β is independent of the molecular constants r_e and D_e, which have known values in molecular spectroscopy. Therefore, an upper bound for β can be estimated by assuming, e.g, that the minimal length corrections in in Eq. (14) are included in the gap between the experimental results and those predicted by the formula (14) in the case case $\beta = 0$. It has been argued in Ref. [2] that a better estimate may be obtained by considering the vibrational ground-state energy E_{00} of the hydrogen molecule (H_2). This led to an upper bound for the minimal length with the value

$$(\Delta X)_{\min} = \hbar\sqrt{5\beta} < 0.01\ \text{Å}. \tag{15}$$

The second view point consists of looking at formula (14) as an energy spectrum of a three-parameter potential, i.e., D_e, r_e, and β; so, the values of β depend now on D_e and r_e. This viewpoint allows for adjusting the three parameters of the "deformed KP" with the spectroscopic data. To this end, we can suggest to extract, from the deformed energy spectrum (14), the spectroscopic constants for diatomic molecules.

The commonly used energy equation in molecular spectroscopy is [32, 34]

$$\begin{aligned} E_{n\ell} = Y_{00} + \omega_e\left(n+\tfrac{1}{2}\right) - \omega_e x_e(n+\tfrac{1}{2})^2 + \omega_e y_e(n+\tfrac{1}{2})^3 \\ + B_e\ell(\ell+1) - \alpha_e\left(n+\tfrac{1}{2}\right)\ell(\ell+1), \end{aligned} \tag{16}$$

where

$$Y_{00} = \tfrac{D_e}{4}\tfrac{\omega_e x_e}{\gamma^2} + \tfrac{\alpha_e\omega_e}{12B_e}\left(1 + \tfrac{\omega_e\omega_e}{12B_e^2}\right). \tag{17}$$

The values of the spectroscopic constants ω_e, $\omega_e x_e$, $\omega_e y_e$, B_e and α_e are determined by fitting the spectroscopic data for any molecules [32].

We write now equation (14) in the form

$$
\begin{aligned}
E_{n\ell} = {} & D_e\Big[-1 + 2(n+\tfrac{1}{2})\tfrac{1}{\gamma} + (\ell+\tfrac{1}{2})^2\tfrac{1}{\gamma^2} - 3(n+\tfrac{1}{2})^2\tfrac{1}{\gamma^2} \\
& +4(n+\tfrac{1}{2})^3\tfrac{1}{\gamma^3} - 3(n+\tfrac{1}{2})(\ell+\tfrac{1}{2})^2\tfrac{1}{\gamma^3}\Big] \\
& + \beta\mu D_e^2\Big[6\left\{(n+\tfrac{1}{2})^2 + \tfrac{1}{4}\right\}\tfrac{1}{\gamma^2} - \tfrac{1}{2}(n+\tfrac{1}{2}) \\
& + 2(n+\tfrac{1}{2})\left\{4(\ell+\tfrac{1}{2})^2 - 15(n+\tfrac{1}{2})^2\right\}\tfrac{1}{\gamma^3}\Big] + \dots \;.
\end{aligned} \tag{18}
$$

The identification of Eqs. (18) and (16) leads to the following deformed spectroscopic constants:

$$
\begin{aligned}
Y_{00} &= \tfrac{1}{4}\tfrac{D_e}{\gamma^2} + \tfrac{3}{2}\beta\mu\tfrac{D_e^2}{\gamma^2}, \quad \omega_e = \tfrac{2D_e}{\gamma} - \tfrac{3}{4}\tfrac{D_e}{\gamma^3} + \tfrac{3}{2}\beta\mu\tfrac{D_e^2}{\gamma^3}, \\
\omega_e x_e &= \tfrac{3D_e}{\gamma^2} - 6\beta\mu\tfrac{D_e^2}{\gamma^2}, \quad \omega_e y_e = \tfrac{4D_e}{\gamma^3} - 30\beta\mu\tfrac{D_e^2}{\gamma^3}, \\
B_e &= \tfrac{D_e}{\gamma^2}, \quad \alpha_e = \tfrac{3D_e}{\gamma^3} - 8\beta\mu\tfrac{D_e^2}{\gamma^3},
\end{aligned} \tag{19}
$$

We observe that the leading corrections of the minimal length are of order $1/\gamma^2$ and concerns the constant of anharmonicity of vibrations $\omega_e x_e$ and the constant Y_{00}, which does not influence the line positions in a spectrum. However, the rotational constant B_e is not affected by this deformed algebra.

As the experimental values of the spectroscopic constants are available for diatomic molecules [32], it follows that the expressions (19) can be used to give values of β for each molecule. This investigation is under consideration in the general case of deformed Heisenberg algebra (4), with $\beta' \neq 2\beta$.

4. Energy spectrum of the pseudoharmonic oscillator with a minimal length

The pseudoharmonic oscillator (PHO) is also one of the molecular interactions, which is used in the study of the vibration-rotation spectra of diatomic molecules. The form of this potential is

$$
V(r) = D_e\left(\frac{r}{r_e} - \frac{r_e}{r}\right)^2, \tag{20}
$$

where D_e is the dissociation energy and r_e is the equilibrium internuclear distance of a given diatomic molecule.

It has been shown in Ref. [1] that the minimal length correction of this potential can also be analytically derived by following the same procedures as in Sec. III. The deformed energy spectrum of the PHO reads [1]

$$
\begin{aligned}
E_{n\ell} = {} & E_{n\ell}^0 + 4\mu\beta\Bigg((E_{n\ell}^0)^2 + 4D_e E_{n\ell}^0 + 6D_e^2 - (4D_e^2 + 2D_e E_{n\ell}^0)\frac{\lambda+2n+1}{\gamma} + \\
& D_e^2\frac{\lambda^2 + (6n+3)\lambda + 6n(n+1) + 2}{\gamma^2} - \gamma\frac{2D_e\left(2D_e + E_{n\ell}^0\right)}{\lambda} + D_e^2\gamma^2\frac{\lambda+2n+1}{\lambda(\lambda^2-1)}\Bigg),
\end{aligned} \tag{21}
$$

where $E_{n\ell}^0$ is the undeformed spectrum, given by

$$
E_{n\ell}^0 = -2D_e\left(1 - \tfrac{1}{\gamma}\left(2n+1+\lambda\right)\right), \tag{22}
$$

with the notations

$$\lambda = \sqrt{\gamma^2 + (\ell + \tfrac{1}{2})^2}, \quad \gamma = \sqrt{\tfrac{2\mu D_e r_e^2}{\hbar^2}}, \tag{23}$$

where, $n = 0, 1, 2, \ldots$, and $\ell = 0, 1, 2, \ldots$are, respectively, the radial (vibrational) and orbital (rotational) quantum numbers, and μ is the reduced mass of the molecule.

The effect of the GUP on the vibration-rotation energy levels of a given diatomic molecule with the PHO interaction was qualitatively investigated in Ref. [1] by using formula (21). Here, we use this deformed spectrum to give the expressions of the molecular constants in the presence of a minimal length.

Spectroscopic constants of diatomic molecules

Formula (21) might be viewed as an energy spectrum of a three-parameter potential function, i.e., D_e, r_e, and β. One can then derive the molecular constants for diatomic molecule by expanding $E_{n\ell}$ in Eq. (21) into powers of $1/\gamma$ ($\gamma \gg 1$); this leads to the expression

$$\begin{aligned} E_{n,\ell} = {} & (\tfrac{1}{4} + 6\mu\beta D_e)\tfrac{D_e}{\gamma^2} + \tfrac{4D_e}{\gamma}\left(1 + \tfrac{3\mu\beta D_e^2}{\gamma^2}\right)(n + \tfrac{1}{2}) + \tfrac{24\mu\beta D_e^2}{\gamma^2}(n + \tfrac{1}{2})^2 \\ & + \tfrac{D_e}{\gamma^2}\ell(\ell + 1) + \tfrac{16\mu\beta D_e^2}{\gamma^3}(n + \tfrac{1}{2})\,\ell(\ell + 1) + \ldots\,, \end{aligned} \tag{24}$$

which is of the form of the spectroscopic formula (16), with the following spectroscopic constants:

$$\begin{aligned} & Y_{00} = (\tfrac{1}{4} + 6\mu\beta D_e)\tfrac{D_e}{\gamma^2}, \quad \omega_e = \tfrac{4D_e}{\gamma}\left(1 + \tfrac{3\mu\beta D_e^2}{\gamma^2}\right) \\ & \omega_e x_e = -\tfrac{24\mu\beta D_e^2}{\gamma^2}, \quad B_e = \tfrac{D_e}{\gamma^2}, \quad \alpha_e = -\tfrac{16\mu\beta D_e^2}{\gamma^3}. \end{aligned} \tag{25}$$

The value of the constant $\omega_e y_e$ is zero in the order $1/\gamma^3$.

In contrast to ordinary quantum mechanics ($\beta = 0$), where the spectrum of the PHO leads to zero values of the spectroscopic constants $\omega_e x_e$ and α_e, in the presence of a minimal length, these constants depend on β and have nonzero values. The basic empirical terms, known in the vibration-rotation energy spectrum of diatomic molecules, are now present in this deformed version of quantum mechanics. However, the sign of $\omega_e x_e$ and α_e is conflicting compared to that of the experimental values, at least for the diatomic molecules listed in Ref. [32].

We can conclude that the PHO in this one-parameter deformed algebra could not be used to fit the deformed spectroscopic constants with the empirical results. It follows that the extension of this work to the general case $\beta' \neq 2\beta$ is interesting; it allows us not only to adjust the parameters D_e, r_e, β and β' but also to establish a constraint between the deformation parameters β and β' by appropriately choosing the sign of the molecular constants. This study is in finalization and would be published else where.

5. Summary and conclusions

We have investigated, in quantum mechanics with a deformed Heisenberg algebra including a minimal length $(\Delta X_i)_{\min} = \hbar\sqrt{5\beta}$, the vibration-rotation energy spectra of diatomic molecules by two molecular interactions: the Kratzer and the PHO potentials. We have discussed how an order of the deformation parameter β can be evaluated from the minimal length corrections and the spectroscopic data of diatomic molecules. With the Kratzer's potential, an upper bound of the minimal length of about 0.01 Å has been obtained by comparing the theoretical and experimental values of the vibrational ground-state energy of the molecule H_2. On the other hand, by supposing that the deformation parameter is a third parameter of the potentials, we have derived the spectroscopic constants of diatomic molecules with both interactions and we have examined the effects of the minimal length on each of these constants. We showed that in the case of the PHO the extension of this study to the general case of a two-parameter deformed

Heisenberg algebra would be mandatory and interesting because it allows to give a physical constraint between β and β'.

This work is supported by the Algerian Ministry of Higher Education and Scientific Research, under the CNEPRU Project No. D01720140007. I am grateful to P. Hajigeorgious for fruitful contact.

References

[1] D. Bouaziz and A. Boukhellout, Mod. Phys. Lett A. **29**, 1450143 (2014).
[2] D. Bouaziz, Ann. Phys. 355, 269 (2015).r
[3] C. A. Mead, Phys. Rev. **135,** 849 (1964).
[4] L. J. Garay, Int. J. Mod. Phys. A **10**, 145 (1995).
[5] C. Rovelli and L. Smolin, Nucl. Phys. B 442, 593 (1995).
[6] D. J. Gross and P. F. Mende, Nucl. Phys. **B 303**, 407 (1988).
[7] D. Amati, M. Ciafaloni, and G. Veneziano, Phys. Lett. B **216**, 41 (1989).
[8] K. Konishi, G. Paffuti, and P. Provero, Phys. Lett. B **234**, 276 (1990).
[9] T. Padmanabhan, Gen. Rel. Grav **17**, 215 (1985).
[10] M. T. Jaekel and S. Reynaud, Phys. Lett. A **185**, 143 (199
[11] M. Magiore, Phys. Lett. B **319**, 83 (1993).
[12] A. Kempf, G. Mangano, and R. B. Mann, Phys. Rev. D **52**, 1108 (1995).
[13] A. Kempf, J. Phys. A. **30**, 2093 (1997).
[14] A. Kempf, J. Math. Phys. **35**, 4483 (1994); A. Kempf, Phys. Rev. D**54**, 5174 (1996); Erratum-ibid. D**55**, 1114 (1997).
[15] A. Kempf, J. Math. Phys. **38**, 1347 (1997); A. Kempf and G. Mangano, Phys. Rev. D **55** , 7909 (1997).
[16] F. Brau, J. Phys. A **32**, 7691 (1999).
[17] L. N. Chang, D. Minic, N. Okamura, and T. Takeuchi, Phys. Rev. D **65**, 125027 (2002).
[18] S. Benczik, L. N. Chang, D. Minic, and T. Takeuchi, Phys. Rev. A **72**, 012104 (2005).
[19] M. M. Stetsko and V. M. Tkachuk, Phys. Rev. A **74**, 012101 (2006); M. M. Stetsko, Phys. Rev. A **74**, 062105 (2006).
[20] L. Menculini, O. Panella, and P. Roy, Phys. Rev. D **87**, 065017 (2013).
[21] Y. Chargui, A. Trabelsi, and L. Chetouani, Phys. Lett. A. **374**, 531 (2010).
[22] C. Quesne and V. M. Tkachuk, J. Phys. A : Math. Gen. **38**, 1747 (2005).
[23] U. Harbach and S. Hossenfelder, Phys. Lett. B **632**, 379 (2006); O. Panella, Phys. Rev. D. **76,** 045012 (2007).
[24] D. Mania, and M. Maziashvili, Phys. Lett. B **705** 521 (2011).
[25] K. Nozari and S. H. Mehdipour, Chaos, Solitons & Fractals, **32**, 1637 (2007); T.V. Fityo, Phys. Lett. A **372**, 5872 (2008); B.Vakili and M. A Gorji, J. Stat. Mech. 10013 (2012).
[26] D. Bouaziz and M. Bawin, Phys. Rev. A. **76**, 032112 (2007); D. Bouaziz and M. Bawin, Phys. Rev. A. **78**, 032110 (2008).
[27] D. Bouaziz and N. Ferkous, Phys. Rev. A. **82**, 022105 (2010).
[28] F. Brau and F. Buisseret, Phys. Rev. D **74**, 036002 (2006).
[29] K. Nozari and P. Pedram, Europhys. Lett. 92, 50013 (2010); P. Pedram, K. Nozari, and S. H. Taheri, J. High. Energy. Phys. **1103**, 093 (2011).
[30] Bibhas Ranjan Majhi, Elias C. Vagenas, Phys. Lett. B **725,** 477 (2013); M. Isi, J. Mureika, P. Nicolini, JHEP. **11**, 139 (2013); M. Sprenger, P. Nicolini, and M. Bleicher, Eur. J. Phys. **33,** 853 (2012).
[31] S. Hossenfelder, Living Rev. Relativity. **16**, 2 (2013); C. Quesne and V. M. Tkachuk, Czech. J. Phys. **56,** 1269 (2006); A. F. Ali , S. Das, and E. C. Vagenas, Phys. Lett. B. **678**, 497 (2009); Das S, E. C. Vagenas, and A. F. Ali, Phys. Lett. B **690**, 407 (2010); K. Nozari and A. Etemadi, Phys. Rev. D **85**, 104029 (2012); P. Pedram, Phys. Lett. B. **714,** 317 (2012).
[32] Karl K. Irikura, J. Phys. Chem. Ref. Data. V. **36,** N. 2, 389 (2007).
[33] S. Flügge, *Practical quantum mechanics*, (Springer-Verlag, 1994), pp. 178-182.
[34] H. Lefebvre-Brion, and R. W. Field, *The spectra and Dynamics of Diatomic Molecules* (Elsevier Academic Press, London, 2004) p. 3.

Journal of Physics: Conference Series **670** (2016) 012015 doi:10.1088/1742-6596/670/1/012015

The Dirac equation in an external electromagnetic field: symmetry algebra and exact integration

A I Breev[1,2] and A V Shapovalov[1,2]

[1] Department of Theoretical Physics, Tomsk State University, 634050 Tomsk, Russia
[2] Department of Higher Mathematics and Mathematical Physics, Tomsk Polytechnic University, 634050 Tomsk, Russia

E-mail: breev@mail.tsu.ru, shpv@phys.tsu.ru

Abstract. Integration of the Dirac equation with an external electromagnetic field is explored in the framework of the method of separation of variables and of the method of noncommutative integration. We have found a new type of solutions that are not obtained by separation of variables for several external electromagnetic fields. We have considered an example of crossed electric and magnetic fields of a special type for which the Dirac equation admits a nonlocal symmetry operator.

Introduction

The Dirac equation for a charge in an external electromagnetic field is the basic equation for relativistic quantum mechanics and quantum electrodynamics. In relativistic quantum mechanics the Dirac equation is interpreted as a single-particle wave equation describing fermions in an external field [1, 2, 3, 4]. In quantum electrodynamics, exact solutions of the Dirac equation are needed to obtain the Furry interaction picture to keep exactly the interaction with the external field and the interaction of photons [5].

To construct exact solutions of the Dirac equation, the separation of variables method (SoV) is commonly used. The external electromagnetic fields admitting SoV in the Dirac and the Klein–Gordon equations have been listed in refs [1, 2].

In refs. [6, 7] a new method, named the noncommutative integration (NI) method, has been proposed to construct basises of exact solutions of linear partial differential equations. The NI method essentially uses a Lie algebra $\mathfrak{g}$ of differential symmetry operators of the first order.

Integration of the free Dirac equation using the NI method was considered in refs. [8, 9, 10]. Note that the method allows one to find exact solutions (NI-solutions) in the cases when the Dirac equation does not allow separation of variables [11, 12].

The paper is organized as follows. In the next section we introduce basic notations used below and briefly describe the SoV method and the NI method applied to the Dirac equation. In Section 2 we apply the NI method for the Dirac equation in a spherically symmetric electromagnetic field. In section 3 we study the case when the magnetic field is uniform and constant. Sections 4 deal with the crossed electromagnetic field of a special type, admitting a nonlocal symmetry operator of the first order. In section 5 we consider integration of the Dirac equation with the electromagnetic field which contracts the symmetry algebra of the free Dirac equation to a commutative subalgebra. Concluding remarks are in Section 6.

1. Symmetry algebra of the Dirac equation and construction of exact solutions

Let us consider the four-dimensional Lorentz manifold M with coordinates x^μ $(\mu, \nu = 1, \ldots, 4)$ having the line element $ds^2 = g_{\mu\nu}(x)dx^\mu dx^\nu$. The Dirac equation for a spinor ψ on M can be written as[1]

$$\hat{H}\psi = m\psi, \quad \hat{H} = \gamma^\mu \hat{\mathcal{P}}_\mu, \quad \hat{\mathcal{P}}_\mu = \hat{p}_\mu - eA_\mu(x), \quad \hat{p}_\mu = i(\partial_\mu + \Gamma_\mu), \quad x \in M. \tag{1}$$

Here, $A_\mu(x)$ is a potential of an external electromagnetic field, e and m are charge and mass of a particle, respectively. The Dirac matrices γ^μ are defined as a solution of the system $\{\gamma^\mu, \gamma^\nu\} = 2g^{\mu\nu}(x)$. The spin connection Γ_μ is defined by [10]

$$\Gamma_\mu = -\frac{1}{4}\gamma_{\nu;\mu}\gamma^\nu, \quad \gamma_{\nu;\mu} = \partial_\mu \gamma_\nu - \Gamma^\rho_{\nu\mu}\gamma_\rho,$$

where $\Gamma^\rho_{\nu\mu}$ are the Christoffel symbols of the symmetric connection compatible with the metric $g_{\mu\nu}(x)$.

The matrices

$$E_4, \quad \gamma^\mu, \quad \gamma^{\mu\nu} = \frac{1}{2}[\gamma^\mu, \gamma^\nu], \quad \gamma = -\frac{1}{4!}e_{\mu\nu\tau\rho}\gamma^\mu\gamma^\nu\gamma^\tau\gamma^\nu, \quad \overset{*}{\gamma}_\mu = -\frac{1}{3!}e_{\mu\nu\tau\rho}\gamma^\nu\gamma^\tau\gamma^\rho,$$

form a basis for 4×4 matrices. Here $e_{\mu\nu\tau\rho} = \sqrt{-\det(g_{\mu\nu}(x)}\varepsilon_{\mu\nu\tau\rho}$ is totally antisymmetric tensor, $(\varepsilon_{1234} = 1)$.

In Minkowski space $\mathcal{M}$ for the Dirac gamma matrices we use the standard representation

$$\gamma^1 = \begin{pmatrix} 1 & 0 \\ 0 & -1 \end{pmatrix}; \quad \gamma^{k+1} = \begin{pmatrix} 0 & \sigma^k \\ -\sigma^k & 0 \end{pmatrix}, \quad \Sigma^k = \begin{pmatrix} \sigma^k & 0 \\ 0 & \sigma^k \end{pmatrix}, \quad k = 1, 2, 3,$$

where σ^k are the usual Pauli matrices.

The matrix differential symmetry operators of the first order for the Dirac equation (1) have been obtained as a linear combination of the following three independent operators (see [13] and [14]):

$$\hat{X} = \xi^\mu(x)\hat{\mathcal{P}}_\mu - \frac{i}{4}\gamma^{\mu\nu}\xi_{\mu;\nu}(x) + \varphi(x), \tag{2}$$

$$\hat{L} = \overset{*}{\gamma}_\mu f^{\nu\mu}(x)\hat{\mathcal{P}}_\nu + \frac{i}{3}\gamma_\mu \tilde{f}^{\nu\mu}_{;\nu}(x), \tag{3}$$

$$\hat{J} = 2\gamma\gamma^{\mu\nu}f_\nu(x)\hat{\mathcal{P}}_\mu + \frac{3i}{4}\gamma f^\mu_{;\mu}(x), \tag{4}$$

where $\tilde{f}_{\mu\nu}(x) = \frac{1}{2}e_{\mu\nu\tau\rho}f^{\tau\rho}(x)$ is the dual tensor of the tensor $f^{\mu\nu}(x)$. We call operators (2) *the Killing symmetry operators* as they are expressed in terms of the Killing vector field $\xi^\mu(x)$,

$$\xi_{\mu;\nu}(x) + \xi_{\nu;\mu}(x) = 0, \quad \partial_\nu\varphi(x) = F_{\nu\mu}(x)\xi^\mu(x).$$

Here $F_{\nu\mu}(x) = \partial_\nu A_\mu(x) - \partial_\mu A_\nu(x)$ is the tensor of an electromagnetic field. The symmetry operators (3) are given by the antisymmetric Yano-Killing tensor field $f_{\mu\nu}(x)$:

$$f_{\mu\nu;\tau}(x) + f_{\mu\tau;\nu}(x) = 0, \quad f_\nu{}^\alpha(x)F_{\alpha\mu}(x) = f_\mu{}^\alpha(x)F_{\alpha\nu}(x).$$

The symmetry operations (4) are determined in terms of the Yano vector field $f_\mu(x)$:

$$f_{\mu;\nu}(x) = \frac{1}{4}g_{\mu\nu}(x)f^\mu{}_{;\mu}(x), \quad F_{\nu\mu}(x)f^\mu(x) = 0.$$

[1] We will use a unit system in which $\hbar = c = 1$ and the Minkowski metric is $\eta_{\mu\nu} = \mathrm{diag}(1, -1, -1, -1)$

We call operators (3) and (4) *the spin symmetry operators.*

A necessary condition for the SoV in the Dirac equation is the existence of three mutually commuting independent symmetry operators of the first order. Under this condition, the problem of finding solutions of the Dirac equation is reduced to the eigenvalue problem

$$\hat{H}\psi = m\psi, \quad \hat{S}_a\psi = \lambda_a\psi, \quad a = 1, \dots, 3, \tag{5}$$

where $\hat{S}_a$ are symmetry operators of the form (2) – (4), and λ_a are essential separation parameters.

The separation of variables method can be applied to the squared Dirac equation [16] which is obtained from the Dirac equation (1) by the substitution $\psi = (\hat{H}+m)\varphi$. Then for φ we obtain *the squared Dirac equation* reads $(\hat{H}^2 - m^2)\varphi = 0$. If there exists a constant matrix Ω such that $(\hat{H}^2 - m^2)\Omega = \Omega\hat{L}_2$ where $\hat{L}_2$ is a diagonal second order differential operator, we say that the squared Dirac equation *admits diagonalization.* Then the complete system of solutions of the Dirac equation can be written as $\psi = (\hat{H} + m)\Omega\Phi$ where Φ is a complete system of solutions for the equation $\hat{L}_2\Phi = 0$. The last one is integrated in the framework of the SoV method.

Let $\mathfrak{g}$ be a Lie algebra of the Killing symmetry operators (2). Denote by $G = \exp(\mathfrak{g})$ a Lie group of motions on M. The spaces $\mathfrak{g}^*$ and $\mathfrak{g}$ are conjugates of each other. Consider a neighborhood U_0 of a regular point $x_0 \in M$ in which we will seek a basis of solutions of the Dirac equation. In this neighborhood we introduce a new coordinate system (r, z) where r are independent invariants of a Lie group G, and z are local coordinates on an orbit of x_0. Denote by $\mathfrak{h}$ a stationary Lie algebra of the point z_0. Without loss of generality, we assume that the Killing symmetry operators depend only on the coordinates z. The method consists in generalization of the eigenvalue problem (5) as follows [6]:

$$\hat{X}_a(z)\psi(x; q, \lambda) = -l_a(q, \partial_q, \lambda)\psi(x; q, \lambda), \quad a = 1, \dots, \dim \mathfrak{g}, \tag{6}$$

where $l_a(q, \partial_q, \lambda) = \alpha_a^\nu(q)\partial_{q^\nu} + i\chi_a(q, \lambda)$ are operators of an irreducible λ-representation of the algebra $\mathfrak{g}$ constructed using a K-orbit $\mathcal{O}_\lambda$ passing through the covector $\lambda \in \mathfrak{g}^*$. The variables q are local coordinates on a Lagrangian submanifold Q of the orbit $\mathcal{O}_\lambda$. If the Lie algebra $\mathfrak{g}$ is Abelian, then $l_a(q, \lambda) = \lambda_a$ and we come to the SoV method. The basis of solutions of equation (6) is represented as $\psi(x; q, \lambda) = \exp(R(x; q, \lambda))\hat{\psi}(u; r, \lambda)$, where $R(x; q, \lambda)$ is a certain function, and $\hat{\psi}(u; r, \lambda)$ is an arbitrary function depending on the characteristics of the system $u = u(q, x)$. Substituting this solution into the Dirac equation (1), we obtain the reduced equation

$$\hat{H}(u, \partial_u, r, \partial_r; \lambda)\hat{\psi}(u; r, \lambda) = m\hat{\psi}(u; r, \lambda). \tag{7}$$

The number of independent variables of the reduced equation (7) is shown to be determined by algebraic properties of the orbit of z_0 [21].

If equation (7) is algebraic or an ordinary differential equation, it is said that the original Dirac equation (1) is *noncommutatively integrable.* Note that the functions $\psi(z; q, \lambda)$ are the eigenfunctions of all nontrivial Casimir operators

$$\hat{K}(z, \partial_z)\psi(z; q, \lambda) = \kappa(\lambda)\psi(z; q, \lambda).$$

The basis of solutions of equation (1) can be represented as a generalized Fourier transform

$$\psi(z; q, \lambda) = \int_U \psi(u; r, \lambda) D_{qu}^\lambda(x) d\mu(u),$$

where the distributions $D_{qu}^\lambda(x)$ have the properties of completeness and orthogonality [21, 19].

Consider the important case in which the Lie group G acts on M simply transitively and the neighborhood U_0 can be identified with a local Lie group G by entering the group coordinates $g = g(z)$. Without loss of generality, we assume that the group acts by right translations. The noncommutative integration method in this case may be formulated using Kirillov's orbit method as follows [19, 9]. The basis of solutions of equation (1) reads

$$\psi(g;q,\lambda) = \int_Q \psi(q';q,\lambda) D^{\lambda}_{qq'}(g^{-1}) d\mu(q'), \quad q, q' \in Q,$$

where $D^{\lambda}_{qq'}(g)$ are generalized kernels of an irreducible unitary representation of the Lie group G constructed on a non-degenerate K-orbit $\mathcal{O}_\lambda$ (see [22]). The representation of the Lie algebra $\mathfrak{g}$ is at the same time the λ-representation of $\mathfrak{g}$. The distributions $D^{\lambda}_{qq'}(g)$ can be found from the following system of equations:

$$\left(\xi_a(g) + \overline{l_a(q,\partial_q)}\right) D^{\lambda}_{qq'}(g) = 0, \quad \left(\eta_a(g) + l_a(q',\partial_{q'})\right) D^{\lambda}_{qq'}(g) = 0. \tag{8}$$

Here $\xi_a(g)$ and $\eta_a(g)$ are the left-invariant and right-invariant vector fields on U_0, respectively. The functions $\psi(q',\lambda)$ satisfy the reduced equation

$$\hat{H}(q',\partial_{q'};q,\lambda)\hat{\psi}(q';q,\lambda) = m\psi(q';q,\lambda)$$

with the number of independent variables q' equal to $\dim Q = (\dim \mathfrak{g} - \operatorname{ind} \mathfrak{g})/2$. Here $\operatorname{ind}\mathfrak{g}$ is the index of the Lie algebra $\mathfrak{g}$ defined as the dimension of the annihilator of a generic covector.

2. The Dirac equation in a spherically symmetric electric field

Let us define a potential $A_\mu(x)$ by the relations

$$A_1 = V(r), \quad A_k = 0, \quad r = \sqrt{x^2+y^2+z^2}, \quad k = 2,\dots,4,$$

where $V(r)$ is a smooth function. The Dirac equation on $\mathcal{M}$ can be written in a Hamiltonian form as $i\partial_t\psi = \hat{H}\psi$. The Hamiltonian $\hat{H}$ in spherical coordinate system (r,θ,ϕ) takes the form:

$$\hat{H} = -i\alpha^1\left(\frac{1}{r} + \partial_r\right) - i\alpha^2\frac{1}{r}\left(\frac{1}{2}\cot\theta + \partial_\theta\right) - i\alpha^3\frac{1}{r\sin\theta}\partial_\phi + \beta m + eV(r),$$

where $\beta = \gamma^1, \alpha^1 = \gamma^1\gamma^2,\ \alpha^2 = \gamma^1\gamma^3,\ \alpha^3 = \gamma^1\gamma^4$. Consider the stationary Dirac equation

$$\hat{H}\psi = E\psi, \quad \psi = \psi(r,\theta,\phi). \tag{9}$$

The Killing symmetry operators corresponding to the generators of the rotation group $SO(3)$ are given by

$$\hat{X}_1(\theta,\phi) = \partial_\phi, \quad \hat{X}_2(\theta,\phi) = -\cot\theta\sin\phi\partial_\phi + \cos\phi\partial_\theta + \frac{i\sin\phi}{2\sin\theta}\Sigma^1,$$
$$\hat{X}_3(\theta,\phi) = -\cot\theta\cos\phi\partial_\phi - \sin\phi\partial_\theta + \frac{i\cos\phi}{2\sin\theta}\Sigma^1.$$

These operators the following realization of the Lie algebra $\mathfrak{so}(3)$:

$$[\hat{X}_1,\hat{X}_2] = \hat{X}_3, \quad [\hat{X}_3,\hat{X}_1] = \hat{X}_2, \quad [\hat{X}_2,\hat{X}_3] = \hat{X}_1.$$

Note that inclusion of the external electromagnetic field does not affect the algebra of symmetry operators $\hat{X}_a(\theta,\phi)$. The Casimir operator of the algebra $\mathfrak{so}(3)$ is the square of the total angular momentum:

$$\mathbf{J}^2 = K(-i\hat{X}), \quad K(f) = f_1^2 + f_2^2 + f_3^2.$$

The spin Yano-Killing symmetry operator reads

$$\hat{S} = -\beta\left(\Sigma^2 \frac{1}{\sin\theta}\partial_\phi - \Sigma^3\left[\frac{1}{2}\cot\theta + \partial_\theta\right]\right), \quad [\hat{S}, X_a] = 0.$$

Consider briefly the separation of variables in equation (9) (for more details see [1] and [17]). There exists a complete set $\{\hat{S}, -iX_3, \mathbf{J}^2\}$ of mutually commuting symmetry operators. The solution of the eigenvalue problem

$$\mathbf{J}^2\psi_{jM\zeta} = j(j+1)\psi_{jM\zeta}, \quad \hat{S}\psi_{jM\zeta} = \zeta\left(j+\frac{1}{2}\right)\psi_{jM\zeta}, \quad -iX_3\psi_{jM\zeta} = M\psi_{jM\zeta} \tag{10}$$

is obtained as

$$\psi_{jM\zeta}(r,\theta,\phi) = \frac{1}{r}\begin{pmatrix} \Omega^j_{M\zeta}(\theta,\phi)f(r) \\ -\zeta\Omega^j_{M-\zeta}(\theta,\phi)g(r), \end{pmatrix}, \quad M = -j\ldots j, \quad \zeta = \pm 1, \quad j = 1,2,3,\ldots, \tag{11}$$

where $\Omega^j_{M\zeta}(\theta,\phi)$ is a spherical spinor [18]. The Dirac equation (9) is reduced to a radial stationary Dirac equation with a fixed angular momentum j and a projection of angular momentum in the direction of M, and a spin quantum number ζ. The radial Dirac equation for $\psi_{jM\zeta}$ is a system of ordinary differential equations for the functions $f(r)$ and $g(r)$:

$$\begin{aligned} f'(r) - \frac{1}{r}\zeta\left(j+\frac{1}{2}\right)f(r) - (E + m - eV(r))g(r) = 0, \\ g'(r) + \frac{1}{r}\zeta\left(j+\frac{1}{2}\right)g(r) + (E - m - eV(r))f(r) = 0. \end{aligned} \tag{12}$$

Let us perform the noncommutative reduction of the Dirac equation (9) with the use of the Lie algebra $\mathfrak{so}(3)$ of symmetry operators of the first order $\hat{X}_a(\theta,\phi)$. Denote an invariant of the Lie group $SO(3)$ as r and let (θ,ϕ) be coordinates on an orbit of the Lie group $SO(3)$. The noncommutative integration consists in solving the system of equations

$$X_a(\theta,\phi)\psi = -l_a(q,\lambda)\psi, \quad \hat{H}\psi = E\psi, \tag{13}$$

where $l_a(q,\lambda)$ are operators of a λ- representation of the Lie algebra $\mathfrak{so}(3)$. The λ- representation is constructed for functions defined on a Lagrangian submanifold Q of orbits of the coadjoint representation (K-orbits) of the Lie algebra $\mathfrak{so}(3)$ [19, 20]. The manifold Q has the topology of the cylinder: $\mathrm{Re}\, q \in [0, 2\pi)$, $\mathrm{Im}\, q \in \mathbb{R}^1$. A covector $\lambda = (j,0,0) \in \mathfrak{so}^*(3)$, $j \in \mathbb{Z}$ parametrizes a nondegenerate K-orbit. The operators $-il_a(q,\lambda)$ are Hermitian with respect to the scalar product

$$(\psi_1(q), \psi_2(q)) = \int_Q \overline{\psi_1(q)}\psi_2(q)d\mu(q), \quad d\mu(q) = \frac{(2j+1)!}{2^j(j!)^2}\frac{dq\wedge d\overline{q}}{(1+\cos(q-\overline{q}))^{j+1}}$$

and have the form

$$l_1(q,\lambda) = -i(\sin(q)\partial_q - j\cos(q)), \quad l_2(q,\lambda) = -i(\cos(q)\partial_q + j\sin(q)), \quad l_3(q,\lambda) = \partial_q.$$

Integrating equations (13), we obtain the basis of solutions

$$\psi_\sigma(r,\theta,\phi) = \frac{1}{r}\begin{pmatrix} D^j_{q\zeta}(\theta,\phi) f_\sigma(r) \\ -\zeta D^j_{q(-\zeta)}(\theta,\phi) g_\sigma(r)\end{pmatrix}, \quad \sigma = (q,j,E,\zeta), \tag{14}$$

where

$$D^j_q(\theta,\phi) = \frac{2^j (j!)^2}{(2j)!}\Big(-i\cos\theta\sin q + \sin\theta(\cos\phi - i\cos q\sin\phi)\Big)^{j-1/2} \times \tag{15}$$

$$\times\sqrt{\cos\theta\sin q + i\cos\phi(\cos q + \sin\theta) + \sin\phi + \cos q\sin\theta\sin\phi}\Bigg[$$

$$\frac{\cos q\cos\theta + i\sin q(\cos\phi + i\sin\theta\sin\phi)}{1+\cos q\sin\theta + \cos\theta\sin q\sin\phi}\sigma^1 + 1\Bigg]\begin{pmatrix} -i\zeta \\ 1\end{pmatrix},$$

and the functions $f_\sigma(r)$ and $g_\sigma(r)$ satisfy the radial equation (12). Solutions (14) are the eigenvalues of the Casimir operator $\mathbf{J}^2$ and of the spin operator $\hat{S}$, but (14) are not eigenfunctions for the operator of angular momentum $-iX_3$:

$$\mathbf{J}^2\psi_\sigma = j(j+1)\psi_\sigma, \quad \hat{S}\psi_\sigma = \zeta\left(j+\frac{1}{2}\right)\psi_\sigma, \quad -iX_3\psi_\sigma = i\partial_q\psi_\sigma. \tag{16}$$

The Fourier transform of functions (15) are the spherical spinors:

$$\Omega^j_{M\zeta}(\theta,\phi) = \int_Q D^j_{q\zeta}(\theta,\phi)e^{iMq}d\mu(q). \tag{17}$$

Indeed, given (16), we obtain that (11) is a family of solutions for the Dirac equation (9) and (11) satisfies equations (10) if the spherical spinors are defined by expression (17). Thus, formula (17) establishes a link between the basis of the solutions obtained using separation of variables and by the method of noncommutative integration. Solution (14) describes a state with definite value of the orbital angular momentum j and spin ζ. Relation (17) allows considering a complex parameter q and a projection of the moment M as dual variables.

Note that (15) is expressed in terms of elementary functions, while the spherical spinors are expressed in terms of special functions.

3. Electromagnetic field extending a subalgebra of symmetry operators

Consider an electromagnetic field with nonzero components $E_z = V'(z)$, $H_z = H = const$ in Cartesian coordinates (x,y,z) and choose a potential in the form

$$A_0 = -V(z), \quad \mathbf{A} = (0, Hx, 0). \tag{18}$$

The Dirac equation on $\mathcal{M}(t,x,y,z)$ with the potential (18) reads

$$i\partial_t\psi = \hat{H}\psi, \quad \hat{H} = -i\alpha^1\partial_x - \alpha^2(i\partial_y + eHx) - i\alpha^3\partial_z - eV(z) + \beta m. \tag{19}$$

The Killing symmetry operators can be written as

$$\hat{X}_0 = ie, \quad \hat{X}_1 = \partial_x - ieHy, \quad \hat{X}_2 = \partial_y, \quad \hat{X}_3 = y\partial_x - x\partial_y - \frac{i}{2}\Sigma^3 + i\frac{eH}{2}(x^2-y^2). \tag{20}$$

Operators (20) form a basis of the Lie algebra $\mathfrak{g}$ with non-trivial brackets:

$$[\hat{X}_1,\hat{X}_3] = -\hat{X}_2, \quad [\hat{X}_2,\hat{X}_3] = \hat{X}_1, \quad [\hat{X}_1,\hat{X}_2] = H\hat{X}_0.$$

In the absence of the electromagnetic field, the symmetry operators (20) form a Lie algebra $\mathfrak{e}(2)$ of the Lie group $E(2)$ acting on the plane (x, y). Inclusion of the external field leads to the fact that the Lie algebra of symmetry operators (20) becomes a central extension of the Lie algebra $\mathfrak{e}(2)$.

We can show that if the symmetry operators of the first order without matrix coefficients at the derivatives form a Lie algebra, then this algebra will be a central extension of the Lie algebra of symmetry operators for the free Dirac equation [15].

Note that in addition to (20), there exists an additional spin symmetry operator

$$\hat{S} = \beta\left(\Sigma^2 \partial_x - \Sigma^1(\partial_y - ieHx)\right), \quad [\hat{S}, \hat{X}_a] = 0. \tag{21}$$

Separation of variables in the Dirac equation is possible if we have a set of operators $\{\hat{H}, \hat{X}_2, \hat{S}\}$ (see [2, 17]). The basis of solutions decreasing at $|\xi| \to \infty$ has the form [17]:

$$\psi_\sigma(x, y, z) = e^{ipy} \begin{pmatrix} -i\frac{\zeta\sqrt{2}}{n} D_k(\xi) f(z) \\ D_{k-1}(\xi) f(z) \\ -i\frac{\zeta\sqrt{2}}{n} D_k(\xi) g(z) \\ -D_{k-1}(\xi) g(z) \end{pmatrix}, \quad \xi = \sqrt{\frac{2}{eH}}(eHx - p), \quad k = -n^2/2, \tag{22}$$

where $D_k(\xi)$ are parabolic cylinder functions. A spinor (22) is a solution of the eigenvalue problem

$$-i\hat{X}_2\psi_\sigma = p\psi_\sigma, \quad \hat{S}\psi_\sigma = \zeta n\sqrt{eH}\psi_\sigma, \quad \sigma = (n, p, \zeta) \quad n = 0, 1, 2, \ldots, \quad \zeta = \pm 1.$$

The functions $f(z)$ and $g(z)$ satisfy the system of ordinary differential equations

$$\begin{aligned} &if'(z) - n\sqrt{eH}\zeta f(z) + (m + E + eV(z))g(z) = 0, \\ &ig'(z) + n\sqrt{eH}\zeta g(z) - (m - E - eV(z))f(z) = 0. \end{aligned} \tag{23}$$

Consider the process of noncommutative reduction. The operators of λ representation of the Lie algebra $\mathfrak{g}$ can be taken as follows:

$$l_0 = -ie, \quad l_1 = -i\left(\frac{1}{2}\frac{\partial}{\partial q} - eHq\right), \quad l_2 = \frac{1}{2}\frac{\partial}{\partial q} + eHq, \quad l_3 = -iq\frac{\partial}{\partial q}, \quad q \in \mathbb{C}. \tag{24}$$

Operators (24) are skew-Hermitian with respect to the measure

$$d\mu(q) = \exp(2eH|q|^2) dq \wedge d\bar{q}.$$

From the system $\hat{X}_a\psi = -l_a\psi$ we have

$$\psi(x, y, z) = \exp\left(eH\left[\frac{x^2 + 2ixy + y^2}{4} - iq(x - iy)\right]\right)\left(\cosh(q') + \Sigma_3 \sinh(q')\right)\Phi(z), \tag{25}$$

where $q' = -\frac{1}{2}\log\left(q + \frac{i}{2}(x + iy)\right)$. Let us substitute (25) into the equation $\hat{H}\psi = E\psi$, then we obtain the reduced equation in the form

$$\hat{H}' = \left(\frac{1}{4} - eH\right)\alpha^1 + i\left(\frac{1}{4} + eH\right)\alpha^2 - i\alpha^3\frac{d}{dz} - eV(z) + \beta m.$$

Let us write the eigenvalue problem $\hat{S}\psi = \zeta\psi$ as follows:

$$\hat{S}'\Phi(z) = \zeta\sqrt{eH}\Phi(z), \quad \hat{S}' = \beta\left(eH(\Sigma^1 - i\Sigma^2) + \frac{1}{4}(\Sigma^1 + i\Sigma^2)\right).$$

Then, we find the basis of solutions parameterized by the spin quantum number ζ and a complex parameter q for the Dirac equation (19):

$$\psi_\sigma(x,y,z) = \begin{pmatrix} D_{q,\zeta}(x,y)f_\sigma(z) \\ D_{q,-\zeta}(x,y)g_\sigma(z) \end{pmatrix}, \quad \sigma = (q,\zeta,E). \tag{26}$$

Here the following notations are used:

$$D_{q,\zeta}(x,y) = \exp\left(eH\left[\frac{1}{4}(x+iy)^2 - iq(x-iy) + \frac{1}{2}y^2\right]\right)\begin{pmatrix} \exp(q') \\ 2\exp(-q')\sqrt{eH}\zeta \end{pmatrix}.$$

The functions $f(z)$ and $g(z)$ are determined by the system of ordinary differential equations (23) for $n = 1$.

The functions of the noncommutative basis of solutions (26) are the eigenfunctions of the spin operator (21) and are not eigenfunctions of the momentum operator $\hat{p}_2 = -iX_2$:

$$\hat{S}\psi_\sigma = \zeta\sqrt{eH}\psi_\sigma, \quad \hat{p}_2\psi_\sigma = il_2(q)\psi_\sigma.$$

Thus, the dependence on the variables (x, y) is expressed in terms of elementary functions. The reduced system (23) is generally different from the reduced system of equations obtained by the noncommutative reduction method, unlike the case of a spherically symmetric field, where the reduced system of equations is the same for the both bases of solutions. Therefore, elucidating the connection between basis (22) obtained by the method of separation of variables and basis (26) obtained by the noncommutative reduction is more difficult problem than in the case of a spherically symmetric potential, and is the subject of a separate study.

4. Electromagnetic field contracting a subalgebra of symmetry operators

Consider an electromagnetic field

$$\mathbf{H} = [\mathbf{n}, \mathbf{E}], \quad (\mathbf{n}, \mathbf{E}) = 0, \quad \mathbf{n} = (0,0,-1), \quad E_x = \frac{\alpha}{t-z}, \quad E_y = \varphi(y), \tag{27}$$

where α is a real parameter and $\varphi(y)$ is a smooth function. The potential of the field (27) can be taken as

$$A_2 = \alpha, \quad A_3 = -\frac{\alpha}{\varepsilon}, \quad A_4 = -A_1 = \frac{\alpha x}{t-z} + \varphi(y), \quad \varepsilon > 0.$$

The Dirac equation on $\mathcal{M}(t,x,y,z)$

$$\hat{H}\psi = m\psi, \quad \hat{H} = \gamma^\mu\hat{\mathcal{P}}_\mu, \quad \hat{\mathcal{P}}_\mu = \hat{p}_\mu - eA_\mu \tag{28}$$

when $e = 0$ admits the Lie algebra $\mathfrak{g}$ of symmetry operators

$$\hat{X}_1 = \hat{L}_{21} + \hat{L}_{24} + \frac{i}{2}(\gamma^{12} + \gamma^{24}), \quad \hat{X}_2 = \hat{L}_{14} - \frac{i}{2}\gamma^{14}, \quad \hat{X}_3 = \hat{\partial}_t + \partial_z, \quad \hat{X}_4 = \hat{\partial}_x + \varepsilon\hat{\partial}_y,$$

with with non-trivial commutators

$$[\hat{X}_1, \hat{X}_2] = \hat{X}_1, \quad [\hat{X}_1, \hat{X}_4] = -\hat{X}_3, \quad [\hat{X}_2, \hat{X}_3] = -\hat{X}_3.$$

Here $\hat{L}_{\mu\nu} = x_\mu\partial_\nu - x_\nu\partial_\mu$, $x_\mu = (t, -x, -y, -z)$. In presence of the external electromagnetic field (27), the Dirac equation admits the Abelian subalgebra $\{\hat{X}_1, \hat{X}_3\}$ of an algebra $\mathfrak{g}$ as the algebra of differential symmetry operators of the first order. However, there are no spin symmetry operators. That is, the inclusion of the electromagnetic field (27) contracts the symmetry algebra of the free Dirac equation to the commutative subalgebra.

A necessary condition for the separation of variables in the Dirac equation is the existence of three commuting symmetry operators of the first order. In our case, this condition does not hold and for the reduction of the Dirac equation to a system of ordinary differential equations one needs to go beyond the classical separation of variables. One way may be the separation of variables in the squared Dirac equation $(\hat{H}^2 - m^2)\phi = 0$ with the use of a complete set of symmetry operators $\{\hat{Y}_1, \hat{Y}_3, \hat{p}_{22}\}$. Then the basis of solutions of the original Dirac equation (28) can be obtained in the form of $\psi = (\hat{H} + m)\phi$.

We show that one can perform the noncommutative reduction of the Dirac equation (28) in the external electromagnetic field (27) using the Lie algebra $\mathfrak{g}$ of symmetry operators of the free Dirac equation.

Let U be a neighborhood of a point with the coordinates $t = x = y = z = 0$. Denote by $g = \{g_1, g_2, g_3, g_4\}$ a curvilinear coordinate system on U,

$$t = -\frac{1}{2}(g_1^2 e^{g_2} + e^{-g_2}), \quad x = g_4 - g_1, \quad y = \varepsilon g_4, \quad z = -\frac{1}{2}(g_1^2 e^{g_2} - e^{-g_2}). \tag{29}$$

The local Lie group $G = \exp(\mathfrak{g})$ acts simply transitively on Minkowski space. We can identify $U(g)$ with the local Lie group $G \simeq U(g)$ by introducing on $U(g)$ the group multiplication law:

$$(g \cdot g')_1 = g_1 + e^{-g_2} g'_1, \quad (g \cdot g')_2 = g_2 + g'_2,$$
$$(g \cdot g')_3 = g'_3 + e^{g'_2}(g_3 + g_4 g'_1), \quad (g \cdot g')_4 = g_4 + g'_4.$$

Note that (29) is the canonical coordinate system of the second kind on the Lie group G. The left-invariant $\xi(g)$ and right-invariant $\eta(g)$ basis vector fields on the Lie group G have the form

$$\xi_1 = e^{-g_2}\partial_1 + g_4\partial_3, \quad \xi_2 = \partial_2 + g_3\partial_3, \quad \xi_3 = \partial_3, \quad \xi_4 = \partial_4,$$
$$\eta_1 = -\partial_1, \quad \eta_2 = g_1\partial_1 - \partial_2, \quad \eta_3 = -e^{g_2}\partial_3, \quad \eta_4 = -(g_1 e^{g_2}\partial_3 + \partial_4).$$

The operators of λ-representation have the form

$$l_1 = iq_1 \exp(-q_2), \quad l_2 = \partial_{q_2}, \quad l_3 = i\exp(-q_2), \quad l_4 = \partial_{q_1}, \quad (q_1, q_2) \in Q = \mathbb{R}^2.$$

We introduce the measure $d\mu(q) = dq_1 dq_2$ in the space Q with respect to which the operators $-il_a(q, \partial_q)$ are Hermitian. Solving the system of equations (8) we obtain

$$D_{qq'}(g^{-1}) = \exp\left(-i\left[g_3 e^{-g_1 - q'_2} + g_1 q'_1 e^{-q'_2}\right] - 2q'_2\right)\delta(g_4 + q'_1 - q_1)\delta(g_2 + q'_2 - q_2).$$

Let us write the Dirac equation in the new coordinate system. The metric in the curvilinear coordinate system (29) is represented as a right-invariant metric on a Lie group G with the metric tensor

$$g^{\mu\nu}(g) = G^{ab}\eta_a^\mu(g)\eta_b^\nu(g), \quad G^{ab} = (G_{ab})^{-1}, \quad G_{ab} = \begin{pmatrix} -1 & 0 & 0 & 1 \\ 0 & 0 & 1 & 0 \\ 0 & 1 & 0 & 0 \\ 1 & 0 & 0 & -(1+\varepsilon^2) \end{pmatrix}.$$

Journal of Physics: Conference Series **670** (2016) 012015 doi:10.1088/1742-6596/670/1/012015

Here, G_{ab} are tetrad components of the metric tensor $g_{\mu\nu}$ with respect to the moving frame of the right-invariant vector fields $\eta_a(g)$. We decompose the Dirac gamma matrices $\gamma^\mu(g)$ in a curvilinear coordinate system $g_\nu = g_\nu(t,x,y,z)$ in the moving frame: $\gamma^\mu(g) = \hat\gamma^a \eta_a^\mu(g)$. The constant gamma matrices $\hat\gamma^a$ are defined as an arbitrary but fixed solution of the equation $\{\hat\gamma^a, \hat\gamma^b\} = 2G^{ab}$. Let us choose a solution of this equation in the form

$$\hat\gamma^1 = \frac{1}{\varepsilon}\gamma^3 - \gamma^4, \quad \hat\gamma^2 = -\frac{1}{2}(\gamma^1 - \gamma^2), \quad \hat\gamma^3 = -(\gamma^1 + \gamma^2), \quad \hat\gamma^4 = \frac{1}{\varepsilon}\gamma^3.$$

The Dirac operator (28) takes the form

$$\hat H = \hat\gamma^a \left(i(\eta_a(g) + \Gamma_a) - eA_a(g)\right), \tag{30}$$

where Γ_a are tetrad components of the spin connection

$$\Gamma_1 = -\frac{1}{2}(\hat\gamma^{12} + \hat\gamma^{24}), \quad \Gamma_2 = -\frac{1}{2}\hat\gamma^{23}, \quad \Gamma_3 = \Gamma_4 = 0.$$

The electromagnetic potential $A_a(g) = A_\mu(g)\eta_a^\mu(g)$ is given by the following tetrad components

$$A_1(g) = \alpha, \quad A_2(g) = \exp(-g_2)\phi(\varepsilon g_4) - \alpha g_4, \quad A_3(g) = A_4(g) = 0.$$

We seek the basis of solutions for the Dirac equation with operator (30) in the form

$$\psi_q(g) = \int_Q \hat\psi(q')D_{qq'}(g^{-1})d\mu(q) = \tag{31}$$
$$= \exp\left(-i\left[g_3 e^{-g_1 - q'_2} + g_1 q'_1 e^{-q'_2}\right] - 2q'_2\right)\hat\psi(q_1 - g_4, q_2 - g_2).$$

Substituting (31) into the Dirac equation $\hat H\psi = m\psi$, we obtain the reduced Dirac equation with the Hamiltonian

$$\hat H = \hat\gamma^1\left[i(l_1(q') + \Gamma_1) - e\alpha\right] + i\hat\gamma^3 l_3(q') + i\hat\gamma^4 l_4(q') +$$
$$+\hat\gamma^2\left[i(l_2(q') + \Gamma_2) - e\left(\varphi(\varepsilon(q_1 - q'_1))e^{q'_2 - q_2} - \alpha(q_1 - q_1')\right)\right].$$

Making a change of variables, $u = q_1 - q_1'$, $v = \exp(q_2' - q_2)$, we can write the reduced Dirac operator as

$$\hat H = -i\hat\gamma^4\partial_u + iv\hat\gamma^2\partial_v - \frac{1}{v}e^{-q_2}\left((q_1 - u)\hat\gamma^1 + \hat\gamma^3\right) - e\alpha\hat\gamma^1 - \tag{32}$$
$$-e\hat\gamma^2\left(v\varphi(\varepsilon u) - \alpha u\right) + \frac{i}{2}\hat\gamma^2\left(\hat\gamma^1\hat\gamma^4 + \frac{1}{\varepsilon^2} + 2\right)$$

The Dirac operator (32) admits the following symmetry operator of the first order:

$$\hat Y = -\partial_v + \frac{1}{2v}\left(\hat\gamma^{23} + (u - q_1)(\hat\gamma^{12} + \gamma^{24}) + \frac{i}{v}e^{-q_2}(u - q_1)^2 + 2ie\alpha q_1(1 - v) - 1.\right)$$

This the symmetry operator allows to perform separation of variables in the reduced Dirac equation. The solution of the equation $-i\hat Y\psi_\kappa(u,v) = \kappa\psi_\kappa(u,v)$ is

$$\psi_\kappa(u,v) = \frac{1}{\sqrt{v}}\exp\left(\frac{1}{2}\left[\hat\gamma^{23} + (u - q_1)(\hat\gamma^{12} + \hat\gamma^{24}) + 2ie\alpha q_1\right]\log v - \tag{33}$$
$$-i(\kappa + e\alpha q_1)v - \frac{i}{2v}e^{-q_2}(u - q_1)^2\right)\Phi_\kappa(u).$$

Substituting (33) into the Dirac equation $\hat{H}\psi_\kappa(u,v) = m\psi_\kappa(u,v)$, we leads to a system of ordinary differential equations:

$$-i\hat{\gamma}^4\Phi'_\kappa(u) + \Big(\hat{\gamma}^1\left((u-q_1)e^{-q_2} - e\alpha\right) - e^{-q_2}\left(\hat{\gamma}^3 + (u-q_1)\hat{\gamma}^4\right) + \tag{34}$$
$$+\hat{\gamma}^2\left(\frac{i}{2}\hat{\gamma}^1\hat{\gamma}^4 - \frac{1}{2}e^{-q_2}(u-q_1)^2 - e\varphi(\varepsilon u) + e\alpha u + \frac{1}{2\varepsilon^2} + \kappa\right) - m\Big)\Phi_\kappa(u) = 0.$$

We explored the integrability features of the Dirac equation with an external electromagnetic field by means of the noncommutative integration method and the method of separation of variables in terms of external fields of special form. Both methods use essentially the Lie algebra of symmetry operators of the Dirac equation. In the examples considered, we studied changes of the subalgebras of symmetry operators that are used to construct exact solutions of the Dirac equation with an external field, compared to the subalgebras of the free Dirac equation.

In the classical problem of integration of the Dirac equation with a spherically symmetric potential, a noncommutative reduction was carried out using the Lie algebra $\mathfrak{so}(3)$ of differential symmetry operators of the first order. It was shown that the reduced Dirac equation obtained by means of the NI method is identical to the similar equation, which occurs in the SoV method.This is due to the fact that in this problem the inclusion of an external field does not change the subalgebra $\mathfrak{so}(3)$ of the Lie symmetry algebra of the free Dirac equation, which is used for constructing solutions. The relationship between the NI-solutions and the separable solutions was found. The angular part of the NI-solution is expressed in terms of elementary functions and these solutions are of interest in the quantum theory of angular momentum. An expression was obtained for the spherical spinors (17) which implies that a continuous parameter q numbering the basis functions of the NI-solutions is a dual value with respect to the projection of the momentum on the preferred direction.

By the example of the Dirac equation in an external electric $E_z = E_z(z)$ and a uniform static magnetic field $H = H_z$, we investigated the situation when an external field leads to a central extension of the Lie algebra of symmetry operators. In this case the reduced Dirac equation, obtained by the noncommutative integration method, is shown to be the same as the similar equation arising in the SoV method only in a special case.

The noncommutative reduction was performed for the Dirac equation in the crossed fields (27). The external field contracts the Lie algebra of symmetry operators to its Abelian subalgebra. In this case the Dirac equation does not admit separation of variables. But, it is possible to perform the noncommutative reduction of the Dirac equation to a system of ordinary differential equations (34) using a subalgebra of symmetry operators for the free Dirac equation.

Acknowledgments

The work was partially supported by Tomsk State University Competitiveness Improvement Program and by program 'Nauka' under contract No. 1.676.2014/ K.

References

[1] Bagrov V G and Gitman D M 1990 *Exact solutions of relativistic wave equations* (Dordrecht: Kluwer);
[2] Bagrov V G and Gitman D M 2014 *The Dirac equation and its Solutions*(Boston: De Gruyter);
[3] Simulik V M and Krivski I Yu 2014 Link between the relativistic canonical quantum mechanics and the Dirac equation *Univ. J. Phys. Appl.* **2** 115
[4] Sitenko Yu A and Yushchenko S A 2014 The Casimir effect with quantized charged scalar matter in background magnetic field *Int. J. Mod. Phys. A* **29** 1450052
[5] Fradkin E S, Gitman D M and Shvartsman Sh M 1991 *Quantum Electrodynamics with Unstable Vacuum* (Springer-Verlag, Berlin Heidelberg New-York London Paris Hong-Kong Barcelona)

[6] Shapovalov A V and Shirokov I V 1995 Noncommutative integration of linear differential equations *Theor. Math. Phys.* **104** 921
[7] Shapovalov A V and Shirokov I V 1996 Noncommutative integration method for linear partial differential equations. Functional algebras and dimensional reduction *Theor. Math. Phys.* **106** 1
[8] Breev A I and Shapovalov A V 2014 Yang-Mills gauge fields conserving the symmetry algebra of the Dirac equation in a homogeneous space *Journal of Physics: Conference Series*, **563** 012004
[9] Breev A I and Shirokov I V 2009 Polarization of a spinor field vacuum on manifolds of the lie groups *Russian Physics Journal* **52** 823
[10] Klishevich V V 2001 Exact solution of Dirac and Klein–Gordon–Fock equations in a curved space admitting a second Dirac operator *Class. Quantum Grav.* **18** 3735
[11] Klishevich V Vand Tyumentsev V A 2005 On the solution of the Dirac equation in de Sitter space *Class. Quantum Grav.* **22** 4263
[12] Varaksin O L and Shirokov I V 1996 Integration of the Dirac equation, which does not presume complete separation of variables, in Stackel spaces *Russian Physics Journal* **39** 27
[13] Shapovalov V N 1975 Symmetry of Dirac-Fock equation *Izv. Vuz. Phys.* **6** 57.
[14] Carter B and McLenaghan R G 1979 Generalized total angular momentum operator for the Dirac equation in curved space-time *Phys. Rev. D* **19** 1093.
[15] Magazev A A 2012 Integration Klein-Gordon-Fock equations in an external electromagnetic field on Lie groups *Theor. Math. Physics* **3** 1654.
[16] Bagrov V G and Obukhov V V 1992 New method of integration for the Dirac equation on a curved space-time *Journal of Mathematical Physics* **33** 2279.
[17] Akhiezer A I and Berestetskii V B 1965 *Quantum Electrodynamics* (New York: Interscience Publishers).
[18] Edmonds A R 1957 *Angular Momentum in Quantum Mechanics* (Princeton: Princeton University Press)
[19] Baranovsky S P, Mikheev V V and Shirokov I V 2001 Quantum Hamiltonian Systems on K-Orbits: Semiclassical Spectrum of the Asymmetric Top *Theor. Math. Phys.* **129** 1311
[20] Gonchapovskii M M and Shirokov I V 2009 An integrable class of differential equations with nonlocal nonlinearity on Lie groups *Theor. Math. Phys.*, **161** 1604.
[21] Breev A I 2014 Scalar field vacuum polarization on homogeneous spaces with an invariant metric *Theor. Math. Phys.* **178** 59
[22] Kirillov A A 1976 *Elements of the Theory of Representation* (New York: Springer)

Journal of Physics: Conference Series **670** (2016) 012016 doi:10.1088/1742-6596/670/1/012016

Unifying Ancient and Modern Geometries Through Octonions

Sultan Catto[1,2,†], Yasemin Gürcan[3], Amish Khalfan[4] and Levent Kurt[3,†]

[1]Physics Department, The Graduate School, City University of New York, New York, NY 10016-4309
[2]Theoretical Physics Group, Rockefeller University, 1230 York Avenue, New York, NY 10021-6399
[3] Department of Science, Borough of Manhattan CC, The City University of NY, New York, NY 10007
[4] Physics Department, LaGuardia CC, The City University of New York, LIC, NY 11101

† Work supported in part by the PSC-CUNY Research Awards, and DOE contracts No. DE-AC-0276 ER 03074 and 03075; and NSF Grant No. DMS-8917754.

Abstract.

We show the first unified description of some of the oldest known geometries such as the Pappus' theorem with more modern ones like Desargues' theorem, Monge's theorem and Ceva's theorem, through octonions, the highest normed division algebra in eight dimensions. We also show important applications in hadronic physics, giving a full description of the algebra of color applicable to quark physics, and comment on further applications.

1. Introduction

In mid sixties Miyazawa, in a series of papers[1], extended the $SU(6)$ group to the supergroup $SU(6/21)$ that could be generated by constituent quarks and diquarks that could be transformed to each other. In particular, he found the following: (a) A general definition of $SU(m/n)$ superalgebras, expressing the symmetry between m bosons and n fermions, with Grassman-valued parameters. (b) A derivation of the super-Jacobi identity. (c) The relation of the baryon mass splitting to the meson mass splitting through the new mass formulae.

This work contained the first classification of superalgebras (later rediscovered by mathematicians in the seventies). Because of the field-theoretic prejudice against $SU(6)$, Miyazawa's work was generally ignored. Supersymmetry was, of course, rediscovered in the seventies within the dual resonance model by Ramond[2], and Neveu and Schwarz[3]. Golfand and Likhtman[4], and independently Volkov and Akulov[5], proposed the extension of the Poincaré group to the super-Poincaré group. Examples of supersymmetric field theories were given and the general method based on the super-Poincaré group was discovered by Wess and Zumino[6]. The super-Poincaré group allowed transformations between fields associated with different spins 0, $\frac{1}{2}$ and 1. The Coleman-Mandula theorem was amended in 1975 by Haag, Lopuszanski and Sohnius to allow super-Poincaré e group $\times G_{int}$ as the maximum symmetry of the S-matrix. Unfortunately, $SU(6)$ symmetry was still forbidden.

2. SU(6) and Hadronic Supersymmetry

How do we interpret the symmetries of the QCD spectrum in this light? In the ultraviolet, the running coupling constant tends to zero and quarks behave like free point particles. Thus an approximate conformal symmetry exists, allowing spin to be conserved separately from orbital angular momentum. Thus spin behaves as an internal quantum number; this makes a $SU(6)$ symmetry possible, since the quarks are almost free Dirac particles. Single vector-gluon exchange breaks this symmetry; thus, as shown by Glashow, Georgi and deRujula[7], the mass-degeneracy of hadrons of different spins is lifted by a hyperfine-interaction term.

Here is the main point. In the infrared we expect confinement to set in. The quark-antiquark potential becomes proportional to the distance. Careful studies of quarkonium spectra and lattice-gauge calculations show that at large separation the quark forces become spin-independent. QCD is also flavor independent. We therefore find approximate spin- and flavor-independent quark binding forces; these are completely consistent with $SU(6)$ symmetry. This is not an exact symmetry, but is a good starting point, before spin and flavor effects are included.

There is good phenomenological evidence that in a rotationally excited baryon a quark-diquark $(q - D)$ structure is favored over a three-quark (qqq) structure[8],[9],[10]. Eguchi[11] had shown that it is energetically favorable for the three quarks in a baryon to form a linear structure with a quark on one end and bilocal structure qq at the other end. Similarly, Bars and Hanson[12], and independently Johnson and Thorn[13] had shown that the string-like hadrons may be pictured as vortices of color flux lines which terminate on concentration of color at the end points. A baryon with three valence quarks would be arranged as a linear chain of molecule where the largest angular momentum for a state of a given mass is expected when two quarks are at one end, and the third is at the other: At large spin, two of the quarks form a diquark at one end of the string, the remaining quark being at the other. Regge trajectories for mesons and baryons are closely parallel; both have a slope of about $0.9(GeV)^{-2}$. If the quarks are light, the underlying quark-diquark symmetry leads to a Miyazawa symmetry between mesons and baryons. Thus we studied QCD with a weakly broken supergroup $SU(6/21)$. Note that the fundamental theory is not supersymmetric. For quarks, the generators of the Poincaré group and those of the color group $SU(3)^c$ commute. It is only the effective Hamiltonian which exhibits an approximate supersymmetry among the bound states $q\bar{q}$ and qD.

Under the color group $SU(3)^c$, meson $q\bar{q}$ and diquark $(D = qq)$ states transform as[10],[14]

$$qq: \ \mathbf{3} \times \mathbf{3} = \bar{\mathbf{3}} + \mathbf{6} \ ; \quad q\bar{q}: \ \mathbf{3} \times \bar{\mathbf{3}} = \mathbf{1} + \mathbf{8} \tag{1}$$

and under the spin flavor $SU(6)$ they transform as

$$qq: \ \mathbf{6} \times \mathbf{6} = \mathbf{15} + \mathbf{21} \ ; \quad q\bar{q}: \ \mathbf{6} \times \bar{\mathbf{6}} = \mathbf{1} + \mathbf{35} \tag{2}$$

Dimensions of internal degrees of quarks and diquarks are shown in the following table:

	$SU_f(3)$	$SU_s(2)$	$dim.$
q	□	$s = 1/2$	$3 \times 2 = 6$
D	□ □	$s = 1$	$6 \times 3 = 18$
	□ □	$s = 0$	$3 \times 1 = 3$

For a more general case of the above table we refer the reader to the table with the flavor-spin content at the end of the paper.

If one writes qqq as qD, then the quantum numbers of D are $\bar{\mathbf{3}}$ for color since when combined with q must give a color singlet, and $\mathbf{21}$ for spin-flavor since combined with color must give antisymmetric wavefunctions. The quantum numbers for $\bar{q}$ are for color, $\bar{\mathbf{3}}$, and for spin-flavor, $\bar{\mathbf{6}}$. Thus $\bar{q}$ and D have the same quantum numbers (color forces can not distinguish between $\bar{q}$ and D). Therefore there is a dynamic supersymmetry in hadrons with supersymmetric partners

$$\psi = \begin{pmatrix} \bar{q} \\ D \end{pmatrix}, \qquad \bar{\psi} = (q \ \ \bar{D}) \tag{3}$$

We can obtain all hadrons by combining ψ and $\bar{\psi}$: mesons are $q\bar{q}$, baryons are qD, and exotics are $D\bar{D}$ states. Inside rotationally excited baryons, QCD leads to the formation of diquarks well separated from the remaining quark. At this separation the scalar, spin-independent, confining part of the effective QCD potential is dominant. Since QCD forces are also flavor-independent, the force between the quark q and the diquark D inside an excited baryon is essentially the same as the one between q and the antiquark $\bar{q}$ inside an excited meson. Thus the approximate spin-flavor independence of hadronic physics expressed by $SU(6)$ symmetry is extended to $SU(6/21)$ supersymmetry[10],[14] through a symmetry between $\bar{q}$ and D, resulting in parallelism of mesonic and baryonic Regge trajectories.

3. Color Algebra and Octonions

We shall now give an algebraic justification to our remarks above. We will find all the answers in an algebra we build in terms of octonions and their split basis. The exact, unbroken color group $SU(3)^c$ is the backbone of the strong interaction. It is worthwhile to understand its role in the diquark picture more clearly.

In what follows we first give a simple description of octonion algebra (also known as Cayley algebra). Later we'll show how to build split octonion algebra that will close into a fermionic Heisenberg algebra. Split octonion algebra will then be shown to produce algebra of color forces in QCD in application to hadronic supersymmetry when the split units and their conjugates become associated with quark and antiquark fields, respectively.

An octonion x is a set of eight real numbers

$$x = (x_0, x_1, \ldots, x_7) = x_0 e_0 + x_1 e_1 + \ldots + x_7 e_7 \tag{4}$$

that are added like vectors and multiplied according to the rules

$$e_0 = 1, \quad e_0 e_i = e_i e_0 = e_i, \quad i = 0, 1, \ldots, 7 \tag{5}$$

$$e_\alpha e_\beta = -\delta_{\alpha\beta} + \epsilon_{\alpha\beta\gamma} e_\gamma. \quad \alpha, \beta, \gamma = 1, 2, \ldots, 7 \tag{6}$$

where e_0 is the multiplicative unit element and e_α's are the imaginary octonion units. The structure constants $\epsilon_{\alpha\beta\gamma}$ are completely antisymmetric and take the value 1 for combinations

$$\epsilon_{\alpha\beta\gamma} = (165), (257), (312), (471), (543), (624), (736) \tag{7}$$

Note that summation convention is used for repeated indices.

The octonion algebra C is an algebra defined over the field $\mathbf{Q}$ of rational numbers, which as a vector space over $\mathbf{Q}$ has dimension 8.

We shall now give reasons for incorporation of the octonion algebra for hadronic physics, showing only they through their split octonionic parts one can provide the correct description of the color algebra in hadrons. Later in another publication we shall show[15] a previously unknown multiplication rules for octonions by producing a wheel that allows generalized multiplication rules for doublets and triplets of octonionic units.

First, the reasons: Two of the colored quarks in the baryon combine into an anti-triplet $\mathbf{3} \times \mathbf{3} = \bar{\mathbf{3}} + (\mathbf{6})$, and in a nucleon $\mathbf{3} \times \bar{\mathbf{3}} = \mathbf{1} + (\mathbf{8})$. The $(\mathbf{6})$ partner of the diquark and the $(\mathbf{8})$ partner of the nucleon do not exist. In hadron dynamics the only color combinations to consider are $\mathbf{3} \times \mathbf{3} \to \bar{\mathbf{3}}$ and $\bar{\mathbf{3}} \times \mathbf{3} \to \mathbf{1}$. These relations imply the existence of split octonion units u_i defined below through a representation of the Grassmann algebra $\{u_i, u_j\} = 0$, $i = 1, 2, 3$. What is a bit strange is that operators u_i, unlike ordinary fermionic operators, are not associative. We also have $\frac{1}{2}[u_i, u_j] = \epsilon_{ijk}\, u_k^*$. The Jacobi identity does not hold since $[u_i, [u_j, u_k]] = -ie_7 \neq 0$, where e_7, anticommute with u_i and u_i^*.

The behavior of various states under the color group are best seen if we use split octonion units defined by[16]

$$u_0 = \frac{1}{2}(1 + ie_7), \qquad u_0^* = \frac{1}{2}(1 - ie_7) \tag{8}$$

$$u_j = \frac{1}{2}(e_j + ie_{j+3}), \quad u_j^* = \frac{1}{2}(e_j - ie_{j+3}), \quad j = 1, 2, 3 \tag{9}$$

The automorphism group of the octonion algebra is the 14-parameter exceptional group G_2. The imaginary octonion units $e_\alpha (\alpha = 1, ..., 7)$ fall into its 7-dimensional representation.

Under the $SU(3)^c$ subgroup of G_2 that leaves e_7 invariant, u_0 and u_0^* are singlets, while u_j and u_j^* correspond, respectively, to the representations **3** and $\bar{\mathbf{3}}$. The multiplication table can now be written in a manifestly $SU(3)^c$ invariant manner (together with the complex conjugate equations):

$$u_0^2 = u_0, \qquad u_0 u_0^* = 0 \tag{10}$$

$$u_0 u_j = u_j u_0^* = u_j, \qquad u_0^* u_j = u_j u_0 = 0 \tag{11}$$

$$u_i u_j = -u_j u_i = \epsilon_{ijk} u_k^* \tag{12}$$

$$u_i u_j^* = -\delta_{ij} u_0 \tag{13}$$

where ϵ_{ijk} is completely antisymmetric with $\epsilon_{ijk} = 1$ for ijk = 123, 246, 435, 651, 572, 714, 367; and zero otherwise. Here, one sees the virtue of octonion multiplication. If we consider the direct products

$$C: \quad \mathbf{3} \otimes \bar{\mathbf{3}} = \mathbf{1} + \mathbf{8} \tag{14}$$

$$G: \quad \mathbf{3} \otimes \mathbf{3} = \bar{\mathbf{3}} + \mathbf{6} \tag{15}$$

for $SU(3)^c$, then these equations show that octonion multiplication gets rid of **8** in $\mathbf{3} \otimes \bar{\mathbf{3}}$, while it gets rid of **6** in $\mathbf{3} \otimes \mathbf{3}$. Combining Eq.(12) and Eq.(13) we find

$$(u_i u_j) u_k = -\epsilon_{ijk} u_0^* \tag{16}$$

Thus the octonion product leaves only the color part in $\mathbf{3} \otimes \bar{\mathbf{3}}$ and $\mathbf{3} \otimes \mathbf{3} \otimes \mathbf{3}$, so that it is a natural algebra for colored quarks.

For convenience we now produce the following multiplication table for the split octonion units:

	u_0	u_0^*	u_k	u_k^*
u_0	u_0	0	u_k	0
u_0^*	0	u_0^*	0	u_k^*
u_j	0	u_j	$\epsilon_{jki} u_i^*$	$-\delta_{jk} u_0$
u_j^*	u_j^*	0	$-\delta_{jk} u_0^*$	$\epsilon_{jki} u_i$

Journal of Physics: Conference Series **670** (2016) 012016 doi:10.1088/1742-6596/670/1/012016

Double and Triple Octonionic Multiplication

1 2 3	2 4 6	4 3 5	3 6 7	6 5 1	5 7 2	7 1 4
4 6 5 7	3 5 7 1	6 7 1 2	5 1 2 4	7 2 4 3	1 4 3 6	2 3 6 5

It is worth noting that u_i and u_j^* behave like fermionic annihilation and creation operators:

$$\{u_i, u_j\} = \{u_i^*, u_j^*\} = 0, \quad \{u_i, u_k^*\} = -\delta_{ik} \tag{17}$$

For more recent reviews on octonions and nonassociative algebras we refer to papers by Okubo[17], Baez[18] and Catto[19].

What we would like to discuss now is a recent unpublished, taking the usual octonionic multiplication rules from doublets of octonions to triplets.

The usual octonion multiplication once sees in literature is a diagram (as shown by, for example, in Baez's paper[18]). That triangle was replaced by Gürsey and Günaydin. We extended and generalized their work by adding the dashed lines shown below in Figure 1. In Double and Triple Octonionic Multiplication table as shown, we describe the rotation of triangle and dashed lines for all octonionic multiplications. We also show by moving to the left instead of the right, we build a completely new multiplication table for doublets and triplets in the Left Handed Double and Triple Octonionic Multiplication and Figure 2. There is a complete correspondence between those diagrams and figures which will be explained in another publication.

We can now show how one can map some of the old and new geometries into each other.

In the well-known Papus' geometry, one can put the *e*'s into a diagram and show how a center line can be mapped into two arbitrary lines outside, above and below as shown Figure 3. One can perform six more distinct copies of these mappings.

In a similar way, one can see all these mappings into the Desargues' theorem in a beautiful generalized way. One can show that these pictures can be generated in two and three dimensions. These will also be published in another publication.

Finally, we want to mention that these mappings described above can be extended into Monge's and Ceva's theorems as shown in the Figures 4 and 5. These have tremendous applications not only in hadronic physics, but also in solutions of problems in astrophysics.

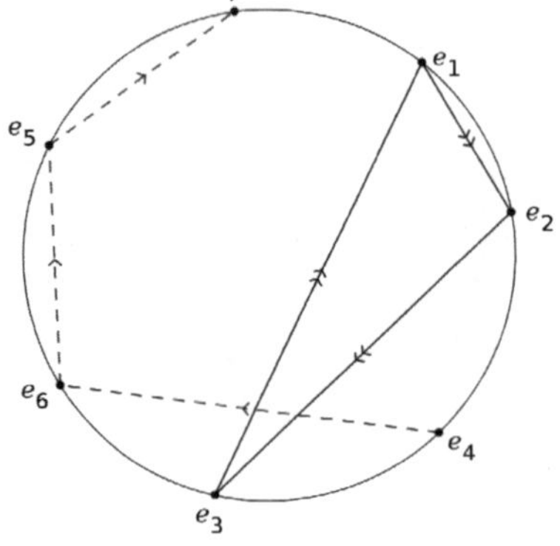

Figure 1.

Left Handed Double and Triple Octonionic Multiplication

1 7 6	7 5 3	5 6 4	6 3 2	3 4 1	4 2 7	2 1 5
5 3 4 2	6 4 2 1	3 2 1 7	4 1 7 5	2 7 5 6	1 5 6 3	7 6 3 4

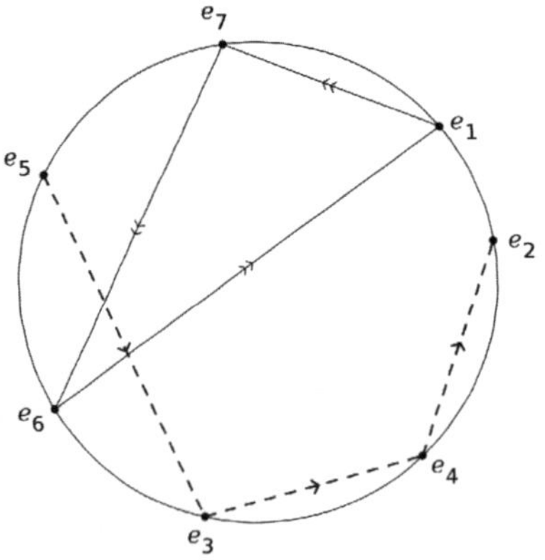

Figure 2.

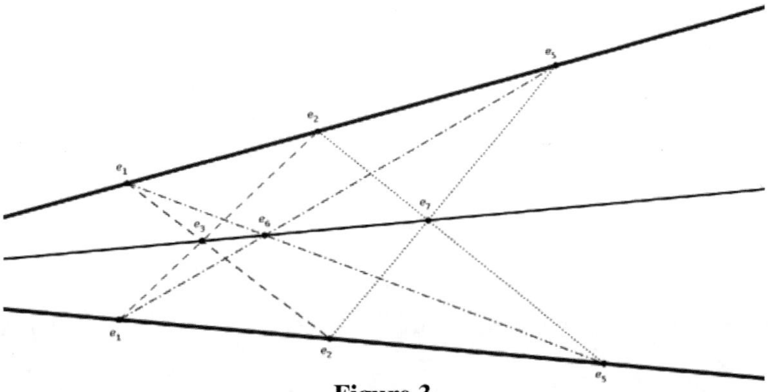

Figure 3.

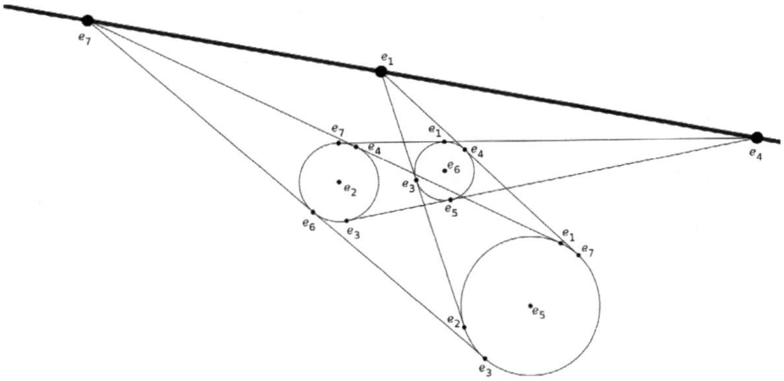

Figure 4.

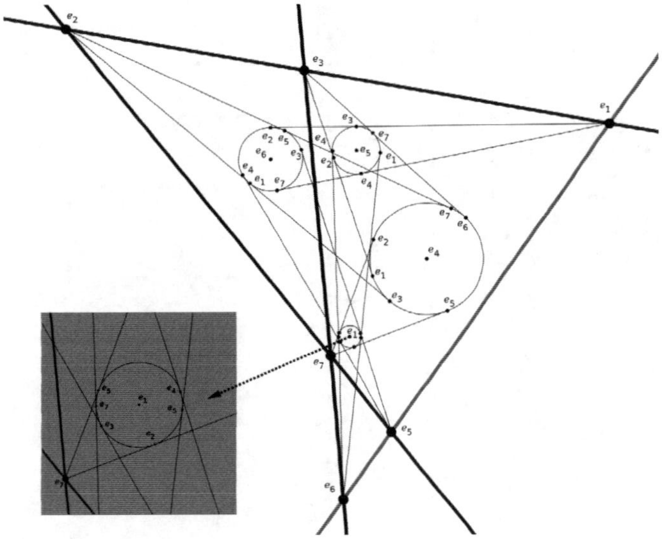

Figure 5.

Journal of Physics: Conference Series **670** (2016) 012016 doi:10.1088/1742-6596/670/1/012016

Table: Flavor-Spin Content for three quark and quark-diquark baryons

Configuration	$SU_c(3)$	$SU_f(3)$	$SU_s(2)$
q	□ $\equiv \underline{3}$	□ $\equiv \underline{3}$	$S = 1/2$
$D \equiv q^2$	□ □ $\equiv \underline{\bar{3}}$	□□ $\equiv \underline{6}$	$S = 0$ not allowed $S = 1$
		□ □ $\equiv \underline{\bar{3}}$	$S = 0$ $S = 1$ not allowed
q^3	□ □ $\equiv \underline{1}$ □	□□□ $\equiv \underline{10}$	$S = 3/2$
		□□ $\equiv \underline{8}$ □	$S = 1/2$
$q - D$	□ □ $\equiv \underline{1}$ □	$\underline{3} \times \underline{\bar{3}}$: □□ $\equiv \underline{8}$ □	$S = 1/2$
		$\underline{3} \times \underline{\bar{3}}$: □ □ $\equiv \underline{1}$ □	$S = 1/2$
		$\underline{3} \times \underline{6}$: □□□ $\equiv \underline{10}$	$S = 1/2$
		$\underline{3} \times \underline{6}$: □□□ $\equiv \underline{10}$	$S = 3/2$
		$\underline{3} \times \underline{6}$: □□ $\equiv \underline{8}$ □	$S = 1/2$
		$\underline{3} \times \underline{6}$: □□ $\equiv \underline{8}$ □	$S = 3/2$

References

[1] H. Miyazawa, *Prog. Theor. Phys.* **36** *(1966) 1266; Phys. Rev.* **170** *(1968) 1586.*
[2] P. Ramond, *Phys. Rev.* **D3** *(1971)2415.*
[3] A. Neveu and J. Schwarz, *Nucl.Phys.* **B31** *(1971) 86.*
[4] T.A. Golfand and E.P. Likhtman, *JETP Lett.***13** *(1971) 323.*
[5] D.V. Volkov and V.P. Akulov, *Phys. Lett.* **46b***(1973) 109.*
[6] J. Wess and B. Zumino, *Nucl.Phys.* **B70** *(1974) 39; Phys. Lett.* **49b** *(1974) 52.*
[7] A. de Rujula, H. Georgi and S.L. Glashow, *Phys. Rev.* **D12** *(1975) 147.*
[8] D.B. Lichtenberg, W. Namgung, E. Predazzi and J.G. Wills, *Phys. Lett.* **48** *(1982) 1653*; W. Namgung and D.B. Lichtenberg, *Lett. Nuovo Cimento* **41** *(1984) 597*; A. Martin, *Z. Phys.* **C32** *(1986) 359* and in "Symmetry in Nature" vol.II, p. 523 (Scuola Normale Superiore, Pisa, 1989).
[9] M. Anselmino, E. Predazzi, S. Ekelin, S. Fredriksson and D.B. Lichtenberg, *Rev. Mod. Phys.* **65** *(1993) 1199.*
[10] S. Catto and F. Gürsey, *Nuovo Cimento* **A 86** *(1985) 201.*
[11] T. Eguchi, *Phys. Lett. B,* **59** *(1975) 457.*
[12] I. Bars and A.J. Hanson, *Phys. Rev.***D13** *(1976) 1744.*
[13] For strings as elongated bags see K. Johnson and C.B. Thorn, *Phys. Rev.* **D13** *(1976) 1934.*
[14] S. Catto and F. Gürsey, *Nuovo Cimento* **A99** *(1988) 685.*
[15] S. Catto, *To be published.*
[16] M. Günaydin and F. Gürsey, *J. Math. Phys.* **14** *(1973) 1651.*
[17] S. Okubo, *"Introduction to Octonion and other Nonassociative Algebras." Montroll Memorial Lecture Series in Mathematical Physics, 2. Cambridge Univ. Press (1995).*
[18] J. Baez, *Bull. Am. Math. Soc.* **39(2)** *(2002) 145-205.*
[19] S. Catto, *"Exceptional Projective Geometries and Internal Symmetries" ArXiv: Hep-th/0302079*

Acknowledgments:

One of us (SC) would like to thank Professor Čestmir Burdik for the invited talk given in Prague and for his kind hospitality. We would like to thank Professors Itzhak Bars and Francesco Iachello for enlightening conversations. This work is supported by DOE contracts No. DE-AC-0276 ER 03074 and 03075, and PSC-CUNY Research Awards.

M. Riesz Theorem on Conjugate Harmonic Functions for Octonion-Valued Monogenic Functions

Sultan Catto[1], **Alexander Kheyfits**[2], **David E. Te er**[3]

[1] Rockefeller University and Baruch College and Graduate Center, CUNY

E-mail: scatto@gc.cuny.edu, scatto@rockefeller.edu

[2] Graduate Center and Bronx Community College City University of New York

E-mail: akheyfits@gc.cuny.edu

[3] Emeritus Baruch College and Graduate Center CUNY and Temple University

E-mail: detepper@temple.edu

Abstract. The classical theorem of M. Riesz about the conjugate harmonic functions is extended onto octonion-valued monogenic functions.

1. Introduction and Statement o the Main Result

Analytic (monogenic) functions of octonion variables, due to their potential, yet not completely fulfilled utility in physics [9], continue to attract the attention of researchers. John C. Baez in his very readable and informative survey formulates the development of an octonionic analogue of the theory of analytic functions as the first item in his list of potential octonion-related problems [3, p. 201]. In this note we continue the study of this topic, see, e.g., [5, 10, 11, 13, 14] and the references therein. For the reader's convenience, in this section we briefly review some basic definitions and then state the principal result, extending the celebrated theorem of M. Riesz on the conjugate harmonic functions onto the monogenic functions. A quaternionic version of this theorem was recently proved by Avetisyan [2]. Let $\sum_{j=0}^{7} x_j e_j$ be a generic octonion, where $e_0 \equiv 1$, and $e_1, \ldots, e_7$ are the basis octonion (imaginary) units; we identify it with a vector $\mathbf{x} = (x_0, \ldots, x_7) \in \mathbf{R}^8$. In notation we follow [5]. Let

$$f(\mathbf{x}) = \sum_{j=0}^{7} e_j f_j(\mathbf{x})$$

be an octonion-valued left-monogenic function in a domain $\subset \mathbf{R}^8$, where $f_0(\mathbf{x}), \ldots, f_7(\mathbf{x})$ are real-valued $\mathcal{C}^1$ functions. That means

$$D[f] = 0 \tag{1.1}$$

where $D = \sum_{k=0}^{7} e_k \frac{\partial}{\partial x_k}$ is the Dirac (or Cauchy-Riemann) operator. It is known that all the components $f_0, \ldots, f_7$ of a left-monogenic function are harmonic functions, that is,

$f_0 = \cdots = f_7 = 0$ in . The equation $Df = 0$ is a system of eight first-order linear partial differential equations with constant coefficients. It can be written as a matrix equation

$$\begin{bmatrix}
\frac{\partial}{\partial x_0} & -\frac{\partial}{\partial x_1} & -\frac{\partial}{\partial x_2} & -\frac{\partial}{\partial x_3} & -\frac{\partial}{\partial x_4} & -\frac{\partial}{\partial x_5} & -\frac{\partial}{\partial x_6} & -\frac{\partial}{\partial x_7} \\
\frac{\partial}{\partial x_1} & \frac{\partial}{\partial x_0} & -\frac{\partial}{\partial x_3} & \frac{\partial}{\partial x_2} & -\frac{\partial}{\partial x_7} & \frac{\partial}{\partial x_6} & -\frac{\partial}{\partial x_5} & \frac{\partial}{\partial x_4} \\
\frac{\partial}{\partial x_2} & \frac{\partial}{\partial x_3} & \frac{\partial}{\partial x_0} & -\frac{\partial}{\partial x_1} & -\frac{\partial}{\partial x_6} & -\frac{\partial}{\partial x_7} & \frac{\partial}{\partial x_4} & \frac{\partial}{\partial x_5} \\
\frac{\partial}{\partial x_3} & -\frac{\partial}{\partial x_2} & \frac{\partial}{\partial x_1} & \frac{\partial}{\partial x_0} & \frac{\partial}{\partial x_5} & -\frac{\partial}{\partial x_4} & -\frac{\partial}{\partial x_7} & \frac{\partial}{\partial x_6} \\
\frac{\partial}{\partial x_4} & \frac{\partial}{\partial x_7} & \frac{\partial}{\partial x_6} & -\frac{\partial}{\partial x_5} & \frac{\partial}{\partial x_0} & \frac{\partial}{\partial x_3} & -\frac{\partial}{\partial x_2} & -\frac{\partial}{\partial x_1} \\
\frac{\partial}{\partial x_5} & -\frac{\partial}{\partial x_6} & \frac{\partial}{\partial x_7} & \frac{\partial}{\partial x_4} & -\frac{\partial}{\partial x_3} & \frac{\partial}{\partial x_0} & \frac{\partial}{\partial x_1} & -\frac{\partial}{\partial x_2} \\
\frac{\partial}{\partial x_6} & \frac{\partial}{\partial x_5} & -\frac{\partial}{\partial x_4} & \frac{\partial}{\partial x_7} & \frac{\partial}{\partial x_2} & -\frac{\partial}{\partial x_1} & \frac{\partial}{\partial x_0} & -\frac{\partial}{\partial x_3} \\
\frac{\partial}{\partial x_7} & -\frac{\partial}{\partial x_4} & -\frac{\partial}{\partial x_5} & -\frac{\partial}{\partial x_6} & \frac{\partial}{\partial x_1} & \frac{\partial}{\partial x_2} & \frac{\partial}{\partial x_3} & \frac{\partial}{\partial x_0}
\end{bmatrix} F = 0,$$

where $F = [f_0, \ldots, f_7]^T$ is the unknown column vector-function. It can also be rewritten as a matrix equation

$$\sum_{j=0}^{7} A_j \frac{\partial F}{\partial x_j} = 0.$$

Here A_0 is the identity matrix of order 8 and the other seven 8×8 antisymmetric matrices $A_1 - A_7$ are given by

$$A_1 = \begin{bmatrix}
0 & -1 & 0 & 0 & 0 & 0 & 0 & 0 \\
1 & 0 & 0 & 0 & 0 & 0 & 0 & 0 \\
0 & 0 & 0 & -1 & 0 & 0 & 0 & 0 \\
0 & 0 & 1 & 0 & 0 & 0 & 0 & 0 \\
0 & 0 & 0 & 0 & 0 & 0 & 0 & -1 \\
0 & 0 & 0 & 0 & 0 & 0 & 1 & 0 \\
0 & 0 & 0 & 0 & 0 & -1 & 0 & 0 \\
0 & 0 & 0 & 0 & 1 & 0 & 0 & 0
\end{bmatrix}$$

$$A_2 = \begin{bmatrix}
0 & 0 & -1 & 0 & 0 & 0 & 0 & 0 \\
0 & 0 & 0 & 1 & 0 & 0 & 0 & 0 \\
1 & 0 & 0 & 0 & 0 & 0 & 0 & 0 \\
0 & -1 & 0 & 0 & 0 & 0 & 0 & 0 \\
0 & 0 & 0 & 0 & 0 & 0 & -1 & 0 \\
0 & 0 & 0 & 0 & 0 & 0 & 0 & -1 \\
0 & 0 & 0 & 0 & 1 & 0 & 0 & 0 \\
0 & 0 & 0 & 0 & 0 & 1 & 0 & 0
\end{bmatrix}$$

$$A_3 = \begin{bmatrix}
0 & 0 & 0 & -1 & 0 & 0 & 0 & 0 \\
0 & 0 & -1 & 0 & 0 & 0 & 0 & 0 \\
0 & 1 & 0 & 0 & 0 & 0 & 0 & 0 \\
1 & 0 & 0 & 0 & 0 & 0 & 0 & 0 \\
0 & 0 & 0 & 0 & 0 & 1 & 0 & 0 \\
0 & 0 & 0 & 0 & -1 & 0 & 0 & 0 \\
0 & 0 & 0 & 0 & 0 & 0 & 0 & -1 \\
0 & 0 & 0 & 0 & 0 & 0 & 1 & 0
\end{bmatrix}$$

$$A_4 = \begin{bmatrix} 0 & 0 & 0 & 0 & -1 & 0 & 0 & 0 \\ 0 & 0 & 0 & 0 & 0 & 0 & 0 & 1 \\ 0 & 0 & 0 & 0 & 0 & 0 & 1 & 0 \\ 0 & 0 & 0 & 0 & 0 & -1 & 0 & 0 \\ 1 & 0 & 0 & 0 & 0 & 0 & 0 & 0 \\ 0 & 0 & 0 & 1 & 0 & 0 & 0 & 0 \\ 0 & 0 & -1 & 0 & 0 & 0 & 0 & 0 \\ 0 & -1 & 0 & 0 & 0 & 0 & 0 & 0 \end{bmatrix}$$

$$A_5 = \begin{bmatrix} 0 & 0 & 0 & 0 & 0 & -1 & 0 & 0 \\ 0 & 0 & 0 & 0 & 0 & 0 & -1 & 0 \\ 0 & 0 & 0 & 0 & 0 & 0 & 0 & 1 \\ 0 & 0 & 0 & 0 & 1 & 0 & 0 & 0 \\ 0 & 0 & 0 & -1 & 0 & 0 & 0 & 0 \\ 1 & 0 & 0 & 0 & 0 & 0 & 0 & 0 \\ 0 & 1 & 0 & 0 & 0 & 0 & 0 & 0 \\ 0 & 0 & -1 & 0 & 0 & 0 & 0 & 0 \end{bmatrix}$$

$$A_6 = \begin{bmatrix} 0 & 0 & 0 & 0 & 0 & 0 & -1 & 0 \\ 0 & 0 & 0 & 0 & 0 & 1 & 0 & 0 \\ 0 & 0 & 0 & 0 & -1 & 0 & 0 & 0 \\ 0 & 0 & 0 & 0 & 0 & 0 & 0 & 1 \\ 0 & 0 & 1 & 0 & 0 & 0 & 0 & 0 \\ 0 & -1 & 0 & 0 & 0 & 0 & 0 & 0 \\ 1 & 0 & 0 & 0 & 0 & 0 & 0 & 0 \\ 0 & 0 & 0 & -1 & 0 & 0 & 0 & 0 \end{bmatrix}$$

$$A_7 = \begin{bmatrix} 0 & 0 & 0 & 0 & 0 & 0 & 0 & -1 \\ 0 & 0 & 0 & 0 & -1 & 0 & 0 & 0 \\ 0 & 0 & 0 & 0 & 0 & -1 & 0 & 0 \\ 0 & 0 & 0 & 0 & 0 & 0 & -1 & 0 \\ 0 & 1 & 0 & 0 & 0 & 0 & 0 & 0 \\ 0 & 0 & 1 & 0 & 0 & 0 & 0 & 0 \\ 0 & 0 & 0 & 1 & 0 & 0 & 0 & 0 \\ 1 & 0 & 0 & 0 & 0 & 0 & 0 & 0 \end{bmatrix}.$$

It should be mentioned that $\det(A_0) = \cdots = \det(A_7) = 1$. The octonion multiplication table can be written in various ways, for example, the table in [3, p. 150] is different from one in [4]. These different tables result in different systems (1.1), though all these systems are clearly equivalent. Solutions of the system $[f]D = 0$ are called right-monogenic functions; functions, which are both left- and right-monogenic, are called monogenic. Hereafter, we always discuss the left-monogenic functions, but the proofs go word-by-word for the right- and monogenic functions.

Systems (1.1), where each component f_j, $0 \le j \le 7$, is harmonic, are called the generalized Cauchy-Riemann systems (GCR) - see Stein and Weiss [15, pp. 260-262]. Systems $\sum_{j=0}^{7} A_j \frac{\partial F}{\partial x_j} + BF = 0$, where B is also a constant matrix, were considered by Evgrafov [8]. Stein and Weiss have proved that for any GCR system there exists a non-egative index $p_0 < 1$ such that $|F|^p$ is a subharmonic function for all $p \ge p_0$. It is known (ibid, p. 262) that for the M. Riesz system in $\mathbf{R}^n$

$$\frac{\partial u_1}{\partial x_1} + \cdots + \frac{\partial u_n}{\partial x_n} = 0,$$

Journal of Physics: Conference Series **670** (2016) 012017 doi:10.1088/1742-6596/670/1/012017

$$\frac{\partial u_i}{\partial x_j} = \frac{\partial u_j}{\partial x_i},\ i, j = 1, \ldots, n,\ i \neq j,$$

the exact value of p_0 is $(n-2)/(n-1)$. Earlier we proved in [10] that the same is valid for system (1.1) in $\mathbf{R}^8$, that is for system (1.1)

$$p_0 = \left.\frac{n-2}{n-1}\right|_{n=8} = 6/7.$$

In this note we use the subharmonicity of powers of octonion-valued monogenic functions to prove for those functions an analog of M. Riesz theorem on the boundedness of the conjugation operator in L^p, $p > 1$. It should be noted that due to the positivity of the subharmonic function $|f(x)|^p$, $p \geq 6/7$, if f is integrable over the boundary of the domain, then f and all its harmonic components f_i, $i = 0, 1, \ldots, 7$, have non-tangential boundary values almost everywhere on the boundary. We let B denote the unit ball and $S = \partial B$, and $d\sigma$ denote surface measure on S. We will let $f_v(x) = \sum_{1=1}^{7} f_i(x) e_i$ denote the imarginary or vector part of f.

Theorem 1 *If f i n enic in e uni ball B and $S = \partial B$, en f r $0 \leq r < 1$*

$$\int_S |f(r\zeta)|^p d\sigma \leq C_p \int_S |f_v(r\zeta)|^p d\sigma \tag{1.2}$$

w ere

$$\begin{aligned} C_p &= \left(\frac{8}{p-1}\right)^{1/p} \quad for\ 1 < p \leq 2 \\ C_p &= \frac{A^{p/2} - (1-\lambda)^p}{|\lambda|^p} \quad for\ 2 < p < \infty \end{aligned} \tag{1.3}$$

$A \geq 8p(p-1)$ and $0 < |\lambda| \leq min\{1, \frac{1}{32(p-2)}\}$

2. Octonionic version o M. Riesz theorem; the roo o Theorem 1.

Let $f(x) = \sum_{j=0}^{7} u_i(x)\mathbf{e_j}$ be a monogenic octonion-valued function, $f_v(x) = f(x) - u_0(x)\mathbf{e_0}$ its imaginary part, and $f_\lambda(x) = \lambda u_0(x) + f_v(x)$

Lemma 2.1 *e func i n $g = A|f_v|^p - |f|^p$ i ub ar nic f r $1 < p \leq 2$ and $A \geq \frac{8}{p-1}$*

Pr f We first assume $f_v(x) \neq 0$ and we show that $g \geq 0$. Letting $J(f)$ denote the Jacobian of $f(x)$, the following follows from (1.1):

$$\begin{aligned} |\nabla(u_0)|^2 &\leq 7|J(f_v)|^2 \\ 8|\nabla(u_0)|^2 &\leq 7|J(f)|^2 \\ |J(f)|^2 &\leq 8|J(f_v)|^2 \end{aligned} \tag{2.1}$$

We will use a well known formula to calculate the laplacian, the following:

$$|f|^p = p|f|^{p-4}[|f^2||J(f)|^2 + (p-2)\sum_{j=0}^{7}(f \cdot f_{x_j})^2.] \tag{2.2}$$

Since $p \leq 2$, from equation (2.2) we immediately have

$$|f|^p \leq p|f|^{p-2}|J(f)|^2$$

Applying the Schwartz inequality to (2.2) gives

$$|f_v|^p \geq p(p-1)|f_v|^{p-2}|J(f_v)|^2$$

Using these inequalities with (2.1), $p-2 \leq 0$, and $|f|^{p-2} \leq |f_v|^{p-2}$ gives

$$\begin{aligned} g &\geq Ap(p-1)|f_v|^p|J(f_v)|^2 - p|f|^{p-2}|J(f)|^2 \\ &\geq p|f|^{p-2}[A(p-1)|J(f_v)|^2 - |J(f)|^2] \\ &\geq p|f|^{p-2}|J(f)|^2\Big[\frac{A}{8}(p-1) - 1\Big] \\ &\geq 0 \end{aligned}$$

The set of zero points of a harmonic function is a polar set, which completes the proof for all x.

In the proof of the next lemma we will need two identities.

$$\begin{aligned} (hg) &= fh \quad g + g \quad h + 2(\nabla g \cdot \nabla h) \\ \nabla|f|^p &= p|f|^{p-2} < f \cdot f_{x_0}, f \cdot f_{x_1}, ..., f \cdot f_{x_7} >= p|f|^{p-2}(f \cdot J(f)). \end{aligned} \tag{2.3}$$

Lemma 2.2 *Le* $h = A|f_v|^2|f_\lambda|^{p-2} - |f_\lambda|^p$ *en i ub ar nic if*

$$\begin{aligned} &p \geq 2 \\ &A \geq 8p(p-1) \\ &|\lambda| \leq min\Big\{1, \frac{1}{32(p-2)}\Big\} \end{aligned} \tag{2.4}$$

Pr f We assume that $f \neq 0$ and $f_v \neq 0$, and use (2.2) to calculate h.

$$\begin{aligned} h &= 2A|f_\lambda|^{p-2}|J(f_v)|^2 \\ &+ A(p-2)|f_v|^2|f_\lambda|^{p-6}(p-4)\sum_{j=0}^{7}(f_\lambda \cdot f_{\lambda xj})^2 \\ &+ A(p-2)|f_v|^2|f_\lambda|^{p-4}|J(f_\lambda)|^2 \\ &+ 4A(p-2)|f_\lambda|^{p-4}\sum_{j=0}^{7}(f_v \cdot f_{vx_j})(f_\lambda \cdot f_{\lambda x_j}) \\ &- \quad |f_\lambda|^p \end{aligned} \tag{2.5}$$

Multiplying through by $|f_\lambda|^{2-p}$, and using $|f_v| \leq |f_\lambda| \leq |f|$ and using $|\lambda||f_{\lambda x_j}|| \leq |f_{\lambda x_j}| \leq |f_{x_j}|$, we obtain

$$\begin{aligned} |f_\lambda|^{2-p} \quad h &\geq 2A|J(f_v)|^2 \\ &- 4A(p-2)|f_\lambda|^{-2}\sum_{j=0}^{7}(f_v \cdot f_{vx_j})^2 \\ &+ 4A(p-2)|f_\lambda|^{-2}\sum_{j=0}^{7}(f_v \cdot f_{vx_j})(f_\lambda \cdot f_{\lambda x_j}) \\ &- |f_\lambda|^{2-p} \quad |f_\lambda|^p. \\ &\geq 2A|J(f_v)|^2 \\ &+ 4A(p-2)|f_\lambda|^{-2}\sum_{j=0}^{7}(f_\lambda \cdot f_{\lambda x_j})[(f_v \cdot f_{vx_j}) - (f_\lambda \cdot f_{\lambda x_j})] \\ &- |f_\lambda|^{2-p} \quad |f_\lambda|^p. \end{aligned} \tag{2.6}$$

Journal of Physics: Conference Series **670** (2016) 012017 doi:10.1088/1742-6596/670/1/012017

Observing that

$$(f_v \cdot f_{vx_j}) - (f_\lambda \cdot f_{\lambda x_j}) = -\lambda^2 u_0 \Big\{ \frac{\partial u_o}{\partial x_j} \Big\}$$

we have

$$\begin{aligned} & |f_\lambda|^{-2} |(f_\lambda \cdot f_{\lambda x_j})| |[(f_v \cdot f_{vx_j}) - (f_\lambda \cdot f_{\lambda x_j})]| \\ & \leq \frac{|\lambda u_0|}{|f_\lambda|} |f_{\lambda x_j}| |\lambda| |\{ \frac{\partial u_o}{\partial x_j} \}| \\ & \leq |f_{\lambda x_j}|^2 |\lambda| \\ & \leq |\lambda| |f_{x_j}|^2 \end{aligned} \tag{2.7}$$

Hence,

$$\begin{aligned} |f_\lambda|^{2-p} \quad h \geq\ & 2A|J(f_v)|^2 \\ & - |f_\lambda|^{2-p} \quad |f_\lambda|^p - 4A|\lambda|(p-2)|J(f)|^2 \\ & = A[|J(f_v)|^2 - 4|\lambda|(p-2)|J(f)|^2] + [A|J(f_v)|^2 - |f_\lambda|^{2-p} \quad |f_\lambda|^p]. \end{aligned}$$

We have

$$|J(f_v)|^2 - 4|\lambda|(p-2)|J(f)|^2 \geq \big[1 - 32|\lambda|(p-2)\big] |J(f_v)|^2 \geq 0$$

and using (2.7) with the Cauchy-Schartz inequality we have

$$\begin{aligned} & A|J(f_v)|^2 - |f_\lambda|^{2-p} \quad |f_\lambda|^p \\ & \geq A|J(f_v)|^2 - p(p-2)|f_\lambda|^{-2} \sum_{j=0}^{7} |f_\lambda|^2 \Big| \Big\{ \frac{\partial f_\lambda}{\partial x_j} \Big\} \Big|^2 - p(J(f_\lambda)|^2 \\ & = A|J(f_v)|^2 - p(p-2)|J(f_\lambda)|^2 - p|J(f_\lambda)|^2 \\ & \geq \big[\frac{A}{8} - p(p-1)\big] |J(f_\lambda)|^2 \geq 0 \end{aligned}$$

since $A \geq 8p(p-1)$. This completes the proof since once again the zero set of a non-constant harmonic function is a polar set.

We now prove Theorem 1.

Pr f We may assume $u_0(0) = 0$. For $1 < p \leq 2$, letting $A = \frac{8}{p-1}$ in Lemma 1, we have

$$g(0) = \frac{8}{p-1} |f_v(0)|^p - |f(0)|^p = \frac{9-p}{p-1} |f(0)|^p \geq 0$$

Hence, by subharmonicity

$$\int_S \Big(g(r\zeta) \Big) d\sigma(\zeta) \geq 0$$

so

$$\int_S |f(r\zeta)|^p d\sigma \leq \Big(\frac{8}{p-1} \Big)^{1/p} \int_S |f_v(r\zeta)|^p d\sigma, \quad 0 \leq r < 1$$

For $2 < p < \infty$ we again assume $u_0(0) = 0$ so $h(x)$ is subharmonic and A and $|\lambda|$ satisfying the conditions of Lemma 2. Since $h(0) \geq 0$, so it follows that

$$\int_S |f_\lambda(r\zeta)|^p d\sigma(\zeta) \leq \int_S |f_v(r\zeta)|^2 |f_\lambda(r\zeta)|^{p-2} d\sigma(\zeta).$$

Journal of Physics: Conference Series **670** (2016) 012017 doi:10.1088/1742-6596/670/1/012017

We now use the Hölder inequality with exponents $p/2$ and $p/(p-2)$ and we have

$$\int_S |f_\lambda(r\zeta)|^p d\sigma(\zeta) \le A\Big(\int_S |f_v(r\zeta)|^p d\sigma(\zeta)\Big)^{\frac{2}{p}}\Big(\int_S |f_\lambda(r\zeta)|^p d\sigma(\zeta)\Big)^{\frac{p-2}{p}}$$

Dividing both sides by $\Big(\int_S |f_\lambda(r\zeta)|^p d\sigma(\zeta)\Big)^{\frac{p-2}{p}}$ we have

$$\int_S |f_\lambda(r\zeta)|^p d\sigma(\zeta) \le A^{p/2}\int_S |f_v(r\zeta)|^p d\sigma(\zeta)$$

We use the inequality $(a+b)^{2/p} \ge a^{2/p} + b^{2/p}$ to $f_\lambda = (1-\lambda)f_v + \lambda f$ and octonion algebra,

$$x = \sum_{j=0}^{7} x_j e_j$$

$$x^* = x_0 e_0 - \sum_{j=1}^{7} x_j e_j$$

$$x x^* = \sum_{j=0}^{7} |x_j|^2$$

From this it follows that $f_v^* = -f_v$ and $f^* = u_0 e_0 - f_v$ so $|f_\lambda|^2 = |(1-\lambda)f_v + \lambda f|^2 = (1-\lambda)^2|f_v|^2 + \lambda^2|f|^2$ and we have—

$$\begin{aligned} A^{p/2}\int_S |f_v(r\zeta)|^p d\sigma(\zeta) &\ge \int_S |f_\lambda|^p d\sigma \\ &= \int_S |f_\lambda|^p d\sigma = \int_S |(1-\lambda)f_v + \lambda f|^p d\sigma \\ &= \int_S ((1-\lambda)^2|f_v|^2 + \lambda^2|f|^2)^{p/2} d\sigma \\ &\ge (1-\lambda)^p \int_S |f_v|^p d\sigma + |\lambda|^p \int_S |f|^p d\sigma \end{aligned}$$

Finally,

$$\int_S |f(r\zeta)|^p d\sigma(\zeta) \le \frac{A^{p/2} - (1-\lambda)^p}{|\lambda|^p}\int_S |f_v(r\zeta)|^p d\sigma(\zeta)$$

which completes the proof.

[1] Armitage D H and Gardiner S J 2001 *Classical Potential Theory* (Springer-Verlag, London Limited)

[2] Avetisyan K 2014 Subharmonicity and a version of Riesz theorem on harmonic conjugates *Adv. Appl. Clifford Algebras* **24** 909

[3] Baez J C 2002 The octonions *Bull. Amer. Math. Soc.* **39** 145

[4] Catto S 2003 Exceptional projective geometries and internal symmetries *Preprint* hep-th 0302079

[5] Catto S and Kheyfits A 2009 Asymptotic behavior of octonion-valued monogenic functions in tube domains, in *Group Theoretical Methodds in Physics, 2006, Proc. XXVI International Colloquium on Group Theoretical Methods in Physics (Group 26)* New York, USA, 26-30 June 2006. (Canopus Publishing Limited) p 140

[6] Coifman RR and Weiss G 1970 On subharmonicity inequalities involving solutions of generalized Cauchy-Riemann equations *Studia Math.* **36** 77

[7] Essén M 1984 A superharmonic proof of the M. Riesz conjugate function theorem *Ark. Mat.* **22** 241

[8] Evgrafov M A 1963 *Generalization of the Phragmén - Lindelöf theorems for analytic functions to solutions of different elliptic systems* (Izv. Akad. Nauk SSSR, Ser. Mat. **27**) p 843 (in Russian)

[9] Gürsey F and Tze C-H 1996 *On the Role of Division, Jordan, and Related Algebras in Particle Physics* (Singapore: World Scientific)

[10] Kheyfits A and Tepper D 2006 Subharmonicity of powers of octonion-valued monogenic functions and some applications *Bull. of the Belgian Math. Soc - Simon Stevin* **13** 609

[11] Kheyfits A 2007 Phragmén-Lindelöf principle for Clifford monogenic functions *Proc. of the* 15th *International Conference on Finite or Infinite-Dimensional Complex Analysis and Applications* (Osaka: Osaka City Univ.) p 237

[12] Kuran Ü 1965 $n-$Dimensional extensions of theorems on conjugate functions *Proc. London Math. Soc.* **15** 713

[13] Li and Peng L 2002 The Cauchy integral formulas on the octonions *Bull. Math. Soc. Belgique* **9** 47

[14] Li and Peng L 2004 Three-line theorem on the octonions *Acta Math. Sinica, English Ser.* **20** 483

[15] Stein E M and Weiss G 1971 *Introduction to Fourier Analysis on Euclidean Spaces* (Princeton: Princeton Univ. Press)

Journal of Physics: Conference Series **670** (2016) 012018 doi:10.1088/1742-6596/670/1/012018

Vibron properties in quasi 1D molecular structures: the case of two parallel unshifted macromolecuar chains

D. Čevizović[1], S. Petković[1], S. Galović[1], A. Reshetnyak[2], A. Chizhov[3,4]

[1]Vinča Institute of Nuclear Sciences, Belgrade, Serbia

[2]Institute of Strength Physics and Materials Science SB RAS, Tomsk, Russia

[3]Bogoliubov Laboratory of Theoretical Physics, Joint Institute for Nuclear Research, 141980 Dubna, Russia

[4]Dubna State University, 141980 Dubna, Russia

E-mail: chizhov@theor.jinr.ru

Abstract. We study the hopping mechanism of the vibron excitation transport in the system of two parallel unshifted 1D macromolecuar chains in the framework of non-adiabatic polaron theory. We suppose that the vibron interaction with thermal oscillations of the macromolecular structural elements will result in vibron self-trapping and the formation of the partial dressed vibron state. We also suppose that quasiparticle motion takes place via a sequence of random sitejumps, in each of which the quasiparticle can migrate either to the first neighbor site of the macromolecular chain. With use of the modified Holstein polaron model, we calculate the vibron effective mass in dependence of the basic system parameters and temperature. Special attention is paid to the influence of interchain coupling on vibron dressing. We find that for certain values of the system parameters the quasiparticle mass abruptly changes.

1. Introduction

The interest for the application of organic macromolecules (e.g., protein macromolecules, DNA, polymers, and other biological macromolecular structures) in construction of nanostructures, such as nanoparticles, nanocrystals, nanowires, and molecular circuits becomes again actual. Its ability in the field of the miniaturization of microelectronic and optoelectronic devices and, at the same time, its self-assembly capabilities makes them as a very promising materials in microelectronic and optoelectronic technology [1, 2, 3]. Efficient application of such materials requires the knowledge of mechanisms that rule the charge and energy migration along the macromolecule spine at such distances which are comparable to the dimension of the macromolecule. Similarly, the processes of energy and charge transport in biological macromolecules that operate inside living cells are not sufficiently elucidated. Unfortunately, formulation of a theoretical model based on the principle of quantum mechanics that might explain such processes on a satisfactory manner is not finished yet. At the present it is believed that the universal source of energy for many physical processes that take place in living cells is the process of the hydrolysis of adenosine triphophate, ATP. The energy quanta that emerge in this process is large enough for the excitation of the intramolecular oscillation (IMO) mode (vibron excitation) of the CO group (Amide I quanta) which is a part of the peptide bond of all protein macromolecules. But the lifetime of Amide I quanta is approximately 1 picosecond, which is too short to explain the process of long-distance energy transport through the MC spine. The inability of a consistent explanation of the above mentioned problems led to a crisis in bioenergetics in the seventies of the last century.

One of the earliest quantum mechanical-based models of the charge and energy transfer in protein molecules was developed by Davydov and his coworkers [4, 5]. This model is founded on the assumption that an external excitation might be captured by MC due to the interaction with thermal oscillations of the MC structure elements. Such a self-trapped (ST) excitation becomes dressed by the phonon cloud and, according to Davydov, it can move in a soliton form through the macromolecular chain. The assumption of the soliton existence might explain the long lifetime of the formed quasiparticle.

However, it turned out that the soliton model can successfully describe the process of ST of the charge carriers and its transport through the macromolecule. As regards the intramolecular vibrational excitations, the Davydov model is inapplicable because the basic energy parameters of the protein macromolecules do not satisfy basic conditions for the soliton formation [6, 7, 8, 9, 10]. After some delay in further progress, in the late eighties it was proposed a new model based on the assumption that in the process of vibron excitation ST might arise a non-adiabatic (small) polaron (SP), rather than a soliton [8]. The new approach to the problem of ST of IMO was originated from the attempt to explain the origin of the anomalous peak in the IR absorption spectra of crystalline acetanilide (ACN).

The non-adiabatic (small) polaron theory is founded on the assumption that the interaction between quasiparticle and phonon subsystem is strong. Unfortunately, the values of system parameters that manage the process of a vibron ST do not belong to the vibron-phonon strong coupling region, but rather correspond to the weak or intermediate coupling strength [10]. This uncertainty stems mostly from the fact that the vibron-phonon coupling strength cannot be directly measured in experiment, but it is estimated according to pre-accepted theoretical model (i.e., it is the model dependent quantity). For that reason, it was necessary to go beyond the standard small polaron theory and to develop a model that will overcome its limitations. One of ways to overcome this problem is to apply the method of the partially dressed vibron quasiparticle. This method was developed in order to intermediate between the exciton-phonon strong coupling and weak coupling interactions. The method of the partially dressed quasiparticle is commonly used in complicated numerical calculations, which does not allow one to have a clear insight into the essence of the problem of quasiparticle dressing. Any analytical model, which will provide "insight" into the heart of physical processes that take place in the quasiparticle self-trapping or quasiparticle migration in the crystal lattice, requires the introduction of various approximations, which however should not neglect the basic processes in the system under consideration. In general, the model of the partially dressed vibron quasiparticle gave satisfactory results, but certain details still did not agree with the experimental results [9]. These details concern especially the problem of the temperature dependence of the anomalous peak position of Amide I quanta in IR absorption spectra of ACN. But all studies devoted to this problem were made under the assumption that ACN can be regarded as a truly 1D system [8, 13, 14, 15, 16, 17], neglecting the interaction among chains, which can affect the SP properties.

The mechanism of IMO migration in the ACN (and other macromolecular structures that are built by two macromolecular chains, like DNA) becomes very important in order to understand the processes of bio-information transmission in the living cells. According to the Holstein MC model, there are two basic mechanisms of polaron motion: a band mechanism, which is dominant at low temperatures, and polaron migration thorough the crystal by the series of successive jumps from one to its neighbor structure element. The latter mechanism dominates at intermediate and high temperatures, including the temperature at which living cells function. In both mechanisms, a key role has the probability of quasiparticle transition from one to some other structure element of a crystal structure. In this paper, the hopping mechanism of vibron exciton transport in the system which consists of two parallel macromolecular chain is investigated in the framework of partially dressed vibron states. It is taken into account the coupling between the chains. Among the typical examples of such a structure we can specify the ACN macromolecule in which its two polypeptide chains are embeded and form a two-chain structure (Fig. 1). We calculate the dependence of the dressing parameter on the system parameters and temperature. Special attention is paid to the vibron properties at room temperatures and for the system parameter values that belong to moderately non-adiabatic and intermediate coupling regions.

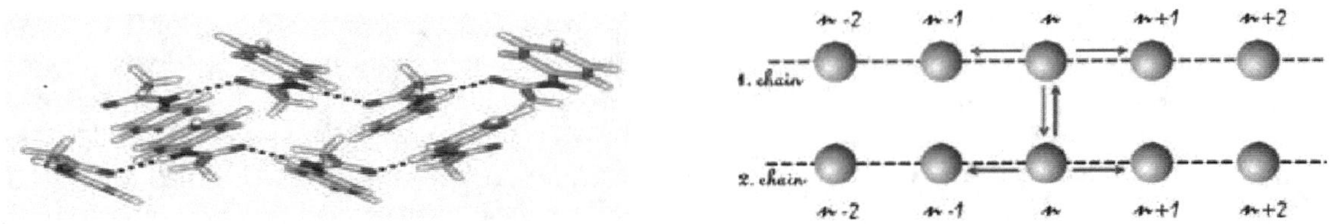

Figure 1. The real structure of the part of crystaline acetanilide macromolecular chains is shown in the left, whereas its schematic structure is demonstrated in the right. The possible vibron hopping transitions are shown by the arrows.

2. Theoretical model

In order to investigate single-vibron properties in the system consisting of two parallel macromolecular chains, we modify the Holstein molecular crystal model [6] and adopt several assumptions: a vibron quasiparticle becomes self-trapped and forms a dressed quasiparticle due to the vibron-phonon interaction with optic phonon modes only; a vibron excited on the particular MC can interact only with phonons that belong to the same MC; a vibron can migrate along the macromolecule spine from the n-th to $(n \pm 1)$-th structural element. The Hamiltonian of the system of two parallel and unstaggered from each other MCs (see right Fig. 1) is defined as

$$\begin{aligned} H &= E_0 \sum_{n,j=1,2} A^\dagger_{j,n} A_{j,n} - J \sum_{n,j=1,2} A^\dagger_{j,n}(A_{j,n-1} + A_{j,n+1}) + L \sum_{n} \left(A^\dagger_{1,n} A_{2,n} + A^\dagger_{2,n} A_{1,n} \right) \\ &+ \sum_{q,j=1,2} \hbar\omega_0 B^\dagger_{j,q} B_{j,q} + \frac{1}{\sqrt{N}} \sum_{n,q,j=1,2} F_q e^{iqnR_0} A^\dagger_{j,n} A_{j,n} (B_{j,q} + B^\dagger_{j,-q}), \end{aligned} \tag{1}$$

where E_0 is the vibron excitation energy in a particular structural element of two N-site MCs, $A^\dagger_{j,n}$ $(A_{j,n})$ are the vibron quasiparticle creation (annihilation) operators in the n-th site of the j-th MC, J and L stand for the energies of the dipole-dipole interaction of neighboring structure elements that belongs to the same chain and to different chains, respectively. The operators $B^\dagger_{j,q}$ $(B_{j,q})$ in Eq. (1) are the creation (annihilation) operators of phonon quanta with frequency $\omega_q = \omega_0$ of the wave number q which belong to the j-th chain. The vibron-phonon coupling parameter is defined as $F_q = F = \chi_{hb}\sqrt{\hbar/2M\omega_0}$ (χ_{hb} is the coupling constant and M is a mass of the molecular group). R_0 is a distance between two neighboring structure elements placed in the same MC. The transition to the polaron picture can be realized by the (modified) Lang-Firsov unitary transformation [9, 11, 18] with the operator $U = U_1 \cdot U_2$, where $U_j = \exp\{-\sum_n \sigma_{j,n} A^\dagger_{j,n} A_{j,n}\}$ and $\sigma_{j,n} = \frac{1}{\sqrt{N}} \sum_q f_{q,j} e^{-iqnR_0}(B_{j,q} - B^\dagger_{j,-q})$. In the standard theory of Lang and Firsov, $f_{q,j}$ are the vibron-phonon coupling parameters. Such an approach corresponds to the standard non-adiabatic polaron (fully dressed vibron quasiparticle). In the case of partially dressed vibron quasiparticles, these parameters can be refered as a set of variational parameters, or they can contain one or more variational parameters. According to [12], we use the variational approach with only one single variational parameter $\delta \in [0, 1]$ introduced as follows $f_{q,j} = \delta \cdot F^*_q/\hbar\omega_0$. The influence of the thermal fluctuations on the vibron properties are accounted by the averaging of the transformed

Hamiltonian over the phonon subsystem [11, 18, 19, 20]. At this manner, we find for the dressed vibron Hamiltonian

$$H_{vib} = \sum_{k,j=1,2} E(k) a^{\dagger}_{j,k} a_{j,k} + \lambda_0 \sum_{k} \left(a^{\dagger}_{1,k} a_{2,k} + a^{\dagger}_{2,k} a_{1,k} \right). \tag{2}$$

Here $a^{\dagger}_{j,k}$ ($a_{j,k}$) denote the creation (annihilation) operators of the dressed vibrons, $\tau = k_B T/\hbar\omega_0$ is the normalized temperature, $\lambda_0 = L \exp\{-\delta^2 S \coth(1/2\tau)\}$, and

$$E(k) = E_0 + \delta(\delta - 2)E_b - 2J\mathrm{e}^{-\delta^2 S \coth(1/2\tau)} \cos(kR_0) \tag{3}$$

is the polaron band energy, $E_b = F^2/\hbar\omega_0$ is the small polaron binding energy, and $S = E_b/\hbar\omega_0$ is the coupling constant. The diagonalization of the Hamiltonian (2) can be done by means of the unitary transformation $\alpha_{\pm,k} = (a_{1,k} \pm a_{2,k})/\sqrt{2}$. As the result, the vibron Hamiltonian takes the following form

$$H_{vib} = \sum_{k} \left[E_+(k) \alpha^{\dagger}_{+,k} \alpha_{+,k} + E_-(k) \alpha^{\dagger}_{-,k} \alpha_{-,k} \right], \tag{4}$$

when the polaron band splits into two subbands:

$$E_{\pm}(k) = E_0 + \delta(\delta - 2)E_b - 2J\mathrm{e}^{-\delta^2 S \coth(1/2\tau)} \cos(kR_0) \pm \lambda_0. \tag{5}$$

The optimal state of the system is determined by the minimum of the free energy of the whole system. Therefore, the value of the variational parameter should be found from the conditions $\partial F_B/\partial\delta = 0$ and $\partial^2 F_B/\partial\delta^2 > 0$, where F_B is the upper bound of the system free energy determined by the Bogoliubov theorem [11, 19].

3. Obtained results and discussion

To study the vibron properties in the system of the molecular structure composed of two unstaggered single MCs, we have considered the dependence of the dressing parameter on coupling constant S, adiabatic parameter $B = 2J/\hbar\omega_0$, and interchain coupling parameter $g = J/L$. Obtained results are presented by the set of iso-lines $\delta = \{0.2, 0.4, 0.6, 0.8, 1.0\}$ of the function $\delta(S, B)$ in the low temperature limit ($T = 100$ K) in Fig. 2 and at room temperature ($T = 300$ K)

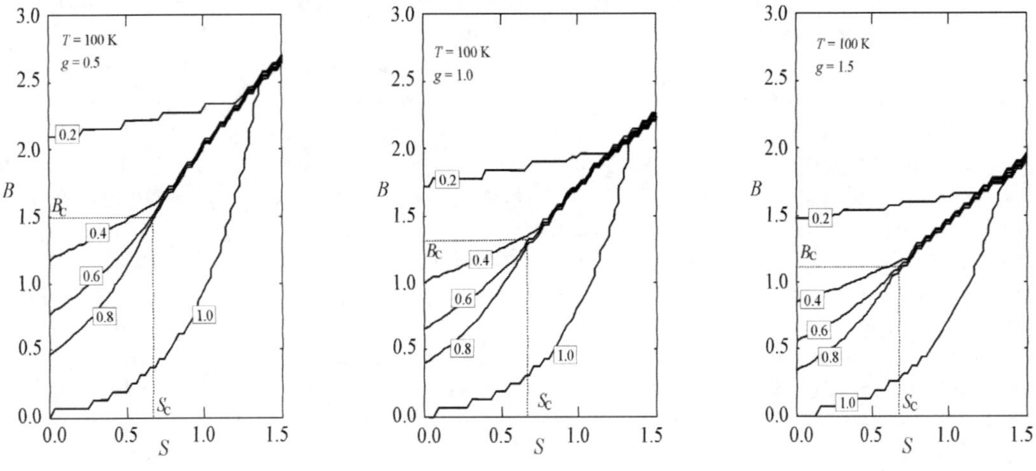

Figure 2. The dependence of the variational parameter δ on the system parameters S, B, g at $T = 100$ K.

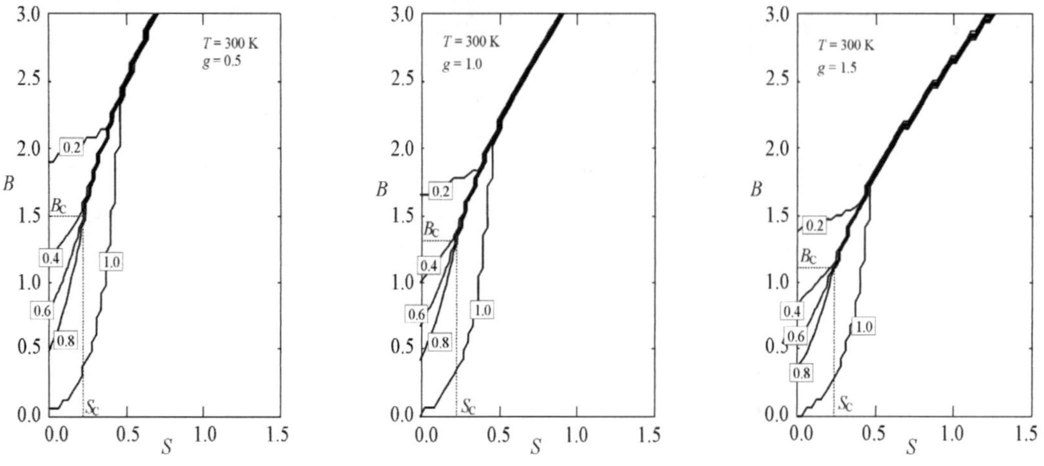

Figure 3. The same as in Fig. 2 at $T = 300$ K.

in Fig. 3.

In these figures one can see that for small values of parameters B and S, variational parameter δ has a relatively high value, which is close to 1. With increasing of S (and for the fixed small value of B) the variational parameter smoothly increases and reaches its maximum value. Consequently, in the strongly non-adiabatic regime the vibron quasiparticle is fully dressed by the phonon cloud and its properties are quite similar to the standard small polaron quasiparticle [6, 8, 18]. On the other hand, with increasing of B and for the fixed (small) value of S, the variational parameter smoothly decreases, and for the moderately adiabatic regime it becomes small (for example, when $g = 0.5$ and $B \approx 2.5$, the variational parameter becomes $\delta \approx 0.1$). Consequently, for the fixed small value of S, the nature of vibron quasiparticle moves toward undressed, practically free vibron state, when the adiabatic parameter increases. Such behavior is fully consistent with the standard polaron theory in the weak coupling, adiabatic limit.

However, there is a region in (S, B)-plane where the dressing parameter abruptly changes its value: for the sufficiently high value of $B = B_c$ with increasing of S, the dressing parameter at first slowly increases, and for $S = S_c$ it suffers an abrupt change to a value which is close to 1. Then its value again slowly decreases with a further increase of B. Such points (S_c, B_c) in the parameter space, where δ suffers a discontinuous change, define so called the "critical area" of parameter space. Since the variational parameter measures the degree of vibron dressing, such a sudden change of the value of δ indicates an abrupt change of the degree of vibron dressing, i.e., a sudden change of its nature.

It is worth noting that the strength of interchain coupling has no significant influence on S_c, but the increase of g may significantly reduce the value of B_c. On the other hand, the increase of temperature significantly decreases the minimum value of S_c, which lies in the weak coupling area at room temperature, while the minimum value of B_c remains practically unchanged.

Obtained results may play an important role in explaining the process of IMO self-trapping in biological macromolecules which are composed of two macromolecular chains, such as DNA and ACN. System parameter values of these structures belong to the non-adiabatic and to weak or moderately strong vibron-phonon interaction. Our analysis indicates that for such structures at room temperatures and for high interchain coupling the vibron quasiparticle must suffer a sharp change in its nature with a slight change in the value of system parameters and (or) temperature. This change of the properties of vibrational energy carriers may be easily verified by the study of temperature dependence of the maxima in the infrared absorption spectrum of these molecules.

Acknowledgments
The work was supported by the Serbian Ministry of Education and Science under Contracts Nos. III-45010, III-45005, and by the Project within the Cooperation Agreement between the JINR, Dubna, Russian Federation and Ministry of Education and Science of Republic of Serbia. The work of A. R. was partially supported by the grant of Leading Scientific Schools of the Russian Federation under Project No. 88.2014.2. A. Ch. is grateful to the Votruba-Blokhintsev Program (JINR, Dubna) for the support of participation in the ISQS-23.

References

[1] Dekker C and Ratner M A 2001 *Phys. World* **14** 29
[2] Convell E 2004 *Topics in Current Chemistry* **237** 73
[3] Mirkin C A, Letsinger R L, Mucic R C and Storhoff J J 1996 *Nature* **382** 607
[4] Davydov A S 1973 *J. Theor. Biol.* **38** 559
[5] Davydov A S 1979 *Phys. Scr.* **20** 387
[6] Holstein T 1959 *Annals of Physics* **8** 325
[7] Rashba E 1982 *Excitons* (Amsterdam: North-Holland Publishing Company)
[8] Alexander D M and Krumhansl J A 1986 *Phys. Rev. B* **33** 7172
[9] Pržulj Ž, Čevizović D, Zeković S and Ivić Z 2008 *Chem. Phys. Lett.* **462** 213
[10] Pouthier V 2010 *J. Chem. Phys.* **132** 035106
[11] Yarkony D and Silbey R 1976 *J. Chem. Phys.* **65** 1042
[12] Brown D W and Ivić Z 1989 *Phys. Rev. B* **40** 9876
[13] Barthes M 1989 *J. Mol. Liq.* **41** 143
[14] Edler J, Hamm P and Scott A C 2002 *Phys. Rev. Lett.* **88** 067403
[15] Hamm P and Edler J 2006 *Phys. Rev. B* **73** 094302
[16] Hamm P and Tsironis G 2007 *Eur. Phys. J. Spec. Top.* **147** 303
[17] Scott A C 1982 *Phys. Rev. A* **26** 578
[18] Lang I G and Firsov Yu A 1962 *Zhurnal. Eksp. Theor. Fiz.* **43** 1843
[19] Čevizović D, Galović S, Reshetnyak A and Ivić Z 2013 *Chin. Phys. B* **22** 060501
[20] Čevizović D, Galović S and Ivić Z 2011 *Phys. Rev. E* **84** 011920

Journal of Physics: Conference Series **670** (2016) 012019 doi:10.1088/1742-6596/670/1/012019

ITP-UU-15/15
KUNS-2579

Towards the gravity/CYBE correspondence beyond integrability – *Yang-Baxter deformations of $T^{1,1}$* –

P. Marcos Crichigno

Institute for Theoretical Physics and Spinoza Institute, Utrecht University, Leuvenlaan 4, 3854 CE Utrecht, The Netherlands.

E-mail: p.m.crichigno@uu.nl

Takuya Matsumoto [1]

Graduate School of Mathematics and Institute for Advanced Research, Nagoya University, Nagoya 464-8602, Japan

E-mail: takuya.matsumoto@math.nagoya-u.ac.jp

Kentaroh Yoshida

Department of Physics, Kyoto University, Kyoto 606-8502, Japan.

E-mail: kyoshida@gauge.scphys.kyoto-u.ac.jp

Abstract. Yang-Baxter sigma models, proposed by Klimcik and Delduc-Magro-Vicedo, have been recognized as a powerful framework for studying integrable deformations of two-dimensional non-linear sigma models. In this short article, as an important generalization, we review a *non-integrable* sigma model in the Yang-Baxter sigma model approach based on [ar iv:1406.2249]. In particular, we discuss a family of deformations of the 5D Sasaki-Einstein manifold $T^{1,1}$, instead of the standard deformations of the 5-sphere S^5. For this purpose, we first describe a novel construction of $T^{1,1}$ as a supercoset, and provide a physical interpretation of this construction from viewpoint of the dual Klebanov-Witten field theory. Secondly, we consider a 3-parameter deformation of $T^{1,1}$ by using classical r-matrices satisfying the classical Yang–Baxter equation (CYBE). The resulting metric and NS-NS two-form completely agree with the ones previously obtained via TsT (T-dual – shift – T-dual) transformations, and contain the Lunin-Maldacena background as a special case. Our result indicates that what we refer to as the *gravity/CYBE(Classical Yang-Baxter Equation) correspondence* can be extended beyond integrable cosets.

1. Introduction and main results

Ten-dimensional supergravity describes the low energy dynamics of string theory and has a rich mathematical structure. Finding explicit solutions to the supergravity field equations is an important task to better understand the low energy dynamics of string theory and to develop

[1] This is a proceeding of "The IIIth International Conference on Integrable Systems and Quantum symmetries (ISQS-23)" held in Prague, Czech Republic, from June 23 till June 27, 2015, based on a talk given by T.M.

new examples of the AdS/CFT correspondence [2, 3, 4], which is often limited to cases with a large amount of symmetry. However, these equations are highly nonlinear and finding explicit solutions without a large amount of symmetry is not an easy task.

One approach that has been quite successful in the past is to use known string theory duality transformations, such as T-duality [5], as as solution-generating technique; starting from a known seed solution one applies this symmetry to generated a new solution. A particular example are the so-called TsT transformations, which are a sequence of T-dualities (T) and coordinate shifts (s) leading to a continuous family of solutions [6, 7].

In a series of works in recent years [8, 9, 10, 1], it has been observed that classical r-matrices, which are solutions to the classical Yang–Baxter equation (*CYBE* in short), characterize various deformations of type IIB supergravity backgrounds. One of the simplest examples is the correspondence between an Abelian r-matrix, consisting of three Cartan generators of $\mathfrak{su}(4)$, and the Lunin-Maldacena-Frolov backgrounds [6, 7, 8] obtained by TsT transformations. The gravity duals of the 4D non-commutative gauge theory [11, 12] can also be obtained by certain r-matrices [9]. These observations have led us to propose a correspondence between classical r-matrices and deformed supergravity solutions, which we refer to as the *ravi /CYBE c rre p ndence* [8].

The reader may wonder how it is that r-matrices, which are purely algebraic objects, can be used to study geometric objects such as gravity backgrounds. The notion which makes this possible is *Yan -Bax er i a del* . These models, originally introduced by Klimcik [13], are integrable deformations of the principal chiral model based on an r-matrix satisfying the *dified* classical Yang-Baxter equation (*CYBE* in short).[2] These were extended to sigma models defined on bosonic symmetric cosets [15], and applied to study an integrable deformation of the $AdS_5 \times S^5$ superstring [16] in [17, 18]. We note that the work [15] is the generalization of *qua ed i a del* defined on the squashed S^3, and their integrable structures have been investigated in [19, 20, 21]. The deformed metric and the NS-NS two-form of the deformed $AdS_5 \times S^5$ superstring were calculated in [22]. However, the deformed background does not seem to be a supergravity solution for a generic value of the deformation parameter [23].

Deformations of the $AdS_5 \times S^5$ superstring by the CYBE (rather than by the mCYBE) were proposed in [24] and the supergravity solution was constructed in [25]. The generic formulation for group manifolds and bosonic cosets was introduced in [26]. These works were inspired by a careful study of the three-dimensional case, namely *Sc r¨din er i a del* [27, 28, 29]. Although the CYBE is a particular case of the mCYBE, their solutions seem to be very different. One of the advantages of using the CYBE is the variety of classical r-matrices and partial deformations of the target space that are possible. In the case of the mCYBE it seems difficult to find solutions, apart from the famous r-matrix of the Drinfeld-Jimbo type [30, 31]. An important aspect of Yang-Baxter deformations is that they preserve the integrability of the system, *i.e.*, if the undeformed model is integrable, so is the deformed model. Thus, it is not surprising that the initial focus was on studying deformations of known integrable models such as $AdS_5 \times S^5$ and four-dimensional Minkowski spacetime [32, 33].[3] The various deformed Lax pairs are summarized in [34] and the expressions in terms of the local coordinates are presented. However, one may wonder whether integrability of the undeformed model is essential in the gravity/CYBE correspondence, or if this correspondence may hold beyond integrability. If so, what are the resulting deformed backgrounds?

This question led us in [1] to consider deformations of the sigma model defined on the five-dimensional Sasaki-Einstein space known as $T^{1,1}$ [38] (see also [39, 40] for relevant work). As shown in [41], due to the appearance of chaotic behavior, this model is non-integrable. Thus, this is a good testing ground for possible generalizations of the gravity/CYBE correspondence in a non-integrable setting. Another important aspect of $T^{1,1}$ is that it admits a Sasaki-Einstein

[2] Quantum aspects of these models are discussed in [14, 56].

[3] Integrable structures in AdS CFT were discovered in [35, 36]. For a comprehensive review see [37].

metric and, as a consequence, $AdS_5 \times T^{1,1}$ is a good supergravity background. In fact, this was an early example [42] of AdS/CFT with reduced supersymmetry. Thus, the study of deformations of $T^{1,1}$ is also interesting from the holographic perspective. See [43] for a two-parameter deformation of metrics on $T^{1,1}$ in terms of gauged linear sigma models, and [44] for a one-parameter deformation and a study of supersymmetric localization of the two-dimensional sigma model on this space.

The first obstacle to apply the Yang-Baxter deformation to $T^{1,1}$ is how to describe the Sasaki-Einstein structure on the manifold by a coset. Although it is well-known that $T^{1,1}$ [38, 42] is given by

$$\frac{SU(2) \times SU(2)}{U(1)_1}\,, \tag{1}$$

the canonical coset metric on the space is not the Sasaki-Einstein metric we are interested in. Although this problem may be circumvented by a (rather artificial) rescaling of the vielbeins (see App. A in [1] for details and references), this is an unsatisfactory resolution from the point of view of the Yang-Baxter deformation; since the coset construction above does not make the full $SU(2) \times SU(2) \times U(1)$ symmetry of the space manifest, it is not clear how one may perform a general (thee-parameter) Yang-Baxter deformation of the space. As shown in [1] a novel construction which solves all these problems is given by the coset

$$T^{1,1} = \frac{SU(2) \times SU(2) \times U(1)_R}{U(1)_1 \times U(1)_2}\,. \tag{2}$$

Although this coset is closely related to the one above, there is an important difference; despite involving only bosonic groups, it is in fact a *uper-c e* in the sense that one must think of these bosonic groups as the bosonic part of a larger supergroup (this will be explained in detail below). Although this construction might seem at first unnatural, a clear advantage is that the full symmetry appears in the numerator of the coset and is therefore manifest. Furthermore, the fact that one should view this as a supercoset has a natural interpretation in terms of the $\mathcal{N} = 1$ superconformal group $PSU(2,2|1)$ of the dual field theory.

Having an appropriate description of the space as a coset, the next step is to apply the Yang-Baxter deformation. As a simple example we consider deformations associated with an Abelian classical r-matrix, given by

$$r = \frac{1}{3}\hat{\gamma}_1 L_3 \wedge M + \frac{1}{3}\hat{\gamma}_2 M \wedge K_3 - \frac{1}{6}\hat{\gamma}_3 K_3 \wedge L_3\,, \tag{3}$$

where K_3 and L_3 are the Cartan generators of two $SU(2)$'s respectively and M is the $U(1)_R$ generator of the numerator in (2) and $\hat{\gamma}_1, \hat{\gamma}_2, \hat{\gamma}_3$ are three continuous parameters. Note that if the $U(1)_R$ was not present in the numerator, only a one-parameter deformation would be possible in this setting. Although it is not guaranteed *a-pri ri* that such a deformation leads to a new supergravity background, it turns out that the deformed metric and the NS-NS two-form agree exactly with the ones obtained by TsT-transformations [45] (in particular for $\hat{\gamma}_1 = \hat{\gamma}_2 = 0$, the deformed background reduces to the one given by Lunin-Maldacena [6]), and is therefore a good supergravity background. Thus, our result indicates that the Yang-Baxter sigma models based on the CYBE can be applied beyond non-integrable cosets and the *ravi /CYBE c rre p ndence* seems to extend to a wider class of supergravity backgrounds than one may have expected.

This article is organized as follows. In Section 2, we discuss the new coset construction of $T^{1,1}$ given in (2). In Section 3, we apply the Yang-Baxter deformation and show that the resulting backgrounds coincide with the TsT-transformed supergravity solutions, which corresponds to a β-deformation for the one-parameter case and to a γ_i-deformation for the three-parameter case, respectively. Section 4 is devoted conclusions and discussions.

2. Coset construction o $T^{1,1}$

§ *e e ric f* $T^{1,1}$

Let us start by recalling some basic elements. The five-dimensional space $T^{1,1}$ admits a Sasaki-Einstein metric, given by [38]

$$ds^2_{T^{1,1}} = \frac{1}{6}(d\theta_1^2 + \sin^2\theta_1 d\phi_1^2) + \frac{1}{6}(d\theta_2^2 + \sin^2\theta_2 d\phi_2^2) + \frac{1}{9}(d\psi + \cos\theta_1 d\phi_1 + \cos\theta_2 d\phi_2)^2\,, \tag{4}$$

where the angle variables $\theta_i \in [0,\pi), \phi_i \in [0,2\pi)$ with $i = 1,2$ are the coordinates of the two two-spheres and the coordinate $\psi \in [0,4\pi)$ is along the $U(1)$-fiber. The isometry of this manifold is $SU(2) \times SU(2) \times U(1)_R$, an important fact when studying general (Abelian) Yang-Baxter deformations below.

§ *Brief review f c e pace*

Before going into the details of the coset description of $T^{1,1}$, let us review some basics of general coset spaces and their metrics. Let $\mathcal{M}$ be a homogeneous space, on which a Lie group G is acting. For a fixed point of $\mathcal{M}$ we take a subgroup $H \subset G$ which preserves the reference point. Conceptually, G is the global symmetry of the space $\mathcal{M}$ and H is the local symmetry around the fixed point. Then, $\mathcal{M}$ corresponds to the coset space

$$\mathcal{M} = \frac{G}{H} = \frac{\text{Global symmetry}}{\text{Local symmetry}}\,. \tag{5}$$

A standard example of a coset space is the two-sphere S^2, whose global symmetry is $SO(3) \simeq SU(2)$. The residual symmetry around the north (or south) pole is $SO(2) \simeq U(1)$. This leads to the well-known coset description of the two-sphere; $\mathrm{S}^2 = SU(2)/U(1)$.

To obtain the metric on $\mathcal{M}$ from the coset (5), we need two ingredients. These are the Maurer-Cartan one-form

$$A = g^{-1}dg \qquad \text{with} \qquad g \in G\,, \tag{6}$$

and the coset projector

$$P : \mathfrak{g} \to \mathfrak{g}/\mathfrak{h}\,, \tag{7}$$

where $\mathfrak{g}$ and $\mathfrak{h}$ are the Lie algebras associated with the Lie group G and H, respectively.

Next, let us introduce the local coordinates $\{\theta^i\}$ through the parameterization of the group element $g \in G$,

$$g = \exp\left(\sum_{i=1}^{\dim\mathfrak{g}} \theta^i T_i\right) \qquad \text{with} \qquad T_i \in \mathfrak{g}\,. \tag{8}$$

Then, a metric g_{ij} on $\mathcal{M}$ is obtained by

$$ds^2_{\mathcal{M}} = \mathrm{Tr}[AP(A)] = \sum_{i,j=1}^{\dim\mathcal{M}} g_{ij}d\theta^i d\theta^j\,, \tag{9}$$

where the trace is taken over a particular representation of $\mathfrak{g}$, which in our case will be the fundamental representation.

Journal of Physics: Conference Series **670** (2016) 012019 doi:10.1088/1742-6596/670/1/012019

§ *Superc e de crip i n f* $T^{1,1}$
Now we discuss the (super)coset construction of $T^{1,1}$ introduced in [1]. Unlike the usual description (1), we take the numerator to be the *full* isometry of the space and divide by an appropriate subgroup:

$$T^{1,1} = \frac{SU(2)\times SU(2)\times U(1)_R}{U(1)_1\times U(1)_2}\,. \tag{10}$$

We denote the generators of the two $SU(2)$'s by K_i, L_i $(i=1,2,3)$, where K_3 and L_3 are the Cartan generators, and we denote the generator of $U(1)_R$ by M. The generators of $U(1)_i$ in the denominator are denoted by T_i $(i=1,2)$, which we choose to be embedded into the numerator by

$$T_1 = K_3 + L_3\,, \qquad T_2 = K_3 - L_3 + 4M\,. \tag{11}$$

We note that T_1 corresponds to the usual $U(1)_1$ in the standard coset description [38, 42], while T_2 has been chosen appropriately in order to obtain the metric we are after.[4]

The Maurer-Cartan one-form in our case is given by

$$A = g^{-1}dg \qquad \text{with} \qquad g\in SU(2)\times SU(2)\times U(1)_R\,, \tag{12}$$

with the parameterization

$$g = \exp\big(\phi_1 K_3 + \phi_2 L_3 + 2\psi M\big)\exp\big(\theta_1 K_2 + (\theta_2+\pi)L_2\big)\,. \tag{13}$$

To define the coset projector explicitly, it is convenient to work with the fundamental representation. Since those of $\mathfrak{su}(2)$ and $\mathfrak{u}(1)$ are 2×2 and 1×1 matrices, respectively, the whole Lie algebra $\mathfrak{su}(2)\oplus\mathfrak{su}(2)\oplus\mathfrak{u}(1)$ is represented by a 5×5 block matrix. We, however, adopt the following $(4|1)\times(4|1)$ *uper*-matrix instead of the 5×5 bosonic block matrix,

$$K_i = -\frac{i}{2}\left(\begin{array}{cc|c}\sigma_i & 0 & 0\\ 0 & 0 & 0\\ \hline 0 & 0 & 0\end{array}\right), \quad L_i = -\frac{i}{2}\left(\begin{array}{cc|c}0 & 0 & 0\\ 0 & \sigma_i & 0\\ \hline 0 & 0 & 0\end{array}\right), \quad M = -\frac{i}{2}\left(\begin{array}{cc|c}0 & 0 & 0\\ 0 & 0 & 0\\ \hline 0 & 0 & 1\end{array}\right), \tag{14}$$

where σ_i $(i=1,2,3)$ are the Pauli matrices.[5] As we shall discuss in the next subsection, the appearance of the supermatrix is rather natural from the perspective of the full ten-dimensional $\mathrm{AdS}_5\times T^{1,1}$ coset space. Since all off-diagonal blocks in (14) are zero, there are no fermionic elements. Thus, the essential difference in considering supermatrices, rather than bosonic ones, is that in computing the metric one must use the *upertrace* STr, rather than the Tr, which is defined by

$$\mathrm{STr}\left(\begin{array}{c|c}A & B\\ \hline C & D\end{array}\right) \equiv \mathrm{Tr}(A) - \mathrm{Tr}(D)\,, \tag{15}$$

where A, D are the bosonic and B, C are the fermionic block matrices. The representation given in (14) has the following normalization:

$$\mathrm{STr}(K_aK_b) = \mathrm{STr}(L_aL_b) = -\frac{1}{2}\delta_{ab}\,, \qquad \mathrm{STr}(MM) = \frac{1}{4}\,. \tag{16}$$

[4] The manifolds $T^{p,q}$ with p, q non-negative integers are defined by the same coset $T^{p,q}\simeq[SU(2)\times SU(2)]/U(1)_1$, but with $T_1 = pK_3 + qL_3$ (see [38] for details). This explains why the manifold is called $T^{1,1}$.

[5] The convention are given by $\sigma_1 = \begin{pmatrix}0 & 1\\ 1 & 0\end{pmatrix}$, $\sigma_2 = \begin{pmatrix}0 & -i\\ i & 0\end{pmatrix}$, $\sigma_3 = \begin{pmatrix}1 & 0\\ 0 & -1\end{pmatrix}$.

Journal of Physics: Conference Series **670** (2016) 012019 doi:10.1088/1742-6596/670/1/012019

Now, we define the coset projector as follows;

$$P(x) = x - \frac{\mathrm{STr}(T_1 x)}{\mathrm{STr}(T_1 T_1)} T_1 - \frac{\mathrm{STr}(T_2 x)}{\mathrm{STr}(T_2 T_2)} T_2\,, \tag{17}$$

for $x \in \mathfrak{su}(2) \oplus \mathfrak{su}(2) \oplus \mathfrak{u}(1)$. Explicit expressions for T_1 and T_2 follow from (11) and (14). Finally, with this prescription, the metric (4) is obtained by computing [1]

$$ds^2_{T^{1,1}} = -\frac{1}{3}\mathrm{STr}[AP(A)]\,. \tag{18}$$

§ *W   Super   a rice ?*
One may wonder about the significance of the appearance the supertrace STr, rather than the usual trace. Although it would be interesting to address this question directly (and study possible generalizations to other spaces), here we offer an explanation of the case at hand based on AdS/CFT. It is well known that backgrounds of the form $\mathrm{AdS}_5 \times X$, with X a 5D Sasaki-Einstein manifold describe the gravity duals of certain $\mathcal{N} = 1$ superconformal field theories. The case $X = T^{1,1}$ was an early example proposed by Klebanov-Witten [42] (see [46] for a review). According to the AdS/CFT dictionary, the isometries of the internal manifold X translate into global symmetries of the dual field theory. In the canonical case of S^5, the $SO(6) \sim SU(4)$ symmetry corresponds to the R-symmetry of the $\mathcal{N} = 4$ dual field theory. In the case of $T^{1,1}$ the $U(1)_R$ part of the isometry corresponds to R-symmetry of the $\mathcal{N} = 1$ dual field theory, while the remaining $SU(2) \times SU(2)$ corresponds to the flavor symmetry of the field theory. Thus, these two symmetries should be physically distinguished. We recall that $U(1)_R$ is included in the bosonic subgroup of the $\mathcal{N} = 1$ superconformal group $PSU(2,2|1)$ as follows,

$$SO(2,4) \times U(1)_R = SU(2,2) \times U(1)_R \overset{\text{bosonic}}{\subset} PSU(2,2|1)\,. \tag{19}$$

Thus, from the perspective of the ten-dimensional space $\mathrm{AdS}_5 \times T^{1,1}$, the internal manifold $T^{1,1}$ in (10) appears naturally as the boconic part of the supercoset[6]

$$\begin{aligned}\mathrm{AdS}_5 \times T^{1,1} \quad &\simeq \quad \frac{SO(2,4)}{SO(1,4)} \times \frac{U(1)_R \times SU(2) \times SU(2)}{U(1)_1 \times U(1)_2} \\ &\overset{\text{bosonic}}{\subset} \frac{PSU(2,2|1) \times SU(2) \times SU(2)}{SO(1,4) \times U(1)_1 \times U(1)_2}\,.\end{aligned} \tag{21}$$

Here we have emphasized the $T^{1,1}$ part by the gray background. Keeping in mind the above supercoset, the $(4|1) \times (4|1)$ supermatrix in (14) may be justified by interpreting the matrix as a part of the bigger $(8|1) \times (8|1)$ supermatrix including the $\mathcal{N} = 1$ superconformal algebra as follows:

$$\left(\begin{array}{cc|c} SU(2) & 0 & 0 \\ 0 & SU(2) & 0 \\ \hline 0 & 0 & U(1)_R \end{array}\right) \hookrightarrow \left(\begin{array}{ccc|c} SU(2,2) & 0 & 0 & \overline{F}^A \\ 0 & SU(2) & 0 & 0 \\ 0 & 0 & SU(2) & 0 \\ \hline F_A & 0 & 0 & U(1)_R \end{array}\right). \tag{22}$$

[6] Here it is worth recalling the more familiar case of $\mathrm{AdS}_5 \times \mathrm{S}^5$, in which case the supercoset reads

$$\mathrm{AdS}_5 \times \mathrm{S}^5 \simeq \frac{SO(2,4)}{SO(1,4)} \times \frac{SU(4)_R}{SO(5)} \overset{\text{bosonic}}{\subset} \frac{PSU(2,2|4)}{SO(1,4) \times SO(5)}\,. \tag{20}$$

In this case, the R-symmetry is $SU(4)_R$ and is a part of the $\mathcal{N} = 4$ superconformal algebra $PSU(2,2|4)$.

In the larger supermatrix on the right of (22) the $\mathcal{N}=1$ superconformal algebra $PSU(2,2|1)$ is sitting at the four corners, together with the eight supercharges $F_A, \overline{F}^A$ $(A=1,\cdots,4)$. From this supermatrix one easily recognizes that the lines separating the $\mathbb{Z}_2$-grading of the $(4|1)\times(4|1)$ supermatrix on the left arises from the structure of the $\mathcal{N}=1$ superconformal algebra $PSU(2,2|1)$ on the right. Thus, the appearance of supermatrices (and with this particular grading) is illuminated by AdS/CFT.

3. De ormations o $T^{1,1}$ as Yang-Baxter sigma models

§ *Yan -Bax er i a del*

Now that we have discussed the coset (10) in detail, we apply the Yang-Baxter sigma model approach to study deformations of this space. We will use a classical r-matrix [26, 24].

The Lagrangian of the Yang-Baxter sigma model on $T^{1,1}$ is defined by

$$L = \frac{1}{3}(\gamma^{\alpha\beta} - \epsilon^{\alpha\beta})\mathrm{STr}\left[A_\alpha P \frac{1}{1-2R_g \circ P} A_\beta\right], \tag{23}$$

where the coset projector P given in (17) and the flat metric on the world-sheet $\gamma^{\alpha\beta}$ and the anti-symmetric tensor $\epsilon^{\alpha\beta}$ are normalized as $\gamma^{\alpha\beta} = \mathrm{diag}(-1,+1)$ and $\epsilon^{\tau\sigma}=1$. The *dre ed* operator R_g with $g \in SU(2)\times SU(2)\times U(1)_R$ is defined by

$$R_g(x) \equiv g^{-1}R(gxg^{-1})g\,, \tag{24}$$

where the linear R-operator satisfies the CYBE (classical Yang-Baxter equation)

$$[R(x),R(y)] - R([R(x),y]+[x,R(y)]) = 0 \tag{25}$$

for $x,y \in \mathfrak{su}(2)\oplus\mathfrak{su}(2)\oplus\mathfrak{u}(1)_R$, rather than the mCYBE (modified classical Yang-Baxter equation) [17].

The deformed projected current in the Lagrangian (23)

$$P(J_\beta) := P\frac{1}{1-2R_g\circ P}A_\beta \tag{26}$$

can be obtained by solving the equation

$$(1-2P\circ R_g)P(J_\beta) = P(A_\beta)\,. \tag{27}$$

§ *Cla ical r- a rix and e linear R- pera r*

Deformations of Yang-Baxter sigma models (23) are characterized by the linear R-operator, and reduces to the undeformed model for $R=0$. In the case of $AdS_5\times S^5$, it was revealed in [8] that the R-operators which non-trivially act on the Cartan generators of $\mathfrak{su}(4)$ yield the β- or $\hat{\gamma}_i$-deformed backgrounds [6, 7]. As a simple generalization of the case S^5 case, here we also consider the Abelian r-matrix consisting of the three Cartan generators,[7]

$$r = \frac{1}{3}\hat{\gamma}_1 L_3\wedge M + \frac{1}{3}\hat{\gamma}_2 M\wedge K_3 - \frac{1}{6}\hat{\gamma}_3 K_3\wedge L_3\,, \tag{28}$$

with three deformation parameters $\hat{\gamma}_i$ $(i=1,2,3)$. The wedge stands for the skew-symmetric tensor product $x\wedge y \equiv x\otimes y - y\otimes x$ and K_3 and L_3 are the Cartan generators of two $\mathfrak{su}(2)$'s and M is the $\mathfrak{u}(1)_R$ generator (14).

[7] In principle, the R-operator does not have to satisfy the (m)CYBE since the undeformed model is no longer integrable.

In general, from a tensorial classical r-matrix such as in (28), the linear R-operator is obtained by tracing out the second entries;

$$R(x) = \mathrm{STr}_2[r(1 \otimes x)] \equiv \sum_i \left(a_i \mathrm{STr}(b_i x) - b_i \mathrm{STr}(a_i x)\right), \tag{29}$$

$$\text{with} \quad r = \sum_i a_i \wedge b_i = \sum_i (a_i \otimes b_i - b_i \otimes a_i).$$

Here it is noted that, as a special feature of $T^{1,1}$, we shall use the supertrace STr in (15) instead of the usual one.

Using the inner products in (16), the explicit action of the linear R-operator associated with the classical r-matrix (28) reads

$$R\begin{pmatrix} K_3 \\ L_3 \\ M \end{pmatrix} = -\frac{1}{12}\begin{pmatrix} 0 & \gamma_3 & 2\gamma_2 \\ -\gamma_3 & 0 & -2\gamma_1 \\ \gamma_2 & -\gamma_1 & 0 \end{pmatrix}\begin{pmatrix} K_3 \\ L_3 \\ M \end{pmatrix}, \qquad R(\text{others}) = 0\,. \tag{30}$$

§ *Def r ed back r und*
Plugging the linear R-operator (30) into the Yang-Baxter sigma model (23), one can read off the deformed metric and the NS-NS two-from [1];

$$\begin{aligned} ds^2 = {} & G(\hat\gamma_1,\hat\gamma_2,\hat\gamma_3)\Big[\frac{1}{6}\sum_{i=1,2}\left(G(\hat\gamma_1,\hat\gamma_2,\hat\gamma_3)^{-1}d\theta_i^2 + \sin^2\theta_i d\phi_i^2\right) \\ & + \frac{1}{9}(d\psi + \cos\theta_1 d\phi_1 + \cos\theta_2 d\phi_2)^2 + \frac{\sin^2\theta_1\sin^2\theta_2}{324}(\hat\gamma_3 d\psi + \hat\gamma_1 d\phi_1 + \hat\gamma_2 d\phi_2)^2\Big], \end{aligned} \tag{31}$$

$$\begin{aligned} B_2 = {} & G(\hat\gamma_1,\hat\gamma_2,\hat\gamma_3)\Big[\Big\{\hat\gamma_3\Big(\frac{\sin^2\theta_1\sin^2\theta_2}{36} + \frac{\cos^2\theta_1\sin^2\theta_2 + \cos^2\theta_2\sin^2\theta_1}{54}\Big) \\ & - \hat\gamma_2\frac{\cos\theta_2\sin^2\theta_1}{54} - \hat\gamma_1\frac{\cos\theta_1\sin^2\theta_2}{54}\Big\}d\phi_1\wedge d\phi_2 \\ & + \frac{(\hat\gamma_3\cos\theta_2 - \hat\gamma_2)\sin^2\theta_1}{54}d\phi_1\wedge d\psi - \frac{(\hat\gamma_3\cos\theta_1 - \hat\gamma_1)\sin^2\theta_2}{54}d\phi_2\wedge d\psi\Big], \end{aligned} \tag{32}$$

where the scalar function G is defined by

$$\begin{aligned} G(\hat\gamma_1,\hat\gamma_2,\hat\gamma_3)^{-1} \equiv {} & 1 + \hat\gamma_3^2\Big(\frac{\sin^2\theta_1\sin^2\theta_2}{36} + \frac{\cos^2\theta_1\sin^2\theta_2 + \cos^2\theta_2\sin^2\theta_1}{54}\Big) + \hat\gamma_2^2\frac{\sin^2\theta_1}{54} \\ & + \hat\gamma_1^2\frac{\sin^2\theta_2}{54} - \hat\gamma_2\hat\gamma_3\frac{\sin^2\theta_1\cos\theta_2}{27} - \hat\gamma_3\hat\gamma_1\frac{\sin^2\theta_2\cos\theta_1}{27}\,. \end{aligned} \tag{33}$$

These expressions for the metric and NS-NS two-form coincide exactly with the ones obtained in [45] via TsT-transformations.[8] Thus, we have shown by means of an explicit example that the *ravi /CYBE c rre p ndence* holds beyond the strict case of integrable backgrounds.

4. Conclusions and discussions

The main question addressed in this short article is whether the *ravi /CYBE c rre p ndence* [8, 9, 10] can be extended beyond the integrable case. The answer seems to be yes, as first realized in [1] and reviewed here.

To recapitulate, we have shown this to be the case by means of an explicit example, namely the 5D Sasaki-Einstein space $T^{1,1}$, which was shown to be non-integrable in [41, 47]. Along the

[8] In the special case $\hat\gamma_1 = \hat\gamma_2 = 0$ and $\hat\gamma_3 \neq 0$ one obtains the Lunin-Maldacena background [6].

way to showing this we had to develop a novel (super)coset construction for this space, namely (10). As discussed, unlike the standard coset construction, this prescription leads directly to the Sasaki-Einstein metric on the space and is specially suited to applying the Yang-Baxter deformation. Although the origin of the supermatrix construction is not entirely clear from a purely mathematical standpoint, it is a rather natural construction from the viewpoint of the $\mathcal{N} = 1$ superconformal symmetry of the dual field theory. It would be interesting to study the mathematical origin of this construction, as well as possible generalizations of it to other coset spaces.

Next, we have applied the Yang-Baxter sigma model deformation with a general Abelian r-matrix (28), showing that the resulting deformed metric and the NS-NS two-form are in complete agreement with the deformed supergravity backgrounds obtained by TsT-transformations, corresponding to β- or $\hat{\gamma}_i$-deformations [6, 45]. This establishes that integrability does not seem to be an essential feature in the gravity/CYBE correspondence. At the moment, it is not clear to the authors if this relation is specific to $T^{1,1}$ or if it may extend to general non-integrable backgrounds. Thus, it would certainly be interesting to explore further examples. One may consider for example deformations of $T^{1,1}$ by non-Abelian r-matrices, or possible extensions to the general family of Sasaki-Einstein manifolds such as $Y^{p,q}$, which are referred as to *c enei* 1.[9] At a more fundamental level, it would be very interesting to have a better understanding in general of the role of the CYBE providing deformations of non-integrable backgrounds.

It would also be interesting to study whether the supercoset construction described here, and its Yang-Baxter deformations, can be realized in terms of generalized $\mathcal{N} = (2, 2)$ gauged linear sigma models, along the lines of [44, 43].

Other deformations of sigma models which have been proposed and are actively studied are [48, 49].[10] One of them is referred as to *λ-def r a i n* [49]. The relation between η-deformations, which are based on classical r-matrices of the Drinfeld-Jimbo type, and λ-deformations was considered by [54, 55, 56], where it was explained that these are *P i n-Lie -dual*. The integrated picture is also captured from the perspective of *e $\mathcal{E}$- del* by [57].

Acknowledgments

We are grateful to Gleb Arutyunov, Riccardo Borsato, Martin Roček and Stefan Vandoren for valuable comments and discussions for the work [1]. P.M.C. is supported by the Netherlands Organization for Scientific Research (NWO) under the VICI Grant 680-47-603. This work is part of the D-ITP consortium, a program of the NWO that is funded by the Dutch Ministry of Education, Culture and Science (OCW). The work of K.Y. is supported by Supporting Program for Interaction-based Initiative Team Studies (SPIRITS) from Kyoto University and by the JSPS Grant-in-Aid for Scientific Research (C) No.15K05051. This work is also supported in part by the JSPS Japan-Russia Research Cooperative Program and the JSPS Japan-Hungary Research Cooperative Program.

Re erences

[1] P. M. Crichigno, T. Matsumoto and K. Yoshida, "Deformations of $T^{1,1}$ as Yang-Baxter sigma models," JHEP **1412** (2014) 085 [ar iv:1406.2249 [hep-th]].

[2] J. M. Maldacena, "The large N limit of superconformal field theories and supergravity," Adv. Theor. Math. Phys. **2** (1998) 231 [Int. J. Theor. Phys. **38** (1999) 1113]. [ar iv:hep-th 9711200].

[3] E. Witten, "Anti-de Sitter space and holography," Adv. Theor. Math. Phys. **2**, 253 (1998) [hep-th 9802150].

[4] S. S. Gubser, I. R. Klebanov and A. M. Polyakov, "Gauge theory correlators from noncritical string theory," Phys. Lett. B **428**, 105 (1998) [hep-th 9802109].

[9] Although this may seem possible conceptually, the general case is technically much more difficult.

[10] For the further references of the Yang-Baxter sigma models, for instance, see [50, 51, 52, 53].

[5] T. Buscher, "A Symmetry of the String Background Field Equations," Phys. Lett. B **184** (1987) 59; "Path-Integral Derivation of Quantum Duality in Non-Linear Sigma Models," Phys. Lett. B **201** (1988) 466.
[6] O. Lunin and J. M. Maldacena, "Deforming field theories with $U(1) \times U(1)$ global symmetry and their gravity duals," JHEP **0505** (2005) 033 [hep-th 0502086].
[7] S. Frolov, "Lax pair for strings in Lunin-Maldacena background," JHEP **0505** (2005) 069 [hep-th 0503201].
[8] T. Matsumoto and K. Yoshida, "Lunin-Maldacena backgrounds from the classical Yang-Baxter equation – Towards the gravity CYBE correspondence," JHEP **1406** (2014) 135 [ar iv:1404.1838 [hep-th]].
[9] T. Matsumoto and K. Yoshida, "Integrability of classical strings dual for noncommutative gauge theories," JHEP **1406** (2014) 163 [ar iv:1404.3657 [hep-th]].
[10] T. Matsumoto and K. Yoshida, "Yang-Baxter deformations and string dualities," JHEP **1503** (2015) 137 [ar iv:1412.3658 [hep-th]].
[11] A. Hashimoto and N. Itzhaki, "Noncommutative Yang-Mills and the AdS CFT correspondence," Phys. Lett. B **465** (1999) 142 [hep-th 9907166].
[12] J. M. Maldacena and J. G. Russo, "Large N limit of noncommutative gauge theories," JHEP **9909** (1999) 025 [hep-th 9908134].
[13] C. Klimcik, "Yang-Baxter sigma models and dS AdS T duality," JHEP **0212** (2002) 051 [hep-th 0210095]; "On integrability of the Yang-Baxter sigma-model," J. Math. Phys. **50** (2009) 043508 [ar iv:0802.3518 [hep-th]]; "Integrability of the bi-Yang-Baxter sigma model," ar iv:1402.2105 [math-ph].
[14] R. Squellari, "Yang-Baxter σ model: Quantum aspects," Nucl. Phys. B **881** (2014) 502 [ar iv:1401.3197 [hep-th]].
[15] F. Delduc, M. Magro and B. Vicedo, "On classical q-deformations of integrable sigma-models," JHEP **1311** (2013) 192 [ar iv:1308.3581 [hep-th]].
[16] R. R. Metsaev and A. A. Tseytlin, "Type IIB superstring action in $AdS_5 \times S^5$ background," Nucl. Phys. B **533** (1998) 109 [hep-th 9805028].
[17] F. Delduc, M. Magro and B. Vicedo, "An integrable deformation of the $AdS_5 \times S^5$ superstring action," Phys. Rev. Lett. **112** (2014) 051601 [ar iv:1309.5850 [hep-th]].
[18] F. Delduc, M. Magro and B. Vicedo, "Derivation of the action and symmetries of the q-deformed AdS_5 x S^5 superstring," ar iv:1406.6286 [hep-th].
[19] I. Kawaguchi and K. Yoshida, "Hybrid classical integrability in squashed sigma models," Phys. Lett. B **705** (2011) 251 [ar iv:1107.3662 [hep-th]]; "Hybrid classical integrable structure of squashed sigma models: A short summary," J. Phys. Conf. Ser. **343** (2012) 012055 [ar iv:1110.6748 [hep-th]].
[20] I. Kawaguchi, T. Matsumoto and K. Yoshida, "The classical origin of quantum affine algebra in squashed sigma models," JHEP **1204** (2012) 115 [ar iv:1201.3058 [hep-th]]; "On the classical equivalence of monodromy matrices in squashed sigma model," JHEP **1206** (2012) 082 [ar iv:1203.3400 [hep-th]].
[21] I. Kawaguchi and K. Yoshida, "Hidden Yangian symmetry in sigma model on squashed sphere," JHEP **1011** (2010) 032 [ar iv:1008.0776 [hep-th]]. I. Kawaguchi, D. Orlando and K. Yoshida, "Yangian symmetry in deformed WZNW models on squashed spheres," Phys. Lett. B **701** (2011) 475 [ar iv:1104.0738 [hep-th]]; I. Kawaguchi and K. Yoshida, "A deformation of quantum affine algebra in squashed WZNW models," J. Math. Phys. **55** (2014) 062302 [ar iv:1311.4696 [hep-th]].
[22] G. Arutyunov, R. Borsato and S. Frolov, "S-matrix for strings on η-deformed $AdS_5 \times S^5$," JHEP **1404** (2014) 002 [ar iv:1312.3542 [hep-th]].
[23] G. Arutyunov, R. Borsato and S. Frolov, "Puzzles of eta-deformed $AdS_5 \times S^5$," ar iv:1507.04239 [hep-th].
[24] I. Kawaguchi, T. Matsumoto and K. Yoshida, "Jordanian deformations of the $AdS_5 \times S^5$ superstring," JHEP **1404** (2014) 153 [ar iv:1401.4855 [hep-th]].
[25] I. Kawaguchi, T. Matsumoto and K. Yoshida, "A Jordanian deformation of AdS space in type IIB supergravity," JHEP **1406** (2014) 146 [ar iv:1402.6147 [hep-th]].
[26] T. Matsumoto and K. Yoshida, "Yang-Baxter sigma models based on the CYBE," Nucl. Phys. B **893** (2015) 287 [ar iv:1501.03665 [hep-th]].
[27] I. Kawaguchi and K. Yoshida, "Classical integrability of Schrödinger sigma models and q-deformed Poincare symmetry," JHEP **1111** (2011) 094 [ar iv:1109.0872 [hep-th]];
[28] I. Kawaguchi and K. Yoshida, "Exotic symmetry and monodromy equivalence in Schrödinger sigma models," JHEP **1302** (2013) 024 [ar iv:1209.4147 [hep-th]].
[29] I. Kawaguchi, T. Matsumoto and K. Yoshida, "Schrödinger sigma models and Jordanian twists," JHEP **1308** (2013) 013 [ar iv:1305.6556 [hep-th]].
[30] V. G. Drinfel'd, "Hopf algebras and the quantum Yang-Baxter equation," Sov. Math. Dokl. **32** (1985) 254; "Quantum groups," J. Sov. Math. **41** (1988) 898 [Zap. Nauchn. Semin. **155**, 18 (1986)].
[31] M. Jimbo, "A q difference analog of $U(g)$ and the Yang-Baxter equation," Lett. Math. Phys. **10** (1985) 63.
[32] T. Matsumoto, D. Orlando, S. Reffert, J. i. Sakamoto and K. Yoshida, "Yang-Baxter deformations of Minkowski spacetime," ar iv:1505.04553 [hep-th].

[33] A. Borowiec, H. Kyono, J. Lukierski, J. Sakamoto and K. Yoshida, "Yang-Baxter sigma models and Lax pairs arising from κ-Poincaré r-matrices," ar iv:1510.03083 [hep-th].
[34] T. Kameyama, H. Kyono, J. i. Sakamoto and K. Yoshida, "Lax pairs on Yang-Baxter deformed backgrounds," ar iv:1509.00173 [hep-th].
[35] J. A. Minahan and K. Zarembo, "The Bethe ansatz for N=4 superYang-Mills," JHEP **0303** (2003) 013 [hep-th 0212208].
[36] I. Bena, J. Polchinski and R. Roiban, "Hidden symmetries of the $AdS_5 \times S^5$ superstring," Phys. Rev. D **69** (2004) 046002 [hep-th 0305116].
[37] N. Beisert *et al.*, "Review of AdS CFT Integrability: An Overview," Lett. Math. Phys. **99** (2012) 3 [ar iv:1012.3982 [hep-th]].
[38] P. Candelas and . C. de la Ossa, "Comments on Conifolds," Nucl. Phys. B **342** (1990) 246.
[39] L. Castellani, L. J. Romans and N. P. Warner, "Symmetries of Coset Spaces and Kaluza-Klein Supergravity," Annals Phys. **157**, 394 (1984).
[40] L. J. Romans, "New Compactifications of Chiral $N = 2d = 10$ Supergravity," Phys. Lett. B **153** (1985) 392.
[41] P. Basu and L. A. Pando Zayas, "Chaos Rules out Integrability of Strings in $AdS_5 \times T^{1,1}$," Phys. Lett. B **700** (2011) 243 [ar iv:1103.4107 [hep-th]].
[42] I. R. Klebanov and E. Witten, "Superconformal field theory on three-branes at a Calabi-Yau singularity," Nucl. Phys. B **536** (1998) 199 [hep-th 9807080].
[43] P. M. Crichigno and M. Roček, "On gauged linear sigma models with torsion," ar iv:1506.00335 [hep-th].
[44] F. Benini, P. M. Crichigno, D. Jain and J. Nian, "Semichiral Fields on S^2 and Generalized Kähler Geometry," ar iv:1505.06207 [hep-th].
[45] A. Catal-Ozer, "Lunin-Maldacena deformations with three parameters," JHEP **0602** (2006) 026 [hep-th 0512290].
[46] C. P. Herzog, I. R. Klebanov and P. Ouyang, "D-branes on the conifold and N=1 gauge gravity dualities," hep-th 0205100.
[47] Y. Asano, D. Kawai, H. Kyono and K. Yoshida, "Chaotic strings in a near Penrose limit of $AdS_5\times T^{1,1}$," JHEP **1508** (2015) 060 [ar iv:1505.07583 [hep-th]].
[48] T. J. Hollowood, J. L. Miramontes and D. M. Schmidtt, "Integrable Deformations of Strings on Symmetric Spaces," ar iv:1407.2840 [hep-th]; T. J. Hollowood, J. L. Miramontes and D. Schmidtt, "An Integrable Deformation of the $AdS_5\times S^5$ Superstring," ar iv:1409.1538 [hep-th].
[49] K. Sfetsos, "Integrable interpolations: From exact CFTs to non-Abelian T-duals," Nucl. Phys. B **880** (2014) 225 [ar iv:1312.4560 [hep-th]]; K. Sfetsos and D. C. Thompson, "Spacetimes for λ-deformations," JHEP **1412** (2014) 164 [ar iv:1410.1886 [hep-th]]; K. Sfetsos and K. Siampos, "The anisotropic λ-deformed SU(2) model is integrable," Phys. Lett. B **743** (2015) 160 [ar iv:1412.5181 [hep-th]]; S. Demulder, K. Sfetsos and D. C. Thompson, "Integrable λ-deformations: Squashing Coset CFTs and $AdS_5 \times S^5$," JHEP **07** (2015) 019 [ar iv:1504.02781 [hep-th]]; G. Georgiou, K. Sfetsos and K. Siampos, "All-loop anomalous dimensions in integrable λ-deformed σ-models," ar iv:1509.02946 [hep-th]; G. Itsios, K. Sfetsos, K. Siampos and A. Torrielli, "The classical Yang-Baxter equation and the associated Yangian symmetry of gauged WZW-type theories," ar iv:1409.0554 [hep-th].
[50] B. Hoare, R. Roiban and A. A. Tseytlin, "On deformations of $AdS_n\times S^n$ supercosets," JHEP **1406** (2014) 002 [ar iv:1403.5517 [hep-th]].
[51] G. Arutyunov and S. J. van Tongeren, "The $AdS_5 \times S^5$ mirror model as a string," ar iv:1406.2304 [hep-th].
[52] G. Arutynov, M. de Leeuw and S. J. van Tongeren, "On the exact spectrum and mirror duality of the $(AdS_5 \times S^5)_\eta$ superstring," ar iv:1403.6104 [hep-th].
[53] S. J. van Tongeren, "On classical Yang-Baxter based deformations of the $AdS_5\times S^5$ superstring," JHEP **1506** (2015) 048 [ar iv:1504.05516 [hep-th]]; S. J. van Tongeren, "Yang-Baxter deformations, AdS CFT, and twist-noncommutative gauge theory," ar iv:1506.01023 [hep-th].
[54] B. Vicedo, "Deformed integrable σ-models, classical R-matrices and classical exchange algebra on Drinfel'd doubles," J. Phys. A **48** (2015) 35, 355203 [ar iv:1504.06303 [hep-th]].
[55] B. Hoare and A. A. Tseytlin, "On integrable deformations of superstring sigma models related to $AdS_n \times S^n$ supercosets," Nucl. Phys. B **897** (2015) 448 [ar iv:1504.07213 [hep-th]].
[56] K. Sfetsos, K. Siampos and D. C. Thompson, "Generalised integrable λ- and η-deformations and their relation," Nucl. Phys. B **899** (2015) 489 [ar iv:1506.05784 [hep-th]].
[57] C. Klimcik, "η and λ deformations as $\mathcal{E}$-models," ar iv:1508.05832 [hep-th].

Hamiltonian action of spinning particle with gravimagnetic moment

Alexei A Deriglazov[1,2]1, **and W Guzmán Ramírez**[1]

[1] Depto. de Matemática, ICE, Universidade Federal de Juiz de Fora, MG, Brazil

[2] Laboratory of Mathematical Physics, Tomsk Polytechnic University, 634050 Tomsk, Lenin Ave. 30, Russian Federation

E-mail: alexei.deriglazov@ufjf.edu.br

Abstract. We develop Hamiltonian variational problem for spinning particle non-minimally interacting with gravity through the gravimagnetic moment κ. For $\kappa = 0$ our model yields Mathisson-Papapetrou-Tulczyjew-Dixon (MPTD) equations, the latter show unsatisfactory behavior of MPTD-particle in ultra-relativistic regime: its longitudinal acceleration increases with velocity. $\kappa = 1$ yields a modification of MPTD-equations with the reasonable behavior: in the homogeneous fields, both longitudinal acceleration and (covariant) precession of spin-tensor vanish as $v \to c$.

Evolution of a rotating body in curved background usually described on the base of equations formulated by Mathisson [1], Papapetrou [2], Tulczyjew [3], and Dixon [4]. In this work we will refer to MPTD-equations as those (6.31)-(6.33) in [4]. Detailed analysis and interpretation of these equations and their generalizations is an actual task since they are widely used now to account spin effects in compact binaries and rotating black holes, see [5, 6] and references therein.

It was recently observed that MPTD-equations have unsatisfactory behavior in ultra-relativistic limit: longitudinal acceleration of the particle grows up to infinity in this limit [7]. Here we briefly describe the origin for such behavior and then obtain Hamiltonian variational problem which yields equations with improved behavior. The improved equations correspond to the spinning particle non-minimally interacting with gravity through the gravimagnetic moment κ. Our main conclusion is that spinning particle with gravimagnetic moment $\kappa = 1$ seems to be more promising for description of a rotating body in a gravitational field than MPTD-equations (the latter correspond to the spinning particle with vanishing gravimagnetic moment).

We first obtain MPTD-equations as the Hamiltonian system starting from the vector model of spin. This allows us to exclude, in unambiguous way, the momenta and to obtain second-order equation for trajectory, which is necessary for analysis of ultra-relativistic behavior of the particle. In the vector model of relativistic spin the configuration-space variables[1] are the position $x^\mu(\tau)$ and non Grassmann vector $\omega^\mu(\tau)$ attached to the point x^μ. The Frenkel spin-tensor [8] in our model is a composite quantity constructed from ω^μ and its conjugated

[1] The dynamical variables are taken in arbitrary parametrization τ, then $\dot{x}^\mu = \frac{dx^\mu}{d\tau}$. Covariant derivative is $\nabla P^\mu = \frac{dP^\mu}{d\tau} + \Gamma^\mu_{\alpha\beta}\dot{x}^\alpha P^\beta$ and curvature is $R^\sigma{}_{\lambda\mu\nu} = \partial_\mu\Gamma^\sigma{}_{\lambda\nu} - \partial_\nu\Gamma^\sigma{}_{\lambda\mu} + \Gamma^\sigma{}_{\beta\mu}\Gamma^\beta{}_{\lambda\nu} - \Gamma^\sigma{}_{\beta\nu}\Gamma^\beta{}_{\lambda\mu}$. The square brackets mean antisymmetrization, $\omega^{[\mu}\pi^{\nu]} = \omega^\mu\pi^\nu - \omega^\nu\pi^\mu$. We use the condensed notation $\dot{x}^\mu G_{\mu\nu}\dot{x}^\nu = \dot{x}G\dot{x}$, $N^\mu{}_\nu\dot{x}^\nu = (N\dot{x})^\mu$, $\omega^2 = g_{\mu\nu}\omega^\mu\omega^\nu$, $\mu, \nu = 0, 1, 2, 3$, and so on. Notation for the scalar functions constructed from second-rank tensors are $\theta S = \theta^{\mu\nu}S_{\mu\nu}$, $S^2 = S^{\mu\nu}S_{\mu\nu}$.

momentum $\pi_\mu = \frac{\partial L}{\partial \dot{\omega}^\mu}$ as follows: $S^{\mu\nu} = 2(\omega^\mu \pi^\nu - \omega^\nu \pi^\mu)$. In contrast to its constituents ω^μ and π^ν, the spin-tensor is invariant under local spin-plane symmetry presented in the model [9], and thus represents an observable quantity. The spinning particle in flat space is described by the Lagrangian action [10, 7]

$$S = -\frac{1}{\sqrt{2}} \int d\tau \sqrt{m^2c^2 - \frac{\alpha}{\omega^2}} \sqrt{-\dot{x}N\dot{x} - \dot{\omega}N\dot{\omega} + \sqrt{[\dot{x}N\dot{x} + \dot{\omega}N\dot{\omega}]^2 - 4(\dot{x}N\dot{\omega})^2}}. \quad (1)$$

The matrix $N_{\mu\nu} = \eta_{\mu\nu} - \frac{\omega_\mu \omega_\nu}{\omega^2}$ is the projector on the plane orthogonal to ω^ν: $N_{\mu\nu}\omega^\nu = 0$, the parameter α determines the value of spin. The value $\alpha = \frac{3\hbar^2}{4}$ corresponds to the elementary spin one-half particle. In the spinless limit, $\alpha = 0$ and $\omega^\mu = 0$, the expression (1) reduces to the standard Lagrangian of relativistic particle, $-mc\sqrt{-\dot{x}^\mu \dot{x}_\mu}$.

The minimal interaction with gravity is achieved by covariantization of the formulation (1), that is we replace $\eta_{\mu\nu} \to g_{\mu\nu}$, and usual derivative by the covariant one, $\dot{\omega}^\mu \to \nabla\omega^\mu = \frac{d\omega^\mu}{d\tau} + \Gamma^\mu_{\alpha\beta}\dot{x}^\alpha \omega^\beta$. With the resulting expression at hands, we pass to the Hamiltonian formulation. This gives the Hamiltonian variational problem which is equivalent to (1)

$$S_H = \int d\tau \; p_\mu \dot{x}^\mu + \pi_\mu \dot{\omega}^\mu - \left[\frac{\lambda_1}{2}\left(P^2 + (mc)^2 + \pi^2 - \frac{\alpha}{\omega^2}\right) + \lambda_2(\omega\pi) + \lambda_3(P\omega) + \lambda_4(P\pi)\right]. \quad (2)$$

By λ_a we denoted the Lagrangian multipliers, they are written in front of the corresponding primary constraints. For the general-covariant and spin-plane invariant phase-space quantities $\dot{x}^\mu$, $P_\mu = \frac{\partial L}{\partial \dot{x}^\mu} - \Gamma^\beta_{\alpha\mu}\omega^\alpha \pi_\beta$ and $S^{\mu\nu}$, the variational problem (1) (or, equivalently, the variational problem (2)) yields the Hamiltonian equations which can be written as follows:

$$P^\mu = \frac{mc}{\sqrt{-\dot{x}G\dot{x}}}\tilde{T}^\mu{}_\nu \dot{x}^\nu, \quad \nabla P^\mu = -\frac{1}{4}\theta^\mu{}_\nu \dot{x}^\nu, \quad \nabla S^{\mu\nu} = 2P^{[\mu}\dot{x}^{\nu]}, \quad (3)$$

as well as the constraints

$$S^{\mu\nu}P_\nu = 0, \quad (4)$$

$$P^2 + (mc)^2 = 0, \qquad S^2 = 8\alpha. \quad (5)$$

We have denoted

$$\theta^\mu{}_\nu \equiv R^\mu{}_{\nu\alpha\beta}S^{\alpha\beta}. \quad (6)$$

In the equations appeared the matrix G constructed with help of "tetrad field" $\tilde{T}$

$$G_{\mu\nu} = g_{\mu\nu} + h_{\mu\nu} \equiv \tilde{T}^\alpha{}_\mu g_{\alpha\beta}\tilde{T}^\beta{}_\nu, \qquad \tilde{T}^\mu{}_\nu = \delta^\mu{}_\nu - \frac{1}{8m^2c^2}S^{\mu\sigma}\theta_{\sigma\nu}. \quad (7)$$

The matrix G is composed from the original metric $g_{\mu\nu}$ plus (spin and field-dependent) contribution $h_{\mu\nu}$. So we call G the effective metric produced along the world-line by interaction of spin with gravity.

Comparing (3)-(5) with MPTD-equations[2], we concluded [7] that all the trajectories of MPTD-equations with given integrals of motion $P^2 = k$ and $S^2 = \beta$ are described by our

[2] In [11] we noticed that MPTD-equations have an expected spinless limit being completed with mass-shell type condition. We also point out that the Lagrangian considered above yields the fixed value of spin, that is this corresponds to an elementary particle. In [11] we presented the modification which leads to the theory with unfixed spin, and, similarly to Hanson-Regge approach [12], with a mass-spin trajectory constraint. The modified model has four physical degrees of freedom in the spin-sector.

spinning particle with mass $m = \frac{k}{c}$ and spin $\alpha = \frac{\beta}{8}$. By the way, we demonstrated that MPTD-equations correspond to the minimal interaction of the spinning particle with gravity.

Inverting the first equation from (3), we can exclude momenta P^μ from the Hamiltonian equations. This yields second-order equation for the particle's position $x^\mu(\tau)$ (we refer the resulting equations as Lagrangian form of MPTD-equations)

$$\nabla\left[\frac{\tilde{T}^\mu{}_\nu\dot{x}^\nu}{\sqrt{-\dot{x}G\dot{x}}}\right] = -\frac{1}{4mc}\theta^\mu{}_\nu\dot{x}^\nu\,, \tag{8}$$

$$\nabla S^{\mu\nu} = \frac{1}{4mc\sqrt{-\dot{x}G\dot{x}}}\dot{x}^{[\mu}(S\theta\dot{x})^{\nu]}\,, \tag{9}$$

$$S^\mu{}_\nu\dot{x}^\nu - \frac{1}{8(mc)^2}(S\theta\dot{x})^\mu = 0\,. \tag{10}$$

The last equation is the Lagrangian counterpart of the Frenkel condition (4).

In the spinless limit the equation (8) turn into the geodesic equation. Spin causes deviations from the geodesic motion due to right hand side of this equation, as well as due to presence of the tetrad field $\tilde{T}$ and the effective metric G in the left hand side.

The equations (8) and (9) became singular at critical velocity which obeys the equation $\dot{x}G\dot{x} = 0$. The singularity determines behavior of the particle in ultra-relativistic limit. To clarify this point, let us consider the standard equations of spinless particle interacting with electromagnetic field in the physical-time parametrization $x^\mu(t) = (ct, \mathbf{x}(t))$

$$\left(\frac{\dot{x}^\mu}{\sqrt{c^2-\mathbf{v}^2}}\right)^{\cdot} = \frac{e}{mc^2}F^\mu{}_\nu\dot{x}^\nu, \tag{11}$$

then the factor is just $c^2 - \mathbf{v}^2$. Rewriting the equations in the form of second law of Newton we find an acceleration. For the case, the longitudinal acceleration reads $a_{||} = \mathbf{va} = \frac{e(c^2-\mathbf{v}^2)^{\frac{3}{2}}}{mc^3}(\mathbf{Ev})$, that is the factor, elevated in some degree, appears on the right hand side of the equation, and thus determines the value of velocity at which the longitudinal acceleration vanishes, $a_{||} \overset{v\to c}{\longrightarrow} 0$. In resume, for the present case[3] the singularity implies that during its evolution in external background, the particle can not exceed the speed of light c.

We can repeat this computation for our case (8)-(10). Due to the dependence of the tetrad field in Eq. (8) on the spin-tensor S, the singularity presented in (9) causes the appearance of the term proportional to $\frac{1}{\sqrt{\dot{x}G\dot{x}}}$ in the expression for longitudinal acceleration. In the result, the acceleration grows up to infinity as the particle's speed approximates to the critical speed $\sim c$, see [7].

We have seen above that in the Lagrangian formalism MPTD-equations correspond to the minimal interaction of spinning particle with gravity. Let us demonstrate that MPTD-particle admits also a non-minimal interaction. We add the term $\frac{\lambda_1}{2}\kappa R_{\alpha\beta\mu\nu}\omega^\alpha\pi^\beta\omega^\mu\pi^\nu \equiv \frac{\lambda_1}{2}\frac{\kappa}{16}\theta S$ into the Hamiltonian action (2). By analogy with the magnetic moment, the interaction constant κ is called gravimagnetic moment [14]. Thus we consider the variational problem

$$S_\kappa = \int d\tau\; p_\mu\dot{x}^\mu + \pi_\mu\dot{\omega}^\mu - \left[\frac{\lambda_1}{2}\left(P^2 + \kappa R_{\alpha\beta\mu\nu}\omega^\alpha\pi^\beta\omega^\mu\pi^\nu + (mc)^2 + \pi^2 - \frac{\alpha}{\omega^2}\right) + \lambda_2(\omega\pi) + \lambda_3(P\omega) + \lambda_4(P\pi)\right]. \tag{12}$$

on the space of independent variables $x^\mu, p_\nu, \omega^\mu, \pi_\nu$ and λ_a. We have denoted $P_\mu \equiv p_\mu - \Gamma^\beta_{\alpha\mu}\omega^\alpha\pi_\beta$, $P^2 = g^{\mu\nu}P_\mu P_\nu$ and so on. Note also that the first two terms can be identically rewritten in the

[3] In a curved space we need to be more careful since the three-dimensional geometry should respect the coordinate independence of the speed of light, see [13, 7].

general-covariant form $p_\mu \dot{x}^\mu + \pi_\mu \dot{\omega}^\mu = P_\mu \dot{x}^\mu + \pi_\mu \nabla\omega^\mu$. Variation of the action with respect to λ_a gives the algebraic equations

$$P^2 + \kappa R_{\alpha\beta\mu\nu}\omega^\alpha\pi^\beta\omega^\mu\pi^\nu + (mc)^2 + \pi^2 - \frac{\alpha}{\omega^2} = 0, \tag{13}$$

$$\omega\pi = 0,, \quad P\omega = 0, \qquad P\pi = 0, \tag{14}$$

while variations with respect to the remaining variables yield dynamical equations which can be written in the covariant form as follows[4]

$$\frac{\partial S_\kappa}{\partial p_\mu} = 0 \quad \Leftrightarrow \quad \dot{x}^\mu = \lambda_1 P^\mu + \lambda_3\omega^\mu + \lambda_4\pi^\mu, \tag{15}$$

$$\frac{\partial S_\kappa}{\partial x^\mu} = 0 \quad \Leftrightarrow \quad \nabla P_\mu = -R_{\mu\nu\alpha\beta}\dot{x}^\nu\omega^\alpha\pi^\beta - \frac{1}{2}\lambda_1\kappa\nabla_\mu R_{\sigma\nu\alpha\beta}\omega^\sigma\pi^\nu\omega^\alpha\pi^\beta, \tag{16}$$

$$\frac{\partial S_\kappa}{\partial \pi_\mu} = 0 \quad \Leftrightarrow \quad \nabla\omega^\mu = \lambda_1\pi^\mu - \lambda_1\kappa R^\mu{}_{\alpha\beta\nu}\omega^\alpha\omega^\beta\pi^\nu + \lambda_2\omega^\mu + \lambda_4 P^\mu, \tag{17}$$

$$\frac{\partial S_\kappa}{\partial \omega^\mu} = 0 \quad \Leftrightarrow \quad \nabla\pi_\mu = -\frac{\lambda_1\alpha}{\omega^4}\omega_\mu - \lambda_1\kappa R_{\mu\nu\alpha\beta}\pi^\nu\omega^\alpha\pi^\beta - \lambda_2\pi_\mu - \lambda_3 P_\mu. \tag{18}$$

Eq. (15) has been repeatedly used to obtain the final form (16)-(18) of the equations $\frac{\partial S_\kappa}{\partial x^\mu} = 0$, $\frac{\partial S_\kappa}{\partial \pi_\mu} = 0$ and $\frac{\partial S_\kappa}{\partial \omega^\mu} = 0$. Computing time-derivative of algebraic equations equations (14) and using (15)-(18) we obtain the consequences

$$\pi^2 - \frac{\alpha}{\omega^2} = 0, \tag{19}$$

$$\begin{aligned}\lambda_3 &= 4a\lambda_1\left[2(1-\kappa)R_{\alpha\beta\mu\nu}\omega^\alpha\pi^\beta\pi^\mu P^\nu + \kappa\pi^\sigma(\nabla_\sigma R_{\mu\nu\alpha\beta})\omega^\mu\pi^\nu\omega^\alpha\pi^\beta\right], \\ \lambda_4 &= -4a\lambda_1\left[2(1-\kappa)R_{\alpha\beta\mu\nu}\omega^\alpha\pi^\beta\omega^\mu P^\nu + \kappa\omega^\sigma(\nabla_\sigma R_{\mu\nu\alpha\beta})\omega^\mu\pi^\nu\omega^\alpha\pi^\beta\right],\end{aligned} \tag{20}$$

Here and below we use the following notation. In the equation which relates velocity and momentum will appear the matrix $\mathcal{T}$

$$\mathcal{T}^\alpha{}_\nu \equiv \delta^\alpha{}_\nu - (\kappa-1)aS^{\alpha\sigma}\theta_{\sigma\nu}, \qquad a = \frac{2}{16m^2c^2 + (\kappa+1)(S\theta)}. \tag{21}$$

The matrix has an inverse given by

$$\tilde{\mathcal{T}}^\alpha{}_\nu \equiv \delta^\alpha{}_\nu + (\kappa-1)bS^{\alpha\sigma}\theta_{\sigma\nu}, \qquad b = \frac{1}{8m^2c^2 + \kappa(S\theta)}. \tag{22}$$

The vector Z^μ is defined by

$$Z^\mu = \frac{b}{8}S^{\mu\sigma}(\nabla_\sigma R_{\alpha\beta\rho\delta})S^{\alpha\beta}S^{\rho\delta} \equiv \frac{b}{8}S^{\mu\sigma}\nabla_\sigma(S\theta). \tag{23}$$

[4] Note that by construction of S_κ, the variational equation $\frac{\partial S_\kappa}{\partial p_\mu} = 0$ is equivalent to $\dot{x}^\mu = \{x^\mu, H\}$, and so on.

This vanishes in a space with homogeneous curvature, $\nabla R = 0$.

The time-derivatives of (13), (19) and (20) do not yield new algebraic equations. Due to (19) we can replace the constraint (13) on $P^2 + \kappa R_{\alpha\beta\mu\nu}\omega^\alpha\pi^\beta\omega^\mu\pi^\nu + (mc)^2 = 0$. The obtained expressions for λ_3 and λ_4 can be used to exclude these variables from the equations (15)-(18).

Neither constraints nor equations of motion do not determine the functions λ_1 and λ_2, that is the non-minimal interaction preserves both reparametrization and spin-plane symmetries of the theory. The presence of λ_1 and λ_2 in the equations of motion implies that evolution of our basic variables is ambiguous. According to general theory [15, 16, 17], variables with ambiguous dynamics do not represent observable quantities, so we need to search for the variables that can be candidates for observables. We note that (17) and (18) imply an equation for $S^{\mu\nu}$ which does not contain λ_2. So we rewrite (15) and (16) in terms of spin-tensor and add to them the equation for $S^{\mu\nu}$, this gives the system

$$\dot{x}^\mu = \lambda_1\left[\mathcal{T}^\mu{}_\nu P^\nu + \kappa\frac{a}{b}Z^\mu\right], \tag{24}$$

$$\nabla P_\mu = -\frac{1}{4}\theta_{\mu\nu}\dot{x}^\nu - \frac{\lambda_1\kappa}{32}\nabla_\mu(S\theta), \tag{25}$$

$$\nabla S^{\mu\nu} = 2P^{[\mu}\dot{x}^{\nu]} - \frac{\lambda_1\kappa}{4}\theta^{[\mu}{}_\sigma S^{\nu]\sigma}. \tag{26}$$

Besides, the constraints (13), (14) and (19) imply

$$P^2 + \frac{\kappa}{16}\theta S + (mc)^2 = 0, \tag{27}$$

$$S^{\mu\nu}P_\nu = 0, \qquad S^{\mu\nu}S_{\mu\nu} = 8\alpha. \tag{28}$$

The equations (28) imply that only two components of spin-tensor are independent, as it should be for spin one-half particle. Eq. (26), contrary to the equations for ω and π, does not depend on λ_2. This proves that the spin-tensor is invariant under local spin-plane symmetry. The remaining ambiguity due to λ_1 is related with reparametrization invariance and disappears when we work with physical dynamical variables $x^i(t)$. Equations (24)-(26), together with (27) and (28), form a closed system which determines evolution of a spinning particle.

The gravimagnetic moment κ can generally take any value. When $\kappa = 0$ we recover the MPTD equations.

Let us exclude P^μ and λ_1 from the equations (25) and (26). Using (22) we solve (24) with respect to P^μ. Using the resulting expression in the constraint (27) we obtain λ_1

$$\lambda_1 = \frac{\sqrt{-\dot{x}\mathcal{G}\dot{x}}}{m_r c}, \quad \text{with} \quad m_r^2 \equiv m^2 + \frac{\kappa}{16c^2}(S\theta) - \left(\frac{\kappa a}{cb}\right)^2 Z^2, \tag{29}$$

where the effective metric now is given by

$$\mathcal{G}_{\mu\nu} = \tilde{\mathcal{T}}^\alpha{}_\mu g_{\alpha\beta}\tilde{\mathcal{T}}^\beta{}_\nu. \tag{30}$$

Then the expression for momentum in terms of velocities implied by (24) is

$$P^\mu = \frac{m_r c}{\sqrt{-\dot{x}\mathcal{G}\dot{x}}}\tilde{\mathcal{T}}^\mu{}_\nu\dot{x}^\nu - \kappa Z^\mu. \tag{31}$$

We substitute this P_μ into (25), (26)

$$\nabla\left[\frac{m_r c}{\sqrt{-\dot{x}\mathcal{G}\dot{x}}}\tilde{\mathcal{T}}_{\mu\nu}\dot{x}^\nu\right] = -\frac{1}{4}\theta_{\mu\nu}\dot{x}^\nu - \kappa\frac{\sqrt{-\dot{x}\mathcal{G}\dot{x}}}{32 m_r c}\nabla^\mu(S\theta) + \kappa\nabla Z^\mu, \tag{32}$$

$$\nabla S^{\mu\nu} = \frac{2m_r c(1-\kappa)b}{\sqrt{-\dot{x}\mathcal{G}\dot{x}}}\dot{x}^{[\mu}S^{\nu]\sigma}\theta_{\sigma\alpha}\dot{x}^\alpha + \frac{\kappa\sqrt{-\dot{x}\mathcal{G}\dot{x}}}{4m_r c}\theta^{[\mu}{}_\alpha S^{\nu]\sigma} - 2\kappa Z^{[\mu}\dot{x}^{\nu]}. \tag{33}$$

Together with (28), this gives us the Lagrangian equations for the spinning particle with gravimagnetic moment.

In resume, we have formulated Hamiltonian variational problem for spinning particle with gravimagnetic moment κ in an arbitrary gravitational background. The model is consistent for any value of κ. When $\kappa = 0$, our equations of motion (24)-(26) coincide with MPTD-equations (3)-(5). The latter also follow from the Lagrangian variational problem without auxiliary variables (1), and correspond to the minimal interaction of our spinning particle with gravity.

Comparing our equations for arbitrary κ to the equations of spinning particle on an arbitrary electromagnetic background (see Eqs. (28)-(29) in [10]), we see that the two systems have the same structure after the identification $\kappa \sim \mu$ and $\theta_{\mu\nu} \equiv R_{\mu\nu\alpha\beta}S^{\alpha\beta} \sim F_{\mu\nu}$, where μ is the magnetic moment. That is curvature influences trajectory of spinning particle in the same way as the electromagnetic field strength $\theta_{\mu\nu}$.

Now we are ready to discuss the main subject of the work. As we have pointed out above, the MPTD-equations, which correspond to the value $\kappa = 0$ of our model (12), have non satisfactory behavior in ultra-relativistic limit. The reason is that variation rate of spin-tensor diverges in this limit and contributes into longitudinal acceleration of the particle through the tetrad field $\tilde{\mathcal{T}}^\mu{}_\nu(S)$, see also [7]. Consider now our spinning particle with gravimagnetic moment $\kappa = 1$. This implies $\tilde{\mathcal{T}}^\mu{}_\nu = \delta^\mu{}_\nu$, $\mathcal{G}_{\mu\nu} = g_{\mu\nu}$, and crucially simplifies the equations of motion. The Hamiltonian equations (24)-(26) read

$$\frac{m_r c}{\sqrt{-\dot{x}g\dot{x}}}\dot{x}^\mu = P^\mu + \frac{a}{b}Z^\mu\,, \tag{34}$$

$$\nabla P_\mu = -\frac{1}{4}\theta_{\mu\nu}\dot{x}^\nu - \frac{\sqrt{-\dot{x}g\dot{x}}}{32m_r c}\nabla_\mu(S\theta)\,, \tag{35}$$

$$\nabla S^{\mu\nu} = 2P^{[\mu}\dot{x}^{\nu]} - \frac{\sqrt{-\dot{x}g\dot{x}}}{4m_r c}\theta^{[\mu}{}_\sigma S^{\nu]\sigma}\,. \tag{36}$$

while the Lagrangian equations are composed now by the equation for trajectory

$$\nabla\left[\frac{m_r c\dot{x}^\mu}{\sqrt{-\dot{x}g\dot{x}}}\right] = -\frac{1}{4}\theta^\mu{}_\nu\dot{x}^\nu - \frac{\sqrt{-\dot{x}g\dot{x}}}{32m_r c}\nabla^\mu(S\theta) + \nabla Z^\mu\,, \tag{37}$$

and by the equation for precession of spin-tensor

$$\nabla S^{\mu\nu} = \frac{\sqrt{-\dot{x}g\dot{x}}}{4m_r c}\theta^{[\mu}{}_\sigma S^{\nu]\sigma} - 2Z^{[\mu}\dot{x}^{\nu]}\,. \tag{38}$$

These equations can be compared with (8)-(9). Even in homogeneous field, $\nabla R = 0$, we have modified dynamics for both x and S. Covariant precession of spin (38) has a smooth behavior $\nabla S \sim (\sqrt{-\dot{x}g\dot{x}})^0$ contrary to $\nabla S \sim \frac{1}{\sqrt{-\dot{x}g\dot{x}}}$ in the equation (9). The equation (34) in space with homogeneous curvature has the structure similar to (11), hence we expect that longitudinal acceleration vanishes as $v \to c$ This has been confirmed by direct computations in [7].

Acknowledgments
The work of AAD has been supported by the Brazilian foundations CNPq (Conselho Nacional de Desenvolvimento Cientfico e Tecnolgico - Brasil) and FAPEMIG (Fundao de Amparo Pesquisa do estado de Minas Gerais- Brasil). WGR thanks CAPES for the financial support (Programm PNPD/2011).

References

[1] Mathisson M 1937 *Acta Phys. Polon.* **6** 163
[2] Papapetrou A 1951 *Proc. Roy. Soc. Lond.* A **209** 248
[3] Tulczyjew W M 1959 *Acta Phys. Polon.* **18** 393
[4] Dixon W G 1964 *Nuovo Cimento* **34** 317
[5] Pomeransky A A, Sen'kov R A and Khriplovich I B 2000 *Physics-Uspekhi* **43 (10)** 1055
[6] BalakinA B and Popov V A 2015 Spin-axion coupling *Preprint* arXiv:1509.06058 (gr-qc)
[7] Deriglazov A A and Ramirez W G 2015 Mathisson-Papapetrou-Tulczyjew-Dixon (MPTD) equations in ultra-relativistic regime and gravimagnetic moment *Preprint* arXiv:1509.05357 (gr-qc)
[8] Frenkel J 1926 *Z. Phys.* **37** 243
[9] Deriglazov A A and Pupasov-Maksimov A M 2014 SIGMA **10** 012
[10] Deriglazov A A 2014 *Phys. Lett.* B **736** 278
[11] Ramirez W G and Deriglazov A A 2015 Lagrangian formulation for Mathisson-Papapetrou-Tulczyjew-Dixon (MPTD) equations *Preprint* arXiv:1509.04926 (gr-qc)
[12] Hanson A J and Regge T 1974 *Annals of Physics* **87 (2)** 498
[13] Deriglazov A A and Ramirez W G 2015 *Mod. Phys. Lett.* A **30** 1550101 (*Preprint* arXiv:1409.4756 (physics.gen-ph))
[14] Khriplovich I B 1989 *Sov. Phys. JETP* **69** 217
[15] Dirac P A M 1964 *Lectures on Quantum Mechanics* (Yeshiva University, New York)
[16] Gitman D M and Tyutin I 1990 *Quantization of fields with constraints* (Springer-Verlag, Berlin)
[17] Deriglazov A 2010 *Classical Mechanics: Hamiltonian and Lagrangian Formalism* (Springer-Verlag)

Contact symmetries of constrained quadratic Lagrangians

N Dimakis[1], Petros A Terzis[2] and T Christodoulakis[2]

[1] Instituto de Ciencias Fisicas y Matematicas,
Universidad Austral de Chile, Valdivia, Chile
[2] Nuclear and Particle Physics Section, Physics Department,
University of Athens, GR 157–71 Athens

E-mail: nsdimakis@gmail.com, pterzis@phys.uoa.gr, tchris@phys.uoa.gr

Abstract. The conditions for the existence of (polynomial in the velocities) contact symmetries of constrained systems that are described by quadratic Lagrangians is presented. These Lagrangians mainly appear in mini-superspace reductions of gravitational plus matter actions. In the literature, one usually adopts a gauge condition (mostly for the lapse N) prior to searching for symmetries. This, however, is an unnecessary restriction which may lead to a loss of symmetries and consequently to the respective integrals of motion. A generalization of the usual procedure rests in the identification of the lapse function N as an equivalent degree of freedom and the according extension of the infinitesimal generator. As a result, conformal Killing tensors (with appropriate conformal factors) can define integrals of motion (instead of just Killing tensors used in the regular gauge fixed case). Additionally, rheonomic integrals of motion - whose existence is unique in this type of singular systems - of various orders in the momenta can be constructed. An example of a relativistic particle in a pp-wave space-time and under the influence of a quadratic potential is illustrated.

1. Introduction

Symmetries have always played a fundamental role in physical theories. In recent years there exists an increased interest on the subject of Noether point and/or generalized symmetries of physical systems and their geometric nature [1]-[7].

In general, one can discern two approaches regarding the treatment of symmetries in mini-superspace systems: The first entails an a priori fixation of the gauge (usualy $N = 1$) at the Lagrangian level and the treatment of the ensuing system as if it were regular ([8]-[12]). The second method ([6], [7], [13], [14]) utilizes the gauge invariance of parametrization invariant systems, resulting in the emergence of larger symmetry groups.

The general form of a mini-superspace Lagrangian is

$$L = \frac{1}{2N}G_{\mu\nu}(q)\dot{q}^\mu\dot{q}^\nu - NV(q), \quad \mu,\nu = 1,...,d, \quad \det(G_{\mu\nu}) \neq 0 \tag{1.1}$$

where $q^\mu(t)$ and $N(t)$ are the $d+1$ degrees of freedom of the system while the dot symbolizes the total time derivative $\frac{d}{dt}$. The Lagrangian denoted by (1.1) is singular, due to the quadratic

Journal of Physics: Conference Series **670** (2016) 012021 doi:10.1088/1742-6596/670/1/012021

constraint equation for N ($\frac{\partial L}{\partial N} = 0$). This constraint can be used to generalize what is considered as an integral of motion, since the time derivative of such a quantity need not be strictly zero, but just a multiple of the constraint.

Let us first review some basic facts from the theory of symmetries of the action for regular systems [16]. A contact transformation whose infinitesimal generator is $X = \Xi^\kappa(t,q,\dot{q})\frac{\partial}{\partial q^\kappa}$, constitutes a symmetry of the action if the condition

$$pr^{(1)}X(L) = \frac{dF}{dt} \tag{1.2}$$

is satisfied, where $pr^{(1)}X$ is the first prolongation of the generator (that expresses the change induced by X in the velocities $\dot{q}$'s) and $F(t,q,\dot{q})$ a gauge function. Given the form of the generator, a conserved quantity can be constructed as $\Xi^\kappa \frac{\partial L}{\partial \dot{q}^\kappa} - F$.

For singular systems it has been shown [6] that the form of X should include dependence in N, since the latter is also a degree of freedom for the system. Thus, one is led to consider

$$X = \Xi^\kappa(t,N,q,\dot{q})\frac{\partial}{\partial q^\kappa} + \Omega(t,N,q,\dot{q})\frac{\partial}{\partial N}. \tag{1.3}$$

Notice that when the additional term $\Omega(t,N,q,\dot{q})\frac{\partial}{\partial N}$ is acting on the Lagrangian through (1.2), creates a multiple of the quadratic constraint. Thus, (1.2) becomes an equation that holds modulo the constraint.

2. General results for polynomial in the velocities contact symmetries

In our case we shall restrict ourselves to the consideration of symmetries with corresponding generators polynomial in the velocities. This means that Ξ^κ in (1.3) is taken as

$$\Xi^\kappa = \xi^\kappa(t,N,q) + S^\kappa{}_{\alpha_1}(t,N,q)\dot{q}^{\alpha_1} + \dots + S^\kappa{}_{\alpha_1\dots\alpha_n}(t,N,q)\dot{q}^{\alpha_1}\dots\dot{q}^{\alpha}_n. \tag{2.1}$$

This automatically restricts the forms of $\Omega(t,N,q,\dot{q})$ and $F(t,N,q,\dot{q})$. The first has to be polynomial of first order in $\dot{N}$ and of n order in $\dot{q}$'s, while the second of zero order in $\dot{N}$ and of $n+1$ order in $\dot{q}$'s. However, the existence of the gauge function can be proven to be trivial [7], so we won't refer to it any longer.

The procedure in order to derive the symmetry generator in these cases is the following: After the application of the symmetry criterion (1.2), the coefficients of terms involving velocities $\dot{N}$ and $\dot{q}^\mu$ of various orders are gathered and set to zero. By solving these partial differential equations, the general form for the symmetry generator is obtained:

$$\begin{aligned} X = &\left(\xi^\kappa(q) + \frac{1}{N}S^\kappa_{\alpha_1}(q)\dot{q}^{\alpha_1} + \frac{1}{N^2}S^\kappa_{\alpha_1\alpha_2}(q)\dot{q}^{\alpha_1}\dot{q}^{\alpha_2} + \dots + \frac{1}{N^n}S^\kappa_{\alpha_1\dots\alpha_n}(q)\dot{q}^{\alpha_1}\dots\dot{q}^{\alpha_n}\right)\frac{\partial}{\partial q^\kappa} \\ &- \left(\frac{N}{V}\tilde{\xi}^\sigma V_{,\sigma} + \frac{2V^{,\sigma}}{V}S_{\sigma\alpha_1}\dot{q}^{\alpha_1} + \frac{n+1}{N^{n-1}}\frac{V^{,\sigma}}{V}S_{\sigma\alpha_1\dots\alpha_n}\dot{q}^{\alpha_1}\dots\dot{q}^{\alpha_n}\right)\frac{\partial}{\partial N} \end{aligned}$$

together with the subsequent conditions:

$$\mathcal{L}_\xi G_{\mu\nu} = -\frac{\mathcal{L}_\xi V}{V}G_{\mu\nu} \tag{2.2a}$$

$$S_{(\mu\alpha_1;\nu)} = -\frac{V^{,\sigma}}{V}S_{\sigma(\alpha_1}G_{\mu\nu)}, \; \dots \;, S_{(\mu\alpha_1\dots\alpha_n;\nu)} = -\frac{n+1}{2}\frac{V^{,\sigma}}{V}S_{\sigma(\alpha_1\dots\alpha_n}G_{\mu\nu)}. \tag{2.2b}$$

From the derived symmetry generator, one can construct integrals of motion of several orders in the momenta. It is clear that equations (2.2) already contain the possibility of ξ (the S's) being a Killing field (Killing tensors) of the metric and the potential ($S^{\sigma}\alpha_1 ... \alpha_n V_{,\sigma} = 0$), which are of course the known cases of symmetries for regular systems. Equations (2.2) indicate that the situation in constraint systems is more general. What is more, it can be easily verified that for systems described by (1.1), rheonomic integrals of motion (conserved quantities that have an explicit time dependence) can also be constructed by any conformal Killing field of $G_{\mu\nu}$ [6]. If there is a conformal Killing vector $\xi(q)$ that satisfies

$$\mathcal{L}_{\xi} G_{\mu\nu} = \omega(q) G_{\mu\nu} \tag{2.3}$$

then the quantity

$$\tilde{Q} = \xi^{\alpha} p_{\alpha} + \int N(t) \left(\omega(q(t)) + f(q(t))\right) V(q(t)) dt, \tag{2.4}$$

where $f = \frac{\mathcal{L}_{\xi} V}{V}$, is an integral of motion of the system, since

$$\frac{dQ}{dt} = \frac{\partial Q}{\partial t} + \{Q, H\} = \omega(q)\mathcal{H} \approx 0. \tag{2.5}$$

In the case where $\omega = -f$ the integral of motion is autonomous and one returns to the case described by (2.2a).

An analogous construction can be made for any conformal Killing tensor of arbitrary order. Let us suppose that

$$S^{(\kappa\alpha_1 ... \alpha_n;\mu)} = \frac{1}{2} \omega^{(\alpha_1 ... \alpha_n} G^{\mu\nu)} \tag{2.6}$$

then one can we can consider the quantity

$$\tilde{Q} = S^{\kappa\alpha_1 ... \alpha_n} p_{\kappa} p_{\alpha_1} ... p_{\alpha_n} + \int N \left(\omega^{\alpha_1 ... \alpha_n} + (n+1) f^{\alpha_1 ... \alpha_n}\right) V \frac{\partial L}{\partial \dot{q}^{\alpha_1}} ... \frac{\partial L}{\partial \dot{q}^{\alpha_n}} dt \tag{2.7}$$

where everything inside the integral is to be regarded strictly as a function of time and $f^{\alpha_1 ... \alpha_n}$ is an n-rank tensor defined as

$$f^{\alpha_1 ... \alpha_n} = S^{\kappa\alpha_1 ... \alpha_n} \frac{V_{,\kappa}}{V}. \tag{2.8}$$

It is easy to check that, in the same manner as previously, the time derivative of $\tilde{Q}$ results in

$$\frac{d\tilde{Q}}{dt} = N\omega^{\alpha_1 ... \alpha_n} p_{\alpha_1} ... p_{\alpha_n} \mathcal{H} \approx 0 \tag{2.9}$$

and thus $\tilde{Q}$ is a rheonomic integral of motion due to the constraint.

Now we have all the tools necessary to apply into a simple example that illustrates the significance of this generalization.

3. Example: A particle in a curved space-time under the influence of a quadratic potential

We have already mentioned that Lagrangians of the form (1.1) are mainly encountered in cosmology. However, here we shall consider an example of a relativistic particle moving in a type Biv pp-wave space-time described by the line element

$$ds^2 = -2dvdu - \frac{2}{z^2} du^2 + dx^2 + dy^2 \tag{3.1}$$

under the influence of a quadratic potential of the form $V(u) = \frac{\mu}{2}u^2$. In the case where $\mu > 0$ the latter can be considered to represent an oscillator in the u direction. The space-time described by (3.1) has been examined in [17] with respect to its conformal algebra and its Killing tensors.

By using (1.1), the Lagrangian of the system reads

$$L = \frac{1}{N}\left(-\dot{u}\,\dot{v} - \frac{\dot{u}^2}{z^2} + \frac{1}{2}\dot{y}^2 + \frac{1}{2}\dot{z}^2\right) - N\frac{\mu\, u^2}{2} \tag{3.2}$$

and, as we can observe, it possesses five degrees of freedom. However, due to its constrained nature, four of them can be expressed as functions of the fifth on the solution space.

There exist three vector fields satisfying condition (2.2a) (which in this case happen also to be Killing fields of both the metric $G_{\mu\nu}$ and the potential $V(u)$). At the same time, there also exists a homothetic vector of both the metric and the potential. With their help the following four linear integrals of motion can be written:

$$Q_1 = -\frac{\dot{u}}{N}, \quad Q_2 = \frac{\dot{y}}{N}, \quad Q_3 = \frac{u\,\dot{y}}{N} - \frac{y\,\dot{u}}{N} \tag{3.3a}$$

$$Q_h = -\frac{2\,u\,\dot{u}}{N\,z^2} - \frac{u\,\dot{v}}{N} + \frac{y\,\dot{y}}{2\,N} + \frac{z\,\dot{z}}{2\,N} + \frac{3}{2}\,\mu \int N\,u^2\,dt. \tag{3.3b}$$

that consequently define the equations $Q_i = \kappa_i$ and $Q_h = \kappa_h$.

It can also be verified that there exist seven second rank non reducible conformal Killing tensors satisfying criterion (2.2b), all of which, albeit one, are also Killing tensors for the space-time under consideration. In our example we are going to make use of the following two

$$S_1^{\mu\nu} = \begin{pmatrix} -\frac{8}{u^2 z^2 \mu^2} & \frac{4}{u^2\mu^2} & 0 & 0 \\ \frac{4}{u^2\mu^2} & 0 & 0 & 0 \\ 0 & 0 & -\frac{4}{u^2\mu^2} & 0 \\ 0 & 0 & 0 & -\frac{4}{u^2\mu^2} \end{pmatrix}, \quad S_2^{\mu\nu} = \begin{pmatrix} \frac{8}{z^2\mu^2} & 0 & 0 & 0 \\ 0 & 0 & 0 & 0 \\ 0 & 0 & 0 & 0 \\ 0 & 0 & 0 & \frac{4}{\mu^2} \end{pmatrix},$$

with $S_1^{\mu\nu}$ being a proper CKT. Thus, we are supplemented with two additional quadratic in the momenta integrals of motion $Q_4 = S_1^{\mu\nu} p_\mu p_\nu = \kappa_4$ and $Q_5 = S_2^{\mu\nu} p_\mu p_\nu = \kappa_5$. We can now proceed with the derivation of the solution.

3.1. Case $\kappa_1 \neq 0$ *and* $\kappa_5 \neq 0$

In order to completely integrate the equations, we choose to write $N(t) = \dot{f}(t)$, where $f(t)$ is considered to be a non-constant (scalar under time re-parameterizations) function. We first integrate equations $Q_1 = \kappa_1$ and $Q_2 = \kappa_2$ with respect to $u(t)$ and $y(t)$ respectively, and get

$$u(t) = c_u - \kappa_1\, f(t) \quad \text{and} \quad y(t) = c_y + \kappa_2\, f(t) \tag{3.4}$$

where c_u and c_y are constants of integration. With the help of the first, equation $Q_5 = \kappa_5$, yields the following quartet of solutions

$$z(t) = z_{\pm\pm}(t) = \pm\frac{\sqrt{\pm 4\, c_z\, \kappa_5^2\, \mu^4\, f(t) + 4\, c_z^2\, \kappa_5^2\, \mu^4 + \kappa_5^2\, \mu^4\, f(t)^2 + 32\, \kappa_1^2}}{2\sqrt{\kappa_5}\, \mu}, \tag{3.5}$$

where c_z is a constant of integration; of course we must also require $\kappa_5 \neq 0$ (the case $\kappa_5 = 0$ leads to a complex solution). At this point, equation $Q_4 = \kappa_4$ can be integrated to give

$$\begin{aligned} v(t) = v_\pm(t) =& c_v - \frac{f(t)\left(3\left(\mu^2\left(c_u^2\kappa_4 + \kappa_5\right) + 4\kappa_2^2\right) - 3c_u\kappa_1\kappa_4\mu^2 f(t) + \kappa_1^2\kappa_4\mu^2 f(t)^2\right)}{24\kappa_1} \\ &+ \sqrt{2}\tan^{-1}\left(\frac{\kappa_5\mu^2(f(t) \pm 2c_z)}{4\sqrt{2}\kappa_1}\right), \end{aligned} \tag{3.6}$$

with c_v being constant and $\kappa_1 \neq 0$; the plus sign $v_+(t)$ corresponds to solutions $z = z_{++}(t)$ and $z = z_{-+}(t)$, while the minus $v_-(t)$ is for the cases $z = z_{--}(t)$ and $z = z_{+-}(t)$.

By using the solutions (3.4), (3.5) and (3.6) in equation $Q_h = \kappa_h$ we see that, in order for Q_h to be constant for any function $f(t)$, one needs to set $\kappa_4 = \frac{4}{\mu}$. It is easy to check that substitution of the same solutions in the equations of motion together with the constraint $\kappa_4 = \frac{4}{\mu}$ satisfies the system. It is noteworthy that, we were able to obtain the general solution without even turning to the Euler-Lagrange equations of the system.

3.2. Case $\kappa_1 = 0$

Equations $Q_1 = 0$, $Q_2 = \kappa_2$ and $Q_5 = \kappa_5$ lead to the same result with (3.4) and (3.5) by simply setting $\kappa_1 = 0$. One can check that all the other equations regarding autonomous integrals of motion reduce to relations between constants. However, the one defined by the rheonomous integral of motion $Q_h = \kappa_h$ yields, upon integration,

$$v(t) = v_\pm(t) = c_v + \frac{\frac{1}{2}f(t)^2\left(12c_u^2\mu + 4\kappa_2^2 + |\kappa_5|\mu^2\right) + 2f(t)\left(2c_y\kappa_2 \pm c_z\sqrt{|\kappa_5|}\mu - 4\kappa_h\right)}{8c_u}, \tag{3.7}$$

where the $v_+(t)$ corresponds to $z_+(t)$, while $v_-(t)$ to $z_-(t)$ and of course in each case $c_u \neq 0$ (the subcase where $c_u = 0$ will be explored later on).

At this point we have used all equations available by the integrals of motion, substitution of their solutions into the Euler - Lagrange equations yields the constraint relation between constants $\kappa_5 = -\frac{4(c_u^2\mu + \kappa_2^2)}{\mu^2}$.

As we previously remarked, apart from this solution we must also check the case where $c_u = 0$. The application of the relations (3.4) and (3.5) under the condition $\kappa_1 = 0$, leads to the observation that Q_h is constant for every $f(t)$, if and only if $\kappa_5 = -\frac{4\kappa_2^2}{\mu^2}$. Substitution of all these results into the equations of motion leads to a final ordinary differential equation for $v(t)$ that by integration yields

$$v(t) = c_1 f(t) + c_2 \tag{3.8}$$

and thus, we have obtained the complete solution space.

4. Discussion

In this work we have considered, polynomial in the velocities, contact variational symmetries of the action. We derived that, for constrained systems described by (1.1), the infinitesimal criterion (1.2) allows for more symmetries than in the case of regular systems. The main reason for this result is the consideration of gauge invariance not as a redundant degree of freedom that must be fixed out, but rather as an essential part of the system under consideration.

The existence of the constraint gives rise to larger symmetry groups. Of course, this does not mean that one must never choose a gauge. The optimum procedure would be to first calculate all symmetries and then apply any gauge fixing, if necessary, in order to simplify the equations. The important thing is not to do so prior to the determination of the integrals of motion, since their existence may depend on the constraint itself.

In order to exhibit the importance of the method, we examined an example of motion of a relativistic particle in a curved space-time. Due to the consideration of conformal Killing tensors instead of just Killing, we were able to obtain the general solution with no need of gauge fixing and without even using the equations of motion themselves.

Acknowledgments

ND acknowledges financial support by FONDECYT postdoctoral grant no. 3150016.

References

[1] Sarlet W and Cantrijn F 1981 *J. Phys. A: Math. Gen.* **14** 479-92
[2] Kalotas T M and Wybourne B G 1982 *J. Phys. A: Math. Gen.* **15** 2077
[3] Rosquist K and Uggla C 1991 *J. Math. Phys* **32** 3412
[4] Tsamparlis M and Paliathanasis A 2010 *Gen. Rel. Grav.* **42** 2957-80
[5] Tsamparlis M and Paliathanasis A 2011 *J. Phys. A: Math. Theor.* **44** 175202
[6] Christodoulakis T, Dimakis N and Terzis P A 2014 *J. Phys. A: Math. Theor.* **47** 095292
[7] Terzis P A, Dimakis N, Christodoulakis T, Paliathanasis A and Tsamparlis M 2015 *Preprint gr-qc/1503.00932*
[8] Vakili B 2008 *Phys. Lett.* B **664** 16-20
[9] Capozziello S, Piedipalumbo E, Rubano c and Scudellaro P 2009 *Phys. Rev.* D **80** 104030
[10] Basilakos S, Tsamparlis M and Paliathanasis A 2011 *Phys. Rev.* D **83** 103512
[11] Paliathanasis A, Tsamparlis M and Basilakos S 2011 *Phys. Rev.* D **84** 123514
[12] Paliathanasis A, Basilakos S, Saridakis E N, Capozziello S, Atazadeh K, Darabi F and Tsamparlis M 2014 *Phys. Rev.* D **89** 104042
[13] Dimakis N, Christodoulakis T and Terzis P A 2014 *J. Geom. and Phys.* **77** 97-112
[14] Terzis P A, Dimakis N and Christodoulakis T 2015 *Phys. Rev.* D **90** 123543
[15] Dirac P A M 1964 *Lectures on Quantum Mechanics* (Yeshiva University, Academic Press, New York)
[16] Olver P J 2000 *Applications of Lie Groups to Differential Equations, Graduate Texts in Mathematics* vol 107 (Springer-Verlag)
[17] Keane A J and Tupper B O J 2010 *Clas. Quant. Grav.* **27** 245011

Journal of Physics: Conference Series **670** (2016) 012022 doi:10.1088/1742-6596/670/1/012022

2d CFT/Gauge/Bethe correspondence and solvable quantum–mechanical systems

F Ferrari[1], M R Piątek[1,3] and A R Pietrykowski[2,3]

[1] Institute of Physics, University of Szczecin, Wielkopolska 15, 70-451 Szczecin, Poland
[2] Institute of Theoretical Physics, University of Wrocław, M. Borna 9, 50-204 Wrocław, Poland
[3] Bogoliubov Laboratory of Theoretical Physics, JINR, 141980 Dubna, Russia

E-mail: `ferrari@fermi.fiz.univ.szczecin.pl`, `piatek@fermi.fiz.univ.szczecin.pl`, `pietrie@theor.jinr.ru`

Abstract. The holomorphic accessory parameters for the Heun (Fuchs with four singularities) and the Lamé equations are obtained by taking the classical limit of some null vector decoupling equations appearing in two-dimensional conformal field theory. In addition, a link between the aforementioned accessory parameters and the solution of some Bethe-like equations which occur in the supersymmetric gauge theories, is established.

1. Introduction

Let $C_{g,n}$ be the Riemann surface with genus g and n punctures. The basic objects of any two-dimensional conformal field theory (2dCFT) living on C_g [1, 2] are the n-point correlation functions of primary physical vertex operators defined on $C_{g,n}$. Any correlation function can be factorized according to the pattern given by the pant decomposition of $C_{g,n}$ and written as a sum (or an integral for theories with a continuous spectrum) which includes the terms consisting of holomorphic and anti-holomorphic conformal blocks times the 3-point functions of the model for each pair of pants. The Virasoro conformal block $\mathcal{F}_{c,\Delta_p}[\Delta_i](\mathsf{Z})$ on $C_{g,n}$ depends on the cross ratios of the vertex operators locations denoted symbolically by Z and on the $3g-3+n$ intermediate conformal weights Δ_p. Moreover, it depends on the n external conformal weights Δ_i and on the central charge c.

Conformal blocks are fully determined by the underlying conformal symmetry. These functions possess an interesting, although not yet completely understood analytic structure. In general, they can be expressed only as a formal power series and no closed formula is known for its coefficients. Among the issues concerning conformal blocks which are still not fully understood there is the problem of their classical limit [3]. This is the limit in which all parameters of the conformal blocks tend to infinity in such a way that their ratios are fixed:

$$\Delta_i,\ \Delta_p,\ c\ \rightarrow\ \infty, \qquad \frac{\Delta_i}{c}\ =\ \frac{\Delta_p}{c}\ =\ \text{const.}\ .$$

For the standard parametrization of the central charge $c = 1 + 6Q^2$, where $Q = b + \frac{1}{b}$ and for 'heavy' weights $(\Delta_p, \Delta_i) = \frac{1}{b^2}(\delta_p, \delta_i)$ with $\delta_p, \delta_i = \mathcal{O}(b^0)$ the classical limit corresponds to $b \to 0$. There exist many convincing arguments that in the classical limit the conformal blocks behave

Journal of Physics: Conference Series **670** (2016) 012022 doi:10.1088/1742-6596/670/1/012022

exponentially with respect to Z:

$$\mathcal{F} \overset{b\to 0}{\sim} \mathrm{e}^{\frac{1}{b^2} f}.$$

The function f is known as the classical conformal block [3, 4].

In 2009 Alday, Gaiotto and Tachikawa (AGT) discovered an amazing relationship [5] which, in particular, identifies the Virasoro conformal blocks $\mathcal{F}_{c,\Delta_p}[\Delta_i](\mathsf{Z})$ on the Riemann surfaces $C_{g,n}$ with the instanton sectors $\mathcal{Z}_{\rm inst}$ of the Nekrasov partition functions [6, 7]:

$$\mathcal{Z}_{\rm Nekrasov}(\tilde{q}, \tilde{a}, \tilde{m}, \epsilon_1, \epsilon_2) = \mathcal{Z}_{\rm class}\mathcal{Z}_{1-\rm loop}\mathcal{Z}_{\rm inst}$$

for certain class $T_{g,n}$ of (Ω-deformed) four-dimensional supersymmetric $\mathcal{N} = 2$ SU(2) quiver gauge theories. The AGT correspondence works provided that certain relations between parameters of the conformal blocks and the Nekrasov functions are assumed. Let us remember that the Nekrasov partition function depends on the following set of parameters: $\tilde{q}$, $\tilde{a}$, $\tilde{m}$, ϵ_1, ϵ_2. The components of $\tilde{q} = \{\exp 2\pi\tau_1, \ldots, \exp 2\pi\tau_{3g-3+n}\}$ are the gluing parameters associated with the pant decomposition of $C_{g,n}$, where the $\tau_p = \theta_p/2\pi + 4\pi i/g_p^2$ are the complexified gauge couplings. The multiplet $\tilde{m} = \{m_1, \ldots, m_n\}$ contains the mass parameters. Moreover, $\tilde{a} = \{a_1, \ldots, a_{3g-3+n}\}$, where the a's are the vacuum expectation values of the scalar fields in the vector multiplets. Finally, ϵ_1, ϵ_2 represent the complex Ω-background parameters.

Soon after its discovery, the AGT conjecture has been extended to the SU(N)-gauge theories/conformal Toda correspondence [8].

On the other hand, at the same time Nekrasov and Shatashvili observed that in the limit where one of the Ω-deformation parameters vanishes, $\epsilon_2 \to 0$ ($\epsilon_1 = \mathrm{const.}$), the $\mathcal{N} = 2$ SU(N) super Yang–Mills theories describe some N-particle quantum integrable systems (Gauge/Bethe correspondence) [9]. The AGT correspondence dictionary says that $b = \sqrt{\epsilon_2/\epsilon_1}$. Therefore, the Nekrasov–Shatashvili limit via the AGT hypothesis corresponds to the classical limit of the conformal blocks. Hence, one gets the triple correspondence relating two-dimensional conformal field theory to supersymmetric gauge theories and then to quantum integrable systems. The latter in the simplest case of SU(2) gauge group are just quantum–mechanical systems. This triple correspondence can be applied, in particular, to study the spectra of some Schrödinger operators [10, 11, 12] or to compute solutions of the monodromy problems for some ordinary differential equations [3, 13].

In the present note we give an overview of some part of our recent research related to the aforementioned dualities, cf. [14, 15, 16, 17, 18]. In particular, we show how it is possible to derive the so-called accessory parameters for the Fuchs (on $C_{0,4}$) and the Lamé equations by taking the classical limit of the so-called null vector decoupling equations, or equivalently, by solving certain Bethe-like equations.[1]

2. Classical conformal blocks vs. effective twisted superpotentials

To begin with, let us recall a definition and basic facts concerning the highest weight representation of the Virasoro algebra. Let $\mathcal{V}^n_{c,\Delta}$ be the free vector space generated by all vectors of the form:

$$|\,\nu^n_{\Delta,X}\,\rangle = L_{-X}|\,\nu_\Delta\,\rangle = L_{-n_k}\ldots L_{-n_2}L_{-n_1}|\,\nu_\Delta\,\rangle, \tag{1}$$

where $X = (n_k \geqslant \ldots \geqslant n_1 \geqslant 1)$ is an ordered set of positive integers of the length $|X| \equiv n_1 + \ldots + n_k = n$. In eq. above L_n's are the Virasoro generators obeying

$$[L_n, L_m] = (n-m)L_{n+m} + \frac{c}{12}(n^3 - n)\delta_{n+m,0} \tag{2}$$

[1] The same can be done for the Mathieu equation, cf. [11, 12] and a talk by A. R. Pietrykowski.

and $|\,\nu_\Delta\,\rangle$ is the highest weight state with the following property:

$$L_0|\,\nu_\Delta\,\rangle = \Delta|\,\nu_\Delta\,\rangle, \qquad L_n|\,\nu_\Delta\,\rangle = 0 \quad \forall\, n > 0. \tag{3}$$

The representation of the Virasoro algebra on the space:

$$\mathcal{V}_{c,\Delta} = \bigoplus_{n=0}^{\infty} \mathcal{V}^n_{c,\Delta}$$

defined by the relations (2) and (3) is called the Verma module with the central charge c and the highest weight Δ. The dimension of the subspace $\mathcal{V}^n_{c,\Delta} \subset \mathcal{V}_{c,\Delta}$ of all homogeneous elements of degree n is given by the number of partitions $p(n)$ ($p(0) = 1$). The subspaces $\mathcal{V}^n_{c,\Delta}$ are eigenspaces of L_0 with the eigenvalues $\Delta + n$. On $\mathcal{V}^n_{c,\Delta}$ there exists the symmetric bilinear form $\langle\,\cdot\,|\,\cdot\,\rangle_{c,\Delta}$ uniquely defined by the relations:

$$\langle\,\nu_\Delta\,|\,\nu_\Delta\,\rangle_{c,\Delta} \;=\; 1, \qquad (L_n)^\dagger \;=\; L_{-n}.$$

The Gram matrix $G_{c,\Delta}$ of the form $\langle\,\cdot\,|\,\cdot\,\rangle_{c,\Delta}$ is block-diagonal in the basis $\{|\,\nu_{\Delta,X}\,\rangle\}$ with blocks:

$$\left[G^n_{c,\Delta}\right]_{XY} \;=\; \langle\,\nu^n_{\Delta,X}\,|\,\nu^n_{\Delta,Y}\,\rangle_{c,\Delta}. \tag{4}$$

The Verma module $\mathcal{V}_{c,\Delta}$ is irreducible if and only if the form $\langle\,\cdot\,|\,\cdot\,\rangle_{c,\Delta}$ is non-degenerate. The criterion for the irreducibility is vanishing of the determinant $\det G^n_{c,\Delta}$ of the Gram matrix, known as the Kac determinant which is given by the formula:

$$\det G^n_{c,\Delta} \;=\; C_n \prod_{\substack{r,s\in\mathbf{N},\\ s\leqslant r,\\ 1\leqslant rs\leqslant n}} \Phi_{rs}(c,\Delta)^{p(n-rs)}. \tag{5}$$

In the equation above C_n is a constant and

$$\Phi_{rs}(c,\Delta) \;=\; \left(\Delta + \tfrac{r^2-1}{24}(c-13) + \tfrac{rs-1}{2}\right)\left(\Delta + \tfrac{s^2-1}{24}(c-13) + \tfrac{rs-1}{2}\right) + \tfrac{(r^2-s^2)^2}{16}.$$

The Kac determinant vanishes for

$$\Delta_{rs}(c) \;=\; \frac{Q^2}{4} - \frac{1}{4}\left(rb + \frac{s}{b}\right)^2, \tag{6}$$

where the central charge is given by $c = 1 + 6Q^2$ with $Q = b + \frac{1}{b}$. For the so-called degenerate conformal weights Δ_{rs} the representations $\mathcal{V}_{c,\Delta_{rs}}$ are reducible.

The non-zero element $|\,\chi_{rs}\,\rangle \in \mathcal{V}_{c,\Delta_{rs}}$ of degree $n = rs$ is called a null vector if

$$L_0\,|\,\chi_{rs}\,\rangle = (\Delta_{rs} + rs)\,|\,\chi_{rs}\,\rangle, \qquad L_k\,|\,\chi_{rs}\,\rangle = 0 \quad \forall\, k > 0.$$

Hence, $|\,\chi_{rs}\,\rangle$ is the highest weight state which generates its own Verma module $\mathcal{V}_{c,\Delta_{rs}+rs}$, which is a submodule of $\mathcal{V}_{c,\Delta_{rs}}$. One can prove that each submodule of the Verma module $\mathcal{V}_{c,\Delta_{rs}}$ is generated by a null vector. Then, the module $\mathcal{V}_{c,\Delta_{rs}}$ is irreducible if and only if it does not contain null vectors with positive degree.

Let $\mathcal{V}_\Delta$ be the Verma module with the highest weight state $|\,\nu_\Delta\,\rangle$. We define the chiral vertex operator (CVO) as the linear map

$$V^{\Delta_3\,\Delta_2\,\Delta_1}_{\infty\;\;z\;\;0} : \mathcal{V}_{\Delta_2} \otimes \mathcal{V}_{\Delta_1} \to \mathcal{V}_{\Delta_3}$$

such that for all $|\,\xi_2\,\rangle \in \mathcal{V}_{\Delta_2}$ the operator

$$V(\xi_2|z) \equiv V^{\Delta_3\,\Delta_2\,\Delta_1}_{\infty\;\,z\;\,0}(|\,\xi_2\,\rangle \otimes \cdot\,) : \mathcal{V}_{\Delta_1} \to \mathcal{V}_{\Delta_3}$$

satisfies the following conditions:

$$[L_n, V(\nu_2|z)] = z^n\left(z\frac{\partial}{\partial z} + (n+1)\Delta_2\right)V(\nu_2|z)\,, \qquad n \in \mathbf{Z} \tag{7}$$

$$V(L_{-1}\xi_2|z) = \frac{\partial}{\partial z}V(\xi_2|z)\,, \tag{8}$$

$$V(L_n\xi_2|z) = \sum_{k=0}^{n+1}\binom{n+1}{k}(-z)^k\,[L_{n-k}, V(\xi_2|z)]\,, \qquad n > -1, \tag{9}$$

$$\begin{aligned} V(L_{-n}\xi_2|z) &= \sum_{k=0}^{\infty}\binom{n-2+k}{n-2} z^k\, L_{-n-k}\,V(\xi_2|z) \\ &+ (-1)^n\sum_{k=0}^{\infty}\binom{n-2+k}{n-2} z^{-n+1-k}\,V(\xi_2|z)\,L_{k-1}, \qquad n > 1 \end{aligned} \tag{10}$$

and $\langle\,\nu_{\Delta_3}\,|V(\nu_{\Delta_2}|z)\,|\,\nu_{\Delta_1}\,\rangle = z^{\Delta_3-\Delta_2-\Delta_1}$. The commutation relation (7) defines the primary vertex operator corresponding to the highest weight state $|\,\nu_2\,\rangle \in \mathcal{V}_{\Delta_2}$. Eqs. (8)–(10) characterize the decendant CVO's.

The matrix elements of the primary chiral vertex operator $V_\Delta(z) \equiv V(\nu_\Delta|z)$ between the basis states (1) are building elements for the conformal blocks. Let us consider for example,

(i) the 4-point block on the sphere:

$$\mathcal{F}_{c,\Delta}\!\left[\begin{smallmatrix}\Delta_3 & \Delta_2\\ \Delta_4 & \Delta_1\end{smallmatrix}\right]\!(x) = x^{\Delta-\Delta_2-\Delta_1}\left(1 + \sum_{n=1}^{\infty}\mathcal{F}^n_{c,\Delta}\!\left[\begin{smallmatrix}\Delta_3 & \Delta_2\\ \Delta_4 & \Delta_1\end{smallmatrix}\right] x^n\right), \tag{11}$$

$$\mathcal{F}^n_{c,\Delta}\!\left[\begin{smallmatrix}\Delta_3 & \Delta_2\\ \Delta_4 & \Delta_1\end{smallmatrix}\right] = \sum_{n=|I|=|J|}\langle\nu_{\Delta_4}, V_{\Delta_3}(1)\nu_{\Delta,I}\rangle\left[G_{c,\Delta}\right]^{IJ}\langle\nu_{\Delta,J}, V_{\Delta_2}(1)\nu_{\Delta_1}\rangle; \tag{12}$$

(ii) the 1-point block on the torus ($q = \mathrm{e}^{2\pi i\tau}$, where τ is the modular parameter of the torus):

$$\mathcal{F}^{\tilde\Delta}_{c,\Delta}(q) = q^{\Delta-\frac{c}{24}}\left(1 + \sum_{n=1}^{\infty}\mathcal{F}^{\tilde\Delta,n}_{c,\Delta}q^n\right), \tag{13}$$

$$\mathcal{F}^{\tilde\Delta,n}_{c,\Delta} = \sum_{n=|I|=|J|}\langle\nu_{\Delta,I}, V_{\tilde\Delta}(1)\nu_{\Delta,J}\rangle\left[G_{c,\Delta}\right]^{IJ}. \tag{14}$$

The conformal blocks coefficients written explicitly in eqs. (12) and (14) are given in terms of the inverse

$$\left[G_{c,\Delta}\right]^{IJ} = \left(\left[G_{c,\Delta}\right]_{IJ}\right)^{-1}$$

of the Gramm matrix (4) and the matrix elements of $V_\Delta(z)$. To compute the latter it is enough to know the covariance properties (7) of the primary CVO w.r.t. the Virasoro algebra.

Due to the AGT correspondence the sphere and torus conformal blocks can be expressed through the SU(2) Nekrasov instanton partition functions respectively for

(i) the theory with four flavors $N_f = 4$:

$$x^{\Delta_1+\Delta_2-\Delta}\,\mathcal{F}_{c,\Delta}\!\left[\begin{smallmatrix}\Delta_3 & \Delta_2\\ \Delta_4 & \Delta_1\end{smallmatrix}\right]\!(x) = \mathcal{Z}_{\text{inst}}^{N_f=4,\mathrm{SU}(2)}(x,a,m_i,\epsilon_1,\epsilon_2) = (1-x)^{-\frac{(m_1+m_2)(m_3+m_4)}{2\epsilon_1\epsilon_2}}\,\mathcal{Z}_{\text{inst}}^{N_f=4,\mathrm{U}(2)}(x,a,m_i,\epsilon_1,\epsilon_2), \tag{15}$$

where the following relations between the conformal block and the instanton function parameters are assumed:

$$c = 1+6\frac{(\epsilon_1+\epsilon_2)^2}{\epsilon_1\epsilon_2} \equiv 1+6Q^2, \qquad \Delta = \frac{(\epsilon_1+\epsilon_2)^2-4a^2}{4\epsilon_1\epsilon_2},$$
$$\Delta_1 = \frac{\frac{1}{4}(\epsilon_1+\epsilon_2)^2-\frac{1}{4}(m_1-m_2)^2}{\epsilon_1\epsilon_2}, \qquad \Delta_2 = \frac{\frac{1}{2}(m_1+m_2)(\epsilon_1+\epsilon_2-\frac{1}{2}(m_1+m_2))}{\epsilon_1\epsilon_2},$$
$$\Delta_3 = \frac{\frac{1}{2}(m_3+m_4)(\epsilon_1+\epsilon_2-\frac{1}{2}(m_3+m_4))}{\epsilon_1\epsilon_2}, \qquad \Delta_4 = \frac{\frac{1}{4}(\epsilon_1+\epsilon_2)^2-\frac{1}{4}(m_3-m_4)^2}{\epsilon_1\epsilon_2};$$

(ii) the theory with one massive hypermultiplet in the adjoint representation — the so-called $\mathcal{N}=2^*$ theory:

$$q^{\frac{c}{24}-\Delta}\mathcal{F}_{c,\Delta}^{\tilde{\Delta}}(q) = \mathcal{Z}_{\text{inst}}^{\mathcal{N}=2^*,\mathrm{SU}(2)}(q,a,m,\epsilon_1,\epsilon_2) = \left[\prod_{n=1}^{\infty}(1-q^n)\right]^{1-2\tilde{\Delta}}\mathcal{Z}_{\text{inst}}^{\mathcal{N}=2^*,\mathrm{U}(2)}(q,a,m;\epsilon_1,\epsilon_2), \tag{16}$$

where $\tilde{\Delta} = m(\epsilon_1+\epsilon_2-m)/(\epsilon_1\epsilon_2)$ and the relations among the central charge c, the intermediate weight Δ, the Ω-background parameters: ϵ_1,ϵ_2 and the v.e.v. a are all the same as in the first example above.

As has been already mentioned, the conformal blocks exponentiate in the classical limit. In particular, in the case of our two canonical examples one can verify that

$$\mathcal{F}_{1+6Q^2,\Delta}\!\left[\begin{smallmatrix}\Delta_3 & \Delta_2\\ \Delta_4 & \Delta_1\end{smallmatrix}\right]\!(x) \overset{b\to 0}{\sim} \exp\left\{\frac{1}{b^2}\,f_\delta\!\left[\begin{smallmatrix}\delta_3 & \delta_2\\ \delta_4 & \delta_1\end{smallmatrix}\right]\!(x)\right\}, \tag{17}$$

$$\mathcal{F}_{1+6Q^2,\Delta}^{\tilde{\Delta}}(q) \overset{b\to 0}{\sim} \exp\left\{\frac{1}{b^2}f_\delta^{\tilde{\delta}}(q)\right\}, \tag{18}$$

where $\Delta = \frac{1}{b^2}\delta$, $\Delta_i = \frac{1}{b^2}\delta_i$ and $\tilde{\Delta} = \frac{1}{b^2}\tilde{\delta}$. In eqs. above the so-called classical 4-point block on the sphere $f_\delta\!\left[\begin{smallmatrix}\delta_3 & \delta_2\\ \delta_4 & \delta_1\end{smallmatrix}\right]\!(x)$ and the classical 1-point block on the torus $f_\delta^{\tilde{\delta}}(q)$ are available as the formal power series

$$f_\delta\!\left[\begin{smallmatrix}\delta_3 & \delta_2\\ \delta_4 & \delta_1\end{smallmatrix}\right]\!(x) = (\delta-\delta_1-\delta_2)\log x + \sum_{n=1}^{\infty} \mathsf{f}_\delta^n\!\left[\begin{smallmatrix}\delta_3 & \delta_2\\ \delta_4 & \delta_1\end{smallmatrix}\right] x^n, \tag{19}$$

$$f_\delta^{\tilde{\delta}}(q) = \left(\delta-\frac{1}{4}\right)\log q + \sum_{n=1}^{\infty}\mathsf{f}_\delta^{\tilde{\delta},n}\,q^n \tag{20}$$

with the following coefficients:

a) in the case of the classical spherical block

$$
\begin{aligned}
\mathsf{f}_{\delta}^{1}\left[\begin{smallmatrix}\delta_3 & \delta_2\\ \delta_4 & \delta_1\end{smallmatrix}\right] &= \frac{(\delta+\delta_3-\delta_4)(\delta+\delta_2-\delta_1)}{2\delta},\\
\mathsf{f}_{\delta}^{2}\left[\begin{smallmatrix}\delta_3 & \delta_2\\ \delta_4 & \delta_1\end{smallmatrix}\right] &= \Big[16\delta^3(4\delta+3)\Big]^{-1}\Big[13\delta^5+\delta^4\left(18\delta_2-14\delta_1+18\delta_3-14\delta_4+9\right)\\
&+ \delta^3\Big(\delta_1^2+\delta_2^2-2\delta_1(\delta_2+6\delta_3-10\delta_4+6)\\
&+ 4\delta_2(5\delta_3-3\delta_4+3)+(\delta_3-\delta_4)(\delta_3-\delta_4+12)\Big)\\
&- 3\delta^2\Big(\delta_1^2(2\delta_3+2\delta_4-1)+2\delta_1(\delta_3^2+\delta_4^2+2\delta_3+\delta_2-2\delta_2\delta_3-2\delta_4(\delta_2+\delta_3+1))\\
&+ \delta_2^2(2\delta_3+2\delta_4-1)+2\delta_2(\delta_3-\delta_4-2)(\delta_3-\delta_4)-(\delta_3-\delta_4)^2\Big)\\
&+ 5\delta(\delta_1-\delta_2)^2(\delta_3-\delta_4)^2-3(\delta_1-\delta_2)^2(\delta_3-\delta_4)^2\Big],\;\ldots\;;
\end{aligned}
$$

b) in the case of the classical torus block

$$
\mathsf{f}_{\delta}^{\tilde{\delta},1}=\frac{\tilde{\delta}^2}{2\delta},\qquad \mathsf{f}_{\delta}^{\tilde{\delta},2}=\frac{\tilde{\delta}^2[24\delta^2\left(4\delta+1\right)+\tilde{\delta}^2\left(5\delta-3\right)-48\delta\tilde{\delta}^2]}{16\delta^3\left(4\delta+3\right)},\;\ldots\,.
$$

The classical limits (17) and (18) are very nontrivial statements concerning the quantum conformal blocks. There is still no rigorous proof of these conjectures. However, there are many convincing arguments for the existence of the classical conformal blocks. In particular, remembering that $b=\sqrt{\epsilon_2/\epsilon_1}$, eqs. (17) and (18) are consistent with the AGT relations (15)-(16) and the Nekrasov–Shatashvili limit [9]:

$$
\mathcal{Z}_{\rm inst}(\,\cdot\,,\epsilon_1,\epsilon_2)\;\overset{\epsilon_2\to 0}{\sim}\;\exp\left\{\frac{1}{\epsilon_2}\mathcal{W}_{\rm inst}(\,\cdot\,,\epsilon_1)\right\} \tag{21}
$$

of the Nekrasov instanton functions. Indeed, the following identities[2]

$$
\begin{aligned}
f_{\delta}\left[\begin{smallmatrix}\delta_3 & \delta_2\\ \delta_4 & \delta_1\end{smallmatrix}\right](x) &= (\delta-\delta_1-\delta_2)\log x-\frac{(m_1+m_2)(m_3+m_4)}{2\epsilon_1^2}\log(1-x)\\
&+ \frac{1}{\epsilon_1}\mathcal{W}_{\rm inst}^{{\rm U}(2),N_f=4}(x,a,m_i,\epsilon_1),
\end{aligned} \tag{22}
$$

$$
f_{\delta}^{\tilde{\delta}}(q) = \left(\delta-\frac{1}{4}\right)\log q-2\tilde{\delta}\log\left(\frac{\eta(q)}{q^{\frac{1}{24}}}\right)+\frac{1}{\epsilon_1}\mathcal{W}_{\rm inst}^{\mathcal{N}=2^*,{\rm U}(2)}(q,a,m,\epsilon_1) \tag{23}
$$

hold for the classical conformal weights δ, δ_i, $\tilde{\delta}$ defined as follows $\delta=\lim_{b\to 0}b^2\Delta=\lim_{\epsilon_2\to 0}\frac{\epsilon_2}{\epsilon_1}\Delta$. The instanton twisted superpotentials:

$$
\mathcal{W}_{\rm inst}^{{\rm U}(2),N_f=4},\qquad \mathcal{W}_{\rm inst}^{\mathcal{N}=2^*,{\rm U}(2)}
$$

in eqs. (22) and (23) are given in terms of the critical value of some functions $H_{\rm inst}^{\rm U(2)}(x_{ui})$. The latter can be derived from the Nekrasov instanton sums within the Nekrasov–Shatashvili limit. Indeed, the Nekrasov instanton partition functions can be approximated in the limit $\epsilon_2\to 0$ by a kind of 'path integral' with the integrand:

$$
\exp\left\{\frac{1}{\epsilon_2}H_{\rm inst}^{\rm U(2)}(x_{ui})\right\}.
$$

[2] $\eta(q)$ is the Dedekind eta function.

Therefore, the saddle point method yields:

$$\mathcal{W}^{\mathrm{U}(2)}_{\mathrm{inst}} \equiv \lim_{\epsilon_2\to 0} \epsilon_2 \log \mathcal{Z}^{\mathrm{U}(2)}_{\mathrm{inst}} = H^{\mathrm{U}(2)}_{\mathrm{inst}}(x^*_{ui}),$$

where x^*_{ui} denote the 'critical configurations' extremizing the 'actions' $H^{\mathrm{U}(2)}_{\mathrm{inst}}$. The extremality conditions (saddle point equations) for the 'actions' $H^{\mathrm{U}(2)}_{\mathrm{inst}}$ take the form of the Bethe-like equations. In particular, one gets

- in the case of the U(2) $N_f = 4$ theory:

$$-q\left[\prod_{v=1}^{2}\prod_{l=1}^{\infty}\frac{(x_{ui}-x_{vl}-\epsilon_1)(x_{ui}-x^0_{vl}+\epsilon_1)}{(x_{ui}-x_{vl}+\epsilon_1)(x_{ui}-x^0_{vl}-\epsilon_1)}\right]\left[\frac{\prod_{\alpha=1}^{4}(x_{ui}+m_\alpha)}{\prod_{v=1}^{2}(x_{ui}-a_v)(x_{ui}-a_v+\epsilon_1)}\right] = 1. \quad (24)$$

- in the case of the U(2) $\mathcal{N} = 2^*$ theory:

$$-q\Bigg(\prod_{v,j}\frac{(x_{ui}-x_{vj}-\epsilon_1)(x_{ui}-x^0_{vj}+\epsilon_1)}{(x_{ui}-x^0_{vj}-\epsilon_1)(x_{ui}-x_{vj}+\epsilon_1)}\frac{(x_{ui}-x_{vj}-m)(x_{ui}-x^0_{vj}+m)}{(x_{ui}-x^0_{vj}-m)(x_{ui}-x_{vj}+m)} \quad (25)$$
$$\times\frac{(x_{ui}-x_{vj}+m+\epsilon_1)(x_{ui}-x^0_{vj}-m-\epsilon_1)}{(x_{ui}-x^0_{vj}+m+\epsilon_1)(x_{ui}-x_{vj}-m-\epsilon_1)}\Bigg)\Bigg(\prod_{v}\frac{(x_{ui}-m-a_v)(x_{ui}+m+\epsilon_1-a_v)}{(x_{ui}-a_v)(x_{ui}+\epsilon_1-a_v)}\Bigg) = 1,$$

where $u, v = 1, 2;\ i, j = 1, \ldots, \infty$.

The solutions $x^*_{ui} \equiv x_{ui}$ to the above equations can be found recursively in terms of certain formal expansions in the parameter q.

The problem of the computation of the Nekrasov–Shatashvili limit of the Nekrasov functions can be also well formulated and studied within the field theory framework, cf. eg. [17, 18].

3. Accessory parameters for the Heun and the Lamé equations

The Hilbert space of (the simplest) two-dimensional conformal field theories consists of the tensor products of the 'left' and 'right' irreducible Verma modules:

$$\mathcal{H} = \bigoplus_{\Delta,\bar\Delta\in\mathbf{S}} \mathcal{V}_\Delta\otimes\mathcal{V}_{\bar\Delta}. \quad (26)$$

The direct sum above is taken over the spectrum $\mathbf{S}$ of the theory, i.e. over the set of the highest weights of the model. In the case of 2d CFT's with the continuous spectrum the direct sum in eq. (26) becomes a direct integral. It is assumed that in the spectrum one can find a pair of weights equal to zero $(\Delta,\bar\Delta) = (0,0)\in\mathbf{S}$. Hence, there exists in $\mathcal{H}$ the sector $\mathcal{V}_0\otimes\mathcal{V}_0$ in which the Verma modules are generated from the highest weight state $|\,0\,\rangle \equiv |\,\nu_0\,\rangle$ with the highest weight $\Delta = 0$.

In $\mathcal{H}$ operate physical fields which are built out of the chiral vertex operators. In particular, the physical primary fields are made of the primary CVO's:

$$\mathsf{V}_{\Delta_2,\bar\Delta_2}(z,\bar z) = \sum_{\Delta_3,\bar\Delta_3,\Delta_1,\bar\Delta_1\in\mathbf{S}} C_{(\Delta_3,\bar\Delta_3),(\Delta_2,\bar\Delta_2),(\Delta_1,\bar\Delta_1)} V_{\Delta_2}(z)\otimes V_{\bar\Delta_2}(\bar z).$$

The symbols $C_{(\Delta_3,\bar\Delta_3),(\Delta_2,\bar\Delta_2),(\Delta_1,\bar\Delta_1)}$ denote the model dependent structure constants:

$$C_{(\Delta_3,\bar\Delta_3),(\Delta_2,\bar\Delta_2),(\Delta_1,\bar\Delta_1)} = \langle\nu_{\Delta_3}\otimes\nu_{\bar\Delta_3}|\mathsf{V}_{\Delta_2,\bar\Delta_2}(1,1)|\nu_{\Delta_1}\otimes\nu_{\bar\Delta_1}\rangle$$

which are nothing but the 3-point functions $\langle 0|\mathsf{V}_3(z_3,\bar z_3)\mathsf{V}_2(z_2,\bar z_2)\mathsf{V}_1(z_1,\bar z_1)|0\rangle$ evaluated at the special locations of the physical primary operators. Let us recall that according to the field–state correspondence the primary fields, acting on the 'vacuum' $|0\rangle \equiv |0\rangle\otimes|0\rangle$, generate the highest weight states in $\mathcal{H}$:

$$|\nu_\Delta\otimes\nu_{\bar\Delta}\rangle = \lim_{z,\bar z\to 0}\mathsf{V}_{\Delta,\bar\Delta}(z,\bar z)|0\rangle, \qquad \langle\nu_\Delta\otimes\nu_{\bar\Delta}| = \lim_{z,\bar z\to\infty} z^{2\Delta}\bar z^{2\bar\Delta}\langle 0|\mathsf{V}_{\Delta,\bar\Delta}(z,\bar z).$$

Furthermore, each state in the Hilbert space corresponds to some local operator which generates this state from the vacuum. In particular, null states correspond to the so-called null fields. Since the physical space of states can not contain null vectors then the correlation functions (the vacuum expectation values) with the null fields must vanish. The last condition, using the conformal Ward identities [1, 2], can be converted to certain partial differential equations obeyed by the correlation functions with the degenerate primary operators. The latter are simply physical primary fields with the degenerate conformal weights (6).

Let us consider the projected 5-point function on the sphere in the diagonal theory ($\Delta_i = \bar\Delta_i$):

$$G_\Delta(z,x) \equiv \Big\langle \mathsf{V}_4(\infty,\infty)\mathsf{V}_3(1,1)\mathbf{1}_{\Delta,\Delta}\mathsf{V}_{-\frac{b}{2}}(z,\bar z)\mathsf{V}_2(x,\bar x)\mathsf{V}_1(0,0)\Big\rangle$$

where $\mathsf{V}_{\alpha=-\frac{b}{2}}$ is the degenerate field with the conformal weight

$$\Delta_{-\frac{b}{2}} = \Delta_{\alpha=-\frac{b}{2}} = \alpha(Q-\alpha) = -\frac{1}{2}-\frac{3}{4}b^2, \qquad Q = b+\frac{1}{b}$$

and V_i's are the four heavy primary operators ($\Delta_i = b^{-2}\,\delta_i$, $\delta_i = \mathcal{O}(1)$). The projected correlation functions are elements for building full physical correlators. Indeed, this can be done by summing or integrating such quantities over the spectrum of the theory. The function $G_\Delta(z,x)$ satisfies the following null vector decoupling (NVD) equation:

$$\left[\frac{\partial^2}{\partial z^2} - b^2\left(\frac{1}{z}-\frac{1}{1-z}\right)\frac{\partial}{\partial z}\right]G_\Delta(z,x) =$$
$$-b^2\left[\frac{\Delta_1}{z^2}+\frac{\Delta_2}{(z-x)^2}+\frac{\Delta_3}{(1-z)^2}+\frac{\Delta_1+\Delta_2+\Delta_3+\Delta_{-\frac{b}{2}}-\Delta_4}{z(1-z)}+\frac{x(1-x)}{z(z-x)(1-z)}\frac{\partial}{\partial x}\right]G_\Delta(z,x).$$

Our aim now is to consider the classical limit of the equation above. The key point here is an observation that in the limit $b\to 0$ only the operator with weight $\Delta_{-\frac{b}{2}}$ remains light ($\Delta_{-\frac{b}{2}} = \mathcal{O}(1)$) and its presence in the correlation function has no influence on the classical dynamics. Then, for $b\to 0$

$$G_\Delta(z,x) \sim \psi(z)\,\mathrm{e}^{-\frac{1}{b^2}\left(S^{\rm cl}(\delta_4,\delta_3,\delta)+S^{\rm cl}(\delta,\delta_2,\delta_1)-f_\delta\left[\begin{smallmatrix}\delta_3 & \delta_2\\ \delta_4 & \delta_1\end{smallmatrix}\right](x)-\bar f_\delta\left[\begin{smallmatrix}\delta_3 & \delta_2\\ \delta_4 & \delta_1\end{smallmatrix}\right](\bar x)\right)}. \tag{27}$$

Indeed, assuming that the light field does not contribute to the classical limit we are left with the projected 4-point function of the heavy operators:

$$\begin{aligned}\langle \mathsf{V}_4(\infty,\infty)\mathsf{V}_3(1,1)\mathbf{1}_{\Delta,\Delta}\mathsf{V}_2(x,\bar x)\mathsf{V}_1(0,0)\rangle &=\\ = C_{\Delta_4,\Delta_3,\Delta}\,C_{\Delta,\Delta_2,\Delta_1}\,&\mathcal{F}_{1+6Q^2,\Delta}\left[\begin{smallmatrix}\Delta_3 & \Delta_2\\ \Delta_4 & \Delta_1\end{smallmatrix}\right](x)\bar{\mathcal{F}}_{1+6Q^2,\Delta}\left[\begin{smallmatrix}\Delta_3 & \Delta_2\\ \Delta_4 & \Delta_1\end{smallmatrix}\right](\bar x)\\ \overset{b\to 0}{\sim}\ &\mathrm{e}^{-\frac{1}{b^2}\left(S^{\rm cl}(\delta_4,\delta_3,\delta)+S^{\rm cl}(\delta,\delta_2,\delta_1)-f_\delta\left[\begin{smallmatrix}\delta_3 & \delta_2\\ \delta_4 & \delta_1\end{smallmatrix}\right](x)-\bar f_\delta\left[\begin{smallmatrix}\delta_3 & \delta_2\\ \delta_4 & \delta_1\end{smallmatrix}\right](\bar x)\right)}.\end{aligned}$$

The quantities $S^{\rm cl}(\delta_3,\delta_2,\delta_1)$ — the so-called classical 3-point actions — are the classical limits of the structure constants. The substitution of eq. (27) into the NVD equation leads, within the limit $b\to 0$, to the Fuchsian differential equation:

$$\partial_z^2\psi(z)+\left[\frac{\delta_1}{z^2}+\frac{\delta_2}{(z-x)^2}+\frac{\delta_3}{(1-z)^2}+\frac{\delta_1+\delta_2+\delta_3-\delta_4}{z(1-z)}+\frac{x(1-x)\mathcal{C}_2(x)}{z(z-x)(1-z)}\right]\psi(z)=0$$

with singularities located at $z_4=\infty, z_3=1, z_2=x, z_1=0$ and the accessory parameter $\mathcal{C}_2(x)$ determined by the classical 4-point block, cf. [13]:

$$\boxed{\mathcal{C}_2(x)=\frac{\partial}{\partial x}\,f_\delta\!\left[\begin{smallmatrix}\delta_3 & \delta_2\\ \delta_4 & \delta_1\end{smallmatrix}\right]\!(x)}\;.$$

On the other hand, it follows from eq. (22) that the parameter $\mathcal{C}_2(x)$ can be found once the solution of the saddle point equation (24) is known, cf. [15].

The above argumentation can be repeated in the case of the projected correlation functions on the torus:

$$\Big\langle\,\cdot\,\Big\rangle_{\tau,\Delta}=\mathrm{Tr}_{\mathcal{V}_\Delta\otimes\mathcal{V}_\Delta}\left(q^{L_0-c/24}\bar{q}^{\bar{L}_0-c/24}\;\cdot\;\right).$$

In particular, considering the NVD equation fulfilled by the projected torus 2-point function with the degenerate operator $\mathsf{V}_{\alpha=-\frac{b}{2}}$, it is possible to obtain in the classical limit $b\to 0$ the Weierstrassian-form Lamé equation:

$$\partial_z^2\psi(z)-\left[\,\kappa\,\wp(z)+\mathcal{B}\,\right]\psi(z)=0$$

with the accessory parameter/eigenvalue:[3]

$$\boxed{\frac{\mathcal{B}(\tau)}{4\pi^2}=q\frac{\partial}{\partial q}\,f_\delta^{\tilde{\delta}}(q)-\frac{\tilde{\delta}}{12}\,\mathrm{E}_2(\tau)}$$

determined by the classical 1-point block on the torus. Analogously, eq. (23) implies that the another way to compute the parameter $\mathcal{B}(\tau)$ is to solve the Bethe-like equation (25), cf. [16].

As a final remark in this section let us stress that the *holomorphic* accessory parameters $\mathcal{C}_2(x)$ and $\mathcal{B}(\tau)$ are closely related to that which emerge in the uniformization theory of the 4-punctured sphere and 1-punctured torus.[4] The accessory parameters for $C_{0,4}$ and $C_{1,1}$ can be found by taking the classical limit of the NVD equations obeyed by the Liouville 5-point function on the sphere and the 2-point function on the torus with $\mathsf{V}_{\alpha=-\frac{b}{2}}$. This means that one has to consider the projected correlation functions discussed above which are integrated over the Liouville field theory spectrum $\Delta=\Delta_\alpha$, $\alpha=\frac{Q}{2}+i\mathbf{R}^+$ with the structure constants $C_{\Delta_3,\Delta_2,\Delta_1}$ identified as the DOZZ Liouville 3-point functions [3, 19].

4. Concluding remarks

The main goal of the present article was to demonstrate that the calculation of the holomorphic accessory parameters for the Fuchs on $C_{0,4}$ and the Lamé equations can be reduced to the problem of computing corresponding classical conformal blocks or solving appropriate Bethe-like saddle point equations. The link between the solutions of certain Bethe-like equations and the classical conformal blocks yields a new method of calculating of the latter. This is of great importance, because the efficient methods of the computation of the classical blocks can be useful in studies in other areas of theoretical physics where classical conformal blocks have applications. Indeed, classical conformal blocks arise in a few other interesting contexts: the entanglement entropy in 2d CFT, 2d CFT/AdS$_3$ holography, matrix models, KdV equation and S-duality in 2d $\mathcal{N}=2$ SYM theories. We plan to examine these contexts soon.

[3] Here $\mathrm{E}_2(\tau)$ is the second Eisenstein series.
[4] Cf. discussion in ref. [13].

References

[1] Belavin A, Polyakov A M and Zamolodchikov A 1984 *Nucl. Phys.* B **241** 333-380
[2] Eguchi T, Ooguri H 1987 *Nucl. Phys.* B **282** 308-328
[3] Zamolodchikov A B, Zamolodchikov A B 1996 *Nucl. Phys.* B **477** 577-605.
[4] Hadasz L, Jaskolski Z and Piatek M 2005 *Nucl. Phys.* B **724** 529554
[5] Alday L F, Gaiotto D and Tachikawa Y 2010 *Lett. Math. Phys.* **91** 167-197
[6] Nekrasov N A 2004 *Adv. Theor. Math. Phys.* **7** 831-864
[7] Nekrasov N, Okounkov A (*Preprint* hep-th/0306238v2)
[8] Wyllard N 2009 *J. High Energy Phys.* JHEP11(2009)002
[9] Nekrasov N A, Shatashvili S L 2009 Quantization of Integrable Systems and Four Dimensional Gauge Theories (*Preprint* 0908.4052v1)
[10] Basar G, Dunne G V 2015 *J. High Energy Phys.* JHEP02(2015)160
[11] Piatek M, Pietrykowski A R 2014 *J. High Energy Phys.* JHEP12(2014)032
[12] Piatek M, Pietrykowski A R 2015 Classical limit of irregular blocks and Mathieu functions (*Preprint* 1509.08164)
[13] Litvinov A, Lukyanov S, Nekrasov N and Zamolodchikov A 2014 *J. High Energy Phys.* JHEP07(2014)144
[14] Piatek M 2011 *J. High Energy Phys.* JHEP06(2011)050
[15] Ferrari F, Piatek M 2012 *J. High Energy Phys.* JHEP05(2012)025
[16] Piatek M 2014 *J. High Energy Phys.* JHEP03(2014)124
[17] Ferrari F, Piatek M 2014 *Can. J. Phys.* **92** 267-270
[18] Ferrari F, Piatek M 2013 *Phys. Lett.* B **718** 1142-1147
[19] Dorn H, Otto H J 1994 *Nucl. Phys.* B **429** 375-388

Physical systems in a space with noncommutativity of coordinates

Kh. P. Gnatenko

Department for Theoretical Physics, Ivan Franko National University of Lviv,
12 Drahomanov St., Lviv, 79005, Ukraine

E-mail: khrystyna.gnatenko@gmail.com

Abstract. We consider a space with canonical noncommutativity of coordinates. The problem of rotational symmetry breaking is studied in this space. To preserve the rotational symmetry we consider the generalization of constant matrix of noncommutativity to a tensor defined with the help of additional coordinates governed by a rotationally symmetric system. The properties of physical systems are examined in the rotationally invariant space with noncommutativity of coordinates. Namely, we consider an effect of coordinate noncommutativity on the energy levels of the hydrogen atom in the rotationally invariant noncommutative space. The motion of a particle in the uniform field is also studied in the noncommutative space with preserved rotational symmetry. On the basis of exact calculations we show that there is an effect of coordinate noncommutativity on the mass of a particle and conclude that noncommutativity causes the anisotropy of mass.

1. Introduction

A lot of attention has been devoted recently to studies of properties of physical systems in a noncommutative space. The idea of noncommutative structure of space was suggested by Heisenberg. Later, Snyder formalized the idea in his paper [1]. In recent years, the interest in noncommutativity is motivated by the development of String Theory and Quantum Gravity (see, for instance, [2, 3]).

Different problems have been studied in a space with canonical noncommutativity of coordinates

$$[X_i, X_j] = i\hbar\theta_{ij}, \tag{1}$$

$$[X_i, P_j] = i\hbar\delta_{ij}, \tag{2}$$

$$[P_i, P_j] = 0, \tag{3}$$

where θ_{ij} is a constant antisymmetric matrix. Among them the hydrogen atom [4–11], the Landau problem (see, for instance, [12–16]), quantum mechanical system in a central potential [17], classical particle in a gravitational potential [18, 19], system of particles in a gravitational field [20], motion of a body in a gravitational field and the equivalence principle [21] and many others.

It is important to note that in the case of canonical noncommutativity of coordinates (1) there is a problem of rotational symmetry breaking [4, 22]. So, in order to preserve the symmetry different classes of noncommutative algebras were considered (see, for example, [23, 24] and references therein).

In previous paper [23] in order to construct rotationally invariant noncommutative algebra we considered the generalization of the constant matrix θ_{ij} to a tensor. We proposed to construct the tensor in the following form

$$\theta_{ij} = \frac{l_0}{\hbar}\varepsilon_{ijk}a_k, \tag{4}$$

where l_0 is a constant with the dimension of length and a_i are additional coordinates which are governed by a rotationally symmetric system. For simplicity, we suppose that coordinates a_i are governed by the harmonic oscillator

$$H_{osc} = \frac{(p^a)^2}{2m_{osc}} + \frac{m_{osc}\omega^2 a^2}{2}. \tag{5}$$

It is generally believed that the parameter of noncommutativity of coordinates is of the order of the Planck scale. So, we put

$$\sqrt{\frac{\hbar}{m\omega}} = l_P, \tag{6}$$

where l_P is the Planck length. We also consider the case when the frequency of harmonic oscillator is very large. Therefore, the distance between the energy levels of harmonic oscillator tends to infinity. So, harmonic oscillator put into the ground state remains in it.

We would like to note that coordinates a_i can be treated as some internal coordinates of a particle. Quantum fluctuations of these coordinates lead effectively to a non-point-like particle, size of which is of the order of the Planck scale.

So, the rotationally invariant noncommutative algebra reads

$$[X_i, X_j] = i\varepsilon_{ijk}l_0 a_k, \tag{7}$$
$$[X_i, P_j] = i\hbar\delta_{ij}, \tag{8}$$
$$[P_i, P_j] = 0. \tag{9}$$

The coordinates a_i and momenta p^a_i satisfy the ordinary commutation relations $[a_i, a_j] = 0$, $[a_i, p^a_j] = i\hbar\delta_{ij}$, $[p^a_i, p^a_j] = 0$. Also, a_i commute with X_i and P_i. As a consequence, tensor of noncommutativity θ_{ij} given by (4) commutes with X_i and P_i too. Therefore, X_i, P_i and θ_{ij} satisfy the same commutation relations as in the case of the canonical version of noncommutativity. Besides, noncommutative algebra (7)-(9) is manifestly rotationally invariant.

It is worth noting that the rotational symmetry is preserved in the case of another way of generalization of the tensor of noncommutativity $\theta_{ij} = \alpha(a_i b_j - a_j b_i)/\hbar$ where a_i, b_i are additional coordinates governed by a rotationally symmetric system and α is a dimensionless constant [23]. In previous papers [23, 25] we studied the hydrogen atom in the rotationally invariant noncommutative space $[X_i, X_j] = i\alpha(a_i b_j - a_j b_i)$, $[X_i, P_j] = i\hbar\delta_{ij}$, $[P_i, P_j] = 0$.

In this paper we consider physical systems in rotationally invariant space with noncommutativity of coordinates (7). We study the motion of a particle in the uniform field in the space. On the basis of exact calculations we show that there is an effect of noncommutativity of coordinates on the mass of a particle and noncommutativity causes the anisotropy of mass. Also we consider the hydrogen atom in rotationally invariant noncommutative space (7)-(9) and study the effect of coordinate noncommutativity on the energy levels of the atom.

The paper is organized as follows. In Section 2, we consider the energy levels of the hydrogen atom in rotationally invariant noncommutative space (7)-(9). In Section 3, the motion of a particle in the uniform field in rotationally invariant noncommutative space is studied. Conclusions are presented in Section 4.

2. Energy levels of hydrogen atom in noncommutative space with preserved rotational symmetry

Let us consider the hydrogen atom in noncommutative space (7)-(9). The Hamiltonian of the hydrogen atom reads

$$H_h = \frac{P^2}{2M} - \frac{e^2}{R}, \tag{10}$$

where $R = \sqrt{\sum_i X_i^2}$ and X_i satisfy (7).

Because of definition of the tensor of noncommutativity (4) in rotationally invariant noncommutative space we have to take into account additional terms that correspond to the harmonic oscillator (5). Therefore, we consider the total Hamiltonian as follows

$$H = H_h + H_{osc}. \tag{11}$$

Let us use the following representation

$$X_i = x_i - \frac{1}{2}\theta_{ij}p_j, \tag{12}$$

$$P_i = p_i, \tag{13}$$

where θ_{ij} is given by (4). Coordinates x_i and momenta p_i satisfy the ordinary commutation relations

$$[x_i, x_j] = 0, \tag{14}$$

$$[p_i, p_j] = 0, \tag{15}$$

$$[x_i, p_j] = i\hbar\delta_{ij}, \tag{16}$$

and commute with a_i, p_i^a, namely $[x_i, a_j] = 0$, $[x_i, p_j^a] = 0$, $[p_i, a_j] = 0$, $[p_i, p_j^a] = 0$. It is worth mentioning that coordinates X_i do not commute with p_j^a. Taking into account (4) and (12), the coordinates X_i can be written as follows

$$X_i = x_i + \frac{l_0}{2\hbar}[\mathbf{a} \times \mathbf{p}]_i. \tag{17}$$

Therefore, we have

$$[X_i, p_j^a] = i\varepsilon_{ijk}\frac{l_0}{2}p_k. \tag{18}$$

Let us write the expansion for H up to the second order in $\boldsymbol{\theta} = l_0\mathbf{a}/\hbar$. Using (17), we have

$$R = \sqrt{\sum_i X_i^2} = \sqrt{r^2 - \frac{l_0}{\hbar}(\mathbf{a} \cdot \mathbf{L}) + \frac{l_0^2}{4\hbar^2}[\mathbf{a} \times \mathbf{p}]^2}, \tag{19}$$

with $r = \sqrt{\sum_i x_i^2}$ and $\mathbf{L} = [\mathbf{r} \times \mathbf{p}]$. It is important to note that the operators under the square root do not commute. Therefore, the expansion for R can be written as follows

$$R = r - \frac{l_0}{2\hbar r}(\mathbf{a} \cdot \mathbf{L}) - \frac{l_0^2}{8\hbar^2 r^3}(\mathbf{a} \cdot \mathbf{L})^2 + \frac{l_0^2}{16\hbar^2}\left(\frac{1}{r}[\mathbf{a} \times \mathbf{p}]^2 + [\mathbf{a} \times \mathbf{p}]^2\frac{1}{r} + a^2 f(\mathbf{r})\right), \tag{20}$$

where $f(\mathbf{r})$ is unknown function. Squaring left- and right-hand sides of equation (20) we obtain

$$\frac{\hbar^2}{r^4}[\mathbf{a}\times\mathbf{r}]^2 - ra^2 f(\mathbf{r}) = 0. \tag{21}$$

Finally, from (21) we have

$$a^2 f(\mathbf{r}) = \frac{\hbar^2}{r^5}[\mathbf{a}\times\mathbf{r}]^2. \tag{22}$$

So, using (20) and (22), it is easy to write expansion for the inverse distance R^{-1}

$$\frac{1}{R} = \frac{1}{r} + \frac{l_0}{2\hbar r^3}(\mathbf{a}\cdot\mathbf{L}) + \frac{3l_0^2}{8\hbar^2 r^5}(\mathbf{a}\cdot\mathbf{L})^2 - \frac{l_0^2}{16\hbar^2}\left(\frac{1}{r^2}[\mathbf{a}\times\mathbf{p}]^2\frac{1}{r} + \frac{1}{r}[\mathbf{a}\times\mathbf{p}]^2\frac{1}{r^2} + \frac{\hbar^2}{r^7}[\mathbf{a}\times\mathbf{r}]^2\right). \tag{23}$$

Therefore, the Hamiltonian (11) can be rewritten as follows

$$H = H_0 + V, \tag{24}$$

with

$$H_0 = H_h^{(0)} + H_{osc}. \tag{25}$$

Here $H_h^{(0)} = \frac{p^2}{2M} - \frac{e^2}{r}$ is the Hamiltonian of the hydrogen atom in the ordinary space and perturbation V caused by the noncommutativity of coordinates is given by

$$V = -\frac{l_0 e^2}{2\hbar r^3}(\mathbf{a}\cdot\mathbf{L}) - \frac{3l_0^2 e^2}{8\hbar^2 r^5}(\mathbf{a}\cdot\mathbf{L})^2 + \frac{l_0^2 e^2}{16\hbar^2}\left(\frac{1}{r^2}[\mathbf{a}\times\mathbf{p}]^2\frac{1}{r} + \frac{1}{r}[\mathbf{a}\times\mathbf{p}]^2\frac{1}{r^2} + \frac{\hbar^2}{r^7}[\mathbf{a}\times\mathbf{r}]^2\right). \tag{26}$$

Let us find the corrections to the energy levels of the hydrogen atom caused by the noncommutativity of coordinates (7). Note that $H_h^{(0)}$ commutes with H_{osc}. So, the eigenvalues and the eigenstates which correspond to H_0 (25) read

$$E^{(0)}_{n,\{n^a\}} = -\frac{e^2}{2a_B n^2} + \hbar\omega\left(n_1^a + n_2^a + n_3^a + \frac{3}{2}\right), \tag{27}$$

$$\psi^{(0)}_{n,l,m,\{n^a\}} = \psi_{n,l,m}\psi^a_{n_1^a,n_2^a,n_3^a}, \tag{28}$$

where $\psi_{n,l,m}$ are the eigenfunctions of the hydrogen atom in the ordinary space, $\psi^a_{n_1^a,n_2^a,n_3^a}$ are the eigenfunctions of three-dimensional harmonic oscillator, and a_B is the Bohr radius. In the case when harmonic oscillator is in the ground state, according to the perturbation theory, in the first order in V we have

$$\Delta E^{(1)}_{n,l} = \langle\psi^{(0)}_{n,l,m,\{0\}}|V|\psi^{(0)}_{n,l,m,\{0\}}\rangle =$$

$$= -\frac{\hbar^2 e^2\langle\theta^2\rangle}{a_B^5 n^5}\left(\frac{1}{6l(l+1)(2l+1)} - \frac{6n^2 - 2l(l+1)}{3l(l+1)(2l+1)(2l+3)(2l-1)} + \right.$$

$$\left. + \frac{5n^2 - 3l(l+1) + 1}{2(l+2)(2l+1)(2l+3)(l-1)(2l-1)} - \frac{5}{6}\frac{5n^2 - 3l(l+1) + 1}{l(l+1)(l+2)(2l+1)(2l+3)(l-1)(2l-1)}\right), \tag{29}$$

where $\langle\theta^2\rangle$ is given by

$$\langle\theta^2\rangle = \frac{l_0^2}{\hbar^2}\langle\psi^a_{0,0,0}|a^2|\psi^a_{0,0,0}\rangle = \frac{3l_0^2}{2\hbar}\left(\frac{1}{m\omega}\right) = \frac{3}{2}\left(\frac{l_0 l_P}{\hbar}\right)^2. \quad (30)$$

The details of calculation of the corresponding integrals can be found in our previous paper [23].

In the second order of the perturbation theory we obtain

$$\Delta E^{(2)}_{n,l,m,\{0\}} = \sum_{n',l',m',\{n^a\}} \frac{\left|\left\langle \psi^{(0)}_{n',l',m',\{n^a\}}\right|V\left|\psi^{(0)}_{n,l,m,\{0\}}\right\rangle\right|^2}{E^{(0)}_n - E^{(0)}_{n'} - \hbar\omega(n^a_1+n^a_2+n^a_3)}, \quad (31)$$

here the set of numbers n', l', m', $\{n^a\}$ does not coincide with the set n, l, m, $\{0\}$, and $E^{(0)}_n = -e^2/(2a_Bn^2)$ is the unperturbed energy of the hydrogen atom. Note that matrix elements $\left\langle \psi^{(0)}_{n',l',m',\{n^a\}}\right|V\left|\psi^{(0)}_{n,l,m,\{0\}}\right\rangle$ do not depend on ω because of our assumption (6). We consider the frequency of the harmonic oscillator ω to be very large. In the case of $\omega\to\infty$ we have

$$\lim_{\omega\to\infty}\Delta E^{(2)}_{n,l,m,\{0\}} = 0. \quad (32)$$

So, we obtain the following corrections up to the second order in the parameter of noncommutativity

$$\Delta E_{n,l} = \Delta E^{(1)}_{n,l}. \quad (33)$$

It is worth noting that in the case of $l=0$ or $l=1$ corrections (33) are divergent. In order to find corrections to the ns energy levels let us write the perturbation V in the following form

$$V = -\frac{e^2}{R}+\frac{e^2}{r} = -\frac{e^2}{\sqrt{r^2-\frac{l_0}{\hbar}(\mathbf{a}\cdot\mathbf{L})+\frac{l_0^2}{4\hbar^2}[\mathbf{a}\times\mathbf{p}]^2}}+\frac{e^2}{r}. \quad (34)$$

Consequently for the corrections to the ns energy levels we have

$$\Delta E_{ns} = \left\langle \psi^{(0)}_{n,0,0,\{0\}}\right|\frac{e^2}{r} - \frac{e^2}{\sqrt{r^2-\frac{l_0}{\hbar}(\mathbf{a}\cdot\mathbf{L})+\frac{l_0^2}{4\hbar^2}[\mathbf{a}\times\mathbf{p}]^2}}\left|\psi^{(0)}_{n,0,0,\{0\}}\right\rangle. \quad (35)$$

It is important that $(\mathbf{a}\cdot\mathbf{L})$ commutes with $[\mathbf{a}\times\mathbf{p}]^2$ and r^2. Also, it can be shown that $(\mathbf{a}\cdot\mathbf{L})\psi^{(0)}_{n,0,0,\{0\}} = 0$. So, the corrections (35) can be written as follows

$$\Delta E_{ns} = \left\langle \psi^{(0)}_{n,0,0,\{0\}}\right|\frac{e^2}{r} - \frac{e^2}{\sqrt{r^2+\frac{l_0^2}{4\hbar^2}[\mathbf{a}\times\mathbf{p}]^2}}\left|\psi^{(0)}_{n,0,0,\{0\}}\right\rangle = \frac{\chi^2e^2}{a_B}I_{ns}(\chi), \quad (36)$$

where we use the following notation

$$I_{ns}(\chi) = \int d\mathbf{a}'\tilde{\psi}^a_{0,0,0}(\mathbf{a}')\int d\mathbf{r}'\tilde{\psi}_{n,0,0}(\chi\mathbf{r}')\left(\frac{1}{r'}-\frac{1}{\sqrt{(r')^2+[\mathbf{a}'\times\mathbf{p}']^2}}\right)\tilde{\psi}_{n,0,0}(\chi\mathbf{r}')\tilde{\psi}^a_{0,0,0}(\mathbf{a}'), \quad (37)$$

with

$$\chi = \sqrt{\frac{l_0 l_P}{2a_B^2}}. \tag{38}$$

Here $\tilde{\psi}_{n,0,0}(\chi\mathbf{r}') = \sqrt{\frac{1}{\pi n^5}} e^{-\frac{\chi r'}{n}} L^1_{n-1}\left(\frac{2\chi r'}{n}\right)$ are the dimensionless eigenfunctions of the hydrogen atom, $L^1_{n-1}\left(\frac{2\chi r'}{n}\right)$ are the generalized Laguerre polynomials, $\tilde{\psi}^a_{0,0,0}(\mathbf{a}') = \pi^{-\frac{3}{4}} e^{-\frac{(a')^2}{2}}$ are the dimensionless eigenfunctions corresponding to the harmonic oscillator, $\mathbf{a}' = \mathbf{a}/l_P$ and $\mathbf{r}' = \mathbf{r}\sqrt{2}/\sqrt{l_0 l_P}$.

It is important to mention that in the case of $\chi = 0$ integral (37) has a finite value. Therefore, for $\chi \to 0$ the asymptotic of ΔE_{ns} can be written as follows

$$\Delta E_{ns} = \frac{\chi^2 e^2}{a_B} I_{ns}(0). \tag{39}$$

In order to find $I_{ns}(0)$, let us first consider the following integral

$$I_{ns}(\chi, \mathbf{a}') = \int d\mathbf{r}' \tilde{\psi}_{n,0,0}(\chi\mathbf{r}') \left(\frac{1}{r'} - \frac{1}{\sqrt{(r')^2 + [\mathbf{a}' \times \mathbf{p}']^2}}\right) \tilde{\psi}_{n,0,0}(\chi\mathbf{r}'). \tag{40}$$

In the case of $\chi = 0$ the integral reads

$$I_{ns}(0, \mathbf{a}') \simeq 1.72 \frac{\pi a'}{4n^3}, \tag{41}$$

here $a' = |\mathbf{a}'|$. The details of calculation of integral (41) can be found in our previous paper [25].

It is clear that

$$I_{ns}(0) = \langle I_{ns}(0, \mathbf{a}')\rangle_{\mathbf{a}'}, \tag{42}$$

where $\langle ... \rangle_{\mathbf{a}'}$ denotes $\langle \tilde{\psi}^a_{0,0,0}(\mathbf{a}')|...|\tilde{\psi}^a_{0,0,0}(\mathbf{a}')\rangle$. So, taking into account (38), (39), (41), (42) and returning to $\mathbf{a} = l_P \mathbf{a}'$, the leading term in the asymptotic expansion of the corrections to the ns energy levels over the small parameter of noncommutativity reads

$$\Delta E_{ns} \simeq 1.72 \frac{\hbar\langle\theta\rangle\pi e^2}{8a_B^3 n^3}, \tag{43}$$

where

$$\langle\theta\rangle = \frac{l_0}{\hbar} \langle \psi^a_{0,0,0}| \sqrt{\sum_i a_i^2} |\psi^a_{0,0,0}\rangle = \frac{2l_0 l_P}{\sqrt{\pi}\hbar}. \tag{44}$$

Note that corrections to the ns energy levels (43) are proportional to $\langle\theta\rangle$. In the case of $l > 1$ we found that corrections (33) are proportional to $\langle\theta^2\rangle$. So, we can conclude that ns energy levels are more sensitive to the noncommutativity of coordinates (7).

3. A particle in the uniform field in rotationally invariant noncommutative space

Let us consider the motion of a particle in the uniform field in rotationally invariant noncommutative space (7)-(9). In the case when the field is pointed in the X_3 direction and is characterized by the factor κ the Hamiltonian of the particle reads

$$H_p = \frac{P^2}{2m} + \kappa X_3, \tag{45}$$

where m is the mass of a particle. For example, in a particular case of motion of a charged particle q in the uniform electric field E directed along the X_3 axis, we have $\kappa = qE$. In the case of motion of a particle of mass m in the uniform gravitational field g directed along the X_3 axis factor κ reads $\kappa = -mg$.

Taking into account the additional terms which correspond to the harmonic oscillator (5), we have

$$H = H_p + H_{osc} = \frac{P^2}{2m} + \kappa X_3 + \frac{(p^a)^2}{2m_{osc}} + \frac{m_{osc}\omega^2 a^2}{2}. \tag{46}$$

It is convenient to use representation (12), (13). Therefore, we can write Hamiltonian (46) in the following form

$$H = \frac{p^2}{2m} + \kappa x_3 + \frac{\kappa l_0}{2\hbar}(a_1 p_2 - a_2 p_1) + \frac{(p^a)^2}{2m_{osc}} + \frac{m_{osc}\omega^2 a^2}{2}. \tag{47}$$

After algebraic transformations, Hamiltonian (47) can be rewritten as

$$H = \left(1 - \frac{\kappa^2 l_0^2 m}{4\hbar^2\omega^2 m_{osc}}\right)\frac{p_1^2}{2m} + \left(1 - \frac{\kappa^2 l_0^2 m}{4\hbar^2\omega^2 m_{osc}}\right)\frac{p_2^2}{2m} + \frac{p_3^2}{2m} + \kappa x_3 + \frac{(p^a)^2}{2m_{osc}}$$
$$+\frac{m_{osc}\omega^2}{2}\left(a_1 + \frac{\kappa l_0}{2\hbar\omega^2 m_{osc}}p_2\right)^2 + \frac{m_{osc}\omega^2}{2}\left(a_2 - \frac{\kappa l_0}{2\hbar\omega^2 m_{osc}}p_1\right)^2 + \frac{m_{osc}\omega^2 a_3^2}{2}. \tag{48}$$

So, we can represent Hamiltonian (48) as follows

$$H = \tilde{H}_p + \tilde{H}_{osc}. \tag{49}$$

Here we use the notations

$$\tilde{H}_p = \frac{p_1^2}{2m_{eff}} + \frac{p_2^2}{2m_{eff}} + \frac{p_3^2}{2m} + \kappa x_3, \tag{50}$$

where m_{eff} is an effective mass which is defined as

$$m_{eff} = m\left(1 - \frac{\kappa^2 l_0^2 m}{4\hbar^2\omega^2 m_{osc}}\right)^{-1}, \tag{51}$$

and

$$\tilde{H}_{osc} = \frac{(p^a)^2}{2m_{osc}} + \frac{m_{osc}\omega^2 q^2}{2}. \tag{52}$$

The components of $\mathbf{q}$ read

$$q_1 = a_1 + \frac{\kappa l_0}{2\hbar\omega^2 m_{osc}}p_2, \tag{53}$$

$$q_2 = a_2 - \frac{\kappa l_0}{2\hbar\omega^2 m_{osc}}p_1, \tag{54}$$

$$q_3 = a_3. \tag{55}$$

Journal of Physics: Conference Series **670** (2016) 012023 doi:10.1088/1742-6596/670/1/012023

It is worth mentioning that q_i satisfy the ordinary commutation relations

$$[q_i, q_j] = 0, \tag{56}$$

$$[q_i, p^a_j] = i\hbar\delta_{ij}. \tag{57}$$

Therefore, Hamiltonian $\tilde{H}_{osc}$ corresponds to the tree-dimensional harmonic oscillator in the ordinary space. Also the following commutation relations are satisfied $[q_i, x_j] = -i\varepsilon_{ij3}\kappa l_0/(2m_{osc}\omega^2)$, $[q_i, p_j] = 0$.

It is important to note that

$$\tilde{H}_1 = \frac{p_1^2}{2m_{eff}}, \tag{58}$$

$$\tilde{H}_2 = \frac{p_2^2}{2m_{eff}}, \tag{59}$$

$$\tilde{H}_3 = \frac{p_3^2}{2m} + \kappa x_3, \tag{60}$$

and $\tilde{H}_{osc}$ which is given by (52) commute with each other. The eigenfunctions of $H = \tilde{H}_1 + \tilde{H}_2 + \tilde{H}_3 + \tilde{H}_{osc}$ (49) can be written as follows

$$\psi(\mathbf{x}, \tilde{\mathbf{q}}) = Ce^{ik_1x_1}e^{ik_2x_2}\psi^{(3)}(x_3)\psi^{\tilde{q}}(\tilde{\mathbf{q}}). \tag{61}$$

where C is a constant, k_1 and k_2 are the components of the wave vector corresponding to the free motion of a particle in the perpendicular directions to the field direction, $\psi^{(3)}(x_3)$ are well known eigenfunctions of $\tilde{H}_3$ which correspond to the motion of a particle in the field direction and can be written in terms of the Airy function, and $\psi^{\tilde{q}}(\tilde{\mathbf{q}})$ are eigenfunctions of tree-dimensional harmonic oscillator with parameters m_{osc} and ω. The components of $\tilde{\mathbf{q}}$ are the following

$$\tilde{q}_1 = a_1 + \frac{\kappa l_0 k_2}{2\omega^2 m_{osc}}, \tag{62}$$

$$\tilde{q}_2 = a_2 - \frac{\kappa l_0 k_1}{2\omega^2 m_{osc}}, \tag{63}$$

$$\tilde{q}_3 = a_3. \tag{64}$$

The eigenvalues of H (49) read

$$E = \frac{\hbar^2 k_1^2}{2m_{eff}} + \frac{\hbar^2 k_2^2}{2m_{eff}} + E_3 + \frac{1}{2}\hbar\omega, \tag{65}$$

where E_3 corresponds to the motion of a particle in the field direction. Here we take into account that the harmonic oscillator is in the ground state.

So, from (65) and (51) we can conclude that there is an effect of noncommutativity (7) on the mass of a particle in the uniform field. Note that the motion of a particle in the field direction described by $\tilde{H}_3$ (60) is the same as in the ordinary space. Noncommutativity has an effect on the motion of a particle in perpendicular directions to the direction of uniform field (see first two terms in (65)). So, noncommutativity of coordinates (7) causes the anisotropy of mass.

In this Section we have considered large but finite limit for ω. Note that in the limit $\omega \to \infty$ effect of noncommutativity on the mass of a particle tends to zero.

It is worth mention that because of the rotationally invariance obtained results can be easy generalized to the case of an arbitrary direction of the uniform field.

4. Conclusion

In this paper we have studied physical systems in rotationally invariant space with noncommutativity of coordinates. We have considered rotationally invariant noncommutative algebra (7)-(9) proposed in [23]. The algebra is constructed with the help of generalization of constant matrix of noncommutativity to the tensor defined by additional coordinates which are governed by harmonic oscillator.

The hydrogen atom has been considered in rotationally invariant noncommutative space (7)-(9). We have studied the corrections to the energy levels of the atom caused by the noncommutativity of coordinates (7). We have found that corrections to the ns energy levels (43) are proportional to $\langle\theta\rangle$, whereas the corrections to the energy levels with $l > 1$ (33) are proportional to $\langle\theta^2\rangle$. Therefore, we have concluded that ns energy levels are more sensitive to the noncommutativity of coordinates (7). The motion of a particle in the uniform field in rotationally invariant noncommutative space (7)-(9) has been also examined. On the basis of exact calculations we have concluded that there is an effect of coordinate noncommutativity on the mass of a particle. The motion of a particle in the perpendicular to the field directions can be described with the help of effective mass whereas the motion of a particle in the field direction is the same as in the ordinary space. So, noncommutativity of coordinates (7) causes the anisotropy of mass.

Acknowledgments

The author thanks Prof. V. M. Tkachuk for his advices and great support during research studies and Dr. A. A. Rovenchak for a careful reading of the manuscript. The author also thanks the Organizers of the XXIIIth International Conference on Integrable Systems and Quantum symmetries (Prague, Czech Republic, June 2015) for the possibility to give a talk.

References

[1] Snyder H 1947 *Phys. Rev.* **71** 38
[2] Seiberg N and Witten E 1999 *J. High Energy Phys.* **9909** 032
[3] Doplicher S, Fredenhagen K and Roberts J E 1994 *Phys. Lett. B* **331** 39
[4] Chaichian M, Sheikh-Jabbari M M and Tureanu A 2001 *Phys. Rev. Lett.* **86** 2716
[5] Ho P M and Kao H C 2002 *Phys. Rev. Lett.* **88** 151602
[6] Chaichian M, Sheikh-Jabbari M M and Tureanu A 2004 *Eur. Phys. J. C* **36** 251
[7] Chair N and Dalabeeh M A 2005 *J. Phys. A, Math. Gen.* **38** 1553
[8] Stern A 2008 *Phys. Rev. Lett.* **100** 061601
[9] Zaim S, Khodja L and Delenda Y 2011 *Int. J. Mod. Phys. A* **26** 4133
[10] Adorno T C, Baldiotti M C, Chaichian M, Gitman D M and Tureanu A 2009 *Phys. Lett. B* **682** 235
[11] Khodja L and Zaim S 2012 *Int. J. Mod. Phys. A* **27** 1250100
[12] Nair V P and Polychronakos A P 2001 *Phys. Lett. B* **505** 267
[13] Bellucci S, Nersessian A and Sochichiu C 2001 *Phys. Lett. B* **522** 345
[14] Dayi O F and Kelleyane L T 2002 *Mod. Phys. Lett. A* **17** 1937
[15] Li K, Cao X H and Wang D Y 2006 *Chin. Phys.* **15** 2236
[16] Dulat S and Li K 2008 *Chin. Phys. C* **32** 92
[17] Gamboa J, Loewe M and Rojas J C 2001 *Phys. Rev. D* **64** 067901
[18] Romero J M and Vergara J D 2003 *Mod. Phys. Lett. A* **18** 1673
[19] Mirza B and Dehghani M 2004 *Commun. Theor. Phys.* **42** 183
[20] Daszkiewicz M and Walczyk C J 2011 *Mod. Phys. Lett. A* **26** 819
[21] Gnatenko K P 2013 *Phys. Lett. A* **377** 3061
[22] Balachandran A P and Padmanabhan P 2010 *J. High Energy Phys.* **1012** 001
[23] Gnatenko K P and Tkachuk V M 2014 *Phys. Lett. A* **378** 3509
[24] Kupriyanov V G 2014 *Fortschr. Phys.* **62** 881
[25] Gnatenko K P, Krynytskyi Y S and Tkachuk V M 2015 *Mod. Phys. Lett. A* **30** 1550033

Family of N-dimensional superintegrable systems and quadratic algebra structures

Md Fazlul Hoque, Ian Marquette and Yao-Zhong Zhang

School of Mathematics and Physics, The University of Queensland, Brisbane, QLD 4072, Australia

E-mail: m.hoque@uq.edu.au; i.marquette@uq.edu.au; yzz@maths.uq.edu.au

Abstract. Classical and quantum superintegrable systems have a long history and they possess more integrals of motion than degrees of freedom. They have many attractive properties, wide applications in modern physics and connection to many domains in pure and applied mathematics. We overview two new families of superintegrable Kepler-Coulomb systems with non-central terms and superintegrable Hamiltonians with double singular oscillators of type $(n, N-n)$ in N-dimensional Euclidean space. We present their quadratic and polynomial algebras involving Casimir operators of $so(N+1)$ Lie algebras that exhibit very interesting decompositions $Q(3) \oplus so(N-1)$, $Q(3) \oplus so(n) \oplus so(N-n)$ and the cubic Casimir operators. The realization of these algebras in terms of deformed oscillator enables the determination of a finite dimensional unitary representation. We present algebraic derivations of the degenerate energy spectra of these systems and relate them with the physical spectra obtained from the separation of variables.

1. Introduction

Algebraic methods are powerful tools in modern physics. Well known examples include the N-dimensional hydrogen atom and harmonic oscillator which were studied using the $so(N+1)$ [1, 2] and $su(N)$ [3, 4] Lie algebras respectively. In particular the spectrum of the 5D hydrogen atom have been calculated using its $so(6)$ Lie algebra and the corresponding Casimir operators of order two, three and four [5]. Superintegrable models are an important class of quantum systems which can be solved using algebraic approaches. An important property of such systems is the existence of non-Abelian symmetry algebras generated by integrals of motion. These symmetry algebras can be embedded in certain non-invariance algebras involving non-commuting operators. Such symmetry algebras are in general finitely generated polynomial algebras and only exceptionally finite dimensional Lie algebras.

Quadratic algebras have been used to obtain energy spectrum of superintegrable systems [6]. The structure of a class of quadratic algebras with only three generators was studied and applied to 2D superintegrable systems in [7]. Many researchers have studied quadratic and polynomial symmetry algebras of superintegrable systems and their representation theory (see e.g. [8, 9, 10, 11, 12, 13, 14, 15]).

We review the results of our recent two papers [16, 17] and present the algebraic derivation of the complete energy spectrum of the N-dimensional superintegrable Kepler-Coulomb system with non-central terms and superintegrable double singular oscillators.

Journal of Physics: Conference Series **670** (2016) 012024 doi:10.1088/1742-6596/670/1/012024

2. The main results

Let us consider the N-dimensional superintegrable Kepler-Coulomb system with non-central terms and superintegrable Hamiltonian with double singular oscillators of type $(n, N-n)$ introduced in [16, 17]

$$H_{KC} = \frac{1}{2}p^2 - \frac{c_0}{r} + c_1\chi_1(r, x_N) + c_2\chi_2(r, x_N), \quad (1)$$

$$H_{dso} = \frac{p^2}{2} + \frac{\omega^2 r^2}{2} + c_1\phi_1(x_1, \ldots, x_n) + c_2\phi_2(x_{n+1}, \ldots, x_N), \quad (2)$$

where $\vec{r} = (x_1, x_2, ..., x_N)$, $\vec{p} = (p_1, p_2, ..., p_N)$, $r^2 = \sum_{i=1}^N x_i^2$, $p_i = -i\hbar\partial_i$, $\chi_1(r, x_N) = \frac{1}{r(r+x_N)}$, $\chi_2(r, x_N) = \frac{1}{r(r-x_N)}$, $\phi_1(x_1, \ldots, x_n) = \frac{1}{x_1^2+\ldots+x_n^2}$, $\phi_2(x_{n+1}, \ldots, x_N) = \frac{1}{x_{n+1}^2+\ldots+x_{N-n}^2}$ and c_0, c_1, c_2 are positive real constants. The system (1) is a generalization of the 3D superintegrable system in E_3 [18]. It includes as a particular 3D case the Hartmann potential which has applications in quantum chemistry and its classical analog possesses closed trajectories and periodic motion [5]. The model (2) is the generalization of the 4D and 8D systems obtained via the Hurwitz transformation for which only the symmetry cases (2,2) and (4,4) were studied [19, 20, 21].

The model (1) is multiseparable and allows separation of variables in hyperspherical and hyperparabolic coordinates. The wave function is

$$\psi(r, \Omega) = R(r)\Theta(\Phi_{N-1})y(\Omega_{N-2}) \quad (3)$$

in terms of special functions

$$R(r) \propto (\varepsilon r)^{l+\frac{\delta_1+\delta_2}{2}} e^{\frac{-\varepsilon r}{2}} {}_1F_1(-n+l+1, 2l+\delta_1+\delta_2+N-1; \varepsilon r), \quad (4)$$

$$\Theta(z) \propto (1+z)^{\frac{(\delta_1+I_{N-2})}{2}}(1-z)^{\frac{(\delta_2+I_{N-2})}{2}} P_{l-I_{N-2}}^{(\delta_2+I_{N-2},\delta_1+I_{N-2})}(z), \quad (5)$$

$$\Lambda^2(N-1)y(\Omega_{N-2}) = I_{N-2}(I_{N-2}+N-3)y(\Omega_{N-2}), \quad (6)$$

where $z = \cos(\Phi_{N-1})$, $I_{N-2}(I_{N-2}+N-3)$ being the general form of the separation constant, $P_\lambda^{(\alpha,\beta)}$ denotes a Jacobi polynomial and $l \in \mathbb{N}$, $\delta_i = \{\sqrt{(I_{N-2}+\frac{N-3}{2})^2 + 4c_i'} - \frac{N-3}{2}\} - I_{N-2}$, $c_i' = \frac{c_i}{\hbar^2}$, $i = 1, 2$. The energy spectrum of the system (1) is [16]

$$E_{KC} \equiv E_n = \frac{-c_0^2}{2\hbar^2\left(n + \frac{\delta_1+\delta_2}{2} + \frac{N-3}{2}\right)^2}, \quad n = 1, 2, 3, \ldots \quad (7)$$

The system (2) is also multiseparable and allows separation of variables in double hyper Eulerian and double hyperspherical coordinates. We can split (2) into the sum of two singular oscillators of dimensions n and $N-n$ as $H_{dso} = H_1 + H_2$, where

$$H_1 = \frac{1}{2}(p_1^2 + \ldots + p_n^2) + \frac{\omega^2 r_1^2}{2} + c_1\phi_1(x_1, \ldots, x_n), \quad (8)$$

$$H_2 = \frac{1}{2}(p_{n+1}^2 + \ldots + p_N^2) + \frac{\omega^2 r_2^2}{2} + c_2\phi_2(x_{n+1}, \ldots, x_N). \quad (9)$$

The wave function $\psi(r_1, \Omega) = R_1(r_1)y(\Omega_{n-1})$ of H_1 in terms of special functions is given by

$$\psi_{n_1 l_n}(u) \propto \frac{a e^{-\frac{u}{2}} u^{(\delta_1+\frac{1}{2}l_n)/2}}{\sqrt{2(\delta_1+\frac{1}{2}l_n+\frac{n}{4})}} \times {}_1F_1\left(-n_1; 2\{\delta_1+\frac{1}{2}l_n+\frac{n}{4}\}; u\right), \quad (10)$$

where $\delta_1 = \left\{\sqrt{(\frac{1}{2}l_n + \frac{n-2}{4})^2 + \frac{1}{2}c_1'} - \frac{n-2}{4}\right\} - \frac{1}{2}l_n$, $n_1 = \frac{E_1'}{2\omega'} - \left(\delta_1 + \frac{1}{2}l_n + \frac{n}{4}\right)$, $c_1' = \frac{c_1}{\hbar^2}$, $\omega' = \frac{\omega}{\hbar}$ and $E_1' = \frac{E_1}{\hbar^2}$. The wave equation of H_2 in $(N-n)$-dimensional hyperspherical coordinates has similar solution. The energy spectrum of the system (2) is [17]

$$E_{dso} = 2\hbar\omega\left(p + 1 + \frac{\alpha_1 + \alpha_2}{2}\right), \tag{11}$$

where $\alpha_1 = 2\delta_1 + l_n + \frac{n-2}{2}$, $\alpha_2 = 2\delta_2 + l_{N-n} + \frac{N-n-2}{2}$ and $p = n_1 + n_2$.

The integrals of motion of (1) are given by

$$A = \sum_{i<j}^{N} L_{ij}^2 + 2r^2[c_1\chi_1(r, x_N) + c_2\chi_2(r, x_N)], \tag{12}$$

$$B = -M_N + c_1(r - x_N)\chi_1(r, x_N) - c_2(r + x_N)\chi_2(r, x_N), \tag{13}$$

$$J^2 = \sum_{i<j}^{N-1} L_{ij}^2, \quad L_{ij} = x_i p_j - x_j p_i, \quad i, j = 1, ..., N-1, \tag{14}$$

and for the model (2) they are

$$\begin{aligned} A &= -\frac{h^2}{4}\left\{\sum_{i,j=1}^{N} x_i^2\partial_{x_j}^2 - \sum_{i,j=1}^{N} x_i x_j \partial_{x_i}\partial_{x_j} - (N-1)\sum_{i=1}^{N} x_i\partial_{x_i}\right\} \\ &+ \frac{c_1\phi_1(x_1,\ldots,x_n) + c_2\phi_2(x_{n+1},\ldots,x_N)}{2}\left[\frac{1}{\phi_1(x_1,\ldots,x_n)} + \frac{1}{\phi_2(x_{n+1},\ldots,x_N)}\right], \end{aligned} \tag{15}$$

$$\begin{aligned} B &= \frac{1}{2}\left\{\sum_{i=1}^{n} p_i^2 - \sum_{i=n+1}^{N} p_i^2\right\} + \frac{\omega^2}{2}\left\{\frac{1}{\phi_1(x_1,\ldots,x_n)} - \frac{1}{\phi_2(x_{n+1},\ldots,x_N)}\right\} \\ &+ c_1\phi_1(x_1,\ldots,x_n) - c_2\phi_2(x_{n+1},\ldots,x_N), \end{aligned} \tag{16}$$

$$J_{(2)} = \sum_{i<j} J_{ij}^2, \quad J_{ij} = x_i p_j - x_j p_i, i, j = 1, 2, ..., n, \tag{17}$$

$$K_{(2)} = \sum_{i<j} K_{ij}^2, \quad K_{ij} = x_i p_j - x_j p_i, i, j = n+1, ..., N, \tag{18}$$

where the Runge-Lenz vector $M_j = \frac{1}{2}\sum_{i=1}^{N}(L_{ji}p_i - p_i L_{ij}) - \frac{c_0 x_j}{r}$. The integrals of motion $\{A, B, C\}$ close to the general form of the quadratic algebra $Q(3)$ [7],

$$[A, B] = C, \tag{19}$$

$$[A, C] = \alpha A^2 + \gamma\{A, B\} + \delta A + \epsilon B + \zeta, \tag{20}$$

$$[B, C] = aA^2 - \gamma B^2 - \alpha\{A, B\} + dA - \delta B + z \tag{21}$$

and the Casimir operator

$$\begin{aligned} K &= C^2 - \alpha\{A^2, B\} - \gamma\{A, B^2\} + (\alpha\gamma - \delta)\{A, B\} + (\gamma^2 - \epsilon)B^2 \\ &+ (\gamma\delta - 2\zeta)B + \frac{2a}{3}A^3 + (d + \frac{a\gamma}{3} + \alpha^2)A^2 + (\frac{a\epsilon}{3} + \alpha\delta + 2z)A. \end{aligned} \tag{22}$$

Here coefficients $\alpha, \gamma, \delta, \epsilon, \zeta, a, d, z$ are given in the following table for models H_{KC} and H_{dso} [16, 17].

	$\mathbf{H_{KC}}$	$\mathbf{H_{dso}}$
α	0	0
γ	$2\hbar^2$	$2\hbar^2$
δ	0	0
ϵ	$(N-1)(N-3)\hbar^4$	$\frac{\hbar^4}{4}N(N-4)$
ζ	$-4(c_1-c_2)\hbar^2 c_0$	$-\hbar^2 J_{(2)}H+\hbar^2 K_{(2)}H-\frac{\hbar^2}{4}\{8c_1-8c_2-(N-4)(N-2n)\hbar^2\}H$
a	0	0
d	$8\hbar^2 H$	$-16\hbar^2\omega^2$
z	$-4\hbar^2 J^2 H+(N-1)^2\hbar^4 H-8\hbar^2(c_1+c_2)H+2\hbar^2 c_0^2$	$2\hbar^2 H^2+4\hbar^2\omega^2 J_{(2)}+4\hbar^2\omega^2 K_{(2)}+8\hbar^2\omega^2\{c_1+c_2-\frac{\hbar^2}{4}n(N-n)\}$

Table 1. Coefficients in $Q(3)$ and K.

By means of the explicit expressions of A, B, C, we can write the Casimir operator in terms only of central elements as

$$\begin{aligned} K_{KC} &= 2(N-3)(N-1)\hbar^4 HJ^2 - 8\hbar^2(c_1-c_2)^2 H + 4(N-3)(N-1)(c_1+c_2)\hbar^4 H \\ &\quad -\hbar^6(N-3)(N-1)^2 H + 4\hbar^2 c_0^2 J^2 + 8\hbar^2(c_1+c_2)c_0^2 - 2(N-3)\hbar^4 c_0^2, \end{aligned} \quad (23)$$

$$\begin{aligned} K_{dso} &= 2\hbar^2 J_{(2)}H^2 + 2\hbar^2 K_{(2)}H^2 + \frac{\hbar^2}{4}\left[16c_1+16c_2-\{4(N-4)-(N-2n)^2\}\hbar^2\right]H^2 \\ &\quad +\hbar^2\omega^2 J_{(2)}^2 + \hbar^2\omega^2 K_{(2)}^2 - 2\hbar^2\omega^2 J_{(2)}K_{(2)} + 4\hbar^2\omega^2\{c_1-c_2-\frac{1}{4}(N-4)(N-n)\hbar^2\}J_{(2)} \\ &\quad -4\hbar^2\omega^2\{c_1-c_2+\frac{1}{4}n(N-4)\hbar^2\}K_{(2)} + 4\hbar^2\omega^2\left[(c_1-c_2)^2-\frac{1}{2}(N-n)(N-4)\hbar^2 c_1\right. \\ &\quad \left.-\frac{1}{2}n(N-4)\hbar^2 c_2 + \frac{1}{4}n(N-n)(N-4)\hbar^4\right]. \end{aligned} \quad (24)$$

The first order integrals L_{ij}, $i,j,k,l=1,..,N-1$ of H_{KC} generate $so(N-1)$ Lie algebra

$$[L_{ij}, L_{kl}] = i(\delta_{ik}L_{jl} + \delta_{jl}L_{ik} - \delta_{il}L_{jk} - \delta_{jk}L_{il})\hbar \quad (25)$$

and $J_{ij} = x_i p_j - x_j p_i$, $i,j = 1,2,....,n$ and $K_{ij} = x_i p_j - x_j p_i$, $i,j = n+1,....,N$ of H_{dso} generate $so(n)$ and $so(N-n)$ Lie algebras

$$\begin{aligned} [J_{ij}, J_{kl}] &= i(\delta_{ik}J_{jl} + \delta_{jl}J_{ik} - \delta_{il}J_{jk} - \delta_{jk}J_{il})\hbar, \quad (26) \\ [K_{ij}, K_{kl}] &= i(\delta_{ik}K_{jl} + \delta_{jl}K_{ik} - \delta_{il}K_{jk} - \delta_{jk}K_{il})\hbar. \quad (27) \end{aligned}$$

Thus the full symmetry algebra is $Q(3) \oplus so(N-1)$ for H_{KC} and $Q(3) \oplus so(n) \oplus so(N-n)$ for H_{dso}. Chains of second order Casimir operators associated with $so(N-1)$, $so(n)$ and $so(N-n)$ components may be used to label quantum states for H_{KC} and H_{dso} respectively.

These quadratic algebras can be realized in terms of the deformed oscillator algebra [7, 22]

$$[\aleph, b^\dagger] = b^\dagger, \quad [\aleph, b] = -b, \quad bb^\dagger = \Phi(\aleph+1), \quad b^\dagger b = \Phi(\aleph), \quad (28)$$

where $\aleph$ is the number operator and $\Phi(x)$ is a well behaved real function satisfying $\Phi(x) > 0$ for all $x > 0$. Then the structure function has the form [16, 17]

$$\Phi(x; u, E) = \nu_0 \prod_{i=1}^{6}[x + u - \nu_i], \quad (29)$$

where ν_i are given in the following table.

	$\mathbf{H_{KC}}$	$\mathbf{H_{dso}}$
ν_0	$6291456E\hbar^{18}$	$-12582912\hbar^{18}\omega^2$
ν_1	$\frac{1}{2}(1+m_1+m_2)$	$\frac{1}{4}(2+m_1'+m_2')$
ν_2	$\frac{1}{2}(1+m_1-m_2)$	$\frac{1}{4}(2+m_1'-m_2')$
ν_3	$\frac{1}{2}(1-m_1+m_2)$	$\frac{1}{4}(2-m_1'+m_2')$
ν_4	$\frac{1}{2}(1-m_1-m_2)$	$\frac{1}{4}(2-m_1'-m_2')$
ν_5	$\frac{1}{2}+\frac{c_0}{\hbar\sqrt{-2E}}$	$\frac{\hbar\omega+H}{2\hbar\omega}$
ν_6	$\frac{1}{2}-\frac{c_0}{\hbar\sqrt{-2E}}$	$\frac{\hbar\omega-H}{2\hbar\omega}$

Table 2. Coefficients ν_i in (29).

In the above table, $\hbar^2 m_{1,2}^2 = 16c_{1,2} + \{4J^2 + (N-3)^2\}\hbar^2$, $\hbar^2 m_1'^2 = 8c_1 + 4J_{(2)} + (n-2)^2\hbar^2$ and $\hbar^2 m_2'^2 = 8c_2 + 4K_{(2)} + (N-n-2)^2\hbar^2$.

For a finite-dimensional unitary representation, we have the following constraints

$$\Phi(p+1; u, E) = 0; \quad \Phi(0; u, E) = 0; \quad \Phi(x) > 0, \quad 0 < x < p+1 \tag{30}$$

on the structure function (29). The solutions of these constraints provide the energy spectra $E_{KC,dso}$ and the arbitrary constant $u_{KC,dso}$ ($\epsilon_1 = \pm 1, \epsilon_2 = \pm 1$) as

$$E_{KC} = \frac{-2c_0^2}{\hbar^2(2+2p+\epsilon_1 m_1+\epsilon_2 m_2)^2}, \qquad u_{KC} = \frac{1}{2} + \frac{c_0}{\hbar\sqrt{-2E}}, \tag{31}$$

$$Case \quad 1: \quad E_{dso} = 2\hbar\omega(p+1+\frac{\epsilon_1 m_1'+\epsilon_2 m_2'}{4}), \qquad u_{dso} = \frac{E+\hbar\omega}{2\hbar\omega}, \tag{32}$$

$$Case \quad 2: \quad E_{dso} = 2\hbar\omega(p+1+\frac{\epsilon_1 m_1'+\epsilon_2 m_2'}{4}), \qquad u_{dso} = \frac{-E+\hbar\omega}{2\hbar\omega}, \tag{33}$$

$$Case \quad 3: \quad E_{dso} = 2\hbar\omega(p+1+\frac{\epsilon_1 m_1'+\epsilon_2 m_2'}{4}), \qquad u_{dso} = \frac{1}{4}(2+\epsilon_1 m_1+\epsilon_2 m_2). \tag{34}$$

We have four possible structure functions for H_{KC} [16] and twenty four for H_{dso} [17]. Making the identification $n_1+n_2+I_{N-2} = n-1$, $p = n_1+n_2$, $m_i = \frac{1}{2}(3-2I_{N-2}-N-2\delta_i)$, $m_i' = 2\alpha_i$, $i = 1, 2$, the energy spectra E_{KC} and E_{dso} coincide (7) and (11) respectively.

3. Conclusion

We have presented some of the results in [16, 17] and shown how the $su(N)$ and $so(N+1)$ symmetry algebras of the N-dimensional Kepler-Coulomb (i.e., $c_1 = c_2 = 0$) and the N-dimensional harmonic oscillator (i.e., $c_1 = c_2 = 0$) are broken to higher rank polynomial algebra of the form $Q(3) \oplus L_1 \oplus L_2 \oplus \ldots$ for non-zero c_1 and c_2, where $L_1, L_2, \ldots,$ are certain Lie algebras. $Q(3)$ is a quadratic algebra involving Casimir operators of the Lie algebras in its structure constants. We have also presented the realizations of these quadratic algebras in terms of deformed oscillator algebras and obtained the finite dimensional unitary representations which yield the energy spectra of these superintegrable systems. In addition, we have compared them with the physical spectra obtained from the separation of variables.

The systems (1) and (2) could be studied using new approaches such as the recurrence method related to special functions and orthogonal polynomials and approaches combining ladder, shift,

Journal of Physics: Conference Series **670** (2016) 012024 doi:10.1088/1742-6596/670/1/012024

intertwining and supercharges operators [20, 23]. The generalizations of these systems to include monopole interactions and their duals [21, 24] could be also investigated.

Acknowledgements: The research of FH was supported by International Postgraduate Research Scholarship and Australian Postgraduate Award. IM was supported by the Australian Research Council through a Discovery Early Career Researcher Award DE 130101067. YZZ was partially supported by the Australian Research Council, Discovery Project DP 140101492.

References

[1] Bargmann 1936 *Zeitschrift fur Physik* **99**, 576.
[2] Louck J D and Galbraith 1972 *Rev..Mod. Phys.* **44**, 540.
[3] Louck J D 1965 *J. Math. Phys.* **6**, 1786.
[4] Hwa R C and Huyts J 1966 *Phys. Rev.* **145**, 1188.
[5] Trunk M 1996 *Int. J. Mod. Phys.* **11**, 2329.
[6] Granovskii Ya I, Zhedanov A S and Lutzenko I M 1991 *J. Phys. A: Math. Theor.* **24**, 3887.
[7] Daskaloyannis C 2001 *J Math. Phys.* **42**, 1100.
[8] Kalnins E G, Kress J M and Miller W Jr 2006 *J. Math. Phys.* **47**, 093501.
[9] Kalnins E G, Miller W Jr and Post S 2008 *SIGMA*, **4**, 008.
[10] Tanoudis Y and Daskaloyannis C 2011 *SIGMA* **7**, 054.
[11] Miller W J, Post S and Winternitz P 2013 *J. Phys. A: Math. Theor.* **46**, 423001.
[12] Isaac P S and Marquette I 2014 *J. Phys. A: Math. and Theor.* **47**, 205203.
[13] Genest V X, Vinet L and Zhedanov A 2014 *J. Phys.: Conf. Ser.* **512**, 012011.
[14] Genest V X, Vinet L and Zhedanov A 2014 *SIGMA* **10**, 038.
[15] Marquette I and Quesne C 2014 *J. Math. Phys.* **55**, 112103.
[16] Hoque M F, Marquette I and Zhang Y-Z 2015 *J. Phys. A : Math. Theor.* **48**, 185201.
[17] Hoque M F, Marquette I and Zhang Y-Z 2015 *J. Phys. A : Math. Theor.* **48** (in press).
[18] Kibler M, Mardoyan L G and Pogosyan G S 1994 *Int. J. Quan. Chem.* **52**, 1301.
[19] Hurwitz A 1993 *Mathematische Werke (Band II Birkhausser, Basel)*, 641.
[20] Marquette I 2010 *J. Math. Phys.* **51**, 102105.
[21] Marquette I 2012 *J. Math. Phys.* **53**, 022103.
[22] Daskaloyannis C 1991 *J. Phys. A: Math. and Gen* **24**, L789.
[23] Kalnins E G, Kress J M and Miller W Jr 2011 *SIGMA*, **7**, 031.
[24] Mardoyan L 2003 *Journal of Mathematical Physics* **44**, 4981-87.

Journal of Physics: Conference Series **670** (2016) 012025 doi:10.1088/1742-6596/670/1/012025

On complex Riemannian foliations

Cristian Ida

Department of Mathematics and Computer Science, University Transilvania of Braşov, 500091, Str. Iuliu Maniu 50, Braşov, România

E-mail: cristian.ida@unitbv.ro

Abstract. In this paper we make some general considerations about the geometry of complex Riemannian foliations, we introduce a leafwise characteristic connection and we write Einstein equations with respect to it. Next, using an one-to-one correspondence between leafwise holomorphic Riemannian metrics and leafwise anti-Kählerian metrics, we focus on the Einstein condition for a leafwise holomorphic Riemannian metric and the associated real leafwise anti-Kählerian metric on a manifold endowed with a complex foliation.

1. Introduction

The importance of (holomorphic) complex Riemannian metrics in mathematical physics is without question (see for instance [3, 16, 20]). We recall that a (holomorphic) complex Riemannian manifold is a complex manifold M, together with a (holomorphic) complex tensor field G that is a complex scalar product (i.e., nondegenerate, symmetric, $\mathbb{C}$-bilinear form) on each holomorphic tangent space of M. The (holomorphic) complex Riemannian geometry possesses an underlying real geometry consisting of a pseudo-Riemannian metric of neutral signature and there is a strong relation between holomorphic Riemannian metrics and anti-Kählerian metrics (also known as Kähler-Norden metrics [10, 18, 19, 21]). In [2], it is proved that an anti-Kählerian metric must be the real part of a holomorphic metric. There is studied the Einstein condition for anti-Kählerian metrics and some generalized Einstein conditions on holomorphic Riemannian manifolds are investigated in [19]. Also, we notice that in [6, 7, 8], a study of connections and curvatures on (holomorphic) complex Riemannian manifolds is given, and the anti-Hermitian metrics which are locally conformal with anti-Kählerian metrics are investigated in [5].

In this paper we work in the category of smooth manifolds endowed with complex foliations, and we check that more of the techniques used in the study of (holomorphic) complex Riemannian geometry can be adjusted to the foliated picture. Thus, we generalize the main notions from the geometry of complex Riemannian manifods to that of complex Riemannian foliations. Incidentally the classical case is contained in the foliation formalism by taking the foliation which consists of the leaf M only.

The structure of the paper is as follows. In the second section, we brief recall the definition of a complex foliation which is used for instance in [4, 13, 14] , we define the leafwsise complex Riemannian metrics and we present some examples. Also, following [2, 7, 11] we generalize some notions from complex Riemannian manifolds to that of complex Riemannian foliations (not necessarily leafwise holomorphic). More exactly we consider the associated leafwise characteristic connection, we study its properties and we write Einstein equations with respect to it. In the last section we study the Einstein condition for leafwise holomorphic Riemannian metrics.

The main methods used here are similar and closely related to those used in the study of complex Riemannian manifolds [7, 8, 11] and anti-Kählerian manifolds [3, 2, 19, 21]. For this reason the most proofs are omitted here.

2. Complex Riemannian foliations

Let M be a smooth $(2m+n)$-dimensional manifold endoweed with a smooth regular foliation $\mathcal{F}$ of codimension n. Then the dimension of the foliation $\mathcal{F}$ is $2m$. We denote by $T\mathcal{F}$ the tangent bundle along the leaves and by $T^*\mathcal{F}$ its dual.

An *almost complex structure* along the leaves of $\mathcal{F}$ is defined as a smooth real vector bundle automorphism $J_{\mathcal{F}}$ of $T\mathcal{F}$ satisfying $J_{\mathcal{F}}^2 = -\mathrm{Id}_{\mathcal{F}}$. Given such a leafwise almost complex structure, we obtain the decomposition $T_{\mathbb{C}}\mathcal{F} = T\mathcal{F} \otimes_{\mathbb{R}} \mathbb{C} = T^{1,0}\mathcal{F} \oplus T^{0,1}\mathcal{F}$, where

$$T^{1,0}\mathcal{F} = \{X - iJ_{\mathcal{F}}X \mid X \in \Gamma(T\mathcal{F})\} \text{ and } T^{1,0}\mathcal{F} = \{X + iJ_{\mathcal{F}}X \mid X \in \Gamma(T\mathcal{F})\}.$$

Moreover, if the Nijenhuis tensor along the leaves

$$N_{J_{\mathcal{F}}} := [J_{\mathcal{F}}X, J_{\mathcal{F}}Y] - J_{\mathcal{F}}[J_{\mathcal{F}}X, Y] - J_{\mathcal{F}}[X, J_{\mathcal{F}}Y] - [X, Y], \text{ for } X, Y \in \Gamma(T\mathcal{F}) \tag{2.1}$$

vanishes, then $J_{\mathcal{F}}$ is called a *complex structure* on $\mathcal{F}$ and $(\mathcal{F}, J_{\mathcal{F}})$ is called a complex foliation on M. In this case, the complex foliation $\mathcal{F}$ can be defined by an open cover $\{U_i\}$, $i \in I$, of M and diffeomorphisms $\phi_i : \Omega_i \times \mathcal{O}_i \to U_i$ (where Ω_i is an open polydisc in $\mathbb{C}^m$ and $\mathcal{O}_i$ is an open ball in $\mathbb{R}^n$) such that, for every pair $(i, j) \in I \times I$ with $U_i \cap U_j \neq \phi$, the coordinate change $\phi_{ij} = \phi_j^{-1} \circ \phi_i : \phi_i^{-1}(U_i \cap U_j) \to \phi_j^{-1}(U_i \cap U_j)$ is of the form $(z', x') = (\phi_{ij}^1(z, x), \phi_{ij}^2(x))$ with $\phi_{ij}^1(z, x)$ holomorphic in z for every x fixed. A such adapted atlas will be called *leafwise complex*.

If we set $z^a = u^a + iu^{m+a}$, $a = 1, \dots, m$, then the complex structure along the leaves $J_{\mathcal{F}} : T\mathcal{F} \to T\mathcal{F}$ is given by

$$J_{\mathcal{F}}\left(\frac{\partial}{\partial u^a}\right) = \frac{\partial}{\partial u^{m+a}},\ J_{\mathcal{F}}\left(\frac{\partial}{\partial u^{m+a}}\right) = -\frac{\partial}{\partial u^a},\ a = 1, \dots, m$$

and its complex extension to $T_{\mathbb{C}}\mathcal{F}$ is given by

$$J_{\mathcal{F}}\left(\frac{\partial}{\partial z^a}\right) = i\frac{\partial}{\partial z^a},\ J_{\mathcal{F}}\left(\frac{\partial}{\partial \overline{z}^a}\right) = -i\frac{\partial}{\partial \overline{z}^a},\ a = 1, \dots, m.$$

Remark 2.1. The above notion of complex foliations is also related to laminations. For instance, a lamination by Riemann surfaces is a topological space locally homeomorphic to a model space of type $D \times T$, where D is the open unit disc in $\mathbb{C}$ and T is a topological space. The transition functions between two such charts is assumed to be of the form $D \times T \to D \times T'$ and defined by $(z, x) \mapsto (f(z, x), g(x))$, where $f(z, x)$ is holomorphic in z and continuous in x, and $g(x)$ is a continuous function of x (see Section 2 in [9]).

We have the following simple examples of complex foliations.

Example 2.1. Any complex manifold M of $\dim_{\mathbb{C}} M = m$ is a complex foliation of complex dimension m and codimension 0.

Example 2.2. Let M be an open set of $\mathbb{C}^m \times N$, where N is an n-dimensional smooth manifold. For every $t \in N$, the set $M^t = \{z \in \mathbb{C}^m \mid (z, t) \in M\}$ is an open set of $\mathbb{C}^m$ called the *section* of M along t. Then, sections of M are leaves of a complex foliation $\mathcal{F}$ of dimension m called the complex *canonical* foliation of M.

Example 2.3. Let F be a complex manifold of $\dim_{\mathbb{C}} F = m$ and N an n-dimensional smooth manifold. Every locally trivial fibration $F \hookrightarrow M \to N$ whose cocycle takes values in the automorphism group $\mathrm{Aut}(F)$ of (the complex manifold) F is a complex foliation, the fibers being the leaves. If the fibration is trivial, that is $M = F \times N$, we say that $\mathcal{F}$ is a *complex product foliation.*

For more examples of complex foliations, see for instance [13, 14] and references therein.

Definition 2.1. A *leafwise complex Riemannian metric* on $(M, \mathcal{F}, J_{\mathcal{F}})$ is a covariant symmetric 2-tensor field $G : \Gamma(T_{\mathbb{C}}\mathcal{F}) \times \Gamma(T_{\mathbb{C}}\mathcal{F}) \to \mathbb{C}$, which is non-degenerate at each point (z, x) of $(M, \mathcal{F})$ and satisfies

$$G(\overline{Z}_1, \overline{Z}_2) = \overline{G(Z_1, Z_2)} \text{ for every } Z_1, Z_2 \in \Gamma(T_{\mathbb{C}}\mathcal{F}), \tag{2.2}$$

$$G(Z_1, Z_2) = 0 \text{ for every } Z_1 \in \Gamma(T^{1,0}\mathcal{F}) \text{ and } Z_2 \in \Gamma(T^{0,1}\mathcal{F}). \tag{2.3}$$

It is easy to see that the relation (2.3) is equivalent to

$$G(J_{\mathcal{F}}Z_1, J_{\mathcal{F}}Z_2) = -G(Z_1, Z_2) \text{ for every } Z_1, Z_2 \in \Gamma(T_{\mathbb{C}}\mathcal{F}), \tag{2.4}$$

where, we have denoted again by $J_{\mathcal{F}}$ the $\mathbb{C}$-linear extension of $J_{\mathcal{F}}$ to $T_{\mathbb{C}}\mathcal{F}$. Thus, a leafwise complex Riemannian metric on $(M, \mathcal{F}, J_{\mathcal{F}})$ is completely determined by its values on $\Gamma(T^{1,0}\mathcal{F})$.

Definition 2.2. The pair $(M, \mathcal{F}, J_{\mathcal{F}}, G)$ consisting by a smooth $(2m+n)$-dimensional manifold M endowed with a complex foliation $(\mathcal{F}, J_{\mathcal{F}})$ and with a leafwise complex Riemannian metric G on $(M, \mathcal{F}, J_{\mathcal{F}})$, will be called a *complex Riemannian foliation.*

If $(z^1, \ldots, z^m, x^1, \ldots, x^n)$ is an adapted local coordinate system (leafwise complex) on $(M, \mathcal{F}, J_{\mathcal{F}})$, such that $\Gamma(T_{\mathbb{C}}\mathcal{F}) = span\{\partial/\partial z^a, \partial/\partial z^{\overline{a}}\}$, we put

$$G_{AB}(z, x) = G\left(\frac{\partial}{\partial z^A}, \frac{\partial}{\partial z^B}\right), \ A, B \in \{1, \ldots, m, \overline{1}, \ldots, \overline{m}\}. \tag{2.5}$$

Then, for a leafwise complex Riemannian metric G, the defining conditions (2.2) and (2.3) can be expressed locally in the form

$$G_{\overline{A}\,\overline{B}} = \overline{G_{AB}} \text{ and } G_{a\overline{b}} = G_{\overline{a}b} = 0. \tag{2.6}$$

Definition 2.3. A leafwise complex Riemannian metric G on $(M, \mathcal{F}, J_{\mathcal{F}})$ is called *leafwise holomorphic Riemannian metric* if the local components $G_{ab}(z, x)$ are leafwise holomorphic functions, that is

$$\frac{\partial G_{ab}}{\partial z^{\overline{c}}} = 0, \text{ for every } c \in \{1 \ldots, m\}. \tag{2.7}$$

As in the case of complex Riemannian manifolds (see [7]) for a given leafwise complex Riemannian metric G on $(M, \mathcal{F}, J_{\mathcal{F}})$, we define the leafwise tensor field $\widetilde{G}$ on $(M, \mathcal{F}, J_{\mathcal{F}})$ by setting

$$\widetilde{G}(Z_1, Z_2) = (G \circ J_{\mathcal{F}})(Z_1, Z_2) := G(J_{\mathcal{F}}Z_1, Z_2) \text{ for every } Z_1, Z_2 \in \Gamma(T_{\mathbb{C}}\mathcal{F}). \tag{2.8}$$

This metric is called *leafwise twin metric*, and locally, it satisfies

$$\widetilde{G}_{ab} = iG_{ab} \text{ and } \widetilde{G}_{\overline{a}\,\overline{b}} = -iG_{\overline{a}\,\overline{b}}. \tag{2.9}$$

Also, we notice that given a leafwise complex Riemannian metric G on $(M, \mathcal{F}, J_{\mathcal{F}})$ it induces a leafwise real Riemannian metric g on the underlying real foliated manifold $(M, \mathcal{F}, J_{\mathcal{F}})$ by setting

$$g(X, Y) = 2Re\, G(\widehat{X}, \widehat{Y}), \ X, Y \in \Gamma(T\mathcal{F}), \tag{2.10}$$

Journal of Physics: Conference Series **670** (2016) 012025 doi:10.1088/1742-6596/670/1/012025

where $\widehat{X} = (1/2)(X - iJ_{\mathcal{F}}X), \widehat{Y} = (1/2)(Y - iJ_{\mathcal{F}}Y) \in \Gamma(T^{1,0}\mathcal{F})$, and this real metric satisfies

$$g(J_{\mathcal{F}}X, J_{\mathcal{F}}Y) = -g(X,Y) \text{ for every } X, Y \in \Gamma(T\mathcal{F}), \tag{2.11}$$

or, equivalently

$$g(J_{\mathcal{F}}X, Y) = g(X, J_{\mathcal{F}}Y) \text{ for every } X, Y \in \Gamma(T\mathcal{F}). \tag{2.12}$$

Such a leafwise real metric will be called a *leafwise anti-Hermitian metric*, or *leafwise Norden metric* and $(M, \mathcal{F}, J_{\mathcal{F}}, g)$ will be called the *realization* of $(M, \mathcal{F}, J_{\mathcal{F}}, G)$.

Conversely, every leafwise anti-Hermitian metric on the underlying real foliated manifold $(M, \mathcal{F}, J_{\mathcal{F}})$ induces a leafwise complex Riemannian metric on the $(M, \mathcal{F}, J_{\mathcal{F}})$ by setting

$$G(\widehat{X}, \widehat{Y}) = \frac{1}{2}\left(g(X,Y) - ig(X, J_{\mathcal{F}}Y)\right), \tag{2.13}$$

where $X, Y \in \Gamma(T\mathcal{F})$ and $\widehat{X} = (1/2)(X - iJ_{\mathcal{F}}X), \widehat{Y} = (1/2)(Y - iJ_{\mathcal{F}}Y) \in \Gamma(T^{1,0}\mathcal{F})$ as above, and next we extend G to have the conditions (2.2) and (2.3), which is possible because of (2.11).

We recall (see [17]) that a leafwise (or tangential) connection ∇ on $(M, \mathcal{F})$ can be seen as a linear map $\nabla : \Gamma(T_{\mathbb{C}}\mathcal{F}) \times \Gamma(T_{\mathbb{C}}\mathcal{F}) \to \Gamma(T_{\mathbb{C}}\mathcal{F})$ with $\nabla_{fX}Y = f\nabla_X Y$ and $\nabla_X(gY) = X(g) \cdot Y + g\nabla_X Y$ for every $f, g \in C^\infty(M) \otimes_{\mathbb{R}} \mathbb{C}$ and $X, Y \in \Gamma(T_{\mathbb{C}}\mathcal{F})$. Given any leafwise connection D on $(M, \mathcal{F}, J_{\mathcal{F}})$, with respect to an adapted coordinate system (z, x) (leafwise complex), we put

$$D_{\frac{\partial}{\partial z^A}} \frac{\partial}{\partial z^B} = L^C_{AB}(z,x) \frac{\partial}{\partial z^C}.$$

We notice that the leafwise covariant differentiation, which is defined for leafwise real vector fields in $\Gamma(T\mathcal{F})$, can be extended by complex linearity on leafwise complex vector fields from $\Gamma(T_{\mathbb{C}}\mathcal{F})$. Then $L^{\overline{C}}_{\overline{A}\overline{B}} = \overline{L^C_{AB}}$, where $\overline{\overline{A}} = A$.

Definition 2.4. A leafwise (real) connection D on $(M, \mathcal{F}, J_{\mathcal{F}})$ is called *leafwise almost complex* if $DJ_{\mathcal{F}} = 0$.

By direct calculus, we easy obtain

Proposition 2.1. *A leafwise connection D on $(M, \mathcal{F}, J_{\mathcal{F}})$ is leafwise almost complex if and only if* $L^{\bar{c}}_{ab} = L^{c}_{a\bar{b}} = 0$

Let us consider the leafwise (or tangential) Levi-Civita connection of a tangential Riemannian metric (see Proposition 5.18 in [17]). We have

Definition 2.5. A leafwise anti-Hermitian metric g on $(M, \mathcal{F}, J_{\mathcal{F}})$ is called *leafwise anti-Kählerian metric* if the leafwise Levi-Civita connection of g is leafwise almost complex.

In the following we present some examples of leafwise anti-(Kählerian) Hermitian metrics on manifolds endowed with complex foliations.

Example 2.4. Let (M, J, g) be a locally conformal anti-Kähler manifold, with a parallel Lee form and a non light-like Lee vector field (see [5]). Then, its vertical foliation (defined by Lee and anti-Lee vector fields) carries a complex structure with respect to which the induced leafwise metric is anti-Hermitian (see Theorem 5 in [5]).

Example 2.5. It is well known [3] that every parallelisable complex manifold G (including complex Lie groups) can be endowed with anti-Kählerian metrics. Thus, it is easy to see that every product complex foliation defined by trivial fibration $M = G \times N \to N$ where G is a parallelisable complex manifold and N is a paracompact smooth manifold, can be endowed with leafwise anti-Kählerian metrics.

Example 2.6. We consider an m-dimensional smooth manifold M and let $\pi : TM \to M$ its tangent bundle with the total space $\mathcal{T}M$, called the *tangent manifold* of M. Let (x^a, y^a), $a = 1, \ldots, m$ the local coordinates on the manifold $\mathcal{T}M$, where (x^a) are the local coordinates on M and (y^a) are the vector coordinates with respect to the basis $\{\frac{\partial}{\partial x^a}\}$. The fibers of TM define the *vertical foliation* $\mathcal{V}$, and we shall denote by $V = T\mathcal{V}$ the tangent bundle along the vertical leaves, which is the vertical bundle. Now, we consider the total space $\mathcal{T}V$ of the vertical bundle $V \to TM$, which is a $3m$-dimensional manifold, called the *vertical tangent manifold* of M. The iterated tangent manifold $\mathcal{T}(\mathcal{T}M)$ has local coordinates $(x^a, y^a, \xi^a, \eta^a)$, where ξ, η are vector coordinates with respect to the natural basis $\{\frac{\partial}{\partial x^a}, \frac{\partial}{\partial y^a}\}$, and the vertical tangent manifold $\mathcal{T}V$ may be seen as the submanifold of $\mathcal{T}(\mathcal{T}M)$ defined by $\xi^a = 0$. For the projection $p : \mathcal{T}V \to M$ given by $p(x, y, \eta) = x$ its kernel is the Whitney sum $W = V_1 \oplus V_2$, where $V_2 \cong V_1 = V$. The vertical bundle $W = span\{\frac{\partial}{\partial y^a}, \frac{\partial}{\partial \eta^a}\}$ is called the *double vertical bundle*, which defines the *double vertical foliation* $\mathcal{W}$, with leaves defined by $x^a = const.$, see [22]. Moreover the following operator

$$J_{\mathcal{W}}(\frac{\partial}{\partial y^a}) = \frac{\partial}{\partial \eta^a}, \; J_{\mathcal{W}}(\frac{\partial}{\partial \eta^a}) = -\frac{\partial}{\partial y^a}, \tag{2.14}$$

defines an almost complex structure on W, which is integrable. Thus $(\mathcal{W}, J_{\mathcal{W}})$ is a complex foliation on $\mathcal{T}V$ and, we can consider the holomorphic coordinates along the leaves of $\mathcal{W}$ by putting $z^a = y^a + i\eta^a$.

Let us consider now a (locally) Lagrange metric $L : TM \to \mathbb{R}$ with the fundamental metric tensor $g_{ab}(x, y) = \partial^2 L/\partial y^a \partial y^b$. It defines a nondegenerate Riemannian metric on the vertical bundle V by formula

$$g_{\mathcal{V}}(X, Y) = g_{ab}(x, y) X^a(x, y) Y^b(x, y), \tag{2.15}$$

for every $X = X^a(x, y)\partial/\partial y^a$, $Y = Y^a(x, y)\partial/\partial y^a \in \Gamma(V)$. This metric descends to a vertical pseudo-Riemannian metric on W, defined by

$$g_{\mathcal{W}}(X, Y) = g_{ab}(x, y) X_1^a(x, y, \eta) Y_1^b(x, y, \eta) - g_{ab}(x, y) X_2^a(x, y, \eta) Y_2^b(x, y, \eta), \tag{2.16}$$

for every $X = X_1^a(x, y, \eta)\partial/\partial y^a + X_2^a(x, y, \eta)\partial/\partial \eta^a$, $Y = Y_1^a(x, y, \eta)\partial/\partial y^a + Y_2^a(x, y, \eta)\partial/\partial \eta^a \in \Gamma(W)$. By direct calculus we get $g_{\mathcal{W}}(J_{\mathcal{W}}(X), J_{\mathcal{W}}(Y)) = -g_{\mathcal{W}}(X, Y)$, for every $X, Y \in \Gamma(W)$ which says that $g_{\mathcal{W}}$ is a leafwise anti-Hermitian metric on the complex foliation $(\mathcal{T}V, \mathcal{W}, J_{\mathcal{W}})$. If we consider ∇ the leafwise Levi-Civita connection associated to this metric, then it is locally given by

$$\nabla_{\frac{\partial}{\partial y^a}} \frac{\partial}{\partial y^b} = \nabla_{\frac{\partial}{\partial \eta^a}} \frac{\partial}{\partial \eta^b} = C_{ab}^c \frac{\partial}{\partial y^c}, \; \nabla_{\frac{\partial}{\partial y^a}} \frac{\partial}{\partial \eta^b} = \nabla_{\frac{\partial}{\partial \eta^b}} \frac{\partial}{\partial y^a} = C_{ab}^c \frac{\partial}{\partial \eta^c}, \tag{2.17}$$

where

$$C_{ab}^c = \frac{1}{2} g^{cd} \frac{\partial g_{bd}}{\partial y^a}. \tag{2.18}$$

We observe that $\nabla_{\frac{\partial}{\partial y^a}} J_{\mathcal{W}}(\frac{\partial}{\partial \eta^b}) = -J_{\mathcal{W}}\left(\nabla_{\frac{\partial}{\partial y^a}} \frac{\partial}{\partial \eta^b}\right)$, which says that $\nabla J_{\mathcal{W}} \neq 0$, so the metric $g_{\mathcal{W}}$ is not leafwise anti-Kählerian.

Example 2.7. *Sasakian-like almost contact complex Riemannian manifolds.* Let $(M, \varphi, \xi, \eta, g)$ be an almost contact complex Riemannian manifold (see [12]), that is M is a $(2m+1)$-dimensional smooth manifold endowed with a quadruple (φ, ξ, η, g) consisting of an endomorphism φ of the tangent bundle, a vector field ξ and its dual 1-form η, and g is a pseudo-Riemannian metric on M of signature $(m+1, m)$, satisfying the conditions

$$\varphi \circ \xi = 0 \,,\, \varphi^2 = -\mathrm{Id} + \eta \otimes \xi \,,\, \eta \circ \varphi = 0 \,,\, \eta(\xi) = 1 \,,\, g(\varphi X, \varphi Y) = -g(X, Y) + \eta(X)\eta(Y). \tag{2.19}$$

The associated twin metric is given by $\widetilde{g}(X,Y) = g(\varphi X, Y) + \eta(X)\eta(Y)$ and, the almost contact complex Riemannian manifold (M,φ,ξ,η,g) is said to be *normal* if the Nijenhuis tensor associated with the endomorphism φ satisfies $N_\varphi(X,Y) + 2d\eta(X,Y)\xi = 0$ for every $X, Y \in \Gamma(TM)$. Also, we consider the structure tensor F of type $(0,3)$ on (M,φ,ξ,η,g) given by $F(X,Y,Z) = g((\nabla_X\varphi)Y, Z)$, and according to [12], the almost contact complex Riemannian manifold (M,φ,ξ,η,g) is called *Sasakian-like almost contact complex Riemannian manifold* if the structure tensors φ,ξ,η,g satisfy

$$F(X,Y,Z) = F(\xi,Y,Z) = F(\xi,\xi,Z) = 0 \text{ and } F(X,Y,\xi) = -g(X,Y),\ X,Y,Z \in \Gamma(TM). \tag{2.20}$$

Moreover, it is proved (see [12]) that if (M,φ,ξ,η,g) is a Sasakian-like almost contact complex Riemannian manifold, then it is normal and the fundamental 1-form η is closed. Then, $H = \ker\eta$ is an integrable $2m$-dimensional distribution which defines a complex foliation $(H, J_H = \varphi|_H)$ on M and the leafwise metric $h = g|_H$, induced on each leaf of H, is anti-Kählerian. Indeed, from the first relation of (2.20) it follows that $h((\nabla^h_X J_H)Y, Z) = F(X,Y,Z) = 0$, $X,Y,Z \in \Gamma(H)$, where ∇^h is the leafwise Levi-Civita connection of h. For more concrete examples of such kind of structures we refer [12].

Similary to the case of complex manifolds [18, 19, 21] (see also, [2, 3]), we have the following one-to-one correspondence between the leafwise anti-Kählerian metrics and leafwise holomorphic Riemannian metrics on $(M,\mathcal{F},J_\mathcal{F})$.

Proposition 2.2. *Let M be a smooth $(2m+n)$-dimensional manifold endowed with a complex foliation $(\mathcal{F},J_\mathcal{F})$. If G is a leafwise holomorphic Riemannian metric $(M,\mathcal{F},J_\mathcal{F})$ then g defined in* (2.10) *is a leafwise anti-Kählerian metric on the realization $(M,\mathcal{F},J_\mathcal{F})$, and conversely if g is a leafwise anti-Kählerian metric on the underlying real foliated manifold $(M,\mathcal{F},J_\mathcal{F})$ then G defined in* (2.13) *is a leafwise holomorphic Riemannian metric on $(M,\mathcal{F},J_\mathcal{F})$.*

Now, let us denote by ∇ and $\widetilde{\nabla}$ the leafwise Levi-Civita connections of G and $\widetilde{G}$, respectively. Then, as usual, the leafwise Christoffel symbols of G are given by

$$\Gamma^C_{AB} = \frac{1}{2}G^{CD}\left(\frac{\partial G_{BD}}{\partial z^A} + \frac{\partial G_{AD}}{\partial z^B} - \frac{\partial G_{AB}}{\partial z^D}\right), \tag{2.21}$$

where $(G^{AB})_{m\times m}$ denotes the inverse matrix of $(G_{AB})_{m\times m}$, and similarly for the leafwise Christoffel symbols $\widetilde{\Gamma}^C_{AB}$ of $\widetilde{G}$.

Taking into account (2.6) and (2.9), we have the following relations which relates the leafwise Christoffel symbols of G and $\widetilde{G}$, respectively

$$\widetilde{\Gamma}^c_{ab} = \Gamma^c_{ab} = \frac{1}{2}G^{cd}\left(\frac{\partial G_{bd}}{\partial z^a} + \frac{\partial G_{ad}}{\partial z^b} - \frac{\partial G_{ab}}{\partial z^d}\right) \tag{2.22}$$

$$\widetilde{\Gamma}^{\overline{c}}_{ab} = -\Gamma^{\overline{c}}_{ab} = \frac{1}{2}G^{\overline{c}\overline{d}}\frac{\partial G_{ab}}{\partial z^{\overline{d}}}\ ,\ \widetilde{\Gamma}^c_{\overline{a}b} = \Gamma^c_{\overline{a}b} = \frac{1}{2}G^{cd}\frac{\partial G_{bd}}{\partial z^{\overline{a}}}. \tag{2.23}$$

By analogy with the case of complex manifolds [7], we define the fundamental leafwise tensor Φ of a leafwise complex Riemannian metric G by setting

$$\Phi(Z_1,Z_2) = \widetilde{\nabla}_{Z_1}Z_2 - \nabla_{Z_1}Z_2\,,\ \text{for every } Z_1,Z_2 \in \Gamma(T_\mathbb{C}\mathcal{F}). \tag{2.24}$$

By this definition, we deduce

$$\Phi(\overline{Z}_1,\overline{Z}_2) = \overline{\Phi(Z_1,Z_2)}\,,\ \text{for every } Z_1,Z_2 \in \Gamma(T_\mathbb{C}\mathcal{F}). \tag{2.25}$$

Using (2.24), (2.22), (2.23) and (2.25), it follows that the nonvanishing components of the fundamental leafwise tensor Φ are given by

$$\Phi^{\bar{c}}_{ab} = G^{\bar{c}\,\bar{d}}\frac{\partial G_{ab}}{\partial z^{\bar{d}}} \text{ and } \Phi^{c}_{\bar{a}\,\bar{b}} = \overline{\Phi^{\bar{c}}_{ab}}. \tag{2.26}$$

Also, from (2.24) and (2.26) we have

Proposition 2.3. *The fundamental leafwise tensor of a complex leafwise Riemannian metric G satisfy*

$$\Phi(Z_1, Z_2) = \Phi(Z_2, Z_1)\,,\ \Phi(J_{\mathcal{F}}Z_1, Z_2) = -J_{\mathcal{F}}\Phi(Z_1, Z_2)\,,\ \forall\, Z_1, Z_2 \in \Gamma(T_{\mathbb{C}}\mathcal{F}). \tag{2.27}$$

Remark 2.2. If $(M, \mathcal{F}, J_{\mathcal{F}}, g)$ is the realization of a complex Riemannian foliation $(M, \mathcal{F}, J_{\mathcal{F}}, G)$ we can define as in (2.24) the fundamental leafwise tensor for real leafwise vector fields, and the property (2.25) of Φ implies that Φ is the complex extension of the real fundamental leafwise tensor on $(M, \mathcal{F}, J_{\mathcal{F}}, g)$.

In the following, we extend the study from [7] to the case of complex Riemannian foliations, and we shall construct a *leafwise characteristic connection* on $(M, \mathcal{F}, J_{\mathcal{F}}, G)$.

We consider the fundamental leafwise tensor of type $(0, 3)$ defined by

$$\Psi(Z_1, Z_2, Z_3) = G(\Phi(Z_1, Z_2), Z_3)\,, \text{ for every } Z_1, Z_2, Z_3 \in \Gamma(T_{\mathbb{C}}\mathcal{F}). \tag{2.28}$$

In an adapted coordinate system (leafwise complex) on $(M, \mathcal{F}, J_{\mathcal{F}})$, we have

$$\Psi_{AB,C} = \Phi^{D}_{AB}G_{DC}, \tag{2.29}$$

and the nonvanishing componets of $\Psi_{AB,C}$ are

$$\Psi_{ab,\bar{c}} = \frac{\partial G_{ab}}{\partial z^{\bar{c}}} \text{ and } \Psi_{\bar{a}\bar{b},c} = \overline{\Psi_{ab,\bar{c}}}. \tag{2.30}$$

We have

Theorem 2.1. *On every complex Riemannian foliation $(M, \mathcal{F}, J_{\mathcal{F}}, G)$ there exists an unique leafwise connection D with local coefficients L^{C}_{AB} such that*

(i) D is symmetric, that is $L^{C}_{AB} = L^{C}_{BA}$;

(ii) D is leafwise almost complex, that is $L^{\bar{c}}_{ab} = L^{c}_{\bar{a}\bar{b}} = 0$;

(iii) The leafwise covariant derivatives $D_aG_{bc} = \partial G_{bc}/\partial z^a - L^{d}_{ab}G_{dc} - L^{d}_{ac}G_{bd}$ vanishes.

Proof. If we define the local coefficients of D by

$$L^{C}_{AB} = \Gamma^{C}_{AB} + \frac{1}{2}\Phi^{C}_{AB} - \frac{1}{2}G^{CD}(\Psi_{DA,B} + \Psi_{DB,A}), \tag{2.31}$$

where Γ^{C}_{AB} are the leafwise complex Christoffel symbols of G, then by direct calculus we obtain that D satisfies the conditions of theorem.

Also, if D' is another leafwise connection with local coefficients L'^{C}_{AB} which satisfy the all conditions of theorem, we denote by $D^{C}_{AB} = L^{C}_{AB} - L'^{C}_{AB}$ the leafwise difference tensor. Then, we easily obtain

$$D^{C}_{AB} = D^{C}_{BA}\,,\ D^{\bar{c}}_{ab} = D^{c}_{a\bar{b}} = 0\,,\ D^{d}_{ab}G_{dc} + D^{d}_{ac}G_{ab} = 0, \tag{2.32}$$

which implies $D^{C}_{AB} = 0$, that is $D = D'$, and the uniqueness then follows. □

The leafwise connection from the above theorem, will be called the *leafwise characteristic connection* of the complex Riemannian foliation $(M, \mathcal{F}, J_{\mathcal{F}}, G)$.

The defining equality (2.31) of the leafwise characteristic connection and the properties of the fundamental leafwise tensor, implies

Corollary 2.1. *On every complex Riemannian foliation $(M, \mathcal{F}, J_{\mathcal{F}}, G)$ there exists an unique leafwise connection D such that*

(i) D is symmetric;

(ii) D is leafwise almost complex;

(iii) $D_A G_{BC} = \Psi_{BC,A}$, i.e the leafwise covariant derivative of the metric G is the fundamental leafwise tensor Ψ.

Remark 2.3. The third condition of Theorem 2.1 says that the nonvanishing components of the leafwise tensor $D_A G_{BC}$ are

$$D_{\overline{a}} G_{bc} = \Psi_{bc,\overline{a}} \text{ and } D_a G_{\overline{b}\,\overline{c}} = \overline{D_{\overline{a}} G_{bc}}. \tag{2.33}$$

On the realization of a complex Riemannian foliation we have

Corollary 2.2. *If $(M, \mathcal{F}, J_{\mathcal{F}}, g)$ is the realization of a complex Riemannian foliation $(M, \mathcal{F}, J_{\mathcal{F}}, G)$, then the leafwise characteristic connection D on $(M, \mathcal{F}, J_{\mathcal{F}}, g)$ is the unique leafwise connection which satisfy the conditions*

(i) D is symmetric;

(ii) D is leafwise almost complex;

(iii) $(D_X g)(Y, Z) = (D_{J_{\mathcal{F}}X} g)(J_{\mathcal{F}}Y, Z)$, for every $X, Y, Z \in \Gamma(T\mathcal{F})$.

The defining equality (2.31) together with (2.30) imply that the nonvanishing coefficients of the leafwise characteristic connection D are

$$L^c_{ab} = \Gamma^c_{ab} \text{ and } L^{\overline{c}}_{\overline{a}\,\overline{b}} = \overline{L^c_{ab}}, \tag{2.34}$$

that is, D is completely determined on $\Gamma(T^{1,0}\mathcal{F})$.

We notice that a leafwise vector field $Z = Z^a(z, x)\partial/\partial z^a \in \Gamma(T^{1,0}\mathcal{F})$ is leafwise holomorphic if Z^a are leafwise holomorphic functions on $(M, \mathcal{F}, J_{\mathcal{F}})$. Also, using the leafwise Cauchy-Riemann equations, it is easy to see that for a given leafwise vector field $X \in \Gamma(T\mathcal{F})$, then $\widehat{X} = (1/2)(X - iJ_{\mathcal{F}}X) \in \Gamma(T^{1,0}\mathcal{F})$ is leafwise holomorphic if and only if

$$(\mathcal{L}_X J_{\mathcal{F}})Y = [X, J_{\mathcal{F}}Y] - J_{\mathcal{F}}[X, Y] = 0\,, \forall Y \in \Gamma(T\mathcal{F}). \tag{2.35}$$

In that follows we denote the set of leafwise holomorphic vector fields on $(M, \mathcal{F}, J_{\mathcal{F}})$ by $\Gamma_{\text{hol}}(T^{1,0}\mathcal{F})$.

Definition 2.6. A leafwise connection D on $(M, \mathcal{F}, J_{\mathcal{F}})$ is called *leafwise holomorphic* if $D_{Z_1} Z_2 \in \Gamma_{\text{hol}}(T^{1,0}\mathcal{F})$ for arbitrary leafwise holomorphic vector fields Z_1, Z_2.

We have

Proposition 2.4. *The leafwise characteristic connection D of a complex Riemannian foliation $(M, \mathcal{F}, J_{\mathcal{F}}, G)$ is leafwise holomorphic if and only if the leafwise complex Christoffel symbols $L^c_{ab} = \Gamma^c_{ab}$ are leafwise holomorphic functions.*

As a direct consequence of (2.30), (2.22), (2.23), Corollary 2.1 and (2.31), we get

Theorem 2.2. *For every complex Riemannian foliation $(M, \mathcal{F}, J_{\mathcal{F}}, G)$, the following assertions are equivalent:*

(i) The fundamental leafwise tensor Φ (or Ψ) is zero;

(ii) The local components G_{ab} of the leafwise metric G are leafwise holomorphic functions;

(iii) The leafwise Levi-Civita connection ∇ of G is leafwise almost complex, that is $\nabla J_{\mathcal{F}} = 0$;

(iv) The leafwise characteristic connection D is metrical with respect to G, that is $DG = 0$;

(v) The leafwise Levi-Civita connection ∇ coincides with the leafwise characteristic connection D.

Let R be the leafwise characteristic curvature tensor of the leafwise characteristic connection D, defined as usual by

$$R(X,Y)Z = [D_X, D_Y]Z - D_{[X,Y]}Z, \text{ for every } X,Y,Z \in \Gamma(T_{\mathbb{C}}\mathcal{F}).$$

The local components of R are given by

$$R\left(\frac{\partial}{\partial z^A}, \frac{\partial}{\partial z^B}\right)\frac{\partial}{\partial z^C} = R^D_{C,AB}\frac{\partial}{\partial z^D}, \tag{2.36}$$

and the nonvanishing components of R are

$$R^d_{c,ab} = \frac{\partial L^d_{cb}}{\partial z^a} - \frac{\partial L^d_{ca}}{\partial z^b} + L^f_{cb}L^d_{fa} - L^f_{ca}L^d_{fb}\,,\; R^{\bar{d}}_{\bar{c},\bar{a}\,\bar{b}} = \overline{R^d_{c,ab}}, \tag{2.37}$$

$$R^d_{c,\bar{a}b} = \frac{\partial L^d_{bc}}{\partial z^{\bar{a}}}\,,\; R^{\bar{d}}_{\bar{c},a\bar{b}} = \overline{R^d_{c,\bar{a}b}}. \tag{2.38}$$

It is easy to see that $R^d_{c,\bar{a}b} = 0$ if and only if D is a leafwise holomorphic connection. Also, the leafwise characteristic Riemann curvature tensor of D is defined as usual by $\mathcal{R}(Z_1,Z_2,Z_3,Z_4) = G(R(Z_1,Z_2)Z_3,Z_4)$ and its local components are $R_{ABCD} = G_{DF}R^F_{C,AB}$. Its nonvanishing components are

$$R_{abcd} = G_{df}R^f_{c,ab} \text{ and } R_{\bar{a}bcd} = G_{df}R^f_{c,\bar{a}b}, \tag{2.39}$$

and their complex conjugates.

Moreover, every nondegenerate 2-plane in $T^{1,0}_{(z,x)}\mathcal{F}$ is called a *leafwise holomorphic* 2-plane, and the *leafwise holomorphic characteristic sectional curvature*, for a given leafwise holomorphic 2-plan $P = span\{Z_1, Z_2\}$, where $Z_1, Z_2 \in \Gamma(T^{1,0}_{(z,x)}\mathcal{F})$, $(z,x) \in M$, is defined by

$$K_{(z,x)}(P) = \frac{\mathcal{R}(Z_1,Z_2,Z_1,Z_2)}{G(Z_1,Z_1)G(Z_2,Z_2) - (G(Z_1,Z_2))^2}. \tag{2.40}$$

Then, the following Schur type theorem holds.

Theorem 2.3. *Let $(M,\mathcal{F},J_{\mathcal{F}})$ be a connected complex foliation with $m \geq 3$ endowed with a leafwise holomorphic metric G. If the leafwise holomorphic sectional curvatures does not depend on the 2-plane P, then it is a basic function $c(x)$.*

In the end of this section we describe the Einstein equations for complex Riemannian foliations. The associated leafwise characteristic Ricci tensor Ric is locally given by

$$\mathrm{Ric}\left(\frac{\partial}{\partial z^C}, \frac{\partial}{\partial z^A}\right) = \mathrm{Ric}_{CA} = R^B_{C,AB}, \tag{2.41}$$

and its nonvanishing components are

$$\mathrm{Ric}_{ca} = R^b_{c,ab}\,,\; \mathrm{Ric}_{c\bar{a}} = R^b_{c,\bar{a}b}\,,\; \mathrm{Ric}_{\bar{c}\bar{a}} = \overline{\mathrm{Ric}_{ca}}\,,\; \mathrm{Ric}_{\bar{c}a} = \overline{\mathrm{Ric}_{c\bar{a}}}. \tag{2.42}$$

Journal of Physics: Conference Series **670** (2016) 012025 doi:10.1088/1742-6596/670/1/012025

The function ρ defined by

$$\rho = G^{CA}\mathrm{Ric}_{CA} = G^{ca}\mathrm{Ric}_{ca} + G^{\bar{c}\bar{a}}\mathrm{Ric}_{\bar{c}\bar{a}} \tag{2.43}$$

is called the *leafwise scalar curvature* of D and it is a real valued function.

The equation

$$\mathrm{Ric} - \frac{\rho}{2}G = 8\pi c T \tag{2.44}$$

is called the *Einstein equation* of the complex Riemannian foliation $(M, \mathcal{F}, J_{\mathcal{F}}, G)$. In the equation (2.44), the left hand side is called the *leafwise Einstein curvature* which is constructed using the leafwise complex Riemannian metric G, while in the right hand side we have a leafwise tensor T called the *leafwise stress-energy-momentum tensor* and represents the matter and energy that generate the gravitational field of potentials (G_{AB}). The constant c is the gravitational constant. Locally, the leafwise Einstein equation is expressed as

$$\mathrm{Ric}_{AB} - \frac{\rho}{2}G_{AB} = 8\pi c T_{AB}. \tag{2.45}$$

Remark 2.4. i) If the Einstein equation holds, then taking into account (2.6) it follows that

$$\mathrm{Ric}_{a\bar{b}} = 8\pi c T_{a\bar{b}}. \tag{2.46}$$

ii) In the empty leave space (no matter, no energy) we have $T_{AB} = 0$, and contracting (2.45) with G^{AB} one gets $\rho = 0$ and so it reduced to

$$\mathrm{Ric}_{AB} = 0. \tag{2.47}$$

Consequently, $\mathrm{Ric}_{ab} = \mathrm{Ric}_{\bar{a}\bar{b}} = 0$.

iii) Letting $E_{AB} = \mathrm{Ric}_{AB} - (\rho/2)G_{AB}$ and $E^A_B = G^{AC}E_{CB}$, the leafwise divergence of E is defined by

$$\operatorname{div} E = E^A_{B|A}, \tag{2.48}$$

where "$|$" denotes the leafwise covariant derivative with respect to ∇ and we have $\operatorname{div} E = 0$. Indeed, the assertion follows using the second leafwise Bianchi identity $\sum_{cycl}(\nabla_X R)(Y, Z) = 0$ written in a local basis $\{\partial/\partial z^A\}$ of $\Gamma(T_{\mathbb{C}}\mathcal{F})$. Assuming the Einstein equation holds, by using $\operatorname{div} E = 0$, we must have

$$\operatorname{div} T = 0, \tag{2.49}$$

which is called the *leafwise continuity condition* for complex Riemannian foliation $(M, \mathcal{F}, J_{\mathcal{F}}, G)$.

Also, by analogy with the case of complex manifolds, see [11], the following result concerning the Einstein condition for complex Riemannian foliations holds.

Definition 2.7. The complex Riemannian foliation $(M, \mathcal{F}, J_{\mathcal{F}}, G)$ is said to be *leafwise characteristic Einstein* if $\mathrm{Ric}_{c\bar{a}} = 0$ and $\mathrm{Ric}_{ca} = fG_{ca}$, where $f = f_1 + if_2$ is a complex valued function on $(M, \mathcal{F})$.

Theorem 2.4. *Let $(M, \mathcal{F}, J_{\mathcal{F}}, G)$ be a leafwise characteristic Einstein complex Riemannian foliation with $m \geq 3$. Then the leafwise characteristic scalar curvature $\rho_0 = G^{ca}\mathrm{Ric}_{ca}$ is a leafwise anti-holomorphic function on $(M, \mathcal{F}, J_{\mathcal{F}})$ and $\mathrm{Ric}_{ca} = (\rho_0/m)G_{ca}$.*

Example 2.8. Let us consider the complex foliation $(\mathcal{W}, J_{\mathcal{W}})$ defined by the double vertical bundle W on the vertical tangent manifold $\mathcal{T}V$ as in the Example 2.6 endowed with leafwise anti-hermitian metric $g_{\mathcal{W}}$ from (2.16). If we consider $z^a = y^a + i\eta^a$, $a = 1, \ldots, m$ the holomorphic coordinates along the leaves of $\mathcal{W}$ and the leafwise complex vector fields

$$\frac{\partial}{\partial z^a} = \frac{1}{2}\left(\frac{\partial}{\partial y^a} - i\frac{\partial}{\partial \eta^a}\right), \ \frac{\partial}{\partial z^{\bar{a}}} = \frac{1}{2}\left(\frac{\partial}{\partial y^a} + i\frac{\partial}{\partial \eta^a}\right), \ a = 1, \ldots, m,$$

then the corresponding leafwise complex Riemannian metric $G_{\mathcal{W}}$ induced by $g_{\mathcal{W}}$ (using (2.13)) have the local components

$$G_{ab} = G_{\mathcal{W}}\left(\frac{\partial}{\partial z^a}, \frac{\partial}{\partial z^b}\right) = \frac{1}{2}g_{ab}\,,\ G_{\bar{a}\bar{b}} = G_{\mathcal{W}}\left(\frac{\partial}{\partial z^{\bar{a}}}, \frac{\partial}{\partial z^{\bar{b}}}\right) = \overline{G_{ab}} = \frac{1}{2}g_{ab} \text{ and } G_{a\bar{b}} = G_{\bar{a}b} = 0.$$

Moreover, the local coefficients of the corresponding leafwise characteristic connection are $L^c_{ab} = (1/8)C^c_{ab}$ and the local components of characteristic curvatures of $G_{\mathcal{W}}$ are given by

$$\begin{aligned} R^d_{c,ab}(G_{\mathcal{W}}) &= \frac{1}{16}\left(\frac{\partial C^d_{cb}}{\partial y^a} - \frac{\partial C^d_{ca}}{\partial y^b}\right) + \frac{1}{64}(C^f_{cb}C^d_{fa} - C^f_{ca}C^d_{fb}) \\ &= \frac{1}{64}R^d_{c,ab}(g_{\mathcal{V}}) + \frac{3}{64}\left(\frac{\partial C^d_{cb}}{\partial y^a} - \frac{\partial C^d_{ca}}{\partial y^b}\right) \end{aligned}$$

and

$$R^d_{c,\bar{a}b}(G_{\mathcal{W}}) = \frac{1}{16}\frac{\partial C^d_{cb}}{\partial y^a},$$

where $R^{\cdot}_{\cdot,\cdot\cdot}(g_{\mathcal{V}})$ denotes the local components of the curvature of vertical metric $g_{\mathcal{V}}$ from (2.15).

The corresponding leafwise Ricci tensors are

$$\mathrm{Ric}_{ca}(G_{\mathcal{W}}) = \frac{1}{64}\mathrm{Ric}_{ca}(g_{\mathcal{V}}) + \frac{3}{64}\left(\frac{\partial C^b_{cb}}{\partial y^a} - \frac{\partial C^b_{ca}}{\partial y^b}\right) \text{ and } \mathrm{Ric}_{c\bar{a}}(G_{\mathcal{W}}) = \frac{1}{16}\frac{\partial C^b_{cb}}{\partial y^a}.$$

Hence, it is easy to see that if $\partial C^b_{cb}/\partial y^a = \partial C^b_{ca}/\partial y^b = 0$ and the vertical metric $g_{\mathcal{V}}$ is Einstein (in the next section, the definition of an Einstein metric is recalled), then $(\mathcal{T}V, \mathcal{W}, J_{\mathcal{W}}, G_{\mathcal{W}})$ is leafwise characteristic Einstein.

3. Leafwise holomorphic Riemannain Einstein metrics

We recall that a (real) metric g on the (real) manifold M is said to be Einsteinian if $\mathrm{Ric}(g) = \lambda g$, where λ is a real constant and $\mathrm{Ric}(g)$ denotes the Ricci tensor of the metric g. By analogy, a (real or complex) leafwise metric G on $(M, \mathcal{F})$ is called *leafwise Einstein metric* if

$$\mathrm{Ric}(G) = \lambda G, \tag{3.1}$$

where λ is a (real or complex) constant and $\mathrm{Ric}(G)$ denotes the Ricci tensor of the leafwise metric G.

The aim of this section, is to point out that by taking the real part of a leafwise holomorphic Einstein metric on a smooth $(2m+n)$-dimensional manifold M endowed with a complex foliation $(\mathcal{F}, J_{\mathcal{F}})$ one gets a real leafwise Einstein metric on the real foliated manifold $(M, \mathcal{F})$ obtaining a result similar to Theorem 5.1 from [2] from the anti-Kählerian manifolds case.

Let $(M, \mathcal{F}, J_{\mathcal{F}}, G)$ be a complex foliation $(\mathcal{F}, J_{\mathcal{F}})$ on M endowed with a leafwise holomorphic Riemannian metric G. Then, as we already noticed in the previous section the relations (2.10) and (2.13) establishes an one-to-one correspondence between the leafwise anti-Kählerian metrics on the (real) foliated manifold $(M, \mathcal{F}, J_{\mathcal{F}})$ and the leafwise holomorphic Riemannian metrics on the complex foliation $(\mathcal{F}, J_{\mathcal{F}})$ on M.

Although we can follow an argument similar from [2, 3], for a better presentation of the notions that we use, in this section we denote the leafwise holomorphic Riemannain metric G by $\widehat{g}$ and we follow an argument similar to [19, 21] for Kähler-Norden manifolds.

Without loss of generality, we consider the real leafwise vector fields $X, Y, \ldots \in \Gamma(T\mathcal{F})$ such that $\widehat{X}, \widehat{Y}, \ldots \in \Gamma_{\rm hol}(T^{1,0}\mathcal{F})$, are leafwise holomorphic vector fields on the complex foliation $(M, \mathcal{F}, J_{\mathcal{F}})$, that is the relation (2.35) holds. Then, we easily obtain

$$[J_{\mathcal{F}}X, Y] = [X, J_{\mathcal{F}}Y] = J_{\mathcal{F}}[X, Y]\,,\; [J_{\mathcal{F}}X, J_{\mathcal{F}}Y] = [X, Y]\,,\; [\widehat{X}, \widehat{Y}] = \widehat{[X, Y]} =: [X, Y]\widehat{\ }. \quad (3.2)$$

Also, by a direct calculation, we have that for every leafwise complex function $f = Re\, f + iIm\, f$ on $(M, \mathcal{F}, J_{\mathcal{F}})$, and every real leafwise vector field $X \in \Gamma(T\mathcal{F})$, the following relation holds

$$f\widehat{X} = ((Re\, f)X + (Im\, f)X)\widehat{\ }, \quad (3.3)$$

and, moreover, if f is leafwise holomorphic, then the leafwise Cauchy-Riemann equations imply

$$X(Re\, f) = (J_{\mathcal{F}}X)(Im\, f)\,,\; (J_{\mathcal{F}}X)(Re\, f) = -X(Im\, f)\,,\; \widehat{X}f = X(Re\, f) + iX(Im\, f). \quad (3.4)$$

Now, for every real tangent space along the leaves $T_{(z,x),\mathbb{R}}\mathcal{F}$, $(z, x) \in M$, we can choose an adapted orthonormal (real) frame $\{e_a, J_{\mathcal{F}}e_a\}$, $a \in \{1, \ldots, m\}$ in $\Gamma(T\mathcal{F})$, such that

$$g(e_a, e_b) = \delta_{ab}\,,\; g(J_{\mathcal{F}}e_a, J_{\mathcal{F}}e_b) = -\delta_{ab}\,,\; g(e_a, J_{\mathcal{F}}e_b) = 0\,,\; a, b \in \{1, \ldots, m\}. \quad (3.5)$$

Then, we obtain an adapted leafwise complex frame $\{\widehat{e_a}\}$, $a \in \{1, \ldots, m\}$, for $\Gamma(T^{1,0}_{(z,x)}\mathcal{F})$, where $\widehat{e_a} = (1/2)(e_a - iJ_{\mathcal{F}}e_a)$ for which $\widehat{g}(\widehat{e_a}, \widehat{e_b}) = (1/2)\delta_{ab}$.

Let ∇ and $\widehat{\nabla}$ be the leafwise Levi-Civita connections of the leafwise anti-Kählerian metric g and of the leafwise holomorphic Riemannian metric $\widehat{g}$, respectively. We also consider the leafwise Ricci tensor fields associated to the leafwise metrics g and $\widehat{g}$, respectively, given by

$${\rm Ric}(g)(X, Y) = {\rm Tr}\{Z \mapsto R(Z, X)Y\} \text{ and } {\rm Ric}(\widehat{g})(\widehat{X}, \widehat{Y}) = {\rm Tr}\{\widehat{Z} \mapsto \widehat{R}(\widehat{Z}, \widehat{X})\widehat{Y}\}, \quad (3.6)$$

and let us denote by Q and $\widehat{Q}$ be the associated leafwise Ricci operators, given by

$$g(QX, Y) = {\rm Ric}(g)(X, Y) \text{ and } \widehat{g}(\widehat{Q}\widehat{X}, \widehat{Y}) = {\rm Ric}(\widehat{g})(\widehat{X}, \widehat{Y}). \quad (3.7)$$

Following step by step the construction from the case of Kähler-Norden manifolds (see [21]), we have the following result which relates the leafwise Ricci tensors ${\rm Ric}(g)$ and ${\rm Ric}(\widehat{g})$.

Proposition 3.1. *The leafwise Ricci tensors* ${\rm Ric}(g)$, ${\rm Ric}(\widehat{g})$ *and the leafwise Ricci operators* Q, $\widehat{Q}$ *satisfy the following relations*

$${\rm Ric}(g)(J_{\mathcal{F}}X, Y) = {\rm Ric}(g)(X, J_{\mathcal{F}}Y)\,,\; {\rm Ric}(g)(J_{\mathcal{F}}X, J_{\mathcal{F}}Y) = -{\rm Ric}(g)(X, Y)\,,\; QJ_{\mathcal{F}} = J_{\mathcal{F}}Q \quad (3.8)$$

and

$${\rm Ric}(\widehat{g})(\widehat{X}, \widehat{Y}) = \frac{1}{2}({\rm Ric}(g)(X, Y) - i{\rm Ric}(g)(X, J_{\mathcal{F}}Y))\,,\; \widehat{Q}\widehat{X} = \widehat{QX}. \quad (3.9)$$

The first relation of (3.9) leads to the announced result, that is

Theorem 3.1. *Let us suppose that g is a leafwise anti-Kählerian metric on $(M, \mathcal{F}, J_{\mathcal{F}})$, that is $(M, \mathcal{F}, J_{\mathcal{F}})$ is endowed with a leafwise holomorphic Riemannian metric $\widehat{g} \equiv (\widehat{g}_{ab}(z, x))$, $a, b \in \{1, \ldots, m\}$ and with a real leafwise metric $g \equiv (g_{\alpha\beta}(u, x))$, $\alpha, \beta \in \{1, \ldots, 2m\}$ given by $g = 2Re\,\widehat{g}$. Then, the leafwise holomorphic metric $\widehat{g}$ is Einstein with the real constant λ if and only if the real leafwise metric g is Einstein with the same constant.*

Journal of Physics: Conference Series **670** (2016) 012025 doi:10.1088/1742-6596/670/1/012025

Remark 3.1. We notice that starting from the original leafwise anti-Kählerian metric g on $(M, \mathcal{F}, J_{\mathcal{F}})$, the real leafwise twin metric $\widetilde{g} = h$ can be considered, that is $h(X,Y) := (g \circ J_{\mathcal{F}})(X,Y) = g(J_{\mathcal{F}}X, Y)$, for every $X, Y \in \Gamma(T\mathcal{F})$. We find

$$h(X,Y) = 2\mathrm{Im}\widehat{g}(\widehat{X},\widehat{Y}), \ \forall X, Y \in \Gamma(T\mathcal{F}). \tag{3.10}$$

Moreover, if we denote by ∇ the covariant differentiation of the Levi-Civita connection associated to the leafwise anti-Kählerian metric g, then we have

$$\nabla h = \nabla g \circ J_{\mathcal{F}} + g \circ \nabla J_{\mathcal{F}} = 0. \tag{3.11}$$

The above relation says that, the leafwise Levi-Civita connection of g coincides with the leafwise Levi-Civita connection of h, thus they have the same real and complex leafwise Riemann and Ricci tensors (see also the discussion from the previous section). In the real case only one of two leafwise twin metrics can be Einsteinian. In the complex case the Einstein condition $\mathrm{Ric}(\widehat{g}) = \lambda\widehat{g}$ implies $\mathrm{Ric}(\widehat{h}) = i\lambda\widehat{h}$, that is, both leafwise holomorphic metrics $\widehat{g}$ and $\widehat{h}$ are Einstein metrics at the same time. We can conclude that the leafwise metric h is an Einstein metric with an imaginary cosmological constant.

Finally, we notice that if the leafwise holomorphic metric $\widehat{g}$ is Einstein with complex constant $\widehat{\lambda}$, that is

$$\mathrm{Ric}(\widehat{g}) = \widehat{\lambda}\widehat{g}, \ \widehat{\lambda} \in \mathbb{C}, \tag{3.12}$$

then, similarly to the Kähler-Norden manifolds case (see [19]), we can describe the following generalization of Theorem 3.1.

Theorem 3.2. *The leafwise holomorphic Riemannian metric $\widehat{g}$ on $(M, \mathcal{F}, J_{\mathcal{F}})$ is leafwise holomorphic Einstein with complex constant $\widehat{\lambda} = \lambda_1 + i\lambda_2$ if and only if*

$$\mathrm{Ric}(g)(X,Y) = \lambda_1 g(X,Y) + \lambda_2 g(X, J_{\mathcal{F}}Y). \tag{3.13}$$

Moreover, in the formula (3.13), *we have $\lambda_1 = K/2m$ and $\lambda_2 = -K^*/2m$, where $K = \mathrm{Tr}Q$ and $K^* = \mathrm{Tr}(J_{\mathcal{F}}Q)$.*

Acknowledgement
This paper is supported by the Sectorial Operational Program Human Resources Development (SOP HRD), financed from the European Social Fund and by the Romanian Government under the Project number POSDRU/159/1.5/S/134378.

References

[1] Anastasiei M and Gîrţu M 2014 Einstein equations in Lie algebroids *Scientific Studies and Research, Series Mathematics and Informatics* **24(1)** 5.
[2] Borowiec A, Francaviglia M and Volovich I 2000 Anti-Kählerian manifolds *Diff Geom. Appl.* **12** (2000) 281.
[3] Borowiec, Ferraris M, Francaviglia M and Volovich I 1999 Almost complex and almost product Einstein manifolds from a variational principle *J. Math. Phys.* **40** 3446.
[4] Deninger C and Singhof W 2002 Real polarizable Hodge structures arising from foliations *Ann. Glob. Anal. Geom.* **21** 377.
[5] Dragomir S and Francaviglia M 2000 On Norden Metrics which Are Locally Conformal to Anti-Kählerian Metrics *Acta Appl. Math.* **60** 115.
[6] Ganchev G and Borisov A 1986 Note on the almost complex manifolds with Norden metric *Compt. rend. l'Acad. Bulg. Sci.* **39(5)** 31.
[7] Ganchev G and Ivanov S 1991 *Connections and curvatures on complex Riemannian manifolds.* (Internal Report, International Centre for Theoretical Physics, Miramare-Trieste, 41).
[8] Ganchev G and Ivanov S 1992 Characteristic curvatures on complex Riemannian manifolds *Rivista di Mat. della Univ. di Parma* **1(5)** 155.

Journal of Physics: Conference Series **670** (2016) 012025 doi:10.1088/1742-6596/670/1/012025

[9] Ghys È, *Laminations par surfaces de Riemann* 1999, Dynamique et Géométrie Complexes (Lyon, 1997) Panormas et Symthèses **8** Soc. Math. France Paris 49.
[10] Iscan M and Salimov A 2009 On Kähler-Norden manifolds *Proc. Indian Acad. Sci. (Math. Sci)* **119(1)** 71.
[11] Ivanov S 1994 Holomorphically projective transformations on complex Riemannian manifold *J. Geom.* **49** 106.
[12] Ivanov S, Manev H and Manev M 2014 Sasaki-like almost contact complex Riemannian manifolds (*Preprint* dg/1402.5426v1)
[13] El Kacimi Alaoui A 2010 The $\overline{\partial}$ operator along the leaves and Guichard's theorem for a complex simple foliation *Math. Ann.* **347(4)** 885.
[14] El Kacimi Alaoui A and Slimène J 2010 Cohomologie de Dolbeault le long des feuilles de certains feuilletages complexes *Ann. Inst. Fourier* **60(1)** 727.
[15] Kobayashi S and Nomizu K 1969 *Foundations of Differential Geometry II* (New York: Interscience).
[16] LeBrun C 1983 Spaces of complex null geodesics in complex-Riemannian geometry *Trans. AMS* **278(1)** 209.
[17] Moore C C and Schochet C 1988 *Global Analysis on Foliated Spaces* MSRI Publications, **9** (New York: Springer-Vedag).
[18] Olszak K 2005 On the Bochner conformal curvature of Kähler-Norden manifolds *Cent. Eur. J. of Math.* **3(2)** 309.
[19] Olszak K and Olszak Z 2006 Generalized Einstein conditions on holomorphic Riemannain manifolds *Acta Math. Hungar.* **113(4)** 345.
[20] Penrose R 1976 The nonlinear gravitons and curved twistor theory *Gen. Rel. Grav.* **7** 31.
[21] Sluka K 2005 On the curvature of Kähler-Norden manifolds *J. Geom. Phys.* **54** 131.
[22] Vaisman I 2014 Hessian geometry on Lagrange spaces *Int. J. of Math. and Math. Sci.*, Art. ID 793473, 9 pp. (Hindawi Publishing Corporation).

Journal of Physics: Conference Series **670** (2016) 012026 doi:10.1088/1742-6596/670/1/012026

SU(2|1) Supersymmetric Mechanics

E Ivanov and S Sidorov

Bogoliubov Laboratory of Theoretical Physics, JINR, 141980, Dubna, Moscow Region, Russia

E-mail: eivanov@theor.jinr.ru, sidorovstepan88@gmail.com

Abstract. We review the salient features of the new kind of $\mathcal{N}=4$ supersymmetric mechanics associated with the supergroup $SU(2|1)$, proceeding from the worldline $SU(2|1)$ superspace formalism.

1. Motivations

Supersymmetric Quantum Mechanics (SQM) [1] is the simplest ($d=1$) supersymmetric theory. It probes the structure of higher-dimensional supersymmetric theories via the dimensional reduction and, in addition, displays a number of distinguishing features on its own. For instance, it provides superextensions of integrable models like Calogero-Moser systems and Landau-type models. An extended $d=1$ supersymmetry exhibits interesting specific properties, such as dualities between various supermultiplets [2], the existence of nonlinear "cousins" of off-shell linear multiplets [3], etc.

Symmetry algebra of the standard "flat" $\mathcal{N}$ extended SQM is

$$\{Q^A, Q^B\} = 2\delta^{AB} H\,, \quad [H, Q^A] = 0\,, \quad A, B = 1\ldots\mathcal{N}\,. \tag{1.1}$$

Recently, a considerable attention was paid to rigid supersymmetric theories based on curved analogs of the Poincaré supergroup in diverse dimensions (see, e.g., [4]). With this in mind, it is of obvious interest to consider "curved" analogs of SQM.

A way to define such generalized SQM models is suggested by $\mathcal{N}=2, d=1$ "Poincaré" superalgebra in the complex notation

$$Q = \frac{1}{\sqrt{2}}(Q^1 + iQ^2)\,,\; \bar{Q} = \frac{1}{\sqrt{2}}(Q^1 - iQ^2)\,,$$
$$\{Q, \bar{Q}\} = 2H\,,\; Q^2 = \bar{Q}^2 = 0,\; [H, Q] = [H, \bar{Q}] = 0\,. \tag{1.2}$$

The relations (1.2) form the superalgebra $su(1|1)$, with H as the central charge generator.

This two-fold interpretation of $\mathcal{N}=2, d=1$ "Poincaré" superalgebra suggests two ways of extending it to higher-rank $d=1$ supersymmetries.

A. The standard extension:

$$(\mathcal{N}=2\,,\; d=1) \quad \Rightarrow \quad (\mathcal{N}>2\,,\; d=1 \text{ "Poincaré"})\,,$$

B. A non-standard extension:

$$(\mathcal{N}=2\,,\; d=1) \equiv u(1|1) \;\subset\; su(2|1) \;\subset\; su(2|2) \;\subset\; \ldots\,.$$

In the links of the chain **B**, the closure of supercharges contains, besides H, also internal symmetry generators. They commute with H, but not with the supercharges.

The subject of the present talk is the simplest version of such SQM based on the supergroup $SU(2|1)$ [5]-[8].

The universal way to construct supersymmetric theories is the *superspace approach*. So our primary aim was to construct the worldline superfield realizations of $SU(2|1)$ and to show that all off-shell multiplets of $\mathcal{N}=4, d=1$ supersymmetry have the well-defined $SU(2|1)$ analogs [5]. It turned out, e.g., that the "weak supersymmetry" models [9] are associated with the $SU(2|1)$ multiplet $(\mathbf{1},\mathbf{4},\mathbf{3})$.

The supergroup $SU(2|1)$ has also invariant chiral subspaces which are carriers of the chiral multiplets $(\mathbf{2},\mathbf{4},\mathbf{2})$ [5, 6]. The relevant component actions naturally provide the bosonic $d=1$ Wess-Zumino-type terms. Besides this, $SU(2|1)$ also admits a supercoset [8] which is an analog of the harmonic analytic superspace of the $\mathcal{N}=4, d=1$ supersymmetry [10]. There, $SU(2|1)$ counterparts of the multiplets $(\mathbf{4},\mathbf{4},\mathbf{0})$ can be defined.

In [7] we demonstrated that the superconformal $D(2,1;\alpha)$ invariant models in the $SU(2|1)$ superspace formulation naturally yield *trigonometric* realization of the $d=1$ conformal group $SO(2,1)$ [11, 12].

2. $SU(2|1)$ superspace

The (central-extended) superalgebra $su(2|1)$ is written as:

$$\{Q^i,\bar{Q}_j\} = 2m\left(I^i_j-\delta^i_jF\right)+2\delta^i_jH\,, \qquad \left[I^i_j,I^k_l\right]=\delta^k_jI^i_l-\delta^i_lI^k_j\,,$$
$$\left[I^i_j,\bar{Q}_l\right]=\frac{1}{2}\delta^i_j\bar{Q}_l-\delta^i_l\bar{Q}_j\,, \qquad \left[I^i_j,Q^k\right]=\delta^k_jQ^i-\frac{1}{2}\delta^i_jQ^k,$$
$$\left[F,\bar{Q}_l\right]=-\frac{1}{2}\bar{Q}_l\,, \qquad \left[F,Q^k\right]=\frac{1}{2}Q^k.$$

The basic $SU(2|1)$ worldline superspace is identified with the supercoset:

$$\frac{SU(2|1)}{SU(2)\times U(1)}\sim\frac{\{Q^i,\bar{Q}_j,H,I^i_j,F\}}{\{I^i_j,F\}}\,.$$

The superspace coordinates $\{t,\theta_i,\bar{\theta}^j\}$ are the parameters associated with the coset generators. The supercoset element can be conveniently parametrized as

$$g=\exp\left(itH+i\tilde{\theta}_iQ^i-i\bar{\tilde{\theta}}^j\bar{Q}_j\right),\quad \tilde{\theta}_i=\left[1-\frac{2m}{3}\left(\bar{\theta}\cdot\theta\right)\right]\theta_i\,.$$

The $SU(2|1)$ superspace coordinates are transformed under the $Q,\bar{Q}$ left shifts as

$$\delta\theta_i=\epsilon_i+2m\left(\bar{\epsilon}\cdot\theta\right)\theta_i\,,\qquad \delta\bar{\theta}^j=\bar{\epsilon}^i-2m\left(\epsilon\cdot\bar{\theta}\right)\bar{\theta}^i,$$
$$\delta t=i\left[\left(\epsilon\cdot\bar{\theta}\right)+\left(\bar{\epsilon}\cdot\theta\right)\right].$$

The $SU(2|1)$ generators in the superspace realization read

$$Q^i=-i\frac{\partial}{\partial\theta_i}+2im\bar{\theta}^i\bar{\theta}^j\frac{\partial}{\partial\bar{\theta}^j}+\bar{\theta}^i\frac{\partial}{\partial t},\quad \bar{Q}_j=i\frac{\partial}{\partial\bar{\theta}^j}+2im\theta_j\theta_k\frac{\partial}{\partial\theta_k}-\theta_j\frac{\partial}{\partial t},$$
$$I^i_j=\left(\bar{\theta}^i\frac{\partial}{\partial\bar{\theta}^j}-\theta_j\frac{\partial}{\partial\theta_i}\right)-\frac{\delta^i_j}{2}\left(\bar{\theta}^k\frac{\partial}{\partial\bar{\theta}^k}-\theta_k\frac{\partial}{\partial\theta_k}\right),$$
$$F=\frac{1}{2}\left(\bar{\theta}^k\frac{\partial}{\partial\bar{\theta}^k}-\theta_k\frac{\partial}{\partial\theta_k}\right),\quad H=i\partial_t.$$

The corresponding covariant derivatives are defined by

$$\mathcal{D}^i = \left[1 + m\left(\bar\theta\cdot\theta\right) - \frac{3m^2}{4}\left(\bar\theta\cdot\theta\right)^2\right]\frac{\partial}{\partial\theta_i} - m\bar\theta^i\theta_j\frac{\partial}{\partial\theta_j} - i\bar\theta^i\frac{\partial}{\partial t} + \ldots,$$
$$\bar{\mathcal{D}}_j = -\left[1 + m\left(\bar\theta\cdot\theta\right) - \frac{3m^2}{4}\left(\bar\theta\cdot\theta\right)^2\right]\frac{\partial}{\partial\bar\theta^j} + m\bar\theta^k\theta_j\frac{\partial}{\partial\bar\theta^k} + i\theta_j\frac{\partial}{\partial t} + \ldots.$$

Here, "dots" stand for matrix $U(2)$ connection parts.The invariant integration measure is defined by the expression $d\zeta = dtd^2\theta d^2\bar\theta(1 + 2m\,\bar\theta\cdot\theta)\,,\quad \delta d\zeta = 0\,.$

3. $(\mathbf{1},\mathbf{4},\mathbf{3})$ multiplet

The $(\mathbf{1},\mathbf{4},\mathbf{3})$ multiplet is described by the real neutral superfield $G(t,\theta,\bar\theta)$ satisfying

$$\varepsilon^{lj}\bar{\mathcal{D}}_l\,\bar{\mathcal{D}}_j G = \varepsilon_{lj}\mathcal{D}^l\,\mathcal{D}^j G = 0 \quad \Rightarrow$$
$$G = x - mx\left(\bar\theta\cdot\theta\right)\left[1 - 2m\left(\bar\theta\cdot\theta\right)\right] + \frac{\ddot x}{2}\left(\bar\theta\cdot\theta\right)^2 - i\left(\bar\theta\cdot\theta\right)\left(\theta_i\,\dot\psi^i + \bar\theta^j\,\dot{\bar\psi}_j\right)$$
$$+ \left[1 - 2m\left(\bar\theta\cdot\theta\right)\right]\left(\theta_i\,\psi^i - \bar\theta^j\,\bar\psi_j\right) + \bar\theta^j\theta_i\,B^i_j\,,\quad B^k_k = 0\,.$$

The irreducible set of off-shell fields is $x(t), \psi^i(t), \bar\psi_i(t), B^i_j(t) (B^k_k = 0)$.

The ϵ transformation properties of the component fields are as follows

$$\delta x = \left(\bar\epsilon\cdot\bar\psi\right) - \left(\epsilon\cdot\psi\right),\qquad \delta\psi^i = i\bar\epsilon^i\dot x - m\bar\epsilon^i x + \bar\epsilon^k B^i_k\,,$$
$$\delta B_{(ij)} = -2i\left[\epsilon_{(i}\dot\psi_{j)} + \bar\epsilon_{(i}\dot{\bar\psi}_{j)}\right] + 2m\left[\bar\epsilon_{(i}\bar\psi_{j)} - \epsilon_{(i}\psi_{j)}\right].$$

The general off-shell Lagrangian is written as

$$\mathcal{L}_{\rm off} = -\int d^2\theta\, d^2\bar\theta\left(1 + 2m\,\bar\theta\cdot\theta\right)\, f(G)\,,\quad S = \int dt\mathcal{L}_{\rm off}\,. \tag{3.1}$$

After doing θ integral and eliminating the auxiliary field, we obtain $\mathcal{L}_{\rm on}$ as

$$\mathcal{L}_{\rm on} \;=\; \dot x^2 g(x) + i\left(\bar\psi_i\dot\psi^i - \dot{\bar\psi}_i\psi^i\right)g(x) - \frac{1}{2}\left(\bar\psi_i\psi^i\right)^2\left[g''(x) - \frac{3\left(g'(x)\right)^2}{2g(x)}\right]$$
$$-m^2x^2g(x) + 2m\bar\psi_i\psi^i g(x) + mx\bar\psi_i\psi^i g'(x)\,,\quad g(x) := f''(x)\,.$$

It can be simplified by passing to the new variables $y(x), \zeta^i$,

$$\dot x^2 g(x) = \frac{1}{2}\dot y^2\,,\quad \Rightarrow\; y'(x) = \sqrt{2g(x)}\,,\; \zeta^i = \psi^i y'(x)\,,$$
$$\mathcal{L}_{\rm on} = \frac{\dot y^2}{2} + \frac{i}{2}\left(\bar\zeta_i\dot\zeta^i - \dot{\bar\zeta}_i\zeta^i\right) - \frac{m^2}{2}V^2(y) + m\bar\zeta_i\zeta^i V'(y)$$
$$-\frac{1}{2}\left(\bar\zeta_i\zeta^i\right)^2\partial_y\left(\frac{V'(y) - 1}{V(y)}\right),\; V(y) := \frac{x(y)}{x'(y)}\,.$$

This Lagrangian defines the SQM model with "weak" $\mathcal{N} = 4$ supersymmetry [9].

3.1 Quantization

The simplest choice in (3.1) is $f(x) = \frac{x^2}{4}$:

$$\mathcal{L}^{(\rm free)}_{\rm on} = \frac{\dot x^2}{2} - \frac{m^2x^2}{2} + \frac{i}{2}\left(\bar\psi_i\dot\psi^i - \dot{\bar\psi}_i\psi^i\right) + m\bar\psi_i\psi^i,$$
$$\delta x = \left(\bar\epsilon\cdot\bar\psi\right) - \left(\epsilon\cdot\psi\right),\qquad \delta\psi^i = i\bar\epsilon^i\dot x - m\bar\epsilon^i x\,. \tag{3.2}$$

The corresponding conserved Noether charges and Hamiltonian are computed to be

$$Q^i = \psi^i\,(p - imx)\,, \qquad \bar{Q}_i = \bar{\psi}_i\,(p + imx)\,,$$
$$H = \frac{p^2}{2} + \frac{m^2x^2}{2} + m\psi^i\bar{\psi}_i\,, \quad F = \frac{1}{2}\psi^k\bar{\psi}_k, \quad I^i_j = \psi^i\bar{\psi}_j - \frac{1}{2}\delta^i_j\psi^k\bar{\psi}_k.$$

We quantize as

$$[\hat{x}, \hat{p}] = i\,, \quad \{\hat{\psi}^i, \hat{\bar{\psi}}_j\} = \delta^i_j\,, \quad \hat{p} = -i\partial_x\,, \ \hat{\bar{\psi}}_j = \partial/\partial\hat{\psi}^j\,.$$

The quantum Hamiltonian and (super)charges which form the superalgebra $su(2|1)$ are

$$\hat{H} = \frac{1}{2}\,(\hat{p} + im\hat{x})\,(\hat{p} - im\hat{x}) + m\hat{\psi}^i\hat{\bar{\psi}}_i, \quad \hat{F} = \frac{1}{2}\hat{\psi}^k\hat{\bar{\psi}}_k, \quad \hat{I}^i_j = \hat{\psi}^i\hat{\bar{\psi}}_j - \frac{1}{2}\delta^i_j\hat{\psi}^k\hat{\bar{\psi}}_k\,,$$
$$\hat{Q}^i = \hat{\psi}^i\,(\hat{p} - im\hat{x})\,, \qquad \hat{\bar{Q}}_i = \hat{\bar{\psi}}_i\,(\hat{p} + im\hat{x})\,.$$

3.2 Spectrum

The super wave-function $\Omega^{(\ell)}$ at the energy level $\ell \geq 2$, reveals the four-fold degeneracy

$$\Omega^{(\ell)} = a^{(\ell)}|\ell\rangle + b^{(\ell)}_i\,\psi^i|\ell - 1\rangle + \frac{1}{2}\,c^{(\ell)}\,\varepsilon_{ij}\psi^i\psi^j|\ell - 2\rangle, \qquad \ell \geq 2\,,$$

where $|\ell\rangle, |\ell - 1\rangle, |\ell - 2\rangle$ are the harmonic oscillator functions. We treat the operators $\hat{p} \pm imx$ in $\hat{H}$ as the creation and annihilation operators and impose the standard conditions

$$\hat{\bar{\psi}}_k|\ell\rangle = 0, \qquad (\hat{p} - im\hat{x})\,|0\rangle = 0, \qquad (\hat{p} + im\hat{x})\,|\ell\rangle = |\ell + 1\rangle.$$

The spectrum of the Hamiltonian is then

$$\hat{H}\,\Omega^{(\ell)} = m\,\ell\,\Omega^{(\ell)}, \qquad m > 0.$$

The ground state ($\ell = 0$) and the first excited states ($\ell = 1$) are special, they encompass non-equal numbers of bosonic and fermionic states:

$$\Omega^{(0)} = a^{(0)}\,|0\rangle\,, \qquad \Omega^{(1)} = a^{(1)}\,|1\rangle + b^{(1)}_i\,\psi^i|0\rangle\,.$$

The ground state is annihilated by all $SU(2|1)$ generators including Q^i and $\bar{Q}_i$, so it is $SU(2|1)$ singlet. The states with $\ell = 1$ form the fundamental $(\mathbf{2}|\mathbf{1})$ representation of $SU(2|1)$. The supercharges act on them as

$$Q^i\,\psi^k|0\rangle = 0\,, \qquad \bar{Q}_i\,\psi^k|0\rangle = \delta^k_i\,|1\rangle\,,$$
$$Q^i\,|1\rangle = 2m\,\psi^i|0\rangle\,, \qquad \bar{Q}_i\,|1\rangle = 0\,.$$

The states with $\ell > 1$ form the representations $(\mathbf{2}|\mathbf{2})$, with equal numbers of bosonic and fermionic states.

For the considered model, the $SU(2|1)$ Casimirs are nicely expressed as

$$m^2C_2 = \hat{H}\left(\hat{H} - m\right), \quad m^3C_3 = \hat{H}\left(\hat{H} - m\right)\left(\hat{H} - \frac{m}{2}\right) \quad \Rightarrow$$
$$C_2(\ell) = (\ell - 1)\,\ell, \quad C_3(\ell) = (\ell - 1/2)\,(\ell - 1)\,\ell.$$

The ground state with $\ell = 0$ is atypical [13], because Casimir operators are zero on it. On the states with $\ell = 1$ both Casimirs as well vanish, so these states also form an atypical (fundamental) $SU(2|1)$ representation. On the $\ell > 1$ states both Casimirs are non-zero, so these states form typical [13] $SU(2|1)$ representations.

4. Chiral multiplet

The $SU(2|1)$ counterpart of the $\mathcal{N}=4, d=1$ chiral multiplet $(\mathbf{2},\mathbf{4},\mathbf{2})$ can also be defined. This is due to the existence of the chiral coset

$$\frac{\{Q^i,\bar{Q}_j,H,I^i_k,F\}}{\{\bar{Q}_j,I^i_k,F\}}\sim(t_L,\theta_i)\,,\qquad t_L=t+\frac{i}{2m}\ln(1+2m\,\bar{\theta}\cdot\theta),$$
$$\delta\theta_i=\epsilon_i+2m\,(\bar{\epsilon}\cdot\theta)\theta_i\,,\qquad \delta t_L=2i\,(\bar{\epsilon}\cdot\theta)\,.$$

The multiplet $(\mathbf{2},\mathbf{4},\mathbf{2})$ is described by the chiral superfield Φ

$$\bar{\mathcal{D}}_j\Phi=0,\qquad \hat{I}^i_j\Phi=0,\qquad \hat{F}\Phi=2\kappa\Phi\,, \tag{4.1}$$

where, in general, $\kappa\neq0$. The solution of this constraint is

$$\Phi=\left[1+m\left(\bar{\theta}\cdot\theta\right)\right]^{-\kappa}\varphi_L(t_L,\theta),\quad \varphi_L(t_L,\theta)=z+\sqrt{2}\,\theta_i\xi^i+\varepsilon^{ij}\theta_i\theta_jF.$$

The superfield φ_L and its components transform as

$$\delta^*\varphi_L=4\kappa m\,(\bar{\epsilon}^j\theta_j)\,\varphi_L\;\Rightarrow$$
$$\delta z=-\sqrt{2}\,\epsilon_i\xi^i,\quad \delta\xi^i=\sqrt{2}\,i\bar{\epsilon}^i\nabla_t z-\sqrt{2}\,\varepsilon^{ik}\epsilon_kF,$$
$$\delta F=-\sqrt{2}\,\varepsilon_{ik}\bar{\epsilon}^k\left[m\xi^i+i\nabla_t\xi^i\right],\quad \nabla_t:=\partial_t+2i\kappa m\,.$$

The general superfield Lagrangian reads

$$\mathcal{L}^{(\kappa)}_{\text{off}}=\frac{1}{4}\int d^2\theta\,d^2\bar{\theta}(1+2m\,\bar{\theta}\cdot\theta)\,f(\Phi,\Phi^\dagger).$$

It results in the following component on-shell Lagrangian, with $g:=f_{z\bar{z}}$,

$$\begin{aligned}\mathcal{L}^{(\kappa)}_{\text{on}}&=g\dot{\bar{z}}\dot{z}+2i\kappa m\,(\dot{\bar{z}}z-\dot{z}\bar{z})\,g-\frac{im}{2}\,(\dot{\bar{z}}f_{\bar{z}}-\dot{z}f_z)-\frac{i}{2}\,(\bar{\xi}\cdot\xi)\,(\dot{\bar{z}}g_{\bar{z}}-\dot{z}g_z)\\&+\frac{i}{2}\left(\bar{\xi}_i\dot{\xi}^i-\dot{\bar{\xi}}_i\xi^i\right)g-m^2V-m\,(\bar{\xi}\cdot\xi)\,U+\frac{1}{2}\,(\bar{\xi}\cdot\xi)^2\,R,\quad R=g_{z\bar{z}}-\frac{g_zg_{\bar{z}}}{g}\,,\\V&=\kappa\,(\bar{z}\partial_{\bar{z}}+z\partial_z)\,f-\kappa^2\,(\bar{z}\partial_{\bar{z}}+z\partial_z)^2\,f,\;U=\kappa\,(\bar{z}\partial_{\bar{z}}+z\partial_z)\,g-(1-2\kappa)\,g\,.\end{aligned}$$

It is invariant, up to a total derivative, under the transformations

$$\delta z=-\sqrt{2}\,\epsilon_i\xi^i,\qquad \delta\xi^i=\sqrt{2}\,i\bar{\epsilon}^i\nabla_t z+\sqrt{2}\,\epsilon_k\xi^k\xi^i\frac{g_z}{g}.$$

The bosonic "core" of this Lagrangian is

$$\mathcal{L}^{(\kappa)}_{\text{bos}}=g\dot{\bar{z}}\dot{z}+2i\kappa m\,(\dot{\bar{z}}z-\dot{z}\bar{z})\,g-\frac{im}{2}\,(\dot{\bar{z}}f_{\bar{z}}-\dot{z}f_z)-m^2V\,,$$
$$V=\kappa\,(\bar{z}\partial_{\bar{z}}+z\partial_z)\,f-\kappa^2\,(\bar{z}\partial_{\bar{z}}+z\partial_z)^2\,f.$$

We observe that the standard $\mathcal{N}=4, d=1$ kinetic term is deformed to a non-trivial Lagrangian with WZ-term, as well as a potential term. The basic novel feature compared to the standard $\mathcal{N}=4$ Kähler sigma model Lagrangian for the multiplet $(\mathbf{2},\mathbf{4},\mathbf{2})$ is just the necessary presence of this WZ term with the strength $\sim m$.

4.1 Model on a complex plane

The simplest Kähler potential is $f\left(\Phi, \Phi^{\dagger}\right) = \Phi\Phi^{\dagger}$, that yields

$$\begin{aligned}\mathcal{L}^{(\kappa)}_{\text{free}} &= \dot{\bar{z}}\dot{z} + im\left(2\kappa - \frac{1}{2}\right)\left(\dot{\bar{z}}z - \dot{z}\bar{z}\right) + \frac{i}{2}\left(\bar{\xi}_i\dot{\xi}^i - \dot{\bar{\xi}}_i\xi^i\right)\\ &+ 2\kappa\left(2\kappa - 1\right)m^2\bar{z}z + \left(1 - 2\kappa\right)m\left(\bar{\xi}\cdot\xi\right).\end{aligned}$$

This Lagrangian is invariant (modulo a total derivative) under the transformations

$$\delta z = -\sqrt{2}\,\epsilon_i\xi^i, \qquad \delta\xi^i = \sqrt{2}\,i\bar{\epsilon}^i\dot{z} - 2\sqrt{2}\,\kappa m\bar{\epsilon}^i z.$$

The relevant quantum Hamiltonian and (super)charges are:

$$\begin{aligned}&\hat{H} = \bar{\nabla}_{\bar{z}}\nabla_z - 2\kappa m\left(\hat{z}\partial_z - \hat{\bar{z}}\partial_{\bar{z}}\right) + m\left(1 - 2\kappa\right)\hat{\eta}^k\hat{\bar{\eta}}_k\\ &\hat{Q}^i = \sqrt{2}\,\hat{\eta}^i\nabla_z\,, \qquad \hat{\bar{Q}}_j = \sqrt{2}\,\hat{\bar{\eta}}_j\bar{\nabla}_{\bar{z}}\,,\\ &\hat{F} = -2\kappa\left(\hat{z}\partial_z - \hat{\bar{z}}\partial_{\bar{z}}\right) - \left(2\kappa - \frac{1}{2}\right)\hat{\eta}^k\hat{\bar{\eta}}_k\,, \qquad \hat{I}^i_j = \hat{\eta}^i\hat{\bar{\eta}}_j - \frac{1}{2}\delta^i_j\hat{\eta}^k\hat{\bar{\eta}}_k\,,\end{aligned}$$

with

$$\nabla_z = -i\partial_z - \frac{i}{2}m\bar{z}\,, \quad \bar{\nabla}_{\bar{z}} = -i\partial_{\bar{z}} + \frac{i}{2}mz\,, \quad \left[\nabla_z, \bar{\nabla}_{\bar{z}}\right] = m.$$

These generators form $su(2|1)$ superalgebra.

4.2 Wave functions and spectrum

We take advantage of the fact that there exists an extra $U(1)$ charge generator commuting with all $SU(2|1)$ generators,

$$\hat{E} = -\left(\hat{z}\partial_z - \hat{\bar{z}}\partial_{\bar{z}}\right) - \hat{\eta}^k\hat{\bar{\eta}}_k\,.$$

Then the relevant wave functions can be constructed in terms of bosonic eigenfunctions of this external generator

$$\Omega^{(\alpha)} = \bar{z}^{\alpha}A\left(z\bar{z}\right), \quad \hat{E}\,\Omega^{(\alpha)} = \alpha\,\Omega^{(\alpha)}\,, \tag{4.2}$$

α being some positive real number. Requiring this set to simultaneously form the full set of the eigenfunctions of the bosonic part of the Hamiltonian yields

$$\begin{aligned}&\Omega^{(\alpha)} \to \Omega^{(\ell;\alpha)}\,, \quad \hat{H}\,\Omega^{(\ell;\alpha)} = m(\ell + 2\kappa\alpha)\,\Omega^{(\ell;\alpha)}\\ &\Omega^{(\ell;\alpha)} = \bar{z}^{\alpha}e^{-\frac{mz\bar{z}}{2}}L^{(\alpha)}_{\ell}\left(m\,z\bar{z}\right),\end{aligned}$$

where $L^{(\alpha)}_{\ell}$ are Laguerre polynomials and ℓ is Landau level.

Acting by supercharges on $\Omega^{(\ell;\alpha)}$ and imposing the obvious vacuum condition,

$$\bar{\eta}_j\,\Omega^{(\ell;\alpha)} = 0 \;\Rightarrow\; \bar{Q}_j\,\Omega^{(\ell;\alpha)} = 0\,,$$

we obtain other eigenstates of $\hat{H}$ and $\hat{E}$:

$$\begin{aligned}\Psi^{(\ell;\alpha)} &= \left[a^{(\ell;\alpha)} + b^{(\ell;\alpha)}_i\eta^i\,\nabla_z + \frac{1}{2}\,c^{(\ell;\alpha)}\,\varepsilon_{ij}\eta^i\eta^j\,\nabla^2_z\right]\Omega^{(\ell;\alpha)}, \quad \ell \geq 2,\\ \Psi^{(1;\alpha)} &= a^{(1;\alpha)}\,\Omega^{(1;\alpha)} + b^{(1;\alpha)}_i\eta^i\,\nabla_z\,\Omega^{(1;\alpha)}, \quad \Psi^{(0;\alpha)} = a^{(0;\alpha)}\,\Omega^{(0;\alpha)}.\end{aligned}$$

The ground state ($\ell = 0$) is $SU(2|1)$ singlet. The wave functions with $\ell = 1$ form the atypical fundamental representation of $SU(2|1)$ (one bosonic and two fermionic states), while those with $\ell \geq 2$ form typical $(\mathbf{2}|\mathbf{2})$ representations.

The Casimir operators in the case at hand are represented as

$$m^2 C_2 = \left(\hat{H} - 2\kappa m\hat{E}\right)\left(\hat{H} - 2\kappa m\hat{E} - m\right),$$
$$m^3 C_3 = \left(\hat{H} - 2\kappa m\hat{E}\right)\left(\hat{H} - 2\kappa m\hat{E} - m\right)\left(\hat{H} - 2\kappa m\hat{E} - \frac{m}{2}\right),$$

whence

$$C_2(\ell) = (\ell - 1)\,\ell, \quad C_3(\ell) = (\ell - 1/2)\,(\ell - 1)\,\ell\,.$$

They are vanishing for the wave functions with $\ell = 0, 1$, confirming the interpretation of the corresponding representations as atypical, and are non-vanishing on the wave functions with $\ell \geq 2$, implying them to form typical representations of $SU(2|1)$.

5. Generalized $SU(2|1)$ chirality

One can choose another $SU(2|1)$ coset as the basic superspace

$$\frac{SU(2|1) \rtimes U(1)_{\text{ext}}}{SU(2) \times U(1)_{\text{ext}}} = \frac{SU(2|1)}{SU(2)} \sim \frac{\{Q^i, \bar{Q}_j, \tilde{H}, I^i_j\}}{\{I^i_j\}}\,.$$

The Hamiltonian is now the full internal $U(1)$ generator $\tilde{H} = H - mF$. The covariant derivatives $\mathcal{D}_i$, $\bar{\mathcal{D}}^i$ are $U(1)$ inert and a generalized chirality condition can be imposed

$$(\cos\lambda\,\bar{\mathcal{D}}_i - \sin\lambda\,\mathcal{D}_i)\Phi = 0\,, \tag{5.1}$$

λ being a new real parameter. Eq. (5.1) is solved by $\Phi = \varphi_L(\hat{t}_L, \hat{\theta}_i)$, where $(\hat{t}_L, \hat{\theta}_i)$ is the proper set of the chiral coordinates. The components of φ_L are transformed as

$$\delta z = -\sqrt{2}\cos\lambda\,(\epsilon\cdot\xi)\,e^{\frac{i}{2}mt} + \sqrt{2}\sin\lambda\;(\bar{\epsilon}\cdot\xi)\,e^{-\frac{i}{2}mt},$$
$$\delta\xi^i = \sqrt{2}\,\bar{\epsilon}^i\,[i\cos\lambda\,\dot{z} - \sin\lambda\,B]\,e^{-\frac{i}{2}mt} - \sqrt{2}\,\epsilon^i\,[i\sin\lambda\,\dot{z} + \cos\lambda\,B]\,e^{\frac{i}{2}mt},$$
$$\delta B = \sqrt{2}\cos\lambda[i(\bar{\epsilon}\cdot\dot{\xi}) + \frac{m}{2}(\bar{\epsilon}\cdot\xi)]e^{-\frac{i}{2}mt} + \sqrt{2}\sin\lambda[i(\epsilon\cdot\dot{\xi}) - \frac{m}{2}(\epsilon\cdot\xi)]e^{\frac{i}{2}mt}.$$

The most general $SU(2|1)$ invariant action of $\varphi^a(\hat{t}_L, \hat{\theta})$, $a = 1, \ldots N$, is

$$S_{\text{kin}} = \int dt\,\mathcal{L}_{\text{kin}} = \frac{1}{4}\int d\zeta\,f(\varphi^a, \bar{\varphi}^{\bar{a}})\,.$$

Its on-shell bosonic core

$$\mathcal{L}^{\text{on}}_{\text{kin}}| = g_{\bar{a}b}\dot{\bar{z}}^{\bar{a}}\dot{z}^b - \frac{i}{2}m\cos 2\lambda(\dot{\bar{z}}^{\bar{a}}f_{\bar{a}} - \dot{z}^a f_a) - \frac{m^2}{4}g^{\bar{a}b}\sin^2 2\lambda\,f_{\bar{a}}f_b$$

is recognized as the Lagrangian of the Kähler oscillator [14] extended by a coupling to an external magnetic field.

The supercharges do not commute with the Hamiltonian $\tilde{H}$, but are still conserved due to their explicit t-dependence

$$\frac{d}{dt}Q^i = \partial_t Q^i + \{Q^i, \tilde{H}\}_{\text{P.B.}} = 0, \qquad \frac{d}{dt}\bar{Q}_j = \partial_t\bar{Q}_j + \{\bar{Q}_j, \tilde{H}\}_{\text{P.B.}} = 0\,.$$

6. Superconformal models

The most general $\mathcal{N}=4, d=1$ superconformal algebra is $D(2,1;\alpha)$ [15]:

$$
\begin{aligned}
&\{Q_{\alpha ii'}, Q_{\beta jj'}\} = 2\left(\epsilon_{ij}\epsilon_{i'j'}T_{\alpha\beta} + \alpha\,\epsilon_{\alpha\beta}\epsilon_{i'j'}J_{ij} - (1+\alpha)\,\epsilon_{\alpha\beta}\epsilon_{ij}L_{i'j'}\right),\\
&\left[T_{\alpha\beta}, Q_{\gamma ii'}\right] = -i\,\epsilon_{\gamma(\alpha}Q_{\beta)ii'}\,, \qquad \left[T_{\alpha\beta}, T_{\gamma\delta}\right] = i\left(\epsilon_{\alpha\gamma}T_{\beta\delta} + \epsilon_{\beta\delta}T_{\alpha\gamma}\right),\\
&\left[J_{ij}, Q_{\alpha ki'}\right] = -i\,\epsilon_{k(i}Q_{\alpha j)i'}\,, \qquad \left[J_{ij}, J_{kl}\right] = i\left(\epsilon_{ik}J_{jl} + \epsilon_{jl}J_{ik}\right),\\
&\left[L_{i'j'}, Q_{\alpha ik'}\right] = -i\,\epsilon_{k'(i'}Q_{\alpha ij')}\,, \qquad \left[L_{i'j'}, L_{k'l'}\right] = i\left(\epsilon_{i'k'}L_{j'l'} + \epsilon_{j'l'}L_{i'k'}\right).
\end{aligned}
$$

Here, $Q_{\alpha ii'}$ are eight supercharges and the bosonic subalgebra is

$$
su(2)\oplus su(2)'\oplus so(2,1) \equiv \{J_{ik}\}\oplus\{L_{i'k'}\}\oplus\{T_{\alpha\beta}\}.
$$

At $\alpha=-1,0,$ $D(2,1;\alpha)$ is reduced to the semi-direct product

$$
D\,(2,1;\alpha) \cong PSU(1,1|2)\rtimes SU(2)_{\rm ext}\,, \quad \alpha = 0,-1\,.
$$

How to implement $D(2,1;\alpha)$ in the $SU(2|1)$ superspaces? The crucial property is the existence of *two* different $su(2|1)\subset D(2,1;\alpha)$, so that the latter is a closure of them:

$$
\begin{aligned}
&(\mathrm{I}).\quad \{Q^i, \bar{Q}_j\} = 2m(\mu)\,I^i_j + 2\delta^i_j[H(\mu) - m(\mu)\,F]\,,\\
&(\mathrm{II}).\quad \{S^i, \bar{S}_j\} = 2m(-\mu)\,I^i_j + 2\delta^i_j[H(-\mu) - m(-\mu)\,F]\,,\\
&m(\mu) := -\alpha\mu\,, \quad H(\mu) := \mathcal{H} + \mu F\,, \quad \mathcal{H} = \hat{H} + \frac{\mu^2}{4}\hat{K}\,,\\
&(\hat{H},\hat{K})\in so(2,1)\,, \quad F\in su(2)'\,, \quad I^i_j \in su(2)\,.
\end{aligned}
$$

The remaining $D\,(2,1;\alpha)$ generators appear in $\{Q,S\}$ and $\{Q,\bar{S}\}$. The supergroup $SU(2|1)$ (I) is identified with the manifest superisometry of the $SU(2|1)$ superspace; then $SU(2|1)$ (II) is realized on the superspace coordinates and superfields as a hidden symmetry. One faces the "*trigonometric*" realization of the $d=1$ conformal generators [11, 12]:

$$
\hat{H} = \frac{i}{2}\left[1+\cos\mu t\,\right]\partial_t\,, \quad \hat{K} = \frac{2i}{\mu^2}\left[1-\cos\mu t\,\right]\partial_t\,, \quad \hat{D} = \frac{i}{\mu}\sin\mu t\,\partial_t\,.
$$

The basic superfield constraints are $D\,(2,1;\alpha)$ covariant, at least for some special values of α. The multiplet $(\mathbf{1},\mathbf{4},\mathbf{3})$ is superconformal for any α, the chiral multiplet admits the superconformal symmetry only for $\alpha=0,-1$. The superconformal subclasses of the general $SU(2|1)$ actions are singled out by requiring them to be *even* functions of μ, in accord with the above structure of $D\,(2,1;\alpha)$ as a closure of its two $SU(2|1)$ subgroups.

6.1 Some examples of superconformal actions

i. The multiplet $(\mathbf{1},\mathbf{4},\mathbf{3})$:

$$
S^{(\alpha)}_{\rm conf} = -\int d\zeta\, f(G)\,, \; f(G) = \begin{cases} \frac{1}{8(\alpha+1)}\,G^{-\frac{1}{\alpha}} & \text{for } \alpha\neq -1,0\,,\\ \frac{1}{8}\,G\ln G & \text{for } \alpha = -1 \end{cases}\,.
$$

The simplest choice is $\alpha=-1/2$ yielding the free Lagrangian

$$
\mathcal{L}^{(\alpha=-1/2)}_{\rm conf} = \frac{\dot{y}^2}{2} + \frac{i}{2}\left(\bar{\zeta}_i\dot{\zeta}^i - \dot{\bar{\zeta}}_i\zeta^i\right) + \tilde{B}^i_j\tilde{B}^j_i - \frac{\mu^2}{8}\,y^2.
$$

The most general G superfield superconformally covariant constraints read

$$\bar{\mathcal{D}}^2\,\tilde{G} = \mathcal{D}^2\,\tilde{G} = 0\,, \quad \left[\mathcal{D},\bar{\mathcal{D}}\right]\tilde{G} = 4m\,\tilde{G} - 4c\,.$$

At $c \neq 0$ they are covariant only under the $\alpha = -1$ supergroup. The bosonic sector of $\mathcal{L}^{(c,\alpha=-1)}_{\rm conf}$ contains the standard conformal potential:

$$\mathcal{L}^{(c,\alpha=-1)}_{\rm conf} \;\Rightarrow\; \frac{\dot{y}^2}{2} - \frac{\mu^2 y^2}{8} - \frac{c^2}{8y^2}\,.$$

ii. The standard chiral multiplet $(\mathbf{2},\mathbf{4},\mathbf{2})$:

$$\bar{\mathcal{D}}_i\Phi = 0\,, \qquad \mathcal{D}^i\bar{\Phi} = 0\,, \quad \tilde{F}\Phi = 2\kappa\Phi\,.$$

The constraints are covariant only for $\alpha = -1$. At $\kappa \neq 0$:

$$S^{(\kappa)}_{\rm conf} = \frac{1}{4}\int d\zeta\, f\left(\Phi,\bar{\Phi}\right), \quad f = \left(\Phi\bar{\Phi}\right)^{\frac{1}{4\kappa}}\,.$$

Neither WZ term, nor standard conformal potential appear, only the oscillator term is present. At $\kappa = 1/4$, we reproduce the free Lagrangian:

$$\mathcal{L}^{(\kappa=1/4)}_{\rm conf} = \dot{\bar{z}}\dot{z} + \frac{i}{2}\left(\bar{\xi}_i\dot{\xi}^i - \dot{\bar{\xi}}_i\xi^i\right) - \frac{\mu^2}{4}\,z\bar{z}\,. \tag{6.1}$$

iii. The generalized chiral multiplet$(\mathbf{2},\mathbf{4},\mathbf{2})$:

$$(\cos\lambda\,\bar{\mathcal{D}}_i - \sin\lambda\,\mathcal{D}_i)\Phi = 0\,.$$

It is also covariant under $D(2,1;\alpha=-1)$ only.

In both cases, the only superconformally invariant superpotential is the holomorphic integral $S^{\rm pot}_{\rm conf} \sim \nu\int d\zeta_L \ln\varphi_L + \text{c.c.}$ yielding the standard conformal potential $\sim -\nu^2|z|^{-2}$. The extra potential $\sim \mu^2 z\bar{z}$ comes always from the sigma-model action. At any κ the conformal action is reduced, by a field redefinition, to the free one (6.1) plus $S^{\rm pot}_{\rm conf}$.

7. Harmonic $SU(2|1)$ superspace

The harmonic extension of the standard $SU(2|1)$ superspace is defined as follows:

$$\begin{aligned}
&\{t,\theta_i,\bar{\theta}^j\} \;\Rightarrow\; \{t,\theta_i,\bar{\theta}^j,w^\pm_i\}, \quad w^{+i}w^-_i = 1\,,\\
&\delta\theta_i = \epsilon_i + 2m\,\bar{\epsilon}^k\,\theta_k\,\theta_i\,,\; \delta\bar{\theta}^j = \bar{\epsilon}^i - 2m\,\epsilon_k\,\bar{\theta}^k\,\bar{\theta}^i,\; \delta t = i\left(\epsilon_k\,\bar{\theta}^k + \bar{\epsilon}^k\,\theta_k\right),\\
&\delta w^+_i = \lambda^{++}w^-_i\,,\; \delta w^-_i = 0\,,\; \lambda^{++} = -\,m\left(1 - m\bar{\theta}^l\theta_l\right)\left(\bar{\theta}^k\epsilon^j + \theta^k\bar{\epsilon}^j\right)w^+_k w^+_j\,.
\end{aligned}$$

Passing to the analytic basis is accomplished by

$$\begin{aligned}
&\{t,\theta_i,\bar{\theta}^j,w^\pm_i\} \;\Rightarrow\; \{t_{(A)},\theta^\pm,\bar{\theta}^\pm,w^\pm_i\}\,,\\
&\delta\theta^+ = \epsilon^+ + m\,\bar{\theta}^+\theta^+\epsilon^-,\; \delta\bar{\theta}^+ = \bar{\epsilon}^+ - m\,\bar{\theta}^+\theta^+\bar{\epsilon}^-,\; \delta t_{(A)} = 2i\left(\epsilon^-\bar{\theta}^+ + \theta^+\bar{\epsilon}^-\right),\\
&\delta\theta^- = \epsilon^- + 2m\,\bar{\epsilon}^-\theta^-\theta^+,\; \delta\bar{\theta}^- = \bar{\epsilon}^- + 2m\,\epsilon^-\bar{\theta}^-\bar{\theta}^+,\\
&\delta w^+_i = -\,m\left(\bar{\theta}^+\epsilon^+ + \theta^+\bar{\epsilon}^+\right)w^-_i\,,\; \delta w^-_i = 0\,,\; \epsilon^\pm = \epsilon^i w^\pm_i\,,\; \bar{\epsilon}^\pm = \bar{\epsilon}^i w^\pm_i\,.
\end{aligned}$$

The analytic subspace closed under the $SU(2|1)$ transformation amounts to the set

$$\zeta_A := \left(t_{(A)},\bar{\theta}^+,\theta^+,w^\pm_i\right).$$

In the analytic basis, the algebra of covariant derivatives (both spinor and harmonic) contains the following closed subalgebra, the so called "CR-structure":

$$\{\mathcal{D}^+,\bar{\mathcal{D}}^+\} = -2m\,\mathcal{D}^{++}, \qquad [\mathcal{D}^{++},\mathcal{D}^+] = [\mathcal{D}^{++},\bar{\mathcal{D}}^+] = 0\,,$$

where

$$\mathcal{D}^+ = \frac{\partial}{\partial\theta^-} + m\,\bar{\theta}^-\mathcal{D}^{++}, \qquad \bar{\mathcal{D}}^+ = -\frac{\partial}{\partial\bar{\theta}^-} - m\,\theta^-\mathcal{D}^{++}$$

and $\mathcal{D}^{++} = w^{+i}\partial_{w^{-i}} + \ldots$ is one of the three harmonic derivatives forming an $su(2)$ algebra. Analytic superfields are defined as :

$$\mathcal{D}^+\varphi^{+q} = \bar{\mathcal{D}}^+\varphi^{+q} = 0\,, \Rightarrow\ \mathcal{D}^{++}\varphi^{+q} = 0\,.$$

The simplest choice is q^{+a}, $\widetilde{q^{+a}} = \varepsilon_{ab}q^{+b}$, describing the off-shell multiplet $(\mathbf{4},\mathbf{4},\mathbf{0})$

$$q^{+a}\left(\zeta_A\right) = x^{ia}w_i^+ + \theta^+\psi^a + \bar{\theta}^+\bar{\psi}^a - 2i\theta^+\bar{\theta}^+\dot{x}^{ia}w_i^-\,.$$

It transforms as

$$\begin{aligned}
\delta q^{+a} &= -\,m\left(\bar{\theta}^+\epsilon^- + \theta^+\bar{\epsilon}^-\right)q^{+a} \quad \Rightarrow \\
\delta x^{ia} &= -\,\epsilon^i\psi^a - \bar{\epsilon}^i\bar{\psi}^a, \quad \delta\bar{\psi}_a = 2i\epsilon_k\dot{x}_a^k - m\,\epsilon_k x_a^k\,, \quad \delta\psi^a = 2i\bar{\epsilon}^k\dot{x}_k^a + m\,\bar{\epsilon}^k x_k^a\,.
\end{aligned}$$

The sigma-model type invariant action is written as:

$$S\left(q^{\pm a}\right) = \int d\zeta_H\, L\left(q^{+a}q_a^-\right), \quad q^{-a} = \mathcal{D}^{--}q^{+a}$$

and, in components, yields:

$$\begin{aligned}
\mathcal{L} \;=\; & G\left[\dot{x}^{ia}\dot{x}_{ia} + \frac{i}{2}\left(\bar{\psi}_a\dot{\psi}^a - \dot{\bar{\psi}}_a\psi^a\right) + \frac{m}{2}\,\psi^a\bar{\psi}_a\right] - \frac{i}{2}\,\dot{x}^{ia}\partial_{ic}G\left(\psi_a\bar{\psi}^c + \psi^c\bar{\psi}_a\right) \\
& - \frac{\Delta_x G}{16}\left(\bar{\psi}\right)^2\left(\psi\right)^2 + \frac{m}{2}\,x^2G'\,\psi^a\bar{\psi}_a - \frac{m^2}{4}\,x^2\,G\,, \quad G := \Delta_x L(x^2) = G(x^2)\,,
\end{aligned}$$

where $\Delta_x = \varepsilon^{ik}\varepsilon^{ab}\partial_{ia}\partial_{kb}$. The simplest Lagrangian corresponds to $L = q^{+a}q_a^+$:

$$\mathcal{L}_{\text{free}} = \dot{x}^{ia}\dot{x}_{ia} + \frac{i}{2}\left(\bar{\psi}_a\dot{\psi}^a - \dot{\bar{\psi}}_a\psi^a\right) + \frac{m}{2}\,\psi^a\bar{\psi}_a - \frac{m^2}{4}\,x^{ia}x_{ia}\,.$$

As distinct from the flat $(\mathbf{4},\mathbf{4},\mathbf{0})$ case [10], no explicit $SU(2|1)$ invariant Wess-Zumino term can be constructed. An internal WZ term, coming from the kinetic sigma-model action, can be defined for $\tilde{F}q^{+a} \neq 0$, where $\tilde{F}$ is the matrix part of the generator F.

7.1 Quantization in the free case

The quantum Hamiltonian and supercharges are given by:

$$\begin{aligned}
H &= -\frac{1}{4}\left(\partial^{ia} - mx^{ia}\right)\left(\partial_{ia} + mx_{ia}\right) + \frac{m}{2}\,\bar{\xi}_a\xi^a\,, \\
Q_i &= -i\xi^a\left(\partial_{ia} - mx_{ia}\right), \quad \bar{Q}_i = -i\bar{\xi}^a\left(\partial_{ia} + mx_{ia}\right), \\
F &= -\frac{1}{2}\,\bar{\xi}_a\xi^a,\ I_{ik} = x_{(i}^a\,\partial_{k)a}\,,\ E_{ab} = x_{(a}^i\,\partial_{ib)} - \bar{\xi}_{(a}\xi_{b)}\,,\ [E_{ab},E_{cd}] = \varepsilon_{cb}E_{ad} - \varepsilon_{ad}E_{cb}\,.
\end{aligned}$$

Here E_{ab} are the extra $SU(2)_{PG}$ generators commuting with $SU(2|1)$. The ground state conditions are imposed as:

$$\xi^a\,|0\rangle = 0\,, \quad (\partial_{ia} + m x_{ia})\,|0\rangle = 0 \;\Rightarrow\; |0\rangle = e^{-\frac{m}{2}x^2}, \quad Q^i\,|0\rangle = \bar{Q}_i\,|0\rangle = 0\,.$$

The general bosonic state $|\ell\,; s\rangle$ is constructed as:

$$|\ell\,; s\rangle = A_{(i_1 i_2 \ldots i_s)(a_1 a_2 \ldots a_s)} \nabla^{i_1 a_1}\nabla^{i_2 a_2}\ldots\nabla^{i_s a_s}\left(\nabla^{ia}\nabla_{ia}\right)^{\ell}|0\rangle, \tag{7.1}$$

where $s/2$ is the highest weight ("isospin") of the irreducible representation of the group $SU(2)_{\rm PG}$ acting on indices a. The spin content with respect to $SU(2)_{\rm int}$ is the same.

The general wave function $\Omega^{(l;s)}$ is a collection of $|\ell\,; s\rangle$ and their fermionic descendants obtained through action of $\bar{Q}_i$ on $|\ell\,; s\rangle$. The spectrum is:

$$H\,\Omega^{(\ell;s)} = \frac{m}{2}\,(2\ell + s)\,\Omega^{(\ell;s)}, \qquad m > 0\,.$$

The Casimirs are given by the following expressions

$$m^2 C_2 = H\,(H + m) - \frac{m^2}{2}\,E^a_b E^b_a\,,\; m^3 C_3 = \left(H + \frac{m}{2}\right) C_2\,,$$
$$\frac{1}{2}E^a_b E^b_a\,\Omega^{(\ell;s)} = \frac{s}{2}(\frac{s}{2} + 1)\,\Omega^{(\ell;s)}.$$

All cases with $\ell \geq 0$ correspond to the typical representations, with the degeneracy $4(s+1)^2$ and equal number of bosonic and fermionic states. The wave function $\Omega^{(0;s)}$ describes atypical representations, with $(s+1)^2$ bosonic and $s(s+1)$ fermionic states and the degeneracy $(2s+1)(s+1)$.

7.2 Mirror $(\mathbf{4},\mathbf{4},\mathbf{0})$ *multiplet*

The standard $(\mathbf{4},\mathbf{4},\mathbf{0})$ multiplet has the content (x^{ia}, ξ^a), i being the doublet index of $SU(2)_{\rm int}$ and a - of $SU(2)_{\rm PG}$; the mirror $(\mathbf{4},\mathbf{4},\mathbf{0})$ multiplet has the content (y^A, ψ^{iA}), A being the doublet index of some other $SU'(2)_{\rm PG}$. In the flat $\mathcal{N} = 4, d = 1$ supersymmetry these multiplets and the relevant SQM models are equivalent, up to switching two $SU(2)$ automorphism algebras. In the $SU(2|1)$ case, the models based on these two multiplets cease to be equivalent.

The salient feature of the mirror multiplet is the existence of the explicit $SU(2|1)$ invariant Wess-Zumino term:

$$\tilde{\mathcal{L}}_{\rm WZ} = 2\gamma\left\{i\,(\dot{y}^A\,\partial_A f - \dot{\bar{y}}^A\,\bar{\partial}_A f) - \frac{m}{2}\,(y^A\,\partial_A f + \bar{y}^A\bar{\partial}_A f) - \frac{1}{2}\,\psi^{iA}\psi^B_i\,\partial_A\bar{\partial}_B f\right\},$$

where $f(y, \bar{y})$ obey the constraints

$$\Delta_y f = 0\,, \quad m\left(y^B\partial_B - \bar{y}^B\bar{\partial}_B\right) f\,(y, \bar{y}) = 0\,.$$

It is interesting to study the corresponding SQM models.

8. Summary and outlook

We reviewed a new type of $\mathcal{N} = 4$ SQM based on the supergroup $SU(2|1)$. It is a deformation of the standard $\mathcal{N} = 4$ SQM by a mass parameter m. We presented the superfield formalism on two different coset supermanifolds of $SU(2|1)$ treated as the real and chiral $SU(2|1), d = 1$ superspaces. They are carriers of the off-shell multiplets $(\mathbf{1},\mathbf{4},\mathbf{3})$ and $(\mathbf{2},\mathbf{4},\mathbf{2})$. There are two non-equivalent types of the $(\mathbf{2},\mathbf{4},\mathbf{2})$ multiplet.

Journal of Physics: Conference Series **670** (2016) 012026 doi:10.1088/1742-6596/670/1/012026

The $SU(2|1)$ SQM models reveal surprising features. For the $(\mathbf{1},\mathbf{4},\mathbf{3})$ multiplet, the oscillator-type potential terms are present in the sigma-model action. For the $(\mathbf{2},\mathbf{4},\mathbf{2})$ models, the kinetic term is accompanied by the $d=1$ WZ term and potential terms.

We also constructed harmonic $SU(2|1)$ superspace and described two sorts of the "root" $(\mathbf{4},\mathbf{4},\mathbf{0})$ multiplets in its framework. The mirror $(\mathbf{4},\mathbf{4},\mathbf{0})$ multiplet admits an explicit $SU(2|1)$ invariant $d=1$ super WZ term.

In all cases the sets of the quantum states reveal deviations from the standard rule of equality of the bosonic and fermionic states, in accordance with the existence of atypical $SU(2|1)$ representations.

Superconformally invariant subclasses of the $SU(2|1)$ actions were constructed, based on the property that the superconformal algebra $D(2,1;\alpha)$ is a closure of its two $su(2|1)$ subalgebras. The $d=1$ conformal generators are always in the trigonometric realization.

In conclusion, we outline some further possible lines of study.

(a). Multi-particle extensions: to take a few superfields of one or different types, to construct the relevant off- and on-shell actions, to quantize, to identify the relevant target bosonic geometries (m-deformed?), etc.

(b). To inquire whether the remaining $\mathcal{N}=4, d=1$ multiplets (e.g. the multiplet $(\mathbf{3},\mathbf{4},\mathbf{1})$) have their $SU(2|1)$ counterparts and to construct the corresponding SQM models. By analogy with the flat $\mathcal{N}=4, d=1$ harmonic superspace approach [10], one can expect that all such multiplets and the associate SQM models follow from the multiplets $(\mathbf{4},\mathbf{4},\mathbf{0})$ and their SQM models through the well defined gauging procedure [16].

(c). To generalize all this to the next in complexity case of the supergroup $SU(2|2)$. It involves **8** supercharges and so can presumably be treated as a deformation of $\mathcal{N}=8, d=1$ supersymmetry (and of $\mathcal{N}=(4,4), d=2$ supersymmetry, in fact).

(d). To establish possible links with the higher-dimensional theories with "curved" supersymmetries in the localization approach [4].

Finally, it is worth to mention a recent application of the $SU(2|1)$ SQM considerations for calculation of the vacuum energy in some $d>1$ supersymmetric models [17, 18].

Acknowledgments

We acknowledge a support from the RFBR grant 15-02-06670. E.I. thanks the organizers of the conference ISQS'2015 for the kind hospitality in Prague.

References

[1] Witten E 1981 *Nucl. Phys.* B **188** 513; 1982 *Nucl. Phys.* B **202** 253
[2] James Gates Jr S and Rana L 1995 *Phys. Lett.* B **342** 132 (*Preprint* hep-th/9410150)
Pashnev A and Toppan F 2001 *J. Math. Phys.* **42** 5257 (*Preprint* hep-th/0010135)
[3] Ivanov E *et al* 2004 *Class. Quant. Grav.* **21** 1031 (*Preprint* hep-th/0310299)
[4] Festuccia G and Seiberg N 2011 *JHEP* **1106** 114 (*Preprint* arXiv:1105.0689 [hep-th])
Dumitrescu T T *et al* 2012 *JHEP* **1208** 141 (*Preprint* arXiv:1205.1115 [hep-th])
[5] Ivanov E and Sidorov S 2014 *Class. Quant. Grav.* **31** 075013 (*Preprint* arXiv:1307.7690 [hep-th])
[6] Ivanov E and Sidorov S 2014 *J. Phys.* A **47** 292002 (*Preprint* arXiv:1312.6821 [hep-th])
[7] Ivanov E *et al* 2015 *Phys. Rev.* D **91** 085032 (*Preprint* arXiv:1501.05622 [hep-th])
[8] Ivanov E and Sidorov S 2015 *SU*(2|1) mechanics and harmonic superspace (*Preprint* arXiv:1507.00987 [hep-th])
[9] Smilga A V 2004 *Phys. Lett.* B **585** 173 (*Preprint* hep-th/0311023)
[10] Ivanov E and Lechtenfeld O 2003 *JHEP* **0309** 073 (*Preprint* hep-th/0307111)
[11] Papadopoulos G 2013 *Class. Quant. Grav.* **30** 075018 (*Preprint* arXiv:1210.1719 [hep-th])

Journal of Physics: Conference Series **670** (2016) 012026 doi:10.1088/1742-6596/670/1/012026

[12] Holanda N L and Toppan F 2014 *J. Math. Phys.* **55** 061703 (*Preprint* arXiv:1402.7298 [hep-th])
[13] Scheunert M *et al* 1977 *J. Math. Phys.* **18** 155
[14] Bellucci S and Nersessian A 2003 *Phys. Rev.* D **67** 065013 (*Preprint* hep-th/0211070) Bellucci S and Nersessian A 2004 Supersymmetric Kähler oscillator in a constant magnetic field (*Preprint* hep-th/0401232)
[15] Frappat L *et al* 2000 *Dictionary on Lie algebras and superalgebras* (Academic Press, New-York U.S.A., *Preprint* hep-th/9607161)
[16] Delduc F and Ivanov E 2006 *Nucl. Phys.* B **753** 211 (*Preprint* hep-th/0605211)
[17] Assel B *et al* 2015 *JHEP* **07** 043 (*Preprint* arXiv:1503.05537 [hep-th])
[18] Ivanov E and Sidorov S 2015 Long multiplets in supersymmetric mechanics (*Preprint* arXiv:1509.05561 [hep-th])

Twist deformations leading to κ-Poincaré Hopf algebra and their application to physics

Tajron Jurić[1], Stjepan Meljanac[2] and Andjelo Samsarov[3,4]

[1,2] Rudjer Bošković Institute, Theoretical Physics Division, Bijenička c.54, HR-10002 Zagreb, Croatia

[3] Dipartimento di Matematica e Informatica, Universita di Cagliari, viale Merello 92, 09123 Cagliari, Italy and INFN, Sezione di Cagliari

E-mail: [1] `tjuric@irb.hr`, [2] `meljanac@irb.hr`, [3] `samsarov@unica.it`

Abstract. We consider two twist operators that lead to kappa-Poincaré Hopf algebra, the first being an Abelian one and the second corresponding to a light-like kappa-deformation of Poincaré algebra. The adventage of the second one is that it is expressed solely in terms of Poincaré generators. In contrast to this, the Abelian twist goes out of the boundaries of Poincaré algebra and runs into envelope of the general linear algebra. Some of the physical applications of these two different twist operators are considered. In particular, we use the Abelian twist to construct the statistics flip operator compatible with the action of deformed symmetry group. Furthermore, we use the light-like twist operator to define a star product and subsequently to formulate a free scalar field theory compatible with kappa-Poincaré Hopf algebra and appropriate for considering the interacting ϕ^4 scalar field model on kappa-deformed space.

1. Introduction

It is known in the literature that the star product realization of the well known κ-deformed Minowski spacetime algebra [1, 2, 3, 4], $[x_0, x_j]_* = iax_j \equiv \frac{i}{\kappa}x_j$, for $j = 1, ..., n-1$ with remaining elements commuting [5], may arise from several different types of Drinfeld's twist operators [5]-[12], the Jordanian twist operator [13, 14, 15, 16] and the Abelian twist operator [17, 18, 19] being the most prominent ones, or at least being among those that have been mostly studied. That being said normaly assumes the existence of the algebra of functions $\mathcal{A}[[a]]$ where the role of the multiplication in the algebra is being played by the twisted star product $*$ given by

$$f * g = \mu \circ \mathcal{F}^{-1}(f \otimes g) = \bar{f}^{\alpha}(f)\bar{f}_{\alpha}(g), \tag{1}$$

for any two functions f and g in the algebra and where the twist operator is being symbolically written as $\mathcal{F} = f^{\alpha} \otimes f_{\alpha}$. [6] As was pointed out in [16], in the case that this twist is either Abelian or Jordanian, the star product (1) leads to the star commutation relation written above.

[4] On the leave of absence from the Rudjer Bošković Institute, Theoretical Physics Division, Zagreb, Croatia

[5] Here a is the formal parameter of deformation and $\kappa = \frac{1}{a}$ has the mass dimension. n is the number of generators of the spacetime algebra.

[6] $\mathcal{F}^{-1} = \bar{f}^{\alpha} \otimes \bar{f}_{\alpha}$,

On the other hand, when the focus is drawn toward κ-Poincaré Hopf algebra, a different algebraic structure, although closely related to κ-deformed Minkowski spacetime algebra (see [4, 20] for the exact relation between the two), then the appearance of this structure from out of the original Poincaré Hopf algebra by means of the twist deformation method does not appear to be paved by as many proposals for the twists accomplishing this task, yet alone not in a format closed under the Poincaré algebra. Indeed, not only that there haven't been many proposals on offer, but also it was believed that the twist which would account for such a deformation does not even exist.

It is though true that some attempts have been made to get κ-Poincaré Hopf algebra from the Poincaré Hopf algebra, but neither of them finished with much of a success. The main problem being that the twist used either gives rise to a coalgebra (or at least the part of a coalgebra) that runs out of the Poincaré algebra and escapes into $igl(n)$ or the twist operator itself is not made of Poincaré generators only, but in building it up one needs to supplement the Poincaré generators by the elements of $igl(n)$.

In this paper we consider two types of Drinfeld's twists that can accomplish the aforementioned deformation that leads from the nondeformed Hopf algebra to the deformed one (κ-Poincaré Hopf algebra). The first one is the Abelian twist

$$\mathcal{F} = \exp\left\{i(\lambda x_k p_k \otimes A - (1-\lambda) A \otimes x_k p_k)\right\}, \tag{2}$$

where $0 \leq \lambda \leq 1$, is a real parameter and $A = ap_0$, and the second one is the so called light-like twist

$$\mathcal{F} = \exp\left\{ i a^\alpha P^\beta \frac{\ln(1 + a\cdot P)}{a \cdot P} \otimes M_{\alpha\beta} \right\}, \tag{3}$$

where $a^2 = 0$ (therefrom the name). It was recently realized that the Abelian twist (2), despite the previous lack of success, may though reproduce the κ-Poincaré Hopf algebra. However, the price paid is that one needs to replace the Hopf algebra framework with a somewhat more general algebraic structure, like e.g. that of the Hopf algebroid [21, 22, 23]. Contrary to this, when the light-like twist (3) is concerned, everything is smooth there. This twist operator can reproduce κ-Poincaré Hopf algebra by staying only within the framework of the Hopf algebra. There is no need to go outside of this framework and to invoke certain more general algebraic structures. In addition to that, there are only Poincaré generators appearing in the twist (3).

In what follows we briefly describe how these twists came up onto the surface and discuss their physical applications. But before that, the notion of κ-Poincaré algebra in a given basis is introduced and the relation between particular basis of κ-Poincaré and κ-Minkowski spacetime realization, the operator ordering prescription and the star product is explained.

2. κ-Poincaré Hopf algebra

We take a brief look at κ-Poincaré algebra, which is one particular algebraic structure obtained by quantizing the usual Poincaré algebra and where the κ is a formal deformation parameter. Our interest shall be primarily focused on its coalgebraic sector, particularly on the coproducts for the Poincaré generators, that is the momentum generators $p_\mu = -i\partial_\mu$,

$$\begin{aligned}
\triangle_\varphi(p_0) &= p_0 \otimes \mathbf{1} + \mathbf{1} \otimes p_0, \\
\triangle_\varphi(p_i) &= \varphi(A \otimes \mathbf{1} + \mathbf{1} \otimes A)\left[\frac{p_i}{\varphi(A)} \otimes \mathbf{1} + e^A \otimes \frac{p_i}{\varphi(A)}\right]
\end{aligned} \tag{4}$$

and the Lorentz generators $M_{\mu\nu}$,

$$\begin{aligned}
\triangle_\varphi(M_{ij}) &= M_{ij} \otimes \mathbf{1} + \mathbf{1} \otimes M_{ij}, \\
\triangle_\varphi(M_{i0}) &= M_{i0} \otimes \mathbf{1} + e^A \otimes M_{i0} - ap_j \frac{1}{\varphi(A)} \otimes M_{ij},
\end{aligned} \tag{5}$$

where $A = -ia_0\partial_0 = ap_0$. Here we chose $a_\mu = (a, 0, ..., 0)$, $a \sim \frac{1}{\kappa}$.

It can be noted that the coalgebra is parametrized by the function $\varphi(A)$. It is a smooth function of A and each choice of this function corresponds to a certain basis of κ-Poincaré algebra. For example, if $\varphi(A) = 1$, then the coalgebra (4),(5) corresponds to bicrossproduct basis of κ-Poincaré.

More direct interpretation of the function $\varphi(A)$ may be acquired if we turn to another interesting object, which is intimately related to κ-Poincaré algebra: κ-Minkowski spacetime,

$$[\hat{x}_i, \hat{x}_j] = 0, \quad [\hat{x}_0, \hat{x}_j] = ia\hat{x}_j \equiv \frac{i}{\kappa}\hat{x}_j, \quad \text{for} \quad j = 1, ..., n-1. \tag{6}$$

In this context, each choice of $\varphi(A)$ corresponds to a particular differential representation of κ-Minkowski spacetime in terms of the generators x_μ, p_ν of the undeformed Heisenberg algebra

$$[x_\mu, x_\nu] = 0, \quad [p_\mu, p_\nu] = 0, \quad [p_\mu, x_\nu] = -i\eta_{\mu\nu}, \tag{7}$$

where $\eta_{\mu\nu} = diag(-1, 1, ..., 1)$. More explicitly, the realizations of κ-Minkowski spacetime that correspond to the coalgebra (4),(5) all belong to the family

$$\hat{x}_i = x_i\varphi(A), \tag{8}$$
$$\hat{x}_0 = x_0\psi(A) - ax_kp_k\gamma(A), \tag{9}$$

with two other functions, $\gamma(A)$ and $\psi(A)$, not being independent of $\varphi(A)$, but rather being related to it through the consistency relation $\frac{\varphi'}{\varphi}\psi = \gamma - 1$.

Furthermore, to each particular realization parametrized by the function $\varphi(A)$ there is a uniquely defined operator ordering prescription, so that for example one has

$$\begin{aligned} :e^{ik_\mu\hat{x}^\mu}:_L &\equiv e^{-ik_0\hat{x}_0}e^{ik_i\hat{x}_i} = e^{-ik_0\hat{x}_0 + ik_i\hat{x}_i\varphi_S(ak_0)e^{ak_0}}, \\ :e^{ik_\mu\hat{x}^\mu}:_R &\equiv e^{ik_i\hat{x}_i}e^{-ik_0\hat{x}_0} = e^{-ik_0\hat{x}_0 + ik_i\hat{x}_i\varphi_S(ak_0)}, \\ :e^{ik_\mu\hat{x}^\mu}:_S &\equiv e^{ik_\mu\hat{x}^\mu}, \end{aligned} \tag{10}$$

for the left, right and totaly symmetric Weyl ordering, respectively. In the above expressions $\varphi_S(A) = \frac{A}{e^A - 1}$. Moreover, for the general ordering prescription labeled by $: \quad :_\varphi$ and corresponding to an arbitrary realization $\varphi(A)$, one can write

$$:e^{ik_\mu\hat{x}^\mu}:_\varphi = e^{-ik_0\hat{x}_0 + ik_i\hat{x}_i\frac{\varphi_S(ak_0)}{\varphi(ak_0)}}. \tag{11}$$

It is readily seen that from the later relation the left, right and Weyl totaly symmetric ordering can be recovered by taking the choices $\varphi = e^{-A}$, $\varphi = 1$, $\varphi = \varphi_S$, respectively. Of interest here will be the following family of operator orderings,

$$:e^{ik_\mu\hat{x}^\mu}:_\lambda = e^{-i\lambda k_0\hat{x}_0}e^{ik_i\hat{x}_i}e^{-i(1-\lambda)k_0\hat{x}_0}, \tag{12}$$

corresponding to the κ-Minkowski spacetime realization $\varphi(A) = e^{-\lambda A}$. It interpolates between the right, time-symmetric and left ordering, corresponding respectively to $\lambda = 0, \frac{1}{2}$ and 1.

The exponentials $:e^{ik_\mu\hat{x}^\mu}:_\varphi$ may be viewed as the plane waves on the κ-Minkowski spacetime. If further they are to be considered as the elements of the Borel group, then the notion of coproduct encodes the group multiplication rule

$$:e^{ik_\mu\hat{x}^\mu}:_\varphi :e^{iq_\mu\hat{x}^\mu}:_\varphi = :e^{i(k\oplus q)_\mu\hat{x}^\mu}:_\varphi = e^{i\mathcal{D}^\varphi(k,q)x}, \tag{13}$$

where the connection between the coproduct and the function $\mathcal{D}^\varphi(k,q)$ is unraveled by the relation $\triangle_\varphi(p_\mu) = \mathcal{D}^\varphi_\mu(p\otimes 1, 1\otimes p)$.

Thus, to conclude, for each $\varphi(A)$ parametrizing the initial coalgebra, there is a unique operator ordering prescription, there is a unique realization of the accompanying structure (κ-Minkowski space) and there is a unique star product and a twist operator, as we shall see shortly.

A comment is needed regarding the algebra sector. In the algebra sector, the Lorentz algebra remains undeformed in this setting, while the commutator $[M_{\mu\nu}, p_\lambda]$ is deformed and depends on the realization $\varphi(A)$. However, it is also possible to formulate the κ-Poincaré algebra in the so called classical basis [20, 24, 25]. Classical here means that the algebraic sector is undeformed and the whole deformation is contained within the coalgebraic sector. This coalgebra is not contained within the coalgebra parametrization (4),(5) and needs to be treated separately. It looks as

$$\triangle(P_\mu) = P_\mu\otimes Z^{-1} + \mathbf{1}\otimes P_\mu - a_\mu(P_\lambda Z)\otimes P^\lambda + \frac{a_\mu}{2}\Box\, Z\otimes aP, \qquad (14)$$

$$\begin{aligned}\triangle(M_{\mu\nu}) &= M_{\mu\nu}\otimes\mathbf{1} + \mathbf{1}\otimes M_{\mu\nu}\\ &+ a_\mu\left(-P^\lambda + \frac{a^\lambda}{2}\Box\right) Z\otimes M_{\lambda\nu} + a_\nu\left(P^\lambda - \frac{a^\lambda}{2}\Box\right) Z\otimes M_{\lambda\mu}, \end{aligned}\qquad (15)$$

where this time a_μ is the fourvector deformation parameter and

$$Z^{-1}(P) = aP + \sqrt{1+a^2P^2}, \qquad \Box(P) = \frac{2}{a^2}(1-\sqrt{1+a^2P^2}), \quad P_\mu = -i\partial_\mu. \qquad (16)$$

Similarly as before, there exists a realization of the accompanying κ-Minkowski space,

$$[\hat{x}_\mu, \hat{x}_\nu] = i(a_\mu\hat{x}_\nu - a_\nu\hat{x}_\mu), \qquad (17)$$

that corresponds to the above coalgebra and it is given by

$$\hat{x}_\mu = x_\mu\left(aP + \sqrt{1+a^2P^2}\right) - (ax)P_\mu. \qquad (18)$$

3. Twist operators and twisting

3.1. *Abelian twist*

Star product for the general parametrization $\varphi(A)$ can be derived [MSSG] and the result is

$$(f\star_\varphi g)(x) = \lim_{\substack{u\to x\\ t\to x}} e^{x_j\partial^u_j\left(\frac{\varphi(A_u+A_t)}{\varphi(A_u)}-1\right)+x_j\partial^t_j\left(\frac{\varphi(A_u+A_t)}{\varphi(A_t)}e^{A_u}-1\right)} f(u)g(t), \qquad (19)$$

where $A_u = -ia_0\partial^u_0$, $A_t = -ia_0\partial^t_0$. It corresponds to the coalgebra (4),(5) and also to the κ-Minkowski spacetime realization (9), both of them being parametrized with $\varphi(A)$. As a next step, we recall the general definition of the star product

$$f * g = \mu_*(f\otimes g) = \mu\circ\mathcal{F}^{-1}(f\otimes g) \qquad (20)$$

and the result which says that for an arbitrary operator f commuting with $D\equiv x_i\partial_i$ (D is the dilatation operator) it holds [26]

$$:e^{Df}: \;=\; e^{D\ln(1+f)}. \qquad (21)$$

Relying on these two, it is possible to extract the twist operator from the above star product,

$$\mathcal{F}_\varphi = \exp\left\{(D\otimes 1)\ln\frac{\varphi(A\otimes 1 + 1\otimes A)}{\varphi(A\otimes 1)} + (1\otimes D)(A\otimes 1 + \ln\frac{\varphi(A\otimes 1+1\otimes A)}{\varphi(1\otimes A)})\right\}. \qquad (22)$$

For $\varphi = e^{-\lambda A}$ this reduces to the family of Abelian twists parametrized with the real number λ,

$$\mathcal{F} = \exp\left\{i(\lambda x_k p_k\otimes A - (1-\lambda)A\otimes x_k p_k)\right\}. \qquad (23)$$

3.2. Light-like twist

There is a straightforward, but tedious and not always feasible procedure for constructing the twist operator from a given coalgebra. The method requires introducing the coordinate generators and considering their coproducts along with the coproducts for the symmetry generators (momenta). It also utilises the perturbative methods which in due course give a twist operator expressed in the form of an infinite series of terms, whose summuation is anything but trivial. However, the ultimate hope is that this series can be resummed and brought into a closed form. The method is the following [27]: Starting from the nontrivial coproducts $\triangle x$ and $\triangle p$, representing a given deformed coalgebra, one seeks for the operator $\mathcal{F}$ which executes the following transformations:

$$\triangle(x_\mu) = \mathcal{F}\triangle_0(x_\mu)\mathcal{F}^{-1}, \quad \triangle(p_\mu) = \mathcal{F}\triangle_0(p_\mu)\mathcal{F}^{-1}. \tag{24}$$

Here

$$\triangle_0(x_\mu) = x_\mu \otimes \mathbf{1}, \quad \triangle_0(p_\mu) = p_\mu \otimes \mathbf{1} + \mathbf{1} \otimes p_\mu, \tag{25}$$

are the primitive coproducts for the coordinate and momentum generators. Next one expresses $\triangle x$ and $\triangle p$ as a power series in the deformation parameter $a \sim \frac{1}{\kappa}$,

$$\triangle(x_\mu) = \sum_{k=0}^{\infty} \triangle_k(x_\mu), \quad \triangle(p_\mu) = \sum_{k=0}^{\infty} \triangle_k(p_\mu), \quad \triangle_k(x_\mu), \triangle_k(p_\mu) \sim a^k. \tag{26}$$

For the twist operator one may take the ansatz $\mathcal{F} = e^f$, where $f = \sum_{k=1}^{\infty} f_k$ and symbolically $f_k \propto a^k x p^{k+1}$ are the objects that need to be found. Then by using (24) and comparing both sides in these expressions, order by order in the deformation parameter, one gets the set of conditions

$$\begin{aligned}
&\Delta_1(x_\mu) = [f_1, \Delta_0(x_\mu)] \\
&\Delta_2(x_\mu) = [f_2, \Delta_0(x_\mu)] + \frac{1}{2}[f_1, [f_1, \Delta_0(x_\mu)]] \\
&\Delta_3(x_\mu) = [f_3, \Delta_0(x_\mu)] + \frac{1}{2}\left([f_1, [f_2, \Delta_0(x_\mu)]] + [f_2, [f_1, \Delta_0(x_\mu)]]\right) + \frac{1}{3!}[f_1, [f_1, [f_1, \Delta_0(x_\mu)]]] \\
&\ldots \\
&\Delta_k(x_\mu) = [f_k, \Delta_0(x_\mu)] + \ldots + \frac{1}{k!}[f_1, [f_1, \ldots [f_1, \Delta_0(x_\mu)]]] \\
&\ldots
\end{aligned} \tag{27}$$

and analogously for $\Delta(p_\mu)$. The interesting thing is that the above described procedure can be carried out in full extent, when applied to the coalgebra (14),(15). Even more, in the special case when $a^2 = 0$, the infinite sum $f = \sum_{k=1}^{\infty} f_k$ can be evaluated and written in a compact form [28]. The result is the twist operator (3).

3.3. Twisting

Taking the Abelian twist (2) and applying it to the primitive coproduct $\triangle_0(p) = p \otimes \mathbf{1} + \mathbf{1} \otimes p$ (by means of the twist transformation (24)), one recovers the momentum part (4) of the coalgebra (4),(5). However, if one repeats the story and applies the Abelian twist (2) to the primitive coproduct of Lorentz generators, he will not get the part (5) of the κ-Poincaré algebra! Nevertheless, despite this inconvenience, the Abelian twist though may be used to reconstruct the full coalgebra (4),(5), as well as the appropriate antipodes, which for simplicity reasons were not written in this letter.) (See [1303]). Though, to accomplish this, one needs to extend the algebraic framework of the Hopf algebra and go beyond this. In this case an appeal to a broader algebraic framework, like for example that of the Hopf algebroid, appears to be necessary. It is also necessary to use certain tensor exchange identities and when twisting, one does not twist the primitive coproduct itself, but the object that is compatible with the homomorphic property of the coproduct. The twist (2) satisfies the cocycle and counital conditions:

$$\begin{aligned} (\mathcal{F}\otimes \mathbf{1})\cdot(\triangle\otimes id)\mathcal{F} &= (\mathbf{1}\otimes\mathcal{F}))\cdot(id\otimes\triangle)\mathcal{F}, \\ \mu\circ(\epsilon\otimes id)\mathcal{F} &= \mathbf{1}=\mu\circ(id\otimes\epsilon)\mathcal{F}. \end{aligned} \tag{28}$$

The twist (3) for the light-like κ-deformation ($a^2=0$), also complies with the above conditions of cocyclicity and normalization. However, the adventage of this twist is that unlike the Abelian one, it is expressed in terms of Poincaré generators only (it doesn't contain the dilatation operator). Also, it is written in a covariant form. And finally, via twist deformation, it gives rise to κ-Poincaré Hopf algebra written in a classical basis,

$$\begin{aligned} &\Delta^{\mathcal{F}}(M_{\mu\nu})=\mathcal{F}\Delta_0(M_{\mu\nu})\mathcal{F}^{-1}=\Delta_0(M_{\mu\nu})+(\delta^{\alpha}_{\mu}a_\nu-\delta^{\alpha}_{\nu}a_\mu)\left(P^\beta+\frac{1}{2}a^\beta P^2\right)Z\otimes M_{\alpha\beta}, \\ &\Delta^{\mathcal{F}}(P_\mu)=\mathcal{F}\Delta_0(P_\mu)\mathcal{F}^{-1}=\Delta_0(P_\mu)+\left[P_\mu a^\alpha-a_\mu\left(P^\alpha+\frac{1}{2}a^\alpha P^2\right)Z\right]\otimes P_\alpha, \\ &S^{\mathcal{F}}(M_{\mu\nu})=\chi S(M_{\mu\nu})\chi^{-1}=-M_{\mu\nu}+(-a_\mu\delta^\beta_\nu+a_\nu\delta^\beta_\mu)\left(P^\alpha+\frac{1}{2}a^\alpha P^2\right)M_{\alpha\beta}, \\ &S^{\mathcal{F}}(P_\mu)=\chi S(P_\mu)\chi^{-1}=\left[-P_\mu-a_\mu\left(P_\alpha+\frac{1}{2}a_\alpha P^2\right)P^\alpha\right]Z, \\ &\epsilon^{\mathcal{F}}(M_{\mu\nu})=0, \quad \epsilon^{\mathcal{F}}(P_\mu)=0, \end{aligned} \tag{29}$$

where $Z=\frac{1}{1+aP}$ and $\chi^{-1}=\mu\circ\left[(S\otimes 1)\mathcal{F}^{-1}\right]$, with S and ϵ denoting the antiopode and counit, respectively. Superscript $\mathcal{F}$ on this quantities refers to their deformed counterparts. That is, it reproduces the coalgebra (14),(15) for the special case when $a^2=0$, that is a deformed Poincaré Hopf algebra corresponding to a light-like deformation. The important point to be emphasized is that to derive κ-Poincaré Hopf algebra by using the twist (3), one doesn't need to go beyond the Hopf algebra setting, everything can be done within that framework.

4. Applications

4.1. Twisted statistics

We use this example to illustrate the application of the Abelian twist (2). Suppose first that we have a standard situation meaning that there are no effects induced by the noncommutativity ($a=0$). Then assume that we have a certain symmetry algebra $\mathcal{G}$ (e.g. κ-Poincaré) under which the system under consideration is invariant. Then the action of this symmetry algebra on the Hilbert space (here denoted by $\mathcal{A}$) of physical states is realized in some representation of $\mathcal{G}$. Therefore, if $|\phi\rangle$ is some state in the Hilbert space $\mathcal{A}$, and Λ is some element in the symmetry algebra, then the action of $\mathcal{G}$ on $\mathcal{A}$ is realized in some representation D of $\mathcal{G}$, $|\phi\rangle\longrightarrow D(\Lambda)\,|\phi\rangle$. If we further want to extend the action of this symmetry group from one-particle to two-particle (and generally many-particle states), we need to use the notion of the coproduct, so that the action of $\mathcal{G}$ on the two-particle states is performed as,

$$|\phi\rangle\otimes|\psi\rangle\longrightarrow(D\otimes D)\Delta_0(\Lambda)|\phi\rangle\otimes|\psi\rangle, \tag{30}$$

where for $a=0$ the coproduct Δ_0 is defined as $\Delta_0\ :\ \Lambda\longrightarrow\Lambda\otimes\mathbf{1}+\mathbf{1}\otimes\Lambda$. Furthermore, this coproduct has to be compatible with the multiplication μ (usual pointwise multiplication) in the algebra $\mathcal{A}$ of physical states. This is achieved by the requirement

$$\mu\left((D\otimes D)\Delta_0(\Lambda)|\phi\rangle\otimes|\psi\rangle\right)=D(\Lambda)\,\mu(|\phi\rangle\otimes|\psi\rangle). \tag{31}$$

When considering quantum mechanics and its premise of indistinguishable particles, the quantum statistics is usually implemented by restricting to subspaces which are composed of either totally symmetric or totally antysymmetric states. This is achieved by considering the so called statistics flip operator τ_0, which on an element $|\phi\rangle \otimes |\psi\rangle$ from $\mathcal{A} \otimes \mathcal{A}$ has the action

$$\tau_0(|\phi\rangle \otimes |\psi\rangle) = |\psi\rangle \otimes |\phi\rangle, \tag{32}$$

and by using the projector $\frac{1}{2}(1 \pm \tau_0)$, with which help the symmetrization or antisymmetrization on a two particle Hilbert space is carried out. If we further want that these subspaces, built out by projectors, remain invariant under the action of the symmetry algebra (meaning that the process of (anty)symmetrization is invariant under the action of the symmetry algebra), we need to require that this τ_0 must commute with Δ_0 coproduct, through which the symmetry is implemented,

$$[\Delta_0(\Lambda), \tau_0] = 0. \tag{33}$$

Physically this means that the process of symmetrization or antisymmetrization is frame independent. The statistics thus remains invariant under the action of the symmetry algebra $\mathcal{G}$.

In the noncommutative case, when $a = \frac{1}{\kappa} \neq 0$, the deformed coproduct $\Delta_\varphi = \mathcal{F}_\varphi{}^{-1}\Delta_0\mathcal{F}_\varphi$, where $\mathcal{F}_\varphi$ is the twist element in (22), will no more be compatible with the multiplication μ. Therefore μ must be replaced with the new multiplication μ_*, so that the compatibility condition be satisfied,

$$\mu_* \left((D \otimes D)\Delta_\varphi(\Lambda)|\phi\rangle \otimes |\psi\rangle\right) = D(\Lambda)\, \mu_*(|\phi\rangle \otimes |\psi\rangle). \tag{34}$$

What about subspaces projected with $\frac{1}{2}(1 \pm \tau_0)$? The subspaces obtained in this way will no more be invariant under the action of the symmetry algebra. To correct this, one introduces a new statistics flip operator by means of the Abelian twist deformation, $\tau_\varphi = \mathcal{F}_\varphi^{-1}\tau_0\mathcal{F}_\varphi$. Unlike τ_0, this new τ_φ will now commute with Δ_φ and consequently may serve to project out subspaces irreducibly invariant under the twisted symmetry algebra. The role of the projector in this case is played by the operator $\frac{1}{2}(1 \pm \tau_\varphi)$. It gives rise to irreducible tensors of certain symmetry type.

Recall now the Abelian twist (22). One can use it to find the twisted statistics flip operator corresponding to the general class of realizations (8),(9) of κ-Minkowski space,

$$\tau_\varphi = e^{(D \otimes A - A \otimes D)}\tau_0, \tag{35}$$

and to find the corresponding universal R-matrix. The result for τ_φ may be used to derive the oscillator algebra of creation and annihilation operators for the scalar field on the noncommutative (NC) κ-deformed spacetime and to investigate the quantum particle statistics on it. This was illustrated in the example of the scalar field probing the background of the NC BTZ []. Deformed oscillator algebras of the similar or slightly different type on κ-deformed spaces are also considered in []. Note that as A is directly proportional to the deformation parameter a, the twisted flip operator τ_φ goes over smoothly to the untwisted flip operator τ_0 as the deformation parameter $a \to 0$.

4.2. Scalar field propagation in the ϕ^4 κ-Minkowski model

We use this example to illustrate the application of the light-like twist (3). Hence we consider the properties of the massive scalar field propagation within the ϕ^4 κ-Minkowski model based on the light-like twist (3). Before writing the appropriate action, it is worthy to note that the star product induced by the twist (3) has the property

$$\int d^n x\, \phi^\dagger \star_h \psi = \int d^n x\, \phi^* \cdot \psi\,, \tag{36}$$

that is, under the integration sign the star product between two functions may simply be droped out and replaced by the pointwise multiplication. As a consequence, the free part (first two terms) of the NC action

$$\begin{aligned} S_n[\phi] &= \int d^n x \, (\partial_\mu \phi)^\dagger \star_h (\partial^\mu \phi) + m^2 \int d^n x \, \phi^\dagger \star_h \phi \\ &+ \frac{\lambda}{4} \int d^n x \, \frac{1}{2} (\phi^\dagger \star_h \phi^\dagger \star_h \phi \star_h \phi + \phi^\dagger \star_h \phi \star_h \phi^\dagger \star_h \phi) , \end{aligned} \tag{37}$$

which is built by using the same star product, will be completely reduced to a usual commutative free scalar field action. The only nontrivial contributions in the action (37) will come from the interaction terms. As for the interaction part of the action, a comment is in order. Actually, there are altogether six interaction terms that could in principle appear in the action (37), corresponding to six possible permutations between the fields $\phi^\dagger$ and ϕ. However, due to the properties of the star product, they reduce to only two mutually nonequivalent combinations and these two are that appearing in (37).

After expanding the star product, the efective action emerges,

$$\begin{aligned} S_n[\phi] &= \int d^n x \, \Big[(\partial_\mu \phi^*)(\partial^\mu \phi) + m^2 \, \phi^* \phi + \frac{\lambda}{4} (\phi^* \phi)^2 \Big] \qquad (38) \\ &+ i \frac{\lambda}{4} \int d^n x \Big[a_\mu \, x^\mu \Big(\phi^{*2} (\partial_\nu \phi) \partial^\nu \phi - \phi^2 (\partial_\nu \phi^*) \partial^\nu \phi^* \Big) \\ &+ a_\nu \, x^\mu \Big(\phi^2 (\partial_\mu \phi^*) \partial^\nu \phi^* - \phi^{*2} (\partial_\mu \phi) \partial^\nu \phi \Big) + \frac{1}{2} a_\nu x^\mu \, \phi^* \phi \Big((\partial_\mu \phi^*) \partial^\nu \phi - (\partial_\mu \phi) \partial^\nu \phi^* \Big) \Big], \end{aligned}$$

which can be treated within the standard framework of perturbative quantum field theory. This means that we may find the Feynman rules resulting from this effective action, by calculating 2-point and 4-point Green's functions. In a momentum space they look as

$$G \equiv G(k_1, k_2) = \frac{i}{k_1^2 + m^2} \delta^{(n)}(k_1 - k_2), \tag{39}$$

and for the vertex function in the momentum space,

$$\tilde{\Gamma}(k_1, k_2, k_3, k_4; a) = i \frac{\delta^4 S[\tilde{\phi}]}{\delta \tilde{\phi}(k_1) \delta \tilde{\phi}(k_2) \delta \tilde{\phi}^*(k_3) \delta \tilde{\phi}^*(k_4)} , \tag{40}$$

amounting to the following expression:

$$\begin{aligned} \tilde{\Gamma}(k_1, k_2, k_3, k_4; a) &= i (2\pi)^n \frac{\lambda}{2} a_\nu \Bigg[\frac{a_\nu}{a^2} + \frac{1}{4} \Big(k_{4\mu} k_{3\nu} + k_{3\mu} k_{4\nu} - 2\delta_{\mu\nu} k_{4\rho} k_{3\rho} \\ &+ \frac{1}{2} (k_{2\mu} k_{4\nu} - k_{4\mu} k_{2\nu} + k_{2\mu} k_{3\nu} - k_{3\mu} k_{2\nu}) \Big) \partial_\mu^{k_1} \\ &+ \frac{1}{4} \Big(k_{4\mu} k_{3\nu} + k_{3\mu} k_{4\nu} - 2\delta_{\mu\nu} k_{4\rho} k_{3\rho} \\ &+ \frac{1}{2} (k_{1\mu} k_{4\nu} - k_{4\mu} k_{1\nu} + k_{1\mu} k_{3\nu} - k_{3\mu} k_{1\nu}) \Big) \partial_\mu^{k_2} \\ &+ \frac{1}{4} \Big(k_{1\mu} k_{2\nu} + k_{2\mu} k_{1\nu} - 2\delta_{\mu\nu} k_{1\rho} k_{2\rho} \Big) (\partial_\mu^{k_3} + \partial_\mu^{k_4}) \Bigg] \delta^{(n)}(k_1 + k_2 - k_3 - k_4), \end{aligned} \tag{41}$$

where we denote $\partial^k_\mu = \frac{\partial}{\partial k_\mu}$, and all four momenta k_i are flowing into the vertex. The coupling λ has to be dimensionally regularized. Thus, while the free propagator remains unchanged, the vertex function gets heavily modified.

There is one more thing one has to take into account when considering the scalar field theory with a deformed underlying symmetry. Namely, due to a deformation of the symmetry which in this case is not the standard Poincaré, but rather its quantum deformed counterpart disguised in the form of κ-Poincaré symmetry, the coalgebraic part of the symmetry will suffer from drastic changes. Since the coalgebra, particularly the coproduct for the momentum generator, is closely related to the momentum addition rule, the connection being disclosed by

$$\Delta^{\mathcal{F}} p_\mu \equiv \mathcal{D}_\mu(p\otimes 1, 1\otimes p), \qquad \mathcal{D}_\mu(p,k) = p\oplus k, \tag{42}$$

it is clear that the symmetry deformation will give rise to a reinterpretation of the energy-momentum conservation, with the actual form of the energy-momentum conservation being dictated by the form of coproduct for the momentum generators. We want to implement this conclusion into our formalism.

Before doing that, recall the coproduct from (29) which is relevant in this case,

$$\Delta^{\mathcal{F}}(P_\mu) = \mathcal{F}\Delta_0(P_\mu)\mathcal{F}^{-1} = \Delta_0(P_\mu) + \left[P_\mu a^\alpha - a_\mu\left(P^\alpha + \frac{1}{2}a^\alpha P^2\right)Z\right]\otimes P_\alpha,$$

with $Z = \frac{1}{1+aP}$. The corresponding expansion up to the first order in a is

$$\mathcal{D}_\mu(p,q) = p\oplus q = p_\mu(1+aq) + q_\mu - a_\mu\frac{pq}{1+ap} - \frac{1}{2}a_\mu(aq)\frac{p^2}{1+ap}. \tag{43}$$

Going back to the vertex function (41), we consistently modify it in accordance with Eq.(43), so as to keep the track with the change in the coalgebra sector of the underlying symmetry. The modification includes the intervention in the argument of the δ-function in (41), resulting with the Feynman rule which respects the κ-deformed momentum addition/subtraction rule:

$$\begin{aligned}\tilde{\Gamma}(k_1,k_2,k_3,k_4;a) \;=\; & i(2\pi)^n\frac{\lambda}{2}a_\nu\Bigg[\frac{a_\nu}{a^2} + \frac{1}{4}\Bigg(k_{4\mu}k_{3\nu} + k_{3\mu}k_{4\nu} - 2\delta_{\mu\nu}k_{4\rho}k_{3\rho} \\ & + \frac{1}{2}(k_{2\mu}k_{4\nu} - k_{4\mu}k_{2\nu} + k_{2\mu}k_{3\nu} - k_{3\mu}k_{2\nu})\Bigg)\partial^{k_1}_\mu \\ & + \frac{1}{4}\Bigg(k_{4\mu}k_{3\nu} + k_{3\mu}k_{4\nu} - 2\delta_{\mu\nu}k_{4\rho}k_{3\rho} \\ & + \frac{1}{2}(k_{1\mu}k_{4\nu} - k_{4\mu}k_{1\nu} + k_{1\mu}k_{3\nu} - k_{3\mu}k_{1\nu})\Bigg)\partial^{k_2}_\mu \\ & + \frac{1}{4}\Bigg(k_{1\mu}k_{2\nu} + k_{2\mu}k_{1\nu} - 2\delta_{\mu\nu}k_{1\rho}k_{2\rho}\Bigg)(\partial^{k_3}_\mu + \partial^{k_4}_\mu)\Bigg] \\ & \times\Bigg[\delta^{(n)}((k_1\oplus k_2)\ominus(k_3\oplus k_4)) + \delta^{(n)}((k_1\oplus k_2)\ominus(k_4\oplus k_3))\Bigg].\end{aligned} \tag{44}$$

For the more detailed explanation, see [].

In order to investigate the properties of the massive scalar field propagation, it is most convenient to consider the connected 2-point Green's function $G^a_{(c,2)}$. The important piece of information for calculating the connected 2-point Green's function is contained in the tad pole diagram contribution to the self-energy of the scalar particle, which is here calculated by using the Feynman rules (39),(44),

$$\Pi_2^a = \int \frac{d^n \ell}{(2\pi)^n} \tilde{\Gamma}(k_1, \ell, \ell, k_4; a, \mu) \frac{i}{\ell^2 + m^2} \, . \tag{45}$$

After regularizing (45) and isolating the divergences, one gets

$$\Pi_2^a = \frac{\lambda m^2}{32\pi^2} \left[(1 + 3ak) \left(\frac{2}{\epsilon} + \psi(2) + \log \frac{4\pi\mu^2}{m^2} \right) - \frac{9}{4} ak \right] . \tag{46}$$

Since Π_2^a is UV divergent, the Green's function $G^a_{(c,2)}$ will also diverge,

$$G^a_{(c,2)}(k_1, k_4) \;\; \propto \;\; \left[\frac{i}{k_1^2 + m^2} + \frac{i}{k_1^2 + m^2} \Pi_2^a \frac{i}{k_1^2 + m^2} + ... \right] ,$$

$$G^a_{(c,2)}(k_1, k_4) \;\; \longrightarrow \;\; (2\pi)^n \delta^{(n)}(k_1 - k_4) \left[\frac{i}{k_1^2 + m^2 - \Pi_2^a} \right] . \tag{47}$$

To remove the infinities from $G^a_{(c,2)}$, that is to renormalize it, one adds a mass counterterm to the Lagrangian. As a consequence, the resulting Green's function $\tilde{G}^a_{(c,2)}$, with the contribution from the mass counterterm included, is finite. More explicitly,

$$\begin{aligned} \tilde{G}^a_{(c,2)}(k_1, k_4) \;\; &= \;\; \left[G^a_{(c,2)}(k_1, k_4) + G^a_{(c,2)}(k_1, k_4)(-\delta m^2) G^a_{(c,2)}(k_1, k_4) + ... \right] \\ &\longrightarrow \;\; (2\pi)^n \delta^{(n)}(k_1 - k_4) \left[\frac{i}{k_1^2 + m^2 + \delta m^2 - \Pi_2^a} \right] , \end{aligned} \tag{48}$$

where $\tilde{G}^a_{(c,2)}$ denotes the Green's function which includes the contribution from the mass counter term. It can be rendered finite if

$$\delta m^2 = \frac{\lambda m^2}{32\pi^2} \left[(1 + 3ak) \frac{2}{\epsilon} + f\left(\frac{4 - \epsilon}{2}, \frac{\mu^2}{m^2}, ak \right) \right] . \tag{49}$$

Therefore, in the appropriate mass counterterm the UV divergent part is scaled with the particle statistics factor, which depends linearly on the noncommutativity parameter a and the ingoing momentum.

Acknowledgments

This article is prepared as the contribution to XXIII ISQS conference on Integrable systems and Quantum Symmetries. A.S. acknowledges support by the European Commission and the Croatian Ministry of Science, Education and Sports through grant project financed under the Marie Curie FP7-PEOPLE-2011-COFUND, project NEWFELPRO for providing the necessary funds for attending the conference and presenting this work.

Journal of Physics: Conference Series **670** (2016) 012027 doi:10.1088/1742-6596/670/1/012027

References

[1] Lukierski J, Nowicki A, Ruegg H and Tolstoy V N 1991 *Phys. Lett.* B **B264** 331
[2] Lukierski J, Nowicki A and Ruegg H 1992 *Phys. Lett.* B **293**, 344
[3] Lukierski J and Ruegg H 1994 *Phys. Lett.* B **329** 189
[4] Majid S and Ruegg H 1994 *Phys. Lett.* B **334** 348
[5] Drinfeld V G 1983 *Soviet Math. Dokl.* **28** 667
[6] Drinfeld V G 1990 *Leningrad Math. J.* **1** 1419
[7] Chaichian M, Kulish P P, Nishijima K and Tureanu A 2004 *Phys. Lett.* B **604** 98
[8] Balachandran A P, Govindarajan T R, Mangano G, Pinzul A, Qureshi B A and Vaidya S 2007 *Phys. Rev.* D **75** 045009
[9] Aschieri P, Lizzi F and Vitale P 2008 *Phys. Rev.* D **77** 025037
[10] Aschieri P 2007 *Preprint* hep-th/0703013
[11] Aschieri P, Dimitrijevic M, Kulish P, Lizzi F and Wess J 2009 *Lecture Notes in Physics* **774** Springer , Chapter 7
[12] Aschieri P 2009 *Russ. J. Math. Phys.* **16** 371
[13] Kulish P P, Lyakhovsky V D and Mudrov A I 1999 *J. Math. Phys.* **40** 4569
[14] Gerstenhaber M, Giaquinto A and Schack S D 1992 *Quantum symmetry, Quantum Groups* (LNM 1510) (ed. P.P. Kulish), Springer Verlag, Berlin, pp. 9-46
[15] Tolstoy V 2004 *Proc. of International Workshop Supersymmetries and Quantum Symmetries (SQS03)* 242 (Russia, Dubna, eds: E. Ivanov and A. Pashnev, Publ. JINR, Dubna) *Preprint* math/0402433
[16] Borowiec A and Pachol A 2009, *Phys. Rev.* D **79** 045012
[17] Reshetikhin N Yu 1990 *Lett. Math. Phys.* **20** 331
[18] Bu Jong-Gepn, Kim Hyeong-Chan, Lee Youngone , Vac Chang Hyon and Yee Jae Hyung 2008 *Phys. Lett.* B **665** 95
[19] Govindarajan T R, Gupta K S, Harikumar E, Meljanac S and Meljanac D 2008 *Phys. Rev.* D **77** 105010
[20] Kowalski-Glikman J and Nowak S 2002 *Phys. Lett.* B **539** 126
[21] Lu J H 1996 *Int. J. Math.* B **7**
[22] Xu P 2001 *Commun. Math. Phys.* **216** 539-581
[23] Bohm G 2008 *Hopf algebroids* (Handbook of algebra) (*Preprint* 0805.3806[math.QA])
[24] Dimitrijevic M, Jonke L, Moller L, Tsouchnika E, Wess J and Wohlgenannt M 2003 *Eur. Phys. J. C* **31** 129
[25] Borowiec A and Pachol A 2010 **J. Phys.** A **43** 045203
[26] Meljanac S, Samsarov A, Stojic M and Gupta K S 2008 *Eur. Phys. J.* C **53** 295
[27] Juric T, Meljanac S and Strajn R 2014 *Int. J. Mod. Phys.* A **29** 5 1450022
[28] Juric T, Meljanac S, Pikutic D and Strajn R 2015 *JHEP* **1507** 055 **

S. Meljanac and A. Samsarov, "Scalar field theory on kappa-Minkowski spacetime and translation and Lorentz invariance," Int. J. Mod. Phys. A **26**, 1439 (2011) [arXiv:1007.3943 [hep-th]]

[29]
[30] S. Meljanac, A. Samsarov, J. Trampetic and M. Wohlgenannt, "Scalar field propagation in the *phi*^{4} kappa-Minkowski model," JHEP **1112**, 010 (2011) [arXiv:1111.5553 [hep-th]]
S. Meljanac, A. Samsarov and R. Strajn, "Kappa-deformation of Heisenberg algebra and coalgebra: generalized Poincare algebras and R-matrix," JHEP **1208**, 127 (2012) [arXiv:1204.4324 [math-ph]]
T. R. Govindarajan, K. S. Gupta, E. Harikumar, S. Meljanac and D. Meljanac, "Deformed Oscillator Algebras and QFT in kappa-Minkowski Spacetime," Phys. Rev. D **80**, 025014 (2009) [arXiv:0903.2355 [hep-th]]

[31] M. Dimitrijevic, F. Meyer, L. Moller and J. Wess, "Gauge theories on the kappa Minkowski space-time," Eur. Phys. J. C **36**, 117 (2004) [hep-th/0310116]
[32] T. Juric, D. Kovacevic and S. Meljanac, "κ-Deformed Phase Space, Hopf Algebroid and Twisting," SIGMA **10**, 106 (2014) [arXiv:1402.0397 [math-ph]]
[33] T. Juric, S. Meljanac and R. Strajn, "Universal κ-Poincar covariant differential calculus over κ-Minkowski space," Int. J. Mod. Phys. A **29**, 1450121 (2014) [arXiv:1312.2751 [hep-th]]
[34]
[35] A. Borowiec and A. Pachol, "Unified description for κ-deformations of orthogonal groups," Eur. Phys. J. C **74**, no. 3, 2812 (2014) [arXiv:1311.4499 [math-ph]]
****

Journal of Physics: Conference Series **670** (2016) 012028 doi:10.1088/1742-6596/670/1/012028

Minimal surfaces in q-deformed $AdS_5 \times S^5$

Takashi Kameyama

Department of Physics, Kyoto University, Kyoto 606-8502, Japan.

E-mail: kame@gauge.scphys.kyoto-u.ac.jp

Abstract. We study minimal surfaces in q-deformed $AdS_5 \times S^5$. For this purpose, it is convenient to introduce a coordinate system which describes the spacetime only inside the singularity surface and treat the singularity surface as the holographic screen. In particular, we consider minimal surfaces whose boundary shapes are 1) a straight line and 2) a circle. In the $q \to 1$ limit, the solutions correspond to a 1/2 BPS straight line Wilson loop and a 1/2 BPS circular one, respectively. A remarkable point is that the classical Euclidean actions have no linear divergence unlike the original ones. This finiteness indicates that the q-deformation may be regarded as a UV regularization.

1. Introduction

The AdS/CFT correspondence is a realization of the equivalence between string theories and gauge theories. The most well-studied example is a duality between type IIB string theory on the $AdS_5 \times S^5$ background and $\mathcal{N} = 4$ super Yang-Mills theory at large N limit [1]. A great discovery in the recent is an integrable structure behind the AdS/CFT [2]. On the string-theory side, the Green-Schwarz string action on $AdS_5 \times S^5$ is constructed based on a supersymmetric coset [3] and its classical integrability has been shown in [4]. Although the essential mechanism of the duality has not been fully understood yet, the integrability has played a crucial role in checking conjectured relations in the AdS/CFT.

To reveal a deeper structure behind gauge/gravity dualities beyond the conformal invariance, it would be worth considering integrable deformations of the AdS/CFT. On the string-theory side, a good way is to employ the Yang-Baxter sigma model approach [5]. This is a systematic way to consider integrable deformations of 2D non-linear sigma models. In this approach, an integrable deformation is specified by picking up a skew-symmetric classical r-matrix satisfying the modified classical Yang-Baxter equation (mCYBE). The deformed action is classically integrable in the sense that a Lax pair exists.

The original argument was restricted to principal chiral models, then it was generalized to the symmetric coset case [6][1]. Based on this generalization, a q-deformed action of the $AdS_5 \times S^5$ superstring has been constructed in [11] by picking up the Drinfeld-Jimbo type r-matrix [12] and the resulting action exhibits a quantum group symmetry. The metric and NS-NS two-form of the deformed background have been found in [13]. In particular, a singularity surface exists in the deformed AdS part. For this deformed string theory, many works have been done. Some special limits of deformed $AdS_n \times S^n$ were studied in [14]. A mirror description were proposed in [15,16].

[1] For earlier developments on q-deformations of $su(2)$ and $sl(2)$, see [7–10].

The fast-moving string limits were considered in [17]. Giant magnon solutions have been argued in [15,18]. The deformed Nuemann models were derived in [19]. A possible holographic setup was proposed in [20] and minimal surfaces have been studied in [20,21]. Two-parameter deformations have been studied in [14, 22]. For some arguments towards the complete supergravity solution, see [23–25]. More recently, another integrable deformation called the λ-deformation has been proposed in [26, 27]. This deformation is closely linked to the Yang-Baxter deformation by a Poisson-Lie duality [26, 28–30].

A generalization of the Yang-Baxter sigma model to the (non-modified) classical Yang-Baxter equation (CYBE) has been considered in [31] . In this reformulation, the Lax pair and the kappa transformation should be reconstructed and this generalization is not so trivial. An advantage in comparison to the mCYBE case is that one may also consider partial deformations of $AdS_5 \times S^5$. So far, in a series of works [32–38] , many kinds of classical r-matrices have been identified with the well-known type IIB supergravity solutions including the γ-deformations of S^5 [39], gravity duals for noncommutative gauge theories [40] and Schrödinger spacetimes [41], in addition to new backgrounds [32]. This identification may be referred to as the gravity/CYBE correspondence (For a short summary, see [42]). The solutions of the CYBE may be regarded as a moduli space of the $AdS_5 \times S^5$ string background. In the recent, this identification has been generalized to integrable deformations of 4D Minkowski spacetime in [43]. Then Lax pairs for string theories on Yang-Baxter deformed backgrounds have been derived explicitly in [44]. Another remarkable feature of the CYBE case is that non-integrable backgrounds can be described. The well-known example is $AdS_5 \times T^{1,1}$ [45] and the non-integrability of this background has been shown by the existence of chaotic string solutions in [46, 47]. It has been found that Yang-Baxter deformations can also reproduce this background in [48]. This result indicates that the gravity/CYBE correspondence is not restricted to a class of integrable backgrounds but also applicable to a wider class of gravity solutions.

In this article, we focus on the q-deformed $AdS_5 \times S^5$ superstring [11]. An interesting issue is to consider a holographic relation in the q-deformed geometry. A proposal is that the singularity surface in the deformed AdS may be regarded as the holographic screen [20]. For this purpose, it is useful to introduce a coordinate system which describes the spacetime enclosed by the singularity surface [20]. With this coordinate system, minimal surfaces whose boundary shapes are a straight line and a circle have been considered in [20, 21]. The solutions are reduced to the well-known results [49–52] in the $q \to 1$ limit. A remarkable feature is that the classical Euclidean actions have no linear divergence, in comparison to the original ones. This result indicates that the q-deformation may be regarded as a UV regularization.

This article is organized as follows. Section 2 gives a brief review of a coordinate system for the q-deformed $AdS_5 \times S^5$ which describes the spacetime only inside the singularity surface and the associated Poincaré coordinates. In section 3, we consider two types of minimal surfaces. The former is a static solution whose boundary is a straight line. The classical action does not have the standard linear divergence but a logarithmic divergence, unlike the usual AdS/CFT case. The solution corresponds to a 1/2 BPS straight line Wilson loop [49,50] in the undeformed limit $q \to 1$. The latter is a circular solution which is constructed by supposing whose boundary shape is a circle. A remarkable point is that in the circular case, the resulting classical action is finite even though there is no UV cut-off. The solution corresponds to a 1/2 BPS circular Wilson loop [51, 52] in the undeformed limit $q \to 1$. Section 4 is devoted to conclusion and discussion.

2. A q-deformed $AdS_5 \times S^5$ background

We consider the bosonic part of the classical action of a q-deformed $AdS_5 \times S^5$ superstring [11]. In the following, we focus on minimal surfaces ending at the singularity surface in the deformed

AdS. For this purpose, it is helpful to employ a coordinate system which describes the spacetime only inside the singularity surface [20]. In this coordinate system, the singularity surface is located at the boundary.

2.1. The bosonic part of the q-deformed action

Let us first introduce the metric part and the WZ term of the bosonic action with the coordinate system [20] , then the associated Poincaré coordinates [21].

The bosonic action (in the conformal gauge) is composed of the metric part and the Wess-Zumino (WZ) term which describes the coupling of string to an NS-NS two-form.

With the coordinate system proposed in [20] , the metric for the deformed AdS and sphere parts are given by, respectively,

$$\begin{aligned} ds^2_{\mathrm{AdS}_5} &= R^2\sqrt{1+C^2}\Big[-\cosh^2\chi\, dt^2 + \frac{d\chi^2}{1+C^2\cosh^2\chi} \\ &\quad + \frac{(1+C^2\cosh^2\chi)\sinh^2\chi}{(1+C^2\cosh^2\chi)^2 + C^2\sinh^4\chi\sin^2\zeta}(d\zeta^2 + \cos^2\zeta\, d\psi_1^2) + \frac{\sinh^2\chi\sin^2\zeta\, d\psi_2^2}{1+C^2\cosh^2\chi}\Big]\,, \end{aligned} \tag{1}$$

$$\begin{aligned} ds^2_{\mathrm{S}^5} &= R^2\sqrt{1+C^2}\Big[\cos^2\gamma\, d\phi^2 + \frac{d\gamma^2}{1+C^2\cos^2\gamma} \\ &\quad + \frac{(1+C^2\cos^2\gamma)\sin^2\gamma}{(1+C^2\cos^2\gamma)^2 + C^2\sin^4\gamma\sin^2\xi}(d\xi^2 + \cos^2\xi\, d\phi_1^2) + \frac{\sin^2\gamma\sin^2\xi\, d\phi_2^2}{1+C^2\cos^2\gamma}\Big]\,. \end{aligned} \tag{2}$$

Here the coordinates $(t\,,\psi_1\,,\psi_2\,,\zeta\,,\chi)$ describe the deformed AdS_5, while the the coordinates $(\phi\,,\phi_1\,,\phi_2\,,\xi\,,\gamma)$ parameterize the deformed S^5. The deformation is characterized by a real parameter $C \in [0,\infty)$. When $C=0$, the geometry is reduced to the undeformed $\mathrm{AdS}_5\times\mathrm{S}^5$ with the curvature radius R.

It should be mentioned that a curvature singularity exists at $\chi=\infty$ in this coordinate system as well as the original one [13]. In [20], it has been shown that it takes infinite affine time for a massless particle to reach the singularity surface, while it does not take infinite time to reach the singularity in the coordinate time. This is the same feature as in the usual AdS space with the global coordinates. Hence it seems likely to treat the singularity surface as the boundary in the holographic setup for the q-deformed geometry.

The WZ term for the AdS part and the sphere part are given by, respectively,

$$\mathcal{L}^{\mathrm{WZ}}_{\mathrm{AdS}} = \frac{\sqrt{\lambda}}{4\pi}C\sqrt{1+C^2}\,\epsilon^{\mu\nu}\frac{r^4\sin 2\zeta}{[z^2+C^2(z^2+r^2)]^2 + C^2r^4\sin^2\zeta}\,\partial_\mu\varphi\,\partial_\nu\zeta\,, \tag{3}$$

$$\mathcal{L}^{\mathrm{WZ}}_{\mathrm{S}} = -\frac{\sqrt{\lambda}}{4\pi}C\sqrt{1+C^2}\,\epsilon^{\mu\nu}\frac{\sin^4\gamma\sin 2\xi}{(1+C^2\cos^2\gamma)^2 + C^2\sin^4\gamma\sin^2\xi}\,\partial_\mu\phi_1\partial_\nu\xi\,. \tag{4}$$

Here the totally anti-symmetric tensor $\epsilon^{\mu\nu}$ is normalized as $\epsilon^{01}=+1$ and the coupling λ is defined as

$$\sqrt{\lambda} \equiv \frac{R^2}{\alpha'}\,. \tag{5}$$

Each component of the WZ term is proportional to C, and hence it vanishes when $C=0$.

2.2. Poincaré coordinates

To consider minimal surfaces in the deformed AdS (1), it is helpful to introduce the associated Poincaré coordinates.

Let us first perform a coordinate transformation,

$$\cosh\chi \equiv \frac{1}{\cos\theta}\,, \tag{6}$$

and then the Wick rotation $t \to i\tau$ to move to the Euclidean signature. Performing the following coordinate transformation,

$$z \equiv \mathrm{e}^{\tau}\cos\theta\,, \qquad r \equiv \mathrm{e}^{\tau}\sin\theta\,, \tag{7}$$

the Poincaré analogue of the deformed Euclidean AdS_5 is obtained as

$$\begin{aligned} ds^2_{\mathrm{AdS}_5} &= R^2\sqrt{1+C^2}\Bigg[\frac{dz^2+dr^2}{z^2+C^2(z^2+r^2)} + \frac{C^2(z\,dz + r\,dr)^2}{z^2(z^2+C^2(z^2+r^2))} \\ &\quad + \frac{(z^2+C^2(z^2+r^2))r^2}{\left(z^2+C^2(z^2+r^2)\right)^2 + C^2 r^4\sin^2\zeta}(d\zeta^2+\cos^2\zeta\, d\psi_1^2) + \frac{r^2\sin^2\zeta\, d\psi_2^2}{z^2+C^2(z^2+r^2)}\Bigg]\,. \end{aligned} \tag{8}$$

When $C=0$, (8) is reduced to the Euclidean AdS_5 metric with the Poincaré coordinates.

After the Wick rotation has been performed, the space-like path is the only sensible measure rather than time-like and null ones. As argued in [21], the space-like proper distance to the singularity surface is finite when $C \neq 0$, unlike the undeformed case with $C=0$. This property might be crucial in the next section.

3. Minimal surfaces

In this section, we consider minimal surfaces in the deformed AdS. The gauge-theory dual has not been unveiled yet, but at least when $C=0$, these solutions should correspond to Wilson loops in the $\mathcal{N}$=4 super Yang-Mills theory. The minimal surfaces may be a good clue to seek for the dual gauge theory.

In the following, we consider minimal surfaces in the deformed AdS_2 subspace for two cases, 1) a straight line solution 2) a circular solution. In the undeformed limit, the solutions are reduced to minimal surfaces dual for a 1/2 BPS straight Wilson loop [49, 50] and a 1/2 BPS circular one [51, 52], respectively.

3.1. A straight line solution

Let us first study a straight string solution. Due to the deformation, it seems difficult to construct a straight line solution with the Poincaré coordinates (8), while it is possible to construct a static configuration of the string world-sheet in the global Lorentzian deformed AdS_5 (1). Then the solution ends with two lines on the boundary.

Let us consider a static configuration of the string world-sheet with the ansatz:

$$\begin{aligned} &t = \kappa\tau\,, \qquad \chi = \chi(\sigma)\,, \qquad \zeta = \psi_1 = \psi_2 = 0\,, \\ &\phi = \phi_1 = \phi_2 = \gamma = \xi = 0\,. \end{aligned} \tag{9}$$

Then the full metric is reduced to that of a deformed AdS_2 subspace,

$$ds^2_{\mathrm{AdS}_2} = R^2\sqrt{1+C^2}\left[-\cosh^2\chi dt^2 + \frac{d\chi^2}{1+C^2\cosh^2\chi}\right]\,. \tag{10}$$

Note that the WZ terms vanish under this ansatz (9). Then the Lagrangian is given by

$$\mathcal{L}_{\mathrm{AdS}_2} = -\frac{\sqrt{\lambda}}{4\pi}\sqrt{1+C^2}\left[\kappa^2\cosh^2\chi + \frac{\chi'^2}{1+C^2\cosh^2\chi}\right]. \quad (11)$$

The Virasoro constraint is

$$0 = -\kappa^2\cosh^2\chi + \frac{\chi'^2}{1+C^2\cosh^2\chi}, \quad (12)$$

and which gives the following equation,

$$\chi' = -\kappa\cosh\chi\sqrt{1+C^2\cosh^2\chi}. \quad (13)$$

By integrating (13) with the boundary condition $\chi(\sigma=0)=\infty$,

$$-\int_\infty^\chi \frac{d\tilde{\chi}}{\cosh\tilde{\chi}\sqrt{1+C^2\cosh^2\tilde{\chi}}} = \kappa\int_0^\sigma d\tilde{\sigma}, \quad (14)$$

the following expression is obtained,

$$\kappa\sigma = \arctan\left[\frac{1}{C}\right] - \arctan\left[\frac{\sinh\chi}{\sqrt{1+C^2\cosh^2\chi}}\right]. \quad (15)$$

By introducing σ_0 as $\sinh\chi(\sigma_0)=0$, $\kappa\sigma_0$ is related to C through,

$$\kappa\sigma_0 = \arctan\left[\frac{1}{C}\right]. \quad (16)$$

Then the solution (15) can be expressed as

$$\cosh^2\chi = \frac{1+\tan^2\left[\arctan[\frac{1}{C}]-\kappa\sigma\right]}{1-C^2\tan^2\left[\arctan[\frac{1}{C}]-\kappa\sigma\right]}. \quad (17)$$

Note that (17) describes a solution which stretches from the boundary to the center of the deformed AdS. The other direction of the AdS part can be described by the negative σ region. In total, the solution stretches from one boundary to the other boundary through the center.

Next, let us convert the solution (17) in the global coordinates into the one in the Poincaré coordinates. Note that it is divided into two sections according to the positive (or negative) σ region. For the positive σ region, the solution is given by

$$\begin{aligned} z^{(+)}(\tau,\sigma) &= \mathrm{e}^{\kappa\tau}\sqrt{\cos^2(\arctan[1/C]-\kappa\sigma) - C^2\sin^2(\arctan[1/C]-\kappa\sigma)}, \\ r^{(+)}(\tau,\sigma) &= \sqrt{1+C^2}\,\mathrm{e}^{\kappa\tau}\sin(\arctan[1/C]-\kappa\sigma), \end{aligned} \quad (18)$$

and, for the negative σ region, the one is

$$\begin{aligned} z^{(-)}(\tau,\sigma) &= \mathrm{e}^{\kappa\tau}\sqrt{\cos^2(\arctan[1/C]+\kappa\sigma) - C^2\sin^2(\arctan[1/C]+\kappa\sigma)}, \\ r^{(-)}(\tau,\sigma) &= -\sqrt{1+C^2}\,\mathrm{e}^{\kappa\tau}\sin(\arctan[1/C]+\kappa\sigma). \end{aligned} \quad (19)$$

Here τ is taken as $-\infty<\tau<\infty$ and the $(\pm)$-signatures correspond to the solution in the positive σ region $(0\le\sigma\le\sigma_0)$ and that of the negative σ region $(-\sigma_0\le\sigma<0)$, respectively.

Journal of Physics: Conference Series **670** (2016) 012028 doi:10.1088/1742-6596/670/1/012028

The positive σ region is mapped to $\theta \equiv \arctan(r/z)$,

$$\theta(\sigma=0)=\frac{\pi}{2}\ \ \text{(boundary)} \quad\longrightarrow\quad \theta(\sigma=\sigma_0)=0\ \ \text{(origin)}\,, \tag{20}$$

while the negative σ region covers,

$$\theta(\sigma=-\sigma_0)=0\ \ \text{(origin)} \quad\longrightarrow\quad \theta(\sigma=0)=-\frac{\pi}{2}\ \ \text{(boundary)}\,. \tag{21}$$

Note that the contribution coming from the negative σ region is the same value as that from the positive σ region.

Then the classical Euclidean action is evaluated as

$$\begin{aligned} S &= \frac{\sqrt{\lambda}}{2\pi}\sqrt{1+C^2}\int_{-\frac{T}{2}}^{\frac{T}{2}}d\tau\int_{\epsilon}^{\sigma_0}d\sigma\,2\kappa^2\Big(\frac{1+\tan^2\big[\arctan[\frac{1}{C}]-\kappa\sigma\big]}{1-C^2\tan^2\big[\arctan[\frac{1}{C}]-\kappa\sigma\big]}\Big) \\ &= \frac{T\sqrt{\lambda}}{\pi}\frac{\sqrt{1+C^2}}{C}\,\kappa\,\text{arctanh}\Big[C\cot\big(\kappa\epsilon+\arctan[C]\big)\Big]\,, \end{aligned} \tag{22}$$

where T is the interval of the world-sheet τ and taken to be large. When $C\neq 0$, the above expression can be expanded in terms of ϵ as

$$S=\frac{T\sqrt{\lambda}}{2\pi}\frac{\sqrt{1+C^2}}{C}\,\kappa\,\log\left[\frac{2C}{(1+C^2)\,\kappa\epsilon}\right]+\mathcal{O}(\epsilon)\,. \tag{23}$$

Note that the divergence becomes logarithmic, unlike the usual AdS/CFT case.

The Legendre transformation It would be worth mentioning about an additional contribution coming from the boundary [52]. In the undeformed case, it is well recognized that the classical action has a linear divergence and it can be removed by considering a Legendre transformation. The origin of this additional contribution is the surface term which appears in taking a variation of the classical action to obtain the equations motion. This is just because the minimal surface has the boundary. Thus it is important to discuss this contribution in the deformed case as well.

By taking account of a Legendre transformation, the total action is written as

$$\tilde{S} = S+S_L\,, \qquad S_L=2\int_{-\frac{T}{2}}^{\frac{T}{2}}d\tau\int_{\epsilon}^{\sigma_0}d\sigma\,\partial_\sigma\Big(z\,\frac{\partial L}{\partial(\partial_\sigma z)}\Big)\,. \tag{24}$$

The term S_L is evaluated as

$$\begin{aligned} S_L &= 2\int_{-\frac{T}{2}}^{\frac{T}{2}}d\tau\Big(z\,\frac{\partial L}{\partial(\partial_\sigma z)}\Big|_{\sigma_0}-z\,\frac{\partial L}{\partial(\partial_\sigma z)}\Big|_{\epsilon}\Big) \\ &= -\frac{T\sqrt{\lambda}}{\pi}\sqrt{1+C^2}\,\kappa\,\cot\big(\kappa\epsilon+\arctan[C]\big)\,. \end{aligned}$$

In the non-zero C case, S_L can be expanded in terms of ϵ as

$$S_L = -\frac{T\sqrt{\lambda}}{\pi}\frac{\sqrt{1+C^2}}{C}\,\kappa+\mathcal{O}(\epsilon)\,. \tag{25}$$

Thus, even though the boundary contribution (25) has been taken into account, the logarithmic divergent term in (23) cannot be canceled out. There might be a proper method to regularize (23), or it may be divergent essentially.

It is worth noting the undeformed limit of the classical action (23). By taking the $C \to 0$ limit (ϵ : fixed), the well-known result in the undeformed case is reproduced,

$$S = \frac{T\sqrt{\lambda}}{\pi}\,\kappa\cot[\kappa\epsilon] = \frac{T\sqrt{\lambda}}{\pi}\frac{1}{\epsilon} + \mathcal{O}(\epsilon)\,. \tag{26}$$

In the undeformed limit $C \to 0$ with ϵ fixed, S_L (25) results in

$$S_L = -\frac{T\sqrt{\lambda}}{\pi}\,\kappa\cot[\kappa\epsilon]\,, \tag{27}$$

and it cancels out the divergent term in (26) as usual.

3.2. A circular solution

Next, we shall consider minimal surfaces which ends up with a circle at the boundary of the q-deformed AdS_5 with the Poincaré coordinates (8) .

Let us consider an ansatz with the conformal gauge :

$$z = \sqrt{a^2 - r^2}\,, \qquad r = r(\sigma)\,, \qquad \psi_1 = \psi_1(\tau)\,, \qquad \psi_2 = \zeta = 0\,, \tag{28}$$

with $0 \le \tau < 2\pi$, $0 \le \sigma < \infty$. Here a is the radius of the circle at the boundary. Note that (28) is a consistent ansatz and the WZ term vanishes under (28) . Then the geometry with the metric (8) is reduced to the following deformed AdS_2 ,

$$ds^2_{\mathrm{AdS}_2} = \frac{R^2\sqrt{1+C^2}\,r^2}{(1+C^2)a^2 - r^2}\Big[\frac{a^2\,dr^2}{r^2(a^2-r^2)} + d\psi_1^2\Big]\,. \tag{29}$$

This metric leads to a solution of the string equation of motion :

$$z = a\tanh\sigma\,, \qquad r = \frac{a}{\cosh\sigma}\,, \qquad \psi_1 = \tau\,. \tag{30}$$

The next task is to evaluate the classical Euclidean action. By following the undeformed case, let us formally introduce a cut-off ϵ for the coordinate z . It may be regarded as a cut-off σ_0 for the string world-sheet coordinate σ through the classical solution (30) for z ,

$$\epsilon = a\tanh\sigma_0\,. \tag{31}$$

Then the classical action is evaluated as

$$\begin{aligned} S &= \frac{\sqrt{\lambda}}{4\pi}\sqrt{1+C^2}\int_0^{2\pi}d\tau\int_{\sigma_0}^{\infty}d\sigma\Big[\frac{2}{\sinh^2\sigma + C^2\cosh^2\sigma}\Big] \\ &= \sqrt{\lambda}\frac{\sqrt{1+C^2}}{C}\Big(\mathrm{arccot}[C] - \mathrm{arccot}\Big[\frac{Ca}{\epsilon}\Big]\Big)\,. \end{aligned} \tag{32}$$

Note that the second term in (32) can be expanded in terms of ϵ as

$$-\sqrt{\lambda}\frac{\sqrt{1+C^2}}{C}\,\mathrm{arccot}\Big[\frac{Ca}{\epsilon}\Big] = -\sqrt{\lambda}\frac{\sqrt{1+C^2}}{C}\Big(\frac{\epsilon}{Ca} + \mathcal{O}(\epsilon^2)\Big)\,, \tag{33}$$

where C has been fixed in this expansion. It is easy to see that the cut-off can be removed for non-vanishing C. By taking $\epsilon \to 0$, the classical action (32) becomes,

$$S = \sqrt{\lambda}\frac{\sqrt{1+C^2}}{C}\,\mathrm{arccot}[C]\,. \tag{34}$$

It should be mentioned that the action (34) is finite even if there is no UV cut-off for the string world-sheet (equivalently for the radial direction of the deformed AdS). The result would come from the finiteness of the space-like proper distance to the singularity surface. Then the deformation parameter C works as a UV regularization and one does not need to introduce ϵ any more in evaluating the classical action. This point becomes clear by taking the $C \to 0$ limit of (34). By expanding (34) in terms of C, we obtain the following expression:

$$S = -\sqrt{\lambda} + \sqrt{\lambda}\frac{\pi}{2C} + \mathcal{O}(C)\,. \tag{35}$$

However one needs to include ϵ so as to reproduce the regularized result in the undeformed limit [51, 52] as shown below. For this purpose, it is helpful to consider the undeformed limit of the classical action (32) by taking $C \to 0$ while keeping ϵ finite. This corresponds to keeping the boundary of the solution away from the singularity surface. Then the result in the undeformed case [51, 52] is reproduced as

$$S = -\sqrt{\lambda} + \sqrt{\lambda}\frac{a}{\epsilon}\,. \tag{36}$$

Note that the linear divergence is also reproduced in the undeformed limit and it can be canceled by taking account of a Legendre transformation as usual.

The Legendre transformation It is of importance to consider the boundary term via a Legendre transformation [52]. In the present case, the total derivative term is given by

$$\tilde{S} = S + S_L\,, \qquad S_L = \int_0^{2\pi} d\tau \int_{\sigma_0}^{\infty} d\sigma\, \partial_\sigma\Big(z\,\frac{\partial L}{\partial(\partial_\sigma z)}\Big)\,. \tag{37}$$

The term S_L is a total derivative and evaluated as

$$\begin{aligned} S_L &= -\int_0^{2\pi} d\tau\, z\,\frac{\partial L}{\partial(\partial_\sigma z)}\bigg|_{\sigma=\sigma_0} = -\sqrt{\lambda}\,\frac{\sqrt{1+C^2}\,\tanh\sigma_0}{\sinh^2\sigma_0 + C^2\cosh^2\sigma_0} \\ &= -\sqrt{\lambda}\,\sqrt{1+C^2}\,\frac{\epsilon\,(a^2-\epsilon^2)}{a(\epsilon^2+C^2a^2)}\,. \end{aligned} \tag{38}$$

A remarkable point is that (38) vanishes in the limit $\epsilon \to 0$ when $C \neq 0$, hence it does not contribute to the final expression of the action.

The next is to consider the undeformed limit of S_L so as to cancel the divergent term in the undeformed limit of S (32). The term S_L in (38) vanishes when $C \neq 0$, hence this is the first place one need to introduce ϵ so as to keep the boundary away from the singularity surface. By taking the $C \to 0$ limit with ϵ fixed, the resulting expression of S_L is given by

$$S_L = -\sqrt{\lambda}\frac{a}{\epsilon} + \mathcal{O}(\epsilon)\,, \tag{39}$$

and it cancels the divergent term in (36) as usual.

It would be worth mentioning about the cut-off ϵ and the deformation parameter C. When $C \neq 0$, there is no strict need to introduce ϵ in evaluating the classical action and the Legendre term vanishes. However, if the limit $\epsilon \to 0$ is taken first, or ϵ is not turned on, the finite result cannot be reproduced correctly in the $C \to 0$ limit. In this scene, there is a subtlety of the order the two limits: 1) $\epsilon \to 0$ and 2) $C \to 0$. At least so far, a possible resolution is to take the limit $C \to 0$ first while keeping ϵ finite.

Journal of Physics: Conference Series **670** (2016) 012028 doi:10.1088/1742-6596/670/1/012028

4. Conclusion and discussion

In this article, we have discussed a q-deformation of the $AdS_5\times S^5$ superstring. It has been conjectured in [20] that the singularity surface may be treated as a holographic screen in the deformed theory. To look for some support for this conjecture, we have further considered minimal surfaces by employing the coordinate system which is enclosed by the singularity surface. A remarkable feature is that the classical Euclidean actions have no linear divergence unlike the original ones. This finiteness comes from the fact that the q-deformation may be regarded as a UV regularization. In the undeformed limit, the linear divergent term is also reproduced. This result may indicate that our conjecture would make sense. To obtain more support, it would be nice to consider other minimal surfaces like cusped solutions by employing the Poincaré coordinates and calculate the corresponding quark-antiquark potential by following [53].

There are a lot of issues to be studied. The most interesting issue is to find out the gauge-theory side dual to the q-deformation of the $AdS_5\times S^5$ superstring. To tackle this issue, it would be useful to look for some clues of the corresponding gauge theory by employing the Poincaré coordinates.

We believe that our results on minimal surfaces can play an important role in unveiling a possible gauge-theory dual.

Acknowledgments

I am grateful to Kentaroh Yoshida for collaboration concerning with [20, 21] and comments on this manuscript. I would also like to thank Ben Hoare, Takuya Matsumoto and Stijn van Tongeren for useful discussions. The work of T.K. was supported by the Japan Society for the Promotion of Science (JSPS). This work is also supported in part by the JSPS Japan-Russia Research Cooperative Program and the JSPS Japan-Hungary Research Cooperative Program.

References

[1] J. M. Maldacena, "The large N limit of superconformal field theories and supergravity," Int. J. Theor. Phys. **38** (1999) 1113 [Adv. Theor. Math. Phys. **2** (1998) 231] [hep-th/9711200].

[2] N. Beisert *et al.*, "Review of AdS/CFT Integrability: An Overview," Lett. Math. Phys. **99** (2012) 3 [arXiv:1012.3982 [hep-th]].

[3] R. R. Metsaev and A. A. Tseytlin, "Type IIB superstring action in $AdS_5\times S^5$ background," Nucl. Phys. B **533** (1998) 109 [hep-th/9805028].

[4] I. Bena, J. Polchinski and R. Roiban, "Hidden symmetries of the $AdS_5\times S^5$ superstring," Phys. Rev. D **69** (2004) 046002 [hep-th/0305116].

[5] C. Klimcik, "Yang-Baxter sigma models and dS/AdS T duality," JHEP **0212** (2002) 051 [hep-th/0210095]; "On integrability of the Yang-Baxter sigma-model," J. Math. Phys. **50** (2009) 043508 [arXiv:0802.3518 [hep-th]].

[6] F. Delduc, M. Magro and B. Vicedo, "On classical q-deformations of integrable σ-models," JHEP **1311** (2013) 192 [arXiv:1308.3581 [hep-th]].

[7] I. Kawaguchi and K. Yoshida, "Hybrid classical integrability in squashed sigma models," Phys. Lett. B **705** (2011) 251 [arXiv:1107.3662 [hep-th]]; "Hybrid classical integrable structure of squashed sigma models: A short summary," J. Phys. Conf. Ser. **343** (2012) 012055 [arXiv:1110.6748 [hep-th]]; "Hidden Yangian symmetry in sigma model on squashed sphere," JHEP **1011** (2010) 032. [arXiv:1008.0776 [hep-th]].

[8] I. Kawaguchi, T. Matsumoto and K. Yoshida, "The classical origin of quantum affine algebra in squashed sigma models," JHEP **1204** (2012) 115 [arXiv:1201.3058 [hep-th]]; "On the classical equivalence of monodromy matrices in squashed sigma model," JHEP **1206** (2012) 082 [arXiv:1203.3400 [hep-th]].

[9] I. Kawaguchi, D. Orlando and K. Yoshida, "Yangian symmetry in deformed WZNW models on squashed spheres," Phys. Lett. B **701** (2011) 475. [arXiv:1104.0738 [hep-th]]; I. Kawaguchi and K. Yoshida, "A deformation of quantum affine algebra in squashed WZNW models,"J. Math. Phys. **55** (2014) 062302 [arXiv:1311.4696 [hep-th]].

[10] I. Kawaguchi and K. Yoshida, "Classical integrability of Schrodinger sigma models and q-deformed Poincare symmetry," JHEP **1111** (2011) 094 [arXiv:1109.0872 [hep-th]]; I. Kawaguchi and K. Yoshida, "Exotic symmetry and monodromy equivalence in Schrodinger sigma models," JHEP **1302** (2013) 024

[arXiv:1209.4147 [hep-th]]; I. Kawaguchi, T. Matsumoto and K. Yoshida, "Schroedinger sigma models and Jordanian twists," JHEP **1308** (2013) 013 [arXiv:1305.6556 [hep-th]].

[11] F. Delduc, M. Magro and B. Vicedo, "An integrable deformation of the $AdS_5 \times S^5$ superstring action," Phys. Rev. Lett. **112** (2014) 051601 [arXiv:1309.5850 [hep-th]]; F. Delduc, M. Magro and B. Vicedo, "Derivation of the action and symmetries of the q-deformed $AdS_5 \times S^5$ superstring," arXiv:1406.6286 [hep-th].

[12] V. G. Drinfel'd, "Hopf algebras and the quantum Yang-Baxter equation," Sov. Math. Dokl. **32** (1985) 254; "Quantum groups," J. Sov. Math. **41** (1988) 898 [Zap. Nauchn. Semin. **155**, 18 (1986)]; M. Jimbo, "A q difference analog of $U(g)$ and the Yang-Baxter equation," Lett. Math. Phys. **10** (1985) 63.

[13] G. Arutyunov, R. Borsato and S. Frolov, "S-matrix for strings on η-deformed $AdS_5 \times S^5$," JHEP **1404** (2014) 002 [arXiv:1312.3542 [hep-th]].

[14] B. Hoare, R. Roiban and A. A. Tseytlin, "On deformations of $AdS_n \times S^n$ supercosets," JHEP **1406** (2014) 002 [arXiv:1403.5517 [hep-th]].

[15] G. Arutynov, M. de Leeuw and S. J. van Tongeren, "The exact spectrum and mirror duality of the $(AdS_5 \times S^5)_\eta$ superstring," Theor. Math. Phys. **182** (2015) 1, 23 [Teor. Mat. Fiz. **182** (2014) 1, 28] [arXiv:1403.6104 [hep-th]].

[16] G. Arutyunov and S. J. van Tongeren, "$AdS_5 \times S^5$ mirror model as a string," Phys. Rev. Lett. **113** (2014) 261605 [arXiv:1406.2304 [hep-th]]; "Double Wick rotating Green-Schwarz strings," JHEP **1505** (2015) 027 [arXiv:1412.5137 [hep-th]].

[17] T. Kameyama and K. Yoshida, "Anisotropic Landau-Lifshitz sigma models from q-deformed $AdS_5 \times S^5$ superstrings," JHEP **1408** (2014) 110 [arXiv:1405.4467 [hep-th]]; "String theories on warped AdS backgrounds and integrable deformations of spin chains," JHEP **1305** (2013) 146 [arXiv:1304.1286 [hep-th]].

[18] M. Khouchen and J. Kluson, "Giant Magnon on Deformed $AdS_3 \times S^3$," Phys. Rev. D **90** (2014) 066001 [arXiv:1405.5017 [hep-th]]; C. Ahn and P. Bozhilov, "Finite-size giant magnons on η-deformed $AdS_5 \times S^5$," Phys. Lett. B **737** (2014) 293 [arXiv:1406.0628 [hep-th]]; A. Banerjee and K. L. Panigrahi, "On the Rotating and Oscillating strings in $(AdS_3 \times S^3)_\varkappa$," JHEP **1409** (2014) 048 [arXiv:1406.3642 [hep-th]];

[19] G. Arutyunov and D. Medina-Rincon, "Deformed Neumann model from spinning strings on $(AdS_5 \times S^5)_\eta$," JHEP **1410** (2014) 50 [arXiv:1406.2536 [hep-th]].

[20] T. Kameyama and K. Yoshida, "A new coordinate system for q-deformed $AdS_5 \times S^5$ and classical string solutions," J. Phys. A **48** (2015) 7, 075401 [arXiv:1408.2189 [hep-th]].

[21] T. Kameyama and K. Yoshida, " Minimal surfaces in q-deformed $AdS_5 \times S^5$ string with Poincare coordinates," J. Phys. A **48** (2015) 24, 245401 [arXiv:1410.5544 [hep-th]].

[22] B. Hoare, "Towards a two-parameter q-deformation of $AdS_3 \times S^3 \times M^4$ superstrings," Nucl. Phys. B **891** (2015) 259 [arXiv:1411.1266 [hep-th]].

[23] O. Lunin, R. Roiban and A. A. Tseytlin, "Supergravity backgrounds for deformations of $AdS_n \times S^n$ supercoset string models," Nucl. Phys. B **891** (2015) 106 [arXiv:1411.1066 [hep-th]].

[24] G. Arutyunov, R. Borsato and S. Frolov, "Puzzles of eta-deformed $AdS_5 \times S^5$," arXiv:1507.04239 [hep-th].

[25] B. Hoare and A. A. Tseytlin, "Type IIB supergravity solution for the T-dual of the eta-deformed $AdS_5 \times S^5$ superstring," arXiv:1508.01150 [hep-th].

[26] K. Sfetsos, "Integrable interpolations: From exact CFTs to non-Abelian T-duals," Nucl. Phys. B **880** (2014) 225 [arXiv:1312.4560 [hep-th]]; K. Sfetsos and D. C. Thompson, "Spacetimes for λ-deformations," JHEP **1412** (2014) 164 [arXiv:1410.1886 [hep-th]]; S. Demulder, K. Sfetsos and D. C. Thompson, "Integrable λ-deformations: Squashing Coset CFTs and $AdS_5 \times S^5$," JHEP **07** (2015) 019 [arXiv:1504.02781 [hep-th]]; "Generalised integrable λ- and η-deformations and their relation," Nucl. Phys. B **899** (2015) 489 [arXiv:1506.05784 [hep-th]]; G. Georgiou, K. Sfetsos and K. Siampos, "All-loop anomalous dimensions in integrable λ-deformed σ-models," arXiv:1509.02946 [hep-th]

[27] T. J. Hollowood, J. L. Miramontes and D. M. Schmidtt, "Integrable Deformations of Strings on Symmetric Spaces," JHEP **1411** (2014) 009 [arXiv:1407.2840 [hep-th]]; T. J. Hollowood, J. L. Miramontes and D. M. Schmidtt, "An Integrable Deformation of the $AdS_5 \times S^5$ Superstring," J. Phys. A **47** (2014) 49, 495402 [arXiv:1409.1538 [hep-th]]; T. J. Hollowood, J. L. Miramontes and D. M. Schmidtt, "S-Matrices and Quantum Group Symmetry of k-Deformed Sigma Models," arXiv:1506.06601 [hep-th]; C. Appadu and T. J. Hollowood, "Beta Function of k Deformed $AdS_5 \times S^5$ String Theory," arXiv:1507.05420 [hep-th].

[28] B. Vicedo, "Deformed integrable σ-models, classical R-matrices and classical exchange algebra on Drinfel'd doubles," J. Phys. A **48** (2015) 35, 355203 [arXiv:1504.06303 [hep-th]].

[29] B. Hoare and A. A. Tseytlin, "On integrable deformations of superstring sigma models related to $AdS_n \times S^n$ supercosets," Nucl. Phys. B **897** (2015) 448 [arXiv:1504.07213 [hep-th]].

[30] C. Klimcik, "η and λ deformations as $\mathcal{E}$-models," arXiv:1508.05832 [hep-th].

[31] I. Kawaguchi, T. Matsumoto and K. Yoshida, "Jordanian deformations of the $AdS_5 \times S^5$ superstring," JHEP **1404** (2014) 153 [arXiv:1401.4855 [hep-th]];

[32] I. Kawaguchi, T. Matsumoto and K. Yoshida, "A Jordanian deformation of AdS space in type IIB supergravity," JHEP **1406** (2014) 146 [arXiv:1402.6147 [hep-th]].
[33] T. Matsumoto and K. Yoshida, "Lunin-Maldacena backgrounds from the classical Yang-Baxter equation – Towards the gravity/CYBE correspondence," JHEP **1406** (2014) 135 [arXiv:1404.1838 [hep-th]].
[34] T. Matsumoto and K. Yoshida, "Integrability of classical strings dual for noncommutative gauge theories," JHEP **1406** (2014) 163 [arXiv:1404.3657 [hep-th]].
[35] T. Matsumoto and K. Yoshida, "Yang-Baxter sigma models based on the CYBE," Nucl. Phys. B **893** (2015) 287 [arXiv:1501.03665 [hep-th]].
[36] T. Matsumoto and K. Yoshida, "Yang-Baxter deformations and string dualities," JHEP **1503** (2015) 137 [arXiv:1412.3658 [hep-th]].
[37] T. Matsumoto and K. Yoshida, "Schrödinger geometries arising from Yang-Baxter deformations," JHEP **1504** (2015) 180 [arXiv:1502.00740 [hep-th]].
[38] S. J. van Tongeren, "On classical Yang-Baxter based deformations of the $AdS_5 \times S^5$ superstring," JHEP **1506** (2015) 048 [arXiv:1504.05516 [hep-th]]; S. J. van Tongeren, "Yang-Baxter deformations, AdS/CFT, and twist-noncommutative gauge theory," arXiv:1506.01023 [hep-th].
[39] O. Lunin and J. M. Maldacena, "Deforming field theories with $U(1) \times U(1)$ global symmetry and their gravity duals," JHEP **0505** (2005) 033 [hep-th/0502086].
[40] A. Hashimoto and N. Itzhaki, "Noncommutative Yang-Mills and the AdS / CFT correspondence," Phys. Lett. B **465** (1999) 142 [hep-th/9907166].
J. M. Maldacena and J. G. Russo, "Large N limit of noncommutative gauge theories," JHEP **9909** (1999) 025 [hep-th/9908134].
[41] C. P. Herzog, M. Rangamani and S. F. Ross, "Heating up Galilean holography," JHEP **0811** (2008) 080 [arXiv:0807.1099 [hep-th]]; J. Maldacena, D. Martelli and Y. Tachikawa, "Comments on string theory backgrounds with non-relativistic conformal symmetry," JHEP **0810** (2008) 072 [arXiv:0807.1100 [hep-th]]; A. Adams, K. Balasubramanian and J. McGreevy, "Hot Spacetimes for Cold Atoms," JHEP **0811** (2008) 059 [arXiv:0807.1111 [hep-th]].
[42] T. Matsumoto and K. Yoshida, "Integrable deformations of the $AdS_5 \times S^5$ superstring and the classical Yang-Baxter equation – Towards the gravity/CYBE correspondence –," J. Phys. Conf. Ser. **563** (2014) 1, 012020 [arXiv:1410.0575 [hep-th]].
[43] T. Matsumoto, D. Orlando, S. Reffert, J. Sakamoto and K. Yoshida, "Yang-Baxter deformations of Minkowski spacetime," arXiv:1505.04553 [hep-th].
[44] T. Kameyama, H. Kyono, J. i. Sakamoto and K. Yoshida, "Lax pairs on Yang-Baxter deformed backgrounds," arXiv:1509.00173 [hep-th].
[45] I. R. Klebanov and E. Witten, "Superconformal field theory on three-branes at a Calabi-Yau singularity," Nucl. Phys. B **536** (1998) 199 [hep-th/9807080].
[46] P. Basu and L. A. Pando Zayas, "Chaos Rules out Integrability of Strings in $AdS_5 \times T^{1,1}$," Phys. Lett. B **700** (2011) 243 [arXiv:1103.4107 [hep-th]]; "Analytic Non-integrability in String Theory," Phys. Rev. D **84** (2011) 046006 [arXiv:1105.2540 [hep-th]].
[47] Y. Asano, D. Kawai, H. Kyono and K. Yoshida, "Chaotic strings in a near Penrose limit of $AdS_5 \times T^{1,1}$," JHEP **1508** (2015) 060 [arXiv:1505.07583 [hep-th]].
[48] P. M. Crichigno, T. Matsumoto and K. Yoshida, "Deformations of $T^{1,1}$ as Yang-Baxter sigma models," JHEP **1412** (2014) 085 [arXiv:1406.2249 [hep-th]].
[49] S. -J. Rey and J. -T. Yee, "Macroscopic strings as heavy quarks in large N gauge theory and anti-de Sitter supergravity," Eur. Phys. J. C **22** (2001) 379 [hep-th/9803001];
J. M. Maldacena, "Wilson loops in large N field theories," Phys. Rev. Lett. **80** (1998) 4859 [hep-th/9803002].
[50] N. Drukker and S. Kawamoto, "Small deformations of supersymmetric Wilson loops and open spin-chains," JHEP **0607** (2006) 024 [hep-th/0604124].
[51] D. E. Berenstein, R. Corrado, W. Fischler and J. M. Maldacena, "The Operator product expansion for Wilson loops and surfaces in the large N limit," Phys. Rev. D **59** (1999) 105023 [hep-th/9809188].
[52] N. Drukker, D. J. Gross and H. Ooguri, "Wilson loops and minimal surfaces," Phys. Rev. D **60** (1999) 125006 [hep-th/9904191].
[53] N. Drukker, S. Giombi, R. Ricci and D. Trancanelli, "Supersymmetric Wilson loops on S^3," JHEP **0805** (2008) 017 [arXiv:0711.3226 [hep-th]]; N. Drukker and V. Forini, "Generalized quark-antiquark potential at weak and strong coupling," JHEP **1106** (2011) 131 [arXiv:1105.5144 [hep-th]].

Journal of Physics: Conference Series **670** (2016) 012029 doi:10.1088/1742-6596/670/1/012029

Yang-Baxter relations with orthogonal or symplectic symmetry

D Karakhanyan[1] **and R Kirschner**[2]

[1] Yerevan Physics Institute, 2 Alikhanyan br., 0036 Yerevan, Armenia
[2] Institut für Theoretische Physik, Universität Leipzig, 04009 Leipzig, Germany

E-mail: `roland.kirschner@itp.uni-leipzig.de`

Abstract. Solutions of the Yang-Baxter relations with orthogonal or symplectic symmetry are studied emphasizing the analogies of both cases. Starting from Jordan-Schwinger representations the distinguishing features of the spinor representation and its symplectic analogon are shown. The corresponding L matrix and the spinorial R operator are discussed.

1. Introduction

The R matrix obeying the Yang-Baxter (YB) relation and representing the orthogonal or symplectic algebras in their fundamental representation is not linear in the spectral parameter [1, 2, 4]. The RLL case of the YB relation involving this fundamental R matrix together with the L matrix operators of the form $L_{a,b}(u) = u\delta_{ab} + M_{ab}$ does not hold for an arbitrary representation with the generators M_{ab}. The spinor representation of the orthogonal algebra with $M_{ab} = \frac{1}{4}[\gamma_a, \gamma_b]$ is a distinguished case, where the RLL relation is obeyed. Also the spinorial R matrix, intertwining two spinor representations, is known [3]. This and other representations of the orthogonal algebra distinguished in this sense and the corresponding R operators have been considered recently and the spinorial R operator has been analysed in detail [6].

We consider the YB relations with symplectic symmetry relying on the known analogies to the orthogonal case. We use the Jordan-Schwinger (JS) type of representations, where the generators M_{ab} are composed of canonical pairs x_a, ∂_a obeying the Heisenberg commutation relation in the usual bosonic or in the fermionic/Grassmann version [5].

2. Jordan-Schwinger representations

From $2n(+1)$ canonical pairs $x_a, \partial_a, a = 1, ..., 2n(+1)$ one can construct operators obeying the $g\ell(2n(+1))$ algebra relations. We consider two versions,

$$E^+_{ab} = x_a \partial_b, \; E^-_{ab} = -\partial_a x_b, \tag{1}$$

related by the elementary canonical transformations

$$\mathcal{C} E^+_{ab} \mathcal{C}^{-1} = E^-_{ab} \tag{2}$$

defined as

$$\mathcal{C}_a \begin{pmatrix} x_a \\ \partial_a \end{pmatrix} \mathcal{C}_a^{-1} = \begin{pmatrix} \partial_a \\ -x_a \end{pmatrix}, \quad \mathcal{C} = \prod \mathcal{C}_a. \tag{3}$$

We consider the action of the constructed generators on functions of x_a. On the linear space spanned by x_a we have the fundamental representations. The two representations are dual to each other in the following sense. If the first set of generators acts on $\mathcal{V}^+$ as infinitesimal transformations and the second one on $\mathcal{V}^-$, both spanned by the x_a as basis vectors, the scalar product of $\mathbf{x}_1 \in \mathcal{V}^+$ and $\mathbf{x}_2 \in \mathcal{V}^-$ defined as $(\mathbf{x}_1\mathbf{x}_2) = \sum_a x_{1,a}x_{2,a}$ is invariant in the sense

$$[E^{(1)+}_{ab} + E^{(2)-}_{ab}, (\mathbf{x}_1\mathbf{x}_2)] = 0. \tag{4}$$

In both cases the algebra commutes with $\sum N_a,\ N_a = x_a\partial_a$.

We need also a fermionic/Grassmann version of the JS representation based on odd canonical pairs $\theta_a, \partial^\theta_a,\ [\theta_a, \partial^\theta_b]_+ = \delta_{ab}$. The relations of $g\ell(2n(+1))$ are fulfilled as well by

$$\mathrm{E}^+_{ab} = \theta_a\partial^\theta_b,\ \ \mathrm{E}^-_{ab} = \partial^\theta_a\theta_b$$

The elementary canonical transformations are now

$$\mathcal{C}^{(\alpha)}_a \begin{pmatrix} \theta_a \\ \partial^\theta_a \end{pmatrix} \mathcal{C}^{(\alpha)}_a = \begin{pmatrix} \alpha\partial^\theta_a \\ \alpha^{-1}\theta_a \end{pmatrix}.$$

We shall use the cases $\alpha = \pm 1$. The bilinear form is the same one with the bosonic coordinates substituted by the fermionic ones.

Now we consider the case that both sets $E^\pm_{ab}$ act on the same space. The sum is invariant with respect to the elementary canonical transformation,

$$\mathcal{C}F_{ab}\mathcal{C}^{-1} = F_{ab},\ F_{ab} = E^+_{ab} + E^-_{ab} + \delta_{ab} = x_a\partial_b - x_b\partial_a,$$

and the scalar product of two elements of this space is invariant,

$$[F_{ab}, (\mathbf{x}_1\mathbf{x}_2)] = 0.$$

In this way in the JS case the subalgebra $so(2n(+1))$ of $g\ell(2n(+1))$ can be considered as the one invariant with respect to the elementary canonical transformation.

In order to have a smooth comparison with the symplectic case we rewrite the bilinear form as a result of relabelling the indices $1, 2, ...2n + (1) \to -n, -n+1, ..., (0), ..., n-1, n$. There is a difference between the even and odd cases. In the latter the index value 0 is included and in the even case it is excluded. The bilinear form reads now

$$(\mathbf{x}_1\mathbf{x}_2) = \sum_{-n}^{n} x_{1,a}x_{2,-a}, \tag{5}$$

and the generators of the $so(2n(+1))$ subalgebra are

$$F_{ab} = E^+_{ab} + E^-_{-a,-b}. \tag{6}$$

They are invariant with respect to the elementary canonical transformation including the index value reflection.

$$\mathcal{C} = \prod \mathcal{C}_a,$$

where in the bosonic case we have

$$\mathcal{C}_a \begin{pmatrix} x_a \\ \partial_a \end{pmatrix} \mathcal{C}^{-1}_a = \begin{pmatrix} \partial_{-a} \\ -x_{-a} \end{pmatrix}, \tag{7}$$

and the fermionic case

$$\mathcal{C}_a^{(\pm)} \begin{pmatrix} \theta_a \\ \partial_a^\theta \end{pmatrix} \mathcal{C}_a^{(\pm)} = \begin{pmatrix} \pm\partial^\theta_{-a} \\ \pm\theta_{-a} \end{pmatrix}.$$

The subalgebra $sp(n)$ of $g\ell(2n)$ is defined as the one commuting with the bilinear form

$$< \mathbf{x}_1 \mathbf{x}_2 >= \sum_{-n}^{n} \varepsilon_a x_{1,a} x_{2,-a}$$

$$[G_{ij}, < \mathbf{x}_1 \mathbf{x}_2 >] = 0, \; G_{ij} = E^+_{ab} + \varepsilon_a \varepsilon_b E^-_{-a,-b}$$

where ε_a denotes the sign of the index a. We have an analogous statement about the characterisation of $sp(n)$ as the subalgebra of $g\ell(2n)$, but the elementary canonical transformations enter with an index sign dependent modification. In the bosonic case we have

$$\tilde{\mathcal{C}} = \prod_1^n \mathcal{C}_a \prod_{-n}^{-1} \mathcal{C}_a^{-1}, \tag{8}$$

and in the fermionic case

$$\tilde{\mathcal{C}} = \prod_1^n \mathcal{C}_a^+ \prod_{-n}^{-1} \mathcal{C}_a^-. \tag{9}$$

Thus the subalgebra $sp(n)$ for JS representations can be regarded as the one invariant with respect to this canonical transformation $\tilde{\mathcal{C}}$.

It is important to notice that there are linear combinations of the underlying Heisenberg generators, invarant (up to sign) under the above elementary canonical transformations.

$$\mathcal{C}(\partial_a \pm \theta_{-a})\mathcal{C} = \pm(\partial_a \pm \theta_{-a}), \tag{10}$$

$$\tilde{\mathcal{C}}(\partial_a \pm \varepsilon_a x_{-a})\tilde{\mathcal{C}}^{-1} = \mp(\partial_a \pm \varepsilon_a x_{-a}).$$

The first relation involving $\mathcal{C}$ is relevant for orthogonal and the second with the index sign dependent $\tilde{\mathcal{C}}$ in the symplectic case. Interchanging the bosonic with the fermionic versions we do not obtain invariance, but the two linear combinations transform into each other:

$$\mathcal{C}(\partial_a \pm x_{-a})\mathcal{C}^{-1} = \pm(\partial_a \mp x_{-a}), \quad \tilde{\mathcal{C}}(\partial_a \pm \varepsilon_a\theta_{-a})\tilde{\mathcal{C}} = \mp(\partial_a \mp \varepsilon_a\theta_{-a}).$$

3. Linear transformations separating generators

We shall study linear transformations of the canonical pairs x_a, ∂_a or $\theta_a, \partial_a^\theta$. Let us write the bosonic case

$$x'_a = A_{ab}x_b + B_{ab}\partial_b, \; \partial'_a = C_{ab}x_b + D_{ab}\partial_b.$$

We may calculate the commutation relations of the new combinations in terms of the transformation matrix blocks A, B, C, D.

$$[x'_a, \partial'_b] = (BC^T - AD^T)_{ab}, \;\; [x'_a, x'_b] = (BA^T - AB^T)_{ab}, \;\; [\partial'_a, \partial_b] = (DC^T - CD^T)_{ab}.$$

Now we have the important case of linear canonical transformations, where the latter two (anti) commutators vanish and the first one is proportional to the unit matrix.

We shall consider the case of the transformation, where the opposite occurs. Namely, the first commutator vanishes and the other two are proportional to the distinguished matrix defining the invariant scalar product. In the orthogonal case it is the unit matrix I or the matrix $\hat{\varepsilon}^+$ with the elements $\varepsilon^+_{ab} = \delta_{a,-b}$ if one prefers to label the components by $-n, ..., +n$. In the symplectic case it is the antisymmetric matrix $\hat{\varepsilon}^-, \varepsilon^-_{ab} = \varepsilon_a \delta_{a,-b}$.

In the orthogonal case this can work only in the fermionic version, since the distinguished matrix is symmetric. In the symplectic case it can work only in the bosonic version since the distinguished matrix is anti-symmetric. In the following we specify the transformations as $A = I, B = -\varepsilon, C = -\varepsilon, D = I$. Then we have for the symplectic case

$$[x'_a, \partial'_b] = 0, \ [x'_a, x'_b] = -2\varepsilon_a \delta_{a,-b}, \ [\partial'_a, \partial'_b] = 2\varepsilon_a \delta_{a,-b},$$

and for the orthogonal case

$$[\theta'_a, \partial^{\theta,\prime}_b]_+ = 0, \ [\theta'_a, \theta'_b]_+ = 2\delta_{a,-b}, \ [\partial'_a, \partial'_b]_+ = -2\delta_{a,-b}.$$

Now let us rewrite the JS generators in the transformed canonical variables. In the orthogonal case we have

$$M_{ab} = \theta_a \partial^\theta_b - \theta_b \partial^\theta_a = M_{\gamma\ ab} - \tilde{M}_{\gamma\ ab}, \tag{11}$$

with the notations

$$\gamma_a = \theta'_a = \theta_a + \partial^\theta_a, \quad \tilde{\gamma}_a = -\partial^{\theta\prime}_a = \theta_a - \partial^\theta_a. \tag{12}$$

$$M_{\gamma\ ab} = \frac{1}{4}[\gamma_a, \gamma_b], \quad \tilde{M}_{\gamma\ ab} = \frac{1}{4}[\tilde{\gamma}_a, \tilde{\gamma}_b]. \tag{13}$$

In particular we have

$$[M_{\gamma\ ab}, \tilde{M}_{\gamma\ cd}] = 0$$

Therefore the two terms in (11) obey the orthogonal algebra relations independently. γ_a may be identified with the conventional gamma matrices. Indeed, they generate the Clifford algebra in $2n(+1)$ dimensions.

Note that the two sets of Clifford generators, γ_a and $\tilde{\gamma}_a$ are just the linear combinations (10) of the fermionic Heisenberg generators that are invariant up to sign with respect to the fermionic elementary canonical transformations (7).

In the symplectic case we have

$$G_{ab} = x_a \partial_b - \varepsilon_a \varepsilon_b x_{-b} \partial_{-a} = G_{\Gamma\ ab} + \tilde{G}_{\Gamma\ ab}, \tag{14}$$

with the notations

$$\tilde{\Gamma}_a = x'_a = x_a + \varepsilon_a \partial_{-a}, \ \Gamma_a = \partial'_a = \partial_a - \varepsilon_a x_{-a}, \tag{15}$$

$$\tilde{G}_{\Gamma\ ab} = \frac{1}{4}\varepsilon_b[\tilde{\Gamma}_a, \tilde{\Gamma}_{-b}]_+, \quad G_{\Gamma\ ab} = \frac{1}{4}\varepsilon_a[\Gamma_{-a}, \Gamma_b]_+. \tag{16}$$

In particular we have in analogy

$$[G_{\Gamma\ ab}, \tilde{G}_{\Gamma\ cd}] = 0. \tag{17}$$

Therefore the two terms in (14) obey the symplectic algebra relations independently. Γ_a generate the symplectic analogon of the Clifford algebra,

$$[\Gamma_a, \Gamma_b] = [\partial'_a, \partial'_b] = 2\varepsilon_a \delta_{a,-b} = 2\epsilon_{ab}. \tag{18}$$

We recall that the two sets of symplectic Clifford generators Γ_a and $\tilde{\Gamma}_a$ are just the linear combinations of the bosonic Heisenberg generators (10) invariant with respect to the sign dependent bosonic elementary canonical transformation (8).

In this way we have seen that the JS generators of $sp(n), so(2n)$ both separate in two terms, commuting and generating the equivalent algebra independently. Each set of such separated generators is built from $2n$ quasi-Clifford generators $\gamma_a, a = -n, ..., +n$.

The latter can also be viewed as n Heisenberg pairs, $\gamma_{-a}, \gamma_a, a = 1, ..., n$, bosonic in the symplectic case and fermionic in the orthogonal case. And these quasi-Clifford generators can be used for building JS representations of $g\ell(n)$.

4. Fundamental YB matrices and L operators

In both orthogonal and symplectic cases there are analoga of Yang's fundamental Yang-Baxter matrix $R_{12} = uI + P_{12}$ obeying the Yang-Baxter relation in the form

$$R^{a_1a_2}_{b_1b_2}(u)R^{b_1a_3}_{c_1b_3}(u+v)R^{b_2b_3}_{c_2c_3}(v) = R^{a_2a_3}_{b_2b_3}(v)R^{a_1b_3}_{b_1c_3}(u+v)R^{b_1b_2}_{c_1c_2}(u). \tag{19}$$

In the orthogonal and symplectic fundamental R matrices the existence of the invariant tensor causes a third term in the corresponding expressions, and related to this the dependence on the spectral parameter involves its second power. The well known results [1, 2, 4] are easily checked starting from the ansatz

$$R^{a_1a_2}_{b_1b_2}(u) = u(u+\beta)\delta^{a_1}_{b_1}\delta^{a_2}_{b_2} + (u+\beta)\delta^{a_1}_{b_2}\delta^{a_2}_{b_1} - u\varepsilon_{a_1}\varepsilon_{b_2}\delta^{a_1,-a_2}\delta^{b_1,-b_2}, \tag{20}$$

where we have denoted $\varepsilon_a = 1$ in the orthogonal case and $\varepsilon_a = sign(a)$ in the symplectic case. One finds that the above YB relation holds if

$$\beta = \beta_\pm = n \mp 1 \tag{21}$$

for the orthogonal or symplectic case, respectively.

Above we have reformulated the orthogonal case to get closer to the symplectic one by choosing the index range $a, b = -n, ...(0)..., n$ and have written the scalar product with the metric $\delta_{a,-b}$. The anti-symmetry relation then reads

$$M_{ab} = -M_{-b-a}, \ a, b = -n, ...(0)..., n \tag{22}$$

If we agree to denote by ε_a in the symplectic case $sign(a)$ and just 1 in the orthogonal case then the anti-symmetry relation in both cases reads

$$G_{ab} = -\varepsilon_a\varepsilon_b G_{-b,-a}, \tag{23}$$

and the invariant tensor defining the scalar product can be written as $\varepsilon_{a,b} = \varepsilon_a\delta_{a,-b}$. It is symmetric in $a \to b$ in the orthogonal case and anti-symmetric in the symplectic case. We have also $\varepsilon_{ab}\varepsilon^{bc} = \delta^c_a, \ \varepsilon^{ab} = \varepsilon_{-a}\delta^{a,-b}$.

Let us define the L matrix as

$$L_{ab} = u\delta_{ab} - G_{ab} \tag{24}$$

where $G_{ab}, a, b = -n, -n+1, ..., -1, (0), 1, ..., n$ obey the Lie algebra relations of $so(2n(+1)), sp(n)$, respectively.

We consider the RLL relation in the form

$$R^{a_1a_2}_{b_1b_2}(u)L^{b_1}_{c_1}(u+v)L^{b_2}_{c_2}(v) = L^{a_2}_{b_2}(v)L^{a_1}_{b_1}(u+v)R^{b_1b_2}_{c_1c_2}(u), \tag{25}$$

and look for the conditions under which this relation is obeyed.

We obtain that the additional condition on the generators for allowing a linear L matrix reads in the uniform notation

$$\varepsilon_{b_1,b_2}[G^{a_1b_1}, G^{a_2,b_2}]_+ = A\varepsilon^{a_1,a_2}. \tag{26}$$

If we choose the generators G^{ab} composed of the Clifford generators γ^a (13, 16) as

$$G^{ab} = \frac{1}{4}\varepsilon_{-a}[\gamma^a, \gamma^{-b}]_\mp \tag{27}$$

the additional condition (26) is fulfilled.

The Clifford algebra relation and its symplectic modification can be written uniformly as

$$[\gamma^a, \gamma^b]_\pm = 2\varepsilon^{ba}, \tag{28}$$

where we can choose the Clifford generators as the combinations of the basic Heisenberg generators, $\gamma^a|_{so} = \partial^\theta_{-a} + \theta_{\bar{a}}, \quad \gamma^a|_{sp} = \partial_a - \varepsilon_a x_{-a}$.

5. The spinorial Yang-Baxter operators

In general, for $L_i(u) = uI - G_i$ with $G^{ab} = \frac{1}{2}\gamma^{[ab]}$ (27), the RLL relation

$$R_{12}(u-v)L_1(u)L_2(v) = L_1(v)L_2(u)R_{12}(u-v) \tag{29}$$

can be considered as the defining relation for $R_{12}(u)$ which acts in the tensor product of two spinor representations. This relation results by separation of the dependence on $u+v$ from $u-v$ in the symmetry condition

$$[G_1 + G_2, R_{12}(u)] = 0,$$

and in the additional defining condition

$$u(R_{12}(u)G_2 - G_1R_{12}(u)) - (R_{12}(u)G_1G_2 - G_1G_2R_{12}(u)) = 0. \tag{30}$$

The symmetry condition implies that the spinorial R operator decomposes into the invariants composed from the Clifford generators in both tensor factors γ_1^a and γ_2^a,

$$\varepsilon_{a_1,a'_1}...\varepsilon_{a_k,a'_k}\mathcal{S}\{\gamma_1^{a_1}...\gamma_1^{a_k}\}\ \mathcal{S}\{\gamma_2^{a'_1}...\gamma_2^{a'_k}\} = \varepsilon_{A_k,A'_k}\gamma_1^{A_k}\ \gamma_2^{A'_k}. \tag{31}$$

The last form defines abbreviations in terms of which we write the ansatz

$$R(u) = \sum_k \varepsilon_{A_k,A_{k'}}\gamma_1^{A_k}\gamma_2^{A'_k}\frac{r_k(u)}{k!}. \tag{32}$$

The product of Clifford generators $\mathcal{S}\{\gamma_1^{a_1}...\gamma_1^{a_k}\}$ is anti-symmetrised in the orthogonal case and symmetrised in the symplectic case. Recall that the generators G_i^{ab} are composed of those γ_i^a in the anti-symmetrised or symmetrised way too (27).

In the defining condition (30) we encounter the multiplication of (anti-)symmetrised products. The transformation of such products into (anti-)symmetrised terms is conveniently done by using generating functions. We use auxiliary variables κ^a and their scalar product with the Clifford generators $(\kappa\gamma) = \varepsilon_{ab}\kappa^a\gamma^b$. They are anti-commuting (Grassmann) ones in the orthogonal case and commuting in the symplectic case.

$$\mathcal{S}\{\gamma^{a_1}...\gamma^{a_k}\} = \partial^{a_k}...\partial^{a_1}\ e^{(\kappa\gamma)}|_{\kappa=0} = \partial^{A_k}\ e^{(\kappa\gamma)}|_{\kappa=0},$$
$$\partial_a\kappa^b = \delta_a^b, \partial^a = \varepsilon^{ab}\partial_b, \partial^a\kappa^b = \varepsilon^{ab}, \varepsilon_{ab}\varepsilon^{bc} = \delta_a^c.$$

We illustrate the computations encountered here in the following example.

$$\gamma^{A_k}\ \gamma^{[ab]} = \partial_1^{A_k}\ e^{(\kappa_1\gamma)}\ \partial_2^b\partial_2^a e^{(\kappa_2\gamma)}|_{\kappa_1=\kappa_2=0} =$$
$$\partial_1^{A_k}\partial_2^b\partial_2^a\ e^{((\kappa_1+\kappa_2)\gamma)}\varepsilon^{(\kappa_1\kappa_2)}|_{\kappa_1=\kappa_2=0}.$$

Here we have used the Baker-Hausdorff formula and the generalised Clifford relation (28),

$$[(\kappa_1\gamma),(\kappa_2\gamma)] = \varepsilon_{ab}\varepsilon_{cd}\kappa_1^a\kappa_2^c[\gamma^b,\gamma^d]_\pm = 2\varepsilon^{db}\varepsilon_{ab}\varepsilon_{cd}\kappa_1^a\kappa_2^c =$$
$$2\varepsilon_{ac}\kappa_1^a\kappa_2^c = 2(\kappa_1\kappa_2).$$

Now we transform κ_1, κ_2 to $\kappa = \kappa_1 + \kappa_2, \kappa_2$. The derivatives with respect to the auxiliary κ variables then transform as $\partial_1 \to \partial,\ \partial_2 \to \partial + \partial_2$. This results in

$$\gamma^{A_k}\ \gamma^{[ab]} = \partial^{A_k}(\partial^b + \partial_2^b)(\partial^a + \partial_2^a)\ e^{(\kappa\gamma)}e^{(\kappa\kappa_2)}|_{\kappa=\kappa_2=0} =$$
$$\partial^{A_k}\{(\partial^b - \kappa^b)(\partial^a - \kappa^a) + \varepsilon^{ba}\}\ e^{(\kappa\gamma)}|_{\kappa=0}.$$

In this way we obtain that the additional defining condition (30) results in the relation for the coefficients of the expansion (32)

$$r_{k+2}(u) = \frac{u+k}{u-k\pm 2\beta_\pm}r_k(u). \tag{33}$$

The iteration goes in steps of 2, thus the spinorial R operator decomposes into the even and the odd parts, both obeying the defining relation independently, $R(u) = R^+(u) + R^-(u)$.

6. Discussion

We have considered the special representations (generators G_{ab}) of the orthogonal and symplectic algebras resulting in L operators of the form $L = uI - G$ obeying the YB relation with the fundamental R matrix. We have discussed the orthogonal and symplectic cases in analogy starting from the JS type representations of the general linear algebra.

In the Jordan-Schwinger case, the reduction of the algebra of $g\ell(2n(+1))$ to $so(2n(+1))$ or $sp(n)$ can be related to the elementary linear canonical transformation. This is the same transformation which connects the two fundamental representations of $g\ell(2n(+1))$ dual to each other.

In the resulting JS formulation of the orthogonal or symplectic generators a linear transformation of the underlying Heisenberg algebra generators leads to a separation in two terms which both obey the orthogonal or symplectic algebra and mutually commute. Thus in this form the algebra decomposes into two subalgebras. The generators of both subalgebras are found to be composed bilinearly of linear combinations of Heisenberg generators obeying the (quasi-) Clifford algebra relation.

These linear combinations of the underlying Heisenberg generators coincide with those obeying the condition of symmetry under the elementary canonical transformations.

The Clifford generators can be regarded equivalently as n Heisenberg pairs (creation and annihilation operators) fermionic in the orthogonal case and bosonic in the symplectic case. In the fermionic case we have the known spinor representation and in the symplectic case we find that the representation generated by n bosonic Heisenberg pairs is the appropriate counterpart of the spinorial one.

There are linear in the spectral parameter L operators in both cases, $L = Iu - G$, where $G^{a,b}$ are the generators of the orthogonal or symplectic algebra and obey additionally the condition (26). The generators bilinear in the Clifford generators obey this additional condition.

The Yang-Baxter R operator intertwining two spinorial representaions can be obtained in both orthogonal and symplectic cases in analogous way.

It can be checked that the fundamental R matrix (quadratic in u) can be reproduced by fusion including projection from the product of the spinorial L with its conjugate. More details about this and further results can be found in [7].

Acknowledgements

The authors are grateful to D Chicherin, S Derkachov and A Isaev for discussions. The work is supported by a Heisenberg-Landau grant (RK) and a Smorodinski-Ter-Antonyan grant (DK).

References

[1] Zamolodchikov A B and Zamolodchikov A B 1979 Factorized s Matrices in Two-Dimensions as the Exact Solutions of Certain Relativistic Quantum Field Models *Annals Phys.* **120** 253
[2] Berg B Karowski M Weisz P and Kurak V 1978 Factorized U(n) Symmetric s Matrices in Two-Dimensions *Nucl. Phys. B* **134** 125
[3] Shankar R and Witten E 1978 The S Matrix of the Kinks of the (ψ^-bar ψ)**2 Model *Nucl. Phys. B* **141** 349 *[Erratum-ibid. B* **148** (1979) 538]
[4] Reshetikhin N Yu 1985 Integrable models of quantum one-dimensional models with $O(n)$ and $Sp(2k)$ symmetry *Theor. Math. Fiz.* **63** 347-366
[5] Karakhanyan D and Kirschner R 2010 Jordan-Schwinger representations and factorised Yang-Baxter operators *SIGMA* **6** 029 [arXiv:0910.5144 [hep-th]]
[6] Chicherin D Derkachov S and Isaev A P 2013 Conformal group: R-matrix and star-triangle relation arXiv:1206.4150 [math-ph] *and* The spinorial R-matrix *J. Phys. A* **46** 485201 arXiv:1303.4929 [math-ph]
[7] Isaev A P Karakhanyan D and Kirschner R *in preparation*

The right-hand side of the Jacobi identity: to be naught or not to be?

Arthemy V Kiselev

Johann Bernoulli Institute for Mathematics and Computer Science, University of Groningen, P.O.Box 407, 9700 AK Groningen, The Netherlands

E-mail: `A.V.Kiselev@rug.nl`

Abstract. The geometric approach to iterated variations of local functionals – e.g., of the (master-)action functional – resulted in an extension of the deformation quantisation technique to the set-up of Poisson models of field theory. It also allowed of a rigorous proof for the main inter-relations between the Batalin–Vilkovisky (BV) Laplacian Δ and variational Schouten bracket $[\![\,,\,]\!]$. The ad hoc use of these relations had been a known analytic difficulty in the BV-formalism for quantisation of gauge systems; now achieved, the proof does actually not require the assumption of graded-commutativity. Explained in our previous work, geometry's self-regularisation is rendered by Gel'fand's calculus of singular linear integral operators supported on the diagonal.

We now illustrate that analytic technique by inspecting the validity mechanism for the graded Jacobi identity which the variational Schouten bracket does satisfy (whence $\Delta^2 = 0$, i.e., the BV-Laplacian is a differential acting in the algebra of local functionals). By using one tuple of three variational multi-vectors twice, we contrast the new logic of iterated variations – when the right-hand side of Jacobi's identity vanishes altogether – with the old method: interlacing its steps and stops, it could produce some non-zero representative of the trivial class in the top-degree horizontal cohomology. But we then show at once by an elementary counterexample why, in the frames of the old approach that did not rely on Gel'fand's calculus, the BV-Laplacian failed to be a graded derivation of the variational Schouten bracket.

Keywords: Variational multi-vectors, Schouten bracket, Jacobi identity, Batalin–Vilkovisky Laplacian, symbolic computations.

1. Introduction

To describe Nature, we need analytic, geometric, and algebraic instruments. Their formalism must simultaneously be mathematically rigorous and yield relevant physical predictions. One of such tools – to work with the geometry of iterated variations of local functionals in field theory models – was designed in [1, 2], see also [3, 4]. The aim of this paper is to illustrate by explicit examples of calculations, commented at every step, how this new technique is used in practice. We refer to the previous report [1] for basic theory outline in §1–2 therein, viewing it as a pre-requisite for the further exercise. Let us keep in mind that the calculus of singular linear integral operators [5] is the working language of that mathematical technique — itself serving the tasks of theoretical physics (cf. [6, 7]). That geometry's and formalism's noncommutative extension has been developed in [4, §1–2] (see also [8] and [9] in a similar context), whereas their implementation in the graded-commutative set-up is performed in [1, §3] and [3, §3]. Besides the proof of theorems in [1, 2, 4], the reader will find there an overview of the history of the problem – e.g., in the context of Batalin–Vilkovisky's formalism for quantisation of gauge systems [10] – accompanied with a list of literature references. The use of new technique in the lift of Kontsevich's deformation quantisation procedure [11] to the geometry of field models is crucial in [3]. We finally recall that the master-functional realisation of zero-curvature representations for kinematically integrable systems was obtained in [12], giving rise to the problem of BV-quantisation in the context of inverse scattering method. That line of research also relies on the technique which is exemplified in this note.

1.1. The geometry of functionals: an overview

Before motivating the choice of a class of examples to analyse, let us first recall an elementary still crucial distinction between the concepts of *functions* on finite-dimensional smooth manifolds and, on the other hand, *functionals* that appear in field theory models.[1] (In retrospect, the examples under study would have motivated the necessity to re-consider the geometry of iterated variations and find the proper analytic apparatus [1] that did resolve the difficulties; the existence of conceptual difficulties in the formalism was of course known long ago, see [7, §15.1].)

Specifically, a function $f\colon N^n \to \Bbbk$ on a finite-dimensional manifold N^n is the *zero function* whenever its value at every point of the domain of definition is the zero $0 \in \Bbbk$. On the other hand, by integrating the physical field values $\phi(\boldsymbol{x}) \in N^n$ over points $\boldsymbol{x} \in M^m$ of the underlying space-time $\bigl(M^m, \mathrm{dvol}(\boldsymbol{x}, \phi(\boldsymbol{x}))\bigr)$, a given field model's functional $F\colon \Gamma(\pi) \to \Bbbk$ is a *zero functional* if its value $F\bigl([\phi]\bigr)$ is $0 \in \Bbbk$ for all configurations $\phi \in \Gamma(\pi)$. Still is it readily seen that, provided that the topology of the bundle π and all the boundary conditions are properly taken into account (typically, either the boundary conditions are periodic or one assumes the sections' rapid decay towards the boundary ∂M^m, if any), a local functional would always attain the zero value at all sections $\phi \in \Gamma(\pi)$ if at least one factor in every integral functionals' formal product is realised by an exact top-degree horizontal form $\mathrm{d}_h(\eta)$, so that $\int \mathrm{d}_h(\eta) \cong \int 0 \in \bar{H}^m(\pi)$. (Whenever the topology of π creeps in, one quotients out the respective invariants of the bundle π; for instance, we have that $\int_{\mathbb{S}^1} \mathrm{d}\alpha - 2\pi \equiv 0$ for all sections $\phi \in \Gamma(\pi)$ of bundles over the circle $M^1 := \mathbb{S}^1$.)

However, we discovered in [1] (cf. [3, §3.4]) that there is a crucial distinction between generic homologically-trivial integral factors $\int \mathrm{d}_h(\eta) \cong \int 0 \in \bar{H}^m(\pi)$ and the zero-density functional $\int 0 \cdot \mathrm{dvol}(\cdot, \phi(\cdot))$. This alternative echoes whenever a functional which manifestly yields zero for every section $\phi \in \Gamma(\pi)$ appears at an intermediate but strictly not the last step of a reasoning. Due to that distinction's mechanism, which will be re-addressed and illustrated in what follows, the *synonyms for zero* $\int \mathrm{d}_h(\eta) \cong \int 0 \in \bar{H}^m(\pi)$ can contribute nontrivially to reasoning's end-product. This paradoxical behaviour was explained within the geometric approach to iterated variations of local functionals [1, 4]; the resolution did allow of a rigorous proof of formalism's main identities (e.g., compare (12) on p. 15, which had been postulated ad hoc, with (14) below).

The geometric mechanism behind that puzzling effect is based on the following fact. Each integral object in a given geometry of fibre (super)bundle π for fields $\phi \in \Gamma(\pi)$ refers to its own copy of the base manifold M^m over which the integration by using the volume element $\mathrm{dvol}(\cdot, \phi(\cdot))$ is performed. Represented by functionals $\Gamma(\pi) \to \Bbbk$, the theory's physical objects interact by serving as the input data for operators and structures such as the Batalin–Vilkovisky Laplacian Δ or variational Schouten bracket $[\![\,,\,]\!]$; in this way, the new object is produced. In the frames of local field theory, the processing of input data entails the

[1] Not only the *integral* functionals are then considered – e.g., such as the action $S = \int L(\boldsymbol{x}, [\boldsymbol{q}]) \cdot \mathrm{dvol}(\boldsymbol{x}) \colon \Gamma(\pi) \to \Bbbk$ that takes the field configurations $\{\boldsymbol{q} = \phi(\boldsymbol{x}),\ \boldsymbol{x} \in M^m\}$ from the set $\Gamma(\pi) \ni \phi$ of sections for the (super)bundle π of a given physical model to the (complex) number field $\Bbbk$. The (sums of) formal products of integral functionals are also introduced, thus extending the vector space $\bar{H}^m(\pi) \ni S$ to the algebra $\overline{\mathfrak{M}}^m(\pi)$ of *local* functionals. For instance, the functional $\exp\bigl(\frac{i}{\hbar} S([\boldsymbol{q}])\bigr) = 1 + \frac{i}{\hbar} S - \frac{1}{2\hbar^2} S \cdot S + \cdots \colon \Gamma(\pi) \to \mathbb{C}[[\hbar]]$, depending on the field configurations $\{\boldsymbol{q} = \phi(\boldsymbol{x})\}$ and possibly, on the sections' derivatives (which is denoted by using the square brackets: $[\ldots]$), serves as a weight factor in the construction of Feynman path integral.

merging of all copies of the constintuent objects' integration domains M^m, so that such functionals' densities – or their derivatives – are then evaluated at the diagonal in the respective products $M^m \times \ldots \times M^m$ of the base copies. (Moreover, the evaluation of all densities belonging to different input objects occurs at the graph of a given field portrait $\{\phi(\boldsymbol{x}),\ \boldsymbol{x} \in M^m\}$, now shared by all the input objects' contributions.) When does the restriction to such diagonals take place? The answer is not obvious. Whereas the full geometric picture keeps track of several copies $M^m \times \ldots \times M^m$ of the integration manifold M^m – in both the functionals and (pairs of) field variations' parity-even and odd components, – too early does the traditional approach merge the integration domains. In effect, the results of two seemingly identical calculations can become unequal as colomology classes[2] — for the same input data!

Clearly, it is desirable that, serving the geometry of field theory, the calculus were consistent, not led to contradictions in the course of verification of formalism's main identities, and that it allowed of maningful predictions in the models at hand. The practical value of this note is that it helps to avoid inconsistencies or brough-in mismatches, hence preventing a necessity of postfactum regularisation by adding the various correction counterterms etc.

1.2. Another look on the variation of functionals

Now let us describe the class of identities which are examined and exemplified in what follows. To this end, we recall that the variational geometry of jet bundles $J^\infty\big(\pi\colon E^{m+(n_0|n_1)} \xrightarrow{N^{(n_0|n_1)}} M^m\big)$ and the calculus of variational multi-vectors on jet spaces enlarge the symplectic geometry for usual (super) manifolds $N^{(n_0|n_1)}$, parametrised locally by using parity-even and odd local coordinates $(\boldsymbol{q}, \boldsymbol{q}^\dagger)$; one can integrate by parts in the new set-up. The classes of highest horizontal cohomology groups $\bar{H}^m(\pi)$ are generated by the lift d_h of the de Rham differential on the base M^m to the total space $J^\infty(\pi)$. Each d_h-cohomology class carries a substantial freedom: a d_h-exact term added to a class representative in the input data, nothing changes in the output of a calculation — provided that a variational derivative acts on that input term *at once.*[3]

In these terms, the variational derivatives $\delta/\delta\boldsymbol{q}$ and $\delta/\delta\boldsymbol{q}^\dagger$ are viewed in the Lagrangian or Hamiltonian formalisms [13] as proper extensions of the partial derivatives $\partial/\partial\boldsymbol{q}$ and $\partial/\partial\boldsymbol{q}^\dagger$ along the fibre $N^{(n_0|n_1)}$ in the (super-)bundle π. These extensions are immediate indeed, meaning that all the integrations by parts over M^m are indivisibly attached to the derivations along the fibre. The two analytic operations are not separated, even though the integrations could be postponed to a later moment. The resulting approach is adequate for the one-step derivation of Euler–Lagrange equations and for verification of the property for some systems of PDEs to be manifestly Euler–Lagrange (by using the Helmholtz criterion, see [14]). The conventional formalism is sufficient also for the construction of variational Poisson brackets of integral functionals.[4]

Strictly speaking, it follows from nowhere why the integrations by parts over the base manifold M^m should be inseparable from the partial derivatives along the fibres in $J^\infty(\pi)$. It could well be that some other modus operandi is preferrable as soon as the transition from the supermanifold $N^{(n|n)}$ standing alone to the super-bundle π with the fibre $N^{(n|n)}$ over each point $\boldsymbol{x} \in M^m$ is accomplished.

An alternative approach has been reported in [1], see also [4, §2.2] and [3, §3.2]. In brief, every calculation is split in two steps. First, all the partial derivatives $\partial/\partial\boldsymbol{q}_\sigma$ and $\partial/\partial\boldsymbol{q}^\dagger_\tau$ act on the input functionals' densities. We remember that each input object brings its own copy of

[2] For instance, the inequality $[\![\cdot,\Delta(\cdot)]\!](F,H) \not\cong [\![\cdot,\cdot]\!](F,\Delta H)$ *can* be valid, cf. conclusive Remark 1 on p. 16.

[3] Let us recall that the identity $\delta/\delta\boldsymbol{q} \circ \mathrm{d}/\mathrm{d}x \equiv 0$ is the primary exercise in the entire theory of variations.

[4] The variational Schouten bracket $[\![\,,\,]\!]$ itself is an example of variational Poisson bracket for the specific $\mathbb{Z}_2$-graded set-up such that the fibres $N^{(n|n)}$ are described by using the parity-even coordinates $\boldsymbol{q}$ and their canonical conjugates $\boldsymbol{q}^\dagger$ of odd parity, cf. [4, §1–2].

the base M^m into the emergent product of bundles (namely, there is one copy of the bundle π for each integral functional and there is a copy of the tangent bundle $T\pi$ for each variation of $\phi \in \Gamma(\pi)$ along the fibres of π). This means that the newly constructed, intermediate objects still retain a kind of memory about the way how they have been produced. Now is the second step of a reasoning: the integration domains M^m in the products $M^m \times \ldots \times M^m$ are merged by the restriction to the diagonal. The integrations by parts then produce the respective total derivatives $\pm \mathrm{d}/\mathrm{d}\boldsymbol{x}^\sigma$ and $\pm \mathrm{d}/\mathrm{d}\boldsymbol{x}^\tau$ along the base $M^m \ni \boldsymbol{x}$ (e.g., contrast (4) on p. 8 with (11) on p. 13 below).

The Batalin–Vilkovisky technique for quantisation of gauge systems [10] puts the mathematical methods of the calculus of variations to a hard test (cf. [3, §3.2.3]). The traditional approach [13] to variation of local functionals is incapable of avoiding contradictions in the calculus (e.g., see a counterexample on p. 15 below; let us repeat that this class of difficulties had been known well in the literature [7, §15.1]). In consequence, the claims in [15, §1.3] about interrelation between the Batalin–Vilkovisky Laplacian Δ and variational Schouten bracket $[\![\,,\,]\!]$ are legitimate; they codify the picture that can be rigorously substantiated in the supergeometry of ordinary supermanifolds, but it is the jet-superbundle set-up in which there were neither proof nor examples (in fact, identity (12) on p. 15 below would have been the sumbling-stone). The effect produced by our revision of action priorities is that the constructions of BV-formalism and the logic of its formulae resume working (in particular, justifying the claims made in [15]).

The right-hand side of Jacobi's identity (6) for the variational Schouten bracket $[\![\,,\,]\!]$ is a convenient indicator of the difference between the two approaches. By following the traditional scheme, one obtains some cohomologically trivial term in the right-hand side of the identity; that representative of the equivalence class of zero functional depends on a choice of representatives for the three inputs of the Jacobiator $\mathrm{Jac}(\cdot,\cdot,\cdot)$ in the left-hand side. But within the geometric picture, the identity's right-hand side is the integral functional the density of which vanishes by construction. (It is the researcher who could then add to it at will any other trivial functional.) The aim of this paper is to illustrate such distinction. For this, we also show that within the same (super)geometry yet in a class of elementary problems just next to the verification of Jacobi's identity, all rigour is destroyed if the old, step-by-step calculation recipe is used. This is confirmed in this paper by using a counterexample (the full matching of the same objects is demonstrated in [1] by using a "counter-counterexample").

The structure of this paper parallels the length of formulae under consideration. Indeed, the three functionals F, G, and H which we use in all the examples in this text are such that the calculation of $\mathrm{Jac}(F,G,H) = 0$ within the true theory of variations is done by hand in §2. The cohomology class estimate $\mathrm{Jac}\bigl(F(x),G(x),H(x)\bigr) \cong 0$ in the frames of traditional approach would be also manageable by hand but hard (see Example 2 in §3), whereas the inspection of validity mechanism for Jacobi's identity via the restriction $\mathrm{Jac}\bigl(F(x),G(y),H(z)\bigr)\Big|_{x=y=z} \cong 0$ is fairly impossible without using proper software for symbolic computations [16, 17], see Example 3 on p. 13. A short Counterexample 4 in §4 takes us back to the main theorem in [1] and its proper illustration (contained on pp. 34–36 therein): the full picture of iterated variations is a key to the resolution of apparent inconsistency.

The notation is standard and as simple as possible. We shall work in the (graded-) commutative set-up; we refer to [4] for its generalisation to a class of non-commutative geometries (see also references therein; the seminal paper is [8]). In all examples, the base manifold $M^m \ni \boldsymbol{x}$ is one-dimensional and boundary terms are always discarded.[5] The two fibre coordinates in the

[5] Whenever the model geometry is taken from the closed string or brane theory, the source manifold M^m is closed so that there is no boundary to have any such terms at (cf. [18]). In the frames of field theory one postulates the rapid decay of every section $\phi \in \Gamma(\pi)$ at the space-time infinity [14, 19], but it would take quite some effort to get

vector superbundle π are the parity-even q and its canonical conjugate, parity-odd $q^\dagger$. The volume form $\mathrm{dvol}(x)$ in the integral functionals is just $\mathrm{d}x$ in the weak-field approximation.

2. Geometric approach to the Jacobi identity

2.1. The variational Schouten bracket

Denote by $\boldsymbol{x}$ a local coordinate on the base M^m of the variational cotangent superbundle π; denote by $\boldsymbol{q}$ and $\boldsymbol{q}^\dagger$ the n-tuples of respective parity-even and parity-odd canonical conjugate variables along the fibres of π. Suppose $\boldsymbol{s} \in \Gamma(\pi)$ is a section. Denote by $\vec{e}_i(\boldsymbol{x}, \boldsymbol{s}(\boldsymbol{x}))$ and $\vec{e}^{\dagger,i}(\boldsymbol{x}, \boldsymbol{s}(\boldsymbol{x}))$ a smooth field of dual bases in the parity-even and odd halves of the spaces $T_{(\boldsymbol{x},\boldsymbol{s}(\boldsymbol{x}))}\pi^{-1}(\boldsymbol{x})$ tangent –at points $(\boldsymbol{x}, \boldsymbol{s}(\boldsymbol{x}))$– to the fibres over points $\boldsymbol{x} \in M^m$. By construction, the two ordered couplings between elements of those bases are normalised: at every value of the index i we have that (no summation!)

$$\underrightarrow{\langle \overset{\text{first}}{\vec{e}_i(\boldsymbol{y}_1, \boldsymbol{s}(\boldsymbol{y}_1))}, \overset{\text{second}}{\vec{e}^{\dagger,i}(\boldsymbol{y}_2, \boldsymbol{s}(\boldsymbol{y}_2))}\rangle}\Big|_{\boldsymbol{y}_1=\boldsymbol{y}_2} = +1, \qquad \underrightarrow{\langle \overset{\text{first}}{\vec{e}^{\dagger,i}(\boldsymbol{y}_2, \boldsymbol{s}(\boldsymbol{y}_2))}, \overset{\text{second}}{\vec{e}_i(\boldsymbol{y}_1, \boldsymbol{s}(\boldsymbol{y}_1))}\rangle}\Big|_{\boldsymbol{y}_1=\boldsymbol{y}_2} = -1. \tag{1}$$

For the sake of brevity, we shall not indicate the basic (co)vectors' dependence on points in the fibres of π, writing only $\vec{e}_i(\boldsymbol{y})$ or $\vec{e}^{\dagger,i}(\boldsymbol{z})$ from now on.

Consider an infinitesimal shift,

$$\big(\delta\boldsymbol{q}(\boldsymbol{y}), \delta\boldsymbol{q}^\dagger(\boldsymbol{y})\big) = \sum_{i=1}^{n}\Big(\delta s^i\big(\boldsymbol{y}, \boldsymbol{s}(\boldsymbol{y})\big)\cdot\vec{e}_i(\boldsymbol{y}) + \delta s_i^\dagger\big(\boldsymbol{y}, \boldsymbol{s}(\boldsymbol{y})\big)\cdot\vec{e}^{\dagger,i}(\boldsymbol{y})\Big),$$

of the bipartite section $\boldsymbol{s} = \big(s^i(\boldsymbol{x}), s_i^\dagger(\boldsymbol{x})\big)$ of the variational cotangent superbundle π. We postulate that for every i and over all points $\boldsymbol{x} \in M^m$ of the support $\operatorname{supp}\delta s^i\big(\boldsymbol{x}, \boldsymbol{s}(\boldsymbol{x})\big)$, the respective components δs^i and $\delta s_i^\dagger$ of the section $(\delta\boldsymbol{q}, \delta\boldsymbol{q}^\dagger) \in \Gamma(T\pi)$ are normalised by using the rule

$$\delta s^i\big(\boldsymbol{x}, \boldsymbol{s}(\boldsymbol{x})\big)\cdot\delta s_i^\dagger\big(\boldsymbol{x}, \boldsymbol{s}(\boldsymbol{x})\big) \equiv 1 \qquad \text{(no summation over } i\text{)}.$$

By convention, the vertical differentials (i.e., those involving only the derivations along the fibres of $J^\infty(\pi)$) of all objects from the variational symplectic supergeometry are always expanded with respect to the elements $+\vec{e}^{\dagger,i}(\cdot)$ and $+\vec{e}_i(\cdot)$ of the mutually dual bases. Namely, such differentials are expressed by

$$\Big(f\big(\boldsymbol{x}_1, [\boldsymbol{q}], [\boldsymbol{q}^\dagger]\big)\Big)\frac{\overleftarrow{\partial}}{\partial q^i}\cdot\vec{e}^{\dagger,i}(\boldsymbol{x}_1) + \ldots \qquad \text{or} \qquad \ldots + \vec{e}_i(\boldsymbol{x}_2)\cdot\frac{\overrightarrow{\partial}}{\partial q_i^\dagger}\Big(g\big(\boldsymbol{x}_2, [\boldsymbol{q}], [\boldsymbol{q}^\dagger]\big)\Big),$$

where f and g are densities of the integral objects to-vary. Supported on the diagonal, the singular linear integral operators for integral objects' variation –in the course of infinitesimal shifts of the sections at which the objects will be evaluated– are such that the couplings in (1) always yield $+1$. Specifically, we have that

$$\overrightarrow{\delta\boldsymbol{s}} = \int \mathrm{d}\boldsymbol{y}\,\Big\langle (\delta s^i)\Big(\frac{\overleftarrow{\partial}}{\partial\boldsymbol{y}}\Big)^{\sigma}(\boldsymbol{y})\cdot\underrightarrow{\overset{\text{first}}{\vec{e}_i(\boldsymbol{y})}\mid\overset{\text{second}}{\vec{e}^{\dagger,i}(\cdot)}}\Big\rangle\frac{\overrightarrow{\partial}}{\partial q^i_\sigma},$$

there and bring back the minus sign from the integration by parts. However, the core idea of fields as excitations of the local degrees of freedom [6, 7, 10] allows us to consider their test shifts $\delta\boldsymbol{q}(\boldsymbol{x}, \phi(\boldsymbol{x}))$ with compact supports concentrated only around a point of M^m, whence improper integrals over the space-time make sense, reducing to proper integrals over such tiny neighbourhoods.

$$\overrightarrow{\delta\boldsymbol{s}^\dagger} = \int \mathrm{d}\boldsymbol{y} \left\langle (\delta s_i^\dagger)\Bigl(\frac{\overleftarrow{\partial}}{\partial\boldsymbol{y}}\Bigr)^\sigma(\boldsymbol{y}) \cdot \underset{\longrightarrow}{\overset{\text{first}\qquad\text{second}}{(-\vec{e}^{\,\dagger,i})(\boldsymbol{y}) \mid \vec{e}_i(\,\cdot\,)}} \right\rangle \frac{\overrightarrow{\partial}}{\partial q^\dagger_{i,\sigma}},$$

$$\overleftarrow{\delta\boldsymbol{s}} = \int \mathrm{d}\boldsymbol{y}\, \frac{\overleftarrow{\partial}}{\partial q^i_\sigma} \left\langle \underset{\longleftarrow}{\overset{\text{second}\qquad\text{first}}{\vec{e}^{\,\dagger,i}(\,\cdot\,) \mid \vec{e}_i(\boldsymbol{y})}} \cdot \Bigl(\frac{\overrightarrow{\partial}}{\partial\boldsymbol{y}}\Bigr)^\sigma (\delta s^i)(\boldsymbol{y}) \right\rangle,$$

$$\overleftarrow{\delta\boldsymbol{s}^\dagger} = \int \mathrm{d}\boldsymbol{y}\, \frac{\overleftarrow{\partial}}{\partial q^\dagger_{i,\sigma}} \left\langle \underset{\longleftarrow}{\overset{\text{second}\qquad\text{first}}{\vec{e}_i(\,\cdot\,) \mid (-\vec{e}^{\,\dagger,i})(\boldsymbol{y})}} \cdot \Bigl(\frac{\overleftarrow{\partial}}{\partial\boldsymbol{y}}\Bigr)^\sigma (\delta s_i^\dagger)(\boldsymbol{y}) \right\rangle,$$

see [1, §2.2–3] for details.[6]

Let $F = \int f(\boldsymbol{x}_1, [\boldsymbol{q}], [\boldsymbol{q}^\dagger])\,\mathrm{dvol}(\boldsymbol{x}_1)$ and $G = \int g(\boldsymbol{x}_2, [\boldsymbol{q}], [\boldsymbol{q}^\dagger])\,\mathrm{dvol}(\boldsymbol{x}_2)$ be two integral functionals in the variational symplectic supergeometry of jet bundle $J^\infty(\pi)$ over π. Consider their formal product $F \cdot G$; let us analyse the construction of response $(\overrightarrow{\delta\boldsymbol{s}} \circ \overrightarrow{\delta\boldsymbol{s}^\dagger})(F \cdot G)$ to consecutive variations of the parity-odd, then parity-even components of a section $\boldsymbol{s} \in \Gamma(\pi)$. (Assume that the infinitesimal shifts $\delta s^i(\boldsymbol{y}, \boldsymbol{s}(\boldsymbol{y}))$ and $\delta s_i^\dagger(\boldsymbol{y}, \boldsymbol{s}(\boldsymbol{y}))$ are given.) By the graded Leibniz rule, we have that

$$\begin{aligned}(\overrightarrow{\delta\boldsymbol{s}} \circ \overrightarrow{\delta\boldsymbol{s}^\dagger})(F \cdot G) &= \overrightarrow{\delta\boldsymbol{s}}\Bigl(\overrightarrow{\delta\boldsymbol{s}^\dagger}(F) \cdot G + (-)^{|F|} F \cdot \overrightarrow{\delta\boldsymbol{s}^\dagger}(G)\Bigr) = \\ &= (\overrightarrow{\delta\boldsymbol{s}} \circ \overrightarrow{\delta\boldsymbol{s}^\dagger})(F) \cdot G + (-)^{|F|}\overrightarrow{\delta\boldsymbol{s}}(F) \cdot \overrightarrow{\delta\boldsymbol{s}^\dagger}(G) + \overrightarrow{\delta\boldsymbol{s}^\dagger}(F) \cdot \overrightarrow{\delta\boldsymbol{s}}(G) + (-)^{|F|} F \cdot (\overrightarrow{\delta\boldsymbol{s}} \circ \overrightarrow{\delta\boldsymbol{s}^\dagger})(G).\end{aligned}$$

Using a slightly counterintuitive lemma[7] $\overrightarrow{\partial}/\partial\boldsymbol{q}^\dagger(F) = (-)^{|F|-1}(F)\overleftarrow{\partial}/\partial\boldsymbol{q}^\dagger$, let us reverse the direction in which the operators $\overrightarrow{\delta\boldsymbol{s}}$ and $\overrightarrow{\delta\boldsymbol{s}^\dagger}$ act on F in the second and third terms of the formula above; this yields[8]

$$= (\overrightarrow{\delta\boldsymbol{s}} \circ \overrightarrow{\delta\boldsymbol{s}^\dagger})(F) \cdot G + (-)^{|F|}\Bigl((F)\overleftarrow{\delta\boldsymbol{s}} \cdot \overrightarrow{\delta\boldsymbol{s}^\dagger}(G) - (F)\overleftarrow{\delta\boldsymbol{s}^\dagger} \cdot \overrightarrow{\delta\boldsymbol{s}}(G)\Bigr) + (-)^{|F|} F \cdot (\overrightarrow{\delta\boldsymbol{s}} \circ \overrightarrow{\delta\boldsymbol{s}^\dagger})(G).$$

Let us have a closer look on the difference of second and third terms: it is

$$\begin{aligned}&\iint \mathrm{d}\boldsymbol{y}_1\,\mathrm{dvol}(\boldsymbol{x}_1)\,(f(\boldsymbol{x}_1, [\boldsymbol{q}], \boldsymbol{q}^\dagger]))\frac{\overleftarrow{\partial}}{\partial q^{i_1}_{\sigma_1}} \left\langle \underset{\longleftarrow}{\overset{\text{second}\qquad\text{first}}{\vec{e}^{\,\dagger,i_1}(\boldsymbol{x}_1), \vec{e}_{i_1}(\boldsymbol{y}_1)}} \cdot \Bigl(\frac{\overrightarrow{\partial}}{\partial\boldsymbol{y}_1}\Bigr)^{\sigma_1}(\delta s^{i_1})(\boldsymbol{y}_1)\right\rangle \cdot \\ &\cdot \iint \mathrm{d}\boldsymbol{y}_2 \left\langle (\delta s^\dagger_{i_2})\Bigl(\frac{\overleftarrow{\partial}}{\partial\boldsymbol{y}_2}\Bigr)^{\sigma_2}(\boldsymbol{y}_2) \cdot \underset{\longrightarrow}{\overset{\text{first}\qquad\text{second}}{(-\vec{e}^{\,\dagger,i_2})(\boldsymbol{y}_2), \vec{e}_{i_2}(\boldsymbol{x}_2)}}\right\rangle \frac{\overrightarrow{\partial}}{\partial q^\dagger_{i_2,\sigma_2}}(g(\boldsymbol{x}_2, [\boldsymbol{q}], [\boldsymbol{q}^\dagger]))\,\mathrm{dvol}(\boldsymbol{x}_2) - \\ &- \iint \mathrm{d}\boldsymbol{y}_1\,\mathrm{dvol}(\boldsymbol{x}_1)\,(f(\boldsymbol{x}_1, [\boldsymbol{q}], \boldsymbol{q}^\dagger]))\frac{\overleftarrow{\partial}}{\partial q^\dagger_{i_1,\sigma_1}} \left\langle \underset{\longleftarrow}{\overset{\text{second}\qquad\text{first}}{\vec{e}_{i_1}(\boldsymbol{x}_1), (-\vec{e}^{\,\dagger,i_1})(\boldsymbol{y}_1)}} \cdot \Bigl(\frac{\overrightarrow{\partial}}{\partial\boldsymbol{y}_1}\Bigr)^{\sigma_1}(\delta s^\dagger_{i_1})(\boldsymbol{y}_1)\right\rangle \cdot \\ &\cdot \iint \mathrm{d}\boldsymbol{y}_2 \left\langle (\delta s^{i_2})\Bigl(\frac{\overleftarrow{\partial}}{\partial\boldsymbol{y}_2}\Bigr)^{\sigma_2}(\boldsymbol{y}_2) \cdot \underset{\longrightarrow}{\overset{\text{first}\qquad\text{second}}{\vec{e}_{i_2}(\boldsymbol{y}_2), \vec{e}^{\,\dagger,i_2}(\boldsymbol{x}_2)}}\right\rangle \frac{\overrightarrow{\partial}}{\partial q^{i_2}_{\sigma_2}}(g(\boldsymbol{x}_2, [\boldsymbol{q}], [\boldsymbol{q}^\dagger]))\,\mathrm{dvol}(\boldsymbol{x}_2).\end{aligned}$$

[6] In particular, the matching of indexes i in the expansions of sections' shifts and objects' differentials reflects the fact that the bases $\vec{e}_i$ and $\vec{e}^{\,\dagger,i}$ are dual. Likewise, the matching of multi-indices σ refers to the definition of vector as equivalence class of trajectories passing through its attachment point.

[7] Not only the parity-odd derivation $\vec{\partial}$ overtakes a graded object F but also, whenever the derivation reaches its final location at the object's right side, the graded operator $\vec{\partial}$ is *destroyed* and replaced by the graded derivation acting in the right-to-left direction. Without such last step, the rule of signs would dictate the formula $\vec{\partial} \circ F = \vec{\partial}(F) + (-)^{|F|\cdot|\partial|} F \circ \vec{\partial}$.

[8] Further processing of the first and last terms in the formula at hand –that is, the on-the-diagonal reconfigurations of couplings and integrations by parts– is completely analogous to the algorithm for dealing with the second and third terms, see below; the result is (3).

As it has been explained in [1] and then in [4, §2.5–6], the conversion of two pairs of variations in $(F)\overleftarrow{\delta s}\cdot\overrightarrow{\delta s^\dagger}(G)-(F)\overleftarrow{\delta s^\dagger}\cdot\overrightarrow{\delta s}(G)$ into one integral object – via integrations by parts on the diagonal $\boldsymbol{x}_1=\boldsymbol{y}_1=\boldsymbol{y}_2=\boldsymbol{x}_2$ through many consecutive reconfigurations of the couplings – determines the functional[9]

$$\begin{aligned}
&\iiiint \mathrm{d}\boldsymbol{x}_1\mathrm{d}\boldsymbol{y}_1\mathrm{d}\boldsymbol{y}_2\,\bigl(f(\boldsymbol{x}_1,[\boldsymbol{q}],\boldsymbol{q}^\dagger])\bigr)\frac{\overleftarrow{\partial}}{\partial q^{i_1}_{\sigma_1}}\ulcorner\Bigl(-\frac{\overleftarrow{\mathrm{d}}}{\mathrm{d}\boldsymbol{y}_1}\Bigr)^{\sigma_1}\urcorner\,\Bigl\langle\underset{}{\overset{\text{first}}{\underline{\vec{e}^{\,\dagger,i_1}(\boldsymbol{x}_1)}}}\Bigr|\cdot\\
&\quad\cdot\Bigl\langle\delta s^{i_1}(\boldsymbol{y}_1)\cdot\overset{\text{first}}{\underline{\underline{\vec{e}_{i_1}(\boldsymbol{y}_1)}}},\overset{\text{second}}{\underline{\underline{(-\vec{e}^{\,\dagger,i_2})(\boldsymbol{y}_2)}}}\cdot\delta s^\dagger_{i_2}(\boldsymbol{y}_2)\Bigr\rangle\cdot\\
&\qquad\cdot\Bigl|\overset{\text{second}}{\underline{\vec{e}_{i_2}(\boldsymbol{x}_2)}}\Bigr\rangle\cdot\ulcorner\Bigl(-\frac{\overrightarrow{\mathrm{d}}}{\mathrm{d}\boldsymbol{y}_2}\Bigr)^{\sigma_2}\urcorner\frac{\overrightarrow{\partial}}{\partial q^\dagger_{i_2,\sigma_2}}\bigl(g(\boldsymbol{x}_2,[\boldsymbol{q}],[\boldsymbol{q}^\dagger])\bigr)\,\mathrm{dvol}(\boldsymbol{x}_2)-\\
&-\iiiint \mathrm{d}\boldsymbol{x}_1\mathrm{d}\boldsymbol{y}_1\mathrm{d}\boldsymbol{y}_2\,\bigl(f(\boldsymbol{x}_1,[\boldsymbol{q}],\boldsymbol{q}^\dagger])\bigr)\frac{\overleftarrow{\partial}}{\partial q^\dagger_{i_1,\sigma_1}}\ulcorner\Bigl(\frac{\overleftarrow{\mathrm{d}}}{\mathrm{d}\boldsymbol{y}_1}\Bigr)^{\sigma_1}\urcorner\,\Bigl\langle\overset{\text{first}}{\underline{\vec{e}_{i_1}(\boldsymbol{x}_1)}}\Bigr|\cdot\\
&\quad\cdot\Bigl\langle\delta s^\dagger_{i_1}(\boldsymbol{y}_1)\overset{\text{first}}{\underline{\underline{(-\vec{e}^{\,\dagger,i_1})(\boldsymbol{y}_1)}}},\overset{\text{second}}{\underline{\underline{\vec{e}_{i_2}(\boldsymbol{y}_2)}}}\cdot\delta s^{i_2}(\boldsymbol{y}_2)\Bigr\rangle\cdot\\
&\qquad\cdot\Bigl|\overset{\text{second}}{\underline{\vec{e}^{\,\dagger,i_2}(\boldsymbol{x}_2)}}\Bigr\rangle\cdot\ulcorner\Bigl(-\frac{\overrightarrow{\mathrm{d}}}{\mathrm{d}\boldsymbol{y}_2}\Bigr)^{\sigma_2}\urcorner\frac{\overrightarrow{\partial}}{\partial q^{i_2}_{\sigma_2}}\bigl(g(\boldsymbol{x}_2,[\boldsymbol{q}],[\boldsymbol{q}^\dagger])\bigr)\,\mathrm{dvol}(\boldsymbol{x}_2).
\end{aligned}\tag{2}$$

Evaluating both couplings in the minuend, we obtain $(-1)\cdot(-1)=+1$; likewise, in the subtrahend we have that $(+1)\cdot(+1)=+1$; at every value of the indexes, the respective shift components contribute with $\delta s^\bullet\cdot\delta s^\dagger_\bullet=1$. In effect, the only minus sign making the *difference* of two terms is determined by the precedence $\boldsymbol{q}\prec\boldsymbol{q}^\dagger$ versus succedence $\boldsymbol{q}^\dagger\succ\boldsymbol{q}$, that is, by the sequential order in which the parity-even and odd partial derivatives along the fibres of $J^\infty(\pi)$ are distributed between the arguments $F\prec G$, which were the ordered pair of input objects. It is now readily seen that the operational algorithm for reconfiguration of normalised couplings in $(\overrightarrow{\delta s}\circ\overrightarrow{\delta s^\dagger})(F\cdot G)$, which we started with, establishes the definition of variational Schouten bracket $[\![\,,\,]\!]$ as the descendent structure to the Batalin–Vilkovisky Laplacian Δ, cf. (13) on p. 15:

$$\Delta(F\cdot G)\stackrel{\text{def}}{=}\Delta F\cdot G+(-)^{|F|}[\![F,G]\!]+(-)^{|F|}F\cdot\Delta G.\tag{3}$$

In effect, the variational Schouten bracket $[\![\,,\,]\!]$ measures the deviation of Δ from being a derivation. Simultaneously, the variational Schouten bracket $[\![\,,\,]\!]$ extends the commutator $[\,,\,]$ of (variational) one-vectors to the space of variational multi-vectors.[10]

However, we emphasize that the expression $[\![F,G]\!]$, which has been constructed by following the couplings' re-attachment mechanism, itself can serve as a constituent part of a larger object. Because the reconfigurations of couplings and integrations by parts occur prior only to the

[9] The remaining volume element can be either $\mathrm{dvol}(\boldsymbol{x}_1)$ or $\mathrm{dvol}(\boldsymbol{x}_2)$; its final location is prescribed by either the right-to-left or left-to-right (which is the case here) direction of couplings in the output. From (1) it is clear that a simultaneous swap "first ⇄ second" in a pair of couplings would give the extra factor $(-1)\cdot(-1)=+1$, so that expression's overall sign does not change.

[10] The sequential order in which the densities of two arguments in $[\![\,,\,]\!]$ are differentiated with respect to the parity-even jet coordinates q^i_σ and parity-odd variables $q^\dagger_{j,\tau}$ is often chosen in such a way that the shifted-graded skew-symmetric Schouten bracket of variational one-vectors $F=\int X^i(\boldsymbol{x},[\boldsymbol{q}])\,q^\dagger_i\,\mathrm{dvol}(\boldsymbol{x})$ and $G=\int Y^j(\boldsymbol{y},[\boldsymbol{q}])\,q^\dagger_j\,\mathrm{dvol}(\boldsymbol{y})$ is determined by *minus* the usual commutator of the respective evolutionary vector fields within the purely even geometry of variables $\boldsymbol{q}$: one has that $[\![F,G]\!]=-\int[X,Y]^k(\boldsymbol{z},[\boldsymbol{q}])\,q^\dagger_k\,\mathrm{dvol}(\boldsymbol{z})$, where $[X,Y]^k=+(X(Y^k)-Y(X^k))$ is the componentwise action; this convention is adopted in [1, 2]. The two conventions for $[\,,\,]$ and $[\![\,,\,]\!]$ coincide if one takes $[X,Y]^k=-(X(Y^k)-Y(X^k))$ as in Lie theory.

Journal of Physics: Conference Series **670** (2016) 012030 doi:10.1088/1742-6596/670/1/012030

restriction of output to the jet of a section $\boldsymbol{s} \in \Gamma(\pi)$, this would mean that the horizontal derivatives along the base M^m have not yet been brought to their anticipated final locations. In the meantime, other partial derivatives can freely overtake them.[11] This is precisely why the total derivatives were embraced by using $\lceil \ldots \rceil$ and why the shifts' own base variables $\boldsymbol{y}_i$ were used instead of $\boldsymbol{x}_i$ from the functionals. The delayed integration by parts guarantees that the iterated variations are *graded-permutable*, stemming from the terms like this:

$$\Big(-\frac{\overrightarrow{\mathrm{d}}}{\mathrm{d}\boldsymbol{x}}\Big)^{\sigma\cup\tau} \circ \frac{\overrightarrow{\partial^2}}{\partial \boldsymbol{q}_\tau \partial \boldsymbol{q}^\dagger_\sigma}, \tag{4}$$

or similar — with any other combination of variables $\boldsymbol{q}$ and $\boldsymbol{q}^\dagger$.

Suppose still that the bracket of functionals F and G is the endpoint of a calculation (that is, the reasoning stops there and the object $[\![F,G]\!]\colon \Gamma(\pi) \to \Bbbk$ is used only for its evaluation at sections but it is *not* contained in any larger formula such as the left-hand side of Jacobi's identity for $[\![\,,\,]\!]$). Should this be known in advance, then one re-derives the familiar provisional formula (in fact, one of many – see [20]),

$$[\![F,G]\!] \text{ ``=''} \int \Big((f)\frac{\overleftarrow{\delta}}{\delta \boldsymbol{q}} \cdot \frac{\overrightarrow{\delta}}{\delta \boldsymbol{q}^\dagger}(g) - (f)\frac{\overleftarrow{\delta}}{\delta \boldsymbol{q}^\dagger} \cdot \frac{\overrightarrow{\delta}}{\delta \boldsymbol{q}}(g) \Big)\, \mathrm{dvol}(\boldsymbol{x}). \tag{5}$$

We recall that a step-by-step construction of objects which are then evaluated at sections is typical in the search for stationary points of action functionals in the Lagrangian formalism. This may not be the case in a larger framework.

2.2. The Jacobi identity for $[\![\,,\,]\!]$

The Jacobi identity for the Schouten bracket $[\![\,,\,]\!]$ can be understood as the graded Leibniz rule

$$[\![F,[\![G,H]\!]]\!] = [\![[\![F,G]\!],H]\!] + (-)^{(|F|-1)(|G|-1)}\, [\![G,[\![F,H]\!]]\!]; \tag{6}$$

the bracket's own grading equals -1, which is responsible for the shifts of grading $|\cdot|$ in the exponent. Equivalently, the Jacobi identity for $[\![\,,\,]\!]$ is the (shifted-)graded commutator of operators $[\![F,\cdot]\!]$ and $[\![G,\cdot]\!]$ acting on a test functional H,

$$[\![F,[\![G,\cdot]\!]]\!](H) - (-)^{(|F|-1)(|G|-1)}\, [\![G,[\![F,\cdot]\!]]\!](H) = [\![[\![F,G]\!],\cdot]\!](H);$$

we refer to [4, §2.6] and [2] for the proof and many essential details.[12]

Example 1. Let us illustrate the validity mechanism for Jacobi's identity (6) by verifying it at three given functionals. For simplicity, let there be just one independent variable x, one parity-even coordinate q and its parity-odd canonical conjugate $q^\dagger$. Set

$$F = \int q^\dagger q q_{x_1x_1}\, \mathrm{dvol}(x_1), \quad G = \int q^\dagger_{x_2} \exp(q_{x_2})\, \mathrm{dvol}(x_2), \quad \text{and } H = \int q^\dagger_{x_3x_3} \cos q\, \mathrm{dvol}(x_3);$$

we note that the functionals F and H re-appear in Counterexample 4 on p. 15 below and in the resolution to the paradox contained in it, see [1, pp. 34–36]. We have $|F| = 1$ and $|G| = 1$, whence $(-)^{(|F|-1)(|G|-1)} = +1$ in (6).

[11] We refer to [2] for a demonstration how these conventions work in the proof of Jacobi's identity for the variational Schouten bracket $[\![\,,\,]\!]$.

[12] The bi-linear, shifted-graded skew-symmetric structure $[\![\,,\,]\!]$ extends via

$$[\![F, G\cdot H]\!] = [\![F,G]\!]\cdot H + (-)^{(|F|-1)\cdot|G|} G\cdot [\![F,H]\!]$$

to the vector space of formal products $H_1 \cdot \ldots \cdot H_\ell \colon \Gamma(\pi) \to \Bbbk$ of integral functionals.

Let $\delta\boldsymbol{s}_1 = (\delta s_1, \delta s_1^\dagger)$ and $\delta\boldsymbol{s}_2 = (\delta s_2, \delta s_2^\dagger)$ be two normalized test shifts, i.e., suppose that $\delta s_\alpha(y) \cdot \delta s_\alpha^\dagger(y) = 1$ at every y for $\alpha = 1, 2$. We recall from Lemma 1 in [1, p. 24] that the values of Schouten brackets in (6) are independent of a concrete choice of the normalized functional coefficients δs_α and $\delta s_\alpha^\dagger$, which implies that the test shifts $\delta\boldsymbol{s}_1$ and $\delta\boldsymbol{s}_2$ in the inner and outer brackets can be swapped (this would amount to relabelling $y \rightleftarrows z$ of their arguments). We shall not write the basic (co)vectors $\vec{e}(y)$ and $\vec{e}^{\,\dagger}(y)$ in expansions of the test shifts and of the differentials of functionals' densities; it is enough to know the contributing values, which are specified by the precedence $\boldsymbol{q} \prec \boldsymbol{q}^\dagger$ or antecedence $\boldsymbol{q}^\dagger \succ \boldsymbol{q}$ and which are equal to ± 1 (see §2.1 above).

We have that[13] $[\![G, H]\!] =$

$$\iiiint \mathrm{d}y_2\, \mathrm{d}y_3\, \mathrm{d}x_2\, \mathrm{dvol}(x_3) \cdot \Big\{ \Big\langle \ulcorner -\tfrac{\mathrm{d}}{\mathrm{d}y_2} \urcorner \underbrace{\big(q^\dagger_{x_2} \exp(q_{x_2})\big)}_{x_2} \cdot \ulcorner \tfrac{\mathrm{d}^2}{\mathrm{d}y_3^2} \urcorner \underbrace{(\cos q)}_{x_3} \Big\rangle \cdot \underbrace{\langle \delta s(y_2), \delta s^\dagger(y_3) \rangle}_{+1} + \\ + \Big\langle \ulcorner -\tfrac{\mathrm{d}}{\mathrm{d}y_2} \urcorner \underbrace{\big(\exp(q_{x_2})\big)}_{x_2} \cdot \underbrace{q^\dagger_{x_3x_3} \cdot (-\sin q)}_{x_3} \Big\rangle \cdot \underbrace{\langle \delta s^\dagger(y_2), \delta s(y_3) \rangle}_{-1} \Big\};$$

the integration variables x_i are displayed under the remnants of respective densities. Next, we obtain that $[\![F, [\![G, H]\!]]\!] =$

$$\int \mathrm{d}z_1 \int \mathrm{d}z_{23} \int \mathrm{d}y_2 \int \mathrm{d}y_3 \int \mathrm{d}x_1 \int \mathrm{d}x_2 \int \mathrm{dvol}(x_3) \cdot \underbrace{\langle \delta s(z_1), \delta s^\dagger(z_{23}) \rangle}_{+1} \cdot \\ \Big\{ \Big\langle \underbrace{\big({}^{\langle 1 \rangle}\, q^\dagger q_{x_1x_1} + {}^{\langle 2 \rangle}\, \tfrac{\mathrm{d}^2}{\mathrm{d}z_1^2}(q^\dagger q)\big)}_{x_1} \cdot \big(-\tfrac{\mathrm{d}}{\mathrm{d}z_{23}}\big)\big(-\tfrac{\mathrm{d}}{\mathrm{d}y_2}\big) \underbrace{\big(\exp(q_{x_2})\big)}_{x_2} \cdot \tfrac{\mathrm{d}^2}{\mathrm{d}y_3^2} \underbrace{(\cos q)}_{x_3} \Big\rangle \cdot \underbrace{\langle \delta s(y_2), \delta s^\dagger(y_3) \rangle}_{+1} + \\ + \Big\langle \underbrace{\big({}^{\langle 3 \rangle}\, q^\dagger q_{x_1x_1} + {}^{\langle 4 \rangle}\, \tfrac{\mathrm{d}^2}{\mathrm{d}z_1^2}(q^\dagger q)\big)}_{x_1} \cdot \big(-\tfrac{\mathrm{d}}{\mathrm{d}y_2}\big) \underbrace{\big(\exp(q_{x_2})\big)}_{x_2} \cdot \tfrac{\mathrm{d}^2}{\mathrm{d}z_{23}^2} \underbrace{(-\sin q)}_{x_3} \Big\rangle \cdot \underbrace{\langle \delta s^\dagger(y_2), \delta s(y_3) \rangle}_{-1} \Big\} + \\ + \int \mathrm{d}z_1 \int \mathrm{d}z_{23} \int \mathrm{d}y_2 \int \mathrm{d}y_3 \int \mathrm{d}x_1 \int \mathrm{d}x_2 \int \mathrm{dvol}(x_3) \cdot \underbrace{\langle \delta s^\dagger(z_1), \delta s(z_{23}) \rangle}_{-1} \cdot \\ \Big\{ \Big\langle \underbrace{\big({}^{\langle 5 \rangle}\, q q_{x_1x_1}\big)}_{x_1} \cdot \big(-\tfrac{\mathrm{d}}{\mathrm{d}z_{23}}\big)\big(-\tfrac{\mathrm{d}}{\mathrm{d}y_2}\big) \underbrace{\big(q^\dagger_{x_2} \exp(q_{x_2})\big)}_{x_2} \cdot \tfrac{\mathrm{d}^2}{\mathrm{d}y_3^2} \underbrace{(\cos q)}_{x_3} + \\ + \underbrace{\big({}^{\langle 6 \rangle}\, q q_{x_1x_1}\big)}_{x_1} \cdot \big(-\tfrac{\mathrm{d}}{\mathrm{d}y_2}\big) \underbrace{\big(q^\dagger_{x_2} \exp(q_{x_2})\big)}_{x_2} \cdot \tfrac{\mathrm{d}^2}{\mathrm{d}y_3^2} \underbrace{(-\sin q)}_{x_3} \Big\rangle \cdot \underbrace{\langle \delta s(y_2), \delta s^\dagger(y_3) \rangle}_{+1} + \\ + \Big\langle \underbrace{\big({}^{\langle 7 \rangle}\, q q_{x_1x_1}\big)}_{x_1} \cdot \big(-\tfrac{\mathrm{d}}{\mathrm{d}z_{23}}\big)\big(-\tfrac{\mathrm{d}}{\mathrm{d}y_2}\big) \underbrace{\big(\exp(q_{x_2})\big)}_{x_2} \cdot \underbrace{\big(q^\dagger_{x_3x_3} \cdot (-\sin q)\big)}_{x_3} + \\ + \underbrace{\big({}^{\langle 8 \rangle}\, q q_{x_1x_1}\big)}_{x_1} \cdot \big(-\tfrac{\mathrm{d}}{\mathrm{d}y_2}\big) \underbrace{\big(\exp(q_{x_2})\big)}_{x_2} \cdot \underbrace{\big(q^\dagger_{x_3x_3} \cdot (-\cos q)\big)}_{x_3} \Big\rangle \cdot \underbrace{\langle \delta s^\dagger(y_2), \delta s(y_3) \rangle}_{-1} \Big\}.$$

[13] Let us repeat that integrations by parts, which cast the derivatives off the test shifts, are performed only when all the objects – such as the l.-h.s. or r.-h.s. of (6) – are fully composed, all partial derivatives of the functionals' densities are calculated, and reconfigurations of the couplings are ready to start. In practice, this means that partial derivatives like $\overrightarrow{\partial}/\partial q_x$ or $\overleftarrow{\partial}/\partial q^\dagger_{xx}$ dive under $\overrightarrow{\mathrm{d}}/\mathrm{d}y$ or $\overleftarrow{\mathrm{d}}/\mathrm{d}z$ because those total derivatives have not yet appeared at the places where we write them ahead of time.

On the other hand, $[\![F,G]\!] =$

$$\iiiint \mathrm{d}y_1\,\mathrm{d}y_2\,\mathrm{d}x_1\ \mathrm{dvol}(x_2)\cdot\Bigl\{\Bigl\langle\underbrace{\bigl(q^\dagger q_{x_1x_1}+\tfrac{\mathrm{d}^2}{\mathrm{d}y_1^2}(q^\dagger q)\bigr)}_{x_1}\cdot\ulcorner-\tfrac{\mathrm{d}}{\mathrm{d}y_2}\urcorner\underbrace{\bigl(\exp(q_{x_2})\bigr)}_{x_2}\Bigr\rangle\cdot\underbrace{\langle\delta s(y_1),\delta s^\dagger(y_2)\rangle}_{+1}+$$
$$+\Bigl\langle\underbrace{\bigl(qq_{x_1x_1}\bigr)}_{x_1}\cdot\ulcorner-\tfrac{\mathrm{d}}{\mathrm{d}y_2}\urcorner\underbrace{\bigl(q^\dagger_{x_2}\exp(q_{x_2})\bigr)}_{x_2}\Bigr\rangle\cdot\underbrace{\langle\delta s^\dagger(y_1),\delta s(y_2)\rangle}_{-1}\Bigr\}.$$

We infer that $[\![[\![F,G]\!],H]\!] =$

$$\int\mathrm{d}z_{12}\int\mathrm{d}z_3\int\mathrm{d}y_1\int\mathrm{d}y_2\int\mathrm{d}x_1\int\mathrm{d}x_2\int\mathrm{dvol}(x_3)\cdot\underbrace{\langle\delta s(z_{12}),\delta s^\dagger(z_3)\rangle}_{+1}\cdot$$
$$\Bigl\{\Bigl\langle\underbrace{\bigl({}^{\langle 9\rangle}\ \tfrac{\mathrm{d}^2}{\mathrm{d}z_{12}^2}(q^\dagger)+{}^{\langle 10\rangle}\ \tfrac{\mathrm{d}^2}{\mathrm{d}y_1^2}(q^\dagger)\bigr)}_{x_1}\cdot\bigl(-\tfrac{\mathrm{d}}{\mathrm{d}y_2}\bigr)\underbrace{\bigl(\exp(q_{x_2})\bigr)}_{x_2}\cdot\tfrac{\mathrm{d}^2}{\mathrm{d}z_3^2}\underbrace{(\cos q)}_{x_3}+$$
$$+\underbrace{\bigl({}^{\langle 1\rangle}\ q^\dagger q_{x_1x_1}+{}^{\langle 2\rangle}\ \tfrac{\mathrm{d}^2}{\mathrm{d}y_1^2}(q^\dagger q)\bigr)}_{x_1}\cdot\bigl(-\tfrac{\mathrm{d}}{\mathrm{d}z_{12}}\bigr)\bigl(-\tfrac{\mathrm{d}}{\mathrm{d}y_2}\bigr)\underbrace{\bigl(\exp(q_{x_2})\bigr)}_{x_2}\cdot\tfrac{\mathrm{d}^2}{\mathrm{d}z_3^2}\underbrace{(\cos q)}_{x_3}\Bigr\rangle\cdot\underbrace{\langle\delta s(y_1),\delta s^\dagger(y_2)\rangle}_{+1}+$$

$$+\Bigl\langle\underbrace{\bigl({}^{\langle 11\rangle}\ q_{x_1x_1}+{}^{\langle 12\rangle}\ \tfrac{\mathrm{d}^2}{\mathrm{d}z_{12}^2}(q)\bigr)}_{x_1}\cdot\bigl(-\tfrac{\mathrm{d}}{\mathrm{d}y_2}\bigr)\underbrace{\bigl(q^\dagger_{x_2}\exp(q_{x_2})\bigr)}_{x_2}\cdot\tfrac{\mathrm{d}^2}{\mathrm{d}z_3^2}\underbrace{(\cos q)}_{x_3}+$$
$$+\underbrace{\bigl({}^{\langle 5\rangle}\ qq_{x_1x_1}\bigr)}_{x_1}\cdot\bigl(-\tfrac{\mathrm{d}}{\mathrm{d}z_{12}}\bigr)\bigl(-\tfrac{\mathrm{d}}{\mathrm{d}y_2}\bigr)\underbrace{\bigl(q^\dagger_{x_2}\exp(q_{x_2})\bigr)}_{x_2}\cdot\tfrac{\mathrm{d}^2}{\mathrm{d}z_3^2}\underbrace{(\cos q)}_{x_3}\Bigr\rangle\cdot\underbrace{\langle\delta s^\dagger(y_1),\delta s(y_2)\rangle}_{-1}\Bigr\}+$$
$$+\int\mathrm{d}z_{12}\int\mathrm{d}z_3\int\mathrm{d}y_1\int\mathrm{d}y_2\int\mathrm{d}x_1\int\mathrm{d}x_2\int\mathrm{dvol}(x_3)\cdot\underbrace{\langle\delta s^\dagger(z_{12}),\delta s(z_3)\rangle}_{-1}\cdot$$
$$\Bigl\{\Bigl\langle\underbrace{\bigl({}^{\langle 13\rangle}\ q_{x_1x_1}+{}^{\langle 14\rangle}\ \tfrac{\mathrm{d}^2}{\mathrm{d}y_1^2}(q)\bigr)}_{x_1}\cdot\bigl(-\tfrac{\mathrm{d}}{\mathrm{d}y_2}\bigr)\underbrace{\bigl(\exp(q_{x_2})\bigr)}_{x_2}\cdot\underbrace{\bigl(q^\dagger_{x_3x_3}\cdot(-\sin q)\bigr)}_{x_3}\Bigr\rangle\cdot\underbrace{\langle\delta s(y_1),\delta s^\dagger(y_2)\rangle}_{+1}+$$
$$+\Bigl\langle\underbrace{\bigl({}^{\langle 7\rangle}\ qq_{x_1x_1}\bigr)}_{x_1}\cdot\bigl(-\tfrac{\mathrm{d}}{\mathrm{d}z_{12}}\bigr)\bigl(-\tfrac{\mathrm{d}}{\mathrm{d}y_2}\bigr)\underbrace{\bigl(\exp(q_{x_2})\bigr)}_{x_2}\cdot\underbrace{\bigl(q^\dagger_{x_3x_3}\cdot(-\sin q)\bigr)}_{x_3}\Bigr\rangle\cdot\underbrace{\langle\delta s^\dagger(y_1),\delta s(y_2)\rangle}_{-1}\Bigr\}.$$

Thirdly, $[\![F,H]\!] =$

$$\iiiint \mathrm{d}y_1\,\mathrm{d}y_3\,\mathrm{d}x_1\ \mathrm{dvol}(x_3)\cdot\Bigl\{\Bigl\langle\underbrace{\bigl(q^\dagger q_{x_1x_1}+\tfrac{\mathrm{d}^2}{\mathrm{d}y_1^2}(q^\dagger q)\bigr)}_{x_1}\cdot\ulcorner\tfrac{\mathrm{d}^2}{\mathrm{d}y_3^2}\urcorner\underbrace{(\cos q)}_{x_3}\Bigr\rangle\cdot\underbrace{\langle\delta s(y_1),\delta s^\dagger(y_3)\rangle}_{+1}+$$
$$+\Bigl\langle\underbrace{\bigl(qq_{x_1x_1}\bigr)}_{x_1}\cdot\underbrace{\bigl(q^\dagger_{x_3x_3}\cdot(-\sin q)\bigr)}_{x_3}\Bigr\rangle\cdot\underbrace{\langle\delta s^\dagger(y_1),\delta s(y_3)\rangle}_{-1}\Bigr\}.$$

In view of the functionals' gradings, we have $+1 \cdot [\![G, [\![F, H]\!]]\!] =$

$$\int \mathrm{d}z_2 \int \mathrm{d}z_{13} \int \mathrm{d}y_1 \int \mathrm{d}y_3 \int \mathrm{d}x_1 \int \mathrm{d}x_2 \int \mathrm{dvol}(x_3) \cdot \underbrace{\langle \delta s(z_2), \delta s^\dagger(z_{13})\rangle}_{+1} \cdot$$

$$\Big\{\Big\langle \big(-\tfrac{\mathrm{d}}{\mathrm{d}z_2}\big)\underbrace{\big(q^\dagger_{x_2}\exp(q_{x_2})\big)}_{x_2} \cdot \underbrace{\big({}^{\langle 11\rangle}\ q_{x_1x_1} + {}^{\langle 12\rangle}\ \tfrac{\mathrm{d}^2}{\mathrm{d}y_1^2}(q)\big)}_{x_1} \cdot \tfrac{\mathrm{d}^2}{\mathrm{d}y_3^2}\underbrace{(\cos q)}_{x_3}\Big\rangle \cdot \underbrace{\langle \delta s(y_1), \delta s^\dagger(y_3)\rangle}_{+1} +$$

$$+ \Big\langle \big(-\tfrac{\mathrm{d}}{\mathrm{d}z_2}\big)\underbrace{\big(q^\dagger_{x_2}\exp(q_{x_2})\big)}_{x_2} \cdot \underbrace{\big({}^{\langle 6\rangle}\ q q_{x_1x_1}\big)}_{x_1} \cdot \tfrac{\mathrm{d}^2}{\mathrm{d}z_{13}^2}\underbrace{(-\sin q)}_{x_3}\Big\rangle \cdot \underbrace{\langle \delta s^\dagger(y_1), \delta s(y_3)\rangle}_{-1}\Big\} +$$

$$+ \int \mathrm{d}z_2 \int \mathrm{d}z_{13} \int \mathrm{d}y_1 \int \mathrm{d}y_3 \int \mathrm{d}x_1 \int \mathrm{d}x_2 \int \mathrm{dvol}(x_3) \cdot \underbrace{\langle \delta s^\dagger(z_2), \delta s(z_{13})\rangle}_{-1} \cdot$$

$$\Big\{\Big\langle \big(-\tfrac{\mathrm{d}}{\mathrm{d}z_2}\big)\underbrace{\big(\exp(q_{x_2})\big)}_{x_2} \cdot \underbrace{\big({}^{\langle 10\rangle}\ \tfrac{\mathrm{d}^2}{\mathrm{d}z_{13}^2}(q^\dagger) + {}^{\langle 9\rangle}\ \tfrac{\mathrm{d}^2}{\mathrm{d}y_1^2}(q^\dagger)\big)}_{x_1} \cdot \tfrac{\mathrm{d}^2}{\mathrm{d}y_3^2}\underbrace{(\cos q)}_{x_3} +$$

$$+ \big(-\tfrac{\mathrm{d}}{\mathrm{d}z_2}\big)\underbrace{\big(\exp(q_{x_2})\big)}_{x_2} \cdot \underbrace{\big({}^{\langle 3\rangle}\ q^\dagger q_{x_1x_1} + {}^{\langle 4\rangle}\ \tfrac{\mathrm{d}^2}{\mathrm{d}y_1^2}(q^\dagger q)\big)}_{x_1} \cdot \tfrac{\mathrm{d}^2}{\mathrm{d}y_3^2}\underbrace{(-\sin q)}_{x_3}\Big\rangle \cdot \underbrace{\langle \delta s(y_1), \delta s^\dagger(y_3)\rangle}_{+1} +$$

$$+ \Big\langle \big(-\tfrac{\mathrm{d}}{\mathrm{d}z_2}\big)\underbrace{\big(\exp(q_{x_2})\big)}_{x_2} \cdot \underbrace{\big({}^{\langle 13\rangle}\ q_{x_1x_1} + {}^{\langle 14\rangle}\ \tfrac{\mathrm{d}^2}{\mathrm{d}z_{13}^2}(q)\big)}_{x_1} \cdot \underbrace{\big(q^\dagger_{x_3x_3} \cdot (-\sin q)\big)}_{x_3} +$$

$$+ \big(-\tfrac{\mathrm{d}}{\mathrm{d}z_2}\big)\underbrace{\big(\exp(q_{x_2})\big)}_{x_2} \cdot \underbrace{\big({}^{\langle 8\rangle}\ q q_{x_1x_1}\big)}_{x_1} \cdot \underbrace{\big(q^\dagger_{x_3x_3} \cdot (-\cos q)\big)}_{x_3}\Big\rangle \cdot \underbrace{\langle \delta s^\dagger(y_1), \delta s(y_3)\rangle}_{-1}\Big\}.$$

Each term $\langle 1\rangle - \langle 8\rangle$ meets its match from the other side of (6), whereas terms $\langle 9\rangle - \langle 14\rangle$ occur in pairs of opposite signs; therefore, they all cancel out in the r.-h.s. of the Jacobi identity.

We conclude that within the true geometry of iterated variations, the Jacobi identity for $[\![\,,\,]\!]$ is

$$\mathrm{Jac}\big(F(x_1), G(x_2), H(x_3)\big) = 0. \tag{7}$$

By construction, its right-hand side is the functional whose density vanishes identically.

3. The old approach to Jacobi's identity

3.1. Cohomologically trivial r.-h.s. with nonzero density

Let us "forget" the operational definition of variational Schouten bracket, that is, the way how the structure $[\![\,,\,]\!]$ is determined by the Batalin–Vilkovisky Laplacian, see Eq. (3), now regarding the operation $[\![\,,\,]\!]$ as if it were introduced by formula (5). At first glance, the geometry becomes very simple: all the intermediate objects are realised in the same way, namely, as integral functionals over the infinite jet superbundle $J^\infty(\pi)$. The price that one pays is the inconsistency of calculus — but let us postpone a counterexample till §4.

Example 2. By taking the three functionals

$$F = \int q^\dagger q q_{xx}\,\mathrm{d}x, \qquad G = \int q^\dagger_y \exp(q_x)\,\mathrm{d}x, \qquad H = \int q^\dagger_{xx}\cos q\,\mathrm{d}x,$$

and first, plugging any two of them into formula (5), one then inserts the output for an argument of the outer bracket in Jacobi's identity (6). In this way one calculates the integral functional[14]

$$[\![G, H]\!] \text{ ``$=$'' } \int \exp(q_x) \cdot \{q^\dagger_x q_x^2 q_{xx}\cos q + q^\dagger_{xx} q_x^2 \cos q + q^\dagger_x q^2_{xx} \sin q\}\,\mathrm{d}x,$$

[14] The reader is invited to ponder whether it would be *these* formulas that he or she is tempted to write.

then making a short break. Resuming the job, one deduces that

$$\begin{aligned}
[\![F,[\![G,H]\!]]\!] \text{ “=” } -\int \exp(q_x)\cdot\Big\{ & q_x^\dagger q q_{xx}^4 \sin q + 2q_x^\dagger q_x^4 q_{xx}\sin q + 2q^\dagger q_x^3 q_{xx}^2 \sin q \\
&+ 4q_{xx}^\dagger q q_{xx}^3 \sin q + 3q_x^\dagger q q_{xx}^3 \cos q + 2q_x^\dagger q_x q_{xx}^3 \sin q - 2q_x^\dagger q_x^2 q_{xx}^2 \cos q \\
&- 2q^\dagger q_x q_{xx}^3 \cos q + 10q^\dagger q_x^2 q_{xx}^2 \sin q + 4q^\dagger q_{xx}^2 q_{xxx} sinq + 2q_{xxx}^\dagger q q_{xx}^2 \sin q \\
&- 4q_{xx}^\dagger q q_{xx}^2 \cos q + q_{xx}^\dagger q q_x^4 \cos q - q_{xx}^\dagger q q_x^3 q_{xx}^2 \sin q - q_x^\dagger q q_x^4 q_{xx} \cos q \\
&+ 3q_x^\dagger q q_x q_{xx}^3 \cos q - 6q_x^\dagger q q_x^2 q_{xx}^2 \sin q + 4q_x^\dagger q q_{xx}^2 q_{xxx} \sin q + q_{xx}^\dagger q q_x q_{xx}^2 \cos q \\
&+ 6q_{xx}^\dagger q q_x^2 q_{xx} \sin q + 6q_{xx}^\dagger q q_{xx} q_{xxx} \sin q - 2q_{xx}^\dagger q q_x q_{xxx} \cos q + 4q_x^\dagger q_x q_{xx} q_{xxx} \sin q \\
&- 4q^\dagger q_x q_{xx} q_{xxx} \cos q - 2q_{xxx}^\dagger q q_x q_{xx} \cos q + 2q_x^\dagger q q_{xx} q_{4x} \sin q + 2q_x^\dagger q_x^5 \cos q \\
&+ 2q^\dagger q_{xx}^4 \sin q - 4q^\dagger q_{xx}^3 \cos q - 4q_x^\dagger q_x^2 q_{xxx} \cos q + 10q_x^\dagger q_x^3 q_{xx} \sin q \\
&+ 2q^\dagger q_x^4 q_{xx} \cos q - 4q_x^\dagger q_x q_{xx}^2 \cos q + 4q_x^\dagger q q_x q_{xx} q_{xxx} \cos q \Big\}\,dx.
\end{aligned}$$

Likewise,

$$[\![F,G]\!] \text{ “=” } \int q_{xx}\exp(q_x)\cdot\big\{q_x^\dagger q q_{xx} - 2q_x^\dagger q_x - 2q^\dagger q_{xx}\big\}\,dx;$$

here one stops for a while. The line of reasoning continues with

$$\begin{aligned}
[\![[\![F,G]\!],H]\!] \text{ “=” } -\int \exp(q_x)\cdot\Big\{ & q_x^\dagger q q_x^2 q_{xx}^3 \cos q + 3q_{xx}^\dagger q q_x^2 q_{xx}^2 \cos q + 4q_{xx}^\dagger q q_x^2 q_{xxx} \cos q \\
&+ 2q_{xxx}^\dagger q q_x^2 q_{xx} \cos q + 2q_x^\dagger q q_x^2 q_{4x} \cos q - 2q_{xxx}^\dagger q_x^3 \cos q - 3q_x^\dagger q_{xx}^3 \sin q \\
&- 2q_{xx}^\dagger q_x q_{xxx} \sin q - 2q_{xxx}^\dagger q_x q_{xx} \sin q - 4q^\dagger q_{xx} q_{4x} \sin q + 3q_x^\dagger q_x^3 q_{xx}^2 \cos q \\
&- 2q^\dagger q_x^2 q_{xx}^3 \cos q + 4q_x^\dagger q_x^3 q_{xxx} \cos q + 2q_{xx}^\dagger q_x^3 q_{xx} \cos q - 4q^\dagger q_x^2 q_{4x} \cos q \\
&+ q_{xx}^\dagger q_x q_{xx}^2 \sin q + q_x^\dagger q q_{xx}^4 \sin q + 4q_{xx}^\dagger q q_{xx}^3 \sin q + 3q_x^\dagger q_x q_{xx}^3 \sin q \\
&- 3q_x^\dagger q_x^2 q_{xx}^2 \cos q - 8q^\dagger q_{xx}^2 q_{xxx} \sin q + 2q_{xxx}^\dagger q q_{xx}^2 \sin q + 4q_x^\dagger q q_{xx}^2 q_{xxx} \sin q \\
&+ 6q_{xx}^\dagger q q_{xx} q_{xxx} \sin q + 4q_x^\dagger q_x q_{xx} q_{xxx} \sin q + 2q_x^\dagger q q_{xx} q_{4x} \sin q - 2q^\dagger q_{xx}^4 \sin q \\
&- 6q_{xx}^\dagger q_{xx}^2 \sin q - 6q_{xx}^\dagger q_x^2 q_{xx} \cos q - 8q_x^\dagger q_x^2 q_{xxx} \cos q - 8q_x^\dagger q_{xx} q_{xxx} \sin q \\
&- 8q^\dagger q_x^2 q_{xx} q_{xxx} \cos q + 4q_x^\dagger q q_x^2 q_{xx} q_{xxx} \cos q \Big\}\,dx.
\end{aligned}$$

Thirdly,

$$[\![F,H]\!] \text{ “=” } -\int \big\{q_{xx}^\dagger q q_x^2 \cos q + 2q_x^\dagger q_x^3 \cos q + 2q^\dagger q_x^2 q_{xx} \cos q + 2q_x^\dagger q_x q_{xx} \sin q + 2q^\dagger q_{xx}^2 \sin q\big\}\,dx,$$

and now is the time for a pause. Finally, one obtains

$$\begin{aligned}
[\![G,[\![F,H]\!]]\!] \text{ “=” } \int \exp(q_x)\cdot\Big\{ & -q_x^\dagger q q_x^4 q_{xx} \cos q - q_{xx}^\dagger q q_x^4 \cos q - 5q_x^\dagger q q_x^2 q_{xx}^2 \sin q \\
&+ 2q_x^\dagger q q_{xx} q_{xxx} \cos q - 6q_{xx}^\dagger q q_x^2 q_{xx} \sin q + 2q_x^\dagger q q_{xx}^3 \cos q - q_x^\dagger q_x^2 q_{xx}^2 \cos q \\
&+ 2q_x^\dagger q_x^3 q_{xx} \sin q + 12q^\dagger q_x^2 q_{xx}^2 \sin q + 2q_{xx}^\dagger q q_x q_{xxx} \cos q - 2q_x^\dagger q_x q_{xx} q_{xxx} \sin q \\
&- 8q^\dagger q_x q_{xx} q_{xxx} \cos q + 2q_{xxx}^\dagger q q_x q_{xx} \cos q + 4q_{xx}^\dagger q q_{xx}^2 \cos q - 6q_x^\dagger q_x q_{xx}^2 \cos q \\
&- 6q^\dagger q_{xx}^3 \cos q - 2q_{xx}^\dagger q_x q_{xxx} \sin q - 8q_x^\dagger q_{xx} q_{xxx} \sin q - 2q_{xxx}^\dagger q_x q_{xx} \sin q \\
&+ 2q^\dagger q_x^4 q_{xx} \cos q - 4q^\dagger q_{xx} q_{4x} \sin q - 6q_{xx}^\dagger q_{xx}^2 \sin q \Big\}\,dx.
\end{aligned}$$

By using the software Jets [16] for symbolic calculations, one verifies that

$$[\![F,[\![G,H]\!]]\!] - \big([\![[\![F,G]\!],H]\!] + (-)^{(|F|-1)(|G|-1)}[\![G,[\![F,H]\!]]\!]\big) \cong 0. \tag{8}$$

Journal of Physics: Conference Series **670** (2016) 012030 doi:10.1088/1742-6596/670/1/012030

3.2. Monitoring each argument's "contribution" to the r.-h.s.

It is easy to explore the structure of cohomologically trivial functional in the right-hand side of the equivalence

$$\mathrm{Jac}\big(F(x), G(x), H(x)\big) \cong 0; \tag{9}$$

for F, G, and H as above, we have that

$$\begin{aligned}\mathrm{Jac}\big(F(x), G(x), H(x)\big) = -\int \mathrm{d}x\, \frac{\mathrm{d}}{\mathrm{d}x}\Big(\exp(q_x)\cdot\Big\{&2q^\dagger q_{xx}^2 q_x^2 \cos q - q_x^\dagger q q_{xx}^2 q_x^2 \cos q - 2q_x^\dagger q q_{xx} q_x^3 \sin q\\
&+ 5q_x^\dagger q q_{xx}^2 q_x \cos q - 2q_x^\dagger q_{xx} q_x^3 \cos q + 4q^\dagger q_{xx} q_x^3 \sin q - 2q_{xx}^\dagger q q_{xx} q_x^2 \cos q + 4q^\dagger q_{xxx} q_x^2 \cos q\\
&+ 2q_x^\dagger q_x^4 \sin q - 2q_x^\dagger q q_x^2 q_{xxx} \cos q + 4q^\dagger q_{xx}^3 \sin q - 10q^\dagger q_{xx}^2 q_x \cos q - q_x^\dagger q_{xx}^2 q_x \sin q + 2q_{xx}^\dagger q_x^3 \cos q\Big\}\Big).\end{aligned} \tag{10}$$

To track where these extra terms in the right-hand side came from, let us notice the following. First, the integrations by parts were always attached to the respective vertical differentials in the fibres of $J^\infty(\pi)$. Consequently, whenever two variations fell on (whatever remained of) one functional, they produced the terms with operators such that the total and partial derivatives are interlaced, e.g., in

$$\Big(-\frac{\overrightarrow{\mathrm{d}}}{\mathrm{d}\boldsymbol{x}}\Big)^\tau \circ \frac{\overrightarrow{\partial}}{\partial \boldsymbol{q}_\tau} \circ \Big(-\frac{\overrightarrow{\mathrm{d}}}{\mathrm{d}\boldsymbol{x}}\Big)^\sigma \circ \frac{\overrightarrow{\partial}}{\partial \boldsymbol{q}^\dagger_\sigma} \tag{11}$$

or in a similar sequence of interlacing total and partial derivatives — with any other combinations of the variables $\boldsymbol{q}$ and $\boldsymbol{q}^\dagger$.

Which is worse, the construction of nested brackets in a term like $[\![F,[\![G,H]\!]]\!]$ according to formula (5) prescribes that, in the course of integration by parts in the bracket of F and entire $[\![G,H]\!]$, whenever a partial derivative $\partial/\partial\boldsymbol{q}_\tau$ or $\partial/\partial\boldsymbol{q}^\dagger_\tau$ with $|\tau|>0$ falls on the density of H, the total derivative $(-\overrightarrow{\mathrm{d}}/\mathrm{d}\boldsymbol{x})^\tau$ does act on G because it spreads over G and H via Newton's binomial formula. Note the cancellation of terms in which the total derivative $(-\overrightarrow{\mathrm{d}}/\mathrm{d}\boldsymbol{x})^\tau$ falls exclusively on the image of its "native" partial derivative $\partial/\partial\boldsymbol{q}_\tau$ or $\partial/\partial\boldsymbol{q}^\dagger_\tau$; this was proved in [2] and illustrated in §2 above.[15]

Example 3. Let us analyse how the remaining cross-terms emerge and contribute to the right-hand side of (9). To visualize their origin from one of the three functionals, we formally denote by x, y, and z the respective base variables so that

$$F = \int q^\dagger q q_{xx}\,\mathrm{d}x, \qquad G = \int q_y^\dagger \exp(q_y)\,\mathrm{d}y, \qquad H = \int q_{zz}^\dagger \cos q\,\mathrm{d}z;$$

the restriction $\mathrm{Jac}\big(F(x), G(y), H(z)\big)\big|_{x=y=z}$ to the diagonal at the end of the day will yield (10) in the right-hand side of (9).

We emphasize that the genuine contribution to the right-hand side of (8) is identically zero; the rest of its density is *naught* (i.e., not a mathematical description of any existing object) which equals

$$\begin{aligned}\exp(q_y)\cdot\Big\{&2q^\dagger q_{xx} q_y q_{yy} q_{zz} \cos q - 8q^\dagger q_{xx} q_y q_z q_{yz} \sin q + q_{xx}^\dagger q q_y q_{yy} q_{zz} \cos q\\
&- 4q_{xx}^\dagger q q_y q_z q_{yz} \sin q + 2q_x^\dagger q_x q_y q_{yy} q_{zz} \cos q - 8q_x^\dagger q_x q_y q_z q_{yz} \sin q + q_y^\dagger q q_{xx} q_y^2 q_z^2 \cos q\\
&+ q_y^\dagger q q_{xx} q_{yy} q_z^2 \sin q - q_y^\dagger q q_{xx} q_{yy} q_{yz}^2 \sin q + q_y^\dagger q q_{xx} q_y^2 q_{zz} \sin q + q_{yy}^\dagger q q_{xx} q_y q_z^2 \sin q\end{aligned}$$

[15] An explanation why the mess of redundant cross-terms remains overall exact is an instructive exercise for the reader.

$$
\begin{aligned}
&-2q^{\dagger}_{yz}qq_{xx}q_{yy}q_{yz}\sin q-2q^{\dagger}_{y}qq_{xx}q_{yy}q_{yzz}\sin q+q^{\dagger}_{zz}qq_{xx}q_{y}q_{yy}\cos q-q^{\dagger}_{y}qq_{xx}q_{yy}q_{zz}\cos q\\
&-2q^{\dagger}_{y}qq_{xx}q_{yz}q_{yyz}\sin q-q^{\dagger}_{yy}qq_{xx}q_{y}q_{zz}\cos q-2q^{\dagger}_{y}qq_{xx}q_{y}q_{yzz}\cos q-2q^{\dagger}_{yy}qq_{xx}q_{yz}q_{z}\cos q\\
&-2q^{\dagger}_{y}qq_{xx}q_{yyz}q_{z}\cos q+q^{\dagger}_{y}qq_{xy}^{2}q_{yy}q_{z}^{2}\cos q+q^{\dagger}_{y}q_{xx}q_{y}q_{yy}q_{z}^{2}\cos q+2q^{\dagger}_{y}q_{x}q_{xy}q_{yy}q_{z}^{2}\cos q\\
&+q^{\dagger}_{y}qq_{xy}^{2}q_{yy}q_{zz}\sin q+2q^{\dagger}_{y}qq_{xxy}q_{yy}q_{z}^{2}\cos q+2q^{\dagger}_{xy}qq_{xy}q_{yy}q_{z}^{2}\cos q+2q^{\dagger}_{y}qq_{xy}q_{xyy}q_{z}^{2}\cos q\\
&+q^{\dagger}_{y}q_{xx}q_{y}q_{yy}q_{zz}\sin q+2q^{\dagger}_{y}q_{x}q_{xy}q_{yy}q_{zz}\sin q+2q^{\dagger}_{y}qq_{xxy}q_{yy}q_{zz}\sin q+2q^{\dagger}_{xy}qq_{xy}q_{yy}q_{zz}\sin q\\
&+2q^{\dagger}_{y}qq_{xy}q_{xyy}q_{zz}\sin q-8q^{\dagger}q_{x}q_{yy}q_{xz}q_{z}\sin q+q^{\dagger}_{y}qq_{x}^{2}q_{yy}q_{z}^{2}\cos q+q^{\dagger}_{y}qq_{x}^{2}q_{yy}q_{zz}\sin q\\
&-2q^{\dagger}_{y}qq_{x}q_{yy}q_{xzz}\cos q+4q^{\dagger}_{yy}qq_{x}q_{xz}q_{z}\sin q-2q^{\dagger}q_{xx}q_{y}q_{yy}q_{z}^{2}\sin q-q^{\dagger}_{xx}qq_{y}q_{yy}q_{z}^{2}\sin q\\
&-2q^{\dagger}_{x}q_{x}q_{y}q_{yy}q_{z}^{2}\sin q-4q^{\dagger}_{xy}q_{xy}q_{zz}\sin q-2q^{\dagger}_{zz}q_{x}^{2}q_{yy}\cos q+4q^{\dagger}q_{yy}q_{xz}^{2}\cos q\\
&+4q^{\dagger}_{x}q_{yy}q_{xzz}\sin q+2q^{\dagger}_{xxz}q_{yy}q_{z}\sin q+4q^{\dagger}_{xz}q_{yy}q_{xz}\sin q+4q^{\dagger}_{z}q_{yy}q_{xxz}\sin q\\
&+4q^{\dagger}q_{yy}q_{xxzz}\sin q-2q^{\dagger}_{zz}q_{xxy}q_{y}\sin q+q^{\dagger}_{yy}q_{xx}q_{z}^{2}\cos q-2q^{\dagger}_{yy}qq_{xz}^{2}\cos q\\
&+2q^{\dagger}_{yy}q_{xxz}q_{z}\sin q+4q^{\dagger}q_{xx}q_{yz}^{2}\cos q+2q^{\dagger}_{xx}qq_{yz}^{2}\cos q+4q^{\dagger}_{x}q_{x}q_{yz}^{2}\cos q\\
&-q^{\dagger}_{xx}q_{yy}q_{z}^{2}\cos q-4q^{\dagger}_{x}q_{xyy}q_{z}^{2}\cos q-2q^{\dagger}_{xxy}q_{y}q_{z}^{2}\cos q-4q^{\dagger}_{y}q_{xxy}q_{z}^{2}\cos q\\
&-4q^{\dagger}q_{xxyy}q_{z}^{2}\cos q-4q^{\dagger}_{xy}q_{xy}q_{z}^{2}\cos q-4q^{\dagger}_{x}q_{xyy}q_{zz}\sin q-2q^{\dagger}_{xxy}q_{y}q_{zz}\sin q\\
&-4q^{\dagger}_{y}q_{xxy}q_{zz}\sin q-4q^{\dagger}q_{xxyy}q_{zz}\sin q+4q^{\dagger}q_{xxz}q_{yy}q_{z}\cos q+q^{\dagger}_{zz}qq_{xy}^{2}q_{yy}\sin q\\
&-q^{\dagger}_{zz}q_{xx}q_{y}q_{yy}\sin q+2q^{\dagger}_{zz}qq_{xy}q_{xyy}\sin q+q^{\dagger}_{yy}qq_{x}^{2}q_{z}^{2}\cos q+q^{\dagger}_{yy}qq_{x}^{2}q_{zz}\sin q\\
&-2q^{\dagger}_{y}qq_{yy}q_{xz}^{2}\cos q-2q^{\dagger}_{yy}qq_{x}q_{xzz}\cos q+2q^{\dagger}_{y}q_{xxz}q_{yy}q_{z}\sin q-4q^{\dagger}_{x}q_{x}q_{yz}q_{yyz}\sin q\\
&+4q^{\dagger}_{x}q_{x}q_{y}q_{yzz}\cos q-q^{\dagger}_{zz}qq_{xx}q_{y}^{2}\sin q-q^{\dagger}_{yy}qq_{xx}q_{yz}^{2}\sin q-2q^{\dagger}_{y}qq_{xx}q_{yz}^{2}\cos q\\
&-2q^{\dagger}_{yyz}qq_{xx}q_{yz}\sin q-2q^{\dagger}_{yy}qq_{xx}q_{yzz}\sin q-2q^{\dagger}_{yz}qq_{xx}q_{yyz}\sin q-2q^{\dagger}_{y}qq_{xx}q_{yyzz}\sin q\\
&+2q^{\dagger}_{yzz}qq_{xx}q_{y}\cos q-2q^{\dagger}q_{yy}q_{xy}^{2}q_{z}^{2}\cos q+q^{\dagger}_{yy}qq_{xy}^{2}q_{z}^{2}\cos q+2q^{\dagger}_{y}q_{xx}q_{yy}q_{z}^{2}\cos q\\
&+q^{\dagger}_{yy}q_{xx}q_{y}q_{z}^{2}\cos q-q^{\dagger}_{xx}q_{y}q_{yy}q_{z}^{2}\cos q-4q^{\dagger}_{x}q_{xy}q_{yy}q_{z}^{2}\cos q+2q^{\dagger}_{yy}q_{xy}q_{x}q_{z}^{2}\cos q\\
&+2q^{\dagger}_{y}q_{x}q_{xyy}q_{z}^{2}\cos q-2q^{\dagger}q_{xy}^{2}q_{yy}q_{zz}\sin q-4q^{\dagger}q_{xxy}q_{yy}q_{z}^{2}\cos q+q^{\dagger}_{yy}qq_{xy}^{2}q_{zz}\sin q\\
&+2q^{\dagger}_{y}q_{xxy}q_{y}q_{z}^{2}\cos q+2q^{\dagger}_{y}q_{x}q_{xyy}q_{zz}\sin q-4q^{\dagger}q_{xxy}q_{yy}q_{zz}\sin q+2q^{\dagger}_{y}q_{xxy}q_{y}q_{zz}\sin q\\
&+2q^{\dagger}_{yy}qq_{xxy}q_{zz}\sin q+2q^{\dagger}_{y}qq_{xxyy}q_{zz}\sin q+2q^{\dagger}_{xyy}qq_{xy}q_{zz}\sin q+2q^{\dagger}_{xy}qq_{xyy}q_{zz}\sin q\\
&-4q^{\dagger}q_{xy}q_{xyy}q_{zz}\sin q-2q^{\dagger}q_{x}^{2}q_{yy}q_{z}^{2}\cos q-2q^{\dagger}q_{xx}q_{yy}q_{z}^{2}\sin q+2q^{\dagger}_{yy}qq_{xxy}q_{z}^{2}\cos q\\
&+2q^{\dagger}_{y}qq_{xxyy}q_{z}^{2}\cos q+2q^{\dagger}_{xyy}qq_{xy}q_{z}^{2}\cos q+2q^{\dagger}_{xy}qq_{xyy}q_{z}^{2}\cos q-4q^{\dagger}q_{xy}q_{xyy}q_{z}^{2}\cos q\\
&+q^{\dagger}_{y}q_{xx}q_{yy}q_{zz}\sin q+q^{\dagger}_{yy}q_{xx}q_{y}q_{zz}\sin q-q^{\dagger}_{xx}q_{yy}q_{y}q_{zz}\sin q-4q^{\dagger}_{x}q_{xy}q_{yy}q_{zz}\sin q\\
&+2q^{\dagger}_{yy}q_{x}q_{xy}q_{zz}\sin q-2q^{\dagger}q_{xx}q_{y}^{2}q_{z}^{2}\cos q-q^{\dagger}_{xx}qq_{y}^{2}q_{z}^{2}\cos q-2q^{\dagger}_{x}q_{x}q_{y}^{2}q_{z}^{2}\cos q\\
&-2q^{\dagger}q_{xx}q_{yy}q_{yz}^{2}\sin q-2q^{\dagger}q_{xx}q_{y}^{2}q_{zz}\sin q-q^{\dagger}_{xx}qq_{yy}q_{yz}^{2}\sin q-q^{\dagger}_{xx}qq_{y}^{2}q_{zz}\sin q\\
&-2q^{\dagger}_{x}q_{x}q_{yy}q_{yz}^{2}\sin q-2q^{\dagger}_{x}q_{x}q_{y}^{2}q_{zz}\sin q-4q^{\dagger}q_{xx}q_{yz}q_{yyz}\sin q+4q^{\dagger}q_{xx}q_{y}q_{yzz}\cos q\\
&-2q^{\dagger}_{xx}qq_{yz}q_{yyz}\sin q+2q^{\dagger}_{xx}qq_{y}q_{yzz}\cos q-2q^{\dagger}_{x}q_{x}q_{yy}q_{z}^{2}\sin q-2q^{\dagger}q_{x}^{2}q_{yy}q_{zz}\sin q\\
&+q^{\dagger}_{zz}qq_{x}^{2}q_{yy}\sin q+2q^{\dagger}q_{xx}q_{yy}q_{zz}\cos q+2q^{\dagger}_{x}q_{x}q_{yy}q_{zz}\cos q+4q^{\dagger}_{x}q_{xz}q_{yy}q_{z}\cos q\\
&+4q^{\dagger}q_{x}q_{yy}q_{xzz}\cos q-2q^{\dagger}_{xzz}qq_{x}q_{yy}\cos q+q^{\dagger}_{y}qq_{xx}q_{y}q_{yy}q_{z}^{2}\sin q-q^{\dagger}_{y}qq_{xx}q_{y}q_{yy}q_{zz}\cos q\\
&-2q^{\dagger}_{y}qq_{xx}q_{yy}q_{yz}q_{z}\cos q+4q^{\dagger}_{y}qq_{xx}q_{y}q_{yz}q_{z}\sin q+4q^{\dagger}_{y}qq_{x}q_{xz}q_{yy}q_{z}\sin q\Big\}.
\end{aligned}
$$

Clearly, it would have been fairly impossible to calculate this quantity without suitable software ([16, 17]). Yet it is this ephemeral fiction that consumed most of the processor time; the genuine right-hand side (which there is none) was calculated by hand, see Eq. (7) on p. 11.

4. A yet another manifestation of old formalism's inconsistency

We agree that the presence of exact terms in the right-hand side of (9) does not discredit any result known from the theories in which all objects are processed only by the Schouten bracket (possibly, in its transcript (5)). For instance, the claims remain true for variational Poisson bi-vectors, the Poisson cohomology groups they give rise to (see [21]), or bi-Hamiltonian cohomology [18]. The contrast between (4) and (11) makes no harm in that narrow sub-class of problems which are posed in the frames of variational symplectic supergeometry of parity-even variables $\boldsymbol{q}$ and their parity-odd canonical conjugates $\boldsymbol{q}^\dagger$ over $\boldsymbol{x} \in M^m$.

Unfortunately, there is much amiss if the superbundle π stays the same but the class of problems to-consider is less narrow (e.g., see [1, 3, 4] and references therein). Namely, the Batalin–Vilkovisky Laplacian Δ stops being a graded derivation of the Schouten bracket,

$$\Delta([\![F,G]\!]) = [\![\Delta F, G]\!] + (-)^{|F|-1}[\![F,\Delta G]\!], \qquad (12)$$

whence the parity-odd linear operator Δ stops being a differential,[16] violating (15). Let us substantiate this claim by a counterexample.

It is often accepted that the Batalin–Vilkovisky Laplacian Δ is no more than the parity-odd linear operator which acts on a given integral functional $H = \int h(\boldsymbol{x},[\boldsymbol{q}],[\boldsymbol{q}^\dagger])\,\mathrm{dvol}(\boldsymbol{x})$ by the formula[17]

$$\Delta(H) \text{ ``=''} \int \frac{\vec{\delta}}{\delta \boldsymbol{q}} \cdot \frac{\vec{\delta}}{\delta \boldsymbol{q}^\dagger}(h)\,(\boldsymbol{x},[\boldsymbol{q}],[\boldsymbol{q}^\dagger])\,\mathrm{dvol}(\boldsymbol{x}) = \\ = \int \sum_{i=1}^{m} \sum_{\substack{|\sigma|\geq 0\\ |\tau|\geq 0}} \left(\Big(-\frac{\vec{\mathrm{d}}}{\mathrm{d}\boldsymbol{x}}\Big)^{\tau} \circ \frac{\vec{\partial}}{\partial q^i_\tau} \circ \Big(-\frac{\vec{\mathrm{d}}}{\mathrm{d}\boldsymbol{x}}\Big)^{\sigma} \circ \frac{\vec{\partial}}{\partial q^\dagger_{i,\sigma}} \right) h(\boldsymbol{x},[\boldsymbol{q}],[\boldsymbol{q}^\dagger])\,\mathrm{dvol}(\boldsymbol{x}). \qquad (13)$$

We now demonstrate[18] that formula (13) is oversimplified to the extent that it is not able to let the Batalin–Vilkovisky Laplacian satisfy important identity (12).

Counterexample 4. Let us denote by f and h the respective integrands in $F = \int q^\dagger q q_{xx}\,\mathrm{d}x$ and $H = \int q^\dagger_{xx} \cos q\,\mathrm{d}x$. One eagerly calculates[19]

$$\frac{\delta f}{\delta q} = q^\dagger q_{xx} + \tfrac{\mathrm{d}^2}{\mathrm{d}x^2}(q^\dagger q) = 2q^\dagger q_{xx} + 2q^\dagger_x q_x + q^\dagger_{xx} q, \qquad \frac{\delta f}{\delta q^\dagger} = q q_{xx},$$

$$\frac{\delta h}{\delta q^\dagger} = \tfrac{\mathrm{d}^2}{\mathrm{d}x^2}(\cos q) = -q_{xx}\sin q - q_x^2 \cos q, \qquad \frac{\delta h}{\delta q} = -q^\dagger_{xx}\sin q.$$

Consider first $\Delta([\![F,H]\!])$. Our new working formula (13), combined with the primary exercise $\delta/\delta q \circ \mathrm{d}/\mathrm{d}x \equiv 0 \equiv \delta/\delta q^\dagger \circ \mathrm{d}/\mathrm{d}x$, suggests that one writes $\Delta(G)$ "$\cong$" $\int \big(\partial/\partial q \circ \partial/\partial q^\dagger\big)(g)\,\mathrm{d}x$ for

[16] Equally bad an option would it be to postulate the validity of main relations between the Batalin–Vilkovisky Laplacian Δ and variational Schouten bracket $[\![\,,\,]\!]$, so that neither their proof nor explicit examples are possible any longer.

[17] The genuine linear operation Δ is extended to the vector space of formal products of integral functionals by the Leibniz rule for $\vec{\partial}/\partial \boldsymbol{q}_\tau$ and $\vec{\partial}/\partial \boldsymbol{q}^\dagger_\sigma$, yielding the operational definition of variational Schouten bracket, see (3).

[18] Let us warn the reader that all claims in Counterexample 4 about any equalities between functionals (specifically, for ΔF and ΔH or for $[\![F,H]\!]$ and $\Delta([\![F,H]\!])$, etc.) should be viewed as the classical parable about a cage which contains an elephant but carries an inscription "BUFFALO" — one may not trust his own eyes. We refer to Remark 1 below and also to Example 2.4 in [1], in which we explicitly calculate the objects ΔF and ΔH or $\Delta([\![F,H]\!])$, confirming that equality (14) *is* valid.

[19] Note that since all the four variational derivatives contain at most one parity-odd $q^\dagger$ or its derivatives, the directions of all the derivations do not actually matter — i.e., reversing their direction would not result in minus signs.

Journal of Physics: Conference Series **670** (2016) 012030 doi:10.1088/1742-6596/670/1/012030

any integral functional $G = \int g\,\mathrm{d}x$. This implies that only those terms survive under Δ in which the density of $[\![F, H]\!]$ carries $q^\dagger$ without derivatives with respect to x. So, one takes into account only

$$[\![F, H]\!] \text{ “=” } \int \big((2q^\dagger q_{xx} + \ldots) \cdot (-q_{xx}\sin q - q_x^2\cos q) + \ldots\big)\,\mathrm{d}x,$$

where the dots indicate the irrelevant terms with $q_x^\dagger$ or $q_{xx}^\dagger$. This produces the functional

$$\begin{aligned}\Delta\big([\![F, H]\!]\big) \text{ “}\cong\text{” } &-2\int \frac{\partial}{\partial q}\frac{\partial}{\partial q^\dagger}\big(q^\dagger q_{xx}^2 \sin q + q^\dagger q_x^2 q_{xx}\cos q\big)\,\mathrm{d}x\\ &= -2\int \frac{\partial}{\partial q}\big(q_{xx}^2\sin q + q_x^2 q_{xx}\cos q\big)\,\mathrm{d}x = -2\int\big(q_{xx}^2\cos q - q_x^2 q_{xx}\sin q\big)\,\mathrm{d}x,\end{aligned}$$

the integrand of which is not cohomogically trivial (as could be readily seen by calculating its variational derivative, which gives nonzero).

On the other hand, ΔF “$\cong$” $\int q_{xx}\,\mathrm{d}x \cong 0$, whence $[\![\Delta F, H]\!]$ “=” $[\![0, H]\!] = 0$. At the same time, the density h contains no $q^\dagger$ but only the second derivative $q_{xx}^\dagger$, which means now that ΔH “$\cong$” 0 as well, whence $[\![F, \Delta H]\!]$ “=” $[\![F, 0]\!] = 0$. In conclusion to this “counterexample,”

$$\Delta\big([\![F, H]\!]\big) \text{ “}\neq\text{” } [\![\Delta F, H]\!] + (-)^{|F|-1}[\![F, \Delta H]\!] = 0. \tag{14}$$

This “contradiction” marks the limits of jet-bundle approach [13] to the BV-geometry via Vinogradov's $\mathcal{C}$-spectral sequence $E_i^{p,q}$ (specifically, by using the upper line $E_1^{n,q}$ of its first term such that $E_1^{n,0} = \bar{H}^n(\pi) \ni F, G, H$, cf. [19]). The resolution of apparent difficulties is achieved in [1, 2], where we rigorously prove the validity of identities (12), (3), and

$$\Delta^2 = 0. \tag{15}$$

In particular, identity (12) does of course hold for the functionals F and H as above — this is confirmed in [1, pp. 34–36].

Remark 1. A very interesting effect which the true theory of iterated variations now offers is the natural existence of synonyms for zero functional; that is, there are objects which would take every section $\boldsymbol{s} \in \Gamma(\pi)$ of the superbundle to $0 \in \Bbbk$ but which, belonging in fact to spaces larger than that for the trivial cohomology class $\int \mathrm{d}_h\eta \in E_1^{n,0}$, can contribute *non*trivially to the output of a calculation. (To comprehend why this is possible, compare (4) with (11) and refer to [1, §1.4] for more explanation.) Such is the object ΔH in (14) for the functional $H = \int q_{xx}^\dagger \cos q\,\mathrm{d}x$, which we used in all the examples here.

This phenomenon manifests the first main slogan in the geometry of iterated variations: **no calculation can be interrupted at any intermediate step**.

Remark 2. The second guiding principle is that the integrations by parts always fall only on the functionals which they stem from but never hit the contributions from other functionals. Therefore, the variation of G or H within $[\![G, H]\!]$ in a term like $[\![F, [\![G, H]\!]]\!]$ is such that the specific choice of H (or, respectively, G) does not matter; the structure $[\![\,,\,]\!]$ is uniquiely defined by (2) – or, equivalently, by (3) – for all pairs of functionals.

Conversely, formula (5) tells us that the bi-linear operation introduced by it makes the derivations which will fall on G or H within $[\![G, H]\!]$ in the future calculation of $[\![F, [\![G, H]\!]]\!]$ explicitly dependent on a choice of the other argument (i.e., H or G, respectively). In other words, formula (5) encodes *infinitely many* structures (roughly speaking, its own structure for each pair of arguments). This is in contrast to the standard idea[10 on p. 7] that the variational Schouten bracket is a *unique* extension of the commutator $[\,,\,]$ of evolutionary vector fields on the jet bundle $J^\infty(\pi)$ to the space of variational multi-vectors on it.

Acknowledgments
The author thanks the Organizing committee of XXIII International conference 'Integrable systems & quantum symmetries' (22 – 28 June 2015; CVUT Prague, Czech Republic) for a warm atmosphere during the meeting. This research was partially supported by JBI RUG project 103511 (Groningen). A part of this work was done while the author was visiting at the IHÉS (Bures-sur-Yvette, France); the hospitality and partial financial support of that institution are gratefully acknowledged.

[1] Kiselev A V 2013 The geometry of variations in Batalin–Vilkovisky formalism *J. Phys.: Conf. Ser.* **474** Paper 012024, 51 (*Preprint* `1312.1262 [math-ph]`)
[2] Kiselev A V 2014 The Jacobi identity for graded-commutative variational Schouten bracket revisited *PEPAN Letters* **11** 1–4 (*Preprint* `1312.4140 [math-ph]`)
[3] Kiselev A V 2015 Deformation approach to quantisation of field models *Preprint* IHÉS/M/15/13
[4] Kiselev A V 2015 The calculus of multivectors on noncommutative jet spaces *Preprint* IHÉS/M/14/39, `1210.0726 v3 [math.DG]`
[5] Gel'fand I M and Shilov G E 1964 *Generalized functions* vol **1** Properties and operations (NY–London: Academic Press)
Gel'fand I M and Shilov G E 1968 *Generalized functions* vol **2** Spaces of fundamental and generalized functions (NY–London: Academic Press)
[6] Gitman D M and Tyutin I V 1990 *Quantization of fields with constraints* (Springer Ser. Nucl. Part. Phys.) (Berlin: Springer-Verlag)
[7] Henneaux M and Teitelboim C 1992 *Quantization of gauge systems* (Princeton, NJ: Princeton Univ. Press)
[8] Kontsevich M 1993 Formal (non)commutative symplectic geometry *The Gel'fand Mathematical Seminars, 1990-1992* eds L Corwin, I Gelfand and J Lepowsky (Boston MA: Birkhäuser) pp 173–87
[9] Olver P J and Sokolov V V 1998 Integrable evolution equations on associative algebras *Comm. Math. Phys.* **193** 245–68
[10] Batalin I A and Vilkovisky G A 1981 Gauge algebra and quantization *Phys. Lett. B* **102** 27–31
Batalin I A and Vilkovisky G A 1983 Quantization of gauge theories with linearly dependent generators *Phys. Rev. D (3)* **28** 2567–82
[11] Kontsevich M 2003 Deformation quantization of Poisson manifolds *Lett. Math. Phys.* **66** 157–216 (*Preprint* `q-alg/9709040`)
[12] Kiselev A V and Krutov A O 2014 Non-Abelian Lie algebroids over jet spaces *J. Nonlin. Math. Phys.* **21** 188–213 (*Preprint* `1305.4598 [math.DG]`)
[13] Kupershmidt B A 1980 Geometry of jet bundles and the structure of Lagrangian and Hamiltonian formalisms *Geometric methods in mathematical physics (Proc. NSF-CBMS Conf., Univ. Lowell, Lowell, Mass., 1979)* (*Lecture Notes in Math.* vol 775) (Springer, Berlin) pp 162–218
[14] Olver P J 1993 *Applications of Lie groups to differential equations* 2nd ed (*Graduate Texts in Mathematics* vol 107) (Springer-Verlag, New York)
[15] Kosmann-Schwarzbach Y 2008 Poisson manifolds, Lie algebroids, modular classes: a survey *SIGMA Symmetry Integrability Geom. Methods Appl.* **4** Paper 005, 30 (*Preprint* `0710.3098 [math.SG]`)
[16] Marvan M 2009 Sufficient set of integrability conditions of an orthonomic system *Found. Comput. Math.* **9** 651–74 (*Preprint* `nlin.SI/0605009`)
[17] Kiselev A V and Wolf T 2007 Classification of integrable super-systems using the SsTools environment *Comput. Phys. Comm.* **177** 315–28 (*Preprint* `nlin.SI/0609065`)
[18] Dubrovin B and Zhang Y 2001 Normal forms of hierarchies of integrable PDEs, Frobenius manifolds and Gromov–Witten invariants (*Preprint* `math.DG/0108160`)
[19] Vinogradov A M 1984 The $\mathcal{C}$-spectral sequence, Lagrangian formalism, and conservation laws. I, II. *J. Math. Anal. Appl.* **100** 1–129
[20] Kiselev A V and Ringers S 2013 A comparison of definitions for the Schouten bracket on jet spaces *Proc. 6th Int. workshop 'Group analysis of differential equations and integrable systems' (June 18–20, 2012; Protaras, Cyprus)* pp 127–41 (*Preprint* `1208.6196 [math.DG]`)
[21] De Sole A and Kac V G 2013 The variational Poisson cohomology *Jpn. J. Math.* **8** 1–145 (*Preprint* `1106.0082 [math-ph]`)

Journal of Physics: Conference Series **670** (2016) 012031 doi:10.1088/1742-6596/670/1/012031

The classical Darboux III oscillator: factorization, Spectrum Generating Algebra and solution to the equations of motion

D Latini[3], O Ragnisco[3], A Ballesteros[1], A Enciso[2], F J Herranz[1], and D Riglioni[3]

[1] Departamento de Física, Universidad de Burgos, E-09001 Burgos, Spain
[2] Instituto de Ciencias Matemáticas, CSIC, Nicolás Cabrera 13-15, E-28049 Madrid, Spain
[3] Dipartimento di Matematica e Fisica, Università di Roma Tre and Istituto Nazionale di Fisica Nucleare sezione di Roma Tre, Via Vasca Navale 84, I-00146 Roma, Italy

E-mail: angelb@ubu.es, aenciso@icmat.es, fjherranz@ubu.es, latini@fis.uniroma3.it, ragnisco@fis.uniroma3.it, riglioni@fis.uniroma3.it

Abstract. In a recent paper the so-called *Spectrum Generating Algebra* (SGA) technique has been applied to the N-dimensional Taub-NUT system, a maximally superintegrable Hamiltonian system which can be interpreted as a one-parameter deformation of the Kepler–Coulomb system. Such a Hamiltonian is associated to a specific Bertrand space of non-constant curvature. The SGA procedure unveils the symmetry algebra underlying the Hamiltonian system and, moreover, enables one to solve the equations of motion. Here we will follow the same path to tackle the Darboux III system, another maximally superintegrable system, which can indeed be viewed as a natural deformation of the isotropic harmonic oscillator where the flat Euclidean space is again replaced by another space of non-constant curvature.

1. Introduction

We consider the classical Hamiltonian in $\mathbb{R}^N$ given by:

$$\mathcal{H}_\lambda(\mathbf{q},\mathbf{p}) = \mathcal{T}_\lambda(\mathbf{q},\mathbf{p}) + \mathcal{U}_\lambda(\mathbf{q}) = \frac{\mathbf{p}^2}{2m(1+\lambda\mathbf{q}^2)} + \frac{m\omega^2\mathbf{q}^2}{2(1+\lambda\mathbf{q}^2)}, \tag{1}$$

where ω and λ are real positive parameters, $\mathbf{q} = (q_1,\ldots,q_N)$, $\mathbf{p} = (p_1,\ldots,p_N) \in \mathbb{R}^N$ are conjugate coordinates and momenta, $\mathbf{q}^2 = \sum_{i=1}^N q_i^2$ and $\mathbf{p}^2 = \sum_{i=1}^N p_i^2$. The kinetic energy $\mathcal{T}_\lambda(\mathbf{q},\mathbf{p})$ can be interpreted as the Lagrangian generating the geodesic motion of a particle of mass m on a conformally flat space known as the Darboux space of type III (D-III). As remarked in [1], such an N-dimensional (ND) curved space is the spherically symmetric generalization of the D-III surface [2, 3, 4], which was constructed in [5, 6]. Moreover, the central potential $\mathcal{U}_\lambda(\mathbf{q})$ was proven to be an 'intrinsic oscillator' potential on that space [6, 7]. We recall that this definition was first proposed in [6] and that a generalization of the Bertrand's Theorem [8] to 3D conformally flat Riemannian spaces has been developed in [9, 10, 11]. In this respect, we shall say that the Hamiltonian (1) determines the D-III oscillator.

We also notice that in [6] the Hamiltonian $\mathcal{H}_\lambda$ has been proven to be maximally superintegrable by taking advantage of its super-separability (like the usual harmonic oscillator, the system turns out to be separable both in Cartesian and (hyper)spherical coordinates). This property has been further analysed in a number of subsequent papers (see, for instance, [11, 12, 13] and references therein), where its algebraic content was explained in terms of the *Demkov–Fradkin tensor* [14, 15] given by

$$\mathcal{I}_{ij} = p_i p_j + m^2 \left(\omega^2 - \frac{2\lambda}{m} \mathcal{H}_\lambda(\mathbf{q}, \mathbf{p}) \right) q_i q_j \,, \qquad i, j = 1, \ldots, N \,.$$

We recall here that the Hamiltonian $\mathcal{H}_\lambda$ can be expressed in terms of hyperspherical coordinates r, θ_j and canonical momenta p_r, p_{θ_j} , $(j = 1, \ldots, N-1)$ defined by

$$q_j = r \cos\theta_j \prod_{k=1}^{j-1} \sin\theta_k \quad (1 \le j \le N-1) \,, \qquad q_N = r \prod_{k=1}^{N-1} \sin\theta_k \,,$$

such that

$$r = |\mathbf{q}| \,, \qquad \mathbf{p}^2 = p_r^2 + \frac{\mathbf{L}^2}{r^2} \,, \quad \text{with} \quad \mathbf{L}^2 = \sum_{j=1}^{N-1} p_{\theta_j}^2 \prod_{k=1}^{j-1} \frac{1}{\sin^2\theta_k} \,,$$

where the radial coordinate r is canonically conjugated to the radial momentum p_r. Then, for a fixed value of $\mathbf{L}^2 = l^2$, the Hamiltonian (1) can be written as a one-degree of freedom radial system.

As a matter of fact the system associated with $\mathcal{H}_\lambda$ can be considered as a genuine (maximally superintegrable) λ-deformation of the usual ND isotropic harmonic oscillator with frequency ω, since the limit $\lambda \to 0$ yields

$$\mathcal{H}_0(\mathbf{p}, \mathbf{q}) = \frac{\mathbf{p}^2}{2m} + \frac{m\omega^2 \mathbf{q}^2}{2} \,.$$

We remind also that both the classical and the quantum problems have been investigated (see [6, 12, 13]) in the special case characterized by positive values of the deformation parameter λ. Also, an in-depth discussion of the $\lambda < 0$ case for the classical system can be found in [13] where, however, the explicit formula of the trajectory has not been derived.

In the present work our scope is rather different from the one pursued in [6]. On one hand, we deal just with the classical system and restrict our considerations to the 3D case, with emphasis on the algebraic side of the problem. In this context we show that the so-called *Spectrum Generating Algebra* (SGA) technique [16, 17] provides us with all the necessary ingredients to achieve its solution, *at least in the* $\lambda > 0$ *case*, in full analogy with the classical Taub-NUT system [18]. In particular, we will derive the solution for both bounded and unbounded motions. On the other hand, through a direct approach, we will be able to handle the $\lambda < 0$ case as well, getting an explicit formula for $t = t(r)$ holding in the *punctured open ball*

$$0 < r < r_s \doteq \frac{1}{\sqrt{|\lambda|}},$$

and thus making some progress with respect to what had been already derived in the literature [6, 12, 13].

The paper is organized as follows:

(i) In section 2 we perform the factorization of the classical Hamiltonian and derive the appropriate Poisson algebra for the ordinary isotropic harmonic oscillator, recovering the well-known solution to the equation of motion.

(ii) In section 3, paraphrasing the procedure followed in [18] we solve the equations of motion in the $\lambda > 0$ case. As it happened for the Taub-NUT system, the solution is obtained in the form $t = t(r)$. As already noticed in [6], in contrast with the usual harmonic oscillator, for sufficiently high energy values there are unbounded orbits (the particle can escape to infinity) which in the quantum setting corresponds to the onset of a continuous spectrum. Coherently with the predictions of the Bertrand's Theorem, the bounded orbits will be closed and there will be stable circular trajectories.

(iii) In section 4 the $\lambda < 0$ case will be tersely investigated. We will get behaviours that are similar to those exhibited by the Taub-NUT system for negative values of the deformation parameter η [18]. Indeed, in a sort of dual way with respect to the Taub-NUT case, the kinetic energy (and thus the metric) remains positive in the bounded region $0 < r < r_s$ while it takes negative values for $r > r_s$. So, in order to get an admissible metric for $r > r_s$, we should there resort to time-reversal symmetry to restore positivity. On the other hand, in that unbounded region the effective potential is a monotonic function of r having a constant limiting value at infinity, implying that no closed orbits can arise.

(iv) The main results and some further perspectives are outlined in the concluding section.

2. The Euclidean case: isotropic harmonic oscillator

Before using the machinery of the SGA approach to tackle the $\lambda \neq 0$ case, we will firstly characterize the Euclidean case in order to check our future results in the (flat) limit $\lambda \to 0$ and to illustrate the usual SGA procedure. Therefore, let us consider the Hamiltonian

$$H_0(r,p) = T_0(r,p) + V_{0,\text{eff}}(r) = \frac{p^2}{2m} + \frac{l^2}{2mr^2} + \frac{1}{2}m\omega^2 r^2 \,, \tag{2}$$

where r is the radial coordinate and $p \equiv p_r$ is the radial momentum related to the particle of mass m, frequency ω and angular momentum l. The effective potential is so given by

$$V_{0,\text{eff}}(r) = \frac{l^2}{2mr^2} + \frac{1}{2}m\omega^2 r^2 \,, \tag{3}$$

Clearly, in order to have closed orbits, the values of the energy have to be confined in the region $V_{0,\text{eff}}(r_{\text{min}}) < E < \infty$, where $V_{0,\text{eff}}(r_{\text{min}}) = l\omega$ is the minimum of the effective potential, i.e. its value calculated at $r_{\text{min}} = \sqrt{\frac{l}{m\omega}}$ (see lhs of figure 1).

Now, in order to apply the SGA technique, we proceed by factorizing the Hamiltonian (2) as follows

$$p^2r^2 + m^2\omega^2r^4 - 2mr^2H_0 = A_0^+A_0^- + \gamma(H_0) = -l^2 \,, \tag{4}$$

where $A_0^\pm$ are unknown functions to be determined. In particular, taking into account (4), we obtain

$$A_0^\pm = \mp irp + m\omega r^2 - \frac{H_0}{\omega} \,, \tag{5}$$

and then

$$\gamma(H_0) = -\frac{H_0^2}{\omega^2} . \tag{6}$$

The Hamiltonian H_0 and $A_0^\pm$ are functions of the canonical coordinates (r,p) and they have to close the following Poisson algebra [16, 17]:

$$\{H_0, A_0^\pm\} = \mp i\alpha(H_0)A_0^\pm \,, \qquad \{A_0^+, A_0^-\} = i\beta(H_0) \,.$$

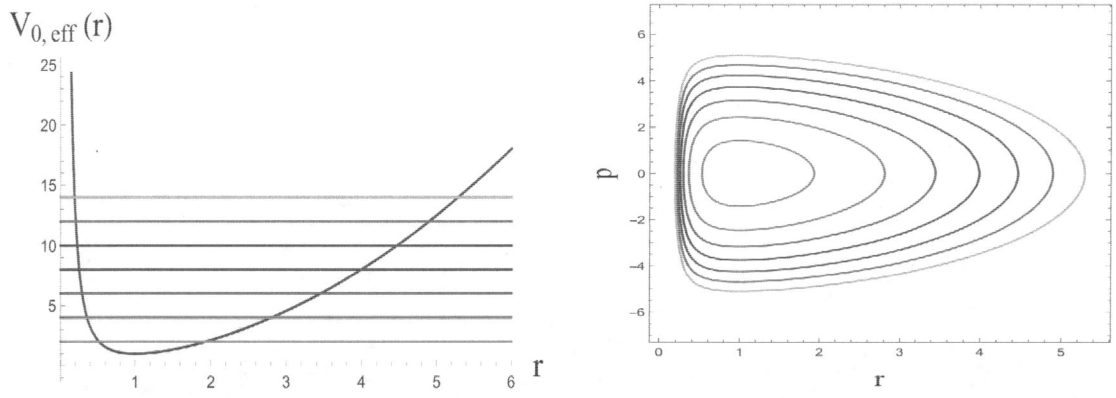

Figure 1. Effective potential $V_{0,\text{eff}}(r)$ (3) and phase plane (r,p) calculated for $m=\omega=l=1$ and $E=2,4,6,8,10,12,14$ (in appropriate units).

In particular, using the expressions (5), we obtain that

$$\alpha(H_0)=2\omega\,,\qquad \beta(H_0)=\frac{4H_0}{\omega}\,.$$

Summarizing, we have the following relations

$$\{H_0,A_0^{\pm}\}=\mp i2\omega A_0^{\pm}\,,\qquad \{A_0^{+},A_0^{-}\}=i\frac{4H_0}{\omega}\,. \tag{7}$$

At this point we can define the *time-dependent constants of motion*

$$Q_0^{\pm}=A_0^{\pm}e^{\mp i\alpha(H_0)t}\,,$$

such that

$$\frac{\mathrm{d}Q_0^{\pm}}{\mathrm{d}t}=\{Q_0^{\pm},H_0\}+\frac{\partial Q_0^{\pm}}{\partial t}=0\,.$$

As we know, those dynamical variables can be written as $Q_0^{\pm}=q_0e^{\pm i\theta_0}$ and allow us to determine the motion [16, 17, 18]. Indeed it holds:

$$\left(\mp irp+m\omega r^2-\frac{H_0}{\omega}\right)e^{\mp 2i\,\omega t}=q_0e^{\pm i\theta_0}\,.$$

By imposing (4) along (6), that is,

$$A_0^{+}A_0^{-}-\frac{H_0^2}{\omega^2}=q_0^2-\frac{H_0^2}{\omega^2}=-l^2\,,$$

we find that

$$\left\{\begin{aligned} -irp+m\omega r^2-\frac{H_0}{\omega}&=q_0e^{i\left(2\,\omega t+\theta_0\right)}\,,\\ irp+m\omega r^2-\frac{H_0}{\omega}&=q_0e^{-i\left(2\,\omega t+\theta_0\right)}\,, \end{aligned}\right. \tag{8}$$

where

$$q_0=\sqrt{-l^2+\frac{H_0^2}{\omega^2}}\,.$$

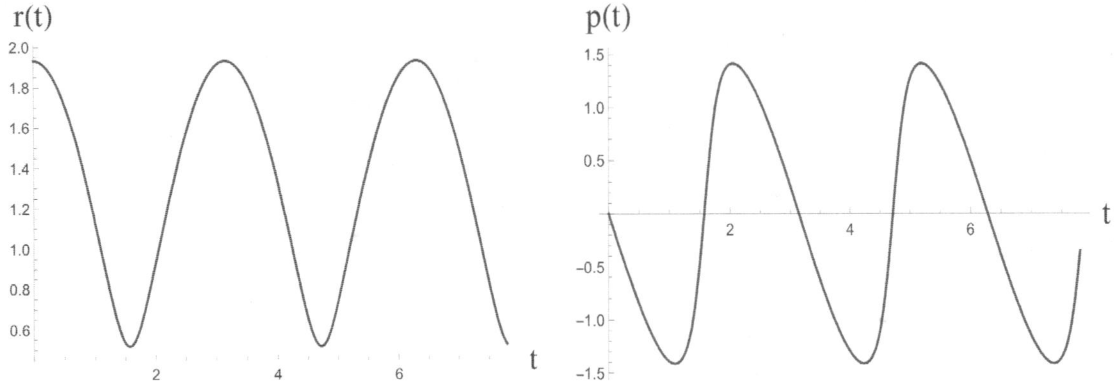

Figure 2. Trajectory $r(t)$ and momentum $p(t)$ calculated for $m = \omega = l = 1$ and $E = 2$.

Now, since the Hamiltonian does not depend explicitly on time, it is a constant of motion, i.e. the energy of the system $H_0 = E$. Therefore, summing and subtracting (8) we obtain (on the level surface $H_0 = E$)

$$\begin{cases} m\omega r^2 - \dfrac{E}{\omega} = q_0 \cos\left(2\,\omega t + \theta_0\right)\,, \\ rp = -q_0 \sin\left(2\,\omega t + \theta_0\right)\,. \end{cases} \tag{9}$$

Finally, using (9) we are able to write r and p as functions of t:

$$\begin{cases} r(t) = \sqrt{\dfrac{E}{m\omega^2} + \dfrac{q_0}{m\omega}\cos\left(2\,\omega t + \theta_0\right)}\,, \\ p(t) = -\dfrac{q_0 \sin\left(2\,\omega t + \theta_0\right)}{\sqrt{\dfrac{E}{m\omega^2} + \dfrac{q_0}{m\omega}\cos\left(2\,\omega t + \theta_0\right)}}\,. \end{cases} \tag{10}$$

In this way we have found $r(t)$ and $p(t)$ and then the motion has been fully determined by means of the SGA procedure. In the rhs of figure 1 the phase plane is depicted, while in figure 2 the curves (10) are represented.

In the following, we shall deal with the deformed case using the same procedure. Clearly, the results just obtained in this section have to be recovered in the Euclidean (i.e. $\lambda \to 0$) limit.

3. The deformed $\lambda > 0$ case: D-III oscillator

In this section we focus our attention on the D-III oscillator Hamiltonian (1) written as a 1D radial system by means of hyperspherical coordinates, namely

$$H(r,p) = T(r,p) + V_{\text{eff}}(r) = \frac{p^2}{2m(1+\lambda r^2)} + \frac{l^2}{2mr^2(1+\lambda r^2)} + \frac{m\omega^2 r^2}{2(1+\lambda r^2)} = \mathcal{F}(r) H_0\,, \tag{11}$$

where m, ω and l are positive constants, λ is the deformation parameter, H_0 is the 'undeformed' isotropic oscillator Hamiltonian (2) and

$$\mathcal{F}(r) \doteq \frac{1}{1+\lambda r^2}\,.$$

Journal of Physics: Conference Series **670** (2016) 012031 doi:10.1088/1742-6596/670/1/012031

Multiplying both sides of (11) by $r^2(1+\lambda r^2)$ we get:

$$r^2(1+\lambda r^2)H = r^2\left(\frac{p^2}{2m}+\frac{l^2}{2mr^2}+\frac{1}{2}m\omega^2 r^2\right) = \frac{1}{2m}(r^2p^2+l^2+m^2\omega^2 r^4). \tag{12}$$

Now, as it has been done in the previous section for the undeformed case, at any r we can factorize (12) as:

$$r^2p^2 + m^2r^4\left(\omega^2-\frac{2\lambda}{m}H\right) - 2mr^2H = A^+A^- + \gamma(H) = -l^2\,, \tag{13}$$

where A^+, A^- are unknown functions of r, p. We make the following *ansatz* for A^+, A^-:

$$A^{\pm} = \left(\mp irp + mr^2\sqrt{\omega^2-\frac{2\lambda}{m}H} - \frac{H}{\sqrt{\omega^2-\frac{2\lambda}{m}H}}\right)e^{\pm f(r,p)}\,.$$

The 'arbitrary function' $f(r,p)$ will be determined by requiring the closure of the Poisson algebra generated by H and $A^{\pm}$:

$$\{H, A^{\pm}\} = \mp i\alpha(H)A^{\pm}\,,$$
$$\{A^+, A^-\} = i\beta(H)\,,$$

where the functions α, β have to be determined. Inserting $A^{\pm}$ in (13) we get

$$\gamma(H) = -\frac{H^2}{\omega^2-\frac{2\lambda}{m}H}\,,$$

and requiring that $A^{\pm}$ obey the proper Poisson brackets we arrive at

$$f(r,p) = -\frac{i\lambda rp\sqrt{\omega^2-\frac{2\lambda}{m}H}}{m(\omega^2-\frac{\lambda}{m}H)}\,,\qquad \alpha(H) = \frac{2(\omega^2-\frac{2\lambda}{m}H)^{\frac{3}{2}}}{\omega^2-\frac{\lambda}{m}H}\,,\qquad \beta(H) = \frac{4H}{\sqrt{\omega^2-\frac{2\lambda}{m}H}}\,.$$

Hence we find that

$$A^{\pm} = \left(\mp irp + mr^2\sqrt{\omega^2-\frac{2\lambda}{m}H} - \frac{H}{\sqrt{\omega^2-\frac{2\lambda}{m}H}}\right)\exp\left\{\mp\frac{i\lambda rp\sqrt{\omega^2-\frac{2\lambda}{m}H}}{m(\omega^2-\frac{\lambda}{m}H)}\right\}\,,$$

$$\{H, A^{\pm}\} = \mp i\,\frac{2(\omega^2-\frac{2\lambda}{m}H)^{\frac{3}{2}}}{\omega^2-\frac{\lambda}{m}H}\,A^{\pm}\,,\qquad \{A^+,A^-\} = i\frac{4H}{\sqrt{\omega^2-\frac{2\lambda}{m}H}}\,.$$

We notice that in the limit $\lambda\to 0$ one gets back the undeformed Poisson algebra (7) as expected. We also point out that the requirement $\omega^2-\frac{2\lambda}{m}H>0$ implies the upper bound $E<\frac{m\omega^2}{2\lambda}$. Moreover, for bounded motion, the energy has to be greater than the minimum of the effective potential $V_{\text{eff}}(r)$. The latter turns out to be

$$V_{\text{eff}}(r_{\text{min}}) = \frac{l^2}{m}\left(\sqrt{\lambda^2+\frac{m^2\omega^2}{l^2}}-\lambda\right)\,,\qquad r^2_{\text{min}} = \frac{l^2}{m^2\omega^2}\left(\lambda+\sqrt{\lambda^2+\frac{m^2\omega^2}{l^2}}\right)\,.$$

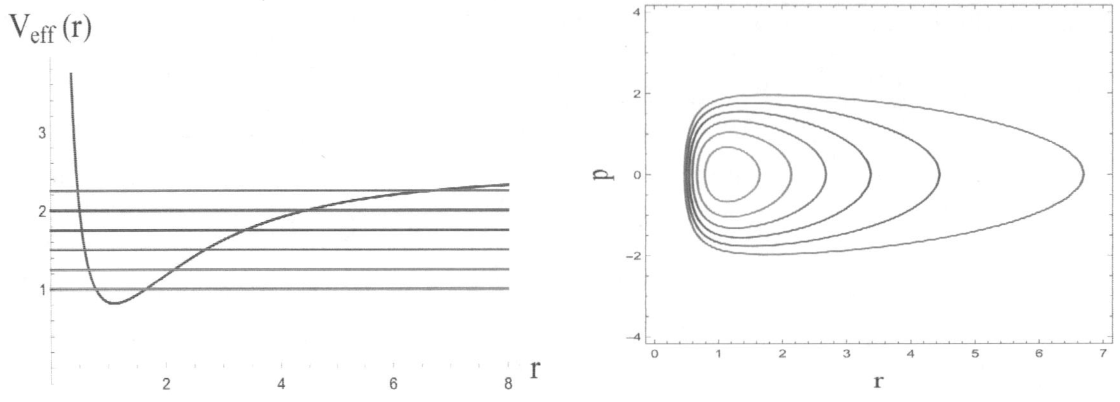

Figure 3. Effective potential $V_{\rm eff}(r)$ (11) and phase plane (r,p) calculated for $m=\omega=l=1$ and $E=1.00, 1.25, 1.50, 1.75, 2.00, 2.25$. The deformation parameter is fixed at the value $\lambda=0.20$.

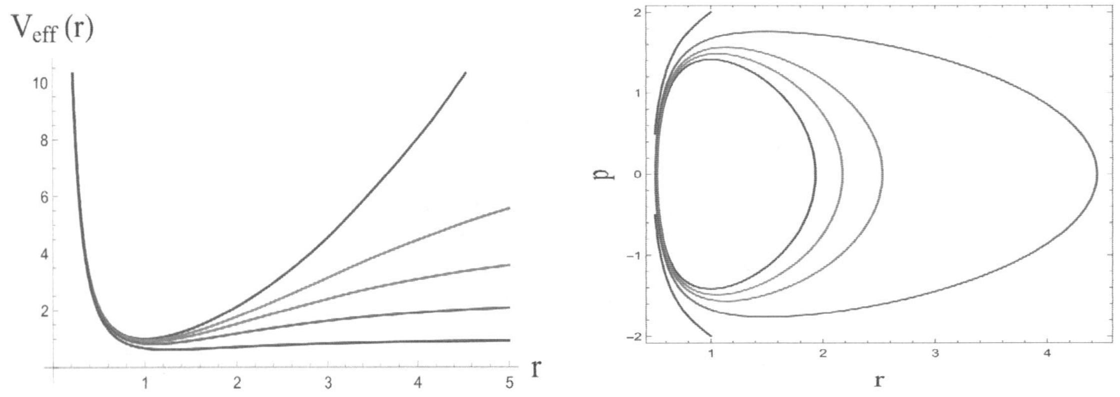

Figure 4. Effective potential $V_{\rm eff}(r)$ (11) and phase plane (r,p) calculated for $m=\omega=l=1$ and $E=2$. The deformation parameter take the values $\lambda=0, 0.05, 0.10, 0.20, 0.50$.

As a consequence the energy belongs to the interval (see lhs of figures 3 and 4)

$$V_{\rm eff}(r_{\rm min}) < E < \frac{m\omega^2}{2\lambda}\,.$$

Next, as usual, in order to find the corresponding trajectory we define the *time-dependent constants of the motion*

$$Q^{\pm} = A^{\pm} e^{\mp i\alpha(H)t}\,,$$

from which we can write (on-shell):

$$\left(\mp irp + mr^2\sqrt{\omega^2-\frac{2\lambda}{m}E} - \frac{E}{\sqrt{\omega^2-\frac{2\lambda}{m}E}}\right)\exp\left\{\mp i\left(\frac{\lambda rp\sqrt{\omega^2-\frac{2\lambda}{m}E}}{m(\omega^2-\frac{\lambda}{m}E)} + \frac{2(\omega^2-\frac{2\lambda}{m}E)^{\frac{3}{2}}}{\omega^2-\frac{\lambda}{m}E}\,t\right)\right\}$$
$$= q\,e^{\pm i\theta}\,,$$

or else

$$\begin{cases} -irp + mr^2\sqrt{\omega^2 - \frac{2\lambda}{m}E} - \dfrac{E}{\sqrt{\omega^2 - \frac{2\lambda}{m}E}} = q\exp\left\{ i\left(\dfrac{\lambda rp\sqrt{\omega^2 - \frac{2\lambda}{m}E}}{m(\omega^2 - \frac{\lambda}{m}E)} + \dfrac{2\left(\omega^2 - \frac{2\lambda}{m}E\right)^{\frac{3}{2}}}{\omega^2 - \frac{\lambda}{m}E} t + \theta \right) \right\}, \\ +irp + mr^2\sqrt{\omega^2 - \frac{2\lambda}{m}E} - \dfrac{E}{\sqrt{\omega^2 - \frac{2\lambda}{m}E}} = q\exp\left\{ -i\left(\dfrac{\lambda rp\sqrt{\omega^2 - \frac{2\lambda}{m}E}}{m(\omega^2 - \frac{\lambda}{m}E)} + \dfrac{2\left(\omega^2 - \frac{2\lambda}{m}E\right)^{\frac{3}{2}}}{\omega^2 - \frac{\lambda}{m}E} t + \theta \right) \right\}, \end{cases} \tag{14}$$

where now

$$q = \sqrt{-l^2 + \frac{E^2}{\omega^2 - \frac{2\lambda}{m}E}}.$$

Once again, summing and subtracting (14) we obtain:

$$\begin{cases} mr^2\sqrt{\omega^2 - \dfrac{2\lambda}{m}E} - \dfrac{E}{\sqrt{\omega^2 - \frac{2\lambda}{m}E}} = q\cos\left(\dfrac{\lambda rp\sqrt{\omega^2 - \frac{2\lambda}{m}E}}{m(\omega^2 - \frac{\lambda}{m}E)} + \dfrac{2(\omega^2 - \frac{2\lambda}{m}E)^{\frac{3}{2}}}{\omega^2 - \frac{\lambda}{m}E} t + \theta \right), \\ rp = -q\sin\left(\dfrac{\lambda rp\sqrt{\omega^2 - \frac{2\lambda}{m}E}}{m(\omega^2 - \frac{\lambda}{m}E)} + \dfrac{2(\omega^2 - \frac{2\lambda}{m}E)^{\frac{3}{2}}}{\omega^2 - \frac{\lambda}{m}E} t + \theta \right). \end{cases}$$

Taking the sum of the square of these two equations we recover (13) restricted to the level surface $H = E$. Finally, thanks to the above relations, we are able to express t as a function of r:

$$t(r) = \frac{1}{\Omega_{[\lambda]}(E)} \left[\arccos\left(\frac{mr^2(\omega^2 - \frac{2\lambda}{m}E) - E}{q\sqrt{\omega^2 - \frac{2\lambda}{m}E}} \right) - \frac{\lambda\sqrt{\omega^2 - \frac{2\lambda}{m}E}}{m(\omega^2 - \frac{\lambda}{m}E)} \sqrt{2mEr^2 - l^2 - m^2r^4(\omega^2 - \frac{2\lambda}{m}E)} - \theta \right], \tag{15}$$

where

$$\Omega_{[\lambda]}(E) = \frac{2(\omega^2 - \frac{2\lambda}{m}E)^{\frac{3}{2}}}{\omega^2 - \frac{\lambda}{m}E} \equiv \alpha(E)$$

is the angular frequency of the motion. In the rhs of figures 3 and 4 some representations of the resulting phase space can be found.

At this point, defining the quantities

$$a^2_{[\lambda]} \doteq \frac{E}{m\left(\omega^2 - \frac{2\lambda}{m}E\right)}, \qquad \epsilon_{[\lambda]} \doteq \sqrt{1 - \frac{\left(\omega^2 - \frac{2\lambda}{m}E\right)l^2}{E^2}},$$

$a^2_{[\lambda]}$ being the square of the major semi-axes of the ellipse and $\epsilon_{[\lambda]}$ a parameter directly related to its eccentricity, equation (15) can be cast into the following form:

$$\Omega_{[\lambda]}(E)t(r) + \theta = \arccos\left[-\frac{1}{\epsilon_{[\lambda]}}\left(1 - \left(\frac{r}{a_{[\lambda]}}\right)^2\right)\right] - \frac{\lambda a^2_{[\lambda]}}{1 + \lambda a^2_{[\lambda]}} \sqrt{\epsilon^2_{[\lambda]} - \left[1 - \left(\frac{r}{a_{[\lambda]}}\right)^2\right]^2}, \tag{16}$$

whose structure is very similar to the one obtained for the classical Taub-NUT system [18]. In the limit $\lambda \to 0$ the equation (16) can be easily inverted to obtain the trajectory $r(t)$ given in (10), showing that it just represents the λ-deformation of the solution of the Euclidean isotropic harmonic oscillator. Then, we can conclude that the motion has been solved by means of the SGA method.

Now we turn to investigate the unbounded motion for positive values of the deformation parameter (see also [6]). The latter will arise for $E > \frac{m\omega^2}{2\lambda}$. In this case

$$\omega^2 - \frac{2\lambda}{m}E = -\left(\frac{2\lambda}{m}E - \omega^2\right) < 0\,,$$

therefore all the previous functions have to be adapted to this situation. In particular, we can write:

$$\tilde{A}^{\pm} = i\left(\mp rp + mr^2\sqrt{\frac{2\lambda}{m}H - \omega^2} + \frac{H}{\sqrt{\frac{2\lambda}{m}H - \omega^2}}\right)\exp\left\{\pm\frac{\lambda rp\sqrt{\frac{2\lambda}{m}H - \omega^2}}{m(\omega^2 - \frac{\lambda}{m}H)}\right\},$$

$$\tilde{\alpha}(H) = -i\,\frac{2(\frac{2\lambda}{m}H - \omega^2)^{\frac{3}{2}}}{\omega^2 - \frac{\lambda}{m}H}\,, \qquad \tilde{\beta}(H) = -i\,\frac{4H}{\sqrt{\frac{2\lambda}{m}H - \omega^2}}\,.$$

All the functions are purely imaginary and the time-dependent constants of motion (also purely imaginary) take the form (see [17]):

$$\tilde{Q}^{\pm} = \tilde{A}^{\pm}e^{\mp i\tilde{\alpha}(H)t} = i\tilde{q}e^{\mp\theta}\,, \qquad \tilde{q} \doteq \sqrt{l^2 + \frac{E^2}{\frac{2\lambda}{m}E - \omega^2}}\,.$$

This leads to the two following equations (again on-shell):

$$\begin{cases} -rp + mr^2\sqrt{\frac{2\lambda}{m}E - \omega^2} + \dfrac{E}{\sqrt{\frac{2\lambda}{m}E - \omega^2}} = \tilde{q}\exp\left\{-\left(\frac{\lambda rp\sqrt{\frac{2\lambda}{m}E-\omega^2}}{m(\omega^2-\frac{\lambda}{m}E)} - \frac{2\left(\frac{2\lambda}{m}E-\omega^2\right)^{\frac{3}{2}}}{\omega^2-\frac{\lambda}{m}E}t + \theta\right)\right\}, \\ +rp + mr^2\sqrt{\frac{2\lambda}{m}E - \omega^2} + \dfrac{E}{\sqrt{\frac{2\lambda}{m}E - \omega^2}} = \tilde{q}\exp\left\{+\left(\frac{\lambda rp\sqrt{\frac{2\lambda}{m}E-\omega^2}}{m(\omega^2-\frac{\lambda}{m}E)} - \frac{2\left(\frac{2\lambda}{m}E-\omega^2\right)^{\frac{3}{2}}}{\omega^2-\frac{\lambda}{m}E}t + \theta\right)\right\}, \end{cases}$$

implying that

$$\begin{cases} mr^2\sqrt{\frac{2\lambda}{m}E - \omega^2} + \dfrac{E}{\sqrt{\frac{2\lambda}{m}E - \omega^2}} = \tilde{q}\cosh\left(\dfrac{\lambda rp\sqrt{\frac{2\lambda}{m}E - \omega^2}}{m(\omega^2 - \frac{\lambda}{m}E)} - \dfrac{2(\frac{2\lambda}{m}E - \omega^2)^{\frac{3}{2}}}{\omega^2 - \frac{\lambda}{m}E}t + \theta\right), \\ rp = \tilde{q}\sinh\left(\dfrac{\lambda rp\sqrt{\frac{2\lambda}{m}E - \omega^2}}{m(\omega^2 - \frac{\lambda}{m}E)} - \dfrac{2(\frac{2\lambda}{m}E - \omega^2)^{\frac{3}{2}}}{\omega^2 - \frac{\lambda}{m}E}t + \theta\right). \end{cases}$$

We notice that the equation (13) (for $H = E$) is now recovered by taking the difference of the square of these two equations. Also in this case, due to the above relations, we are able to obtain

t as a function of r, namely:

$$t(r) = \frac{1}{\zeta_{[\lambda]}(E)}\left[\operatorname{arccosh}\left(\frac{mr^2(\frac{2\lambda}{m}E-\omega^2)+E}{\tilde{q}\sqrt{\frac{2\lambda}{m}E-\omega^2}}\right) + \frac{\lambda\sqrt{\frac{2\lambda}{m}E-\omega^2}}{m(\frac{\lambda}{m}E-\omega^2)}\sqrt{2mEr^2 - l^2 + m^2r^4(\frac{2\lambda}{m}E-\omega^2)} - \theta\right], \tag{17}$$

where we defined the function

$$\zeta_{[\lambda]}(E) \doteq \frac{2(\frac{2\lambda}{m}E-\omega^2)^{\frac{3}{2}}}{\frac{\lambda}{m}E-\omega^2}.$$

Similarly to the case of bounded motion, by defining the quantities

$$\tilde{a}^2_{[\lambda]} \doteq \frac{E}{m(\frac{2\lambda E}{m}-\omega^2)}, \qquad \tilde{\epsilon}_{[\lambda]} \doteq \sqrt{1 + \frac{(\frac{2\lambda E}{m}-\omega^2)l^2}{E^2}},$$

the expression (17) can be written more elegantly as:

$$\zeta_{[\lambda]}(E)t(r) + \theta = \operatorname{arccosh}\left[\frac{1}{\tilde{\epsilon}_{[\lambda]}}\left(1 + \left(\frac{r}{\tilde{a}_{[\lambda]}}\right)^2\right)\right] + \frac{\lambda\tilde{a}^2_{[\lambda]}}{1-\lambda\tilde{a}^2_{[\lambda]}}\sqrt{\left[1+\left(\frac{r}{\tilde{a}_{[\lambda]}}\right)^2\right]^2 - \tilde{\epsilon}^2_{[\lambda]}}. \tag{18}$$

Equation (18) might be compared with the formula (15) given in [6], having in mind the relation existing between the Hamiltonian H defined in the present paper and the one defined in the aforementioned reference, here denoted as $\bar{H}$, calculated for $\kappa = \frac{1}{\lambda}$ and $b_j = 0$ in [6]. Such relation is given by $\bar{H} = \lambda H - \frac{m\omega^2}{2}$. In [6], the equation of motion has been solved assuming that $\bar{E} := 2\bar{H} = 2\lambda E - m\omega^2 > 0$.

4. The deformed $\lambda < 0$ case

In this case the deformation parameter is negative and the Hamiltonian can be written as follows

$$H(r,p) = T(r,p) + V_{\text{eff}}(r) = \frac{p^2}{2m(1-|\lambda|r^2)} + \frac{l^2}{2mr^2(1-|\lambda|r^2)} + \frac{m\omega^2r^2}{2(1-|\lambda|r^2)}. \tag{19}$$

It is clear that it presents a singularity at the point $r_s = \frac{1}{\sqrt{|\lambda|}}$.

As a matter of fact, owing to the above singularity, the domain of definition of the effective potential splits in two subdomains. The first one, corresponding to the punctured open ball $0 < r < r_s$, is characterized by a positive kinetic energy term in the Hamiltonian and the effective potential exhibits a typical confining shape. Viceversa, the second one, $r_s < r < \infty$, is characterized by a negative kinetic energy and the effective potential has no critical points (see figure 5 and lhs of figure 6).

The interesting region is the open set $0 < r < r_s$. In fact, in this case there will be only closed orbits for energy values $V_{\text{eff}}(r_{\text{min}}) < E < \infty$, where

$$V_{\text{eff}}(r_{\text{min}}) = \frac{l^2}{m}\left(\sqrt{\lambda^2 + \frac{m^2\omega^2}{l^2}} + |\lambda|\right), \qquad r^2_{\text{min}} = \frac{l^2}{m^2\omega^2}\left(-|\lambda| + \sqrt{\lambda^2 + \frac{m^2\omega^2}{l^2}}\right).$$

In that bounded region the solution can be immediately obtained by means of the SGA technique by letting in (16) $\lambda \to -|\lambda|$. Clearly, since the term $\omega^2 - \frac{2\lambda}{m}E = \omega^2 + \frac{2|\lambda|}{m}E$ is always positive, no

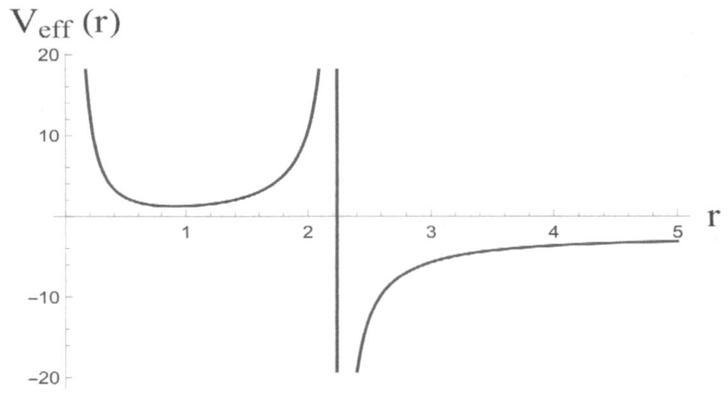

Figure 5. Effective potential $V_{\text{eff}}(r)$ (19) calculated for $m = \omega = l = 1$ and $\lambda = -0.20$. The singularity is located at the point $r_s = \sqrt{5} \simeq 2.24$.

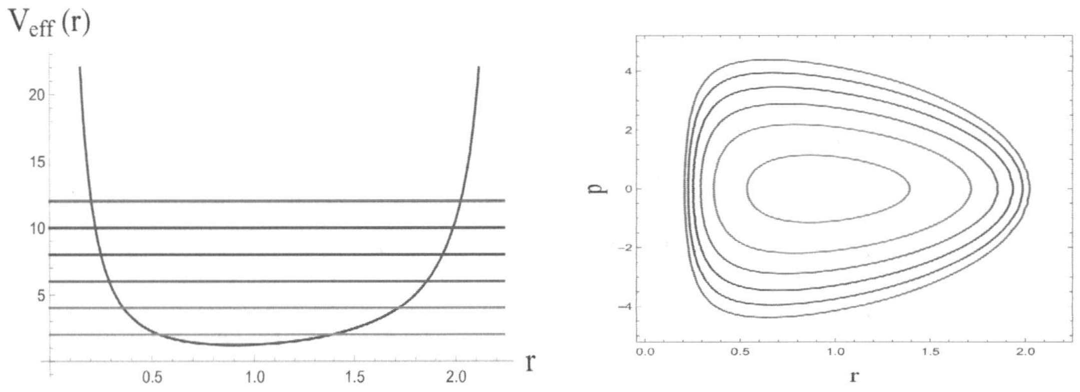

Figure 6. Effective potential $V_{\text{eff}}(r)$ (19) and phase plane (r, p) calculated for $m = \omega = l = 1$ and $E = 2, 4, 6, 8, 10, 12$ in the region $0 < r < r_s$. The deformation parameter is fixed at the value $\lambda = -0.2$.

further restrictions on the energy will apply, i.e. the energy has no upper bound, as expected. For this case the phase plane is depicted in the rhs of figure 6.

Viceversa, in order to obtain the solution in the unbounded region $r > r_s$, as we already noticed in the case of the Taub-NUT system [18], we have to factorise the new Hamiltonian $\tilde{H} = -H$, the latter being defined with an overall change of sign. This allows us to restore a positive kinetic energy and then to obtain a physically meaningful Hamiltonian (see also [13]). However, this case is not so interesting since no closed orbits are allowed.

5. Conclusions

To summarize the key points of our paper, we can say that we have successfully applied the SGA approach to the classical D-III oscillator, identifying the structure of its Poisson algebra, and solving the equations of motion by finding an explicit expression in the form $t = t(r)$. In particular, after having calculated the solution to the equation of motion for the usual isotropic harmonic oscillator (which in our framework just represents the Euclidean limit of the D-III system), we have found the trajectory $t = t(r)$ of the D-III oscillator, for positive values of the deformation parameter λ, both in the case of bounded and unbounded motion. Furthermore, the case of λ negative has also been analyzed. For the latter situation, as we have already seen in the classical Taub-NUT system [18], the most interesting features of the dynamics are somehow hidden 'beyond the singularity', in the open set $0 < r < r_s$, where for suitable initial conditions the particle remains trapped during its whole life. As a matter of fact, in that region the effective potential shows a typical confining shape and the kinetic energy is always positive definite. We were then enabled to recover an explicit formula for the trajectory, restricted to that domain, simply by taking $t = t(r)$ for $\lambda = -|\lambda|$.

In the quantum case we expect similar features, however *in that punctured open ball* we might miss the algebraic nature of the spectrum and, accordingly, an analytic expression for the wave functions might not be available.

Acknowledgments

The research related to this paper has been developed in the framework of the PRIN prot. n. 2010JJ4KPA_004. It has been partially supported by the Spanish Ministerio de Economía y Competitividad (MINECO) under grant MTM2013-43820-P and by the Spanish Junta de Castilla y León under grant BU278U14.

References

[1] Ballesteros A, Enciso A, Herranz F J, Ragnisco O and Riglioni D 2011 *Ann. Phys.* **326** 2053
[2] Koenigs G 1972 *Leçons sur la thèorie gènèrale des surfaces* vol 4 ed G Darboux (New York: Chelsea) p 368
[3] Kalnins E G, Kress J M, Miller W Jr and Winternitz P 2003 *J. Math. Phys.* **44** 5811
[4] Grosche C, Pogosyan G S and Sissakian A N 2007 *Phys. Part. Nuclei* **38** 525
[5] Ballesteros A, Enciso A, Herranz F J and Ragnisco O 2007 *Phys. Lett. B* **652** 376
[6] Ballesteros A, Enciso A, Herranz F J and Ragnisco O 2008 *Physica D* **237** 505
[7] Ballesteros A, Enciso A, Herranz F J and Ragnisco O 2009 *Ann. Phys.* **324** 1219
[8] Bertrand J 1873 *C.R. Math. Acad. Sci. Paris* **77** 849
[9] Perlick V 1992 *Class. Quantum Grav.* **9** 1009
[10] Ballesteros A, Enciso A, Herranz F J and Ragnisco O 2008 *Class. Quantum Grav.* **25** 165005
[11] Ballesteros A, Enciso A, Herranz F J and Ragnisco O 2009 *Commun. Math. Phys.* **290** 1033
[12] Ballesteros A, Enciso A, Herranz F J, Ragnisco O and Riglioni D 2011 *Phys. Lett. A* **375** 1431
[13] Ballesteros A, Enciso A, Herranz F J, Ragnisco O and Riglioni D 2011 *Int. J. Theor. Phys.* **50** 2268
[14] Demkov Y N 1959 *Soviet Phys. JETP* **9** 63
[15] Fradkin D M 1965 *Amer. J. Phys.* **33** 207
[16] Kuru S and Negro J 2008 *Ann. Phys.* **323** 413
[17] Kuru S and Negro J 2012 *J. Phys.: Conf. Ser.* **343** 012063
[18] Latini D and Ragnisco O 2015 *J. Phys. A: Math. Teor.* **48** 175201

Journal of Physics: Conference Series **670** (2016) 012032 doi:10.1088/1742-6596/670/1/012032

Biquaternion Construction of SL(2,C) Yang-Mills Instantons

Jen-Chi Lee

Department of Electrophysics, National Chiao-Tung University, Hsinchu, Taiwan, R.O.C.

E-mail: jcclee@cc.nctu.edu.tw

Abstract. We use biquaternion to construct SL(2,C) ADHM Yang-Mills instantons. The solutions contain 16k-6 moduli parameters for the kth homotopy class, and include as a subset the SL(2,C) (M,N) instanton solutions constructed previously. In constrast to the SU(2) instantons, the SL(2,C) instantons inhereit jumping lines or singulariries which are not gauge artifacts and can not be gauged away.

1. Introduction

The classical exact solutions of Euclidean $SU(2)$ (anti)self-dual Yang-Mills (SDYM) equation were intensively studied by pure mathematicians and theoretical physicists in 1970s. The first BPST 1-instanton solution [1] with 5 moduli parameters was found in 1975. The CFTW k-instanton solutions [2] with $5k$ moduli parameters were soon constructed, and then the number of moduli parameters of the solutions for each homotopy class k was extended to $5k+4$ (5,13 for $k=$ 1,2) [3] based on the conformal symmetry of massless pure YM equation. The complete solutions with $8k-3$ moduli parameters for each k-th homotopy class were finally worked out in 1978 by mathematicians ADHM [4] using theory in algebraic geometry. Through an one to one correspondence between anti-self-dual SU(2)-connections on S^4 and holomorphic vector bundles on CP^3, ADHM converted the highly nontrivial anti-SDYM equations into a much more simpler system of quadratic algebraic equations in quaternions. The explicit closed form of the complete solutions for $k=2,3$ had been worked out [5].

There are many important applications of instantons to algebraic geometry and quantum field theory. One important application of instantons in algebraic geometry was the classification of four-manifolds [6]. On the physics side, the non-perturbative instanton effect in QCD resolved the $U(1)_A$ problem [7]. Another important application of YM instantons in quantum field theory was the introduction of θ- vacua [8] in nonperturbative QCD, which created the strong CP problem.

In addition to $SU(2)$, the ADHM construction has been generalized to the cases of $SU(N)$ SDYM and many other SDYM theories with compact Lie groups [5, 9]. In this talk we are going to consider the classical solutions of non-compact $SL(2,C)$ SDYM system. YM theory based on $SL(2,C)$ was first discussed in 1970s [10, 11]. It was found that the complex $SU(2)$ YM field configurations can be interpreted as the real field configurations in $SL(2,C)$ YM theory. However, due to the non-compactness of $SL(2,C)$, the Cartan-Killing form or group metric of $SL(2,C)$ is not positive definite. Thus the action integral and the Hamiltonian of non-compact $SL(2,C)$ YM theory may not be positve. Nevertheless, there are still important motivations to

Journal of Physics: Conference Series **670** (2016) 012032 doi:10.1088/1742-6596/670/1/012032

study $SL(2,C)$ SDYM theory. For example, it was shown that the $4D$ $SL(2,C)$ SDYM equation can be dimensionally reduced to many important $1+1$ dimensional integrable systems [12], such as the KdV equation and the nonlinear Schrodinger equation.

2. SL(2,C) SDYM E uation

We first briefly review the $SL(2,C)$ YM theory. It was shown that [10] there are two linearly independent choices of $SL(2,C)$ group metric

$$g^a = \begin{pmatrix} I & 0 \\ 0 & -I \end{pmatrix}, g^b = \begin{pmatrix} 0 & I \\ I & 0 \end{pmatrix} \tag{2.1}$$

where I is the 3×3 unit matrix. In general, we can choose

$$g = \cos\theta g^a + \sin\theta g^b \tag{2.2}$$

where $\theta =$ real constant. Note that the metric is not positive definite due to the non-compactness of $SL(2,C)$. On the other hand, it was shown that $SL(2,C)$ group can be decomposed such that [13]

$$SL(2,C) = SU(2)\cdot P, P \in H \tag{2.3}$$

where $SU(2)$ is the maximal compact subgroup of $SL(2,C)$, $P \in H$ (not a group) and $H = \{P|P$ is Hermitain, positive definite, and $detP = 1\}$. The parameter space of H is a noncompact space R^3. The third homotopy group is thus [13]

$$\pi_3[SL(2,C)] = \pi_3[S^3 \times R^3] = \pi_3(S^3)\cdot \pi_3(R^3) = Z\cdot I = Z \tag{2.4}$$

where I is the identity group, and Z is the integer group.

On the other hand, Wu and Yang [10] have shown that a complex $SU(2)$ gauge field is related to a real $SL(2,C)$ gauge field. Starting from $SU(2)$ complex gauge field formalism, we can write down all the $SL(2,C)$ field equations. Let

$$G^a_\mu = A^a_\mu + iB^a_\mu \tag{2.5}$$

and, for convenience, we set the coupling constant $g = 1$. The complex field strength is defined as

$$F^a_{\mu\nu} \equiv H^a_{\mu\nu} + iM^a_{\mu\nu}, a, b, c = 1,2,3 \tag{2.6}$$

where

$$\begin{aligned} H^a_{\mu\nu} &= \partial_\mu A^a_\nu - \partial_\nu A^a_\mu + \epsilon^{abc}(A^b_\mu A^c_\nu - B^b_\mu B^c_\nu), \\ M^a_{\mu\nu} &= \partial_\mu B^a_\nu - \partial_\nu B^a_\mu + \epsilon^{abc}(A^b_\mu B^c_\nu - A^b_\mu B^c_\nu), \end{aligned} \tag{2.7}$$

then $SL(2,C)$ Yang-Mills equation can be written as

$$\begin{aligned} \partial_\mu H^a_{\mu\nu} + \epsilon^{abc}(A^b_\mu H^c_{\mu\nu} - B^b_\mu M^c_{\mu\nu}) = 0, \\ \partial_\mu M^a_{\mu\nu} + \epsilon^{abc}(A^b_\mu M^c_{\mu\nu} - B^b_\mu H^c_{\mu\nu}) = 0. \end{aligned} \tag{2.8}$$

The $SL(2,C)$ SDYM equations are

$$\begin{aligned} H^a_{\mu\nu} &= \frac{1}{2}\epsilon_{\mu\nu\alpha\beta}H_{\alpha\beta}, \\ M^a_{\mu\nu} &= \frac{1}{2}\epsilon_{\mu\nu\alpha\beta}M_{\alpha\beta}. \end{aligned} \tag{2.9}$$

The Yang-Mills Equation above can be derived from the following Lagrangian

$$L_\theta = \frac{1}{4}[F^i_{\mu\nu}]^T g_{ij}[F^j_{\mu\nu}] = \cos\theta(\frac{1}{4}H^a_{\mu\nu}H^a_{\mu\nu} - \frac{1}{4}M^a_{\mu\nu}M^a_{\mu\nu}) + \sin\theta(\frac{1}{2}H^a_{\mu\nu}M^a_{\mu\nu}) \tag{2.10}$$

where $F^k_{\mu\nu} = H^k_{\mu\nu}$ and $F^{3+k}_{\mu\nu} = M^k_{\mu\nu}$ for $k = 1, 2, 3$. Note that L_θ is indefinite for any real value θ. We shall only consider the particular case for $\theta = 0$ in this talk, i.e.

$$L = \frac{1}{4}(H^a_{\mu\nu}H^a_{\mu\nu} - M^a_{\mu\nu}M^a_{\mu\nu}), \tag{2.11}$$

for the action density in discussing the homotopic classifications of our solutions.

3. Bi uaternion construction o $SL(2, C)$ YM Instantons

Instead of quaternion in the $Sp(1)$ ($= SU(2)$) ADHM construction, we will use *biqua erni n* to construct $SL(2, C)$ SDYM instantons. A quaternion x can be written as

$$x = x_\mu e_\mu, \ \ x_\mu \in R, \ \ e_0 = 1, e_1 = i, e_2 = j, e_3 = k \tag{3.12}$$

where e_1, e_2 and e_3 anticommute and obey

$$e_i \cdot e_j = -e_j \cdot e_i = \epsilon_{ijk} e_k; \ \ i, j, k = 1, 2, 3, \tag{3.13}$$

$$e_1^2 = -1, e_2^2 = -1, e_3^2 = -1. \tag{3.14}$$

A (ordinary) biquaternion (or complex-quaternion) z can be written as

$$z = z_\mu e_\mu, \ \ z_\mu \in C, \tag{3.15}$$

which will be used in this talk. Occasionally z can be written as

$$z = x + yi \tag{3.16}$$

where x and y are quaternions and $i = \sqrt{-1}$, not to be confused with e_1 in Eq.(3.12). For biquaternion, the biconjugation [14]

$$z^{\circledast} = z_\mu e^\dagger_\mu = z_0 e_0 - z_1 e_1 - z_2 e_2 - z_3 e_3 = x^\dagger + y^\dagger i, \tag{3.17}$$

will be heavily used in this talk. In contrast to the real number norm square of a quaternion, the norm square of a biquarternion used in this talk is defined to be

$$|z|^2_c = z^{\circledast} z = (z_0)^2 + (z_1)^2 + (z_2)^2 + (z_3)^2 \tag{3.18}$$

which is a c *plex* number in general as a subscript c is used in the norm.

We are now ready to proceed the construction of $SL(2, C)$ instantons. We begin by introducing the $(k + 1) \times k$ biquarternion matrix $(x) = a + bx$

$$(x)_{ab} = a_{ab} + b_{ab} x, \ \ a_{ab} = a^\mu_{ab} e_\mu, b_{ab} = b^\mu_{ab} e_\mu \tag{3.19}$$

where a^μ_{ab} and b^μ_{ab} are complex numbers, and a_{ab} and b_{ab} are biquarternions. The biconjugation of the (x) matrix is defined to be

$$(x)^{\circledast}_{ab} = \ (x)^\mu_{ba} e^\dagger_\mu = \ (x)^0_{ba} e_0 - \ (x)^1_{ba} e_1 - \ (x)^2_{ba} e_2 - \ (x)^3_{ba} e_3. \tag{3.20}$$

In contrast to the of $SU(2)$ instantons, the quadratic condition of $SL(2, C)$ instantons reads

$$(x)^{\circledast} \quad (x) = f^{-1} = \text{symmetric, non-singular } k \times k \text{ matrix for } x \notin J, \tag{3.21}$$

from which we can deduce that $a^{\circledast}a, b^{\circledast}a, a^{\circledast}b$ and $b^{\circledast}b$ are all symmetric matrices. We stress here that it will turn out the choice of *bic nju a i n* operation is crucial for the follow-up discussion in this work. On the other hand, for $x \in J$, det $(x)^{\circledast} \quad (x) = 0$. The set J is called singular locus or jumping lines in the mathematical literatures and was discussed in [15]. In contrast to the $SL(2,C)$ instantons, there are no jumping lines for the case of $SU(2)$ instantons. In the $Sp(1)$ quaternion case, the symmetric condition on f^{-1} means f^{-1} is real. For the $SL(2,C)$ biquaternion case, however, it can be shown that symmetric condition on f^{-1} implies f^{-1} is *c plex*.

To construct the self-dual gauge field, we introduce a $(k+1) \times 1$ dimensional biquaternion vector $v(x)$ satisfying the following two conditions

$$v^{\circledast}(x) \quad (x) = 0, \tag{3.22}$$
$$v^{\circledast}(x)v(x) = 1. \tag{3.23}$$

Note that $v(x)$ is fixed up to a $SL(2,C)$ gauge transformation

$$v(x) \longrightarrow v(x)g(x), \quad g(x) \in \ 1 \times 1 \text{ Biquaternion}. \tag{3.24}$$

Note also that in general a $SL(2,C)$ matrix can be written in terms of a 1×1 biquaternion as

$$g = \frac{q_\mu e_\mu}{\sqrt{q^{\circledast} q}} = \frac{q_\mu e_\mu}{|q|_c}. \tag{3.25}$$

The next step is to define the gauge field

$$G_\mu(x) = v^{\circledast}(x)\partial_\mu v(x), \tag{3.26}$$

which is a 1×1 biquaternion. Note that, unlike the case for $Sp(1)$, $G_\mu(x)$ needs not to be anti-Hermitian.

We can now define the $SL(2,C)$ field strength

$$F_{\mu\nu} = \partial_\mu G_\nu(x) + G_\mu(x)G_\nu(x) - [\mu \longleftrightarrow \nu]. \tag{3.27}$$

To show that $F_{\mu\nu}$ is self-dual, one first show that the operator

$$P = 1 - v(x)v^{\circledast}(x) \tag{3.28}$$

is a projection operator $P^2 = P$, and can be written in terms of as

$$P = \quad (x) f \quad ^{\circledast}(x). \tag{3.29}$$

The self-duality of $F_{\mu\nu}$ can now be proved as following

$$\begin{aligned} F_{\mu\nu} &= \partial_\mu (v^{\circledast}(x)\partial_\nu v(x)) + v^{\circledast}(x)\partial_\mu v(x) v^{\circledast}(x)\partial_\nu v(x) - [\mu \longleftrightarrow \nu] \\ &= v^{\circledast}(x) b (e_\mu e_\nu^\dagger - e_\nu e_\mu^\dagger) f b^{\circledast} v(x) \end{aligned} \tag{3.30}$$

where we have used Eqs.(3.19),(3.22) and (3.29). Finally the factor $(e_\mu e_\nu^\dagger - e_\nu e_\mu^\dagger)$ above can be shown to be self-dual

$$\sigma_{\mu\nu} \equiv \frac{1}{4i}(e_\mu e_\nu^\dagger - e_\nu e_\mu^\dagger) = \frac{1}{2}\epsilon_{\mu\nu\alpha\beta}\sigma_{\alpha\beta}, \tag{3.31}$$

$$\bar{\sigma}_{\mu\nu} = \frac{1}{4i}(e_\mu^\dagger e_\nu - e_\nu^\dagger e_\mu) = -\frac{1}{2}\epsilon_{\mu\nu\alpha\beta}\bar{\sigma}_{\alpha\beta}. \tag{3.32}$$

This proves the self-duality of $F_{\mu\nu}$. We thus have constructed many $SL(2,C)$ SDYM field configurations.

To count the number of moduli parameters for the $SL(2,C)$ k-instantons we have constructed , one uses transformations which preserve conditions Eq.(3.21), Eq.(3.22) and Eq.(3.23), and the definition of G_μ in Eq.(3.26) to bring b and a in Eq.(3.19) into a simple canonical form

$$b = \begin{bmatrix} 0_{1\times k} \\ I_{k\times k} \end{bmatrix}, \tag{3.33}$$

$$a = \begin{bmatrix} \lambda_{1\times k} \\ -y_{k\times k} \end{bmatrix} \tag{3.34}$$

where λ and y are biquaternion matrices with orders $1\times k$ and $k\times k$ respectively, and y is symmetric

$$y = y^T. \tag{3.35}$$

The constraints for the moduli parameters are

$$a_{ci}^{\circledast} a_{cj} = 0, i \neq j, \text{ and } y_{ij} = y_{ji}. \tag{3.36}$$

The total number of moduli parameters for k-instanton can be calculated through Eq.(3.36) to be

$$\#\text{ of moduli for } SL(2,C)\ k\text{-instantons} = 16k - 6, \tag{3.37}$$

which is twice of that of the case of $Sp(1)$. Roughly speaking, there are $8k$ parameters for instanton biquaternion positions and $8k$ parameters for instanton sizes . Finally one has to subtract an overall $SL(2,C)$ gauge group degree of freesom 6. This picture will become more clear when we give examples of explicit constructions of $SL(2,C)$ instantons in the next section.

4. Exam les o $SL(2,C)$ instantons and Jum ing lines

In this section, we will explicitly construct examples of $SL(2,C)$ YM instantons to illustrate our prescription given in the last section. Example of $SL(2,C)$ instantons with jumping lines will also be given.

e $SL(2,C)$ (M,N) In an n

In this first example, we will reproduce from the ADHM construction the $SL(2,C)$ (M,N) instanton solutions constructed in [13]. We choose the biquaternion λ_j in Eq.(3.34) to be $\lambda_j e_0$ with λ_j a c plex number, and choose $y_{ij} = y_j\delta_{ij}$ to be a diagonal matrix with $y_j = y_{j\mu}e_\mu$ a quaternion. That is

$$(x) = \begin{bmatrix} \lambda_1 & \lambda_2 & \dots & \lambda_k \\ x - y_1 & 0 & \dots & 0 \\ 0 & x - y_2 & \dots & 0 \\ . & \dots & \dots & \dots \\ 0 & 0 & \dots & x - y_k \end{bmatrix}, \tag{4.38}$$

which satisfies the constraint in Eq.(3.36). One can calculate the gauge potential as

$$
\begin{aligned}
G_\mu = v^{\circledast}\partial_\mu v &= \frac{1}{4}[e_\mu^\dagger e_\nu - e_\nu^\dagger e_\mu]\partial_\nu \ln(1 + \frac{\lambda_1^2}{|x-y_1|^2} + ... + \frac{\lambda_k^2}{|x-y_k|^2}) \\
&= \frac{1}{4}[e_\mu^\dagger e_\nu - e_\nu^\dagger e_\mu]\partial_\nu \ln(\phi)
\end{aligned}
\qquad (4.39)
$$

where

$$
\phi = 1 + \frac{\lambda_1^2}{|x-y_1|^2} + ... + \frac{\lambda_k^2}{|x-y_k|^2}. \qquad (4.40)
$$

For the case of $Sp(1)$, λ_j is a real number and $\lambda_j \lambda_j^\dagger = \lambda_j^2$ is a real number. So ϕ in Eq.(4.40) is a complex-valued function in general. If we choose $k = 1$ and define $\lambda_1^2 = \frac{\alpha_1^2}{1+i}$, then

$$
\phi = 1 + \frac{\frac{\alpha_1^2}{1+i}}{|x-y_1|^2}. \qquad (4.41)
$$

The gauge potential is

$$
\begin{aligned}
G_\mu &= \frac{1}{4}[e_\mu^\dagger e_\nu - e_\nu^\dagger e_\mu]\partial_\nu \ln(1 + \frac{\frac{\alpha_1^2}{1+i}}{|x-y_1|^2}) = \frac{1}{4}[e_\mu^\dagger e_\nu - e_\nu^\dagger e_\mu]\partial_\nu \ln(1 + \frac{\alpha_1^2}{|x-y_1|^2} + i) \\
&= \frac{1}{2}[e_\mu^\dagger e_\nu - e_\nu^\dagger e_\mu]\frac{-\alpha_1^2(x-y_1)_\nu}{|x-y_1|^4 + (|x-y_1|^2 + \alpha_1^2)^2}[\frac{|x-y_1|^2 + \alpha_1^2}{|x-y_1|^2} - i]
\end{aligned}
\qquad (4.42)
$$

which reproduces the $SL(2,C)$ $(M,N) = (1,0)$ solution calculated in [13]. It is easy to generalize the above calculations to the general (M,N) cases, and it can be shown that the topological charge of these field configuations is $k = M + N$ [13].

SL(2, C) CF W k-in an n and ju pin line

For another subset of k-instanton field configurations, one chooses $\lambda_i = \lambda_i e_0$ (with λ_i a *c plex* number) and y_i to be a *biqua erni n* in Eq.(4.38). It is important to note that for these choices, the constraints in Eq.(3.36) are still satisfied *wi u* turning on the off-diagonal elements y_{ij} in Eq.(3.34). It can be shown that, for these field configurations, there are non-removable singularities which are zeros ($x \in J$) of

$$
\phi = 1 + \frac{\lambda_1\lambda_1^{\circledast}}{|x-y_1|_c^2} + ... + \frac{\lambda_k\lambda_k^{\circledast}}{|x-y_k|_c^2}, \qquad (4.43)
$$

or

$$
\det \quad (x)^{\circledast} \quad (x) = |x-y_1|_c^2|x-y_2|_c^2 \cdots |x-y_k|_c^2\phi = P_{2k}(x) + iP_{2k-1}(x) = 0. \qquad (4.44)
$$

For the k-instanton case, one encounters intersections of zeros of $P_{2k}(x)$ and $P_{2k-1}(x)$ polynomials with degrees $2k$ and $2k-1$ respectively

$$
P_{2k}(x) = 0, \quad P_{2k-1}(x) = 0. \qquad (4.45)
$$

These new singularities can not be gauged away and do not show up in the field configurations of $SU(2)$ k-instantons. Mathematically, the existence of singular structures of the non-compact $SL(2,C)$ SDYM field configurations is consistent with the inclusion of sheaves by Frenkel-Jardim [16] recently, rather than just the restricted notion of vector bundles , in the one to one correspondence between ASDYM and certain algebraic geometric objects.

5. Acknowledgments

This talk is based on a collaboration paper with S.H. Lai and I.H. Tsai. I thank the financial support of MoST, Taiwan.

[1] A. Belavin, A. Polyakov, A. Schwartz, Y. Tyupkin, "Pseudo-particle solutions of the Yang-Mills equations", Phys. Lett. B 59 (1975) 85.
[2] E.F. Corrigan, D.B. Fairlie, Phys. Lett. 67B (1977)69; G. 'tHooft, Phys. Rev. Lett., 37 (1976) 8; F. Wilczek, in "Quark Confinement and Field Theory", Ed. D.Stump and D. Weingarten, John Wiley and Sons, New York (1977).
[3] R. Jackiw, C. Rebbi, "Conformal properties of a Yang-Mills pseudoparticle", Phys. Rev. D 14 (1976) 517; R. Jackiw, C. Nohl and C. Rebbi, "Conformal properties of pseu-doparticle con gurations", Phys. Rev. D 15 (1977) 1642.
[4] M. Atiyah, V. Drinfeld, N. Hitchin, Yu. Manin, "Construction of instantons", Phys. Lett. A 65 (1978) 185.
[5] N. H. Christ, E. J. Weinberg and N. K. Stanton, "General Self-Dual Yang-Mills Solutions", Phys. Rev. D 18 (1978) 2013. V. Korepin and S. Shatashvili, "Rational parametrization of the three instanton solutions of the Yang-Mills equations", Math. USSR Izversiya 24 (1985) 307.
[6] S.K. Donaldson and P.B. Kronheimer, "The Geometry of Four Manifolds", Oxford University Press (1990).
[7] G. 't Hooft, "Computation of the quantum effects due to a four-dimensional pseudoparticle", Phys. Rev. D 14 (1976) 3432. G. 't Hooft, "Symmetry breaking through Bell-Jackiw anomalies", Phys. Rev. Lett. 37 (1976) 8.
[8] C. Callan Jr., R. Dashen, D. Gross, "The structure of the gauge theory vacuum", Phys. Lett. B 63 (1976) 334; "Toward a theory of the strong interactions", Phys. Rev. D 17 (1978) 2717. R. Jackiw, C. Rebbi, "Vacuum periodicity in a Yang-Mills quantum theory", Phys. Rev. Lett. 37 (1976) 172.
[9] R. Jackwi and C. Rebbi, Phys. Lett. 67B (1977) 189. C. W. Bernard, , N. H. Christ, A. H. Guth and E. J. Weinberg, Phys. Rev. D16 (1977) 2967.
[10] Tai Tsun Wu and Chen Ning Yang, Phys. Rev. D12, 3843 (1975); Phys. Rev.D13, (1976) 3233.
[11] J. P. Hsu and E. Mac, J. Math. Phys. 18 (1977) 1377.
[12] L. J. Mason and G. A. J. Sparling, "Nonlinear Schrodinger and Korteweg-de Vries are reductions of self-dual Yang-Mills," Phys. Lett. A 137, 29–33 (1989).
[13] K. L. Chang and J. C. Lee, "On solutions of self-dual SL(2,C) gauge theory", Chinese Journal of Phys. Vol. 44, No.4 (1984) 59. J.C. Lee and K. L. Chang, "SL(2,C) Yang-Mills Instantons", Proc. Natl. Sci. Counc. ROC (A), Vol 9, No 4 (1985) 296.
[14] W. R. Hamilton, "Lectures on Quaternions", Macmillan & Co, Cornell University Library (1853).
[15] Sheng-Hong Lai, Jen-Chi Lee and I-Hsun Tsai, "Biquaternions and ADHM Construction of Non-Compact SL(2,C) Yang-Mills Instantons", Ann. Phys. 361,(2015) 14-32.
[16] I. Frenkel, M. Jardim, "Complex ADHM equations and sheaves on P^3", Journal of Algebra 319 (2008) 2913-2937. J. Madore, J.L. Richard and R. Stora, "An Introduction to the Twistor Programme", Phys. Rept. 49, No. 2 (1979) 113-130.

Journal of Physics: Conference Series **670** (2016) 012033 doi:10.1088/1742-6596/670/1/012033

Towards the gravity/CYBE correspondence
– *the current status* –

Takuya Matsumoto
Graduate School of Mathematics and Institute for Advanced Research, Nagoya University, Nagoya 464-8602, Japan
E-mail: takuya.matsumoto@math.nagoya-u.ac.jp

Kentaroh Yoshida
Department of Physics, Kyoto University, Kitashirakawa, Kyoto 606-8502, Japan
E-mail: kyoshida@gauge.scphys.kyoto-u.ac.jp

Abstract. We give a brief summary of Yang-Baxter deformations of the $\mathrm{AdS}_5\times\mathrm{S}^5$ superstring by focusing upon four examples, 1) gravity duals for noncommutative gauge theories, 2) γ-deformations of S^5, 3) Schrödinger spacetimes and 4) abelian twists of the global AdS_5 .

1. Introduction

The most prototypical example of AdS/CFT correspondences [1] is the conjectured equivalence between type IIB superstring on the $\mathrm{AdS}_5\times\mathrm{S}^5$ background and the four-dimensional $\mathcal{N}=4$ $SU(N)$ super Yang-Mills (SYM) theory in the large N limit. A great progress on this subject is that an integrable structure behind this duality was unveiled (For a comprehensive review, see [2]). Motivated by this discovery, we are interested in the associated classical integrability on the string-theory side. The classical string action on the $\mathrm{AdS}_5\times\mathrm{S}^5$ background was constructed by adopting the Green-Schwarz formulation [3] with the supercoset representation

$$\text{super } \mathrm{AdS}_5\times\mathrm{S}^5 = \frac{PSU(2,2|4)}{SO(1,4)\times SO(5)}\,. \tag{1}$$

This supercoset enjoys the $\mathbb{Z}_4$-grading property, which ensures the classical integrability [4] in the sense of kinematical integrability that means the existence of Lax pairs (For excellent reviews, see [5, 6]). However, it should be remarked that the complete integrability in the sense of Liouville has not been shown due to the ambiguity stemming from non-ultra local terms in the Poisson structure.

As the next issue, it is intriguing to study integrable deformations of the $\mathrm{AdS}_5\times\mathrm{S}^5$ superstring. In fact, there is a long history along this direction and a vast number of the deformations are already known. Thus it is desirable to study them with the most systematic and efficient way. A promising way is to follow the Yang-Baxter sigma model description proposed by Klimcik [7]. This is such a way to study integrable deformations of 2D non-linear sigma models. By following this approach, the integrable deformations are specified by skew-symmetric linear R-operators which satisfy the modified classical Yang-Baxter equation (mCYBE). The original work [7] was invented for principal chiral models, but it was recently generalized to the symmetric cosets by Delduc-Magro-Vicedo [8][1]. Then it was straightforwardly applied to construct a q-deformed action of the $AdS_5 \times S^5$ superstring [14] with the classical r-matrix of Drinfeld-Jimbo type [15]. The metric in the string frame and NS-NS two-form have been obtained in [16] (Here "NS" is an abbreviation of Neveu-Schwarz). For generalizations to $AdS_n \times S^n$, see [17]. For the recent progress towards the full solution, see [18–20]. This issue is closely related to our interest here.

As a possible generalization, Jordanian deformations of the $AdS_5 \times S^5$ superstring action were presented in [21] by adopting the homogeneous classical Yang-Baxter equation (CYBE). Here the Lax pair and kappa transformation are different from the ones of the work [14] and this generalization is not so difficult but not so obvious. One of the advantages of this formulation is that partial deformations of $AdS_5 \times S^5$ are possible, because the zero map $R = 0$ is allowed as a solution of the CYBE, but not of the mCYBE. Then, many skew-symmetric solutions of the CYBE have been identified with well-known solutions of type IIB supergravity in a series of papers [22–30]. The list of the solutions includes γ-deformations of S^5 [31, 32], gravity duals for noncommutative (NC) gauge theories [33], Schrödinger spacetimes [34], abelian twists of the global AdS_5 [35] and further new backgrounds [22]. These identifications may be regarded as a new perspective of Yang-Baxter deformations. The conjectured relation between solutions of type IIB supergravity and classical r-matrices are called the gravity/CYBE correspondence [23] (For a short summary of the works in 2014, see [36]. This review is the update of [36] in 2015).

To establish this correspondence, it is necessary to do much effort. However, if it has been established, then it indicates that the moduli space of a certain class of solutions of type IIB supergravity can be described by the CYBE. Due to the fact that the size of classical r-matrices is finite and then the number of the solutions is also finite, the number of the associated solutions of type IIB supergravity should also be finite. It is also analogous to the bubbling scenario [37], where 1/2 BPS solutions of type IIB supergravity preserving a certain symmetry are specified by droplet configurations in a free fermion system.

Recently, Yang-Baxter deformations are further generalized to 4D Minkowski spacetime [38]. In this case, there is an obstacle that the inner product entering into the YB sigma model action is degenerate. A possible way around is to employ an embedding of 4D Minkowski spacetime into the bulk AdS_5 space. By adopting this resolution, a Yang-Baxter sigma model providing deformations of 4D Minkowski spacetime was proposed in [38]. Then classical r-matrices have been identified with a lot of gravity solutions such as Melvin backgrounds [39–42], pp-wave backgrounds [43], Hashimoto-Sethi backgrounds [44] and Spradlin-Takayanagi-Volovich backgrounds [45]. More interestingly, T-duals of dS_4 and AdS_4 also have been reproduced. In addition, new backgrounds generated by the standard q-deformation were also presented. Furthermore, the above result has an intimate connection with kappa-Minkowski spacetime [46] via preceding works e.g., [47]. For an argument with gravity duals, see [29].

It is worth noting that the gravity/CYBE correspondence may work beyond the integrability. There are many examples of non-integrable AdS/CFT correspondences. A landmark example is the $AdS_5 \times T^{1,1}$ background [48], for which the complete integrability is broken because chaotic string solutions appear on the $R \times T^{1,1}$ geometry [49, 50]. Hence TsT transformations of $T^{1,1}$ [31, 51] give rise to non-integrable deformations. Interestingly enough, these deformations can be reproduced as Yang-Baxter deformations with abelian classical r-matrices [52]. This result indicates that such Yang-Baxter deformations work well even for non-integrable backgrounds of type IIB supergravity.

[1] For earlier arguments concerned with this generalization, see [9–13].

In this article, we will give a summary for a series of our works with some updates by focusing upon four examples of classical r-matrices. The explanation includes Lax pairs as well as the metric and NS-NS two-form.

This article is organized as follows. Section 2 is a brief review of Yang-Baxter deformations of the $AdS_5 \times S^5$ superstring. In the subsequent sections, we explain four examples of classical r-matrices: (i) gravity duals of non-commutative gauge theories [in sec. 3] , (ii) γ-deformations of S^5 [in sec. 4] , (iii) Schrödinger spacetimes [in sec. 5] , and (iv) abelian twists of the global AdS_5 [in sec. 6] . Section 7 is summary and outlooks. In Appendix A, our notation and convention are summarized.

2. Jordanian deformations of the $AdS_5 \times S^5$ superstring

We give a brief review of Yang-Baxter deformations of the $AdS_5 \times S^5$ superstring with the CYBE [21]. The deformations are often called Jordanian deformations.

First of all, the deformed classical action of the $AdS_5 \times S^5$ superstring is given by

$$S = -\frac{\sqrt{\lambda_c}}{4} \int_{-\infty}^{\infty} d\tau \int_0^{2\pi} d\sigma \, (\gamma^{\alpha\beta} - \epsilon^{\alpha\beta}) \mathrm{STr}\Big[A_\alpha \, d \circ \frac{1}{1 - \eta R_g \circ d}(A_\beta)\Big] . \tag{2}$$

The left-invariant one-form A_α is defined as

$$A_\alpha \equiv g^{-1} \partial_\alpha g \,, \qquad g \in SU(2,2|4) \,. \tag{3}$$

Here $\alpha = (\tau, \sigma)$ is the world-sheet index. Then the world-sheet metric is taken as

$$\gamma^{\alpha\beta} = \mathrm{diag}(-1, +1) \tag{4}$$

with the conformal gauge, in which there is no coupling of the dilaton to the world-sheet scalar curvature. The anti-symmetric tensor $\epsilon^{\alpha\beta}$ is normalized as $\epsilon^{\tau\sigma} = +1$. The constant parameter λ_c in front of the action (2) is defined as

$$\sqrt{\lambda_c} \equiv 2\pi L^2 T = \frac{L^2}{\alpha'} \,, \qquad T = \frac{1}{2\pi\alpha'} \,,$$

where T is the string tension and L is the curvature radius of the undeformed AdS_5 . Note that η is a deformation parameter and hence the undeformed action [3] is reproduced when $\eta = 0$.

A key ingredient appearing in (2) is a chain of operations R_g , which is defined as

$$R_g(X) \equiv g^{-1} R(g X g^{-1}) g \,, \qquad X \in \mathfrak{su}(2,2|4) \,. \tag{5}$$

A linear operator $R : \mathfrak{su}(2,2|4) \to \mathfrak{su}(2,2|4)$ is a solution of the CYBE[2],

$$[R(X), R(Y)] - R([R(X), Y] + [X, R(Y)]) = 0 \,. \tag{6}$$

This R-operator is related to a skew-symmetric classical r-matrix in the tensorial notation through the following supertrace operation on the second site:

$$R(X) = \mathrm{STr}_2[r(1 \otimes X)] = \sum_i \left(a_i \, \mathrm{STr}[b_i X] - b_i \, \mathrm{STr}[a_i X] \right) \,. \tag{7}$$

[2] In the original work [21], a wider class of R-operators whose image is given by $\mathfrak{gl}(4|4)$ has been proposed. The $\mathfrak{gl}(4|4)$ image is restricted on $\mathfrak{su}(2,2|4)$ in essential under the coset projection d, as pointed out in [28]. We will concentrate here on a restricted class in which the image is $\mathfrak{su}(2,2|4)$ from the beginning, so as to deal with pre-projected quantities like the deformed current J itself, without introducing extra generators. For general cases argued in [22, 25], a more detailed study would be necessary.

Here the classical r-matrix is represented by

$$r = \sum_i a_i \wedge b_i \equiv \sum_i (a_i \otimes b_i - b_i \otimes a_i) \qquad \text{with} \qquad a_i, b_i \in \mathfrak{su}(2,2|4)\,. \tag{8}$$

If a_i and b_j (do not) commute with each other, then r is called (non-)abelian.

The coset (1) enjoys the $\mathbb{Z}_4$-grading property with the projections $P_i\,(i = 0,1,2,3)$ to the graded components of the decomposition

$$\mathfrak{su}(2,2|4) = \mathfrak{su}(2,2|4)_{(0)} \oplus \mathfrak{su}(2,2|4)_{(1)} \oplus \mathfrak{su}(2,2|4)_{(2)} \oplus \mathfrak{su}(2,2|4)_{(3)}\,, \tag{9}$$

where $P_i(\mathfrak{su}(2,2|4)) \equiv \mathfrak{su}(2,2|4)_{(i)}$. Note that

$$\mathfrak{su}(2,2|4)_{(0)} = \mathfrak{so}(1,4) \oplus \mathfrak{so}(5) \tag{10}$$

is a local symmetry of the classical action (2). Then d is defined as a linear combination of P_1 , P_2 and P_3 like

$$d \;\equiv\; P_1 + 2P_2 - P_3\,. \tag{11}$$

The numerical coefficients have been fixed by requiring the kappa-symmetry [21]. This d is the same in the undeformed case [3], while it depends on η in the case of the mCYBE [14].

In the following analysis, it is convenient to introduce the light-cone expression of A_α on the world-sheet like

$$A_\pm \equiv A_\tau \pm A_\sigma\,. \tag{12}$$

In particular, it makes the expression of Lax pair much simpler.

2.1. *The bosonic part of the Lagrangian*

Le us study the bosonic part of the deformed action (2) , which can be rewritten as

$$L = \frac{\sqrt{\lambda_c}}{2}\,\mathrm{STr}(A_-\,P_2(J_+))\,. \tag{13}$$

Here $J_\pm$ is a deformed current defined as

$$J_\pm \equiv \frac{1}{1 \mp 2\eta R_g \circ P_2} A_\pm\,. \tag{14}$$

Note here that the factor 2 in front of η comes from the projection operator d given in (11) . The deformed current $J_\pm$ is determined by solving the following equations:[3]

$$(1 \mp 2\eta R_g \circ P_2)\, J_\pm = A_\pm\,. \tag{15}$$

Then the metric in the string frame and NS-NS two-form can be evaluated from the Lagrangian (13) .

[3] In order to derive the metric and NS-NS two-form, it is enough to determine the projected current $P_2(J_\pm)$ by solving the projected conditions

$$(1 \mp 2\eta P_2 \circ R_g)\, P_2(J_\pm) = P_2(A_\pm)\,.$$

On the other hand, the unprojected current $J_\pm$ itself is necessary to derive the explicit form of Lax pair.

Taking a variation of the Lagrangian (13), the equation of motion is obtained as

$$\mathcal{E} \equiv \partial_+ P_2(J_-) + \partial_- P_2(J_+) + [J_+, P_2(J_-)] + [J_-, P_2(J_+)] = 0\,. \tag{16}$$

By definition, the undeformed current $A_\pm$ satisfies the flatness condition,

$$\mathcal{Z} \equiv \partial_+ A_- - \partial_- A_+ + [A_+, A_-] = 0\,. \tag{17}$$

Then, in terms of the deformed current $J_\pm$, this condition can be rewritten as

$$\partial_+ J_- - \partial_- J_+ + [J_+, J_-] + 2\eta\, R_g(\mathcal{E}) + 4\eta^2\, \mathrm{CYBE}_g(P_2(J_+), P_2(J_-)) = 0\,, \tag{18}$$

where we have introduced a new quantity defined as

$$\mathrm{CYBE}_g(X, Y) \equiv [R_g(X), R_g(Y)] - R_g([R_g(X), Y] + [X, R_g(Y)])\,. \tag{19}$$

Note that $\mathrm{CYBE}_g(X, Y)$ vanishes if the R-operator satisfies the CYBE in (6). Then the relation (18) indicates that $J_\pm$ also satisfies the flatness condition with the equation of motion $\mathcal{E} = 0$. That is, $J_\pm$ is the on-shell flat current, while $A_\pm$ is the off-shell flat one.

It is helpful to decompose $J_\pm$ with the projections P_0 and P_2 like

$$J_\pm = P_0(J_\pm) + P_2(J_\pm) \equiv J_\pm^{(0)} + J_\pm^{(2)}\,, \tag{20}$$

with the completeness condition $P_0 + P_2 = 1$. The concrete expressions of the projections are given in Appendix A. Then the equation of motion (16) can be rewritten as

$$\mathcal{E} = \partial_+ J_-^{(2)} + \partial_- J_+^{(2)} + [J_+^{(0)}, J_-^{(2)}] + [J_-^{(0)}, J_+^{(2)}] = 0\,. \tag{21}$$

The flatness condition (17) can also be rewritten in a similar way:

$$\mathcal{Z} = P_0(\mathcal{Z}) + P_2(\mathcal{Z}) = 0\,. \tag{22}$$

With the help of the linear independence of the grade 0 and grade 2 parts, one can obtain the following two conditions:

$$\begin{aligned} P_0(\mathcal{Z}) &= \partial_+ J_-^{(0)} - \partial_- J_+^{(0)} + [J_+^{(0)}, J_-^{(0)}] + [J_+^{(2)}, J_-^{(2)}] + 2\eta\, P_0(R_g(\mathcal{E})) = 0\,, \\ P_2(\mathcal{Z}) &= \partial_+ J_-^{(2)} - \partial_- J_+^{(2)} + [J_+^{(0)}, J_-^{(2)}] + [J_+^{(2)}, J_-^{(0)}] + 2\eta\, P_2(R_g(\mathcal{E})) = 0\,. \end{aligned} \tag{23}$$

Note here that the terms proportional to η vanish under the equation of motion, i.e., $\mathcal{E} = 0$.

Then the three conditions in (21) and (23) can be recast into the following set of the equations $\mathcal{C}_i = 0$ $(i = 1, 2, 3)$:

$$\begin{aligned} \mathcal{C}_1 &\equiv \partial_- J_+^{(2)} - [J_+^{(2)}, J_-^{(0)}]\,, \\ \mathcal{C}_2 &\equiv \partial_+ J_-^{(2)} + [J_+^{(0)}, J_-^{(2)}]\,, \\ \mathcal{C}_3 &\equiv \partial_+ J_-^{(0)} - \partial_- J_+^{(0)} + [J_+^{(0)}, J_-^{(0)}] + [J_+^{(2)}, J_-^{(2)}]\,. \end{aligned} \tag{24}$$

Namely, the relations $\mathcal{C}_i = 0$ $(i = 1, 2, 3)$ are equivalent to the equation of motion (16) and the flatness condition (17).

2.2. Lax pair

Finally, let us introduce a Lax pair for the deformed Lagrangian (13), which is given by

$$\mathcal{L}_\pm = J_\pm^{(0)} + \lambda^{\pm 1} J_\pm^{(2)} \tag{25}$$

with a spectral parameter $\lambda \in \mathbb{C}$. Note that the existence of the Lax pair (25) is based on the $\mathbb{Z}_2$-grading property of the bosonic $\mathrm{AdS}_5 \times \mathrm{S}^5$ group manifold.

As usual, the flatness condition of $\mathcal{L}_\pm$

$$0 \;=\; \partial_+ \mathcal{L}_- - \partial_- \mathcal{L}_+ + [\mathcal{L}_+, \mathcal{L}_-] \tag{26}$$

is equivalent to the equation of motion $\mathcal{E} = 0$ [in (16)] and the flatness condition $\mathcal{Z} = 0$ [in (17)]. In order to confirm the equivalence, it is helpful to notice that the right-hand side of (26) can be rewritten in terms of $\mathcal{C}_i$ as follows:

$$\partial_+ \mathcal{L}_- - \partial_- \mathcal{L}_+ + [\mathcal{L}_+, \mathcal{L}_-] = -\lambda\, \mathcal{C}_1 + \frac{1}{\lambda}\mathcal{C}_2 + \mathcal{C}_3 \,. \tag{27}$$

Thus the equivalence is now obvious.

In the following sections, we will explain classical r-matrices associated with four backgrounds, 1) gravity duals for noncommutative gauge theories, 2) γ-deformations of S^5, 3) Schrödinger spacetimes, and 4) abelian twists of the global AdS_5. For each of the examples, we show the metric (in the string frame) and NS-NS two-form, and present an explicit form of the associated Lax pair (25). For the conventions of the generators, see Appendix A.

3. Gravity duals of noncommutative gauge theories

Firstly, we will describe gravity duals of noncommutative gauge theories [33] from the viewpoint of Yang-Baxter deformations.

The backgrounds are associated with abelian Jordanian classical r-matrices [24]

$$r = c_1\, p_2 \wedge p_3 + c_2\, p_0 \wedge p_1 \,, \tag{28}$$

which consist of the translation generators p_μ in $\mathfrak{su}(2,2)$. The constant parameters c_1 and c_2 are related to magnetic and electric NS-NS two-forms in the solutions [33], respectively.

Note that p_μ's commute with each other and $p_\mu p_\nu = 0$ $(\mu, \nu = 0,1,2,3)$. Then the square of the associated R-operator vanishes and the classical r-matrices (28) are of Jordanian type. Since the r-matrices include no $\mathfrak{su}(4)$ generator, only the AdS_5 part is deformed. Hence, we will concentrate on the AdS_5 part below, while ignoring the S^5 part.

3.1. The deformed metric and NS-NS two-form

Let us derive the metric and NS-NS two-form from the Lagrangian (13).

We first introduce a coordinate system through a parametrization of an $SU(2,2)$ element,

$$g_a(\tau,\sigma) = \exp\Big[p_0\, x^0 + p_1\, x^1 + p_2\, x^2 + p_3\, x^3\Big] \,\exp\left[\gamma_5^a\, \frac{1}{2}\log z\right] \quad \in SU(2,2)\,. \tag{29}$$

By solving the relations in (15), the deformed current J_α is determined as

$$J_\pm \;=\; \frac{z}{z^4 - 4{c_2}^2\eta^2}\Big[(z^2\partial_\pm x^0 \pm 2c_2\eta\partial_\pm x^1)\, p_0 + (z^2\partial_\pm x^1 \pm 2c_2\eta\partial_\pm x^0)\, p_1\Big]$$

$$
\begin{aligned}
&+\frac{z}{z^4+4{c_1}^2\eta^2}\Big[(z^2\partial_\pm x^2 \pm 2c_1\eta\partial_\pm x^3)\,p_2 + (z^2\partial_\pm x^3 \mp 2c_1\eta\partial_\pm x^2)\,p_3\Big] \\
&+\frac{1}{2z}\partial_\pm z\,\gamma_5^a\,. \qquad (30)
\end{aligned}
$$

Then the resulting metric and NS-NS two-form are given by

$$
\begin{aligned}
ds^2 &= \frac{z^2[-(dx^0)^2+(dx^1)^2]}{z^4-4c_2^2\eta^2} + \frac{z^2[(dx^2)^2+(dx^3)^2]}{z^4+4c_1^2\eta^2} + \frac{dz^2}{z^2}\,, \\
B &= -\frac{2c_2\eta}{z^4-4c_2^2\eta^2}dx^0\wedge dx^1 + \frac{2c_1\eta}{z^4+4c_1^2\eta^2}dx^2\wedge dx^3\,. \qquad (31)
\end{aligned}
$$

This result exactly agrees with the solutions in [33]. When $c_1 = c_2 = 0$, the Poincaré AdS_5 is reproduced.

The result in (31) can also be reproduced as a special limit of the q-deformed $\mathrm{AdS}_5\times\mathrm{S}^5$ [19]. This limit has been further confirmed at the level of Lax pairs [30].

3.2. Lax pair

Then let us see the associated Lax pair $\mathcal{L}_\pm^{\mathrm{NC}}$. It is explicitly given by [24, 30]

$$
\begin{aligned}
\mathcal{L}_\pm^{\mathrm{NC}} &= \frac{z}{z^4-4c_2^2\eta^2}\Big[(z^2\partial_\pm x^0 \pm 2c_2\eta\partial_\pm x^1)\Big(\frac{\lambda^{\pm1}}{2}\gamma_0^a - n_{05}^a\Big) \\
&\qquad +(z^2\partial_\pm x^1 \pm 2c_2\eta\partial_\pm x^0)\Big(\frac{\lambda^{\pm1}}{2}\gamma_1^a - n_{15}^a\Big)\Big] \\
&\quad +\frac{z}{z^4+4c_1^2\eta^2}\Big[(z^2\partial_\pm x^2 \pm 2c_1\eta\partial_\pm x^3)\Big(\frac{\lambda^{\pm1}}{2}\gamma_2^a - n_{25}^a\Big) \\
&\qquad + (z^2\partial_\pm x^3 \mp 2c_1\eta\partial_\pm x^2)\Big(\frac{\lambda^{\pm1}}{2}\gamma_3^a - n_{35}^a\Big)\Big] + \frac{\lambda^{\pm1}\partial_\pm z}{2z}\gamma_5^a\,. \qquad (32)
\end{aligned}
$$

In the undeformed limit $c_1,\, c_2 \to 0$, the above expression is reduced to

$$
\mathcal{L}_\pm^{\mathrm{PAdS}_5} = \frac{\partial_\pm x^\mu}{z}\Big(\frac{\lambda^{\pm1}}{2}\gamma_\mu^a - n_{\mu5}^a\Big) + \frac{\lambda^{\pm1}\partial_\pm z}{2z}\gamma_5^a\,. \qquad (33)
$$

This is nothing but a Lax pair for the Poincaré AdS_5.

3.3. Twisted boundary condition

In fact, the deformations with the r-matrices (28) can be regarded as twisted boundary conditions with the undeformed $\mathrm{AdS}_5\times\mathrm{S}^5$, as argued in [32].

For simplicity, suppose $c_1 \neq 0$ and $c_2 = 0$. The analysis for the case with $c_2 \neq 0$ is quite similar, though there is a subtlety for the signature of the metric (For the detail, see [33]).

After performing the Yang-Baxter deformation (equivalently the associated TsT transformation), the original coordinates $\tilde{x}^2$ and $\tilde{x}^3$ for the undeformed $\mathrm{AdS}_5\times\mathrm{S}^5$ are mapped to x^2 and x^3. Then the relations between the coordinates are given by

$$
\begin{aligned}
\frac{1}{z^2}\,\partial_\pm\tilde{x}^2 &= \frac{z^2}{z^4+4{c_1}^2\eta^2}\left[\partial_\pm x^2 \pm \frac{2c_1\eta}{z^2}\partial_\pm x^3\right], \\
\frac{1}{z^2}\,\partial_\pm\tilde{x}^3 &= \frac{z^2}{z^4+4{c_1}^2\eta^2}\left[\partial_\pm x^3 \mp \frac{2c_1\eta}{z^2}\partial_\pm x^2\right]. \qquad (34)
\end{aligned}
$$

These relations indicate the following equivalence of Noether currents

$$\tilde{P}_i^\alpha = P_i^\alpha \qquad (i = 2, 3)\,. \tag{35}$$

Here P_i^α and $\tilde{P}_i^\alpha$ are conserved currents associated with translation invariance for the x^i and $\tilde{x}^i$ directions, respectively. The τ-component of (35) indicates that the momentum $p_i \equiv P_i^\tau$ is identical to $\tilde{p}_i \equiv \tilde{P}_i^\tau$, namely $p_i = \tilde{p}_i$. Then the σ-component of (35) leads to the relations:

$$\partial_\sigma \tilde{x}^2 = \partial_\sigma x^2 + \frac{2c_1\eta}{\sqrt{\lambda_c}}\, p_3\,, \qquad \partial_\sigma \tilde{x}^3 = \partial_\sigma x^3 - \frac{2c_1\eta}{\sqrt{\lambda_c}}\, p_2\,. \tag{36}$$

By integrating both relations, one can realize that the deformed backgrounds with the periodic boundary condition are equivalent to the undeformed $AdS_5 \times S^5$ with twisted boundary conditions:

$$\tilde{x}^2(\sigma = 2\pi) = \tilde{x}^2(\sigma = 0) + \frac{2c_1\eta}{\sqrt{\lambda_c}}\, P_3\,, \qquad \tilde{x}^3(\sigma = 2\pi) = \tilde{x}^3(\sigma = 0) - \frac{2c_1\eta}{\sqrt{\lambda_c}}\, P_2\,. \tag{37}$$

Here P_i are Noether charges obtained by integrating p_i .

Thus the Yang-Baxter deformations with the classical r-matrices (28) can be reinterpreted as twisted boundary conditions with the usual $AdS_5 \times S^5$ background.

4. γ-deformations of S^5

An example of marginal deformations of $\mathcal{N} = 4$ SYM, which preserves an $\mathcal{N} = 1$ superconformal symmetry, is called the β-deformation [53]. The gravity dual for the β-deformation was constructed in [31], and then it was generalized to three-parameter cases called the γ-deformations [32]. The AdS_5 part is not deformed due to the conformal symmetry, while the S^5 part is deformed due to the reduced supersymmetry.

Let us study here the γ-deformations [31,32] from the viewpoint of Yang-Baxter deformations. In the following, the AdS_5 part will be dropped off.

The backgrounds are associated with abelian classical r-matrices [23] ,

$$r = \mu_3\, h_4 \wedge h_5 + \mu_1\, h_5 \wedge h_6 + \mu_2\, h_6 \wedge h_4\,. \tag{38}$$

Here h_4 , h_5 and h_6 are the three Cartan generators of $\mathfrak{su}(4)$, and μ_i $(i = 1, 2, 3)$ are the deformation parameters.

4.1. The deformed metric and NS-NS two-form

It is useful to employ the following parametrization of a group element g_s of $SU(4)$,

$$g_s(\tau, \sigma) = \exp\left[\frac{i}{2}(\phi_1\, h_4 + \phi_2\, h_5 + \phi_3\, h_6)\right] \exp\left[-\zeta n_{13}^{s}\right] \exp\left[-\frac{i}{2}\, r\, \gamma_1^{s}\right]. \tag{39}$$

In the undeformed case, this describes the round S^5 with the coordinates r , ζ and ϕ_i $(i = 1, 2, 3)$.

By solving the equations in (15) , the deformed current $J_\alpha^{\hat{\gamma}_1, \hat{\gamma}_2, \hat{\gamma}_3}$ is determined as

$$J_\pm^{\hat{\gamma}_1, \hat{\gamma}_2, \hat{\gamma}_3} \;=\; -i\partial_\pm r\, \frac{\gamma_1^{s}}{2} - \partial_\pm \zeta \left[i \sin r\, \frac{\gamma_3^{s}}{2} + \cos r\, n_{13}^{s}\right]$$

$$
\begin{aligned}
&-G(\hat{\gamma}_i)\Big[\partial_\pm\phi_1 \pm (\hat{\gamma}_3 \sin^2 r \sin^2\zeta\partial_\pm\phi_2 - \hat{\gamma}_2\cos^2 r\partial_\pm\phi_3) \\
&\qquad\qquad +\hat{\gamma}_1 \sin^2 r\cos^2 r\sin^2\zeta\sum_{i=1}^{3}\hat{\gamma}_i\partial_\pm\phi_i\Big] \\
&\qquad\qquad \times\left[\cos\zeta(i\sin r\,\frac{\gamma_2^s}{2} + \cos r\, n_{12}^s) + \sin\zeta\, n_{23}^s\right] \\
&-G(\hat{\gamma}_i)\Big[\partial_\pm\phi_2 \pm (\hat{\gamma}_1 \cos^2 r\partial_\pm\phi_3 - \hat{\gamma}_3\sin^2 r\cos^2\zeta\partial_\pm\phi_1) \\
&\qquad\qquad +\hat{\gamma}_2 \sin^2 r\cos^2 r\cos^2\zeta\sum_{i=1}^{3}\hat{\gamma}_i\partial_\pm\phi_i\Big] \\
&\qquad\qquad \times\left[\sin\zeta(i\,\sin r\,\frac{1}{2}\gamma_4^s + \cos r\, n_{14}^s) + \cos\zeta\, n_{34}^s\right] \\
&+G(\hat{\gamma}_i)\Big[\partial_\pm\phi_3 \pm (\hat{\gamma}_2 \sin^2 r \cos^2\zeta\partial_\pm\phi_1 - \hat{\gamma}_1\sin^2 r\sin^2\zeta\partial_\pm\phi_2) \\
&\qquad\qquad +\hat{\gamma}_3 \sin^4 r\sin^2\zeta\cos^2\zeta\sum_{i=1}^{3}\hat{\gamma}_i\partial_\pm\phi_i\Big] \\
&\qquad\qquad \times\left[i\cos r\,\frac{\gamma_5^s}{2} - \sin r\, n_{15}^s\right] .
\end{aligned} \tag{40}
$$

Here the constant parameters $\hat{\gamma}_i$ are defined as

$$\hat{\gamma}_i \equiv 8\,\eta\mu_i\,, \tag{41}$$

and the scalar function $G(\hat{\gamma}_i)$ is given by

$$G^{-1}(\hat{\gamma}_i) \equiv 1 + \sin^2 r(\hat{\gamma}_1^2\cos^2 r\sin^2\zeta + \hat{\gamma}_2^2\cos^2 r\cos^2\zeta + \hat{\gamma}_3^2\sin^2 r\sin^2\zeta\cos^2\zeta)\,. \tag{42}$$

By substituting (40) into the Lagrangian (13), the deformed background is obtained as

$$
\begin{aligned}
ds^2 &= \sum_{i=1}^{3}\left(d\rho_i{}^2 + G(\hat{\gamma}_i)\rho_i{}^2 d\phi_i{}^2\right) + G(\hat{\gamma}_i)\rho_1{}^2\rho_2{}^2\rho_3{}^2\left(\sum_{i=1}^{3}\hat{\gamma}_i d\phi_i\right)^2 , \\
B_2 &= G(\hat{\gamma}_i)\left(\hat{\gamma}_3\rho_1{}^2\rho_2{}^2 d\phi_1\wedge d\phi_2 + \hat{\gamma}_1\rho_2{}^2\rho_3{}^2 d\phi_2\wedge d\phi_3 + \hat{\gamma}_2\rho_3{}^2\rho_1{}^2 d\phi_3\wedge d\phi_1\right).
\end{aligned} \tag{43}
$$

Here new coordinates ρ_i $(i = 1, 2, 3)$ are defined as

$$\rho_1 \equiv \sin r\cos\zeta\,, \qquad \rho_2 \equiv \sin r\sin\zeta\,, \qquad \rho_3 \equiv \cos r\,. \tag{44}$$

The metric and NS-NS two-form in (43) agree with the ones of γ-deformations of S^5 [32].

A particular one-parameter case with

$$\hat{\gamma}_1 = \hat{\gamma}_2 = \hat{\gamma}_3 \equiv \hat{\gamma} \tag{45}$$

corresponds to the metric and NS-NS two-form of the Lunin-Maldacena solution [31],

$$
\begin{aligned}
ds^2 &= \sum_{i=1}^{3}\left(d\rho_i{}^2 + G\rho_i{}^2 d\phi_i{}^2\right) + G\hat{\gamma}^2\rho_1{}^2\rho_2{}^2\rho_3{}^2\left(\sum_{i=1}^{3} d\phi_i\right)^2 , \\
B_2 &= G\,\hat{\gamma}(\rho_1{}^2\rho_2{}^2 d\phi_1\wedge d\phi_2 + \rho_2{}^2\rho_3{}^2 d\phi_2\wedge d\phi_3 + \rho_3{}^2\rho_1{}^2 d\phi_3\wedge d\phi_1)\,,
\end{aligned} \tag{46}
$$

where the scalar function $G(\hat{\gamma})$ is defined as

$$G^{-1}(\hat{\gamma}) \equiv 1 + \frac{\hat{\gamma}^2}{4}(\sin^2 2r + \sin^4 r\sin^2 2\zeta)\,. \tag{47}$$

This background is a holographic dual of the β-deformation of the $\mathcal{N} = 4$ SYM [53].

4.2. Lax pair

Next, the explicit form of the Lax pair is given by [23, 30]

$$\begin{aligned}
\mathcal{L}_{\pm}^{\hat{\gamma}_1,\hat{\gamma}_2,\hat{\gamma}_3} &= -i\frac{\lambda^{\pm 1}}{2}\partial_{\pm}r\,\gamma_1^s - \partial_{\pm}\zeta\left[i\sin r\,\frac{\lambda^{\pm 1}}{2}\gamma_3^s + \cos r\,n_{13}^s\right] \\
&\quad -G(\hat{\gamma}_i)\Big[\partial_{\pm}\phi_1 \pm (\hat{\gamma}_3\sin^2 r\sin^2\zeta\partial_{\pm}\phi_2 - \hat{\gamma}_2\cos^2 r\partial_{\pm}\phi_3) \\
&\qquad\qquad +\hat{\gamma}_1\sin^2 r\cos^2 r\sin^2\zeta\sum_{i=1}^{3}\hat{\gamma}_i\partial_{\pm}\phi_i\Big] \\
&\qquad\times\left[\cos\zeta(i\sin r\,\frac{\lambda^{\pm 1}}{2}\gamma_2^s + \cos r\,n_{12}^s) + \sin\zeta\,n_{23}^s\right] \\
&\quad -G(\hat{\gamma}_i)\Big[\partial_{\pm}\phi_2 \pm (\hat{\gamma}_1\cos^2 r\partial_{\pm}\phi_3 - \hat{\gamma}_3\sin^2 r\cos^2\zeta\partial_{\pm}\phi_1) \\
&\qquad\qquad +\hat{\gamma}_2\sin^2 r\cos^2 r\cos^2\zeta\sum_{i=1}^{3}\hat{\gamma}_i\partial_{\pm}\phi_i\Big] \\
&\qquad\times\left[\sin\zeta(i\,\sin r\,\frac{\lambda^{\pm 1}}{2}\gamma_4^s + \cos r\,n_{14}^s) + \cos\zeta\,n_{34}^s\right] \\
&\quad +G(\hat{\gamma}_i)\Big[\partial_{\pm}\phi_3 \pm (\hat{\gamma}_2\sin^2 r\cos^2\zeta\partial_{\pm}\phi_1 - \hat{\gamma}_1\sin^2 r\sin^2\zeta\partial_{\pm}\phi_2) \\
&\qquad\qquad +\hat{\gamma}_3\sin^4 r\sin^2\zeta\cos^2\zeta\sum_{i=1}^{3}\hat{\gamma}_i\partial_{\pm}\phi_i\Big] \\
&\qquad\times\left[i\cos r\,\frac{\lambda^{\pm 1}}{2}\gamma_5^s - \sin r\,n_{15}^s\right]. \qquad (48)
\end{aligned}$$

This result is equivalent to the Lax pair obtained in [32]. Note here that, in the undeformed limit $\hat{\gamma}_i \to 0$, $\mathcal{L}_{\pm}^{\hat{\gamma}_1,\hat{\gamma}_2,\hat{\gamma}_3}$ is reduced to

$$\begin{aligned}
\mathcal{L}_{\pm}^{S} &= -i\,\partial_{\pm}r\frac{\lambda^{\pm 1}}{2}\gamma_1^s - \partial_{\pm}\zeta\left[i\sin r\,\frac{\lambda^{\pm 1}}{2}\gamma_3^s + \cos r\,n_{13}^s\right] \\
&\quad -\partial_{\pm}\phi_1\left[\cos\zeta(i\sin r\,\frac{\lambda^{\pm 1}}{2}\gamma_2^s + \cos r\,n_{12}^s) + \sin\zeta\,n_{23}^s\right] \\
&\quad -\partial_{\pm}\phi_2\left[\sin\zeta(i\,\sin r\,\frac{\lambda^{\pm 1}}{2}\gamma_4^s + \cos r\,n_{14}^s) + \cos\zeta\,n_{34}^s\right] \\
&\quad +\partial_{\pm}\phi_3\left[i\cos r\,\frac{\lambda^{\pm 1}}{2}\gamma_5^s - \sin r\,n_{15}^s\right]. \qquad (49)
\end{aligned}$$

This is just a Lax pair for the round S^5.

4.3. Twisted boundary condition

Again, the deformations can be regarded as the undeformed $AdS_5\times S^5$ with twisted boundary conditions [32]. The twisted boundary conditions are given by

$$\begin{aligned}
\tilde{\phi}_1(\sigma = 2\pi) &= \tilde{\phi}_1(\sigma = 0) + \gamma_3 J_2 - \gamma_2 J_3 + 2\pi n_1\,, \\
\tilde{\phi}_2(\sigma = 2\pi) &= \tilde{\phi}_2(\sigma = 0) + \gamma_1 J_3 - \gamma_3 J_1 + 2\pi n_2\,, \\
\tilde{\phi}_3(\sigma = 2\pi) &= \tilde{\phi}_3(\sigma = 0) + \gamma_2 J_1 - \gamma_1 J_2 + 2\pi n_3\,, \qquad (50)
\end{aligned}$$

with $\gamma_i \equiv \hat{\gamma}/\sqrt{\lambda_c}$. Here J_i are the Noether charges for rotation invariance in the ϕ_i directions. The integers n_i are winding numbers along the ϕ_i directions.

5. Schrödinger spacetimes

We are concerned with Schrödinger spacetimes here. Originally, 3D Schrödinger spacetimes were introduced as light-like deformations of AdS_3 [55]. Higher-dimensional Schrödinger spacetimes were constructed as holographic duals for non-relativistic conformal field theories [56]. For the coset construction of the metrics, see [54]. Then the backgrounds have been embedded into type IIB supergravity [34][4].

In the following, let us study Schrödinger spacetimes embedded in type IIB supergravity [34] from the viewpoint of Yang-Baxter deformations. Then the backgrounds are associated with the following abelian r-matrix [27]:

$$r = \frac{i}{4\sqrt{2}}\,(p_0 - p_3)\wedge(h_4 + h_5 + h_6)\,. \tag{51}$$

It contains generators of both $\mathfrak{su}(2,2)$ and $\mathfrak{su}(4)$ and hence deforms both AdS_5 and S^5 parts.

5.1. The deformed metric and NS-NS two-form

Let us parametrize group elements of $SU(2,2)$ and $SU(4)$ like

$$\begin{aligned} g_a(\tau,\sigma) &= \exp\left[x^0 p_0 + x^1 p_1 + x^2 p_2 + x^3 p_3\right]\exp\left[\gamma_5^a\,\frac{1}{2}\,\log z\right] \quad \in\ SU(2,2)\,, \\ g_s(\tau,\sigma) &= \exp\left[\frac{i}{2}(\psi_1 h_4 + \psi_2 h_5 + \psi_3 h_6)\right]\exp\left[-\zeta n_{13}^s\right]\exp\left[-\frac{i}{2}r\,\gamma_1^s\right] \quad \in\ SU(4)\,. \end{aligned} \tag{52}$$

The deformed current $J_\pm$ can be expanded in terms of the generators of $\mathfrak{su}(2,2)\oplus\mathfrak{su}(4)$. Then, by solving the equations in (15) , $J_\pm$ is determined as

$$\begin{aligned} J_\pm^a &= \frac{1}{z}\partial_\pm x^1 p_1 + \frac{1}{z}\partial_\pm x^2 p_2 + \frac{1}{2z}\partial_\pm z\,\gamma_5^a \\ &\quad + \frac{1}{\sqrt{2}\,z}\partial_\pm x^+\,(p_0 + p_3) \\ &\quad + \frac{1}{\sqrt{2}\,z}\left[\partial_\pm x^- \pm \eta\partial_\pm\chi \pm \frac{\eta}{2}\sin^2\mu(\partial_\pm\psi + \cos\theta\partial_\pm\phi) + \frac{\eta^2}{z^2}\partial_\pm x^+\right](p_0 - p_3)\,, \\ J_\pm^s &= -\frac{i}{2}\partial_\pm\mu\,\gamma_1^s - \frac{1}{2}\partial_\pm\theta\left[\frac{i}{2}\sin\mu\,\gamma_3^s + \cos\mu\,n_{13}^s\right] \\ &\quad - \left[\partial_\pm\chi \pm \frac{\eta\,\partial_\pm x^+}{z^2}\right]\left[\sin\frac{\theta}{2}\left(\frac{i}{2}\sin\mu\,\gamma_4^s + \cos\mu\,n_{14}^s + n_{23}^s\right)\right. \\ &\qquad \left. + \cos\frac{\theta}{2}\left(\frac{i}{2}\sin\mu\,\gamma_2^s + \cos\mu\,n_{12}^s + n_{34}^s\right) - \frac{i}{2}\cos\mu\,\gamma_5^s + \sin\mu\,n_{15}^s\right] \\ &\quad + \frac{1}{2}\partial_\pm\phi\left[\sin\frac{\theta}{2}\left(\frac{i}{2}\sin\mu\,\gamma_4^s + \cos\mu\,n_{14}^s - n_{23}^s\right)\right. \\ &\qquad \left. - \cos\frac{\theta}{2}\left(\frac{i}{2}\sin\mu\,\gamma_2^s + \cos\mu\,n_{12}^s - n_{34}^s\right)\right] \\ &\quad - \frac{1}{2}\partial_\pm\psi\left[\sin\frac{\theta}{2}\left(\frac{i}{2}\sin\mu\,\gamma_4^s + \cos\mu\,n_{14}^s + n_{23}^s\right)\right. \\ &\qquad \left. + \cos\frac{\theta}{2}\left(\frac{i}{2}\sin\mu\,\gamma_2^s + \cos\mu\,n_{12}^s + n_{34}^s\right)\right]\,. \end{aligned} \tag{53}$$

[4] For another embedding into type IIB supergravity, see [57].

Here we have performed a coordinate transformation,

$$
\begin{aligned}
x^{\pm} &= \frac{x^0 \pm x^3}{\sqrt{2}}\,, \\
r &= \mu\,, \quad \zeta = \tfrac{1}{2}\theta\,, \quad \psi_1 = \chi + \tfrac{1}{2}(\psi+\phi)\,, \quad \psi_2 = \chi + \tfrac{1}{2}(\psi-\phi)\,, \quad \psi_3 = \chi\,.
\end{aligned}
$$

With the deformed current (53) , the resulting background is given by

$$
\begin{aligned}
ds^2 &= \frac{-2dx^+dx^- + (dx^1)^2 + (dx^2)^2 + dz^2}{z^2} - \eta^2 \frac{(dx^+)^2}{z^4} + ds^2_{\mathrm{S}^5}\,, \\
B_2 &= \frac{\eta}{z^2}\, dx^+ \wedge (d\chi + \omega)\,.
\end{aligned}
\tag{54}
$$

Here the S^5 metric is written as an S^1-fibration over $\mathbb{CP}^2$,

$$
\begin{aligned}
ds^2_{\mathrm{S}^5} &= (d\chi + \omega)^2 + ds^2_{\mathbb{CP}^2}\,, \\
ds^2_{\mathbb{CP}^2} &= d\mu^2 + \sin^2\mu \left(\Sigma_1^2 + \Sigma_2^2 + \cos^2\mu\, \Sigma_3^2\right).
\end{aligned}
\tag{55}
$$

Now χ is the fiber coordinate and ω is a one-form potential of the Kähler form on $\mathbb{CP}^2$. The symbols Σ_i $(i = 1, 2, 3)$ and ω are defined as

$$
\begin{aligned}
\Sigma_1 &\equiv \tfrac{1}{2}(\cos\psi\, d\theta + \sin\psi \sin\theta\, d\phi)\,, \\
\Sigma_2 &\equiv \tfrac{1}{2}(\sin\psi\, d\theta - \cos\psi \sin\theta\, d\phi)\,, \\
\Sigma_3 &\equiv \tfrac{1}{2}(d\psi + \cos\theta\, d\phi)\,, \qquad \omega \equiv \sin^2\mu\, \Sigma_3\,.
\end{aligned}
\tag{56}
$$

It is remarkable that only the AdS_5 metric is deformed while the S^5 part is not, in spite of the expression of the classical r-matrix (51) . On the other hand, the NS-NS two-form carries two indices, one of which is from AdS_5 and the other is S^5 .

So far, we have considered the one-parameter deformation, but it is easy to reproduce three-parameter deformations of [58] and non-Cartan deformations of [59] (For the detail, see [27]) .

5.2. Lax pair

In the present case, the associated Lax pair is a bit messy but given by [27, 30]

$$
\begin{aligned}
\mathcal{L}^{\mathrm{Sch}}_{\pm} &= \frac{1}{z}\partial_{\pm}x^1 \left[\frac{\lambda^{\pm1}}{2}\gamma_1^a - n_{15}^a\right] + \frac{1}{z}\partial_{\pm}x^2 \left[\frac{\lambda^{\pm1}}{2}\gamma_2^a - n_{25}^a\right] + \frac{\lambda^{\pm1}}{2z}\partial_{\pm}z\,\gamma_5^a \\
&\quad + \frac{1}{\sqrt{2}\,z}\partial_{\pm}x^+ \left[\frac{\lambda^{\pm1}}{2}\gamma_0^a + \frac{\lambda^{\pm1}}{2}\gamma_3^a - n_{05}^a - n_{35}^a\right] \\
&\quad + \frac{1}{\sqrt{2}\,z}\left[\partial_{\pm}x^- \pm \eta\partial_{\pm}\chi \pm \frac{\eta}{2}\sin^2\mu(\partial_{\pm}\psi + \cos\theta\partial_{\pm}\phi) + \frac{\eta^2}{z^2}\partial_{\pm}x^+\right] \\
&\qquad \times \left[\frac{\lambda^{\pm1}}{2}\gamma_0^a - \frac{\lambda^{\pm1}}{2}\gamma_3^a - n_{05}^a + n_{35}^a\right] \\
&\quad - \frac{i\lambda^{\pm1}}{2}\partial_{\pm}\mu\,\gamma_1^s - \frac{1}{2}\partial_{\pm}\theta\left[\frac{i\lambda^{\pm1}}{2}\sin\mu\,\gamma_3^s + \cos\mu\, n_{13}^s\right] \\
&\quad - \left[\partial_{\pm}\chi \pm \frac{\eta\,\partial_{\pm}x^+}{z^2}\right]\left[\sin\frac{\theta}{2}\left(\frac{i\lambda^{\pm1}}{2}\sin\mu\,\gamma_4^s + \cos\mu\, n_{14}^s + n_{23}^s\right)\right. \\
&\qquad \left. + \cos\frac{\theta}{2}\left(\frac{i\lambda^{\pm1}}{2}\sin\mu\,\gamma_2^s + \cos\mu\, n_{12}^s + n_{34}^s\right) - \frac{i\lambda^{\pm1}}{2}\cos\mu\,\gamma_5^s + \sin\mu\, n_{15}^s\right]
\end{aligned}
$$

$$
\begin{aligned}
&+\frac{1}{2}\partial_\pm\phi\Bigg[\sin\frac{\theta}{2}\left(\frac{i\lambda^{\pm 1}}{2}\sin\mu\,\gamma_4^s+\cos\mu\,n_{14}^s-n_{23}^s\right)\\
&\qquad\qquad -\cos\frac{\theta}{2}\left(\frac{i\lambda^{\pm 1}}{2}\sin\mu\,\gamma_2^s+\cos\mu\,n_{12}^s-n_{34}^s\right)\Bigg]\\
&-\frac{1}{2}\partial_\pm\psi\Bigg[\sin\frac{\theta}{2}\left(\frac{i\lambda^{\pm 1}}{2}\sin\mu\,\gamma_4^s+\cos\mu\,n_{14}^s+n_{23}^s\right)\\
&\qquad\qquad +\cos\frac{\theta}{2}\left(\frac{i\lambda^{\pm 1}}{2}\sin\mu\,\gamma_2^s+\cos\mu\,n_{12}^s+n_{34}^s\right)\Bigg]\,. \qquad (57)
\end{aligned}
$$

As $\eta\to 0$, the above Lax pair $\mathcal{L}_\pm^{\rm Sch}$ is reduced to the following:

$$
\begin{aligned}
\mathcal{L}_\pm &= \frac{1}{z}\partial_\pm x^1\left[\frac{\lambda^{\pm 1}}{2}\gamma_1^a-n_{15}^a\right]+\frac{1}{z}\partial_\pm x^2\left[\frac{\lambda^{\pm 1}}{2}\gamma_2^a-n_{25}^a\right]+\frac{\lambda^{\pm 1}}{2z}\partial_\pm z\,\gamma_5^a\\
&+\frac{1}{\sqrt{2}\,z}\partial_\pm x^+\left[\frac{\lambda^{\pm 1}}{2}\gamma_0^a+\frac{\lambda^{\pm 1}}{2}\gamma_3^a-n_{05}^a-n_{35}^a\right]\\
&+\frac{1}{\sqrt{2}\,z}\partial_\pm x^-\left[\frac{\lambda^{\pm 1}}{2}\gamma_0^a-\frac{\lambda^{\pm 1}}{2}\gamma_3^a-n_{05}^a+n_{35}^a\right]\\
&-\frac{i\lambda^{\pm 1}}{2}\partial_\pm\mu\,\gamma_1^s-\frac{1}{2}\partial_\pm\theta\left[\frac{i\lambda^{\pm 1}}{2}\sin\mu\,\gamma_3^s+\cos\mu\,n_{13}^s\right]\\
&-\partial_\pm\chi\Bigg[\sin\frac{\theta}{2}\left(\frac{i\lambda^{\pm 1}}{2}\sin\mu\,\gamma_4^s+\cos\mu\,n_{14}^s+n_{23}^s\right)\\
&\qquad +\cos\frac{\theta}{2}\left(\frac{i\lambda^{\pm 1}}{2}\sin\mu\,\gamma_2^s+\cos\mu\,n_{12}^s+n_{34}^s\right)-\frac{i\lambda^{\pm 1}}{2}\cos\mu\,\gamma_5^s+\sin\mu\,n_{15}^s\Bigg]\\
&+\frac{1}{2}\partial_\pm\phi\Bigg[\sin\frac{\theta}{2}\left(\frac{i\lambda^{\pm 1}}{2}\sin\mu\,\gamma_4^s+\cos\mu\,n_{14}^s-n_{23}^s\right)\\
&\qquad\qquad -\cos\frac{\theta}{2}\left(\frac{i\lambda^{\pm 1}}{2}\sin\mu\,\gamma_2^s+\cos\mu\,n_{12}^s-n_{34}^s\right)\Bigg]\\
&-\frac{1}{2}\partial_\pm\psi\Bigg[\sin\frac{\theta}{2}\left(\frac{i\lambda^{\pm 1}}{2}\sin\mu\,\gamma_4^s+\cos\mu\,n_{14}^s+n_{23}^s\right)\\
&\qquad\qquad +\cos\frac{\theta}{2}\left(\frac{i\lambda^{\pm 1}}{2}\sin\mu\,\gamma_2^s+\cos\mu\,n_{12}^s+n_{34}^s\right)\Bigg]\,. \qquad (58)
\end{aligned}
$$

5.3. Twisted boundary condition

The deformations can be reinterpreted as twisted boundary conditions, again. The following twisted boundary conditions

$$
\begin{aligned}
\tilde{x}^-(\sigma=2\pi) &= \tilde{x}^-(\sigma=0)+\frac{\eta}{\sqrt{\lambda_{\rm c}}}J_\chi+2\pi n_\chi\,,\\
\tilde{\chi}(\sigma=2\pi) &= \tilde{\chi}(\sigma=0)-\frac{\eta}{\sqrt{\lambda_{\rm c}}}P_- \qquad (59)
\end{aligned}
$$

with the undeformed $\mathrm{AdS}_5\times\mathrm{S}^5$ is equivalent to the deformed background with the periodic boundary condition. Here P_- and J_χ are the Noether charges for translation and rotation invariance for the x^- and χ directions, respectively. The integer n_χ is a winding number for the χ direction. It may be interesting to consider a relation between the above argument and the symmetric two-form studied in [54].

6. Abelian twists of the global AdS$_5$

In section 3, we have considered gravity duals for gauge theories on noncommutative planes. Thus the deformed backgrounds have been described as deformations of the Poincaré AdS$_5$.

On the other hand, one may consider noncommutative deformations of the global AdS$_5$ [35], in which the dual gauge theories are living on $R\times$ deformed S^3. The deformed backgrounds can be realized by performing abelian twists (equivalently TsT transformations) as in section 4 for the global AdS$_5$.

Here, let us consider abelian twists of the global AdS$_5$ as Yang-Baxter deformations. The twists are associated with the classical r-matrix [24],

$$r=-\frac{i}{2}n_{12}^{a}\wedge n_{03}^{a}\,. \tag{60}$$

This r-matrix is composed of two Cartan generators of $\mathfrak{su}(2,2)$ and deforms only the AdS$_5$ part. Hence we will omit the S^5 part hereafter.

6.1. The deformed metric and NS-NS two-form

Let us consider the following parameterization of a group element of $SU(2,2)$:

$$g_a(\tau,\sigma)=\exp\left[\frac{i}{2}(\phi_1\,h_1+\phi_2\,h_2+\tau\,h_3)\right]\exp\left[-\theta\,n_{13}^{a}\right]\exp\left[-\rho\,\frac{\gamma_1^{a}}{2}\right]\quad\in SU(2,2)\,. \tag{61}$$

The deformed current $J_\pm$ is expanded in terms of the basis of $\mathfrak{su}(2,2)$. Then, by solving the equations in (15), $J_\pm$ can be determined as

$$\begin{aligned}J_\pm\ =\ &-\partial_\pm\rho\frac{1}{2}\gamma_1^{a}-\partial_\pm\theta\left[\frac{1}{2}\sinh\rho\,\gamma_3^{a}+\cosh\rho\,n_{13}^{a}\right]\\&+i\partial_\pm\tau\left[\frac{1}{2}\cosh\rho\,\gamma_5^{a}+\sinh\rho\,n_{15}^{a}\right]\\&-\hat{G}\left(\partial_\pm\phi_1\mp\eta\sin^2\theta\sinh^2\rho\,\partial_\pm\phi_2\right)\\&\qquad\times\left[\cos\theta\left(\frac{1}{2}\sinh\rho\,\gamma_2^{a}+\cosh\rho\,n_{12}^{a}\right)+\sin\theta\,n_{23}^{a}\right]\\&+i\hat{G}\left(\partial_\pm\phi_2\pm\eta\cos^2\theta\sinh^2\rho\,\partial_\pm\phi_1\right)\\&\qquad\times\left[\sin\theta\left(\frac{1}{2}\sinh\rho\,\gamma_0^{a}-\cosh\rho\,n_{01}^{a}\right)-\cos\theta\,n_{03}^{a}\right],\end{aligned} \tag{62}$$

where $\hat{G}(\eta)$ is a scalar function defined as

$$\hat{G}^{-1}(\eta)\equiv 1+\eta^2\sin^2\theta\cos^2\theta\sinh^2\rho\,. \tag{63}$$

By using the current (62), the deformed metric and NS-NS two-form are obtained as

$$\begin{aligned}ds^2\ &=\ -\cosh^2\rho\,d\tau^2+d\rho^2+\sinh^2\rho\left(d\theta^2+\frac{\cos^2\theta\,d\phi_1^2+\sin^2\theta\,d\phi_2^2}{1+\eta^2\sin^2\theta\cos^2\theta\sinh^4\rho}\right),\\B_2\ &=\ -\frac{\eta\sin^2\theta\cos^2\theta\sinh^4\rho}{1+\eta^2\sin^2\theta\cos^2\theta\sinh^4\rho}d\phi_1\wedge d\phi_2\,.\end{aligned} \tag{64}$$

This result precisely agrees with the ones of abelian twists of the global AdS$_5$ [35]. Incidentally, three-parameter cases have been discussed in [24].

6.2. Lax pair

The next is to determine the associated Lax pair. By using the deformed current (62), the Lax pair can be explicitly evaluated. The resulting expression is given by [24, 30]

$$\begin{aligned}
\mathcal{L}_{\pm}^{\rm AT} &= -\partial_{\pm}\rho\frac{\lambda^{\pm 1}}{2}\gamma_1^a - \partial_{\pm}\theta\left[\frac{\lambda^{\pm 1}}{2}\sinh\rho\,\gamma_3^a + \cosh\rho\, n_{13}^a\right] \\
&\quad + i\partial_{\pm}\tau\left[\frac{\lambda^{\pm 1}}{2}\cosh\rho\,\gamma_5^a + \sinh\rho\, n_{15}^a\right] \\
&\quad - \hat{G}\left(\partial_{\pm}\phi_1 \mp \eta\sin^2\theta\sinh^2\rho\partial_{\pm}\phi_2\right) \\
&\qquad\qquad \times\left[\cos\theta\left(\frac{\lambda^{\pm 1}}{2}\sinh\rho\,\gamma_2^a + \cosh\rho\, n_{12}^a\right) + \sin\theta\, n_{23}^a\right] \\
&\quad + i\hat{G}\left(\partial_{\pm}\phi_2 \pm \eta\cos^2\theta\sinh^2\rho\partial_{\pm}\phi_1\right) \\
&\qquad\qquad \times\left[\sin\theta\left(\frac{\lambda^{\pm 1}}{2}\sinh\rho\,\gamma_0^a - \cosh\rho\, n_{01}^a\right) - \cos\theta\, n_{03}^a\right]. \qquad (65)
\end{aligned}$$

The existence of the Lax pair was anticipated in [35], but the explicit form has been derived in [24, 30]. In the $\eta \to 0$ limit, the above Lax pair is reduced to the following form:

$$\begin{aligned}
\mathcal{L}_{\pm}^{\rm GAdS_5} &= -\partial_{\pm}\rho\frac{\lambda^{\pm 1}}{2}\gamma_1^a - \partial_{\pm}\theta\left[\frac{\lambda^{\pm 1}}{2}\sinh\rho\,\gamma_3^a + \cosh\rho\, n_{13}^a\right] \\
&\quad + i\partial_{\pm}\tau\left[\frac{\lambda^{\pm 1}}{2}\cosh\rho\,\gamma_5^a + \sinh\rho\, n_{15}^a\right] \\
&\quad - \partial_{\pm}\phi_1\left[\cos\theta\left(\frac{\lambda^{\pm 1}}{2}\sinh\rho\,\gamma_2^a + \cosh\rho\, n_{12}^a\right) + \sin\theta\, n_{23}^a\right] \\
&\quad + i\partial_{\pm}\phi_2\left[\sin\theta\left(\frac{\lambda^{\pm 1}}{2}\sinh\rho\,\gamma_0^a - \cosh\rho\, n_{01}^a\right) - \cos\theta\, n_{03}^a\right]. \qquad (66)
\end{aligned}$$

This is nothing but a Lax pair for the global AdS_5. It may be possible to consider the three-parameter generalization, up to a subtlety of the signature for the time direction.

6.3. Twisted boundary condition

As in the previous three examples, the deformations can be reinterpreted as twisted boundary conditions again. After performing a similar analysis, one can see that the deformed backgrounds with the periodic boundary condition is equivalent to the undeformed theory with the following twisted boundary conditions:

$$\begin{aligned}
\tilde{\phi}_1(\sigma = 2\pi) &= \tilde{\phi}_1(\sigma = 0) + \frac{\eta}{\sqrt{\lambda_{\rm c}}}J_2 + 2\pi n_1\,, \\
\tilde{\phi}_2(\sigma = 2\pi) &= \tilde{\phi}_2(\sigma = 0) - \frac{\eta}{\sqrt{\lambda_{\rm c}}}J_1 + 2\pi n_2\,. \qquad (67)
\end{aligned}$$

Here J_i are the Noether charges for rotation invariance in the ϕ_i directions. The integers n_i are winding numbers along the ϕ_i directions.

Journal of Physics: Conference Series **670** (2016) 012033 doi:10.1088/1742-6596/670/1/012033

7. Summary and Outlooks

In this article, we have given a short summary of Yang-Baxter deformations of the $AdS_5 \times S^5$ superstring by focusing upon four type IIB supergravity backgrounds, 1) gravity duals for noncommutative gauge theories, 2) γ-deformations of S^5, 3) Schrödinger spacetimes and 4) abelian twists of the global AdS_5 . All of them are associated with abelian classical r-matrices.

For all of the examples presented here, it has been shown that the deformed backgrounds with the periodic boundary condition are equivalent with the undeformed one with twisted boundary conditions. This result has been anticipated by the preceding works [32, 60], because these backgrounds can also be realized as TsT transformations of the $AdS_5 \times S^5$. Thus it seems likely that the abelian classical r-matrices can be seen as abelian twists (equivalently TsT transformations). In fact, there are general arguments supporting this statement [29, 61]. This would be the case even in non-integrable cases like $T^{1,1}$ [52].

However, in non-abelian cases, it would not be the case. Then non-local gauge transformations should enter into the argument and one needs to take account of the global structure like an integration constant matrix as argued in the case of 3D Schrödinger spacetime [12]. In particular, periodic boundary conditions are not suitable for non-local gauge fields and further the undeformed geometry cannot be reproduced even after undoing the twist for the affine extended algebra. The resulting geometry is described in terms of the dipole-like coordinates [12].

More interesting observation is that the gravity/CYBE correspondence may contain S-duality as argued in [25]. It is quite natural that the integrability survives T-dualities, but it is not certain for S-dualities. It would be very nice to reveal a relation between the integrability and S-dualities from the viewpoint of Yang-Baxter deformations.

Thus there are a lot of open problems in the case of non-abelian classical r-matrices. As a matter of course, it is of significance to study Jordanian deformations of the $AdS_5 \times S^5$ superstring. However, as a simpler exercise, it may be interesting to consider non-abelian classical r-matrices in the context of Yang-Baxter-deformations of 4D Minkowski spacetime [38]. An intriguing example is the κ-deformations of the Poincaré algebra. For the progress along this line, see the upcoming work [62].

So far, we have concentrated on the metric and NS-NS two-form. One of the most important issues to be confirmed is the Ramond-Ramond (R-R) sector and the dilaton. These can be investigated by including the fermionic sector and then performing supercoset construction. The κ-symmetry is preserved even after the deformations have been performed. Hence, one may expect that the R-R sector and the dilaton would be reproduced as well. However, it is still necessary to be confirmed directly[5]. A nice candidate is the Schrödinger spacetimes, where the dilaton is constant and the R-R sector is the same as the undeformed $AdS_5 \times S^5$ background.

Still, one need to make much effort towards establishing the gravity/CYBE correspondence.

Acknowledgments

We are very grateful to Io Kawaguchi for collaboration at an early stage of this work. We also appreciate Ben Hoare, Takashi Kameyama, Hideki Kyono, Jerzy Lukierski, Jun-ichi Sakamoto and Stijn J. van Tongeren for useful discussions. The work of K.Y. is supported by Supporting Program for Interaction-based Initiative Team Studies (SPIRITS) from Kyoto University and by the JSPS Grant-in-Aid for Scientific Research (C) No.15K05051. This work is also supported in part by the JSPS Japan-Russia Research Cooperative Program and the JSPS Japan-Hungary Research Cooperative Program.

[5] At the present moment, there is a puzzle in the case of the q-deformation [19, 20].

Appendix A. Notation and convention

We summarize here our notation and convention of the $\mathfrak{su}(2,2)$ and $\mathfrak{su}(4)$ generators.

The gamma matrices

Let us first introduce the following gamma matrices:

$$\gamma_1 = \begin{pmatrix} 0 & 0 & 0 & -1 \\ 0 & 0 & 1 & 0 \\ 0 & 1 & 0 & 0 \\ -1 & 0 & 0 & 0 \end{pmatrix}, \quad \gamma_2 = \begin{pmatrix} 0 & 0 & 0 & i \\ 0 & 0 & i & 0 \\ 0 & -i & 0 & 0 \\ -i & 0 & 0 & 0 \end{pmatrix}, \quad \gamma_3 = \begin{pmatrix} 0 & 0 & 1 & 0 \\ 0 & 0 & 0 & 1 \\ 1 & 0 & 0 & 0 \\ 0 & 1 & 0 & 0 \end{pmatrix},$$

$$\gamma_0 = i\gamma_4 = \begin{pmatrix} 0 & 0 & 1 & 0 \\ 0 & 0 & 0 & -1 \\ -1 & 0 & 0 & 0 \\ 0 & 1 & 0 & 0 \end{pmatrix}, \quad \gamma_5 = i\gamma_1\gamma_2\gamma_3\gamma_0 = \begin{pmatrix} 1 & 0 & 0 & 0 \\ 0 & 1 & 0 & 0 \\ 0 & 0 & -1 & 0 \\ 0 & 0 & 0 & -1 \end{pmatrix}. \qquad (A.1)$$

To embed $\mathfrak{su}(2,2)$ and $\mathfrak{su}(4)$ into $\mathfrak{su}(2,2|4)$, we follow an 8×8 matrix representation as

$$\begin{aligned} \gamma^a_\mu &= \begin{pmatrix} \gamma_\mu & 0 \\ 0 & 0 \end{pmatrix}, \quad \gamma^a_5 = \begin{pmatrix} \gamma_5 & 0 \\ 0 & 0 \end{pmatrix} \qquad \text{with} \qquad \mu = 0,1,2,3\,, \\ \gamma^s_i &= \begin{pmatrix} 0 & 0 \\ 0 & \gamma_i \end{pmatrix}, \quad \gamma^s_5 = \begin{pmatrix} 0 & 0 \\ 0 & \gamma_5 \end{pmatrix} \qquad \text{with} \qquad i = 1,2,3,4\,. \end{aligned} \qquad (A.2)$$

Note that each block of the matrices is a 4×4 matrix.

The $\mathfrak{su}(2,2)$ and $\mathfrak{su}(4)$ generators

The Lie algebras $\mathfrak{su}(2,2) \sim \mathfrak{so}(2,4)$ and $\mathfrak{su}(4) \sim \mathfrak{so}(6)$ are spanned as follows:

$$\begin{aligned} \mathfrak{su}(2,2) &= \text{span}_{\mathbb{R}}\left\{ \frac{1}{2}\gamma^a_\mu\,, \frac{1}{2}\gamma^a_5\,, n^a_{\mu\nu} = \frac{1}{4}[\gamma^a_\mu\,, \gamma^a_\nu]\,, n^a_{\mu 5} = \frac{1}{4}[\gamma^a_\mu\,, \gamma^a_5] \mid \mu\,,\nu = 0,1,2,3 \right\}, \\ \mathfrak{su}(4) &= \text{span}_{\mathbb{R}}\left\{ \frac{1}{2}\gamma^s_i\,, \frac{1}{2}\gamma^s_5\,, n^s_{ij} = \frac{1}{4}[\gamma^s_i\,, \gamma^s_j]\,, n^s_{i5} = \frac{1}{4}[\gamma^s_i\,, \gamma^s_5] \mid i,j = 1,2,3,4 \right\}. \end{aligned} \qquad (A.3)$$

The subalgebras $\mathfrak{so}(1,4)$ and $\mathfrak{so}(5)$ in the spinor representation are formed as

$$\begin{aligned} \mathfrak{so}(1,4) &= \text{span}_{\mathbb{R}}\{ n^a_{\mu\nu}\,, n^a_{\mu 5} \mid \mu\,,\nu = 0,1,2,3 \}, \\ \mathfrak{so}(5) &= \text{span}_{\mathbb{R}}\{ n^s_{ij}\,, n^s_{i5} \mid i,j = 1,2,3,4 \}. \end{aligned} \qquad (A.4)$$

For a coset construction of Poincaré AdS_5, it is useful to employ the following basis:

$$\mathfrak{su}(2,2) = \text{span}_{\mathbb{R}}\{ p_\mu\,, k_\mu\,, h_1\,, h_2\,, h_3\,, n^a_{13}\,, n^a_{10}\,, n^a_{23}\,, n^a_{20} \mid \mu = 0,1,2,3 \}. \qquad (A.5)$$

Here the generators p_μ, k_μ and the Cartan generators $h_1\,, h_2\,, h_3$ are defined as

$$\begin{aligned} p_\mu &\equiv \frac{1}{2}\gamma^a_\mu - n^a_{\mu 5}\,, \qquad k_\mu \equiv \frac{1}{2}\gamma^a_\mu + n^a_{\mu 5}\,, \\ h_1 &\equiv 2i\,n^a_{12} = \text{diag}(-1,1,-1,1,0,0,0,0)\,, \\ h_2 &\equiv 2n^a_{30} = \text{diag}(-1,1,1,-1,0,0,0,0)\,, \end{aligned}$$

$$h_3 \equiv \gamma_5^a = \text{diag}(1,1,-1,-1,0,0,0,0)\,.$$

Note that the generators p_μ and k_μ commute each other,

$$[p_\mu\,, p_\nu] = [k_\mu\,, k_\nu] = [p_\mu\,, k_\nu] = 0 \qquad \text{for} \qquad \mu\,, \nu = 0,1,2,3\,. \tag{A.6}$$

For the S^5 part, the Cartan generators $h_4\,, h_5\,, h_6$ of $\mathfrak{su}(4)$ are given by

$$\begin{aligned} h_4 &\equiv 2i\, n_{12}^s = \text{diag}(0,0,0,0,-1,1,-1,1)\,, \\ h_5 &\equiv 2i\, n_{34}^s = \text{diag}(0,0,0,0,-1,1,1,-1)\,, \\ h_6 &\equiv \gamma_5^s = \text{diag}(0,0,0,0,1,1,-1,-1)\,. \end{aligned} \tag{A.7}$$

Since non-Cartan generators of $\mathfrak{su}(4)$ are not used in our analysis here, we will not write them down explicitly.

The bosonic coset projectors

In deriving the bosonic part of Lax pairs, it is necessary to employ the coset projectors P_0 and P_2 regarding the $\mathbb{Z}_2$-grading property. The projectors P_0 and P_2 are decomposed into the AdS_5 part and the S^5 part like

$$P_0(x) = P_0^a(x) + P_0^s(x)\,, \qquad P_2(x) = P_2^a(x) + P_2^s(x)\,, \tag{A.8}$$

where $P_0^{a,s}$ and $P_2^{a,s}$ are the following coset projectors for $\mathfrak{so}(2,4)$ and $\mathfrak{su}(4)$,

$$\begin{aligned} &P_0^a : \quad \mathfrak{su}(2,2) \longrightarrow \mathfrak{so}(1,4)\,, \qquad P_2^a : \quad \mathfrak{su}(2,2) \longrightarrow \frac{\mathfrak{su}(2,2)}{\mathfrak{so}(1,4)}\,, \\ &P_0^s : \quad \mathfrak{su}(4) \longrightarrow \mathfrak{so}(5)\,, \qquad P_2^s : \quad \mathfrak{su}(4) \longrightarrow \frac{\mathfrak{su}(4)}{\mathfrak{so}(5)}\,. \end{aligned} \tag{A.9}$$

These coset projectors can be represented by the $\mathfrak{su}(2,2)$ and $\mathfrak{su}(4)$ generators as follows:

$$\begin{aligned} P_0^a(x) &= \frac{1}{2}\sum_{\mu,\nu=0}^{3} \frac{\text{Tr}[n_{\mu\nu}^a x]}{\text{Tr}[n_{\mu\nu}^a n_{\mu\nu}^a]} n_{\mu\nu}^a + \sum_{\mu=0}^{3} \frac{\text{Tr}[n_{\mu 5}^a x]}{\text{Tr}[n_{\mu 5}^a n_{\mu 5}^a]} n_{\mu 5}^a\,, \\ P_2^a(x) &= \sum_{\mu=0}^{3} \frac{\text{Tr}[\gamma_\mu^a x]}{\text{Tr}[\gamma_\mu^a \gamma_\mu^a]}\gamma_\mu^a + \frac{\text{Tr}[\gamma_5^a x]}{\text{Tr}[\gamma_5^a \gamma_5^a]}\gamma_5^a\,, \\ P_0^s(x) &= \frac{1}{2}\sum_{\mu,\nu=1}^{4} \frac{\text{Tr}[n_{\mu\nu}^s x]}{\text{Tr}[n_{\mu\nu}^s n_{\mu\nu}]} n_{\mu\nu}^s + \sum_{\mu=1}^{4} \frac{\text{Tr}[n_{\mu 5}^s x]}{\text{Tr}[n_{\mu 5}^s n_{\mu 5}^s]} n_{\mu 5}^s\,, \\ P_2^s(x) &= \sum_{\mu=1}^{4} \frac{\text{Tr}[\gamma_\mu^s x]}{\text{Tr}[\gamma_\mu^s \gamma_\mu^s]}\gamma_\mu^s + \frac{\text{Tr}[\gamma_5^s x]}{\text{Tr}[\gamma_5^s \gamma_5^s]}\gamma_5^s\,. \end{aligned} \tag{A.10}$$

The projectors are utilized in evaluating the deformed metric, NS-NS two-form and Lax pair.

References

[1] J. M. Maldacena, "The large N limit of superconformal field theories and supergravity," Int. J. Theor. Phys. **38** (1999) 1113 [Adv. Theor. Math. Phys. **2** (1998) 231] [hep-th/9711200].

[2] N. Beisert *et al.*, "Review of AdS/CFT Integrability: An Overview," Lett. Math. Phys. **99** (2012) 3 [arXiv:1012.3982 [hep-th]].

[3] R. R. Metsaev and A. A. Tseytlin, "Type IIB superstring action in $AdS_5 \times S^5$ background," Nucl. Phys. B **533** (1998) 109 [hep-th/9805028].

[4] I. Bena, J. Polchinski and R. Roiban, "Hidden symmetries of the $AdS_5 \times S^5$ superstring," Phys. Rev. D **69** (2004) 046002 [hep-th/0305116].

[5] G. Arutyunov and S. Frolov, "Foundations of the $AdS_5 \times S^5$ Superstring. Part I," J. Phys. A **42** (2009) 254003 [arXiv:0901.4937 [hep-th]].

[6] M. Heinze, "Spectrum and Quantum Symmetries of the $AdS_5 \times S^5$ Superstring," arXiv:1507.03005 [hep-th].

[7] C. Klimcik, "Yang-Baxter sigma models and dS/AdS T duality," JHEP **0212** (2002) 051 [hep-th/0210095]; "On integrability of the Yang-Baxter sigma-model," J. Math. Phys. **50** (2009) 043508 [arXiv:0802.3518 [hep-th]].

[8] F. Delduc, M. Magro and B. Vicedo, "On classical q-deformations of integrable σ-models," JHEP **1311** (2013) 192 [arXiv:1308.3581 [hep-th]].

[9] I. Kawaguchi and K. Yoshida, "Hybrid classical integrability in squashed sigma models," Phys. Lett. B **705** (2011) 251 [arXiv:1107.3662 [hep-th]]; "Hybrid classical integrable structure of squashed sigma models: A short summary," J. Phys. Conf. Ser. **343** (2012) 012055 [arXiv:1110.6748 [hep-th]]; "Hidden Yangian symmetry in sigma model on squashed sphere," JHEP **1011** (2010) 032. [arXiv:1008.0776 [hep-th]].

[10] I. Kawaguchi, T. Matsumoto and K. Yoshida, "The classical origin of quantum affine algebra in squashed sigma models," JHEP **1204** (2012) 115 [arXiv:1201.3058 [hep-th]]; "On the classical equivalence of monodromy matrices in squashed sigma model," JHEP **1206** (2012) 082 [arXiv:1203.3400 [hep-th]].

[11] I. Kawaguchi, D. Orlando and K. Yoshida, "Yangian symmetry in deformed WZNW models on squashed spheres," Phys. Lett. B **701** (2011) 475. [arXiv:1104.0738 [hep-th]]; I. Kawaguchi and K. Yoshida, "A deformation of quantum affine algebra in squashed WZNW models," J. Math. Phys. **55** (2014) 062302 [arXiv:1311.4696 [hep-th]].

[12] I. Kawaguchi and K. Yoshida, "Classical integrability of Schrodinger sigma models and q-deformed Poincare symmetry," JHEP **1111** (2011) 094 [arXiv:1109.0872 [hep-th]]; "Exotic symmetry and monodromy equivalence in Schrodinger sigma models," JHEP **1302** (2013) 024 [arXiv:1209.4147 [hep-th]]; I. Kawaguchi, T. Matsumoto and K. Yoshida, "Schroedinger sigma models and Jordanian twists," JHEP **1308** (2013) 013 [arXiv:1305.6556 [hep-th]].

[13] T. Kameyama and K. Yoshida, "String theories on warped AdS backgrounds and integrable deformations of spin chains," JHEP **1305** (2013) 146 [arXiv:1304.1286 [hep-th]]; "Anisotropic Landau-Lifshitz sigma models from q-deformed $AdS_5 \times S^5$ superstrings," JHEP **1408** (2014) 110 [arXiv:1405.4467 [hep-th]].

[14] F. Delduc, M. Magro and B. Vicedo, "An integrable deformation of the $AdS_5 \times S^5$ superstring action," Phys. Rev. Lett. **112** (2014) 051601 [arXiv:1309.5850 [hep-th]]; F. Delduc, M. Magro and B. Vicedo, "Derivation of the action and symmetries of the q-deformed $AdS_5 \times S^5$ superstring," JHEP **1410**, 132 (2014) [arXiv:1406.6286 [hep-th]].

[15] V. G. Drinfel'd, "Hopf algebras and the quantum Yang-Baxter equation," Sov. Math. Dokl. **32** (1985) 254; "Quantum groups," J. Sov. Math. **41** (1988) 898 [Zap. Nauchn. Semin. **155**, 18 (1986)]; M. Jimbo, "A q difference analog of $U(g)$ and the Yang-Baxter equation," Lett. Math. Phys. **10** (1985) 63.

[16] G. Arutyunov, R. Borsato and S. Frolov, "S-matrix for strings on η-deformed $AdS_5 \times S^5$," JHEP **1404** (2014) 002 [arXiv:1312.3542 [hep-th]].

[17] B. Hoare, R. Roiban and A. A. Tseytlin, "On deformations of $AdS_n \times S^n$ supercosets," JHEP **1406** (2014) 002 [arXiv:1403.5517 [hep-th]]; B. Hoare and A. A. Tseytlin, "On integrable deformations of superstring sigma models related to $AdS_n \times S^n$ supercosets," Nucl. Phys. B **897** (2015) 448 [arXiv:1504.07213 [hep-th]].

[18] O. Lunin, R. Roiban and A. A. Tseytlin, "Supergravity backgrounds for deformations of $AdS_n \times S^n$ supercoset string models," Nucl. Phys. B **891** (2015) 106 [arXiv:1411.1066 [hep-th]].

[19] G. Arutyunov, R. Borsato and S. Frolov, "Puzzles of eta-deformed $AdS_5 \times S^5$," arXiv:1507.04239 [hep-th].

[20] B. Hoare and A. A. Tseytlin, "Type IIB supergravity solution for the T-dual of the eta-deformed $AdS_5 \times S^5$ superstring," arXiv:1508.01150 [hep-th].

[21] I. Kawaguchi, T. Matsumoto and K. Yoshida, "Jordanian deformations of the $AdS_5 \times S^5$ superstring," JHEP **1404** (2014) 153 [arXiv:1401.4855 [hep-th]];

[22] I. Kawaguchi, T. Matsumoto and K. Yoshida, "A Jordanian deformation of AdS space in type IIB supergravity," JHEP **1406** (2014) 146 [arXiv:1402.6147 [hep-th]].

[23] T. Matsumoto and K. Yoshida, "Lunin-Maldacena backgrounds from the classical Yang-Baxter equation – Towards the gravity/CYBE correspondence," JHEP **1406** (2014) 135 [arXiv:1404.1838 [hep-th]].

[24] T. Matsumoto and K. Yoshida, "Integrability of classical strings dual for noncommutative gauge theories," JHEP **1406** (2014) 163 [arXiv:1404.3657 [hep-th]].
[25] T. Matsumoto and K. Yoshida, "Yang-Baxter deformations and string dualities," JHEP **1503** (2015) 137 [arXiv:1412.3658 [hep-th]].
[26] T. Matsumoto and K. Yoshida, "Yang-Baxter sigma models based on the CYBE," Nucl. Phys. B **893** (2015) 287 [arXiv:1501.03665 [hep-th]].
[27] T. Matsumoto and K. Yoshida, "Schrödinger geometries arising from Yang-Baxter deformations," JHEP **1504** (2015) 180 [arXiv:1502.00740 [hep-th]].
[28] S. J. van Tongeren, "On classical Yang-Baxter based deformations of the $AdS_5 \times S^5$ superstring," JHEP **1506** (2015) 048 [arXiv:1504.05516 [hep-th]].
[29] S. J. van Tongeren, "Yang-Baxter deformations, AdS/CFT, and twist-noncommutative gauge theory," arXiv:1506.01023 [hep-th].
[30] T. Kameyama, H. Kyono, J. i. Sakamoto and K. Yoshida, "Lax pairs on Yang-Baxter deformed backgrounds," arXiv:1509.00173 [hep-th].
[31] O. Lunin and J. M. Maldacena, "Deforming field theories with $U(1) \times U(1)$ global symmetry and their gravity duals," JHEP **0505** (2005) 033 [hep-th/0502086].
[32] S. Frolov, "Lax pair for strings in Lunin-Maldacena background," JHEP **0505** (2005) 069 [hep-th/0503201].
[33] A. Hashimoto and N. Itzhaki, "Noncommutative Yang-Mills and the AdS / CFT correspondence," Phys. Lett. B **465** (1999) 142 [hep-th/9907166].
J. M. Maldacena and J. G. Russo, "Large N limit of noncommutative gauge theories," JHEP **9909** (1999) 025 [hep-th/9908134].
[34] C. P. Herzog, M. Rangamani and S. F. Ross, "Heating up Galilean holography," JHEP **0811** (2008) 080 [arXiv:0807.1099 [hep-th]]; J. Maldacena, D. Martelli and Y. Tachikawa, "Comments on string theory backgrounds with non-relativistic conformal symmetry," JHEP **0810** (2008) 072 [arXiv:0807.1100 [hep-th]]; A. Adams, K. Balasubramanian and J. McGreevy, "Hot Spacetimes for Cold Atoms," JHEP **0811** (2008) 059 [arXiv:0807.1111 [hep-th]].
[35] D. Dhokarh, S. S. Haque and A. Hashimoto, "Melvin Twists of global $AdS_5 \times S^5$ and their Non-Commutative Field Theory Dual," JHEP **0808** (2008) 084 [arXiv:0801.3812 [hep-th]]; T. McLoughlin and I. Swanson, "Integrable twists in AdS/CFT," JHEP **0608** (2006) 084 [hep-th/0605018].
[36] T. Matsumoto and K. Yoshida, "Integrable deformations of the $AdS_5 \times S^5$ superstring and the classical Yang-Baxter equation – Towards the gravity/CYBE correspondence –," J. Phys. Conf. Ser. **563** (2014) 1, 012020 [arXiv:1410.0575 [hep-th]].
[37] H. Lin, O. Lunin and J. M. Maldacena, "Bubbling AdS space and 1/2 BPS geometries," JHEP **0410** (2004) 025 [hep-th/0409174].
[38] T. Matsumoto, D. Orlando, S. Reffert, J. Sakamoto and K. Yoshida, "Yang-Baxter deformations of Minkowski spacetime," arXiv:1505.04553 [hep-th].
[39] M. A. Melvin, "Pure Magnetic And Electric Geons," Phys. Lett. **8** (1964) 65.
[40] G. W. Gibbons and K. i. Maeda, "Black Holes and Membranes in Higher Dimensional Theories with Dilaton Fields," Nucl. Phys. B **298** (1988) 741.
[41] A. A. Tseytlin, "Melvin solution in string theory," Phys. Lett. B **346** (1995) 55 [hep-th/9411198].
[42] A. Hashimoto and K. Thomas, "Dualities, twists, and gauge theories with non-constant non-commutativity," JHEP **0501** (2005) 033 [hep-th/0410123].
[43] C. R. Nappi and E. Witten, "A WZW model based on a nonsemisimple group," Phys. Rev. Lett. **71** (1993) 3751 [hep-th/9310112].
[44] A. Hashimoto and S. Sethi, "Holography and string dynamics in time dependent backgrounds," Phys. Rev. Lett. **89** (2002) 261601 [hep-th/0208126].
[45] M. Spradlin, T. Takayanagi and A. Volovich, "String theory in beta deformed spacetimes," JHEP **0511** (2005) 039 [hep-th/0509036].
[46] J. Lukierski, H. Ruegg, A. Nowicki and V. N. Tolstoy, "Q deformation of Poincare algebra," Phys. Lett. B **264** (1991) 331.
J. Lukierski and A. Nowicki, "Quantum deformations of D = 4 Poincare and Weyl algebra from Q deformed D = 4 conformal algebra," Phys. Lett. B **279** (1992) 299.
[47] A. Ballesteros, F. J. Herranz and N. R. Bruno, "Quantum (anti)de Sitter algebras and generalizations of the kappa-Minkowski space," hep-th/0409295.
[48] I. R. Klebanov and E. Witten, "Superconformal field theory on three-branes at a Calabi-Yau singularity," Nucl. Phys. B **536** (1998) 199 [hep-th/9807080].
[49] P. Basu and L. A. Pando Zayas, "Chaos Rules out Integrability of Strings in $AdS_5 \times T^{1,1}$," Phys. Lett. B **700** (2011) 243 [arXiv:1103.4107 [hep-th]]; "Analytic Non-integrability in String Theory," Phys. Rev. D **84** (2011) 046006 [arXiv:1105.2540 [hep-th]].

[50] Y. Asano, D. Kawai, H. Kyono and K. Yoshida, "Chaotic strings in a near Penrose limit of $AdS_5 \times T^{1,1}$," JHEP **1508** (2015) 060 [arXiv:1505.07583 [hep-th]].
[51] A. Catal-Ozer, "Lunin-Maldacena deformations with three parameters," JHEP **0602** (2006) 026 [hep-th/0512290].
[52] P. M. Crichigno, T. Matsumoto and K. Yoshida, "Deformations of $T^{1,1}$ as Yang-Baxter sigma models," JHEP **1412** (2014) 085 [arXiv:1406.2249 [hep-th]].
[53] R. G. Leigh and M. J. Strassler, "Exactly marginal operators and duality in four-dimensional N=1 supersymmetric gauge theory," Nucl. Phys. B **447** (1995) 95 [arXiv:hep-th/9503121].
[54] S. Schafer-Nameki, M. Yamazaki and K. Yoshida, "Coset Construction for Duals of Non-relativistic CFTs," JHEP **0905** (2009) 038 [arXiv:0903.4245 [hep-th]].
[55] D. Israel, C. Kounnas, D. Orlando and P. M. Petropoulos, "Electric/magnetic deformations of S^3 and AdS_3, and geometric cosets," Fortsch. Phys. **53** (2005) 73 [hep-th/0405213].
[56] D. T. Son, "Toward an AdS/cold atoms correspondence: A Geometric realization of the Schrödinger symmetry," Phys. Rev. D **78** (2008) 046003 [arXiv:0804.3972 [hep-th]];
K. Balasubramanian and J. McGreevy, "Gravity duals for non-relativistic CFTs," Phys. Rev. Lett. **101** (2008) 061601 [arXiv:0804.4053 [hep-th]].
[57] S. A. Hartnoll and K. Yoshida, "Families of IIB duals for nonrelativistic CFTs," JHEP **0812** (2008) 071 [arXiv:0810.0298 [hep-th]].
[58] N. Bobev and A. Kundu, "Deformations of Holographic Duals to Non-Relativistic CFTs," JHEP **0907** (2009) 098 [arXiv:0904.2873 [hep-th]].
[59] N. Bobev, A. Kundu and K. Pilch, "Supersymmetric IIB Solutions with Schrödinger Symmetry," JHEP **0907** (2009) 107 [arXiv:0905.0673 [hep-th]].
[60] L. F. Alday, G. Arutyunov and S. Frolov, "Green-Schwarz strings in TsT-transformed backgrounds," JHEP **0606** (2006) 018 [hep-th/0512253].
[61] B. Vicedo, "Deformed integrable σ-models, classical R-matrices and classical exchange algebra on Drinfel'd doubles," J. Phys. A **48** (2015) 35, 355203 [arXiv:1504.06303 [hep-th]].
[62] A. Borowiec, H. Kyono, J. Likierski, J. Sakamoto and K. Yoshida, "Yang-Baxter sigma models and Lax pairs arising from κ-Poincaré r-matrices," to appear.

Journal of Physics: Conference Series **670** (2016) 012034 doi:10.1088/1742-6596/670/1/012034

A holographic realization of light dilatons

Eugenio Megías

Max-Planck-Institut für Physik (Werner-Heisenberg-Institut), Föhringer Ring 6, D-80805, Munich, Germany

E-mail: emegias@mppmu.mpg.de

Abstract. We study a realization of light dilatons by considering Conformal Field Theories (CFT) deformed by a nearly marginal operator $\mathcal{O}$. This is discussed in the holographic setup consisting of Renormalization Group (RG) flows that end on a soft wall. We obtain a mass formula for the dilaton as an average along the RG flow. Finally we discuss the holographic method to compute the condensate $\langle \mathcal{O} \rangle$.

1. Introduction

Spontaneous breaking of conformal invariance (SBCI) is nowadays a scenario that is extensively studied in the context of the most fundamental problems in physics. On the one hand, this symmetry breaking may give rise to a Goldstone boson, the *dilaton*, which might have important consequences in cosmology and particle physics. In particular, some interesting beyond the Standard Model (SM) realizations assume that the SM can be part of a nearly-conformal sector. The electroweak scale would arise from the SBCI, and a naturally light dilaton would appear in the spectrum [1, 2, 3]. On the other hand, the effective theory for the dilaton presents a toy version of the Cosmological Constant (CC) problem [4]. Both applications are connected with the concept of naturalness of physical theories.

Some significant progress on these issues have been achieved recently thanks to the observation by Contino, Pomarol and Rattazzi (CPR) [5, 6, 7] that a light dilaton should occur naturally in certain deformations of CFTs. While the original holographic implementation of the CPR proposal is dual to a CFT that contains at least two scalar operators, in this manuscript we try to test it in slightly simpler and perhaps more realistic models, involving a single nearly marginal operator $\mathcal{O}$. Under a relevant deformation one generally expects that the operator develops a condensate $\langle \mathcal{O} \rangle$, and the fluctuation of $\langle \mathcal{O} \rangle$ can already play the role of a dilaton. This will be tested using the techniques developed in the context of holographic RG flows [8, 9, 10, 11, 12, 13] and soft-wall models [14, 15].

Let us include a brief note on related literature. The interest for the dilaton as the pseudo-Goldstone boson of SBCI has a long story [16]. More recently there has been a renewed interest mostly triggered by LHC phenomenology, see e.g. [17, 18]. The identification of a dilaton in holographic RG flows goes back to [19], though some of the models involve what in the present manuscript we classify as tuned SBCI [20]. There is also a considerable amount of literature on string-theoretic embeddings of walking dynamics and light dilatons, see e.g. [21] and references therein. However, in these scenarios it is hard to disentangle whether the dilaton lightness stems from SBCI or from SUSY. In the present work, we consider non-supersymmetric flows.

2. Spontaneous breaking of conformal invariance and dilatons

Physics at low energies can be captured by effective field theory methods. We will address in this section some basic features of the effective theory for a dilaton, and discuss the mechanism to get a naturally light dilaton recently proposed in [5].

2.1. Spontaneous breaking of conformal invariance

A scale transformation is defined as $x^\mu \to \frac{1}{b}x^\mu$, where b is the scaling factor. The starting point to compute the effective theory of a dilaton is to construct the most general Lagrangian invariant under this transformation. The scale transformation acts like $\chi \to b^{\frac{d-2}{2}}\chi$, in $d > 2$ space-time dimensions, where χ is the canonically normalized dilaton field. Therefore, the effective Lagrangian compatible with this symmetry allows a potential of the form

$$V_{\rm dil}^{\rm CFT}(\chi) = U_0\, \chi^{\frac{2d}{d-2}} \,. \tag{1}$$

The value of the "quartic" coupling U_0, which is the analogue of the CC, is arbitrary. SBCI means the appearance of a nonzero value for the condensate $\langle\chi\rangle \neq 0$. For $U_0 > 0$ the minimum of the potential is at $\langle\chi\rangle = 0$, while for $U_0 < 0$ there is no minimum and the theory is unstable. So, SBCI can only happen in a CFT if $U_0 = 0$, and then the dilaton is massless and it is identified as the corresponding Goldstone boson. However, this can only occur if either i) the theory is fine-tuned, or ii) the theory enjoys additional symmetries (SUSY) that enforce $U_0 = 0$. This picture is illustrated in Fig. 1. As we mention in the Introduction, in the following we will consider non-supersymmetric theories.

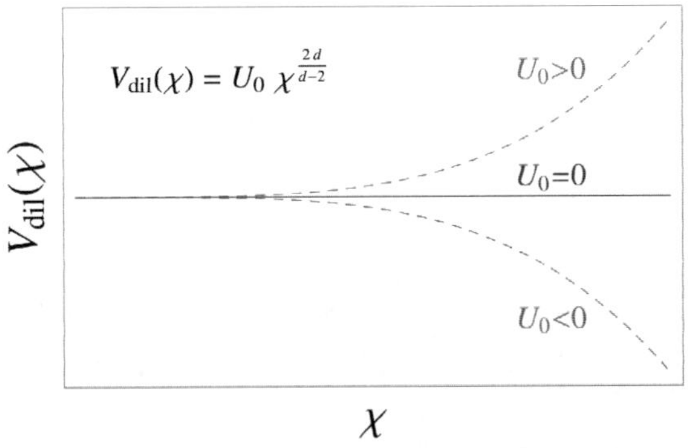

Figure 1. Dilaton potential in a CFT for different values of the quartic coupling U_0. SBCI is only possible when $U_0 = 0$. These curves illustrate the results for a potential of the form Eq. (1).

2.2. Naturally light dilatons

After the discussion above, a logical question arises: can a naturally light dilaton appear in Quantum Field Theories that are close to conformal invariance (CI)? We will study in the following the simplest situation, consisting of a CFT deformed by a single trace operator $\mathcal{O}$, i.e.

$$\mathcal{L} = \mathcal{L}^{\rm CFT} + \delta\mathcal{L} \,, \qquad \text{with} \qquad \delta\mathcal{L} = -\lambda\,\mathcal{O} \,. \tag{2}$$

The deformation $\delta\mathcal{L}$ introduces a small explicit breaking of CI if the operator is relevant and nearly marginal, i.e.

$$\text{Dim}(\mathcal{O}) = d - \Delta \qquad \text{with} \qquad \Delta \ll 1 \,. \tag{3}$$

The dimension of the Lagrangian and the coupling constant are $\text{Dim}(\mathcal{L}) = d$ and $\text{Dim}(\lambda) = \Delta$, respectively. This is the starting point of the CPR mechanism proposed in Refs. [5, 6, 7].

Let us study now whether SBCI can occur in addition to the explicit breaking of CI. RG computations capture the χ dependence of the potential, as the dilaton shifts like a scale transformation, so that the effective potential is obtained by evaluating the running coupling at the scale $\mu = \chi^{\frac{2}{d-2}}$. This generally takes the form

$$V_{\rm dil}^{\rm CFT+def}(\chi) = \chi^{\frac{2d}{d-2}} U(\lambda(\mu))\Big|_{\mu=\chi^{\frac{2}{d-2}}}, \qquad (4)$$

where $\lambda(\mu)$ is the running coupling constant. The appearance of SBCI amounts to having a nontrivial minimum of the potential. For later convenience, we will refer to this minimum as the infrared (IR) scale $\chi_{\rm IR}$. Since $V_{\rm dil}^{\rm CFT+def\prime}(\chi_{\rm IR}) = \frac{2}{d-2}\chi_{\rm IR}^{\frac{d+2}{d-2}}(dU(\lambda) + \beta U'(\lambda))$, where we have defined the beta function as $\beta = \frac{d\lambda}{d\log\mu}$, it follows that the dilaton mass is small if β and U are small at the minimum. Indeed

$$V_{\rm dil}(\chi_{\rm IR}) = U_{\rm IR}\, \chi_{\rm IR}^{\frac{2d}{d-2}} \sim \beta_{\rm IR}\, \chi_{\rm IR}^{\frac{2d}{d-2}} \qquad \text{and} \qquad V_{\rm dil}''(\chi_{\rm IR}) = m_{\rm dil}^2 \sim \beta_{\rm IR}\, \chi_{\rm IR}^{\frac{4}{d-2}}, \qquad (5)$$

are generically suppressed by one power of $\beta_{\rm IR}$. This picture is illustrated in Fig. 2. In the rest

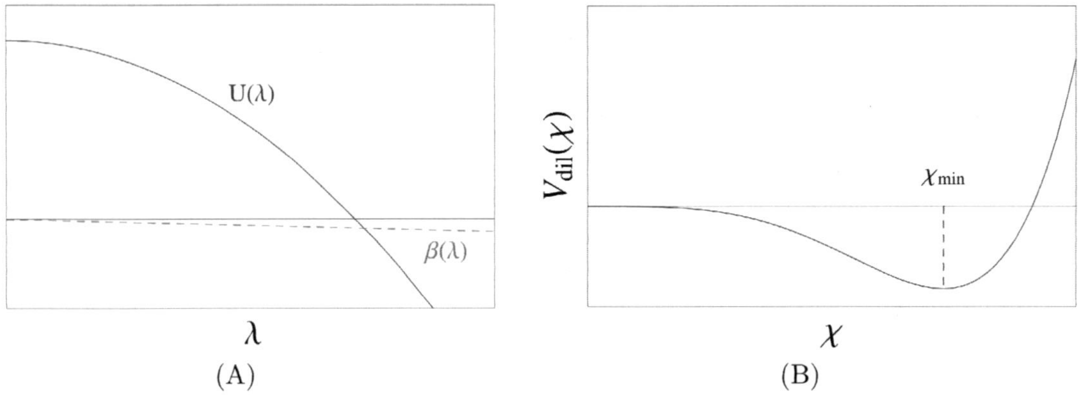

Figure 2. (A) Quartic coupling and beta function as a function of the coupling constant. (B) Dilaton potential as a function of the dilaton field. These curves illustrate the results for a potential of the form Eq. (4).

of this manuscript we will present an explicit holographic realization of the CPR proposal.

3. Holographic description of CFT deformations

In this section we review the most important ingredients for the holographic description of CFT deformations. In particular, we stress the importance of the holographic beta function to describe how confinement and condensation of operators arise in the holographic RG flows.

3.1. Holographic model

Let us consider a dilaton-gravity model in (d+1) dim given by

$$S = \frac{M^{d-1}}{2}\int d^{d+1}x\sqrt{-g}\left\{R - (\partial\phi)^2 - 2V(\phi)\right\}. \qquad (6)$$

Holographically, a CFT deformed by a scalar operator $\mathcal{O}$ as presented in Eq. (2), is realized by a domain wall geometry in which the scalar field ϕ develops a profile in the extra dimension y,

$$ds^2 = dy^2 + a^2(y)dx^\mu dx^\nu\eta_{\mu\nu}, \qquad \phi = \phi(y). \qquad (7)$$

The equations of motion for this model can be written as a system of first order equations

$$\frac{d\phi}{dy} = -W'(\phi)\,, \qquad \frac{1}{a}\frac{da}{dy} = \frac{W(\phi)}{d-1}\,, \tag{8}$$

where the superpotential W obeys the equation $W'(\phi)^2 = 2V(\phi) + \frac{d}{d-1}W^2(\phi)$. In a near boundary expansion the scalar field contains two modes,

$$\phi(y) = \lambda\, e^{-\Delta y} + \langle \mathcal{O}\rangle\, e^{-(d-\Delta)y} + \cdots\,, \qquad y \to +\infty\,. \tag{9}$$

Following the holographic dictionary, the leading mode is identified with the deformation parameter λ, while the sub-leading one corresponds to the condensate $\langle \mathcal{O}\rangle$. These two quantities are eventually related to the two integration constants needed to solve the equations of motion.

3.2. Holographic beta function

The warp factor a plays the role holographically of the renormalization scale μ. In addition, for a nearly marginal deformation the scalar field close to the boundary behaves as $\phi(y) = \lambda\, e^{-\Delta y} + \cdots$ with $\Delta \ll 1$, so that it can be approximately identified as the coupling strength of the deformation. Then one immediately identifies the holographic version of the beta function [9, 10] as

$$\beta(\phi) := a\frac{\partial\phi}{\partial a} = -(d-1)\frac{W'(\phi)}{W(\phi)}\,. \tag{10}$$

In the second equality we have made use of the equations of motion Eq. (8). This definition implies the existence of a 1 to 1 relation between the beta function and the superpotential. It follows also that β obeys the following integral equation

$$\left(\beta^2 - d(d-1)\right)\frac{W^2}{2(d-1)^2} = V\,. \tag{11}$$

Since $V(\phi) < 0$, this implies that β is bounded everywhere along the flow, i.e. $|\beta| < \sqrt{d(d-1)}$. Finally, by using Eq. (11) and its derivative, one can obtain that $\beta(\phi)$ obeys the first order differential equation

$$\beta\,\beta' = \left(\beta^2 - d(d-1)\right)\left(\frac{\beta}{d-1} + \frac{1}{2}\frac{V'}{V}\right)\,. \tag{12}$$

We can easily derive from Eq. (12) some interesting properties of the beta function. In particular: i) existence of an integration constant, whose physical meaning will be explained in Sec. 3.3; ii) fixed points $\beta = 0$ map to extrema of V, i.e. $V' = 0$; iii) existence of three IR attractors:

$$\beta'(\phi) = 0 \quad \Longrightarrow \quad \begin{cases} \beta = \beta_b \equiv \pm\sqrt{d(d-1)} \\ \beta \simeq \beta_g \equiv -\frac{(d-1)}{2}\frac{V'}{V} \end{cases}\,. \tag{13}$$

In the second line of this equation we have assumed that V'/V is smooth. The attractors $\beta = \beta_b$ and $\beta = \beta_g$ give rise to singularities of Gubser's bad and good type respectively [14], provided that V'/V is not too large. The bad (good) attractor is attractive towards large (small) ϕ. The behavior of $\beta(\phi)$ and the location of the attractors are displayed in Fig. 3. Note that $\beta \simeq$ cte implies $V(\phi) \sim e^{\nu\phi}$, so that it is natural to consider an exponential behavior for the scalar potential in the IR.

Journal of Physics: Conference Series **670** (2016) 012034 doi:10.1088/1742-6596/670/1/012034

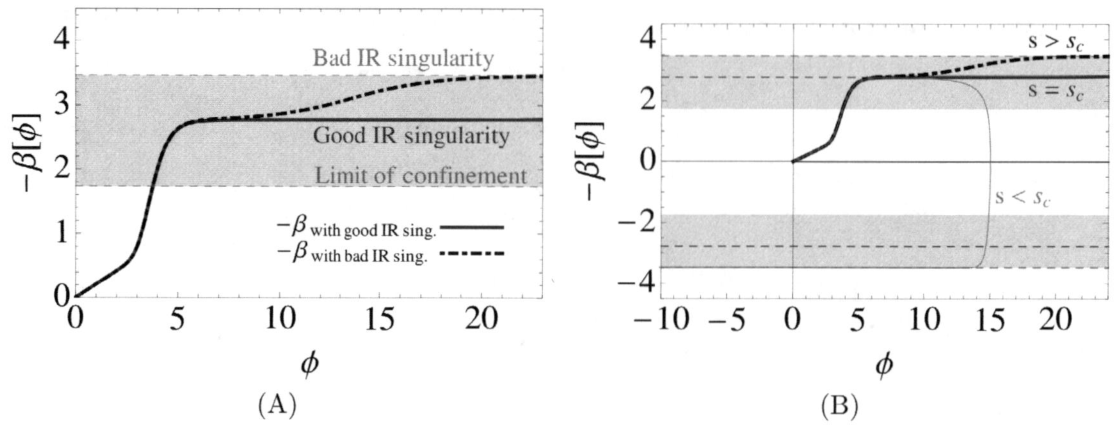

Figure 3. (A) Behavior of $-\beta$ as a function of ϕ. It is shown two different flows approaching the IR attractors of Eq. (13). The shaded band corresponds to the bounds of Eq. (15). (B) RG flow for several values of the integration constant of Eq. (19): $s > s_c$, $s = s_c$ and $s < s_c$. In these plots we have considered $\Delta_- = 0.21$.

3.3. CFTs with confining deformations

In the following we will concentrate on flows from close to a CFT in the ultraviolet (UV) down to a confining IR. Some of the main points presented here have been derived previously in e.g. [13, 22, 23, 24].

Confinement leads to the generation of a universal mass gap in the spectrum of excitations, and the dilaton is going to be easily identifiable as a low-lying mass-eigenvalue. A simple holographic criterion for confinement is that the conformal coordinate, defined as $dz = -dy/a$, has a finite range. Placing the AdS boundary at $z = 0$, the IR singularity is at $z = z_s$ with

$$z_s = \int_0^\infty \frac{d\phi}{a(\phi)W'(\phi)} < \infty\,, \tag{14}$$

and the mass gap is identified as $\Lambda_{\rm IR} = 1/z_s$. This leads to the following bounds for the beta function in the IR,

$$\beta(\phi) \to \beta_\infty \qquad \text{with} \qquad \sqrt{d-1} < -\beta_\infty < \sqrt{d(d-1)}\,. \tag{15}$$

The upper bound corresponds to limit of good IR singularity solutions, see Fig. 3(A). In the following we will restrict to scalar potentials that are analytic in ϕ and even under $\phi \to -\phi$, so that they admit the small ϕ expansion

$$V(\phi) = -\frac{d(d-1)}{2\ell^2} + \frac{M_\phi^2}{2}\phi^2 + \mathcal{O}(\phi^4)\,. \tag{16}$$

The solutions of Eq. (12) contain generically two contributions that we call perturbative and non-perturbative parts,

$$\beta(\phi) = \beta_{\rm P}(\phi) + \beta_{\rm NP}(\phi)\,. \tag{17}$$

$\beta_{\rm P}$ is analytic as it is completely fixed by V. In particular at first order $\beta_{\rm P} = -\Delta\phi + \ldots$, and from Eq. (16) one gets that Δ is set to either of $\Delta_\pm$ with

$$\Delta_\pm = \frac{d}{2} \pm \sqrt{\left(\frac{d}{2}\right)^2 + M_\phi^2\,\ell^2}\,. \tag{18}$$

Then one finds that there are two types of flows with behavior $\beta_\pm \simeq -\Delta_\pm \phi$. In the case of β_+ flows, the scalar field behaves as $\phi \simeq 0 + \langle \mathcal{O} \rangle\, e^{-\Delta_+ y}$. Then one concludes that they correspond to a CFT with SBCI but no explicit breaking. As discussed in Sec. 2, one expects that this situation is either fine-tuned or supersymmetric. On the other hand, the β_- flows lead to the general expansion of Eq. (9), and so they correspond to a genuine deformation of a CFT with explicit and SBCI.

$\beta_{\rm NP}$ is a non-analytical function of ϕ^2 that is controlled by an integration constant. When $\beta_{\rm NP}$ is a small correction to $\beta_{\rm P}$, one can express this solution as [13]

$$\beta_{\rm NP}(\phi) \propto s \cdot \exp\left(- \int_0^\phi d\bar{\phi}\, \frac{d + \beta_{\rm P}'(\bar{\phi})}{\beta_{\rm P}(\bar{\phi})} \right) \sim s \cdot \phi^{\frac{d}{\Delta_\pm} - 1} , \tag{19}$$

where s is the integration constant. For β_+ flows the only choice that does not break the AdS asymptotic behavior near the boundary is $s = 0$. The fine-tuned feature of this kind of flows is then evident from the particular value of the integration constant. On the other hand, β_- flows can have in principle any value of s, as $\phi^{\frac{d}{\Delta_-} - 1}$ is sub-leading to ϕ.

In order to discuss the physical meaning of the integration constant s, let us compute the near boundary expansion of the scalar field for the flow $\beta_-(\phi) = -\Delta_- \phi + \ldots - s\, \phi^{\frac{d}{\Delta_-} - 1} + \ldots$. From Eqs. (8) and (10) one gets

$$\phi(y) = \lambda\, e^{-\Delta_- y} (1 + \ldots) + s\, \frac{\lambda^{\frac{\Delta_+}{\Delta_-}}}{\Delta_+ - \Delta_-}\, e^{-\Delta_+ y}\, (1 + \ldots) . \tag{20}$$

By comparison with Eq. (9) we conclude that s is none other than the value of the condensate, i.e. $s \sim \langle \mathcal{O} \rangle$. In order to determine $\langle \mathcal{O} \rangle$, the main idea is that s should be adjusted in the UV so that the IR part of the flow is as regular as possible. This selects uniquely a critical value for the integration constant $s = s_c$ which maps to the physical value of the condensate, and this corresponds to a flow falling in the good IR singularity of Eq. (13), i.e. $\beta \to -\frac{(d-1)}{2} \frac{V'}{V}$. In effect, as it is shown in Fig. 3(B), if one over-shoots (under-shoots) from the UV by taking $s > s_c$ ($s < s_c$), then the solution falls on the bad attractors $\beta \to \mp\sqrt{d(d-1)}$. Then we conclude that there are always explicit and SBCI flows with $\langle \mathcal{O} \rangle \neq 0$. The procedure to find s_c numerically can be made systematic and efficient by shooting instead from the IR, since then the bad attractor is in fact repulsive.

Excitations around the regular flow determines the dilaton mass. We will shortly obtain that it is positive in a large class of models, i.e. $m^2_{\rm dil} > 0$, thus fulfilling the *Minimum Energy Theorems* of scalar-gravity systems dual to a CFT with a scalar operator [25, 26, 27]. Then, the flow $s_c \sim \langle \mathcal{O} \rangle_{\rm phys}$ corresponds to the ground state of the deformed CFT.

3.4. Models

Up to now we have presented all the background material on holographic RG flows. We are now ready to specify two concrete holographic models for CFTs with confining deformations to illustrate the physics of light dilatons. Model A corresponds to a non-zero value of the condensate $\langle \mathcal{O} \rangle \neq 0$, and it is defined in terms of the scalar potential as

$$\frac{V'(\phi)}{V(\phi)} = \begin{cases} \frac{2\Delta_-\Delta_+}{d(d-1)}\phi & \text{for} \quad \phi < \phi_{conf} \\ 2\nu\sqrt{\frac{d}{d-1}} & \text{for} \quad \phi > \phi_{conf} \end{cases} . \tag{21}$$

The behavior of the beta function is then obtained by solving Eq. (12), and the integration constant s is fixed as explained in Sec. 3.3. The resulting RG flow is displayed in Fig. 4(A).

In addition to the confinement scale a_{conf}, we observe the appearance of a condensation scale a_{cond}. The regime $\phi_{cond} < \phi < \phi_{conf}$ is a condensate-dominated region in which $\beta(\phi)$ is well approximated by a condensate type flow, i.e. $\beta_+ \simeq -\Delta_+(\phi - \phi_{cond})$.

It is also possible to design a model with no condensate, i.e. $\langle \mathcal{O} \rangle = 0$, just by providing an analytical expression for $\beta(\phi)$, so that $\beta_{\rm NP}(\phi) = 0$ in Eq. (17). We will refer it as model B, and it is defined as

$$\beta(\phi) = \beta_{\rm P}(\phi) = \begin{cases} -\Delta_- \phi & \text{for} \quad \phi < \phi_{conf} \\ -\nu\sqrt{d(d-1)} & \text{for} \quad \phi > \phi_{conf} \end{cases} . \tag{22}$$

The oscillating behavior of $\frac{V'}{V}$ shown in Fig. 4(B) is a manifestation that the potential needs to be carefully adjusted to avoid the formation of a condensate. As we will explain in Sec. 4 both models lead to a light dilaton, and the main difference is the fine-tuning character of model B which is manifest in the precise value of the condensate.

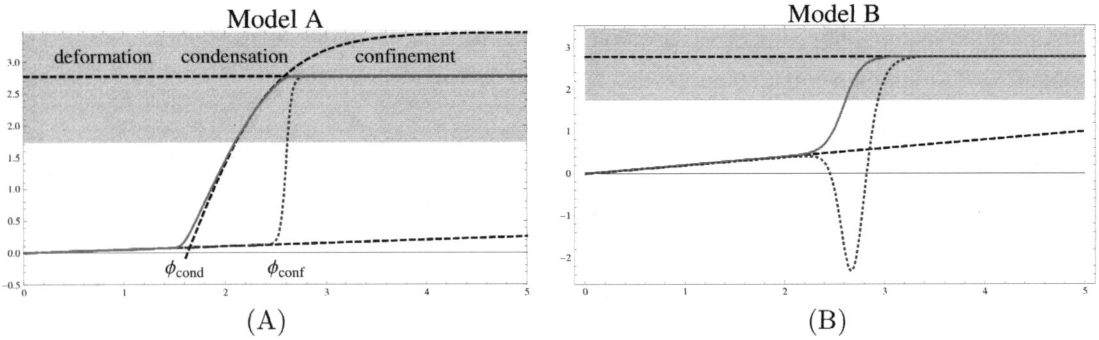

Figure 4. (A) $\frac{(d-1)}{2}\frac{V'}{V}$ (blue dotted) and $-\beta(\phi)$ (continuous red) as a function of ϕ for the model A defined in Eq. (21). The three regimes (deformation, condensation and confinement) are clearly distinguishable in $\beta(\phi)$. (B) The same as (A) for model B defined in Eq. (22).

4. Spectra and mass formula: the dilaton

In order to compute the scalar spectrum and in particular its lightest mode, we have to consider a scalar perturbation of the background flow solution. The most general one is given by allowing

$$\phi = \phi_0(y) + \delta\phi \qquad \text{and} \qquad ds^2 = N^2 dy^2 + g_{\mu\nu}(dx^\mu + N^\mu dy)(dx^\nu + N^\nu dy) \,, \tag{23}$$

with $N = 1 + \delta N$, $N_\mu = \partial_\mu \psi$, $g_{\mu\nu} = a^2(y)\, e^{2\xi}\, \eta_{\mu\nu}$. [1] In the unitary gauge, $\delta\phi = 0$, one finds the following reduced action for ξ

$$S_{red} = \frac{M^{d-1}}{2} \int d^d x dy\, a^d \beta^2 \left[\frac{(\partial_\mu \xi)^2}{a^2} - (\partial_y \xi)^2 \right] , \tag{24}$$

in which we have used the background equations of motion, and integrated by parts a number of times. Then the spectrum can be obtained from the 2nd order differential equation

$$\frac{1}{a^d \beta^2} \partial_y \left[a^d \beta^2 \partial_y \xi_n \right] + \frac{m_n^2 \xi_n}{a^2} = 0 \,. \tag{25}$$

[1] We refer the reader to Refs. [28, 2] for details on the computation.

To get this equation we have considered a Fourier decomposition in the boundary coordinates $\xi = \xi_n e^{-ik_\mu x^\mu}$, with the mode masses $k_\mu k^\mu = -m_n^2$. Since $\xi + \delta\phi/\beta$ is a gauge invariant variable, one can translate from the ξ variable into a $\delta\phi$ perturbation by adding a factor β , i.e. $\delta\phi = \beta\xi$. In the following we will work in the standard quantization, so that we will require that $\delta\phi$ contains only the sub-leading mode, i.e. $\delta\phi \propto e^{-\Delta_+ y}$. [2] This, in addition to the requirement of regularity in the IR, leads to a discrete spectrum m_n.

Firstly, we will make an analytical study of the fluctuations for massless and light dilatons, and then present the full numerical computation with Eq. (25).

4.1. Massless dilaton

By setting $m_n = 0$ in Eq. (25) one finds the following exact solution for the eigenvalue problem

$$\xi_{m^2=0} = c_1 + c_2 \int_0^z dz' \frac{1}{a^{d-1}\beta^2} \,. \tag{26}$$

The mode c_2 diverges at the IR singularity, so that $c_2 = 0$. Note however that the solution $\xi_{m^2=0} = c_1$ leads to $\delta\phi = c_1\beta \sim z^{\Delta_-}$, so that it does not satisfy the standard quantization prescription unless $c_1 = 0$. This is interpreted as the impossibility of getting a massless dilaton in CFT deformations, a feature already anticipated in Sec. 2.

4.2. Light dilaton

Let us now study under which circumstances a light dilaton can appear in the spectrum, which means the existence of a mode with mass $m_{\rm dil} \ll \Lambda_{\rm IR}$. This problem allows for an analytical treatment based on the method of matched asymptotic expansions [20, 30, 2]. Treating m^2 as a small parameter in Eq. (25), one can obtain the following UV and IR expansions of the solution

$$\xi^\alpha_{\rm dil} = \left(1 - m^2 \mathcal{I}_\alpha + \ldots\right) \left[c_1^\alpha + c_2^\alpha \int_{z_\alpha}^z dz' \frac{1}{a^{d-1}\,\beta^2}\right] , \qquad \text{for} \qquad \alpha = \mathrm{UV}, \mathrm{IR}\,, \tag{27}$$

where $z_{\rm UV} = 0$ and $z_{\rm IR} = z_s$. The integral operators act to the right, and they are defined as

$$\mathcal{I}_\alpha = \int_{z_\alpha}^z dz' \frac{1}{a^{d-1}\beta^2} \int_{z_\alpha}^{z'} dz'' a^{d-1}\beta^2 \,. \tag{28}$$

Regularity in the IR and standard quantization select $c_2^{\rm IR} = 0$ and $c_1^{\rm UV} = 0$ respectively. One expects the existence of an overlap region, $\frac{z_s}{2} \lesssim z \ll z_s$, where $\xi^{\rm UV}_{\rm dil}(z) \approx \xi^{\rm IR}_{\rm dil}(z)$. Then one can impose the matching conditions

$$\xi^{\rm UV}_{\rm dil}(z_{\rm M}) = \xi^{\rm IR}_{\rm dil}(z_{\rm M}) \qquad \text{and} \qquad \xi^{\rm UV\,\prime}_{\rm dil}(z_{\rm M}) = \xi^{\rm IR\,\prime}_{\rm dil}(z_{\rm M})\,, \tag{29}$$

at any point $z_{\rm M}$ in this region. Finally from Eq. (29) one can easily obtain the solution

$$\frac{c_2^{\rm UV}}{c_1^{\rm IR}} \simeq -m^2_{\rm dil} \int_0^{z_s} dz\, a^{d-1}\,\beta^2\,, \qquad \text{and} \qquad m^2_{\rm dil} \simeq \left[\int_0^{z_s} dz\, \frac{1}{a^{d-1}\beta^2} \int_z^{z_s} dz' a^{d-1}\beta^2\right]^{-1} . \tag{30}$$

In these expressions we have made the choice $z_{\rm M} = 0$, which in fact it has been found to be a good approximation, see Ref. [2]. The mass formula given by Eq. (30) constitutes one of the main results of this work. Note that in the absence of fixed points apart from the UV one, i.e. $\beta(\phi) \neq 0$, this formula predicts $m^2_{\rm dil} > 0$, leading to stability of the theory.

We show in Fig. 5 the wave function and the spectrum obtained with a numerical solution of the equation of motion for the fluctuation, Eq. (25). It is obvious from the figure that there

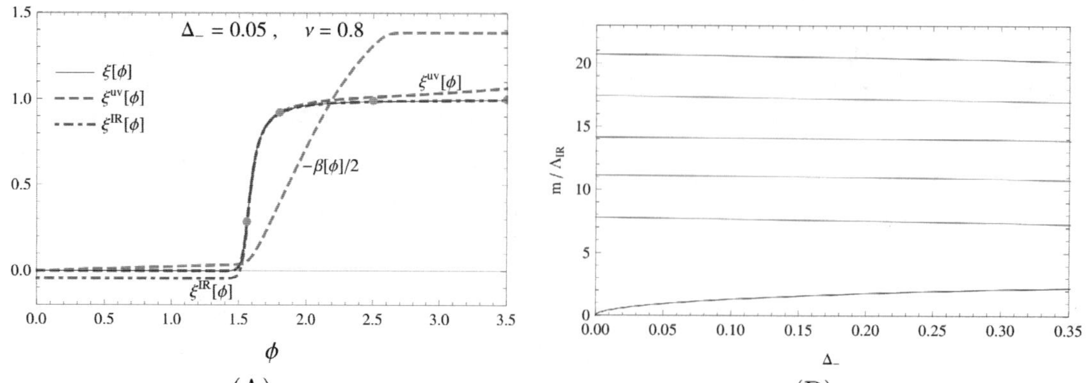

(A) (B)

Figure 5. (A) Dilaton wave function ξ as a function of ϕ, obtained by numerically solving Eq. (25) (continuous line) and the UV- and IR-approximations of Eq. (27) (dashed lines). It is also displayed the background beta function. (B) Spectrum as a function of Δ_- from a numerical computation of Eq. (25). The first 6 modes are depicted. We have used in both figures the model A defined in Eq. (21). Similar results are obtained for model B defined in Eq. (22), except for the dilaton mass scaling with Δ_-.

is a light state, which in fact is well described by Eq. (30). We find that the lightest mode – the dilaton – scales like $m^2_{\rm dil}/\Lambda^2_{\rm IR} \sim \Delta_-$ and $\sim \Delta_-^2$ for model A and B respectively, so that it becomes light for nearly marginal deformations, i.e. $\Delta_- \ll 1$.

It is possible to compute the integral of Eq. (30) analytically in a toy model which captures the main features of the flow. By extending the model B of Eq. (22) with a condensation region, i.e. $\beta(\phi) = -\widetilde{\Delta}_+(\phi - \phi_{cond}) - \Delta_-\phi_{cond}$ for $\phi_{cond} < \phi < \phi_{conf}$, one finds the following result for small Δ_- [2]

$$m^2_{\rm dil}/\Lambda^2_{\rm IR} \sim \beta_{cond}^{2-d/\widetilde{\Delta}_+} . \tag{31}$$

The dilaton mass is suppressed so long as $\beta_{cond} \ll 1$ and $\widetilde{\Delta}_+ > \frac{d}{2}$. In particular, this model reproduces the suppression of $m_{\rm dil}$ for model A when $\widetilde{\Delta}_+ = \Delta_+$, and model B when $\widetilde{\Delta}_+ \gg 1$. Eq. (31) illustrates the main result of this work, i.e. that the dilaton is light whenever i) the beta function is small in the IR (more precisely at the condensation scale) and ii) the rise in β towards confinement ($\beta_{conf} \sim 1$) is fast enough. The natural way that the beta function experiences a fast rise is by the condensation of $\langle \mathcal{O} \rangle$ itself.

5. Discussion and conclusions

In this work we have studied the realization of a light dilaton by considering CFTs deformed by a single nearly-marginal scalar operator $\mathcal{O}$. Such deformation induces generically a nonzero condensate $\langle \mathcal{O} \rangle$, which we have called models of type A. In addition, we have found a family of fine-tuned models, called models of type B, with a vanishing value for the condensate $\langle \mathcal{O} \rangle = 0$. The dilaton mode is identifiable with the fluctuation of the condensate, regardless of whether $\langle \mathcal{O} \rangle$ vanishes or not. In both cases we find that the dilaton mass is controlled by the value of the beta function in the IR as conjectured in Ref. [5], i.e. $m^2_{\rm dil} \sim \beta_{\rm IR}$. More precisely, the dilaton is light when $\beta_{\rm IR}$ is small and the rise towards confinement is fast enough.

It has been found a rather simple mass formula for the dilaton mass involving an integral over the flow, cf. Eq. (30), which can be used as a diagnostic for the presence of a light dilaton in the

[2] A study in other quantizations will be presented in a forthcoming publication [29].

spectrum. This formula is explicitly positive, and this can be seen as a perturbative stability result. This allows to identify the regular RG flow as the ground state of the CFT deformation.

Finally we have studied the method to compute the condensate $\langle \mathcal{O} \rangle$ resulting from the CFT deformation. Holographically, this is based on the solution of a first-order differential equation for the beta function, supplemented with a regularity condition in the IR. This uniquely determines the condensate once the holographic model is specified.

There remain some open questions, in particular i) the generalization of these results to other quantizations like alternative and double trace deformations, and ii) the study of stability and the possible existence of tachyons in more general backgrounds. There are in the literature some computations of the effective potential in related theories, see e.g. [31], and this can be used as well to shed some light on these and other issues. A study of these topics is currently in progress [29].

Concerning possible applications with a naturally light dilaton, one of the most obvious is adopting the light dilaton as the Higgs boson found at the LHC. Within the present mechanism this would allow to describe the light mass of the Higgs without fine-tunings. To test this possibility it should be studied the coupling of the dilaton with SM matter fields

$$\mathcal{L}_{\phi-T_{\mu\nu}} = \frac{1}{f_d}\phi T^{\mu}_{\mu} \,, \tag{32}$$

and compare the strength of the couplings with the ones predicted for the Higgs by the SM [32, 17]. It is important to stress that even if the dilaton were not a Higgs *imposter*, it could affect its phenomenology in such a way that it might be detected from deviations of the SM predictions at the LHC or future colliders. In any case, this could be the starting point for beyond the SM phenomenology: extra-dimensions and/or strongly coupled sector; and the dilaton would be naturally the lightest state of this sector of new physics. This study is currently in progress [33].

Other target is to study the dilaton as a (light) dark matter candidate. In this case it could happen that the coupling of the dilaton to matter fields [34] would require the assumption of additional symmetries in order to get stability at cosmological scales. Finally, it deserves to be investigated the CC problem in light with a mechanism similar to the one presented in this work, see e.g. [4].

Acknowledgments

This work is based on Ref. [2], co-authored with O. Pujolàs. We thank M. Baggioli, B. Bellazzini, A. Pineda, J. Serra, W. Skiba, M. Quirós and especially A. Pomarol for valuable discussions. We acknowledge support from the European Union under a Marie Curie Intra-European fellowship (FP7-PEOPLE-2013-IEF) with project number PIEF-GA-2013-623006.

References

[1] Coradeschi F, Lodone P, Pappadopulo D, Rattazzi R and Vitale L 2013 *JHEP* **11** 057 (*Preprint* 1306.4601)
[2] Megias E and Pujolas O 2014 *JHEP* **08** 081 (*Preprint* 1401.4998)
[3] Cox P and Gherghetta T 2015 *JHEP* **02** 006 (*Preprint* 1411.1732)
[4] Bellazzini B, Csaki C, Hubisz J, Serra J and Terning J 2014 *Eur. Phys. J.* **C74** 2790 (*Preprint* 1305.3919)
[5] Contino R, Pomarol A and Rattazzi R , *unpublished work (2010).*
[6] Rattazzi R , *The naturally light dilaton, talk given at Planck (2010). From the Planck Scale to the ElectroWeak Scale, CERN, Geneva, 31 May - 4 June 2010. [http://indico.cern.ch/event/75810/].*
[7] Pomarol A , *Elementary or Composite: The particle physics dilemma, talk at The XVI IFT UAM/CSIC Xmas Workshop, Madrid, Spain, 15 - 17 December 2010. [http://www2.ift.uam-csic.es/workshops/Xmas10/doc/pomarol.pdf].*
[8] DeWolfe O, Freedman D Z, Gubser S S and Karch A 2000 *Phys. Rev.* **D62** 046008 (*Preprint* hep-th/9909134)
[9] de Boer J, Verlinde E P and Verlinde H L 2000 *JHEP* **08** 003 (*Preprint* hep-th/9912012)
[10] Anselmi D, Girardello L, Porrati M and Zaffaroni A 2000 *Phys. Lett.* **B481** 346–352 (*Preprint* hep-th/0002066)

[11] Papadimitriou I and Skenderis K 2004 73–101 (*Preprint* hep-th/0404176)
[12] Papadimitriou I and Skenderis K 2004 *JHEP* **10** 075 (*Preprint* hep-th/0407071)
[13] Papadimitriou I 2007 *JHEP* **05** 075 (*Preprint* hep-th/0703152)
[14] Gubser S S 2000 *Adv. Theor. Math. Phys.* **4** 679–745 (*Preprint* hep-th/0002160)
[15] Karch A, Katz E, Son D T and Stephanov M A 2006 *Phys. Rev.* **D74** 015005 (*Preprint* hep-ph/0602229)
[16] Gildener E and Weinberg S 1976 *Phys. Rev.* **D13** 3333
[17] Goldberger W D, Grinstein B and Skiba W 2008 *Phys. Rev. Lett.* **100** 111802 (*Preprint* 0708.1463)
[18] Vecchi L 2010 *Phys. Rev.* **D82** 076009 (*Preprint* 1002.1721)
[19] Bianchi M, Freedman D Z and Skenderis K 2001 *JHEP* **08** 041 (*Preprint* hep-th/0105276)
[20] Hoyos C, Kol U, Sonnenschein J and Yankielowicz S 2013 *JHEP* **10** 181 (*Preprint* 1307.2572)
[21] Elander D, Faedo A F, Hoyos C, Mateos D and Piai M 2014 *JHEP* **05** 003 (*Preprint* 1312.7160)
[22] Galow B, Megias E, Nian J and Pirner H J 2010 *Nucl. Phys.* **B834** 330–362 (*Preprint* 0911.0627)
[23] Bourdier J and Kiritsis E 2014 *Class. Quant. Grav.* **31** 035011 (*Preprint* 1310.0858)
[24] Cabrer J A, von Gersdorff G and Quiros M 2010 *New J. Phys.* **12** 075012 (*Preprint* 0907.5361)
[25] Amsel A J and Marolf D 2006 *Phys. Rev.* **D74** 064006 [Erratum: Phys. Rev.D75,029901(2007)] (*Preprint* hep-th/0605101)
[26] Faulkner T, Horowitz G T and Roberts M M 2010 *Class. Quant. Grav.* **27** 205007 (*Preprint* 1006.2387)
[27] Elder B, Joyce A, Khoury J and Tolley A J 2015 *Phys. Rev.* **D91** 064002 (*Preprint* 1405.7696)
[28] Maldacena J M 2003 *JHEP* **05** 013 (*Preprint* astro-ph/0210603)
[29] Megias E, Pujolas O and Yang K , *work in progress (2015).*
[30] Bednik G, Pujolas O and Sibiryakov S 2013 *JHEP* **11** 064 (*Preprint* 1305.0011)
[31] Kiritsis E, Li W and Nitti F 2015 *JHEP* **04** 125 (*Preprint* 1410.1091)
[32] Csaki C, Hubisz J and Lee S J 2007 *Phys. Rev.* **D76** 125015 (*Preprint* 0705.3844)
[33] Megias E, Pujolas O and Quiros M , *work in progress (2015).*
[34] Cembranos J A R 2009 *Phys. Rev. Lett.* **102** 141301 (*Preprint* 0809.1653)

First results on the representation theory of the Ultrahyperbolic BMS group $\mathcal{UHB}(2,2)$

Evangelos Melas
University of Athens, Sofokleous 1, Athens, Greece
E-mail: emelas@econ.uoa.gr

Abstract. The Bondi–Metzner–Sachs (BMS) group B is the common asymptotic group of all asymptotically flat (lorentzian) space–times, and is the best candidate for the universal symmetry group of General Relativity (G.R.). B admits generalizations to real space–times of any signature, to complex space–times, and supersymmetric generalizations for any space–time dimension. With this motivation McCarthy constructed the strongly continuous unitary irreducible representations (IRs) of B some time ago, and he identified $B(2,2)$ as the generalization of B appropriate to the to the ultrahyperbolic signature (+,+,−,−) and asymptotic flatness in null directions. We continue this programme by introducing a new group $\mathcal{UHB}(2,2)$ in the group theoretical study of ultrahyperbolic G.R. which happens to be a proper subgroup of $B(2,2)$. We report on the first general results on the representation theory of $\mathcal{UHB}(2,2)$. In particular the main general results are that the all little groups of $\mathcal{UHB}(2,2)$ are compact and that the Wigner–Mackeys inducing construction is exhaustive despite the fact that $\mathcal{UHB}(2,2)$ is not locally compact in the employed Hilbert topology.

1. Introduction

In 1939 Wigner laid the foundations of special relativistic quantum mechanics and relativistic quantum field theory by constructing the Hilbert space strongly continuous unitary irreducible representations (IRs) of the (universal cover) of the Poincare group P. The Bondi–Metzner–Sachs (BMS) group B is the common asymptotic group of all curved real lorentzian space–times which are asymptotically flat in future null directions [1, 2], and is the best candidate for the universal symmetry group of G.R.. In a quantum setting the universal property of B for G.R. make it reasonable to attempt to lay a similarly firm foundation for quantum gravity by following through the analogue of Wigner's programme with B replacing P. Some years ago McCarthy constructed explicitly [3, 4, 5, 6, 7, 8, 9, 10] the IRs of B for exactly this purpose. This work was based on G.W.Mackey's pioneering work on group representations [11, 12, 13, 14, 15]; in particular an extension to the relevant infinite–dimensional case of his semi–direct product theory.

It is difficult to overemphasize the importance of Piard's results [16, 17] who soon afterwards proved that *all* the IRs of B, when this is equipped with the Hibert topology, are derivable by the inducing construction. This proves the exhaustivity of McCarthy's list of representations and renders his results even more important.

However, in quantum gravity, complexified or euclidean versions of G.R. are frequently considered and the question arises: Are there similar symmetry groups for these versions of the theory? McCarthy constructed [18], in abstract form, all possible analogues of B, both real and in any signature, or complex, with all possible notions of asymptotic flatness 'near infinity'. There are, in fact, forty−one such groups. These abstract constructions were given in a quantum setting; the paper was concerned with finding the IRs of the groups $\mathcal{G}$ in Hilbert spaces (especially for the complexification $\mathcal{CB}$ of B itself). It was argued that these Hilbert space representations were related to elementary particles and quantum gravity (via gravitational instantons).

Let $B(2,2)$ be the BMS group $\mathcal{G}$ appropriate to the 'ultrahyperbolic signature' and asymptotic flatness in null directions. $B(2,2)$, is like B itself, based on a null cone [18], and it is given by

$$B(2,2) = C_e^\infty(T^2, R) \circledS_T G^2 \tag{1}$$

,i.e., it is the semi−direct product of the group G^2, where $G = SL(2,R)$, times the abelian normal subgroup $C_e^\infty(T^2,R)$ of so called supertranslations; $C_e^\infty(T^2,R)$ is the set of *even* real−valued infinitely−differentiable functions defined on the 2−Torus $T^2 = S^1 \times S^1$, S^1 being the set of vectors of unit length in $R^2 - \{0\}$. That is the functions $\alpha(m,n) \in C_e^\infty(T^2,R)$ satisfy the even−ness condition

$$\alpha(m,n) = \alpha(-m,-n),$$

where, $m = \frac{x}{|x|}, x = (x_1, x_2) \in R^2 - \{0\}$, and, similarly, $n = \frac{y}{|y|}, y = (y_1, y_2) \in R^2 - \{0\}$. The representation theory of $B(2,2)$ has been initiated elsewhere [19, 20].

The present paper reports the first general results on the representation theory of

$$\mathcal{UHB}(2,2) = C^\infty(P_1(R) \times P_1(R), R) \circledS_T G^2, \tag{2}$$

$P_1(R) = S^1/Z_2$ being the one−dimensional real projective space (the circle quotient the antipodal map). $\mathcal{UHB}(2,2)$ arises naturally in the construction of the generalizations of B given in [18] but it remained unnoticed in [18]. The crucial difference between $\mathcal{UHB}(2,2)$ and $B(2,2)$ is that for $\mathcal{UHB}(2,2)$ the supertranslations are *c ple el unc n rained,* whereas, for $B(2,2)$ they are described by *even* functions on the torus T^2. The representation theory of $\mathcal{UHB}(2,2)$ has been initiated in [21, 22, 23].

2. The grou $\mathcal{UHB}(2,2)$

Recall that the ultrahyperbolic version of Minkowski space is the vector space R^4 of row vectors with 4 real components, with scalar product defined as follows. Let $x, y \in R^4$ have components x^μ and y^μ respectively, where $\mu = 0,1,2,3$. Define the scalar product $x.y$ between x and y by

$$x.y = x^0y^0 + x^2y^2 - x^1y^1 - x^3y^3. \tag{3}$$

Then the ultrahyperbolic version of Minkowski space, sometimes written $R^{2,2}$, is just R^4 with this scalar product.

In [21] it was shown that

Theorem 1 *e r up* $\mathcal{UHB}(2,2)$ *can be reali ed a*

$$\mathcal{UHB}(2,2) = L^2(\mathcal{P}, \lambda, R) \circledS_T G^2 \tag{4}$$

wi e i−direc pr duc pecified b

$$(T(g,h)\alpha)(x,y) = k_g(x)s_g(x)k_h(w)s_h(w)\alpha(xg, yh), \tag{5}$$

where $\alpha \in L^2(\mathcal{P},\lambda,R)$ and $(x,y) \in \mathcal{P}$. For ease of notation, we write $\mathcal{P}$ for the torus $\mathrm{T} \simeq P_1(R) \times P_1(R)$, $P_1(R)$ is the one−dimensional real projective space, and $\mathcal{G}$ for $G \times G$, $G = SL(2,R)$. In analogy to B, it is natural to choose a measure λ on $\mathcal{P}$ which is invariant under the maximal compact subgroup SO(2) × SO(2) of $\mathcal{G}$. $L^2(\mathcal{P}, \lambda, R)$ is the separable Hilbert space of real−valued functions defined on $\mathcal{P}$.

Moreover, if $g \in G$ is

$$\begin{bmatrix} a & b \\ c & d \end{bmatrix}, \tag{6}$$

then the components x_1, x_2 of $\mathbf{x} \in R^2$ transform linearly, so that the ratio $x = x_1/x_2$ transforms fraction linearly. Writing xg for the transformed ratio,

$$xg = \frac{(\mathbf{x}g)_1}{(\mathbf{x}g)_2} = \frac{x_1 a + x_2 c}{x_1 b + x_2 d} = \frac{xa + c}{xb + d}. \tag{7}$$

The factors $k_g(x)$ and $s_g(x)$ on the right hand side of (5) are defined by

$$k_g(x) = \left\{ \frac{(xb+d)^2 + (xa+c)^2}{1+x^2} \right\}^{\frac{1}{2}}, \tag{8}$$

$$s_g(x) = \frac{xb+d}{|xb+d|}, \tag{9}$$

with similar formulae for yh, $k_h(y)$ and $s_h(y)$.

It is well known that the topological dual of a Hilbert space can be identified with the Hilbert space itself, so that we have $L^{2'}(\mathcal{P}, \lambda, R) \simeq L^2(\mathcal{P}, \lambda, R)$. In fact, given a continuous linear functional $\phi \in L^{2'}(\mathcal{P}, \lambda, R)$, we can write, for $\alpha \in L^2(\mathcal{P}, \lambda, R)$

$$(\phi, \alpha) =< \phi, \alpha > \tag{10}$$

where the function $\phi \in L^2(\mathcal{P}, \lambda, R)$ on the right is uniquely determined by (and denoted by the same symbol as) the linear functional $\phi \in L^{2'}(\mathcal{P}, \lambda, R)$ on the left. The representation theory of $\mathcal{UHB}(2, 2)$ is governed by the dual action T' of $\mathcal{G}$ on the topological dual $L^{2'}(\mathcal{P}, \lambda, R)$ of $L^2(\mathcal{P}, \lambda, R)$. The dual action T' is defined by:

$$< T'(g,h)\phi, \alpha >=< \phi, T(g^{-1}, h^{-1})\alpha > \cdot \tag{11}$$

A short calculation gives

$$(T'(g,h)\phi)(x,y) = k_g^{-3}(x) s_g(x) k_h^{-3}(y) s_h(y) \phi(xg, yh). \tag{12}$$

Now, this action T' of $\mathcal{G}$ on $L^{2'}(\mathcal{P}, \lambda, R)$, given explicitly above, is like the action T of $\mathcal{G}$ on $L^2(\mathcal{P}, \lambda, R)$, continuous. The 'little group' L_ϕ of any $\phi \in L^{2'}(\mathcal{P}, \lambda, R)$ is the stabilizer

$$L_\phi = \{(g,h) \in \mathcal{G} \mid T'(g,h)\phi = \phi\}. \tag{13}$$

By continuity, $L_\phi \subset \mathcal{G}$ is a closed subgroup.

3. Re resentation theory

Let A and $\mathcal{G}$ be topological groups, and let T be a given homomorphism from $\mathcal{G}$ into the group of automorphisms Aut(A) of A. Suppose A is abelian and $\mathcal{H} = A \circledS_T \mathcal{G}$ is the semi−direct product of A and $\mathcal{G}$, specified by the continuous action $T : \mathcal{G} \longrightarrow$ Aut(A). In the product topology of $A \times \mathcal{G}$, $\mathcal{H}$ then becomes a topological group. It is assumed that it becomes a separable locally compact topological group.

In order to give the operators of the induced representations explicitly it is necessary ([11], [12], [13], [14], [15] and references therein) to give the following information

Journal of Physics: Conference Series **670** (2016) 012035 doi:10.1088/1742-6596/670/1/012035

(i) An irreducible unitary representation U of L_{ϕ_o} on a Hilbert space D for each L_{ϕ_o}.
(ii) A $\mathcal{G}$ −quasi−invariant measure μ on each orbit $\mathcal{G}\phi \approx \mathcal{G}/L_{\phi_o}$; where L_{ϕ_o} denotes the little group of the base point $\phi_o \in A^{'}$ of the orbit $\mathcal{G}\phi_o$; $A^{'}$ is the topological dual of A.

Let D_μ be the space of functions $\psi : \mathcal{G} \to D$ which satisfy the conditions

$$(a) \qquad \psi(gl) = U(l^{-1})\psi(g) \qquad (g \in \mathcal{G},\ l \in L_\phi)$$

$$(b) \quad \int_{\mathcal{G}\phi_o} < \psi(q), \psi(q) > \mathrm{d}\mu(q) < \infty,$$

where the scalar product under the integral sign is that of D. Note, that the constraint (a) implies that $< \psi(gl), \psi(gl) >=< \psi(g), \psi(g) >$, and therefore the inner product $< \psi(g), \psi(g) >$, $g \in \mathcal{G}$, is constant along every element q of the coset space $\mathcal{G}/L_{\phi_o} \approx \mathcal{G}\phi_o$. This allows to assign a meaning to $< \psi(q), \psi(q) >$, where $q = gL_{\phi_o}$, by defining $< \psi(q), \psi(q) >:=< \psi(g), \psi(g) >$. Thus the integrand in (b) becomes meaningful due to the condition (a). A pre−Hilbert space structure can now be given to D_μ by defining the scalar product

$$< \psi_1, \psi_2 >= \int_{\mathcal{G}\phi_o} < \psi_1(q), \psi_2(q) > \mathrm{d}\mu(q), \tag{14}$$

where $\psi_1, \psi_2 \in D_\mu$. It is convenient to complete the space D_μ with respect to the norm defined by the scalar product (14). In the resulting Hilbert space, functions are identified whenever they differ, at most, on a set of μ−measure zero. Thus our Hilbert space is

$$D_\mu = L^2(\mathcal{G}\phi_o, \mu, D). \tag{15}$$

Define now an action of $\mathcal{H} = A \circledS_T \mathcal{G}$ on D_μ by

$$(g_o\psi)(q) = \sqrt{\frac{\mathrm{d}\mu_{g_o}}{\mathrm{d}\mu}(q)}\psi(g_o^{-1}q), \tag{16}$$

$$\alpha\psi(q) = e^{i<g_o\phi_o,\alpha>}\psi(q) \tag{17}$$

where, $g_o \in \mathcal{G}$, $q \in \mathcal{G}\phi_o$, and $\alpha \in A$. Eqs. (16) and (17) define the IRs of $\mathcal{HB}$ induced for *eac* $\phi_o \in A^{'}$ and *eac* irreducible representation U of L_{ϕ_o}. The 'Jacobian' $\frac{\mathrm{d}\mu_{g_o}}{\mathrm{d}\mu}$ of the group transformation is known as the *Rad n−Nik d* derivative of μ_{g_o} with respect to μ and ensures that the resulting IRs of $\mathcal{HB}$ are unitary.
The central results of induced representation theory ([11], [12], [13], [14], [15] and references therein) are the following

(i) Given the topological restrictions on $\mathcal{H} = A \circledS_T \mathcal{G}$ (separability and local compactness), any representation of $\mathcal{H}$, constructed by the method above, is irreducible if the representation U of L_{ϕ_o} on D is irreducible. Thus an irreducible representation of $\mathcal{H}$ is obtained for *eac* $\phi_o \in A^{'}$ and *eac* irreducible representation U of L_{ϕ_o}.
(ii) If $\mathcal{H} = A \circledS_T \mathcal{G}$ is a regular semi−direct product (i.e., $A^{'}$ contains a Borel subset which meets each orbit in $A^{'}$ under $\mathcal{H}$ in just one point) then *all* of its irreducible representations can be obtained in this way.

4. Obstructions and resolutions

Two remarks are in order regarding the representations of $\mathcal{UHB}(2,2)$ obtained by the above construction

(i) As it is explained in [21] the subgroup $L^2(\mathcal{P},\lambda,R)$ of $\mathcal{UHB}(2,2) = L^2(\mathcal{P},\lambda,R) \circledS_T \mathcal{G}$ is topologised as a (pre) Hilbert space by using a natural measure on $\mathcal{P} = P_1(R) \times P_1(R)$ and by introducing a scalar product into $L^2(\mathcal{P},\lambda,R)$. If R^8 is endowed with the natural metric topology then the group $\mathcal{G} = SL(2,R) \times SL(2,R)$, considered as a subset of R^8, inherits the induced topology on $\mathcal{G}$. In the product topology of $L^2(\mathcal{P},\lambda,R) \times \mathcal{G}$ $\mathcal{UHB}(2,2)$ is a non−locally compact group (the proof follows without substantial change Cantoni's proof [24], see also [3]). (In fact the subgroup $L^2(\mathcal{P},\lambda,R)$, and therefore the group $\mathcal{UHB}(2,2)$ can be employed with many different topologies. The Hilbert type topology employed here appears to describe quantum mechanical systems in asymptotically flat space−times [9]). Since in the Hilbert type topology $\mathcal{UHB}(2,2) = L^2(\mathcal{P},\lambda,R) \circledS_T \mathcal{G}$ is not locally compact the theorems dealing with the irreducibility of the representations obtained by the above construction no longer apply (see e.g. [13]). However, it can be proved that the induced representations obtained above *are* irreducible. The proof follows very closely the one given in [6] for the case of the original BMS group B.

(ii) Here it is assumed that $\mathcal{UHB}(2,2)$ is equipped with the Hilbert topology. It is of outmost significance that it can be proved [21] that in this topology $\mathcal{UHB}(2,2)$ is a regular semi−direct−product. The proof follows the corresponding proof [16, 17] for the group B. Regularity amounts to the fact that [12] $L^{2'}(\mathcal{P},\lambda,R)$ can have no equivalent classes of quasi−invariant measures μ such that the action of $\mathcal{G}$ is strictly ergodic with respect to μ. When such measures μ do exist it can be proved [12] that an irreducible representation of the group, with the semi−direct−product structure at hand, may be associated with each that is not equivalent to any of the IRs constructed by the Wigner−Mackey's inducing method. In a different topology it is not known if $\mathcal{UHB}(2,2)$ is a regular or irregular semi−direct−product. Irregularity of $\mathcal{UHB}(2,2)$ in a topology different from the Hilbert topology would imply that there are IRs of $\mathcal{UHB}(2,2)$ that are not not equivalent to any of the IRs obtained above by the inducing construction. Strictly ergodic actions are notoriously hard to deal with even in the locally compact case. Indeed, for locally compact non−regular semi−direct products, there is no known example for which all inequivalent irreducibles arising from strictly ergodic actions have been found. For the other 41 groups defined in [18] regularity has only been proved for B [16, 17] when B is equipped with the Hilbert topology. Similar remarks apply to all of them regarding IRs arising from strictly ergodic actions in a given topology.

5. Results

The new results are: A new group $\mathcal{UHB}(2,2)$ is introduced for the group theoretical study of ultrahyperbolic G.R.. $\mathcal{UHB}(2,2)$ is a proper subgroup of the group $B(2,2)$ initially proposed in [18] as appropriate to the 'ultrahyperbolic signature' and asymptotic flatness in null directions. Both $\mathcal{UHB}(2,2)$ and $B(2,2)$ are based on the null cone $\mathcal{N}$ of $R^{2,2}$. The crucial difference between $\mathcal{UHB}(2,2)$ and $B(2,2)$ is that the supertranslations in the case of $\mathcal{UHB}(2,2)$ are *free* functions defined on $P_1(R) \times P_1(R)$, whereas in the case of $B(2,2)$ the supertranslations are *even* functions defined on $S^1 \times S^1$. Remarkably, it is the nature of the supertranslations of $\mathcal{UHB}(2,2)$ which allows to establish contact [21] of the representation theory of $\mathcal{UHB}(2,2)$ with standard representation theory; something which is not feasible for the representation theory of $B(2,2)$. $\mathcal{UHB}(2,2)$ captures more efficiently, via its subgroup $L^2(\mathcal{P},\lambda,R)$, the fundamental characteristic of $R^{2,2}$, namely, that there is no clear−cut distinction of past and future in $R^{2,2}$. In [21] it is proved that when $\mathcal{UHB}(2,2)$ is employed with the Hilbert topology *all* little groups of $\mathcal{UHB}(2,2)$ are compact. Moreover in [21] it is shown that the Wigner–Mackey's inducing construction is exhaustive despite the fact that $\mathcal{UHB}(2,2)$ is not locally compact in the employed Hilbert topology. This result is rather important because other group theoretical approaches to quantum gravity which invoke Wigner–Mackey's inducing construction (see e.g. [25]) are typically plagued by the non–exhaustiveness of the inducing construction which results precisely from the fact that the group in question is not locally compact in the prescribed topology. Exhaustiveness is not just a mathematical nicety: If the inducing construction is not exhaustive one cannot simply know if the most interesting information or part of it is coded in the irreducibles which cannot be found by the Wigner–Mackey's inducing procedure. These results, compactness of the little groups and exhaustiveness of the inducing construction, not only are they significant for the group theoretical approach to quantum gravity advocated here, but also they have repercussions [21] for the other approaches to quantum gravity.

In comparing the representation theory of $\mathcal{UHB}(2,2)$ [21, 22, 23] with the representation theory of $B(2,2)$ [19, 20], we find both similarities and differences. The key difference between $\mathcal{UHB}(2,2)$ and $B(2,2)$, the supertranslations in the case of $\mathcal{UHB}(2,2)$ are *free* functions defined on $P_1(R) \times P_1(R)$, whereas in the case of $B(2,2)$ the supertranslations are *even* functions defined on $S^1 \times S^1$, leaves its trace on the representation theory: the proof [21] of compactness for little groups of $\mathcal{UHB}(2,2)$ is similar to, but subtly different from, the corresponding proof [19] for $B(2,2)$. On the other hand, an interesting similarity between $\mathcal{UHB}(2,2)$ and $B(2,2)$ lies in the structure of their little groups: Their one–dimensional little groups form an unexpected family of continuous/discrete groups (with many connected components). Also, their finite little groups involve subgroups of direct products of the symmetry groups of the regular polygons only; the regular polyhedra do not appear at all here. The regular polyhedra appear [5] in the representation theory of the ordinary Bondi–Metzner–Sachs group.

Re erences

[1] Bondi H, Van Der Berg M G J and Metzner A W K 1962 *Proc. R. Soc. Lond. A* **269** 21
[2] Sachs R K 1962 *Proc. R. Soc. Lond. A* **270** 103
[3] McCarthy P J 1972 *Proc. R. Soc. Lond. A.* **330** 517
[4] McCarthy P J 1972 *J. Math. Phys.* **13** 1837
[5] McCarthy P J 1973 *Proc. R. Soc. Lond. A* **333** 317
[6] McCarthy P J and Crampin M 1973 *Proc. R. Soc. Lond. A.* **335** 301
[7] McCarthy P J 1975 *Proc. R. Soc. Lond. A* **343** 489
[8] McCarthy P J 1978 *Proc. R. Soc. Lond. A* **358** 141
[9] McCarthy P J and Crampin M 1974 *Phys. Rev. Lett.* **33** 547
[10] McCarthy P J 1972 *Phys. Rev. Lett.* **29** 817
[11] Wigner E 1939 *Annals of Mathematics* **40** 149
[12] Mackey G W 1968 *Induced representations of groups and quantum mechanics* Benjamin
[13] Mackey G W 1955 *The theory of group representations* The University of Chicago Press
[14] Simms D J 1968 *Lie groups and quantum mechanics* (Bonn notes) Berlin: Springer
[15] Isham C J 1984 Topological and quantum aspects of quantum theory *Relativity groups and topology* DeWitt B S and Stora R eds
[16] Piard A 1977 *Rep. Math. Phys.* **11** 259
[17] Piard A 1977 *Rep. Math. Phys.* **11** 279
[18] McCarthy P J 1992 *Soc. Lond. A.* **338** 271
[19] McCarthy P J and Melas E 2003 *Nucl. Phys. B.* **653** 369
[20] Melas E 2006 *J. Phys. A : Math. Gen.* **39** 3341
[21] Melas E Representations of the Ultrahyperbolic BMS group $\mathcal{UHB}(2,2)$. I. General Results (unpublished)
[22] Melas E Representations of the ultrahyperbolic BMS group $\mathcal{HB}$. II. Determination of the representations induced from infinite little groups *Preprint* ar iv:1312.0532v1
[23] Melas E Representations of the ultrahyperbolic BMS group $\mathcal{HB}$. II. Determination of the representations induced from finite little groups *Preprint* ar iv:1402.1428v1

[24] Cantoni V 1967 On the representations of the Bondi–Metzner–Sachs group Ph.D. Thesis, University of London
[25] Isham C J and Kakas A C 1984 *Class. Quantum Grav.* **1** 633

Journal of Physics: Conference Series **670** (2016) 012036 doi:10.1088/1742-6596/670/1/012036

Dispersionless 2D Toda hierarchy, Hurwitz numbers and Riemann theorem

Sergey M.Natanzon

National Research University Higher School of Economics, 20 Myasnitskaya street, Moscow 101000, Russia

Institute for Theoretical and Experimental Physics, 25 Bolshaya Cheremushkinskaya street, Moscow 117218, Russia

Laboratory of Quantum Topology, Chelyabinsk State University, 129 Brat'ev Kashirinykh street, Chelyabinsk 454001, Russia

E-mail: natanzons@mail.ru

Abstract. We describe all formal symmetric solutions of dispersionless 2D Toda hierarchy. This classification we use for solving of two classical problems: 1) The calculation of conformal mapping of an arbitrary simply connected domain to the standard disk; 2) Calculation of 2-Hurwitz numbers of genus 0.

Explain from the beginning of these 2 problems are reduced to integrable systems.

Calculation of conformal mapping

Let Q be a simply connected domain on the Riemann sphere without two points. According to Riemann theorem, there exists a biholomorphic function into a standard disk

$$w_Q(z) : Q \to \Lambda.$$

The Riemann theorem does not give algorithm for calculation of the function. But this function is very important for applications in gas dynamics, hydrodynamics, the oil industry ets.

After conformal transformation of the Riemann sphere, we may assume that

$$0 \notin Q, \Lambda \ni \infty.$$

Let $\mathcal{H}$ be the set of all such Q with smooth boundary $\gamma = \partial \mathrm{Q}$. The set of moments

$$t_0(Q) = \frac{1}{\pi} \iint\limits_{\overline{\mathbb{C}} \setminus Q} dx\, dy \quad t_k(Q) = -\frac{1}{\pi k} \iint\limits_{Q} z^{-k}\, dx\, dy\,, \quad k \geq 1$$

forms local coordinates $\mathbf{t}(Q) = (t_1(Q), t_2(Q), \ldots,)$ on $\mathcal{H}$ (1992, P.Etingof, A.Varchenko [1]). Moreover, $\mathbf{t}(Q)$ are global coordinates for stars domains Q (1938, P.S.Novikov [2]). Put $\bar{\mathbf{t}}(Q) = (\bar{t}_1(Q), \bar{t}_2(Q)\ , \ldots,)$.

The conformal function $w_Q(z) : Q \to \Lambda$ is single if

$$w_Q(\infty) = \infty, \quad \text{and} \quad w'_Q(\infty) \quad \text{is real positive.}$$

Journal of Physics: Conference Series **670** (2016) 012036 doi:10.1088/1742-6596/670/1/012036

P.Wiegmann and A.Zabrodin (2000, [3]) proved, that there is a not dependent from Q **symmetric solution of dispersionless 2D Toda hierarchy** $F_{con}(t_0, \mathbf{t}, \bar{\mathbf{t}})$, such that

$$w_Q(z) = z \exp\left(\left(-\frac{1}{2}\frac{\partial^2}{\partial t_0^2} - \frac{\partial}{\partial t_0}\sum_{k\geq 1}\frac{z^{-k}}{k}\frac{\partial}{\partial t_k}\right)F_{con}\right)(t_0(Q), \mathbf{t}(Q), \bar{\mathbf{t}}(Q)).$$

Thus, in order to find a conformal mapping for all Q we have to find the universal function $F_{con}(t_0, \mathbf{t}, \bar{\mathbf{t}})$.

Hurwitz numbers

Hurwitz numbers are appeared in the 19th century. They describe a numbers of coverings of surfaces. In recent years they have been actively used in moduli spaces of algebraic curves and mathematical physics. We consider only 2-Hurwitz numbers of genus 0. That is, the number of different rational functions of general position with fixed types divisor of poles and zeros.

In more detail. Fix Young diagrams $\Delta = [\mu_1, \ldots, \mu_\ell]$, $\bar{\Delta} = [\bar{\mu}_1, \ldots, \bar{\mu}_{\bar{\ell}}]$ of degrees $|\Delta| = |\bar{\Delta}| = d$ and different points $z_1, \ldots, z_k \in \mathbb{C} \setminus 0$, where $k = \ell + \bar{\ell} - 2$. Let us consider rational functions $\varphi : \overline{\mathbb{C}} \to \overline{\mathbb{C}}$ of degree d with degrees of its zeros $\Delta = [\mu_1, \ldots, \mu_\ell]$, degrees of its poles $\bar{\Delta} = [\bar{\mu}_1, \ldots, \bar{\mu}_{\bar{\ell}}]$ and simple critical values $z_1, \ldots, z_k$.

Functions φ and $\tilde{\varphi}$ we consider as equivalent if $\tilde{\varphi}(z) = \varphi(hz)$, where h is an automorphism of Riemann sphere. The number $H_0(\Delta, \bar{\Delta})$ of equivalence classes is finite and does not depend on $z_1, \ldots, z_k$. For $k > 0$ the number $H_0(\Delta, \bar{\Delta})$ is called a *double Hurwitz number* of genus 0. For $k = 0$ the double Hurwitz number is $\frac{1}{d}$.

Let us look the generating function

$$F(t_0, \mathbf{t}, \bar{\mathbf{t}}) = \frac{t_0^3}{6} + \sum_{|\Delta|=|\bar{\Delta}|} \frac{e^{|\Delta| t_0} H_0(\Delta, \bar{\Delta})}{(\ell(\Delta)+\ell(\bar{\Delta})-2)!} \prod_{i=1}^{\ell(\Delta)} \mu_i t_{\mu_i} \prod_{i=1}^{\ell(\bar{\Delta})} \bar{\mu}_i \bar{t}_{\bar{\mu}_i},$$

where $t = \{t_1, t_2, \ldots\}$, $\bar{t} = \{\bar{t}_1, \bar{t}_2, \ldots\}$, $\Delta = [\mu_1, \ldots, \mu_{\ell(\Delta)}]$, $\bar{\Delta} = [\bar{\mu}_1, \ldots, \bar{\mu}_{\ell(\bar{\Delta})}]$.

A.Okounkov proved (2000, [4]), that $F(t_0, \mathbf{t}, \bar{\mathbf{t}})$ is a **symmetric solution of dispersionless 2D Toda hierarchy.**

Dispersionless 2D Toda hierarchy

Dispersionless 2D Toda hierarchy was defined by K.Takasaki and T.Takebe (1994, [5]) for application in mathematical physics. In the future, its applications have been found in the topology, complex analysis, and algebraic geometry.

Dispersionless 2D Toda hierarchy are equations on a function $F = F(t_0, \mathbf{t}, \bar{\mathbf{t}})$ from a variable t_0 and two infinite sets of variables , $\mathbf{t} = \{t_1, t_2, \ldots\}$, $\bar{\mathbf{t}} = \{\bar{t}_1, \bar{t}_2, \ldots\}$. To determine the hierarchy is convenient to introduce the auxiliary formal complex variables z, $\bar{z}$, ξ, $\bar{\xi}$.

Let us put

$$\partial_0 = \partial/\partial t_0, \quad \partial_k = \partial/\partial t_k, \quad \bar{\partial}_k = \partial/\partial \bar{t}_k$$

and

$$D(z) = \sum_{k\geq 1}\frac{z^{-k}}{k}\,\partial_k\,, \qquad \bar{D}(\bar{z}) = \sum_{k\geq 1}\frac{\bar{z}^{-k}}{k}\,\bar{\partial}_k\,,$$

The hierarchy is presented by equations

$$(z-\xi)\exp\big(D(z)D(\xi)F\big) = z\exp\big(-\partial_0 D(z)F\big) - \xi\exp\big(-\partial_0 D(\xi)F\big), \tag{1}$$

$$(\bar z-\bar\xi)\exp\big(\bar D(\bar z)\bar D(\bar\xi)F\big) = \bar z\exp\big(-\partial_0 \bar D(\bar z)F\big) - \bar\xi\exp\big(-\partial_0 \bar D(\bar\xi)F\big), \tag{2}$$

$$1-\exp\big(-D(z)\bar D(\bar\xi)F\big) = \frac{1}{z\bar\xi}\exp\big(\partial_0(\partial_0 + D(z)+\bar D(\bar\xi))F\big). \tag{3}$$

Differential equations arise, if we expand the right-hand and left-hand sides in the formal Laurent series and equate the coefficients of like powers.

The coefficient for $z^{-i}\xi^{-j}$ in (1) for positive i,j gives, for example an equation

$$\partial_i\partial_j F = \sum_{m=1}^{\infty}\sum_{\substack{s_1+\dots+s_m=\\ i+j;\ s_k>1}} \widetilde{P}_{ij}(s_1,\dots,s_m)\partial_1\partial_{s_1-1}F\dots\partial_1\partial_{s_m-1}F.$$

Formal solutions

We consider *formal solutions* i.e. solutions in form of formal Taylor series

$$F(t_0,\mathbf{t},\bar{\mathbf{t}}) =$$

$$\sum_{\substack{\mu_1\ge\mu_2\ge\dots\ge\mu_\ell>0\\ \bar\mu_1\ge\bar\mu_2\ge\dots\ge\bar\mu_{\bar\ell}>0}} F(\mu_1,\mu_2,\dots,\mu_\ell|\bar\mu_1,\bar\mu_2,\dots,\bar\mu_{\bar\ell}|t_0)t_{\mu_1}t_{\mu_2}\dots t_{\mu_k}\bar t_{\bar\mu_1}\bar t_{\bar\mu_2}\dots\bar t_{\mu_{\bar\ell}}.$$

The indexes $\mu_1\ge\mu_2\ge\dots\ge\mu_\ell>0$ and $\bar\mu_1\ge\bar\mu_2\ge\dots\ge\bar\mu_{\bar\ell}>0$ form Young diagrammes Δ and $\bar\Delta$.

Denote by $t_\Delta = t_{\mu_1}t_{\mu_2}\dots t_{\mu_k}$, and $t_{\bar\Delta} = \bar t_{\bar\mu_1}\bar t_{\bar\mu_2}\dots\bar t_{\mu_{\bar\ell}}$. Then

$$F(t_0,\mathbf{t},\bar{\mathbf{t}}) = F(\emptyset|\emptyset|t_0) + \sum_{\Delta}F(\Delta|\emptyset|t_0)t_\Delta + \sum_{\bar\Delta}F(\emptyset|\bar\Delta|t_0)\bar t_{\bar\Delta}$$

$$+\sum_{\Delta,\bar\Delta}F(\Delta|\bar\Delta|t_0)\,t_\Delta\bar t_{\bar\Delta}.$$

Main Theorem

We say that a function $F(t_0,\mathbf{t},\bar{\mathbf{t}})$ is *symmetric* if

$$\partial_k F(t_0,0,0) = \bar\partial_k F(t_0,0,0) = 0 \quad \text{for} \quad k>0.$$

Theorem 1 (2015, S.Natanzon, A.Zabrodin [6]). Any formal symmetric solution of 2D dispersionless Toda hierarchy F is defined by any formal function $\Phi(t_0)$ and it is equal

$$F = \Phi(t_0) + \sum_{i>0}^{\infty} i f^i t_i\bar t_i +$$

$$\sum_{|\Delta|=|\bar\Delta|}\ \sum_{\substack{s_1+\dots+s_m=|\Delta|\\ r_1+\dots+r_m=\ell(\Delta)+\ell(\bar\Delta)-2>0}} N_{(\Delta|\bar\Delta)}\begin{pmatrix}s_1\dots s_m\\ r_1\dots r_m\end{pmatrix}\partial_0^{r_1}(f^{s_1})\dots\partial_0^{r_m}(f^{s_m})\,t_\Delta\bar t_{\bar\Delta}, \tag{4}$$

where $f(t_0) = \exp(\Phi'')(t_0)$ and the sum is given by all Young diagrams Δ, $\bar\Delta$ and all positive integer indexes.

In Theorem 1 and in Theorem 2 we are correcting misprints in [6] in indexes of sums. The correct formula given also in S.Natanzon, A.Zabrodin arxiv.org/1302.7288v4 .

Calculation of $N_{(\Delta|\bar{\Delta})}\begin{pmatrix} s_1 \dots s_m \\ r_1 \dots r_m \end{pmatrix}$

We find also recursion algorithm for calculation of $N_{(\Delta|\bar{\Delta})}\begin{pmatrix} s_1 \dots s_m \\ r_1 \dots r_m \end{pmatrix}$. It consists of several steps.

1. **Denote** by $P_{ij}(r_1, \dots, r_m)$ is the number of sequences of positive integers $(i_1, \dots, i_m)$, $(j_1, \dots, j_m)$ such that $i_1 + \dots + i_m = i$, $j_1 + \dots + j_m = j$ and $r_k = i_k + j_k$.

Put $T_{ij}(p_1, \dots, p_m) = \sum\limits_{\substack{k>0,\, n_i>0 \\ n_1+\dots+n_k=m}} \frac{(-1)^{m+1}}{k\, n_1! \dots n_k!} P_{ij}\left(\sum\limits_{i=1}^{n_1} p_i, \sum\limits_{i=n_1+1}^{n_1+n_2} p_i, \dots, \sum\limits_{i=n_1+\dots+n_{k-1}+1}^{m} p_i \right)$.

2. **Define**

$$T_{i_1 i_2}\begin{pmatrix} s_1 \dots s_m \\ \ell_1 \dots \ell_m \end{pmatrix} = \begin{cases} T_{i_1 i_2}(s_1, \dots, s_m), & \text{if } \ell_1 = \dots = \ell_m = 1 \\ 0 & \text{otherwise,} \end{cases}$$

and

$$T_{i_1 \dots i_k}\begin{pmatrix} s_1 \dots s_m \\ \ell_1 \dots \ell_m \end{pmatrix} = \sum_{1 \leqslant i \leqslant j \leqslant m} \frac{\ell!}{(\ell_i - 1)! \dots (\ell_j - 1)!} \times$$

$$T_{i_1 \dots i_{k-1}}\begin{pmatrix} s_1 \dots s_{i-1}\, s\, s_{j+1} \dots s_m \\ \ell_1 \dots \ell_{i-1}\, \ell\, \ell_{j+1} \dots \ell_m \end{pmatrix} T_{s, i_k}(s_i, s_{i+1}, \dots, s_j),$$

where

$$s = s_i + s_{i+1} + \dots + s_j - i_k > 0, \quad \ell = (\ell_i - 1) + \dots + (\ell_j - 1) > 0.$$

3. **Consider** $\tilde{N}_{\begin{pmatrix} i_1 \dots i_k \\ \bar{i}_1 \dots \bar{i}_{\bar{k}} \end{pmatrix}}\begin{pmatrix} s_1 \dots s_m \\ r_1 \dots r_m \end{pmatrix} =$

$$\frac{i_1 \dots i_k\, \bar{i}_1 \dots \bar{i}_{\bar{k}}}{s_1 \dots s_m} \sum T_{i_1 \dots i_k}\begin{pmatrix} s_1 & \dots & s_m \\ r_1 - n_1 + 1 & \dots & r_m - n_m + 1 \end{pmatrix},$$

where the summation is carried over all representation of the set $\{\bar{i}_1, ..., \bar{i}_{\bar{k}}\}$ as a union of non-intersecting non-empty subsequences $\{b_1^j, ..., b_{n_j}^j\} \subset \{\bar{i}_1, ..., \bar{i}_{\bar{k}}\}$ such that $b_1^j + \dots + b_{n_j}^j = s_j$ and $j = 1, \dots, m$.

Put

$$N_{(\Delta|\bar{\Delta})}\begin{pmatrix} s_1 \dots s_m \\ r_1 \dots r_m \end{pmatrix} = \frac{1}{\sigma(\Delta)\sigma(\bar{\Delta})} \tilde{N}_{\begin{pmatrix} \mu_1 \dots \mu_k \\ \bar{\mu}_1 \dots \bar{\mu}_{\bar{k}} \end{pmatrix}}\begin{pmatrix} s_1 \dots s_m \\ r_1 \dots r_m \end{pmatrix}, \tag{5}$$

where Δ, $\bar{\Delta}$ are Young diagrams with rows $[\bar{\mu}_1, \dots, \bar{\mu}_\ell]$ and $[\mu_1, \dots, \mu_\ell]$ and $\sigma(\Delta)$, $\sigma(\bar{\Delta})$ are orders of the automorphism groups of rows for the Young diagrams Δ, $\bar{\Delta}$.

Example 1. If $\Delta = [\mu_1, \dots, \mu_l]$, $\bar{\Delta} = [\bar{\mu}_1, \dots, \bar{\mu}_{\bar{l}}]$ and $d = |\Delta| = |\bar{\Delta}|$, then

$$N_{(\Delta|\bar{\Delta})}\begin{pmatrix} d \\ \ell(\Delta) + \ell(\bar{\Delta}) - 2 \end{pmatrix} = \frac{\rho(\Delta)\rho(\bar{\Delta})}{d\sigma(\Delta)\sigma(\bar{\Delta})},$$

where $\rho(\Delta) = \mu_1 \cdots \mu_l$. For other cases $N_{(\Delta|\bar{\Delta})}\begin{pmatrix} s \\ r \end{pmatrix} = 0$.

Journal of Physics: Conference Series **670** (2016) 012036 doi:10.1088/1742-6596/670/1/012036

Example 2.

$$N_{\binom{i_1 i_2}{\bar{i}_1 \bar{i}_2}} \binom{\bar{i}_1 \bar{i}_2}{11} = N_{\binom{i_1 i_2}{\bar{i}_1 \bar{i}_2}} \binom{\bar{i}_2 \bar{i}_1}{11} = -\frac{i_1 i_2}{2\sigma([i_1, i_2])\sigma([\bar{i}_1, \bar{i}_2])} \min\{i_1, i_2, \bar{i}_1, \bar{i}_2\}).$$

For other cases $N_{\binom{i_1 i_2}{\bar{i}_1 \bar{i}_2}} \binom{s_1 s_2}{r_1 r_2} = 0$.

Application to Hurwitz numbers

Okounkov's theorem says that

$$F(t_0, \mathbf{t}, \bar{\mathbf{t}}) = \frac{t_0^3}{6} + \sum_{|\Delta|=|\bar{\Delta}|} \frac{e^{|\Delta| t_0} H_0(\Delta, \bar{\Delta})}{(\ell(\Delta)+\ell(\bar{\Delta})-2)!} \prod_{i=1}^{\ell(\Delta)} \mu_i t_{\mu_i} \prod_{i=1}^{\ell(\bar{\Delta})} \bar{\mu}_i \bar{t}_{\bar{\mu}_i}$$

is a symmetric formal solutions of dispersionless 2D Toda hierarchy with $\Phi(t_0) = \frac{t_0^3}{6}$ and therefore $f(t_0) = \exp(t_0)$. Thus, Theorem 1 gives

Theorem 2 (2015, S.Natanzon, A.Zabrodin [6]). Double Hurwitz numbers of genus 0 for $\ell(\Delta) + \ell(\bar{\Delta}) > 2$ are

$$H_0(\Delta|\bar{\Delta}) = \frac{(\ell(\Delta) + \ell(\bar{\Delta})-2)!}{\rho(\Delta)\,\rho(\bar{\Delta})} \sum s_1^{r_1} \dots s_m^{r_m} N_{(\Delta|\bar{\Delta})} \binom{s_1 \dots s_m}{r_1 \dots r_m},$$

where the sum is carried over all matrixes $\binom{s_1 \dots s_m}{r_1 \dots r_m}$ such that $s_1 + \dots + s_m = |\Delta|$ and $r_1 + \dots + r_m = \ell(\Delta) + \ell(\bar{\Delta}) - 2$.

In particulary examples 1 and 2 give

Example 3.

The double Hurwitz numbers of polynomials for $\ell(\Delta) > 1$ are

$$H_0(\Delta|[n]) = \frac{(\ell(\Delta) - 1)!}{\sigma(\Delta)} n^{\ell(\Delta)-2}.$$

Example 4.

The double Hurwitz numbers of simplest Laurent polynomials are

$$H_0([i_1, i_2]|[\bar{i}_1, \bar{i}_2]) = 2\frac{d - \min\{i_1, i_2, \bar{i}_1, \bar{i}_2\}}{(1 + \delta_{i_1 i_2})(1 + \delta_{\bar{i}_1 \bar{i}_2})}.$$

These nice formulas were at first found by integration on compactification of moduli spaces of algebraic curves (2002, S.Lando [7]; 2005, I.Goulden, D.Jackson, Vakil [8]; 2008, S.Shadrin, M.Shapiro, A.Vainshtein [9]). Theorem 2 enables us to find a lot of other formulas of this type.

Application to Riemann theorem

Wiegmann-Zabrodin function F_{con} is a symmetric solutions to dispersionless 2D Toda hierarchy, $\Phi(t_0) = F_{con}(t_0, 0, 0) = \frac{1}{2}t_0^2 \ln t_0 - \frac{3}{4}t_0^2$ and therefore $f(x) = t_0$. Thus, theorem 1 gives

Theorem 3 (2005, S.Natanzon [10]). The conformal map $w_Q : Q \to \Lambda$ is

$$w_Q(z) = z \exp\left(\left(-\frac{1}{2}\frac{\partial^2}{\partial t_0^2} - \frac{\partial}{\partial t_0}\sum_{k\geq 1}\frac{z^{-k}}{k}\frac{\partial}{\partial t_k}\right)F_{con}\right)(t_0(Q), \mathbf{t}(Q), \bar{\mathbf{t}}(Q)),$$

where

$$F_{con}(t_0, \mathbf{t}, \bar{\mathbf{t}}) = \frac{1}{2}t_0^2 \ln t_0 - \frac{3}{4}t_0^2 + \sum_{i>0}^{\infty} i t_0^i t_i \bar{t}_i +$$

$$\sum_{|\Delta|=|\bar{\Delta}|} \sum_{\substack{s_1+\ldots+s_m=|\Delta| \\ r_1+\ldots+r_m=\ell(\Delta)+\ell(\bar{\Delta})-2>0}} N_{(\Delta|\bar{\Delta})}\begin{pmatrix} s_1 \ldots s_m \\ r_1 \ldots r_m \end{pmatrix} \partial_0^{r_1}(t_0^{s_1})\ldots\partial_0^{r_m}(t_0^{s_m})\, t_\Delta \bar{t}_{\bar{\Delta}}.$$

Acknowledgments

The author was partially supported in part by RFBR grant 13-01-00755 and partially supported by Laboratory of Quantum Topology of Chelyabinsk State University (Russian Federation government grant 14.Z50.31.0020).

References

[1] Varchenko A and Etingof P 1992 Why the boundary of a round drop becomes a curve of order four *University Lecture Series, AMS Povidence RI* **3** 72.
[2] Novikov P.S 1938 On uniqueness of the solution of the inverse problem for a potential *Soviet Math.Dok.* **18:3** 165-168.
[3] Wiegmann P and Zabrodin A 2000 Conformal maps and integrable hierarchies *Commun. Math. Phys.* **213** 523-538.
[4] Okounkov A 2000 Toda equations for Hurwitz numbers *Math. Res. Lett.* **7** 447-453.
[5] Takasaki K and Takebe T 1991 SDiff(2) Toda equation - hierachy, tau function and symmetries *Lett. Math. Phys.* **23** 205-214.
[6] Natanzon S and Zabrodin A 2015 Simmetric solutions to dispersionless 2D Toda hierachy, Hurwitz numbers and conformal dynamics *Int. Math. Res. Notices* **2015** 2082-2110.
[7] Lando S 2002 Ramifield coverings of the two-dimensional sphere and the intersection theory in spaces of meromorphic functions on algebraic curves *Russ. Math. Surv.* **57:3** 463-533.
[8] Gulden I, Jeckson D and Vakil R 2005 Towards the geometry of double Hurwitz numbers *Adv. Math.* **198** 43-92.
[9] Shadrin S, Shapiro M and Wainshtein A 2008 Chamber behevior of double Hurwitz numbers in genus 0 *Adv. Math.* **217** 79-96.
[10] Natanzon S 2005 Towards an effectivisation of the Riemann theorm *Ann. Global Anal. Geom.* **28** 233-255.

Journal of Physics: Conference Series **670** (2016) 012037 doi:10.1088/1742-6596/670/1/012037

Quantum affine algebras and universal functional relations

Kh S Nirov[1,2] **and A V Razumov**[3]

[1] Institute for Nuclear Research of the Russian Academy of Sciences, 60th October Ave 7a, 117312 Moscow, Russia
[2] Fachbereich C – Physik, Bergische Universität Wuppertal, 42097 Wuppertal, Germany
[3] Institute for High Energy Physics, NRC "Kurchatov Institute", 142281 Protvino, Moscow region, Russia

E-mail: nirov@uni-wuppertal.de and Alexander.Razumov@ihep.ru

Abstract. By the universal integrability objects we mean certain monodromy-type and transfer-type operators, where the representation in the auxiliary space is properly fixed, while the representation in the quantum space is not. This notion is actually determined by the structure of the universal R-matrix. We call functional relations between such universal integrability objects, and so, being independent of the representation in the quantum space, the universal functional relations. We present a short review of the universal functional relations for the quantum integrable systems associated with the quantum groups of loop Lie algebras.

1. Introduction

The discovery of the so-called *Hidden Grassmann Structure* in the state space of the XXZ model [1–5] opened principally novel perspectives in the study of correlation functions in quantum integrable systems. At the same time, it raised fundamental questions about the algebraic origin of the fermionic basis in the XXZ model and its possible generalizations. Trying to give any adequate response to this and other related problems, we came to the conclusion that we need a better understanding of Baxter's TQ-relations [6, 7], with all their ingredients and possible generalizations. Then, at certain point of our study, we came across an earlier series of remarkable papers on integrable structure of conformal field theory [8–11]. For us, the main idea of these pioneer works was that the objects describing the model and related to its integrability should be constructed on the basis of the *universal R-matrix*, which is an element of the tensor product of two copies of the quantum group [12]. Traditionally, the first factor in this tensor product is regarded as the *auxiliary space*, and the second one as the *quantum space*. But their roles can certainly be interchanged, giving rise to various interesting relations between different integrable structures. The functional relations, supposed to be a substitute and generalization of the famous Bethe Ansatz [13], should follow from the characteristics of the appropriate representations of the underlying quantum group. Altogether, our deliberations over the contents of these papers made clear that, in the theory of quantum integrable systems, the functional relations are too important to treat them superficially. And so, we have carefully reconsidered the whole quantum group approach to quantum integrable systems. Essential part of this revision is actually what we are presenting here in a short review based mainly on our publications [14–19].

By $\mathcal{L}(\mathfrak{g})$ we denote the loop Lie algebra of a finite-dimensional Lie algebra $\mathfrak{g}$ and by $\widetilde{\mathcal{L}}(\mathfrak{g})$ its standard central extension [20]. The corresponding quantum group is denoted by adding the symbol U_q, as appropriate for the q-deformation of the universal enveloping algebra of a given Lie algebra. It is assumed here that q is not a root of unity. To construct integrability objects one uses spectral parameters. They are introduced by means of a $\mathbb{Z}$-gradation of the quantum group under consideration. In the case of our interest, the quantum group $U_q(\mathcal{L}(\mathfrak{sl}_n))$, a $\mathbb{Z}$-gradation is determined by n integers s_i, $i = 0, \ldots, n-1$, and we use the notation $s = s_0 + \ldots + s_{n-1}$ for their total sum. Hence, the functional relations are certain difference equations for the integrability objects considered as functions of the spectral parameters. When the Bethe Ansatz is not applicable anymore, the functional relations can serve to diagonalize the transfer matrix of such models. However, they can also be studied independently to elucidate certain algebraic structures of quantum integrable systems. In particular, they can help in understanding the group-theoretic background of the Hidden Grassmann Structure and see ways to extend it to other models. It turns out here that it is justified to fix the auxiliary space by an appropriate representation, while keeping the quantum space free. Possible advantage of such an approach was mentioned earlier in different contexts, see, for example, [21, 22]. In our recent works [16–19], we have developed and elaborated this idea and have given the complete proof of the functional relations in the form independent of the representation of the quantum group in the quantum space. We have also described the specialization of the universal functional relations to the case when the quantum space is the state space of a discrete spin chain. In this presentation we consider only the first part of the program, the universal integrability objects and the corresponding universal functional relations without their specialization in the quantum space.

We use the notation $\kappa_q = q - q^{-1}$, so that the definition of the q-number can be written as

$$[\nu]_q = \frac{q^\nu - q^{-\nu}}{q - q^{-1}} = \kappa_q^{-1}(q^\nu - q^{-\nu}), \qquad \nu \in \mathbb{C}.$$

2. Quantum groups in general

Here we recollect basic definitions of a well-known object and fix the corresponding notations. The object in question, denoted here by $\mathcal{A}$, is nothing but the *quantum group*, being actually the main tool for the whole construction. We generally follow the definitions proposed by Drinfeld [23, 24] and Jimbo [25]; for a review see [26]. Hence, it is a unital associative algebra obtained by a deformation of the universal enveloping algebra of a given Lie algebra. The quantum group is thus an algebra, and if the initial Lie algebra is an affine Kac–Moody Lie algebra, then one deals with a *quantum affine algebra*. This is the case of our consideration.

It should be noted here that, depending on the nature of the deformation parameter q, the quantum group can be interpreted differently. If $q = \exp \hbar$, where $\hbar$ is an indeterminate, then the quantum group is a $\mathbb{C}[[\hbar]]$-algebra. If q itself is an indeterminate, then the quantum group is a $\mathbb{C}(q)$-algebra. If $q = \exp \hbar$, where $\hbar$ is a complex number, then the quantum group is a $\mathbb{C}$-algebra. We deal with the quantum group defined according to the last way in order to have meaningful traces, see, for example, [27, 28].

The quantum group is a Hopf algebra with respect to appropriate co-multiplication, antipode and co-unit. Moreover, as any Hopf algebras, it also has an opposite co-multiplication Δ^{op}, defined by the initial co-multiplication Δ as

$$\Delta^{\mathrm{op}} = \Pi \circ \Delta, \qquad \Pi(a \otimes b) = b \otimes a, \qquad a, b \in \mathcal{A}.$$

Now, assuming that there exists an invertible element $\mathcal{R}$ of the tensor product of two copies of the quantum group, such that

$$\Delta^{\mathrm{op}}(a) = \mathcal{R}\, \Delta(a) \mathcal{R}^{-1}, \qquad \mathcal{R} \in \mathcal{A} \otimes \mathcal{A}, \tag{2.1}$$

for any element a of the quantum group, and the following relations

$$(\Delta \otimes \mathrm{id})(\mathcal{R}) = \mathcal{R}^{13}\mathcal{R}^{23}, \qquad (\mathrm{id} \otimes \Delta)(\mathcal{R}) = \mathcal{R}^{13}\mathcal{R}^{12}, \tag{2.2}$$

with the standard convention about the indices, are fulfilled, we restrict ourselves by the so-called *quasi-triangular* Hopf algebras.[1] The element $\mathcal{R}$ is called the *universal R-matrix*. It follows from (2.1) and (2.2) that it satisfies the *master equation*

$$\mathcal{R}^{12}\mathcal{R}^{13}\mathcal{R}^{23} = \mathcal{R}^{23}\mathcal{R}^{13}\mathcal{R}^{12} \tag{2.3}$$

defined on the tensor product of three copies of the quantum group. However, it is important to note that in the case under consideration the universal R-matrix belongs actually to a smaller subset of the tensor square of the full quantum group; to be precise, to the tensor product of the Borel subalgebras $\mathcal{B}_+$ and $\mathcal{B}_-$ of the quantum group $\mathcal{A}$, i. e. $\mathcal{R} \in \mathcal{B}_+ \otimes \mathcal{B}_- \subset \mathcal{A} \otimes \mathcal{A}$. Therefore, and in particular, the master equation is defined on the tensor product $\mathcal{B}_+ \otimes \mathcal{A} \otimes \mathcal{B}_-$.

3. Universal integrability objects

Let us introduce a group-like element, that is an invertible element of the quantum group with the co-multiplication property given by

$$\Delta(t) = t \otimes t, \qquad t \in \mathcal{A}.$$

With the help of the universal R-matrix we define the *universal integrability objects*. These are the monodromy- and transfer-type operators defined as follows.

First, let φ be a representation of the quantum group $\mathcal{A}$ in a vector space V[2]

$$\varphi : \mathcal{A} \to \mathrm{End}(V).$$

Then, the corresponding universal monodromy operator $\mathcal{M}_\varphi$ is defined as

$$\mathcal{M}_\varphi = (\varphi \otimes \mathrm{id})(\mathcal{R}) \in \mathrm{End}(V) \otimes \mathcal{A}. \tag{3.1}$$

The corresponding universal transfer operator is defined as

$$\mathcal{T}_\varphi = (\mathrm{tr}_V \otimes \mathrm{id})(\mathcal{M}_\varphi(\varphi(t) \otimes 1)) = ((\mathrm{tr}_V \circ \varphi) \otimes \mathrm{id})(\mathcal{R}(t \otimes 1)), \tag{3.2}$$

where 1 means the unit element of $\mathcal{A}$, and now t makes sense as a *twist element*. Using appropriate t, in general, we can obtain meaningful traces over infinite-dimensional representation spaces, see, for example, [17]. The representation φ can depend on certain parameters, which we collectively denote by ζ. In such a case we write φ_ζ and $\mathcal{M}_\varphi(\zeta)$, $\mathcal{T}_\varphi(\zeta)$, respectively.

The other universal integrability objects are introduced in a similar way, but the representation ρ used for them is essentially different from the representation φ used earlier. Indeed, here we recall that, in the case under consideration, the universal R-matrix is an element of the tensor product of two principal Borel subalgebras of $\mathcal{A}$, that is $\mathcal{R} \in \mathcal{B}_+ \otimes \mathcal{B}_-$. Then, let ρ be such a representation of $\mathcal{B}_+$ that it cannot be extended to a representation of the whole quantum group $\mathcal{A}$. Therefore, it cannot be obtained simply by the restriction of φ (from $\mathcal{A}$ to $\mathcal{B}_+$). We should understand, however, that to produce nontrivial relations between the

[1] In fact, there is a subtle conflict between the definition of the quantum group as a $\mathbb{C}$-algebra and the last requirement. However, this problem is resolvable in our case, see, for example, the discussions in [17–19,29].
[2] In other words, V is an $\mathcal{A}$-module.

integrability objects, these representations, φ and ρ, should be somehow connected. Now, with such a representation of $\mathcal{B}_+$ in a vector space W,[3]

$$\rho : \mathcal{B}_+ \to \operatorname{End}(W),$$

we introduce the universal L-operator

$$\mathcal{L}_\rho = (\rho \otimes \mathrm{id})(\mathcal{R}) \in \operatorname{End}(W) \otimes \mathcal{B}_- \tag{3.3}$$

and the corresponding universal Q-operator

$$\mathcal{Q}_\rho = (\mathrm{tr}_W \otimes \mathrm{id})(\mathcal{L}_\rho(\rho(t) \otimes 1)) = ((\mathrm{tr}_W \circ \rho) \otimes \mathrm{id})(\mathcal{R}(t \otimes 1)), \tag{3.4}$$

where the group-like element t is the same as for the universal transfer-operator, and 1 again means the unit of the quantum group.

Further, we can rewrite the master equation (2.3) as follows:

$$(\mathcal{R}^{13}t^1)(\mathcal{R}^{23}t^2) = (\mathcal{R}^{12})^{-1}(\mathcal{R}^{23}t^2)(\mathcal{R}^{13}t^1)\mathcal{R}^{12}. \tag{3.5}$$

Using the tensor products of the maps $(\mathrm{tr} \circ \varphi_\zeta)$, $(\mathrm{tr} \circ \rho_\zeta)$ for various φ and ζ in the providently rewritten master equation (3.5), we obtain the first functional relations. Applying $(\mathrm{tr} \circ \varphi_{1\zeta_1}) \otimes (\mathrm{tr} \circ \varphi_{2\zeta_2})$ to both sides of (3.5), we obtain

$$\mathcal{T}_{\varphi_1}(\zeta_1)\mathcal{T}_{\varphi_2}(\zeta_2) = \mathcal{T}_{\varphi_2}(\zeta_2)\mathcal{T}_{\varphi_1}(\zeta_1). \tag{3.6}$$

Applying $(\mathrm{tr} \circ \rho_{\zeta_1}) \otimes (\mathrm{tr} \circ \varphi_{\zeta_2})$ to both sides of (3.5), we come to the relation

$$\mathcal{Q}_\rho(\zeta_1)\mathcal{T}_\varphi(\zeta_2) = \mathcal{T}_\varphi(\zeta_2)\mathcal{Q}_\rho(\zeta_1). \tag{3.7}$$

These are the commutativity relations for the universal integrability objects.

The commutativity between two universal Q-operators is not obtained in this way due to the specific nature of the representation ρ, as just described above. Nevertheless, we have a useful formula for the product of two (and more) universal Q-operators presented here:

$$\mathcal{Q}_{\rho_1}(\zeta_1)\mathcal{Q}_{\rho_2}(\zeta_2) = ((\mathrm{tr}_{W_1 \otimes W_2} \circ (\rho_{1\zeta_1} \otimes_\Delta \rho_{2\zeta_2}) \otimes \mathrm{id})(\mathcal{R}(t \otimes 1)), \tag{3.8}$$

where we have used the equation

$$\mathcal{R}^{13}t^1\mathcal{R}^{23}t^2 = [(\Delta \otimes \mathrm{id})(\mathcal{R})][(\Delta \otimes \mathrm{id})(t \otimes 1)] = (\Delta \otimes \mathrm{id})(\mathcal{R}(t \otimes 1)). \tag{3.9}$$

Equation (3.8) proves to be useful in establishing the commutativity between the universal Q-operators. However, the very proof of this fact, and also further functional relations, require more details on the quantum group and its representations. This is the subject of the next section.

[3] Similarly as before, W is a $\mathcal{B}_+$-module.

4. Quantum groups $U_q(\mathfrak{gl}_n)$ and $U_q(\mathcal{L}(\mathfrak{sl}_n))$

We need two quantum groups, the quantum group of the general linear Lie group $\mathfrak{gl}_n$, and the quantum group of the loop Lie algebra $\mathcal{L}(\mathfrak{sl}_n)$, the latter justifying the term *quantum affine algebra*.[4]

Recall that there are $2(n-1)$ generators E_i, F_i, $i = 1, \ldots, n-1$, of the Lie algebra $\mathfrak{gl}_n$, and n basis elements G_i, $i = 1, \ldots, n$, of its standard Cartan subalgebra $\mathfrak{k}_n$. They are subject to well-known commutation relations and Serre relations. Then, the quantum group $U_q(\mathfrak{gl}_n)$ is generated by the elements[5]

$$E_i, \quad F_i, \quad i = 1, \ldots, n-1, \qquad q^X, \quad X \in \mathfrak{k}_n,$$

satisfying the corresponding q-deformed defining relations

$$q^0 = 1, \qquad q^{X_1} q^{X_2} = q^{X_1 + X_2}, \tag{4.1}$$

$$q^X E_i q^{-X} = q^{\alpha_i(X)} E_i, \qquad q^X F_i q^{-X} = q^{-\alpha_i(X)} F_i, \tag{4.2}$$

$$[E_i, F_j] = \delta_{ij} \kappa_q^{-1} (q^{G_i - G_{i+1}} - q^{G_{i+1} - G_i}), \tag{4.3}$$

where $\alpha_i \in \mathfrak{k}_n^*$ denote the respective simple positive roots, such that $\alpha_j(G_i) = c_{ij}$, where c_{ij} are the entries of an $n \times (n-1)$ matrix with $c_{ii} = 1$, $c_{i+1,i} = -1$, $i = 1, \ldots, n-1$, and $c_{ij} = 0$ otherwise (i. e. for $|j - i| \geq 2$). It means, in particular, that $\alpha_j(G_i) - \alpha_j(G_{i+1}) = a_{ij}$ are the entries of the Cartan matrix of the Lie algebra $\mathfrak{sl}_n$. Besides, we have the q-deformed Serre relations

$$E_i E_j = E_j E_i, \qquad F_i F_j = F_j F_i, \qquad |i - j| \geq 2, \tag{4.4}$$

$$E_i^2 \, E_{i\pm1} - [2]_q E_i \, E_{i\pm1} \, E_i + E_{i\pm1} \, E_i^2 = 0, \tag{4.5}$$

$$F_i^2 \, F_{i\pm1} - [2]_q F_i \, F_{i\pm1} \, F_i + F_{i\pm1} \, F_i^2 = 0. \tag{4.6}$$

Remember that the deformation parameter q here is the exponential of a complex number $\hbar$, therefore, the quantum group is just a complex algebra. Here we assume that

$$q^{X+\nu} = q^\nu q^X,$$

$$[X + \nu]_q = \kappa_q^{-1} (q^{X+\nu} - q^{-X-\nu}) = \kappa_q^{-1} (q^\nu q^X - q^{-\nu} q^{-X})$$

for any complex number ν and element X of $\mathfrak{k}_n$. It is important that $U_q(\mathfrak{gl}_n)$ is a Hopf algebra with respect to appropriately defined co-multiplication, antipode and counit; their explicit form can be omitted, though.

The generators E_i and F_i, $i = 1, \ldots, n-1$, are certainly the root vectors corresponding to the simple positive and simple negative roots α_i and $-\alpha_i$. There are also non-simple roots for $n > 2$, and one constructs more root vectors corresponding to these non-simple roots. All these root vectors are used in constructing the basis vectors of the highest-weight $U_q(\mathfrak{gl}_n)$-modules (see Appendix). Actually, we will need infinite-dimensional highest-weight representations $\widetilde{\pi}^\lambda$ defined by

$$E_i v^\lambda = 0, \quad i = 1, \ldots, n-1, \qquad q^X v^\lambda = q^{\lambda(X)} v^\lambda, \quad X \in \mathfrak{k}_n, \quad \lambda \in \mathfrak{k}_n^*, \tag{4.7}$$

where v^λ is the highest-weight vector with the weight λ which can be seen in terms of its n components $\lambda_i = \lambda(G_i)$. In particular, denoting the highest-weight vector by v_0, one usually chooses the basis vectors for $n = 2$ as

$$v_k = F^k v_0, \qquad E v_0 = 0, \qquad q^{\nu G_i} v_0 = q^{\nu \lambda_i} v_0, \quad i = 1, 2,$$

[4] The term *quantum loop algebra* is also used.

[5] We use the same notation for the generators of the quantum group as for the corresponding Lie algebra. The notation q^X, $X \in \mathfrak{k}_n$, is used to emphasize that $\mathfrak{k}_n$ parametrizes the corresponding set of elements of $U_q(\mathfrak{gl}_n)$.

and for $n = 3$ one chooses

$$v_k = F_1^{k_1} F_3^{k_3} F_2^{k_2} v_0, \qquad E_i v_0 = 0, \qquad q^{\nu G_i} v_0 = q^{\lambda_i} v_0, \quad i = 1,2,3,$$

where $F_3 = F_2 F_1 - q F_1 F_2$ is the root vector corresponding to the root $-\alpha_1 - \alpha_2$. The infinite-dimensional $U_q(\mathfrak{gl}_n)$-modules corresponding to the representations $\widetilde{\pi}^\lambda$ are denoted by $\widetilde{V}^\lambda$. We will also need the corresponding finite-dimensional representations π^λ that can be obtained from $\widetilde{\pi}^\lambda$ as the quotient representations over infinite-dimensional maximal sub-representations in the case when all the differences $\lambda_i - \lambda_{i+1}$ are non-negative integers. The finite-dimensional $U_q(\mathfrak{gl}_n)$-modules corresponding to the representations π^λ are denoted by V^λ.

The next object, the quantum affine algebra $U_q(\mathcal{L}(\mathfrak{sl}_n))$, is more complicated, because, unlike the preceding case, it is the quantum group of an infinite-dimensional Lie algebra. Again, it is a unital associative complex algebra which can be defined in terms of its generators and Cartan subalgebra. We start with a reminder that the Lie algebra $\widetilde{\mathcal{L}}(\mathfrak{sl}_n)$ has $2n$ generators e_i, f_i, $i = 0,1,\ldots,n-1$, its Cartan subalgebra $\widetilde{\mathfrak{h}}_n$ can be described by n basis elements h_i, $i = 0,1,\ldots,n-1$, and there is a nontrivial center generated by the element $c = \sum_{i=0}^{n-1} h_i$. All these elements are subject to well-known commutation relations and Serre relations. The quantum group $U_q(\widetilde{\mathcal{L}}(\mathfrak{sl}_n))$ is generated by the elements

$$e_i, \quad f_i, \quad i = 0,1,\ldots,n-1, \qquad q^x, \quad x \in \widetilde{\mathfrak{h}}_n,$$

satisfying certain defining relations. These are the following q-deformed commutation

$$q^0 = 1, \qquad q^{x_1} q^{x_2} = q^{x_1 + x_2}, \tag{4.8}$$
$$q^x e_i q^{-x} = q^{\alpha_i(x)} e_i, \qquad q^x f_i q^{-x} = q^{-\alpha_i(x)} f_i, \tag{4.9}$$
$$[e_i, f_j] = \delta_{ij} \kappa_q^{-1} (q^{h_i} - q^{-h_i}) \tag{4.10}$$

and Serre relations

$$\sum_{k=0}^{1-\widetilde{a}_{ij}} (-1)^k \begin{bmatrix} 1-\widetilde{a}_{ij} \\ k \end{bmatrix}_q (e_i)^{1-\widetilde{a}_{ij}-k} e_j (e_i)^k = 0, \tag{4.11}$$
$$\sum_{k=0}^{1-\widetilde{a}_{ij}} (-1)^k \begin{bmatrix} 1-\widetilde{a}_{ij} \\ k \end{bmatrix}_q (f_i)^{1-\widetilde{a}_{ij}-k} f_j (f_i)^k = 0, \tag{4.12}$$

where $\widetilde{a}_{ij}$ are the entries of the generalized Cartan matrix of type $A_{n-1}^{(1)}$, and the set of simple positive roots is extended by the additional root α_0.

Since we need finite-dimensional representations, we should first note that there is no finite-dimensional representation of $\widetilde{\mathcal{L}}(\mathfrak{sl}_n)$ with $c \neq 0$, hence we consider the loop Lie algebra $\mathcal{L}(\mathfrak{sl}_n)$ defined as the quotient

$$\mathcal{L}(\mathfrak{sl}_n) = \widetilde{\mathcal{L}}(\mathfrak{sl}_n)/\mathbb{C}c.$$

Moreover, also the quantum group $U_q(\widetilde{\mathcal{L}}(\mathfrak{sl}_n))$ does not have any finite-dimensional representations with $q^{\nu c} \neq 1$ for any $\nu \in \mathbb{C}^\times$. Therefore, we consider the quantum group $U_q(\mathcal{L}(\mathfrak{sl}_n))$ defined as the quotient

$$U_q(\mathcal{L}(\mathfrak{sl}_n)) = U_q(\widetilde{\mathcal{L}}(\mathfrak{sl}_n))/\langle q^{\nu c} - 1 \rangle_{\nu \in \mathbb{C}}.$$

The quantum group $U_q(\mathcal{L}(\mathfrak{sl}_n))$ can be considered in terms of the same generators and defining relations as $U_q(\widetilde{\mathcal{L}}(\mathfrak{sl}_n))$, where the additional relation $q^{\nu c} = 1$, $\nu \in \mathbb{C}$, is taken into account.

Journal of Physics: Conference Series **670** (2016) 012037 doi:10.1088/1742-6596/670/1/012037

The restriction $q^{\nu c} = 1$ makes it possible to construct finite-dimensional representations of the quantum affine algebras under consideration.

The Hopf algebra structure of $U_q(\mathcal{L}(\mathfrak{sl}_n))$ can be defined by the relations with the comultiplication Δ,

$$\Delta(q^x) = q^x \otimes q^x, \qquad \Delta(e_i) = e_i \otimes 1 + q^{-h_i} \otimes e_i, \qquad \Delta(f_i) = f_i \otimes q^{h_i} + 1 \otimes f_i, \tag{4.13}$$

antipode S,

$$S(q^x) = q^{-x}, \qquad S(e_i) = -q^{h_i} e_i, \qquad S(f_i) = -f_i \, q^{-h_i}, \tag{4.14}$$

and counit ε,

$$\varepsilon(q^x) = 1, \qquad \varepsilon(e_i) = 0, \qquad \varepsilon(f_i) = 0, \tag{4.15}$$

acting on the generators explicitly as given.

5. Jimbo's homomorphism and representations of $U_q(\mathcal{L}(\mathfrak{sl}_n))$

As the representation φ mentioned in section 3 we use a representation $\widetilde{\varphi}^{\lambda}_{\zeta}$ constructed as follows.[6] We first define a general grading of the quantum affine algebra $U_q(\mathcal{L}(\mathfrak{sl}_n))$ with the help of a map Γ_ζ which acts on the generators as

$$\Gamma_\zeta(q^x) = q^x, \quad x \in \widetilde{\mathfrak{h}}_n, \qquad \Gamma_\zeta(e_i) = \zeta^{s_i} e_i, \qquad \Gamma_\zeta(f_i) = \zeta^{-s_i} f_i, \tag{5.1}$$

where $\zeta \in \mathbb{C}^\times$ is called the *spectral parameter*, and s_i are arbitrary integers. We denote the total sum of these integers by s. Secondly, we use Jimbo's homomorphism from the quantum affine algebra $U_q(\mathcal{L}(\mathfrak{sl}_n))$ to the quantum group $U_q(\mathfrak{gl}_n)$ [30],

$$\varphi : U_q(\mathcal{L}(\mathfrak{sl}_n)) \to U_q(\mathfrak{gl}_n). \tag{5.2}$$

For some values of n of our specific interest, we write it down explicitly. Thus, Jimbo's homomorphism $\varphi : U_q(\mathcal{L}(\mathfrak{sl}_2)) \to U_q(\mathfrak{gl}_2)$ is given by the relations

$$\varphi(q^{\nu h_0}) = q^{\nu(G_2 - G_1)}, \qquad \varphi(q^{\nu h_1}) = q^{\nu(G_1 - G_2)}, \tag{5.3}$$
$$\varphi(e_0) = F \, q^{-G_1 - G_2}, \qquad \varphi(e_1) = E, \tag{5.4}$$
$$\varphi(f_0) = E \, q^{G_1 + G_2}, \qquad \varphi(f_1) = F, \tag{5.5}$$

and for $n = 3$, we have $\varphi : U_q(\mathcal{L}(\mathfrak{sl}_3)) \to U_q(\mathfrak{gl}_3)$ defined as

$$\varphi(q^{\nu h_0}) = q^{\nu(G_3 - G_1)}, \qquad \varphi(q^{\nu h_1}) = q^{\nu(G_1 - G_2)}, \qquad \varphi(q^{\nu h_2}) = q^{\nu(G_2 - G_3)}, \tag{5.6}$$
$$\varphi(e_0) = F_3 \, q^{-G_1 - G_3}, \qquad \varphi(e_1) = E_1, \qquad \varphi(e_2) = E_2, \tag{5.7}$$
$$\varphi(f_0) = E_3 \, q^{G_1 + G_3}, \qquad \varphi(f_1) = F_1, \qquad \varphi(f_2) = F_2, \tag{5.8}$$

where $E_3 = E_1 E_2 - q^{-1} E_2 E_1$ and $F_3 = F_2 F_1 - q F_1 F_2$. Thirdly, and finally, we use the highest-weight representation $\widetilde{\pi}^{\lambda}$ of $U_q(\mathfrak{gl}_n)$ briefly described in section 4, see equation (4.7) and around. We understand that Jimbo's homomorphism allows us to use representations of $U_q(\mathfrak{gl}_n)$ to construct representations of $U_q(\mathcal{L}(\mathfrak{sl}_n))$. Hence, our basic representation $\widetilde{\varphi}^{\lambda}_{\zeta}$ is constructed as the superposition of the above three maps,

$$\widetilde{\varphi}^{\lambda}_{\zeta} = \widetilde{\pi}^{\lambda} \circ \varphi \circ \Gamma_\zeta.$$

[6] In our notation, the tilde corresponds to infinite-dimensional representations; for the corresponding finite-dimensional representations we use the same symbols simply omitting the tilde.

Here, using the finite-dimensional representations π^λ of $U_q(\mathfrak{gl}_n)$, we will obtain finite-dimensional representations of $U_q(\mathcal{L}(\mathfrak{sl}_n))$ with $\varphi^\lambda_\zeta = \pi^\lambda \circ \varphi \circ \Gamma_\zeta$.

Recall again that the universal R-matrix is an element of the tensor product of two Borel subalgebras of the quantum group, in the case under consideration, $\mathcal{R} \in U_q(\mathfrak{b}_+) \otimes U_q(\mathfrak{b}_-)$, where the Borel subalgebra $U_q(\mathfrak{b}_+)$ of $U_q(\mathcal{L}(\mathfrak{sl}_n))$ is generated by the elements e_i, $i = 0, 1, \ldots, n-1$, and q^x, and the Borel subalgebra $U_q(\mathfrak{b}_-)$ of $U_q(\mathcal{L}(\mathfrak{sl}_n))$ is generated by the elements f_i, $i = 0, 1, \ldots, n-1$, and q^x. One can try to define the representation ρ of $U_q(\mathfrak{b}_+)$ introduced in section 3 in the following way. Let $\widetilde{\varphi}^\lambda_\zeta$ be a representation of $U_q(\mathcal{L}(\mathfrak{sl}_n))$ constructed in accordance with the above prescription, and $\xi \in \widetilde{\mathfrak{h}}^*_n$. Then the relations

$$\widetilde{\varphi}^\lambda_\zeta[\xi](e_i) = \widetilde{\varphi}^\lambda_\zeta(e_i), \qquad \widetilde{\varphi}^\lambda_\zeta[\xi](q^x) = q^{\xi(x)} \widetilde{\varphi}^\lambda_\zeta(q^x) \tag{5.9}$$

define a representation of $U_q(\mathfrak{b}_+)$ called a *shifted representation*. We see that the only difference between the shifted and initial representations appears simply in the factor $q^{\xi(x)}$ in the second relation of (5.9). Moreover, choosing the twist element t explicitly in the form

$$t = q^{\sum_{i=0}^{n-1} \phi_i h_i / n}$$

for some complex numbers ϕ_i, which are subject to the condition $\sum_i \phi_i = 0$ to respect the restriction $q^{\nu c} = 1$, we obtain that the universal integrability objects based on the representation $\widetilde{\varphi}^\lambda_\zeta$ are related to the universal integrability objects based on its shifted counterpart $\widetilde{\varphi}^\lambda_\zeta[\xi]$ simply as

$$\mathcal{T}_{\widetilde{\varphi}^\lambda[\xi]}(\zeta) = \mathcal{T}_{\widetilde{\varphi}^\lambda}(\zeta)\, q^{\sum_{i=0}^{n-1} \xi(h_i) h'_i / n}, \tag{5.10}$$

where we use the notation $h'_i = h_i + \phi_i$.

It is not difficult to see that the shifted representation $\widetilde{\varphi}^\lambda_\zeta[\xi]$ for a nonzero ξ cannot be extended to a representation of the full quantum affine algebra. Therefore, the shifted representation is what we actually need to construct analogs of the representation ρ introduced in section 3. It is clear that the universal Q-operators constructed with the help of the shifted representations $\widetilde{\varphi}^\lambda_\zeta[\xi]$ will be connected with the corresponding universal transfer operators based on $\widetilde{\varphi}^\lambda_\zeta$ according to (5.10). Hence, we do not obtain a really new object. However, considering all the $U_q(\mathfrak{b}_+)$-modules with general nonzero shifts ξ, we can demonstrate that there are interesting limits of the corresponding representations when the differences $\lambda_i - \lambda_{i+1} = \mu_i$, $i = 1, \ldots, n-1$, go to infinity and ξ is chosen appropriately. To be precise, we note that for $n > 2$ the obtained representation is reducible and we take the corresponding irreducible subrepresentation. The final formulas for the Q-operators look for $n = 2$ as

$$\mathcal{Q}(\zeta) = \lim_{\mu \to \infty} \left(\widetilde{\mathcal{T}}^{(\mu,0)}(q^{-1/s}\zeta)\, q^{(\mu h'_0 - \mu h'_1)/2} \right),$$

and for $n = 3$ as

$$\mathcal{Q}(\zeta) = (1 - q^{(h'_0 - 2h'_1 + h'_2)/3}) \lim_{\mu_1, \mu_2 \to \infty} \left(\widetilde{\mathcal{T}}^{(\mu_1+\mu_2, \mu_2, 0)}(q^{-2/s}\zeta)\, q^{((\mu_1+\mu_2)h'_0 - \mu_1 h'_1 - \mu_2 h'_2)/3} \right).$$

Here, we use the notation $\widetilde{\mathcal{T}}^\lambda(\zeta)$ for $\mathcal{T}_{\widetilde{\varphi}^\lambda}(\zeta)$, and we do not write explicitly the representation ρ_ζ in $\mathcal{Q}(\zeta)$, simply keeping in mind that it is obtained from $\widetilde{\varphi}^\lambda_\zeta[\xi]$ by the procedure shortly described above.

The above two (lower and higher rank) basic examples can be directly generalized to the case of general n. Remarkably, it follows from the limit relation between the universal integrability objects that also the universal Q-operators commute as well as the universal transfer operators in (3.6) and (3.7),

$$\mathcal{Q}_{\rho_1}(\zeta_1)\mathcal{Q}_{\rho_2}(\zeta_2) = \mathcal{Q}_{\rho_2}(\zeta_2)\mathcal{Q}_{\rho_1}(\zeta_1). \tag{5.11}$$

The representations ρ_ζ have a useful interpretation in terms of the so-called *q-oscillators*, see e. g. [9–11, 31]. We define them by a natural deformation of the usual oscillators with the deformation parameter $q = \exp \hbar$, where $\hbar$ is a complex number, such that q is not a root of unity. The q-oscillator algebra Osc_q is a unital associative $\mathbb{C}$-algebra with generators $b^\dagger$, b, $q^{\nu N}$, $\nu \in \mathbb{C}$, and relations

$$q^0 = 1, \qquad q^{\nu_1 N} q^{\nu_2 N} = q^{(\nu_1+\nu_2)N},$$
$$q^{\nu N} b^\dagger q^{-\nu N} = q^\nu b^\dagger, \qquad q^{\nu N} b q^{-\nu N} = q^{-\nu} b,$$
$$b^\dagger b = [N]_q, \qquad b b^\dagger = [N+1]_q.$$

The monomials $(b^\dagger)^{k+1} q^{\nu N}$, $b^{k+1} q^{\nu N}$ and $q^{\nu N}$ for $k \in \mathbb{Z}_+$ and $\nu \in \mathbb{C}$ form a basis of Osc_q.

The representations of the q-oscillator algebra are constructed as follows. One can see that the relations

$$q^{\nu N} v_k = q^{\nu k} v_k, \tag{5.12}$$
$$b^\dagger v_k = v_{k+1}, \qquad b\, v_k = [k]_q v_{k-1}, \tag{5.13}$$

supplied with the assumption $v_{-1} = 0$, endow free vector space generated by the set $\{v_0, v_1, \ldots\}$ with the structure of an Osc_q-module. We denote the corresponding representations by χ.[7]

In the case of $n = 2$ for the representation ρ_ζ we have

$$q^{\nu h_0} v_k = q^{2\nu k} v_k, \qquad q^{\nu h_1} v_k = q^{-2\nu k} v_k,$$
$$e_0 v_k = v_{k+1}, \qquad e_1 v_k = \kappa_q^{-1} q^{-k} [k]_q v_{k-1}.$$

Here v_k, $k \in \mathbb{Z}_+$, are the basis vectors in the representation space. Comparing these relations with (5.12) and (5.13), we see that it is natural to define the mapping θ from $\mathrm{U}_q(\mathfrak{b}_+)$ to Osc_q as

$$\theta(q^{\nu h_0}) = q^{2\nu N}, \qquad \theta(q^{\nu h_1}) = q^{-2\nu N},$$
$$\theta(e_0) = b^\dagger, \qquad \theta(e_1) = \kappa_q^{-1} b\, q^{-N}.$$

It can be shown that θ is a homomorphism. Now, for the representation ρ_ζ we have

$$\rho_\zeta = \chi \circ \theta \circ \Gamma_\zeta. \tag{5.14}$$

In the case of $n = 3$ we need two copies of the algebra Osc_q. The corresponding homomorphism from $\mathrm{U}_q(\mathfrak{b}_+)$ to $\mathrm{Osc}_q \otimes \mathrm{Osc}_q$ has here the form

$$\theta(q^{\nu h_0}) = q^{\nu(2N_1+N_2)}, \qquad \theta(q^{\nu h_1}) = q^{\nu(-N_1+N_2)}, \qquad \theta(q^{\nu h_2}) = q^{\nu(-N_1-2N_2)},$$
$$\theta(e_0) = b_1^\dagger q^{-N_2}, \qquad \theta(e_1) = -b_1 b_2^\dagger q^{-N_1+N_2+1}, \qquad \theta(e_2) = \kappa_q^{-1} b_2 q^{-N_2}.$$

One can produce more universal integrability objects with the help of the automorphisms of the quantum group $U_q(\mathcal{L}(\mathfrak{sl}_n))$. These are the automorphism σ corresponding to the cyclic

[7] There is one more useful representation, but we do not use it here.

Journal of Physics: Conference Series **670** (2016) 012037 doi:10.1088/1742-6596/670/1/012037

permutations of the Dynkin diagram of the Kac–Moody Lie algebra of type $A^{(1)}_{n-1}$ transforming the simple positive roots as $\alpha_i \to \alpha_{i+1}$, and the automorphism τ acting as $\alpha_i \to \alpha_{n-i}$ while leaving α_0 alone. Here, $\sigma^n = \mathrm{id}$ and $\tau^2 = \mathrm{id}$. If $\theta : U_q(\mathfrak{b}_+) \to \underbrace{\mathrm{Osc}_q \otimes \cdots \otimes \mathrm{Osc}_q}_{n-1}$ is the initial homomorphism for the basic representation ρ_ζ, we define

$$\theta_i = \theta \circ \sigma^{-i}, \qquad \overline{\theta}_i = \theta \circ \tau \circ \sigma^{-i+1}, \qquad i = 1, \ldots, n,$$

and use the formulas of type (5.14) to define the representations $\rho_{i\zeta}$ and $\overline{\rho}_{i\zeta}$ and then the set of $2n$ universal Q-operators. Similarly, one can produce more representations $\widetilde{\varphi}^{\lambda}_{i\zeta}$ with the help of the automorphisms σ and τ starting from a basic one, and construct the respective universal transfer operators. However, not all of them will be independent. In fact, there is only one universal transfer operator for $n = 2$, and there are two independent universal transfer operators in the higher-rank case.

6. The universal functional relations

6.1. The key relations

Thus, we have $2n$ different universal Q-operators.[8] Specifying the irreducible representations ρ_ζ and constructing the corresponding universal Q-operators, we can use the formula for their products given in section 3. We see here that to analyze these products, we have to consider tensor products $(\rho_{\ell\zeta_\ell} \otimes_\Delta \cdots \otimes_\Delta \rho_{1\zeta_1})$ of the representations $\rho_{i\zeta}$, with $\ell = 2, 3, \ldots, n$, and go similarly with $\overline{\rho}_{i\zeta}$. Then, choosing appropriate bases for the corresponding $U_q(\mathfrak{b}_+)$-modules $(W_\ell)_{\zeta_\ell} \otimes_\Delta \cdots \otimes_\Delta (W_1)_{\zeta_1}$, we should consider their defining module relations. In this way we obtain the whole set of the universal functional relations.

In particular, putting $\ell = 2$ in a higher-rank case ($n = 3$), we see that the $U_q(\mathfrak{b}_+)$-module $(W_2)_{\zeta_2} \otimes_\Delta (W_1)_{\zeta_1}$ for some special choice of ζ_1 and ζ_2 is reducible and the corresponding quotient module is isomorphic to $(\overline{W}_3)_\zeta[\xi]$ with a new spectral parameter ζ expressed in terms of ζ_1 and ζ_2, and a certain shift ξ of the corresponding representation $\overline{\rho}_{3\zeta}$. Similarly, one can obtain the $U_q(\mathfrak{b}_+)$-module $(W_3)_\zeta[\xi]$ for some ζ and ξ as a quotient of $(\overline{W}_2)_{\zeta_2} \otimes_\Delta (\overline{W}_1)_{\zeta_1}$. This consideration allows us to write down the universal double-Q functional relations in the determinant form,

$$\mathcal{C}_i \overline{\mathcal{Q}}_i(\zeta) = \mathcal{Q}_j(q^{1/s}\zeta)\mathcal{Q}_k(q^{-1/s}\zeta) - \mathcal{Q}_j(q^{-1/s}\zeta)\mathcal{Q}_k(q^{1/s}\zeta), \tag{6.1}$$

$$\mathcal{C}_i \mathcal{Q}_i(\zeta) = \overline{\mathcal{Q}}_j(q^{-1/s}\zeta)\overline{\mathcal{Q}}_k(q^{1/s}\zeta) - \overline{\mathcal{Q}}_j(q^{1/s}\zeta)\overline{\mathcal{Q}}_k(q^{-1/s}\zeta), \tag{6.2}$$

where (i, j, k) runs over all cyclic permutations of the set $(1, 2, 3)$, and we use the notation

$$\mathcal{C}_i = q^{-\mathcal{D}_i/s}(q^{2\mathcal{D}_j/s} - q^{2\mathcal{D}_k/s})^{-1}, \qquad \mathcal{D}_i = (h'_{i-1} - h'_i)s/6. \tag{6.3}$$

Note that such relations are absent in the lower-rank case.

To derive the major functional relations, we have to put $\ell = n$ and analyze the n-tuple product representation $(\rho_{n\zeta_n} \otimes_\Delta \cdots \otimes_\Delta \rho_{1\zeta_1})$ corresponding to the tensor product of n $U_q(\mathfrak{b}_+)$-modules $(W_n)_{\zeta_n} \otimes_\Delta \cdots \otimes_\Delta (W_1)_{\zeta_1}$. It allows one to obtain the key relation between the universal transfer operator and universal Q-operators,

$$\mathcal{C}\widetilde{\mathcal{T}}^\lambda(\zeta) = \mathcal{Q}_1(q^{-2(\lambda+\rho)_1/s}\zeta) \cdots \mathcal{Q}_n(q^{-2(\lambda+\rho)_n/s}\zeta), \qquad \mathcal{C} = \mathcal{C}_1 \cdots \mathcal{C}_n, \tag{6.4}$$

where $\rho \in \mathfrak{k}^*_n$ is the half-sum of the positive roots,

$$\rho = ((n-1)/2, (n-3)/2, \ldots, -(n-1)/2).$$

[8] One has only two universal Q-operators for the lower-rank case (for $n = 2$).

This is the central equation from which follow all other functional relations. One can see that relation (6.4) contains only the universal transfer operator for the infinite-dimensional representation $\widetilde{\varphi}^{\lambda}_{\zeta}$. To obtain the corresponding relations for the universal transfer operator based on the finite-dimensional representation $\varphi^{\lambda}_{\zeta}$, one uses the quantum version of the so-called *B.G.G. resolution* [18, 19, 32], which implies an exact sequence of $U_q(\mathfrak{gl}_n)$-modules and $U_q(\mathfrak{gl}_n)$-homomorphisms,

$$U_k = \bigoplus_{\substack{w \in W \\ \ell(w)=k}} \widetilde{V}^{w\cdot\lambda}, \qquad w\cdot\lambda = w(\lambda+\rho)-\rho,$$

$$0 \longrightarrow U_n \xrightarrow{\varphi_n} U_{n-1} \xrightarrow{\varphi_{n-1}} \dots \xrightarrow{\varphi_1} U_0 \xrightarrow{\varphi_0} U_{-1} \longrightarrow 0, \qquad U_{-1} = V^{\lambda},$$

where w means any element of the Weyl group W of the root system of $\mathfrak{gl}_n$, with $\ell(w)$ being its length, and $w\cdot\lambda$ stands for the affine action of w defined explicitly as above. The B.G.G. resolution allows one to express the trace over the finite-dimensional $U_q(\mathfrak{gl}_n)$-module V^{λ} as certain linear combination of traces over the infinite-dimensional $U_q(\mathfrak{gl}_n)$-modules $\widetilde{V}^{w\cdot\lambda}$,

$$\mathrm{tr}^{\lambda} = \sum_{w\in W} (-1)^{\ell(w)} \widetilde{\mathrm{tr}}^{w\cdot\lambda}. \tag{6.5}$$

Then, using our definition of the universal transfer operator (3.2), we immediately come to a remarkable relation between the universal transfer operator based on the finite-dimensional representation $\varphi^{\lambda}_{\zeta}$ and universal transfer operators based on the infinite-dimensional representations $\widetilde{\varphi}^{w\cdot\lambda}_{\zeta}$,

$$\mathcal{T}^{\lambda}(\zeta) = \sum_{p\in S_n} \mathrm{sgn}(p)\, \widetilde{\mathcal{T}}^{p(\lambda+\rho)-\rho}(\zeta). \tag{6.6}$$

It is taken into account here that W can be identified with the symmetric group S_n, and so, p is an element of S_n acting on $\mathfrak{k}_n$ by appropriate permutations. Relation (6.6) leads to the following representation of the universal transfer operator in terms of the universal Q-operators:

$$\mathcal{C}\,\mathcal{T}^{\lambda-\rho}(\zeta) = \det\left(\mathcal{Q}_i(q^{-2\lambda_j/s}\zeta)\right)_{i,j=1,\dots,n}. \tag{6.7}$$

Similar consideration holds as well when also the automorphism τ is involved giving rise to the barred universal integrability objects. In this case we come to the determinant representation

$$\mathcal{C}\,\overline{\mathcal{T}}^{\lambda-\rho}(\zeta) = \det\left(\overline{\mathcal{Q}}_i(q^{2\lambda_j/s}\zeta)\right)_{i,j=1,\dots,n} \tag{6.8}$$

It is worthwhile noting that for the lower-rank case, $n = 2$, we would be able to obtain the universal functional relations in the determinant form even if we started with Jimbo's homomorphism from $U_q(\mathcal{L}(\mathfrak{sl}_2))$ to $U_q(\mathfrak{sl}_2)$. However, for higher-rank cases, $n \geq 3$, it is most convenient to use Jimbo's homomorphism from $U_q(\mathcal{L}(\mathfrak{sl}_n))$ to $U_q(\mathfrak{gl}_n)$.

6.2. Universal TQ- and TT-relations

Further universal functional relations can be obtained from the vanishing of the determinants of certain $(n+1)\times(n+1)$ matrices with one dependent row using equations (6.7), (6.8). The universal TQ-relations follow from the identity

$$\det \begin{pmatrix} \mathcal{Q}_1(q^{-2\lambda_1/s}\zeta) & \cdots & \mathcal{Q}_1(q^{-2\lambda_n/s}\zeta) & \mathcal{Q}_1(q^{-2\lambda_{n+1}/s}\zeta) \\ \mathcal{Q}_2(q^{-2\lambda_1/s}\zeta) & \cdots & \mathcal{Q}_2(q^{-2\lambda_n/s}\zeta) & \mathcal{Q}_2(q^{-2\lambda_{n+1}/s}\zeta) \\ \vdots & \ddots & \vdots & \vdots \\ \mathcal{Q}_n(q^{-2\lambda_1/s}\zeta) & \cdots & \mathcal{Q}_n(q^{-2\lambda_n/s}\zeta) & \mathcal{Q}_n(q^{-2\lambda_{n+1}/s}\zeta) \\ \mathcal{Q}_j(q^{-2\lambda_1/s}\zeta) & \cdots & \mathcal{Q}_j(q^{-2\lambda_n/s}\zeta) & \mathcal{Q}_j(q^{-2\lambda_{n+1}/s}\zeta) \end{pmatrix} = 0$$

obviously satisfied for any $j = 1, \ldots, n$. Such matrices can be constructed also for the barred universal integrability objects. In the lower-rank case, putting $n = 2$, we obtain

$$\mathcal{T}^{(\lambda_1-1/2,\lambda_2+1/2)}(\zeta)\mathcal{Q}_j(q^{-2\lambda_3/s}\zeta) - \mathcal{T}^{(\lambda_1-1/2,\lambda_3+1/2)}(\zeta)\mathcal{Q}_j(q^{-2\lambda_2/s}\zeta) + \mathcal{T}^{(\lambda_2-1/2,\lambda_3+1/2)}(\zeta)\mathcal{Q}_j(q^{-2\lambda_1/s}\zeta) = 0. \quad (6.9)$$

We call this equation the universal TQ-relation. For example, choosing in (6.9) the components of λ as $\lambda_1 = 1$, $\lambda_2 = 0$, $\lambda_3 = -1$, after some transformations based on obvious symmetry properties of $\mathcal{T}^{(\lambda_1,\lambda_2)}(\zeta)$, we derive

$$\mathcal{T}^{(1/2,-1/2)}(\zeta)\mathcal{Q}_j(\zeta) = \mathcal{Q}_j(q^{2/s}\zeta) + \mathcal{Q}_j(q^{-2/s}\zeta).$$

This is the universal analog of the famous Baxter's TQ-relation, that is given in the form independent of the representation of the quantum group in the quantum space.

In a higher-rank case, $n = 3$, we obtain the $U_q(\mathcal{L}(\mathfrak{sl}_3))$ universal TQ-relations

$$\mathcal{T}^{(\lambda_1-1,\lambda_2,\lambda_3+1)}(\zeta)\mathcal{Q}_j(q^{-2\lambda_4/s}\zeta) - \mathcal{T}^{(\lambda_1-1,\lambda_2,\lambda_4+1)}(\zeta)\mathcal{Q}_j(q^{-2\lambda_3/s}\zeta) + \mathcal{T}^{(\lambda_1-1,\lambda_3,\lambda_4+1)}(\zeta)\mathcal{Q}_j(q^{-2\lambda_2/s}\zeta) - \mathcal{T}^{(\lambda_2-1,\lambda_3,\lambda_4+1)}(\zeta)\mathcal{Q}_j(q^{-2\lambda_1/s}\zeta) = 0.$$

Now, with the choice $\lambda_1 = 2$, $\lambda_2 = 1$, $\lambda_3 = 0$, $\lambda_4 = -1$, also using symmetry properties of $\mathcal{T}^{(\lambda_1,\lambda_2,\lambda_3)}(\zeta)$, we derive

$$\mathcal{T}^{(1,1,0)}(\zeta)\mathcal{Q}_j(\zeta) - \mathcal{T}^{(1,0,0)}(\zeta)\mathcal{Q}_j(q^{-2/s}\zeta) = \mathcal{Q}_j(q^{2/s}\zeta) - \mathcal{Q}_j(q^{-4/s}\zeta). \quad (6.10)$$

In a similar way we obtain

$$\overline{\mathcal{T}}^{(1,1,0)}(\zeta)\overline{\mathcal{Q}}_j(\zeta) - \overline{\mathcal{T}}^{(1,0,0)}(\zeta)\overline{\mathcal{Q}}_j(q^{2/s}\zeta) = \overline{\mathcal{Q}}_j(q^{-2/s}\zeta) - \overline{\mathcal{Q}}_j(q^{4/s}\zeta). \quad (6.11)$$

Unlike the lower-rank case, each of the above equations involves different universal transfer operators. However, one can derive functional relations containing only one universal transfer operator, $\mathcal{T}^{(1,0,0)}(\zeta)$ or $\mathcal{T}^{(1,1,0)}(\zeta)$, or their barred analog, but having mixed $\mathcal{Q}_i$ and $\overline{\mathcal{Q}}_j$ for distinct i and j in one same equation. To this end, one can use the Jacoby identity for determinants [33]. Then, in the simplest higher-rank case $n = 3$, we obtain the corresponding $TQ\overline{Q}$-relations

$$\mathcal{T}^{(1,0,0)}(\zeta)\mathcal{Q}_i(q^{-2/s}\zeta)\overline{\mathcal{Q}}_j(q^{-1/s}\zeta) = \mathcal{Q}_i(q^{-4/s}\zeta)\overline{\mathcal{Q}}_j(q^{-1/s}\zeta) + \mathcal{Q}_i(\zeta)\overline{\mathcal{Q}}_j(q^{-3/s}\zeta) + \mathcal{Q}_i(q^{-2/s}\zeta)\overline{\mathcal{Q}}_j(q^{1/s}\zeta)$$

and

$$\mathcal{T}^{(1,1,0)}(\zeta)\mathcal{Q}_i(\zeta)\overline{\mathcal{Q}}_j(q^{-1/s}\zeta) = \mathcal{Q}_i(q^{2/s}\zeta)\overline{\mathcal{Q}}_j(q^{-1/s}\zeta) + \mathcal{Q}_i(q^{-2/s}\zeta)\overline{\mathcal{Q}}_j(q^{1/s}\zeta) + \mathcal{Q}_i(\zeta)\overline{\mathcal{Q}}_j(q^{-3/s}\zeta).$$

The universal TT-relations can be derived from the equation

$$\det \begin{pmatrix} \mathcal{Q}_1(q^{-2\lambda_1/s}\zeta) & \cdots & \mathcal{Q}_1(q^{-2\lambda_n/s}\zeta) & \mathcal{Q}_1(q^{-2\lambda_{n+1}/s}\zeta) \\ \mathcal{Q}_2(q^{-2\lambda_1/s}\zeta) & \cdots & \mathcal{Q}_2(q^{-2\lambda_n/s}\zeta) & \mathcal{Q}_2(q^{-2\lambda_{n+1}/s}\zeta) \\ \vdots & \ddots & \vdots & \vdots \\ \mathcal{Q}_n(q^{-2\lambda_1/s}\zeta) & \cdots & \mathcal{Q}_n(q^{-2\lambda_n/s}\zeta) & \mathcal{Q}_n(q^{-2\lambda_{n+1}/s}\zeta) \\ \mathcal{T}^{(\lambda'_1,\lambda'_{n+2},\ldots,\lambda'_{2n})}(\zeta) & \cdots & \mathcal{T}^{(\lambda'_n,\lambda'_{n+2},\ldots,\lambda'_{2n})}(\zeta) & \mathcal{T}^{(\lambda'_{n+1},\lambda'_{n+2},\ldots,\lambda'_{2n})}(\zeta) \end{pmatrix} = 0,$$

where $\lambda'_j = \lambda_j - (n-1)/2$, $j = 1, \ldots, n+1$, and $\lambda'_{n+1+k} = \lambda_{n+1+k} - (n-2k-1)/2$, $k = 1, \ldots, n-1$. Using equation (6.7) to express the operators $\mathcal{T}^{(\lambda'_j,\lambda'_{n+2},\ldots,\lambda'_{2n})}(\zeta)$ in terms of the universal Q-operators for all $j = 1, \ldots, n+1$, one can see that the last row of the above $(n+1) \times (n+1)$ matrix is a linear combination of the first n rows. Therefore, its determinant identically vanishes. Expanding this determinant over the last row and using again (6.7), we obtain the universal TT-relations. The case with the barred universal integrability objects can be considered in the same way. Actually, the universal TT-relations for the barred integrability objects can be obtained directly from the relations for the unbarred quantities changing q by q^{-1} there [18].

In the simplest lower-rank case we obtain

$$\mathcal{T}^{(\lambda_1-1/2,\lambda_2+1/2)}(\zeta)\mathcal{T}^{(\lambda_3-1/2,\lambda_4+1/2)}(\zeta) - \mathcal{T}^{(\lambda_1-1/2,\lambda_3+1/2)}(\zeta)\mathcal{T}^{(\lambda_2-1/2,\lambda_4+1/2)}(\zeta) + \mathcal{T}^{(\lambda_2-1/2,\lambda_3+1/2)}(\zeta)\mathcal{T}^{(\lambda_1-1/2,\lambda_4+1/2)}(\zeta) = 0. \quad (6.12)$$

Choosing the weights subsequently as $\lambda_1 = \ell+1$, $\lambda_2 = \ell$, $\lambda_3 = 0$, $\lambda_4 = -1$ and $\lambda_1 = \ell+1$, $\lambda_2 = \ell$, $\lambda_3 = \ell-1$, $\lambda_4 = -1$, we derive the universal analog of the TT-relations of particular interest,

$$\mathcal{T}^{(\ell,0)}(q^{-1/s}\zeta)\mathcal{T}^{(\ell,0)}(q^{1/s}\zeta) = 1 + \mathcal{T}^{(\ell-1,0)}(q^{-1/s}\zeta)\mathcal{T}^{(\ell+1,0)}(q^{1/s}\zeta) \quad (6.13)$$

and

$$\mathcal{T}^{(1,0)}(q^{-2\ell/s}\zeta)\mathcal{T}^{(\ell,0)}(\zeta) = \mathcal{T}^{(\ell+1,0)}(\zeta) + \mathcal{T}^{(\ell-1,0)}(\zeta), \quad (6.14)$$

respectively.

For a more complicated higher-rank case, $n = 3$, we obtain

$$\mathcal{T}^{(\lambda_1-1,\lambda_2,\lambda_3+1)}(\zeta)\mathcal{T}^{(\lambda_4-1,\lambda_5,\lambda_6+1)}(\zeta) - \mathcal{T}^{(\lambda_1-1,\lambda_2,\lambda_4+1)}(\zeta)\mathcal{T}^{(\lambda_3-1,\lambda_5,\lambda_6+1)}(\zeta) + \mathcal{T}^{(\lambda_1-1,\lambda_3,\lambda_4+1)}(\zeta)\mathcal{T}^{(\lambda_2-1,\lambda_5,\lambda_6+1)}(\zeta) - \mathcal{T}^{(\lambda_2-1,\lambda_3,\lambda_4+1)}(\zeta)\mathcal{T}^{(\lambda_1-1,\lambda_5,\lambda_6+1)}(\zeta) = 0. \quad (6.15)$$

Putting here $\lambda_1 = \ell+2$, $\lambda_2 = \ell+1$, $\lambda_3 = 1$, $\lambda_4 = 0$, $\lambda_5 = 0$, $\lambda_6 = -1$, we derive one TT-relation of particular interest in the universal form,

$$\mathcal{T}^{(\ell-1,0,0)}(\zeta)\mathcal{T}^{(\ell+1,0,0)}(q^{2/s}\zeta) = \mathcal{T}^{(\ell,0,0)}(\zeta)\mathcal{T}^{(\ell,0,0)}(q^{2/s}\zeta) - \mathcal{T}^{(\ell,\ell,0)}(\zeta), \quad (6.16)$$

while the choice $\lambda_1 = \ell+2$, $\lambda_2 = \ell$, $\lambda_3 = 0$, $\lambda_4 = \ell+1$, $\lambda_5 = \ell+1$, $\lambda_6 = -1$ produces another interesting universal TT-relation,

$$\mathcal{T}^{(\ell-1,\ell-1,0)}(q^{-2/s}\zeta)\mathcal{T}^{(\ell+1,\ell+1,0)}(\zeta) = \mathcal{T}^{(\ell,\ell,0)}(q^{-2/s}\zeta)\mathcal{T}^{(\ell,\ell,0)}(\zeta) - \mathcal{T}^{(\ell,0,0)}(\zeta). \quad (6.17)$$

Relations of the type (6.13), (6.14) and (6.16), (6.17) are usually called fusion relations, see [11, 34, 35]. They are given here in the form independent of the representation of the quantum group in the quantum space, that is, they are the *universal fusion relations*.

Journal of Physics: Conference Series **670** (2016) 012037 doi:10.1088/1742-6596/670/1/012037

6.3. *Quantum Jacobi–Trudi identity*

Equation (6.14) allows one to express $\mathcal{T}^{(\ell,0)}$ through $\mathcal{T}^{(1,0)}$. In the higher-rank case the situation is more intricate. Considering $n = 3$, we first note that equation (6.16) can be generalized as follows. Let ℓ_1, ℓ_2 be positive integers, such that $\ell_1 \geq \ell_2$. Choosing the weight components λ_i in (6.15) as $\lambda_1 = \ell_1 + 2$, $\lambda_2 = \ell_2 + 1$, $\lambda_3 = 1$, $\lambda_4 = 0$, $\lambda_5 = 0$, $\lambda_6 = -1$, we obtain

$$\mathcal{T}^{(\ell_1,\ell_2,0)}(\zeta) = \mathcal{T}^{(\ell_1,0,0)}(\zeta)\mathcal{T}^{(\ell_2,0,0)}(q^{2/s}\zeta) - \mathcal{T}^{(\ell_1+1,0,0)}(q^{2/s}\zeta)\mathcal{T}^{(\ell_2-1,0,0)}(\zeta). \tag{6.18}$$

Equation (6.18) implies that $\mathcal{T}^{(\ell_1,\ell_2,0)}(\zeta)$ can be expressed via $\mathcal{T}^{(\ell,0,0)}(\zeta)$, and subsequently via $\mathcal{T}^{(1,0,0)}(\zeta)$, $\mathcal{T}^{(1,1,0)}(\zeta)$. Explicitly, the result of this procedure is given by the determinant

$$\mathcal{T}^{(\ell_1,\ell_2,0)}(\zeta) = \det\left(\mathcal{E}_{\ell_i^t-i+j}(q^{-2(j-1)/s}\zeta)\right)_{1\leq i,j\leq \ell_1}, \tag{6.19}$$

where we use the notation

$$\mathcal{E}_0(\zeta) = \mathcal{E}_3(\zeta) = 1, \qquad \mathcal{E}_1(\zeta) = \mathcal{T}^{(1,0,0)}(\zeta), \qquad \mathcal{E}_2(\zeta) = \mathcal{T}^{(1,1,0)}(\zeta),$$
$$\mathcal{E}_k(\zeta) = 0 \qquad \forall\, k < 0,\ k > 3,$$
$$\ell_i^t = 2, \quad 1 \leq i \leq \ell_2, \qquad \ell_i^t = 1, \quad \ell_2 < i \leq \ell_1.$$

Here, one connects the integers ℓ_1 and ℓ_2 with the Young diagram with the rows of the length ℓ_1 and ℓ_2, then the numbers ℓ_i^t describe the rows of the transposed diagram. Equation (6.19) can thus be regarded as the universal quantum analog of the Jacobi–Trudi identity from the theory of symmetric polynomials [11, 36, 37]. To prove this identity in the universal form, we use $(n+1)$-term universal functional relations and certain symmetries of the universal transfer operators [18].

Remarkably, also the barred universal transfer operators $\overline{\mathcal{T}}^{(\ell_1,\ell_2,0)}(\zeta)$ can be expressed through the same basic universal transfer operators $\mathcal{T}^{(1,0,0)}(\zeta)$ and $\mathcal{T}^{(1,1,0)}(\zeta)$. To come to this conclusion, one can write the Jacobi–Trudi identity for $\overline{\mathcal{T}}^{(\ell_1,\ell_2,0)}(\zeta)$ down and use the relations

$$\overline{\mathcal{T}}^{(1,0,0)}(\zeta) = \mathcal{T}^{(1,0,0)}(q^{3/s}\zeta), \qquad \overline{\mathcal{T}}^{(1,1,0)}(\zeta) = \mathcal{T}^{(1,1,0)}(q^{1/s}\zeta),$$

following actually from the comparison of the representations φ_ζ^λ and $\overline{\varphi}_\zeta^\lambda$.

7. Conclusions

We have presented a short review of the universal functional TQ- and TT-relations for the quantum integrable systems associated with the quantum affine algebra $U_q(\mathcal{L}(\mathfrak{sl}_n))$, emphasizing the lower-rank ($n = 2$) and a higher-rank ($n = 3$) cases as basic examples. We have also given the quantum analog of the Jacobi–Trudi identity allowing one to express 'higher-weight' universal transfer operators by means of two non-trivial basic operators with lowest weight. Here, the representation of the quantum group in the auxiliary space is properly specified, giving rise to this or another universal integrability object, while the representation in the quantum space is not.

To consider concrete physical models, one should make the corresponding specialization of the quantum group also in the quantum space. Thus, choosing an appropriate infinite- or finite-dimensional representation of the quantum group in the quantum space, one can consider either a low-dimensional quantum field theory or a lattice model with the corresponding spin-chain counterpart. Upon such a specialization, one can use the remarkable Khoroshkin–Tolstoy formula [38–41] for the universal R-matrix to carry out explicit calculations of the monodromy operators [19, 42], R-operators [14, 15, 19, 43–46] and L-operators [14, 15]. Another possibility is offered by a $3D$ approach to the Yang–Baxter equation [47].

Acknowledgments
This work was supported in part by the German Research Foundation grant KL 645/10-1, the Russian Foundation for Basic Research grants # 13-01-00217, # 14-01-91335, and by the Volkswagen Foundation. We thank our colleagues and coauthors H. Boos, F. Göhmann and A. Klümper for discussions. Kh.S.N. thanks the organizers of the ISQS'23 in Prague (June 23–27, 2015) for invitation and hospitality.

Appendix. Defining $U_q(\mathfrak{gl}_n)$-module relations
The root system of type A_{n-1} consists of $n(n-1)$ roots. We introduce for the system of positive roots the *normal ordering* [48,49]

$$(\alpha_1), (\alpha_1+\alpha_2, \alpha_2), (\alpha_1+\alpha_2+\alpha_3, \alpha_2+\alpha_3, \alpha_3), \ldots$$
$$\ldots(\alpha_1+\alpha_2+\ldots+\alpha_i, \alpha_2+\ldots+\alpha_i, \ldots, \alpha_i), \ldots$$
$$\ldots(\alpha_1+\alpha_2+\ldots+\alpha_{n-1}, \alpha_2+\ldots+\alpha_{n-1}, \ldots, \alpha_{n-1}),$$

where α_j, $j = 1, \ldots, n-1$ are the simple positive roots. As usual, we define the root vectors corresponding to the negative composite roots as follows. Let $\gamma = \alpha + \beta$ be a composite positive root. Then the relation

$$F_\gamma = F_\beta F_\alpha - q F_\alpha F_\beta$$

gives the root vector F_γ corresponding to the negative root $-\gamma$. The whole set of negative root vectors ordered in accordance with the above given normal ordering of the roots is used to define the basis vectors of the $U_q(\mathfrak{gl}_n)$ modules. We write

$$v_k = (F_1^{k_1})(F_{12}^{k_{12}} F_2^{k_2})(F_{13}^{k_{13}} F_{23}^{k_{23}} F_3^{k_3}) \cdots (F_{1i}^{k_{1i}} F_{2i}^{k_{2i}} \cdots F_i^{k_i}) \cdots (F_{1,n-1}^{k_{1,n-1}} F_{2,n-1}^{k_{2,n-1}} \cdots F_{n-1}^{k_{n-1}}) v_0$$

where k means the set of non-negative integers $(k_1;\ k_{12}, k_2;\ k_{13}, k_{23}, k_3;\ \ldots\ k_{1i}, k_{2i}, \ldots, k_i;\ \ldots\ k_{1,n-1}, k_{2,n-1}, \ldots, k_{n-1})$ being the powers of the root vectors acting on the highest-weight vector v_0. Here we use the notation F_i for the root vector corresponding to the simple negative root $-\alpha_i$, and F_{ij} means the root vector corresponding to the composite negative root $-\alpha_i - \ldots - \alpha_j$ which is defined according to the above relation.

Acting on the basis vectors v_k by the generators $q^{\nu G_i}$, E_i and F_i of $U_q(\mathfrak{gl}_n)$, we obtain the defining relations of the $U_q(\mathfrak{gl}_n)$ modules. For the simplest lower-rank case $n = 2$ the basis vectors are $v_k = F^k v_0$, and we obtain

$$q^{\nu G_1} v_k = q^{\nu(\lambda_1 - k)} v_k, \qquad q^{\nu G_2} v_k = q^{\nu(\lambda_2 + k)} v_k,$$
$$F v_k = v_{k+1}, \qquad E v_k = [k]_q [\lambda_1 - \lambda_2 - k + 1]_q v_{k-1}.$$

For a more complicated and rich higher-rank case $n = 3$ we act on $v_k = F_1^{k_1} F_{12}^{k_{12}} F_2^{k_2} v_0$ and obtain

$$q^{\nu G_1} v_k = q^{\nu(\lambda_1 - k_1 - k_3)} v_k, \quad q^{\nu G_2} v_k = q^{\nu(\lambda_2 + k_1 - k_2)} v_k, \quad q^{\nu G_3} v_k = q^{\nu(\lambda_3 + k_2 + k_3)} v_k,$$
$$F_1 v_k = v_{k+\varepsilon_1}, \quad F_2 v_k = q^{k_1 - k_3} v_{k+\varepsilon_3} + [k_1]_q v_{k-\varepsilon_1+\varepsilon_2}, \quad F_3 v_k = q^{-k_1} v_{k+\varepsilon_2},$$
$$E_1 v_k = [\lambda_1 - \lambda_2 - k_1 + k_2 - k_3 + 1]_q [k_1]_q v_{k-\varepsilon_1} - q^{\lambda_1 - \lambda_2 + k_2 - k_3 + 2} [k_2]_q v_{k-\varepsilon_2+\varepsilon_3},$$
$$E_2 v_k = [\lambda_2 - \lambda_3 - k_2 + 1]_q [k_2]_q v_{k-\varepsilon_3} + q^{-(\lambda_2 - \lambda_3) + 2k_2} [k_3]_q v_{k+\varepsilon_1-\varepsilon_2},$$
$$E_3 v_k = q^{k_1} [\lambda_1 - \lambda_3 - k_1 - k_2 - k_3 + 1]_q [k_3]_q v_{k-\varepsilon_2},$$
$$- q^{-(\lambda_1 - \lambda_2) + k_1 - k_2 + k_3 - 1} [\lambda_2 - \lambda_3 - k_2 + 1]_q [k_1]_q [k_2]_q v_{k-\varepsilon_1-\varepsilon_3}.$$

Here, for brevity, we have denoted F_{12} and k_{12} by F_3 and k_3, respectively, and use the notation $k - \varepsilon_i$, $i = 1, 2, 3$, for the sets of three integers $(k_1 - 1, k_3, k_2)$, $(k_1, k_3 - 1, k_2)$ and $(k_1, k_3, k_2 - 1)$, respectively.

Thorough consideration of these module relations and appropriate interpretation of their consequences is the actual basis for the complete proof of the universal functional relations.

References

[1] Boos H, Jimbo M, Miwa T, Smirnov F and Takeyama Y 2007 Hidden Grassmann structure in the XXZ model *Commun. Math. Phys.* **272** 263–281

[2] Boos H, Jimbo M, Miwa T, Smirnov F and Takeyama Y 2009 Hidden Grassmann structure in the XXZ model II: Creation operators *Commun. Math. Phys.* **286** 875–932

[3] Jimbo M, TMiwa and Smirnov F 2009 Hidden Grassmann structure in the XXZ model III: Introducing the Matsubara direction *J. Phys. A: Math. Theor.* **42** 304018 (31pp)

[4] Boos H, Jimbo M, Miwa T and Smirnov F 2010 Hidden Grassmann structure in the XXZ model IV: CFT limit *Commun. Math. Phys.* **299** 825–866

[5] Jimbo M, Miwa T and Smirnov F 2011 Hidden Grassmann structure in the XXZ model v: sine-Gordon model *Lett. Math. Phys.* **96** 325–365

[6] Baxter R J 1982 *Exactly Solved Models in Statistical Mechanics* (London: Academic Press)

[7] Baxter R J 2004 The six and eight-vertex models revisited *J. Stat. Phys.* **116** 43–66

[8] Bazhanov V V, Lukyanov S L and Zamolodchikov A B 1996 Integrable structure of conformal field theory, quantum KdV theory and thermodynamic Bethe ansatz *Commun. Math. Phys.* **177** 381–398

[9] Bazhanov V V, Lukyanov S L and Zamolodchikov A B 1997 Integrable structure of conformal field theory II. Q-operator and DDV equation *Commun. Math. Phys.* **190** 247–278

[10] Bazhanov V V, Lukyanov S L and Zamolodchikov A B 1999 Integrable structure of conformal field theory III. The Yang–Baxter relation *Commun. Math. Phys.* **200** 297–324

[11] Bazhanov V V, Hibberd A N and Khoroshkin S M 2002 Integrable structure of $\mathcal{W}_3$ conformal field theory, quantum Boussinesq theory and boundary affine Toda theory *Nucl. Phys. B* **622** 475–574

[12] Chari V and Pressley A 1991 Quantum affine algebras *Commun. Math. Phys.* **142** 261–283

[13] Bethe H 1931 Zur Theorie der Metalle I. Eigenwerte und Eigenfunktionen der linearen Atomkette. *Z. Phys.* **71** 205–231

[14] Boos H, Göhmann F, Klümper A, Nirov K S and Razumov A V 2010 Exercises with the universal R-matrix *J. Phys. A: Math. Theor.* **43** 415208 (35pp)

[15] Boos H, Göhmann F, Klümper A, Nirov K S and Razumov A V 2011 On the universal R-matrix for the Izergin–Korepin model *J. Phys. A: Math. Theor.* **44** 355202 (25pp)

[16] Boos H, Göhmann F, Klümper A, Nirov K S and Razumov A V 2013 Universal integrability objects *Theor. Math. Phys.* **174** 21–39

[17] Boos H, Göhmann F, Klümper A, Nirov K S and Razumov A V 2014 Universal R-matrix and functional relations *Rev. Math. Phys.* **26** 1430005 (66pp)

[18] Boos H, Göhmann F, Klümper A, Nirov K S and Razumov A V 2014 Quantum groups and functional relations for higher rank *J. Phys. A: Math. Theor.* **47** 275201 (47pp)

[19] Nirov K S and Razumov A V 2014 Quantum groups and functional relations for lower rank *Preprint* arXiv:1412.7342 [math-ph]

[20] Kac V 1990 *Infinite-Dimensional Lie algebras* (Cambridge: Cambridge University Press)

[21] Antonov A and Feigin B 1997 Quantum group representations and the Baxter equation *Phys. Lett. B* **392** 115–122

[22] Bazhanov V V and Tsuboi Z 2008 Baxter's Q-operators for supersymmetric spin chains *Nucl. Phys. B* **805** 451–516

[23] Drinfeld V G 1985 Hopf algebras and quantum Yang–Baxter equation *Soviet Math. Dokl.* **32** 254–258

[24] Drinfeld V G 1987 *Proceedings of the International Congress of Mathematicians, Berkeley, 1986* vol 1 ed Gleason A E (Providence: American Mathematical Society) pp 798–820

[25] Jimbo M 1985 A q-difference analogue of $\mathrm{U}(\mathfrak{g})$ and the Yang-Baxter equation *Lett. Math. Phys.* **10** 63–69

[26] Chari V and Pressley A 1994 *A Guide to Quantum Groups* (Cambridge: Cambridge University Press)

[27] Jimbo M and Miwa T 1995 *Algebraic Analysis of Solvable Lattice Models* (*Regional Conference Series in Mathematics* no 85) (Providence: American Mathematical Society)

[28] Etingof P, Frenkel B and Kirillov A A 1998 *Lectures on Representation Theory and Knizhnik–Zamolodchikov Equations* (*Mathematical Surveys and Monographs* vol 58) (Providence: American Mathematical Society)

[29] Tanisaki T 1992 *Infinite Analyis* (*Advanced Series in Mathematical Physics* vol 16) ed Tsuchiya A, Eguchi T and Jimbo M (Singapore: World Scientific) pp 941–962

[30] Jimbo M 1986 A q-analogue of $\mathrm{U}(\mathfrak{gl}(N+1))$, Hecke algebra, and the Yang–Baxter equation *Lett. Math. Phys.* **11** 247–252

[31] Klimyk A and Schmüdgen K 1997 *Quantum groups and their representations* Texts and Monographs in Physics (Heidelberg: Springer)

[32] Rosso M 1991 An analogue of B.G.G. resolution for the quantum $SL(N)$-group *Symplectic geometry and mathematical physics* (*Progress in Mathematics* vol 99) ed Donato P, Duval C, Elhadad J and Tynman G M (Aix-en-Provence, 1990: Birkhäuser, Boston) pp 422–432

[33] Hirota R 2004 *The direct method in soliton theory* (Cambridge: Cambridge University Press)

[34] Klümper A and Pearce P A 1992 Conformal weights of RSOS lattice models and their fusion hierarchies *Physica A* **183** 304–350

[35] Kuniba A, Nakanishi T and Suzuki J 1994 Functional relations in solvable lattice models. I. Functional relations and representation theory *Int. J. Mod. Phys. A* **9** 5215–5266

[36] Cherednik I V 1987 An analogue of the character formula for Hecke algebras *Funct. Anal. Appl.* **21** 172–174

[37] Bazhanov V V and Reshetikhin N 1990 Restricted solid-on-solid models connected with simply laced algebras and conformal field theory *J. Phys. A: Math. Gen.* **23** 1477–1492

[38] Khoroshkin S M and Tolstoy V N 1992 The uniqueness theorem for the universal R-matrix *Lett. Math. Phys.* **24** 231–244

[39] Tolstoy V N and Khoroshkin S M 1992 The universal R-matrix for quantum untwisted affine Lie algebras *Funct. Anal. Appl.* **26** 69–71

[40] Khoroshkin S M and Tolstoy V N 1993 On Drinfeld's realization of quantum affine algebras *J. Geom. Phys.* **11** 445–452

[41] Khoroshkin S M, Stolin A A and Tolstoy V N 1995 Gauss decomposition of trigonometric R-matrices *Mod. Phys. Lett. A* **10** 1375–1392

[42] Razumov A V 2013 Monodromy operators for higher rank *J. Phys. A: Math. Theor.* **46** 385201 (24pp)

[43] Levenderovskiĭ S, Soibelman Y and Stukopin V 1993 The quantum Weyl group and the universal quantum R-matrix for affine Lie algebra $A_1^{(1)}$ *Lett. Math. Phys.* **27** 253–264

[44] Zhang Y Z and Gould M D 1994 Quantum affine algebras and universal R-matrix with spectral parameter *Lett. Math. Phys.* **31** 101–110

[45] Bracken A J, Gould M D, Zhang Y Z and Delius G W 1994 Infinite families of gauge-equivalent R-matrices and gradations of quantized affine algebras *Int. J. Mod. Phys. B* **8** 3679–3691

[46] Bracken A J, Gould M D and Zhang Y Z 1995 Quantised affine algebras and parameter-dependent R-matrices *Bull. Austral. Math. Soc.* **51** 177–194

[47] Mangazeev V V 2014 On the Yang-Baxter equation for the six-vertex model *Nucl. Phys. B* **882** 70–96

[48] Asherova R M, Smirnov Y F and Tolstoy V N 1979 Description of a class of projection operators for semisimple complex lie algebras *Math. Notes* **26** 499–504

[49] Tolstoy V N 1990 Extremal projectors for quantized Kac–Moody superalgebras and some of their applications *Quantum Groups* (*Lecture Notes in Physics* vol 370) ed Doebner H D and Hennig J D pp 118–125

Journal of Physics: Conference Series **670** (2016) 012038 doi:10.1088/1742-6596/670/1/012038

Exact scattering eigenstates in double quantum-dot systems with an interdot Coulomb interaction

Akinori Nishino[1], Naomichi Hatano[2] and Gonzalo Ordonez[3]

[1]Faculty of Engineering, Kanagawa University, 3-27-1, Rokkakubashi, Kanagawa-ku, Yokohama-shi, Kanagawa, 221-8686, Japan

[2]Institute of Industrial Science, The University of Tokyo, 4-6-1 Komaba, Meguro-ku, Tokyo, 153-8505, Japan

[3]Department of Physics and Astronomy, Butler University, 4600 Sunset Ave., Indianapolis, IN 46208, USA

E-mail: `nishino@kanagawa-u.ac.jp`

Abstract. We study a double quantum-dot system which consists of two leads of noninteracting electrons and two quantum dots with an interdot Coulomb interaction. We assume spinless electrons and consider arbitrary complex values of all lead-dot couplings and an interdot coupling. We construct exact many-electron scattering eigenstates whose incident states are free-electronic plane waves in the leads. Due to the interaction, some of the incident plane waves are scattered to two-body and three-body bound states. The binding strength of the many-body bound states is affected by the arrangement of the two quantum dots, by which we observe an interplay of the Coulomb interaction and quantum interference. We can understand the many-body bound states in terms of many-body resonances.

1. Introduction

Electronic transport in nanoscale systems is a rapidly developing research field of condensed matter physics. In particular, quantum-dot (QD) systems have been studied intensively in the last two decades. One of the significant experimental results is the measurement of a zero-bias peak of the electrical conductance across a QD at low temperatures [1–3], which is described by the Kondo effect in the Anderson model theoretically [4–6]. Recently, the subjects of interest have shifted to double quantum-dot (DQD) systems. The Kondo states realized in each QD are affected by the arrangement of the two QDs due to quantum interference [7–13] and, furthermore, the interdot Coulomb interaction yields the orbital Kondo effect [14–22]. For the interdot interaction larger than a critical value, correlation-induced resonances, which are different from the Kondo resonance, are observed in the conductance profile as an interplay of the Coulomb interaction and quantum interference [16, 23, 24].

Several methods have been proposed to compute the electrical conductance across nanoscale scatterers. The Landauer formula enables us to calculate it by using scattering eigenstates of open quantum systems [25, 26]. However, the application of the formula has been restricted to noninteracting cases. In previous papers [27–29], we obtained *exact* many-electron scattering eigenstates for the QD system described by the two-lead interacting resonant-level models (IRLM). By applying them in order to extend the Landauer formula, we analytically calculated the average electric current for the system under bias voltages in the first order of the interaction parameters. The universal electric current that we obtained through a renormalized-group technique [30, 31] is consistent with those obtained by other approaches; specifically, we observed *negative differential conductance* for large bias voltages [30–35].

It is remarkable that, due to the interactions, some of the incident plane waves in our scattering eigenstates are scattered to *two-body bound states* that exponentially decay with the distance of the two electrons [27–29]. The two-body bound states promote the reflection of electrons at the QD and result in negative differential conductance in the universal electric current. We can understand the appearance of the two-body bound states in terms of one-body *resonant states* [36–41]. Due to the Coulomb interactions, some of the incident electrons are scattered to a resonant state and the corresponding anti-resonant state to form a two-body bound state. The resonant energy thereby determines the binding strength of the exponential decay of the two-body bound state. We remark that such many-body bound states in exact scattering eigenstates have been found in the Anderson model with spin degrees of freedom [42] and in optical systems of waveguide QED [43].

In this paper, we study many-electron scattering eigenstates for a DQD system with an interdot Coulomb interaction. We neglect spin degrees of freedom and assume a single energy level, called a gate energy, on each QD. We consider arbitrary complex values of all lead-dot couplings and dot-dot coupling [12, 16]. We construct exact many-electron scattering eigenstates with plane-wave incident states. Due to the Coulomb interaction, some of the incident plane waves are scattered to two-body and three-body bound states. We can understand the two-body bound states in terms of the one-body resonant states, as was done in the IRLM cases, while we can understand the three-body bound states in terms of the one-body and two-body resonant states. We also find that the many-body bound states are affected by the quantum interference brought by the arrangement of the two QDs. Indeed, the binding strength as a function of the detuning of the two gate energies behaves differently for parallel, serial and T-shaped DQD systems. Thus we observe an interplay of the Coulomb interactions and quantum interference in the many-body bound states.

The paper is organized as follows: In Section 2, the DQD system with an interdot Coulomb interaction is introduced as an open quantum system. In Section 3, the exact scattering eigenstates are presented for one, two and three electrons. The binding strength of the two-body bound states as a function of the detuning of the gate energies is investigated for parallel, serial and T-shaped DQD systems. In Section 4, the construction of the two- and three-electron scattering eigenstates is described in detail. Section 5 is devoted to a summary.

2. Double quantum-dot systems

We introduce a DQD system which consists of two QDs connected to two infinite leads of noninteracting spinless electrons. Figure 1 illustrates the arrangement of the two QDs and the two leads. The Hamiltonian of the DQD system is given by

$$H = \sum_{\ell=1,2} \int_{-\infty}^{\infty} dx\, c_\ell^\dagger(x) \frac{1}{\mathrm{i}} \frac{d}{dx} c_\ell(x) + \sum_{\ell=1,2} \sum_{\alpha=1,2} \left(t_{\ell\alpha} c_\ell^\dagger(0) d_\alpha + t_{\ell\alpha}^* d_\alpha^\dagger c_\ell(0) \right) \\ + t' d_1^\dagger d_2 + t'^* d_2^\dagger d_1 + \sum_{\alpha=1,2} \epsilon_{\mathrm{d}\alpha} n_{\mathrm{d}\alpha} + U n_{\mathrm{d}1} n_{\mathrm{d}2}. \quad (1)$$

Here, $c_\ell^\dagger(x)$ and $c_\ell(x)$ are the creation and the annihilation operators of an electron at a point x in the lead $\ell(=1,2)$ while $d_\alpha^\dagger$ and d_α are those on the QD $\alpha(=1,2)$. We have also defined the number operator $n_{\mathrm{d}\alpha} = d_\alpha^\dagger d_\alpha$ of electrons on the QD α. We consider all the lead-dot couplings denoted by the parameters $t_{\ell\alpha}(\in \mathbb{C})$ and the dot-dot coupling denoted by $t'(\in \mathbb{C})$. We assume a single energy level $\epsilon_{\mathrm{d}\alpha}(\in \mathbb{R})$ on each QD. The parameter $U(>0)$ expresses the strength of the interdot Coulomb repulsion. The first term of the Hamiltonian corresponds to the kinetic energy of electrons in the leads. Here we focus on the electrons with positive velocities and linearize the local dispersion relations of the leads in the vicinity of the Fermi energy under the assumption that other parameters are small compared with the Fermi energy.

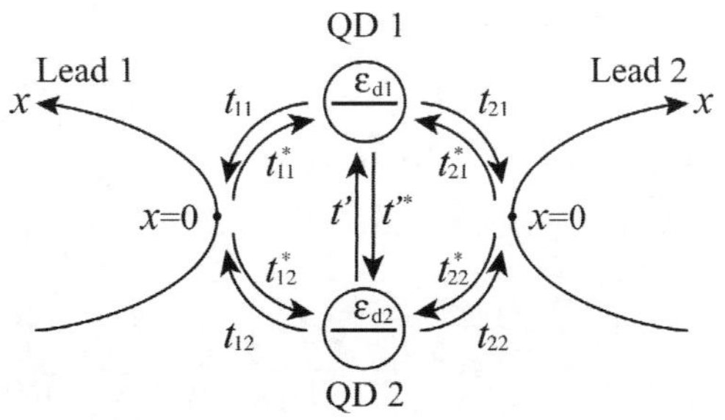

Figure 1. A schematic diagram of the DQD system.

The DQD system described by Eq. (1) includes various DQD systems such as a parallel DQD system ($t' = 0$), a serial DQD system ($t_{12} = t_{21} = 0$) and a T-shaped DQD system ($t_{12} = t_{22} = 0$) [12, 16]. In the previous work [44] studying a parallel DQD system, we employed an even-odd transformation that maps the two-lead system to decoupled two single-lead systems. The even-odd transformation, however, works only for the cases with $t_{11}t_{22} = t_{12}t_{21}$, which do not include the serial DQD system.

3. Exact scattering eigenstates

In the present section, we present exact solutions of one-, two- and three-electron scattering eigenstates. Details of the construction of two- and three-electron eigenstates will be given in the next section.

3.1. One-electron case

First, we construct the one-electron scattering eigenstates. The one-electron eigenstates are given in the form

$$|E\rangle = \Big(\int dx \sum_m g_m(x) c_m^\dagger(x) + \sum_\alpha e_\alpha d_\alpha^\dagger\Big)|0\rangle. \tag{2}$$

Here $|0\rangle$ is the vacuum state satisfying $c_m(x)|0\rangle = d_\alpha|0\rangle = 0$, ($m = 1, 2$, $\alpha = 1, 2$), and $g_m(x)$ and e_α are the eigenfunctions to be determined by the Schrödinger equations

$$\Big(\frac{1}{\mathrm{i}}\frac{d}{dx} - E\Big) g_m(x) + \delta(x) \sum_\alpha t_{m\alpha} e_\alpha = 0, \tag{3a}$$

$$(\epsilon_{d\alpha} - E)e_\alpha + \sum_m t_{m\alpha}^* g_m(0) + t'_\alpha e_{\bar\alpha} = 0, \tag{3b}$$

where we put $t'_1 = t'$, $t'_2 = t'^*$ and $\bar\alpha = 3 - \alpha$.

We construct the one-electron scattering eigenfunctions $g_m(x)$ and e_α in the following way. We find that the eigenfunctions $g_m(x)$ are discontinuous at $x = 0$ due to the δ-function term in Eq. (3a). In fact, by integrating the Schrödinger equation (3a) around $x = 0$, we obtain the matching condition

$$g_m(0+) = g_m(0-) - \mathrm{i} \sum_\alpha t_{m\alpha} e_\alpha. \tag{4}$$

Since the Schrödinger equations do not determine the value of the eigenfunctions at the discontinuous point, we assume

$$g_m(0) = \frac{1}{2}\big(g_m(0+) + g_m(0-)\big). \tag{5}$$

By applying Eqs. (4) and (5) to the Schrödinger equation (3b), we have

$$(\epsilon_{\mathrm{d}\alpha} - \mathrm{i}\Gamma_{\alpha\alpha} - E)e_\alpha + (t'_\alpha - \mathrm{i}\Gamma_{\alpha\bar{\alpha}})e_{\bar{\alpha}} = -\sum_m t^*_{m\alpha} g_m(0-), \tag{6}$$

where $\Gamma_{\alpha\beta} = \sum_\ell t^*_{\ell\alpha} t_{\ell\beta}/2$ is the level-width matrix of the QDs. By solving Eq. (6) as coupled equations for e_1 and e_2, we obtain

$$e_\alpha = \frac{\sum_m \big((E-\epsilon_{\mathrm{d}\bar{\alpha}}+\mathrm{i}\Gamma_{\bar{\alpha}\bar{\alpha}})t^*_{m\alpha} - (\mathrm{i}\Gamma_{\alpha\bar{\alpha}}-t'_\alpha)t^*_{m\bar{\alpha}}\big)g_m(0-)}{(E-\epsilon_{\mathrm{d}1}+\mathrm{i}\Gamma_{11})(E-\epsilon_{\mathrm{d}2}+\mathrm{i}\Gamma_{22}) - (\mathrm{i}\Gamma_{12}-t'_1)(\mathrm{i}\Gamma_{21}-t'_2)}. \tag{7}$$

Then, for given $g_m(0-)$, the values e_α and $g_l(0+)$ are determined by Eq. (7) and Eq. (4), respectively.

In order to study the electron transport by the Landauer formula [25, 26], we need the scattering eigenstates whose incident states are plane waves in the leads. We note that the incident plane waves come from the part $x < 0$ of the leads since we focus on the electrons with positive velocities. Then we consider the eigenstate whose incident state is a plane wave with wave number k in the lead ℓ as follows:

$$g^{(\ell)}_{m,k}(x) = \frac{1}{\sqrt{2\pi}}\delta_{m\ell}\mathrm{e}^{\mathrm{i}kx}, \quad (x<0), \tag{8}$$

which we call scattering boundary conditions. The reflected and transmitted waves are determined by the matching conditions (4) with Eq. (7). As a result, we obtain the exact one-electron scattering eigenfunctions satisfying the scattering boundary conditions (8) as

$$g^{(\ell)}_{m,k}(x) = \frac{1}{\sqrt{2\pi}}\Big(\delta_{m\ell} - \sqrt{2\pi}\mathrm{i}\sum_\alpha t_{m\alpha}e^{(\ell)}_{\alpha,k}\theta(x)\Big)\mathrm{e}^{\mathrm{i}kx}, \tag{9a}$$

$$e^{(\ell)}_{\alpha,k} = \frac{1}{\sqrt{2\pi}}\frac{(k-\epsilon_{\mathrm{d}\bar{\alpha}}+\mathrm{i}\Gamma_{\bar{\alpha}\bar{\alpha}})t^*_{\ell\alpha} - (\mathrm{i}\Gamma_{\alpha\bar{\alpha}} - t'_\alpha)t^*_{\ell\bar{\alpha}}}{(k-\epsilon_{\mathrm{d}1}+\mathrm{i}\Gamma_{11})(k-\epsilon_{\mathrm{d}2}+\mathrm{i}\Gamma_{22}) - (\mathrm{i}\Gamma_{12} - t'_1)(\mathrm{i}\Gamma_{21} - t'_2)}, \tag{9b}$$

where $\theta(x)$ is a step function: $\theta(x) = 1$ if $x > 0$, $\theta(0) = 1/2$ and $\theta(x) = 0$ if $x < 0$. The energy eigenvalue is given by $E = k$.

Let us rewrite the one-electron eigenfunction $e^{(\ell)}_{\alpha,k}$ in Eq. (9b) as follows:

$$e^{(\ell)}_{\alpha,k} = \frac{1}{\sqrt{2\pi}}\frac{(k-\epsilon_{\mathrm{d}\bar{\alpha}}+\mathrm{i}\Gamma_{\bar{\alpha}\bar{\alpha}})t^*_{\ell\alpha} - (\mathrm{i}\Gamma_{\alpha\bar{\alpha}} - t'_\alpha)t^*_{\ell\bar{\alpha}}}{(k-\bar{\epsilon}_{\mathrm{d}}+\mathrm{i}\bar{\Gamma}-\frac{\mathrm{i}}{2}\eta)(k-\bar{\epsilon}_{\mathrm{d}}+\mathrm{i}\bar{\Gamma}+\frac{\mathrm{i}}{2}\eta)}, \tag{10}$$

where we define $\bar{\epsilon}_{\mathrm{d}} = (\epsilon_{\mathrm{d}1} + \epsilon_{\mathrm{d}2})/2$, $\bar{\Gamma} = (\Gamma_{11} + \Gamma_{22})/2$, $\Delta\Gamma_\alpha = \Gamma_{\alpha\alpha} - \Gamma_{\bar{\alpha}\bar{\alpha}}$, $\Delta\epsilon_{\mathrm{d}\alpha} = \epsilon_{\mathrm{d}\alpha} - \epsilon_{\mathrm{d}\bar{\alpha}}$ and

$$\eta = \sqrt{(\mathrm{i}\Delta\epsilon_{\mathrm{d}\alpha} + \Delta\Gamma_\alpha)^2 - 4(t' - \mathrm{i}\Gamma_{12})(t'^* - \mathrm{i}\Gamma_{21})}. \tag{11}$$

From this expression, we find resonant poles of the one-electron scattering at $k = \bar{\epsilon}_{\mathrm{d}} - \mathrm{i}\bar{\Gamma} \pm \mathrm{i}\eta/2$ on the complex-k plane. "Eigenstates" with resonant energies are called *resonant states* [36–41]. It is known that the resonant states are characterized as the eigenstates with no incoming waves, which is called the Siegert boundary conditions [37]. Indeed, the resonant energy $k = \bar{\epsilon}_{\mathrm{d}} - \mathrm{i}\bar{\Gamma} \pm \mathrm{i}\eta/2$ is determined by Eq. (6) with the Siegert boundary conditions $g_m(0-) = 0$. While the resonant states are unphysical in the framework of the standard quantum mechanics, we shall see in the two- and three-electron cases that such resonant states "emerge" as many-body bound states in the scattering eigenstates.

3.2. Two-electron case

Second, we consider the two-electron case, in which an effect of the interaction appears. The two-electron eigenstates are given in the form

$$|E\rangle = \Big(\sum_{l,m}\int_{x_1<x_2} dx_1 dx_2\, g_{lm}(x_1,x_2)c_l^\dagger(x_1)c_m^\dagger(x_2) + \sum_{l,\alpha}\int dx_1 e_{l\alpha}(x_1)c_l^\dagger(x_1)d_\alpha^\dagger + f_{12}d_1^\dagger d_2^\dagger\Big)|0\rangle. \tag{12}$$

Here we assume the anti-symmetry relations $g_{lm}(x_1,x_2) = -g_{ml}(x_2,x_1)$ and $f_{12} = -f_{21}$. The Schrödinger equations are given by

$$\Big(\frac{1}{\mathrm{i}}(\partial_1+\partial_2) - E\Big)g_{lm}(x_1,x_2) + \sum_\alpha \big(t_{m\alpha}e_{l\alpha}(x_1)\delta(x_2) - t_{l\alpha}\delta(x_1)e_{m\alpha}(x_2)\big) = 0, \tag{13a}$$

$$\Big(\frac{1}{\mathrm{i}}\frac{d}{dx_1} + \epsilon_{\mathrm{d}\alpha} - E\Big)e_{l\alpha}(x_1) + \sum_m t_{m\alpha}^* g_{lm}(x_1,0) + t_{\bar\alpha}^{\prime *} e_{l\bar\alpha}(x_1) - t_{l\bar\alpha}\delta(x_1)f_{\alpha\bar\alpha} = 0, \tag{13b}$$

$$(\epsilon_{\mathrm{d}1} + \epsilon_{\mathrm{d}2} + U - E)f_{\alpha\bar\alpha} + \sum_l \big(t_{l\alpha}^* e_{l\bar\alpha}(0) - t_{l\bar\alpha}^* e_{l\alpha}(0)\big) = 0, \tag{13c}$$

where $\partial_i = \partial/\partial x_i$.

In a way similar to the one-electron case, we consider the two-electron scattering eigenstates satisfying the following scattering boundary conditions: the incident states are free-electronic plane waves with wave number k_1 in the lead ℓ_1 and with wave number k_2 in the lead ℓ_2,

$$g_{lm,k_1k_2}^{(\ell_1\ell_2)}(x_1,x_2) = \frac{1}{2\pi}\sum_P \mathrm{sgn}(P)\delta_{l\ell_{P_1}}\delta_{m\ell_{P_2}}\mathrm{e}^{\mathrm{i}(k_{P_1}x_1+k_{P_2}x_2)}, \qquad (x_1,x_2<0), \tag{14}$$

where $P = (P_1,P_2)$ denotes a permutation of $(1,2)$. We shall present how to construct the two-electron scattering eigenstates in the next section. As a result, we obtain the exact two-electron scattering eigenfunctions with energy eigenvalue $E = k_1 + k_2$ as

$$\begin{aligned} g_{m_1m_2,k_1k_2}^{(\ell_1\ell_2)}(x_1,x_2) &= \sum_P \mathrm{sgn}(P) g_{m_1,k_{P_1}}^{(\ell_{P_1})}(x_1) g_{m_2,k_{P_2}}^{(\ell_{P_2})}(x_2) \\ &\quad - \sum_{Q,\alpha,s}\mathrm{sgn}(Q)\mathrm{i}t_{m_{Q_1}\alpha}y_{m_{Q_2}\alpha,s}Z_{\alpha,k_1k_2,s}^{(\ell_1\ell_2)}(x_{Q_1Q_2})\mathrm{e}^{\mathrm{i}Ex_{Q_2}}\theta(x_{Q_1}), \end{aligned} \tag{15a}$$

$$e_{m_1\alpha,k_1k_2}^{(\ell_1\ell_2)}(x_1) = \sum_P \mathrm{sgn}(P) g_{m_1,k_{P_1}}^{(\ell_{P_1})}(x_1) e_{\alpha,k_{P_2}}^{(\ell_{P_2})} - \sum_s y_{m_1\alpha,s}Z_{\alpha,k_1k_2,s}^{(\ell_1\ell_2)}(-x_1)\mathrm{e}^{\mathrm{i}Ex_1}, \tag{15b}$$

$$f_{\alpha\bar\alpha,k_1k_2}^{(\ell_1\ell_2)} = \sum_P \mathrm{sgn}(P) e_{\alpha,k_{P_1}}^{(\ell_{P_1})} e_{\bar\alpha,k_{P_2}}^{(\ell_{P_2})} + Z_{\alpha,k_1k_2}^{(\ell_1\ell_2)}. \tag{15c}$$

Here $x_{12} = x_1 - x_2$, $s = +$ or $-$, $Q = (Q_1,Q_2)$ is a permutation of $(1,2)$ and

$$y_{m\alpha,s} = \frac{1}{2\eta_s}((-\Delta\epsilon_{\mathrm{d}\alpha} + \mathrm{i}\Delta\Gamma_\alpha - \mathrm{i}\eta_s)t_{m\bar\alpha} + 2(t_\alpha' - \mathrm{i}\Gamma_{\alpha\bar\alpha})t_{m\alpha}), \tag{16}$$

$$Z_{\alpha,k_1k_2}^{(\ell_1\ell_2)} = \frac{U}{k_1+k_2-2\bar\epsilon_{\mathrm{d}}-U+2\mathrm{i}\bar\Gamma}(e_{\alpha,k_1}^{(\ell_1)}e_{\bar\alpha,k_2}^{(\ell_2)} - e_{\alpha,k_2}^{(\ell_2)}e_{\bar\alpha,k_1}^{(\ell_1)}), \tag{17}$$

$$Z_{\alpha,k_1k_2,s}^{(\ell_1\ell_2)}(x) = Z_{\alpha,k_1k_2}^{(\ell_1\ell_2)}\mathrm{e}^{\mathrm{i}(\bar\epsilon_{\mathrm{d}}-\mathrm{i}\bar\Gamma+\frac{\mathrm{i}}{2}\eta_s)x}\theta(-x) \tag{18}$$

with $\eta_\pm = \pm\eta$, where η is given in Eq. (11).

The first term of each eigenfunction in Eqs. (15) is the Slater determinant of the one-electron scattering eigenfunctions, which corresponds to the noninteracting case. The second term of the eigenfunction $g^{(\ell_1\ell_2)}_{m_1m_2,k_1k_2}(x_1,x_2)$ corresponds to *two-body bound states* since the term exponentially decays as $\mathrm{e}^{-\kappa_\pm|x_1-x_2|}$ with the distance $|x_1-x_2|$ between the two electrons. Here $\kappa_\pm = \bar{\Gamma} \mp \mathrm{Re}(\eta)/2$ is the binding strength of the two-body bound states. We note that the complete eigenstate in Eq. (12) is stationary in time since the total energy eigenvalue E is real. Similarly, the second term of the eigenfunction $e^{(\ell_1\ell_2)}_{m_1\alpha,k_1k_2}(x_1)$ corresponds to two-body bound states of two electrons one of which is on the QDs.

We observe an effect of quantum interference in the binding strength $\kappa_\pm$ of the two-body bound states. Let us consider the following three cases: a parallel DQD system ($t_{11}=t_{12}=t_{21}=t_{22}=t$, $t'=0$) [44], a serial DQD system ($t_{11}=t_{22}=\sqrt{2}t$, $t_{12}=t_{21}=0$) and a T-shaped DQD system ($t_{11}=t_{21}=\sqrt{2}t$, $t_{12}=t_{22}=0$). The binding strength $\kappa_\pm$ is explicitly given by

$$\kappa_\pm = \begin{cases} \Gamma \mp \frac{1}{2}\mathrm{Re}\left(\sqrt{-\Delta\epsilon_\mathrm{d}^2+4\Gamma^2}\right) & \text{for the parallel DQD system,} \\ \Gamma \mp \frac{1}{2}\mathrm{Re}\left(\sqrt{-\Delta\epsilon_\mathrm{d}^2-4|t'|^2}\right) = \Gamma & \text{for the serial DQD system,} \\ \Gamma \mp \frac{1}{2}\mathrm{Re}\left(\sqrt{(\mathrm{i}\Delta\epsilon_\mathrm{d}+2\Gamma)^2-4|t'|^2}\right) & \text{for the T-shaped DQD system,} \end{cases} \tag{19}$$

where $\Delta\epsilon_\mathrm{d} = \epsilon_{\mathrm{d}1} - \epsilon_{\mathrm{d}2}$ and $\Gamma = |t|^2$. The behavior of the binding strength $\kappa_\pm$ as a function in the detuning $\Delta\epsilon_\mathrm{d}$ in each case is illustrated in Fig. 2.

For the parallel DQD system, two binding strengths appear in the regime $0 < \Delta\epsilon_\mathrm{d} < 2\Gamma$ but merge into $\kappa_\pm = \Gamma$ at $\Delta\epsilon_\mathrm{d} = 2\Gamma$. For the T-shaped DQD system, two binding strengths appear for $\Delta\epsilon_\mathrm{d} > 0$ and each monotonically increases/decreases with the detuning $\Delta\epsilon_\mathrm{d}$. We can regard the existence of two different binding strengths as an effect of the quantum interference brought by the two paths connecting the left and the right leads. For the serial DQD system, on the other hand, only one binding strength appears, which takes a constant value $\kappa_\pm = \Gamma$. In this case, there is indeed only one path connecting the left and the right leads.

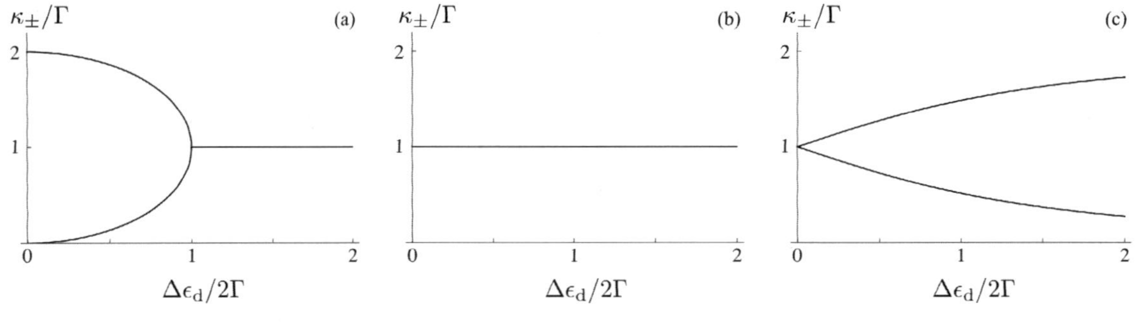

Figure 2. Binding strength $\kappa_\pm$ of the two-body bound states as a function in the detuning $\Delta\epsilon_\mathrm{d}$: (a) Parallel DQD system ($t_{11}=t_{12}=t_{21}=t_{22}=1$, $t'=0$); (b) Serial DQD system ($t_{11}=t_{22}=t'=\sqrt{2}$, $t_{12}=t_{21}=0$); (c) T-shaped DQD system ($t_{11}=t_{21}=t'=\sqrt{2}$, $t_{12}=t_{22}=0$).

We can understand the appearance of the two-body bound states as a consequence of the resonances of the one-body scattering. Let us inspect the set of wave numbers characterizing each term of the eigenfunctions in Eqs. (15). Due to the interaction, the wave-number set $\{k_1,k_2\}$ of the incident plane waves is not conserved and, as is expressed in the second terms of Eq. (15a), is scattered to the set $\{\bar{\epsilon}_\mathrm{d} - \mathrm{i}\bar{\Gamma} + \mathrm{i}\eta_s/2, E - \bar{\epsilon}_\mathrm{d} + \mathrm{i}\bar{\Gamma} - \mathrm{i}\eta_s/2\}$, which is a pair of the resonant energy and the corresponding anti-resonant energy. Thus the unphysical resonant states "emerge" as physical two-body bound states. We note that such imaginary parts of the wave numbers are allowed since the total energy eigenvalue E is real.

We also find a resonant pole at $E = 2\bar{\epsilon}_\mathrm{d} + U - 2\mathrm{i}\bar{\Gamma}$ in the amplitude $Z^{(\ell_1\ell_2)}_{\alpha,k_1k_2}$ of the terms of the two-body bound states in Eqs. (15). We call it a *two-body resonance* since the resonant energy $E = 2\bar{\epsilon}_\mathrm{d} + U - 2\mathrm{i}\bar{\Gamma}$ cannot be expressed by the sum of the one-body resonant energies. The two-body resonant energy corresponds to the "eigenstate" with no incoming states (see Eq. (31) below), which is an extension of the Siegert boundary conditions [37]. The two-body resonance yields three-body bound states in the three-electron scattering eigenstates, which shall be described in the next subsection.

3.3. Three-electron case

Let us turn to the three-electron case. The three-electron eigenstates are given in the form

$$|E\rangle = \Big(\sum_{l,m,n} \int_{x_1<x_2<x_3} dx_1 dx_2 dx_3\, g_{lmn}(x_1,x_2,x_3) c_l^\dagger(x_1) c_m^\dagger(x_2) c_n^\dagger(x_3) + \sum_{l,m,\alpha} \int_{x_1<x_2} dx_1 dx_2\, e_{lm\alpha}(x_1,x_2) c_l^\dagger(x_1) c_m^\dagger(x_2) d_\alpha^\dagger + \sum_m \int dx_1 f_{m12}(x_1) c_m^\dagger(x_1) d_1^\dagger d_2^\dagger \Big)|0\rangle. \quad (20)$$

Here we assume the anti-symmetry

$$g_{lmn}(x_1,x_2,x_3) = -g_{mln}(x_2,x_1,x_3), \qquad g_{lmn}(x_1,x_2,x_3) = -g_{lnm}(x_1,x_3,x_2), \quad (21a)$$

$$e_{lm\alpha}(x_1,x_2) = -e_{ml\alpha}(x_2,x_1), \qquad f_{m12}(x_1) = -f_{m21}(x_1). \quad (21b)$$

The Schrödinger equations are given by

$$\Big(\frac{1}{\mathrm{i}}(\partial_1+\partial_2+\partial_3) - E\Big) g_{lmn}(x_1,x_2,x_3) + \sum_\alpha \big(t_{l\alpha}\delta(x_1)e_{mn\alpha}(x_2,x_3) - t_{m\alpha}\delta(x_2)e_{ln\alpha}(x_1,x_3) + t_{n\alpha}\delta(x_3)e_{lm\alpha}(x_1,x_2)\big) = 0, \quad (22a)$$

$$\Big(\frac{1}{\mathrm{i}}(\partial_1+\partial_2) + \epsilon_{\mathrm{d}\alpha} - E\Big) e_{lm\alpha}(x_1,x_2) + \sum_n t^*_{n\alpha} g_{lmn}(x_1,x_2,0) + t'_\alpha e_{lm\bar{\alpha}}(x_1,x_2) + t_{l\bar{\alpha}}\delta(x_1) f_{m\alpha\bar{\alpha}}(x_2) - t_{m\bar{\alpha}}\delta(x_2) f_{l\alpha\bar{\alpha}}(x_1) = 0, \quad (22b)$$

$$\Big(\frac{1}{\mathrm{i}}\frac{d}{dx_1} + 2\bar{\epsilon}_\mathrm{d} + U - E\Big) f_{l\alpha\bar{\alpha}}(x_1) + \sum_m \big(t^*_{m\alpha} e_{lm\bar{\alpha}}(x_1,0) - t^*_{m\bar{\alpha}} e_{lm\alpha}(x_1,0)\big) = 0. \quad (22c)$$

We consider the three-electron scattering eigenstates satisfying the scattering boundary conditions:

$$g^{(\ell_1\ell_2\ell_3)}_{lmn,k_1k_2k_3}(x_1,x_2,x_3) = \frac{1}{(2\pi)^{\frac{3}{2}}} \sum_P \mathrm{sgn}(P)\delta_{l\ell_{P_1}}\delta_{m\ell_{P_2}}\delta_{n\ell_{P_3}} \mathrm{e}^{\mathrm{i}\sum_i k_{P_i}x_i}, \quad (x_1,x_2,x_3<0), \quad (23)$$

where $P = (P_1,P_2,P_3)$ denotes a permutation of $(1,2,3)$. The exact three-electron scattering

Journal of Physics: Conference Series **670** (2016) 012038 doi:10.1088/1742-6596/670/1/012038

eigenfunctions that shall be constructed in the next section are given by

$$
\begin{aligned}
g^{(\ell_1\ell_2\ell_3)}_{m_1m_2m_3,k_1k_2k_3}(x_1,x_2,x_3) &= \sum_P \operatorname{sgn}(P) g^{(\ell_{P_1})}_{m_1,k_{P_1}}(x_1) g^{(\ell_{P_2})}_{m_2,k_{P_2}}(x_2) g^{(\ell_{P_3})}_{m_3,k_{P_3}}(x_3) \\
&- \frac{1}{2}\sum_{\substack{P,Q\\ \alpha,s}} \operatorname{sgn}(PQ)\mathrm{i}t_{m_{Q_2}\alpha} y_{m_{Q_3}\alpha,s} g^{(\ell_{P_1})}_{m_{Q_1},k_{P_1}}(x_{Q_1}) Z^{(\ell_{P_2}\ell_{P_3})}_{\alpha,k_{P_2}k_{P_3},s}(x_{Q_2Q_3}) \mathrm{e}^{\mathrm{i}(k_{P_2}+k_{P_3})x_{Q_3}}\theta(x_{Q_2}) \\
&+ \frac{U}{2\sqrt{2\pi}} \sum_{\substack{P,Q\\ \alpha,s,r}} \operatorname{sgn}(PQ)\mathrm{i}t_{m_{Q_1}\alpha} y_{m_{Q_2}\alpha,s} \frac{t^*_{\ell_{P_1}\alpha} y_{m_{Q_3}\bar{\alpha},r} + t^*_{\ell_{P_1}\bar{\alpha}} y_{m_{Q_3}\alpha,r}}{(k_{P_1}-\bar{\epsilon}_{\mathrm{d}}+\mathrm{i}\Gamma-\frac{\mathrm{i}}{2}\eta_{\bar{r}})(k_{P_1}-\bar{\epsilon}_{\mathrm{d}}-U+\mathrm{i}\Gamma-\frac{\mathrm{i}}{2}\eta_{\bar{r}})} \\
&\times Z^{(\ell_{P_2}\ell_{P_3})}_{\alpha,k_{P_2}k_{P_3},sr}(x_{Q_1Q_2}, x_{Q_2Q_3}) \mathrm{e}^{\mathrm{i}(k_{P_1}x_{Q_2}+(k_{P_2}+k_{P_3})x_{Q_3})}\theta(x_{Q_1}) \\
&- \frac{U}{2\sqrt{2\pi}} \sum_{\substack{P,Q\\ \alpha,s,r}} \operatorname{sgn}(PQ)\mathrm{i}t_{m_{Q_1}\alpha} y_{m_{Q_2}\alpha,s} \frac{t^*_{\ell_{P_1}\alpha} y_{m_{Q_3}\bar{\alpha},r} + t^*_{\ell_{P_1}\bar{\alpha}} y_{m_{Q_3}\alpha,r}}{(k_{P_1}-\bar{\epsilon}_{\mathrm{d}}+\mathrm{i}\Gamma-\frac{\mathrm{i}}{2}\eta_{\bar{r}})(k_{P_1}-\bar{\epsilon}_{\mathrm{d}}-U+\mathrm{i}\Gamma-\frac{\mathrm{i}}{2}\eta_{\bar{r}})} \\
&\times \tilde{Z}^{(\ell_{P_2}\ell_{P_3})}_{\alpha,k_{P_2}k_{P_3},s}(x_{Q_1Q_2}, x_{Q_2Q_3}) \mathrm{e}^{\mathrm{i}Ex_{Q_3}}\theta(x_{Q_1}),
\end{aligned} \tag{24a}
$$

$$
\begin{aligned}
e^{(\ell_1\ell_2\ell_3)}_{m_1m_2\alpha,k_1k_2k_3}(x_1,x_2) &= \sum_P \operatorname{sgn}(P) g^{(\ell_{P_1})}_{m_1,k_{P_1}}(x_1) g^{(\ell_{P_2})}_{m_2,k_{P_2}}(x_2) e^{(\ell_{P_3})}_{\alpha,k_{P_3}} \\
&- \frac{1}{2}\sum_{P,Q,s} \operatorname{sgn}(PQ) y_{m_{Q_2}\alpha,s} g^{(\ell_{P_1})}_{m_{Q_1},k_{P_1}}(x_{Q_1}) Z^{(\ell_{P_2}\ell_{P_3})}_{\alpha,k_{P_2}k_{P_3},s}(-x_{Q_2}) \mathrm{e}^{\mathrm{i}(k_{P_2}+k_{P_3})x_{Q_2}} \\
&- \frac{1}{2}\sum_{\substack{P,Q\\ \beta,s}} \operatorname{sgn}(PQ)\mathrm{i}t_{m_{Q_1}\beta} y_{m_{Q_2}\beta,s} e^{(\ell_{P_1})}_{\alpha,k_{P_1}} Z^{(\ell_{P_2}\ell_{P_3})}_{\beta,k_{P_2}k_{P_3},s}(x_{Q_1Q_2}) \mathrm{e}^{\mathrm{i}(k_{P_2}+k_{P_3})x_{Q_2}}\theta(x_{Q_1}) \\
&- \frac{U}{2\sqrt{2\pi}} \sum_{P,Q,s,r} \operatorname{sgn}(PQ) y_{m_{Q_1}\alpha,s} \frac{t^*_{\ell_{P_1}\alpha} y_{m_{Q_2}\bar{\alpha},s} + t^*_{\ell_{P_1}\bar{\alpha}} y_{m_{Q_2}\alpha,s}}{(k_{P_1}-\bar{\epsilon}_{\mathrm{d}}+\mathrm{i}\bar{\Gamma}-\frac{\mathrm{i}}{2}\eta_{\bar{r}})(k_{P_1}-\bar{\epsilon}_{\mathrm{d}}-U+\mathrm{i}\bar{\Gamma}-\frac{\mathrm{i}}{2}\eta_{\bar{r}})} \\
&\times Z^{(\ell_{P_2}\ell_{P_3})}_{\alpha,k_{P_2}k_{P_3},sr}(-x_{Q_1}, x_{Q_1Q_2}) \mathrm{e}^{\mathrm{i}(k_{P_1}x_{Q_1}+(k_{P_2}+k_{P_3})x_{Q_2})} \\
&+ \frac{U}{2\sqrt{2\pi}} \sum_{P,Q,s,r} \operatorname{sgn}(PQ) y_{m_{Q_1}\alpha,s} \frac{t^*_{\ell_{P_1}\alpha} y_{m_{Q_2}\bar{\alpha},r} + t^*_{\ell_{P_1}\bar{\alpha}} y_{m_{Q_2}\alpha,r}}{(k_{P_1}-\bar{\epsilon}_{\mathrm{d}}+\mathrm{i}\bar{\Gamma}-\frac{\mathrm{i}}{2}\eta_{\bar{r}})(k_{P_1}-\bar{\epsilon}_{\mathrm{d}}-U+\mathrm{i}\bar{\Gamma}-\frac{\mathrm{i}}{2}\eta_{\bar{r}})} \\
&\times \tilde{Z}^{(\ell_{P_2}\ell_{P_3})}_{\alpha,k_{P_2}k_{P_3},s}(-x_{Q_1}, x_{Q_1Q_2}) \mathrm{e}^{\mathrm{i}Ex_{Q_2}},
\end{aligned} \tag{24b}
$$

$$
\begin{aligned}
f^{(\ell_1\ell_2\ell_3)}_{m_1\alpha\bar{\alpha},k_1k_2k_3}(x_1) &= \sum_P \operatorname{sgn}(P) g^{(\ell_{P_1})}_{m_1,k_{P_1}}(x_1) e^{(\ell_{P_2})}_{\alpha,k_{P_2}} e^{(\ell_{P_3})}_{\bar{\alpha},k_{P_3}} + \frac{1}{2}\sum_P \operatorname{sgn}(P) g^{(\ell_{P_1})}_{m_1,k_{P_1}}(x_1) Z^{(\ell_{P_2}\ell_{P_3})}_{\alpha,k_{P_2}k_{P_3}} \\
&- \frac{1}{2\sqrt{2\pi}} \sum_{P,s} \operatorname{sgn}(P) \frac{t^*_{\ell_{P_1}\alpha} y_{m_1\bar{\alpha},s} + t^*_{\ell_{P_1}\bar{\alpha}} y_{m_1\alpha,s}}{k_{P_1}-\bar{\epsilon}_{\mathrm{d}}-U+\mathrm{i}\bar{\Gamma}-\frac{\mathrm{i}}{2}\eta_{\bar{s}}} Z^{(\ell_{P_2}\ell_{P_3})}_{\alpha,k_{P_2}k_{P_3},s}(-x_1) \mathrm{e}^{\mathrm{i}(k_{P_2}+k_{P_3})x_1} \\
&+ \frac{U}{2\sqrt{2\pi}} \sum_{P,s} \operatorname{sgn}(P) \frac{t^*_{\ell_{P_1}\alpha} y_{m_1\bar{\alpha},s} + t^*_{\ell_{P_1}\bar{\alpha}} y_{m_1\alpha,s}}{(k_{P_1}-\bar{\epsilon}_{\mathrm{d}}+\mathrm{i}\bar{\Gamma}-\frac{\mathrm{i}}{2}\eta_{\bar{s}})(k_{P_1}-\bar{\epsilon}_{\mathrm{d}}-U+\mathrm{i}\bar{\Gamma}-\frac{\mathrm{i}}{2}\eta_{\bar{s}})} \\
&\times Z^{(\ell_{P_2}\ell_{P_3})}_{\alpha,k_{P_2}k_{P_3}} \mathrm{e}^{\mathrm{i}(E-2\bar{\epsilon}_{\mathrm{d}}-U+2\mathrm{i}\bar{\Gamma})x_1}\theta(x_1),
\end{aligned} \tag{24c}
$$

where

$$y_{m\alpha,s} = \frac{1}{2\eta_s}((-\Delta\epsilon_{\mathrm{d}\alpha} + \mathrm{i}\Delta\Gamma_\alpha - \mathrm{i}\eta_s)t_{m\bar{\alpha}} + 2(t'_\alpha - \mathrm{i}\Gamma_{\alpha\bar{\alpha}})t_{m\alpha}), \tag{25a}$$

$$Z^{(\ell_1\ell_2)}_{\alpha,k_1k_2} = \frac{U}{k_1 + k_2 - 2\bar{\epsilon}_\mathrm{d} - U + 2\mathrm{i}\bar{\Gamma}}(e^{(\ell_1)}_{\alpha,k_1}e^{(\ell_2)}_{\bar{\alpha},k_2} - e^{(\ell_2)}_{\alpha,k_2}e^{(\ell_1)}_{\bar{\alpha},k_1}), \tag{25b}$$

$$Z^{(\ell_1\ell_2)}_{\alpha,k_1k_2,s}(x) = Z^{(\ell_1\ell_2)}_{\alpha,k_1k_2}\mathrm{e}^{\mathrm{i}(\bar{\epsilon}_\mathrm{d} - \mathrm{i}\bar{\Gamma} + \frac{\mathrm{i}}{2}\eta_s)x}\theta(-x), \tag{25c}$$

$$Z^{(\ell_1\ell_2)}_{\alpha,k_1k_2,sr}(x,y) = Z^{(\ell_1\ell_2)}_{\alpha,k_1k_2}\mathrm{e}^{\mathrm{i}((\bar{\epsilon}_\mathrm{d} - \mathrm{i}\bar{\Gamma} + \frac{\mathrm{i}}{2}\eta_s)x + (\bar{\epsilon}_\mathrm{d} - \mathrm{i}\bar{\Gamma} + \frac{\mathrm{i}}{2}\eta_r)y)}\theta(-x)\theta(-y), \tag{25d}$$

$$\tilde{Z}^{(\ell_1\ell_2)}_{\alpha,k_1k_2,s}(x,y) = Z^{(\ell_1\ell_2)}_{\alpha,k_1k_2}\mathrm{e}^{\mathrm{i}((\bar{\epsilon}_\mathrm{d} - \mathrm{i}\bar{\Gamma} + \frac{\mathrm{i}}{2}\eta_s)x + (2\bar{\epsilon}_\mathrm{d} + U - 2\mathrm{i}\bar{\Gamma})y)}\theta(-x)\theta(-y). \tag{25e}$$

The first term of each eigenfunction in Eqs. (24) corresponds to the noninteracting case. The second term of the eigenfunction $g^{(\ell_1\ell_2\ell_3)}_{m_1m_2m_3,k_1k_2k_3}(x_1,x_2,x_3)$ in Eq. (24a) is the two-body bound states which are the same as those in the two-electron case. We interpret both the third and fourth terms of the eigenfunction $g^{(\ell_1\ell_2\ell_3)}_{m_1m_2m_3,k_1k_2k_3}(x_1,x_2,x_3)$ as *three-body bound states* since, for the distance $|x_i - x_j|$ between any two of the three electrons, the third terms exponentially decay as $\mathrm{e}^{-\kappa_\pm|x_{Q_1} - x_{Q_2}| - \kappa_\pm|x_{Q_2} - x_{Q_3}|}$ and the fourth terms also decay as $\mathrm{e}^{-\kappa_\pm|x_{Q_1} - x_{Q_2}| - 2\bar{\Gamma}|x_{Q_2} - x_{Q_3}|}$. Similarly, three-body bound states of the three electrons one of which is on the QDs appear in the fourth and fifth terms of the eigenfunction $e^{(\ell_1\ell_2\ell_3)}_{m_1m_2\alpha,k_1k_2k_3}(x_1,x_2)$ in Eq. (24b).

In a way similar to the two-body bound states, we can also understand the three-body bound states in terms of resonances of the scattering. The wave-number set $\{k_1,k_2,k_3\}$ of the incident plane waves is scattered to the set $\{\bar{\epsilon}_\mathrm{d} - \mathrm{i}\bar{\Gamma} + \mathrm{i}\eta_s/2, k_{P_1} - \mathrm{i}(\eta_s - \eta_r)/2, k_{P_2} + k_{P_3} - \bar{\epsilon}_\mathrm{d} + \mathrm{i}\bar{\Gamma} - \mathrm{i}\eta_r/2\}$ in the third term in Eq. (24a) and to the set $\{\bar{\epsilon}_\mathrm{d} - \mathrm{i}\bar{\Gamma} + \mathrm{i}\eta_s/2, U + \bar{\epsilon}_\mathrm{d} - \mathrm{i}\bar{\Gamma} - \mathrm{i}\eta_s/2, E - 2\bar{\epsilon}_\mathrm{d} - U + 2\mathrm{i}\bar{\Gamma}\}$ in the fourth term in Eq. (24a). We interpret the third term in Eq. (24a) as "non-pure" three-body bound states since the resonant energies appearing in the former wave-number set are not the two-body resonant energy but the one-body (anti-)resonant energies. Here the one-body resonant energy $\bar{\epsilon}_\mathrm{d} - \mathrm{i}\bar{\Gamma} + \mathrm{i}\eta_s/2$ does not pair with the one-body anti-resonant energy $k_{P_2} + k_{P_3} - \bar{\epsilon}_\mathrm{d} + \mathrm{i}\bar{\Gamma} - \mathrm{i}\eta_r/2$ for $r \neq s$, which is a difference from the two-electron case. On the other hand, we interpret the fourth term in Eq. (24a) as "pure" three-body bound states since the two-body anti-resonant energy $E - 2\bar{\epsilon}_\mathrm{d} - U + 2\mathrm{i}\bar{\Gamma}$ appears in the latter wave-number set. We remark that three-body resonances do not appear in the DQD system because only two electrons can interact at a time through the Coulomb interaction.

4. Construction of scattering eigenstates

4.1. Two-electron case

We demonstrate how to construct the two-electron scattering eigenstates in Eqs. (15). The eigenfunction $g_{lm}(x_1,x_2)$ is discontinuous at $x_1 = 0$ and $x_2 = 0$, and $e_{l\alpha}(x)$ is discontinuous at $x = 0$. In fact, by integrating the Schrödinger equations (13a) and (13b) around the discontinuous points, we obtain the matching conditions

$$g_{lm}(x,0+) = g_{lm}(x,0-) - \mathrm{i}\sum_\alpha t_{m\alpha}e_{l\alpha}(x), \tag{26a}$$

$$e_{l\alpha}(0+) = e_{l\alpha}(0-) + \mathrm{i}t_{l\bar{\alpha}}f_{\alpha\bar{\alpha}} = 0. \tag{26b}$$

Since the Schrödinger equations (13) do not determine the values of the eigenfunctions at the discontinuous points, we assume

$$g_{lm}(x,0) = \frac{1}{2}\big(g_{lm}(x,0+) + g_{lm}(x,0-)\big), \tag{27a}$$

$$e_{l\alpha}(0) = \frac{1}{2}\big(e_{l\alpha}(0+) + e_{l\alpha}(0-)\big). \tag{27b}$$

Journal of Physics: Conference Series **670** (2016) 012038 doi:10.1088/1742-6596/670/1/012038

For $x \neq 0$, by using Eqs. (26a) and (27a), the coupled first-order differential equations (13b) for the eigenfunctions $e_{l\alpha}(x)$ and $e_{l\bar{\alpha}}(x)$ give the second-order differential equation

$$\partial_x^2 \tilde{e}_{l\alpha}(x) - \mathrm{i}(\Delta\epsilon_{\mathrm{d}\alpha} - \mathrm{i}\Delta\Gamma_\alpha)\partial_x \tilde{e}_{l\alpha}(x) + (t'^*_{\bar{\alpha}} - \mathrm{i}\Gamma_{\alpha\bar{\alpha}})(t'^*_{\alpha} - \mathrm{i}\Gamma_{\bar{\alpha}\alpha})\tilde{e}_{l\alpha}(x) = G_{l\alpha}(x) \tag{28}$$

for $\tilde{e}_{l\alpha}(x) = \mathrm{e}^{-\mathrm{i}(E-\epsilon_{\mathrm{d}\alpha}+\mathrm{i}\Gamma_{\alpha\alpha})x} e_{l\alpha}(x)$. Here the inhomogeneous term is given by

$$G_{l\alpha}(x) = -\sum_m \mathrm{e}^{-\mathrm{i}(E-\epsilon_{\mathrm{d}\alpha}+\mathrm{i}\Gamma_{\alpha\alpha})x}\big((E-\epsilon_{\mathrm{d}\bar{\alpha}}+\mathrm{i}\Gamma_{\bar{\alpha}\bar{\alpha}}+\mathrm{i}\partial_x)t^*_{m\alpha} + (t'^*_{\bar{\alpha}} - \mathrm{i}\Gamma_{\alpha\bar{\alpha}})t^*_{m\bar{\alpha}}\big) g_{lm}(x, 0-). \tag{29}$$

General solutions for the function $e_{l\alpha}(x)$ are given by

$$e_{l\alpha}(x) = \sum_{s=\pm} C_{l\alpha,s}\mathrm{e}^{\mathrm{i}(E-\epsilon_{\mathrm{d}\alpha}+\mathrm{i}\Gamma_{\alpha\alpha}-\mathrm{i}\lambda_{\alpha,s})x} + \sum_{s=\pm} \frac{1}{\lambda_{\alpha,s}-\lambda_{\alpha,\bar{s}}} \mathrm{e}^{\mathrm{i}(E-\epsilon_{\mathrm{d}\alpha}+\mathrm{i}\Gamma_{\alpha\alpha})x} \int_{x_0}^{x} dz\, \mathrm{e}^{\lambda_{\alpha,s}(x-z)} G_{l\alpha}(z). \tag{30}$$

Here $C_{l\alpha,\pm}$ is an integration constant, $x_0 = -\infty$ if $x < 0$ and $x_0 = 0$ if $x > 0$, and $\lambda_{\alpha,\pm} = (\mathrm{i}\Delta\epsilon_{\mathrm{d}\alpha} + \Delta\Gamma_\alpha + \eta_\pm)/2$ are the roots of the characteristic equation associated with the second-order differential equation (28). By using Eqs. (26b) and (27b), we obtain

$$(E - 2\bar{\epsilon}_{\mathrm{d}} - U + 2\mathrm{i}\bar{\Gamma}) f_{\alpha\bar{\alpha}} = \sum_l \big(t^*_{l\alpha} e_{l\bar{\alpha}}(0-) - t^*_{l\bar{\alpha}} e_{l\alpha}(0-)\big), \tag{31}$$

which gives

$$f_{\alpha\bar{\alpha}} = \frac{1}{E - 2\bar{\epsilon}_{\mathrm{d}} - U + 2\mathrm{i}\bar{\Gamma}} \sum_l \big(t^*_{l\alpha} e_{l\bar{\alpha}}(0-) - t^*_{l\bar{\alpha}} e_{l\alpha}(0-)\big). \tag{32}$$

By using the matching conditions (26a) and (26b) and the solutions (30) and (32), we obtain the eigenfunctions satisfying arbitrary scattering boundary conditions. The construction scheme is indicated in Fig. 3.

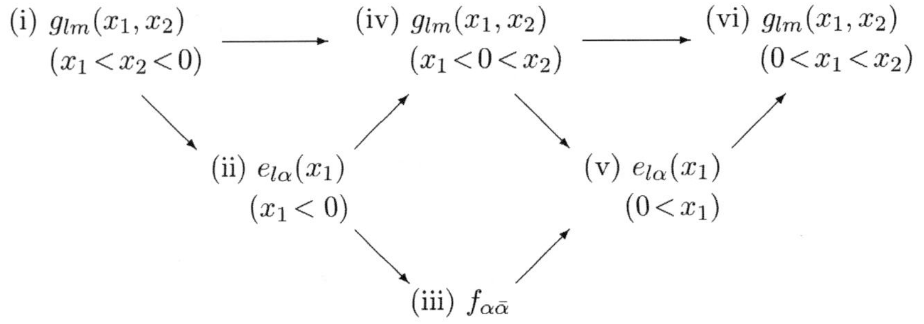

Figure 3. Construction scheme for the two-electron scattering eigenfunctions.

(i)→(ii): For $x < 0$, the scattering boundary conditions (14) read

$$g_{lm}(x, 0-) = \frac{1}{2\pi}\sum_P A_{lm,P}\,\mathrm{e}^{\mathrm{i}k_{P_1}x}, \qquad A_{lm,P} = \delta_{l\ell_{P_1}}\delta_{m\ell_{P_2}}\mathrm{sgn}(P). \tag{33}$$

Since the inhomogeneous term of the second-order differential equation (28) is given by

$$G_{l\alpha}(x) = -\frac{1}{2\pi}\sum_{P,m}\big((k_{P_2} - \epsilon_{\mathrm{d}\bar{\alpha}} + \mathrm{i}\Gamma_{\bar{\alpha}\bar{\alpha}})t^*_{m\alpha} + (t'^*_{\bar{\alpha}} - \mathrm{i}\Gamma_{\alpha\bar{\alpha}})t^*_{m\bar{\alpha}}\big) A_{lm,P}\mathrm{e}^{\mathrm{i}(-k_{P_2}+\epsilon_{\mathrm{d}\alpha}-\mathrm{i}\Gamma_{\alpha\alpha})x}, \tag{34}$$

the general solution (30) is

$$e_{l\alpha}(x) = \sum_s C_{l\alpha,s} e^{\mathrm{i}(E-\epsilon_{\mathrm{d}\alpha}+\mathrm{i}\Gamma_{\alpha\alpha}-\mathrm{i}\lambda_{\alpha,s})x} + \frac{1}{\sqrt{2\pi}} \sum_{P,m} A_{lm,P} e^{(m)}_{\alpha,k_{P_2}} e^{\mathrm{i}k_{P_1}x}. \tag{35}$$

To avoid the divergences in $x \to -\infty$, we take $C_{l\alpha,\pm} = 0$,

$$e_{l\alpha}(x) = \frac{1}{\sqrt{2\pi}} \sum_{P,m} A_{lm,P} e^{(m)}_{\alpha,k_{P_2}} e^{\mathrm{i}k_{P_1}x}, \qquad (x<0). \tag{36}$$

(ii)→(iii): By using Eq. (32), we obtain

$$f_{\alpha\bar{\alpha}} = \frac{E-2\bar{\epsilon}_{\mathrm{d}}+2\mathrm{i}\bar{\Gamma}}{E-2\bar{\epsilon}_{\mathrm{d}}-U+2\mathrm{i}\bar{\Gamma}} \sum_{P,l,m} A_{lm,P} e^{(l)}_{\alpha,k_{P_1}} e^{(m)}_{\bar{\alpha},k_{P_2}}. \tag{37}$$

(i), (ii)→(iv): For $x<0$, we have

$$g_{lm}(x,0+) = g_{lm}(x,0-) - \mathrm{i}\sum_\alpha t_{m\alpha} e_{l\alpha}(x) = \frac{1}{\sqrt{2\pi}} \sum_{P,n} A_{ln,P} e^{\mathrm{i}k_{P_1}x} g^{(n)}_{m,k_{P_2}}(0+). \tag{38}$$

Then, for $x_1 < 0 < x_2$, we obtain

$$g_{lm}(x_1,x_2) = \frac{1}{\sqrt{2\pi}} \sum_{P,n} A_{ln,P} e^{\mathrm{i}k_{P_1}x_1} g^{(n)}_{m,k_{P_2}}(x_2). \tag{39}$$

(iv), (iii)→(v): For $x>0$, we have

$$g_{lm}(0-,x) = \frac{1}{\sqrt{2\pi}} \sum_{P,n} A_{ln,P} g^{(n)}_{m,k_{P_2}}(x). \tag{40}$$

The general solution (30) is

$$e_{l\alpha}(x) = \sum_{P,m,n} A_{mn,P} g^{(m)}_{l,k_{P_1}}(x) e^{(n)}_{\alpha,k_{P_2}} + \sum_s \tilde{C}_{l\alpha,s} e^{\mathrm{i}(E-\epsilon_{\mathrm{d}\alpha}+\mathrm{i}\Gamma_{\alpha\alpha}-\mathrm{i}\lambda_{\alpha,s})x}. \tag{41}$$

We keep the integration constant $\tilde{C}_{l\alpha,\pm}$ since the second term is not divergent in $x \to \infty$. The integration constants $\tilde{C}_{l\bar{\alpha},\pm}$ satisfy the relation

$$\tilde{C}_{l\bar{\alpha},\pm} = -\frac{\lambda_{\alpha,\pm}\tilde{C}_{l\alpha,\pm}}{\mathrm{i}(t'^{*}_{\bar{\alpha}} - \mathrm{i}\Gamma_{\alpha\bar{\alpha}})}. \tag{42}$$

Through the matching condition in Eq. (26b) for $e_{l\alpha}(x)$, we obtain

$$\sum_s \tilde{C}_{l\alpha,s} = \frac{\mathrm{i}t_{l\bar{\alpha}}U}{E-2\bar{\epsilon}_{\mathrm{d}}-U+2\mathrm{i}\bar{\Gamma}} \sum_{P,m,n} A_{mn,P} e^{(m)}_{\alpha,k_{P_1}} e^{(n)}_{\bar{\alpha},k_{P_2}}. \tag{43}$$

By replacing α with $\bar{\alpha}$, we have

$$\sum_s \tilde{C}_{l\bar{\alpha},s} = -\sum_s \frac{\lambda_{\alpha,s}\tilde{C}_{l\alpha,s}}{\mathrm{i}(t'^{*}_{\bar{\alpha}} - \mathrm{i}\Gamma_{\alpha\bar{\alpha}})} = \frac{\mathrm{i}t_{l\alpha}U}{E-2\bar{\epsilon}_{\mathrm{d}}-U+2\mathrm{i}\bar{\Gamma}} \sum_{P,m,n} A_{mn,P} e^{(m)}_{\bar{\alpha},k_{P_1}} e^{(n)}_{\alpha,k_{P_2}}. \tag{44}$$

By solving the coupled equations (43) and (44) for $\tilde{C}_{l\alpha,+}$ and $\tilde{C}_{l\alpha,-}$, we obtain

$$\tilde{C}_{l\alpha,s}=\frac{1}{\eta_s}\frac{U}{E-2\bar{\epsilon}_{\rm d}-U+2{\rm i}\bar{\Gamma}}({\rm i}\lambda_{\bar{\alpha},s}t_{l\bar{\alpha}}-(t'_\alpha-{\rm i}\Gamma_{\alpha\bar{\alpha}})t_{l\alpha})\sum_{P,m,n}A_{mn,P}e^{(m)}_{\alpha,k_{P_1}}e^{(n)}_{\bar{\alpha},k_{P_2}}. \tag{45}$$

Hence, for $x>0$, we have

$$\begin{aligned}e_{l\alpha}(x)&=\sum_{P,m,n}A_{mn,P}g^{(m)}_{l,k_{P_1}}(x)e^{(n)}_{\alpha,k_{P_2}}\\&+\frac{U}{E-2\bar{\epsilon}_{\rm d}-U+2{\rm i}\bar{\Gamma}}\sum_s\frac{1}{\eta_s}({\rm i}\lambda_{\bar{\alpha},s}t_{l\bar{\alpha}}-(t'_\alpha-{\rm i}\Gamma_{\alpha\bar{\alpha}})t_{l\alpha})\sum_{P,m,n}A_{mn,P}e^{(m)}_{\alpha,k_{P_1}}e^{(n)}_{\bar{\alpha},k_{P_2}}{\rm e}^{{\rm i}(E-\epsilon_{{\rm d}\alpha}+{\rm i}\Gamma_{\alpha\alpha}-{\rm i}\lambda_{\alpha,s})x}.\end{aligned} \tag{46}$$

(iv), (v)→(vi): For $x>0$, we have

$$\begin{aligned}g_{lm}(x,0+)&=g_{lm}(x,0-)-{\rm i}\sum_\alpha t_{m\alpha}e_{l\alpha}(x)\\&=-\sum_{P,r,n}A_{rn,P}g^{(r)}_{m,k_{P_1}}(0+)g^{(n)}_{l,k_{P_2}}(x)-{\rm i}\sum_{\alpha,s=\pm}t_{m\alpha}\tilde{C}_{l\alpha,s}{\rm e}^{{\rm i}(E-\epsilon_{{\rm d}\alpha}+{\rm i}\Gamma_{\alpha\alpha}-{\rm i}\lambda_{\alpha,s})x}.\end{aligned} \tag{47}$$

Then, for $0<x_1<x_2$, we obtain

$$\begin{aligned}g_{lm}(x_1,x_2)&=\sum_{P,r,n}A_{rn,P}g^{(r)}_{l,k_{P_1}}(x_1)g^{(n)}_{m,k_{P_2}}(x_2)\\&+{\rm i}\sum_{\alpha,s}t_{l\alpha}\tilde{C}_{m\alpha,s}{\rm e}^{{\rm i}((\epsilon_{{\rm d}\alpha}-{\rm i}\Gamma_{\alpha\alpha}+{\rm i}\lambda_{\alpha,s})x_1+(E-\epsilon_{{\rm d}\alpha}+{\rm i}\Gamma_{\alpha\alpha}-{\rm i}\lambda_{\alpha,s})x_2)}.\end{aligned} \tag{48}$$

By inserting $A_{lm,P}=\delta_{l\ell_{P_1}}\delta_{m\ell_{P_2}}{\rm sgn}(P)$ and $\tilde{C}_{m\alpha,s}$ in Eq. (45), we obtain

$$\begin{aligned}g_{lm}(x_1,x_2)&=\sum_P{\rm sgn}(P)g^{(\ell_{P_1})}_{l,k_{P_1}}(x_1)g^{(\ell_{P_2})}_{m,k_{P_2}}(x_2)\\&+\frac{{\rm i}U}{E-2\bar{\epsilon}_{\rm d}-U+2{\rm i}\bar{\Gamma}}\sum_P{\rm sgn}(P)\sum_{\alpha,s}\frac{1}{\eta_s}t_{l\alpha}({\rm i}\lambda_{\bar{\alpha},s}t_{m\bar{\alpha}}-(t'_\alpha-{\rm i}\Gamma_{\alpha\bar{\alpha}})t_{m\alpha})e^{(\ell_{P_1})}_{\alpha,k_{P_1}}e^{(\ell_{P_2})}_{\bar{\alpha},k_{P_2}}\\&\times{\rm e}^{{\rm i}((\epsilon_{{\rm d}\alpha}-{\rm i}\Gamma_{\alpha\alpha}+{\rm i}\lambda_{\alpha,s})x_1+(E-\epsilon_{{\rm d}\alpha}+{\rm i}\Gamma_{\alpha\alpha}-{\rm i}\lambda_{\alpha,s})x_2)}.\end{aligned} \tag{49}$$

Through the anti-symmetrization with respect to the variables x_1 and x_2, we obtain the exact two-electron scattering eigenfunctions in Eqs. (15).

4.2. Three-electron case

We now describe an outline of the construction of the three-electron scattering eigenstates in Eqs. (24). It is clear from the Schrödinger equations (22) that the eigenfunction $g_{lmn}(x_1,x_2,x_3)$ is discontinuous at $x_i=0$, $(i=1,2,3)$ and $e_{lm\alpha}(x_1,x_2)$ is also discontinuous at $x_i=0$, $(i=1,2)$. Then, in a way similar to the one- and two-electron cases, we assume

$$g_{lmn}(x_1,x_2,0)=\frac{1}{2}(g_{lmn}(x_1,x_2,0+)+g_{lmn}(x_1,x_2,0-)), \tag{50a}$$

$$e_{lm\alpha}(x_1,0)=\frac{1}{2}(e_{lm\alpha}(x_1,0+)+e_{lm\alpha}(x_1,0-)). \tag{50b}$$

We note that the function $f_{l\alpha\bar{\alpha}}(x_1)$ is a continuous function. The matching conditions for the functions $g_{lmn}(x_1, x_2, x_3)$ and $e_{lm\alpha}(x_1, x_2)$ are

$$g_{lmn}(x_1, x_2, 0+) - g_{lmn}(x_1, x_2, 0-) + \mathrm{i}\sum_{\alpha} t_{n\alpha} e_{lm}(x_1, x_2) = 0, \tag{51a}$$

$$e_{lm\alpha}(x_1, 0+) - e_{lm\alpha}(x_1, 0-) - \mathrm{i}t_{m\bar{\alpha}} f_{l\alpha\bar{\alpha}}(x_1) = 0. \tag{51b}$$

General solutions for the functions $e_{lm\alpha}(x_1, x_2)$ and $f_{l\alpha\bar{\alpha}}(x_1)$ are given by

$$\begin{aligned} e_{lm\alpha}(x_1, x_2) &= \sum_{s=\pm} C_{lm\alpha,s}(x_{12}) \mathrm{e}^{\mathrm{i}(E-\epsilon_{\mathrm{d}\alpha}+\mathrm{i}\Gamma_{\alpha\alpha}-\mathrm{i}\lambda_{\alpha,s})\bar{x}} \\ &+ \sum_{s=\pm} \frac{1}{\lambda_{\alpha,s} - \lambda_{\alpha,\bar{s}}} \mathrm{e}^{\mathrm{i}(E-\epsilon_{\mathrm{d}\alpha}+\mathrm{i}\Gamma_{\alpha\alpha})\bar{x}} \int_{x_0}^{\bar{x}} \mathrm{e}^{\lambda_{\alpha,s}(\bar{x}-z)} G_{lm\alpha}\Big(z+\frac{x_{12}}{2}, z-\frac{x_{12}}{2}\Big) dz, \end{aligned} \tag{52}$$

$$\begin{aligned} f_{l\alpha\bar{\alpha}}(x_1) &= C_{l\alpha\bar{\alpha}} \mathrm{e}^{\mathrm{i}(E-2\bar{\epsilon}_{\mathrm{d}}-U+2\mathrm{i}\bar{\Gamma})x_1} \\ &- \mathrm{i}\sum_{m} \int_{x_0}^{x_1} dz\, \mathrm{e}^{\mathrm{i}(E-2\bar{\epsilon}_{\mathrm{d}}-U+2\mathrm{i}\bar{\Gamma})(x_1-z)} \big(t^*_{m\alpha} e_{lm\bar{\alpha}}(z, 0-) - t^*_{m\bar{\alpha}} e_{lm\alpha}(z, 0-)\big). \end{aligned} \tag{53}$$

Here we put $\bar{x} = (x_1 + x_2)/2$ and $x_{12} = x_1 - x_2$ and $x_0 = -\infty$ if $x < 0$ and $x_0 = 0$ if $x > 0$. The function $C_{l\alpha,\pm}(x_{12})$ is a function in the variable x_{12} and $C_{l\alpha\bar{\alpha}}$ is an integration constant. We also introduce

$$\begin{aligned} &G_{lm\alpha}(z_1, z_2) \\ &= -\sum_{n} \mathrm{e}^{\mathrm{i}(-E+\epsilon_{\mathrm{d}\alpha}-\mathrm{i}\Gamma_{\alpha\alpha})\bar{z}} \big(t^*_{n\alpha}(E - \epsilon_{\mathrm{d}\bar{\alpha}} + \mathrm{i}\Gamma_{\bar{\alpha}\bar{\alpha}} + \mathrm{i}\partial_{\bar{z}}) + t^*_{n\bar{\alpha}}(t'_{\alpha} - \mathrm{i}\Gamma_{\alpha\bar{\alpha}})\big) g_{lmn}(z_1, z_2, 0-), \end{aligned} \tag{54}$$

where $\bar{z} = (z_1 + z_2)/2$. In a way similar to the two-electron case, we construct the three-electron scattering eigenstates. The construction scheme is indicated in Fig. 4.

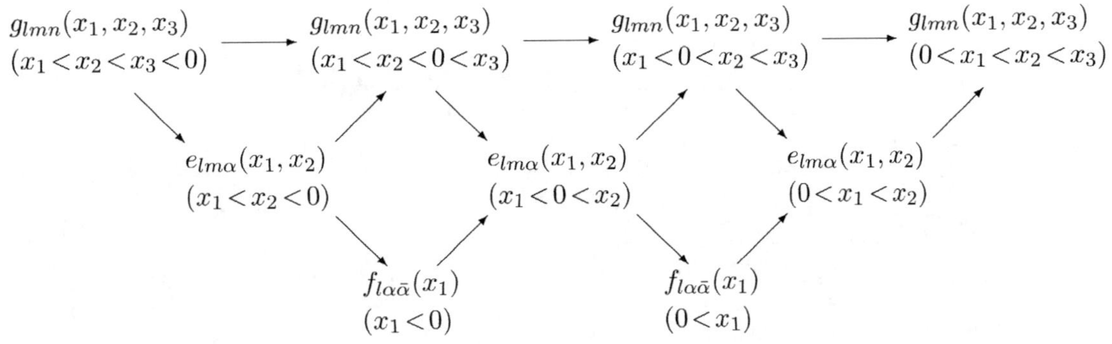

Figure 4. Construction scheme for three-electron scattering eigenfunctions.

5. Summary

We have constructed exact two- and three-electron scattering eigenstates for the DQD system with an interdot Coulomb interaction. The construction can be extended to the N-electron case straightforwardly. We have found that, due to the interdot Coulomb interaction, some of the incident plane waves are scattered to two- and three-body bound states. The binding strength of the many-body bound states is affected by the arrangement of the two QDs, which is an effect of the quantum interference between different electron pathways. Thus we have observed an interplay of the interaction and quantum interference in the many-body bound states.

Let us summarize the relation of the many-body resonances and the many-body bound states in the DQD system. As is indicated in Table 1, we find the one- and two-body resonant poles in the scattering eigenstates, which produce the two- and three-body bound states. We also find that there is no three-body resonances in the three-electron scattering eigenstates, which implies that pure four-body bound states that are not a combination of the two- and three-body bound states do not appear in the scattering eigenstates for electrons more than three.

Table 1. Relation of many-body resonances and many-body bound states in the N-electron scattering eigenstates

N	Positions of resonant poles	Terms of many-body bound states
1	$k_1 = \bar{\epsilon}_{\rm d} - {\rm i}\bar{\Gamma} + \frac{\rm i}{2}\eta_s$	
2	$k_i = \bar{\epsilon}_{\rm d} - {\rm i}\bar{\Gamma} + \frac{\rm i}{2}\eta_s$	${\rm e}^{(-\bar{\Gamma}+\frac{1}{2}{\rm Re}(\eta_s))\|x_1-x_2\|}$
	$k_1 + k_2 = 2\bar{\epsilon}_{\rm d} + U - 2{\rm i}\bar{\Gamma}$	
3	$k_i = \bar{\epsilon}_{\rm d} - {\rm i}\bar{\Gamma} + \frac{\rm i}{2}\eta_s$	${\rm e}^{(-\bar{\Gamma}+\frac{1}{2}{\rm Re}(\eta_s))\|x_1-x_2\|}$
	$k_i = \bar{\epsilon}_{\rm d} + U - {\rm i}\bar{\Gamma} + \frac{\rm i}{2}\eta_s$	${\rm e}^{(-\bar{\Gamma}+\frac{1}{2}{\rm Re}(\eta_s))\|x_i-x_j\|+(-\bar{\Gamma}+\frac{1}{2}{\rm Re}(\eta_r))\|x_j-x_l\|}$
	$k_i + k_j = 2\bar{\epsilon}_{\rm d} + U - 2{\rm i}\bar{\Gamma}$	${\rm e}^{(-\bar{\Gamma}+\frac{1}{2}{\rm Re}(\eta_s))\|x_i-x_j\|-2\bar{\Gamma}\|x_j-x_l\|}$

Acknowledgments
This work was supported by JSPS KAKENHI Grants Numbers 26400409, 15K05200 and 15K05207.

References

[1] Cronenwett S M, Oosterkamp T H and Kouwenhoven L P 1998 *Science* **281** 540
[2] Goldhaber-Gordon D, Shtrikman H, Mahalu D, Abusch-Magder D, Meirav U and Kastner M A 1998 *Nature (London)* **391** 156
[3] W G van der Wiel, S De Franceschi, T Fujisawa, J M Elzerman, S Tarucha, and L P Kouwenhoven 2000 *Science* **289** 2105
[4] Ng T K and Lee P A 1988 *Phys. Rev. Lett.* **61** 1768
[5] Wingreen N S and Meir Y 1994 *Phys. Rev. B* **49** 11040
[6] Costi T A, Hewson A C and Zlatić V 1994 *J. Phys. Condens. Matter* **6** 2519
[7] Aono T, Eto M and Kawamura K 1998 *J. Phys. Soc. Jpn.* **67** 1860–1863
[8] Georges A and Meir Y 1999 *Phys. Rev. Lett.* **82** 3508–3511
[9] López R, Aguado R and Platero G 2002 *Phys. Rev. Lett.* **89** 136802
[10] Kim T S and Hershfield S 2001 *Phys. Rev. B* **63** 245326
[11] Zhang G M, Lü R, Liu Z R and Yu L 2005 *Phys. Rev. B* **72** 073308
[12] Tanaka Y and Kawakami N 2005 *Phys. Rev. B* **72** 085304
[13] Cornaglia P S and Grempel D R 2005 *Phys. Rev. B* **71** 075305
[14] Boese D, Hofstetter W and Schoeller H 2002 *Phys. Rev. B* **66** 125315
[15] Karrasch C, Enss T and Meden V 2006 *Phys. Rev. B* **73** 235337
[16] Kashcheyevs V, Schiller A, Aharony A and Entin-Wohlman O 2007 *Phys. Rev. B* **75** 115313
[17] Hübel A, Held K, Weis J and v Klitzing K 2008 *Phys. Rev. Lett.* **101** 186804
[18] Okazaki Y, Sasaki S and Muraki K 2011 *Phys. Rev. B* **84** 161305
[19] Büsser C A, Feiguin A E and Martins G B 2012 *Phys. Rev. B* **85** 241310(R)

[20] Amasha S, Keller A J, Rau I G, Carmi A, Katine J A, Shtrikman H, Oreg Y and Goldhaber-Gordon D 2013 *Phys. Rev. Lett.* **110** 046406
[21] Tosi L, Roura-Bas P and Aligia A A 2013 *Phys. Rev. B* **88** 235427
[22] Nishikawa Y, Hewson A C, Crow D J G and Bauer J 2013 *Phys. Rev. B* **88** 245130
[23] Meden V and Marquardt F 2006 *Phys. Rev. Lett.* **96** 146801
[24] Nilsson H A, Karlström O, Larsson M, Caroff P, Pedersen J N, Samuelson L, Wacker A, Wernersson L E and Xu H Q 2010 *Phys. Rev. Lett.* **104** 186804
[25] Imry Y 2002 *Introduction to Mesoscopic Physics* 2nd ed (Oxford Univ. Pr., Oxford)
[26] Datta S 1995 *Electronic Transport in Mesoscopic Systems* (Cambridge University Press, Cambridge)
[27] Nishino A, Imamura T and Hatano N 2009 *Phys. Rev. Lett.* **102** 146803
[28] Nishino A, Imamura T and Hatano N 2011 *Phys. Rev. B* **83** 035306
[29] Nishino A, Hatano N and Ordoez G 2015 *Phys. Rev. B* **91** 045140
[30] Doyon B 2007 *Phys. Rev. Lett.* **99** 076806
[31] Golub A 2007 *Phys. Rev. B* **76** 193307
[32] Mehta P 2006 Ph.D. thesis Rutgers University
[33] Boulat E, Saleur H and Schmitteckert P 2008 *Phys. Rev. Lett.* **101** 140601
[34] C Karrasch, S Andergassen, M Pletyukhov, D Schuricht, L Borda, V Meden, and H Schoeller 2010 *Eur. Phys. Lett.* **90** 30003
[35] S Andergassen, M Pletyukhov, D Schuricht, H Schoeller, and L Borda 2011 *Phys. Rev. B* **83** 205103
[36] Gamow G 1928 *Z. Phys. A* **51** 204
[37] Siegert A J F 1939 *Phys. Rev.* **56** 750
[38] Peierls R E 1959 *Proc. Roy. Soc. London A* **253** 16
[39] Brändas E E and Elandar N 1989 *Resonances* (Springer-Verlag, Berlin, Heidelberg)
[40] Kukulin V I, Krasnopol'sky V M and Horáček J 1989 *Theory of Resonances: Principles and Applications* (Kluwer Academic Publishers, Dordrecht)
[41] Hatano N, Sasada K, Nakamura H and Petrosky T 2008 *Prog. Theor. Phys.* **119** 187
[42] Imamura T, Nishino A and Hatano N 2009 *Phys. Rev. B* **80** 245323
[43] Zheng H, Gauthier D J and Baranger H U 2010 *Phys. Rev. A* **82** 063816
[44] Nishino A, Imamura T and Hatano N 2012 *J. Phys.: Conf. Ser.* **343** 012087

Journal of Physics: Conference Series **670** (2016) 012039 doi:10.1088/1742-6596/670/1/012039

On the group theoretical approach to the Quantum Theory of an interacting spin-0 particle

Giuseppe Nisticò

Dipartimento di Matematica e Informatica, Università della Calabria, Italy, and INFN gr. Collegato di Cosenza, Italy

E-mail: giuseppe.nistico@unical.it

Abstract. We point out a difficulty that arises in extending the group theoretical approach that deductively establish the quantum theory of a free particle to the case of an interacting particle. Then we develop an approach which overcomes this difficulty. The result is a theory of an interacting particle where the standard theory is characterized by specific covariance properties related to the interaction.

1. Introduction

A very satisfactory approach to the Quantum Theory of a *free* particle can be accomplished by following group theoretical methods, developed in particular by E.P. Wigner and G.W. Mackey. The key starting point is that Galilei's group $\mathcal{G}$, within a non-relativistic theory, is a group of *symmetry* transformations for a free particle; then Wigner's theorem [1],[2] on the representation of symmetries and Mackey's imprimitivity theorem [3],[4],[5] apply. In so doing the specific Quantum Theory of a free particle is explicitly obtained [3],[6],[7],[8],[9] without invoking canonical quantization, but, instead, by means of a mathematical deduction based on symmetry principles. We recall the main mathematical tools that allow for such an approach in section 2, while the necessary formal implications of the concept of Galileian transformations in the Quantum Theory of a particle are outlined in section 3.

But the *Galileian transformations are not symmetries for an interacting particle*, so that an obstacle is found in extending the approach to this case. To overcome this difficulty, the approaches that in the literature extend the cited group theoretical method to the interacting case (e.g., [7],[8],[9]) adopt, among others, the following assumption.

($\mathcal{P}$) *Each Galileian transformation $g \in \mathcal{G}$ is assigned a unitary operator U_g which realizes the corresponding quantum transformations of states and observables according to $\rho \xrightarrow{g} U_g \rho U_g^{-1}$ and $A \xrightarrow{g} U_g A U_g^{-1}$; the correspondence $g \to U_g$ is a projective representation.*

Now, in section 4 we show that ($\mathcal{P}$) forces the hamiltonian operator into the form $H = -\frac{1}{2\mu}\left(\frac{\partial^2}{\partial x_1^2} + \frac{\partial^2}{\partial x_2^2} + \frac{\partial^2}{\partial x_3^2}\right) + \Phi(x_1, x_2, x_3)$. Therefore a Quantum Theory where ($\mathcal{P}$) is assumed cannot describe all empirically known interactions; in particular it excludes electromagnetic interactions. An empirically more adequate approach must abandon assumption ($\mathcal{P}$).

Then in section 5 we undertake the task of establishing the Quantum Theory of an interacting particle through a group theoretical approach, but without assumption ($\mathcal{P}$). The difficulties raised by the loss of symmetry are addressed through a theoretical development based on the *passive interpretation* of the Galileian transformations and on the concept of *σ-conversion*. We find that the form of the dynamical equation is constrained by the covariance properties preserved by the σ-conversion admitted by the interaction; in particular, we exactly identify which covariance properties characterize electromagnetic interaction among all possible ones.

2. Group theoretical key results

Here we recall the main mathematical tools of the cited group theoretical approach.

2.1. Basic achievements

Given the Hilbert space $\mathcal{H}$ of the Quantum Theory of a physical system, we shall make use of the following operator's structures.

- The set $\Omega(\mathcal{H})$ of all self-adjoint operators of $\mathcal{H}$, which represent *quantum observables*.
- The complete, ortho-complemented lattice $\Pi(\mathcal{H})$ of all projections operators of $\mathcal{H}$, i.e. quantum observables with possible outcomes in $\{0,1\}$.
- The set $\Pi_1(\mathcal{H})$ of all rank one orthogonal projections of $\mathcal{H}$.
- The set $\mathcal{S}(\mathcal{H})$ of all density operators of $\mathcal{H}$, which represent *quantum states*.
- The set $\mathcal{U}(\mathcal{H})$ of all unitary operators of the Hilbert space $\mathcal{H}$.

The main mathematical tools will be used in this work are the *imprimitivity theorem* of Mackey and Wigner's *representation theorem*.

Mackey's *imprimitivity* theorem is a representation theorem for *imprimitivity* systems relative to projective representations [4]. The following definition recalls the notion of projective representation.

Definition 2.1. *Let G be a separable, locally compact group with identity element e. A correspondence $U : G \to \mathcal{U}(\mathcal{H})$, $g \to U_g$, with $U_e = 1\!\!I$, is a projective representation of G if the following conditions hold.*

i) *A complex function $\sigma : \mathcal{G} \times \mathcal{G} \to \mathbf{C}$ such that $|\sigma(g_1, g_2)| = 1$ for all $g_1, g_2 \in G$, called multiplier, exists such that $U_{g_1 g_2} = \sigma(g_1, g_2) U_{g_1} U_{g_2}$;*

ii) *for all $\phi, \psi \in \mathcal{H}$, the mapping $g \to \langle U_g \phi \mid \psi \rangle$ is a Borel function in g.*

A projective representation with multiplier σ is called σ-representation.

A projective representation is said to be continuous if for any fixed $\psi \in \mathcal{H}$ the mapping $g \to U_g \psi$ from $\mathcal{G}$ into $\mathcal{H}$ is continuous with respect to g.

Let $\mathcal{E}$ be the Euclidean group, i.e. the semi-direct product $\mathcal{E} = \mathbf{R}^3 \circledS SO(3)$ between the group of spatial translations $\mathbf{R}^3$ and the group of spatial proper rotations $SO(3)$; each transformation $g \in \mathcal{E}$ bi-univocally corresponds to the pair $(\mathbf{a}, R) \in \mathbf{R}^3 \times SO(3)$ such that $R\mathbf{x} + \mathbf{a} \equiv g(\mathbf{x})$ is the result of the transformation of the spatial point $\mathbf{x} = (x_1, x_2, x_3)$ by g. The general imprimitivity theorem is an advanced mathematical result, but its formulation becomes very simple when adapted to the case of the Euclidean group $\mathcal{E}$, that is the case we are interested to. Then we introduce the concept of *imprimitivity system* and the theorem for this simple case.

Definition 2.2. *Let $\mathcal{H}$ be the Hilbert space of a σ-representation $g \to U_g$ of the Euclidean group $\mathcal{E}$. A projection valued (PV) measure $E : \mathcal{B}(\mathbf{R}^3) \to \Pi(\mathcal{H})$, $\Delta \to E(\Delta)$ is an imprimitivity system for the σ-representation $g \to U_g$ if the relation*

$$U_g E(\Delta) U_g^{-1} = E(g(\Delta)) \equiv E(R(\Delta) + \mathbf{a})$$

holds for all $(\mathbf{a}, R) \in \mathcal{E}$.

Mackey's theorem of imprimitivity. *If a PV measure $E : \mathcal{B}(\mathbf{R}^3) \to \Pi(\mathcal{H})$ is an imprimitivity system for a continuous σ-representation $g \to U_g$ of the Euclidean group $\mathcal{E}$, then a σ-representation $L : SO(3) \to \mathcal{U}(\mathcal{H}_0)$ exists such that, modulo a unitary isomorphism,*

(M.1) $\mathcal{H} = L_2(\mathbf{R}^3, \mathcal{H}_0)$,

(M.2) $(E(\Delta)\psi)(\mathbf{x}) = \chi_\Delta(x)\psi(\mathbf{x})$, *where* χ_Δ *is the characteristic functional of* Δ,

(M.3) $(U_g\psi)(\mathbf{x}) = L_R\psi(g^{-1}(\mathbf{x})) \equiv L_R\psi(R^{-1}\mathbf{x} - R^{-1}\mathbf{a})$, *for every* $g = (\mathbf{a}, R) \in \mathcal{E}$.

In the literature different equivalent formulations of Wigner's theorem have been proved [2],[11]. Here we formulate those "Wigner's' theorems" we need for our work.

Wigner's theorem 1. *If $S^{(1)} : \mathcal{S}(\mathcal{H}) \to \mathcal{S}(\mathcal{H})$ and $S^{(2)} : \Omega(\mathcal{H}) \to \Omega(\mathcal{H})$ are bijective mappings such that*

$$Tr(S^{(1)}[\rho]S^{(2)}[A]) = Tr[\rho A] \tag{1}$$

holds for all $(\rho, A) \in \mathcal{S}(\mathcal{H}) \times \Omega(\mathcal{H})$ for which $Tr(\rho A)$ exists, then either a unitary operator or an anti-unitary operator U of $\mathcal{H}$ exists such that $S^{(1)}[\rho] = U\rho U^*$ and $S^{(2)}[A] = UAU^*$ for all $(\rho, A) \in \mathcal{S}(\mathcal{H}) \times \Omega(\mathcal{H})$, unique up a phase factor.

Wigner's theorem 2. *If $S : \Pi(\mathcal{H}) \to \Pi(\mathcal{H})$ is an automorphism of $\Pi(\mathcal{H})$, i.e. if it is a bijective mapping such that*

$$E_1 \leq E_2 \Leftrightarrow S[E_1] \leq S[E_2] \quad \text{and} \quad S[E^\perp] = (S[E])^\perp, \quad \forall E_1, E_2, E \in \Pi(\mathcal{H}),$$

then either a unitary operator or an anti-unitary operator U of $\mathcal{H}$ exists such that $S(E) = UEU^*$ for all $E \in \Pi(\mathcal{H})$, unique up a phase factor.

2.2. Quantum theoretical implications

Wigner's theorem has very important implications in Quantum Theory. We outline those of interest for our aims.

Let $g \to \hat{U}_g$ be every continuous projective representation of Galilei's group $\mathcal{G}$, i.e. the group generated by $\mathcal{E}$ and by Galileian velocity boosts. Let $\mathcal{T}_1 \subseteq \mathcal{G}$ be the abelian sub-group of spatial translations along the x_1 axis. If $\tau_1(a)$ denotes the translation $(x_1, x_2, x_3) \to (x_1 + a, x_2, x_3)$, then we have $\tau_1(a)\tau_1(b) = \tau_1(a+b)$, so that $\hat{U}_{\tau_1(a+b)} = \sigma(\tau_1(a), \tau_1(b))\,\hat{U}_{\tau_1(a)}\hat{U}_{\tau_1(b)}$. The arbitrary phase factor of each $\hat{U}_{\tau_1(a)}$ can be chosen so that, according to Stone's theorem, $\hat{U}_{\tau_1(a)} = e^{-i\hat{P}_1 a}$, where $\hat{P}_1$ is a self-adjoint operator called the *hermitean generator* of the one parameter unitary sub-group $\{e^{-i\hat{P}_1 a}, a \in \mathbf{R}\} \subseteq \mathcal{U}(\mathcal{H})$. Now, the one-parameter abelian sub-groups $\mathcal{T}_\alpha, \mathcal{R}_\alpha$, $\mathcal{B}_\alpha$ of spatial translation, spatial rotations and Galileian boosts, relative to axis x_α, are all additive; then the argument used for $\mathcal{T}_1$ can be repeated to establish the existence of nine hermitean generators $\hat{P}_\alpha$, $\hat{J}_\alpha$, $\hat{G}_\alpha$, $\alpha \in \{1, 2, 3\}$, of the nine one-parameter unitary subgroups $\{e^{-i\hat{P}_\alpha a_\alpha}, a \in \mathbf{R}\}$, $\{e^{-i\hat{J}_\alpha\theta_\alpha}, \theta_\alpha \in \mathbf{R}\}$, $\{e^{-i\hat{G}_\alpha u_\alpha}, u_\alpha \in \mathbf{R}\}$ representing the sub-groups $\mathcal{T}_\alpha, \mathcal{R}_\alpha$, $\mathcal{B}_\alpha$ according to the projective representation $g \to \hat{U}_g$ of the Galilei's group $\mathcal{G}$. The structural properties of $\mathcal{G}$ as a Lie group imply the validity of the following commutation relations [9].

$$\begin{aligned}&\text{(i) } [\hat{P}_\alpha, \hat{P}_\beta] = \mathbf{0}, \quad \text{(ii) } [\hat{J}_\alpha, \hat{P}_\beta] = i\hat{\epsilon}_{\alpha\beta\gamma}\hat{P}_\gamma, \quad \text{(iii) } [\hat{J}_\alpha, \hat{J}_\beta] = i\hat{\epsilon}_{\alpha\beta\gamma}\hat{J}_\gamma,\\ &\text{(iv) } [\hat{J}_\alpha, \hat{G}_\beta] = i\hat{\epsilon}_{\alpha\beta\gamma}\hat{G}_\gamma, \quad \text{(v) } [\hat{G}_\alpha, \hat{G}_\beta] = \mathbf{0}, \quad \text{(vi) } [\hat{G}_\alpha, \hat{P}_\beta] = i\delta_{\alpha\beta}\mu\mathbb{I},\end{aligned} \tag{2}$$

where $\hat{\epsilon}_{\alpha,\beta,\gamma}$ is the Levi-civita symbol $\epsilon_{\alpha,\beta,\gamma}$ restricted by the condition $\alpha \neq \gamma \neq \beta$, while μ is a non-zero real number which characterizes the projective representation.

Another important achievement implied by Wigner's theorem is the general evolution law of Quantum Theory [5],[6] with respect to a homogeneous time:

$$\rho_t = U_t = e^{-iHt}\rho e^{iHt} \quad \text{and} \quad \frac{d}{dt}A^{(t)} \equiv \dot{A}^{(t)} = i[H, A^{(t)}]. \tag{3}$$

3. Galileian quantum transformations

In this section we recall the concept of *quantum transformation* corresponding to a Galileian transformation $g \in \mathcal{G}$ according to the *active* interpretation.

In correspondence with each Galileian transformation $g \in \mathcal{G}$ we can conceive *two* mappings $R_g^{(1)}$ and $R_g^{(2)}$ which transform quantum states and quantum observables, respectively, according to the active interpretation of the transformation:

$$\rho \to R_g^{(1)}[\rho], \qquad A \to R_g^{(2)}[A]. \tag{4}$$

For instance, if $h \in \mathcal{G}$ is a pure velocity boost characterized by velocity parameters $(u_1, u_2, u_3) \equiv \mathbf{u}$, then $R_h^{(2)}[A]$ is meant to be the observable whose measurement apparatus ${}^h\mathcal{A}$ is identical to an apparatus $\mathcal{A}$ that measures A, also for its location, but it is endowed with a constant velocity $\mathbf{u}$ with respect to $\mathcal{A}$.

If the theory is specializaed to a localizable particle, then three commuting quantum observables $\mathbf{Q} = (Q_1, Q_2, Q_3)$, the *position* observables, exist such that $R_g^{(2)}[\mathbf{Q}] = g^{-1}(\mathbf{Q})$, where the function $g : \mathbf{R}^3 \to \mathbf{R}^3$, $\mathbf{x} \to g(\mathbf{x})$ defines how $g \in \mathcal{G}$ transforms spatial points $\mathbf{x}$.

Remark 3.1. The transformation of the position at time t, i.e. of $\mathbf{Q}^{(t)} = e^{iHt}\mathbf{Q}e^{-iHt}$, requires a function g_t different[1] from g; in general, the following covariance relations hold.

$$(i) \quad R_g^{(2)}[\mathbf{Q}^{(t)}] = g_t^{-1}(\mathbf{Q}^{(t)}); \quad (ii) \text{ in particular, } R_g^{(2)}[\mathbf{Q}] = g^{-1}(\mathbf{Q}). \tag{5}$$

If all transformations $g \in \mathcal{G}$ are *symmetry transformations* for the quantum system, further conditions can be implied in its Quantum Theory. First of all, $R_g^{(1)}$ and $R_g^{(2)}$ have to be bijective on $\mathcal{S}(\mathcal{H})$ and $\Omega(\mathcal{H})$, respectively. Furthermore, the following statement holds.

$$R_{g_1g_2}^{(1)} = R_{g_1}^{(1)} \circ R_{g_2}^{(1)} \text{ and } R_{g_1g_2}^{(2)} = R_{g_1}^{(2)} \circ R_{g_2}^{(2)}, \text{ for all } g_1g_2 \in \mathcal{G}. \tag{6}$$

The symmetry's character of g entails that the expected values of observables do not change if both the quantum state and the observable are transformed by the same symmetry transformation; then, whenever $Tr(\rho A)$ exists, the following statement holds for every $g \in \mathcal{G}$.

$$Tr(R_g^{(1)}[\rho]R_g^{(2)}[A]) = Tr(\rho A). \tag{7}$$

4. The interacting particle problem

For a free particle, the validity of (7) makes possible to apply Wigner's theorem 1, so that for every $g \in \mathcal{G}$ an operator U_g exists such that $R_g^{(1)}[\rho] = U_g \rho U_g^*$ and $R_g^{(2)}[A] = U_g A U_g^*$; together with (6), these relations imply that the restriction to $\mathcal{E}$ of $g \to U_g$ is a projective representation [5]. Then, the relation $U_g \mathbf{Q} U_g^{-1} = g^{-1}(\mathbf{Q})$ follows from (5.ii); it entails that the spectral PV measure of $\mathbf{Q}$ is an imprimitivity system [5]; therefore we can apply Mackey's theorem. In so doing, to each choice of the inducing representation L in Mackey's theorem and of μ in (2), there corresponds a different theory. The simplest choice, i.e. $L : SO(3) \to \mathbf{C}$, $L(R) = 1$, identifies $\mathcal{H} = L_2(\mathbf{R}^3)$ as the Hilbert space of the theory, and the position operators as $(Q_\alpha \psi) = x_\alpha \psi(\mathbf{x})$; moreover, by making use of Galileian invariance, valid for a free particle, it can be proved that the form of the hamiltonian operator must be $H = -\frac{1}{2\mu}\sum_{\alpha=1}^{3} \frac{\partial^2}{\partial x_\alpha^2}$. Hence, the simplest theory

[1] For instance, the pure boost $h \in \mathcal{G}$ characterized by a velocity $\mathbf{u} = (u, 0, 0)$, does not change the position at all; hence $h(\mathbf{x}) = \mathbf{x}$, i.e. $R_h^{(2)}(\mathbf{Q}) = h^{-1}(\mathbf{Q}) = \mathbf{Q}$. But $\mathbf{Q}^{(t)}$ represents the position measured with a delay t, therefore $R_h^{(2)}[\mathbf{Q}^{(t)}] = (Q_1^{(t)} - ut, Q_2, Q_3) \equiv h_t(\mathbf{Q}^{(t)})$, where $h_t(\mathbf{x}) = (x_1 - ut, x_2, x_3)$.

deduced from symmetry principles by the group theoretical approach is the standard Quantum Theory of a spin-0 free particle.

If the system is an *interacting* particle, then the *group of Galileian transformations is not a group of symmetry transformations*, so that conditions (6) and (7) fail to hold and we find an obstacle in extending the group theoretical approach to the interacting particle. However, in the literature several proposals can be found [5],[7],[9] where the group theoretical methods are extended to the interacting case. The aforesaid obstacles are overcome by assuming, more or less implicitly, that the following statement holds (e.g., see [5], page 201 and [7], page 236).

($\mathcal{P}$) *Each Galileian transformation g is represented in the formalism of the Quantum Theory by a unitary or anti-unitary operator U_g in such a way that*

i) $R_g^{(2)}[A] = U_g A U_g^*$ *is the result of the active transformation of the observable A by g;*

ii) $R_{g_1 g_2}^{(2)} = R_{g_1}^{(2)} \circ R_{g_2}^{(2)}$, *for all $g_1, g_2 \in \mathcal{G}$.*

By making use of these assumptions, the cited approaches deduce that in the Quantum Theory of a spin-0 particle undergoing an interaction homogeneous in time the hamiltonian operator H must have the following form, able to describe also interactions with electromagnetic fields.

$$H = \frac{1}{2\mu}\sum_{\alpha=1}^{3}\left(-i\frac{\partial}{\partial x_\alpha} - a_\alpha(\mathbf{Q})\right)^2 + \Phi(\mathbf{x}). \tag{8}$$

Now we prove, instead, the following statement.

($\mathcal{R}$) *Assumption ($\mathcal{P}$) forces the hamiltonian of the Quantum Theory of a spin-0 particle undergoing an interaction homogeneous in time into the form*

$$H = \frac{1}{2\mu}\sum_{\alpha=1}^{3}\left(-i\frac{\partial}{\partial x_\alpha}\right)^2 + \Phi(\mathbf{x}).$$

To prove ($\mathcal{R}$), first we imply from ($\mathcal{P}$) that $g \to U_g$ is a projective representation of the Galileian group $\mathcal{G}$. Then, according to section 2.2, the sub-groups $\mathcal{T}_\alpha$, $\mathcal{R}_\alpha$, $\mathcal{B}_\alpha$ can be represented by the one parameter unitary sub-groups $\{e^{-iP_\alpha a}\}_{a\in\mathbf{R}}$, $\{e^{-iJ_\alpha\theta}\}_{\theta\in\mathbf{R}}$, $\{e^{iG_\alpha u}\}_{u\in\mathbf{R}}$, in such a way that the hermitean generators $P_\alpha, J_\alpha, G_\alpha$ satisfy (2). Once defined the self-adjoint operators $F_\alpha = \frac{G_\alpha}{\mu}$, it can be proved that relations (2) imply that

$$U_g \mathbf{F} U_g^{-1} = g^{-1}(\mathbf{F}). \tag{9}$$

Since the F_α's commute with each other, according to spectral theory, there is a unique PV measure $P : \mathcal{B}(\mathbf{R}^3) \to \Pi(\mathcal{H})$ such that $F_\alpha = \int \lambda dE_\lambda^{(\alpha)}$, where $E_\lambda^{(1)} = P((-\infty,\lambda]\times\mathbf{R}^2)$, $E_\lambda^{(2)} = P(\mathbf{R}\times(-\infty,\lambda]\times\mathbf{R})$, $E_\lambda^{(3)} = P(\mathbf{R}^2\times(-\infty,\lambda])$. Then (9) easily implies that $\Delta \to P(\Delta)$ is an imprimitivity system for the restriction to $\mathcal{E}$ of $g \to U_g$; therefore Mackey's theorem applies. In so doing, the simplest choice for $\mathcal{H}_0$, i.e. $\mathcal{H}_0 = \mathbf{C}$, leads to identify $\mathcal{H}$, F_α, P_α, $\hat{U}_g$ as

$$\mathcal{H} = L_2(\mathbf{R}^3), \quad (F_\alpha\psi)(\mathbf{x}) = x_\alpha\psi(\mathbf{x}), \quad P_\alpha = -i\frac{\partial}{\partial x_\alpha}, \quad (U_g\psi)(\mathbf{x}) = \psi\left(g^{-1}(\mathbf{x})\right). \qquad (10).$$

Both $\mathbf{F}$ and $\mathbf{P}$ in (10) are *complete* systems of operators in $L_2(\mathbf{R}^3)$, and $(\mathbf{F},\mathbf{P})$ is an *irreducible* system of operators [5]. Then we can easily prove the following proposition.

Proposition 4.1. *If ($\mathcal{P}$) holds, then in the simplest Quantum Theory of a localizable interacting particle the equality $\mathbf{F} = \mathbf{Q}$ holds.*

Proof. If $g \in \mathcal{T}_1$ and $(\mathcal{P})$ holds, then $U_g = e^{-iP_\beta a} = \mathbb{I} - iP_\beta a + o_1(a)$, where $o_1(a)$ is an infinitesimal operator of order greater than 1 with respect to a; relation (5.ii) implies $[Q_\alpha, P_\beta] = i\delta_{\alpha\beta}\mathbb{I}$. So, by using (2.vi) we obtain $[F_\alpha - Q_\alpha, P_\beta] = \mathbf{0}$; but (5.ii) for $U_g = e^{iG_\beta u} = \mathbb{I} + iG_\beta u + o_2(u)$ implies also $[F_\alpha - Q_\alpha, F_\beta] = \mathbf{0}$, i.e. $F_\alpha - Q_\alpha = c_\alpha\mathbb{I}$ ≡constant for the irreducibility of $(\mathbf{F}, \mathbf{P})$; on the other hand, $(\mathcal{P}.\text{i})$ together with (2.iv) and (5.ii) for $U_g = e^{-iJ_\alpha\theta}$ imply $[J_\alpha, F_\beta - Q_\beta] = i\hat{\epsilon}_{\alpha,\beta,\gamma}(F_\gamma - Q_\gamma) = [J_\alpha, c_\beta\mathbb{I}] = \mathbf{0}$; thus, $F_\alpha - Q_\alpha = \mathbf{0}$. •

Prop. 4.1 together with (5.i) is sufficient to determine the form of the hamiltonian operator H consistent with $(\mathcal{P})$. First, we determine $[G_\alpha, \dot{Q}_\beta]$. Let us start with

$$e^{iG_\alpha u}\dot{Q}_\beta e^{-iG_\alpha u} = \dot{Q}_\beta + i[G_\alpha, \dot{Q}_\beta]u + o(u). \tag{11}$$

By making use of $\dot{Q}_\beta = i[H, Q_\beta] = \lim_{t\to 0}\frac{Q_\beta^{(t)} - Q_\beta}{t}$ and of $e^{iG_\alpha u}Q_\beta^{(t)}e^{-iG_\alpha u} = Q_\beta^{(t)} - \delta_{\alpha\beta}ut\mathbb{I}$, implied by (5.i), we also find

$$e^{iG_\alpha u}\dot{Q}_\beta e^{-iG_\alpha u} = \lim_{t\to 0} e^{iG_\alpha u}\frac{Q_\beta^{(t)} - Q_\beta}{t}e^{-iG_\alpha u} = \lim_{t\to 0}\frac{Q_\beta^{(t)} - \delta_{\alpha\beta}ut\mathbb{I} - Q_\beta}{t} = \dot{Q}_\beta - \delta_{\alpha\beta}u\mathbb{I}. \tag{12}$$

The comparison between (11) and (12) yields

$$[G_\alpha, \dot{Q}_\beta] = [Q_\alpha, \mu\dot{Q}_\beta] = i\delta_{\alpha\beta}\mathbb{I}, \text{ which implies } [F_\alpha, \mu\dot{Q}_\beta - P_\beta] = \mathbf{0}. \qquad (13).$$

By using $U_g = e^{-iP_\alpha a}$ and $U_g = e^{-iJ_\alpha\theta}$ instead of $e^{iG_\alpha u}$ in this last argument[2], we obtain $[P_\alpha, \mu\dot{Q}_\beta - P_\beta] = \mathbf{0}$ and $[J_\alpha, \mu\dot{Q}_\beta] = i\hat{\epsilon}_{\alpha,\beta,\gamma}\mu\dot{Q}_\gamma$. These two relations, together with (13), imply $\dot{Q}_\beta = (1/\mu)P_\beta$. To determine H, by making use of (2.vi) we obtain

$$i[H, Q_\beta] = \dot{Q}_\beta = \frac{1}{\mu}P_\beta = i\left[(1/2\mu)(P_1^2 + P_2^2 + P_3^2), \frac{G_\alpha}{\mu}\right] \equiv i[(1/2\mu)(P_1^2 + P_2^2 + P_3^2), Q_\alpha]. \tag{14}$$

Then the completeness of $\mathbf{Q}$ implies that $H - (1/2\mu)(P_1^2 + P_2^2 + P_3^2)$ is a function of $\mathbf{Q}$. Thus

$$H = -\frac{1}{2\mu}\left(\frac{\partial^2}{\partial x_1^2} + \frac{\partial^2}{\partial x_2^2} + \frac{\partial^2}{\partial x_3^2}\right) + \Phi(\mathbf{Q}). \tag{15}$$

Thus, assumption $(\mathcal{P})$ forbids the description of electro-magnetic interactions.

5. Quantum Theory of an interacting particle

According to the conclusion of the last section, to develop a Quantum Theory of a particle, able to describe also electromagnetic interactions, assumption $(\mathcal{P})$ must be abandoned besides the symmetry's implications (6) and (7). In this section we undertake such a development.

First, in sect. 5.1, we replace the active transformation $A \to R_g^{(2)}[A]$ of observables by the *passive* transformation $A \to S_g[A]$. This allows us to prove, without making use of (7), that a unitary or an anti-unitary operator U_g exists such that $S_g[A] = U_g A U_g^*$, for every $g \in \mathcal{G}$.

Another problematic task is to exclude anti-unitary operators U_g without the condition that $g \to U_g$ is a projective representation, condition necessary in the known arguments that obtain such an exclusion [3],[5], [10]. Such a further problem is addressed and solved in sect. 5.2.

However, our correspondence $U : \mathcal{G} \to \mathcal{U}(\mathcal{H})$, $S_g[A] = U_g A U_g^{-1}$ is yet not a projective representation, so that Mackey's theorem cannot apply. In section 5.3 we argue how $g \to U_g$

[2] It is just by extending the argument to translations ($U_g = e^{-iP_\alpha a}$) and rotations ($U_g = e^{-iJ_\alpha\theta}$) that we can prove $(\mathcal{R})$. If the argument is limited to boosts ($U_g = e^{iG_\alpha a}$) [5],[7],[9] only the loser form (8) of H is obtained.

can be changed into a projective representation by means of a *σ-conversion*. We show that the operators representing position are identifiable with the multiplication operators in the case that the interaction admits a σ-conversion that leaves unaltered the covariance properties of the position with respect to $\mathcal{G}$. Therefore our approach explains such an identification by the existence of a determined covariance property related to the interaction.

In section 5.4 we determine the dynamical law for this class of interactions, whose hamiltonian operator turns out to be more general than the known hamiltonian of a particle in an electromagnetic field. However, in section 5.5 we completely characterize the electromagnetic interaction, among all possible ones, in terms of covariance properties preserved by the σ-conversion.

5.1. Passive interpretation of quantum transformations

To recall the meaning of the passive interpretation we make use of the Quantum Theory of a free particle, where $\mathcal{H} = L_2(\mathbf{R}^3)$ and the position operators Q_α are the multiplication operators defined by $(Q_\alpha\psi)(\mathbf{x}) = x_\alpha\psi(\mathbf{x})$. In fact, these operators Q_α represent the position observables of the particle just as referred to a pre-specified *spatial reference frame* Σ_0; if Σ_g is the frame related to such a pre-fixed Σ_0 just by g, then the position operators with respect to Σ_g must be $g^{-1}(\mathbf{Q})$, in general different from $\mathbf{Q}$. Such a dependence on the reference frame is not an exclusive feature of the position observables; for each $g \in \mathcal{G}$, we can introduce a mapping

$$S_g : \Omega(\mathcal{H}) \to \Omega(\mathcal{H}), \quad A \to S_g[A] \tag{16}$$

with the following interpretation.

(PI) *If the operator A represents a given observable with respect to the pre-fixed reference frame Σ_0, the operator $S_g[A]$ represents that observable with respect to Σ_g.*

Statement (PI) expresses the *passive* interpretation of the transformation g. For instance, since remark 3.1 extends to the mapping S_g, then the operators $\mathbf{Q}^{(t)}$ transform according to

$$S_g[\mathbf{Q}^{(t)}] = g_t^{-1}(\mathbf{Q}^{(t)}), \quad \text{and, as a particular case, } S_g[\mathbf{Q}] = g^{-1}(\mathbf{Q}). \tag{17}$$

The following statements are conceptually implied by the meaning of the passive quantum transformations.

(S.1) For every $g \in \mathcal{G}$, the mapping S_g is bijective.

(S.2) Let f be any fixed real Borel function such that if A is a self-adjoint operator, then $B = f(A)$ is a self-adjoint operator too. Since in Quantum Theory a measurement of the quantum observable $f(A)$ can be performed by measuring A and then transforming the obtained outcome a by the purely mathematical transformation f into the outcome $b = f(a)$ of $f(A)$, the following equality has to hold:

$$f(S_g[A]) = S_g[f(A)]. \tag{18}$$

Conditions (S.1) and (S.2) are sufficient to show further properties of the mappings S_g, according to the following propositions.

Proposition 5.1. *Let $S : \Omega(\mathcal{H}) \to \Omega(\mathcal{H})$ be a bijective mapping such that $S[f(A)] = f(S[A])$, for every function as in (S2). Then the following statements hold.*

i) *If $E \in \Pi(\mathcal{H})$ then $S[E] \in \Pi[\mathcal{H}]$, i.e., the mapping S is an extension of a bijection of $\Pi(\mathcal{H}]$.*

ii) *If $A, B \in \Omega(\mathcal{H})$ and $A + B \in \Omega(\mathcal{H})$, then $[A, B] = \mathbf{0}$ implies $S[A + B] = S[A] + S[B]$. This partial additivity immediately implies $S[A] = \mathbf{0}$ if and only if $A = \mathbf{0}$.*

iii) *For all* $E, F \in \Pi(\mathcal{H})$, $EF = \mathbf{0}$ *implies* $S[E+F] = S[E] + S[F] \in \Pi(\mathcal{H})$; *as a consequence,* $E \leq F$ *if and only if* $S[E] \leq S[F]$.

iv) $S[P] \in \Pi_1(\mathcal{H})$ *if and only if* $P \in \Pi_1(\mathcal{H})$.

Proof. (i) If $E \in \Pi(\mathcal{H})$ and $f(\lambda) = \lambda^2$ then $f(E) = E$ holds; so $S[f(E)] = f(S[E])$ implies $(S_g[E])^2 \equiv f(S[E]) = S[E^2] \equiv S[E]$, i.e. $S^2[E] = S[E]$.

(ii) If $[A, B] = \mathbf{0}$ then $C \in \Omega(\mathcal{H})$ and two functions f_a, f_b exist so that $A = f_a(C)$ and $B = f_b(C)$; once defined the function $f = f_a + f_b$, we have $S[A+B] \equiv S[f(C)] = f(S[C]) = f_a(S[C]) + f_b(S[C]) = S[f_a(C)] + S[f_b(C)] \equiv S[A] + S[C]$.

(iii) If $EF = \mathbf{0}$, then $[E, F] = \mathbf{0}$ and $(E+F) \in \Pi(\mathcal{H})$ hold. Statements (i) and (ii) imply $S[E+F] = S[E] + S[F] \in \Pi(\mathcal{H}]$.

(iv) If $P \in \Pi_1(\mathcal{H})$ then $S[P] \in \Pi(\mathcal{H})$ by (i). If $Q \in \Pi_1(\mathcal{H})$ and $Q \leq S([P]$ then $P_0 \equiv S^{-1}[Q] \leq P$ by (iii); but P is rank 1, therefore $P_0 = P$ and $Q = S[P]$. •

Corollary 5.1. *From prop. 5.1 immediately follows that the restriction of* S *to* $\Pi(\mathcal{H})$ *is a bijection that also satisfies* $S[\mathbf{0}] = \mathbf{0}$, $S[\mathbb{1}] = \mathbb{1}$, $E \leq F$ *iff* $S[E] \leq S[F]$, $S[E^\perp] = (S[E])^\perp$.

In virtue of corollary 5.1, by Wigner's Theorem 2 the following proposition is easily proved.

Proposition 5.2. *If a mapping* S *satisfies the hypothesis of Prop. 5.1, then a unitary or an anti-unitary operator exists such that* $S[A] = UAU^*$ *for every* $A \in \Omega(\mathcal{H})$; *if another unitary or anti-unitary operator* V *satisfies* $S[A] = VAV^*$ *for every* $A \in \Omega$, *then* $V = e^{i\theta}U$ *with* $\theta \in \mathbf{R}$.

5.2. Further implications

Propositions 5.1 and 5.2 imply that for every transformation $g \in \mathcal{G}$, the corresponding *passive* quantum transformation is given by $A \longrightarrow S_g[A] = U_g A U_g^*$. If the correspondence $g \longrightarrow S_g$ satisfied $S_{g_1 g_2} = S_{g_1} \circ S_{g_2}$, so that $g \longrightarrow U_g$ would be a projective representation, then it could be easily proved, according to [3],[5],[10],[12], that every U_g must be unitary. But in presence of interaction $g \longrightarrow U_g$ in general *is not a projective representation.* Now we prove that anti-unitary U_g can be excluded under the only hypothesis that the correspondence $g \longrightarrow S_g$ is continuous according to the following continuity notion given by Bargmann [12].

Definition 5.1. *Given* $P_1, P_2 \in \Pi_1(\mathcal{H})$, *the distance* $d(P_1, P_2)$ *is defined as the minimal distance* $\|\psi_1 - \psi_2\|$ *between vectors* ψ_1, ψ_2 *such that* $P_1 = |\psi_1\rangle\langle\psi_1|$ *and* $P_2 = |\psi_2\rangle\langle\psi_2|$, *i.e.,* $d(P_1, P_2) = [2(1 - |\langle\psi_1 \mid \psi_2\rangle|]^{1/2}$.

A correspondence $g \longrightarrow T_g$ *from* $\mathcal{G}$ *into the set of all automorphims of* $\Pi(\mathcal{H})$ *is continuous if for any fixed* $P \in \Pi_1(\mathcal{H})$ *the mapping from* $\mathcal{G}$ *into* $\Pi_1(\mathcal{H})$, $g \longrightarrow T_g[P]$ *is continuous in* g *with respect the distance* d *above defined on* $\Pi_1(\mathcal{H})$.

Bargmann proved that if a correspondence $g \longrightarrow T_g$ from $\mathcal{G}$ into the set of all automorphisms of $\Pi(\mathcal{H})$ is continuous then the arbitrary phase factor of the operator U_g for which $T_g[A] = U_g A U_g^*$ can be chosen so that $U_g\psi$ is continuous with respect to g in the topology of $\mathcal{H}$ [12].

Proposition 5.3. *If the correspondence* $g \longrightarrow S_g$ *in (16) is continuous according to def. 5.1, then every operator* U_g *such that* $S_g[A] = U_g A U_g^*$ *for all* $A \in \Omega(\mathcal{H})$ *is unitary.*

Proof. Since $S_e[A] = A = \mathbb{1}A\mathbb{1}^*$, we can choose $U_e = \mathbb{1}$ which is unitary. Hence, because of the continuity of $g \longrightarrow U_g$, a maximal neighborhood K_e of e must exist so that U_g is unitary for all $g \in K_e$; otherwise a sequence $g_n \longrightarrow e$ would exist with U_{g_n} anti-unitary and so $\langle\psi \mid \varphi\rangle = \langle U_{g_n}\varphi \mid U_{g_n}\psi\rangle$, and then $\langle\psi \mid \varphi\rangle = \lim_{n\to\infty}\langle U_{g_n}\varphi \mid U_{g_n}\psi\rangle = \langle U_e\varphi \mid U_e\psi\rangle = \langle\varphi \mid \psi\rangle$ which cannot hold for all ψ, φ unless $\mathcal{H}$ is real. Now, such a neighborhood K_e has no boundary, and since $\mathcal{G}$ is a connected group, $K_e = \mathcal{G}$; indeed, if $g_0 \in \partial K_e$, two sequences $g_n \longrightarrow g_0$ and $h_n \longrightarrow g_0$ would exist with U_{g_n} unitary and U_{h_n} anti-unitary; therefore, the continuity of U_g would imply that U_{g_0} should simultaneously be unitary and anti-unitary. •

5.3. σ-conversions

By switching to the passive interpretation, under a continuity condition we were able to establish the existence of a continuous correspondence $U : \mathcal{G} \to \mathcal{U}(\mathcal{H})$, $g \to U_g$, such that $S_g[A] = U_g A U_g^{-1}$. The argument of section 4, which shows that assumption $(\mathcal{P})$ is empirically inadequate, can be easily adapted to show that such a correspondence is not a σ-representation, in general, even if the passive interpretation is adopted. Without such a representation property we cannot apply Mackey's theorem to proceed with our approach. Now we address this obstacle.

The correspondence $g \to U_g$, can be *converted* into a σ-representation if we multiply each operator U_g by a suitable unitary operator V_g of $\mathcal{H}$; namely, V_g is a unitary operator such that the correspondence $g \to \hat{U}_g = V_g U_g$ turns out to be a σ-representation. The transition from $g \to U_g$ to $g \to \hat{U}_g = V_g U_g$ will be called *σ-conversion*; the mapping $V : \mathcal{G} \to \mathcal{U}(\mathcal{H})$, $g \to V_g$ that realizes the σ-conversion will be called *σ-conversion mapping*. If $g \to V_g$ is a σ-conversion mapping for $g \to U_g$ and $\theta : \mathcal{G} \to \mathbf{R}$ is a real function, then also $g \to e^{i\theta(g)} V_g$ is a σ-conversion mapping, provided that $e^{i\theta(e)} = 1$. In any case, $V_e = \mathbb{1}$ must hold.

If $g \to V_g$ is a σ-conversion mapping for U_g then, according to sect. 2.2, the σ-representation $g \to \hat{U}_g = V_g U_g$ has nine hermitean generators $\hat{P}_\alpha$, $\hat{J}_\alpha$, $\hat{G}_\alpha$ for which (2) hold. Moreover, the common spectral measure of the triple $\mathbf{F} = \hat{\mathbf{G}}/\mu$ is an imprimitivity system for the restriction of $g \to \hat{U}_g$ to $\mathcal{E}$. Then Mackey's theorem allows us to identify $\mathcal{H}$ with $L_2(\mathbf{R}^3, \mathcal{H}_0)$ and explicitly indicates the concrete form of $\mathbf{F}$ and of $\hat{P}_\alpha$, $\hat{J}_\alpha$, $\hat{G}_\alpha$. In the following, we restrict ourselves to the case $\mathcal{H}_0 = \mathbf{C}$; is so doing we can take $\mathcal{H} = L_2(\mathbf{R}^3)$, $(F_\alpha \psi)(\mathbf{x}) = x_\alpha \psi(\mathbf{x})$, $\hat{P}_\alpha = -i\frac{\partial}{\partial x_\alpha}$, $\hat{J}_\alpha = F_\beta \hat{P}_\gamma - F_\gamma \hat{P}_\beta$ where (α, β, γ) is a cyclic permutation of $(1, 2, 3)$, and $\hat{G}_\alpha = \mu F_\alpha$.

But $\hat{U}_g$ *is not* the unitary operator which realizes the *passive* transformation: $S_g[A] \neq \hat{U}_g A \hat{U}_g^{-1}$. Moreover, $\mathbf{F}$ *is not* the triple $\mathbf{Q}$ representing the position. So, our explicit realization of the mathematical formalism of the theory is, in general, devoid of physical significance. But the approach can go on if we restrict our investigation to those interactions which admit σ-conversions $U_g \to \hat{U}_g = V_g U_g$ which are *Q-covariant*, i.e. σ-conversions that *leave unaltered the covariance properties of the position operators* $\mathbf{Q}$, i.e. such that

$$\hat{U}_g \mathbf{Q} \hat{U}_g^{-1} = g^{-1}(\mathbf{Q}) \ \forall g \in \mathcal{G}. \tag{19}$$

The proof of Prop. 4.1 can be easily adapted to show that if the interaction admits a Q-covariant σ-conversion then $\mathbf{F} = \mathbf{Q}$ in the present case too. Hence, the operator which represents the position in the mathematical formalism of the theory is the multiplication operator: $Q_\alpha \psi(\mathbf{x}) = x_\alpha \psi(\mathbf{x})$. So, in the present approach the identification of the multiplication operator with the position operator is implied by the possibility that the interaction admits a σ-conversion which preserves the covariance properties of the position operators. The existence of interaction which are not Q-covariant is not excluded, of course.

However, the operators $\hat{U}_g$ continue *to be not* the representative of the transformations of $\mathcal{G}$, i.e., $S_g[A] = \hat{U}_g A \hat{U}_g^{-1}$ does not hold. The following proposition specify how $\hat{U}_g$ relates to the unitary operator U_g that realizes the transformation g according to the passive interpretation.

Proposition 5.4. *The σ-conversion mapping of a Q-covariant σ-conversion has the form* $g \to V_g = e^{i\theta(g,\mathbf{Q})}$, *for some function* θ.

Proof. The relation $V_g U_g \mathbf{Q} U_g^{-1} V_g^{-1} = g^{-1}(\mathbf{Q})$ implied by (19) and (17) imply $V_g(g^{-1}(\mathbf{Q}))V_g^{-1} = g^{-1}(\mathbf{Q})$, i.e. $[V_g, g^{-1}(\mathbf{Q})] = \mathbf{0}$. Then $[V_g, f(g^{-1}(\mathbf{Q}))] = \mathbf{0}$ for every sufficiently regular function f; by taking $f = g$ we have $[V_g, \mathbf{Q}] = \mathbf{0}$. The thesis follows from the completeness of $\mathbf{Q}$ and the unitary character of V_g. •

5.4. Dynamical equation for Q-covariant interactions

Let us consider a velocity boost $g \in \mathcal{G}$ such that $\hat{U}_g = e^{i\hat{G}_\alpha u}$, for an interaction with Q-covariant σ-conversion. Since $\hat{G}_\alpha = \mu F_\alpha = \mu Q_\alpha$, we can write $\hat{U}_g = e^{i\mu Q_\alpha u}$; therefore

$$\hat{U}_g \dot{Q}_\beta \hat{U}_g^{-1} = \dot{Q}_\beta + i\mu[Q_\alpha, \dot{Q}_\beta]u + o_1(u). \tag{20}$$

On the other hand

$$\hat{U}_g \dot{Q}_\beta \hat{U}_g^{-1} = \lim_{t\to 0} V_g U_g \frac{(Q_\beta^{(t)} - Q_\beta)}{t} U_g^{-1} V_g^{-1}. \tag{21}$$

By making use of $U_g Q_\beta^{(t)} U_g^{-1} = Q_\beta^{(t)} - \delta_{\alpha\beta} ut \mathbb{I}$, implied by (17), in (21) and then comparing with (20) we obtain

$$\hat{U}_g \dot{Q}_\beta \hat{U}_g^{-1} = V_g \dot{Q}_\beta V_g^{-1} - \delta_{\alpha\beta} u \mathbb{I} = \dot{Q}_\beta + i\mu[Q_\alpha, \dot{Q}_\beta]u + o_1(u). \tag{22}$$

But Prop. 5.4 implies that $V_g = e^{i\varsigma_\alpha(u,\mathbf{Q})}$, for some fuction ς_α; replacing in (22) we obtain

$$\dot{Q}_\beta + i[\varsigma_\alpha(u, \mathbf{Q}), \dot{Q}_\beta] + o_2(u) - \delta_{\alpha\beta} u \mathbb{I} = \dot{Q}_\beta + i\mu[Q_\alpha, \dot{Q}_\beta]u + o_1(u). \tag{23}$$

Since $e^{i\varsigma_\alpha(0,\mathbf{Q})} = \mathbb{I}$, the expansion of ς_α with respect to u yields $\varsigma_\alpha(u,\mathbf{Q}) = \frac{\partial \varsigma_\alpha}{\partial u}(0,\mathbf{Q})u + o_3(u)$; by replacing this last relation in (23) we obtain $\mu[Q_\alpha, \dot{Q}_\beta] = [\eta_\alpha(\mathbf{Q}), \dot{Q}_\beta] + i\delta_{\alpha\beta}\mathbb{I}$, where $\eta_\alpha(\mathbf{Q}) = \frac{\partial \varsigma_\alpha}{\partial u}(0,\mathbf{Q})$. By replacing $\dot{Q}_\beta = i[H, Q_\beta]$ in this equation we can apply Jacobi's identity, and in so doing we obtain $[Q_\beta, \mu\dot{Q}_\alpha] = [Q_\beta, \dot{\eta}_\alpha(\mathbf{Q})] + i\delta_{\alpha\beta}\mathbb{I}$, i.e.

$$[Q_\beta, \dot{\eta}_\alpha(\mathbf{Q}) - \mu\dot{Q}_\alpha] = -i\delta_{\alpha\beta}\mathbb{I} = [Q_\beta, -\hat{P}_\alpha]. \tag{24}$$

By the completeness of $\mathbf{Q}$, from (24) we imply that an operator $f_\alpha(\mathbf{Q})$, function of $\mathbf{Q}$, must exist such that the equation $\dot{\eta}(\mathbf{Q}) - \mu\dot{Q}_\alpha + \hat{P}_\alpha = f_\alpha(\mathbf{Q})$ holds; then we can rewrite (24) as

$$i[H, \mu Q_\alpha - \eta_\alpha(\mathbf{Q})] = \hat{P}_\alpha - f_\alpha(\mathbf{Q}). \tag{25}$$

This is a general dynamical equation for a localizable particle whose interaction admits Q-covariant σ-conversions.

5.5. Characterization of the electromagnetic interaction

Once derived the general dynamical law (25) for a localizable particle with Q-covariant and homogeneous in time interaction, it is worth to re-discover the equation currently assumed in quantum physics as a particular case of the general equation (25). The nowadays adopted Schroedinger equation for a spin-0 particle has the form

$$i\frac{d}{dt}\psi_t = \left\{ \frac{1}{2m} \sum_{\alpha=1}^{3} [\hat{P}_\alpha - a_\alpha(\mathbf{Q})]^2 + \Phi(\mathbf{Q}) \right\} \psi_t,$$

i.e. the Hamiltonian operator is $H = (1/2\mu)\sum_{\alpha=1}^{3}\{\hat{P}_\alpha - a_\alpha(\mathbf{x})\}^2 + \Phi(\mathbf{Q})$. The consequent Quantum Theory corresponds to the case that the function η_α in the general law (25) is a constant function. Let us show this result.

Proposition 5.5. *The hamiltonian operator H of an interacting particle which admits Q-covariant σ-conversion has the form $H = (1/2\mu)\sum_{\alpha=1}^{3}\{\hat{P}_\alpha - a_\alpha(\mathbf{x})\}^2 + \Phi(\mathbf{Q})$ if and only if the functions η_α in (25) are constant functions. In this case $a_\alpha = f_\alpha$.*

Proof. If η_α is a constant function, then (25) transforms into $i[H,\mu Q_\alpha] = \hat{P}_\alpha - f_\alpha(\mathbf{Q})$ which holds if $H_0 = \frac{1}{2\mu}\sum_{\alpha=1}^{3}[\hat{P}_\alpha - f_\alpha(\mathbf{Q})]^2$ replaces H. Hence the operator $H - H_0$ must be a function Φ of $\mathbf{Q}$ because of the completeness of $\mathbf{Q}$. Then $\eta_\alpha(\mathbf{Q}) = c_\alpha \mathbb{I}$ implies $H = \frac{1}{2\mu}\sum_{\alpha=1}^{3}[\hat{P}_\alpha - f_\alpha(\mathbf{Q})]^2 + \Phi(\mathbf{Q})$.

Now we prove the converse. Let us suppose that $H = \frac{1}{2\mu}\sum_{\alpha=1}^{3}\{\hat{P}_\alpha - a_\alpha(\mathbf{Q})\}^2 + \Phi(\mathbf{Q})$; by replacing this H in (25) to obtain

$$
\begin{aligned}
&i[H,\mu Q_\alpha - \eta_\alpha(\mathbf{Q})] = \hat{P}_\alpha - f_\alpha(\mathbf{Q}) = \\
&\quad= \frac{i}{2\mu}\sum_\beta[P_\beta^2,\mu Q_\alpha] - \frac{i}{2\mu}\sum_\beta[a_\beta\hat{P}_\beta,\mu Q_\alpha] - \frac{i}{2\mu}\sum_\beta[\hat{P}_\beta a_\beta,\mu Q_\alpha] + \frac{i}{2\mu}\sum_\beta[a_\beta^2,\mu Q_\alpha] + \\
&\quad -\frac{i}{2\mu}\sum_\beta[P_\beta^2,\eta_\alpha] + \frac{i}{2\mu}\sum_\beta[a_\beta\hat{P}_\beta,\eta_\alpha] + \frac{i}{2\mu}\sum_\beta[\hat{P}_\beta a_\beta,\eta_\alpha] - \frac{i}{2\mu}\sum_\beta[a_\beta^2,\eta_\alpha] + \\
&\quad +i[\Phi(\mathbf{Q}),\mu Q_\alpha - \eta_\alpha].
\end{aligned}
$$

The last term and the fourth terms of the second and third lines are zero. Then we have

$$
\begin{aligned}
&i[H,\mu Q_\alpha - \eta_\alpha(\mathbf{Q})] = \hat{P}_\alpha - f_\alpha(\mathbf{Q}) \qquad (26)\\
&= \hat{P}_\alpha - \frac{i}{2}\sum_\beta(a_\beta\hat{P}_\beta Q_\alpha - Q_\alpha a_\beta\hat{P}_\beta + \hat{P}_\beta a_\beta Q_\alpha - Q_\alpha\hat{P}_\beta a_\beta) + \\
&\quad -\frac{i}{2\mu}\sum_\beta[P_\beta^2,\eta_\alpha] + \frac{i}{2\mu}\sum_\beta(a_\beta\hat{P}_\beta\eta_\alpha - \eta_\alpha a_\beta\hat{P}_\beta + \hat{P}_\beta a_\beta\eta_\alpha - \eta_\alpha\hat{P}_\beta a_\beta) \\
&= \hat{P}_\alpha - \frac{i}{2}\sum_\beta(a_\beta[\hat{P}_\beta,Q_\alpha] + [\hat{P}_\beta,Q_\alpha]a_\beta) - \frac{i}{2\mu}\sum_\beta[P_\beta^2,\eta_\alpha] + \frac{i}{2\mu}\sum_\beta(a_\beta[\hat{P}_\beta,\eta_\alpha] + [\hat{P}_\beta,\eta_\alpha]a_\beta) \\
&= \hat{P}_\alpha - \frac{i}{2}(-2ia_\alpha) - \frac{i}{2\mu}\sum_\beta[P_\beta^2,\eta_\alpha] + \frac{i}{2\mu}\sum_\beta\left(-2i\,a_\beta\frac{\partial\eta_\alpha}{\partial q_\beta}\right) \\
&= \hat{P}_\alpha - a_\alpha + \frac{1}{\mu}\sum_\beta a_\beta\frac{\partial\eta_\alpha}{\partial q_\beta} - \frac{i}{2\mu}\sum[P_\beta^2,\eta_\alpha].
\end{aligned}
$$

From the second and last members of this equations' chain we obtain $f_\alpha(\mathbf{Q}) = a_\alpha - \frac{1}{\mu}\sum_\beta a_\beta\frac{\partial\eta_\alpha}{\partial q_\beta} + \frac{i}{2\mu}\sum[P_\beta^2,\eta_\alpha]$, which implies that $\sum_\beta[P_\beta^2,\eta_\alpha]$ is a function of $\mathbf{Q}$. Therefore we have

$$
\begin{aligned}
\sum_\beta[P_\beta^2,\eta_\alpha] = \phi_\alpha(\mathbf{Q}) &= \sum_\beta(\hat{P}_\beta[\hat{P}_\beta,\eta_\alpha] + [\hat{P}_\beta,\eta_\alpha]\hat{P}_\beta) = (-i)\sum_\beta\left(\hat{P}_\beta\frac{\partial\eta_\alpha}{\partial q_\beta} + \frac{\partial\eta_\alpha}{\partial q_\beta}\hat{P}_\beta\right) \\
&= (-i)\sum_\beta\left(\left[\hat{P}_\beta,\frac{\partial\eta_\alpha}{\partial q_\beta}\right] + 2\frac{\partial\eta_\alpha}{\partial q_\beta}\hat{P}_\beta\right) = (-i)\sum_\beta\left((-i)\frac{\partial^2\eta_\alpha}{\partial q_\beta^2} + 2\frac{\partial\eta_\alpha}{\partial q_\beta}\hat{P}_\beta\right).
\end{aligned}
$$

Then $\sum_\beta\frac{\partial\eta_\alpha}{\partial q_\beta}\hat{P}_\beta$ is a function of $\mathbf{Q}$, and this implies $\sum_\beta\left[Q_\gamma,\frac{\partial\eta_\alpha}{\partial q_\beta}\hat{P}_\beta\right] = \mathbf{0} = \frac{\partial\eta_\alpha}{\partial q_\gamma}[Q_\gamma,\hat{P}_\gamma] = i\frac{\partial\eta_\alpha}{\partial q_\gamma}$ for every γ; therefore $\frac{\partial\eta_\alpha}{\partial q_\gamma} = \mathbf{0}$; thus η_α is a constant function. By using this result in the equality between the second and the last members of (26) we obtain $a_\alpha = f_\alpha$. •

Now we show that the case η_α =constant is *completely* characterized in terms of covariance properties tied to the interaction.

Journal of Physics: Conference Series **670** (2016) 012039 doi:10.1088/1742-6596/670/1/012039

Proposition 5.6. *The Q-covariant σ-conversions for which* $\eta_\alpha(\mathbf{Q})$ *= constant are those and only those which leave unaltered the covariant properties of* $\mathbf{Q}^{(t)}$ *with respect to the Galileian boosts* $g \in \mathcal{G}$*, at the first order in the boost's velocity, i.e. the σ-conversions such that*

$$\hat{U}_g \mathbf{Q}^{(t)} \hat{U}_g^{-1} = g_t^{-1}(\mathbf{Q}^{(t)}) + o(u), \quad \textit{where } \hat{U}_g = e^{i\hat{G}_\alpha u} \tag{27}$$

at the first order in u.

Proof. Let $\hat{U}_g = e^{i\hat{G}_\alpha u} = V_g U_g$ be the σ-converted unitary operator associated with the Galileian boost g, where $V_g = e^{i\varsigma_\alpha(u,\mathbf{Q})}$ according to Prop.2. By starting from (27) and by expanding $e^{\pm i\varsigma_\alpha(u,\mathbf{Q})}$ with respect to u we obtain

$$e^{i\hat{G}_\alpha u} Q_\beta^{(t)} e^{-i\hat{G}_\alpha u} = V_g U_g Q_\beta^{(t)} U_g^{-1} V_g^{-1} = Q_\beta^{(t)} + i[\eta_\alpha(\mathbf{Q}), Q_\beta^{(t)}]u - \delta_{\alpha\beta} u t \mathbb{I} + o_1(u). \tag{28}$$

The covariance properties of $\mathbf{Q}^{(t)}$ with respect to our Galileian boost g are expressed by $S_g[Q_\beta^{(t)}] = Q_\beta^{(t)} - \delta_{\alpha\beta} u t \mathbb{I}$. Hence the equation

$$e^{i\hat{G}_\alpha u} Q_\beta^{(t)} e^{-i\hat{G}_\alpha u} = Q_\beta^{(t)} - \delta_{\alpha\beta} u t \mathbb{I} + o_2(u) \tag{29}$$

is the condition in order that the σ-conversion leave unaltered the covariance properties of $\mathbf{Q}^{(t)}$ with respect to the Galileian boosts, at the first order in u; the comparison between (28) and (29) implies that such a condition holds if and only if

$$[\eta_\alpha(\mathbf{Q}), Q_\beta^{(t)}] = \mathbf{0}. \tag{30}$$

If η_α =constant then (30) holds, of course. Therefore, in order to prove the proposition, it is sufficient to prove the inverse implication. Hence we suppose that (30) holds. By expanding $e^{\pm i\hat{G}_\alpha u}$ with respect to u we find $e^{i\hat{G}_\alpha u} Q_\beta^{(t)} e^{-i\hat{G}_\alpha u} = Q_\beta^{(t)} + i[\hat{G}_\alpha, Q_\beta^{(t)}]u + o_3(u)$, so that (29) holds if and only if $i[\hat{G}_\alpha, Q_\beta^{(t)}] = -\delta_{\alpha\beta} t \mathbb{I}$, i.e. if and only if $[Q_\alpha, (\mu Q_\beta^{(t)})/t] = i\delta_{\alpha\beta} \equiv [Q_\alpha, \hat{P}_\beta]$; then the completeness of $\mathbf{Q}$ implies that a function φ_β must exist such that $(\mu Q_\beta^{(t)})/t - \hat{P}_\beta = \varphi_\beta(\mathbf{Q})$ and hence the following equation holds.

$$Q_\beta^{(t)} = \frac{t}{\mu}(\varphi_\beta(\mathbf{Q}) + \hat{P}_\beta). \tag{31}$$

Now, the condition $[\eta_\alpha(\mathbf{Q}), Q_\alpha] = \mathbf{0}$ is obviously satisfied for all α; but it and the necessary and sufficient condition (30), where (31) is used, imply also $[\eta_\alpha(\mathbf{Q}), \hat{P}_\beta] = \mathbf{0}$; therefore $\eta_\alpha(\mathbf{Q})$ is a constant operator $c_\alpha \mathbb{I}$. •

[1] Wigner E P 1959 *Group Theory and its Applications to the Quantum Theory of Atomic Spectra* (Boston: Academic Press)
[2] Cassinelli G, De Vito E, Lahti P and Levrero A 1997 *Rev. Math. Phys.* **8** 921
[3] Mackey G W 1968 *Induced Representations of Group and Quantum Mechanics* (New York: W.A Benjamin, Inc)
[4] Mackey G W 1958 *Acta Mathematica* **99** 265
[5] Mackey G W 1989 *Unitary Representations in Physics, Probability and Number Theory* (Reading: Addison Wesley Pub Co)
[6] Varadarjan V S 1968 *Geometry of Quantum Theory* (Berlin: Springer)
[7] Jauch J M 1968 *Foundations of Quantum Mechanics* (Reading: Addison Wesley Pub Co)
[8] Mackey G W 1993 *Int. J. Th. Phys* **32** 1643
[9] Ballentine L E 2006 *Quantum Mechanics - A Modern Development* (Singapore: World Scientific)
[10] Simon B 1976 in *Studies in Mathematical Physics: Essays in Honor of Valentine Bargmann* ed Lieb E H, Simon B and Wightman A S (Princeton: Princeton Un. Press) pp. 327-350
[11] Molnár L 2007 *Lecture Notes in Matemathics* **1895** (Berlin: Springer)
[12] Bargmann V 1954 *Ann. of Math.* **59** 1

Journal of Physics: Conference Series **670** (2016) 012040 doi:10.1088/1742-6596/670/1/012040

Re-entrant phase transitions in non-commutative quantum mechanics

Orlando Panella$^{(1)}$ and Pinaki Roy$^{(2)}$

$^{(1)}$ INFN, Sezione di Perugia, Via A. Pascoli I-06123, Perugia, Italy
$^{(2)}$ Physics and Applied Mathematics Unit, Indian Statistical Institute, Kolkata, India

E-mail: orlando.panella@pg.infn.it

Abstract. We discuss the (2+1)-dimensional Dirac oscillator in a magnetic field in non commutative quantum mechanics. The system is known to be characterised by a left right-chiral phase transition in ordinary Quantum mechanics. We show that the momentum non-commutativity shifts the know phase transition while the space non commutativity introduces a new right-left chiral quantum phase transition giving rise to the intriguing phenomenon of re-entrant phase transition observed in condensed matter as well as in black hole physics.

1. Introduction

On the basis that both string theory and quantum gravity indicate that space-time can be non-commutative [1–4] several quantum mechanical systems have been studied by a number of authors to determine the role of noncommutativity parameters on a variety of physical observables [5–14].

The Dirac oscillator on the other hand is one of the very few relativistic systems which is exactly solvable [15–22]. In mathematical physics the Dirac oscillator has become a paradigm for the realization of covariant quantum models and it has found applications both in nuclear [23–25] and subnuclear [26, 27] physics as well as in quantum optics [28–30]. Very recently a one dimensional version of the Dirac oscillator has been realised experimentally [31] for the first time with realistic prospects to realise in the near future the two dimensional version of the Dirac oscillator which may be feasible using networks of microwave coaxial cables [32–34]. It is interesting to note that a further interaction in the form of a homogeneous magnetic field can still be incorporated in the Dirac oscillator keeping the system still exactly solvable [28, 35–38] and this combined system has quite interesting properties. In some recent papers it has been shown that for this combined system there is a chirality phase transition if the magnitude of the magnetic field either exceeds or is less than a critical value $B_{\rm cr}$ (which also depends on the oscillator strength) [39, 40]. A consequence of this chirality phase transition is that the spectrum is different for $B > B_{\rm cr}$ and $B < B_{\rm cr}$, B being the magnetic field strength.

Our objective here is to analyse the same system, a 2D Dirac oscillator within a constant magnetic field, but in the framework of non-commutative space and momentum coordinates. It will be shown that in this case the critical value of the magnetic field depends not only on the oscillator strength but on the non-commutativity parameters as well. The interesting feature which we would like to emphasise, is that beside the two left- and right-chiral phases present also in the commutative case, in the non-commutative scenario there appear also a *new* third left phase and so a second quantum phase transition (right-left) which both disappear as the space non-commutativity parameter vanishes [41]. This is the well known phenomenon of re-entrant phase transitions which has been pointed out in condensed matter as well as in black hole physics [42, 43]. While the details of the full calculation can be found in [41] we illustrate here the main results. We shall discuss the spectrum and the degeneracy of the various energy levels in all three phases. Another point which we address is how to characterise the phase transition(s). It will be shown that such quantum phase transitions can be described in terms of the magnetisation of the system.

2. Formulation of the problem

To begin with we note that the Hamiltonian for the $(2+1)$ dimensional Dirac oscillator in the noncommutative plane in the presence of a homogeneous magnetic field is given by

$$H = c\boldsymbol{\sigma}.(\hat{\boldsymbol{p}} - im\omega\beta\hat{\boldsymbol{x}} + \frac{e}{c}\hat{\boldsymbol{A}}) + \beta mc^2\,, \tag{1}$$

where c is the velocity of light and $\boldsymbol{\sigma}$, $\beta = \sigma_z$ denote Pauli matrices. We now choose the vector potential as

$$\hat{\boldsymbol{A}} = (-B\hat{y}/2, B\hat{x}/2, 0)\,. \tag{2}$$

The commutation relation between coordinates and momenta are given by

$$[\hat{x},\hat{y}] = i\theta, \qquad [\hat{p}_x,\hat{p}_y] = i\eta, \qquad [\hat{x}_i,\hat{p}_j] = i\hbar(1+\frac{\theta\eta}{4\hbar^2})\delta_{ij}, \qquad \theta,\eta\in\mathbb{R}\,. \tag{3}$$

Written in full the above Hamiltonian reads

$$H = c\begin{pmatrix} mc & \hat{\Pi}_- \\ \hat{\Pi}_+ & -mc \end{pmatrix}, \tag{4}$$

where $\Pi_\pm$ are given by

$$\begin{aligned}\hat{\Pi}_- &= \hat{p}_x - i\hat{p}_y - i\left(\frac{eB}{2c} - m\omega\right)\hat{x} - \left(\frac{eB}{2c} - m\omega\right)\hat{y}\,;\\ \hat{\Pi}_+ &= \hat{p}_x + i\hat{p}_y + i\left(\frac{eB}{2c} - m\omega\right)\hat{x} - \left(\frac{eB}{2c} - m\omega\right)\hat{y}\,.\end{aligned} \tag{5}$$

It is now necessary to express the noncommuting coordinates and momenta in terms of commuting ones. This can be achieved using the Seiberg-Witten map and are given by

$$\begin{aligned}\hat{x} &= x - \frac{\theta}{2\hbar}p_y, & \hat{p}_x &= p_x + \frac{\eta}{2\hbar}y\,,\\ \hat{y} &= y + \frac{\theta}{2\hbar}p_x, & \hat{p}_y &= p_y - \frac{\eta}{2\hbar}x\,,\end{aligned} \tag{6}$$

where (x, y) and (p_x, p_y) denote commuting coordinates and momenta. Now using the relation (6) the Hamiltonian (4) can be written as

$$H = c\begin{pmatrix} mc & \hat{\Pi}_- \\ \hat{\Pi}_+ & -mc \end{pmatrix}, \tag{7}$$

where $\hat{\Pi}_\pm$ are more conveniently written after introducing the following frequencies out of the parameters of the problem:

$$\tilde{\omega} = \frac{eB}{2mc}, \qquad \omega_\theta = \frac{2\,\hbar}{m\theta}, \qquad \omega_\eta = \frac{\eta}{2\hbar m}. \tag{8}$$

Then we find:

$$\hat{\Pi}_{\pm} = \left(1 - \frac{\tilde{\omega} - \omega}{\omega_\theta}\right)(p_x \pm i\, p_y) \pm i\, m\, (\tilde{\omega} - \omega - \omega_\eta)(x \pm iy)\,. \tag{9}$$

From the above relations one finds that there is a critical value of the magnetic field ($B_{\rm cr}$)

$$B_{\rm cr} = \frac{2mc}{e}(\omega + \omega_\eta) = \frac{2c}{e}(m\omega + \frac{2\eta}{\hbar})\,, \tag{10}$$

such that for $B = B_{\rm cr}$ (or $\tilde{\omega} = \omega + \omega_\eta$) there are no interactions in the model and the Hamiltonian represents a free particle (only kinetic energy). Note that the critical magnetic field $B = B_{\rm cr}$ actually depends on the oscillator frequency as well as the momentum non-commutativity parameter (η or ω_η). We see therefore that the momentum non-commutativity parameter (η) shifts the value of the critical filed relative to the value in the commutative case, where the system is well known to undergo a quantum phase transition.

More importantly we note that space non commutativity ($\theta \neq 0$) introduces an additional critical value for the magnetic field ($B^*_{\rm cr}$) which corresponds to the the condition $\tilde{\omega} = \omega + \omega_\theta$:

$$B^*_{\rm cr} = \frac{2mc}{e}(\omega + \omega_\theta) = \frac{2c}{e}(m\omega + \frac{\hbar}{\theta})\,. \tag{11}$$

At this value of the magnetic field the kinetic part of the Hamiltonian will vanish. Physically we can visualise the system at this critical value as being made up by *only potential energy.* This second critical point signals a new quantum phase transition which is *absent* in the commutative limit, whereas the quantum phase transition at $B = B_{\rm cr}$ is only shifted by the momentum non commutativity parameter η (and this shift goes smoothly to zero when $\eta \to 0$).

In the following section we shall analyse the spectrum as a function of the magnetic field separately in each of the three regions: when $B < B_{\rm cr}$, $B_{\rm cr} < B < B^*_{\rm cr}$ and $B > B^*_{\rm cr}$.

We shall discuss the problem according to the magnitude of the magnetic field (or the cyclotron frequency $\tilde{\omega}$) relative to the other parameters (Dirac oscillator frequency ω and NC frequencies $\omega_\eta, \omega_\theta$). In this regard we note that as the non commutative (NC) parameters $\eta, \theta \to 0$ (the commutative limit) the frequency ω_η vanishes, while $\omega_\theta \to \infty$. Therefore without any loss of generality we may assume the following relation between ω_η, and ω_θ:

$$\omega_\eta < \omega_\theta\,. \tag{12}$$

3. Weak Magnetic Field: $B < B_{\rm cr}$ ($\tilde{\omega} < \omega + \omega_\eta$)

In this case we find from Eq. (9):

$$\hat{\Pi}_{\pm} = \lambda(p_x \pm i\, p_y) \mp i\nu(x \pm y)\,, \tag{13}$$

$$\lambda = 1 - \frac{\tilde{\omega} - \omega}{\omega_\theta} > 0\,, \qquad \nu = m(\omega + \omega_\eta - \tilde{\omega}) > 0\,. \tag{14}$$

To this end let us define the following set of creation and annihilation operators:

$$\begin{aligned} a_x &= \frac{i}{\sqrt{2\nu\lambda\hbar}}(\lambda p_x - i\nu x), \quad & a_x^\dagger &= \frac{-i}{\sqrt{2\lambda\nu\hbar}}(\lambda p_x + i\nu x)\,, \\ a_y &= \frac{i}{\sqrt{2\nu\lambda\hbar}}(\lambda p_y - i\nu y), \quad & a_y^\dagger &= \frac{-i}{\sqrt{2\lambda\nu\hbar}}(\lambda p_y + i\nu y)\,, \end{aligned} \tag{15}$$

and then the so called circular (or chiral) annihilation and creation operators:

$$a_L = \frac{1}{\sqrt{2}}(a_x + i a_y), \qquad a_L^\dagger = \frac{1}{\sqrt{2}}(a_x^\dagger - i a_y^\dagger), \\ a_R = \frac{1}{\sqrt{2}}(a_x - i a_y), \qquad a_R^\dagger = \frac{1}{\sqrt{2}}(a_x^\dagger + i a_y^\dagger). \tag{16}$$

It can easily be verified that the operators in (16) satisfy the following commutation relations

$$[a_R, a_R^\dagger] = [a_L, a_L^\dagger] = 1, \qquad [a_R, a_L] = [a_R, a_L^\dagger] = [a_R^\dagger, a_L] = [a_R^\dagger, a_L^\dagger] = 0. \tag{17}$$

Thus $(a_{R,L}^\dagger, a_{R,L})$ represent creation and annihilation operators for a pair of independent harmonic oscillators. We shall now analyse the spectrum using the creation and annihilation operators defined in (Eq. 16).

As a consequence we obtain:

$$\hat{\Pi}_+ = -2\,i\,\sqrt{\nu\lambda\hbar}\,a_L, \qquad \hat{\Pi}_- = +2\,i\,\sqrt{\nu\lambda\hbar}\,a_L^\dagger. \tag{18}$$

The eigenvalue equation can be written as

$$+\,2\,i\,\sqrt{\nu\lambda\hbar}\,a_L^\dagger\,\psi_L^{(2)} = \epsilon_-\psi_L^{(1)}, \qquad -\,2\,i\,\sqrt{\nu\lambda\hbar}\,a_L\,\psi_L^{(1)} = \epsilon_+\psi_L^{(2)}, \qquad \epsilon_\pm = \frac{E \pm mc^2}{c}. \tag{19}$$

Let us now examine whether there are solutions with energies $E = \pm mc^2$ i.e, $\epsilon_\pm = 0$. It turns out to be convenient to work in standard polar coordinates:

$$r = \sqrt{x^2 + y^2}, \qquad \varphi = \arctan\left(\frac{y}{x}\right), \qquad \rightarrow \qquad (x, y) = (r\cos\varphi, r\sin\varphi),$$

and then we can easily find the form of the annihilation and creation operators:

$$a_L = \frac{e^{i\varphi}}{2\sqrt{\nu\lambda\hbar}}\left\{\lambda\hbar\left(\partial_r + \frac{i}{r}\,\partial_\varphi\right) + \nu r\right\}, \\ a_L^\dagger = \frac{e^{-i\varphi}}{2\sqrt{\nu\lambda\hbar}}\left\{-\lambda\hbar\left(\partial_r - \frac{i}{r}\,\partial_\varphi\right) + \nu r\right\},$$

$$a_L\,\psi_L^{(1)} = 0, \quad \rightarrow \quad \{\partial_r + \tfrac{i}{r}\,\partial_\varphi + \tfrac{\nu}{\lambda\hbar}r\}\,\psi_L^{(1)}(r,\varphi) = 0, \\ a_L^\dagger\,\psi_L^{(2)} = 0, \quad \rightarrow \quad \{\partial_r - \tfrac{i}{r}\,\partial_\varphi - \tfrac{\nu}{\lambda\hbar}r\}\,\psi_L^{(2)}(r,\varphi) = 0. \tag{20}$$

Let us first consider the first of Eqs. (20). This is a first order partial differential equation which can be easily solved by the method of separation of the variables. For the wave function we make the ansatz $\psi_L^{(2)} = u(r)Y(\varphi)$ and find with straightforward calculations that $u(r) = r^M e^{-\frac{\nu}{2\lambda\hbar}r^2} + c_1$ and $Y(\varphi) = e^{iM\varphi} + c_2$ with M, c_1 and c_2 constants. Physical boundary conditions are imposed on the wave function and single-valuedness implies that M can only take integer values $M = 0, \pm1, \pm2, \cdots$, but normalisability further restrict M to non-negative values only ($M \geq 0$). Therefore the normalised solutions of the equation $a_L\,\psi_L^{(1)} = 0$ are:

$$\psi_L^{(1)} = u_{0,M} = C\,r^M\,e^{-\frac{\nu}{2\lambda\hbar}r^2}\,e^{iM\varphi}, \qquad M = 0, 1, 2, \cdots. \tag{21}$$

with $C = (\nu/(\lambda\hbar))^{\frac{M+1}{2}}/(\sqrt{\pi}\,\Gamma(M+1))$.

A spinor solution of the Hamiltonian equation with the upper component $\psi_{L,0}^{(1)}$ given by Eq. (21) can be associated with a null lower component ($\psi_{L,0}^{(2)} = 0$) to build a (singlet like) spinor state solution of the full Hamiltonian eigenvalue equation $H\psi = E\psi$ with energy $E = +mc^2$ for which again *both* intertwining solutions are satisfied:

$$(\epsilon_- = 0) \qquad E = +mc^2, \qquad \psi_{L,0} = \begin{pmatrix} u_{0,M} \\ 0 \end{pmatrix} \qquad M = 0, 1, 2, \cdots . \quad \text{(solution of Eqs. (19))}\,. \tag{22}$$

This is an acceptable (normalized) solution of our eigenvalue problem. Note that this zero mode has an infinite degeneracy with respect to the positive values of M, – also note that here, and thereafter, by zero modes we mean states that are annihilated by the corresponding annihilation operators entering the problem, and singlet (doublet) are referred to spinors with only one (both) component(s) non-vanishing–.

The second of Eqs. (20) can be solved in a similar way and we do not give the details here. It is easily found that the solution $\psi_L^{(2)} = C\, r^{-M}\, e^{+\frac{\nu}{2\lambda\hbar} r^2}\, e^{iM\varphi}$ can be made regular in the origin by choosing negative values of M but it is always diverging at large radial distances ($r \to \infty$). While one would still be able to construct a spinor (singlet) solution of the hamiltonian eigenvalue equations, c.f. Eqs. (20), corresponding to $E = -mc^2$ or ($\epsilon_+ = 0$) this would be un-normalizable for any value of M and hence unphysical and is therefore discarded.

We now turn to a detailed discussion of the excited states and their energy eigen-values which are best discussed by considering the second order eigenvalue equation $H^2\psi_L = E^2\psi_L$ which allows to disentangle the two components $\psi_L^{(1)}$ and $\psi_L^{(2)}$. Then form Eq. (7) we obtain the decoupled equations:

$$\hat{\Pi}_-\, \hat{\Pi}_+\, \psi_L^{(1)} = \epsilon_+\epsilon_-\, \psi_L^{(1)}\,, \tag{23a}$$

$$\hat{\Pi}_+\, \hat{\Pi}_-\, \psi_L^{(2)} = \epsilon_+\epsilon_-\, \psi_L^{(2)}\,, \tag{23b}$$

where $\epsilon_+\epsilon_- = (E^2 - m^2c^4)/c^2$. By using the explicit form of the creation and annihilation operators given in Eqs. (15),(16) we find:

$$a_L^\dagger\, a_L = \frac{\hat{\Pi}_-\, \hat{\Pi}_+}{4\nu\lambda\hbar} = \frac{1}{2\hbar}\left[H_\circ^{2D} - L_z - \hbar\right]\,, \tag{24a}$$

$$a_L\, a_L^\dagger = \frac{\hat{\Pi}_+\, \hat{\Pi}_-}{4\nu\lambda\hbar} = \frac{1}{2\hbar}\left[H_\circ^{2D} - L_z + \hbar\right]\,, \tag{24b}$$

where we have indicated the angular momentum in the two dimensional commuting plane (x, y) by the operator $L_z = x\, p_y - y\, p_x$ and:

$$H_\circ^{2D} = \frac{\lambda}{2\nu}\left(p_x^2 + p_y^2\right) + \frac{\nu}{2\lambda}(x^2 + y^2)\,. \tag{25}$$

We see therefore that the second order Hamiltonians in Eqs. (23)[1] can be related to $H_\circ^{2D}$, the Hamiltonian of a well known *and exactly solvable* non-relativistic system –that of a two dimensional isotropic (or circular) harmonic oscillator (of unit frequency and mass ν/λ)–. The eigen-functions and eigenvalues of this system are well known [52] and can be readily used to solve the second order Hamiltonians of Eqs. (23) since the angular momentum operator L_z commutes with $H_\circ^{2D}$. The complete set of eigenfunctions of $H_\circ^{2D}$ and the corresponding

[1] Eqs. (24) prove indirectly the relation $[a_L\,, a_L^\dagger] = 1$.

eigenvalues are conveniently derived using the polar coordinates of the (x, y) plane and they are identified by a radial quantum number $n_r = 0, 1, 2, \cdots$ and the angular momentum quantum number $M = 0, \pm 1, \pm 2, \cdots$ [52]:

$$H_\circ^{2D}\, u_{n_r,M}(r,\varphi) = \varepsilon\, u_{n_r,M}(r,\varphi)\,, \tag{26}$$

whose eigenstates and eigenvalues are given by:

$$\varepsilon = \hbar\,(|M| + 1 + 2n_r)\,, \tag{27a}$$

$$u_{n_r,M}(r,\varphi) = C_{n_r,M}\, r^{|M|}\, e^{-\frac{\nu}{2\lambda\hbar}r^2}\; {}_1F_1(-n_r, |M|+1; \frac{\nu}{\lambda\hbar} r^2)\, e^{iM\varphi}\,, \tag{27b}$$

where $C_{n_r,M}$ are normalisation constants that can be easily computed as:

$$C_{n_r,M} = \frac{1}{\sqrt{\pi}} \left(\frac{\nu}{\lambda\hbar}\right)^{\frac{|M|+1}{2}} \frac{\sqrt{\Gamma(|M|+1+n_r)}}{\Gamma(|M|+1)\sqrt{\Gamma(n_r+1)}}\,. \tag{28}$$

The relativistic eigenvalues of the second order equations are then easily found. The solution of Eq. (23a) is obtained with the help of Eq. (24a) and is found to be:

$$\psi_L^{(1)}(r,\varphi) = C_1\, u_{n_r,M}(r,\varphi)\,, \tag{29}$$

$$\epsilon_+\epsilon_- = 4\nu\lambda\hbar\, \frac{1}{2\hbar}\, [\hbar\,(|M| + 1 + 2n_r) - \hbar M - \hbar]\,, \tag{30}$$

where C_1 is a spinor normalisation constant. From the above equation the excited states eigenvalues ($n_r \geq 1$) of the original Dirac equation can be extracted:

$$E_N^{\pm} = \pm mc^2 \sqrt{1 + 4\frac{\lambda\nu\hbar}{m^2c^2} N} = \pm mc^2\sqrt{1 + \zeta_L\, N}\,, \quad N = n_r + \frac{|M| - M}{2} \quad N = 1, 2, ,\cdots, \tag{31}$$

and the quantity ζ_L is defined as:

$$\zeta_L = \zeta_L(\tilde\omega - \omega; \omega_\eta, \omega_\theta) = \frac{4\lambda\nu\hbar}{m^2c^2} = -4\frac{\hbar}{mc^2}\left(1 - \frac{\tilde\omega - \omega}{\omega_\theta}\right)(\tilde\omega - \omega - \omega_\eta)\,. \tag{32}$$

We see that every energy level E_N is highly degenerate. In particular every level has an infinite degeneracy with respect to the non-negative values of M, ($M \geq 0$), while there is a finite degeneracy $D = N + 1$ with respect to the negative values of M.

The lower component $\psi_L^{(2)}$ of the eigen-solution of the Dirac Hamiltonian is found by using the intertwining relation in Eq. (19):

$$\psi_L^{(2)}(r,\varphi) = -2i\frac{\sqrt{\lambda\nu\hbar}}{\epsilon_+}\, a_L\, C_1\, u_{n_r,M}(r,\varphi)\,,$$

$$= -i\frac{C_1}{\epsilon_+}\, e^{i\varphi} \left\{\lambda\hbar\left(\partial_r + \frac{i}{r}\partial_\varphi\right) + \nu r\right\} u_{n_r,M}(r,\varphi)\,. \tag{33}$$

We find that the sign of the angular quantum number (M) identifies two classes of eigen-states which we discuss separately.

$M \geq 0$ (inifinite degeneracy):

In this case using the explicit expression of the wave functions $u_{n_r,M}$ and with the aid of the recurrence relation [53]:

$$\frac{d}{dz}\, {}_1F_1(a, b; z) = \frac{a}{b}\, {}_1F_1(a+1, b+1; z)\,, \tag{34}$$

and after taking into account that for $M \geq 0$ the normalisation constants in Eq. (28) satisfy $C_{n_r,M} = C_{n_r-1,M+1}\sqrt{\lambda\hbar/\nu}\,(M+1)/\sqrt{n_r}$ we find explicitly:

$$\psi_L^{(2)} = +i\,C_1\frac{\sqrt{4\nu\lambda\hbar n_r}}{\epsilon_+}\,u_{n_r-1,M+1} = +i\,C_1\,\frac{\sqrt{\epsilon_-\epsilon_+}}{\epsilon_+}\,u_{n_r-1,M+1}\,. \tag{35}$$

Finally the normalised spinor solution for positive values of M can be put in the form:

$$D=\infty, \qquad \psi_L^{(\pm,n_r,M)} = \frac{1}{\sqrt{2}}\begin{pmatrix} \sqrt{1+\frac{mc^2}{E_N^\pm}}\,u_{n_r,M} \\ \pm i\sqrt{1-\frac{mc^2}{E_N^\pm}}\,u_{n_r-1,M+1}\end{pmatrix}, \qquad N=n_r=1,2,\cdots, \tag{36}$$

where the upper (lower) sign corresponds respectively to the positive (negative) branch of the spectrum, and the normalisation condition is, here and thereafter, the usual one for two component spinors, i.e. $\langle\psi|\psi\rangle = \langle\psi_1|\psi_1\rangle + \langle\psi_2|\psi_2\rangle = 1$.

$M < 0$ (finite degeneracy):

In this case when solving for the lower component of the spinor solution in Eq. (33) one has to use a different recurrence relation satisfied by the confluent hypergeometric function [53]:

$$z\frac{d}{dz}\,{}_1F_1(a,b;z) = (b-1)\left[\,{}_1F_1(a,b-1;z) - \,{}_1F_1(a,b;z)\right], \tag{37}$$

and after taking into account that for $M < 0$ the normalisation constants in Eq. (28) satisfy $C_{n_r,M} = -C_{n_r,M+1}\sqrt{\nu\,(n_r-M)/(\lambda\hbar)}\,/M$ we find explicitly:

$$\psi_L^{(2)} = -i\,C_1\frac{\sqrt{4\nu\lambda\hbar(n_r-M)}}{\epsilon_+}\,u_{n_r,M+1} = -i\,C_1\,\frac{\sqrt{\epsilon_-\epsilon_+}}{\epsilon_+}\,u_{n_r,M+1}\,, \tag{38}$$

and the final expression of the spinor solution for the N-th energy level in the case of negative values of M is:

$$D=N+1, \qquad \psi_L^{(\pm,n_r,M)} = \frac{1}{\sqrt{2}}\begin{pmatrix} \sqrt{1+\frac{mc^2}{E_N^\pm}}\,u_{n_r,M} \\ \mp i\sqrt{1-\frac{mc^2}{E_N^\pm}}\,u_{n_r,M+1}\end{pmatrix}, \qquad N=n_r-M=1,2,\cdots, \tag{39}$$

where again the upper (lower) sign correspond to the positive (negative) branch of the spectrum.

Also we note the energy gap is dependent on the commutative parameters θ,η.

4. Intermediate Magnetic Field: $B_{\mathrm{cr}} < B < B^*_{\mathrm{cr}}$ $(\omega_\eta < \tilde\omega - \omega < \omega_\theta)$

In this case we find from Eq. (9):

$$\hat\Pi_\pm = \lambda(p_x \pm i\,p_y) \pm i\mu(x \pm i\,y)\,, \tag{40}$$

$$\lambda = 1 - \frac{\tilde\omega-\omega}{\omega_\theta} > 0\,, \qquad \mu = m(\tilde\omega - \omega - \omega_\eta) > 0\,. \tag{41}$$

The Dirac Hamiltonian is in this case expressed solely in terms of the right handed creation and annihilation operators so in this region of parameters the system is said to be in the right chiral phase.

As regards the zero modes in the right phase we find that the normalisable solution is given in terms of the eigenfunction $\tilde u_{0,M} = C\,r^{-M}\,e^{-\frac{\mu}{2\lambda\hbar}r^2}$ and now it is infinitely degenerate with respect to the non-positive values of M, ($M \leq 0$), and is a spin down singlet with $E = -mc^2$ (negative branch of the spectrum):

$$(\epsilon_+ = 0) \quad E = -mc^2, \quad \psi_{R,0} = \begin{pmatrix} 0 \\ \tilde{u}_{0,M} \end{pmatrix}, \quad M = 0, -1, -2 \cdots, \quad \text{(solution of Eqs. (??))}, \tag{42}$$

which is normalized to unity if the constant C is chosen as:
$C = (\mu/(\lambda\hbar))^{\frac{|M|+1}{2}}/(\sqrt{\pi}\,\Gamma(|M|+1))$.

The excited levels ($n_r \geq 1$) are then derived from the second order equations c.f. Eq (23) similarly to what has been done in the left chiral phase. We find it convenient to solve the lower component $\psi_R^{(2)}$ from Eq. (23b) and then compute the upper component $\psi_R^{(1)}$ with the first of the intertwining relations. We find:

$$E_N^{\pm} = \pm mc^2 \sqrt{1 + 4\,\frac{\lambda\mu\hbar}{m^2c^2}\,N} = \pm mc^2\sqrt{1 + \zeta_R\,N} \qquad N = n_r + \frac{M + |M|}{2} \qquad N = 1, 2\cdots \tag{43}$$

and the quantity ζ_R is defined as:

$$\zeta_R = \zeta_R(\tilde{\omega} - \omega; \omega_\eta, \omega_\theta) = 4\,\frac{\mu\lambda\hbar}{m^2c^2} = 4\,\frac{\hbar}{mc^2}\left(1 - \frac{\tilde{\omega} - \omega}{\omega_\theta}\right)(\tilde{\omega} - \omega - \omega_\eta)\,. \tag{44}$$

We see again that every energy level E_N is highly degenerate and, as expected, there is an infinite degeneracy with respect to the non-positive values of M, ($M \leq 0$), while the degeneracy is finite, $D = N+1$, with respect to the positive values of M. The corresponding eigen-solutions are:

<u>$\boldsymbol{M > 0}$ (finite degeneracy):</u>

In this case using the explicit expression of the wave functions $\tilde{u}_{n_r,M}$ we find the normalised spinor solution for positive values of M can be put in the form:

$$D = N + 1, \quad \psi_R^{(\pm,n_r,M)} = \frac{1}{\sqrt{2}}\begin{pmatrix} \mp i\sqrt{1 + \frac{mc^2}{E_N^{\pm}}}\,\tilde{u}_{n_r,M-1} \\ \sqrt{1 - \frac{mc^2}{E_N^{\pm}}}\,\tilde{u}_{n_r,M} \end{pmatrix}, \quad N = n_r + M = 1, 2, \cdots, \tag{45}$$

where the upper (lower) sign corresponds respectively to the positive (negative) branch of the spectrum.

<u>$\boldsymbol{M \leq 0}$ (infinite degeneracy):</u>

$$D = \infty, \quad \psi_R^{(\pm,n_r,M)} = \frac{1}{\sqrt{2}}\begin{pmatrix} \pm i\sqrt{1 + \frac{mc^2}{E_N^{\pm}}}\,\tilde{u}_{n_r-1,M-1} \\ \sqrt{1 - \frac{mc^2}{E_N^{\pm}}}\,\tilde{u}_{n_r,M} \end{pmatrix}, \quad N = n_r = 1, 2, \cdots. \tag{46}$$

Thus for $B_{\text{cr}} < B < B^*_{\text{cr}}$, the $E = -mc^2$ state is a singlet while for the other energy values the positive and negative energy states are paired in doublets. It may also be noticed that the energy gap is not $2mc^2$ but depends on the magnetic field intensity B as well as on the non-commutativity parameters η and θ.

5. Strong Magnetic Field: $B > B^*_{cr}$ $(\tilde{\omega} - \omega > \omega_\theta)$.

We see that relative to case 2 (see Eq. (40)) now the quantity $\lambda = 1 - (\tilde{\omega} - \omega)/\omega_\theta$ becomes negative and therefore the operators $\hat{\Pi}_\pm$ are given by

$$\hat{\Pi}_\pm = -\left[\delta(p_x \pm i\, p_y) \mp i\mu(x \pm i\, y)\right] , \tag{47}$$

where:

$$\delta = \frac{\tilde{\omega} - \omega}{\omega_\theta} - 1 > 0 , \tag{48}$$

and μ is given as in Eq. 41 and we see that $\Pi_\pm$ in Eq. (47) differ by an overall negative sign with the operators in Eq. (13), of case 1 (the region with a weak magnetic field), –the so called left chiral phase– with the substitutions: $\lambda \to \delta, \nu \to \mu$. We expect therefore this region of very strong magnetic field $(\tilde{\omega} - \omega > \omega_\theta)$ to be described as left chiral phase.

We examine first whether there are solutions with energies $E = \pm mc^2$ i.e, $\epsilon_\pm = 0$, and find now that the normalisable solution is given in terms of the eigenfunction $u'_{0,M} = C'\, r^M\, e^{-\frac{\mu}{2\delta\hbar} r^2}$, infinitely degenerate with respect to the non-negative values of M, $(M \geq 0)$, and is a spin-up singlet with $E = +mc^2$ (positive branch of the spectrum):

$$(\epsilon_- = 0) \qquad E = +mc^2, \qquad \psi_{R,0} = \begin{pmatrix} u'_{0,M} \\ 0 \end{pmatrix}, \quad M = 0, 1, 2 \cdots , \quad \text{(solution of Eqs. (??))} \tag{49}$$

which is normalized to unity if the constant C' is chosen as:
$C' = (\mu/(\delta\hbar))^{\frac{M+1}{2}} / (\sqrt{\pi}\, \Gamma(M + 1))$.

The excited levels $(n_r \geq 1)$ are then derived from the second order equations c.f. Eq (23) in exactly the same way as done in the left chiral phase (case 1).

The spectrum and the eigenfunctions are then:

$$E^\pm_N = \pm mc^2 \sqrt{1 + 4\, \frac{\delta\mu\hbar}{m^2c^2}\, N} \; = \pm mc^2 \sqrt{1 + \zeta'_L\, N} \qquad N = n_r + \frac{|M| - M}{2} \qquad N = 1, 2, \cdots \tag{50}$$

and the quantity ζ'_L is defined by:

$$\zeta'_L = \zeta'_L(\tilde{\omega} - \omega; \omega_\eta, \omega_\theta) \; = \frac{4\, \delta\, \mu\, \hbar}{m^2c^2} = 4\, \frac{\hbar}{mc^2} \left(\frac{\tilde{\omega} - \omega}{\omega_\theta} - 1 \right) (\tilde{\omega} - \omega - \omega_\eta) . \tag{51}$$

$\boldsymbol{M \geq 0}$ (infinite degeneracy):
The normalised spinor solution for positive values of M can be put in the form:

$$D = \infty, \qquad \psi_L^{(\pm, n_r, M)} = \frac{1}{\sqrt{2}} \begin{pmatrix} \sqrt{1 + \frac{mc^2}{E^\pm_N}}\; u_{n_r, M} \\ \mp i \sqrt{1 - \frac{mc^2}{E^\pm_N}}\; u_{n_r - 1, M+1} \end{pmatrix}, \qquad N = n_r = 1, 2, \cdots , \tag{52}$$

where the upper (lower) sign corresponds respectively to the positive (negative) branch of the spectrum.

$\boldsymbol{M < 0}$ (finite degeneracy):
The final expression of the spinor solution for the N-th energy level in the case of negative values of M is:

$$D = N + 1, \qquad \psi_L^{(\pm, n_r, M)} = \frac{1}{\sqrt{2}} \begin{pmatrix} \sqrt{1 + \frac{mc^2}{E^\pm_N}}\; u_{n_r, M} \\ \pm i \sqrt{1 - \frac{mc^2}{E^\pm_N}}\; u_{n_r, M+1} \end{pmatrix}, \qquad N = n_r - M = 1, 2, \cdots , \tag{53}$$

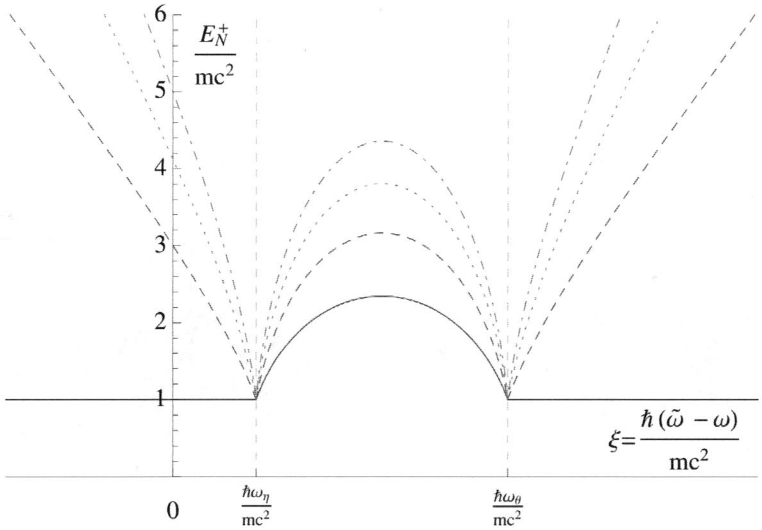

Figure 1. Energy eigenvalues of the positive branch E_N^+, in units of mc^2, as a function of the dimensionless ratio $\xi = \hbar(\tilde{\omega} - \omega)/(mc^2)$. Shown are the ground state, solid line (blue) and of the first few excited states, dashed (red), dotted (light green) and dot-dashed (green). There are two values of the magnetic field (or $\tilde{\omega}$) for which there is a *quantum* phase transition (a point of non analyticity in the energy eigenvalues). In the commutative limit ($\eta, \theta \to 0$) $\omega_\eta \to 0$ and $\omega_\theta \to \infty$ and the spectrum goes smoothly into the one discussed in [39] with only one phase transition at $\xi = 0$.

where again the upper (lower) sign correspond to the positive (negative) branch of the spectrum.

Note that the spinor structure of the eigenstates of the excited levels in this new left chiral phase is identical to that of the initial left phase except for the relative sign between the lower and upper components (in addition of course to the replacement of the parameters).

6. Discussion

In Fig. 1 we show the positive branch of the spectrum for the first few energy levels as function of the dimensionless quantity $\xi = \hbar(\tilde{\omega} - \omega)/(mc^2)$ which is directly related to the intensity of the external magnetic field through $\tilde{\omega} = eB/(2mc)$. Note that while in the two left phases the positive branch of the spectrum starts with the singlet zero mode at $E = +mc^2$ in the right phase the lowest positive energy state (the ground state) is a doublet with $E = +mc^2\sqrt{1+\zeta_R}$. In Fig. 1 one clearly makes out the two points of non-analyticity, $\xi = \hbar\omega_\eta/(mc^2)$ and $\xi = \hbar\omega_\theta/(mc^2)$, which signal the two quantum phase transitions. The same behaviour is of course shown by all the excited states. In the commutative limit ($\eta, \theta \to 0$) we have that $\omega_\eta \to 0$ and $\omega_\theta \to \infty$ and therefore we obtain that while the first phase transition (left-right) is shifted towards the commutative critical value $\xi = 0$ ($\tilde{\omega} = \omega$), the second phase transition (right-left) disappears since it is moved to $\xi = \infty$ ($\tilde{\omega} = \infty$).

We now would like to discuss a physical observable, closely related to the spectrum, the magnetisation, which has the potential of characterising the quantum phase transitions of the system. Indeed for every energy level $E_N^{(i)}$, $i = L, R, L'$, we can define the magnetisation $M_N^{(i)}$ by:

$$M_N^{(i)} = -\frac{\partial E_N^{(i)}}{\partial B}. \tag{54}$$

In each of the quantum phases that we have discussed previously the energy eigenvalues can be written as $E_N^{(i)} = mc^2\sqrt{1+\zeta_{(i)}N}$ with the functions $\zeta_{(i)}$ as in Eqs. (31,44,51) and the

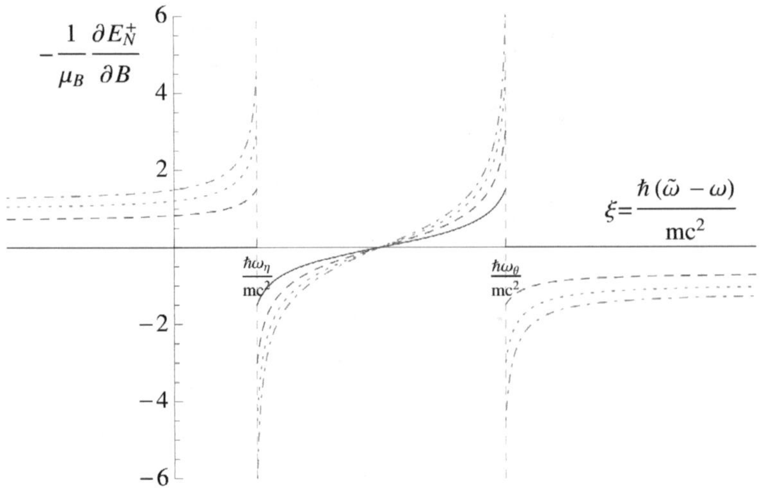

Figure 2. Magnetisation of the energy levels of the positive branch of the spectrum E_N^+ in units of the Bohr magneton $\mu_B = \frac{e\hbar}{2m_e c}$ (and assuming $m_e/m = 1$), as a function of the dimensionless ratio $\xi = \hbar(\tilde{\omega} - \omega)/(mc^2)$. Shown are the ground state, solid line (blue) and first few excited energy levels, dashed (red), dotted (light green) and dot-dashed (green). At the two critical points ($\xi = \frac{\hbar\omega_\eta}{mc^2}$ and $\xi = \frac{\hbar\omega_\eta}{mc^2}$) the magnetisation presents a discontinuity for each energy level.

magnetisation is then easily found to be given by:

$$M_N^{(i)} = -\frac{mc^2}{2\sqrt{1+N\zeta_{(i)}}} N \left(\frac{\partial \zeta_{(i)}}{\partial B}\right)$$
$$= -2\,\mu_B \frac{m_e}{m} \frac{N}{\sqrt{1+N\zeta_{(i)}}} \left(1 + \frac{\omega_\eta}{\omega_\theta} - 2\frac{\tilde{\omega}-\omega}{\omega_\theta}\right). \tag{55}$$

where m_e is the electron mass and $\mu_B = \frac{e\hbar}{2m_e c}$ is the Bohr magneton.

In Fig. 2 we show the magnetisation of our system assuming $m_e = m$. We find that for every level the magnetisation has a finite discontinuity at the two critical points. We point out a very interesting feature which is closely tied to the presence of non-commutativity. We clearly see that in the right chiral phase there is a fixed point of vanishing magnetisation ($M_N^{(R)} = 0$) at $\tilde{\omega} - \omega = (\omega_\eta + \omega_\theta)/2$, exactly the midpoint of the two critical points. Clearly this will remain so even when computing a thermal average of the magnetisation at a given temperature. Since $M_N^{(R)} = 0$ for every level its average value, at any given temperature, will also vanish. So this can be taken as a prediction of our model for the system under consideration: *in the presence of non-commutativity there is a fixed point with vanishing magnetisation in the right chiral phase.*

7. Conclusions

Within a non-commutativity scenario both in the space and momentum coordinates we have studied the (2+1) dimensional Dirac oscillator in a constant homogenous magnetic field. The one dimensional version of the Dirac oscillator has been recently experimentally realised and observed in the laboratory [31] with realistic prospects of realising soon a 2-dimensional version. Here we have solved exactly the eigenvalue equations according to the strength of the magnetic field. The system in the absence of non-commutativity is known to have a left-right chiral phase transition for $B = B_{\text{cr}} = \frac{2mc}{e}\omega$. We find that the non-commutativity of the momentum coordinates ($\eta-$parameter, see Eq. (3)) simply shifts the left-right chiral phase transition. The presence of non-commutativity in the space coordinates ($\theta-$parameter, see Eq. (3)) changes the picture quite dramatically, because it introduces a new right-left phase transition at $B = B_{\text{cr}}^* = \frac{2mc}{e}(\omega + \omega_\theta)$ which is absent in the commutative limit (as $\theta \to 0$, $\omega_\theta \to \infty$). Therefore the system is shown to exibit the interesting phenomenon of reentrant phase transitions.

We have also discussed the magnetisation of the energy levels of the system and we have found that this physical observable has the potential of being able to characterise the two quantum phase transitions since it offers a finite discontinuity (for every energy level) at the two critical points. A rather peculiar prediction for this observable is the existence of a *fixed point* with vanishing magnetisation, for every energy level, exactly located at the midpoint of the two critical points. Work is in progress to present a full thermodynamic treatment of the system based on the solution presented here.

References

[1] Connes A, Douglas M R and Schwarz A 1998 *J. High Energy Phys.* **9802** 003
[2] Seiberg N and Witten E 1999 *J. High Energy Phys.* **9909** 032
[3] Douglas M R and Nekrasov N A 2001 *Rev. Mod. Phys.* **73**(4) 977–1029
[4] Szabo R J 2003 *Phys. Rept.* 0370-1573
[5] Duval C and Horváthy P 2000 *Phys. Lett. B* 0370-2693
[6] Gamboa J, Loewe M and Rojas J C 2001 *Phys. Rev. D* **64**(6) 067901
[7] Gamboa J, Mendez F, Loewe M and Rojas J C 2001 *Mod. Phys. Lett. A* **16** 2075–2078
[8] Nair V and Polychronakos A 2001 *Phys. Lett. B* 0370-2693
[9] Gamboa J, Méndez F, Loewe M and Rojas J C 2002 *Int. J. Mod. Phys. A* **17** 2555–2565
[10] Smailagic A and Spallucci E 2002 *Phys. Rev. D* **65**(10) 107701
[11] Smailagic A and Spallucci E 2002 *J. Phys. A: Math. Gen.* **35** L363
[12] Bellucci S, Nersessian A and Sochichiu C 2001 *Phys. Lett. B* 0370-2693
[13] Bertolami O, Rosa J G, de Aragão C M L, Castorina P and Zappalà D 2005 *Phys. Rev. D* **72**(2) 025010
[14] Horváthy P, Martina L and Stichel P 2010 *SIGMA* **6** 060 (*Preprint* 1002.4772)
[15] Ito D, Mori K and Carriere E 1967 *Nuovo Cimento A* **51** 1119
[16] Cook P 1971 *Lett. Nuovo Cimento* 0375-930X
[17] Moshinsky M and Szczepaniak A 1989 *J. Phys. A: Math. Gen.* **22** L817
[18] Benitez J, Martinez y Romero R P, Nunez-Yepez H N and Salas-Brito A L 1990 *Phys. Rev. Lett.* **64**(14) 1643–1645
[19] Rozmej P and Arvieu R 1999 *J. Phys. A: Math. Gen.* **32** 5367
[20] Mirza B and Mohadesi M 2004 *Comm. Th. Phys.* **42** 664
[21] Sadurni E, Torres J M and Seligman T H 2010 *J. Phys. A: Math. Gen.* **43** 285204
[22] Boumali A and Hassanabadi H 2013 *Eur. Phys. J. Plus* **128** 124
[23] Faessler A, Kukulin V and Shikhalev M 2005 *Annals Phys.* 0003-4916
[24] Grineviciute J and Halderson D 2012 *Phys. Rev. C* **85**(5) 054617
[25] Munarriz J, Dominguez-Adame F and Lima R 2012 *Phys. Lett. A* 0375-9601
[26] Romera E 2011 *Phys. Rev. A* **84**(5) 052102
[27] Wang Y, Cao J and Xiong S 2012 *Eur. Phys. J. B* 1434-6028
[28] Bermudez A, Martin-Delgado M A and Solano E 2007 *Phys. Rev. Lett.* **99**(12) 123602
[29] Dodonov V V 2002 *J. Opt. B: Quantum S. O.* **4** R1
[30] Lamata L, León J, Schätz T and Solano E 2007 *Phys. Rev. Lett.* **98**(25) 253005
[31] Franco-Villafane J A, Sadurni E, Barkhofen S, Kuhl U, Mortessagne F and Seligman T H 2013 *Phys. Rev. Lett.* **111**(17) 170405
[32] Hul O, Bauch S, Pakoński P, Savytskyy N, Życzkowski K and Sirko L 2004 *Phys. Rev. E* **69**(5) 056205

[33] Sadurni E, Seligman T H and Mortessagne F 2010 *New J. Phys.* **12** 053014

[34] Hul O, Tymoshchuk O, Bauch S, Koch P M and Sirko L 2005 *J. Phys. A: Math. Gen.* **38** 10489

[35] Bermudez A, Martin-Delgado M A and Solano E 2007 *Phys. Rev. A* **76**(4) 041801

[36] Bermudez A, Martin-Delgado M A and Luis A 2008 *Phys. Rev. A* **77**(3) 033832

[37] Mandal B P and Verma S 2010 *Phys. Lett. A* 0375-9601

[38] Mandal B P and Rai S K 2012 *Phys. Lett. A* 0375-9601

[39] Bermudez A, Martin-Delgado M A and Luis A 2008 *Phys. Rev. A* **77**(6) 063815

[40] Quimbay C and Strange P 2013 (*Preprint* 1312.5251)

[41] Panella O and Roy P 2014 *Phys. Rev. A* **90**(4) 042111

[42] Altamirano N, Kubizňák D and Mann R B 2013 *Phys. Rev. D* **88**(10) 101502

[43] Frassino A, Kubizňák D, Mann R and Simovic F 2014 *Journal of High Energy Physics* **2014** 80

[44] Geim A K and Novoselov K S 2007 *Nat Mater* **6** 183–191

[45] Castro Neto A H, Guinea F, Peres N M R, Novoselov K S and Geim A K 2009 *Rev. Mod. Phys.* **81**(1) 109–162

[46] Semenoff G W 1984 *Phys. Rev. Lett.* **53**(26) 2449–2452

[47] Lalmi B, Oughaddou H, Enriquez H, Kara A, Vizzini S, Ealet B and Aufray B 2010 *Appl. Phys. Lett.* **97** 223109

[48] Vogt P, De Padova P, Quaresima C, Avila J, Frantzeskakis E, Asensio M C, Resta A, Ealet B and Le Lay G 2012 *Phys. Rev. Lett.* **108**(15) 155501

[49] Fleurence A, Friedlein R, Ozaki T, Kawai H, Wang Y and Yamada-Takamura Y 2012 *Phys. Rev. Lett.* **108**(24) 245501

[50] Lin C L, Arafune R, Kawahara K, Tsukahara N, Minamitani E, Kim Y, Takagi N and Kawai M 2012 *Appl. Phys. Express* **5** 045802

[51] Quimbay C and Strange P 2013 (*Preprint* 1311.2021)

[52] Flügge S 1974 *Practical quantum mechanics. I, II* (Berlin: Springer-Verlag)

[53] Abramowitz M and Stegun I A 1964 *Handbook of Mathematical Functions with Formulas, Graphs, and Mathematical Tables* ninth Dover printing, tenth printing of U.S. Governement Printing Office ed (New York: Dover)

Journal of Physics: Conference Series **670** (2016) 012041 doi:10.1088/1742-6596/670/1/012041

Irregular blocks, $\mathcal{N} = 2$ gauge theory and Mathieu system

M R Piatek[1,3], **A R Pietrykowski**[2,3]

[1] Institute of Physics, University of Szczecin, ul. Wielkopolska 15, 70-451 Szczecin, Poland
[2] Institute of Theoretical Physics, University of Wroclaw, M. Borna 9, 50-204 Wroclaw, Poland
[3] Bogoliubov Laboratory of Theoretical Physics, JINR, 141980 Dubna, Russia

E-mail: piatek@fermi.fiz.univ.szczecin.pl, heartie@ift.uni.wroc.pl

Abstract. The Alday-Gayotto-Tachikawa (AGT) conjecture relates 4d $\mathcal{N} = 2$, $SU(2)$ SYM theories with N_f matter hypermultiplets to 2d CFT. In case of pure 4d $\mathcal{N} = 2$, $SU(2)$ SYM there is a corresponding irregular conformal block in 2d CFT. The AGT correspondence may be extended within a certain limit (the Nekrasov-Shataschvili limit) to the correspondence between an effective twisted superpotentials of 2d $\mathcal{N} = 2$ SUSY and the Zamolodchikov's "classical" conformal blocks. When narrowed to the pure 4d $\mathcal{N} = 2$ SYM case its limit is related to an irregular classical conformal block. It will be shown that according to the triple correspondence (2dCFT/Gauge/Bethe - c.f. Piatek's talk) the irregular classical conformal block yields spectrum of Mathieu operator. The latter can be obtained as a "classical" limit of the null vector decoupling equation for three-point degenerate irregular block. It will also be shown that the Mathieu spectrum can be also obtained from the limit of the pure gauge theory as a solution of the saddle point equation as well as from the Bohr-Sommerfeld quantization of the Seiberg-Witten theory.

1. Introduction

The Mathieu equation and its solution proved useful in many fields of physics ranging from solid state physics to astrophysics and cosmology. It appeared as a result of study of the elliptic membrane oscillations [1]. The equation that describes mathematically this process is a second order elliptic partial differential equation which splits into two independent ordinary differential equations termed the Mathieu and the modified Mathieu equation both with real coefficients. For definiteness let us focus on the former, which assumes the form

$$\left(-\frac{\mathrm{d}^2}{\mathrm{d}x^2} + 2h^2\cos 2x\right)\psi(x) = \lambda\,\psi(x). \tag{1}$$

λ and h are parameters of the equation to whose we further refer as the spectrum of Mathieu operator and the coupling constant respectively. The solution of eq. (1) has the following property

$$\psi(x+\pi) \;=\; \mathrm{e}^{i\pi\nu}\psi(x).$$

The parameter ν is termed Floque exponent. It may assume integer and non-integer values. In what follows we are concerned with the non-integer one. For $h \ll 1$, i.e., in the week coupling

Journal of Physics: Conference Series **670** (2016) 012041 doi:10.1088/1742-6596/670/1/012041

regime the spectrum of Mathieu operator λ can be developed in terms of h and ν, namely

$$\lambda = \nu^2 + \frac{h^4}{2(\nu^2-1)} + \frac{(5\nu^2+7)\,h^8}{32(\nu^2-4)(\nu^2-1)^3} + \frac{(9\nu^4+58\nu^2+29)\,h^{12}}{64(\nu^2-9)(\nu^2-4)(\nu^2-1)^5} + \mathcal{O}\left(h^{16}\right). \quad (2)$$

The Mathieu equation also emerged in the context of recent studies of interrelationship between Quantum Integrable Systems (QIS), 2d Conformal Field Theories ($2d$ CFT) and $4d$ $\mathcal{N}=2$ super-Young-Mills field theories (SYM) [2]. As it turned out the Nekrasov partition function for $4d$ $\mathcal{N}=2$, $SU(N)$ SYM that depend on two regularization parameters ϵ_1, ϵ_2 [3], when the two parameters tend to zero ϵ_1, $\epsilon_2 \to 0$, approaches the exponential with an exponent having a divergent part proportional to a prepotential (for $N=2$ this is the Seiberg-Witten prepotential [4]). The latter is uniquely determined by N moduli and their duals as well as a certain elliptic curve which is topologically equivalent to a genus $N-1$ Riemann surface [5]. The moduli are related to $N-1$ periods of the elliptic curve. The latter, in turn, were revealed to be related to a classical integrable system, namely, N-particle periodic Toda chain [6, 7]. Its quantization corresponds to taking the zero limit in the Nekrasov partition function only in one parameter ϵ_2, called *the Nekrasov-Shataschvili limit* (NS limit). The divergent term in the exponent of thus obtained expression is proportional to the twisted superpotential $\mathcal{W}$ for $2d$ SYM. This twisted superpotential as proposed in ref. [2] plays a role of the Young-Young function, which is a statement of *Bethe gauge correspondence.* This can be summarized by the following diagram:

$$\mathcal{Z}(\hat\Lambda, \mathbf{a}, \epsilon_1, \epsilon_2) \xrightarrow[\text{NS limit}]{\epsilon_2\to 0} \exp\left\{\frac{1}{\epsilon_2}\mathcal{W}(\hat\Lambda/\epsilon_1, \mathbf{a})\right\} \xrightarrow{(\epsilon_1,\epsilon_2)\to(0,0)} \exp\left\{\frac{1}{\epsilon_1\epsilon_2}\mathcal{F}_{\rm SW}(\hat\Lambda, \mathbf{a})\right\}. \quad (3)$$

In particular, for $N=2$, a case we are the most concerned with in what follows, a moduli and its dual are determined by two periods i.e., $2\pi a = \Pi(A)$, $2\pi a_D = \Pi(B)$ that take the form of the action of the classical sine-Gordon model or 2-particle periodic Toda chain

$$\Pi(\Gamma) \equiv \oint_\Gamma P_0(\varphi)\,\mathrm{d}\varphi = \oint_\Gamma \sqrt{2(u-\hat\Lambda^2\cos\varphi)}\,\mathrm{d}\varphi, \quad \Gamma = A, B.$$

Its quantization leads to the Schrödinger equation for the sine-Gordon/2p Toda model

$$\left(-\frac{\epsilon_1^2}{2}\frac{\mathrm{d}^2}{\mathrm{d}\varphi^2} + \hat\Lambda^2\cos\varphi\right)\psi(\varphi) = u\,\psi(\varphi).$$

WKB solution to the above equation [8] provides a spectrum which coincides the one in eq. (1) (see also appendix C of [9] for direct computations)

$$\Pi(\Gamma) \xrightarrow{\epsilon_1=\text{`}\hbar\text{'}} \tilde\Pi(\Gamma) \equiv \oint_\Gamma P(\varphi,\epsilon_1)\,\mathrm{d}\varphi, \qquad \psi_{\rm WKB}(\varphi) = \exp\left\{\frac{i}{\epsilon_1}\int^{\varphi} P(\rho,\epsilon_1)\,\mathrm{d}\rho\right\}.$$

It also corresponds to the quantization of moduli $\mathsf{a}(a)$, which entails the quantization of Seiberg-Witten curves [10]. The twisted superpotential from middle exponent in eq. (3) is found due to the duality relation $\mathsf{a}_D(a) = (\partial\mathcal{W}/\partial\mathsf{a})(a)$.

The spectrum of the Mathieu operator can be also found from the deformed critical colored Young diagram [10, 9]. It has been shown that the Nekrasov partition function for $N_f \geqslant 0$ $\mathcal{N}=2$ $SU(N)$ SYM can be rewritten as a sum over profiles $f_{\mathbf{a},\mathbf{k}}(x|\epsilon_1,\epsilon_2)$ of the deformed, colored Young diagrams [11]. For the case in question $N_f=0, SU(2)$ it takes the form

$$\mathcal{Z}(\hat\Lambda, \mathbf{a}, \epsilon_1, \epsilon_2) = \sum_{f_{\mathbf{a},\mathbf{k}}} \exp\left\{-\tfrac{1}{4}\iint \mathrm{d}x\mathrm{d}y\, f''_{\mathbf{a},\mathbf{k}}(x|\epsilon_1,\epsilon_2)\gamma_{\epsilon_1,\epsilon_2}(x-y;\hat\Lambda) f''_{\mathbf{a},\mathbf{k}}(y|\epsilon_1,\epsilon_2)\right\}, \quad (4)$$

where $\mathbf{k} \equiv \{k_{\alpha,i}\}$, $\alpha = 1,2$; $i \in \mathbb{N}$ and $i > j \Rightarrow k_{\alpha,j} \geqslant k_{\alpha,i} \geqslant 0$ are colored partitions (Young diagrams) with the number of boxes $|\mathbf{k}| = \sum k_{\alpha,i}$. Making use of the argument in ref. [11] within the NS limit $\epsilon_2 \to 0$, $\mathbf{k} \to \infty$ and $\epsilon_2 \mathbf{k} = \boldsymbol{\omega} < \infty$ such that the entire sum in eq. (4) can be approximated by the path integral over sequences of nonpositive real numbers $\omega_{\alpha,i}$ [9]

$$\mathcal{Z}_{\text{inst}}(\Lambda, \mathbf{a}, \epsilon_1, \epsilon_2) \sim \int_{\mathsf{c}_0(\mathbb{R}_{\leqslant 0})^2} \prod_{\alpha,i} \mathrm{d}\omega_{\alpha,i} \, \exp\left\{-\frac{1}{\epsilon_2}\mathcal{H}[\rho_{\mathbf{a},\boldsymbol{\omega}}]\right\}, \qquad \rho_{\mathbf{a},\mathbf{k}} := f_{\mathbf{a},\mathbf{k}} - f_{\mathbf{a},\varnothing}.$$

where $\mathsf{c}_0(\mathbb{R}_{\leqslant 0})^2$ is a space of such a sequences. This path integral is dominated by the term that fulfills the saddle point condition

$$\left(\delta\mathcal{H}[\rho_{\mathbf{a},\boldsymbol{\omega}*}]/\delta\rho^{\mathrm{i}}_{\mathbf{a},\boldsymbol{\omega}}\right)\delta\rho^{\mathrm{i}}_{\mathbf{a},\boldsymbol{\omega}} = 0 \quad \Rightarrow \quad \hat{\Lambda}^4 \prod_{\beta=1}^{2}\prod_{j\geqslant 1} \frac{x_{\alpha,i} - x_{\beta,j} - \epsilon_1}{x_{\alpha,i} - x_{\beta,j} + \epsilon_1} = -1,$$

where $x_{\alpha,i} := a_\alpha + \epsilon_1(i-1) + \omega_{\alpha,i}$, $\omega_{\alpha,i} = k_{\alpha,i}\epsilon_2$. The iterative solution to the above saddle point equation [10] yields columns $\overset{*}{\omega}_{\alpha,i}$ of the critical Young diagram $\boldsymbol{\omega}_*$. The twisted $SU(2)$ superpotential in terms of columns of the critical colored Young diagram reads

$$\mathcal{W}_{\text{inst}}\left(\hat{\Lambda}, \mathbf{a}, \epsilon_1\right) = \mathcal{H}[\rho_{\mathbf{a},\boldsymbol{\omega}*}]\left(\hat{\Lambda}, \epsilon_1\right) = \sum_{i\geqslant 1}(\hat{\Lambda}/\epsilon_1)^{4i}\, W_i(a,\epsilon_1)/i\,, \quad W_i = -\sum_{\alpha=1}^{2}\sum_{j=1}^{i} \overset{*}{\omega}_{\alpha,j,i}\,.$$

By means of the Bethe/gauge correspondence

$$\mathcal{E}_k = \langle \mathrm{tr}\phi^k \rangle_{\mathbf{a}}\big|_{\epsilon_2=0} = \tfrac{1}{2}\int \mathrm{d}x\, x^k f''_{\mathbf{a},\boldsymbol{\omega}*}(x|\epsilon_1),$$

one finds the spectrum for the two particle periodic Toda chain

$$\mathcal{E}_1 = 0, \quad \mathrm{E} = 2\mathcal{E}_2 = \epsilon_1\hat{\Lambda}\partial_{\hat{\Lambda}}\mathcal{W}(\hat{\Lambda}, \mathbf{a}, \epsilon_1) = \epsilon_1^2\lambda,$$

with λ being a spectrum of Mathieu operator given in eq. (2).

2. Nonconformal limit of AGT relation and Mathieu equation

Alday, Gaiotto and Tachikawa have discovered the correspondence between fourdimensional $\mathcal{N} = 2, N_f = 4$ $SU(2)$ field theories and the twodimensional Liouville field theory [12]. This henceforth called AGT correspondence enabled Gaiotto the discovery of entirely new objects within the twodimensional Conformal Field Theory (2d CFT) [13]. These new states – the Gaiotto states – turned out to be the non-conformal limit of the AGT correspondence [14]. The norm of the Gaiotto state with only one parameter Λ corresponds to the Nekrasov partition function for pure gauge ($N_f = 0$) $\mathcal{N} = 2$ SYM with $SU(2)$ symmetry.

2.1. Quantum irregular conformal block

Let us consider the four-point conformal block on twodimensional Riemann sphere.[1] It takes the form

$$\mathcal{F}_{c,\Delta}\left[\begin{smallmatrix}\Delta_3 & \Delta_2\\ \Delta_4 & \Delta_1\end{smallmatrix}\right](x) = x^{\Delta-\Delta_2-\Delta_1}\tilde{\mathcal{F}}_{c,\Delta}\left[\begin{smallmatrix}\Delta_3 & \Delta_2\\ \Delta_4 & \Delta_1\end{smallmatrix}\right](x),$$

[1] The extended exposition of the reasoning in this subsection can be found in ref. [14].

where Δ_i, Δ for $i = 1,\ldots,4$ are four external and one intermediate weight respectively. c denotes the central charge of the Virasoro algebra $\mathfrak{Vir}$. The 4pt conformal block has the following expansion in terms of location x not fixed by $SL(2,\mathbb{C})$ symmetry

$$\tilde{\mathcal{F}}_{c,\Delta}\begin{bmatrix}\Delta_3 & \Delta_2\\ \Delta_4 & \Delta_1\end{bmatrix}(x) := 1 + \sum_{n\geqslant 1} x^n \mathcal{F}^{(n)}_{c,\Delta}\begin{bmatrix}\Delta_3 & \Delta_2\\ \Delta_4 & \Delta_1\end{bmatrix}, \quad \mathcal{F}^{(n)}_{c,\Delta}\begin{bmatrix}\Delta_3 & \Delta_2\\ \Delta_4 & \Delta_1\end{bmatrix} := \sum_{I,J\vdash n} \gamma_\Delta\left[{}^{\Delta_3}_{\Delta_4}\right]_I \left[G_{c,\Delta}\right]^{IJ} \gamma_\Delta\left[{}^{\Delta_2}_{\Delta_1}\right]_J,$$

where $I, J \vdash n$ i.e., are partitions of n. $G_{c,\Delta}$ denotes the Gram matrix of scalar products between vectors in the highest weight representation of Virasoro algebra of generators $L_i \in \mathfrak{Vir}$, $L_{-i} = L_i^\dagger$, $L_I \equiv L_{k_1}\cdots L_{k_{\ell(I)}}$, $\mathcal{V}^{(n)}_{c,\Delta} = \mathrm{span}\{L_{-I}|\Delta\rangle;\ I\vdash n\}$

$$\left[G^{(n)}_{c,\Delta}\right]_{IJ} := \langle\Delta|L_I L_{-J}|\Delta\rangle, \qquad \left[G^{(n)}_{c,\Delta}\right]_{IK}\left[G^{(n)}_{c,\Delta}\right]^{KJ} = \delta^J_K. \tag{5}$$

The object γ_I – the vertex vector – takes the form

$$\gamma_\Delta\left[{}^{\Delta_a}_{\Delta_b}\right]_I \equiv \prod_{i=1}^{\ell(I)}\left(\Delta + k_i\Delta_b - \Delta_a + \sum_{i<j}^{\ell(I)} k_j\right), \quad I = (k_1,\ldots,k_{\ell(I)},0,\ldots). \tag{6}$$

According to AGT correspondence the relationship between the parameters of the two theories for conformal weights parametrized as

$$\Delta_i = (Q^2 + p_i^2)/4,\ \Delta = (Q^2+p^2)/4 \qquad \text{with} \qquad c = 1+6Q^2,\ Q = b + b^{-1} \tag{7}$$

reads

$$p_1 \overset{\text{AGT}}{=} \mu_1 - \mu_2, \quad p_2 \overset{\text{AGT}}{=} \epsilon_1+\epsilon_2 - (\mu_1+\mu_2)/2, \quad p_3 \overset{\text{AGT}}{=} \epsilon_1+\epsilon_2-(\mu_3+\mu_4)/2, \quad p_4 \overset{\text{AGT}}{=} \mu_3-\mu_4,$$
$$p \overset{\text{AGT}}{=} \sqrt{-1}a/(\epsilon_1\epsilon_2), \quad b \overset{\text{AGT}}{=} \sqrt{-\epsilon_2/\epsilon_1}.$$

Let us consider the so-called *mass decoupling limit.* This is the case when all the four masses related to the four external weights by the above relation tend to infinity $\mu_i \to \infty \Rightarrow \Delta_i \to \infty$, $\Delta < \infty$ while $x\mu_1\mu_2\mu_3\mu_4 = \hat{\Lambda}^4$. In this case one obtains

$$\gamma_\Delta\left[{}^{\Delta_a}_{\Delta_b}\right]_I \overset{\mu_a,\mu_b\gg 1}{\sim} \left(-\frac{\mu_a\mu_b}{\epsilon_1\epsilon_2}\right)^{\sum m_i} \leqslant \left(-\frac{\mu_a\mu_b}{\epsilon_1\epsilon_2}\right)^{|I|}\delta_{I,(1^{|I|})},$$

where $I = \{k_i\}_{i\in\mathbb{N}} = \{i^{m_i}\}_{i\in\mathbb{N}}$, $I\vdash n$ and m_i is a multiplicity of i in partition I. Noting, that $|I| = n = \sum k_i = \sum i m_i$ and that $\ell(I) = \sum m_i \leqslant \sum i m_i = |I|$ we conclude that the above inequality is saturated provided $\ell(I) = \ell_{\max}(I) = |I|$. That is, within this limit only those vertex vectors contribute that have maximal lengths. The partition with maximal length is $(1^{|I|})$. Within the mass decoupling limit 4pt conformal block takes the form

$$\tilde{\mathcal{F}}_{c,\Delta}\begin{bmatrix}\Delta_3 & \Delta_2\\ \Delta_4 & \Delta_1\end{bmatrix}(x) \xrightarrow[x\mu_1\mu_2\mu_3\mu_4=\hat{\Lambda}^4]{\mu_1,\mu_2,\mu_3,\mu_4\to\infty} \mathcal{F}_{c,\Delta}(\Lambda) := \sum_{n\geqslant 0}\Lambda^{4n}\left[G_{c,\Delta}\right]^{(1^n)(1^n)}, \quad \Lambda = \frac{\hat{\Lambda}}{\epsilon_1\epsilon_2}. \tag{8}$$

$\mathcal{F}_{c,\Delta}(\Lambda)$ is *irregular pure gauge conformal block.* This conformal block can be built out of the Gaiotto states that are defined as

$$\mathcal{F}_{c,\Delta}(\Lambda) = \langle\Delta,\Lambda^2|\Delta,\Lambda^2\rangle, \quad |\Delta,\Lambda^2\rangle := \sum_{n\geqslant 0}\Lambda^{2n}\sum_{|I|=n}\left[G_{c,\Delta}\right]^{(1^n)I} L_{-I}|\Delta\rangle. \tag{9}$$

The action of the generators that form $\mathfrak{Vir}$ on the Gaiotto state is

$$L_0|\Delta,\Lambda^2\rangle = (\Delta + (\Lambda/2)\partial_\Lambda)|\Delta,\Lambda^2\rangle, \quad L_1|\Delta,\Lambda^2\rangle = \Lambda^2|\Delta,\Lambda^2\rangle, \quad L_n|\Delta,\Lambda^2\rangle = 0,\ n\geqslant 2.$$

The nonconformal limit of AGT relation relates, therefore, the pure gauge $4d$ $\mathcal{N}=2$ SYM with $SU(2)$ symmetry to the irregular pure gauge conformal block within $2d$ CFT.

2.2. Classical irregular conformal block

The classical limit within the $2d$ CFT Liouville theory consists of taking the limit $b \to 0$ in the correlation function or the quantum conformal block in which it can be decomposed. The dependence of 4pt conformal block on b is clear in parametrization (7). Within this limit

$$\Delta, c \to \infty, \ \Delta/c = \text{const.} \quad \Rightarrow \quad \Delta \sim \delta b^{-2}, \ c \sim 6b^{-2}. \tag{10}$$

The irregular conformal block within this limit is expected to behave as

$$\mathcal{F}_{c,\Delta}(\Lambda) = \sum_{n\geqslant 0} \left(\frac{\hat{\Lambda}}{\epsilon_1 b}\right)^{4n} [G_{c,\Delta}]^{(1^n)(1^n)} \overset{b\to 0}{\sim} \exp\left\{\frac{1}{b^2} f_\delta\left(\frac{\hat{\Lambda}}{\epsilon_1}\right)\right\}, \tag{11}$$

where $f_\delta(\hat{\Lambda}/\epsilon_1)$ is *the classical pure gauge irregular block.* This behavior is further corroborated by the leading order analysis in b^{-1} performed in [15]. Making use of the formula

$$f_\delta\left(\frac{\hat{\Lambda}}{\epsilon_1}\right) = \lim_{b\to 0} b^2 \log \mathcal{F}_{c,\Delta}\left(\frac{\hat{\Lambda}}{\epsilon_1 b}\right) = \sum_{n\geqslant 1} \left(\frac{\hat{\Lambda}}{\epsilon_1}\right)^{4n} f_\delta^{(n)},$$

and our *Mathematica* code we were able to find up to 7 of its coefficients five of which we present below

$$\begin{gathered}
f_\delta^{(1)} = \frac{1}{2\delta}, \qquad f_\delta^{(2)} = \frac{5\delta - 3}{16\delta^3(4\delta+3)}, \qquad f_\delta^{(3)} = \frac{9\delta^2 - 19\delta + 6}{48\delta^5(4\delta+3)(\delta+2)}, \\
f_\delta^{(4)} = \frac{5876\delta^5 - 16489\delta^4 - 22272\delta^3 + 17955\delta^2 + 9045\delta - 4050}{512\delta^7(\delta+2)(4\delta+3)^3(4\delta+15)}, \\
f_\delta^{(5)} = \frac{17884\delta^6 - 96187\delta^5 - 156432\delta^4 + 388737\delta^3 - 7317\delta^2 - 138348\delta + 34020}{1280\delta^9(\delta+2)(\delta+6)(4\delta+3)^3(4\delta+15)}, \quad \ldots
\end{gathered} \tag{12}$$

It was also possible to establish the relation between the twisted superpotential and the classical irregular block which reads ($\epsilon_2 \to 0, b \equiv \sqrt{-\epsilon_2/\epsilon_1} \Rightarrow b \to 0$)

$$f_\delta(\hat{\Lambda}/\epsilon_1) = \frac{1}{\epsilon_1} \mathcal{W}_{\text{inst}}(\hat{\Lambda}, a, \epsilon_1), \qquad \frac{1}{n\epsilon_1} W_n(a, \epsilon_1) = -f_\delta^{(n)} \quad \text{for} \quad \delta = \frac{1}{4} - \left(\frac{a}{\epsilon_1}\right)^2.$$

3. Mathieu equation from classical limit of null vector decoupling equation

It is possible to obtain the Mathieu equation directly from the 2d CFT without any reference to the AGT relation. It should be stressed that this is possible due to the discovery of Gaiotto states within $2d$ CFT. The Mathieu equation can be obtained by computing the matrix element of a chiral null field built out of a vertex operator with degenerated weight and inserted between the two different Gaiotto states and using *the null vector decoupling condition.* However, the null vector decoupling condition can be applied to the physical correlation functions with vertex operators having weights from the Kac table, while we use matrix element of the null field between the Gaiotto states which is purely chiral object. There is a variant of the null vector decoupling theorem that applies to the matrix elements with null chiral vertex operator, namely, the Feigin-Fuchs theorem

Theorem 1 (Feigin-Fuchs). *Let $i, j, k \in \{1, 2, 3\}$ be chosen such that $j \neq i$, $k \neq i$, $j \neq k$. Let us assume that*

(i) $\Delta_i = \Delta_{r,s} := \frac{Q^2}{4} - \frac{1}{4}\left(rb + sb^{-1}\right)^2, \quad r, s \in \mathbb{N};$

(ii) the vector $|\xi_i\rangle$ lies in the singular submodule generated by the null vector $|\chi_{rs}\rangle$, i.e.: $|\xi_i\rangle \in \mathcal{V}_{c,\Delta_{rs}(c)+rs} \subset \mathcal{V}_{c,\Delta_{r,s}(c)}$.

Then,

$$\langle\xi_1|V^{\Delta_3,\Delta_2,\Delta_1}_{z_3,\ z_2,\ z_1}(\xi_2|z_2)|\xi_3\rangle = 0,$$

iff $\Delta_j = \Delta\left(\sqrt{-1}\beta_j\right) := \frac{Q^2}{4} - \frac{1}{4}\beta_j^2$ *and* $\Delta_k = \Delta\left(\sqrt{-1}\beta_k\right) := \frac{Q^2}{4} - \frac{1}{4}\beta_k^2$ *satisfy the fusion rules*

$$\beta_j - \beta_k = pb + qb^{-1},$$

where $p \in \{1-r, 3-r, \ldots, r-1\}$ *and* $q \in \{1-s, 3-s, \ldots, s-1\}$.

Thus, the above theorem says that the matrix element with the null field insertion vanishes provided the weights of states from domain and target space of the null vertex operator fulfill appropriate fusion rules.

3.1. Null vector decoupling equation

Having Feigin-Fuchs theorem at hand we can proceed to the derivation of the Mathieu equation through the following construction. First, let us choose the vertex operator to have a weight $\Delta_{r,s}$ from the Kac table (see item *(i)* of Feigin-Fuchs theorem 1 for definition of $\Delta_{r,s}$ and eq. (7) for definition of Q) at the second level i.e., $rs = 2$. There are two weights from which we take

$$\Delta_+ \equiv \Delta_{2,1} = -\frac{1}{2} - \frac{3}{4}b^2.$$

The reason for this choice is that this weight is 'light', that is, in the classical limit $b \to 0$ it becomes constant (it is opposite to the 'heavy' weight $\Delta_{1,2}$ which diverges to infinity within this limit). It is expected that in this limit the light field insertions in the physical correlation function factorize from the heavy ones. The same should be true for the matrix elements with light field insertions. Hence, the degenerate primary chiral vertex operator takes the form

$$V_{\Delta_+}(z) \equiv V_+(z), \qquad V_+(z) := V^{\Delta'\,\Delta_+\,\tilde{\Delta}}_{\infty\ \ z\ \ 0}(|\,\Delta_+\rangle \otimes \cdot\,) : \mathcal{V}_{\tilde{\Delta}} \to \mathcal{V}_{\Delta'}, \quad |\Delta_+\rangle \in \mathcal{V}_{\Delta_+}.$$

Now we can take the null field as a second level descendant of $V_+(z)$, namely

$$\chi_+(z) = \left(\hat{L}_{-2}(z) - \frac{3}{2(2\Delta_+ + 1)}\hat{L}^2_{-1}(z)\right)V_+(z), \quad \hat{L}_{-k}(z) := \frac{1}{2\pi i}\oint\limits_{C_z} dw(w-z)^{1-k}T(w).$$

Moreover, we take the weights Δ', $\tilde{\Delta}$ to fulfill the following fusion rules

$$\Delta(\sigma) := \frac{Q^2}{4} - \sigma^2 \quad \Rightarrow \quad \tilde{\Delta} = \Delta\left(\sigma - \frac{b}{4}\right), \quad \Delta' = \Delta\left(\sigma + \frac{b}{4}\right).$$

Hence, according to the Feigin-Fuchs theorem 1 we obtain the null vector decoupling equation

$$\langle\Delta', \Lambda^2|\chi_+(z)|\tilde{\Delta}, \Lambda^2\rangle = \langle\Delta', \Lambda^2|\hat{L}_{-2}(z)V_+(z)|\tilde{\Delta}, \Lambda^2\rangle + \frac{1}{b^2}\langle\Delta', \Lambda^2|\hat{L}^2_{-1}(z)V_+(z)|\tilde{\Delta}, \Lambda^2\rangle = 0,$$

which, after the use of Conformal Ward Identities assumes the form

$$\left[\frac{1}{b^2}z^2\frac{\partial^2}{\partial z^2} - \frac{3z}{2}\frac{\partial}{\partial z} + \Lambda^2\left(z + \frac{1}{z}\right) + \frac{\Lambda}{4}\frac{\partial}{\partial\Lambda} + \frac{\tilde{\Delta} + \Delta' - \Delta_+}{2}\right]\Psi(\Lambda, z) \;=\; 0, \tag{13a}$$

$$\Psi(\Lambda, z) := \langle\Delta', \Lambda^2|V_+(z)|\tilde{\Delta}, \Lambda^2\rangle. \tag{13b}$$

Making use of the definition of the Gaiotto states (9) we can expand the function $\Psi(\Lambda, z)$ and examine its structure with respect to z dependence

$$\begin{aligned}\Psi(\Lambda, z) =& z^{\Delta'-\Delta_+-\tilde{\Delta}} \sum_{m,n\geqslant 0} \Lambda^{2(m+n)} z^{m-n} \sum_{\substack{|I|=m\\|J|=n}} \left[G^m_{c,\Delta'}\right]^{(1^m)I} \langle \Delta', I \,|\, V_+(1)\,|\,\tilde{\Delta}, J\rangle \left[G^n_{c,\tilde{\Delta}}\right]^{J(1^n)} \\ :=& z^{\kappa}\Phi(\Lambda, z), \qquad \kappa \equiv \Delta' - \Delta_+ - \tilde{\Delta}\,. \end{aligned} \tag{14}$$

Let us note that $\Phi(\Lambda, z)$ when separated into the 'diagonal' and 'off-diagonal' parts in m, n the former does not depend on z leaving the total z dependence to the latter. Indeed, for

$$\Phi(\Lambda, z) = \Phi^{(m=n)}(\Lambda) + \Phi^{(m\neq n)}(\Lambda, z), \tag{15}$$

we obtain

$$\begin{aligned}\Phi^{(m=n)}(\Lambda) &= \sum_{n\geqslant 0} \Lambda^{4n} \sum_{I\vdash n} \left[G^n_{c,\Delta'}\right]^{(1^n)I} \langle\Delta'|L_I V_+(1) L_{-I}|\tilde{\Delta}\rangle \left[G^n_{c,\tilde{\Delta}}\right]^{I\,(1^n)}, \\ \Phi^{(m\neq n)}(\Lambda, z) &= \sum_{\substack{m\neq n\\ m,n\geqslant 0}} \Lambda^{2(m+n)} z^{m-n} \sum_{\substack{I\vdash m\\ J\vdash n}} \left[G^m_{c,\Delta'}\right]^{(1^m)\,I} \langle\Delta'|L_I V_+(1) L_{-J}|\tilde{\Delta}\rangle \left[G^n_{c,\tilde{\Delta}}\right]^{J\,(1^n)}. \end{aligned}$$

Let us note that $\Phi(\Lambda, z)$ can also be written in the factor form, namely

$$\Psi(\Lambda, z) = z^{\kappa} \exp\left\{\log \Phi^{(m=n)}(\Lambda)\right\} \left(1 + \frac{\Phi^{(m\neq n)}(\Lambda, z)}{\Phi^{(m=n)}(\Lambda)}\right) \equiv z^{\kappa}\, \mathrm{e}^{\phi(\Lambda)}\, \zeta(\Lambda, z). \tag{16}$$

This observation is crucial for the 'light' field factorization phenomenon in the classical limit which we consider below.

3.2. Classical limit of Null Vector Decoupling Equation

The Null Vector Decoupling equation, after we express $\Psi(\Lambda, z)$ in terms of $\Phi(\Lambda, z)$ as in eq. (14), assumes the form (recall that $\Lambda = \hat{\Lambda}/\epsilon_1 b$)

$$\begin{aligned}&\left[\frac{1}{b^2} z^2\partial_z^2 + \left(\frac{2\kappa}{b^2} - \frac{3}{2}\right) z\partial_z + \frac{\hat{\Lambda}}{4}\partial_{\hat{\Lambda}} \right.\\ &\left. \qquad + \frac{\kappa(\kappa-1)}{b^2} - \frac{3\kappa}{2} + \frac{\hat{\Lambda}^2}{b^2\epsilon_1^2}\left(z + \frac{1}{z}\right) + \frac{\tilde{\Delta} + \Delta' - \Delta_+}{2}\right] \Phi\left(\frac{\hat{\Lambda}}{\epsilon_1 b}, z\right) = 0. \end{aligned} \tag{17}$$

The parameters in the above equations have the following behavior in the classical limit $b \to 0$:

$$\begin{aligned}&\sigma = \xi/b \quad \Rightarrow \quad \Delta', \tilde{\Delta} \overset{b\to 0}{\sim} \frac{1}{b^2}\delta \qquad \tilde{\Delta} + \Delta' - \Delta_+ \overset{b\to 0}{\sim} \frac{1}{b^2} 2\delta, \quad \text{where} \quad \delta = \frac{1}{4} - \xi^2, \\ &\kappa \xrightarrow{b\to 0} \frac{1}{2} - \xi, \qquad \kappa\,(\kappa - 1) \xrightarrow{b\to 0} -\left(\frac{1}{4} - \xi^2\right) = -\delta \qquad \Delta_+ \xrightarrow{b\to 0} -\frac{1}{2}. \end{aligned} \tag{18}$$

The function Φ also depends on b in its first argument. As we have already mentioned in the beginning of subsection 3.1 it is expected that the physical correlation function that has 'light' fields and 'heavy' fields, in the classical limit, factorizes into some function that depends on locations of the 'light' fields and exponentially divergent reminder of the 'heavy' fields. This factorization phenomenon, sometimes termed the "Zamolodchikov's ansatz", is believed to carry over to case of purely chiral objects like the matrix element defining Ψ function in eq. (13b). Taking into account our earlier observation phrased in eq. (16) the chiral counterpart of the Zamolodchikov's ansatz can be formulated as follows

$$\Phi(\Lambda,z) = \mathrm{e}^{\phi(\Lambda)}\,\zeta(\Lambda,z) = z^{-\kappa}\langle\,\Delta',\Lambda^2\,|\,V_+(z)\,|\,\tilde{\Delta},\Lambda^2\,\rangle \overset{b\to 0}{\sim} v(\hat{\Lambda}/\epsilon_1,z)\,\exp\left\{\frac{1}{b^2}f_\delta(\hat{\Lambda}/\epsilon_1)\right\}, \tag{19a}$$

where

$$f_\delta(\hat{\Lambda}/\epsilon_1) = \lim_{b\to 0} b^2\phi\left(\frac{\hat{\Lambda}}{\epsilon_1 b}\right), \quad v(\hat{\Lambda}/\epsilon_1,z) = \lim_{b\to 0}\zeta\left(\frac{\hat{\Lambda}}{\epsilon_1 b},z\right). \tag{19b}$$

The factorization of matrix element in the form of eq. (19a) has been recently proved up to the leading order in b^{-1} in our paper [15]. Using the factorization ansatz (19b) and asymptotic form of parameters given in eq. (18) the Null Vector Decoupling equation (17) takes its classical form independent of b

$$\left[\,z^2\partial_z^2 + 2\left(\tfrac{1}{2}-\xi\right)z\partial_z + \frac{\hat{\Lambda}^2}{\epsilon_1^2}\left(z+\frac{1}{z}\right) + \frac{\hat{\Lambda}}{4}\partial_{\hat{\Lambda}}f_\delta(\hat{\Lambda}/\epsilon_1)\right]v(\hat{\Lambda}/\epsilon_1,z) \;=\; 0\,.$$

Substituting $v(\hat{\Lambda}/\epsilon_1,z) = z^\xi\,\psi(\hat{\Lambda}/\epsilon_1,z)$ and subsequently changing variable $z = \mathrm{e}^{2ix}$ we arrive at the sought form of classical limit of the Null Vector Decoupling equation, that can be identified with the Mathieu equation (1)

$$\left[-\frac{\mathrm{d}^2}{\mathrm{d}x^2} + 8\frac{\hat{\Lambda}^2}{\epsilon_1^2}\cos 2x + \hat{\Lambda}\,\partial_{\hat{\Lambda}}f_\delta(\hat{\Lambda}/\epsilon_1) - 4\xi^2\right]\psi\left(\hat{\Lambda}/\epsilon_1,\mathrm{e}^{2ix}\right) \;=\; 0. \tag{20}$$

This identification becomes clear when we relate the Mathieu parameters with the parameters of the above equation,

$$\lambda \;=\; -\hat{\Lambda}\,\partial_{\hat{\Lambda}}f_\delta\left(\hat{\Lambda}/\epsilon_1\right) + 4\xi^2, \qquad h \;=\; \pm\frac{2\hat{\Lambda}}{\epsilon_1}, \qquad \xi \;=\; \frac{\nu}{2}.$$

Using coefficients of classical irregular pure gauge block expansion found by means of our *Mathematica* code and presented in eq. (12) one finds the coincidence with h expansion of the spectrum of Mathieu operator λ given in eq. (2),

$$\begin{aligned}
\lambda &= -\,\hat{\Lambda}\,\partial_{\hat{\Lambda}}\left[\sum_{n=1}\left(\hat{\Lambda}/\epsilon_1\right)^{4n} f_\delta^n\right] + 4\xi^2 \\
&= -\,\frac{4h^4}{16}\,f^1_{\frac{1}{4}-\frac{\nu^2}{4}} - \frac{8h^8}{256}\,f^2_{\frac{1}{4}-\frac{\nu^2}{4}} - \frac{12h^{12}}{4096}\,f^3_{\frac{1}{4}-\frac{\nu^2}{4}} - \ldots + 4\left(\frac{\nu^2}{4}\right) \\
&= \nu^2 + \frac{h^4}{2\left(\nu^2-1\right)} + \frac{\left(5\nu^2+7\right)h^8}{32\left(\nu^2-4\right)\left(\nu^2-1\right)^3} + \frac{\left(9\nu^4+58\nu^2+29\right)h^{12}}{64\left(\nu^2-9\right)\left(\nu^2-4\right)\left(\nu^2-1\right)^5} + \ldots\,.
\end{aligned}$$

3.3. Mathieu functions

The direct computations of the limit

$$v(\hat{\Lambda}/\epsilon_1,z) = \lim_{b\to 0}\left(1 + \frac{\Phi^{(m\neq n)}(\Lambda,z)}{\Phi^{(m=n)}(\Lambda)}\right), \qquad z = \mathrm{e}^{2ix},$$

by means of our *Mathematica* code confirmed its finiteness up to $(\hat{\Lambda}/\epsilon_1)^{12}$ order. The result of the computations up to $(\hat{\Lambda}/\epsilon_1)^6$ order reads

$$\begin{aligned} v(\hat{\Lambda}/\epsilon_1, \mathrm{e}^{2ix}) &= 1 + \frac{\hat{\Lambda}^2}{\epsilon_1^2}\left(\frac{\mathrm{e}^{-2ix}}{2\xi-1} - \frac{\mathrm{e}^{2ix}}{2\xi+1}\right) + \frac{\hat{\Lambda}^4}{\epsilon_1^4}\left(\frac{e^{4ix}}{4(\xi+1)(2\xi+1)} + \frac{e^{-4ix}}{4(\xi-1)(2\xi-1)}\right) \\ &\frac{\hat{\Lambda}^6}{\epsilon_1^6}\left(\frac{e^{-2ix}\left(4\xi^2-8\xi+7\right)}{4(\xi-1)(2\xi-1)^3(2\xi+1)} - \frac{e^{2ix}\left(4\xi^2+8\xi+7\right)}{4(\xi+1)(2\xi-1)(2\xi+1)^3}\right. \\ &\left. + \frac{e^{-6ix}}{12(\xi-1)(2\xi-3)(2\xi-1)} - \frac{e^{6ix}}{12(\xi+1)(2\xi+1)(2\xi+3)}\right) + \mathcal{O}\left(\frac{\hat{\Lambda}^8}{\epsilon_1^8}\right). \end{aligned}$$

The relationship between v the Mathieu exponential me_ν is found to be

$$\psi(h/2, z) = \mathrm{e}^{i\nu x} v(h/2, z) = \mathrm{me}_\nu(x, h),$$

where

$$\begin{aligned} \mathrm{me}_\nu(x,h) &= \mathrm{e}^{i\nu x} + \frac{h^2}{4}\left(\frac{\mathrm{e}^{(\nu-2)ix}}{\nu-1} - \frac{\mathrm{e}^{(\nu+2)ix}}{\nu+1}\right) + \frac{h^4}{32}\left(\frac{\mathrm{e}^{(\nu+4)ix}}{(\nu+1)(\nu+2)} + \frac{\mathrm{e}^{(\nu-4)ix}}{(\nu-2)(\nu-1)}\right) \\ &+ \frac{h^6}{128}\left(\frac{\left(\nu^2-4\nu+7\right)\mathrm{e}^{(\nu-2)ix}}{(\nu-2)(\nu-1)^3(\nu+1)} - \frac{\left(\nu^2+4\nu+7\right)\mathrm{e}^{(\nu+2)ix}}{(\nu-1)(\nu+1)^3(\nu+2)}\right. \\ &\left. + \frac{\mathrm{e}^{(\nu-6)ix}}{3(\nu-3)(\nu-2)(\nu-1)} - \frac{\mathrm{e}^{(\nu+6)ix}}{3(\nu+1)(\nu+2)(\nu+3)}\right) + \mathcal{O}(h^8). \qquad (21) \end{aligned}$$

Interestingly, the comparison of the above result with the one in literature (NIST Digital Library of Mathematical Functions http://dlmf.nist.gov/28.15.E3) reveals the discrepancy between the two formulas at h^2 order of h expansion of Mathieu exponential ($q \equiv h^2$),

$$\begin{aligned} \mathrm{me}_\nu(z,h) &= e^{i\nu z} + \frac{q}{4}\left(\frac{e^{i(\nu-2)z}}{\nu-1} - \frac{e^{i(\nu+2)z}}{\nu+1}\right) \\ &+ \frac{q^2}{32}\left(\frac{e^{i(\nu+4)z}}{(\nu+1)(\nu+2)} + \frac{e^{i(\nu-4)z}}{(\nu-1)(\nu-2)} \boxed{-\frac{2(\nu^2+1)}{(\nu^2-1)^2}e^{i\nu z}}\right) + \mathcal{O}(q^4). \qquad (22) \end{aligned}$$

We believe that the above formula in eq. (22) taken from NIST Digital Library of Mathematical Functions is incorrect as it does not obey the formula for the Mathieu sinus function (http://dlmf.nist.gov/28.12.iii)

$$\mathrm{se}_\nu(x,h^2) = \tfrac{i}{2}\left(\mathrm{me}_\nu(x,h^2) - \mathrm{me}_\nu(-x,h^2)\right).$$

4. Conclusions and outlook

The study presented above shows that the classical irregular conformal block is related to the spectrum of the Mathieu operator. Due to the discovery of the Gaiotto states it is possible to derive the Mathieu equation entirely within 2*d* CFT, without any reference to AGT relation. Moreover, taking into account the studies of WKB quantization of Seiber-Witten curves [8, 10] as it was mentioned in the Introduction we have also found a kind of classical AGT relation between the classical irregular conformal block and 2d twisted superpotential. This in turn constitutes the explicit test of the triple correspondence: 2dCFT/$\mathcal{N} = 2$, $SU(2)$ gauge/2-particle QIS.

Journal of Physics: Conference Series **670** (2016) 012041 doi:10.1088/1742-6596/670/1/012041

Furthermore, the quantum pure gauge irregular block and its classical counterpart studied here are merely an examples of much broader branch of possibilities to find a new equations along with their solutions as there are more irregular blocks for $N_f > 0$. As an example let us quote our last result for $N_f = 2$ case. The procedure analogous to the one described above one can find a generalized Mathieu equation,

$$\left[-\frac{\mathrm{d}^2}{\mathrm{d}x^2} + \frac{1}{2}h^2\cos 4x + 4h\mu\cos 2x\right]\psi^{\mathbf{2}} \;=\; \lambda_2\,\psi^{\mathbf{2}}\,,$$

along with its spectrum λ_2 given by the formula

$$\lambda_2 = \nu^2 - 2h\frac{\partial}{\partial h}f_\delta^{\mathbf{2}}\left(\tfrac{1}{2}h,\mu,\mu\right)\,,$$

where $f_\delta^{\mathbf{2}}$ is $N_f = 2$ classical irregular block. This work is in progress.

References

[1] Mathieu E 1868 *Jour. de Math. Pures et Appliquees (Jour. de Liouville)* **13** 137
[2] Nekrasov N A and Shatashvili S L 2009 Quantization of Integrable Systems and Four Dimensional Gauge Theories (*Preprint* 0908.4052v1) URL http://arxiv.org/abs/0908.4052v1
[3] Nekrasov N A 2004 *Adv.Theor.Math.Phys.* **7** 831–864
[4] Seiberg N and Witten E 1994 *Nucl.Phys.* **B426** 19–52 (*Preprint* hep-th/9407087)
[5] Argyres P C and Faraggi A E 1995 *Phys. Rev. Lett.* **74** 3931–3934 (*Preprint* hep-th/9411057)
[6] Gorsky A, Krichever I, Marshakov A, Mironov A and Morozov A 1995 *Phys.Lett.* **B355** 466–474 (*Preprint* hep-th/9505035)
[7] Itoyama H and Morozov A 1996 *Nucl.Phys.* **B477** 855–877 (*Preprint* hep-th/9511126)
[8] Mironov A and Morozov A 2010 *JHEP* **1004** 040 (*Preprint* 0910.5670)
[9] Piatek M and Pietrykowski A R 2014 *JHEP* **12** 032 (*Preprint* 1407.0305)
[10] Poghossian R 2011 *JHEP* **1104** 033 (*Preprint* 1006.4822)
[11] Nekrasov N and Okounkov A (*Preprint* hep-th/0306238v2)
[12] Alday L F, Gaiotto D and Tachikawa Y 2010 *Lett.Math.Phys.* **91** 167–197 (*Preprint* 0906.3219)
[13] Gaiotto D 2009 (*Preprint* 0908.0307)
[14] Marshakov A, Mironov A and Morozov A 2009 *Phys. Lett.* **B682** 125–129 (*Preprint* 0909.2052)
[15] Piatek M and Pietrykowski A R 2015 (*Preprint* 1509.08164)

Journal of Physics: Conference Series **670** (2016) 012042 doi:10.1088/1742-6596/670/1/012042

Analytical simulations of double-well, triple-well and multi-well dynamics via rationally extended Harmonic oscillator

Andrey M. Pupasov-Maksimov
Depto. de Matemática, ICE,
Universidade Federal de Juiz de Fora, MG, Brasil
E-mail: pupasov.maksimov@ufjf.edu.br

Abstract. Quantum tunnelling in two-well, triple-well and multi-well potentials is studied analytically with the aid of exact propagators $K^\sigma(x, y; t)$ corresponding to rational extensions $H^\sigma = H_{\rm osc} + \Delta V^\sigma$ of the Harmonic oscillator.

Introduction

One of the main reason to develop new sophisticated analytical models is to use them as tests for the various approximations or as the starting point for perturbation methods. Due to the tunnelling, quantum evolution in the multi-well potentials becomes very complicated. Moreover, it is difficult to use perturbation theory or WKB approach. For instance, calculating path integral for a symmetric two-well potential one has to take into account instanton contributions [1].

The aim of this paper is to present the well-known rationally extended Harmonic oscillators as suitable analytical models for the description of quantum evolution in multi-well potentials. In this respect we follow the idea of works [2, 3], where wave packet dynamics and tunnelling for the oscillator's polynomial SUSY partners were studied with the aid of SUSY QM relations between propagators. We will show that propagators of monodromy free rational extensions of the Harmonic oscillator are expressed in the elementary functions only.

An important property of rational extensions is related with the analytical structure of wave-functions in the complex plane. There is a fundamental theorem which states that all monodromy free rational extensions of the Harmonic oscillator are constructed by a finite chain of level-removing Darboux transformations [4]. Hence, the general form of the corresponding Hamiltonian reads

$$H^\sigma = -\partial^2_{xx} + \frac{x^2}{4} - 2\partial^2_{xx}(\ln \mathrm{Wr}[\mathrm{He}_\sigma(x), x]) + 2M\,, \tag{1}$$

where σ is a string of increasing integers which label removed energy levels and

$$\mathrm{He}_\sigma(x) = \mathrm{He}_{\{n_1,n_2,\ldots,n_{2M}\}}(x) = \{\mathrm{He}_{n_1}(x), \mathrm{He}_{n_2}(x), \ldots, \mathrm{He}_{n_{2M}}(x)\}\,.$$

Our suggestion is based on the following analytical structure of the corresponding propagators

$$K^\sigma(x,y;t) = K_{\rm osc}(x,y;t)\frac{\sum\limits_{k=0}^{\sigma[[-1]]+1} Q_k^\sigma(x,y)\mathrm{e}^{-\mathrm{i}kt}}{\sum\limits_{k=0}^{\sigma[[-1]]+1} Q_k^\sigma(x,y)} \tag{2}$$

where symmetric polynomials $Q_k^\sigma(x,y) = Q_k^\sigma(y,x)$ can be determined iteratively by a finite number of steps. These polynomials provide a non-linear connection between Hermite and exceptional Hermite polynomials.

Despite to the fact that rational extensions of harmonic oscillator are well studied, some important questions still wait an answers. For instance, it worth to know how the shape of a potential (the number and the structure of wells and barriers) depends on the parameters of extension (i.e. on the sequence σ of deleted levels). Regarding the shape of potentials V^σ, it is known that $V^{\{k,k+1\}}$ is a k-well potential [5]. This "landscape" problem for the rational extensions of the Harmonic oscillator is closely related to the pole configurations of the corresponding potentials, which are defined by zeros of wronskians $\mathrm{Wr}[\mathrm{He}_\sigma(x), x]$. As it was shown in [6] these zeros obey the Calogero relations. Some recent advantages in this direction come from studies of "locus" (the term introduced by Airault, McKean and Moser) [7, 8]. In particular, for a certain class of sequences σ, a simple qualitative relation is observed between the shape of the corresponding Young diagram and the pattern of zeros of the Wronskian of the corresponding Hermite polynomials. It is interesting that zeros of exceptional Hermite polynomials define vortex configurations for quadrupole background flow [8].

The paper is organized as follows. In the first section we present a non-linear connection formula between eigen-projectors of the Harmonic oscillator and its rational extensions. Applying this formula to the propagator expansion we obtain analytical expressions for propagators K^σ. In the next sections we will analyse two-well and triple-well potentials as well the corresponding propagators. In a conclusion possible applications are discussed.

1. Rationally extended propagators

1.1. Non-linear connection coefficients

Consider Harmonic oscillator and its (monodromy free) rational extension

$$H_{\rm osc}\psi_n(x) = \left(n+\frac{1}{2}\right)\psi_n(x)\,, \qquad H^\sigma\psi_n^\sigma(x) = \left(n+\frac{1}{2}\right)\psi_n^\sigma(x)\,. \tag{3}$$

First of all, two Hamiltonians are almost isospectral, that is $\mathrm{spec}H^\sigma \subset \mathrm{spec}H_{\rm osc}$ and $\mathrm{spec}H_{\rm osc}\setminus\mathrm{spec}H^\sigma = \sigma$.

It is well known that bound state wave functions for the rationally extended Harmonic oscillator

$$\psi_n^\sigma(x) = \frac{N_n\mathrm{He}_n^\sigma(x)}{W(x)}\mathrm{e}^{\frac{-x^2}{4}} = \frac{h_n^\sigma(x)}{W(x)}\mathrm{e}^{\frac{-x^2}{4}}\,, \tag{4}$$

are related with the oscillator bound state wave functions

$$\psi_n(x) = p_n\mathrm{He}_n(x)\mathrm{e}^{-\frac{x^2}{4}}\,, \qquad p_n = \left(n!\sqrt{2\pi}\right)^{-\frac{1}{2}}\,, \tag{5}$$

by a Darboux transformation. Here and in what follows, together with standard (probabilistic) Hermite polynomials $\mathrm{He}_n(x)$ and exceptional Hermite polynomials

$$\mathrm{He}_n^\sigma(x) = \mathrm{Wr}[\mathrm{He}_{\sigma\cup\{n\}}(x), x]\,, \tag{6}$$

we will use their normalized versions $h_n(x)$ and $h_n\sigma(x)$

$$h_n(x) = p_n \mathrm{He}_n(x)\,, \tag{7}$$
$$h_n^\sigma(x) = N_n \mathrm{Wr}[h_{\sigma\cup\{n\}}(x), x] \tag{8}$$

where a normalization factor reads

$$N_n = \begin{cases} \left(\prod_{j=1}^{2M}(n - \sigma[[j]])\right)^{-\frac{1}{2}}, & n \notin \sigma\,, \\ 0\,, & n \in \sigma\,. \end{cases} \tag{9}$$

Finally, the compact notations for the following Wronskians are in order

$$W(x) = \mathrm{Wr}[\mathrm{He}_\sigma(x), x]\,, \qquad w(x) = \mathrm{Wr}[\mathrm{h}_\sigma(x), x]\,. \tag{10}$$

Recently we discovered an alternative possibility to express exceptional Hermite polynomials from standard ones using

Non-linear connection Lemma [9]. Given a Krein-Adler sequence $\sigma = \{k_1, k_1+1\ldots, k_M, k_M+1\}$, the corresponding family of (formally normalized) exceptional Hermite polynomials $h_n^\sigma(x)$ obeys the following relation

$$\sum_{k=0}^{\sigma[[-1]]+1} h_{m-k}(x)h_{m-k}(y)Q_k^\sigma(x,y) = h_m^\sigma(x)h_m^\sigma(y)\,, \tag{11}$$

where symmetric connection polynomials Q_k^σ are given by

$$Q_k^\sigma(x,y) = \frac{1}{h_0(x)h_0(y)}\left(h_k^\sigma(x)h_k^\sigma(y) - \sum_{j=1}^{k} Q_{k-j}^\sigma(x,y)h_j(x)h_j(y)\right), \quad 0 \le k \le \sigma[[-1]]+1\,, \tag{12}$$

and sum of connection polynomials is equal to product of Wronskians $w(x)w(y)$ defined above in (10)

$$\sum_{k=0}^{\sigma[[-1]]+1} Q_k^\sigma(x,y) = w(x)w(y)\,. \tag{13}$$

The proof of this lemma was given in [9].

It is also convenient to rewrite these relations in terms of the corresponding wave functions. Introducing projectors

$$\Pi_n(x,y) = \psi_n(x)\psi_n(y)\,, \qquad \Pi_n^\sigma(x,y) = \psi_n^\sigma(x)\psi_n^\sigma(y)\,, \tag{14}$$

we write (11) as follows

$$\frac{\sum_{k=0}^{\sigma[[-1]]+1} \Pi_{m-k}(x,y)Q_k^\sigma(x,y)}{w(x)w(y)} = \Pi_m^\sigma(x,y)\,. \tag{15}$$

There are also several equivalent representations of non-linear connection lemma in a form of some matrix equations. For instance, $\sigma[[-1]]+2$ dimensional vector $\vec{Q}^\sigma$ can be defined as a solution of the following matrix equation

$$\begin{pmatrix} \Pi_0 & 0 & 0 & \dots & 0 \\ \Pi_1 & \Pi_0 & 0 & \dots & 0 \\ \Pi_2 & \Pi_1 & \Pi_0 & 0 & 0 \\ \dots & \dots & \dots & \dots & 0 \\ \Pi_{2M+1} & \Pi_{2M} & \dots & \dots & \Pi_0 \end{pmatrix} \begin{bmatrix} Q_0^\sigma \\ Q_1^\sigma \\ Q_2^\sigma \\ \dots \\ Q_{2M+1}^\sigma \end{bmatrix} = w(x)w(y) \begin{bmatrix} \Pi_0^\sigma \\ \Pi_1^\sigma \\ \Pi_2^\sigma \\ \dots \\ \Pi_{2M+1}^\sigma \end{bmatrix}, \tag{16}$$

defined by a Toeplitz matrix $\hat{\Pi}_t$ and a vector $\vec{\Pi}^\sigma$ with $2M$ vanishing components. That is, the vector $\vec{Q}^\sigma$ yields

$$\vec{Q}^\sigma = w(x)w(y)\hat{\Pi}_t^{-1}\vec{\Pi}^\sigma .$$

From another hand, we have the following (semi-infinite) matrix equation

$$\begin{pmatrix} Q_0^\sigma & 0 & 0 & 0 & 0 & \dots \\ Q_1^\sigma & Q_0^\sigma & 0 & 0 & 0 & \dots \\ Q_2^\sigma & Q_1^\sigma & Q_0^\sigma & 0 & 0 & \dots \\ \dots & \dots & \dots & \dots & 0 & \dots \\ Q_{\sigma[[-1]]+1}^\sigma & Q_{\sigma[[-1]]}^\sigma & \dots & \dots & Q_0^\sigma & \dots \\ 0 & Q_{\sigma[[-1]]+1}^\sigma & Q_{\sigma[[-1]]}^\sigma & \dots & \dots & \dots \\ \dots & \dots & \dots & \dots & \dots & \dots \end{pmatrix} \begin{bmatrix} \Pi_0 \\ \Pi_1 \\ \Pi_2 \\ \dots \\ \dots \\ \dots \\ \dots \\ \dots \end{bmatrix} = w(x)w(y) \begin{bmatrix} \Pi_0^\sigma \\ \Pi_1^\sigma \\ \Pi_2^\sigma \\ \dots \\ \dots \\ \dots \\ \dots \\ \dots \end{bmatrix}, \tag{17}$$

1.2. Rational ansatz for the propagators

Combining non-linear connection (15) between projectors and expansion of propagators

$$K_{\rm osc}(x,y;t) = \lambda^{\frac{1}{2}} \sum_{n=0}^{\infty} \lambda^n \Pi_n(x,y) , \qquad \lambda = {\rm e}^{-{\rm i}t} , \tag{18}$$

$$K^\sigma(x,y;t) = \lambda^{\frac{1}{2}} \sum_{n\in\mathbb{N}\setminus\sigma} \lambda^n \Pi_n^\sigma(x,y) , \tag{19}$$

one can obtain closed expression for the propagator K^σ

$$K^\sigma(x,y;t) = K_{\rm osc}(x,y;t) \frac{\sum\limits_{k=0}^{\sigma[[-1]]+1} Q_k^\sigma(x,y){\rm e}^{-{\rm i}kt}}{\sum\limits_{k=0}^{\sigma[[-1]]+1} Q_k^\sigma(x,y)} \tag{20}$$

in terms of elementary functions.

2. Exact propagators for two-well rational extensions of the Harmonic oscillator

The simplest two-well rational extension of the Harmonic oscillator reads (see also Fig. 1)

$$V^{\{2,3\}}(x) = \frac{x^2}{4} + 2\left(1 + 4x^2 \frac{x^4-9}{(x^4+3)^2}\right) . \tag{21}$$

This is a so-called "shallow" two-well potential since it holds only one pair of bound states below $V^{\{2,3\}}(0)$.

Journal of Physics: Conference Series **670** (2016) 012042 doi:10.1088/1742-6596/670/1/012042

k	$Q_k^{\{2,3\}}(x,y)$
0	$\frac{(1+x^2)(1+y^2)}{4\pi}$
1	$-\frac{xy(-3+x^2y^2)}{6\pi}$
2	$\frac{-3-12x^2y^2+3y^4+3x^4+x^4y^4}{24\pi}$
3	$\frac{xy(-3+x^2y^2)}{6\pi}$
4	$\frac{(x^2-1)(y^2-1)}{4\pi}$

Table 1. Sequence of connection polynomials $Q_k^{\{2,3\}}(x,y)$

Explicit expressions of connection polynomials $Q_k^{\{2,3\}}(x,y)$ are given in Table 2. Hamiltonian $H^{\{2,3\}}$ has the quasi-equidistant spectrum, $E_n = n + \frac{1}{2}$, $n \in \mathbb{N}_0 \setminus \{2,3\}$. That is the ground and the first excited states are separated from the equidistant part of the spectrum by a gap energy $\Delta E_g = 3$. Using connection polynomials we can construct propagator $K^{\{2,3\}}(x,y;t)$ by rational ansatz (20). It reads explicitly

$$K^{\{2,3\}} = \mathrm{e}^{-2\mathrm{i}t} K_{\mathrm{osc}}(x,y;t)\left(1 - \frac{8\mathrm{i}\sin t\left[xy(x^2y^2-3) - 3(x^2+y^2)\cos t - 3\mathrm{i}(x^2y^2+1)\sin t\right]}{(3+x^4)(3+y^4)}\right).$$

Propagator $K^{\{2,3\}}(x,y;t)$ completely describes wave packet dynamics in the potential (21).

In general a wide class of two-well extensions (see Fig. 1) can be classified by two parameters. The first parameter is the splitting $\Delta E_s = 2s+1$, $s = 0,1,\dots$ of the ground state and the first excited state (it is always an odd integer). The second parameter is the gap energy $\Delta E_g = 2g+1$, $g = 1,2,\dots$ between the first excited state and the second excited state. For instance, when $\sigma = \{2,3\}$ we have $\Delta E_s = 1$, $s = 0$, and $\Delta E_g = 3$, $g = 1$. Numerical studies suggest that when

$$s \geq 2g\,,$$

two minima will collapse. An example of collapsing wells when g is fixed and s is increased can be seen in the first column of Fig. 1.

A two-well rational extension of the Harmonic oscillator which corresponds to $s = 1$, $g = 1$ gives the following potential

$$V^{\{1,2,4,5\}}(x) = \frac{x^2}{4} + 4 - \frac{80(5+10x^2+7x^4)}{(5+5x^2+5x^4+x^6)^2} + \frac{12(5+x^4)}{5+5x^2+5x^4+x^6}\,. \tag{22}$$

The corresponding connection polynomials which allow to calculate the propagator $K^{\{1,2,4,5\}}$ are listed in the Table 2.

It should be noted that the parametrization of rationally extended Harmonic oscillators by the splitting energy ΔE_s and the gap energy ΔE_g suggests a possible application to an analytical description of flux's qubit time evolution. Indeed, in general a Josephson qubit behaves like a nonlinear resonator formed from the Josephson inductance and its junction capacitance [10]. The system has many energy levels, nevertheless the operating space of the qubit must contain only the two lowest states. As a result, [11] transition frequencies between the qubit ground

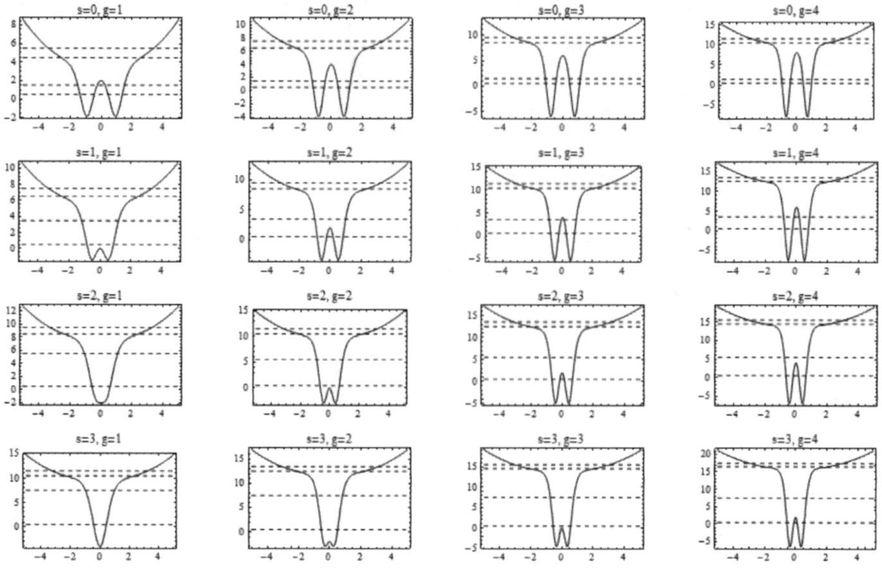

Figure 1. The rationally extended Harmonic oscillators with the splitting energy $\Delta E_s = 2s+1$ and gap energy $\Delta E_g = 2g+1$. Energies of first 4 bound states are plotted by dashed horizontal lines. The spectrum is equidistant starting from the second excited state.

k	$Q_k^{\{1,2,4,5\}}(x,y)$
0	$\frac{9(1+x^2)(1+y^2)}{\pi^2}$
1	$-\frac{9xy(1+x^2)(1+y^2)}{\pi^2}$
2	$-\frac{9(1+x^2)(1+y^2)(y^2+x^2+y^2x^2-1)}{2\pi^2}$
3	$\frac{9xy(-65-30(x^2+y^2)+5(x^4+y^4)+10(x^4y^2+x^2y^4)+3x^4y^4)}{20\pi^2}$
4	$\frac{9(-5+5(x^2+y^2)+15(x^4+y^4)+5(x^6+y^6)-95x^2y^2-35(x^4y^2+x^2y^4)+5(x^6y^2+x^2y^6)-5x^4y^4+5(x^6y^4+x^4y^6)+x^6y^6)}{40\pi^2}$
5	$\frac{9xy(-45-10(x^2+y^2)+5(x^4+y^4)+20x^2y^2+10(x^4y^2+x^2y^4)+3x^4y^4)}{20\pi^2}$
6	$\frac{9(x^4+2x^2-1)(y^4+2y^2-1)}{\pi^2}$

Table 2. Sequence of connection polynomials $Q_k^{\{1,2,4,5\}}(x,y)$

and first excited states should differ from the frequency of transition between the qubit first and second excited states. The Hamiltonian which describes flux qubit reads

$$H_{fq}(p,\delta) = \frac{p^2}{2C} - \frac{I_0\Phi_0}{2\pi}\cos\delta + \frac{1}{2L}\left(\Phi - \frac{\Phi_0}{2\pi}\delta\right)^2, \qquad \Phi_0 = \frac{h}{2e}. \tag{23}$$

The corresponding potential can be fitted at low energies by tuning the splitting and the gap energy of rational extensions of Harmonic oscillator. With the analytical model at hands

Journal of Physics: Conference Series **670** (2016) 012042 doi:10.1088/1742-6596/670/1/012042

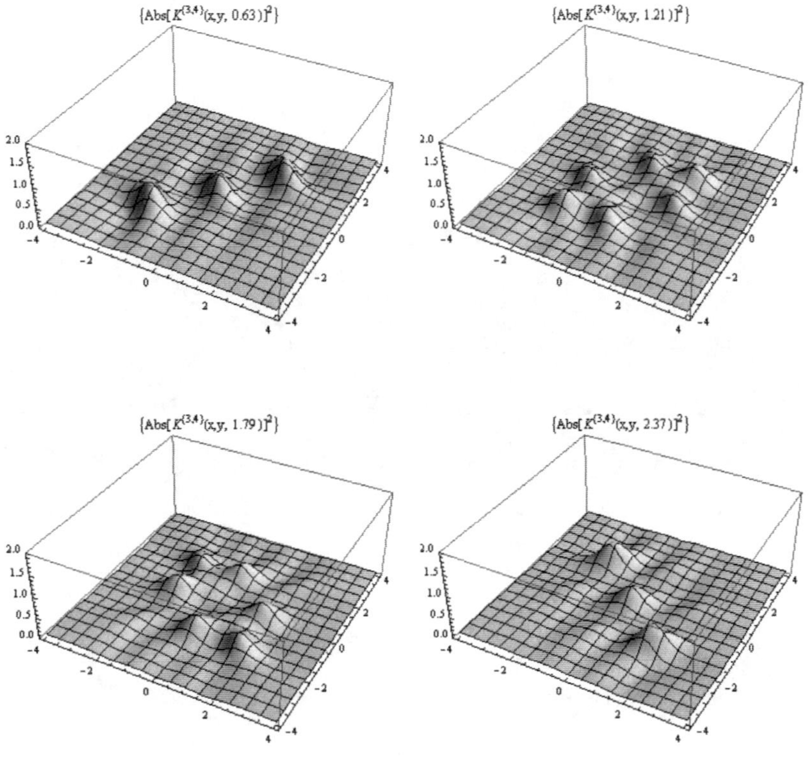

Figure 2. Transition probabilities defined by $|K^{\{3,4\}}(x,y;t)|^2$.

which completely describes all energies and evolution of arbitrary initial state one can optimize efficiency of a flux qubit.

3. Exact propagators for three-well rational extensions of the Harmonic oscillator

The simplest the three-well rational extension of the Harmonic oscillator reads

$$V^{\{3,4\}}(x) = \frac{x^2}{4} + 2 - \frac{1296(x^4 - 2x^2 - 1)}{(9 + 9x^2 - 3x^4 + x^6)^2} + \frac{12(x^4 - 15)}{9 + 9x^2 - 3x^4 + x^6} \tag{24}$$

Explicit expressions of connection polynomials $Q_k^{\{3,4\}}(x,y)$ are given in Table 3. Hamiltonian $H^{\{3,4\}}$ has the quasi-equidistant spectrum, $E_n = n + \frac{1}{2}$, $n \in \mathbb{N}_0 \setminus \{3,4\}$. That is there exists a group of three levels (ground, first and second excited states) separated from the equidistant part of the spectrum by a gap energy $\Delta E_g = 3$.

Using connection polynomials we can construct propagator $K^{\{3,4\}}(x,y;t)$ by rational ansatz (20). In Fig. 2 we visualize $|K^{\{3,4\}}(x,y;t)|^2$ at different moments of time to show tunnelling pattern.

In general a wide class of three-well extensions (see Fig. 3) can be classified by three parameters. There are two splitting energies $\Delta E_s = E_1^{\sigma} - E_0^{\sigma} = 2s + 1$, and $\Delta E_h = E_2^{\sigma} - E_1^{\sigma} = 2h + 1$ $s, h = 0, 1, \ldots$, and the gap energy $\Delta E_g = E_3^{\sigma} - E_2^{\sigma} 2g + 1$, $g = 1, 2, \ldots$. For instance, when $\sigma = \{3,4\}$ we have $\Delta E_s = \Delta E_h = 1$, $s == h = 0$, and $\Delta E_g = 3$, $g = 1$. One can see in Fig. 3 that when g is fixed, wells tend to collapse with increasing parameters s and h. Unfortunately we didn't find any simple criteria to distinguish degenerate and non-degenerate cases.

k	$Q_k^{\{3,4\}}(x,y)$
0	$\frac{(3+x^4)(3+y^4)}{24\pi}$
1	$-\frac{xy(9-6y^2+3y^4+x^4(3-2y^2+y^4)-2x^2(3+2y^2+y^4))}{48\pi}$
2	$\frac{27-27x^2-45x^4+9x^6-27y^2+81x^2y^2+9x^4y^2+9x^6y^2-45y^4+9x^2y^4-21x^4y^4-3x^6y^4+9y^6+9x^2y^6-3x^4y^6+x^6y^6}{288\pi}$
3	$\frac{xy(-27+6x^2+3x^4+6y^2-8x^2y^2-2x^4y^2+3y^4-2x^2y^4+x^4y^4)}{48\pi}$
4	$-\frac{3-6x^2+x^4-6y^2+2x^4y^2+y^4+2x^2y^4-x^4y^4}{16\pi}$
5	$\frac{xy(x^2-3)(y^2-3)}{12\pi}$

Table 3. Sequence of connection polynomials $Q_k^{\{3,4\}}(x,y)$

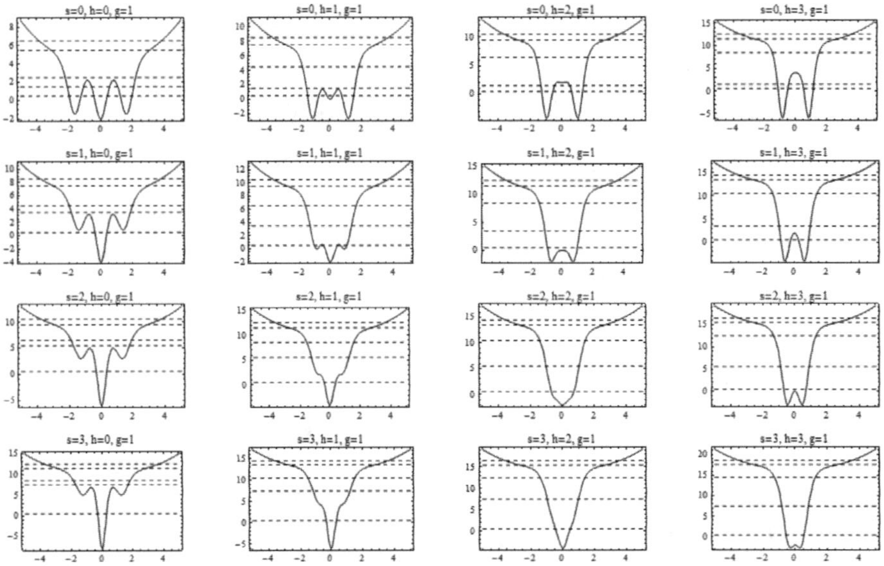

Figure 3. The rationally extended Harmonic oscillators with the splitting energies $\Delta E_s = 2s+1$, $\Delta E_h = 2h+1$, and gap energy $\Delta E_g = 2g+1$, $(g=1)$. Energies of first 5 bound states are plotted by dashed horizontal lines. The spectrum is equidistant starting from the third excited state.

4. Conclusions

Using the rational form of the propagators

$$K^\sigma(x,y;t) = K_{\rm osc}(x,y;t)\frac{\sum\limits_{k=0}^{\sigma[[-1]]+1} Q_k^\sigma(x,y)e^{-ikt}}{\sum\limits_{k=0}^{\sigma[[-1]]+1} Q_k^\sigma(x,y)} \tag{25}$$

we present a new example of Feynman path integrals that can be calculated analytically [12].

Given a sequence σ of levels deleted by Darboux transformation, the rationally extended oscillator potential V^σ appears to be a multi-well potential with N_w minima, where

$$N_w \leq \sigma[[-1]] + 1 - |\sigma| \,. \tag{26}$$

The number N_w of classical vacua (stable equilibrium of V^σ) is related with the structure of non-equidistant part of the spectrum. We established that two-level non-equidistant part of spectrum corresponds to the two well potential when the splitting parameter s, $\Delta E_s = 2s = 1$ is smaller than the doubled gap parameter g, $\Delta E_g = 2g + 1$, that is when $s < 2g$. The resulting two-well potentials and corresponding propagators can be used to model dynamics of a flux qubit [13].

Analytical propagators for the two-well and triple-well potentials may be applied also to studies of oscillations and self-trapping of BEC or to dynamics of a single trapped ion [14, 15, 16, 17, 18].

Acknowledgements

AMPM thanks Brazilian foundation FAPEMIG for the financial support.

References

[1] Guillou J C L and Zinn-Justin J 2012 *Large-order behaviour of perturbation theory* (Elsevier)
[2] Berezovoj V P and Konchatnij M I 2012 *Journal of Physics A: Mathematical and Theoretical* **45** 225302
[3] Berezovoj V P, Konchatnij M I and Nurmagambetov A J 2013 *Journal of Physics A: Mathematical and Theoretical* **46** 065302
[4] Oblomkov A A 1999 *Theoretical and Mathematical Physics* **121** 1574–1584
[5] Bagrov V G and Samsonov B F 1997 *Physics of Particles and Nuclei* **28** 374–397
[6] Veselov A P 2001 *Journal of Physics A: Mathematical and General* **34** 3511
[7] Felder G, Hemery A D and Veselov A P 2012 *Physica D: Nonlinear Phenomena* **241** 2131–2137
[8] Clarkson P A 2009 *Studies in Applied Mathematics* **123** 37–62
[9] Pupasov-Maksimov A M 2015 Propagators of isochronous an-harmonic oscillators and Mehler formula for the x-Hermite polynomials (*Preprint* `math-ph/1502.01778`)
[10] Martinis J M and Osborne K 2004 *arXiv preprint cond-mat/0402415*
[11] Steffen M, Martinis J M and Chuang I L 2003 *Physical Review B* **68** 224518
[12] Grosche C and Steiner F 1998 *Springer tracts in modern physics* **145** 1
[13] Rad A I, Zandi H and Fardmanesh M 2014 *arXiv preprint arXiv:1403.5762*
[14] Milburn G, Corney J, Wright E and Walls D 1997 *Physical Review A* **55** 4318
[15] Leibfried D, Blatt R, Monroe C and Wineland D 2003 *Reviews of Modern Physics* **75** 281
[16] Salgueiro A N, de Toledo P, Lemos G B, Drumond R, Nemes M C and Weidemüller M 2007 *The European Physical Journal D* **44** 537–540
[17] Liu B, Fu L B, Yang S P and Liu J 2007 *Phys. Rev. A* **75**(3) 033601
[18] Feijoo D, Paredes A and Michinel H 2013 *Physical Review A* **87** 063619

Journal of Physics: Conference Series **670** (2016) 012043 doi:10.1088/1742-6596/670/1/012043

Yang-Baxter deformations of Minkowski spacetime

Jun-ichi Sakamoto

Department of Physics, Kyoto University, Kyoto 606-8502, Japan

E-mail: sakajun@gauge.scphys.kyoto-u.ac.jp

Abstract. We discuss Yang–Baxter sigma deformations of 4D Minkowski spacetime proposed recently. To avoid the degeneracy of the standard bilinear form associated with the familiar coset $ISO(1,3)/SO(1,3)$, we consider a slice of AdS_5 in Poincaré coordinates by embedding the 4D Poincaé group into the 4D conformal group $SO(2,4)$. With this procedure we present the metrics and B-fields as Yang–Baxter deformations which correspond to well-known backgrounds such as T–duals of Melvin backgrounds, Hashimoto–Sethi and Spradlin–Takayanagi–Volovich backgrounds, pp-waves, and T–duals of dS_4 and AdS_4. Finally we consider a deformation with a classical r-matrix of Drinfeld–Jimbo type and explicitly derive the associated metric and B-field.

1. Introduction

Yang–Baxter sigma models are known to describe systematic integrable deformations of two–dimensional non-linear sigma models. They have been proposed in [2] and their classical integrability is shown by constructing the universal expression for the Lax pair [3]. This Klimcik's original work [2] [3] employs deformations of principal chiral models on a Lie group G based on a modified classical Yang–Baxter equation (mCYBE). It has been extended to any symmetric coset spaces [4] and the classical Yang–Baxter equation (CYBE) [5].

An important application of Yang–Baxter sigma models are integrable deformations of the superstring on $\mathrm{AdS}_5 \times S^5$ spacetime, which is abbreviated as $\mathrm{AdS}_5 \times S^5$ superstring. The integrable structure of the $\mathrm{AdS}_5 \times S^5$ superstring has been discovered in [1] in the sense of the existence of a Lax pair. A remarkable deformation that preserves the integrability of the $\mathrm{AdS}_5 \times S^5$ superstring is the q–deformed $\mathrm{AdS}_5 \times S^5$ superstring [6]. The deformation is based on a mCYBE and the associated background including the fermionic sector are given by [42] [43]. Also the Maldacena–Russo background [11] and the Lax pair are obtained from q–deformed $\mathrm{AdS}_5 \times S^5$ superstrings by taking a scaling limit [43] [44].

After that, the above models have been generalized to deformations of the $\mathrm{AdS}_5 \times S^5$ superstring based on a classical Yang–Baxter equation [7]. A CYBE has many (skew-symmetric) classical r–matrices as solutions, in contrast to the mCYBE. Yang–Baxter deformed $\mathrm{AdS}_5 \times S^5$ backgrounds for these classical r–matrices can be identified with the well-known γ–deformation of S^5 [8, 9], gravity duals of non-commutative gauge theories [10] [11], Schrödinger spacetimes [12–14] and gravity duals for dipole theories [15–19], as shown in a series of works [20–24]. Remarkably, these deformations can also be applied to non-integrable backgrounds such as $\mathrm{AdS}_5 \times T^{1,1}$ [25]. The deformations of this background [8] [26] can be reproduced as Yang–Baxter deformations [27] . This correspondence between classical r-matrices and deformed backgrounds is called as gravity/CYBE correspondence (for a brief review, see [28]).

Journal of Physics: Conference Series **670** (2016) 012043 doi:10.1088/1742-6596/670/1/012043

Instead of curved spaces, we have generalized this method to flat space in order to study to the applicability of this correspondence [29]. For flat space, there is the obvious problem that the standard bilinear form degenerates if we employ the familiar coset Poincaré group/Lorentz group. A possible resolution is that we consider a conformal embedding of 4D Minkowski spacetime into AdS_5 spacetime in the Poincaré coordinates. This method seems work well because we can reproduce the metric and B–field of well-known backgrounds such as T–duals of Melvin backgrounds [30, 32, 33], Hashimoto–Sethi backgrounds [34], time-dependent backgrounds of Spradlin–Takayanagi–Volovich [35], pp-wave backgrounds, and T–duals of dS_4 and AdS_4. The purpose of this article is to give a short review of this work.

The classical r-matrices considered in this note can be divided into two classes. Yang–Baxter deformations for classical r-matrices of the first class can be interpreted as backgrounds generated by applying a TsT–transformation to 4D Minkowski spacetime in the string theory sense. Since T–duality preserves classical integrability of two dimensional sigma models [36–39], Yang–Baxter sigma models in this class are also classical integrable. In fact, we can construct a general Lax pair in this case [48]. Another class of classical r-matrices containing the dilatation generator $\hat{d}$ correspond to non-twisted backgrounds, such as the T–duals of dS_4 and AdS_4. Although these backgrounds are classically integrable, we have not succeeded in obtaining a formal proof. Hence, it is a conjecture that our Yang–Baxter sigma models for this class have classical integrability.

The structure of this article is as follows. In Section 2, we explain the coset construction of AdS_5 spacetime in Poincaré coordinates. After that we will realize flat space as a slice of the Poincaré AdS_5. In Section 3, we introduce Yang–Baxter deformations of 4D Minkowski spacetime. In Sections 4, we will provide some examples of classical r-matrices and their associated metrics and two-form B–fields. In Section 5, we extend the formulation from the CYBE to the mCYBE and then study a deformation with a classical r-matrix of Drinfeld–Jimbo type. Section 6 is devoted to conclusions and discussion.

2. A coset construction o Minkowski s acetime

Yang–Baxter sigma models need to a non-degenerate bilinear form for their respective group or coset manifolds. The Killing form of the familiar coset space $ISO(1,3)/SO(1,3)$ is however degenerate because the Poincaré algebra is not a semi-simple algebra. A possible resolution is to consider instead an embedding of 4D Minkowski spacetime into the Poincaré AdS_5.

C e c n ruc i n f P incaré AdS_5 *revi i ed*

Let us consider AdS_5 spacetime in Poincaré coordinates. For this purpose, it is helpful to use the conformal basis for $\mathfrak{so}(2,4)$:

$$\mathfrak{so}(2,4) = \text{span}_{\mathbb{R}}\{\, p_\mu \,, n_{\mu\nu} \,, \hat{d} \,, k_\mu \mid \mu,\nu = 0,1,2,3 \,\} \,, \tag{1}$$

where the translation generator p_μ , the Lorentz generators $n_{\mu\nu}$, the dilatation $\hat{d}$ and the special conformal generator k_μ are represented by, respectively,

$$p_\mu \equiv \frac{1}{2}(\gamma_\mu - 2n_{\mu 5}) \,, \quad n_{\mu\nu} \equiv \frac{1}{4}[\gamma_\mu, \gamma_\nu] \,, \quad \hat{d} \equiv \frac{1}{2}\gamma_5 \,, \quad k_\mu \equiv \frac{1}{2}(\gamma_\mu + 2n_{\mu 5}) \,. \tag{2}$$

Here we introduced the gamma matrices γ_μ , $\gamma_5 = -i\gamma_0\gamma_1\gamma_2\gamma_3$ and $n_{\mu 5}$ defined by

$$n_{\mu 5} \equiv \tfrac{1}{4}[\gamma_\mu, \gamma_5] \,. \tag{3}$$

The non-vanishing commutation relations are

$$\begin{aligned}
[p_\mu, k_\nu] &= 2(n_{\mu\nu} + \eta_{\mu\nu}\,\hat{d}\,) \,, \quad [\hat{d}, p_\mu] = p_\mu \,, \quad [\hat{d}, k_\mu] = -k_\mu \,, \\
[p_\mu, n_{\nu\rho}] &= \eta_{\mu\nu}\, p_\rho - \eta_{\mu\rho}\, p_\nu \,, \quad [k_\mu, n_{\nu\rho}] = \eta_{\mu\nu}\, k_\rho - \eta_{\mu\rho}\, k_\nu \,, \\
[n_{\mu\nu}, n_{\rho\sigma}] &= \eta_{\mu\sigma}\, n_{\nu\rho} + \eta_{\nu\rho}\, n_{\mu\sigma} - \eta_{\mu\rho}\, n_{\nu\sigma} - \eta_{\nu\sigma}\, n_{\mu\rho} \,.
\end{aligned} \tag{4}$$

It well known that AdS_5 spacetime is described by a symmetric coset

$$\mathrm{AdS}_5 = \frac{SO(2,4)}{SO(1,4)} . \tag{5}$$

Here the Lie algebra $\mathfrak{so}(1,4)$ of the stabilizer group $SO(1,4)$ is given by

$$\mathfrak{so}(1,4) = \mathrm{span}_{\mathbb{R}}\{ \ \frac{1}{2}(p_\mu - k_\mu) = -n_{\mu 5} \ | \ \ \mu = 0,1,2,3 \ \} . \tag{6}$$

So the coset generators are spanned by $\frac{1}{2}(p_\mu + k_\mu) \sim \gamma_\mu$ and $\hat{d} \sim \gamma_5$. A convenient representative of the coset (5) is given by

$$g = \exp\big[\, p_\mu \, x^\mu \, \big] \ \exp\Big[\hat{d} \log z \Big] \, , \tag{7}$$

and makes manifest the $\mathbb{R}^4$ symmetry in the boundary of AdS_5 in Poincaré coordinates. Then, we can easily get the AdS_5 metric in Poincaré coordinates,

$$ds^2 = \mathrm{Tr}(A\overline{P}(A)) = \frac{-(dx^0)^2 + \sum_{i=1}^3 (dx^i)^2 + dz^2}{z^2} \, , \tag{8}$$

where $\overline{P}$ is a coset projector from $\mathfrak{so}(2,4)$ to $\mathfrak{so}(2,4)/\mathfrak{so}(1,4)$ and is defined as

$$\overline{P}(x) \ \equiv \ \gamma_0 \frac{\mathrm{Tr}(\gamma_0 x)}{\mathrm{Tr}(\gamma_0^2)} + \sum_{i=1}^{3} \gamma_i \frac{\mathrm{Tr}(\gamma_i x)}{\mathrm{Tr}(\gamma_i^2)} + \gamma_5 \frac{\mathrm{Tr}(\gamma_5 x)}{\mathrm{Tr}(\gamma_5^2)} \tag{9}$$

$$= \ \frac{1}{4}\Big[-\gamma_0 \ \mathrm{Tr}(\gamma_0 x) + \sum_{i=1}^{3} \gamma_i \ \mathrm{Tr}(\gamma_i x) + \gamma_5 \ \mathrm{Tr}(\gamma_5 x) \Big] \qquad \text{for } x \in \mathfrak{so}(2,4) \, . \tag{10}$$

Note that the AdS radius is set to 1 . This coset projector respects the $\mathbb{Z}_2$–grading structure of the coset space $SO(2,4)/SO(1,4)$.

Finally, let us comment on a relation between the projection $\overline{P}$ and the standard method. In the usual way, we expand A as

$$A = g^{-1}dg = e^\mu \frac{1}{2}(p_\mu + k_\mu) + e^5 \hat{d} + \omega^\mu \frac{1}{2}(p_\mu - k_\mu) \, , \tag{11}$$

where the vielbein and spin connections are

$$e^\mu = \frac{dx^\mu}{z} \, , \qquad e^5 = \frac{dz}{z} \, , \qquad \omega^\mu = \frac{dx^\mu}{z} \, . \tag{12}$$

Then it is easy to get the metric on AdS_5 spacetime by computing

$$ds^2 = \eta_{\mu\nu} e^\mu e^\nu + e^5 e^5 \, . \tag{13}$$

Now using the relations

$$e^\mu = \frac{1}{2}\mathrm{Tr}(\gamma^\mu A) \, , \quad e^5 = \mathrm{Tr}(\hat{d} A) \, , \tag{14}$$

this expression can be rewritten as

$$\begin{aligned} ds^2 \ &= \ \frac{1}{4}\eta_{\mu\nu}\mathrm{Tr}(\gamma^\mu A)\mathrm{Tr}(\gamma^\nu A) + \mathrm{Tr}(\hat{d}A)\mathrm{Tr}(\hat{d}A) \\ &= \ \mathrm{Tr}\left[A \ \frac{1}{4}\eta^{\mu\nu}\gamma_\mu \mathrm{Tr}(\gamma_\nu \, A) \right] + \mathrm{Tr}\Big[A \, \hat{d}\mathrm{Tr}(\hat{d}A) \Big] \\ &= \ \mathrm{Tr}(A\overline{P}(A)) \, . \end{aligned} \tag{15}$$

Thus our method using the coset projector is equivalent to the standard manner.

A c nf r al e beddin f D Mink w ki pace i e

Now we can get the coset space for 4D Minkowski spacetime by embedding $ISO(1,3)/SO(1,3)$ into the Poincaré AdS$_5$. A representative of the group element g is

$$g = \exp\big[\, p_\mu\, x^\mu \,\big]\,. \tag{16}$$

Unlike the coset representative for the Poincaré AdS$_5$ (7), the radial coordinate z does not appear in the above expression. It implies that we realize 4D Minkowski spacetime as a slice of $z = 1$ of the Poincaré AdS$_5$.

Note here that the 4D Poincaré algebra $\mathfrak{iso}(1,3)$ and the 4D Lorentz algebra $\mathfrak{so}(1,3)$ are generated by the following generators, respectively,

$$\begin{aligned} \mathfrak{iso}(1,3) &= \text{span}_{\mathbb{R}}\{\; n_{\mu\nu}\,, p_\mu \mid \mu,\nu = 0,1,2,3\;\}\,, \\ \mathfrak{so}(1,3) &= \text{span}_{\mathbb{R}}\{\; n_{\mu\nu} \mid \mu,\nu = 0,1,2,3\;\}\,. \end{aligned} \tag{17}$$

Thus it makes sense to use the generators p_μ to parameterize the coset representative of $ISO(1,3)/SO(1,3)$ as (16). Eventually the left-invariant one-form $A = g^{-1}dg$ is written as a linear combination of p_μ.

To avoid the problem of the degeneracy of the Poincaré group $ISO(1,3)$, we can use the projector for 4D Minkowski spacetime

$$P(x) = \frac{1}{4}\Big[-\gamma_0\, \text{Tr}(\gamma_0 x) + \sum_{i=1}^{3} \gamma_i\, \text{Tr}(\gamma_i x)\Big] \qquad \text{for} \quad x \in \mathfrak{so}(2,4)\,. \tag{18}$$

Here due to the dimension of the coset space, we have dropped $\gamma_5 \sim \hat{d}$ from the projector $\overline{P}$. Because the projected one-form $P(A)$ is expanded in terms of $\gamma_\mu (\mu = 0,1,2,3)$, we can obtain the flat metric

$$ds^2 = \text{Tr}(AP(A)) = -(dx^0)^2 + \sum_{i=1}^{3} (dx^i)^2\,. \tag{19}$$

Note that this computation is equivalent to the standard method using the vierbein like in the Poincaré AdS$_5$ case. This result is the starting point of our argument in the following.

3. Yang–Baxter sigma models or 4D Minkowski s acetime

In this section, we consider Yang–Baxter deformations of two-dimensional sigma models with 4D Minkowski spacetime as target space. Next we present a classification of classical r-matrices with values in $\mathfrak{so}(2,4) \otimes \mathfrak{so}(2,4)$.

Def r ed ac i n

Following the above coset construction of flat space, we introduce Yang–Baxter sigma models for 4D Minkowski spacetime.

The deformed action is given by[1]

$$S = -\frac{1}{2}\int_{-\infty}^{\infty} d\tau \int_0^{2\pi} d\sigma\, (\gamma^{\alpha\beta} - \epsilon^{\alpha\beta})\, \text{Tr}\left[A_\alpha P \circ \frac{1}{1 - 2\eta R_g \circ P}(A_\beta)\right], \tag{20}$$

[1] Here the string tension $T = \frac{1}{2\pi\alpha'}$ is set to 1, and the conformal gauge is taken so as to drop the dilaton coupling to the world-sheet scalar curvature.

where $A_\alpha = g^{-1}\partial_\alpha g$ and g is given in Eq. (16). Here η is a deformation parameter and the action (20) is reduced to the undeformed one for $\eta = 0$. $\gamma_{\alpha\beta} = \mathrm{diag}(-1,1)$ and $\epsilon^{\alpha\beta}$ are the world sheet metric and the anti-symmetric tensor that is normalized as $\epsilon^{\tau\sigma} = 1$, respectively. The operator R_g is defined as

$$R_g \equiv g^{-1}R(gXg^{-1})g\,, \tag{21}$$

where a linear operator $R : \mathfrak{so}(2,4) \to \mathfrak{so}(2,4)$ is a solution of the CYBE,

$$\big[R(M), R(N)\big] - R\left([R(M),N] + [M,R(N)]\right) = 0\,, \qquad M,N \in \mathfrak{so}(2,4)\,. \tag{22}$$

The R-operator is related to the *kew- e ric* classical r-matrix in the tensorial notation through

$$R(X) = \mathrm{Tr}_2[r(1\otimes X)] = \sum_i (a_i\,\mathrm{Tr}(b_iX) - b_i\,\mathrm{Tr}(a_iX))\,, \tag{23}$$

where the classical r-matrix is given by

$$r = \sum_i a_i \wedge b_i \equiv \sum_i (a_i\otimes b_i - b_i \otimes a_i) \tag{24}$$

satisfying the CYBE,

$$[r_{12}, r_{13}] + [r_{12}, r_{23}] + [r_{13}, r_{23}] = 0\,. \tag{25}$$

The generators a_i, b_i are elements of $\mathfrak{so}(2,4)$.

To obtain deformed backgrounds, it is convenient to rewrite the deformed Lagrangian as

$$L = -\frac{1}{2}\,\mathrm{Tr}\left[A_- P(J_+)\right]\,, \tag{26}$$

where $J_\pm$ is a deformed current defined as

$$J_\pm \equiv \frac{1}{1\mp 2\eta R_g\circ P}A_\pm\,, \qquad A_\pm = A_\tau \pm A_\sigma\,. \tag{27}$$

By solving the equations

$$(1\mp 2\eta P\circ R_g)P(J_\pm) = P(A_\pm)\,, \tag{28}$$

the projected deformed current $P(J_\pm)$ is determined. Then the metric and NS-NS two-form are given by the symmetric and skew-symmetric parts regarding to worldsheet coordinates in the deformed Lagrangian, respectively.

Cla ifica i n f cla ical r- a rice

One can organize the classical r-matrices as follows:

(a) $r =$ Poincaré $\wedge$ Poincaré
  1. abelian e.g., $r \sim p_3 \wedge n_{12}$, 2. non-abelian e.g., $r \sim p_1 \wedge n_{12}$,

(b) $r =$ Poincaré $\wedge$ non-Poincaré
  1. abelian e.g., $r \sim n_{12} \wedge \hat{d}$, 2. non-abelian e.g., $r \sim p_0 \wedge \hat{d}$,

(c) $r =$ non-Poincaré $\wedge$ non-Poincaré
  1. abelian e.g., $r \sim k_1 \wedge k_2$, 2. non-abelian e.g., $r \sim k_0 \wedge \hat{d}$.

Here, given a classical r-matrix $r = a \wedge b$, the word (non-)abelian means that a and b (do not) commute each other.

We consider Yang–Baxter deformations of 4D Minkowski spacetime for the classes (a)-1 and (b)-2 in the following. The class (a)-1 corresponds to classical r-matrices associated with various kinds of TsT–transformations. Let α and β be dual coordinates for the generators a and b in the sense of the Lie algebra. Concretely for $r \sim p_3 \wedge n_{12}$, we take $\alpha = x^3$ and $\beta = \theta$ is the angular coordinate in the x^1-x^2 plane. Then the corresponding TsT–transformation is the following sequence of T–dualities and shifts: 1. T–dualize along α to α; 2. Replace α with $\alpha + \eta\,\beta$; 3. Finally, T–dualize along the α direction. Of course the coordinates α and β are compactified with periodic boundary conditions under this sequence. Since we have succeeded to construct Lax pairs for these r-matrices, Yang–Baxter sigma models for the class (a)-1 certainly describe integrable deformations of 4D-Minkowski spacetime.

In the case of the class (b)-2, T-duals of dS_4 and AdS_4 are realized from $r \sim p_0 \wedge \hat{d}$ and $r \sim p_3 \wedge \hat{d}$, respectively. These backgrounds cannot be reproduced by applying TsT–transformations to flat space like for the class (a)-1. Although two-dimensional non-linear sigma models with these target spaces are classically integrable, we do not have a formal proof or explicit construction of the Lax pairs. Hence it is a conjecture that Yang–Baxter sigma models in the class (b)-2 are classically integrable. For a more comprehensive list of classical r-matrices, see [47].

4. De ormed backgrounds

TsT–dual f fla pace

In this subsection, we present deformed backgrounds associated with classical r-matrices in the class (a)-1. Here we assume the extra six dimensions to be flat (as resulting from a T^6-compactification) and we introduce only a dilaton field. All backgrounds are identified with a TsT–transformation of flat space.

-dual f Melvin back r und First, let us consider the r-matrix

$$r = \frac{1}{2} p_3 \wedge n_{12}\,. \tag{29}$$

The associated deformed metric and B–field are T–dual to a Melvin background:

$$\begin{aligned} ds^2 &= -(dx^0)^2 + dr^2 + \frac{r^2 d\theta^2 + (dx^3)^2}{1+\eta^2 r^2}\,, \\ B &= \frac{\eta r^2}{1+\eta^2 r^2}\, d\theta \wedge dx^3\,, \end{aligned} \tag{30}$$

where we have performed a coordinate transformation,

$$x^1 = r\,\cos\theta\,, \qquad\qquad x^2 = r\,\sin\theta\,. \tag{31}$$

It should be remarked that the Yang–Baxter deformations cannot reproduce the associated dilaton, although it may be possible to perform a supercoset construction in principle. However we can embed the background into string theory by observing that the one-loop beta function vanishes when adding the dilaton [31–33]

$$\Phi = -\frac{1}{2}\log(1+\eta^2 r^2)\,. \tag{32}$$

Journal of Physics: Conference Series **670** (2016) 012043 doi:10.1088/1742-6596/670/1/012043

This background can also be constructed by applying a sequence of a TsT–transformation to flat space as noted above. More concretely, the sequence in this case is 1.T–duality in x^3; 2. shift $\theta \to \theta + \eta\, x^3$; 3.T–duality in $\hat{x}^3$. Consistency requires x^3 to be periodic with period $x^3 \simeq x^3 + 2\pi/\eta$ and x^3 with period[2] $x^3 \simeq x^3 + \alpha'\eta/(2\pi)$.

pp-wave back r und Let us consider the classical r-matrix

$$r = \frac{1}{2\sqrt{2}}(p_0 - p_3) \wedge n_{12}\,. \tag{33}$$

The associated deformed background is

$$\begin{aligned} ds^2 &= -2dx^+dx^- - \eta^2 r^2 (dx^+)^2 + (dr)^2 + r^2 d\theta^2\,, \\ B &= \eta r^2\, dx^+ \wedge d\theta\,, \end{aligned} \tag{34}$$

where we have introduced the polar coordinate system for x^1 and x^2 given by

$$x^1 = r\cos\theta\,, \qquad x^2 = r\sin\theta\,. \tag{35}$$

This is a pp-wave background which can also be understood as a generalization of a (null) TsT transformation obtained as 1. a T–duality from θ to θ, followed by 2. the shifts $x^0 \to \eta\theta + x^0$, $x^3 \to -\eta\theta + x^3$ and the final 3. T–duality from θ to θ.

Note that this requires the identifications $x^0 \simeq x^0 + \alpha'\eta/(2\pi)$ and $x^3 \simeq x^3 + \alpha'\eta/(2\pi)$.

Ha i –Se i back r und Next we consider the following abelian classical r-matrix

$$r = \frac{1}{2\sqrt{2}} p_2 \wedge (n_{01} + n_{13})\,. \tag{36}$$

The resulting metric and B–field are given by

$$\begin{aligned} ds^2 &= -2dx^-dx^+ + \frac{1}{1+\eta^2(x^+)^2}[(dx^1)^2 + (dx^2)^2 + \eta^2 x^1 dx^+(2x^+dx^1 - x^1 dx^+)]\,, \\ B &= \frac{\eta}{1+\eta^2(x^+)^2}(x^1dx^+ - x^+dx^1) \wedge dx^2\,. \end{aligned} \tag{37}$$

Note that this background depends on the light-cone time x^+ explicitly. The associated dilaton to complete the string embedding is taken to be

$$\Phi = -\frac{1}{2}\log(1 + \eta^2(x^+)^2)\,. \tag{38}$$

The metric and B–field (37) agree with those of the Hashimoto–Sethi background. To show this agreement, one has to introduce new coordinates,

$$x^+ = y^+\,, x^1 = y^+ y\,, x^- = y^- + \frac{1}{2}y^+(y)^2, x^2 = -z\,. \tag{39}$$

Under this coordinate transformation, the above deformed background (37) becomes

$$\begin{aligned} ds^2 &= -2dy^+dy^- + \frac{(y^+)^2(dy)^2 + dz^2}{1+\eta^2(y^+)^2}\,, \\ B &= \frac{\eta\,(y^+)^2}{1+\eta^2(y^+)^2} dy \wedge dz\,. \end{aligned} \tag{40}$$

This reproduces the expression in Eq. (25) of [34] where the background is shown to be the result of a TsT–transformation for $U(1) \times U(1)$ isometries which are shifts of the y and z directions.

[2] We reintroduce the explicit parameter α' in the identifications to manifestly illustrate the dimensions of the variables.

Spradlin–V l v ic – aka ana i back r und Let us here consider the classical r-matrix

$$r = \frac{1}{2} n_{12} \wedge n_{03} \,. \tag{41}$$

Then the associated metric and B–field are given by

$$\begin{aligned} ds^2 &= -dt^2 + dr^2 + \frac{r^2 d\theta^2 + t^2 d\phi^2}{1+\eta^2 r^2 t^2} \,, \\ B &= \frac{\eta r^2 t^2}{1+\eta^2 r^2 t^2} d\phi \wedge d\theta \,, \end{aligned} \tag{42}$$

where we have introduced new coordinates

$$x^0 = t \cosh\phi \,, \qquad x^3 = t \sinh\phi \,. \tag{43}$$

Note that the coordinates in (43) do not cover the whole x^0-x^3 plane and the background (42) contains no singularity. Then the metric and B–field in (42) agree with those of (6.1) in [35]. This is a time-dependent background realized by a TsT–transformation of Minkowski spacetime on the torus generated by ϕ and θ. The associated dilaton is

$$\Phi = -\frac{1}{2} \log(1 + \eta^2 r^2 t^2) \,. \tag{44}$$

N n- wi ca e

We consider Yang-Baxter deformations of type (b)-1 in the following. In this article, we present two examples, the T–duals of dS_4 and AdS_4, associated with classical r-matrices that contain the dilation generator $\hat{d}$. One can find more examples for this type in [29].

–dual f dS_4 Let us consider the non-abelian r-matrix

$$r = \frac{1}{2} p_0 \wedge \hat{d} \quad \text{with} \quad [\hat{d}\,, p_0] = p_0 \,. \tag{45}$$

This r-matrix was used to deform the conformal algebra $\mathfrak{so}(2,4)$ in [41]. The resulting background is

$$\begin{aligned} ds^2 &= \frac{-(dx^0)^2 + dr^2}{1-\eta^2 r^2} + r^2(d\theta^2 + \sin^2\theta\, d\phi^2) \,, \\ B &= \frac{\eta r}{1-\eta^2 r^2} dx^0 \wedge dr \,, \end{aligned} \tag{46}$$

where we have introduced new coordinates r, θ and ϕ through

$$x^1 = r\cos\phi\, \sin\theta \,, \qquad x^2 = r \sin\phi \sin\theta \,, \qquad x^3 = r\cos\theta \,. \tag{47}$$

Note here that the above B–field can be rewritten in the form of a total derivative. One can understand this background as T–dual of four dimensional de Sitter space by following transformation. Firstly, we take a timelike T–duality along the x^0 direction [40]. The deformed metric becomes

$$ds^2 = (dr + \eta r\, dx^0)^2 - (dx^0)^2 + r^2(d\theta^2 + \sin^2\theta\, d\phi^2) \,. \tag{48}$$

Note that the B–field has disappeared due to the T–duality along the x^0 direction. Next we introduce the new coordinate

$$x^0 = t - \frac{1}{2\eta} \log(\eta^2 r^2 - 1) \,. \tag{49}$$

In this way, we get the well-known metric of dS_4 in static coordinates,

$$ds^2 = -(1-\eta^2 r^2) dt^2 + \frac{dr^2}{1-\eta^2 r^2} + r^2(d\theta^2 + \sin^2\theta d\phi^2). \tag{50}$$

Note that there is a cosmological horizon at $r = 1/\eta$.

–dual f AdS_4 As another example, let us consider the classical r-matrix

$$r = \frac{1}{2}\hat{d} \wedge p_1 \quad \text{with} \quad [\hat{d}\,, p_1] = p_1\,. \tag{51}$$

This r-matrix also contains the dilatation $\hat{d}$. The associated metric and B–field are given by

$$\begin{aligned} ds^2 &= \frac{dt^2 + (dx^1)^2}{1+\eta^2 t^2} + t^2\cosh^2\phi d\theta^2 - t^2 d\phi^2\,, \\ B &= \frac{\eta t}{1+\eta^2 t^2} dt \wedge dx^1\,, \end{aligned} \tag{52}$$

where we have introduced new coordinates t, θ and ϕ through

$$x^0 = t\,\sinh\phi\,, \qquad x^2 = t\,\cos\theta\,\cosh\phi\,, \qquad x^3 = t\,\sin\theta\,\cosh\phi\,. \tag{53}$$

Note here that the B–field can be recast as a total derivative.

As in the previous case, it is nice to perform a T–duality along the x^1-direction. Then the resulting background is given by[3]

$$ds^2 = (dt - \eta t\,dx^1)^2 + (dx^1)^2 + t^2(-d\phi^2 + \cosh^2\phi\,d\theta^2)\,. \tag{54}$$

Now the B–field has disappeared. Let us perform a coordinate transformation,

$$x^1 = y + \frac{1}{2\eta}\log(\eta^2 t^2 + 1)\,. \tag{55}$$

Then the resulting metric is given by

$$ds^2 = (1+\eta^2 t^2)dy^2 + \frac{dt^2}{1+\eta^2 t^2} + t^2(-d\phi^2 + \cosh^2\phi\,d\theta^2)\,. \tag{56}$$

By replacing the coordinates (with a double Wick rotation) by

$$y \to it\,, \quad t \to r\,, \quad \phi \to i\theta\,, \quad \theta \to \phi\,, \tag{57}$$

one can obtain the standard metric of AdS_4 with global coordinates

$$ds^2 = -(1+\eta^2 r^2)dt^2 + \frac{dr^2}{1+\eta^2 r^2} + r^2(d\theta^2 + \cos^2\theta\,d\phi^2)\,. \tag{58}$$

Note that η^2 measures the curvature.

5. A Yang–Baxter de ormation with mCYBE

So far, we have considered classical r-matrices satisfying the classical Yang–Baxter equation (25) (or (22)). Here, as an exceptional case, let us consider a classical r-matrix of Drinfeld–Jimbo (DJ) type,

$$r_{\rm DJ} = -i \sum_{1\le i<j\le 4} E_{ij} \wedge E_{ji}\,, \qquad (E_{ij})_{kl} \equiv \delta_{ik}\delta_{jl}\,, \tag{59}$$

[3] Note that, at this stage, one can see that this metric describes AdS_4 by explicitly computing the scalar curvature and the Ricci tensor.

which satisfies the modified Yang–Baxter equation,

$$\big[R(M), R(N)\big] - R\left([R(M), N] + [M, R(N)]\right) = [M, N]\,, \tag{60}$$

where $M, N \in \mathfrak{so}(2,4)$. In comparison to the CYBE in (22), the right-hand side of (60) is modified. The overall factor of the classical r-matrix is determined by a reality condition of the deformed B–field. The r-matrix (59) was used to construct an integrable deformation of the $\mathrm{AdS}_5 \times S^5$ superstring [6]. The deformed metric and B–field were explicitly computed in [42].

Since we are only interested in the associated background, the computation scheme is identical to the one of the CYBE case.

The resulting metric and B–field are given by

$$\begin{aligned} ds^2 &= -r^2 \sin^2\theta\, dt^2 + dr^2 + \frac{r^2}{1+\eta^2 r^4 \sin^2\theta}\left(d\theta^2 + \cos^2\theta\, d\phi^2\right)\,, \\ B &= -\frac{\eta\, r^4 \sin\theta \cos\theta}{1+\eta^2 r^4 \sin^2\theta}\, d\theta \wedge d\phi\,. \end{aligned} \tag{61}$$

Here we have performed a coordinate transformation,

$$\begin{aligned} x^0 &= r \sin\theta\, \sinh t\,, & x^1 &= r\cos\theta\, \cos\phi\,, \\ x^2 &= r\cos\theta\, \sin\phi\,, & x^3 &= r\, \sin\theta\, \cosh t\,, \end{aligned} \tag{62}$$

and rescaled $\eta \to \eta/2$. It is worth noting that the metric in (61) is be regular as opposed to the case of the q-deformed AdS_5 superstring. The scalar curvature has no singularity. The singular surface of the metric identified in [42] has not appeared due to the fact that we are now working on a slice of the Poincaré AdS_5 at $z = 1$.

6. Conclusion and Discussion

In this article, we have presented a brief review of Yang–Baxter deformations of 4D Minkowski spacetime. The essential point is to realize 4D Minkowski spacetime by embedding the Poincaré AdS_5 to avoid the degeneracy of the Killing form for a standard coset $ISO(1,3)/SO(1,3)$. Following this prescription, we can reproduce the metric and B–field of well-known backgrounds such as T–duals of Melvin backgrounds, Hashimoto–Sethi backgrounds, time-dependent backgrounds of Spradlin–Takayanagi–Volovich, pp-wave backgrounds, and T–duals of dS_4 and AdS_4. Finally, we have considered the Yang–Baxter deformation for Drinfeld–Jimbo r-matrix and present the associated deformed background explicitly. The list of correspondences between deformed backgrounds and classical r-matices is given in Table 1.

There are many open questions. Since the Lax pairs associated with classical r-matrices for the class (a)-1 have been constructed explicitly, our Yang–Baxter sigma model describe integrable deformations in this class at least. However we have not yet constructed Lax pairs for the class (b)-2, although the sigma models with the resulting deformed backgrounds are classically integrable. It would be interesting to study the form of the Lax pairs and the associated deformed algebra for these cases.

In particular, deformed Poincaré algebras are studied in [45] in terms of classical r-matrices. These r-matrices are contained in the class (a). As a first step to clarify the relations between the list [47] and our results, the associated backgrounds and corresponding Lax pairs for special cases of the class (a)-2 that describe so-called κ-deformations of the 4D Poincaré algebra are given in a forthcoming paper [49].

r-matrix	Type of Twist	Background
$p_i \wedge p_j \ (i,j=1,2,3)$	Melvin Shift Twist	Seiberg-Witten
$p_0 \wedge p_i$	Melvin Shift Twist	NCOS
$(p_0 + p_i) \wedge p_j \ (i \neq j)$	Null Melvin Shift Twist	light-like NC
$\frac{1}{2} p_3 \wedge n_{12}$	Melvin Twist	T–dual Melvin
$\frac{1}{2\sqrt{2}} p_2 \wedge (n_{01} + n_{13})$	Melvin Null Twist	Hashimoto–Sethi
$\frac{1}{2} n_{12} \wedge n_{03}$	R Melvin R Twist	Spradlin–Takayanagi–Volovich
$\frac{1}{2} p_1 \wedge n_{03}$	Melvin Boost Twist	T–dual of Grant space
$\frac{1}{2\sqrt{2}} (p_0 - p_3) \wedge n_{12}$	Null Melvin Twist	pp-wave
$\frac{1}{2\sqrt{2}} (\hat{d} - n_{03}) \wedge (p_0 - p_3)$	Non-Twist	pp-wave
$\frac{1}{2} \hat{d} \wedge p_0$	Non-Twist	T–dual of dS_4
$\frac{1}{2} \hat{d} \wedge p_1$	Non-Twist	T–dual of AdS_4
DJ-type (mCYBE)	Non-Twist	q-deformation (?)

Table 1. A catalog of classical r-matrices and the associated backgrounds.

It would be most important to generalize our argument to 10D Minkowski spacetime in order to extend our argument to a consistent string theory. For this purpose, we have to consider the 10D conformal group $SO(2,10)$ and realize 10D Minkowski spacetime as a slice of 11D AdS space in Poincaré coordinates. We expect that in this case, 10D supersymmetric configurations like the fluxtrap backgrounds [46] could be reproduced as Yang–Baxter deformations.

Our discussion has been constrained to the bosonic sector. As opposed to deformations of $AdS_5 \times S^5$, it is an easier task to generalize our computations to a supercoset construction than the $AdS_5 \times S^5$ case. In recent work, deformations of the $AdS_5 \times S^5$ superstring for the class (a)-1 have been reinterpreted as twisted boundary conditions of the undeformed $AdS_5 \times S^5$ [44]. We can apply a similar analysis to the flat space case. Thus since for the class (a)-1 deformed sigma models on Yang–Baxter deformed Minkowski spacetime are equivalent to undeformed sigma models up to boundary conditions, one can quantize it and get its quantum spectrum. It would be very interesting to investigate the relation between the string spectrum and Yang–Baxter deformations including the fermionic sector.

Acknowledgments
We are thankful to T. Matsumoto, D. Orlando, S. Reffert, and K. Yoshida for their supports. Also we are grateful to D. Orlando and S. Reffert for correcting the errors in the previous version of this article.

Re erences

[1] I. Bena, J. Polchinski and R. Roiban, "Hidden symmetries of the $AdS_5 \times S^5$ superstring," Phys. Rev. D **69** (2004) 046002 [hep-th 0305116].
[2] C. Klimcik, "Yang-Baxter sigma models and dS AdS T duality," JHEP **0212** (2002) 051 [hep-th 0210095]
[3] C. Klimcik, "On integrability of the Yang-Baxter sigma-model," J. Math. Phys. **50** (2009) 043508 [ar iv:0802.3518 [hep-th]]; "Integrability of the bi-Yang-Baxter sigma model," Lett. Math. Phys. **104** (2014) 1095 [ar iv:1402.2105 [math-ph]].
[4] F. Delduc, M. Magro and B. Vicedo, "On classical q-deformations of integrable sigma-models," JHEP **1311** (2013) 192 [ar iv:1308.3581 [hep-th]].
[5] T. Matsumoto and K. Yoshida, "Yang-Baxter sigma models based on the CYBE," Nucl. Phys. B **893** (2015) 287 [ar iv:1501.03665 [hep-th]].

[6] F. Delduc, M. Magro and B. Vicedo, "An integrable deformation of the $AdS_5 \times S^5$ superstring action," Phys. Rev. Lett. **112** (2014) 051601 [ar iv:1309.5850 [hep-th]]; "Derivation of the action and symmetries of the q-deformed $AdS_5 \times S^5$ superstring," JHEP **1410** (2014) 132 [ar iv:1406.6286 [hep-th]].

[7] I. Kawaguchi, T. Matsumoto and K. Yoshida, "Jordanian deformations of the $AdS_5 \times S^5$ superstring," JHEP **1404** (2014) 153 [ar iv:1401.4855 [hep-th]].

[8] O. Lunin and J. M. Maldacena, "Deforming field theories with $U(1) \times U(1)$ global symmetry and their gravity duals," JHEP **0505** (2005) 033 [hep-th 0502086].

[9] S. Frolov, "Lax pair for strings in Lunin-Maldacena background," JHEP **0505** (2005) 069 [hep-th 0503201].

[10] A. Hashimoto and N. Itzhaki, "Noncommutative Yang-Mills and the AdS CFT correspondence," Phys. Lett. B **465** (1999) 142 [hep-th 9907166].

[11] J. M. Maldacena and J. G. Russo, "Large N limit of noncommutative gauge theories," JHEP **9909** (1999) 025 [hep-th 9908134].

[12] C. P. Herzog, M. Rangamani and S. F. Ross, "Heating up Galilean holography," JHEP **0811** (2008) 080 [ar iv:0807.1099 [hep-th]]; J. Maldacena, D. Martelli and Y. Tachikawa, "Comments on string theory backgrounds with non-relativistic conformal symmetry," JHEP **0810** (2008) 072 [ar iv:0807.1100 [hep-th]]; A. Adams, K. Balasubramanian and J. McGreevy, "Hot Spacetimes for Cold Atoms," JHEP **0811** (2008) 059 [ar iv:0807.1111 [hep-th]].

[13] N. Bobev and A. Kundu, "Deformations of Holographic Duals to Non-Relativistic CFTs," JHEP **0907** (2009) 098 [ar iv:0904.2873 [hep-th]].

[14] N. Bobev, A. Kundu and K. Pilch, "Supersymmetric IIB Solutions with Schrödinger Symmetry," JHEP **0907** (2009) 107 [ar iv:0905.0673 [hep-th]].

[15] A. Bergman and O. J. Ganor, "Dipoles, twists and noncommutative gauge theory," JHEP **0010** (2000) 018 [hep-th 0008030].

[16] K. Dasgupta, O. J. Ganor and G. Rajesh, "Vector deformations of N=4 superYang-Mills theory, pinned branes, and arched strings," JHEP **0104** (2001) 034 [hep-th 0010072].

[17] A. Bergman, K. Dasgupta, O. J. Ganor, J. L. Karczmarek and G. Rajesh, "Nonlocal field theories and their gravity duals," Phys. Rev. D **65** (2002) 066005 [hep-th 0103090].

[18] M. Alishahiha and O. J. Ganor, "Twisted backgrounds, PP waves and nonlocal field theories," JHEP **0303** (2003) 006 [hep-th 0301080].

[19] E. Imeroni, "On deformed gauge theories and their string M-theory duals," JHEP **0810** (2008) 026 [ar iv:0808.1271 [hep-th]].

[20] T. Matsumoto and K. Yoshida, "Lunin-Maldacena backgrounds from the classical Yang-Baxter equation - towards the gravity CYBE correspondence," JHEP **1406** (2014) 135 [ar iv:1404.1838 [hep-th]].

[21] T. Matsumoto and K. Yoshida, "Integrability of classical strings dual for noncommutative gauge theories," JHEP **1406** (2014) 163 [ar iv:1404.3657 [hep-th]].

[22] T. Matsumoto and K. Yoshida, "Schrödinger geometries arising from Yang-Baxter deformations," JHEP **1504** (2015) 180 [ar iv:1502.00740 [hep-th]].

[23] I. Kawaguchi, T. Matsumoto and K. Yoshida, "A Jordanian deformation of AdS space in type IIB supergravity," JHEP **1406** (2014) 146 [ar iv:1402.6147 [hep-th]].

[24] T. Matsumoto and K. Yoshida, "Yang-Baxter deformations and string dualities," JHEP **1503** (2015) 137 [ar iv:1412.3658 [hep-th]].

[25] P. Basu and L. A. Pando Zayas, "Chaos Rules out Integrability of Strings in $AdS_5 \times T^{1,1}$," Phys. Lett. B **700** (2011) 243 [ar iv:1103.4107 [hep-th]].

[26] A. Catal-Ozer, "Lunin-Maldacena deformations with three parameters," JHEP **0602** (2006) 026 [hep-th 0512290].

[27] P. M. Crichigno, T. Matsumoto and K. Yoshida, "Deformations of $T^{1,1}$ as Yang-Baxter sigma models," JHEP **1412** (2014) 085 [ar iv:1406.2249 [hep-th]].

[28] T. Matsumoto and K. Yoshida, "Integrable deformations of the $AdS_5 \times S^5$ superstring and the classical Yang-Baxter equation -*Towards the gravity/CYBE correspondence*-," J. Phys. Conf. Ser. **563** (2014) 1, 012020 [ar iv:1410.0575 [hep-th]].

[29] T. Matsumoto, D. Orlando, S. Reffert, J. Sakamoto and K. Yoshida, "Yang-Baxter deformations of Minkowski spacetime," ar iv:1505.04553 [hep-th].

[30] M. A. Melvin, "Pure Magnetic And Electric Geons," Phys. Lett. **8** (1964) 65.

[31] G. W. Gibbons and K. i. Maeda, "Black Holes and Membranes in Higher Dimensional Theories with Dilaton Fields," Nucl. Phys. B **298** (1988) 741.

[32] A. A. Tseytlin, "Melvin solution in string theory," Phys. Lett. B **346** (1995) 55 [hep-th 9411198].

[33] A. Hashimoto and K. Thomas, "Dualities, twists, and gauge theories with non-constant non-commutativity," JHEP **0501** (2005) 033 [hep-th 0410123].

[34] A. Hashimoto and S. Sethi, "Holography and string dynamics in time dependent backgrounds," Phys. Rev.

Lett. **89** (2002) 261601 [hep-th 0208126].

[35] M. Spradlin, T. Takayanagi and A. Volovich, "String theory in beta deformed spacetimes," JHEP **0511** (2005) 039 [hep-th 0509036].

[36] R. Ricci, A. A. Tseytlin and M. Wolf, "On T-Duality and Integrability for Strings on AdS Backgrounds," JHEP **0712** (2007) 082 [ar iv:0711.0707 [hep-th]].

[37] N. Mohammedi, "On the geometry of classically integrable two-dimensional non-linear sigma models," Nucl. Phys. B **839** (2010) 420 [ar iv:0806.0550 [hep-th]].

[38] N. Beisert, R. Ricci, A. A. Tseytlin and M. Wolf, "Dual Superconformal Symmetry from $AdS_5 \times S^5$ Superstring Integrability," Phys. Rev. D **78** (2008) 126004 [ar iv:0807.3228 [hep-th]].

[39] D. Orlando and L. I. Uruchurtu, "Integrable Superstrings on the Squashed Three-sphere," JHEP **1210**, 007 (2012) [ar iv:1208.3680 [hep-th]].

[40] C. M. Hull, "Timelike T duality, de Sitter space, large N gauge theories and topological field theory," JHEP **9807** (1998) 021 [hep-th 9806146].

[41] A. Ballesteros, F. J. Herranz and N. R. Bruno, "Quantum (anti)de Sitter algebras and generalizations of the kappa-Minkowski space," hep-th 0409295.

[42] G. Arutyunov, R. Borsato and S. Frolov, "S-matrix for strings on η-deformed $AdS_5 \times S^5$," JHEP **1404** (2014) 002 [ar iv:1312.3542 [hep-th]].

[43] G. Arutyunov, R. Borsato and S. Frolov, "Puzzles of eta-deformed $AdS_5 \times S^5$," [ar iv:1507.04239 [hep-th]].

[44] T. Kameyama, H. Kyono, J. Sakamoto and K. Yoshida "Lax pairs on Yang-Baxter deformed backgrounds," [ar iv:1509.00173 [hep-th]]

[45] S. Zakrzewski, "Poisson structures on the Poincaré group," Comm. Math. Phys. **185** (1997) 285 [q-alg 9602001]; J. Lukierski and M. Woronowicz, "New Lie-algebraic and quadratic deformations of Minkowski space from twisted Poincare symmetries," Phys. Lett. B **633** (2006) 116 [hep-th 0508083]; A. Borowiec, J. Lukierski, M. n. Mozrzymas and V. N. Tolstoy, "N=1 2 Deformations of Chiral Superspaces from New Twisted Poincare and Euclidean Superalgebras," JHEP **1206** (2012) 154 [ar iv:1112.1936 [hep-th]].

[46] S. Hellerman, D. Orlando and S. Reffert, "String theory of the Omega deformation," JHEP **1201** (2012) 148 [ar iv:1106.0279 [hep-th]]; "The Omega Deformation From String and M-Theory," JHEP **1207** (2012) 061 [ar iv:1204.4192 [hep-th]].
D. Orlando and S. Reffert, "Deformed supersymmetric gauge theories from the fluxtrap background," Int. J. Mod. Phys. A **28** (2013) 1330044 [ar iv:1309.7350 [hep-th]].
N. Lambert, D. Orlando and S. Reffert, "Alpha- and Omega-Deformations from fluxes in M-Theory," JHEP **1411** (2014) 162 [ar iv:1409.1219 [hep-th]].

[47] V. N. Tolstoy, "Twisted Quantum Deformations of Lorentz and Poincaré algebras," ar iv:0712.3962[math QA].

[48] in preparation

[49] A. Borowiec, H. Kyono, J. Lukierski, J. Sakamoto and K. Yoshida, in preparation

Three-particle hyper-spherical harmonics and quark bound states

Igor Salom, and V. Dmitrašinović

Institute of Physics, Belgrade University, Pregrevica 118, Zemun,
P.O.Box 57, 11080 Beograd, Serbia

E-mail: `isalom@ipb.ac.rs`

Abstract. We construct the three-body permutation symmetric hyperspherical harmonics based on the subgroup chain $S_3 \otimes SO(3)_{rot} \subset O(2) \otimes SO(3)_{rot} \subset O(6)$ (and the subalgebra chain $u(1) \otimes so(3)_{rot} \subset u(3) \subset so(6)$). These hyperspherical harmonics represent a natural basis for solving non-relativistic three-body Schrödinger equation in three spatial dimensions. In particular, we apply the calculated three-particle harmonics to the three-quark bound state problem. We consider confining Δ- and Y-string three-quark effective potentials, and then calculate the spectrum of low-lying (K $\leq$ 4) bound states.

1. Introduction

The quantum-mechanical three-body bound-state problem has been addressed by a huge literature, in which the hyperspherical harmonics, Refs. [1, 2, 3, 4], provide one of the most firmly established theoretical tools. Nevertheless, little is known about the general structure of the three-body bound-state spectrum, such as the ordering of states, even in the (simplest) case of three indentical particles. In comparison, the two-body bound state problem is well understood, see Refs. [5, 6, 7, 8], where theorems controlling the ordering of bound states in convex two-body potentials were proven more than 30 years ago. In this paper we make the first significant advance in the three-body problem after the 1990 paper by Taxil & Richard, Ref. [9].

The basic difficulty lay in the absence of a systematic construction of permutation-symmetric three-body wave functions. Classification of wave functions into distinct classes under permutation symmetry in the three-body system, should be a matter of course, and yet permutation symmetric three-body hyperspherical harmonics in three dimensions were known explicitly only in a few special cases, such as those with total orbital angular momentum $L = 0$, see Refs. [3, 10] before the recent progress made in Ref. [11]. In this paper we confine ourselves to the study of factorizable (in the hyper-radius and hyper-angles) three-body potentials for technical reasons: For this class of potentials our method allows closed-form (analytical) results, at sufficiently small values of the grand angular momentum K (i.e. up to, and including the K $\leq$ 8 shell). Factorizable potentials include homogenous potentials, which in turn include pair-wise sums of two-body power-law potentials, such as the linear (confining) -string , and the Coulomb ones, as well as the genuine three-body Y-string potential [12, 13].

In this paper, we shall: 1) show how the Schrödinger equation for three particles in a homogenous/factorizable potential can be reduced to a single differential equation and an algebraic/numerical problem for their coupling strengths; 2) use this result to explicitly confirm

Journal of Physics: Conference Series **670** (2016) 012044 doi:10.1088/1742-6596/670/1/012044

Richard and Taxil's results, [9], for the ordering of K = 3 shell three-quark states, and thus resolve the controversy with [14]; 3) calculate the K = 4 shell's (purported universal) spectral splittings in terms of four parameters (lowest hyperspherical harmonics expansion coefficients) that characterize the three-body potential. 4) show that the first manifest differences in the ordering of states in the Y- and -string potentials appear in the K = 3 shell, and then reappear more emphatically in the K = 4 shell.

Our work is based on the recent advances in the construction of three-body wave functions with well-defined permutation symmetry, see Ref. [11].

2. Three-body roblem in hy er-s herical coordinates

The three-body wave function $\Psi(\boldsymbol{\rho}, \boldsymbol{\lambda})$ can be transcribed from the Euclidean relative position (Jacobi) vectors $\boldsymbol{\rho} = \frac{1}{\sqrt{2}}(\mathbf{x_1} - \mathbf{x_2})$, $\boldsymbol{\lambda} = \frac{1}{\sqrt{6}}(\mathbf{x_1} + \mathbf{x_2} - 2\mathbf{x_3})$, into hyper-spherical coordinates as $\Psi(R, \ _5)$, where $R = \sqrt{\boldsymbol{\rho}^2 + \boldsymbol{\lambda}^2}$ is the hyper-radius, and five angles $\ _5$ that parametrize a hyper-sphere in the six-dimensional Euclidean space. Three (Φ_i; $i = 1, 2, 3$) of these five angles ($\ _5$) are just the Euler angles associated with the orientation in a three-dimensional space of a spatial reference frame defined by the (plane of) three bodies; the remaining two hyper-angles describe the shape of the triangle subtended by three bodies; they are functions of three independent scalar three-body variables, e.g. $\boldsymbol{\rho} \cdot \boldsymbol{\lambda}$, $\boldsymbol{\rho}^2$, and $\boldsymbol{\lambda}^2$. As we saw above, one linear combination of the two variables $\boldsymbol{\rho}^2$, and $\boldsymbol{\lambda}^2$, is already taken by the hyper-radius R, so the shape-space is two-dimensional, and topologically equivalent to the surface of a three-dimensional sphere.

There are two traditional ways of parameterizing this sphere: 1) the standard Delves choice, [1], of hyper-angles (χ, θ), that somewhat obscures the full S_3 permutation symmetry of the problem; 2) the Iwai, Ref. [4], hyper-angles (α, ϕ): $(\sin\alpha)^2 = 1 - \left(\frac{2\boldsymbol{\rho}\times\boldsymbol{\lambda}}{R^2}\right)^2$, $\tan\phi = \left(\frac{2\boldsymbol{\rho}\cdot\boldsymbol{\lambda}}{\boldsymbol{\rho}^2 - \boldsymbol{\lambda}^2}\right)$, reveal the full S_3 permutation symmetry of the problem: the angle α does not change under permutations, so that all permutation properties are encoded in the ϕ-dependence of the wave functions. We shall use the latter choice, as it leads to permutation-symmetric hyperspherical harmonics, see Ref. [11].

We expand the wave function $\Psi(R, \ _5)$ in terms of hyper-spherical harmonics $\mathcal{Y}^{\mathrm{K}}_{[\mathrm{m}]}(\ _5)$, $\Psi(R, \ _5) = \sum_{\mathrm{K},[\mathrm{m}]} \psi^{\mathrm{K}}_{[\mathrm{m}]}(R) \mathcal{Y}^{\mathrm{K}}_{[\mathrm{m}]}(\ _5)$, where K together with $[\mathrm{m}] = [Q, \nu, L, L_z = m]$ constitute the complete set of hyperspherical quantum numbers: K is the hyper-spherical angular momentum, L is the (total orbital) angular momentum, $L_z = m$ its projection on the z-axis, Q is the Abelian quantum number conjugated with the Iwai angle ϕ, and ν is the multiplicity label that distinguishes between hyperspherical harmonics with remaining four quantum numbers that are identical.

The hyper-spherical harmonics turn the Schrödinger equation into a set of (infinitely) many coupled equations,

$$
\begin{aligned}
&- \frac{1}{2\mu}\left[\frac{d^2}{dR^2} + \frac{5}{R}\frac{d}{dR} - \frac{\mathrm{K}(\mathrm{K}+4)}{R^2} + 2\mu E\right] \psi^{\mathrm{K}}_{[\mathrm{m}]}(R) \\
&+ \ V_{\mathrm{eff.}}(R) \sum_{\mathrm{K}',[\mathrm{m}']} C^{\mathrm{K}\ \mathrm{K}'}_{[\mathrm{m}][\mathrm{m}']} \psi^{\mathrm{K}'}_{[\mathrm{m}']}(R) = 0 \qquad (1)
\end{aligned}
$$

with a hyper-angular coupling coefficients matrix $C^{\mathrm{K}\ \mathrm{K}'}_{[\mathrm{m}][\mathrm{m}']}$ defined by

$$
\begin{aligned}
V_{\mathrm{eff.}}(R) C^{\mathrm{K}'\ \mathrm{K}}_{[\mathrm{m}'][\mathrm{m}]} &= \langle \mathcal{Y}^{\mathrm{K}'}_{[\mathrm{m}']}(\ _5) | V(R, \alpha, \phi) | \mathcal{Y}^{\mathrm{K}}_{[\mathrm{m}]}(\ _5) \rangle \\
&= V(R) \langle \mathcal{Y}^{\mathrm{K}'}_{[\mathrm{m}']}(\ _5) | V(\alpha, \phi) | \mathcal{Y}^{\mathrm{K}}_{[\mathrm{m}]}(\ _5) \rangle. \qquad (2)
\end{aligned}
$$

In Eq. (1) we used the factorizability of the potential $V(R, \alpha, \phi) = V(R) V(\alpha, \phi)$ to reduce this set to one (common) hyper-radial Schrödinger equation. The hyper-angular part $V(\alpha, \phi)$ can be expanded in terms of O(6) hyper-spherical harmonics with zero angular momenta $L = m = 0$ (due to the rotational invariance of the potential),

$$V(\alpha, \phi) = \sum_{\mathrm{K},Q}^{\infty} v_{\mathrm{K},Q}^{3-\mathrm{body}} \mathcal{Y}_{00}^{\mathrm{K}Q\nu}(\alpha, \phi) \tag{3}$$

where $v_{\mathrm{K},Q}^{3-\mathrm{body}} = \int \mathcal{Y}_{00}^{\mathrm{K}Q\nu *}(\ _5)\, V(\alpha, \phi)\, d\ _{(5)}$ leading to

$$\begin{aligned} V_{\mathrm{eff.}}(R) C_{[\mathrm{m}''][\mathrm{m}']}^{\mathrm{K}''\ \mathrm{K}'} = & \quad V(R) \sum_{\mathrm{K},Q}^{\infty} v_{\mathrm{K},Q}^{3-\mathrm{body}} \\ \langle \mathcal{Y}_{[\mathrm{m}'']}^{\mathrm{K}''}(\ _5)| & \quad \mathcal{Y}_{00}^{\mathrm{K}Q\nu}(\alpha, \phi) | \mathcal{Y}_{[\mathrm{m}']}^{\mathrm{K}'}(\ _5)\rangle \end{aligned} \tag{4}$$

There is no summation over the multiplicity index in Eq. (3), because no multiplicity arises for harmonics with $L < 2$. Here we separate out the $\mathrm{K} = 0$ term and absorb the factor $\frac{v_{00}^{3-\mathrm{body}}}{\pi\sqrt{\pi}}$ into the definition of $V_{\mathrm{eff.}}(R) = \frac{v_{00}^{3-\mathrm{body}}}{\pi\sqrt{\pi}} V(R)$ to find

$$\begin{aligned} C_{[\mathrm{m}''][\mathrm{m}']}^{\mathrm{K}''\ \mathrm{K}'} = \delta_{\mathrm{K}'',\mathrm{K}'} \delta_{[\mathrm{m}''],[\mathrm{m}']} + \pi\sqrt{\pi} \textstyle\sum_{\mathrm{K}>0,Q}^{\infty} \frac{v_{\mathrm{K},Q}^{3-\mathrm{body}}}{v_{00}^{3-\mathrm{body}}} \\ \times \langle \mathcal{Y}_{[\mathrm{m}'']}^{\mathrm{K}''}(\ _5) | \mathcal{Y}_{00}^{\mathrm{K}Q\nu}(\alpha, \phi) | \mathcal{Y}_{[\mathrm{m}']}^{\mathrm{K}'}(\ _5)\rangle. \end{aligned} \tag{5}$$

Homogenous potentials, such as the and Y-string ones, which are linear in R, and the Coulomb one, have first coefficients $v_{00}^{3-\mathrm{body}}$ in the h.s. expansion that are one order of magnitude larger than the rest $v_{\mathrm{K}>0,Q}^{3-\mathrm{body}}$. This reflects the fact that, on the average, these potential energies depend more on the overall size of the system than on its shape, thus justifying the perturbative approach taken in Ref. [9], with the first term in Eq. (5) taken as the zeroth-order approximation.[1]

In such cases Eqs. (1) decouple, leading to zeroth order solutions for $\psi_{0[\mathrm{m}]}^{\mathrm{K}}(R)$ that are independent of $[m]$ and thus have equal energies within the same K shell, and different energies in different K shells. Two known exceptions are potentials with the homogeneity degree $\mathrm{k} = -1, 2$, that lead to accidental degeneracies and have to be treated separately.

The first-order corrections are obtained by diagonalization of the block matrices $C_{[\mathrm{m}][\mathrm{m}']}^{\mathrm{K}\ \mathrm{K}}$, $\mathrm{K} = 1, 2, ...$, while the off-diagonal couplings $C_{[\mathrm{m}][\mathrm{m}']}^{\mathrm{K}\ \mathrm{K}'}, \mathrm{K} \neq \mathrm{K}'$ appear only in the second-order corrections. Rather than calculating perturbative first-order energy shifts, a better approximation is obtained when the diagonalized block matrices are plugged back into Eq. (1), which equations then decouple into a set of (separate) individual ODEs in one variable, that differ only in the value of the effective coupling constant:

$$\left[\frac{d^2}{dR^2} + \frac{5}{R}\frac{d}{dR} - \frac{\mathrm{K}(\mathrm{K}+4)}{R^2} + 2\mu(E - V_{[\mathrm{m_d}]}^{\mathrm{K}}(R))\right] \psi_{[\mathrm{m_d}]}^{\mathrm{K}}(R) = 0, \tag{6}$$

where $V_{[\mathrm{m_d}]}^{\mathrm{K}}(R) = C_{[\mathrm{m_d}]}^{\mathrm{K}} V_{\mathrm{eff.}}(R)$, with $C_{[\mathrm{m_d}]}^{\mathrm{K}}$ being the eigenvalues of matrix $C_{[\mathrm{m}][\mathrm{m}']}^{\mathrm{K}\ \mathrm{K}}$.

The spectrum of three-body systems in homogenous potentials is now reduced to finding the eigenvalues of a single differential operator, just as in the two-body problem with a radial potential. The matrix elements in Eq. (5) can be readily evaluated using the permutation-symmetric O(6) hyper-spherical harmonics and the integrals that are spelled out in Ref. [11].

[1] (note that the h.s. matrix elements $\langle \mathcal{Y}_{[\mathrm{m}'']}^{\mathrm{K}''}(\Omega_5) | \mathcal{Y}_{00}^{\mathrm{K}Q\nu}(\alpha, \phi) | \mathcal{Y}_{[\mathrm{m}']}^{\mathrm{K}'}(\Omega_5)\rangle$ under the sum are always less than $\frac{1}{\pi\sqrt{\pi}}$).

Journal of Physics: Conference Series **670** (2016) 012044 doi:10.1088/1742-6596/670/1/012044

This is our main (algebraic) result: combined with the hyperspherical harmonics recently obtained in Ref. [11], it allows one to evaluate the discrete part of the (energy) spectrum of a three-body potential as a function of its shape-sphere harmonic expansion coefficients $v^{3-\text{body}}_{\text{K},Q}$. Generally, these matrix elements obey selection rules: they are subject to the triangular conditions $\text{K}' + \text{K}'' \geq \text{K} \geq |\text{K}' - \text{K}''|$ plus the condition that $\text{K}' + \text{K}'' + \text{K} = 0, 2, 4, \ldots$, and the angular momenta satisfy the selection rules: $L' = L''$, $m' = m''$. Moreover, Q is an Abelian (i.e. additive) quantum number that satisfies the simple selection rule: $Q'' = Q' + Q$. That reduces the sum in Eq. (5) to a finite one, that depends on a finite number of coefficients $v^{3-\text{body}}_{\text{K},Q}$; for small values of K, this number is also small.

A matrix such as that in Eq. (5) is generally sparse in the permutation-symmetric basis, so its diagonalization is not a serious problem, and, for sufficiently small K values it can even be accomplished in closed form: for example, for $\text{K} \leq 5$, all results depend only on four coefficients $(v_{00}, v_{40}, v_{6\pm6}, v_{80})$, and there is at most three-state mixing, so the eigenvalue equations are at most cubic ones, with well-known solutions. For brevity's sake we confine ourselves to $\text{K} \leq 4$ states here.

3. Results

1) In the $\text{K} = 2$ band/shell of the three-body energy spectrum the eigen-energies depend on two coefficients (v_{00}, v_{40}), and the splittings among various levels depend only on the (generally small, see Table 1) ratio v_{40}/v_{00}. This means that the eigen-energies form a fixed pattern (ordering) that does not depend on the shape of the three-body potential. The actual size of the $\text{K} = 2$ shell energy splitting depends on the small parameter v_{40}/v_{00}, provided that the potential is permutation symmetric. This fact was noticed almost 40 years ago, Refs. [15, 16], and it suggested that similar patterns might exist in higher-K shells.

The advantage of permutation-symmetric hyperspherical harmonics over the conventional ones is perhaps best illustrated here: the $\text{K} = 2$ shell splittings in the Y- and -string potentials were obtained, after some complicated calculations using conventional hyperspherical harmonics in Ref. [17], whereas here they follow from the calculation of four (simple) hyper-angular matrix elements.

2) Historically, extensions of this kind of calculations to higher $(\text{K} \geq 3)$ bands, for general three-body potentials turned out more difficult than expected: Bowler et. al, Ref. [14], published a set of predictions for the $\text{K} = 3, 4$ bands, which were later questioned by Richard and Taxil's [9], $\text{K} = 3$ hyperspherical harmonic calculation; see also Refs. [18]. This controversy had not been resolved to the present day, to our knowledge, so we address that problem first: In the $\text{K} = 3$ case the energies depend on three coefficients $(v_{00}, v_{40}, v_{6\pm6})$, and there is no mixing of multiplets, so all eigen-energies can be expressed in simple closed form that agrees with Ref. [9] and depends on two small parameters $v_{40}/v_{00}, v_{6\pm6}/v_{00}$.

Note that the third coefficient $v_{6\pm6}$ vanishes in the simplified Y-string potential without two-body terms and thus causes the first observable difference between Y- and -string potentials: the splittings between $[20, 1^-]$, and $[56, 1^-]$, as well as between $[20, 3^-]$, and $[56, 3^-]$. The vanishing of $v_{6\pm6}$ implies that the Y-string potential is independent of the Iwai angle ϕ, and that consequently there is a (new) dynamical kinematic rotations/democracy transformations O(2) symmetry, [12, 13] associated with it.

3) In the $\text{K} = 4$ band $SU(6)$, or S_3 multiplets have one of the following 12 values of the diagonalized C-matrix $C^{\text{K}}_{[\text{m}_\text{d}]} \times \frac{v_{00}}{\pi\sqrt{\pi}}$, from which one can evaluate the eigen-energies. We use the baryon-spectroscopic notation: $[\text{dim.}, L^P]$, where dim. is the dimension of the $SU_{\text{FS}}(6)$ representation and the correspondence with the representations of the permutation group S_3

is given as $70 \leftrightarrow M$, $20 \leftrightarrow A$, $56 \leftrightarrow S$.

$$[70,0^+]: \frac{1}{\pi\sqrt{\pi}}\left(v_{00} + \frac{\sqrt{3}}{2}v_{40} + \frac{1}{2\sqrt{5}}v_{80}\right)$$

$$[56,0^+]: \frac{1}{\pi\sqrt{\pi}}\left(v_{00} \qquad + \frac{2}{\sqrt{5}}v_{80}\right)$$

$$[70,1^+]: \frac{1}{\pi\sqrt{\pi}}\left(v_{00} \qquad - \frac{1}{\sqrt{5}}v_{80}\right)$$

$$[70,2^+]: \frac{1}{\pi\sqrt{\pi}}\left(v_{00} + \frac{1}{35}\left(7\sqrt{3}v_{40} + 2\sqrt{5}v_{80}\right.\right.$$

$$\left.\left.-3\sqrt{3v_{40}^2 - 2\sqrt{15}v_{40}v_{80} + 5v_{80}^2 + 120v_{6\pm6}^2}\right)\right)$$

$$[70',2^+]: \frac{1}{\pi\sqrt{\pi}}\left(v_{00} + \frac{1}{35}\left(7\sqrt{3}v_{40} + 2\sqrt{5}v_{80}\right.\right.$$

$$\left.\left.+3\sqrt{3v_{40}^2 - 2\sqrt{15}v_{40}v_{80} + 5v_{80}^2 + 120v_{6\pm6}^2}\right)\right)$$

$$[56,2^+]: \frac{1}{\pi\sqrt{\pi}}\left(v_{00} - \frac{12\sqrt{3}}{35}v_{40} + \frac{\sqrt{5}}{7}v_{80}\right)$$

$$[20,2^+]: \frac{1}{\pi\sqrt{\pi}}\left(v_{00} \qquad - \frac{1}{\sqrt{5}}v_{80}\right)$$

$$[20,3^+]: \frac{1}{\pi\sqrt{\pi}}\left(v_{00} - \frac{3\sqrt{3}}{14}v_{40} - \frac{\sqrt{5}}{14}v_{80}\right)$$

$$[70,3^+]: \frac{1}{\pi\sqrt{\pi}}\left(v_{00} - \frac{5\sqrt{3}}{14}v_{40} + \frac{1}{14\sqrt{5}}v_{80}\right)$$

$$[56,4^+]: \frac{1}{\pi\sqrt{\pi}}\left(v_{00} + \frac{5\sqrt{3}}{14}v_{40} + \frac{3}{14\sqrt{5}}v_{80}\right)$$

$$[70,4^+]: \frac{1}{\pi\sqrt{\pi}}\left(v_{00} + \frac{1}{42\sqrt{5}}\left(-2v_{80}\right.\right.$$

$$\left.\left.-\sqrt{1215v_{40}^2 - 54\sqrt{15}v_{40}v_{80} + 9v_{80}^2 + 1280v_{6\pm6}^2}\right)\right)$$

$$[70',4^+]: \frac{1}{\pi\sqrt{\pi}}\left(v_{00} + \frac{1}{42\sqrt{5}}\left(-2v_{80}\right.\right.$$

$$\left.\left.+\sqrt{1215v_{40}^2 - 54\sqrt{15}v_{40}v_{80} + 9v_{80}^2 + 1280v_{6\pm6}^2}\right)\right).$$

Table 1 shows that the ordering of K = 4 states is not universally valid even for these two convex confining potentials. This, of course, is a consequence of different ratios v_{40}/v_{00}, $v_{6\pm6}/v_{00}$ and v_{80}/v_{00}. That goes to show that one cannot expect strongly restrictive ordering theorems

Table 1. Expansion coefficients v_{KQ} of the Y- and -string potentials in terms of O(6) hyper-spherical harmonics $\mathcal{Y}_{0,0}^{K,0,0}$, for K = 0, 4, 8, respectively, and of the hyper-spherical harmonics $\mathcal{Y}_{0,0}^{6,\pm6,0}$.

(K,Q)	(0,0)	(4,0)	(6,±6)	(8,0)
v_{KQ}(Y)	8.18	-0.44	0	-0.09
v_{KQ}( )	16.04	-0.44	-0.14	-0.06

to hold for three-body systems in general, the way they hold in the two-body problem, Ref. [8]. Nevertheless, even the present results can be useful, as they indicate that certain groups of multiplets are jointly lifted, or depressed in the spectrum, subject to the value of the ratio v_{40}/v_{00}, with ordering within each group being subject to the finer structure of the potential, i.e., to higher coefficients ratios $v_{6\pm6}/v_{00}$ and v_{80}/v_{00}.

Of course, similar conclusions hold also for K = 3 spectrum splitting, but are less pronounced, as that shell depends only on two numbers: the ratios v_{40}/v_{00} and $v_{6\pm6}/v_{00}$. As the difference between and Y-string potentials is most pronounced in the value of $v_{6\pm6}$, that is the case where the distinction between these two potentials is most clearly seen.

4. Summary and Conclusions

We have reduced the non-relativistic (quantum) three-identical-body problem to a single ordinary differential equation for the hyper-radial wave function with coefficients multiplying the linear hyper-radial potential determined by O(6) group-theoretical arguments. That equation can be solved in the same way as the radial Schrödinger equation in 3D. The breaking of the O(6) symmetry by the three-body potential determines the ordering of states in the spectrum.

In three dimensions (3D) the hyper-spherical symmetry group is O(6), and the residual dynamical symmetry of the potential is $S_3 \otimes SO(3)_{rot} \subset O(2) \otimes SO(3)_{rot} \subset O(6)$, where $SO(3)_{rot}$ is the rotational symmetry associated with the (total orbital) angular momentum L. We showed how the energy eigenvalues can be calculated in terms of the three-body potential's (hyper-)spherical harmonics expansion coefficients v_{KQ}.

The ordering of bound states has its most immediate application in the physics of three confined quarks, where the question was raised originally, Refs. [9, 14, 15, 16, 17]. We have used these results to calculate the energy splittings of various SU(6)/S_3 multiplets in the K $\leq$ 4 shells of the Y- and -string potential spectra. The dynamical $O(2)$ dynamical symmetry of the Y-string potential was discovered in Ref. [12], with the permutation group $S_3 \subset O(2)$ as its subgroup. The existence of an additional dynamical symmetry strongly suggested an algebraic approach to this problem, such as that used in two-dimensional space, in Ref. [13]. We have shown that the first clear difference between the spectra of these two models of confinement appears in the K $\geq$ 3 shell. That is also the first explicit consequence of the dynamical O(2) symmetry of the Y-string potential. We stress the analytical nature of our results, in contrast to the numerical results of Refs. [18].

The next step would be to apply the method to linear combinations of homogenous potentials, which can only be done numerically, however. Several realistic two-body potentials, such as the Lennard-Jones inter-atomic one, as well as the Coulomb + linear quark-quark one, are simple linear combinations of (only) two homogenous potentials.

Acknowledgments

This work was financed by the Serbian Ministry of Science and Technological Development under grant numbers OI 171031, OI 171037 and III 41011.

Re erences

[1] L. M. Delves, *Nucl. Phys.* **9**, 391 (1958); ibid. **20**, 275 (1960).

[2] F. T. Smith, *J. Chem. Phys.* **31**, 1352 (1959); F. T. Smith, *Phys. Rev.* **120**, 1058 (1960); F. T. Smith, *J. Math. Phys.* **3**, 735 (1962); R. C. Whitten and F. T. Smith, *J. Math. Phys.* **9**, 1103 (1968).

[3] Yu. A. Simonov, *Sov. J. Nucl. Phys.* **3**, 461 (1966) [*Yad. Fiz.* **3**, 630 (1966)].

[4] T. Iwai, *J. Math. Phys.* **28**, 964, 1315 (1987).

[5] H. Grosse and A. Martin, *Phys. Rept.* **60**, 341 (1980).

[6] B. Baumgartner, H. Grosse and A. Martin, *Nucl. Phys.* B **254**, 528 (1985).

[7] A. Martin, J. M. Richard and P. Taxil, *Nucl. Phys.* B **329**, 327 (1990).

[8] H. Grosse and A. Martin, *Particle physics and the Schrödinger equation*, Cambridge University Press, (1997).

[9] J. -M. Richard and P. Taxil, *Nucl. Phys.* B **329**, 310 (1990).

[10] N. Barnea and V. B. Mandelzweig, *Phys. Rev.* A **41**, 5209 (1990).

[11] Igor Salom and V. Dmitrašinović, "Permutation-symmetric three-particle hyper-spherical harmonics based on the $S_3 \otimes SO(3)_{rot} \subset U(1) \otimes SO(3)_{rot} \subset U(3) \subset O(6)$ chain of algebras", in preparation (2015).

[12] V. Dmitrašinović, T. Sato and M. Šuvakov , *Phys. Rev.* D **80**, 054501 (2009).

[13] V. Dmitrašinović and Igor Salom, *J. Math. Phys.* **55**, 082105 (16) (2014).

[14] K. C. Bowler, P. J. Corvi, A. J. G. Hey, P. D. Jarvis and R. C. King, *Phys. Rev.* D **24**, 197 (1981); K. C. Bowler and B. F. Tynemouth, *Phys. Rev.* D **27**, 662 (1983).

[15] D. Gromes and I. O. Stamatescu, *Nucl. Phys.* B **112**, 213 (1976); *Z. Phys.* **C 3**, 43 (1979).

[16] N. Isgur and G. Karl, *Phys. Rev.* D **19**, 2653 (1979).

[17] V. Dmitrašinović, T. Sato and M. Šuvakov, *Eur. Phys. J.* C **62**, 383 (2009).

[18] F. Stancu and P. Stassart, *Phys. Lett.* B **269**, 243 (1991).

Journal of Physics: Conference Series **670** (2016) 012045 doi:10.1088/1742-6596/670/1/012045

Dimensional Regularization of the Spatial wave function for a singular contact interaction in the Relativistic Schrodinger Equation

A. M. Shalaby

Department of Mathematics, Statistics, and Physics, Qatar University, Doha 2713, Qatar

E-mail: amshalab@qu.edu.qa

Abstract. Based on our previous work in PRD 89, 125023 (2014), we stress here (for the first time) the regularization of the spatial wave function for the δ-contact interaction within the relativistic Schrodinger equation. The D-dimensional inverse Fourier transform has been utilized to map the momentum-space wave function to the spatial one. To regularize the logarithmic blow up of the wave function as $x \to 0$, we employed the dimensional regularization technique. We assert that although the technique has been used here for the bound state only, the form of the scattering states in momentum space assures the reliability of the same technique to regularize the spatial scattering wave functions.

In particle physics, the asymptotic freedom property plays a vital role in describing strong interactions. This property is responsible for the confinement of quarks inside protons (for instance). In fact, the need for a theory with such property was based on the experimental results from SLAC [1] where it has been discovered that fast electrons scatter from protons as if they have scattered from quasi-free, point-like constituents inside the protons. At that time there were no known theory that can describe the scattering pattern as **QED** shows an opposite behavior. This leaded to the formulation of quantum chromodynamics, a non-abelian gauge field theory with the group structure $SU(3)$. In **QCD**, the number of colors is equal to or greater than three to secure the existence of the asymptotic freedom [6, 8, 7]. In 1975, Symanzik [2] advocated the simpler one-component $(-\phi^4)$ scalar field theory and later the asymptotic freedom of this theory has been stressed [3, 4, 5]. Other studies also stressed the existence of quantum field-like properties in non-relativistic quantum mechanical theories [9, 10, 11, 12, 13, 14, 15, 16, 17, 18]. Very recently, we discovered the existence of such property in a relativistic Schrodinger equation with contact interactions [19, 20].

In the relativistic Schrodinger equation where the kinetic term takes the form $\sqrt{p^2+m^2}$, certain precautions should be taken into account. According to Leutwyler's non-interaction theorem [21], the potential term to be added to that kinetic term should not let the theory violating the principles of the special theory of relativity. In fact, the no-go theorem of Leutwyler admits the singular contact interactions and thus there is no logic reason of excluding them from relativistic quantum mechanical studies. The study of the singular contact interactions within the context of relativistic Schrodinger equation enables us to avoid certain problems that do exist in either Klein-Gordon or Dirac equations. For instance, the existence of Ghost states in Klein-Gordon equation and the existence of holes in Dirac equation represent main problems in these theories when considering them as relativistic quantum mechanical theories. These theories are in fact successful when considered as quantum field ones.

In Refs.[19, 20] we studied two types of one dimensional contact interactions within the relativistic Schrodinger equation. Although the two potential terms are different, a type of universality has been shown up. The universality characteristic is manifested in finding that the two theories shares the same properties. However, the coupling in the two theories are of different mass dimensions which reflects the need for using two different regularization techniques. For the $\delta-$ potential, the coupling is dimensionless and thus the theory contains only logarithmic type of divergences. On the other hand, the coupling in the $\delta'-$ potential has a length dimension and the theory contains logarithmic, linear and quadratic type of divergences. For logarithmic divergences both dimensional and cut off regularization lead to the same results but for other type of divergences, dimensional regularization may not reflect the correct scale dependance (for instance for the scalar ϕ^4 field theory in $2+1$ space-time dimensions, linear divergences will not appear in using dimensional regularization). In our work in Refs. [19, 20], the two theories have been regularized in momentum space and the spatial wave function has shown up a logarithmic blow up as x approaches zero. In this work, we stress this point and show that using dimensional regularization, one can show that the blow up of the wave function as $x \to 0$ can be regularized using dimensional regularization. This point has not been stressed before and represents the main point of our work here.

To start, consider the relativistic Schrodinger equation in D dimensions of the form:

$$\sqrt{p^2+m^2}\Psi(x_1,x_2,....,x_D) + \lambda\delta(x_1,x_2,....,x_D)\Psi(x_1,x_2,....,x_D) = E\Psi(x_1,x_2,....,x_D). \tag{1}$$

We have in momentum space the following quantities:

$$\Psi(x_1,x_2,....,x_D) = \frac{1}{(2\pi)^D}\int d^Dp\,\widetilde{\Psi}(p)\exp(i\vec{p}\cdot\vec{x}), \tag{2}$$

$$\Psi(0,0,...0) = \frac{1}{(2\pi)^D}\int d^Dp\,\widetilde{\Psi}(p), \tag{3}$$

$$\widetilde{\Psi}(p) = \frac{\lambda\Psi(0,0,...0)}{\left(E_B-\sqrt{p^2+m^2}\right)}. \tag{4}$$

Here E_B represents the bound state energy. Since $\Psi(0,0,...0)$ is a quantity in position space, it would be more plausible to express $\widetilde{\Psi}_B(p)$ in terms of the momentum representations of all quantities. So we put:

$$\widetilde{\Psi}(p) = \frac{\lambda}{2\pi}\frac{\int d^Dp'\widetilde{\Psi}(p')}{\left(E_B-\sqrt{p^2+m^2}\right)}. \tag{5}$$

In fact, this is a simple integral equation for which $\widetilde{\Psi}(p)$ given by:

$$\widetilde{\Psi}(p) = \frac{A\lambda}{2\pi\left(E_B-\sqrt{p^2+m^2}\right)} \tag{6}$$

represents a solution, where A is a normalization constant. This solution will satisfy the integral equation since when substituted back into integral equation we get:

$$\widetilde{\Psi}(p) = \lambda A\,\frac{\int d^Dp'\frac{\lambda}{2\pi\left(E_B-\sqrt{(p')^2+m^2}\right)}}{2\pi\left(E_B-\sqrt{p^2+m^2}\right)}, \tag{7}$$

we can employ the gap equation in Ref.[19] that reads:

$$\int \frac{dp'}{2\pi\left(E_B - \sqrt{(p')^2 + m^2}\right)} = \frac{1}{\lambda}. \tag{8}$$

Thus the form $\widetilde{\Psi}(p)$ in Eq.(6) represents a solution of the integral equation. This form of $\widetilde{\Psi}(p)$ suggests that the regularization of λ will certainly regularize the wave function in momentum space. Using dimensional regularization we get:

$$\begin{aligned}
\frac{m^{D-1}}{\lambda(\varepsilon)} &= \frac{1}{(2\pi)^D}\int d^Dp\,\frac{1}{E_B - \sqrt{p^2+m^2}} = I(E_B) \\
&= -\frac{1}{(2\pi)^D}\int \frac{d^Dp}{\sqrt{p^2+m^2}} - \frac{1}{(2\pi)^D}\int \sum_{n=1}^{\infty} d^Dp\frac{(E_B)^n}{(p^2+m^2)^{\frac{n+1}{2}}} \\
&= -\frac{1}{(2\pi)^D}\int \frac{d^Dp}{\sqrt{p^2+m^2}} - \frac{E_B}{2\pi\sqrt{m^2-E_B^2}}\left(\pi + 2\arcsin\frac{E_B}{m}\right) \\
&= -\frac{1}{(4\pi)^{\frac{D}{2}}}\frac{\Gamma\left(\frac{1-D}{2}\right)}{\Gamma\left(\frac{1}{2}\right)}\left(\frac{1}{m^2}\right)^{\frac{1-D}{2}} - \frac{E_B}{2\pi\sqrt{m^2-E_B^2}}\left(\pi + 2\arcsin\frac{E_B}{m}\right) \\
&\propto \frac{1}{\epsilon} + (\text{finite terms}),\ \epsilon = D-1.
\end{aligned} \tag{9}$$

The UV divergences in momentum space will have its manifestation in position-space as small-x divergences. The position-space wave function is obtained via Fourier inverse transformation of $\widetilde{\Psi}(p)$. Since the inverse Fourier transform is an integration over momentum, one can aim to employ dimensional regularization to get a regularized wave function; a wave function that is finite everywhere but scale dependent on the other hand. To do this let us consider the following:

$$\begin{aligned}
\widetilde{\Psi}(p) &= \frac{A\lambda}{2\pi\left(E_B - \sqrt{p^2+m^2}\right)} \\
&= -\frac{1}{(2\pi)^D}\frac{A\lambda}{\sqrt{p^2+m^2}} - \frac{A\lambda}{(2\pi)^D}\sum_{n=1}^{\infty}\frac{(E_B)^n}{(p^2+m^2)^{\frac{n+1}{2}}}.
\end{aligned} \tag{10}$$

In 1-dimension ($D = 1$), the spatial wave function takes the form:

$$\begin{aligned}
\Psi(x) &= \frac{1}{(2\pi)^D}\int d^Dp\frac{\exp\left(-i\vec{x}\cdot\vec{p}\right)}{E - \sqrt{p^2+m^2}} \\
&= -\frac{1}{(2\pi)^D}\int d^Dp\frac{\exp\left(-i\vec{x}\cdot\vec{p}\right)}{(p^2+m^2)^{\frac{1}{2}}} \\
&\quad -\frac{A\lambda}{\sqrt{\pi}}\sum_{n=1}^{\infty}\left(\frac{E_B}{m}\right)^n\left(\frac{m|x|}{2}\right)^{n/2}\frac{K_{n/2}(m|x|)}{\Gamma\left(\frac{n+1}{2}\right)},
\end{aligned} \tag{11}$$

where $K_{n/2}$ are the modified Bessel functions of second kind. In fact, every term in the summation above is finite in the limit $D \to 1$ but the integral of the form:

$$T_0(x) = -\frac{1}{(2\pi)^D}\int d^Dp\frac{\exp\left(-i\vec{x}\cdot\vec{p}\right)}{(p^2+m^2)^{\frac{1}{2}}} \tag{12}$$

is logarithmic divergent by power counting and thus needs to be regularized. To do this let us consider:

$$\vec{x}\cdot\vec{p} = |x|\,|p|\cos(\theta)\,. \tag{13}$$

In D dimensions, the volume element is

$$\begin{aligned} d^D p &= p^{D-1}\sin^{D-2}(\phi_1)\sin^{D-3}(\phi_2)\cdots\sin(\phi_{D-2})\,dr\,d\phi_1\,d\phi_2\cdots d\phi_{D-1} \\ \text{where } \theta &\equiv \phi_1. \end{aligned} \tag{14}$$

Then,

$$\begin{aligned} & \frac{1}{(2\pi)^D}\int d^D p\frac{\exp(-i\vec{x}\cdot\vec{p})}{(p^2+m^2)^{\frac{1}{2}}} \\ = & \frac{1}{(2\pi)^D}\int d^D p\frac{\exp(-i|x|\,|p|\cos(\phi_{D-2}))}{(p^2+m^2)^{\frac{1}{2}}} \\ = & \frac{1}{(2\pi)^D}\int_{\phi_{D-1}=0}^{2\pi}\int_{\phi_{D-2}=0}^{\pi}\cdots\int_{\phi_1=0}^{\pi}\int_0^{\infty} p^{D-1}\sin^{D-2}(\phi_1)\sin^{D-3}(\phi_2)\cdots\sin(\phi_{D-2})\,\times \\ & \times\frac{\exp(-i|x|\,|p|\cos(\phi_{D-2}))}{(p^2+m^2)^{\frac{1}{2}}}dpdd\phi_1\,d\phi_2\cdots\phi_{D-2}d\phi_{D-1} \end{aligned} \tag{15}$$

Now, the integral I_n over the angles gives:

$$\begin{aligned} I_n &= \int_{\phi=0}^{\pi}\sin^n(\phi)d\phi = \sqrt{\pi}\frac{\Gamma(\frac{1+n}{2})}{\Gamma(\frac{n}{2}+1)}, \\ I_n I_{n-1} &= \frac{2\pi}{n}, \\ I_{n-2}I_{n-3} &= \frac{2\pi}{n-2}. \end{aligned} \tag{16}$$

Or in general, the inverse Fourier transform of the wave function in D-dimensions results in the following form:

$$\Psi(x) = Am^{1-D}\lambda\,(2\pi)^{\frac{D}{2}}\sum_{n=1}^{\infty}E_B^{n-1}\frac{2^{-n/2}n\,|x|^{-D/2}\left(\frac{1}{m^2}\right)^{\frac{1}{4}(-D-n)}\left(\frac{m^2}{|x|}\right)^{-n/2}K_{\frac{D-n}{2}}\left(\frac{|x|}{\sqrt{\frac{1}{m^2}}}\right)}{\Gamma\left(\frac{n}{2}+1\right)}. \tag{17}$$

This sum is rabidly convergent as it can be realized from Fig.1. In fact, this form represents the solution of the original relativistic Schrodinger equation in D-dimensions which has not been obtained before. In $D=1$, the wave function is logarithmically divergent at the origin while for the $D=2$ case the wave function possesses logarithmic as well as linear type of divergences (in momentum space but $1/x$ in position space). In $D=3$ an extra quadratic type of divergences appears. The divergences in higher dimensions (higher than 1) need more studies which will be done in another work.

For $|x|=0$, one can re-sum the series to get:

$$\Psi(0) = Am^{1-D}\lambda\frac{\pi^{\frac{D-1}{2}}\left(\frac{1}{m^2}\right)^{-D/2}\left(\begin{array}{c} -E_B^2\Gamma\left(\frac{1-D}{2}\right){}_2F_1\left(1,\frac{1}{2}-\frac{D}{2};\frac{1}{2};\frac{E_B^2}{m^2}\right) \\ +m^2\Gamma\left(\frac{1-D}{2}\right){}_2F_1\left(1,\frac{1}{2}-\frac{D}{2};\frac{1}{2};\frac{E_B^2}{m^2}\right) \\ +\sqrt{\pi}E_B m\Gamma\left(1-\frac{D}{2}\right)\left(1-\frac{E_B^2}{m^2}\right)^{D/2} \end{array}\right)}{\sqrt{m^2}\left(m^2-E_B^2\right)}. \tag{18}$$

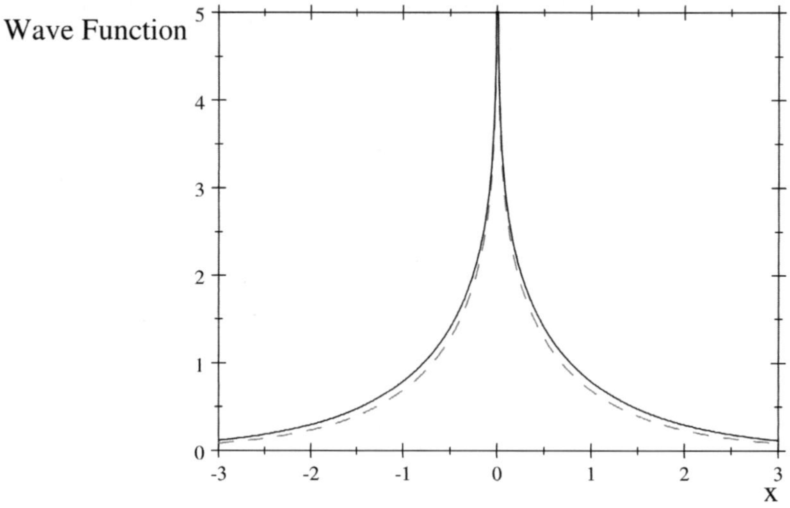

Figure 1. The bound state wave function for the summation of the first three terms (red) compared to the first seven terms (black) for $E_B = 0.5$ and $m = 1$. The figure shows a rapid convergence of the series.

Expanding in powers of $\epsilon = D - 1$, we obtain:

$$\psi(0) \propto A\lambda\left(\frac{1}{\epsilon} + \text{finite terms}\right). \tag{19}$$

Since in Eq.(9) $\frac{1}{\lambda} \propto \left(\frac{1}{\epsilon} + finite\right)$, then as $\epsilon \to 0$, $\psi(0) \propto A$. This means that the dimensional regularization method regularizes the wave function in position space and no need for extra renormalization conditions.

In fact, the logarithmic behavior of the wave function for small $|x|$ values for the unregularized wave function in our paper in Ref. [19] has been reflected here as $\ln\frac{1}{m}$ in the regularized wave function (position has an m^{-1} dimension in natural units). This result is very important as one is not in a need for other renormalization conditions and the coupling renormalization condition used in our paper in Ref. [19] is the only needed one to turn all physical quantities finite. Although in the above discussion we stressed the regularization of the position-space bound state wave function, the same argument can be extended for the scattering states wave functions. In fact, the scattering state wave function in momentum space is given by [19]:

$$\begin{aligned} \widetilde{\Psi}_E(p) &= \delta(p - \sqrt{E^2 - m^2}) + \delta(p + \sqrt{E^2 - m^2}) + \widetilde{\Phi}_E(p), \\ \widetilde{\Phi}_E(p) &= \frac{\lambda}{\pi}\frac{1 + \pi\Phi_E(0)}{E - \sqrt{p^2 + m^2}}. \end{aligned} \tag{20}$$

While the first two terms in $\widetilde{\Psi}_E(p)$ result in two plane waves in position space, the third term has a similar behavior as the bound state. Accordingly, the inverse Fourier transform of the $\widetilde{\Phi}_E(p)$ is logarithmic divergent and can thus be regularized in a similar way.

To conclude, we considered the singular contact interaction within the frame work of the relativistic Schrodinger equation. The theory shows up rigorous quantum-field-like properties such as asymptotic freedom and dimensional transmutation. Since the theory bears divergences in some physical quantities, dimensional regularization has been employed to regularize the theory. In regularizing the theory in momentum space, the coupling renormalization condition has been utilized. To obtain the spatial wave functions, a direct inverse Fourier transom can be applied but then the position-space wave functions below up as $|x| \to 0$. This might lead to the conclusion that the renormalization condition used for the coupling in momentum space is not sufficient to turn all quantities finite in position space. However, based on the fact that position space is totally equivalent to the momentum space, one should not obtain divergent quantities in position space from already regularized quantities in momentum space. To lift this conflict, we employed the D-dimensional inverse Fourier transform to map quantities from momentum space to position space. In doing this, we found that the regularized wave functions in position space are finite but rather scale dependent.

Acknowledgments

This publication was made possible by the NPRP grant # NPRP 5 - 261-1-054 from the Qatar National Research Fund (a member of the Qatar Foundation). The statements made herein are solely the responsibility of the author.

Re erences

[1] E.D. Bloom et al., Phys. Rev. Lett. 23, 930 and 935 (1969)
[2] K. Symanzik, Commun. Math. Phys. 45, 79 (1975)
[3] C. M. Bender, K. A. Milton, and V. M. Savage, Phys. Rev. D 62, 85001 (2000)
[4] Frieder Kleefeld, J. Phys. A: Math. Gen. 39 L9–L15 (2006)
[5] Abouzeid M. Shalaby and Suleiman S. Al-Thoyaib, Phys. Rev. D 82, 085013 (2010)
[6] H. Fritzsch, M. Gell-Mann, and H. Leutwyler, Phys. Lett. B 47 (1973) 365
[7] H. D. Politzer, Phys. Rev. Lett. 30 (1973) 1346
[8] D. Gross and F. Wilczek, Phys. Rev. Lett. 30 (1973) 1343
[9] D. J. E. Callaway, Phys. Rep. 167 (1988) 241
[10] C. Thorn, Phys. Rev. D19 (1979) 639
[11] M. A. B. Beg and R. C. Furlong, Phys. Rev. D31 (1985) 1370
[12] C. R. Hagen, Phys. Rev. Lett. 64 (1990) 503
[13] R. Jackiw, M. A. B. Beg Memorial Volume, A. Ali and P. Hoodbhoy, Eds., World Scientific, Singapore (1991)
[14] J. Fernando Perez and F. A. B. Coutinho, Am. J. Phys. 59 (1991) 52
[15] P. Gosdzinsky and R. Tarrach, Am. J. Phys. 59 (1991) 70
[16] L. R. Mead and J. Godines, Am. J. Phys. 59 (1991) 935
[17] C. Manuel and R. Tarrach, Phys. Lett. B328 (1994) 113
[18] D. R. Phillips, S. R. Beane, and T. D. Cohen, Ann. Phys. 263 (1998) 255
[19] M. H. Al-Hashimi, A. M. Shalaby, and U.-J. Wiese, Phys. Rev.D 89, 125023 (2014)
[20] M. H. Al-Hashimi and A. M. Shalaby, Phys. Rev.D 92, 025043 (2015)
[21] H. Leutwyler, Nuovo Cim. 37 (1965) 543

Symmetry operators of the two-component Gross–Pitaevskii equation with a Manakov-type nonlocal nonlinearity

A V Shapovalov[1,2], A Yu Trifonov[1,2], and A L Lisok[2]

[1] Tomsk State University, Lenin ave., 36, Tomsk, Russia, 634050
[2] Tomsk Polytechnic University, Lenin ave., 30, Tomsk, Russia, 634034

Abstract.
We consider an integro-differential 2-component multidimensional Gross–Pitaevskii equation with a Manakov-type cubic nonlocal nonlinearity. In the framework of the WKB–Maslov semiclassical formalism, we obtain a semiclassically reduced 2-component nonlocal Gross–Pitaevskii equation determining the leading term of the semiclassical asymptotic solution. For the reduced Gross–Pitaevskii equation we construct symmetry operators which transform arbitrary solution of the equation into another solution. Constructing the symmetry operator is based on the Cauchy problem solution technique and uses an intertwining operator which connects two solutions of the reduced Gross–Pitaevskii equation. General structure of the symmetry operator is illustrated with a 1D case for which a family of symmetry operators is found explicitly and a set of exact solutions is generated.

1. Introduction

Symmetry properties of an equation are related to analytical methods of exact solution finding to the equation under consideration. Symmetries are usually can be found explicitly for rather narrow sets of equations. A way for extending classes of equations with symmetries is to use methods of symmetry analysis along with approximate asymptotic methods. For instance, in [1–3] a theory of approximate symmetries was developed for partial differential equations (PDEs) in the framework of the perturbation theory where solutions of an equation and its symmetries were presented as regular asymptotic power series in a formal small parameter.

Another way to relate ideas of symmetry and approximate methods was explored in [4, 5] where WKB–Maslov semiclassical approximation method [6, 7] as well as symmetry properties of nonlocal and nonlinear equations with partial derivatives were considered. Note that in contrast to the perturbation method, the semiclassical asymptotics depends singularly on a small asymptotic parameter.

To specify the problem statement, consider a nonlinear operator $\hat{F}$ acting in a functional space $\mathcal{L}$, $\hat{F}:\mathcal{L}\to\mathcal{L}$. By definition, an operator $\hat{A}:\mathcal{L}\to\mathcal{L}$ is a symmetry operator for $\hat{F}$ if

$$\hat{F}(u)=0\to\hat{F}(\hat{A}(u))=0, \tag{1}$$

or

$$\hat{F}(\hat{A}(u))=\hat{\Phi}(u,\hat{F}(u)),\quad \hat{\Phi}(u,0)=0,\quad \forall u\in\mathcal{L}, \tag{2}$$

where $\hat{\Phi}$ is an operator Lagrangian multiplier.

Symmetry operators can be used directly, for example, for generation of exact solutions from a known particular solution of the equation. Important application of symmetry operators is related to the Lie groups. Let $\hat{A}(\alpha)$ be a Lie group of symmetry operators with a generator $\hat{X}$ and a group parameter α, $\hat{X} = \frac{d\hat{A}(\alpha)}{d\alpha}|_{\alpha=0}$. From (2) one can obtain

$$\hat{F}'(u)\hat{X}(u) = \hat{R}(u)\hat{F}(u), \quad \forall u \in \mathcal{L}. \tag{3}$$

Here, $\hat{R}(u)$ for a given u is a linear operator acting on $\hat{F}(u)$, $\hat{R}$ plays the role of an operator Lagrangian multiplier, $\hat{F}'$ is the derivative of $\hat{F}$.

Unlike (2), equation (3) is the linear operator equation determining an operator $\hat{X}$. For partial differential operators $\hat{F}$ and $\hat{X}$ the determining equation (3) can be solved effectively for many important equations of mathematical physics. $\hat{X}$ is called a symmetry for $\hat{F}$ and $\hat{X}$ is the main object of study in the group analysis of PDEs [8–10].

Finding the symmetry operators $\hat{A}$ for an operator $\hat{F}$ of general form is ill-posed problem on account of complexity of the determining equation (2). Therefore, symmetry operators for nonlinear equations are explored a little. Nevertheless, the problem of symmetry operators finding can be not unreasonable for nonlinear equations of a special form.

In the present work we consider a 2-component generalized multidimensional Gross–Pitaevskii equation (GPE) with a Manakov-type nonlocal nonlinearity. The GPE is the basic model equation in the theory of the Bose–Einstein condensates (see, e.g., the review paper [11]). Nonlocal generalizations of the GPE are introduced to avoid the problem of the wave function collapse and for some other problems [12,13].

Our results use as a main tool the WKB–Maslov formalism of semiclassical asymptotics (e.g., [6]). In terms of this formalism the leading term of the semiclassical asymptotic solution is determined by a reduced 2-component nonlocal GPE (R2GPE).

Here, we construct a class of symmetry operators for the R2GPE using the method developed in [5] for solving the Cauchy problem for the 1-component nonlocal GPE.

This article is structured as follows. In Section 2, following [5], we briefly describe the method of the Cauchy problem solution for the 2-component multidimensional GPE with a Manakov-type nonlocal nonlinearity in the semiclassical approximation. In the framework of the semiclassical approximation, we obtain the reduced 2-component nonlocal GPE which determines the leading term of the semiclassical asymptotics. In Section 3, based on the results of Section 2, we formulate a general structure describing a class of symmetry operators for the R2GPE. The core of this structure is an intertwining operator that connects two linear operators associated with the R2GPE in the framework of the semiclassical formalism. Section 4 contains an illustrative example where a set of symmetry operators is obtained in explicit form for a particular case of 1D R2GPE. A family of exact solutions is generated with the symmetry operators obtained. Concluding remarks are in Section 5.

2. Solution of the Cauchy problem

Consider the generalized 2-component Gross–Pitaevskii equation with the nonlocal Manakov-type interaction term:

$$\hat{F}(\Psi)(\vec{x},t) = \{-i\hbar\partial_t\mathbb{I} + \hat{H}_0(t)\mathbb{I} + \hbar(\vec{\sigma},\vec{H}(t)) + \varkappa\mathbb{I}\hat{V}(\Psi)(t)\}\Psi(\vec{x},t) = 0, \tag{4}$$

$$\hat{V}(\Psi)(t) = V(\Psi)(\hat{z},t) = \int_{\mathbb{R}^n} d\vec{y}\,\Psi^+(\vec{y},t)V(\hat{z},\hat{w},t)\Psi(\vec{y},t), \tag{5}$$

where $\partial_t = \partial/\partial t$, $\Psi(\vec{x},t) = (\Psi_1(\vec{x},t),\Psi_2(\vec{x},t))^\intercal$, $\Psi^+(\vec{x},t) = (\Psi_1^*(\vec{x},t),\Psi_2^*(\vec{x},t))$. Here $\Psi_1(\vec{x},t)$ and $\Psi_2(\vec{x},t)$ are smooth complex scalar functions which belong to a complex Schwartz space

$\mathbb{S}$ in the space variable $\vec{x} \in \mathbb{R}^n$ at each time t, Ψ_1^* and Ψ_2^* denote the complex conjugate to Ψ_1 and Ψ_2, respectively, $\Psi^\intercal$ denotes the transpose of Ψ; $\mathbb{I} = \mathbb{I}_{2\times 2}$ is an 2×2 identity matrix; $\vec{H}(t) = (H_1(t), H_2(t), H_3(t))^\intercal$ is a vector function where $H_k(t)$, $k = 1, 2, 3$, are smooth real functions of t, and $\vec{\sigma} = (\sigma_1, \sigma_2, \sigma_3)^\intercal$ is the column vector where σ_k are the Pauli matrices; $(\vec{\sigma}, \vec{H}(t)) = \sigma_1 H_1(t) + \sigma_2 H_2(t) + \sigma_3 H_3(t)$. We call $\hat{V}(\Psi)(t)$ in (5) the nonlocal Manakov-type nonlinear term. The linear operators $\hat{H}_0(t) = H_0(\hat{z}, t)$ and $V(\hat{z}, \hat{w}, t)$ in (4) and (5) are Hermitian Weyl-ordered functions [14] of time t and of noncommuting operators

$$\hat{z} = (\hat{\vec{p}}, \vec{x}) = (-i\hbar\partial/\partial\vec{x}, \vec{x}), \quad \hat{w} = (-i\hbar\partial/\partial\vec{y}, \vec{y}), \quad \vec{x}, \vec{y} \in \mathbb{R}^n, \tag{6}$$

with commutators $[\hat{z}_k, \hat{z}_j]_- = [\hat{w}_k, \hat{w}_j]_- = i\hbar J_{kj}$, $[\hat{z}_k, \hat{w}_j]_- = 0$, $k, j = \overline{1, 2n}$, where $[\hat{A}, \hat{B}]_- = \hat{A}\hat{B} - \hat{B}\hat{A}$, $J = \|J_{kj}\|_{2n\times 2n}$ is a unit symplectic matrix: $J = \begin{pmatrix} 0 & -I \\ I & 0 \end{pmatrix}_{2n\times 2n}$, and $I = I_{n\times n}$ is an $n \times n$ identity matrix. The space $\mathbb{S}$ is used to provide for existence of the moments of $\Psi(\vec{x}, t)$ and convergence of the integral in (5). In what follows, we use the norm $\|\Psi\|$, $\Psi \in \mathbb{S}(\mathbb{R}^n_x, \mathbb{C}^2)$, of the space $L_2(\mathbb{R}^n_x, \mathbb{C}^2)$, i.e., $\|\Psi\| = \sqrt{(\Psi, \Psi)}$, $(\Phi, \Psi) = \int_{\mathbb{R}^n} d\vec{x}\, \Phi^+(\vec{x})\Psi(\vec{x})$ denotes the Hermitian inner product of the functions $\Phi, \Psi \in \mathbb{S}(\mathbb{R}^n_x, \mathbb{C}^2)$.

From equation (4) it follows that the squared norm of a solution $\Psi(\vec{x}, t)$ is conserved, $\|\Psi(t)\|^2 = \|\Psi(0)\|^2 = \text{const}$.

Equation (4) with variable coefficients of general form cannot be integrated exactly by known methods. Therefore, analytical solutions to this equation we will construct approximately using the method of semiclassical asymptotics as $\hbar \to 0$. The semiclassical solutions of the 2-component nonlocal GPE (4) can be constructed using an appropriate complex WKB–Maslov ansatz [6, 7].

Following [5], we introduce a class $\mathcal{K}^t_\hbar$ of two-component functions

$$\mathcal{K}^t_\hbar = \Big\{\Phi : \Phi(\vec{x}, t, \hbar) = \sum_{k=0}^{\infty} \hbar^{k/2} u^{(k)}(t)\widetilde{\Phi}^{(k)}(\vec{x}, t, \hbar)\Big]\Big\}, \tag{7}$$

where $u^{(k)}(t)$ is a two-component vector-function smoothly dependent on t; $\tilde{\Phi}^{(k)}(\vec{x}, t, \hbar) \in \mathcal{C}^t_\hbar$, $k = \overline{1, \infty}$. Here

$$\mathcal{C}^t_\hbar = \Big\{\Phi : \Phi(\vec{x}, t) = \varphi\Big(\frac{\Delta\vec{x}}{\sqrt{\hbar}}, t\Big)\mathcal{R}\big(S(t, \hbar), \vec{P}(t, \hbar), \Delta\vec{x}\big)\Big\}, \quad \mathcal{R}(S, \vec{\xi}, \vec{\zeta}) = \exp\Big[\frac{i}{\hbar}(S + \langle\vec{\xi}, \vec{\zeta}\rangle)\Big],$$

where $\vec{\xi}, \vec{\zeta} \in \mathbb{R}^n$, the angle brackets $\langle ., . \rangle$ denote the Euclidean inner product of vectors: $\langle\vec{\xi}, \vec{\zeta}\rangle = \sum_{j=1}^{n} \xi_j\zeta_j$; the function $\varphi(\vec{\xi}, t, h)$ belongs to the Schwarz space $\mathbb{S}$ in the variable $\vec{\xi}$ and smoothly depends on t. Here $\Delta\vec{x} = \vec{x} - \vec{X}(t, \hbar)$, and the real function $S(t, \hbar)$ and the $2n$-dimensional vector function $Z(t, \hbar) = (\vec{P}(t, \hbar), \vec{X}(t, \hbar))$, which characterize the class $\mathcal{C}^t_\hbar$, regularly depend on $\sqrt{\hbar}$ in the neighborhood of $\hbar = 0$ and *are to be determined*. Note that $\mathcal{C}^t_\hbar \subset \mathbb{S}$.

In accordance with [5], we call the functions of the class $\mathcal{K}^t_\hbar$ *two-component trajectory-concentrated functions*. The functions belonging to the class $\mathcal{C}^t_\hbar$ at any fixed time $t \in \mathbb{R}^1$ are *concentrated*, as $\hbar \to 0$, in the neighborhood of a point lying on the phase curve $z = Z(t, 0)$, $t \in \mathbb{R}^1$ [15]. This property motivates the term "trajectory-concentrated functions".

In the next section 3 we construct symmetry operators using the complex WKB–Maslov asymptotic solutions of the Cauchy problem to equation (4) with the initial condition:

$$\Psi(\vec{x}, t)\big|_{t=s} = \psi(\vec{x}) = u_0 \cdot \phi(\vec{x}), \quad \phi(\vec{x}) \in \mathcal{C}^0_\hbar, \tag{8}$$

where $u_0 = (u_0^1, u_0^2)^\intercal$ is a constant 2-component vector and $\phi(\vec{x})$ is a scalar complex function,

$$\mathcal{C}_\hbar^0 = \Big\{\phi : \phi(\vec{x},\hbar) = \varphi\Big(\frac{\Delta\vec{x}}{\sqrt{\hbar}}\Big)\mathcal{R}\big(S_0(\hbar), \vec{P}_0(\hbar), \Delta\vec{x}_0\big)\Big\}, \quad \Delta\vec{x}_0 = \vec{x} - \vec{X}_0(\hbar).$$

Let us use $\widehat{O}(\hbar^\nu)$ to designate an operator $\hat{F}$ such that for any function $\Phi \in \mathcal{K}_\hbar^t$ the following asymptotic estimate is valid:

$$\frac{\|\hat{F}\Phi\|}{\|\Phi\|} = O(\hbar^\nu), \qquad \hbar \to 0. \tag{9}$$

In accordance with (9), it may be shown (see [15, 16]) that

$$\Delta\hat{z} = \hat{z} - Z(t,\hbar) = \widehat{O}(\hbar^{1/2}), \quad \hbar \to 0. \tag{10}$$

Let us expand the operators $\hat{H}_0(t) = H_0(\hat{z},t)$ and $V(\hat{z},\hat{w},t)$ in (4) and (5) as Taylor series in the operators $\Delta\hat{z} = \hat{z} - Z(t,\hbar)$ and $\Delta\hat{w} = \hat{w} - Z(t,\hbar)$, respectively, and restrict ourselves to quadratic terms. Then, in view of (10), the solution of the Cauchy problem (4) and (8) asymptotic in a formal small parameter $\hbar$ ($\hbar \to 0$) can be constructed accurate to $O(\hbar^{3/2})$ (see [5, 16]). The leading-order term of the asymptotics can be found by reducing the generalized GPE (4) to an equation with a quadratic nonlocal operator.

The higher-order corrections to the leading-order term can be found using perturbation theory [16]. Thus the study of GPEs with a quadratic nonlocal operator is crucial for the construction of semiclassical asymptotics for the GPE of the form (4) in the class of trajectory concentrated functions (7). Without loss of generality, we consider a reduced 2-component nonlocal GPE of the form

$$\Big\{-i\hbar\partial_t\mathbb{I} + H_{\rm qu}(\hat{z},t)\mathbb{I} + \hbar(\vec{\sigma},\vec{H}(t)) + \varkappa\mathbb{I}\int\limits_{\mathbb{R}^n} d\vec{y}\,\Psi^+(\vec{y},t)V_{\rm qu}(\hat{z},\hat{w},t)\Psi(\vec{y},t)\Big\}\Psi(\vec{x},t) = 0, \tag{11}$$

where the linear operators $H_{\rm qu}(\hat{z},t)$ and $V_{\rm qu}(\hat{z},\hat{w},t)$ are Hermitian and quadratic in $\hat{z}$, $\hat{w}$, respectively:

$$H_{\rm qu}(\hat{z},t) = \frac{1}{2}\langle\hat{z},\mathcal{H}_{zz}(t)\hat{z}\rangle + \langle\mathcal{H}_z(t),\hat{z}\rangle, \tag{12}$$

$$V_{\rm qu}(\hat{z},\hat{w},t) = \frac{1}{2}\langle\hat{z},W_{zz}(t)\hat{z}\rangle + \langle\hat{z},W_{zw}(t)\hat{w}\rangle + \frac{1}{2}\langle\hat{w},W_{ww}(t)\hat{w}\rangle. \tag{13}$$

Here, $\mathcal{H}_{zz}(t)$, $W_{zz}(t)$, $W_{zw}(t)$, and $W_{ww}(t)$ are $2n\times 2n$ matrices; $\mathcal{H}_z(t)$ is a $2n$ vector; $\langle z,w\rangle = \sum\limits_{j=1}^{2n} z_j w_j$; $z, w \in \mathbb{R}^{2n}$.

We call equation (11) with the linear operators $H_{\rm qu}$ and $V_{\rm qu}$ given by (12) and (13), respectively, *a reduced 2-component nonlocal Gross–Pitaevskii equation* (R2GPE).

The R2GPE can be integrated explicitly similar to [5] and its symmetries are of interest in the symmetry analysis of integro-differential equations (IDEs).

In study of symmetries of nonlocal equations the problem arises how to choose a structure of symmetries for non-differential equations. The R2GPE provides an example of IDE, when symmetry operators can be found explicitly.

Let us consider briefly a method for solving the Cauchy problem (8) for the R2GPE (11), following the scheme described elsewhere [5, 16].

We denote the Weyl-ordered symbol of an operator $\hat{A}(t) = A(\hat{z}, t)$ by $A(z,t)$ and define the expectation value for $\hat{A}(t)$ over the state $\Psi(\vec{x}, t)$ as

$$A_\Psi(t) = \frac{1}{\|\Psi\|^2}(\Psi, \hat{A}(t)\Psi) = \frac{1}{\|\Psi\|^2}\int\limits_{\mathbb{R}^n} d\vec{x}\, \Psi^+(\vec{x}, t)\hat{A}(t)\Psi(\vec{x}, t).$$

As $\|\Psi\|^2$ does not depend on time, we have from (11),(12), and (13)

$$\dot{A}_\Psi(t) = \frac{1}{\|\Psi\|^2}\int\limits_{\mathbb{R}^n} d\vec{x}\, \Psi^+(\vec{x}, t)\Big\{\frac{\partial \hat{A}(t)}{\partial t} + \frac{i}{\hbar}[\mathbb{I}H_{\rm qu}(\hat{z}, t), \hat{A}(t)]_- + i[(\vec{\sigma}, \vec{H}(t)), \hat{A}(t)]_- +$$
$$+ \frac{i\tilde{\varkappa}}{\hbar}\int\limits_{\mathbb{R}^n} d\vec{y}\, \Psi^+(\vec{y}, t)[\mathbb{I}V_{\rm qu}(\hat{z}, \hat{w}, t), \hat{A}(t)]_- \Psi(\vec{y}, t)\Big\}\Psi(\vec{x}, t), \tag{14}$$

where $\dot{A}_\Psi(t) = dA_\Psi(t)/dt$ and $\tilde{\varkappa} = \varkappa\|\Psi\|^2 = \varkappa\|\psi\|^2$.

Let $z_\Psi(t) = \big(z_{\Psi\, l}(t)\big)$, $\vec{\eta}_\Psi(t)$ and $\Delta^{(2)}_\Psi(t) = \big(\Delta^{(2)}_{\Psi\, kl}(t)\big)$ denote the expectation values over $\Psi(\vec{x}, t)$ for the operators

$$\hat{z}_l\mathbb{I}, \quad \vec{\sigma}, \quad \hat{\Delta}^{(2)}_{kl}\mathbb{I} = \frac{1}{2}\Big(\mathbb{I}\Delta\hat{z}_k\Delta\hat{z}_l + \mathbb{I}\Delta\hat{z}_l\Delta\hat{z}_k,\Big), \quad k, l = \overline{1, 2n},$$

respectively. Here, $\Delta\hat{z}_l = \hat{z}_l - (z_\Psi)_l(t)$, $\hat{z}_l$ are given in (6), and we set $Z(t, \hbar) = (z_\Psi)(t)$. We call $z_\Psi(t)$ the first moments and $\Delta^{(2)}_\Psi(t)$ do the second centered moments of $\Psi(\vec{x}, t)$.

From (11), (12), (13), and (14) we immediately obtain a dynamical system in matrix notation:

$$\begin{cases} \dot{z}_\Psi = J\{\mathcal{H}_z(t) + [\mathcal{H}_{zz}(t) + \tilde{\varkappa}(W_{zz}(t) + W_{zw}(t))]z_\Psi\}, \\ \dot{\Delta}^{(2)}_\Psi = J[\mathcal{H}_{zz}(t) + \tilde{\varkappa}W_{zz}(t)]\Delta^{(2)}_\Psi - \Delta^{(2)}_\Psi[\mathcal{H}_{zz}(t) + \tilde{\varkappa}W_{zz}(t)]J, \\ \dot{\vec{\eta}}_\Psi(t) = 2\vec{H}(t) \times \vec{\eta}_\Psi(t). \end{cases} \tag{15}$$

Following [5], we call (15) *the Hamilton–Ehrenfest system* (HES) of the second order for the R2GPE (11).

The HES is of the second order, as equations (15) contain the first and second moments. For brevity, we use a shorthand notation for the total set of the first and second moments of $\Psi(\vec{x}, t)$:

$$\mathfrak{g}_\Psi(t) = \big(z_\Psi(t), \vec{\eta}_\Psi(t), \Delta^{(2)}_\Psi(t)\big). \tag{16}$$

Then the R2GPE (11) can be written equivalently as

$$\hat{L}(t, \mathfrak{g}_\Psi(t))\Psi(\vec{x}, t) = \Big\{ - i\hbar\partial_t\mathbb{I} + \hat{H}_q(t, \mathfrak{g}_\Psi(t))\mathbb{I} + \hbar(\vec{\sigma}, \vec{H}(t))\Big\}\Psi(\vec{x}, t) = 0, \tag{17}$$

$$\hat{H}_q(t, \mathfrak{g}_\Psi(t)) = \frac{1}{2}\langle\hat{z}, \mathcal{H}_{zz}(t)\hat{z}\rangle + \langle\mathcal{H}_z(t), \hat{z}\rangle + \frac{\tilde{\varkappa}}{2}\langle\hat{z}, W_{zz}(t)\hat{z}\rangle +$$
$$+ \frac{\tilde{\varkappa}}{2}\langle z_\Psi(t), W_{ww}(t)z_\Psi(t)\rangle + \tilde{\varkappa}\langle\hat{z}, W_{zw}(t)z_\Psi(t)\rangle + \frac{\tilde{\varkappa}}{2}\,\mathrm{Sp}\,\Big[W_{ww}(t)\Delta^{(2)}_\Psi(t)\Big]. \tag{18}$$

The Cauchy problem (17) and (8) for the R2GPE gives the Cauchy problem for the HES (15) which can be written in a concise form as

$$\dot{\mathfrak{g}}_\Psi(t) = \Gamma(t, \mathfrak{g}_\Psi(t)), \tag{19}$$

$$\mathfrak{g}_\Psi(t)\Big|_{t=s} = \mathfrak{g}_\psi, \tag{20}$$

where $\Gamma\big(t, \mathfrak{g}_\Psi(t)\big)$ designates the rhs of (15).

The system (17), (19) allows us to reduce the Cauchy problem for the R2GPE (17) to the Cauchy problem for a linear PDE because the Cauchy problem (20) for HES (19) can be solved independently of equation (17).

Let

$$\mathfrak{g}(t, \mathbf{C}) = \big(Z(t, \mathbf{C}), \vec{\eta}(t, \mathbf{C}), \Delta^{(2)}(t, \mathbf{C})\big) \tag{21}$$

be the general solution of the HES (19) and $\mathbf{C} = (C_1, C_2, \dots, C_N)$ denote the set of integration constants.

Consider a linear PDE with coefficients depending on the parameters $\mathbf{C}$:

$$\hat{L}(t, \mathbf{C})\Phi(\vec{x}, t, \mathbf{C}) = \big\{ - i\hbar\partial_t\mathbb{I} + \hat{H}_q(t, \mathbf{C})\mathbb{I} + \hbar(\vec{\sigma}, \vec{H}(t))\big\}\Phi(\vec{x}, t, \mathbf{C}) = 0, \tag{22}$$

where

$$\hat{H}_q(t, \mathbf{C}) = \frac{1}{2}\langle\hat{z}, \mathcal{H}_{zz}(t)\hat{z}\rangle + \langle\mathcal{H}_z(t), \hat{z}\rangle + \frac{\tilde{\varkappa}}{2}\langle\hat{z}, W_{zz}(t)\hat{z}\rangle + \tilde{\varkappa}\langle\hat{z}, W_{zw}(t)Z(t, \mathbf{C})\rangle +$$
$$+ \frac{\tilde{\varkappa}}{2}\langle Z(t, \mathbf{C}), W_{ww}(t)Z(t, \mathbf{C})\rangle + \frac{\tilde{\varkappa}}{2}\operatorname{Sp}\Big[W_{ww}(t)\Delta^{(2)}(t, \mathbf{C})\Big]. \tag{23}$$

The operator $\hat{H}_q(t, \mathbf{C})$ (23) of (22) is obtained from (18) where the general solution $\mathfrak{g}(t, \mathbf{C})$ of the HES (19) stands for the moments $\mathfrak{g}_\Psi(t)$. We call (22) *the associated linear equation* (ALE) for the R2GPE (17).

A solution of the ALE (22) can be found in the form

$$\Phi(\vec{x}, t, \mathbf{C}) = \phi(\vec{x}, t, \mathbf{C}) \cdot v(t, \mathbf{C}), \tag{24}$$

where a scalar complex function $\phi(\vec{x}, t, \mathbf{C})$ and a 2-component complex function $v(t, \mathbf{C})$ satisfy the equations

$$\big\{ - i\hbar\partial_t + \hat{H}_q(t, \mathbf{C})\big\}\phi(\vec{x}, t, \mathbf{C}) = 0, \tag{25}$$
$$\big\{ - i\mathbb{I}\frac{d}{dt} + (\vec{\sigma}, \vec{H}(t))\big\}v(t, \mathbf{C}) = 0. \tag{26}$$

Let $\Phi(\vec{x}, t, \mathbf{C}[\psi])$ denote the solution of the Cauchy problem for the ALE (22) with the initial condition (8):

$$\Phi(\vec{x}, t, \mathbf{C}[\psi])\Big|_{t=s} = \psi(\vec{x}) = u_0\phi(\vec{x}). \tag{27}$$

Then for (24), (25), and (26) we have, respectively,

$$\phi(\vec{x}, t, \mathbf{C}[\psi])\Big|_{t=s} = \phi(\vec{x}), \qquad v(t, \mathbf{C}[\psi])\Big|_{t=s} = u_0. \tag{28}$$

Here, the integration constants $\mathbf{C}$ in (25) and (26) have been replaced by the functionals $\mathbf{C} = \mathbf{C}[\psi]$ determined from the algebraic conditions

$$\mathfrak{g}(t, \mathbf{C})\Big|_{t=s} = \mathfrak{g}_\psi. \tag{29}$$

Then the solution of the Cauchy problem (8) and (17), (18) for the R2GPE (see [5, 16] for details) is

$$\Psi(\vec{x}, t) = \Phi(\vec{x}, t, \mathbf{C}[\psi]) = \varphi(\vec{x}, t) \cdot u(t), \tag{30}$$

where $\varphi(\vec{x},t) = \phi(\vec{x},t,\mathbf{C}[\psi])$ is the scalar complex function and $u(t) = v(t,\mathbf{C}[\psi])$.

Define $\mathbf{C}[\Psi](t)$ by the algebraic condition

$$\mathfrak{g}(t,\mathbf{C}[\Psi](t)) = \mathfrak{g}_\Psi(t). \tag{31}$$

From the uniqueness of the solution of the Cauchy problem for the HES (20) it follows that

$$\mathfrak{g}(t,\mathbf{C}[\Psi](t)) = \mathfrak{g}(t,\mathbf{C}[\psi])$$

and, hence,

$$\mathbf{C}[\Psi](t) = \mathbf{C}[\psi],$$

i.e., the functionals $\mathbf{C}[\Psi](t)$ are the integrals of (4).

Also, we have

$$\mathfrak{g}(t,\mathbf{C}[\psi]) = \mathfrak{g}_\psi(t), \tag{32}$$

where $\mathfrak{g}_\psi(t)$ is the solution of the HES (19) with the initial condition (29).

Let us now turn to the construction of symmetry operators for the R2GPE (11). To do this, we use an operator intertwining a pair of ALEs of the form (22).

3. Symmetry operators

In this section, we apply the approach developed in [5] for the 1-component nonlocal GPE to find symmetry operators for the reduced 2-component nonlocal Gross–Pitaevskii equation (11) or, equivalently, (17).

According to definition (1), the nonlinear symmetry operator $\hat{A}(t)$ maps any solution $\Psi(\vec{x},t)$ of equation (17) into its another solution:

$$\Psi_A(\vec{x},t) = (\hat{A}(t)\Psi)(\vec{x},t). \tag{33}$$

The main idea of the symmetry operator construction is as follows. Let us take an operator $\hat{a}$ acting on the initial function $\psi(\vec{x})$ of the Cauchy problem (8) and set

$$\psi_a(\vec{x}) = \hat{a}\psi(\vec{x}) = \Psi_A(t)\big|_{t=s}, \quad \hat{A}(t)\big|_{t=s} = \hat{a}. \tag{34}$$

Define the functions $\Psi(\vec{x},t)$ and $\Psi_A(\vec{x},t)$ of the form (30) in (33) as two solutions of the Cauchy problem for the R2GPE (11) with two initial functions $\psi(\vec{x})$ and $\psi_a(\vec{x})$, respectively. Then the desired symmetry operator $\hat{A}(t)$ should relate $\Psi(\vec{x},t)$ and $\Psi_A(\vec{x},t)$.

Introduce the notation $\mathfrak{g}_{\psi_a}$ for the first and second moments of $\psi_a(\vec{x})$ similar to (16). From the solution of the Cauchy problem for the HES (20) with the initial condition $\mathfrak{g}_\Psi(t)\Big|_{t=s} = \mathfrak{g}_{\psi_a}$, by analogy with (32), we have

$$\mathfrak{g}(t,\mathbf{C}[\psi_a]) = \mathfrak{g}_{\psi_a}(t). \tag{35}$$

According to (30), the solutions $\Psi(\vec{x},t)$ and $\Psi_A(\vec{x},t)$ of the R2GPE (17) are found as

$$\Psi(\vec{x},t) = \Phi(\vec{x},t,\mathbf{C})\big|_{\mathbf{C}=\mathbf{C}[\psi]} \tag{36}$$

and

$$\Psi_A(\vec{x},t) = \Phi(\vec{x},t,\mathbf{C}')\big|_{\mathbf{C}'=\mathbf{C}'[\psi_a]}, \tag{37}$$

where $\Phi(\vec{x},t,\mathbf{C})$ and $\Phi(\vec{x},t,\mathbf{C}')$ are the solutions of two ALEs of the form (22) with two different sets of integration constants $\mathbf{C}$ and $\mathbf{C}'$, respectively, and the corresponding linear operators $\hat{L}(t,\mathbf{C})$ and $\hat{L}(t,\mathbf{C}')$.

The essential point in the construction of the symmetry operator $\hat{A}(t)$ in (33) is a linear operator $\hat{M}(t,s,\mathbf{C}',\mathbf{C})$ intertwining the operators $\hat{L}(t,\mathbf{C}')$ and $\hat{L}(t,\mathbf{C})$:

$$\hat{L}(t,\mathbf{C}')\hat{M}(t,s,\mathbf{C}',\mathbf{C}) = \hat{R}(t,s,\mathbf{C}',\mathbf{C})\hat{L}(t,\mathbf{C}). \tag{38}$$

Here, the linear operator $\hat{R}(t,s,\mathbf{C}',\mathbf{C})$ is a Lagrangian multiplier, and the initial condition is $\hat{M}(t,s,\mathbf{C}',\mathbf{C})|_{t=s} = \hat{a}$.

From (38) we have that $\Phi(\vec{x},t,\mathbf{C}') = \hat{M}(t,s,\mathbf{C}',\mathbf{C})\Phi(\vec{x},t,\mathbf{C})$ for two arbitrary sets of constants $\mathbf{C}'$ and $\mathbf{C}$, and this is especially true for $\Phi(\vec{x},t,\mathbf{C}'[\psi_a])$ and $\Phi(\vec{x},t,\mathbf{C}[\psi])$ with the constants $\mathbf{C}'[\psi_a]$ and $\mathbf{C}[\psi]$.

The intertwining operator $\hat{M}(t,s,\mathbf{C}',\mathbf{C})$ relates the functions $\Phi(\vec{x},t,\mathbf{C}'[\psi_a])$ and $\Phi(\vec{x},t,\mathbf{C}[\psi])$ that in turn relates $\Psi_A(\vec{x},t)$ and $\Psi(\vec{x},t)$ according to (37) and (36).

To find the operator $\hat{M}(t,s,\mathbf{C}',\mathbf{C})$, we consider a linear intertwining operator $\widehat{\mathcal{D}}(t,s,\mathbf{C}',\mathbf{C})$ for $\hat{L}(t,\mathbf{C}')$ and $\hat{L}(t,\mathbf{C})$ satisfying the conditions

$$\hat{L}(t,\mathbf{C}')\widehat{\mathcal{D}}(t,s,\mathbf{C}',\mathbf{C}) = \widehat{\mathcal{D}}(t,s,\mathbf{C}',\mathbf{C})\hat{L}(t,\mathbf{C}), \tag{39}$$

$$\widehat{\mathcal{D}}(t,s,\mathbf{C}',\mathbf{C})\Big|_{t=s} = \hat{\mathbb{I}}, \tag{40}$$

where $\hat{\mathbb{I}}$ is an identity operator. We call $\widehat{\mathcal{D}}(t,s,\mathbf{C}',\mathbf{C})$ *the fundamental intertwining operator* for $\hat{L}(t,\mathbf{C}')$ and $\hat{L}(t,\mathbf{C})$.

With the use of $\widehat{\mathcal{D}}(t,s,\mathbf{C}',\mathbf{C})$, the operator $\hat{M}(t,s,\mathbf{C}',\mathbf{C})$ entering into (38) can be presented as

$$\hat{M}(t,s,\mathbf{C}',\mathbf{C}) = \widehat{\mathcal{D}}(t,s,\mathbf{C}',\mathbf{C})\widehat{B}(t,\mathbf{C}).$$

Here, $\widehat{B}(t,\mathbf{C}) \in \mathcal{B}$ is the linear symmetry operator of ALE (22) satisfying the conditions

$$[\hat{L}(t,\mathbf{C}),\widehat{B}(t,\mathbf{C})]_- = 0, \quad \widehat{B}(t,\mathbf{C})|_{t=s} = \hat{a}, \tag{41}$$

and $\mathcal{B}$ is the family of linear symmetry operators of the ALE (41).

Hence, given the operator $\widehat{\mathcal{D}}(t,s,\mathbf{C}',\mathbf{C})$ of (39) and the family $\mathcal{B}$ of linear symmetry operators of the ALE (41), we can construct the family of nonlinear symmetry operators for the GPE (4) as follows:

Let $\widehat{\mathcal{D}}(t,s,\mathbf{C}',\mathbf{C})$ be the fundamental intertwining operator (39) for $\hat{L}(t,\mathbf{C}')$ and $\hat{L}(t,\mathbf{C})$ and let $\widehat{B}(t,\mathbf{C})$ be the linear symmetry operator of the ALE (22) satisfying the conditions (41), then

$$(\hat{A}(t)\Psi)(\vec{x},t) = \widehat{\mathcal{D}}(t,s,\mathbf{C}'[\hat{a}\psi],\mathbf{C}[\Psi](t))\widehat{B}(t,\mathbf{C}[\Psi](t))\Psi(\vec{x},t) \tag{42}$$

defines the family of nonlinear symmetry operators for the GPE (11).

Here, $\mathbf{C}'[\hat{a}\psi]$ and $\mathbf{C}[\Psi](=\mathbf{C}[\psi])$ can be found from (32) and (31), respectively, and $\widehat{B} \in \mathcal{B}$.

Note that the symmetry operator $\hat{A}(t)$ from (42) is nonlinear, as the operators $\widehat{\mathcal{D}}$ and $\widehat{B}$ depend on the parameters $\mathbf{C}$ being functionals of the function Ψ.

4. Symmetry operators in the 1D case

As an illustration of the general formula (42), we will explicitly construct here the symmetry operators for the 1D R2GPE (11) of the form

$$\hat{F}(\Psi)(x,t) = \{-i\hbar\partial_t\mathbb{I} + H_{\rm qu}(\hat{p},x)\mathbb{I} + \hbar(\vec{\sigma},\vec{\mathcal{B}}) + \varkappa\mathbb{I}\hat{V}_{\rm qu}(\Psi)\}\Psi(x,t) = 0, \tag{43}$$

where $H_{\rm qu}(\hat{p},x) = \dfrac{1}{2}\left(\mu\hat{p}^2 + \rho(x\hat{p}+\hat{p}x) + \omega x^2\right)$, $\hat{V}_{\rm qu}(\Psi) = \dfrac{1}{2}\int\limits_{-\infty}^{+\infty} dy\,\left(ax^2 + 2bxy + cy^2\right)|\Psi(y)|^2$, $\hat{p} = -i\hbar\partial/\partial x$; a, b, and c are the real parameters of the nonlocal operator $\hat{V}_{\rm qu}(\Psi)$; μ, ω, and ρ

are the parameters of the linear operator $H_{\text{qu}}(\hat{p}, x)$; $x, y \in \mathbb{R}^1$, $\vec{\mathcal{B}}$ is a real constant vector. The initial condition (8) is written as

$$\Psi(x,t)\big|_{t=0} = \psi(x) = \phi(x)\cdot u_0, \tag{44}$$

where $\phi(x)$ is a scalar function and $u_0 = (u_0^1, u_0^2)^\intercal$ is a constant 2-component vector.

The Hamilton–Ehrenfest system (19) for the first-order moments becomes

$$\begin{cases} \dot{p} = -\rho p - \omega_0 x, \\ \dot{x} = \mu p + \rho x, \end{cases} \tag{45}$$

for the vector $\vec{\eta}(t) = \langle\Psi|\vec{\sigma}|\Psi\rangle$ we similarly obtain

$$\dot{\vec{\eta}}(t) = 2\vec{\mathcal{B}} \times \vec{\eta}(t), \quad \vec{\eta}^2(t) = 1, \tag{46}$$

and for the second-order moments with $\Delta_{21}^{(2)} = \Delta_{12}^{(2)}$ we have

$$\begin{cases} \dot{\Delta}_{11}^{(2)} = -2\rho\Delta_{11}^{(2)} - 2\tilde{\omega}\Delta_{21}^{(2)}, \\ \dot{\Delta}_{21}^{(2)} = \mu\Delta_{11}^{(2)} - \tilde{\omega}\Delta_{22}^{(2)}, \\ \dot{\Delta}_{22}^{(2)} = 2\mu\Delta_{21}^{(2)} + 2\rho\Delta_{22}^{(2)}, \end{cases} \tag{47}$$

where $\omega_0 = \omega + \tilde{\varkappa}(a+b)$, $\tilde{\omega} = \omega + \tilde{\varkappa}a$. We assume that $\bar{\Omega}^2 = \omega_0\mu - \rho^2 > 0$, $\Omega^2 = \tilde{\omega}\mu - \rho^2 > 0$ and introduce the notations

$$\bar{\Omega} = \sqrt{\omega_0\mu - \rho^2}, \qquad \Omega = \sqrt{\tilde{\omega}\mu - \rho^2}. \tag{48}$$

Then the general solution of system (45) is

$$\begin{aligned} X(t,\mathbf{C}) &= C_1 \sin\bar{\Omega}t + C_2\cos\bar{\Omega}t, \\ P(t,\mathbf{C}) &= \frac{1}{\mu}(\bar{\Omega}C_1 - \rho C_2)\cos\bar{\Omega}t - \frac{1}{\mu}(\bar{\Omega}C_2 + \rho C_1)\sin\bar{\Omega}t, \end{aligned} \tag{49}$$

and for system (47) we have

$$\begin{aligned} \Delta_{22}^{(2)}(t,\mathbf{C}) &= C_3\sin 2\Omega t + C_4\cos 2\Omega t + C_5, \\ \Delta_{21}^{(2)}(t,\mathbf{C}) &= \frac{1}{\mu}(\Omega C_3 - \rho C_4)\cos 2\Omega t - \frac{1}{\mu}(\Omega C_4 + \rho C_3)\sin 2\Omega t - \frac{\rho}{\mu}C_5, \\ \Delta_{11}^{(2)}(t,\mathbf{C}) &= \frac{1}{\mu^2}\big((\rho^2 - \Omega^2)C_3 + 2\rho\Omega C_4\big)\sin 2\Omega t + \\ &\quad + \frac{1}{\mu^2}\big((\rho^2 - \Omega^2)C_4 - 2\rho\Omega C_3\big)\cos 2\Omega t + \frac{\tilde{\omega}}{\mu}C_5. \end{aligned} \tag{50}$$

Note that all solutions of system (49) and (47) are localized. Similarly, solution of system (46) subject to the normalization condition $\vec{\eta}^2(t) = 1$ reads

$$\vec{\eta}(t,\mathbf{C}) = \vec{n}\cos C_6 + \vec{e}_\theta \sin C_6 \cos(2\mathcal{B}t + C_7) + \vec{e}_\varphi \sin C_6 \sin(2\mathcal{B}t + C_7). \tag{51}$$

Here, we use the notations: $\mathbf{C} = (C_1, \ldots, C_7)$ are arbitrary integration constants, $\mathcal{B} = |\vec{\mathcal{B}}|$, and

$$\begin{aligned} &\vec{n} = \frac{\vec{\mathcal{B}}}{\mathcal{B}} = (\sin\theta\cos\varphi,\ \sin\theta\sin\varphi,\ \cos\theta), \\ &\vec{e}_\theta = (\cos\theta\cos\varphi,\ \cos\theta\sin\varphi,\ -\sin\theta), \\ &\vec{e}_\varphi = (\sin\varphi,\ -\cos\varphi,\ 0), \\ &\vec{n}\times\vec{e}_\theta = -\vec{e}_\varphi, \quad \vec{n}\times\vec{e}_\varphi = \vec{e}_\theta, \quad \vec{e}_\varphi\times\vec{e}_\theta = \vec{n}. \end{aligned} \tag{52}$$

The 1D associated linear equation (22) for the 1D R2GPE (43) takes the form

$$\hat{L}(t,\mathbf{C})\Phi(\vec{x},t,\mathbf{C}) = \big\{ -i\hbar\partial_t\mathbb{I} + \hat{H}_q(t,\mathbf{C})\mathbb{I} + \hbar(\vec{\sigma},\vec{\mathcal{B}})\big\}\Phi(\vec{x},t,\mathbf{C}) = 0, \tag{53}$$
$$\hat{H}_q(t,\mathbf{C}) = \frac{\mu\hat{p}^2}{2} + \frac{\tilde{\omega}x^2}{2} + \frac{\rho(x\hat{p}+\hat{p}x)}{2} + \tilde{\varkappa}bxX(t,\mathbf{C}) + \tilde{\varkappa}\frac{c}{2}\big[X^2(t,\mathbf{C}) + \Delta_{22}^{(2)}(t,\mathbf{C})\big].$$

Denote by $\widehat{\mathcal{D}}(t,\mathbf{C}',\mathbf{C})$ the fundamental intertwining operator in (39), (40) as $s=0$. To find the operator $\widehat{\mathcal{D}}(t,\mathbf{C}',\mathbf{C})$, we introduce a function $\Upsilon(\vec{x},t,\mathbf{C})$ by the conditions

$$\Phi(x,t,\mathbf{C}) = \hat{\mathcal{K}}(x,t,\mathbf{C})\Upsilon(\vec{x},t,\mathbf{C}), \tag{54}$$
$$\hat{\mathcal{K}}(x,t,\mathbf{C}) = \hat{K}(x,t,\mathbf{C})\mathcal{G}(t,\mathbf{C}),$$
$$\hat{K}(x,t,\mathbf{C}) = \exp[-X(t,\mathbf{C})\partial_x]\exp\Big\{\frac{i}{\hbar}[S(t,\mathbf{C}) + P(t,\mathbf{C})x]\Big\}, \tag{55}$$

where

$$S(t,\mathbf{C}) = \int_0^t \Big\{P(t,\mathbf{C})\dot{X}(t) - H_{\varkappa}(t,\mathbf{C})\Big\}dt, \tag{56}$$
$$H_{\varkappa}(t,\mathbf{C}) = \frac{\mu}{2}P^2(t,\mathbf{C}) + \frac{1}{2}X^2(t,\mathbf{C})\big[\omega_0 + \tilde{\varkappa}(b+c)\big] + \rho P(t,\mathbf{C})X(t,\mathbf{C}) + \tilde{\varkappa}\frac{c}{2}\Delta_{22}^{(2)}(t,\mathbf{C}),$$

and $\mathcal{G}(t)$ is the Cauchy matrix of the system

$$\Big\{-i\mathbb{I}\frac{d}{dt} + (\vec{\sigma},\vec{\mathcal{B}})\Big\}\mathcal{G} = 0, \quad \mathcal{G}\big|_{t=0} = \mathbb{I}. \tag{57}$$

From (57) we have

$$\mathcal{G}(t) = \exp[-i(\vec{\sigma},\vec{\mathcal{B}})t] = \mathbb{I}\cos(\mathcal{B}t) - \frac{i}{\mathcal{B}}(\vec{\sigma},\vec{\mathcal{B}})\sin(\mathcal{B}t), \tag{58}$$

where $\mathcal{B} = |\vec{\mathcal{B}}|$. In the issue relation (58) can be written as

$$\mathcal{G}(t) = \frac{1}{\mathcal{B}}\begin{pmatrix} \mathcal{B}\cos\mathcal{B}t - i\mathcal{B}_3\sin\mathcal{B}t & -(i\mathcal{B}_1+\mathcal{B}_2)\sin\mathcal{B}t \\ -(i\mathcal{B}_1-\mathcal{B}_2)\sin\mathcal{B}t & \mathcal{B}\cos\mathcal{B}t + i\mathcal{B}_3\sin\mathcal{B}t \end{pmatrix}. \tag{59}$$

For $\Upsilon(x,t,\mathbf{C})$ we have from (22) and (54)

$$\hat{L}_0(x,t)\mathbb{I}\Upsilon = 0, \quad \Upsilon(\vec{x},t,\mathbf{C}) = \phi(x,t,\mathbf{C})u_0, \tag{60}$$
$$\hat{L}_0(x,t)\mathbb{I} = \hat{\mathcal{K}}^{-1}(x,t,\mathbf{C})\hat{L}(x,t,\mathbf{C})\hat{\mathcal{K}}(x,t,\mathbf{C}) =$$
$$= \big(-i\hbar\partial_t + \frac{\mu\hat{p}^2}{2} + \frac{\tilde{\omega}x^2}{2} + \frac{\rho(x\hat{p}+\hat{p}x)}{2}\big)\mathbb{I}, \tag{61}$$
$$\hat{L}_0(x,t)\phi(x,t,\mathbf{C}) = 0. \tag{62}$$

For the associated linear equation (62) we can construct the following set of symmetry operators linear in x and $\hat{p}$:

$$\hat{a}(t,\mathbf{C}) = \frac{1}{\sqrt{2\hbar}}\big[C(t)\hat{p} - B(t)x\big], \tag{63}$$
$$\hat{a}^+(t,\mathbf{C}) = \frac{1}{\sqrt{2\hbar}}\big[C^*(t)\hat{p} - B^*(t)x\big]. \tag{64}$$

Here the functions $B(t)$ and $C(t)$ are solutions of the linear Hamiltonian system

$$\begin{cases} \dot{B} = -\rho B - \tilde{\omega} C, \\ \dot{C} = \mu B + \rho C. \end{cases} \tag{65}$$

The Cauchy matrix $\mathcal{X}(t)$ for system (65) can easily be found as

$$\mathcal{X}(t) = \begin{pmatrix} \cos\Omega t - \dfrac{\rho}{\Omega}\sin\Omega t & -\dfrac{\tilde{\omega}}{\Omega}\sin\Omega t \\ \dfrac{\mu}{\Omega}\sin\Omega t & \cos\Omega t + \dfrac{\rho}{\Omega}\sin\Omega t \end{pmatrix}, \quad \mathcal{X}(t)\Big|_{t=0} = \mathbb{I}_{2\times 2}. \tag{66}$$

The set of solutions normalized by the condition [6]

$$B(t)C^*(t) - C(t)B^*(t) = 2i \tag{67}$$

can be written as

$$B(t) = e^{i\Omega t}\frac{(-\rho + i\Omega)}{\sqrt{\Omega\mu}}, \qquad C(t) = e^{i\Omega t}\sqrt{\frac{\mu}{\Omega}}.$$

Equation (67) results in the following commutation relations for the symmetry operators (63) and (64):

$$\big[\hat{a}(t,\mathbf{C}), \hat{a}^+(t,\mathbf{C})\big]_- = 1.$$

Then the symmetry operator $\hat{A}(t)$ in (42) for equation (43) with the initial condition (44) can be presented as

$$(\hat{A}(t)\Psi)(x,t) = \widehat{\mathcal{D}}(t, \mathbf{C}[\hat{a}\psi], \mathbf{C}[\Psi](t))\widehat{B}(t, \mathbf{C}[\Psi](t))\Psi(x,t). \tag{68}$$

Here, $\widehat{B}(t,\mathbf{C})$ is the symmetry operator of the associated linear equation (53), $\widehat{B}(t,\mathbf{C})|_{t=0} = \hat{a}$, and the fundamental intertwining operator $\widehat{\mathcal{D}}(t,\mathbf{C}',\mathbf{C})$ reads

$$\begin{aligned} \widehat{\mathcal{D}}(t,\mathbf{C}',\mathbf{C}) = \exp\Big\{i\frac{C_2' - C_2}{2\hbar\mu}\Big(\bar{\Omega}(C_1' - C_1) - \rho(C_2' - C_2)\Big)\Big\} \times \\ \times \hat{\mathcal{K}}^{-1}(x,t,\mathbf{C}')\exp\Big\{\frac{i}{\hbar}\hat{b}(t,\mathbf{C}',\mathbf{C})\Big\}\hat{\mathcal{K}}(x,t,\mathbf{C}), \end{aligned} \tag{69}$$

where

$$\hat{b}(t,\mathbf{C}',\mathbf{C}) = b_x(t,\mathbf{C}',\mathbf{C})\hat{p} - b_p(t,\mathbf{C}',\mathbf{C})x = \langle b(t,\mathbf{C}',\mathbf{C}), J\hat{z}\rangle,$$

and

$$b(t,\mathbf{C}',\mathbf{C}) = \begin{pmatrix} b_p(t,\mathbf{C}',\mathbf{C}) \\ b_x(t,\mathbf{C}',\mathbf{C}) \end{pmatrix} = \frac{1}{\mu}\mathcal{X}(t)\begin{pmatrix} \bar{\Omega}(C_1' - C_1) - \rho(C_2' - C_2) \\ \mu(C_2' - C_2) \end{pmatrix}.$$

The matrix $\mathcal{X}(t)$ is given by (66).

The symmetry operator $\hat{A}(t)$ in (68) for the 1D R2GPE (43) has the structure of a linear pseudodifferential operator whose parameters are functionals of the function $\Psi(x,t)$ on which the operator acts. Therefore, the operator $\hat{A}(t)$ is nonlinear.

Note that for some values of the parameters the pseudodifferential operator can become a differential one.

The symmetry operator $\hat{A}(t)$ depends on the symmetry operator $\widehat{B}(t,\mathbf{C})$ of the associated linear equation (53). Consider

$$\widehat{B}(t,\mathbf{C}) = \widehat{B}_\nu(t,\mathbf{C}) = \frac{1}{\sqrt{\nu!}}\Big[\hat{a}^+(t,\mathbf{C})\Big]^\nu\mathbb{I}, \quad \nu \in \mathbb{Z}_+, \tag{70}$$

where the operators $\hat{a}^{+}(t,\mathbf{C})$ are defined in (64).

Substituting (70) and (69) in (68), we obtain the symmetry operator for the 1D R2GPE (43), which we denote by $\hat{A}_\nu(t)$.

Following calculations given in section 4 of Ref. [5], we can generate a countable family of exact solutions for the 1D R2GPE (43) with the use of the family of the symmetry operators $\hat{A}_\nu(t)$, $\nu\in\mathbb{Z}_+$ and a starting solution $\Psi_0(x,t)$ of equation (43). Since the construction of these solutions for the 1D R2GPE (43) is similar to the same problem for the reduced one-component GPE [5], we present here only the solutions themselves without detail calculations.

We find the starting solution $\Psi_0(x,t)$ using a stationary solution of the of the Hamilton–Ehrenfest system (45), (47) that can be obtained from the general solution (49), (50) if we take integration constants as $\mathbf{C}=\mathbf{C}^0=(C_1^0,\dots,C_7^0)$, where we set $C_1^0=C_2^0=C_3^0=C_4^0=C_7^0=0$, $\sin C_6^0=0$, and C_5^0 is an arbitrary real constant. The stationary solution is

$$\begin{aligned}
&X(t,\mathbf{C})=P(t,\mathbf{C})=0,\\
&\Delta_{22}^{(2)}(t,\mathbf{C})=C_5^0,\quad \Delta_{21}^{(2)}(t,\mathbf{C})=-\frac{\rho}{\mu}C_5^0,\quad \Delta_{11}^{(2)}(t,\mathbf{C})=\frac{\tilde{\omega}}{\mu}C_5^0,\\
&\vec{\eta}(t,\mathbf{C})=\zeta\vec{n},\quad \zeta=\pm1.
\end{aligned}\tag{71}$$

The function

$$\Phi_0(x,t,\mathbf{C}^0)=\Big(\frac{\Omega}{\pi\hbar\,\mu}\Big)^{1/4}\exp\Big\{-\frac{i}{2\hbar}\frac{\rho}{\mu}x^2-\frac{1}{2\hbar}\frac{\Omega}{\mu}x^2\Big\}\exp\Big\{-\frac{i}{2}\Omega t-\frac{i}{2\hbar}\varkappa c C_5^0 t\Big\}v_\zeta(t)\tag{72}$$

is a particular solution of the corresponding associated linear equation (53). Here, $v_\zeta(t)=\exp\big(-i\zeta\mathcal{B}t\big)f_\zeta$, and f_ζ is given by the system

$$(\vec{\sigma},\vec{\mathcal{B}})f_\zeta=\zeta\mathcal{B}f_\zeta\tag{73}$$

and has the form

$$f_\zeta=\frac{e^{i\alpha}}{\sqrt{2}}\begin{pmatrix}\zeta\sqrt{1+\zeta\,\cos\theta}\,\exp(-i\varphi/2)\\ \sqrt{1-\zeta\,\cos\theta}\exp(i\varphi/2)\end{pmatrix},\quad \alpha=\text{const},\ \zeta=\pm1.\tag{74}$$

Taking $C_5^0=(\hbar\mu/2\Omega)$ in (72), we find the desired starting solution $\Psi_0(x,t)$ of the 1D R2GPE (43):

$$\begin{aligned}
\Psi_0(x,t)=\Phi_0(x,t,\mathbf{C}^0)\Big|_{C_5^0=(\hbar\mu/2\Omega)}&=\Big(\frac{\Omega}{\pi\hbar\,\mu}\Big)^{1/4}\exp\Big\{-\frac{i}{2\hbar}\frac{\rho}{\mu}x^2-\frac{1}{2\hbar}\frac{\Omega}{\mu}x^2\Big\}\times\\
&\times\exp\Big\{-\frac{i}{2}\Omega t-\frac{i\mu}{4\Omega}\varkappa c t\Big\}v_\zeta(t).
\end{aligned}\tag{75}$$

Then the symmetry operator $\hat{A}_\nu(t)$ determined by (68) transforms the solution $\Psi_0(x,t)$ of (75) into a solution $\Psi_\nu(x,t)$ of the 1D R2GPE (43) according to the following relation:

$$\begin{aligned}
\Psi_\nu(x,t)=(\hat{A}_\nu(t)\Psi_0)(x,t)&=\Big(\frac{\Omega}{\pi\hbar\,\mu}\Big)^{1/4}\Big(\frac{1}{\sqrt{2}}\Big)^{\nu}\exp\Big\{-\frac{i}{2\hbar}\frac{\rho}{\mu}x^2-\frac{1}{2\hbar}\frac{\Omega}{\mu}x^2\Big\}\times\\
&\times H_\nu\Big(\sqrt{\frac{\Omega}{\hbar\mu}}x\Big)\exp\Big\{-i\Big(\nu+\frac{1}{2}\Big)\Big(\frac{\varkappa c\mu}{2\Omega}+\Omega\Big)t\Big\}v_\zeta(t),
\end{aligned}\tag{76}$$

where $H_\nu(\zeta)$ are the Hermite polynomials [17] The functions (76) constitute a countable set of particular solutions to equation (43) which are generated from $\Psi_0(x,t)$ by the nonlinear symmetry operator $\hat{A}_\nu(t)$.

5. Concluding remarks

In this paper, which is a sequel to [4] and [5], we continued our study of symmetry properties of nonlinear integro-differential equations using WKB-Maslov semiclassical approximation.

In [5] we considered the 1-component multidimensional Gross–Pitaevskii equation with nonlocal nonlinear term and developed an approach to construct a class of symmetry operators for the semiclassically reduced Gross–Pitaevskii equation. The approach is based on our method proposed to construct the semiclassical asymptotic solution of the Cauchy problem to the nonlocal GPE.

In this work we apply the approach of Ref. [5] to the 2-component Gross–Pitaevskii equation with the nonlocal Manakov-type nonlinear interaction term (4), (5). In the semiclassical approximation the leading term of the asymptotic expansion is determined by the R2GPE (11) (or (17)) which contains nonlocal terms as a finite number of moments of the unknown function $\Psi(\vec{x}, t)$. We assign the R2GPE to the class of nearly linear equations according to the definition of Ref. [4] similar to the semiclassically reduced GPE considered in [5].

The general expression for the symmetry operator (42) is similar to the one obtained in [5] for the scalar (1-component) nonlocal GPE. To explore how the symmetry operators for a system (multi-component equation) transform as compared to a scalar equation, we have considered here the simplest case of the 2-component nonlocal GPE when the vector $\vec{H}(t)$ in (4) depends on t and does not depend on $\vec{x}$. For this case solution (8) can be presented as the product of a 2-component vector $u(t)$ and a scalar function $\varphi(\vec{x}, t)$ that leads to slight modification of the symmetry operator. Finding such symmetry operators in explicit form and generating exact solutions are illustrated with an example of 1D R2GPE considered in section 4.

Acknowledgments

The work was supported in part by Tomsk State University International Competitiveness Improvement Programm and by program "Nauka" No 1.676.2014/K.

References

[1] Baikov V A, Gazizov R K and Ibragimov N Kh 1989 *Current Problems in Mathematics: Newest Results* **34** 85 [English transl.: 1991 *J. Soviet Math.* **55** 1450]
[2] Fushchych W I, Shtelen W M and Serov N I 1989 *Symmetry Analysis and Exact Solutions of Nonlinear Equations of Mathematical Physics* (Kyiv: Naukova Dumka) [English transl.: 1993 (Dordrecht: Kluwer)]
[3] Baikov V A and Kordyukova S A 2003 *Quaest. Math.* **26** 1
[4] Levchenko E A, Shapovalov A V and Trifonov A Yu 2012 *J. Math. Analysis and Applications* **395** 716
[5] Lisok A L, Trifonov A Yu and Shapovalov A V 2013 *Symmetry, Integrability and Geometry: Methods and Applications (SIGMA)* **9**(2013)066 (*Preprint* math-ph/1302.3326)
[6] Maslov V P 1994 *The Complex WKB Method for Nonlinear Equations. I. Linear Theory* (Basel, Boston, Berlin: Birkhauser Verlag)
[7] Belov V V and Dobrokhotov S Yu 1992 *Theor. Math. Phys.* **92** 843
[8] Ovsjannikov L V 1982 *Group Analysis of Differential Equations* (New York: Academic Press)
[9] Olver P J 1986 *Application of Lie Groups to Differential Equations, Graduate Texts in Mathematics* vol 107 (New York: Springer–Verlag)
[10] Bluman G W, Cheviakov A F and Anco S C 2010 *Applications of Symmetry Methods to Partial Differential Equations, Applied Mathematical Sciences* vol 168 (New York: Springer)
[11] Dalfovo F, Giorgini S, Pitaevskii L and Stringari S 1999 *Rev. Mod. Phys.* **71** 463
[12] Lushnikov P M 2002 *Phys.Rev.* A **66** 051601(R)
[13] Frantzeskakis D J 2010 *J. Phys. A: Math. Theor.* **43** 213001
[14] Karasev M V and Maslov V P 1993 *Nonlinear Poisson Brackets: Geometry and Quantization* (Providence, RI:Amer. Math. Soc.) *Traslations of Mathematical Monographs* vol 119.
[15] Bagrov V G, Belov V V and Trifonov A Yu 1996 *Ann. of Phys.* (NY) **246** 231
[16] Belov V V, Trifonov A Yu, and Shapovalov A V 2002 *Int. J. Math. and Math. Sci.* **32** 325
[17] Bateman H and Erdelyi A 1953 *Higher Transcedental Functions* vol 2 (London: McGraw-Hill)

Journal of Physics: Conference Series **670** (2016) 012047 doi:10.1088/1742-6596/670/1/012047

Link between the relativistic canonical quantum mechanics of arbitrary spin and the corresponding field theory

Volodimir Simulik

Principal research associate, Institute of Electron Physics, National Academy of Sciences of Ukraine, 21 Universitetska Str., Uzhgorod 88000, Ukraine

E-mail: `vsimulik@gmail.com`

Abstract. The new relativistic equations of motion for the particles with arbitrary spin and nonzero mass have been introduced. The axiomatic level description of the relativistic canonical quantum mechanics of the arbitrary mass and spin has been given. The 64-dimensional $\mathrm{Cl}^{\mathrm{R}}(0,6)$ algebra in terms of Dirac gamma matrices has been suggested. The link between the relativistic canonical quantum mechanics of the arbitrary spin and the covariant local field theory has been found. Different methods of the Dirac equation derivation have been reviewed. The manifestly covariant field equations for an arbitrary spin that follow from the quantum mechanical equations have been considered. The covariant local field theory equations for spin s = (1,1) particle-antiparticle doublet, spin s = (1,0,1,0) particle antiparticle multiplet, spin s = (3 2,3 2) particle-antiparticle doublet, spin s = (2,2) particle-antiparticle doublet, spin s = (2,0,2,0) particle-antiparticle multiplet and spin s = (2,1,2,1) particle-antiparticle multiplet have been introduced. The Maxwell-like equations for the boson with spin s = 1 and nonzero mass have been introduced as well.

1. Introduction

Recently in [1, 2] the interesting results in the area of relativistic quantum mechanics and quantum field theory have been presented. This article contains brief exposition of the results [1, 2], which were reported on the conference. The general forms of quantum-mechanical and covariant equations for arbitrary spin are presented. The corresponding relativistic quantum mechanics of arbitrary spin is given as the brief version of the system of axioms. The partial cases of the spin s=(0,0) and spin s=(3/2,3/2) particle-antiparticle doublets are considered in explicit forms. The brief review of the different investigations in the area of relativistic canonical quantum mechanics (RCQM) is given and the brief analysis of the existing approaches to the field theory of arbitrary spin is initiated.

Note that in the Dirac model [3, 4] the quantum-mechanical interpretation is not evident. It has been demonstrated in [1, 2, 5, 6] that the quantum-mechanical interpretation is much more clear in the Foldy–Wouthuysen (FW) model [5, 6]. Nevertheless, the complete quantum-mechanical picture is possible only in the framework of RCQM. This assertion is one of the main conclusions proved in [1, 2].

The relativistic quantum mechanics under consideration is called *can nical* due to three main reasons. (i) The model under consideration has direct link with nonrelativistic quantum

Journal of Physics: Conference Series **670** (2016) 012047 doi:10.1088/1742-6596/670/1/012047

mechanics based on nonrelativistic Schrödinger equation. The principles of heredity and correspondence with other models of physical reality leads directly to nonrelativistic Schrödinger quantum mechanics. (ii) The FW model is already called by many authors as the canonical representation of the Dirac equation or a canonical field model, see, e. g., the paper [6]. And the difference between the field model given by FW and the RCQM is minimal – in corresponding equations it is only the presence and absence of beta matrix. (iii) The list of relativistic quantum-mechanical models is long. The Dirac model and the FW model are called by the 'old ' physicists as the relativistic quantum mechanics as well (one of my tasks in this paper is to show in visual and demonstrative way that these models have only weak quantum-mechanical interpretation). Further, the fractional relativistic quantum mechanics and the proper-time relativistic quantum mechanics can be listed (recall matrix formulation by W. Heisenberg, Feynman's sum over path's quantum theory, many-worlds interpretation by H. Everett), etc. Therefore, in order to avoid a confusion the model under consideration must have its proper name. Due to the reasons (i)–(iii) the best name for it is RCQM.

The general and fundamental goals here are as follows: (i) visual and demonstrative generalization of existing RCQM for the case of arbitrary spin, (ii) more complete formulation of this model on axiomatic level (on the test example of spin s=(1/2,1/2) particle-antiparticle doublet), (iii) vertical and horizontal links between the three different models of physical reality: relativistic quantum mechanics of arbitrary spin in canonical form, canonical (FW type) field theory of any spin, locally covariant (Dirac and Maxwell type) field theory of arbitrary spin.

2. Conce ts, definitions and notations

The concepts, definitions and notations here are the same as in [1, 2]. For example, in the Minkowski space-time

$$\mathrm{M}(1,3) = \{x \equiv (x^{\mu}) = (x^0 = t,\ \overrightarrow{x} \equiv (x^j))\}; \quad \mu = \overline{0,3},\ j = 1,2,3, \tag{1}$$

x^{μ} denotes the Cartesian (covariant) coordinates of the points of the physical space-time in the arbitrary-fixed inertial reference frame (IRF). We use the system of units with $\hbar = c = 1$. The metric tensor is given by

$$g^{\mu\nu} = g_{\mu\nu} = g^{\mu}_{\nu},\ (g^{\mu}_{\nu}) = \mathrm{diag}\,(1,-1,-1,-1)\,; \quad x_{\mu} = g_{\mu\nu}x^{\mu}, \tag{2}$$

and summation over the twice repeated indices is implied.

Note that the square-root operator equation $i\partial_t f(x) = \sqrt{m^2 - \quad} f(x)$, which is the main equation of RCQM, has been rejected by Dirac in his consideration in [4] (chapter 11, section 67), see [2] for the details. Nevertheless, today, contrary to the year 1928, the definition of the pseudo-differential (non-local) operator

$$\widehat{\omega} \equiv \sqrt{\widehat{\overrightarrow{p}}^{2} + m^2} = \sqrt{- \quad + m^2} \geq m > 0, \quad \widehat{\overrightarrow{p}} \equiv (\widehat{p}^j) = -i\nabla, \quad \nabla \equiv (\partial_{\ell}), \tag{3}$$

is well known. The action of the operator (3) in the coordinate representation (see, e. g. [7]) is given by

$$\widehat{\omega} f(t,\overrightarrow{x}) = \int d^3y K(\overrightarrow{x} - \overrightarrow{y}) f(t,\overrightarrow{y}), \quad f \in \mathrm{H}^{3,\mathrm{N}}, \tag{4}$$

where the function $K(\overrightarrow{x} - \overrightarrow{y})$ has the form $K(\overrightarrow{x} - \overrightarrow{y}) = -\frac{2m^2 K_2(m|\overrightarrow{x} - \overrightarrow{y}|)}{(2\pi)^2 |\overrightarrow{x} - \overrightarrow{y}|^2}$ and $K_{\nu}(z)$ is the modified Bessel function (Macdonald function), $|\overrightarrow{a}|$ designates the norm of the vector $\overrightarrow{a}$, $\mathrm{H}^{3,\mathrm{N}}$ is the Hilbert space of N-component functions.

Further, the following integral form

$$(\widehat{\omega} f)(t,\overrightarrow{x}) = \frac{1}{(2\pi)^{\frac{3}{2}}}\int d^3k e^{i\overrightarrow{k}\overrightarrow{x}}\widetilde{\omega}\widetilde{f}(t,\overrightarrow{k}), \quad \widetilde{\omega}\equiv\sqrt{\overrightarrow{k}^2+m^2}, \quad \widetilde{f}\in\widetilde{\mathrm{H}}^{3,\mathrm{N}}, \tag{5}$$

of the operator $\widehat{\omega}$ is used often, see, e. g., [6, 8], where f and $\widetilde{f}$ are linked by the 3-dimensional Fourier transformations

$$f(t,\overrightarrow{x}) = \frac{1}{(2\pi)^{\frac{3}{2}}}\int d^3k e^{i\overrightarrow{k}\overrightarrow{x}}\widetilde{f}(t,\overrightarrow{k}) \Leftrightarrow \widetilde{f}(t,\overrightarrow{k}) = \frac{1}{(2\pi)^{\frac{3}{2}}}\int d^3k e^{-i\overrightarrow{k}\overrightarrow{x}}\widetilde{f}(t,\overrightarrow{x}), \tag{6}$$

(in (6) $\overrightarrow{k}$ belongs to the spectrum $\mathrm{R}^3_{\overrightarrow{k}}$ of the operator $\widehat{\overrightarrow{p}}$, and the parameter $t\in(-\infty,\infty)\subset$ M(1,3)).

Note that the space of states $\mathrm{H}^{3,\mathrm{N}}$ is invariant with respect to the Fourier transformation (6). Therefore, both $\overrightarrow{x}$-realization $\mathrm{H}^{3,\mathrm{N}}$ and $\overrightarrow{k}$-realization $\widetilde{\mathrm{H}}^{3,\mathrm{N}}$ of the space of states are suitable for the purposes of our consideration. In the $\overrightarrow{k}$-realization the Schrödinger–Foldy equation has the algebraic-differential form

$$i\partial_t\widetilde{f}(t,\overrightarrow{k}) = \sqrt{\overrightarrow{k}^2+m^2}\widetilde{f}(t,\overrightarrow{k}); \quad \overrightarrow{k}\in\mathrm{R}^3_{\overrightarrow{k}}, \quad \widetilde{f}\in\widetilde{\mathrm{H}}^{3,\mathrm{N}}. \tag{7}$$

Below in the places, where misunderstanding is impossible, the symbol tilde is omitted.

Thus, today on the basis of above given definitions the difficulties, which stopped Dirac in 1928, can be overcome.

The name of the person, whose contribution in the theoretical model based on the square-root operator equation was decisive, is Leslie Lawrance Foldy (1919–2001). His interesting biography is presented in [9]. In our investigations we always marked the role of L. Foldy. Taking into account the L. Foldy's contribution in the construction of RCQM and his proof of the principle of correspondence between RCQM and non-relativistic quantum mechanics, we propose [10, 11] and [1, 2] *call e N-c p nen quare-r pera r equa i n* $i\partial_t f(x) = \sqrt{m^2 - \quad} f(x)$ *a e Sc rödin er–F ld equa i n*. Note here that this equation, which is a direct sum of one component spinless Salpeter equations [12], has been introduced in the formula (21) of [6]. Furthermore, note here that the nonlocal Poincaré group representation generators are known from the formulae (B-25)–(B-28) of the L. Foldy's paper [6].

Contrary to the times of papers [3, 5, 6, 12], the RCQM today is enough approbated and generally accepted theory. The spinless Salpeter equation has been introduced in [12]. The allusion on the RCQM and the first steps are given in [6], where the Salpeter equation for the 2s+1-component wave function was considered and the cases of s=1/2, s=1 were presented as an examples. In [13] Foldy continued his investigations [6] by the consideration of the relativistic particle systems with interaction. The interaction was introduced by the specific group-theoretical method.

After that in the RCQM were developed both the construction of mathematical foundations and the solution of concrete quantum-mechanical problems for different potentials. The brief review of 24 articles devoted to contemporary RCQM has been given in [2]. In the papers [10, 11], where we started our investigations in RCQM, this relativistic model for the test case of the spin s=(1/2,1/2) particle-antiparticle doublet is formulated. In [10], this model is considered as the system of the axioms on the level of the von Neumann monograph [14], where the mathematically well-defined consideration of the nonrelativistic quantum mechanics was given. Furthermore, in [10, 11] the operator link between the spin s=(1/2,1/2) particle-antiparticle doublet RCQM and the Dirac theory is given and Foldy's synthesis of *c varian par icle equa i n* is extended to the start from the RCQM of the spin s=(1/2,1/2) particle-antiparticle doublet. In [1] the same

Journal of Physics: Conference Series **670** (2016) 012047 doi:10.1088/1742-6596/670/1/012047

procedure is fulfilled for the spin s=(1,1), s=(1,0,1,0), s=(3/2,3/2), s=(2,2), s=(2,0,2,0) and spin s=(2,1,2,1) RCQM. The corresponding equations, which follow from the RCQM for the covariant local field theory, are introduced.

Taking into account the 24 RCQM results reviewed in [2] the following conclusion is given. Here and in [1, 2] I am not going to formulate a new relativistic quantum mechanics! The foundations of RCQM based on the spinless Salpeter equation are already formulated in [6, 7] and in Refs. [9–35] given in [2].

3. Brie analysis o the covariant e uations or an arbitrary s in

One of the goals of [1] is the link between the RCQM of an arbitrary spin and the different approaches to the covariant local field theory of an arbitrary spin. Surely, at least the brief analysis of the existing covariant equations for an arbitrary spin should be presented.

Note that in [1] and here only the first-order particle and the field equations (together with their canonical nonlocal pseudo-differential representations) are considered. The second order equations (like the Klein–Gordon–Fock equation) are not the subject of this investigation.

Different approaches to the description of the field theory of an arbitrary spin can be found in [6, 15–24]. Here and in [1, 2] only the approach started in [6] is the basis for further application. Other results given in [15–24] are not used here.

Note only some general deficiencies of the known equations for arbitrary spin. The consideration of the partial cases, when the substitution of the fixed value of spin is fulfilled, is not successful in all cases. For example, for the spin $s > 1$ existing equations have the redundant components and should be complemented by some additional conditions. Indeed, the known equations [25, 26] for the spin s=3/2 (and their confirmation in [27]) should be essentially complemented by the additional conditions. The main difficulty in the models of an arbitrary spin is the interaction between the fields of higher-spin. Even the quantization of higher-spin fields generated the questions. These and other deficiencies of the known equations for higher-spin are considered in Refs. [50–61] given in [2] (a brief review of deficiencies see in [28]).

Equations suggested in [1] and here are free of these deficiencies. The start of such consideration is taken from [6], where the main foundations of the RCQM are formulated. In the text of [1, 2] and here the results of [6] are generalized and extended. The operator link between the results of [5] and [6] (between the canonical FW type field theory and the RCQM) is suggested. Note that the cases s=3/2 and s=2 are not presented in [6], especially in explicit demonstrative forms. The results of [1, 2] are closest to the given in [29, 30]. The difference is explained in the section 5 below after the presentation of the results [1, 2].

Even this brief analysis makes us sure in the prospects of the investigations started in [1]. The successful description of the arbitrary spin field models is not the solved problem today.

4. Axioms o the relativistic canonical uantum mechanics o an arbitrary s in

The RCQM of the arbitrary spin given in sections 2 and 18 of [1] can be formulated at the level of von Neumann's consideration [14]. The difference with [14] is only in relativistic invariance and in the consideration of multicomponent and multidimensional objects.

The partial case of axiomatic formulation is already given in section 7 of [1] at the example of spin s=1/2 particle-antiparticle doublet. The RCQM of the arbitrary spin particle-antiparticle doublet (or particle singlet) can be formulated similarly as the corresponding generalization of this partial case.

Below the brief presentation of the *li f e axi* is given. Note that some particular content of these axioms is already given in section 2 of [1], where the RCQM of the arbitrary spin particle singlet has been formulated.

On e pace f a e
The space of states of isolated arbitrary spin particle singlet in an arbitrarily-fixed inertial frame of reference (IFR) in its $\overrightarrow{x}$-realization is the Hilbert space

$$\mathrm{H}^{3,\mathrm{N}} = \mathrm{L}_2(\mathrm{R}^3) \otimes \mathrm{C}^{\otimes \mathrm{N}} = \{f = (f^{\mathrm{N}}) : \mathrm{R}^3 \rightarrow \mathrm{C}^{\otimes \mathrm{N}}; \quad \int d^3x |f(t, \overrightarrow{x})|^2 < \infty\}, \quad \mathrm{N} = 2s + 1, \quad (8)$$

of complex-valued N-component square-integrable functions of $x \in \mathrm{R}^3 \subset \mathrm{M}(1,3)$ (similarly, in momentum, $\overrightarrow{p}$-realization). In (8) d^3x is the Lebesgue measure in the space $\mathrm{R}^3 \subset \mathrm{M}(1,3)$ of the eigenvalues of the position operator $\overrightarrow{x}$ of the Cartesian coordinate of the particle in an arbitrary-fixed IFR. Further, $\overrightarrow{x}$ and $\overrightarrow{p}$ are the operators of canonically conjugated dynamical variables of the spin s=(1/2,1/2) particle-antiparticle doublet, and the vectors f, f in $\overrightarrow{x}$- and $\overrightarrow{p}$-realizations are linked by the 3-dimensional Fourier transformation (the variable t is the parameter of time-evolution).

e a e a ical c rrec ne f e c n idera i n demands the application of the rigged Hilbert space

$$\mathrm{S}^{3,\mathrm{N}} \equiv \mathrm{S}(\mathrm{R}^3) \times \mathrm{C}^{\mathrm{N}} \subset \mathrm{H}^{3,\mathrm{N}} \subset \mathrm{S}^{3,\mathrm{N}*}. \quad (9)$$

where the Schwartz test function space $\mathrm{S}^{3,\mathrm{N}}$ is the core (i. e., it is dense both in $\mathrm{H}^{3,\mathrm{N}}$ and in the space $\mathrm{S}^{3,\mathrm{N}*}$ of the N-component Schwartz generalized functions). The space $\mathrm{S}^{3,\mathrm{N}*}$ is conjugated to that of the Schwartz test functions $\mathrm{S}^{3,\mathrm{N}}$ by the corresponding topology (see, e. g. [31]).

Strictly speaking, the mathematical correctness of consideration demands to make the calculations in the space $\mathrm{S}^{3,\mathrm{N}*}$ of generalized functions, i. e. with the application of cumbersome functional analysis (see, e. g. [32]). Nevertheless, one can take into account the properties of the Schwartz test function space $\mathrm{S}^{3,\mathrm{N}}$ in the triple (9). The space $\mathrm{S}^{3,\mathrm{N}}$ is dense both in quantum-mechanical space $\mathrm{H}^{3,\mathrm{N}}$ and in the space of generalized functions $\mathrm{S}^{3,\mathrm{N}*}$. Therefore, any physical state $f \in \mathrm{H}^{3,\mathrm{N}}$ can be approximated with an arbitrary precision by the corresponding elements of the Cauchy sequence in $\mathrm{S}^{3,\mathrm{N}}$, which converges to the given $f \in \mathrm{H}^{3,\mathrm{N}}$. Further, taking into account the requirement to measure the arbitrary value of the quantum-mechanical model with non-absolute precision, it means that all concrete calculations can be fulfilled within the Schwartz test function space $\mathrm{S}^{3,\mathrm{N}}$. Thus, such consideration allows us to perform, without any loss of generality, all necessary calculations in the space $\mathrm{S}^{3,\mathrm{N}}$ at the level of correct differential and integral calculus. More detailed consideration see in [1, 2].

Note finally that in the case of arbitrary spin particle-antiparticle doublet the dimension of spaces (8), (9) is M=2N=2(2s+1).

On e i e ev lu i n f e a e vec r
The time dependence of the state vectors $f \in \mathrm{H}^{3,\mathrm{N}}$ (time t is the parameter of evolution) is given either in the integral form by the unitary operator

$$u\,(t_0, t) = \exp\left[-i\widehat{\omega}(t - t_0)\right]; \quad \widehat{\omega} \equiv \sqrt{- \quad + m^2}, \quad (10)$$

(below $t_0 = t$ is put), or in the differential form by the Schrödinger–Foldy equation of motion

$$(i\partial_0 - \widehat{\omega})f(x) = 0. \quad (11)$$

with the wave function

$$f \equiv \mathrm{column}(f^1, f^2, ..., f^{\mathrm{N}}), \quad \mathrm{N} = 2s + 1, \quad f \in \mathrm{H}^{3,\mathrm{N}}. \quad (12)$$

Note that here the operator $\widehat{\omega} \equiv \sqrt{- \quad + m^2}$ is the relativistic analog of the energy operator (Hamiltonian) of nonrelativistic quantum mechanics. The Minkowski space-time M(1,3) is pseudo Euclidean with metric $g = \text{diag}(+1, -1, -1, -1)$. The step from the particle singlet of arbitrary spin to the corresponding particle-antiparticle doublet is evident.

Thus, for the arbitrary spin particle-antiparticle doublet the system of two N-component equations $(i\partial_0 - \widehat{\omega})f(x) = 0$ and $(i\partial_0 - \widehat{\omega})f(x) = 0$ is used. Therefore, the corresponding Schrödinger–Foldy equation is given by (11), where the 2N-component wave function is the direct sum of the particle and antiparticle wave functions, respectively. Due to the historical tradition of the physicists the antiparticle wave function is put in the down part of the 2N-column.

The general solution of the Schrödinger–Foldy equation of motion (11) (in the case of particle-antiparticle arbitrary spin doublet) has the form

$$f(x) = \frac{1}{(2\pi)^{\frac{3}{2}}} \int d^3k e^{-ikx} a^{2\text{N}}\left(\overrightarrow{k}\right) \text{d}_{2\text{N}}, \quad kx \equiv \omega t - \overrightarrow{k}\overrightarrow{x}, \quad \omega \equiv \sqrt{\overrightarrow{k}^2 + m^2}, \tag{13}$$

where the orts of the N-dimensional Cartesian basis can be found in [1] in formulae (10).

The action of the pseudo-differential (non-local) operator $\widehat{\omega} \equiv \sqrt{- \quad + m^2}$ is explained in (4), (5).

On e funda en al d na ical variable

The dynamical variable $\overrightarrow{x} \in \text{R}^3 \subset$M(1,3) (as well as the variable $\overrightarrow{k} \in \text{R}^3_{\overrightarrow{k}}$) represents the external degrees of freedom of the arbitrary spin particle-antiparticle doublet. The spin $\overrightarrow{s}$ of the particle-antiparticle doublet is the first in the list of the carriers of the internal degrees of freedom. Taking into account the Pauli principle and the fact that experimentally an antiparticle is observed as the mirror reflection of a particle, the operators of the charge sign and the spin of the arbitrary particle-antiparticle doublet are taken in the form

$$g \equiv -\Gamma^0_{2\text{N}} \equiv -\sigma^3_{2\text{N}} = \begin{vmatrix} -\text{I}_\text{N} & 0 \\ 0 & \text{I}_\text{N} \end{vmatrix}, \quad \overrightarrow{s}_{2\text{N}} = \begin{vmatrix} \overrightarrow{s}_\text{N} & 0 \\ 0 & -\hat{C}\overrightarrow{s}_\text{N}\hat{C} \end{vmatrix}, \quad \text{N} = 2s + 1, \tag{14}$$

where $\Gamma^0_{2\text{N}}$ is the 2N$\times$2N Dirac Γ^0 matrix, $\sigma^3_{2\text{N}}$ is the 2N$\times$2N Pauli σ^3 matrix, $\hat{C}$ is the operator of complex conjugation in the form of N$\times$N diagonal matrix, the operator of involution in $\text{H}^{3,2\text{N}}$, and I_N is N $\times$ N unit matrix. Thus, the spin is given by the generators of SU(2) algebra! The spin matrices $\overrightarrow{s}_{2\text{N}}$ (14) satisfy the commutation relations

$$\left[s^j_{2\text{N}}, s^\ell_{2\text{N}}\right] = i\varepsilon^{j\ell n} s^n_{2\text{N}}, \quad \varepsilon^{123} = +1, \tag{15}$$

of the algebra of SU(2) group, where $\varepsilon^{j\ell n}$ is the Levi-Civita tensor and $s^j = \varepsilon^{j\ell n} s_{\ell n}$ are the Hermitian 2N $\times$ 2N matrices (14) – the generators of a 2N-dimensional reducible representation of the spin group SU(2) (universal covering of the SO(3)$\subset$SO(1,3) group).

On e rela ivi ic invariance f e e r

The relativistic invariance of the model under consideration (the relativistic invariance of the Schrödinger–Foldy equation (11)) requires, as a first step, consideration of its invariance with respect to the proper ortochronous Lorentz $\text{L}^\uparrow_+ = \text{SO(1,3)} = \{\Lambda = (\Lambda^\mu_\nu)\}$ and Poincaré $\text{P}^\uparrow_+ = \text{T(4)}\times)\text{L}^\uparrow_+ \supset \text{L}^\uparrow_+$ groups. This invariance in an arbitrary relativistic model is the implementation of the Einstein's relativity principle in the special relativity form. Note that the mathematical correctness requires the invariance mentioned above to be considered as the invariance with respect to the *univer al c verin* $\mathcal{L} = \text{SL(2,C)}$ and $\mathcal{P} \supset \mathcal{L}$ of the groups $\text{L}^\uparrow_+$ and $\text{P}^\uparrow_+$, respectively.

For the group $\mathcal{P}$ we choose real parameters $a = (a^{\mu}) \in$M(1,3) and $\varpi \equiv (\varpi^{\mu\nu} = -\varpi^{\nu\mu})$ with well-known physical meaning. For the standard $\mathcal{P}$ generators $(p_{\mu}, j_{\mu\nu})$ we use commutation relations in the manifestly covariant form

$$[p_{\mu}, p_{\nu}] = 0, \ [p_{\mu}, j_{\rho\sigma}] = ig_{\mu\rho}p_{\sigma} - ig_{\mu\sigma}p_{\rho}, \tag{16}$$

$$[j_{\mu\nu}, j_{\rho\sigma}] = -i\left(g_{\mu\rho}j_{\nu\sigma} + g_{\rho\nu}j_{\sigma\mu} + g_{\nu\sigma}j_{\mu\rho} + g_{\sigma\mu}j_{\rho\nu}\right).$$

The following assertion should be noted. Not a matter of fact that *n n-c varian* objects such as the Lebesgue measure d^3x and *n n-c varian (n n-Lie)* generators of algebras are explored, the model of RCQM of arbitrary spin is a relativistic invariant in the following sense. The Schrödinger–Foldy equation (11) and the set of its solution $\{f\}$ (13) are invariant with respect to the irreducible unitary representation of the group $\mathcal{P}$, the N×N matrix-differential generators of which are given by the following nonlocal operators

$$\widehat{p}_0 = \widehat{\omega} \equiv \sqrt{- \quad + m^2}, \quad \widehat{p}_{\ell} = i\partial_{\ell}, \tag{17}$$

$$\widehat{j}_{\ell n} = x_{\ell}\widehat{p}_n - x_n\widehat{p}_{\ell} + s_{ln} \equiv \widehat{m}_{\ell n} + s_{\ell n},$$

$$\widehat{j}_{0\ell} = -\widehat{j}_{\ell 0} = t\widehat{p}_{\ell} - \frac{1}{2}\{x_{\ell}, \widehat{\omega}\} - \left(\frac{s_{\ell n}\widehat{p}_n}{\widehat{\omega} + m} \equiv \breve{s}_{\ell}\right), \tag{18}$$

where the orbital parts of the generators are not changed under the transition from one spin to another. Under such transitions only the spin parts (14), (15) of the expressions (17), (18) are changed. Indeed, the direct calculations visualize that generators (17), (18) commute with the operator of equation (11) and satisfy the commutation relations (16) of the Lie algebra of the Poincaré group $\mathcal{P}$. In formulae (17), (18), the SU(2)-spin generators $s^{\ell n}$ have particular specific forms for each representation of the SU(2) group (see the list of examples in [1]).

Note that the generators (17), (18) are known from the formulae (B-25) (B-28) of the paper [6].

Note also that together with the generators (17), (18) another set of 10 operators commutes with the operator of equation (11), satisfies the commutation relations (16) of the Lie algebra of Poincaré group $\mathcal{P}$, and, therefore, can be chosen as the Poincaré symmetry of the model under consideration. This second set is given by the generators $\widehat{p}^0$, $\widehat{p}^{\ell}$ from (17) together with the orbital parts of the generators $\widehat{j}^{\ell n}$, $\widehat{j}^{0\ell}$ from (18).

Thus, the irreducible unitary representation of the Poincaré group $\mathcal{P}$ in the space (9), with respect to which the Schrödinger-Foldy equation (11) and the set of its solution $\{f\}$ (13) are invariant, is given by a series converges in this space

$$(a, \varpi) \rightarrow U(a, \varpi) = \exp(-ia^0\widehat{p}_0 - i\overrightarrow{a}\overrightarrow{\widehat{p}} - \frac{i}{2}\varpi^{\mu\nu}\widehat{j}_{\mu\nu}) \tag{19}$$

where the generators $(\widehat{p}^{\mu}, \widehat{j}^{\mu\nu})$ are given in (17), (18) with the arbitrary values of the SU(2) spins $\overrightarrow{s} = (s^{\ell n})$ (14), (15).

The validity of this assertion is verified by the following three steps. (i) The calculation that the $\mathcal{P}$-generators (17), (18) commute with the operator $i\partial_0 - \widehat{\omega}$ of the Schrödinger–Foldy equation (11). (ii) The verification that the $\mathcal{P}$-generators (17), (18) satisfy the commutation relations (16) of the Lie algebra of the Poincaré group $\mathcal{P}$. (iii) The proof that generators (17), (18) realize the spin s(s+1) representation of this group. Therefore, the Bargman–Wigner classification on the basis of the corresponding Casimir operators calculation should be given. These three steps can be made by direct and non-cumbersome calculations.

The expression (19) is well known, but rather formal. In fact the transition from a Lie algebra to a finite group transformations is a rather non-trivial action. The mathematical justification of (19) can be fulfilled in the framework of Schwartz test function space and will be given in next special publication.

The corresponding Casimir operators have the form

$$p^2 = \widehat{p}^{\mu}\widehat{p}_{\mu} = m^2 \mathrm{I_N}, \tag{20}$$

$$W = w^{\mu}w_{\mu} = m^2 \overrightarrow{s}^2 = \mathrm{s}(\mathrm{s}+1)m^2\mathrm{I_N}, \tag{21}$$

where $\mathrm{I_N}$ is the $\mathrm{N} \times \mathrm{N}$ unit matrix and s =1/2, 1, 3/2, 2, ...

Note that together with the generators (17), (18) another set of 10 operators commutes with the operator of equation (11), satisfies the commutation relations (16) of the Lie algebra of Poincaré group $\mathcal{P}$, and, therefore, can be chosen as the Poincaré symmetry of the model under consideration. This second set is given by the generators $\widehat{p}^0$, $\widehat{p}^{\ell}$ from (17) together with the orbital parts of the generators $\widehat{j}^{\ell n}$, $\widehat{j}^{0\ell}$ from (18). Thus, this second set of Poincaré generators is given by

$$\widehat{p}_0 = \widehat{\omega} \equiv \sqrt{- \quad + m^2}, \quad \widehat{p}_{\ell} = i\partial_{\ell}, \quad \widehat{m}_{\ell n} = x_{\ell}\widehat{p}_n - x_n\widehat{p}_{\ell}, \quad \widehat{m}_{0\ell} = -\widehat{m}_{\ell 0} = t\widehat{p}_{\ell} - \frac{1}{2}\left\{x_{\ell}, \widehat{\omega}\right\}. \tag{22}$$

Note that in the case s=0 only generators (22) form the Poincaré symmetry.

Note that the modern definition of $\mathcal{P}$ invariance (or $\mathcal{P}$ symmetry) of the equation of motion (11) in $\mathrm{H}^{3,\mathrm{N}}$ is given by the following assertion, see, e. g. [33]. *e e* $\mathrm{F} \equiv \{f\}$ *f all p ible lu i n f e equa i n ( ) i invarian wi re pec e* $\mathcal{P}^{\mathrm{f}}$*-repre en a i n f e r up* $\mathcal{P}$*, if f r arbi rar lu i n f and arbi raril -fixed para e er* (a, ϖ) *e a er i n*

$$(a, \varpi) \to U(a, \varpi)\{f\} = \{f\} \equiv \mathrm{F} \tag{23}$$

i valid. In [2] this axiom is considered together with very useful axiom *n e d na ic and kine a ic a pec f e rela ivi ic invariance.*

5 On e Cliff rd–Dirac al ebra

This axiom is additional and is not necessary. Nevertheless, such axiom is very useful for the dimensions, where the Γ matrices exist. Application of the Clifford–Dirac algebra is the useful method of calculations in RCQM. Three different definitions of the Clifford algebra and their equivalence are considered in [34]. In different approaches to the relativistic quantum mechanics the matrix representation of the Clifford algebra in terms of the Dirac gamma matrices is used. This representation is called the Clifford–Dirac algebra.

For our purposes the anticommutation relations of the Clifford–Dirac algebra generators are taken in the general form

$$\Gamma^{\bar{\mu}}_{2\mathrm{N}}\Gamma^{\bar{\nu}}_{2\mathrm{N}} + \Gamma^{\bar{\nu}}_{2\mathrm{N}}\Gamma^{\bar{\mu}}_{2\mathrm{N}} = 2g^{\bar{\mu}\bar{\nu}}; \quad \bar{\mu}, \bar{\nu} = \overline{0,4}, \quad (g^{\bar{\mu}\bar{\nu}}) = (+----), \tag{24}$$

where $\Gamma^{\bar{\mu}}_{2\mathrm{N}}$ are the $2\mathrm{N} \times 2\mathrm{N}$ Dirac $\Gamma^{\bar{\mu}}$ matrices ($2\mathrm{N} \times 2\mathrm{N}$ generalization of the Dirac 4×4 γ matrices), $\Gamma^4_{2\mathrm{N}} \equiv \Gamma^0_{2\mathrm{N}}\Gamma^1_{2\mathrm{N}}\Gamma^2_{2\mathrm{N}}\Gamma^3_{2\mathrm{N}}$. Here and in our publications (see, e. g. the last years articles [35–39]) we use the $\gamma^4 \equiv \gamma^0\gamma^1\gamma^2\gamma^3$ matrix instead of the γ^5 matrix of other authors. Our γ^4 is equal to $i\gamma^5_{\mathrm{standard}}$. Notation γ^5 is used in [35–39] for a completely different matrix $\gamma^5 \equiv \gamma^1\gamma^3\hat{C}$. As well as the element $\Gamma^4_{2\mathrm{N}} \equiv \Gamma^0_{2\mathrm{N}}\Gamma^1_{2\mathrm{N}}\Gamma^2_{2\mathrm{N}}\Gamma^3_{2\mathrm{N}}$ of (24) is dependent the algebra basis is formed by 4=1+3 independent elements. Therefore, such Clifford algebra over the field of complex numbers is denoted $\mathrm{Cl}^{\mathrm{C}}(1,3)$ and the dimension of the algebra is $2^4 = 16$.

The best consideration of this axiom is given in [2], where the complete analysis is presented. Moreover, in [2] new Clifford–Dirac algebra over the field of real numbers is introduced. Two important representations of this algebra are defined as $\mathrm{Cl}^{\mathrm{R}}(4,2)$, $\mathrm{Cl}^{\mathrm{R}}(0,6)$ and the dimension of this algebra is $2^6 = 64$.

6 Briefl n er axi

Other axioms of arbitrary spin RCQM are given in [2]. The list of these axioms is as follows. *On e ex ernal and in ernal de ree f freed , n e al ebra f b ervable , n e ain and addi i nal c n erva i n law , n e a i nar c ple e e f pera r , n e lu i n f e Sc rödin er-F ld equa i n, n e ean value f e pera r f b ervable , n e principle f eredi and e c rre p ndence, n e ec nd quan i a i n (ex ernal axi), n e p ical in erpre a i n* All axioms of this section eventually need to be reconciled with three levels of description used in this paper: RCQM, canonical FW and Dirac models. Nevertheless, this interesting problem cannot be considered in few pages. The readers of this paper can compare the axioms of RCQM given above with the main principles of the Dirac model given in B. Thaller's monograph [40] on the high mathematical level.

5. General descri tion o the arbitrary s in field theory

The step by step consideration of the different partial examples in [1] (sections 21–27) enabled us to rewrite them in the general form, which is valid for arbitrary spin. Therefore, the generalization of the consideration given in [1] leads to the general formalism of the arbitrary spin fields.

The formalism presented below in this section is valid for an arbitrary particle-antiparticle multiplet in general and for the particle-antiparticle doublet in particular.

5 e can nical (FW pe) del f e arbi rar pin par icle-an ipar icle field

The operator, which transform the RCQM of the arbitrary spin particle-antiparticle multiplet into the corresponding canonical particle-antiparticle field, is given by

$$v_{2\mathrm{N}} = \left| \begin{array}{cc} \mathrm{I_N} & 0 \\ 0 & \hat{C}\mathrm{I_N} \end{array} \right|, \quad v_{2\mathrm{N}}^{-1} = v_{2\mathrm{N}}^{\dagger} = v_{2\mathrm{N}}, \quad v_{2\mathrm{N}}v_{2\mathrm{N}} = \mathrm{I_{2N}}, \quad \mathrm{N} = 2s+1, \tag{25}$$

where $\hat{C}\mathrm{I_N}$ is the $\mathrm{N} \times \mathrm{N}$ operator of complex conjugation. Indeed, the operator (25) translates any operator from canonical field FW representation into the RCQM representation and vice versa:

$$v_{2\mathrm{N}}\hat{q}_{\mathrm{cf}}^{\mathrm{anti-Herm}}v_{2\mathrm{N}} = \hat{q}_{\mathrm{qm}}^{\mathrm{anti-Herm}}, \quad v_{2\mathrm{N}}\hat{q}_{\mathrm{qm}}^{\mathrm{anti-Herm}}v_{2\mathrm{N}} = \hat{q}_{\mathrm{cf}}^{\mathrm{anti-Herm}}. \tag{26}$$

Here $\hat{q}_{\mathrm{qm}}^{\mathrm{anti-Herm}}$ is an arbitrary operator from the RCQM of the 2N-component particle-antiparticle doublet in the anti-Hermitian form, e. g., the operator $(\partial_0 + i\widehat{\omega})$ of equation of motion (11), the operator of spin $\overrightarrow{s}_{2\mathrm{N}}$ (14) taken in anti-Hermitian form, etc., $\hat{q}_{\mathrm{cf}}^{\mathrm{anti-Herm}}$ is an arbitrary operator from the canonical field theory of the 2N-component particle-antiparticle doublet in the anti-Hermitian form. Thus, the only warning is that operators here must be taken in anti-Hermitian form, see section 9 in [1] for the details and see [41, 42] for the mathematical correctness of anti-Hermitian operators application.

Further, the operator (25) translates

$$\phi = v_{2\mathrm{N}}f, \quad f = v_{2\mathrm{N}}\phi, \tag{27}$$

the solution (13) of the Schrödinger–Foldy equation (11) into the solution

$$\phi(x) = \frac{1}{(2\pi)^{\frac{3}{2}}} \int d^3k \left[e^{-ikx}a^{\mathrm{N}}(\overrightarrow{k})\mathrm{d_N} + e^{ikx}a^{*\check{\mathrm{N}}}(\overrightarrow{k})\mathrm{d}_{\check{\mathrm{N}}} \right], \tag{28}$$

$\mathrm{N} = 1, 2, ..., \mathrm{N}, \quad \check{\mathrm{N}} = \mathrm{N}+1, \mathrm{N}+2, ..., 2\mathrm{N}$, of the FW equation

$$(i\partial_0 - \Gamma_{2\mathrm{N}}^0\widehat{\omega})\phi(x) = 0, \quad \Gamma_{2\mathrm{N}}^0 \equiv \sigma_{2\mathrm{N}}^3 = \left| \begin{array}{cc} \mathrm{I_N} & 0 \\ 0 & -\mathrm{I_N} \end{array} \right|, \tag{29}$$

$\widehat{\omega} \equiv \sqrt{- \quad + m^2}$, $\mathrm{N} = 2s + 1$, and *vice ver a*. Thus, the transformation (25), (26) translates the matrices $\Gamma^0_{2\mathrm{N}}$ and

$$\Gamma^j_{2\mathrm{N}} = \begin{vmatrix} 0 & \Sigma^j_{\mathrm{N}} \\ -\Sigma^j_{\mathrm{N}} & 0 \end{vmatrix}, \quad j = 1, 2, 3, \tag{30}$$

into the RCQM representation and *vice ver a*

$$\bar{\Gamma}^{\bar{\mu}}_{2\mathrm{N}} = v_{2\mathrm{N}} \Gamma^{\bar{\mu}}_{2\mathrm{N}} v_{2\mathrm{N}}, \quad \Gamma^{\bar{\mu}}_{2\mathrm{N}} = v_{2\mathrm{N}} \bar{\Gamma}^{\bar{\mu}}_{2\mathrm{N}} v_{2\mathrm{N}}. \tag{31}$$

In (30) Σ^j_{N} are the $\mathrm{N} \times \mathrm{N}$ Pauli matrices. Matrices $\Gamma^\mu_{2\mathrm{N}}$ (29), (30) together with matrix $\Gamma^4_{2\mathrm{N}} \equiv \Gamma^0_{2\mathrm{N}}\Gamma^1_{2\mathrm{N}}\Gamma^2_{2\mathrm{N}}\Gamma^3_{2\mathrm{N}}$ satisfy the anticommutation relations (24) of the Clifford–Dirac algebra.

The formulas mentioned below are found from the corresponding formulas of RCQM with the help of the operator (25) on the basis of its properties (26), (27). Thus, for the general form of arbitrary spin canonical particle-antiparticle field the equation of motion of the FW type is given by (29). The general solution has the form (28), where $a^{\mathrm{N}}(\overrightarrow{k})$ are the quantum-mechanical momentum-spin amplitudes of the particle and $a^{\check{\mathrm{N}}}(\overrightarrow{k})$ are the quantum-mechanical momentum-spin amplitudes of the antiparticle, $\{\mathrm{d}\}$ is 2N-component Cartesian basis.

It is evident from (28) that the model under consideration is not quantum mechanics. Indeed, contrary to (13) the solution (28) contains positive and negative frequency terms and, as a consequence, equation (29) is dealing with positive and negative energies (contrary to equation (11)).

The spin operator, which follows from (14), has the form

$$\overrightarrow{s}_{2\mathrm{N}} = \begin{vmatrix} \overrightarrow{s}_{\mathrm{N}} & 0 \\ 0 & \overrightarrow{s}_{\mathrm{N}} \end{vmatrix}, \quad \mathrm{N} = 2s + 1, \tag{32}$$

where $\overrightarrow{s}_{\mathrm{N}}$ are $\mathrm{N} \times \mathrm{N}$ generators of arbitrary spin irreducible representations of SU(2) algebra, which satisfy the commutation relations $\left[s^j_{\mathrm{N}}, s^\ell_{\mathrm{N}}\right] = i\varepsilon^{j\ell n} s^n_{\mathrm{N}}, \quad \varepsilon^{123} = +1$.

The generators of the reducible unitary representation of the Poincaré group $\mathcal{P}$, with respect to which the canonical field equation (29) and the set $\{\phi\}$ of its solutions (28) are invariant, are given by

$$\widehat{p}^0 = \Gamma^0_{2\mathrm{N}} \widehat{\omega} \equiv \Gamma^0_{2\mathrm{N}} \sqrt{- \quad + m^2}, \quad \widehat{p}^\ell = -i\partial_\ell, \quad \widehat{j}^{\ell n} = x^\ell \widehat{p}^n - x^n \widehat{p}^\ell + s^{\ell n}_{2\mathrm{N}} \equiv \widehat{m}^{\ell n} + s^{\ell n}_{2\mathrm{N}}, \tag{33}$$

$$\widehat{j}^{0\ell} = -\widehat{j}^{\ell 0} = x^0 \widehat{p}^\ell - \frac{1}{2}\Gamma^0_{2\mathrm{N}}\left\{x^\ell, \widehat{\omega}\right\} + \Gamma^0_{2\mathrm{N}} \frac{(\overrightarrow{s}_{2\mathrm{N}} \times \overrightarrow{p})^\ell}{\widehat{\omega} + m}, \tag{34}$$

where arbitrary spin SU(2) generators $\overrightarrow{s}_{2\mathrm{N}} = (s^{\ell n}_{2\mathrm{N}})$ have the form (32), $\Gamma^0_{2\mathrm{N}}$ is given in (29).

Note that together with the generators (33), (34) another set of 10 operators commutes with the operator of equation (29), satisfies the commutation relations (16) of the Lie algebra of Poincaré group $\mathcal{P}$, and, therefore, can be chosen as the Poincaré symmetry of the model under consideration. This second set is given by the generators $\widehat{p}^0$, $\widehat{p}^\ell$ from (33) together with the orbital parts of the generators $\widehat{j}^{\ell n}$, $\widehat{j}^{0\ell}$ from (33), (34), respectively. In another way this set follows from the set (22) after the transformation (25), (26).

The calculation of the Casimir operators $p^2 = \widehat{p}^\mu \widehat{p}_\mu$, $W = w^\mu w_\mu$ (w^μ is the Pauli–Lubanski pseudovector) for the fixed value of spin completes the brief description of the model.

5 e l call c varian del f e arbi rar pin par icle-an ipar icle field

The operator, which transform the canonical (FW type) model of the arbitrary spin particle-antiparticle field into the corresponding locally covariant particle-antiparticle field, is the generalized FW operator and is given by

$$V^{\mp}=\frac{\mp\overrightarrow{\Gamma}_{2\mathrm{N}}\cdot\overrightarrow{p}+\widehat{\omega}+m}{\sqrt{2\widehat{\omega}(\widehat{\omega}+m)}},\quad V^{-}=(V^{+})^{\dagger},\quad V^{-}V^{+}=V^{+}V^{-}=\mathrm{I}_{2\mathrm{N}},\quad \mathrm{N}=2s+1, \tag{35}$$

where $\Gamma^{j}_{2\mathrm{N}}$ are known from (30) and Σ^{j}_{N} are the $\mathrm{N}\times\mathrm{N}$ Pauli matrices.

Note that in formulas (35) and in all formulas before the end of the subsection the values of N are only even. Therefore, the canonical field equation (29) describes the larger number of multiplets then the generalized Dirac equation (36) given below.

The formulas (36)–(41) below are found from the corresponding formulas (28), (29), (32)–(34) of canonical field model on the basis of the operator (35).

For the general form of arbitrary spin locally covariant particle-antiparticle field the Dirac-like equation of motion follows from the equation (29) after the transformation (35) and is given by

$$\left[i\partial_{0}-\Gamma^{0}_{2\mathrm{N}}(\overrightarrow{\Gamma}_{2\mathrm{N}}\cdot\overrightarrow{p}+m)\right]\psi(x)=0. \tag{36}$$

The general solution has the form

$$\psi(x)=V^{-}\phi(x)=\frac{1}{(2\pi)^{\frac{3}{2}}}\int d^{3}k\left[e^{-ikx}a^{\mathrm{N}}(\overrightarrow{k})\mathrm{v}^{-}_{\mathrm{N}}(\overrightarrow{k})+e^{ikx}a^{*\breve{\mathrm{N}}}(\overrightarrow{k})\mathrm{v}^{+}_{\breve{\mathrm{N}}}(\overrightarrow{k})\right], \tag{37}$$

where amplitudes and notation $\breve{\mathrm{N}}$ are the same as in (28); $\left\{\mathrm{v}^{-}_{\mathrm{N}}(\overrightarrow{k}),\ \mathrm{v}^{+}_{\breve{\mathrm{N}}}(\overrightarrow{k})\right\}$ are 2N-component Dirac basis spinors with properties of orthonormalisation and completeness similar to 4-component Dirac spinors from [43].

The spin operator is given by

$$\overrightarrow{s}_{\mathrm{D}}=V^{-}\overrightarrow{s}_{2\mathrm{N}}V^{+}, \tag{38}$$

where operator $\overrightarrow{s}_{2\mathrm{N}}$ is known from (32). The explicit forms of few partial cases of spin operators (38) are given in formulae (259)–(261), (284)–(286), (359) of [1] for the particle-antiparticle multiplets s=(1,0,1,0), s=(3/2,3/2), s=(2,1,2,1), respectively.

The generators of the reducible unitary representation of the Poincaré group $\mathcal{P}$, with respect to which the covariant field equation (36) and the set $\{\psi\}$ of its solutions (37) are invariant, have the form

$$\widehat{p}^{0}=\Gamma^{0}_{2\mathrm{N}}(\overrightarrow{\Gamma}_{2\mathrm{N}}\cdot\overrightarrow{p}+m),\quad \widehat{p}^{\ell}=-i\partial_{\ell},\quad \widehat{j}^{\ell n}=x^{\ell}_{\mathrm{D}}\widehat{p}^{n}-x^{n}_{\mathrm{D}}\widehat{p}^{\ell}+s^{\ell n}_{\mathrm{D}}\equiv\widehat{m}^{\ell n}+s^{\ell n}_{\mathrm{D}}, \tag{39}$$

$$\widehat{j}^{0\ell}=-\widehat{j}^{\ell 0}=x^{0}\widehat{p}^{\ell}-\frac{1}{2}\left\{x^{\ell}_{\mathrm{D}},\widehat{p}^{0}\right\}+\frac{\widehat{p}^{0}(\overrightarrow{s}_{\mathrm{D}}\times\overrightarrow{p})^{\ell}}{\widehat{\omega}(\widehat{\omega}+m)}, \tag{40}$$

where the spin matrices $\overrightarrow{s}_{\mathrm{D}}=(s^{\ell n}_{\mathrm{D}})$ are given in (38) and the operator $\overrightarrow{x}_{\mathrm{D}}$ has the form

$$\overrightarrow{x}_{\mathrm{D}}=\overrightarrow{x}+\frac{i\overrightarrow{\Gamma}_{2\mathrm{N}}}{2\widehat{\omega}}-\frac{\overrightarrow{s}^{\Gamma}_{2\mathrm{N}}\times\overrightarrow{p}}{\widehat{\omega}(\widehat{\omega}+m)}-\frac{i\overrightarrow{p}(\overrightarrow{\Gamma}_{2\mathrm{N}}\cdot\overrightarrow{p})}{2\widehat{\omega}^{2}(\widehat{\omega}+m)}, \tag{41}$$

where specific spin matrices $\overrightarrow{s}^{\Gamma}_{2\mathrm{N}}$ are given by

$$\overrightarrow{s}^{\Gamma}_{2\mathrm{N}}\equiv\overrightarrow{s}_{\mathrm{FW}}=\left(s^{1}_{2\mathrm{N}},s^{2}_{2\mathrm{N}},s^{3}_{2\mathrm{N}}\right)=\frac{i}{2}(\Gamma^{2}_{2\mathrm{N}}\Gamma^{3}_{2\mathrm{N}},\ \Gamma^{3}_{2\mathrm{N}}\Gamma^{1}_{2\mathrm{N}},\ \Gamma^{1}_{2\mathrm{N}}\Gamma^{2}_{2\mathrm{N}}). \tag{42}$$

Journal of Physics: Conference Series **670** (2016) 012047 doi:10.1088/1742-6596/670/1/012047

Note that for corresponding partial cases of $\vec{x}_{\mathrm{D}}$ in [1] ((267) for s=(1,0,1,0), once more (267) but for s=(3/2,3/2), (333) for s=(2,0,2,0), (363) for s=(2,1,2,1)) the corresponding $\vec{s}^{\,\Gamma}_{2\mathrm{N}}$ are given by

$$\mathrm{s} = (1,0,1,0) \, \text{and} \, \mathrm{s} = (3/2,3/2): \quad \vec{s}^{\,\Gamma}_{8} \equiv \left(s^1_8, s^2_8, s^3_8\right) = \frac{i}{2}(\Gamma^2_8\Gamma^3_8,\, \Gamma^3_8\Gamma^1_8,\, \Gamma^1_8\Gamma^2_8), \tag{43}$$

for $\vec{x}_{\mathrm{D}}$ [1] (267), where the Γ^j_8 matrices are given in (253),

$$\mathrm{s} = (2,0,2,0): \quad \vec{s}^{\,\Gamma}_{12} \equiv \left(s^1_{12}, s^2_{12}, s^3_{12}\right) = \frac{i}{2}(\Gamma^2_{12}\Gamma^3_{12},\, \Gamma^3_{12}\Gamma^1_{12},\, \Gamma^1_{12}\Gamma^2_{12}), \tag{44}$$

for $\vec{x}_{\mathrm{D}}$ [1] (333), where the Γ^j_{12} matrices are given in (324),

$$\mathrm{s} = (2,1,2,1): \quad \vec{s}^{\,\Gamma}_{16} \equiv \left(s^1_{16}, s^2_{16}, s^3_{16}\right) = \frac{i}{2}(\Gamma^2_{16}\Gamma^3_{16},\, \Gamma^3_{16}\Gamma^1_{16},\, \Gamma^1_{16}\Gamma^2_{16}), \tag{45}$$

for $\vec{x}_{\mathrm{D}}$ [1] (363), where the Γ^j_{16} matrices are given in (353).

It is easy to verify that the generators (39), (40) for any N commute with the operator of equation (36), and satisfy the commutation relations (16) of the Lie algebra of the Poincaré group. The last step in the brief description of the model is the calculation of the Casimir operators $p^2 = \widehat{p}^{\mu}\widehat{p}_{\mu}$, $W = w^{\mu}w_{\mu}$ (w^{μ} is the Pauli–Lubanski pseudovector) for the fixed value of spin.

5 e exa ple f pin =(0,0) par icle-an ipar icle d uble

The completeness of simplest spin multiplets and doublets consideration of [1] is achieved by the supplementation of this example. The formalism follows from the general formalism of arbitrary spin after the substitution s=0.

The Schrödinger–Foldy equation of RCQM is given by (11) for N=1, i. e. it is 2-component equation. The solution is given by (13) for N=1. The Poincaré group $\mathcal{P}$ generators, with respect to which the equation (11) for s=(0,0) is invariant, are given by (17), (18) taken in the form of 2×2 matrices with spin terms equal to zero, i. e. the corresponding generators are given by 2×2 matrices (22).

The corresponding FW type equation of canonical field theory is given by

$$(i\partial_0 - \sigma^3\widehat{\omega})\phi(x) = 0, \quad \sigma^3 = \begin{vmatrix} 1 & 0 \\ 0 & -1 \end{vmatrix}, \quad \widehat{\omega} \equiv \sqrt{- \quad + m^2}. \tag{46}$$

The general solution is given by

$$\phi(x) = \frac{1}{(2\pi)^{\frac{3}{2}}}\int d^3k \left[e^{-ikx}a^1(\vec{k})\mathrm{d}_1 + e^{ikx}a^{*2}(\vec{k})\mathrm{d}_2\right]. \tag{47}$$

The Poincaré group $\mathcal{P}$ generators, with respect to which the equation (45) and the set $\{\phi\}$ of its solutions (47) are invariant, have the form

$$\widehat{p}^0 = \sigma^3\widehat{\omega} \equiv \sigma^3\sqrt{- \quad + m^2}, \quad \widehat{p}^{\ell} = -i\partial_{\ell}, \quad \widehat{j}^{\ell n} = x^{\ell}\widehat{p}^n - x^n\widehat{p}^{\ell}, \tag{48}$$

$$\widehat{j}^{0\ell} = -\widehat{j}^{\ell 0} = x^0\widehat{p}^{\ell} - \frac{1}{2}\sigma^3\left\{x^{\ell}, \widehat{\omega}\right\}. \tag{49}$$

Generators (48), (49) are the partial 2×2 matrix form of operators (33), (34) taken with the spin terms equal to zero.

5 *e exa ple f c varian field equa i n f r pin =(/ , /) par icle-an ipar icle d uble*

Consider the nontrivial partial example of covariant field equation for arbitrary spin. Such example is given by covariant field equation for spin s=(3/2,3/2) particle-antiparticle doublet. This case presents the demonstrative example how new equations can be derived by the developed in [1] and here methods.

Now contrary to [1] equation for spin s=3/2 is found as the simple partial case of general equation (36):

$$\left[i\partial_0 - \Gamma_8^0(\overrightarrow{\Gamma}_8 \cdot \overrightarrow{p} + m)\right]\psi(x) = 0. \tag{50}$$

Here the Γ_8^μ matrices are given by

$$\Gamma_8^0 = \left|\begin{array}{cc} I_4 & 0 \\ 0 & -I_4 \end{array}\right|, \quad \Gamma_8^j = \left|\begin{array}{cc} 0 & \Sigma^j \\ -\Sigma^j & 0 \end{array}\right|, \tag{51}$$

where Σ^j are the 4×4 Pauli matrices

$$\Sigma^j = \left|\begin{array}{cc} \sigma^j & 0 \\ 0 & \sigma^j \end{array}\right|, \tag{52}$$

and σ^j are the standard 2×2 Pauli matrices. The matrices Σ^j satisfy the similar commutation relations as the standard 2×2 Pauli matrices and have other similar properties. The matrices Γ_8^μ (51) satisfy the anticommutation relations of the Clifford–Dirac algebra in the form (24) with N=4.

Note that equation (50) is not the ordinary direct sum of the two Dirac equations. Therefore, it is not the complex Dirac–Kahler equation [44]. Moreover, it is not the standard 16-component Dirac–Kahler equation [45]. Furthermore, for the same reason it is not the spin 3/2 equation from [29, 30].

The solution of equation (50) is derived as a partial case from the solution (37) of the general equation (36) and is given by

$$\psi(x) = V_8^- \phi(x) = \frac{1}{(2\pi)^{\frac{3}{2}}} \int d^3k \left[e^{-ikx} b^{\mathrm{A}}(\overrightarrow{k}) \mathrm{v}_{\mathrm{A}}^-(\overrightarrow{k}) + e^{ikx} b^{*\mathrm{B}}(\overrightarrow{k}) \mathrm{v}_{\mathrm{B}}^+(\overrightarrow{k})\right], \tag{53}$$

where $\mathrm{A} = \overline{1,4}$, $\mathrm{B} = \overline{5,8}$ and the 8-component spinors $(\mathrm{v}_{\mathrm{A}}^-(\overrightarrow{k}), \mathrm{v}_{\mathrm{B}}^+(\overrightarrow{k}))$ are given by (257) in [1].

The spinors $(\mathrm{v}_{\mathrm{A}}^-(\overrightarrow{k}), \mathrm{v}_{\mathrm{B}}^+(\overrightarrow{k}))$ satisfy the relations of the orthonormalization and completeness similar to the corresponding relations for the standard 4-component Dirac spinors, see, e. g., [43].

In the covariant local field theory, the operators of the SU(2) spin, which satisfy the corresponding commutation relations $\left[s_{8\mathrm{D}}^j, s_{8\mathrm{D}}^\ell\right] = i\varepsilon^{j\ell n} s_{8\mathrm{D}}^n$ and commute with the operator $\left[i\partial_0 - \Gamma_8^0(\overrightarrow{\Gamma}_8 \cdot \overrightarrow{p} + m)\right]$ of equation (50), are derived from the pure matrix operators (279) of [1] with the help of transition operator V_8^-: $\overrightarrow{s}_{8\mathrm{D}} = V_8^- \overrightarrow{s}_8 V_8^+$. The explicit form of the transition operator $V_8^\mp$ is given in (249)–(251) of [1]. The explicit form of these s=(3/2,3/2) SU(2) generators was given already by formulae (284)–(287) in [1].

The equations on eigenvectors and eigenvalues of the operator $s_{8\mathrm{D}}^3$ (286) in [1] follow from the equations (280) of [1] and the transformation V_8^-. In addition to it, the action of the operator $s_{8\mathrm{D}}^3$ (286) in [1] on the spinors $(\mathrm{v}_{\mathrm{A}}^-(\overrightarrow{k}), \mathrm{v}_{\mathrm{B}}^+(\overrightarrow{k}))$ (257) in [1] also leads to the result

$$s_{8\mathrm{D}}^3 \mathrm{v}_1^-(\overrightarrow{k}) = \tfrac{3}{2}\mathrm{v}_1^-(\overrightarrow{k}),\ s_{8\mathrm{D}}^3 \mathrm{v}_2^-(\overrightarrow{k}) = \tfrac{1}{2}\mathrm{v}_2^-(\overrightarrow{k}),\ s_{8\mathrm{D}}^3 \mathrm{v}_3^-(\overrightarrow{k}) = -\tfrac{1}{2}\mathrm{v}_3^-(\overrightarrow{k}),\ s_{8\mathrm{D}}^3 \mathrm{v}_4^-(\overrightarrow{k}) = -\tfrac{3}{2}\mathrm{v}_4^-(\overrightarrow{k}),$$

$$s_{8\mathrm{D}}^3 \mathrm{v}_5^+(\overrightarrow{k}) = \frac{3}{2}\mathrm{v}_5^+(\overrightarrow{k}),\ s_{8\mathrm{D}}^3 \mathrm{v}_6^+(\overrightarrow{k}) = \frac{1}{2}\mathrm{v}_6^+(\overrightarrow{k}),\ s_{8\mathrm{D}}^3 \mathrm{v}_7^+(\overrightarrow{k}) = -\frac{1}{2}\mathrm{v}_7^+(\overrightarrow{k}),\ s_{8\mathrm{D}}^3 \mathrm{v}_8^+(\overrightarrow{k}) = -\frac{3}{2}\mathrm{v}_8^+(\overrightarrow{k}). \tag{54}$$

In order to verify equations (54) the identity $(\omega+m)^2+(\overrightarrow{k})^2=2\omega(\omega+m)$ is used. In the case $\mathrm{v}_{\mathrm{B}}^{+}(\overrightarrow{k})$ in the expression $s_{8\mathrm{D}}^{3}(\overrightarrow{k})$ (286) of [1] the substitution $\overrightarrow{k}\rightarrow-\overrightarrow{k}$ is made.

The equations (54) determine the interpretation of the amplitudes in solution (53). Nevertheless, the direct quantum-mechanical interpretation of the amplitudes should be made in the framework of the RCQM and is already given in [1] (section 14 in paragraph after equations (183)).

The explicit form of the $\mathcal{P}$-generators of the fermionic representation of the Poincaré group $\mathcal{P}$, with respect to which the covariant equation (50) and the set $\{\psi\}$ of its solutions (53) are invariant, is derived as a partial case from the generators (39), (40). The corresponding generators are given by

$$\widehat{p}^0=\Gamma_8^0(\overrightarrow{\Gamma}_8\cdot\overrightarrow{p}+m),\quad \widehat{p}^\ell=-i\partial_\ell,\quad \widehat{j}^{\ell n}=x_{\mathrm{D}}^\ell\widehat{p}^n-x_{\mathrm{D}}^n\widehat{p}^\ell+s_{8\mathrm{D}}^{\ell n}\equiv\widehat{m}^{\ell n}+s_{8\mathrm{D}}^{\ell n}, \tag{55}$$

$$\widehat{j}^{0\ell}=-\widehat{j}^{\ell 0}=x^0\widehat{p}^\ell-\frac{1}{2}\left\{x_{\mathrm{D}}^\ell,\widehat{p}^0\right\}+\frac{\widehat{p}^0(\overrightarrow{s}_{8\mathrm{D}}\times\overrightarrow{p})^\ell}{\widehat{\omega}(\widehat{\omega}+m)}, \tag{56}$$

where the spin matrices $\overrightarrow{s}_{8\mathrm{D}}=(s_{8\mathrm{D}}^{\ell n})$ are given by (284)–(286) in [1] and the operator $\overrightarrow{x}_{\mathrm{D}}$ has the form

$$\overrightarrow{x}_{\mathrm{D}}=\overrightarrow{x}+\frac{i\overrightarrow{\Gamma}_8}{2\widehat{\omega}}-\frac{\overrightarrow{s}_8^{\Gamma}\times\overrightarrow{p}}{\widehat{\omega}(\widehat{\omega}+m)}-\frac{i\overrightarrow{p}(\overrightarrow{\Gamma}_8\cdot\overrightarrow{p})}{2\widehat{\omega}^2(\widehat{\omega}+m)}, \tag{57}$$

with specific spin matrices $\overrightarrow{s}_8^{\Gamma}$ given in (43).

It is easy to verify that the generators (55), (56) with SU(2) spin (284)–(286) from [1] commute with the operator $\left[i\partial_0-\Gamma_8^0(\overrightarrow{\Gamma}_8\cdot\overrightarrow{p}+m)\right]$ of equation (50), satisfy the commutation relations (16) of the Lie algebra of the Poincaré group and the corresponding Casimir operators are given by $p^2=\widehat{p}^\mu\widehat{p}_\mu=m^2\mathrm{I}_8$, $W=w^\mu w_\mu=m^2\overrightarrow{s}_{8\mathrm{D}}^2=\frac{3}{2}\left(\frac{3}{2}+1\right)m^2\mathrm{I}_8$.

The conclusion that equation (50) describes the local field of fermionic particle-antiparticle doublet of the spin s=(3/2,3/2) and mass $m>0$ (and its solution (53) is the local fermionic field of the above mentioned spin and nonzero mass) follows from the analysis of equations (50) and the above given calculation of the Casimir operators p^2, $W=w^\mu w_\mu$.

6. Brie conclusions

Hence, the equation (50) describes the spin s=(3/2,3/2) particle-antiparticle doublet on the same level, on which the standard 4-component Dirac equation describes the spin s=(1/2,1/2) particle-antiparticle doublet. Moreover, the external argument in the validity of such interpretation is the link with the corresponding RCQM of spin s=(3/2,3/2) particle-antiparticle doublet, where the quantum-mechanical interpretation is direct and evident. Therefore, the fermionic spin s=(3/2,3/2) properties of equation (50) are proved.

Contrary to the bosonic spin s=(1,0,1,0) properties of the equation (50) found in [1] (section 22), the fermionic spin s=(1/2,1/2,1/2,1/2) properties of this equation are evident. The fact that equation (50) describes the multiplet of two fermions with the spin s=1/2 and two antifermions with that spin can be proved much more easier then the above given consideration. The proof is similar to that given in the standard 4-component Dirac model. The detailed consideration can be found in sections 7, 9, 10 of [1]. Therefore, equation (50) has more extended property of the Fermi–Bose duality then the standard Dirac equation [35–39]. This equation has the property of the Fermi–Bose triality. The property of the Fermi–Bose triality of the manifestly covariant equation (50) means that this equation describes on equal level (i) the spin s=(1/2,1/2,1/2,1/2) multiplet of two spin s=(1/2,1/2) fermions and two spin s=(1/2,1/2) antifermions, (ii) the spin s=(1,0,1,0) multiplet of the vector and scalar bosons together with their antiparticles, (iii) the spin s=(3/2,3/2) particle-antiparticle doublet.

It is evident that equation (50) is new in comparison with the Pauli–Fierz [25], Rarita–Schwinger [26] and confirmed by Davydov [27] equations for the spin s=3/2 particle. Contrary to 16-component equations from [25–27] equation (50) is 8-component and does not need any additional condition. Formally equation (50) looks like to have some similar features with the Bargman–Wigner equation [29] for arbitrary spin, when the spin value is taken 3/2. The transformation $V_8^{\mp} = \frac{\mp\overrightarrow{\Gamma}_8\cdot\overrightarrow{p}+\widehat{\omega}+m}{\sqrt{2\widehat{\omega}(\widehat{\omega}+m)}}$ looks like the transformation of Pursey [30] in the case of s=3/2. Nevertheless, the difference is clear. Equation (50) is not the ordinary direct sum of the two Dirac equations. Furthermore, the given here model is derived from the first principles of RCQM (not from the FW type representation of the canonical field theory). Our consideration is original and new. The link with corresponding RCQM, the proof of the symmetry properties and relativistic invariance, the well defined spin operator (284)–(286) in [1], the features of the Fermi–Bose duality (triality) of the equation (50), the interaction with electromagnetic field and many other characteristics are suggested firstly.

Interaction, quantization and Lagrange approach in the above given spin s=(3/2,3/2) model are completely similar to the Dirac 4-component theory and standard quantum electrodynamics. For example, the Lagrange function of the system of interacting 8-component spinor and electromagnetic fields (in the terms of 4-vector potential $A^\mu(x)$) is given by

$$L = -\frac{1}{4}F^{\mu\nu}F_{\mu\nu}+\frac{i}{2}\left(\overline{\psi}(x)\Gamma_8^\mu\frac{\partial\psi(x)}{\partial x^\mu}-\frac{\partial\overline{\psi}(x)}{\partial x^\mu}\Gamma_8^\mu\psi(x)\right)-m\overline{\psi}(x)\psi(x)+q\overline{\psi}(x)\Gamma_8^\mu\psi(x)A_\mu(x), \quad (58)$$

where $\overline{\psi}(x)$ is the independent Lagrange variable and $\overline{\psi} = \psi^\dagger\Gamma_8^0$ in the space of solutions $\{\psi\}$. In Lagrangian (101) $F_{\mu\nu} = \partial_\mu A_\nu - \partial_\nu A_\mu$ is the electromagnetic field tensor in the terms of potentials, which play the role of variational variables in this Lagrange approach. Thus, the difficulties mentioned in [28] are absent here.

Therefore, the covariant local quantum field theory model for the interacting particles with spin s=3/2 and photons can be constructed in complete analogy to the construction of the modern quantum electrodynamics. This model can be useful for the investigations of processes with interacting hyperons and photons.

Other results mentioned in abstract are proved similarly. The consideration of these results is presented in [1] and [2].

Re erences

[1] Simulik V M 2014 Covariant local field theory equations following from the relativistic canonical quantum mechanics of arbitrary spin *arXiv:* 1409.2766v2 [quant-ph, hep-th]
[2] Simulik V M 2015 General form of the covariant field equations of arbitrary spin and the relativistic canonical quantum mechanics *arXiv:* 1509.04630 [quant-ph, hep-th]
[3] Dirac P A M 1928 *Proc. R. Soc. Lond.* A **117** 610
[4] Dirac P A M 1958 *The principles of quantum mechanics, 4-th edition* (Oxford: Clarendon Press)
[5] Foldy L L and Wouthuysen S A 1950 *Phys. Rev.* **78** 29
[6] Foldy L L 1956 *Phys. Rev.* **102** 568
[7] Kowalski K and Rembielinski J 2011 *Phys. Rev.* A **84** 012108
[8] Simulik V M, Krivsky I Y and Lamer I L 2013 *Int. J. Theor. Math. Phys.* **3** 109
[9] Brown R W, Krauss L M and Taylor P L 2001 *Phys. Today* **54** 75
[10] Simulik V M and Krivsky I Y 2013 *Ukr. J. Phys.* **58** 1192
[11] Simulik V M and Krivsky I Y 2014 *Univ. J. Phys. Appl.* **2** 115
[12] Salpeter E E 1952 *Phys. Rev.* **87** 328
[13] Foldy L L 1961 *Phys. Rev.* **122** 275
[14] von Neumann J 1932 *Mathematische Grundlagen der Quantenmechanik* (Berlin: Verlag von Julius Springer)
[15] Dirac P A M 1936 *Proc. R. Soc. Lond.* A **155** 447
[16] Bhabha H J 1945 *Rev. Mod. Phys.* **17** 200
[17] Weaver D L, Hammer C L and Good R H Jr. 1964 *Phys. Rev.* **135** B241

[18] Pursey D L 1965 *Ann. Phys. (N.Y.)* **32** 157
[19] Mathews P M 1966 *Phys. Rev.* **143** 978
[20] Nelson T J and Good R H Jr. 1968 *Rev. Mod. Phys.* **40** 508
[21] Guertin R F 1974 *Ann. Phys. (N.Y.)* **88** 504
[22] Kraicik R A and Nieto M M 1977 *Phys. Rev.* D **15** 433
[23] Loide R-K, Ots I and Saar R 1997 *J. Phys.* A **30** 4005
[24] Gitman D M and Shelepin A L 2001 *Int. J. Theor, Phys.* **40** 603
[25] Fierz M and Pauli W 1939 *Proc. R. Soc. Lond.* A **173** 211
[26] Rarita W and Schwinger J 1941 *Phys. Rev.* **60** 61
[27] Davydov A.S. 1943 *J. Exper. Theor. Phys.* **13** 313 (in Russian)
[28] Capri A Z and Kobes R L 1980 *Phys. Rev.* D **22** 1967
[29] Bargman V and Wigner E P 1948 *Proc. Nat Acad. Sci. USA* **34** 211
[30] Pursey D L 1964 *Nucl. Phys.* **53** 174
[31] Vladimirov V S 2002 *Methods of the theory of generalized functions* (London: Taylor and Francis)
[32] Bogoliubov N N, Logunov A A and Todorov I T 1975 *Introduction to axiomatic quantum field theory* (Florida: W.A. Benjamin, Inc.)
[33] Ibragimov N H 1995 *CRC Handbook of Lie group analysis of differential equations, vol 2* (Florida: CRC Press)
[34] Lounesto P 2001 *Clifford algebras and spinors, 2-nd edition* (Cambridge: Cambridge University Press)
[35] Simulik V M and Krivsky I Y 2010 *Reports of the National Academy of Sciences of Ukraine* **5** 82 (in Ukrainian)
[36] Simulik V M and Krivsky I Y 2011 *Phys. Lett.* A **375** 2479
[37] Simulik V M, Krivsky I Y and Lamer I L 2012 *Cond. Matt. Phys.* **15** 43101
[38] Simulik V M, Krivsky I Y and Lamer I L 2013 *TWMS J. App. Eng. Math.* **3** 46
[39] Simulik V M, Krivsky I Y and Lamer I L 2013 *Ukr. J. Phys.* **58** 523
[40] Thaller B 1992 *The Dirac equation* (Berlin: Springer)
[41] Elliott J and Dawber P 1979 *Symmetry in physics, vol 1* (London: Macmillian Press)
[42] Wybourne B 1974 *Classical groups for Physicists* (New York: John Wiley and sons)
[43] Bogoliubov N N and Shirkov D V 1980 *Introduction to the theory of quantized fields* (New York: John Wiley and Sons)
[44] Krivsky I Y, Lompay R R and Simulik V M 2005 *Theor. Math. Phys.* **143** 541
[45] Kruglov S I 2002 *Int. J. Theor. Phys.* **41** 653

Journal of Physics: Conference Series **670** (2016) 012048 doi:10.1088/1742-6596/670/1/012048

Self-adjointness and the Casimir effect with confined quantized spinor matter

Yurii A Sitenko[1,2]

[1] Bogolyubov Institute for Theoretical Physics, National Academy of Sciences of Ukraine, 14-b Metrologichna Str., 03680 Kyiv, Ukraine

[2] Institute for Theoretical Physics, University of Bern, Sidlerstrasse 5, CH-3012 Bern, Switzerland

E-mail: yusitenko@bitp.kiev.ua

Abstract. A generalization of the MIT bag boundary condition for spinor matter is proposed basing on the requirement that the Dirac hamiltonian operator be self-adjoint. An influence of a background magnetic field on the vacuum of charged spinor matter confined between two parallel material plates is studied. Employing the most general set of boundary conditions at the plates in the case of the uniform magnetic field directed orthogonally to the plates, we find the pressure from the vacuum onto the plates. In physically plausible situations, the Casimir effect is shown to be repulsive, independently of a choice of boundary conditions and of a distance between the plates.

1. Introduction

The self-adjointness of operators of physical observables in quantum mechanics is required by general principles of comprehensibility and mathematical consistency, see, e.g., [1, 2]. To put it simply, a multiple action is well defined for a self-adjoint operator only, allowing for the construction of functions of the operator, such as resolvent, evolution, heat kernel and zeta-function operators, with further implications upon second quantization.

The mathematical demand for the self-adjointness of a basic operator acting on functions defined in a bounded spatial region is a somewhat more general than the physical demand for the confinement of quantized matter fields inside this region. The concept of confined matter fields is quite familiar in the context of condensed matter physics: collective excitations (e.g., spin waves and phonons) exist only inside material objects and do not spread outside. In the context of quantum electrodynamics, if one is interested in the effect of a classical background magnetic field on the vacuum of the quantized electron-positron matter, then the latter should be considered as confined to the spatial region between the sources of the magnetic field, as long as collective quasielectronic excitations inside a magnetized material differ from electronic excitations in the vacuum. It should be noted in this respect that the effect of the background electromagnetic field on the vacuum of quantized charged matter was studied for eight decades almost, see [3, 4] and a review in [5]. However, the concern was for the case of a background field filling the whole (infinite) space, that is hard to be regarded as realistic. The case of both the background and quantized fields confined to a bounded spatial region with boundaries serving as sources of the background field looks much more physically plausible, it can even be regarded as realizable in laboratory. Moreover, there is no way to detect the energy density that is induced

Journal of Physics: Conference Series **670** (2016) 012048 doi:10.1088/1742-6596/670/1/012048

in the vacuum in the first case, whereas the pressure from the vacuum onto the boundaries, resulting in the second case, is in principle detectable.

In view of the above, an issue of a choice of boundary conditions ensuring the confinement of the quantized spinor matter gains a crucial significance. It seems that a quest for such boundary conditions was initiated in the context of a model description of hadrons as composites of quarks and gluons [6, 7]. If an hadron is an extended object occupying spatial region bounded by surface ∂ , then the condition that the quark matter field be confined inside the hadron is formulated as

$$\mathbf{n}\cdot\mathbf{J}(\mathbf{r})|_{\mathbf{r}\in\partial\Omega}=0, \tag{1}$$

where $\mathbf{n}$ is the unit normal to the boundary surface, and $\mathbf{J}(\mathbf{r})=\psi^\dagger(\mathbf{r})\boldsymbol{\alpha}\psi(\mathbf{r})$, $\mathbf{r}\in$, with $\psi(\mathbf{r})$ being the quark matter field (an appropriate condition is also formulated for the gluon matter field). Condition (1) should be resolved to take the form of a boundary condition that is linear in $\psi(\mathbf{r})$, and an immediate way of such a resolution is known as the MIT bag boundary condition [8],

$$[I+\mathrm{i}\beta(\mathbf{n}\cdot\boldsymbol{\alpha})]\psi(\mathbf{r})|_{\mathbf{r}\in\partial\Omega}=0 \tag{2}$$

(α^1, α^2, α^3 and β are the generating elements of the Dirac-Clifford algebra), but it is needless to say that this way is not a unique one. The most general boundary condition that is linear in $\psi(\mathbf{r})$ in the case of a simply-connected boundary involves four arbitrary parameters [9], and the explicit form of this boundary condition has been given [10]; the condition is compatible with the self-adjointness of the Dirac hamiltonian operator, and its four parameters can be interpreted as the self-adjoint extension parameters. In the present work, I follow the lines of works [9, 10] by proposing a different, embracing more cases, form of the four-parameter generalization of the MIT bag boundary condition.

Thus, let us consider in general the quantized spinor matter field that is confined to the three-dimensional spatial region bounded by the two-dimensional surface ∂ . To study a response of the vacuum to the background magnetic field, we restrict ourselves to the case of the boundary consisting of two parallel planes; the magnetic field is assumed to be uniform and orthogonal to the planes. Such a spatial geometry is typical for the remarkable macroscopic quantum phenomenon which yields the attraction (negative pressure) between two neutral plates and which is known as the Casimir effect [11], see review in [12]. The conventional Casimir effect is due to vacuum fluctuations of the quantized electromagnetic field obeying certain boundary conditions at the bounding plates, and a choice of boundary conditions is physically motivated by material properties of the plates (for instance, metallic or dielectric, see, e.g., [12]). Such a motivation is lacking for the case of vacuum fluctuations of the quantized spinor matter field. That is why there is a necessity in the last case to start from the most general set of mathematically acceptable (i.e. compatible with the self-adjointness) boundary conditions. Further follow physical constraints that the spinor matter be confined within the plates and that the spectrum of the wave number vector in the direction which is orthogonal to the plates be unambiguously (although implicitly) determined. Employing these mathematical and physical restrictions, I consider the generalized Casimir effect which is due to vacuum fluctuations of the quantized spinor matter field in the presence of the background magnetic field; the pressure from the vacuum onto the bounding plates will be found.

2. Sel -adjointness and boundary conditions

Defining a scalar product as $(\tilde{\chi},\chi)=\int_\Omega \mathrm{d}^3r\,\tilde{\chi}^\dagger\chi$, we get, using integration by parts,

$$(\tilde{\chi},H\chi)=(H^\dagger\tilde{\chi},\chi)-\mathrm{i}\int_{\partial\Omega}\mathrm{d}\mathbf{s}\cdot\tilde{\chi}^\dagger\boldsymbol{\alpha}\chi, \tag{3}$$

where
$$H = H^\dagger = -\mathrm{i}\boldsymbol{\alpha}\cdot\boldsymbol{\nabla} + \beta m \tag{4}$$
is the formal expression for the Dirac hamiltonian operator and $\boldsymbol{\nabla}$ is the covariant derivative involving both the affine and bundle connections (natural units $\hbar = c = 1$ are used). Operator H is Hermitian (or symmetric in mathematical parlance),
$$(\chi, H\chi) = (H^\dagger\chi, \chi), \tag{5}$$
if
$$\int_{\partial\Omega} \mathrm{d}\mathbf{s}\cdot\chi^\dagger\boldsymbol{\alpha}\chi = 0. \tag{6}$$
The latter condition can be satisfied in various ways by imposing different boundary conditions for χ and χ. However, among the whole variety, there may exist a possibility that a boundary condition for χ is the same as that for χ; then the domain of definition of $H^\dagger$ (set of functions χ) coincides with that of H (set of functions χ), and operator H is self-adjoint. The action of a self-adjoint operator results in functions belonging to its domain of definition only, and, therefore, a multiple action and functions of such an operator can be consistently defined.

Condition (6) is certainly fulfilled when the integrand in (6) vanishes, i.e.
$$\chi^\dagger(\mathbf{n}\cdot\boldsymbol{\alpha})\chi|_{\mathbf{r}\in\partial\Omega} = 0. \tag{7}$$
To fulfill the latter condition, we impose the same boundary condition for χ and χ in the form
$$\chi|_{\mathbf{r}\in\partial\Omega} = K\chi|_{\mathbf{r}\in\partial\Omega}, \quad \chi|_{\mathbf{r}\in\partial\Omega} = K\chi|_{\mathbf{r}\in\partial\Omega}, \tag{8}$$
where K is a matrix (element of the Dirac-Clifford algebra) which is determined by two conditions:
$$K^2 = I \tag{9}$$
and
$$K^\dagger(\mathbf{n}\cdot\boldsymbol{\alpha})K = -\mathbf{n}\cdot\boldsymbol{\alpha}. \tag{10}$$
It should be noted that, in addition to (7), the following combination of χ and χ is also vanishing at the boundary:
$$\chi^\dagger(\mathbf{n}\cdot\boldsymbol{\alpha})K\chi|_{\mathbf{r}\in\partial\Omega} = \chi^\dagger K^\dagger(\mathbf{n}\cdot\boldsymbol{\alpha})\chi|_{\mathbf{r}\in\partial\Omega} = 0. \tag{11}$$
Using the standard representation for the Dirac matrices,
$$\beta = \begin{pmatrix} I & 0 \\ 0 & -I \end{pmatrix}, \qquad \boldsymbol{\alpha} = \begin{pmatrix} 0 & \boldsymbol{\sigma} \\ \boldsymbol{\sigma} & 0 \end{pmatrix} \tag{12}$$
(σ^1, σ^2 and σ^3 are the Pauli matrices), one can get
$$K = \begin{pmatrix} 0 & \varrho^{-1} \\ \varrho & 0 \end{pmatrix}, \tag{13}$$
where condition
$$(\mathbf{n}\cdot\boldsymbol{\sigma})\varrho = -\varrho^\dagger(\mathbf{n}\cdot\boldsymbol{\sigma}) \tag{14}$$
defines ϱ as a rank-2 matrix depending on four arbitrary parameters [9]. An explicit form for matrix K is [10]
$$K = \frac{(1+u^2-v^2-\mathbf{t}^2)\beta + (1-u^2+v^2+\mathbf{t}^2)I}{2\mathrm{i}(u^2-v^2-\mathbf{t}^2)}(u\mathbf{n}\cdot\boldsymbol{\alpha} + v\beta\gamma^5 - \mathrm{i}\mathbf{t}\cdot\boldsymbol{\alpha}), \tag{15}$$

where $\gamma^5 = \mathrm{i}\alpha^1\alpha^2\alpha^3$, and $\mathbf{t} = (t^1, t^2)$ is a two-dimensional vector which is tangential to the boundary, $\mathbf{t}\cdot\mathbf{n} = 0$. Matrix K is Hermitian in two cases only when it takes forms

$$K_+ = -\mathrm{i}\beta(\mathbf{n}\cdot\boldsymbol{\alpha}) \quad (u = 1, \quad v = 0, \quad \mathbf{t} = 0) \tag{16}$$

and

$$K_- = \mathrm{i}v\beta\gamma^5 + \mathbf{t}\cdot\boldsymbol{\alpha} \quad (u = 0, \quad v^2 + \mathbf{t}^2 = 1). \tag{17}$$

Matrix K_+ (16) corresponds to the choice of the standard MIT bag boundary condition [8], cf. (2),

$$(I - K_+)\chi|_{\mathbf{r}\in\partial\Omega} = (I - K_+)\chi|_{\mathbf{r}\in\partial\Omega} = 0, \tag{18}$$

when relation (11) takes form

$$\chi^\dagger\beta\chi|_{\mathbf{r}\in\partial\Omega} = 0. \tag{19}$$

It is instructive to go over from off-diagonal matrix K (15) to Hermitian matrix K, presenting boundary condition (8) as

$$\chi|_{\mathbf{r}\in\partial\Omega} = K\chi|_{\mathbf{r}\in\partial\Omega}, \quad \chi|_{\mathbf{r}\in\partial\Omega} = K\chi|_{\mathbf{r}\in\partial\Omega}, \tag{20}$$

with $K = K^\dagger$ determined by conditions

$$K^2 = I \tag{21}$$

and

$$[K, \mathbf{n}\cdot\boldsymbol{\alpha}]_+ = 0. \tag{22}$$

This transition is implemented with the use of the block-diagonal Hermitian matrix,

$$N = \begin{pmatrix} \nu_1 & 0 \\ 0 & \nu_2 \end{pmatrix}, \qquad \nu_1^\dagger = \nu_1, \qquad \nu_2^\dagger = \nu_2, \tag{23}$$

obeying condition

$$(I - N)K = K^\dagger(I - N); \tag{24}$$

namely, the result is

$$K = (I - N)K + N. \tag{25}$$

Using parametrization

$$u = -\frac{\sin\varphi}{\cos\varphi\cos\theta + \cos\varphi}, \qquad v = \frac{\sin\varphi\cos\theta}{\cos\varphi\cos\theta + \cos\varphi},$$
$$t^1 = \frac{\sin\theta\cos\eta}{\cos\varphi\cos\theta + \cos\varphi}, \qquad t^2 = \frac{\sin\theta\sin\eta}{\cos\varphi\cos\theta + \cos\varphi},$$
$$-\pi/2 < \varphi \le \pi/2, \quad -\pi/2 \le \varphi < \pi/2, \quad 0 \le \theta < \pi, \quad 0 \le \eta < 2\pi, \tag{26}$$

one gets

$$K = \mathrm{i}\frac{\beta\cos\varphi\cos\theta + I\cos\varphi}{\cos^2\varphi\cos^2\theta - \cos^2\varphi}[\mathbf{n}\cdot\boldsymbol{\alpha}\sin\varphi - \beta\gamma^5\sin\varphi\cos\theta + \mathrm{i}(\alpha^1\cos\eta + \alpha^2\sin\eta)\sin\theta)], \tag{27}$$

where

$$[\mathbf{n}\cdot\boldsymbol{\alpha},\, \alpha^1]_+ = [\mathbf{n}\cdot\boldsymbol{\alpha},\, \alpha^2]_+ = [\alpha^1, \alpha^2]_+ = 0. \tag{28}$$

Then matrix N takes form

$$N = \beta \cos\varphi \cos\varphi \cos\theta - \beta\gamma^5(\mathbf{n}\cdot\boldsymbol{\alpha}) \sin\varphi \sin\varphi \cos\theta + \mathrm{i}(\alpha^1 \cos\eta + \alpha^2 \sin\eta)(\mathbf{n}\cdot\boldsymbol{\alpha}) \sin\varphi \sin\theta, \quad (29)$$

and one gets

$$K = [\beta \mathrm{e}^{\mathrm{i}\varphi\gamma^5} \cos\theta + (\alpha^1 \cos\eta + \alpha^2 \sin\eta) \sin\theta] \mathrm{e}^{\mathrm{i}\tilde{\varphi}\mathbf{n}\cdot\boldsymbol{\alpha}}; \quad (30)$$

in particular,

$$K_+ = K|_{\varphi=0,\,\tilde{\varphi}=-\pi/2,\,\theta=0}, \quad K_- = K|_{\varphi=\pi/2,\,\tilde{\varphi}=0}. \quad (31)$$

Thus, the explicit form of the boundary condition ensuring the self-adjointness of operator H (4) is

$$\left\{I - [\beta \mathrm{e}^{\mathrm{i}\varphi\gamma^5} \cos\theta + (\alpha^1 \cos\eta + \alpha^2 \sin\eta) \sin\theta] \mathrm{e}^{\mathrm{i}\tilde{\varphi}\mathbf{n}\cdot\boldsymbol{\alpha}}\right\} \chi|_{\mathbf{r}\in\partial\Omega} = 0 \quad (32)$$

(the same condition is for χ), and relation (11) takes form

$$\chi^\dagger [\beta \mathrm{e}^{\mathrm{i}\varphi\gamma^5} \cos\theta + (\alpha^1 \cos\eta + \alpha^2 \sin\eta) \sin\theta] \mathrm{e}^{\mathrm{i}(\tilde{\varphi}+\pi/2)\mathbf{n}\cdot\boldsymbol{\alpha}} \chi|_{\mathbf{r}\in\partial\Omega} = 0. \quad (33)$$

Four parameters of boundary condition (32), φ, φ, θ and η, can be interpreted as the self-adjoint extension parameters. It should be emphasized that the values of these parameters vary in general from point to point of the boundary. In this respect the number of self-adjoint extension parameters is in fact infinite, moreover, it is not countable but is of power of a continuum. This distinguishes the case of an extended boundary from the case of an excluded point (contact interaction), when the number of self-adjoint extension parameters is finite, being equal to n^2 for the deficiency index equal to $\{n, n\}$ (see, e.g., [1]).

In the context of the Casimir effect, one usually considers spatial region with a disconnected boundary consisting of two connected components, $\partial \ = \partial \ ^{(+)} \bigcup \partial \ ^{(-)}$. Choosing coordinates $\mathbf{r} = (x, y, z)$ in such a way that x and y are tangential to the boundary, while z is normal to it, we identify the position of $\partial \ ^{(\pm)}$ with, say, $z = \pm a/2$. In general, there are 8 self-adjoint extension parameters: φ_+, φ_+, θ_+ and η_+ corresponding to $\partial \ ^{(+)}$ and φ_-, φ_-, θ_- and η_- corresponding to $\partial \ ^{(-)}$. However, if some symmetry is present, then the number of self-adjoint extension parameters is diminished. For instance, if the boundary consists of two parallel planes, then the cases differing by the values of η_+ or η_- are physically indistinguishable, since they are related by a rotation around a normal to the boundary. To avoid this unphysical degeneracy, one has to fix

$$\theta_+ = \theta_- = 0, \quad (34)$$

and there remains 4 self-adjoint extension parameters: φ_+, φ_+, φ_- and φ_-. Operator H (4) acting on functions which are defined in the region bounded by two parallel planes is self-adjoint, if the following condition holds:

$$\left\{I - \beta \exp[\mathrm{i}(\varphi_\pm\gamma^5 \pm \varphi_\pm\alpha^z)]\right\} \chi|_{z=\pm a/2} = 0 \quad (35)$$

(the same condition holds for χ). The latter ensures the fulfilment of constraints

$$\chi^\dagger \alpha^z \chi|_{z=\pm a/2} = 0 \quad (36)$$

and

$$\chi^\dagger \beta \exp\left\{\mathrm{i}[\varphi_\pm\gamma^5 \pm (\varphi_\pm + \pi/2)\alpha^z]\right\} \chi|_{z=\pm a/2} = 0. \quad (37)$$

3. Induced vacuum energy in the magnetic field background

The operator of a spinor field which is quantized in an ultrastatic background is presented in the form

$$\hat{\Psi}(t,\mathbf{r}) = \sum\!\!\!\!\!\!\int_{E_\lambda>0} \mathrm{e}^{-\mathrm{i}E_\lambda t}\psi_\lambda(\mathbf{r})\hat{a}_\lambda + \sum\!\!\!\!\!\!\int_{E_\lambda<0} \mathrm{e}^{-\mathrm{i}E_\lambda t}\psi_\lambda(\mathbf{r})\hat{b}^\dagger_\lambda, \tag{38}$$

where $\hat{a}^\dagger_\lambda$ and $\hat{a}_\lambda$ ($\hat{b}^\dagger_\lambda$ and $\hat{b}_\lambda$) are the spinor particle (antiparticle) creation and destruction operators, satisfying anticommutation relations $[\hat{a}_\lambda, \hat{a}^\dagger_{\lambda'}]_+ = [\hat{b}_\lambda, \hat{b}^\dagger_{\lambda'}]_+ = \langle\lambda|\lambda'\rangle$, wave functions $\psi_\lambda(\mathbf{r})$ form a complete set of solutions to the stationary Dirac equation

$$H\psi_\lambda(\mathbf{r}) = E_\lambda\psi_\lambda(\mathbf{r}); \tag{39}$$

λ is the set of parameters (quantum numbers) specifying a one-particle state with energy E_λ; symbol $\sum\!\!\!\!\!\!\int$ denotes summation over discrete and integration (with a certain measure) over continuous values of λ. Ground state $|\mathtt{vac}>$ is defined by condition $\hat{a}_\lambda|\mathtt{vac}>= \hat{b}_\lambda|\mathtt{vac}>= 0$. The temporal component of the operator of the energy-momentum tensor is given by expression

$$\hat{T}^{00} = \frac{\mathrm{i}}{4}[\hat{\Psi}^\dagger(\partial_0\hat{\Psi}) - (\partial_0\hat{\Psi}^T)\hat{\Psi}^{\dagger T} - (\partial_0\hat{\Psi}^\dagger)\hat{\Psi} + \hat{\Psi}^T(\partial_0\hat{\Psi}^{\dagger T})], \tag{40}$$

where superscript T denotes a transposed spinor. Consequently, the formal expression for the vacuum expectation value of the energy density is

$$\varepsilon =< \mathtt{vac}|\hat{T}^{00}|\mathtt{vac}>= -\frac{1}{2}\sum\!\!\!\!\!\!\int|E_\lambda|\psi^\dagger_\lambda(\mathbf{r})\psi_\lambda(\mathbf{r}). \tag{41}$$

Let us consider the quantized charged massive spinor field in the background of a static uniform magnetic field; then $\boldsymbol{\nabla} = \boldsymbol{\partial} - \mathrm{i}e\mathbf{A}$ and the connection can be chosen as $\mathbf{A} = (-yB, 0, 0)$, where B is the value of the magnetic field strength which is directed along the z-axis in Cartesian coordinates $\mathbf{r} = (x, y, z)$, $\mathbf{B} = (0, 0, B)$. The one-particle energy spectrum is

$$E_{nk} = \pm\omega_{nk}, \tag{42}$$

where

$$\omega_{nk} = \sqrt{2n|eB| + k^2 + m^2},\ -\infty < k < \infty,\ n = 0, 1, 2, \dots, \tag{43}$$

k is the value of the wave number vector along the z-axis, and n labels the Landau levels. Using the explicit form of the complete set of solutions to the Dirac equation, one can get that expression (41) takes form

$$\varepsilon^\infty = -\frac{|eB|}{2\pi^2}\int\limits_{-\infty}^{\infty} \mathrm{d}k \sum_{n=0}^{\infty} \iota_n\omega_{nk}, \tag{44}$$

where the superscript on the left-hand side indicates that the magnetic field fills the whole (infinite) space; the appearance of factor $\iota_n = 1 - \frac{1}{2}\delta_{n0}$ on the right-hand side is due to the fact that there is one solution for the lowest Landau level, $\psi^{(0)}_{q0k}(\mathbf{r})$ (q is the value of the wave number vector along the x-axis, $-\infty < q < \infty$), and there are two solutions otherwise, $\psi^{(j)}_{qnk}(\mathbf{r})$ ($j = 1, 2$), $n \geq 1$. The integral and the sum in (44) are divergent and require regularization and renormalization. This problem has been solved long ago by Heisenberg and Euler [3] (see also [4]), and we just list here their result

$$\varepsilon^\infty_{\mathrm{ren}} = \frac{1}{8\pi^2}\int\limits_0^\infty \frac{\mathrm{d}\tau}{\tau}\mathrm{e}^{-\tau}\left[\frac{eBm^2}{\tau}\coth\left(\frac{eB\tau}{m^2}\right) - \frac{m^4}{\tau^2} - \frac{1}{3}e^2B^2\right]; \tag{45}$$

note that the renormalization procedure involves subtraction at $B = 0$ and renormalization of the charge.

Let us turn now to the quantized charged massive spinor field in the background of a static uniform magnetic field in spatial region bounded by two parallel planes $\partial^{(+)}$ and $\partial^{(-)}$; the position of $\partial^{(\pm)}$ is identified with $z = \pm a/2$, and the magnetic field is orthogonal to the boundary. The solution to (39) in region is chosen as a superposition of two plane waves propagating in opposite directions along the z-axis,

$$\psi^{(j)}_{qnl}(\mathbf{r}) = \psi^{(j)}_{qnk_l}(\mathbf{r}) + \psi^{(j)}_{qn-k_l}(\mathbf{r}), \quad j = 0, 1, 2, \tag{46}$$

where the values of wave number vector k_l ($l = 0, \pm1, \pm2, ...$) are determined from the boundary condition, see (35):

$$\left\{I - \beta \exp[\mathrm{i}(\varphi_\pm \gamma^5 \pm \varphi_\pm \alpha^z)]\right\} \psi^{(j)}_{qnl}(\mathbf{r})|_{z=\pm a/2} = 0, \quad (j = 1, 2) \quad n \geq 1 \tag{47}$$

and

$$\left[I + \frac{\beta}{2}\left(\pm\alpha^z\gamma^5 - 1\right)\mathrm{e}^{\mathrm{i}(\varphi_\pm - \tilde{\varphi}_\pm)\gamma^5}\Theta(\pm eB) - \frac{\beta}{2}\left(\pm\alpha^z\gamma^5 + 1\right)\mathrm{e}^{\mathrm{i}(\varphi_\pm + \tilde{\varphi}_\pm)\gamma^5}\Theta(\mp eB)\right]\psi^{(0)}_{q0l}(\mathbf{r})|_{z=\pm a/2} = 0; \tag{48}$$

the step function is introduced as $\Theta(u) = 1$ at $u > 0$ and $\Theta(u) = 0$ at $u < 0$. This boundary condition ensures that the normal component of current $\mathbf{J}_{qnlj}(\mathbf{r}) = \psi^{(j)\dagger}_{qnl}(\mathbf{r})\boldsymbol{\alpha}\psi^{(j)}_{qnl}(\mathbf{r})$ ($j = 0, 1, 2$) vanishes at the boundary, see (36),

$$J^z_{qnlj}(\mathbf{r})|_{z=\pm a/2} = 0, \tag{49}$$

which, cf. (1), signifies that the quantized matter is confined within the boundaries.

The boundary condition depends on four self-adjoint extension parameters, $\varphi_+, \varphi_+, \varphi_-$ and φ_-, in the case of $n \geq 1$, see (47), and on two self-adjoint extension parameters, $\varphi_+ - \varphi_+$ and $\varphi_- + \varphi_-$ ($eB > 0$), or $\varphi_+ + \varphi_+$ and $\varphi_- - \varphi_-$ ($eB < 0$), in the case of $n = 0$, see (48). As was mentioned in the previous section, the values of these self-adjoint extension parameters may vary arbitrarily from point to point of the bounding planes. However, in the context of the Casimir effect, such a generality seems to be excessive, lacking physical motivation and, moreover, being impermissible, as long as boundary condition (47)-(48) is to be regarded as the one determining the spectrum of the wave number vector in the z-direction. Therefore, we shall assume in the following that the self-adjoint extension parameters are independent of coordinates x and y.

It should be noted that value $k_l = 0$ is allowed for special cases only. Really, we have in the case of $k_l = 0$:

$$\psi^{(j)}_{qnl}(\mathbf{r})|_{z=a/2} = \psi^{(j)}_{qnl}(\mathbf{r})|_{z=-a/2}, \tag{50}$$

and boundary condition (47)-(48) can be presented in the form

$$R\,\psi^{(j)}_{qnl}(\mathbf{r})|_{k_l=0} = 0, \tag{51}$$

where

$$\left\{\begin{array}{llll} R_{11} = \sin\frac{\varphi_+ - \tilde{\varphi}_+}{2}, & R_{12} = 0, & R_{13} = \mathrm{i}\cos\frac{\varphi_+ - \tilde{\varphi}_+}{2}, & R_{14} = 0, \\ R_{21} = 0, & R_{22} = \sin\frac{\varphi_+ + \tilde{\varphi}_+}{2}, & R_{23} = 0, & R_{24} = \mathrm{i}\cos\frac{\varphi_+ + \tilde{\varphi}_+}{2}, \\ R_{31} = \sin\frac{\varphi_- + \tilde{\varphi}_-}{2}, & R_{32} = 0, & R_{33} = \mathrm{i}\cos\frac{\varphi_- + \tilde{\varphi}_-}{2}, & R_{34} = 0, \\ R_{41} = 0, & R_{42} = \sin\frac{\varphi_- - \tilde{\varphi}_-}{2}, & R_{43} = 0, & R_{44} = \mathrm{i}\cos\frac{\varphi_- - \tilde{\varphi}_-}{2} \end{array}\right\}. \tag{52}$$

The determinant of matrix R is:

$$\det R = -\sin\frac{\varphi_+ - \varphi_- + \tilde{\varphi}_+ + \tilde{\varphi}_-}{2}\sin\frac{\varphi_+ - \varphi_- - \tilde{\varphi}_+ - \tilde{\varphi}_-}{2}. \tag{53}$$

The necessary and sufficient condition for value $k_l = 0$ to be admissible is $\det R = 0$. Otherwise, $\det R \neq 0$ and value $k_l = 0$ is excluded from the spectrum, because equation (51) then allows for the trivial solution only, $\psi^{(j)}_{qnl}(\mathbf{r})|_{k_l=0} \equiv 0$.

The spectrum of k_l is determined from a transcendental equation which in general possesses two branches and allows for complex values of k_l (details will be published elsewhere). It is not clear which of the branches should be chosen, and, therefore, we restrict ourselves to boundary conditions corresponding to the case of a single branch. The latter is ensured by imposing constraint

$$\varphi_+ = \varphi_- = \varphi, \quad \varphi_+ = \varphi_- = \varphi. \tag{54}$$

Then relations (35) and (37) take forms

$$\left\{I - \beta \exp[\mathrm{i}(\varphi\gamma^5 \pm \varphi\alpha^z)]\right\} \chi|_{z=\pm a/2} = 0 \tag{55}$$

and

$$\chi^\dagger \beta \exp\left\{\mathrm{i}[\varphi\gamma^5 \pm (\varphi + \pi/2)\alpha^z]\right\} \chi|_{z=\pm a/2} = 0. \tag{56}$$

respectively, while the equation determining the spectrum of k_l takes form

$$\cos(k_l a) + \frac{\omega_{nk_l}\,\mathrm{sgn}(E_{nk_l}) \cos\varphi - m\cos\varphi}{k_l \sin\varphi} \sin(k_l a) = 0, \tag{57}$$

where $\mathrm{sgn}(u) = \Theta(u) - \Theta(-u)$ is the sign function; note that the spectrum is real, consisting of values of the same sign, say, $k_l > 0$ (values of the opposite sign ($k_l < 0$) should be excluded to avoid double counting).

In the case of $\varphi = -\pi/2$, the spectrum of k_l is independent of the number of the Landau level, n, and of the sign of the one-particle energy, $\mathrm{sgn}(E_{nk_l})$; it is determined from equation

$$\cos(k_l a) + \frac{m\cos\varphi}{k_l} \sin(k_l a) = 0. \tag{58}$$

In the case of $\varphi = 0$, the k_l-spectrum is also independent of n and of $\mathrm{sgn}(E_{nk_l})$; moreover, it is independent of φ, since the determining equation takes form

$$\sin(k_l a) = 0; \tag{59}$$

note that value $k_l = 0$ is admissible in this case, see (53)-(54). In what follows, we shall consider the most general case of two self-adjoint extension parameters, φ and φ, when the k_l-spectrum depends on n and on $\mathrm{sgn}(E_{nk_l})$, see (57).

Wave functions $\psi^{(j)}_{qnl}(\mathbf{r})$ $(j = 0, 1, 2)$ satisfy the requirements of orthonormality

$$\int\limits_\Omega \mathrm{d}^3r\, \psi^{(j)\dagger}_{qnl}(\mathbf{r}) \psi^{(j')}_{q'n'l'}(\mathbf{r}) = \delta_{jj'}\delta_{nn'}\delta_{ll'}\delta(q - q'), \quad j, j' = 0, 1, 2 \tag{60}$$

and completeness

$$\sum_{\mathrm{sgn}(E_{nk_l})} \int\limits_{-\infty}^{\infty} \mathrm{d}q \sum_l \left[\psi^{(0)}_{q0l}(\mathbf{r})\psi^{(0)\dagger}_{q0l}(\mathbf{r}') + \sum_{n=1}^{\infty}\sum_{j=1,2} \psi^{(j)}_{qnl}(\mathbf{r})\psi^{(j)\dagger}_{qnl}(\mathbf{r}')\right] = I\delta(\mathbf{r} - \mathbf{r}'). \tag{61}$$

Consequently, we obtain the following expression for the vacuum expectation value of the energy per unit area of the boundary surface

$$\frac{E}{S} \equiv \int\limits_{-a/2}^{a/2} \mathrm{d}z\, \varepsilon = -\frac{|eB|}{2\pi} \sum_{\mathrm{sgn}(E_{nk_l})} \sum_l \sum_{n=0}^{\infty} \iota_n \omega_{nk_l}. \tag{62}$$

4. Casimir energy and orce

Expression (62) can be regarded as purely formal, since it is ill-defined due to the divergence of infinite sums over l and n. To tame the divergence, a factor containing a regularization parameter should be inserted in (62). A summation over values $k_l \geq 0$, which are determined by (57), can be performed with the use of the Abel-Plana formula and its generalizations. In the cases of $\varphi = 0$ and of $\varphi = -\varphi = \pi/2$, the well-known versions of the Abel-Plana formula (see, e.g., [12]),

$$\sum_{\mathrm{sgn}(E_{nk_l})} \sum_{k_l \geq 0} f(k_l^2) \Bigg|_{\sin(k_l a)=0} = \frac{a}{\pi} \int_{-\infty}^{\infty} \mathrm{d}k f(k^2) - \frac{2\mathrm{i}a}{\pi} \int_{0}^{\infty} \mathrm{d}\kappa \frac{f[(-\mathrm{i}\kappa)^2] - f[(\mathrm{i}\kappa)^2]}{\mathrm{e}^{2\kappa a} - 1} + f(0) \qquad (63)$$

and

$$\sum_{\mathrm{sgn}(E_{nk_l})} \sum_{k_l > 0} f(k_l^2) \Bigg|_{\cos(k_l a)=0} = \frac{a}{\pi} \int_{-\infty}^{\infty} \mathrm{d}k f(k^2) + \frac{2\mathrm{i}a}{\pi} \int_{0}^{\infty} \mathrm{d}\kappa \frac{f[(-\mathrm{i}\kappa)^2] - f[(\mathrm{i}\kappa)^2]}{\mathrm{e}^{2\kappa a} + 1}, \qquad (64)$$

are used, respectively. Otherwise, we use the following version of the Abel-Plana formula:

$$\sum_{\mathrm{sgn}(E_{nk_l})} \sum_{k_l > 0} f(k_l^2) = \frac{a}{\pi} \int_{-\infty}^{\infty} \mathrm{d}k f(k^2) + \frac{2\mathrm{i}a}{\pi} \int_{0}^{\infty} \mathrm{d}\kappa \Lambda(\kappa) \{f[(-\mathrm{i}\kappa)^2] - f[(\mathrm{i}\kappa)^2]\}$$

$$-f(0) - \frac{1}{\pi} \int_{-\infty}^{\infty} \mathrm{d}k f(k^2) \frac{m \cos\varphi \sin\varphi [k^2 - \mu_n(\varphi, \varphi)]}{[k^2 + \mu_n(\varphi, \varphi)]^2 + 4k^2 m^2 \cos^2\varphi \sin^2\varphi}, \qquad (65)$$

where

$$\Lambda(\kappa) = \Bigg(-[\kappa^2 \cos 2\varphi - \mu_n(\varphi, \varphi)] \mathrm{e}^{2\kappa a} + \kappa^2 + 2\kappa m \cos\varphi \sin\varphi - \mu_n(\varphi, \varphi)$$

$$+ \frac{\sin\varphi}{a} \left\{ -\kappa^2 m \cos\varphi (\cos 2\varphi \mathrm{e}^{2\kappa a} - 1) + [(2\kappa \sin\varphi - m\cos\varphi) \mathrm{e}^{2\kappa a} \right.$$

$$\left. + m\cos\varphi] \mu_n(\varphi, \varphi) \right\} [\kappa^2 - 2\kappa m \cos\varphi \sin\varphi - \mu_n(\varphi, \varphi)]^{-1} \Bigg)$$

$$\times \Big\{ [\kappa^2 - 2\kappa m \cos\varphi \sin\varphi - \mu_n(\varphi, \varphi)] \mathrm{e}^{4\kappa a}$$

$$-2[\kappa^2 \cos 2\varphi - \mu_n(\varphi, \varphi)] \mathrm{e}^{2\kappa a} + \kappa^2 + 2\kappa m \cos\varphi \sin\varphi - \mu_n(\varphi, \varphi) \Big\}^{-1} \qquad (66)$$

and

$$\mu_n(\varphi, \varphi) = 2n|eB| \cos^2\varphi + m^2 \sin(\varphi + \varphi) \sin(\varphi - \varphi). \qquad (67)$$

In (63)-(65), $f(u^2)$ as a function of complex variable u is assumed to decrease sufficiently fast at large distances from the origin of the complex u-plane, and this decrease is due to the use of some kind of regularization for (62). However, the regularization in the second integral on the right-hand side of (63)-(65) can be removed; then

$$\mathrm{i}\{f[(-\mathrm{i}\kappa)^2] - f[(\mathrm{i}\kappa)^2]\} = -\frac{|eB|}{\pi} \sum_{n=0}^{\infty} \iota_n \sqrt{\kappa^2 - \omega_{n0}^2} \qquad (68)$$

with the range of κ restricted to $\kappa > \omega_{n0}$ for the corresponding terms; here, recalling (43), $\omega_{n0} = \sqrt{2n|eB| + m^2}$. As to the first integral on the right-hand side of (63)-(65), one immediately recognizes that it is equal to ε^{∞} (44) multiplied by a. Hence, if one ignores for a moment the last term of (63), as well as the terms in the last line of (65), then the problem of regularization and removal of the divergency in expression (62) is the same as that in the case of no boundaries, when the magnetic field fills the whole space. Defining the Casimir energy as the vacuum energy per unit area of the boundary surface, which is renormalized in the same way as in the case of no boundaries, we obtain

$$\frac{E_{\rm ren}}{S} = a\varepsilon_{\rm ren}^{\infty} - \frac{2|eB|}{\pi^2} a \sum_{n=0}^{\infty} \iota_n \int_{\omega_{n0}}^{\infty} d\kappa \Lambda(\kappa)\sqrt{\kappa^2 - \omega_{n0}^2}$$

$$+\frac{|eB|}{2\pi}\sum_{n=0}^{\infty}\iota_n\omega_{n0} + \frac{|eB|}{2\pi^2}\int_{-\infty}^{\infty} dk \sum_{n=0}^{\infty}\iota_n\sqrt{k^2+\omega_{n0}^2}\frac{m\cos\varphi\sin\varphi[k^2-\mu_n(\varphi,\varphi)]}{[k^2+\mu_n(\varphi,\varphi)]^2+4k^2m^2\cos^2\varphi\sin^2\varphi}, \quad (69)$$

$\varepsilon_{\rm ren}^{\infty}$ is given by (45). The sums and the integral in the last line of (69) (which are due to the terms in the last line of (65) and which can be interpreted as describing the proper energies of the boundary planes containing the sources of the magnetic field) are divergent, but this divergency is of no concern for us, because it has no physical consequences. Rather than the Casimir energy, a physically relevant quantity is the Casimir force per unit area of the boundary surface, i.e. pressure, which is defined as

$$F = -\frac{\partial}{\partial a}\frac{E_{\rm ren}}{S}, \quad (70)$$

and which is free from divergencies. We obtain

$$F = -\varepsilon_{\rm ren}^{\infty} - \frac{2|eB|}{\pi^2}\sum_{n=0}^{\infty}\iota_n\int_{\omega_{n0}}^{\infty} d\kappa \Upsilon(\kappa)\sqrt{\kappa^2-\omega_{n0}^2}, \quad (71)$$

where

$$\Upsilon(\kappa) \equiv -\frac{\partial}{\partial a} a\Lambda(\kappa) = \left[v_1(\kappa)e^{6\kappa a} + v_2(\kappa)e^{4\kappa a} + v_3(\kappa)e^{2\kappa a} + v_4(\kappa)\right]$$
$$\times\Big\{[\kappa^2 - 2\kappa m\cos\varphi\sin\varphi - \mu_n(\varphi,\varphi)]e^{4\kappa a}$$
$$-2[\kappa^2\cos 2\varphi - \mu_n(\varphi,\varphi)]e^{2\kappa a} + \kappa^2 + 2\kappa m\cos\varphi\sin\varphi - \mu_n(\varphi,\varphi)\Big\}^{-2} \quad (72)$$

and

$$v_1(\kappa) = -(2\kappa a - 1)[\kappa^2 - 2\kappa m\cos\varphi\sin\varphi - \mu_n(\varphi,\varphi)][\kappa^2\cos 2\varphi - \mu_n(\varphi,\varphi)]$$
$$-2[\kappa^2 m\cos\varphi\cos 2\varphi - (2\kappa\sin\varphi - m\cos\varphi)\mu_n(\varphi,\varphi)]\kappa\sin\varphi, \quad (73)$$
$$v_2(\kappa) = (4\kappa a - 3)\Big\{[\kappa^2 - \mu_n(\varphi,\varphi)]^2 - 4\kappa^2 m^2\cos^2\varphi\sin^2\varphi\Big\}$$
$$+8\kappa^2[\kappa^2\cos^2\varphi - m^2\cos^2\varphi - \mu_n(\varphi,\varphi)]\sin^2\varphi + 4[\kappa^2 + \mu_n(\varphi,\varphi)]\kappa m\cos\varphi\sin\varphi, \quad (74)$$
$$v_3(\kappa) = -(2\kappa a - 3)[\kappa^2 + 2\kappa m\cos\varphi\sin\varphi - \mu_n(\varphi,\varphi)][\kappa^2\cos 2\varphi - \mu_n(\varphi,\varphi)]$$
$$-2[\kappa^2 m\cos\varphi\cos 2\varphi + (2\kappa\sin\varphi + m\cos\varphi)\mu_n(\varphi,\varphi)]\kappa\sin\varphi, \quad (75)$$
$$v_4(\kappa) = -[\kappa^2 + 2\kappa m\cos\varphi\sin\varphi - \mu_n(\varphi,\varphi)]^2. \quad (76)$$

In the case of $\varphi = -\pi/2$, relations (72)-(76) are simplified:

$$\Upsilon(\kappa)|_{\tilde{\varphi}=-\pi/2} = \frac{\left[(2\kappa a - 1)\left(\kappa^2 - m^2\cos^2\varphi\right) - 2\kappa m\cos\varphi\right]e^{2\kappa a} - (\kappa - m\cos\varphi)^2}{\left[(\kappa + m\cos\varphi)\,e^{2\kappa a} + \kappa - m\cos\varphi\right]^2}. \quad (77)$$

This case was exhaustively studied earlier [10] in a different parametrization corresponding to substitution $\cos\varphi \to 1/\cosh\vartheta$; we remind here that the case of the MIT bag boundary condition is obtainable at $\varphi = 0$ or $\vartheta = 0$, respectively.

In the case of $\varphi = 0$, we obtain

$$\Upsilon(\kappa)|_{\tilde{\varphi}=0} = -\frac{(2\kappa a - 1)\,\mathrm{e}^{2\kappa a} + 1}{(\mathrm{e}^{2\kappa a} - 1)^2}. \tag{78}$$

The spectrum of k_l in this case is explicitly given by $k_l = \frac{\pi}{a} l$ $(l = 0, 1, 2, ...)$, and the Casimir pressure takes form

$$F|_{\tilde{\varphi}=0} = -\varepsilon_{\text{ren}}^{\infty} + \frac{2|eB|}{\pi^2} \sum_{n=0}^{\infty} \iota_n \int_{\omega_{n0}}^{\infty} \frac{\mathrm{d}\kappa}{\mathrm{e}^{2\kappa a} - 1} \frac{\kappa^2}{\sqrt{\kappa^2 - \omega_{n0}^2}}, \tag{79}$$

or, in the alternative representation,

$$F|_{\tilde{\varphi}=0} = -\varepsilon_{\text{ren}}^{\infty} + \frac{2|eB|}{\pi^2} \sum_{n=0}^{\infty} \iota_n \omega_{n0}^2 \sum_{j=1}^{\infty} \left[K_0(2j\omega_{n0}a) + \frac{1}{2j\omega_{n0}a} K_1(2j\omega_{n0}a) \right], \tag{80}$$

where $K_\rho(s)$ is the Macdonald function of order ρ.

It is instructive to consider also the case of $\varphi = \pi/2$, when relations (72)-(76) are reduced to

$$\begin{aligned} \Upsilon(\kappa)|_{\varphi=\pi/2} = -\Bigg\{ & \left[(2\kappa a - 1) \left(1 - \frac{2\kappa^2 \sin^2\varphi}{\kappa^2 - \omega_{n0}^2 \cos^2\varphi} \right) - \frac{\kappa^2 \omega_{n0}^2 \sin^2 2\varphi}{(\kappa^2 - \omega_{n0}^2 \cos^2\varphi)^2} \right] \mathrm{e}^{6\kappa a} \\ & - \left[4\kappa a - 3 + \frac{2\kappa^2 (\kappa^2 - \omega_{n0}^2) \sin^2 2\varphi}{(\kappa^2 - \omega_{n0}^2 \cos^2\varphi)^2} \right] \mathrm{e}^{4\kappa a} \\ & + \left[(2\kappa a - 3) \left(1 - \frac{2\kappa^2 \sin^2\varphi}{\kappa^2 - \omega_{n0}^2 \cos^2\varphi} \right) + \frac{\kappa^2 \omega_{n0}^2 \sin^2 2\varphi}{(\kappa^2 - \omega_{n0}^2 \cos^2\varphi)^2} \right] \mathrm{e}^{2\kappa a} + 1 \Bigg\} \\ & \times \left[\left(\mathrm{e}^{2\kappa a} - 1 \right)^2 + \frac{4\kappa^2 \sin^2\varphi}{\kappa^2 - \omega_{n0}^2 \cos^2\varphi} \mathrm{e}^{2\kappa a} \right]^{-2}. \end{aligned} \tag{81}$$

This case interpolates between the case of spectrum $k_l = \frac{\pi}{a} l$ $(l = 0, 1, 2, ...)$, see (78)-(80), and the case of spectrum $k_l = \frac{\pi}{a}(l + \frac{1}{2})$ $(l = 0, 1, 2, ...)$ with

$$\Upsilon(\kappa)|_{\varphi=-\tilde{\varphi}=\pi/2} = \frac{(2\kappa a - 1)\,\mathrm{e}^{2\kappa a} - 1}{(\mathrm{e}^{2\kappa a} + 1)^2} \tag{82}$$

and

$$F|_{\varphi=-\tilde{\varphi}=\pi/2} = -\varepsilon_{\text{ren}}^{\infty} - \frac{2|eB|}{\pi^2} \sum_{n=0}^{\infty} \iota_n \int_{\omega_{n0}}^{\infty} \frac{\mathrm{d}\kappa}{\mathrm{e}^{2\kappa a} + 1} \frac{\kappa^2}{\sqrt{\kappa^2 - \omega_{n0}^2}}, \tag{83}$$

or, alternatively,

$$F|_{\varphi=-\tilde{\varphi}=\pi/2} = -\varepsilon_{\text{ren}}^{\infty} - \frac{2|eB|}{\pi^2} \sum_{n=0}^{\infty} \iota_n \omega_{n0}^2 \sum_{j=1}^{\infty} (-1)^{j-1} \left[K_0(2j\omega_{n0}a) + \frac{1}{2j\omega_{n0}a} K_1(2j\omega_{n0}a) \right]. \tag{84}$$

5. Asym totics at small and large se arations o lates

The expression for the Casimir pressure, see (71), can be presented as

$$F = -\varepsilon_{\rm ren}^{\infty} + \quad {}_{\varphi,\tilde{\varphi}}(a), \tag{85}$$

where the first term is equal to minus the vacuum energy density which is induced by the magnetic field in unbounded space, see (45), whereas the second term which is given by the sum over n and the integral over κ in (71) depends on the distance between bounding plates and on a choice of boundary conditions at the plates.

In the case of a weak magnetic field, $B \ll m^2|e|^{-1}$, substituting the sum by integral $\int\limits_0^\infty \mathrm{d}n$ and changing the integration variable, we get

$${}_{\varphi,\tilde{\varphi}}(a) = -\frac{1}{\pi^2}\int\limits_m^\infty \mathrm{d}\kappa(\kappa^2 - m^2)^{3/2}\int\limits_0^1 \mathrm{d}v\sqrt{1-v}\,\Upsilon(\kappa, v), \quad |eB| \ll m^2, \tag{86}$$

where $\Upsilon(\kappa, v)$ is obtained from $\Upsilon(\kappa)$ (72) by substitution $\mu_n(\varphi,\varphi) \to \mu_{v,\kappa^2}(\varphi,\varphi)$ with

$$\mu_{v,\kappa^2}(\varphi,\varphi) = v(\kappa^2 - m^2)\cos^2\varphi + m^2\sin(\varphi+\varphi)\sin(\varphi-\varphi). \tag{87}$$

In the limit of small distances between the plates, $ma \ll 1$, (86) becomes independent of the φ-parameter:

$$\begin{aligned}{}_{\varphi,\tilde{\varphi}}(a) = \frac{1}{4a^4}\Bigg\{\frac{\pi^2}{30} - \int\limits_0^1 \mathrm{d}v\,\rho_{\tilde{\varphi}}(v)\left(1 - \frac{|\rho_{\tilde{\varphi}}(v)|}{\pi}\right)\Bigg[\frac{3}{2}\sqrt{1-v}\,\rho_{\tilde{\varphi}}(v)\left(1 - \frac{|\rho_{\tilde{\varphi}}(v)|}{\pi}\right) \\ + \frac{v\sin 2\varphi}{1 - v\cos^2\varphi}\left(\frac{1}{2} - \frac{|\rho_{\tilde{\varphi}}(v)|}{\pi}\right)\Bigg]\Bigg\}, \quad \sqrt{|eB|}a \ll ma \ll 1,\end{aligned} \tag{88}$$

where

$$\rho_{\tilde{\varphi}}(v) = \arcsin\left(\frac{\sin\varphi}{\sqrt{1 - v\cos^2\varphi}}\right). \tag{89}$$

Thus, ${}_{\varphi,\tilde{\varphi}}(a)$ in this case is power-dependent on the distance between the plates as a^{-4} with the dimensionless constant of proportionality, either positive or negative, depending on the value of the φ-parameter. In particular, we get

$${}_{\varphi,0}(a) = \frac{\pi^2}{120}\frac{1}{a^4}, \quad \sqrt{|eB|}a \ll ma \ll 1 \tag{90}$$

and

$${}_{\varphi,-\pi/2}(a) = -\frac{7}{8}\frac{\pi^2}{120}\frac{1}{a^4}, \quad \sqrt{|eB|}a \ll ma \ll 1. \tag{91}$$

In the limit of large distances between the plates, $ma \gg 1$, ${}_{\varphi,\tilde{\varphi}}(a)$ (86) takes form

$$\begin{aligned}{}_{\varphi,\tilde{\varphi}}(a) = \frac{2}{\pi^2}\int\limits_m^\infty \mathrm{d}\kappa\kappa(\kappa^2 - m^2)^{3/2}\mathrm{e}^{-2\kappa a}\int\limits_0^1 \mathrm{d}v\sqrt{1-v}\Bigg\{a\frac{\kappa^2\cos 2\varphi - \mu_{v,\kappa^2}(\varphi,\varphi)}{\kappa^2 - 2\kappa m\cos\varphi\sin\varphi - \mu_{v,\kappa^2}(\varphi,\varphi)} \\ - \frac{(2\kappa\sin\varphi - m\cos\varphi)\mu_{v,\kappa^2}(\varphi,\varphi) - \kappa^2 m\cos\varphi\cos 2\varphi}{[\kappa^2 - 2\kappa m\cos\varphi\sin\varphi - \mu_{v,\kappa^2}(\varphi,\varphi)]^2}\sin\varphi\Bigg\}, \quad |eB| \ll m^2, \quad ma \gg 1.\end{aligned} \tag{92}$$

Clearly, (92) is suppressed as $\exp(-2ma)$. In particular, we get

$$ {}_{\varphi,0}(a) = \frac{1}{2\pi^{3/2}}\frac{m^{5/2}}{a^{3/2}}\mathrm{e}^{-2ma}\left[1 + O\left(\frac{1}{ma}\right)\right], \quad |eB| \ll m^2, \quad ma \gg 1 \tag{93}$$

and

$$ {}_{\varphi,-\pi/2}(a) = \left\{ \begin{array}{ll} -\frac{3}{16\pi^{3/2}}\frac{m^{3/2}}{a^{5/2}}\mathrm{e}^{-2ma}[1+O(\frac{1}{ma})], & \varphi = 0 \\ \\ -\frac{\tan^2(\varphi/2)}{2\pi^{3/2}}\frac{m^{5/2}}{a^{3/2}}\mathrm{e}^{-2ma}[1+O(\frac{1}{ma})], & \varphi \neq 0 \end{array} \right\}, \quad |eB| \ll m^2, \quad ma \gg 1. \tag{94}$$

In the case of a strong magnetic field, $|B| \gg m^2|e|^{-1}$, one has

$$ {}_{\varphi,\tilde{\varphi}}(a) = -\frac{|eB|}{\pi^2}\left[\int\limits_m^\infty d\kappa\sqrt{\kappa^2 - m^2}\Upsilon(\kappa)|_{n=0} \right.$$

$$\left. +2\sum_{n=1}^{\infty}\int\limits_{\sqrt{2n|eB|}}^{\infty} d\kappa\sqrt{\kappa^2 - 2n|eB|}\Upsilon(\kappa)|_{m=0}\right], \quad |eB| \gg m^2. \tag{95}$$

In the limit of extremely small distances between the plates, $ma \ll \sqrt{|eB|}a \ll 1$, the analysis is similar to that of the limit of $\sqrt{|eB|}a \ll ma \ll 1$, yielding the same results as (88)-(91). Otherwise, in the limit of $\sqrt{|eB|}a \gg 1$, only the first term in square brackets on the right-hand side of (95) matters. In the limit of small distances between the plates this term becomes φ-independent, yielding

$$ {}_{\varphi,\tilde{\varphi}}(a) = \frac{|eB|}{4a^2}\left[\frac{1}{6} - \frac{|\varphi|}{\pi}\left(1 - \frac{|\varphi|}{\pi}\right)\right], \quad \sqrt{|eB|}a \gg 1, \quad ma \ll 1. \tag{96}$$

In particular, we get

$$ {}_{\varphi,0}(a) = \frac{|eB|}{24a^2}, \quad \sqrt{|eB|}a \gg 1, \quad ma \ll 1, \tag{97}$$

$$ {}_{\varphi,\pm\pi/4}(a) = -\frac{|eB|}{192a^2}, \quad \sqrt{|eB|}a \gg 1, \quad ma \ll 1 \tag{98}$$

and

$$ {}_{\varphi,-\pi/2}(a) = -\frac{|eB|}{48a^2}, \quad \sqrt{|eB|}a \gg 1, \quad ma \ll 1. \tag{99}$$

In the limit of large distances between the plates, the first term in square brackets on the right-hand side of (95) yields

$$ {}_{\varphi,\tilde{\varphi}}(a) = \frac{2|eB|}{\pi^2}\int\limits_m^\infty d\kappa\kappa(\kappa^2 - m^2)^{1/2}\mathrm{e}^{-2\kappa a}\left\{a\frac{\kappa^2\cos 2\varphi - m^2\sin(\varphi+\varphi)\sin(\varphi-\varphi)}{\kappa^2 - 2\kappa m\cos\varphi\sin\varphi - m^2\sin(\varphi+\varphi)\sin(\varphi-\varphi)}\right.$$

$$\left. + \frac{\kappa^2 m\cos\varphi\cos 2\varphi - (2\kappa\sin\varphi - m\cos\varphi)m^2\sin(\varphi+\varphi)\sin(\varphi-\varphi)}{[\kappa^2 - 2\kappa m\cos\varphi\sin\varphi - m^2\sin(\varphi+\varphi)\sin(\varphi-\varphi)]^2}\sin\varphi\right\}, \quad \sqrt{|eB|}a \gg ma \gg 1, \tag{100}$$

which is obviously suppressed as $\exp(-2ma)$. In particular, we get

$$ {}_{\varphi,0}(a) = \frac{|eB|}{2\pi^{3/2}}\frac{m^{3/2}}{a^{1/2}}\mathrm{e}^{-2ma}\left[1 + O\left(\frac{1}{ma}\right)\right], \quad \sqrt{|eB|}a \gg ma \gg 1 \tag{101}$$

and

$$
{}_{\varphi,-\pi/2}(a) = \left\{ \begin{array}{ll} -\frac{|eB|}{16\pi^{3/2}}\frac{m^{1/2}}{a^{3/2}}\mathrm{e}^{-2ma}[1+O(\frac{1}{ma})], & \varphi = 0 \\ \\ -\frac{|eB|\tan^2(\varphi/2)}{2\pi^{3/2}}\frac{m^{3/2}}{a^{1/2}}\mathrm{e}^{-2ma}[1+O(\frac{1}{ma})], & \varphi \neq 0 \end{array} \right\}, \quad \sqrt{|eB|}a \gg ma \gg 1. \tag{102}
$$

It is appropriate in this section to consider also the limiting case of $m \to 0$. In view of the asymptotical behaviour of the boundary-independent piece of the Casimir pressure,

$$
-\varepsilon^{\infty}_{\rm ren} = \frac{e^2B^2}{24\pi^2}\ln\frac{2|eB|}{m^2}, \quad m^2 \ll |eB| \tag{103}
$$

and

$$
-\varepsilon^{\infty}_{\rm ren} = \frac{1}{360\pi^2}\frac{e^4B^4}{m^4}, \quad m^2 \gg |eB|, \tag{104}
$$

namely asymptotics (88) is relevant for this case, and the pressure from the vacuum of a confined massless spinor matter field is given by expression

$$
F|_{m=0,\,B=0} = \frac{1}{8a^4}\left\{\frac{\pi^2}{30} - \int\limits_0^1 \mathrm{d}v\,\rho_{\tilde\varphi}(v)\left(1-\frac{|\rho_{\tilde\varphi}(v)|}{\pi}\right)\right.
$$

$$
\left.\times\left[\frac{3}{2}\sqrt{1-v}\,\rho_{\tilde\varphi}(v)\left(1-\frac{|\rho_{\tilde\varphi}(v)|}{\pi}\right)+\frac{v\sin 2\varphi}{1-v\cos^2\varphi}\left(\frac{1}{2}-\frac{|\rho_{\tilde\varphi}(v)|}{\pi}\right)\right]\right\}, \tag{105}
$$

which is bounded from above and below by values

$$
F|_{m=0,\,B=0,\,\tilde\varphi=0} = \frac{\pi^2}{240}\frac{1}{a^4} \tag{106}
$$

and

$$
F|_{m=0,\,B=0,\,\tilde\varphi=-\pi/2} = -\frac{7}{8}\frac{\pi^2}{240}\frac{1}{a^4}, \tag{107}
$$

respectively; here, an additional factor of 1/2 has appeared due to diminishment of the number of degrees of freedom (a massless spinor can be either left or right).

6. Summary and discussion

We study an influence of a background (classical) magnetic field on the vacuum of a quantized charged spinor matter field which is confined to a bounded region of space; the sources of the magnetic field are outside of the bounded region, and the magnetic field strength lines are assumed to be orthogonal to a boundary. The confinement of the matter field (i.e. absence of the matter flux across the boundary) is ensured by boundary condition (32) which is compatible with the self-adjointness of the Dirac hamiltonan operator and which generalizes the well-known MIT bag boundary condition to the most extent. In the case which is relevant to the geometry of the Casimir effect (i.e. the spatial region bounded by two parallel planes separated by distance a) and the uniform magnetic field orthogonal to the planes, the most general extension of the MIT bag boundary condition is given by (55); the spectrum of the wave number vector along the magnetic field in this case depends on the number of the Landau level and on the sign of the one-particle energy, see (57). The Casimir pressure is shown to take the form of (71), where $\varepsilon^{\infty}_{\rm ren}$ is given by (45) and $\Upsilon(\kappa)$ is given by (72)-(76); the result for the case of the MIT bag boundary condition is obtained from (71) at $\varphi = 0$, $\varphi = -\pi/2$.

The Casimir effect is usually validated in experiments with nearly parallel plates separated by a distance of order $10^{-8} - 10^{-5}$ m, see, e.g., [12]. The pressure from the vacuum of neutral massless spinor matter onto the bounding plates is given by a^{-4} times a constant of proportionality, either positive or negative, depending on a choice of boundary conditions, see (105)-(107). The situation with charged massive spinor matter is quite different. The Compton wavelength of the lightest charged particle, electron, is $m^{-1} \sim 10^{-12}$ m, thus $ma \gg 1$ and, as has been shown in the preceding section, all the dependence of the Casimir pressure on the distance between the plates and a choice of boundary conditions at the plates is suppressed by factor $\exp(-2ma)$, see (92)-(94) and (100)-(102). Hence, the pressure from the electron-positron vacuum onto the plates separated by distance $a > 10^{-10}$ m is well approximated by $F \approx -\varepsilon_{\rm ren}^{\infty}$, where $\varepsilon_{\rm ren}^{\infty}$ (45) is negative, i.e. the pressure is positive and the plates are repelled. Some possibilities to detect this new-type Casimir effect were pointed out in [10].

Acknowledgments

I would like to thank the Organizers of the XXIII International Conference on Integrable Systems and Quantum Symmetries for kind hospitality during this interesting and inspiring meeting. The work was supported by the National Academy of Sciences of Ukraine (project No.0112U000054), the Program of Fundamental Research of the Department of Physics and Astronomy of the National Academy of Sciences of Ukraine (project No.0112U000056) and the ICTP – SEENET-MTP grant PRJ-09 Strings and Cosmology .

Re erences

[1] Akhiezer N I and Glazman I M 1981 *Theory of Linear Operators in Hilbert Space* (Boston: Pitman)
[2] Naimark M A 1968 *Linear Differential Operators* (New York: Frederick Ungar)
[3] Heisenberg W and Euler H 1936 *Z. Phys.* **98** 714
[4] Schwinger J 1951 *Phys. Rev.* **82** 662
[5] Dittrich W and Reuter M 1985 *Effective Lagrangians in Quantum Electrodynamics* (Berlin: Springer)
[6] Chodos A, Jaffe R L, Johnson K, Thorn C B and Weisskopf V 1974 *Phys. Rev.* D **9** 3471
[7] Chodos A, Jaffe R L, Johnson K and Thorn C B 1974 *Phys. Rev.* D **10** 2599
[8] Johnson K 1975 *Acta Phys. Polon.* B **6** 865
[9] Al-Hashimi M H and Wiese U-J 2012 *Annals Phys.* **327** 1
[10] Sitenko Yu A 2015 *Phys. Rev.* D **91** 085012
[11] Casimir H B G 1948 *Proc. Kon. Ned. Akad. Wetenschap* B **51** 793
[12] M.Bordag, Klimchitskaya G L, Mohideen U and Mostepanenko V M 2009 *Advances in the Casimir Effect* (Oxford: Oxford University)

Fusing defect for the $\mathcal{N} = 2$ super sinh-Gordon model

N I Spano, A R Aguirre, J F Gomes and A H Zimerman

Rua Dr. Bento Teobaldo Ferraz 271, Block 2, 01140-070, São Paulo, Brazil

E-mail: natyspano@ift.unesp.br

Abstract. In this paper we derive the type-II integrable defect for the $\mathcal{N} = 2$ supersymmetric sinh-Gordon (sshG) model by using the fusing procedure. In particular, we show explicitly the conservation of the modified energy, momentum and supercharges.

1. Introduction

An interesting topic in the study of integrable systems is the analysis of their integrability properties in the presence of impurities or defects. Accordingly, defects are introduced in two-dimensional integrable field theories as internal boundary conditions located at a fixed point in the x-axis, which connect two different field theories of both sides of it. In particular, after the introduction of the defect, the spatial translation invariance is broken since some constraints are imposed to be satisfied at a particular space point, and hence it would be expected a violation of momentum conservation. However, it was verified in [1]-[4], that in order to preserve the integrability, the fields of the theory must satisfy a kind of Bäcklund transformation frozen at the defect point.

This kind of integrable defects can be classified into two classes: *type-I*, if the fields on both sides only interact with each other at the defect point, and *type-II* if they interact through additional degrees of freedom present only at the defect point [5]. The type-II formulation proved to be suitable not only for describing defects within the Tzitzéica-Bullough-Dodd ($a_2^{(2)}$-Toda) model, which had been excluded from the type-I setting, but it also provided additional types of defects for the sine-Gordon (sG) and others affine Toda field theories (ATFT) [6]. Interestingly, for the sG model [5, 7], and in general for $a_r^{(1)}$-ATFT [8] and the $\mathcal{N} = 1$ sshG models [9], the type-II defects can be regarded as fused pairs of type-I defects previously placed at different points in space. However, the type-II defects can be allowed in models that cannot support type-I defects, as it was shown for the $a_2^{(2)}$-Toda model [5].

On the other hand, the presence of integrable defects in the $\mathcal{N} = 1$ sshG model has been already discussed in [10, 11]. However, the kind of defect introduced in those papers can be regarded as a partial type-II defect since only auxiliary fermionic fields appear in the defect Lagrangian, and consequently it reduces to type-I defect for sinh-Gordon model in the bosonic limit. The proper supersymmetric extension of the type-II defect for the $\mathcal{N} = 1$ sshG model was recently proposed in [9], by using two methods: the generalization of the super-Bäcklund transformations, and the fusing procedure.

The purpose of this paper is to derive type-II defects for the $\mathcal{N} = 2$ sshG equation by fusing defects of the kind already known in literature [12]. The explicit form of the type-II Bäcklund transformations for the $\mathcal{N} = 2$ sshG model will be presented. We will also compute its modified conserved energy, momentum and supercharges. Finally, by introducing appropriate field transformations, the **PT** symmetry of the bulk and the defect theories will be discussed.

2. $\mathcal{N}=2$ super sinh-Gordon model

The action for the bulk $\mathcal{N}=2$ sshG model is given by,

$$S_{bulk}=\int_{-\infty}^{\infty}dt\int_{-\infty}^{\infty}dx\ \mathcal{L}_{bulk}, \tag{1}$$

with the bulk Lagrangian density,

$$\begin{aligned}\mathcal{L}_{bulk} &= \frac{1}{2}(\partial_x\phi)^2-\frac{1}{2}(\partial_t\phi)^2-\frac{1}{2}(\partial_x\varphi)^2+\frac{1}{2}(\partial_t\varphi)^2-i\psi(\partial_x-\partial_t)\psi+i\bar{\psi}(\partial_x+\partial_t)\bar{\psi}\\ &+i\chi(\partial_x-\partial_t)\chi-i\bar{\chi}(\partial_x+\partial_t)\bar{\chi}+m^2\big[\cosh(2\phi)-\cosh(2\varphi)\big]\\ &+4im(\bar{\psi}\psi+\bar{\chi}\chi)\cosh\phi\cosh\varphi-4im(\bar{\psi}\chi+\bar{\chi}\psi)\sinh\phi\sinh\varphi,\end{aligned} \tag{2}$$

where ϕ, φ are bosonic fields, ψ, $\bar{\psi}$, χ, $\bar{\chi}$ are fermionic fields. Then, the bulk field equations are,

$$\begin{aligned}(\partial_x^2-\partial_t^2)\phi &= 2m^2\sinh(2\phi)+4im(\bar{\psi}\psi+\bar{\chi}\chi)\sinh\phi\cosh\varphi\\ &\quad -4im(\bar{\psi}\chi+\bar{\chi}\psi)\cosh\phi\sinh\varphi,\\ (\partial_x^2-\partial_t^2)\varphi &= 2m^2\sinh(2\varphi)-4im(\bar{\psi}\psi+\bar{\chi}\chi)\cosh\phi\sinh\varphi\\ &\quad +4im(\bar{\psi}\chi+\bar{\chi}\psi)\sinh\phi\cosh\varphi,\\ (\partial_x-\partial_t)\psi &= -2m\left[\bar{\psi}\cosh\phi\cosh\varphi-\bar{\chi}\sinh\phi\sinh\varphi\right],\\ (\partial_x+\partial_t)\bar{\psi} &= -2m\left[\psi\cosh\phi\cosh\varphi-\chi\sinh\phi\sinh\varphi\right],\\ (\partial_x-\partial_t)\chi &= 2m\left[\bar{\chi}\cosh\phi\cosh\varphi-\bar{\psi}\sinh\phi\sinh\varphi\right],\\ (\partial_x+\partial_t)\bar{\chi} &= 2m\left[\chi\cosh\phi\cosh\varphi-\psi\sinh\phi\sinh\varphi\right],\end{aligned} \tag{3}$$

The bulk action and the equation of motion have on-shell $\mathcal{N}=2$ supersymmetry (susy). The susy transformation is given by,

$$\begin{aligned}\delta\phi &= i(\epsilon_1\psi+\bar{\epsilon}_1\bar{\psi})-i(\epsilon_2\chi+\bar{\epsilon}_2\bar{\chi}),\\ \delta\varphi &= i(\epsilon_2\psi+\bar{\epsilon}_2\bar{\psi})-i(\epsilon_1\chi+\bar{\epsilon}_1\bar{\chi}),\\ \delta\psi &= (\epsilon_1\partial_+\phi+\bar{\epsilon}_1 m\sinh\phi\cosh\varphi)-(\epsilon_2\partial_+\varphi+\bar{\epsilon}_2 m\sinh\varphi\cosh\phi),\\ \delta\chi &= (\epsilon_2\partial_+\phi-\bar{\epsilon}_2 m\sinh\phi\cosh\varphi)-(\epsilon_1\partial_+\varphi-\bar{\epsilon}_1 m\sinh\varphi\cosh\phi),\\ \delta\bar{\psi} &= (\bar{\epsilon}_2\partial_-\varphi+\epsilon_2 m\sinh\varphi\cosh\phi)-(\bar{\epsilon}_1\partial_-\phi+\epsilon_1 m\sinh\phi\cosh\varphi),\\ \delta\bar{\chi} &= (\bar{\epsilon}_1\partial_-\varphi-\epsilon_1 m\sinh\varphi\cosh\phi)-(\bar{\epsilon}_2\partial_-\phi-\epsilon_2 m\sinh\phi\cosh\varphi),\end{aligned} \tag{4}$$

where ϵ_k and $\bar{\epsilon}_k$, with $k=1,2$, are fermionic parameters, and the light-cone notation $x_\pm=x\pm t$, and $\partial_\pm=\frac{1}{2}(\partial_x\pm\partial_t)$ has been used. It can be easily verified that the equations of motions are invariant under these transformations. For simplicity, we will focus on the ϵ_1-projection of the susy transformation (4), which will be denoted δ_1, and then we will compute the associated supercharge $Q_{\epsilon_1}\equiv Q_1$.

Under a not-rigid susy transformation, i.e with parameters $\epsilon(x,t)$ and $\bar{\epsilon}(x,t)$, $\mathcal{L}_{bulk}$ changes by a total derivative

$$\begin{aligned}\delta_1\mathcal{L}_{bulk} &= \partial_x\Big[i\epsilon_1\Big(\psi\,(\partial_-\phi+2\partial_+\phi)+\chi\,(\partial_-\varphi+2\partial_+\varphi)+m\bar{\psi}\sinh\phi\cosh\varphi-m\bar{\chi}\sinh\varphi\cosh\phi\Big)\Big]\\ &+\partial_t\Big[i\epsilon_1\Big(\psi\,(\partial_-\phi-2\partial_+\phi)+\chi\,(\partial_-\varphi+2\partial_+\varphi)+m\bar{\psi}\sinh\phi\cosh\varphi-m\bar{\chi}\sinh\varphi\cosh\phi\Big)\Big]\\ &+\epsilon_1\Big[\partial_t\Big(2i\psi\partial_+\phi+2i\chi\partial_+\varphi-2im\bar{\psi}\sinh\phi\cosh\varphi+2im\bar{\chi}\sinh\varphi\cosh\phi\Big)\\ &\quad-\partial_x\Big(2i\psi\partial_+\phi+2i\chi\partial_+\varphi+2im\bar{\psi}\sinh\phi\cosh\varphi-2im\bar{\chi}\sinh\varphi\cosh\phi\Big)\Big],\end{aligned} \tag{5}$$

Journal of Physics: Conference Series **670** (2016) 012049 doi:10.1088/1742-6596/670/1/012049

if the conservation law inside the last square-bracket in (5) is hold. Then, the associated bulk supercharges Q_1 is given by an integral of the fermionic density, namely

$$Q_1 = \int_{-\infty}^{\infty} dx \Big[i\psi(\partial_x + \partial_t)\phi + i\chi(\partial_x + \partial_t)\varphi - 2im\bar{\psi}\sinh\phi\cosh\varphi + 2im\bar{\chi}\sinh\varphi\cosh\phi \Big]. \quad (6)$$

The derivation of the remaining supercharges follows the same line of reasoning. Their explicit form is given by the following expressions,

$$\overline{Q}_1 = \int_{-\infty}^{\infty} dx \Big[i\bar{\psi}(\partial_x - \partial_t)\phi + i\bar{\chi}(\partial_x - \partial_t)\varphi - 2im\psi\sinh\phi\cosh\varphi + 2im\chi\sinh\varphi\cosh\phi \Big], \quad (7)$$

$$Q_2 = \int_{-\infty}^{\infty} dx \Big[i\chi(\partial_x + \partial_t)\phi + i\psi(\partial_x + \partial_t)\varphi + 2im\bar{\chi}\sinh\phi\cosh\varphi - 2im\bar{\psi}\sinh\varphi\cosh\phi \Big], \quad (8)$$

$$\overline{Q}_2 = \int_{-\infty}^{\infty} dx \Big[i\bar{\psi}(\partial_x - \partial_t)\varphi + i\bar{\chi}(\partial_x - \partial_t)\phi + 2im\chi\sinh\phi\cosh\varphi - 2im\psi\sinh\varphi\cosh\phi \Big]. \quad (9)$$

In next section we introduce the Lagrangian description of type-I defects in $\mathcal{N} = 2$ sshG model.

3. Type-I defect for $\mathcal{N} = 2$ sshG model

We consider a defect placed in $x = 0$ connecting two field theories Φ_1 in the region $x < 0$ and Φ_2 in the region $x > 0$, First of all, let us consider a Lagrangian density for the region $x < 0$

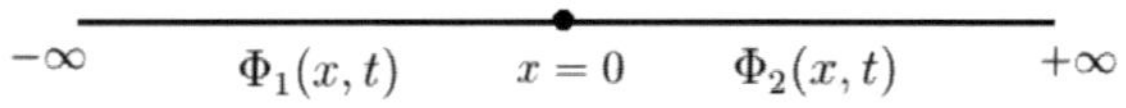

Figure 1. Defect Representation.

describing the set of fields $\Phi_1(\phi_1, \psi_1, \bar{\psi}_1, \varphi_1, \chi_1, \bar{\chi}_1)$ and correspondingly $\Phi_2(\phi_2, \psi_2, \bar{\psi}_2, \varphi_2, \chi_2, \bar{\chi}_2)$ in the region $x > 0$, and a defect located at $x = 0$, in the following way

$$\mathcal{L} = \theta(-x)\mathcal{L}_1 + \theta(x)\mathcal{L}_2 + \delta(x)\mathcal{L}_D, \quad (10)$$

where $\mathcal{L}_1$ and $\mathcal{L}_2$ are the bulk Lagrangian densities corresponding to $x < 0$ and $x > 0$ regions, respectively, and the defect Lagrangian density $\mathcal{L}_D$ is given by

$$\begin{aligned} \mathcal{L}_D = {} & \frac{1}{2}(\phi_2\partial_t\phi_1 - \phi_1\partial_t\phi_2) - \frac{1}{2}(\varphi_2\partial_t\varphi_1 - \varphi_1\partial_t\varphi_2) + B_0(\phi_1, \phi_2, \varphi_1, \varphi_2) \\ & -i(\bar{\psi}_1\bar{\psi}_2 + \psi_1\psi_2) + i(\bar{\chi}_1\bar{\chi}_2 + \chi_1\chi_2) + \frac{i}{2}(f\partial_t g + g\partial_t f) \\ & +B_1(\phi_1, \phi_2, \varphi_1, \varphi_2, \psi_1, \psi_2, \bar{\psi}_1, \bar{\psi}_2, \chi_1, \chi_2, \bar{\chi}_1, \bar{\chi}_2, f, g) \end{aligned} \quad (11)$$

where f, g are fermionic auxiliary fields. For $x = 0$, we obtain the following defect conditions,

$$\begin{aligned} \partial_x\phi_1 - \partial_t\phi_2 &= -\partial_{\phi_1}(B_0 + B_1), & \partial_x\varphi_1 - \partial_t\varphi_2 &= \partial_{\varphi_1}(B_0 + B_1), \\ \partial_x\phi_2 - \partial_t\phi_1 &= \partial_{\phi_2}(B_0 + B_1), & \partial_x\varphi_2 - \partial_t\varphi_1 &= -\partial_{\varphi_2}(B_0 + B_1), \\ i(\psi_1 - \psi_2) &= -\partial_{\psi_1}B_1 = -\partial_{\psi_2}B_1, & i(\chi_1 - \chi_2) &= \partial_{\chi_1}B_1 = \partial_{\chi_2}B_1, \\ i(\bar{\psi}_1 + \bar{\psi}_2) &= \partial_{\bar{\psi}_1}B_1 = -\partial_{\bar{\psi}_2}B_1, & i(\bar{\chi}_1 + \bar{\chi}_2) &= -\partial_{\bar{\chi}_1}B_1 = \partial_{\bar{\chi}_2}B_1, \\ i\partial_t f &= -\partial_g B_1, & i\partial_t g &= -\partial_f B_1, \end{aligned} \quad (12)$$

where the defect potentials B_0 and B_1 are given given by [12],

$$B_0 = m\sigma \left[\cosh \phi_+ - \cosh \varphi_+\right] + \frac{m}{\sigma}\left[\cosh \phi_- - \cosh \varphi_-\right], \tag{13}$$

$$\begin{aligned} B_1 &= i\sqrt{\frac{m\sigma}{2}}\left[\cosh\left(\frac{\phi_+ + \varphi_+}{2}\right) f(\psi_+ - \chi_+) + \cosh\left(\frac{\phi_+ - \varphi_+}{2}\right) g(\psi_+ + \chi_+)\right] \\ &-i\sqrt{\frac{m}{2\sigma}}\left[\cosh\left(\frac{\phi_- - \varphi_-}{2}\right) f(\bar{\psi}_- + \bar{\chi}_-) + \cosh\left(\frac{\phi_- + \varphi_-}{2}\right) g(\bar{\psi}_- - \bar{\chi}_-)\right], \end{aligned} \tag{14}$$

where we have denoted $\phi_\pm = \phi_1 \pm \phi_2$, $\psi_\pm = \psi_1 \pm \psi_2$, $\chi_\pm = \chi_1 \pm \chi_2$ for the others fields the notation is similar, and σ is a free parameter associated with the defect.
In next section, we will perform the fusing of two type-I defects placed at different points in order to construct a type-II defect for the $\mathcal{N} = 2$ sshG model.

4. Fusing Defects

Let us introduce two type-I defects in the $\mathcal{N} = 2$ sshG model, one located at $x = 0$, and a second one located at $x = x_0$ where $\Phi_1(\phi_1, \psi_1, \bar{\psi}_1, \varphi_1, \chi_1, \bar{\chi}_1)$ is a set of fields in the region

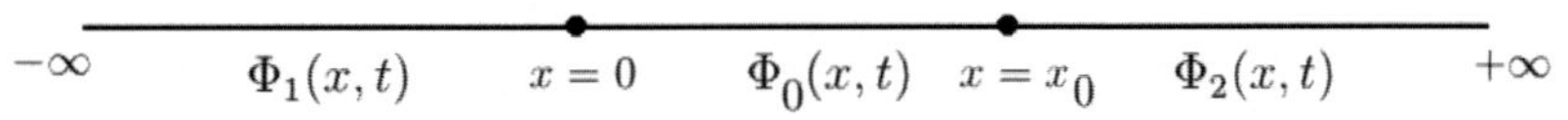

Figure 2. Fusing defects.

$x < 0$, $\Phi_0(\phi_0, \psi_0, \bar{\psi}_0, \varphi_0, \chi_0, \bar{\chi}_0)$ is the correspondingly set of fields for the region $0 < x < x_0$ and $\Phi_2(\phi_2, \psi_2, \bar{\psi}_2, \varphi_2, \chi_2, \bar{\chi}_2)$ in the $x > 0$.
Then, the Lagrangian density describing this system can be written as,

$$\mathcal{L} = \theta(-x)\mathcal{L}_1 + \delta(x)\mathcal{L}_{D_1} + \theta(x)\theta(x_0 - x)\mathcal{L}_0 - \delta(x - x_0)\mathcal{L}_{D_2} + \theta(x - x_0)\mathcal{L}_2, \tag{15}$$

where the two type-I defect Lagrangian densities $\mathcal{L}_{D_k}$ at $x = 0$ $(k = 1)$, and $x = x_0$ $(k = 2)$, are given by

$$\begin{aligned} \mathcal{L}_{D_k} &= \frac{1}{2}[\phi_0\partial_t\phi_k - \phi_k\partial_t\phi_0] - \frac{1}{2}[\varphi_0\partial_t\varphi_k - \varphi_k\partial_t\varphi_0] + i(\bar{\chi}_k\bar{\chi}_0 + \chi_k\chi_0) \\ &-i(\bar{\psi}_k\bar{\psi}_0 + \psi_k\psi_0) - (-1)^k\left(\frac{i}{2}f_k\partial_t g_k + \frac{i}{2}g_k\partial_t f_k + B_1^{(k)} + B_0^{(k)}\right), \end{aligned} \tag{16}$$

with the defect potentials

$$B_0^{(k)} = m\sigma_k\left[\cosh(\phi_0 + \phi_k) - \cosh(\varphi_0 + \varphi_k)\right] + \frac{m}{\sigma_k}\left[\cosh(\phi_0 - \phi_k) - \cosh(\varphi_0 - \varphi_k)\right], \tag{17}$$

$$\begin{aligned} B_1^{(k)} &= i\sqrt{\frac{m\sigma_k}{2}}\cosh\left(\frac{\phi_k + \phi_0 + \varphi_0 + \varphi_k}{2}\right) f_k(\psi_0 + \psi_k - \chi_0 - \chi_k) \\ &+i\sqrt{\frac{m\sigma_k}{2}}\cosh\left(\frac{\phi_k + \phi_0 - \varphi_k - \varphi_0}{2}\right) g_k(\psi_k + \psi_0 + \chi_0 + \chi_k) \\ &-i(-1)^k\sqrt{\frac{m}{2\sigma_k}}\cosh\left(\frac{\phi_0 - \phi_k + \varphi_k - \varphi_0}{2}\right) f_k(\bar{\psi}_0 - \bar{\psi}_k + \bar{\chi}_0 - \bar{\chi}_k) \\ &-i(-1)^k\sqrt{\frac{m}{2\sigma_k}}\cosh\left(\frac{\phi_0 - \phi_k + \varphi_0 - \varphi_k}{2}\right) g_k(\bar{\psi}_0 - \bar{\psi}_k - \bar{\chi}_0 + \bar{\chi}_k), \end{aligned} \tag{18}$$

where σ_k, with $k = 1, 2$ are two free parameters associated two each defect. Thus for each defect we can write the following equations of motion
for $x = 0$:

$$\begin{aligned}
\partial_x\phi_1 - \partial_t\phi_0 &= -\partial_{\phi_1}(B_0^{(1)} + B_1^{(1)}), & \partial_x\varphi_1 - \partial_t\varphi_0 &= \partial_{\varphi_1}(B_0^{(1)} + B_1^{(1)}),\\
\partial_x\phi_0 - \partial_t\phi_1 &= \partial_{\phi_0}(B_0^{(1)} + B_1^{(1)}), & \partial_x\varphi_0 - \partial_t\varphi_1 &= -\partial_{\varphi_0}(B_0^{(1)} + B_1^{(1)}),\\
i(\psi_1 - \psi_0) &= -\partial_{\psi_1}B_1^{(1)} = -\partial_{\psi_0}B_1^{(1)}, & i(\chi_1 - \chi_0) &= \partial_{\chi_1}B_1^{(1)} = \partial_{\chi_0}B_1^{(1)},\\
i(\bar\psi_1 + \bar\psi_0) &= \partial_{\bar\psi_1}B_1^{(1)} = -\partial_{\bar\psi_0}B_1^{(1)}, & i(\bar\chi_1 + \bar\chi_0) &= -\partial_{\bar\chi_1}B_1^{(1)} = \partial_{\bar\chi_0}B_1^{(1)},\\
i\partial_t f_1 &= -\partial_{g_1}B_1^{(1)}, & i\partial_t g_1 &= -\partial_{f_1}B_1^{(1)},
\end{aligned} \tag{19}$$

and for $x = x_0$:

$$\begin{aligned}
\partial_x\phi_0 - \partial_t\phi_2 &= -\partial_{\phi_0}(B_0^{(2)} + B_1^{(2)}), & \partial_x\varphi_0 - \partial_t\varphi_2 &= \partial_{\varphi_0}(B_0^{(2)} + B_1^{(2)}),\\
\partial_x\phi_2 - \partial_t\phi_0 &= \partial_{\phi_2}(B_0^{(2)} + B_1^{(2)}), & \partial_x\varphi_2 - \partial_t\varphi_0 &= -\partial_{\varphi_2}(B_0^{(2)} + B_1^{(2)}),\\
i(\psi_0 - \psi_2) &= -\partial_{\psi_0}B_1^{(2)} = -\partial_{\psi_2}B_1^{(2)}, & i(\chi_0 - \chi_2) &= \partial_{\chi_0}B_1^{(2)} = \partial_{\chi_2}B_1^{(2)},\\
i(\bar\psi_2 + \bar\psi_0) &= \partial_{\bar\psi_0}B_1^{(2)} = -\partial_{\bar\psi_2}B_1^{(2)}, & i(\bar\chi_2 + \bar\chi_0) &= -\partial_{\bar\chi_0}B_1^{(2)} = \partial_{\bar\chi_2}B_1^{(2)},\\
i\partial_t f_2 &= -\partial_{g_2}B_1^{(2)}, & i\partial_t g_2 &= -\partial_{f_2}B_1^{(2)},
\end{aligned} \tag{20}$$

Now taking the limit $x_0 \to 0$ in the Lagrangian density (15), the bulk Langrangian term $\mathcal{L}_0$ vanishes, and then the resulting Lagrangian density for fused defect becomes of the form of eq. (10) with $\mathcal{L}_D = \mathcal{L}_{D1} - \mathcal{L}_{D2}$, namely,

$$\begin{aligned}
\mathcal{L}_D &= \frac{1}{2}(\phi_0\partial_t\phi_- - \phi_-\partial_t\phi_0) - i(\bar\psi_-\bar\psi_0 + \psi_-\psi_0) + \frac{i}{2}(f_1\partial_t g_1 + f_2\partial_t g_2 + g_1\partial_t f_1 + g_2\partial_t f_2)\\
&\quad -\frac{1}{2}(\varphi_0\partial_t\varphi_- - \varphi_-\partial_t\varphi_0) + i(\bar\chi_-\bar\chi_0 + \chi_-\chi_0) + B_0^{(1)} + B_0^{(2)} + B_1^{(1)} + B_1^{(2)}.
\end{aligned} \tag{21}$$

We note that the fields of the bulk Langrangian term $\mathcal{L}_0$ only contribute to the total defect Lagrangian at $x = 0$, and become auxiliary fields.
The fused bosonic potential for $\mathcal{N} = 2$ sshG model $B_0 = B_0^{(1)} + B_0^{(2)} = B_0^+ + B_0^-$, is a combination of two $\mathcal{N} = 1$ potentials previously obtained in [9], and it can be written explicitly as follows,

$$\begin{aligned}
B_0^+ &= \frac{m}{2}\Big[e^{\left(\frac{\phi_+}{2}+\phi_0\right)}\left(\sigma_1 e^{\frac{\phi_-}{2}} + \sigma_2 e^{-\frac{\phi_-}{2}}\right) + e^{-\left(\frac{\phi_+}{2}+\phi_0\right)}\left(\sigma_1 e^{-\frac{\phi_-}{2}} + \sigma_2 e^{\frac{\phi_-}{2}}\right)\\
&\quad - e^{\left(\frac{\varphi_+}{2}+\varphi_0\right)}\left(\sigma_1 e^{\frac{\varphi_-}{2}} + \sigma_2 e^{-\frac{\varphi_-}{2}}\right) - e^{-\left(\frac{\varphi_+}{2}+\varphi_0\right)}\left(\sigma_1 e^{-\frac{\varphi_-}{2}} + \sigma_2 e^{\frac{\varphi_-}{2}}\right)\Big],
\end{aligned} \tag{22}$$

$$\begin{aligned}
B_0^- &= \frac{m}{2}\Big[e^{\left(\frac{\phi_+}{2}-\phi_0\right)}\left(\frac{1}{\sigma_1} e^{\frac{\phi_-}{2}} + \frac{1}{\sigma_2} e^{-\frac{\phi_-}{2}}\right) + e^{-\left(\frac{\phi_+}{2}-\phi_0\right)}\left(\frac{1}{\sigma_1} e^{-\frac{\phi_-}{2}} + \frac{1}{\sigma_2} e^{\frac{\phi_-}{2}}\right)\\
&\quad - e^{\left(\frac{\varphi_+}{2}-\varphi_0\right)}\left(\frac{1}{\sigma_1} e^{\frac{\varphi_-}{2}} + \frac{1}{\sigma_2} e^{-\frac{\varphi_-}{2}}\right) - e^{-\left(\frac{\varphi_+}{2}-\varphi_0\right)}\left(\frac{1}{\sigma_1} e^{-\frac{\varphi_-}{2}} + \frac{1}{\sigma_2} e^{\frac{\varphi_-}{2}}\right)\Big].
\end{aligned} \tag{23}$$

For the fermionic part we need to use the equations of motion (12) for each region, in order to eliminate the auxiliary fields $\bar{\psi}_0, \psi_0, \chi_0, \bar{\chi}_0$, we get

$$\psi_0 = \frac{\psi_+}{2} - \sqrt{\frac{m\sigma_1}{2}}\left[\partial_{\phi_1}u_1^+ f_1 + \partial_{\phi_1}u_1^- g_1\right] + \sqrt{\frac{m\sigma_2}{2}}\left[\partial_{\phi_2}u_2^+ f_2 + \partial_{\phi_2}u_2^- g_2\right] \tag{24}$$

$$\chi_0 = \frac{\chi_+}{2} - \sqrt{\frac{m\sigma_1}{2}}\left[\partial_{\phi_1}u_1^+ f_1 - \partial_{\phi_1}u_1^- g_1\right] + \sqrt{\frac{m\sigma_2}{2}}\left[\partial_{\phi_2}u_2^+ f_2 - \partial_{\phi_2}u_2^- g_2\right] \tag{25}$$

$$\bar{\psi}_0 = -\frac{\bar{\psi}_+}{2} + \sqrt{\frac{m}{2\sigma_1}}\left[\partial_{\phi_1}v_1^- f_1 + \partial_{\phi_1}v_1^+ g_1\right] + \sqrt{\frac{m}{2\sigma_2}}\left[\partial_{\phi_2}v_2^- f_2 + \partial_{\phi_2}v_2^+ g_2\right] \tag{26}$$

$$\bar{\chi}_0 = -\frac{\bar{\chi}_+}{2} - \sqrt{\frac{m}{2\sigma_1}}\left[\partial_{\phi_1}v_1^- f_1 - \partial_{\phi_1}v_1^+ g_1\right] - \sqrt{\frac{m}{2\sigma_2}}\left[\partial_{\phi_2}v_2^- f_2 - \partial_{\phi_2}v_2^+ g_2\right] \tag{27}$$

where we define the functions

$$u_k^{\pm} = \sinh\left(\frac{(\phi_k+\phi_0)\pm(\varphi_k+\varphi_0)}{2}\right), \quad v_k^{\pm} = \sinh\left(\frac{(\phi_k-\phi_0)\pm(\varphi_k-\varphi_0)}{2}\right) \tag{28}$$

Then noting that

$$\begin{aligned} i(\chi_-\chi_0 - \psi_-\psi_0) &= \frac{i}{2}(\chi_-\chi_+ - \psi_-\psi_+) - im\sqrt{\sigma_1\sigma_2}\left[\partial_{\phi_1}u_1^+\partial_{\phi_2}u_2^- f_1 g_2 + \partial_{\phi_1}u_1^-\partial_{\phi_2}u_2^+ g_1 f_2\right], \\ i(\bar{\chi}_-\bar{\chi}_0 - \bar{\psi}_-\bar{\psi}_0) &= -\frac{i}{2}(\bar{\chi}_-\bar{\chi}_+ - \bar{\psi}_-\bar{\psi}_+) - \frac{im}{\sqrt{\sigma_1\sigma_2}}\left[\partial_{\phi_1}v_1^-\partial_{\phi_2}v_2^+ f_1 g_2 + \partial_{\phi_1}v_1^+\partial_{\phi_2}v_2^- g_1 f_2\right], \end{aligned} \tag{29}$$

we find that the fermionic part of the fused defect Lagrangian is given by,

$$\begin{aligned} \mathcal{L}_D\Big|_{fermion} &= i(\bar{\psi}_1\bar{\psi}_2 - \psi_1\psi_2) - i(\bar{\chi}_1\bar{\chi}_2 - \chi_1\chi_2) + \frac{i}{2}(f_1\partial_t g_1 + f_2\partial_t g_2 + g_1\partial_t f_1 + g_2\partial_t f_2) \\ &\quad + B_1^+ + B_1^-, \end{aligned} \tag{30}$$

where

$$\begin{aligned} B_1^+ &= \frac{i}{2}\sqrt{\frac{m}{2}}\Bigg[e^{-\left(\frac{\phi_++\varphi_+}{4}+\frac{\phi_0+\varphi_0}{2}\right)}\left(\sqrt{\sigma_2}\,e^{\frac{\phi_-+\varphi_-}{4}}f_2 + \sqrt{\sigma_1}\,e^{-\left(\frac{\phi_-+\varphi_-}{4}\right)}f_1\right) \\ &\qquad + e^{\left(\frac{\phi_++\varphi_+}{4}+\frac{\phi_0+\varphi_0}{2}\right)}\left(\sqrt{\sigma_2}\,e^{-\left(\frac{\phi_-+\varphi_-}{4}\right)}f_2 + \sqrt{\sigma_1}\,e^{\frac{\phi_-+\varphi_-}{4}}f_1\right)\Bigg](\psi_+ - \chi_+) \\ &+ \frac{i}{2}\sqrt{\frac{m}{2}}\Bigg[e^{-\left(\frac{\phi_+-\varphi_+}{4}+\frac{\phi_0-\varphi_0}{2}\right)}\left(\sqrt{\sigma_1}\,e^{-\left(\frac{\phi_--\varphi_-}{4}\right)}g_1 + \sqrt{\sigma_2}\,e^{\frac{\phi_--\varphi_-}{4}}g_2\right) \\ &\qquad + e^{\left(\frac{\phi_+-\varphi_+}{4}+\frac{\phi_0-\varphi_0}{2}\right)}\left(\sqrt{\sigma_1}\,e^{\frac{\phi_--\varphi_-}{4}}g_1 + \sqrt{\sigma_2}\,e^{-\left(\frac{\phi_--\varphi_-}{4}\right)}g_2\right)\Bigg](\psi_+ + \chi_+) \\ &+ \frac{im}{2}\sqrt{\sigma_1\sigma_2}\left(\cosh\left(\phi_0 + \frac{\phi_+}{2} + \frac{\varphi_-}{2}\right) + \cosh\left(\varphi_0 + \frac{\varphi_+}{2} + \frac{\phi_-}{2}\right)\right)f_1 g_2 \\ &+ \frac{im}{2}\sqrt{\sigma_1\sigma_2}\left(\cosh\left(\phi_0 + \frac{\phi_+}{2} - \frac{\varphi_-}{2}\right) + \cosh\left(\varphi_0 + \frac{\varphi_+}{2} - \frac{\phi_-}{2}\right)\right)g_1 f_2, \end{aligned} \tag{31}$$

and

$$
\begin{aligned}
B_1^- &= -\frac{i}{2}\sqrt{\frac{m}{2}}\Bigg[e^{-\left(\frac{\phi_++\varphi_+}{4}-\frac{\phi_0+\varphi_0}{2}\right)}\left(\frac{1}{\sqrt{\sigma_1}}e^{-\left(\frac{\phi_-+\varphi_-}{4}\right)}g_1-\frac{1}{\sqrt{\sigma_2}}e^{\frac{\phi_-+\varphi_-}{4}}g_2\right) \\
&\quad + e^{\left(\frac{\phi_++\varphi_+}{4}-\frac{\phi_0+\varphi_0}{2}\right)}\left(\frac{1}{\sqrt{\sigma_1}}e^{\frac{\phi_-+\varphi_-}{4}}g_1-\frac{1}{\sqrt{\sigma_2}}e^{-\left(\frac{\phi_-+\varphi_-}{4}\right)}g_2\right)\Bigg](\bar{\psi}_+-\bar{\chi}_+) \\
&\quad -\frac{i}{2}\sqrt{\frac{m}{2}}\Bigg[e^{-\left(\frac{\phi_+-\varphi_+}{4}-\frac{\phi_0-\varphi_0}{2}\right)}\left(\frac{1}{\sqrt{\sigma_1}}e^{-\left(\frac{\phi_--\varphi_-}{4}\right)}f_1-\frac{1}{\sqrt{\sigma_2}}e^{\frac{\phi_--\varphi_-}{4}}f_2\right) \\
&\quad + e^{\left(\frac{\phi_+-\varphi_+}{4}-\frac{\phi_0-\varphi_0}{2}\right)}\left(\frac{1}{\sqrt{\sigma_1}}e^{\frac{\phi_--\varphi_-}{4}}f_1-\frac{1}{\sqrt{\sigma_2}}e^{-\left(\frac{\phi_--\varphi_-}{4}\right)}f_2\right)\Bigg](\bar{\psi}_++\bar{\chi}_+) \\
&\quad +\frac{im}{2\sqrt{\sigma_1\sigma_2}}\left(\cosh\left(\phi_0-\frac{\phi_+}{2}+\frac{\varphi_-}{2}\right)+\cosh\left(\varphi_0-\frac{\varphi_+}{2}+\frac{\phi_-}{2}\right)\right)f_1g_2 \\
&\quad +\frac{im}{2\sqrt{\sigma_1\sigma_2}}\left(\cosh\left(\phi_0-\frac{\phi_+}{2}-\frac{\varphi_-}{2}\right)+\cosh\left(\varphi_0-\frac{\varphi_+}{2}-\frac{\phi_-}{2}\right)\right)g_1f_2. \qquad (32)
\end{aligned}
$$

Finally the type-II defect Lagrangian density can be expressed as follows,

$$
\begin{aligned}
\mathcal{L}_D &= \frac{1}{2}(\phi_0\partial_t\phi_- - \phi_-\partial_t\phi_0) - \frac{1}{2}(\varphi_0\partial_t\varphi_- - \varphi_-\partial_t\varphi_0) + i(\bar{\psi}_1\bar{\psi}_2 - \psi_1\psi_2) - i(\bar{\chi}_1\bar{\chi}_2 - \chi_1\chi_2) \\
&\quad +\frac{i}{2}(f_1\partial_t g_1 + f_2\partial_t g_2 + g_1\partial_t f_1 + g_2\partial_t f_2) + B_0^+ + B_0^- + B_1^+ + B_1^-. \qquad (33)
\end{aligned}
$$

Then, the corresponding type-II defects conditions for the $\mathcal{N}=2$ sshG model at $x=0$ are,

$$
\begin{aligned}
\partial_x\phi_1 - \partial_t\phi_0 &= -\partial_{\phi_1}(B_0+B_1), & \partial_x\phi_2 - \partial_t\phi_0 &= \partial_{\phi_2}(B_0+B_1), \\
\partial_x\varphi_1 - \partial_t\varphi_0 &= \partial_{\varphi_1}(B_0+B_1), & \partial_x\varphi_2 - \partial_t\varphi_0 &= -\partial_{\varphi_2}(B_0+B_1), \\
\partial_t(\phi_1-\phi_2) &= -\partial_{\phi_0}(B_0+B_1), & \partial_t(\varphi_1-\varphi_2) &= \partial_{\varphi_0}(B_0+B_1), \\
i(\psi_1-\psi_2) &= -\partial_{\psi_1}B_1 = -\partial_{\psi_2}B_1, & i(\bar{\psi}_1-\bar{\psi}_2) &= \partial_{\bar{\psi}_1}B_1 = \partial_{\bar{\psi}_2}B_1, \qquad (34) \\
i(\chi_1-\chi_2) &= \partial_{\chi_1}B_1 = \partial_{\chi_2}B_1, & i(\bar{\chi}_1-\bar{\chi}_2) &= -\partial_{\bar{\chi}_1}B_1 = -\partial_{\bar{\chi}_2}B_1, \\
i\partial_t g_1 &= -\partial_{f_1}B_1, & i\partial_t f_1 &= -\partial_{g_1}B_1, \\
i\partial_t g_2 &= -\partial_{f_2}B_1, & i\partial_t f_2 &= -\partial_{g_2}B_1.
\end{aligned}
$$

The explicit form of the Bäcklund transformation for $\mathcal{N}=2$ sshG model is presented in appendix A.

5. Conservation of the momentum and energy

In this section, we will discuss the modified conserved momentum and energy. Let us consider first the total canonical momentum, which is given by the following contributions

$$P = \int_{-\infty}^{0} dx\,\mathcal{P}_1 + \int_{0}^{+\infty} dx\,\mathcal{P}_2, \qquad (35)$$

with

$$\mathcal{P}_p = \partial_t\phi_p\partial_x\phi_p - \partial_t\varphi_p\partial_x\varphi_p - i(\psi_p\partial_x\psi_p + \bar{\psi}_p\partial_x\bar{\psi}_p) + i(\chi_p\partial_x\chi_p + \bar{\chi}_p\partial_x\bar{\chi}_p), \quad p=1,2. \qquad (36)$$

Using the bulk equations (3), we can write the time derivative of momentum as

$$
\begin{aligned}
\frac{dP}{dt} &= \Bigg[\frac{1}{2}(\partial_x\phi_1)^2 + \frac{1}{2}(\partial_t\phi_1)^2 - \frac{1}{2}(\partial_x\varphi_1)^2 - \frac{1}{2}(\partial_t\varphi_1)^2 - i(\psi_1\partial_t\psi_1 + \bar{\psi}_1\partial_t\bar{\psi}_1) \\
&\quad + i(\chi_1\partial_t\chi_1 + \bar{\chi}_1\partial_t\bar{\chi}_1) - \frac{1}{2}(\partial_x\phi_2)^2 - \frac{1}{2}(\partial_t\phi_2)^2 + \frac{1}{2}(\partial_x\varphi_2)^2 + \frac{1}{2}(\partial_t\varphi_2)^2 \\
&\quad + i(\psi_2\partial_t\psi_2 + \bar{\psi}_2\partial_t\bar{\psi}_2) - i(\chi_2\partial_t\chi_2 + \bar{\chi}_2\partial_t\bar{\chi}_2) - V_1 + V_2 - W_1 + W_2\Bigg]_{x=0}. \qquad (37)
\end{aligned}
$$

Now, from the explicit form of the defect potentials $B_0 = B_0^+ + B_0^-$, $B_1 = B_1^+ + B_1^-$ given in eqs. (31)–(23), and the defect conditions (34), we find the following set of relations,

$$\begin{aligned} \partial_{\psi_-} B_1 &= \partial_{\bar{\psi}_-} B_1 = \partial_{\chi_-} B_1 = \partial_{\bar{\chi}_-} B_1 = 0, \\ \partial_{\psi_+} B_1^- &= \partial_{\bar{\psi}_+} B_1^+ = \partial_{\chi_+} B_1^- = \partial_{\bar{\chi}_+} B_1^+ = 0, \end{aligned} \tag{38}$$

and

$$\begin{aligned} \partial_{\phi_0} B_0^+ &= 2\partial_{\phi_+} B_0^+, & \partial_{\phi_0} B_1^+ &= 2\partial_{\phi_+} B_1^+, \\ \partial_{\phi_0} B_0^- &= -2\partial_{\phi_+} B_0^-, & \partial_{\phi_0} B_1^- &= -2\partial_{\phi_+} B_1^-, \\ \partial_{\varphi_0} B_0^+ &= 2\partial_{\varphi_+} B_0^+, & \partial_{\varphi_0} B_1^+ &= 2\partial_{\varphi_+} B_1^+, \\ \partial_{\varphi_0} B_0^- &= -2\partial_{\varphi_+} B_0^-, & \partial_{\varphi_0} B_1^- &= -2\partial_{\varphi_+} B_1^-. \end{aligned} \tag{39}$$

Then, by using the above relations and the defect conditions (34), the equation (37) takes the following form,

$$\begin{aligned} \frac{dP}{dt} = &\Big[2\partial_{\phi_+} B_0 \partial_{\phi_-} B_0 - 2\partial_{\varphi_+} B_0 \partial_{\varphi_-} B_0 + 2\partial_{\phi_+} B_1 \partial_{\phi_-} B_1 - 2\partial_{\varphi_+} B_1 \partial_{\varphi_-} B_1 + 2\partial_{\phi_-} B_0 \partial_{\phi_+} B_1 \\ &+2\partial_{\phi_+} B_0 \partial_{\phi_-} B_1 - 2\partial_{\varphi_-} B_0 \partial_{\varphi_+} B_1 - 2\partial_{\varphi_+} B_0 \partial_{\varphi_-} B_1 - 2\partial_t \phi_0 \partial_{\phi_+}(B_0 + B_1) \\ &-2\partial_t \varphi_0 \partial_{\varphi_+}(B_0 + B_1) - \frac{1}{2}\partial_t \phi_+ \partial_{\phi_0}(B_0 + B_1) - \frac{1}{2}\partial_t \varphi_+ \partial_{\varphi_0}(B_0 + B_1) \\ &-\partial_t \psi_+ \partial_{\psi_+} B_1 + \partial_t \bar{\psi}_+ \partial_{\bar{\psi}_+} B_1 - \partial_t \chi_+ \partial_{\chi_+} B_1 + \partial_t \bar{\chi}_+ \partial_{\bar{\chi}_+} B_1 - V_1 + V_2 - W_1 + W_2 \\ &+i\partial_t(\psi_1\psi_2 + \bar{\psi}_1\bar{\psi}_2 - \chi_1\chi_2 - \bar{\chi}_1\bar{\chi}_2)\Big]_{x=0}, \end{aligned} \tag{40}$$

where the right-hand-side of the above equation becomes a total time derivative since the defect potentials $B_0^{\pm}$ and $B_1^{\pm}$ satisfy the following Poisson-bracket-like relations,

$$V_1 - V_2 = 2(\partial_{\phi_0} B_0^+ \partial_{\phi_-} B_0^- - \partial_{\phi_0} B_0^- \partial_{\phi_-} B_0^+ - \partial_{\varphi_0} B_0^+ \partial_{\varphi_-} B_0^- + \partial_{\varphi_0} B_0^- \partial_{\varphi_-} B_0^+), \tag{41}$$

$$\begin{aligned} W_1 - W_2 = &\ 2(\partial_{\phi_0} B_1^+ \partial_{\phi_-} B_0^- - \partial_{\phi_0} B_1^- \partial_{\phi_-} B_0^+ + \partial_{\phi_0} B_0^+ \partial_{\phi_-} B_1^- - \partial_{\phi_0} B_0^- \partial_{\phi_-} B_1^+) \\ &-2(\partial_{\varphi_0} B_1^+ \partial_{\varphi_-} B_0^- - \partial_{\varphi_0} B_1^- \partial_{\varphi_-} B_0^+ + \partial_{\varphi_0} B_0^+ \partial_{\varphi_-} B_1^- - \partial_{\varphi_0} B_0^- \partial_{\varphi_-} B_1^+) \\ &+2i(\partial_{f_1} B_1^- \partial_{g_1} B_1^+ - \partial_{f_1} B_1^+ \partial_{g_1} B_1^- + \partial_{f_2} B_1^- \partial_{g_2} B_1^+ - \partial_{f_2} B_1^+ \partial_{g_2} B_1^-), \end{aligned} \tag{42}$$

together with the constraint,

$$\partial_{\phi_0} B_1^+ \partial_{\phi_-} B_1^- - \partial_{\phi_0} B_1^- \partial_{\phi_-} B_1^+ - \partial_{\varphi_0} B_1^+ \partial_{\varphi_-} B_1^- + \partial_{\varphi_0} B_1^- \partial_{\varphi_-} B_1^+ = 0. \tag{43}$$

Then, we get that the modified conserved momentum can be written in a simple form,

$$\mathcal{P} = P + \Big[B_1^+ + B_0^+ - B_1^- - B_0^- + i(-\psi_1\psi_2 - \bar{\psi}_1\bar{\psi}_2 + \chi_1\chi_2 + \bar{\chi}_1\bar{\chi}_2)\Big]_{x=0}. \tag{44}$$

Now, let us consider the total energy

$$E = \int_{-\infty}^{0} dx\, \mathcal{E}_1 + \int_{0}^{+\infty} dx\, \mathcal{E}_2, \tag{45}$$

where

$$\begin{aligned} \mathcal{E}_p = &\ \frac{1}{2}(\partial_x \phi_p)^2 + \frac{1}{2}(\partial_t \phi_p)^2 - \frac{1}{2}(\partial_x \varphi_p)^2 - \frac{1}{2}(\partial_t \varphi_p)^2 - i(\psi_p \partial_x \psi_p - \bar{\psi}_p \partial_x \bar{\psi}_p) \\ &+i(\chi_p \partial_x \chi_p - \bar{\chi}_p \partial_x \bar{\chi}_p) + V_p + W_p. \end{aligned} \tag{46}$$

We can find its time-derivative in the same way as before by using the bulk equations (3). The result reads

$$\begin{aligned}\frac{dE}{dt} &= \Big[\partial_t\phi_1\partial_x\phi_1 - \partial_t\varphi_1\partial_x\varphi_1 - i(\psi_1\partial_t\psi_1 - \bar{\psi}_1\partial_t\bar{\psi}_1) + i(\chi_1\partial_t\chi_1 - \bar{\chi}_1\partial_t\bar{\chi}_1) \\ &\quad -\partial_t\phi_2\partial_x\phi_2 + \partial_t\varphi_2\partial_x\varphi_2 + i(\psi_2\partial_t\psi_2 - \bar{\psi}_2\partial_t\bar{\psi}_2) - i(\chi_2\partial_t\chi_2 - \bar{\chi}_2\partial_t\bar{\chi}_2)\Big]_{x=0}. \qquad (47)\end{aligned}$$

Then using the defect conditions (34) and the defect potentials (31)–(23), we find that the modified conserved energy is given by

$$\mathcal{E} = E + \Big[B_0 + B_1 + i(\bar{\psi}_1\bar{\psi}_2 - \psi_1\psi_2 + \chi_1\chi_2 - \bar{\chi}_1\bar{\chi}_2)\Big]_{x=0}. \qquad (48)$$

6. Modified conserved supercharges

We have seen that the bulk theory action is invariant under susy transformation (4), and it was explicitly shown for δ_1 projection. However, this is not necessarily true for the defect theory, and therefore we should show that the presence of the defect will not destroy the supersymmetry of the bulk theory. Let us compute the defect contribution for Q_1. By introducing the defect at $x = 0$, we have

$$\begin{aligned}Q_1 &= \int_{-\infty}^{0} dx\Big[2i\psi_1\partial_+\phi_1 + 2i\chi_1\partial_+\varphi_1 - 2im\bar{\psi}_1\sinh\phi_1\cosh\varphi_1 + 2im\bar{\chi}_1\sinh\varphi_1\cosh\phi_1\Big] \\ &\quad + \int_{0}^{\infty} dx\Big[2i\psi_2\partial_+\phi_2 + i\chi_2\partial_+\varphi_2 - 2im\bar{\psi}_2\sinh\phi_2\cosh\varphi_2 + 2im\bar{\chi}_2\sinh\varphi_2\cosh\phi_2\Big] \qquad (49)\end{aligned}$$

Now, by taking the time-derivative respectively, we get

$$\begin{aligned}\frac{dQ_1}{dt} &= \Big[2i\psi_1\partial_+\phi_1 + 2i\chi_1\partial_+\varphi_1 + 2im\bar{\psi}_1\sinh\phi_1\cosh\varphi_1 - 2im\bar{\chi}_1\sinh\varphi_1\cosh\phi_1\Big]_{x=0} \\ &\quad -\Big[2i\psi_2\partial_+\phi_2 + 2i\chi_2\partial_+\varphi_2 + 2im\bar{\psi}_2\sinh\phi_2\cosh\varphi_2 - 2im\bar{\chi}_2\sinh\varphi_2\cosh\phi_2\Big]_{x=0}. \qquad (50)\end{aligned}$$

Using the defect conditions (34), we get

$$\begin{aligned}\frac{dQ_1}{dt} &= \Bigg[\; i\psi_-\partial_t\left(\frac{\phi_+}{2} + \phi_0\right) - i\psi_-\,\partial_{\phi_-}(B_0 + B_1) - i\psi_+\,\partial_{\phi_0}(B_0^+ + B_1^+) \\ &\quad +i\chi_-\,\partial_t\left(\frac{\varphi_+}{2} + \varphi_0\right) + i\chi_-\partial_{\varphi_-}(B_0 + B_1) + i\chi_+\,\partial_{\varphi_0}(B_0^+ + B_1^+) \\ &\quad -im(\bar{\psi}_+ + \bar{\psi}_-)\sinh\left(\frac{\phi_+ + \phi_-}{2}\right)\cosh\left(\frac{\varphi_+ + \varphi_-}{2}\right) \\ &\quad +im(\bar{\psi}_+ - \bar{\psi}_-)\sinh\left(\frac{\phi_+ - \phi_-}{2}\right)\cosh\left(\frac{\varphi_+ - \varphi_-}{2}\right) \\ &\quad +im(\bar{\chi}_+ + \bar{\chi}_-)\sinh\left(\frac{\varphi_+ + \varphi_-}{2}\right)\cosh\left(\frac{\phi_+ + \phi_-}{2}\right) \\ &\quad -im(\bar{\chi}_+ - \bar{\chi}_-)\sinh\left(\frac{\varphi_+ - \varphi_-}{2}\right)\cosh\left(\frac{\phi_+ - \phi_-}{2}\right)\Bigg]_{x=0}. \qquad (51)\end{aligned}$$

Now, by making use of the defect conditions intensively, we find after some algebra that the right-hand-side of the equation becomes a total time-derivative, and then the modified conserved

Journal of Physics: Conference Series **670** (2016) 012049 doi:10.1088/1742-6596/670/1/012049

supercharge can be written in $\mathcal{Q}_1 = Q_1 + Q_{D_1}$, with the defect contribution given by the following expression,

$$Q_{D_1} = \sum_{k=1}^{2} -i\sqrt{2m\sigma_k}\left(u_k^+ f_k + u_k^- g_k\right)_{x=0}. \tag{52}$$

where we have introduced the function,

$$u_k^{\pm} = \sinh\left(\frac{(\phi_k + \phi_0) \pm (\varphi_k + \varphi_0)}{2}\right). \tag{53}$$

Analogously, we can find the remaining modified conserved supercharges,

$$\mathcal{Q}_2 = Q_2 + Q_{D_2}, \qquad \overline{\mathcal{Q}}_1 = \overline{Q}_1 + \overline{Q}_{D_1}, \qquad \overline{\mathcal{Q}}_2 = \overline{Q}_2 + \overline{Q}_{D_2}, \tag{54}$$

with the corresponding defect contributions given by,

$$Q_{D_2} = \sum_{k=1}^{2} -i\sqrt{2m\sigma_k}\left(u_k^+ f_k - u_k^- g_k\right)_{x=0}, \tag{55}$$

$$\overline{Q}_{D_1} = \sum_{k=1}^{2} \frac{i\sqrt{2m}(-1)^k}{\sqrt{\sigma_k}}\left(v_k^- f_k + v_k^+ g_k\right)_{x=0}, \tag{56}$$

$$\overline{Q}_{D_2} = \sum_{k=1}^{2} \frac{i\sqrt{2m}(-1)^{k-1}}{\sqrt{\sigma_k}}\left(v_k^- f_k - v_k^+ g_k\right)_{x=0}, \tag{57}$$

where the functions $v_k^{\pm}$ are defined to be,

$$v_k^{\pm} = \sinh\left(\frac{(\phi_k - \phi_0) \pm (\varphi_k - \varphi_0)}{2}\right). \tag{58}$$

The derivation of the exact form of the all modified conserved, together with the modified conserved energy and momentum, provides a strong evidence indicating the classical integrability of the fused defect for the $\mathcal{N} = 2$ sshG model. A more rigorous analysis should require the derivation of the generating function of an infinite set of modified conserved quantities. That can be done following the on-shell Lax approach to derive the corresponding type-II defect $\mathcal{K}$-matrix for the model. From its explicit form is possible to derive an infinite set of modified conserved quantities. Alternative approaches can also be used in order to prove the involutivity of the charges, for instance the off-shell r-matrix and the multisymplectic approach as well. Some of these issues will be considered in future investigations.

7. PT symmetry

First of all, it can be shown that the bulk Lagrangian density and fields equations are invariant under the simultaneous transformations of parity transformation (**P**), and time reversal (**T**), namely $(x,t) \to (-x,-t)$, if the fields transform as follows,

$$\begin{aligned} \phi(x,t) &\longrightarrow \phi(-x,-t), & \varphi(x,t) &\longrightarrow \varphi(-x,-t), \\ \psi(x,t) &\longleftrightarrow -\chi(-x,-t), & \bar{\psi}(x,t) &\longleftrightarrow -\bar{\chi}(-x,-t), \end{aligned}$$

In addition, we notice that applying this **PT** transformation, the ϵ_1-projection of the susy transformation maps to the ϵ_2-projection. Then, as it can be verified, by applying the **PT**

Journal of Physics: Conference Series **670** (2016) 012049 doi:10.1088/1742-6596/670/1/012049

transformation over Q_1 we get the second supercharge, namely $Q_2 = \mathbf{PT}Q_1$. Analogously, it happens with $\overline{Q}_2 = \mathbf{PT}\overline{Q}_1$.

Now, in the presence of a type-I defect the **PT** transformation relates the fields on the left-hand side to the ones on the right-hand side, and conversely. Then, to preserve the invariance under PT symmetry the fields in the respective bulk Lagrangian densities $\mathcal{L}_p$ should transform in the following way,

$$\begin{aligned} \phi_1(x,t) &\longleftrightarrow \phi_2(-x,-t), & \varphi_1(x,t) &\longleftrightarrow \varphi_2(-x,-t), \\ \psi_1(x,t) &\longleftrightarrow -\chi_2(-x,-t), & \bar{\psi}_1(x,t) &\longleftrightarrow -\bar{\chi}_2(-x,-t). \end{aligned}$$

Consequently, the auxiliary fermionic fields f, g in the defect Lagrangian should transform as,

$$f(t) \longrightarrow f(-t), \qquad g(t) \longrightarrow -g(-t).$$

Under these field transformations it can be shown that the type-I defect equations are invariant. On the other hand, the type-II defect Lagrangian density for the $\mathcal{N} = 2$ sshG model is invariant under **PT** transformation, if the corresponding auxiliary fields transform in the following way,

$$\begin{aligned} \phi_0(t) &\longleftrightarrow \phi_0(-t), & \varphi_0(t) &\longleftrightarrow \varphi_0(-t), \\ f_1(t) &\longleftrightarrow f_2(-t), & g_1(t) &\longleftrightarrow -g_2(-t). \end{aligned} \tag{59}$$

In this case, we can verified that if this **PT** transformation is applied over Q_{D_1} we obtain the defect contribution to the second supercharge, namely $Q_{D_2} = \mathbf{PT}Q_{D_1}$. The same is valid for $\overline{Q}_{D_2} = \mathbf{PT}\overline{Q}_{D_1}$.

The invariance of the $\mathcal{N} = 2$ sshG model under **PT** symmetry is strongly related with the description of the equation of motion in the superspace formalism. In such language, the fields appear as components of two $\mathcal{N} = 2$ superfields, one of them being a chiral superfield, while the other one is anti-chiral (For more details see [12]). The fact that the **PT** symmetry is preserved in the presence of the type-II defect, somehow suggests the possibility of describing the defect conditions in terms of superfields. In other words, there should exists a type-II Bäcklund transformation for the $\mathcal{N} = 2$ sshG equation consistent with the defect conditions of the fused defect. As it was shown for the $\mathcal{N} = 1$ sshG equations, the type-II defect conditions are equivalent to "frozen" type-II Bäcklund transformation of the model (see appendix A).

8. Final Remarks

In this paper, we have derived a type-II integrable defect for the $\mathcal{N} = 2$ sshG model by using the fusing procedure. At the Lagrangian level, we have shown that the type-II defect for this supersymmetric model can be also obtained by fusing two type-I defects located initially at different points in the x-axis. We have shown the conservation of the modified quantities of the energy, momentum and supercharges. Moreover, the invariance under **PT** symmetry was verified.

From the results obtained in this paper and those previously found in [9], it would be interesting to explore the possibility of finding new integrable boundary conditions for the $\mathcal{N} = 1$ and $\mathcal{N} = 2$ sshG models, by performing a consistent half-line limit, specially following the reasoning of [13]–[15].

There are many others algebraic aspects related to type-II defects that have not been addressed in this work, like the Lax representation, the involutivity of the charges via the r-matrix approach, and the construction of the soliton solutions. These issues are expected to be addressed in future investigations.

Acknowledgments
The authors are very grateful to the organisers of the XXIIIth International Conference on Integrable Systems and Quantum Symmetries (ISQS-23) for the opportunity to present this work. NIS, JFG and AHZ would like to thank CNPq-Brasil for financial support. ARA is supported by FAPESP grant 2012/13866-3.

Appendix A. Type-II Bäcklund transformations for $\mathcal{N}=2$ sshG model

$$\psi_- = \sqrt{2m}\left[\sqrt{\sigma_1}(\partial_{\phi_0}u_1^+ f_1 + \partial_{\phi_0}u_1^- g_1) + \sqrt{\sigma_2}(\partial_{\phi_0}u_2^+ f_2 + \partial_{\phi_0}u_2^- g_2)\right] \tag{A.1}$$

$$\bar{\psi}_- = -\sqrt{2m}\left[\frac{1}{\sqrt{\sigma_1}}(\partial_{\phi_0}v_1^+ g_1 + \partial_{\phi_0}v_1^- f_1) - \frac{1}{\sqrt{\sigma_2}}(\partial_{\phi_0}v_2^+ g_2 + \partial_{\phi_0}v_2^- f_2)\right] \tag{A.2}$$

$$\chi_- = \sqrt{2m}\left[\sqrt{\sigma_1}(\partial_{\phi_0}u_1^+ f_1 - \partial_{\phi_0}u_1^- g_1) + \sqrt{\sigma_2}(\partial_{\phi_0}u_2^+ f_2 - \partial_{\phi_0}u_2^- g_2)\right] \tag{A.3}$$

$$\bar{\chi}_- = -\sqrt{2m}\left[\frac{1}{\sqrt{\sigma_1}}(\partial_{\phi_0}v_1^+ g_1 - \partial_{\phi_0}v_1^- f_1) - \frac{1}{\sqrt{\sigma_2}}(\partial_{\phi_0}v_2^+ g_2 - \partial_{\phi_0}v_2^- f_2)\right] \tag{A.4}$$

$$\begin{aligned}\partial_- g_1 &= \frac{m}{2\sqrt{\sigma_1\sigma_2}}\Big(\cosh\Big(\phi_0 - \frac{\phi_+}{2} + \frac{\varphi_-}{2}\Big) + \cosh\Big(\varphi_0 - \frac{\varphi_+}{2} + \frac{\phi_-}{2}\Big)\Big)g_2 \\ &+ \sqrt{\frac{2m}{\sigma_1}}\partial_{\phi_0}v_1^-(\bar{\psi}_+ + \bar{\chi}_+)\end{aligned} \tag{A.5}$$

$$\begin{aligned}\partial_+ g_1 &= -\frac{m}{2}\sqrt{\sigma_1\sigma_2}\Big(\cosh\Big(\phi_0 + \frac{\phi_+}{2} + \frac{\varphi_-}{2}\Big) + \cosh\Big(\varphi_0 + \frac{\varphi_+}{2} + \frac{\phi_-}{2}\Big)\Big)g_2 \\ &- \sqrt{2m\sigma_1}\partial_{\phi_0}u_1^+(\psi_+ - \chi_+)\end{aligned} \tag{A.6}$$

$$\begin{aligned}\partial_- f_1 &= \frac{m}{2\sqrt{\sigma_1\sigma_2}}\Big(\cosh\Big(\phi_0 - \frac{\phi_+}{2} - \frac{\varphi_-}{2}\Big) + \cosh\Big(\varphi_0 - \frac{\varphi_+}{2} - \frac{\phi_-}{2}\Big)\Big)f_2 \\ &+ \sqrt{\frac{2m}{\sigma_1}}\partial_{\phi_0}v_1^+(\bar{\psi}_+ - \bar{\chi}_+)\end{aligned} \tag{A.7}$$

$$\begin{aligned}\partial_+ f_1 &= -\frac{m}{2}\sqrt{\sigma_1\sigma_2}\Big(\cosh\Big(\phi_0 + \frac{\phi_+}{2} - \frac{\varphi_-}{2}\Big) + \cosh\Big(\varphi_0 + \frac{\varphi_+}{2} - \frac{\phi_-}{2}\Big)\Big)f_2 \\ &- \sqrt{2m\sigma_1}\partial_{\phi_0}u_1^-(\psi_+ + \chi_+)\end{aligned} \tag{A.8}$$

$$\begin{aligned}\partial_- g_2 &= -\frac{m}{2\sqrt{\sigma_1\sigma_2}}\Big(\cosh\Big(\phi_0 - \frac{\phi_+}{2} - \frac{\varphi_-}{2}\Big) + \cosh\Big(\varphi_0 - \frac{\varphi_+}{2} - \frac{\phi_-}{2}\Big)\Big)g_1 \\ &- \sqrt{\frac{2m}{\sigma_2}}\partial_{\phi_0}v_2^-(\bar{\psi}_+ + \bar{\chi}_+)\end{aligned} \tag{A.9}$$

$$\begin{aligned}\partial_+ g_2 &= \frac{m}{2}\sqrt{\sigma_1\sigma_2}\Big(\cosh\Big(\phi_0 + \frac{\phi_+}{2} - \frac{\varphi_-}{2}\Big) + \cosh\Big(\varphi_0 + \frac{\varphi_+}{2} - \frac{\phi_-}{2}\Big)\Big)g_1 \\ &- \sqrt{2m\sigma_2}\partial_{\phi_0}u_2^+(\psi_+ - \chi_+)\end{aligned} \tag{A.10}$$

$$\begin{aligned}\partial_- f_2 &= -\frac{m}{2\sqrt{\sigma_1\sigma_2}}\Big(\cosh\Big(\phi_0 - \frac{\phi_+}{2} + \frac{\varphi_-}{2}\Big) + \cosh\Big(\varphi_0 - \frac{\varphi_+}{2} + \frac{\phi_-}{2}\Big)\Big)f_1 \\ &- \sqrt{\frac{2m}{\sigma_2}}\partial_{\phi_0}v_2^+(\bar{\psi}_+ - \bar{\chi}_+)\end{aligned} \tag{A.11}$$

$$\begin{aligned}\partial_+ f_2 &= \frac{m}{2}\sqrt{\sigma_1\sigma_2}\Big(\cosh\Big(\phi_0 + \frac{\phi_+}{2} + \frac{\varphi_-}{2}\Big) + \cosh\Big(\varphi_0 + \frac{\varphi_+}{2} + \frac{\phi_-}{2}\Big)\Big)f_1 \\ &- \sqrt{2m\sigma_2}\partial_{\phi_0}u_2^-(\psi_+ + \chi_+)\end{aligned} \tag{A.12}$$

Journal of Physics: Conference Series **670** (2016) 012049 doi:10.1088/1742-6596/670/1/012049

$$
\begin{aligned}
\partial_+\phi_- &= -\frac{m}{2}\left[e^{\left(\frac{\phi_+}{2}+\phi_0\right)}\left(\sigma_1 e^{\frac{\phi_-}{2}}+\sigma_2 e^{-\frac{\phi_-}{2}}\right)-e^{-\left(\frac{\phi_+}{2}+\phi_0\right)}\left(\sigma_1 e^{-\frac{\phi_-}{2}}+\sigma_2 e^{\frac{\phi_-}{2}}\right)\right] \\
&- \frac{i}{2}\sqrt{\frac{m}{2}}\left[(\sqrt{\sigma_1}u_1^+ f_1+\sqrt{\sigma_2}u_2^+ f_2)(\psi_+-\chi_+)+(\sqrt{\sigma_1}u_1^- g_1+\sqrt{\sigma_2}u_2^- g_2)(\psi_++\chi_+)\right] \\
&- \frac{im}{2}\sqrt{\sigma_1\sigma_2}\left[\sinh\left(\phi_0+\frac{\phi_+}{2}+\frac{\varphi_-}{2}\right)f_1g_2+\sinh\left(\phi_0+\frac{\phi_+}{2}-\frac{\varphi_-}{2}\right)g_1f_2\right] \qquad \text{(A.13)}\\
\partial_-\phi_- &= -\frac{m}{2}\left[e^{\left(\frac{\phi_+}{2}-\phi_0\right)}\left(\frac{1}{\sigma_1}e^{\frac{\phi_-}{2}}+\frac{1}{\sigma_2}e^{-\frac{\phi_-}{2}}\right)-e^{-\left(\frac{\phi_+}{2}-\phi_0\right)}\left(\frac{1}{\sigma_1}e^{-\frac{\phi_-}{2}}+\frac{1}{\sigma_2}e^{\frac{\phi_-}{2}}\right)\right] \\
&+ \frac{i}{2}\sqrt{\frac{m}{2}}\left[\left(\frac{1}{\sqrt{\sigma_1}}v_1^+ g_1-\frac{1}{\sqrt{\sigma_2}}v_2^+ g_2\right)(\bar\psi_+-\bar\chi_+)+\left(\frac{1}{\sqrt{\sigma_1}}v_1^- f_1-\frac{1}{\sqrt{\sigma_2}}v_2^- f_2\right)(\bar\psi_++\bar\chi_+)\right] \\
&+ \frac{im}{2\sqrt{\sigma_1\sigma_2}}\left[\sinh\left(\phi_0-\frac{\phi_+}{2}+\frac{\varphi_-}{2}\right)f_1g_2+\sinh\left(\phi_0-\frac{\phi_+}{2}-\frac{\varphi_-}{2}\right)g_1f_2\right] \qquad \text{(A.14)}\\
\partial_-(\phi_++2\phi_0) &= -\frac{m}{2}\left[e^{\left(\frac{\phi_+}{2}-\phi_0\right)}\left(\frac{1}{\sigma_1}e^{\frac{\phi_-}{2}}-\frac{1}{\sigma_2}e^{-\frac{\phi_-}{2}}\right)-e^{-\left(\frac{\phi_+}{2}-\phi_0\right)}\left(\frac{1}{\sigma_1}e^{-\frac{\phi_-}{2}}-\frac{1}{\sigma_2}e^{\frac{\phi_-}{2}}\right)\right] \\
&+ \frac{i}{2}\sqrt{\frac{m}{2}}\left[\left(\frac{1}{\sqrt{\sigma_1}}v_1^+ g_1+\frac{1}{\sqrt{\sigma_2}}v_2^+ g_2\right)(\bar\psi_+-\bar\chi_+)+\left(\frac{1}{\sqrt{\sigma_1}}v_1^- f_1+\frac{1}{\sqrt{\sigma_2}}v_2^- f_2\right)(\bar\psi_++\bar\chi_+)\right] \\
&- \frac{im}{2\sqrt{\sigma_1\sigma_2}}\left[\sinh\left(\varphi_0-\frac{\varphi_+}{2}+\frac{\phi_-}{2}\right)f_1g_2-\sinh\left(\varphi_0-\frac{\varphi_+}{2}-\frac{\phi_-}{2}\right)g_1f_2\right] \qquad \text{(A.15)}\\
\partial_+(\phi_+-2\phi_0) &= -\frac{m}{2}\left[e^{\left(\frac{\phi_+}{2}+\phi_0\right)}\left(\sigma_1 e^{\frac{\phi_-}{2}}-\sigma_2 e^{-\frac{\phi_-}{2}}\right)-e^{-\left(\frac{\phi_+}{2}+\phi_0\right)}\left(\sigma_1 e^{-\frac{\phi_-}{2}}-\sigma_2 e^{\frac{\phi_-}{2}}\right)\right] \\
&- \frac{i}{2}\sqrt{\frac{m}{2}}\left[(\sqrt{\sigma_1}u_1^+ f_1-\sqrt{\sigma_2}u_2^+ f_2)(\psi_+-\chi_+)+(\sqrt{\sigma_1}u_1^- g_1-\sqrt{\sigma_2}u_2^- g_2)(\psi_++\chi_+)\right] \\
&- \frac{im}{2}\sqrt{\sigma_1\sigma_2}\left[\sinh\left(\varphi_0+\frac{\varphi_+}{2}+\frac{\phi_-}{2}\right)f_1g_2-\sinh\left(\varphi_0+\frac{\varphi_+}{2}-\frac{\phi_-}{2}\right)g_1f_2\right] \qquad \text{(A.16)}\\
\partial_+\varphi_- &= -\frac{m}{2}\left[e^{\left(\frac{\varphi_+}{2}+\varphi_0\right)}\left(\sigma_1 e^{\frac{\varphi_-}{2}}+\sigma_2 e^{-\frac{\varphi_-}{2}}\right)-e^{-\left(\frac{\varphi_+}{2}+\varphi_0\right)}\left(\sigma_1 e^{-\frac{\varphi_-}{2}}+\sigma_2 e^{\frac{\varphi_-}{2}}\right)\right] \\
&+ \frac{i}{2}\sqrt{\frac{m}{2}}\left[(\sqrt{\sigma_1}u_1^+ f_1+\sqrt{\sigma_2}u_2^+ f_2)(\psi_+-\chi_+)-(\sqrt{\sigma_1}u_1^- g_1+\sqrt{\sigma_2}u_2^- g_2)(\psi_++\chi_+)\right] \\
&+ \frac{im}{2}\sqrt{\sigma_1\sigma_2}\left[\sinh\left(\varphi_0+\frac{\varphi_+}{2}+\frac{\phi_-}{2}\right)f_1g_2+\sinh\left(\varphi_0+\frac{\varphi_+}{2}-\frac{\phi_-}{2}\right)g_1f_2\right] \qquad \text{(A.17)}\\
\partial_-\varphi_- &= -\frac{m}{2}\left[e^{\left(\frac{\varphi_+}{2}-\varphi_0\right)}\left(\frac{1}{\sigma_1}e^{\frac{\varphi_-}{2}}+\frac{1}{\sigma_2}e^{-\frac{\varphi_-}{2}}\right)-e^{-\left(\frac{\varphi_+}{2}-\varphi_0\right)}\left(\frac{1}{\sigma_1}e^{-\frac{\varphi_-}{2}}+\frac{1}{\sigma_2}e^{\frac{\varphi_-}{2}}\right)\right] \\
&- \frac{i}{2}\sqrt{\frac{m}{2}}\left[\left(\frac{1}{\sqrt{\sigma_1}}v_1^+ g_1-\frac{1}{\sqrt{\sigma_2}}v_2^+ g_2\right)(\bar\psi_+-\bar\chi_+)-\left(\frac{1}{\sqrt{\sigma_1}}v_1^- f_1-\frac{1}{\sqrt{\sigma_2}}v_2^- f_2\right)(\bar\psi_++\bar\chi_+)\right] \\
&- \frac{im}{2\sqrt{\sigma_1\sigma_2}}\left[\sinh\left(\varphi_0-\frac{\varphi_+}{2}+\frac{\phi_-}{2}\right)f_1g_2+\sinh\left(\varphi_0-\frac{\varphi_+}{2}-\frac{\phi_-}{2}\right)g_1f_2\right] \qquad \text{(A.18)}\\
\partial_-(\varphi_++2\varphi_0) &= -\frac{m}{2}\left[e^{\left(\frac{\varphi_+}{2}-\varphi_0\right)}\left(\frac{1}{\sigma_1}e^{\frac{\varphi_-}{2}}-\frac{1}{\sigma_2}e^{-\frac{\varphi_-}{2}}\right)-e^{-\left(\frac{\varphi_+}{2}-\varphi_0\right)}\left(\frac{1}{\sigma_1}e^{-\frac{\varphi_-}{2}}-\frac{1}{\sigma_2}e^{\frac{\varphi_-}{2}}\right)\right] \\
&- \frac{i}{2}\sqrt{\frac{m}{2}}\left[\left(\frac{1}{\sqrt{\sigma_1}}v_1^+ g_1+\frac{1}{\sqrt{\sigma_2}}v_2^+ g_2\right)(\bar\psi_+-\bar\chi_+)-\left(\frac{1}{\sqrt{\sigma_1}}v_1^- f_1+\frac{1}{\sqrt{\sigma_2}}v_2^- f_2\right)(\bar\psi_++\bar\chi_+)\right] \\
&+ \frac{im}{2\sqrt{\sigma_1\sigma_2}}\left[\sinh\left(\phi_0-\frac{\phi_+}{2}+\frac{\varphi_-}{2}\right)f_1g_2-\sinh\left(\phi_0-\frac{\phi_+}{2}-\frac{\varphi_-}{2}\right)g_1f_2\right] \qquad \text{(A.19)}
\end{aligned}
$$

$$\begin{aligned}\partial_+(\varphi_+ - 2\varphi_0) &= -\frac{m}{2}\left[e^{\left(\frac{\varphi_+}{2}+\varphi_0\right)}\left(\sigma_1 e^{\frac{\varphi_-}{2}} - \sigma_2 e^{-\frac{\varphi_-}{2}}\right) - e^{-\left(\frac{\varphi_+}{2}+\varphi_0\right)}\left(\sigma_1 e^{-\frac{\varphi_-}{2}} - \sigma_2 e^{\frac{\varphi_-}{2}}\right)\right] \\ &+ \frac{i}{2}\sqrt{\frac{m}{2}}\left[(\sqrt{\sigma_1}u_1^+ f_1 - \sqrt{\sigma_2}u_2^+ f_2)(\psi_+ - \chi_+) - (\sqrt{\sigma_1}u_1^- g_1 - \sqrt{\sigma_2}u_2^- g_2)(\psi_+ + \chi_+)\right] \\ &+ \frac{im}{2}\sqrt{\sigma_1\sigma_2}\left[\sinh\left(\phi_0 + \frac{\phi_+}{2} + \frac{\varphi_-}{2}\right) f_1 g_2 - \sinh\left(\phi_0 + \frac{\phi_+}{2} - \frac{\varphi_-}{2}\right) g_1 f_2\right] \qquad (A.20)\end{aligned}$$

It was verified that these Bäcklund transformations correspond to the equations (3) for each fields.

References

[1] P. Bowcock, E. Corrigan and C. Zambon, *Classically integrable field theories with defects*, *Int. J. Mod. Phys.* **A19** (2004) 82 [hep-th/0305022].

[2] P. Bowcock, E. Corrigan and C. Zambon, *Affine Toda field theories with defects*, *JHEP* **01** (2004) 056 [hep-th/0401020].

[3] E. Corrigan and C. Zambon, *Jump-defects in the nonlinear Schrödinger model and other non-relativistic field theories*, *Nonlinearity* **19** (2006) 1447 [nlin/0512038].

[4] P. Bowcock and J. M. Umpleby, *Quantum complex sine-Gordon dressed boundaries*, *JHEP* **38** (2008) 0811 [hep-th/0809.0661]

[5] E. Corrigan and C. Zambon, *A new class of integrable defects*, *J. Phys.* **A42** (2009) 475203 [hep-th/0908.3126].

[6] E. Corrigan and C. Zambon, *Integrable defects in affine Toda field theory and infinite dimensional representations of quantum groups*, *Nucl. Phys* **B 848** (2011) 545 [arXiv:1012.4186].

[7] E. Corrigan and C. Zambon, *A transmission matrix for a fused pair of integrable defects in the sine-Gordon model*, *J. Phys.* **A 43** (2010) 345201 [arXiv:1006.0939].

[8] C. Robertson, *Folding defect affine Toda field theories*, *J. Phys.* **A 47** (2014) 185201 [arXiv:1304.3129].

[9] A.R Aguirre, J.F Gomes, N.I. Spano and A.H. Zimerman, *Type-II super Baäcklund transformation and integrable defects for the $N = 1$ super sinh-Gordon model*, **JHEP 06** (2015) 125 [arXiv:1504.07978].

[10] J.F. Gomes, L.H. Ymai, and A.H. Zimerman, *Classical integrable super sinh-Gordon equation with defects*, *J. Phys.* **A 39** (2006) 7471 [hep-th/0601014].

[11] A.R. Aguirre, J.F. Gomes, N.I. Spano and A.H. Zimerman, *N=1 super sinh-Gordon model with defects revisited*, **JHEP** 02 (2015) 175 [arXiv:1412.2579].

[12] J.F. Gomes, L.H. Ymai and A.H. Zimerman, *Integrability of a classical $N = 2$ super sinh-Gordon model with jump defects*, *JHEP* **03** (2008) 001 [hep-th/0710.1391].

[13] E. Corrigan and C. Zambon, *Infinite dimension reflection matrices in the sine-Gordon model with a boundary*, *JHEP* **06** (2012) 050 [arXiv:1202.6016].

[14] A.R. Aguirre, J.F. Gomes, L.H. Ymai and A.H. Zimerman, *N=1 super sinh-Gordon model in the half line: Breather solutions*, *JHEP* **04** (2013) 136 [arXiv:1304.4582].

[15] C. Zambon, *The classical nonlinear Schrödinger model with a new integrable boundary*, *JHEP* **08** (2014) 036 [arXiv:1405.0967].

Journal of Physics: Conference Series **670** (2016) 012050 doi:10.1088/1742-6596/670/1/012050

Can Spectral Action be a Window to Very High Energies?

Dmitri Vassilevich

CMCC, Universiade Federal do ABC, Santo André, SP, Brazil

E-mail: dvassil@gmail.com

Abstract. The principles of noncommutative geometry impose severe restrictions on the structure of (almost) commutative field theories. The Standard Model fits surprisingly well into the noncommutative framework. Here we overview some universal predictions of the spectral action principle for the behavior of bosonic theories at very high energies.

1. Introduction

Nowadays, no one doubts that quantum theory is an adequate instrument to describe the real world. There are strong theoretical arguments that the space-time itself can be quantum. Noncommutative Geometry provides a consistent mathematical framework to deal with quantum spaces. This framework appears to be quite restrictive, so that it yields testable predictions even on commutative spaces. In particular, the principles of Noncommutative Geometry have very interesting consequences for particle physics and gravity that will be the subject of present contribution.

For the readers' convenience we start with some basic references. A fairly complete introduction to Noncommutative Geometry is the monograph [1]. More recent developments including the applications to particle physics can be found in [2] and [3].

Geometry of noncommutative spaces is defined through Spectral Triples $(\mathcal{A}, \mathcal{H}, \mathcal{D})$. Here $\mathcal{A}$ is an associative algebra that acts by bounded operators on some Hilbert space $\mathcal{H}$, while $\mathcal{D}$ is a Dirac operator on $\mathcal{H}$. The first two entries of Spectral Triple remind us of the Gelfand-Naimark theory. Any commutative C^* algebra is isomorphic to the algebra of continuous functions on some topological space. In this sense, a noncommutative associative C^* algebra can be thought of as the algebra of functions on a noncommutative space. Abstract algebras are hard to deal with, one has to realize them on some space. The Gelfand-Naimark-Segal construction provides any C^* algebra with a Hilbert space where this algebra acts by bounded operators. Thus, $\mathcal{A}$ and $\mathcal{H}$ serve to describe topology of (possibly) noncommutative spaces. However, just $\mathcal{A}$ and $\mathcal{H}$ are not enough to describe geometries. One needs a scale to measure the distances, which is supplied by the third member of Connes' Spectral Triple – the Dirac operator $\mathcal{D}$. Spectral Triples should satisfy a number of Axioms that we do not write down here.

The Dirac operator may contain a lot of important data as, e.g., the Riemann metric, connections, etc. It is natural to assume that fundamental physics is a geometric theory, i.e., it is fully based on the Spectral Triple. One needs therefore to construct an action principle basing on the structures available in the Spectral Triple only. This construction is called the Spectral

Journal of Physics: Conference Series **670** (2016) 012050 doi:10.1088/1742-6596/670/1/012050

Action principle [4]. The action for spinor fields that belong to the Hilbert space $\mathcal{H}$ may be constructed with the help of inner product $\langle\ ,\ \rangle$ and the Dirac operator

$$\mathcal{S}_{\mathrm{f}} = \langle \psi, \mathcal{D}\psi \rangle \tag{1}$$

This action has the form of the usual fermionic action. The part of the action that does not contain fermions should be constructed from $\mathcal{D}$ only. To get a number out of an operator one can take the trace, so that the bosonic part of the spectral action becomes

$$\mathcal{S}_{\mathrm{b}} = \mathrm{Tr}_{\mathcal{H}} \left(\chi(\mathcal{D}^2/\Lambda^2) \right) \tag{2}$$

Here χ is any function that decays fast enough at infinity to ensure the existence of trace. Λ is a scale parameter.

Spectral triples also include the operators of chirality and real structure. These operators will not be considered in this contribution even though they play an important role.

Apart from the Spectral Action, spectral triples allow to define and compute many other important characteristics of noncommutative spaces (as the distance, e.g.). In particular, any abstract spectral triple with a commutative algebra $\mathcal{A}$ allows to reconstruct the corresponding commutative manifold by what is called the Connes Reconstruction Theorem. More generally, one may be interested in how much the spectrum of natural operators tells about the geometry. This is one of the subjects of Spectral Geometry, that has many applications to Quantum Field Theory [5]. Many notions and methods of spectral geometry may be extended even to differential operators on infinite rank bundles [6].

2. Low and moderate energies

Let us see how the Standard Model of elementary particles looks from the point of view of Noncommutative Geometry. Following the approach of [7], let us take $\mathcal{A} = C^\infty(\mathbb{R}^4) \otimes \mathcal{A}_{\mathrm{SM}}$, where $\mathcal{A}_{\mathrm{SM}} = \mathbb{C} \oplus \mathbb{H} \oplus \mathrm{Mat}_3(\mathbb{C})$ is finite dimensional. Here $\mathbb{H}$ is the algebra of quaternions. The corresponding noncommutative geometry is called almost commutative since the whole noncommutativity is in the finite-dimensional algebra $\mathcal{A}_{\mathrm{SM}}$. This finite-dimensional piece defines the gauge group which appears to be that of the Standard Model. The Hilbert space $\mathcal{H}$ consists of square integrable spinors that carry a representation of the associative algebra $\mathcal{A}_{\mathrm{SM}}$. There are not that many representations of associative algebras (as compared to the Lie algebras). Therefore, the last condition restricts the spectrum of fermions essentially to trivial or fundamental representations of the $SU(2)$ and $SU(3)$ factors in the gauge group. Although the gauge group itself is defined by $\mathcal{A}_{\mathrm{SM}}$ that and may vary in particular realizations of almost commutative geometries, the spectrum of fermionic representations is fixed to the one that we do see in the Standard Model. Note, that the Higgs field appears very natural in this approach as fluctuation of the connection in internal noncommutative directions.

Let us see how much one could say about the dynamics. The fermionic action of the Standard Model defines through (1) the Dirac operator (almost) uniquely. The Dirac operator defines in turn the bosonic action (2) that depends also on an arbitrary function χ. It happens, however, that in the low-energy limit the whole dependence of χ may be reduced to a few constants. To demonstrate this, we need the heat kernel expansion for $\mathcal{D}^2$:

$$K(\mathcal{D}^2, t) \equiv \mathrm{Tr}\left[e^{-t\mathcal{D}^2} \right] \simeq \sum_{p=0}^{\infty} t^{-2+p} a_{2p}(D^2)\,, \qquad t \to +0, \tag{3}$$

Suppose that χ is a Laplace transform,

$$\chi(z) = \int_0^\infty dt\, e^{-tz} \tilde{\chi}(t)\,. \tag{4}$$

Then, by substituting (4) in (2) and using there the expansion (3), one may show that

$$S_{\mathrm{b}}\left(\mathcal{D}\right) \sim \sum_{p=0} \Lambda^{4-2p} \chi_{2p} a_{2p}\left(D^2\right), \tag{5}$$

where

$$\chi_{2p} = \int_0^\infty dt\, t^{-2+p} \tilde{\chi}\left(t\right). \tag{6}$$

To discuss the structure of expansion (5) let us take the Dirac operator in the following simple but sufficiently general form

$$D = i\gamma^\mu(\nabla^{\mathrm{LC}}_\mu + iA_\mu) + \gamma_5 \phi\,, \tag{7}$$

where ∇^{LC} is the Levi-Civita covariant derivative acting on spinors, A_μ is an abelian gauge field, and ϕ is a scalar field that represents the Higgs field. The chirality matrix γ_5 appears due to peculiarities of the Wick rotation, see [8, 9] for the discussion. The presence of of γ_5 does not imply that ϕ is a pseudoscalar.

To compute the heat kernel coefficients (see, e.g., [10]), one first brings D^2 to the standard form

$$\mathcal{D}^2 = -(\nabla^2 + E)\,, \tag{8}$$

where [11]

$$E = -i\gamma^\mu\gamma_5(\partial_\mu\phi) - \phi^2 - \tfrac{1}{4}R + \tfrac{i}{4}[\gamma^\mu, \gamma^\nu]F_{\mu\nu} \tag{9}$$

with $F_{\mu\nu} \equiv \partial_\mu A_\nu - \partial_\nu A_\mu$. The connection $\nabla = \nabla^{\mathrm{LC}} + iA_\mu$ has the curvature

$$\Omega_{\mu\nu} \equiv [\nabla_\mu, \nabla_\nu] = iF_{\mu\nu} - \tfrac{1}{4}\gamma^\sigma\gamma^\rho R_{\sigma\rho\mu\nu}\,. \tag{10}$$

For any operator of the form (8), the heat kernel coefficients read

$$\begin{aligned}
a_0 &= \frac{1}{16\pi^2}\int d^4x\sqrt{g}\mathrm{tr}\left(1\right)\\
a_2 &= \frac{1}{96\pi^2}\int d^4x\sqrt{g}\mathrm{tr}\left(6E + R\right)\\
a_4 &= \frac{1}{182\pi^2}\int d^4x\sqrt{g}\mathrm{tr}\left(\Omega_{\mu\nu}\Omega^{\mu\nu} + 6E^2 + 2ER + R^2\text{–terms}\right)
\end{aligned} \tag{11}$$

By using these formulas, one obtains the following expansion for the bosonic spectral action

$$\begin{aligned}
S_{\mathrm{b}}\left(\mathcal{D}\right) \quad \sim \quad & \frac{1}{48\pi^2}\int d^4x\sqrt{g}\big[12\Lambda^4\chi_0 + \Lambda^2\chi_2(-2\phi^2 - R)\\
& +\chi_4(2F_{\mu\nu}F^{\mu\nu} + 6\phi^4 + 6(\partial_\mu\phi)^2 + \text{higher curvature terms})\big] + \mathcal{O}(\Lambda^{-2})
\end{aligned} \tag{12}$$

The action (12) contains the terms of all types that one expects from a bosonic action in four dimensions. By taking two constants, χ_2 and χ_4 positive, we can ensure correct signs of the Einstein term, Maxwell term, the Higgs kinetic energy, mass and selfcoupling, which is already a miracle. If one uses the standard model Dirac operator instead of (7) the similarities between the spectral action and the bosonic part of Standard Model become even more striking. Almost all terms come out correctly, except for a few problems:

(i) One needs a not very natural renormalization procedure to get all Λ^2 terms right.

(ii) The Higgs mass appeared to be a bit too high, at about 170 GeV.

(iii) The Spectral Action predicts a unification point at some energy where all gauge coupling constants become equal – the feature that contradicts the behavior of running couplings in the Standard Model.

I have nothing new to say about the first point above. The Higgs mass at 125 GeV may be very naturally achieved by including a new scalar field, called σ, that has the meaning of fluctuating Majorana mass term. One may also consult the papers [12, 13, 14] for various approaches to the Higgs mass and Standard Model within noncommutative geometry.

The expansion (5) is an asymptotic $1/\Lambda$ expansion organized in accordance to the canonical mass dimension of the field polynomials. One assumes that the fields and their derivatives are small compared to the scale Λ. There is no fundamental reason to terminate the expansion (5) at the third term, as in (12). One can show [15] that inclusion of higher order terms in the expansion of the Yang-Mills spectral action makes the theory superrenormalizable. The a_6-term in the expansion of the Standard Model spectral action modifies the RG running and allows to achieve the unification of couplings [16], thus solving one of the problems mentioned above.

3. Very high energies

The expansion (5) for the spectral action is a low-energy asymptotic expansion. There are examples when it has a vanishing radius of convergence. Therefore, it is desirable to construct another expansion for the spectral action that would be (i) convergent and (ii) valid for high energies. A suitable expansion of the spectral action was proposed in [17], some mathematical aspects were developed further in [18], while physical consequences were discussed in [11]. Below we include a very brief overview of these works.

To simplify the discussion we take the cut-off function

$$\chi(z) = e^{-z} \tag{13}$$

so that the spectral action becomes equal to the heat trace (3) of D^2 with $t = 1/\Lambda^2$. Let us assume, that

$$\mathcal{D}^2 = T + B \tag{14}$$

where T is regarded as the "main" part, while T is a perturbation. More precisely, we need B to be relatively bounded with respect to T. Consider the heat semigroup that is defined as $G(L,t) \equiv e^{-tL}$ for some operator L. There is an expansion of the heat semigroup (called the Dyson-Phillips or Duhamel expansion):

$$G(T+B,t) = \sum_{n=0}^{\infty} G_n(t) \tag{15}$$

where

$$G_0(t) = G(T,t), \qquad G_{n+1}(t) = -\int_0^t G(T,t-s)BG_n(t) \tag{16}$$

This expansion is trace-norm convergent and generates an expansion of the heat trace since $K(L,t) = \operatorname{Tr} G(L,t)$. In applications, it is essential to have a sufficiently simple operator T, so that the integrals in (16) may be computed. On an asymptotically flat space with $\mathcal{D}$ given by (7) it is natural to take $T = \mathcal{D}_0^2$, where $\mathcal{D}_0$ is the free Dirac operator on flat $\mathbb{R}^4$. Then one arrives at the Barvinsky-Vilkovisky expansion [19]

$$\begin{aligned} K(L,t) \ \simeq \ & \frac{1}{(4\pi t)^2} \int d^4x g^{\frac{1}{2}} \operatorname{tr} \left[1 + tP + t^2 (R_{\mu\nu} f_1(-t\partial^2) R^{\mu\nu} + R f_2(-t\partial^2) R \right. \\ & \left. + P f_3(-t\partial^2) R + P f_4(-t\partial^2) P + \Omega_{\mu\nu} f_5(-t\partial^2) \Omega^{\mu\nu}) \right] + \dots \end{aligned} \tag{17}$$

Journal of Physics: Conference Series **670** (2016) 012050 doi:10.1088/1742-6596/670/1/012050

where $P \equiv E + \frac{1}{6}R$. The form-factors $f_{1,\dots,5}$ are expressed through the function

$$h(z) := \int_0^1 d\alpha \, e^{-\alpha(1-\alpha)\, z} \,. \tag{18}$$

For example,

$$f_1(\xi) = \frac{h(\xi) - 1 + \frac{1}{6}\xi}{\xi^2}$$

while other form-factors may be found in [19].

In this expansion we assume that E, R and Ω are small, but do not restrict their derivatives. In other words, (17) is an expansion in fields, that is exact in the derivatives. This sounds better than the usual heat kernel expansion, but there is a price to pay: the expansion in non-universal, depends in an essential way on the topology, and is much more complicated. In the low-energy limit $-t\partial^2 \to 0$ on reproduces (11). Let us study the high energy limit, when the $-t\partial^2$ is supposed to be large. Let us expand the metric as $g_{\mu\nu} = \delta_{\mu\nu} + h_{\mu\nu}$, assuming that the fluctuations are gravitons, $\partial_\mu h_{\mu\nu} = 0$ and $h_{\mu\mu} = 0$. After some calculations, one obtains the quadratic in fluctuations part of (17) keeping the leading terms in the $-\partial^2/\Lambda^2 \to \infty$ limit only:

$$K(\mathcal{D}^2, t = 1/\Lambda^2) \simeq \frac{\Lambda^4}{(4\pi)^2} \int d^4x \left[-\tfrac{3}{2} h_{\mu\nu} h_{\mu\nu} + 8\phi \frac{1}{-\partial^2} \phi + 8F_{\mu\nu} \frac{1}{(-\partial^2)^2} F_{\mu\nu} \right] \tag{19}$$

We see, that the expression (19) for the spectral action corresponding to the exponential cut-off (13) depends on zero or negative powers of the momenta. This means, that the propagators contain zero or positive powers of the momenta or that they are *local.* In other words, high-momenta bosonic fields do not propagate!

It is interesting to compare the high energy expansion of the heat kernel for $\mathcal{D}^2$ to that of a generic Laplace type operator. It appears that due to some miraculous cancellations the heat kernel for $\mathcal{D}^2$ falls off at large momenta faster than that for a generic Laplacian.

If one takes a generic cut-off function instead of the exponent, the leading term in the high energy expansion of the bosonic spectral action differs by an overall coefficient from (19). All our conclusions remain unchanged.

4. Conclusions and discussion

Noncommutative geometry provides a unified framework for the description of both noncommutative and commutative manifolds. These scheme appears to be very restrictive, so that it excludes many field-theoretical models of particle physics. However, the standard model is consistent with the axioms of noncommutative geometry and with the low energy expansion of the spectral action. We saw, that at high energies the spectral action principle yields some universal predictions. The spectrum of bosonic particle should change dramatically at very high energies. In particular, this implies that there are no high momentum gravitons, and the problem of perturbative nonrenormalizability of quantum gravity (as we know it) is no longer relevant. Unfortunately, the calculation made in the previous section does not tell what the correct high energy gravity looks like. Therefore, it is a bit too early to claim that the renormalization problem of quantum gravity is solved by the spectral action.

The characteristic scale Λ that separates high energies is expected to be somewhere in the range of $10^{14} - 10^{16}$GeV. There is a gap of a few orders of magnitude between Λ and the Planck scale. It is quite possible therefore that the high energy behavior of the spectral action will have some testable predictions, e.g. in Cosmology.

To conclude, I like to mention several relatively recent developments of the spectral action approach to particle physics. These are the Pati-Salam unification [20], the anomaly approach to the spectral action [21], non-associative geometry approach to the standard model [22], and a new zeta-function definition of the spectral action [23].

Acknowledgments

This contribution is based on joint papers with Agostino Devastato, Bruno Iochum, Max Kurkov, Cyril Levy, Fedele Lizzi and Carlos Valcárcel. I am grateful to all of them for collaboration. This work was supported in part by the CNPq project 306208/2013-0 and by the FAPESP project 2012/00333-7.

References

[1] Gracia-Bondía J M, Várilly J C and Figueroa H 2001 *Elements of Noncommutative Geometry* (Boston: Birkhäuser)
[2] Connes A, Marcolli M 2008 *Noncommutative Geometry, Quantum Fields and Motives* (New Dehli: Hindustan Book Agency & AMS)
[3] van Suijlekom W D 2015 *Noncommutative Geometry and Particle Physics* (Dordrecht: Springer)
[4] Chamseddine A H and Connes A 1997 *Commun. Math. Phys.* **186** 731 [hep-th/9606001]
[5] Fursaev D and Vassilevich D 2011 *Operators, Geometry and Quanta : Methods of spectral geometry in quantum field theory* (Dordrecht: Springer)
[6] Vassilevich D 2015 Spectral geometry of symplectic spinors *Preprint* arXiv:1508.03684 [math-ph]
[7] Chamseddine A H, Connes A and Marcolli M 2007 *Adv. Theor. Math. Phys.* **11**, 991 [hep-th/0610241]
[8] Lizzi F, Mangano G, Miele G and Sparano G 1997 *Phys. Rev.* D **55** 6357 [hep-th/9610035]
[9] Gracia-Bondía J M, Iochum B and Schucker T 1998 *Phys. Lett.* B **416** 123 [hep-th/9709145]
[10] Vassilevich D V 2003 *Phys. Rept.* **388** 279 [hep-th/0306138].
[11] Kurkov M, Lizzi F and Vassilevich D 2014 *Phys. Lett.* B **731** 311 [arXiv:1312.2235 [hep-th]]
[12] Stephan C A 2009 *Phys. Rev.* D **79** 065013 [arXiv:0901.4676 [hep-th]]
[13] Devastato A, Lizzi F and Martinetti P 2014 *JHEP* **1401** 042 [arXiv:1304.0415 [hep-th]]
[14] Chamseddine A H, Connes A and van Suijlekom W D 2013 *J. Geom. Phys.* **73** 222 [arXiv:1304.7583 [math-ph]]
[15] van Suijlekom W D 2012 *Commun. Math. Phys.* **312** 883 [arXiv:1104.5199 [math-ph]]
[16] Devastato A, Lizzi F, Valcárcel Flores C E and Vassilevich D V 2015 *Int. J. Mod. Phys.* A **30** 1550033 [arXiv:1410.6624 [hep-ph]]
[17] Iochum B, Levy C and Vassilevich D 2012 *Commun. Math. Phys.* **316** 595 [arXiv:1108.3749 [hep-th]]
[18] Iochum B, Levy C and Vassilevich D 2012 *J. Phys.* A **45** 374020 [arXiv:1201.6637 [math-ph]]
[19] Barvinsky A O and Vilkovisky G A 1990 Nucl. Phys. B **333** 471
[20] Chamseddine A H, Connes A and van Suijlekom W D 2013 *JHEP* **1311** 132 [arXiv:1304.8050 [hep-th]]
[21] Andrianov A A, Kurkov M A and Lizzi F 2011 *JHEP* **1110** 001 [arXiv:1106.3263 [hep-th]]
[22] Boyle L and Farnsworth S 2014 *New J. Phys.* **16** 123027 [arXiv:1401.5083 [hep-th]]
[23] Kurkov M A, Lizzi F, Sakellariadou M and Watcharangkool A 2015 *Phys. Rev.* D **91** 065013 [arXiv:1412.4669 [hep-th]]

Journal of Physics: Conference Series **670** (2016) 012051 doi:10.1088/1742-6596/670/1/012051

Complete integrability of geodesics in toric Sasaki-Einstein spaces

Mihai Visinescu

Department of Theoretical Physics, National Institute for Physics and Nuclear Engineering, Magurele, P.O.Box M.G.-6, Romania

E-mail: mvisin@theory.nipne.ro

Abstract. We describe a method for constructing Killing-Yano tensors on toric Sasaki-Einstein manifolds using their geometrical properties. We take advantage of the fact that the metric cones of these spaces are Calabi-Yau manifolds. The complete list of special Killing forms can be extracted making use of the description of the Calabi-Yau manifolds in terms of toric data. This general procedure for toric Sasaki-Einstein manifolds is exemplified in the case of the 5-dimensional spaces $Y^{p,q}$ and $T^{1,1}$. Finally we discuss the integrability of geodesic motion in these spaces.

1. Introduction

Higher order symmetries associated with Killing-Yano (KY) tensors (or Killing forms) and Stäckel-Killing (SK) tensors play an important role in modern gravitational and mathematical physics. Contrary to Killing vectors, these geometrical objects do not have clear meaning as they do not describe continuous symmetries of the space. They are considered as symmetries of the whole phase-space and consequently they are often called *dynamical* or *hidden symmetries*. Their corresponding conserved quantities are polynomial in momenta and are involved in the study of integrability properties of geodesic motions and separation of variables for the Hamilton-Jacobi or quantum Klein-Gordon, Dirac equations.

The generalized Killing equations for SK and KY tensors are quite intricate and to get their solutions by solving the corresponding differential equations is a very difficult task. From this point of view it is of particular importance to find alternative ways for constructing the higher order Killing tensors.

In this paper we investigate the KY and SK tensors on toric Sasaki-Einstein (SE) spaces using their geometrical properties. Recently SE geometry has become of significant interest in various developments in mathematics and theoretical physics [1]. These geometries are believed to play a significant role in studies of consistent string theory compactification and in the context of the AdS/CFT correspondence. CFT duals to the SE manifolds are $\mathcal{N} = 1$ superconformal quiver gauge theories in $3 + 1$ dimensions. Both the geometry and the gauge theory can be characterized by a common data, called a *toric data*. In the case of toric SE manifolds there is an algorithm which constructs the gauge theory from the toric data of the Calabi-Yau (CY) singularity.

The aim of this paper is to present a method to find the conserved quantities for geodesic motions on toric SE spaces. For this purpose we construct all special KY forms on the SE spaces

Journal of Physics: Conference Series **670** (2016) 012051 doi:10.1088/1742-6596/670/1/012051

using the fact that these forms are exactly those forms which translate into parallel forms on the CY metric cone [2]. In turn the KY forms on the metric cone are obtained from the toric data of the CY manifolds using the Delzant approach [3] to toric geometries.

The general procedure to get KY tensors on toric SE spaces is exemplified in the case of the 5-dimensional spaces $Y^{p,q}$ and $T^{1,1}$. Having the explicit SK tensors on these spaces we discuss the integrability of geodesic motions in these spaces and show that the systems are completely integrable.

The paper is organized as follows. In the next Section we review the basic definitions of SK and KY tensors. In Section 3 we present some well-known results concerning the special KY forms on SE manifolds and their relations with the parallel forms of the corresponding CY metric cones. In Section 4 it is described the construction of KY parallel forms on CY spaces using the toric data. In Section 5 we exemplify the general scheme in the case of the 5-dimensional spaces $Y^{p,q}$ and $T^{1,1}$ and prove the complete integrability of geodesic motions. The paper ends with conclusions in Section 6.

2. Killing tensors

Let (M, g) be an n-dimensional Riemannian manifold with the metric g and let ∇ be its Levi-Civita connection. A natural generalization of Killing vectors is given by SK tensors of rank $r > 1$ defined as totally symmetric tensor fields $K_{\mu_1\cdots\mu_r}$ satisfying the generalized Killing equation

$$\nabla_{(\mu}K_{\nu_1\cdots\nu_r)} = 0\,, \tag{1}$$

where the round brackets indicate symmetrization over the indices within.

In the presence of a SK tensor the system of a free particle with the Hamiltonian

$$H = \frac{1}{2}g^{\mu\nu}p_\mu p_\nu\,, \tag{2}$$

admits the conserved quantity

$$K = K^{\mu_1\cdots\mu_r}p_{\mu_1}\cdots p_{\mu_r}\,, \tag{3}$$

commuting with Hamiltonian (2) in the sense of Poisson brackets. Here p_μ are canonical momenta conjugate to the coordinates x^μ, $p_\mu = g_{\mu\nu}\dot{x}^\nu$ with overdot denoting proper time derivative.

Another important generalization of Killing vector fields is represented by antisymmetric KY tensors which are r-forms obeying the equation [4]

$$\nabla_{(\mu}\Psi_{\nu_1)\nu_2\ldots\nu_r} = 0\,. \tag{4}$$

It turns out that the most part of known interesting KY tensors are the so called special Killing forms satisfying for some constant c the additional equation [2]

$$\nabla_X(d\Psi) = cX^* \wedge \Psi\,, \tag{5}$$

for any vector field X on M, X^* being the 1-form dual to the vector field X.

There is an important connection between these two generalizations of the Killing vectors. To wit, the partially contracted product of two KY tensors $\Psi^{i_1,\ldots,i_r}$ and $\Sigma^{i_1,\ldots,i_r}$ generates a SK tensor of rank 2:

$$K_{ij}^{(\Psi,\Sigma)} = \Psi_{ii_2\ldots i_r}\Sigma_j{}^{i_2\ldots i_r} + \Sigma_{ii_2\ldots i_r}\Psi_j{}^{i_2\ldots i_r}\,. \tag{6}$$

This fact offers a method to generate higher order integrals of motion by identifying the complete set of Killing-Yano tensors.

Let us note that trying to solve straightforwardly the generalized Killing equations (1) or (4) turns out to be very difficult. For this reason in what follows we shall present an approachable way for finding hidden symmetries on SE spaces.

3. Special KY forms on SE spaces

A $(2n-1)$-dimensional manifold M is a *contact manifold* if there exists a 1-form η (called a *contact* 1-*form*) on M such that:

$$\eta \wedge (d\eta)^{n-1} \neq 0\,. \tag{7}$$

For every choice of contact 1-form η there exists a unique vector field K_η, called the *Reeb vector field*, which satisfies:

$$\eta(K_\eta) = 1 \quad \text{and} \quad K_\eta \lrcorner\, d\eta = 0\,, \tag{8}$$

where $\lrcorner$ is the operator dual to the wedge product.

Let us introduce the metric cone $C(M)$ of the manifold M as the product manifold $M \times \mathbb{R}_{>0}$, with $\dim C(M) = 2n$, endowed with the warped metric

$$\bar{g} := dr^2 + r^2 g\,. \tag{9}$$

Here $r \in (0, \infty)$ may be considered as a coordinate on the positive real line $\mathbb{R}_+$.

A compact Riemannian manifold (M, g) is Sasakian if and only if its metric cone $C(M)$ is Kähler and its Kähler 2-form is given by

$$\omega = \frac{1}{2} d(r^2 \eta)\,. \tag{10}$$

If the Sasaki manifold M is Einstein

$$\mathrm{Ric}_g = 2(n-1)g\,, \tag{11}$$

then the Kähler metric cone is Ricci-flat ($\mathrm{Ric}_{\bar{g}} = 0$), i.e. CY manifold.

A first set of KY tensors of a SE manifold is given by the special Killing forms

$$\Psi_k = \eta \wedge (d\eta)^k \quad , \quad k = 0, 1, \cdots, n-1\,. \tag{12}$$

These tensors do not exhaust the complete set of KY tensors on SE spaces. In order to find all special KY tensors we must resort to the CY metric cone of a SE manifold. To write explicitly two additional Killing forms we shall express the volume form of the metric cone in terms of the Kähler form ω:

$$d\mathcal{V} = \frac{1}{n!}\omega^n\,. \tag{13}$$

The volume form of a Kähler manifold can be also written as [2, 5]:

$$d\mathcal{V} = \frac{i^n}{2^n}(-1)^{n(n-1)/2}\Omega \wedge \overline{\Omega}, \tag{14}$$

where Ω is the volume holomorphic $(n, 0)$ form of $C(M)$. The additional Killing forms on $C(M)$ are given by the real respectively the imaginary part of the complex volume form.

The last step to finding all special KY tensors of the SE manifold M makes use of the fact that special Killing forms on Sasaki manifolds are exactly those forms which translate into parallel forms on the metric cone. More precisely, for any p-form Ψ on the space M we can define an associated $(p+1)$-form Ψ^C on the cone $C(M)$:

$$\Psi^C := r^p dr \wedge \Psi + \frac{r^{p+1}}{p+1} d\Psi\,. \tag{15}$$

Ψ^C is parallel if and only if Ψ is a special Killing-Yano tensor (5) with constant $c = -(p+1)$ [2].

Therefore the complete list if special KY forms on SE spaces are given by the Killing forms (12) and two additional KY tensors connected with the real and imaginary parts of the complex volume $(n, 0)$ form of $C(M)$ and extracted as in (15).

Journal of Physics: Conference Series **670** (2016) 012051 doi:10.1088/1742-6596/670/1/012051

4. Complex volume form of CY toric manifold

In order to write the complex volume form we need the complex coordinates of the CY cone manifold.

Let us consider that $C(M)$ is toric, the standard n-torus $\mathbb{T}^n = \mathbb{R}^n/2\pi\mathbb{Z}^n$ acting effectively on $C(M)$

$$\tau : \mathbb{T}^n \to \mathrm{Diff}(C(M), \omega), \tag{16}$$

preserving the Kähler form ω. Let us denote by Φ^i the angular coordinates along the torus action generated by $\partial/\partial\Phi_i$ and write the Reeb vector field as

$$K_\eta = b_i \frac{\partial}{\partial \Phi_i} . \tag{17}$$

Associated to $(C(M), \omega, \tau)$ there is a moment map

$$\mu = \frac{1}{2} r^2 \eta , \tag{18}$$

and the action coordinates are

$$y^i = \mu \left(\frac{\partial}{\partial \Phi_i} \right) . \tag{19}$$

The $\mathbb{T}^n$-invariant Kähler metric on $C(M)$ in the symplectic coordinates (y, Φ) is

$$ds^2 = G_{ij} dy^i dy^j + G^{ij} d\Phi_i d\Phi_j, \tag{20}$$

where G_{ij} is the Hessian of the symplectic potential $G(y)$ in the y coordinates

$$G_{ij} = \frac{\partial^2 G}{\partial y^i \partial y^j} \quad , \quad 1 \le i, j \le n, \tag{21}$$

and $G^{ij} = (G_{ij})^{-1}$.

For what follows we need the evaluation of the symplectic potential G. This object is obtained using the Delzant construction [3]. A *Delzant polytope* is a covex polytope such that there are n edges meeting at each vertex, each edge meeting at the vertex is of the form $1 + tu_i$ where $u_i \in \mathbb{Z}^n$, and $\{u_i\}$ can be chosen to form basis in $\mathbb{Z}^n$. This polytope can be described by the inequalities

$$l_a(y) := (y, v_a) \ge 0, \quad \text{for} \quad 1 \le a \le d , \tag{22}$$

where v_a are the inward pointing normal vectors to the d facets of the polyhedral cone. The set of vectors $\{v_a\}$

$$v_A = v_a^i \left(\frac{\partial}{\partial \Phi_i} \right) , \tag{23}$$

is called a *toric data.*

Using the Delzant construction the general symplectic potential has the following form in terms of the toric data [6, 7]:

$$G = G^{can} + G_b + h, \tag{24}$$

where

$$G^{can} = \frac{1}{2} \sum_a l_a(y) \log l_a(y), \tag{25}$$

$$G_b = \frac{1}{2} l_b(y) \log l_b(y) - \frac{1}{2} l_\infty(y) \log l_\infty(y), \tag{26}$$

with $l_b(y) = (b, y)\,,\, l_\infty(y) = \sum_a (v_a, y)$ and h is a homogeneous degree one function of variables y^i

$$h = \lambda_i y^i + t, \tag{27}$$

λ_i, t being some constants.

The standard complex coordinates of $C(M)$ are w_i on $\mathbb{C}\backslash\{0\}$. Log complex coordinates are $z_i = \log w_i = x_i + \mathrm{i}\Phi_i$, with

$$x_i = \frac{\partial G}{\partial y_i}\,. \tag{28}$$

The final form of the $(n, 0)$ holomorphic form of the Ricci-flat metric on the CY cone turns out to be [8]:

$$\Omega = \exp\left(x_1 + \mathrm{i}\Phi_1\right) dz_1 \wedge \cdots \wedge dz_n\,. \tag{29}$$

5. Applications

We shall exemplify the general procedure in the case of the 5-dimensional spaces $Y^{p,q}$ and $T^{1,1}$. The complete list of special KY tensors allows us to prove the complete integrability of geodesic motions on these spaces.

5.1. $Y^{p,q}$ spaces

A particular interesting class of toric contact structures on $S^2 \times S^3$ have been studied by physicists [9, 10] and denoted by $Y^{p,q}$ where p, q are relative prime integers satisfying $0 \leq q \leq p$.

The explicit local metric of the 5-dimensional $Y^{p,q}$ manifold is given by the line element [11]

$$\begin{aligned} ds^2 = & \frac{1 - cy}{6}(d\theta^2 + \sin^2\theta\, d\phi^2) + \frac{1}{w(y)q(y)}dy^2 + \frac{q(y)}{9}(d\psi - \cos\theta\, d\phi)^2 \\ & + w(y)\left[d\alpha + \frac{ac - 2y + cy^2}{6(a - y^2)}[d\psi - \cos\theta\, d\phi]\right]^2, \end{aligned} \tag{30}$$

where

$$\begin{aligned} w(y) &= \frac{2(a - y^2)}{1 - cy}\,, \\ q(y) &= \frac{a - 3y^2 + 2cy^3}{a - y^2}\,. \end{aligned} \tag{31}$$

The constant c can be rescaled by a diffeomorphism, so in what follows we take $c = 1$. For $0 < \alpha < 1$ we can take the range of the angular coordinates (θ, ϕ, ψ) to be $0 \leq \theta \leq 2\pi\,, 0 \leq \phi \leq 2\pi\,, 0 \leq \psi \leq 2\pi$. Choosing $0 < a < 1$, the range of the coordinate y is taken between the negative and the smallest positive roots of the cubic equation

$$a - 3y^2 + 2y^3 = 0\,. \tag{32}$$

The Sasakian 1-form η is

$$\eta = -2y d\alpha + \frac{1 - y}{3}(d\psi - \cos\theta d\phi)\,, \tag{33}$$

and the Reeb vector field is [11]

$$K_\eta = 3\frac{\partial}{\partial\psi} - \frac{1}{2}\frac{\partial}{\partial\alpha}\,. \tag{34}$$

The explicit form of the KY tensor Ψ_1 (12) is

$$\begin{aligned}\Psi_1 &= (1-y)^2 \sin\theta\, d\theta \wedge d\phi \wedge d\psi - 6 dy \wedge d\alpha \wedge d\psi \\ &+ 6\cos\theta\, d\phi \wedge dy \wedge d\alpha - 6(1-y)y \sin\theta\, d\theta \wedge d\phi \wedge d\alpha\,.\end{aligned} \tag{35}$$

For the additional KY tensors we shall construct the symplectic potential G (24) using the toric data for $Y^{p,q}$ [11, 12] and finally the holomorphic form Ω (29). From (15) in the case of the 5-dimensional $Y^{p,q}$ spaces we have

$$\Omega := r^3 dr \wedge \Psi + \frac{r^3}{3} d\Psi\,. \tag{36}$$

Decomposing Ψ into its real ($\Re\Psi$) and imaginary ($\Im\Psi$) parts, and ignoring the multiplicative constants, we get the following special real Killing forms [13, 14, 15]:

$$\begin{aligned}\Re\Psi = \sqrt{\frac{1-y}{p(y)}}\Big(&\cos\psi\Big[d\theta \wedge dy + 6p(y)\sin\theta\, d\phi \wedge d\alpha + p(y)\sin\theta\, d\phi \wedge d\psi\Big] \\ &-\sin\psi\Big[\sin\theta\, d\phi \wedge dy - 6p(y) d\theta \wedge d\alpha - p(y) d\theta \wedge d\psi \\ &+ p(y)\cos\theta\, d\theta \wedge d\phi\Big]\Big)\,,\end{aligned} \tag{37}$$

$$\begin{aligned}\Im\Psi = \sqrt{\frac{1-y}{p(y)}}\Big(&\sin\psi\Big[d\theta \wedge dy + 6p(y)\sin\theta\, d\phi \wedge d\alpha + p(y)\sin\theta\, d\phi \wedge d\psi\Big] \\ &+\cos\psi\Big[\sin\theta\, d\phi \wedge dy - 6p(y) d\theta \wedge d\alpha - p(y) d\theta \wedge d\psi \\ &+ p(y)\cos\theta\, d\theta \wedge d\phi\Big]\Big)\,.\end{aligned} \tag{38}$$

For $Y^{p,q}$ spaces the conjugate momenta to the coordinates $(\theta, \phi, y, \alpha, \psi)$ are [16]:

$$\begin{aligned} p_\theta &= \frac{1-y}{6}\dot\theta\,, \\ p_\phi + \cos\theta p_\psi &= \frac{1-y}{6}\sin^2\theta\dot\phi\,, \\ p_y &= \frac{1}{6p(y)}\dot y\,, \\ p_\alpha &= w(y)\left(\dot\alpha + f(y)\left(\dot\psi - \cos\theta\dot\phi\right)\right)\,, \\ p_\psi &= w(y)f(y)\dot\alpha + \left[\frac{q(y)}{9} + w(y)f^2(y)\right]\left(\dot\psi - \cos\theta\dot\phi\right)\,. \end{aligned} \tag{39}$$

From the isometry $SU(2)\times U(1)\times U(1)$ of the metric (30) we have that the momenta p_ϕ, p_ψ and p_α are conserved. p_ϕ is the third component of the $SU(2)$ angular momentum and p_ψ, p_α are associated to the $U(1)$ factors. In addition, the total $SU(2)$ angular momentum

$$\vec{J}^{\,2} = p_\theta^2 + \frac{1}{\sin^2\theta}\left(p_\phi + \cos\theta p_\psi\right)^2 + p_\psi^2\,, \tag{40}$$

is also conserved [16, 17].

The next conserved quantities, quadratic in momenta, will be expressed in terms of Stäckel-Killing tensors as in (3). The Stäckel-Killing tensors of rank two on $Y^{p,q}$ will be constructed from Killing-Yano tensors according to (6). For this purpose we shall use the Killing 3-form Ψ_1 (35) and the additional 2-forms (37) and (38).

The first Stäckel-Killing tensor $K^{(1)}_{\mu\nu}$ is constructed according to (6) using the real part of the Killing form Ψ (37):

$$K^{(1)}_{\mu\nu} = (\Re\Psi)_{\mu\lambda}(\Re\Psi)^{\lambda}{}_{\nu}\,. \tag{41}$$

Let us denote by $K^{(1)}$ the corresponding quantity constructed from $K^{(1)}_{\mu\nu}$.

The next SK tensor can be constructed using the imaginary part $\Im\Psi$, but we find that this SK tensor produces the same conserved quantity $K^{(1)}$.

In principle a mixed combination of KY tensors $\Re\Psi$ and $\Im\Psi$ produces a SK tensor (6), but it proves that all components of this tensor are zero.

Finally we construct the Stäckel-Killing tensor from the Killing form Ψ_1:

$$K^{(2)}_{\mu\nu} = (\Psi_1)_{\mu\lambda\sigma}(\Psi_1)^{\lambda\sigma}{}_{\nu}\,, \tag{42}$$

and let us denote by $K^{(2)}$ the corresponding conserved quantity.

In conclusion we have a set of 7 conserved quantities $H, p_\phi, p_\psi, p_\alpha, \vec{J}^{\,2}, K^{(1)}, K^{(2)}$. We can examine if this set constitutes a functionally independent set of constants of motion for the geodesics of $Y^{p,q}$ constructing the Jacobian:

$$\mathcal{J} = \frac{\partial(H, p_\phi, p_\psi, p_\alpha, \vec{J}^{\,2}, K^{(1)}, K^{(2)})}{\partial(\theta, \phi, y, \alpha, \psi, \dot\theta, \dot\phi, \dot y, \dot\alpha, \dot\psi)}\,. \tag{43}$$

The rank of this Jacobian is 5, exactly the number of the degrees of freedom, which means that the system is completely integrable [18] in contrast to the assertion made in [17]. In spite of the presence of the Stäckel-Killing tensors $K^{(1)}$ and $K^{(2)}$, the system is not superintegrable, $K^{(1)}$ and $K^{(2)}$ being a combination of the first integrals $H, p_\phi, p_\psi, p_\alpha, \vec{J}^{\,2}$. Therefore the toric Sasaki-Einstein spaces $Y^{p,q}$ spaces possess several Killing-Yano tensors, but these Killing forms do not generate new Stäckel-Killing tensors, i.e. genuine conserved quantities.

5.2. $T^{1,1}$ *space*

The homogeneous Sasaki-Einstein metric on $S^2 \times S^3$ is usually referred to as $T^{1,1}$. The $T^{1,1}$ space was considered as the first example of toric Sasaki-Einstein/quiver duality [19].

The isometries of $T^{1,1}$ form the group $SU(2)\times SU(2)\times U(1)$ and the metric of this space may be written down explicitly by utilizing the fact that it is a $U(1)$ bundle over $S^2\times S^2$. Let us denote by (θ_1, ϕ_1) and (θ_2, ϕ_2) the coordinates which parametrize the two sphere in a conventional way, and the angle $\psi \in [0, 4\pi)$ to parametrize the $U(1)$ fiber. Using these parametrizations the $T^{1,1}$ metric may be written as [20, 11]:

$$\begin{aligned} ds^2(T^{1,1}) =& \frac{1}{6}(d\theta_1^2 + \sin^2\theta_1 d\phi_1^2 + d\theta_2^2 + \sin^2\theta_2 d\phi_2^2) \\ &+ \frac{1}{9}(d\psi + \cos\theta_1 d\phi_1 + \cos\theta_2 d\phi_2)^2\,. \end{aligned} \tag{44}$$

The globally defined contact 1-form η is:

$$\eta = \frac{1}{3}(d\psi + \cos\theta_1 d\phi_1 + \cos\theta_2 d\phi_2)\,, \tag{45}$$

and the Reeb vector field K_η has the form:

$$K_\eta = 3\frac{\partial}{\partial\psi}\,. \tag{46}$$

From the contact form η (45), using (12) we construct the Killing 3-form

$$\begin{aligned}\Psi_1 =&\frac{1}{9}(\sin\theta_1 d\psi\wedge d\theta_1\wedge d\phi_1+\sin\theta_2 d\psi\wedge d\theta_2\wedge d\phi_2\\ &-\cos\theta_1\sin\theta_2 d\theta_2\wedge d\phi_1\wedge d\phi_2\\ &+\cos\theta_2\sin\theta_1 d\theta_1\wedge d\phi_1\wedge d\phi_2)\,.\end{aligned} \tag{47}$$

Again, using the toric data for $T^{1,1}$ we evaluate the symplectic potential G and the complex coordinates on the CY cone manifold. The additional real Killing forms extracted from the holomorphic volume $(3,0)$ form are [21]

$$\begin{aligned}\Re\Psi =&\cos\psi\, d\theta_1\wedge d\theta_2+\sin\theta_2\sin\psi\, d\theta_1\wedge d\phi_2\\ &-\sin\theta_1\sin\psi\, d\theta_2\wedge d\phi_1\\ &-\sin\theta_1\sin\theta_2\cos\psi\, d\phi_1\wedge d\phi_2\,,\end{aligned} \tag{48}$$

$$\begin{aligned}\Im\Psi =&\sin\psi\, d\theta_1\wedge d\theta_2-\sin\theta_2\cos\psi\, d\theta_1\wedge d\phi_2\\ &+\sin\theta_1\cos\psi\, d\theta_2\wedge d\phi_1\\ &-\sin\theta_1\sin\theta_2\sin\psi\, d\phi_1\wedge d\phi_2\,.\end{aligned} \tag{49}$$

The conjugate momenta to the coordinates $(\theta_1,\theta_2,\phi_1,\phi_2,\psi)$ are:

$$\begin{aligned}p_{\theta_1} &= \frac{1}{6}\dot\theta_1\,,\\ p_{\theta_2} &= \frac{1}{6}\dot\theta_2\,,\\ p_{\phi_1} &= \frac{1}{6}\sin^2\theta_1\,\dot\phi_1+\frac{1}{9}\cos^2\theta_1\,\dot\phi_1+\frac{1}{9}\cos\theta_1\,\dot\psi+\frac{1}{9}\cos\theta_1\cos\theta_2\,\dot\phi_2\,,\\ p_{\phi_2} &= \frac{1}{6}\sin^2\theta_2\,\dot\phi_2+\frac{1}{9}\cos^2\theta_2\,\dot\phi_2+\frac{1}{9}\cos\theta_2\,\dot\psi+\frac{1}{9}\cos\theta_1\cos\theta_2\,\dot\phi_1\,,\\ p_\psi &= \frac{1}{9}\dot\psi+\frac{1}{9}\cos\theta_1\,\dot\phi_1+\frac{1}{9}\cos\theta_2\,\dot\phi_2\,.\end{aligned} \tag{50}$$

Taking into account the isometries of $T^{1,1}$, the momenta p_{ϕ_1}, p_{ϕ_2} and p_ψ are conserved. On the other hand two total $SU(2)$ angular momenta are also conserved:

$$\begin{aligned}\vec{J}_1^{\,2} =&p_{\theta_1}^2+\frac{1}{\sin^2\theta_1}(p_{\phi_1}-\cos\theta_1 p_\psi)^2+p_\psi^2\\ =&\frac{1}{36}\left[\dot\theta_1^2+\sin^2\theta_1\dot\phi_1^2\right]+\frac{1}{81}\Big[\dot\psi^2+\cos^2\theta_1\dot\phi_1^2+\cos^2\theta_2\dot\phi_2^2\\ &+2\cos\theta_1\,\dot\phi_1\dot\psi+2\cos\theta_2\dot\phi_2\dot\psi+2\cos\theta_1\cos\theta_2\dot\phi_1\dot\phi_2\Big]\,,\\ \vec{J}_2^{\,2} =&p_{\theta_2}^2+\frac{1}{\sin^2\theta_2}(p_{\phi_2}-\cos\theta_2 p_\psi)^2+p_\psi^2\\ =&\frac{1}{36}\left[\dot\theta_2^2+\sin^2\theta_2\dot\phi_2^2\right]+\frac{1}{81}\Big[\dot\psi^2+\cos^2\theta_1\dot\phi_1^2+\cos^2\theta_2\dot\phi_2^2\\ &+2\cos\theta_1\,\dot\phi_1\dot\psi+2\cos\theta_2\dot\phi_2\dot\psi+2\cos\theta_1\cos\theta_2\dot\phi_1\dot\phi_2\Big]\,.\end{aligned} \tag{51}$$

As for the $Y^{p,q}$ spaces we construct the conserved quantities associated with the Killing forms (47), (48), (49) and find again that we have only two nontrivial conserved quantities $K^{(1)}$ and $K^{(2)}$ as in the previous case. In fact metric (44) of the $T^{1,1}$ space can be obtained from metric (30) of the $Y^{p,q}$ spaces taking $c = 0$, setting $a = 3$ and making some changes of coordinates [9].

Finally we investigate the integrability of the geodesic motion on $T^{1,1}$ and for this purpose we construct the Jacobian:

$$\mathcal{J} = \frac{\partial(H, p_{\phi_1}, p_{\phi_2}, p_\psi, \vec{J}_1^{\,2}, \vec{J}_2^{\,2}, K^{(1)}, K^{(2)})}{\partial(\theta_1, \theta_2, \phi_1, \phi_2, \psi, \dot{\theta}_1, \dot{\theta}_2, \dot{\phi}_1, \dot{\phi}_2, \dot{\psi})} . \tag{52}$$

As in the case of $Y^{p,q}$ spaces, the rank of this Jacobian is 5 implying the complete integrability of the geodesic motion on $T^{1,1}$. Because not all aforesaid constants of motion are functionally independent, from 8 constants we can choose a subset of 5 constants as functionally independent. For $T^{1,1}$ space the expressions of the constants of motion are simpler than in the case of $Y^{p,q}$ spaces and we can write explicitly the relations between them. For example, selecting the subset $(H, p_{\phi_1}, p_{\phi_2}, p_\psi, \vec{J}_1^{\,2})$ as functionally independent constants of motion, the constants $\vec{J}_2^{\,2}, K^{(1)}$ and $K^{(2)}$ can be expressed in terms of the chosen subset:

$$\begin{aligned} \vec{J}_2^{\,2} &= \frac{1}{3}H + \frac{1}{2}p_\psi^2 - \vec{J}_1^{\,2} , \\ K^{(1)} &= 72H - 324p_\psi^2 , \\ K^{(2)} &= 16H + 72p_\psi^2 . \end{aligned} \tag{53}$$

6. Conclusions

An important point of interest in physics is to identify constants of motion or conserved quantities of a system.

KY tensors are known to be highly relevant for both mathematics and physics as they can be used to construct higher order integrals of motion. In this paper we present a method to extract Killing forms on a toric SE space using successively its contact form, the symplectic, complex coordinates and holomorphic volume form of its CY metric cone.

A detailed analysis of the KY tensors on 5-dimensional $Y^{p,q}$ and $T^{1,1}$ spaces shows that the number of functionally independent constants of motion is five implying the complete integrability.

Using toric geometry, many other SE manifolds can be investigated and these spaces are a good testing ground for the predictions of the AdS/CFT correspondence.

Acknowledgments

The work of M. V. has been partly supported by the joint Romanian-LIT, JINR, Dubna Research Project, theme no. 05-6-1119-2014/2016 and partly by the program NUCLEU PN-09370102, Romania.

References

[1] Sparks J 2011 Sasaki-Einstein manifolds *Surv. Diff. Geom.* **16** 265 (*Preprint* 1004.2461)
[2] Semmelmann U 2003 Conformal Killing forms on Riemannian manifolds *Math. Z.* **245** 503 (*Preprint* math/0206117)
[3] Delzant T 1988 Hamiltoniens périodiques et images convexes de l'application moment *Bull. Soc. Math. France* **116** 315
[4] Yano K 1952 Some remarks on tensor fields and curvature *Ann. of Math.* **55** 328
[5] Visinescu M and Vîlcu G E 2012 Hidden symmetries of Euclideanised Kerr-NUT-(A)dS metrics in certain scaling limit *SIGMA* **8** 058 (*Preprint* 1205.6036)
[6] Guillemin V 1994 Kähler structures on toric varieties *J. Diff. Geom.* **40** 285

[7] Abreu M 2003 Kähler geometry of toric manifolds in symplectic coordinates *Fields Institute Commun.* **35** 1
[8] Martelli D, Sparks J and Yau S T 2005 The geometric dual of a-maximisation for toric Sasaki-Einstein manifolds *Commun. Math. Phys.* **268** 39 (*Preprint* hep-th/0503183)
[9] Gauntlett J P, Martelli D, Sparks J and Waldram D 2004 Sasaki-Einstein metrics on $S^2 \times S^3$ *Adv. Theor. Math. Phys.* **8** 711 (*Preprint* hep-th/0403002)
[10] Martelli D and Sparks J 2008 Notes on toric Sasaki-Einstein seven-manifolds and AdS_4/CFT_3 *J. High Energy Phys.* **11** 016 (*Preprint* 0808.0904)
[11] Martelli D and Sparks J 2006 Toric geometry, Sasaki-Einstein manifolds and a new infinite class of AdS/CFT duals *Commun. Math. Phys.* **262** 51 (*Preprint* hep-th/0411238)
[12] Oota Y and Yasui Y 2006 Toric Sasaki-Einstein manifolds and Heun equations *Nucl. Phys.* **B742** 275 (*Preprint* hep-th/0512124)
[13] Visinescu M 2012 Killing forms on the five-dimensional Einstein-Sasaki Y(p,q) spaces *Mod. Phys. Lett. A* **27** 1250217 (*Preprint* 1207.2581)
[14] Slesar V, Visinescu M and Vîlcu G E 2014 Special Killing forms on Sasaki-Einstein manifolds *Phys. Scripta* **89** 125205 (*Preprint* 1403.1015)
[15] Slesar V, Visinescu M and Vîlcu G E 2015 Toric data, Killing and complete integrability of geodesics in Sasaki-Einstein spaces $Y^{p,q}$ *Annals of Physics* **361** 548
[16] Benvenuti S and Kruczenski M 2006 Semiclassical strings in Sasaki-Einstein manifolds and long operators in $\mathcal{N} = 1$ gauge theories *JHEP* **0610** 051 (*Preprint* hep-th/0505046)
[17] Rubín de Celis E and O. P. Santillán O P 2012 Massless geodesics in $AdS_5 \times Y(p,q)$ as a superintegrable system *JHEP* **09** 032 (*Preprint* 1205.3256)
[18] Babalic E M and Visinescu M 2015 Complete integrability of geodesic motion in Sasaki-Einstein toric $Y^{p,q}$ spaces *Mod. Phys. Lett. A* **30** 1550180 (*Preprint* 1505.03976)
[19] Klebanov I R and Witten E 1998 Superconformal field theory on threebranes at a Calabi-Yau singularity *Nucl. Phys.* **B536** 199
[20] Candelas P and de la Ossa X C 1990 Comments on conifolds *Nucl. Phys.* **B342** 246
[21] Slesar V, Visinescu M and Vîlcu G E 2015 Hidden symmetries on toric Sasaki-Einstein spaces *EPL* **110** 31001

XXIII International Conference on Integrable Systems and Quantum Symmetries (ISQS-23) IOP Publishing
Journal of Physics: Conference Series **670** (2016) 012052 doi:10.1088/1742-6596/670/1/012052

Geometric structures, fractal self-similarity, squeezed coherent states and electrodynamics

Giuseppe Vitiello

Dipartimento di Fisica "E. R. Caianiello", Universitá di Salerno
and INFN, Sezione di Napoli, Gruppo Collegato Salerno
Fisciano (Salerno) 84084, Italy

E-mail: `vitiello@sa.infn.it`

Abstract. The formalism describing fractal self-similar structures has been shown in recent works to be isomorphic to the one of a system of damped/amplified oscillators, which is a prototype of a dissipative system and of the environment in which it is embedded, and to squeezed coherent states. Fractal-like structures appear to be generated by coherent quantum condensation processes, and thus they appear as macroscopic quantum systems, as it happens with crystals, ferromagnets, superconductors and like systems characterized by ordered patterns. In this report, by resorting to such results it is shown that in space-time regions where the magnetic field may be approximated to be constant and the electric field is derivable from a harmonic potential, the isomorphism also exists between electrodynamics and fractal-like structures. A link is thus established between self-similarity, dissipation, coherent states and electrodynamics. The relation between quantum dissipation and non-commutative geometry in the plane is also commented upon. The macroscopic appearances (forms) of the fractals seem to emerge out of a process of morphogenesis as the macroscopic manifestation of the underlying dissipative, coherent quantum dynamics.

1. Introduction

In recent works [1]-[5] it has been shown that an isomorphism exists between the formalism describing fractal self-similar structures and squeezed generalized SU(1,1) coherent states in quantum field theory (QFT). This result can be framed in the research line which has led to recognize that systems characterized by observable ordered patterns, such as crystals, superconductors, ferromagnets, provide examples of "macroscopic quantum systems" arising as the effect of homogeneous, coherent boson condensation in the system ground state [6, 7]. These systems are quantum systems not in the trivial sense that they are made of atoms, molecules and other quantum components, as every system is made of, but in the sense that their observable physical properties, such as electrical conductivity, magnetization, crystal structure, critical temperature behavior, etc. cannot be explained without recourse to the study of the dynamics of their quantum components. Similarly, the appearance in these systems, especially in the course of phase transitions, of topologically non trivial 'extended objects', such as crystal dislocations, vortices, domain wall, kinks and other kind of 'defects', is found to be the manifestation of non-homogeneous boson condensation at quantum dynamical level [6, 7]. These results, which rest on a mathematically sound formulation and wide experimental confirmation, have suggested the possibility that also the self-similarity properties of fractal-like structures might arise from the

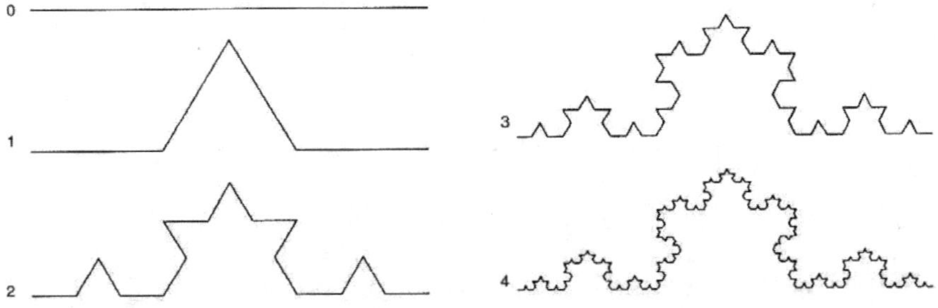

Figure 1. The first five stages of the Koch curve

underlying dynamics of their quantum components involving as well the mechanism of coherent boson condensation in the system ground state. Recent observations seem to confirm such a view since they show that when a crystal is submitted to deforming stress actions the induced deformations in the lattice structure form, at low temperature, self-similar fractal patterns and provide an example of "emergence of fractal dislocation structures" [8] in dissipative, non-equilibrium systems, thus appearing as the result of non-homogeneous coherent phonon condensation. A first conjecture of the relation between fractal self-similarity and coherent states was proposed in [9] and then the study of fractals in the frame of the entire analytical functions was presented in [1] - [5], also with application to the observed brain functional activity [1], [10] - [13], to water molecular structure in the presence of nafion and filtering [14] and in a numerical simulation of pancreatic beta cell clusters (Lagerhans islets) [15]. More recently, it has been shown [16] that the isomorphism between the logarithmic spiral and other fractal structures and the squeezed coherent states representing the system of damped/amplified oscillators, a prototype of a dissipative system and the environment in which it is embedded, may be extended to electrodynamics in space-time regions where the magnetic field may be approximated to be constant and the electric field is derivable from a harmonic potential. The plan of the report is the following. In Section 2 a summary is presented of the isomorphism between the Koch curve, the logarithmic spiral and squeezed coherent states. Section 3 summarizes the isomorphism with electrodynamics. Section 5 is devoted to the conclusions.

2. Geometric structures emerging from coherent dynamics

In order to show how the dynamical description of the geometrical feature of fractal self-similarity is obtained, in this Section I briefly summarize the examples of the Koch curve (Figure 1) and of the logarithmic spiral (Figure 2) [17, 18]. The results can be extended to fractals which are generated iteratively according to a prescribed recipe (deterministic fractals), such as the Sierpinski gasket and carpet, the Cantor set, etc. [17, 19]. The results also extend to the golden spiral and its relation with Fibonacci progression [4].

Denote by $u_0 = 1$ the starting stage and by $u_{n,q}(\alpha)$ the n-th stage of the Koch curve construction, with $\alpha = 4$ and $q = 1/3^d$. One has [1, 2]

$$u_{n,q}(\alpha) = (q\,\alpha)^n = 1, \qquad \text{for any } n. \tag{1}$$

The fractal or self-similarity dimension [17] $d = \ln 4/\ln 3 \approx 1.2619$, is thus obtained. Notice that self-similarity is properly defined only in the $n \to \infty$ limit.

In full generality one may consider the complex α-plane. Then the study of the fractal properties is carried on in the space $\mathcal{F}$ of the entire analytic functions. One restricts at the end the conclusions to real $q\,\alpha$, $q\,\alpha \to Re(q\,\alpha)$. Putting $q = e^{-d\,\theta}$, Eq. (1) is written as $d\,\theta = \ln\,\alpha$ and the functions $u_{n,q}(\alpha)$ are, apart the normalization factor $1/\sqrt{n!}$, nothing but the restriction

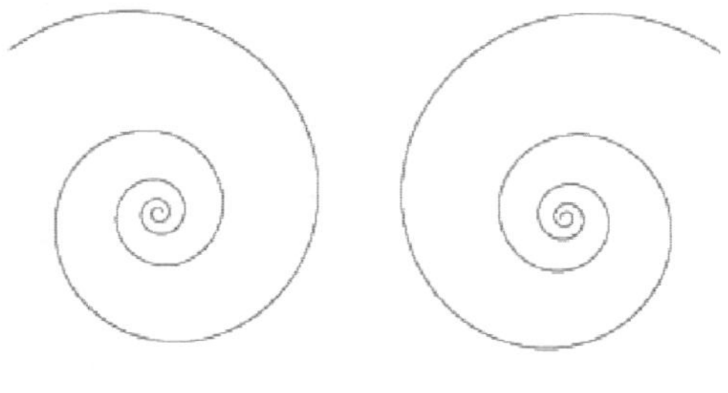

Figure 2. The anti-clockwise and the clockwise logarithmic spiral.

to real $q\,\alpha$ of the functions which form indeed a basis in the space $\mathcal{F}$ of the entire analytic functions:

$$u_{n,q}(\alpha) = \frac{(q\,\alpha)^n}{\sqrt{n!}}\;, \qquad n \in \mathcal{N}_+\;, \quad q\,\alpha \in \mathcal{C}\;, \tag{2}$$

One then realizes that $\mathcal{F}$ is the so-called Fock-Bargmann representation of the Weyl–Heisenberg algebra [20] where the (Glauber) coherent states are described. The fractal self-similarity properties and coherent states are thus readily recognized to be related. One introduces then the finite difference operator $\mathcal{D}_q$, called the q-derivative operator [21, 22, 23] and the q-deformed algebraic structure, with $q = e^{\zeta}$, $\zeta \in \mathcal{C}$, is obtained. Denote the coherent state by $|\alpha\rangle$, with $a|\alpha\rangle = \alpha|\alpha\rangle$, and a the annihilator operator. Application of q^N to $|\alpha\rangle$, $N \equiv \alpha\, d/d\alpha$, gives the q-deformed coherent state

$$q^N|\alpha\rangle = |q\alpha\rangle = \exp\Big(-\frac{|q\alpha|^2}{2}\Big)\sum_{n=0}^{\infty}\frac{(q\alpha)^n}{\sqrt{n!}}\;|n\rangle\;. \tag{3}$$

$|q\alpha\rangle$ is a squeezed coherent state [21], with $\zeta = \ln q$ the squeezing parameter and q^N acts in $\mathcal{F}$ as the squeezing operator. By applying $(a)^n$ to $|q\alpha\rangle$ and restricting to real $q\alpha$, the n-th iteration stage of the fractal is obtained

$$\langle q\alpha|(a)^n|q\alpha\rangle = (q\alpha)^n = u_{n,q}(\alpha), \qquad q\alpha \rightarrow Re(q\alpha). \tag{4}$$

Thus, the n-th term, $n = 0, 1, 2, .., \infty$, in the coherent state series Eq. (3) represents, in a one-to-one correspondence, the fractal n-th stage of iteration. $(a)^n$ acts as a "magnifying" lens [1, 2, 19]. q^N is called *the fractal operator* [1, 2].

The self-similarity properties of the Koch curve can be thus described in terms of the coherent state squeezing transformation.

I consider now the logarithmic spiral. It is represented in polar coordinates (r, θ) by

$$r = r_0\, e^{d\,\theta}, \tag{5}$$

with r_0 and d arbitrary real constants and $r_0 > 0$. Eq. (5) is represented by the straight line of slope d in a log-log plot with abscissa $\theta = \ln e^{\theta}$: $d\,\theta = \ln(r/r_0)$.

The self-similarity property consists in the constancy of the angular coefficient $\tan^{-1} d$ and the rescaling $\theta \rightarrow n\,\theta$ affects r/r_0 by the power $(r/r_0)^n$.

The associated parametric equations are:

$$\xi = r(\theta)\,\cos\theta = r_0\, e^{d\,\theta}\,\cos\theta \tag{6}$$
$$\eta = r(\theta)\,\sin\theta = r_0\, e^{d\,\theta}\,\sin\theta \tag{7}$$

The point $z = \xi + i\,\eta = r_0\, e^{d\,\theta}\, e^{i\,\theta}$ on the spiral is fully specified in the complex z-plane by the sign of $d\,\theta$. Due to the completeness of the (hyperbolic) basis $\{e^{-\,d\,\theta},\, e^{+\,d\,\theta}\}$ one needs to consider both the points $z_1 = r_0\, e^{-\,d\,\theta}\, e^{-\,i\,\theta}$ and $z_2 = r_0\, e^{+\,d\,\theta}\, e^{+\,i\,\theta}$. Opposite signs for the imaginary exponent $\pm i\,\theta$ are also considered for convenience. z_1 and z_2 solve the equations

$$m\,\ddot{z}_1 + \gamma\,\dot{z}_1 + \kappa\,z_1 = 0 \tag{8}$$

$$m\,\ddot{z}_2 - \gamma\,\dot{z}_2 + \kappa\,z_2 = 0 \tag{9}$$

respectively, where the parameter t has been introduced, $\theta = \theta(t)$; "dot" denotes derivative with respect to t, and, up to an arbitrary additive constant, it is put $\theta(t) = \Gamma t$, with $\Gamma \equiv \gamma/2m$. m, γ and κ are positive real constants. Thus, $z_1(t) = r_0\; e^{-\,i\,\Omega\,t}\, e^{-\Gamma\,t}$ and $z_2(t) = r_0\; e^{+\,i\,\Omega\,t}\, e^{+\Gamma\,t}$ solutions of Eqs. (8) and (9) describe the logarithmic spirals and the parameter t can be interpreted as the time parameter.

$|\,d\theta/dt\,| = |\,\Gamma/d\,|$ denotes the spiral "angular velocity". The notation $\Omega^2 = (1/m)(\kappa - \gamma^2/4m) = \Gamma^2/d^2$, with $\kappa > \gamma^2/4m$, is used. Note that the time-reversed, *but distinct*, image of the right-handed chirality spiral (indirect spiral, $q \equiv e^{-d\theta} < 1$) is the left-handed chirality spiral (direct spiral, $q > 1$).

By putting $[z_1(t) + z_2^*(-t)]/2 = x(t)$ and $[z_1^*(-t) + z_2(t)]/2 = y(t)$, Eqs. (8) and (9) reduce to

$$m\ddot{x} + \gamma\dot{x} + kx = 0, \tag{10}$$

$$m\ddot{y} - \gamma\dot{y} + ky = 0. \tag{11}$$

which by using [16, 24] $x_1 \equiv (x + y)/\sqrt{2}$ and $x_2 \equiv (x - y)/\sqrt{2}$ become

$$m\ddot{x}_1 + \gamma\dot{x}_2 + kx_1 = 0, \tag{12}$$

$$m\ddot{x}_2 + \gamma\dot{x}_1 + kx_2 = 0. \tag{13}$$

In passing, I observe that such a system of damped/amplified oscillators belongs to the class of deterministic systems à la 't Hooft [25]-[29], which remain deterministic even when described by means of Hilbert space techniques. These oscillators have a quantum representation in terms of squeezed $SU(1,1)$ coherent states. The QFT quantization procedure can be summarized as follows.

One assumes, as customary, the commutators $[\,x, p_x\,] = i\,\hbar = [\,y, p_y\,]$, $[\,x, y\,] = 0 = [\,p_x, p_y\,]$ and the annihilation and creation operators are introduced:

$$a \equiv \left(\frac{1}{2\hbar\Omega}\right)^{\frac{1}{2}} \left(\frac{p_x}{\sqrt{m}} - i\sqrt{m}\Omega x\right); \quad a^\dagger \equiv \left(\frac{1}{2\hbar\Omega}\right)^{\frac{1}{2}} \left(\frac{p_x}{\sqrt{m}} + i\sqrt{m}\Omega x\right) \tag{14}$$

$$b \equiv \left(\frac{1}{2\hbar\Omega}\right)^{\frac{1}{2}} \left(\frac{p_y}{\sqrt{m}} - i\sqrt{m}\Omega y\right); \quad b^\dagger \equiv \left(\frac{1}{2\hbar\Omega}\right)^{\frac{1}{2}} \left(\frac{p_y}{\sqrt{m}} + i\sqrt{m}\Omega y\right) \tag{15}$$

satisfying the commutation relations $[\,a, a^\dagger\,] = 1 = [\,b, b^\dagger\,]$, $[\,a, b\,] = 0 = [\,a, b^\dagger\,]$. Then the Hamiltonian H for the quantum damped/amplified oscillator system is obtained [30] (see also [7, 31, 32])

$$H = H_0 + H_I \tag{16}$$

$$H_0 = \hbar\Omega(A^\dagger A - B^\dagger B), \qquad H_I = i\hbar\Gamma(A^\dagger B^\dagger - AB) \tag{17}$$

where $A \equiv (1/\sqrt{2})(a + b)$, $B \equiv (1/\sqrt{2})(a - b)$.

The vacuum state is $|0\rangle \equiv |n_A = 0, n_B = 0\rangle = |0\rangle \otimes\; |0\rangle$, where n_A and n_B denote the number of A's and B's and $(A\otimes\; 1)|0\rangle \otimes\; |0\rangle \equiv A|0\rangle = 0$; $(1\otimes\; B)|0\rangle \otimes\; |0\rangle \equiv B|0\rangle = 0$. The vacuum time

Journal of Physics: Conference Series **670** (2016) 012052 doi:10.1088/1742-6596/670/1/012052

evolution is controlled by H_I: $|0(t)\rangle = e^{-it\frac{H}{\hbar}}|0\rangle = e^{-it\frac{H_I}{\hbar}}|0\rangle$, with $\langle 0(t)|0(t)\rangle = 1, \quad \forall t$. $|0(t)\rangle$ is explicitly given by

$$|0(t)\rangle = \prod_{\kappa} \frac{1}{\cosh(\Gamma_\kappa t)} \exp\left(\tanh(\Gamma_\kappa t) A_\kappa^\dagger B_\kappa^\dagger\right)|0\rangle \tag{18}$$

We have $\lim_{t\to\infty}\langle 0(t)|0\rangle \propto \lim_{t\to\infty} \exp(-t\Gamma) = 0$. In the infinite volume limit, for $\int d^3\kappa\, \Gamma_\kappa$ finite and positive,

$$\langle 0(t)|0\rangle \to 0 \quad \text{as} \quad V \to \infty \quad \forall\, t \tag{19}$$

and $\langle 0(t)|0(t')\rangle \to 0$ as $V \to \infty \quad \forall\, t$ and t', $\quad t' \neq t$. The meaning of these relations is that a representation $\{|0(t)\rangle\}$ of the canonical commutation relations (CCR) is defined at each time t and is unitarily inequivalent to any other representation $\{|0(t')\rangle, \ \forall t' \neq t\}$ in the infinite volume limit. The system thus evolves in time through unitarily inequivalent representations of CCR [30]. The number of modes A_κ (or B_κ) condensed in $|0(t)\rangle$ is given by

$$\mathcal{N}_{A_\kappa}(t) = \langle 0(t)|A_\kappa^\dagger A_\kappa|0(t)\rangle = \sinh^2 \Gamma_\kappa t \tag{20}$$

The states generated by $B^\dagger$ represent the sink where the energy dissipated by the quantum damped oscillator flows, or, in other words, the B-oscillator represents the reservoir coupled to the A-oscillator [30].

The two mode realization of the algebra $su(1,1)$ is obtained by defining $J_+ \equiv A^\dagger B^\dagger$, $J_- \equiv J_+^\dagger \equiv AB$, $\quad J_3 \equiv (1/2)(A^\dagger A + B^\dagger B + 1)$, then $\quad [\,J_+, J_-\,] = -2J_3, \quad [\,J_3, J_\pm\,] = \pm J_\pm$, with the $SU(1,1)$ Casimir operator $\mathcal{C}$ given by $\mathcal{C}^2 = (1/4)(A^\dagger A - B^\dagger B)^2$. The commutation relation $[\,H_0, H_I\,] = 0$ guarantees that the initial condition of positiveness for the eigenvalues of H_0 is protected against transitions to negative energy states.

As said, $|0(t)\rangle$ is thus a two-mode time dependent generalized $SU(1,1)$ coherent state [7, 20, 30, 33]. $|0(t)\rangle$ is also recognized to be a squeezed coherent state characterized by the q-deformation of Lie-Hopf algebra [7, 21, 22, 23]. I also remark that A and B are entangled modes. This entanglement cannot be destroyed by the action of any unitary operator, a feature absent in quantum mechanics.

Finally, I observe that the variations in time of the number of particles condensed in the vacuum gives heat dissipation $dQ = \frac{1}{\beta}dS$ and time evolution is controlled by entropy variations [30, 34], which is consistent with the fact that dissipation implies breaking of time-reversal invariance (*the arrow of time*). The Hamiltonian H turns out to be actually the *fractal free energy* for the coherent boson condensation process out of which the fractal is formed. The system temperature $T = \hbar\,\Gamma$ is proportional to the background zero point energy: $\hbar\,\Gamma \propto \hbar\,\Omega/2$ [7, 27, 28, 29].

It can be shown [4] that also in the case of the Koch curve and other fractals the isomorphism with the system of damped/amplified oscillators and squeezed SU(1,1) coherent states can be established. For brevity, here I do not report the derivation of these results.

In closing this Section, I observe that letting $p_{z_\pm}$ denote the momenta and $v_\pm = \dot{z}_\pm$ the forward in time and backward in time velocities, with the notation $+ \equiv 1$ and $- \equiv 2$, it is:

$$v_\pm = \pm\frac{1}{m}\left(p_{z_\pm} \mp (1/2)\gamma z_\mp\right), \qquad \text{with} \qquad [v_+, v_-] = i\,\frac{\gamma}{m^2} \tag{21}$$

Conjugate position coordinates (ξ_+, ξ_-) can be defined by putting $\xi_\pm = \mp(m/\gamma)v_\mp$, with

$$[\xi_+, \xi_-] = i\,\frac{1}{\gamma}. \tag{22}$$

This suggests that the relation between dissipation and noncommutative geometry in the plane can be shown to exist. For brevity I will not discuss further such an issue and the role of the deformed Hopf algebra [16, 35, 36, 37].

A final remark is that the continuous time evolution includes the discrete group of transformations $z_1(m) = r_0(e^{-2\pi d})^m \to z_1(m+1) = r_0(e^{-2\pi d})^{(m+1)} = z_1(m)(e^{-2\pi d})$, with integer m = 1, 2, 3..., due to the T integer multiplicity, $\theta(T) = 2\pi$ at $T = 2\pi d/\Gamma$ and, at $t = mT$, $z_1 = r_0\,(e^{-2\pi d})^m$, $z_2 = r_0\,(e^{2\pi d})^m$. The isomorphism seems thus to appear as a homomorphism.

3. Self-similarity and electrodynamics

In the previous Section we have established the isomorphism between (fractal) self-similarity and the squeezed SU(1,1) coherent states associated to the damped/amplified oscillator system Eqs. (8) and (9) (or, equivalently Eqs. (10) and (11) or Eqs. (12) and (13)). We now discuss the relation with electrodynamics. Consider the vector potential given by $\vec{A} = (1/2)\vec{B} \times \vec{r}$, $\vec{r} = (x_1, x_2, x_3)$. It is $\vec{B} = \vec{\nabla} \times \vec{A}$, $\vec{\nabla} \cdot \vec{A} = 0$. Choose the reference frame so that the magnetic field $\vec{B}$ be a constant vector, $\vec{B} = \vec{\nabla} \times \vec{A} = -B\hat{\mathbf{3}}$. Then, $A_3 = 0$ and

$$A_i = \frac{B}{2}\epsilon_{ij}x_j\,, \qquad i,j = 1,2. \tag{23}$$

where $\epsilon_{12} = -\epsilon_{21} = 1$; $\epsilon_{ii} = 0$. Let $B \equiv \gamma/e$. The third component, $i = 3$, of $(\vec{v} \times \vec{B})$, $v_i = \dot{x}_i$, vanishes. Moreover, assume that $\vec{E}$ is given by the gradient of the harmonic potential $\Phi \equiv \frac{k}{2e}({x_1}^2 - {x_2}^2) \equiv \Phi_1 - \Phi_2$, $\vec{E} = -\vec{\nabla}\Phi$; and $E_3 = 0$. Eqs. (12) and (13) are now rewritten as

$$F_e^i = e\,E^i + e\,(\vec{v} \times \vec{B})^i \tag{24}$$

$$F_{-e}^i = -e\,E^i - e\,(\vec{v} \times \vec{B})^i \tag{25}$$

namely, the Lorentz forces $\vec{F}_e$ and $\vec{F}_{-e}$, acting on two opposite charges with same velocity $\vec{v}$ in the same electric and magnetic fields, $\vec{E}$ and $\vec{B}$. Use of $i = 1$ in Eq. (24) and $i = 2$ in Eq. (25) gives indeed Eqs. (12) and (13). Considering $i = 2$ in Eq. (24) and $i = 1$ in Eq. (25) leads to similar result.

One can show [24] that by using Eqs. (23), Eqs. (12) and (13) can be derived from the Lagrangian

$$L = \frac{1}{2m}(m\dot{x}_1 + e_1 A_1)^2 - \frac{1}{2m}(m\dot{x}_2 + e_2 A_2)^2 - \frac{e^2}{2m}({A_1}^2 - {A_2}^2) - e\Phi, \tag{26}$$

and the Hamiltonian is

$$H = H_1 - H_2 = \frac{1}{2m}(p_1 - e_1 A_1)^2 + e_1\Phi_1 - \frac{1}{2m}(p_2 + e_2 A_2)^2 + e_2\Phi_2 \tag{27}$$

In the least energy state (where $H = 0$, $H_1 = H_2$) the respective contributions to the energy compensate each other. One of the charged particles (one of the oscillators) may be considered to represent the em field in which the other one is embedded and vice-versa.

I remark that Eqs. (24) and (25) (i.e. Eqs. (12) and (13)) are derived [16] by usual integration over the volume of the equations for the matter part $T_m^{\mu\nu}$ and the em part $T_\gamma^{\mu\nu}$ of the total energy-momentum tensor $T^{\mu\nu}$ in electrodynamics

$$\partial_\mu T_m^{\mu\nu} = e\,F^{\alpha\nu}J_\alpha \tag{28}$$

$$\partial_\mu T_\gamma^{\mu\nu} = -e\,F^{\alpha\nu}J_\alpha \tag{29}$$

where J_α denotes the current and as usual $F^{\alpha\beta} = \partial^\beta A^\alpha - \partial^\alpha A^\beta$ ($g^{\mu\nu} = (1,-1,-1,-1)$, $\mu = 0,1,2,3$; $\hbar = 1 = c$). We see that the non-vanishing divergences of $T_m^{\mu\nu}$ and $T_\gamma^{\mu\nu}$

compensate each other, $\partial_\mu T_m^{\mu\nu} = -\partial_\mu T_\gamma^{\mu\nu}$, so that the conservation of the total $T^{\mu\nu}$ holds: $\partial_\mu T^{\mu\nu} = \partial_\mu(T_m^{\mu\nu} + T_\gamma^{\mu\nu}) = 0$.

For $\nu = 0$, volume integration of Eqs. (28) and (29) gives the rate of changes in time of the energy of the matter field and em field, $\mathcal{E}_m$ and $\mathcal{E}_\gamma$, respectively:

$$\partial_0 \mathcal{E}_m = e\,\vec{E}\cdot\vec{v} = -\partial_0 \mathcal{E}_\gamma \tag{30}$$

For $\nu = i = 1, 2, 3$, integration of Eqs. (28) and (29) over the volume gives, as said, Eqs. (24) and (25).

A representation of the content of Maxwell equations and the associated conservation laws is thus provided, under the conditions specified above, by the considered system of damped/amplified oscillators. As seen, their realization in terms of squeezed coherent states turns out to be isomorphic to fractal self-similarity properties. This establishes the link between electrodynamics and fractal self-similarity.

4. Conclusions

In this paper I have summarized results showing that an isomorphism exists between the formalism describing fractal self-similarity properties and squeezed coherent states in QFT representing a system of damped/amplified oscillators. I have also shown that the isomorphism extends to electrodynamics involving the basic conservation laws implied by Maxwell equation. These conclusions may turn out to be of interest in view of the widely diffused presence in Nature of fractal-like structures and of the ubiquitous occurrence of power laws in physics, in biology and neuroscience. Fractals appear to be macroscopic quantum systems, in the sense specified in the Introduction, arising from a process of morphogenesis as the manifestation of deformations (squeezing) of coherent dissipative quantum dynamics. Such results lead to an integrated vision of Nature resting on the paradigm of coherence and dissipation. Nature appears to be modulated by coherence, rather than being hierarchically layered in isolated compartments, in collections of isolated systems and phenomena.

References

[1] Vitiello, G 2009a *New Math. Nat. Comput.* **5** 245
[2] Vitiello, G 2009b *Quantum Interaction; Lecture Notes in Artificial Intelligence* ed W Lawless, K van Rijsbergen, M Klusch, (Berlin/Heidelberg: Springer-Verlag) p 6
[3] Vitiello, G 2011 *Vision of Oneness* ed I Licata, A J Sakaji (Roma: Aracne Edizioni) p 155
[4] Vitiello, G 2012a *Phys. Lett.* A **376** 2527
[5] Vitiello, G 2012b *J. Phys.: Conf. Series* **380** 012021
[6] Umezawa, H 1999 *Advanced Field Theory: Micro, Macro and Thermal Concepts* (Melville, NY: American Institute of Physics)
[7] Blasone, M, Jizba, P, Vitiello, G 2011 *Quantum Field Theory and Its Macroscopic Manifestations* (London: Imperial College Press)
[8] Chen, Y S, Choi, W, Papanikolaou, S, Sethna, J P 2010 *Phys. Rev. Lett.* **105** 105501
[9] Vitiello G 2005 *Brazilian J. of Physics* **35** 351
[10] Boudreau, J C and Freeman, W J 1963 *Exp. Neurol.* **8** 423
[11] Freeman, W J 1975 *Mass Action in the Nervous System* (New York: AcademicPress)
[12] Bressler, S L 1987 *Brain. Res.* **409** 285
[13] Beggs, J M, Plenz, D 2003 *J. Neurosc.* **23** (35) 11167
[14] Capolupo, A et al. 2014 *Int. J. Mod. Phys.* B **28** 1450007
[15] Loppini, A et al. 2014, *Phys. Lett.* A **378** 3210
[16] Vitiello G 2014 *Systems* **2** 203
[17] Peitgen, H O, Jürgens, H, Saupe, D 1986 *Chaos and Fractals. New Frontiers of Science* (Berlin: Springer-Verlag)
[18] Andronov, Vitt, A A, Khaikin, S E 1966 *Theory of Oscillators* (Mineola, NY: Dover Publications).
[19] Bunde, A, Havlin S (Eds.) 1995 *Fractals in Science* (Berlin: Springer-Verlag)
[20] Perelomov, A 1986 *Generalized Coherent States and Their Applications* (Berlin: Springer-Verlag)

Journal of Physics: Conference Series **670** (2016) 012052 doi:10.1088/1742-6596/670/1/012052

[21] Celeghini, E, De Martino, S, De Siena, S, Rasetti, M, Vitiello, G 1995 *Ann. Phys.* **241** 50
[22] Celeghini, E, Rasetti, M, Tarlini, M, Vitiello, G 1989 *Mod. Phys. Lett.* **B3** 1213
[23] Celeghini, E, Rasetti, M, Vitiello, G 1991 *Phys. Rev. Lett.* **66** 2056
[24] Blasone, M, Graziano, E, Pashaev, O K, Vitiello, G 1996 *Ann. Phys.* **252** 115
[25] 't Hooft, G 1999 *Classical Quant. Grav.* **16** 3263
[26] 't Hooft, G 2007 *J. Phys.: Conf. Series* **67** 012015
[27] Blasone, M, Jizba, P, Vitiello, G 2001 *Phys. Lett.* **A287** 205
[28] Blasone, M, Celeghini, E, Jizba, P, Vitiello, G 2003 *Phys. Lett.* **A310** 393
[29] Blasone, M, Jizba, P, Scardigli, F, Vitiello, G 2009 *Phys. Lett.* **A373** 4106
[30] Celeghini, E, Rasetti, M, Vitiello, G 1992 *Ann. Phys.* **215** 156
[31] Blasone, M, Srivastava, Y N, Vitiello, G, Widom, A 1998 *Ann. Phys.* **267** 61
[32] Srivastava, Y N, Vitiello, G, Widom, A 1995 *Ann. Phys.* **238** 200
[33] Klauder, J R, Skagerstam, B 1985 *Coherent States* (Singapore: World Scientific)
[34] De Filippo, S, Vitiello, G 1977 *Lett. Nuovo Cim.* **19** 92
[35] Sivasubramanian, S, Srivastava, Y N, Vitiello, G, Widom, A 2003 *Phys. Lett.* **A311** 97
[36] Celeghini, E, De Martino, S, De Siena, S, Iorio, A, Rasetti, M, Vitiello, G 1998 *Phys. Lett.* **A244** 455
[37] Vitiello, G 2007 *Quantum Analogues: From Phase Transitions to Black Holes and Cosmology*; Lectures Notes in Physics 718; ed W.G. Unruh, R. Schuetzhold (Berlin: Springer) p 165

Journal of Physics: Conference Series **670** (2016) 012053 doi:10.1088/1742-6596/670/1/012053

Representations of quantum SU(2) operators on a local chart

Elmar Wagner

Instituto de Física y Matemáticas, Universidad Michoacana, Edificio C-3, Cd. Universitaria, 58040 Morelia, Mich., Mexico

E-mail: elmar@ifm.umich.mx

Abstract. Hilbert space representations of quantum SU(2) by multiplication operators on a local chart are constructed, where the local chart is given by tensor products of square integrable functions on a quantum disc and on the classical unit circle. The actions of generators of quantum SU(2), generators of the opposite algebra, and noncommutative partial derivatives are computed on a Hilbert space basis.

1. Introduction

In [4], a twisted Dirac operator for quantum SU(2) is constructed by using so-called disc coordinates. The starting point of this construction is a representation of quantum SU(2) by multiplication operators on the tensor product of $L_2(\mathbb{S}^1)$ with the quantum disc algebra, where the inner product on the latter space is defined by a positive weighted trace. The construction of the Dirac operator uses noncommutative first order differential operators satisfying a twisted Leibniz rule. Only this twisted Leibniz rule is needed to prove that the Dirac operator has bounded twisted commutators with differentiable functions, thus avoiding an explicit description of the Hilbert space actions of the involved operators.

On the other hand, an explicit description of the Hilbert space representation of the quantum algebra of differentiable functions and the action of the Dirac operator is crucial for determining the analytic properties of the twisted spectral triple, for instance the spectrum of the Dirac operator, its K-homology class, its eigenvectors, and the C*-closure of the quantum algebra. The purpose of the present work is to describe the actions of all operators occurring in [4] on a Hilbert space basis. Unfortunately, a detailed analysis of the twisted spectral triple from [4] is behind the scope of this paper.

2. Quantum SU(2) in disc coordinates

Let $q \in (0, 1)$. The coordinate ring $\mathcal{O}(\mathrm{SU}_q(2))$ of quantum SU(2) is the *-algebra generated by c and d satisfying

$$cd = qdc, \qquad c^*d = qdc^*, \qquad cc^* = c^*c, \tag{1}$$

$$d^*d + q^2cc^* = 1, \qquad dd^* + cc^* = 1. \tag{2}$$

Classically, i. e. $q = 1$, the universal C*-algebra generated by c and d is isomorphic to $C(\mathbb{S}^3)$. The universal C*-closure $C(\mathrm{SU}_q(2))$ of $\mathcal{O}(\mathrm{SU}_q(2))$ has been studied in [6] and [8]. Here we will use the fact that it is generated by the operators $c, d \in B(\ell_2(\mathbb{N}) \bar{\otimes} \ell_2(\mathbb{Z}))$ given by

$$c(e_n \otimes b_k) = q^n \, e_n \otimes b_{k+1}\,, \qquad d(e_n \otimes b_k) = \sqrt{1 - q^{2(n+1)}}\, e_{n+1} \otimes b_k\,, \tag{3}$$

where $\{e_n\}_{n\in\mathbb{N}}$ and $\{b_k\}_{k\in\mathbb{Z}}$ are orthonormal bases for $\ell_2(\mathbb{N})$ and $\ell_2(\mathbb{Z})$, respectively. Note that c and d from (3) do indeed satisfy the relations (1) and (2). Moreover, the representation of $\mathcal{O}(\mathrm{SU}_q(2))$ defined by (3) is faithful.

Let $u : \mathbb{S}^1 \longrightarrow \mathbb{S}^1$, $u(\mathrm{e}^{\mathrm{i}t}) = \mathrm{e}^{\mathrm{i}t}$, denote the unitary generator of $C(\mathbb{S}^1)$. On the orthonormal basis $\{b_k := \frac{1}{\sqrt{2\pi}}\mathrm{e}^{\mathrm{i}kt} : k \in \mathbb{Z}\}$ for $L_2(\mathbb{S}^1) \cong \ell_2(\mathbb{Z})$, multiplication by u becomes the bilateral shift

$$u b_k = b_{k+1}\,, \qquad k \in \mathbb{Z}. \tag{4}$$

Consider the bounded operators z and z^* on $\ell_2(\mathbb{N})$ given by

$$z e_n := \sqrt{1 - q^{2(n+1)}}\, e_{n+1}\,, \qquad z^* e_n := \sqrt{1 - q^{2n}}\, e_{n-1}\,. \tag{5}$$

Obviously, z and z^* satisfy the quantum disc relation

$$z^* z - q^2 z z^* = 1 - q^2. \tag{6}$$

The polynomial *-algebra generated by z and z^* will be denoted by $\mathcal{O}(\mathrm{D}_q)$. Here and subsequently, $\mathrm{D} := \{z \in \mathbb{C} : |z| < 1\}$ and $\bar{\mathrm{D}} := \{z \in \mathbb{C} : |z| \leq 1\}$ stand for the open and closed unit discs, respectively, and $\mathcal{O}(\mathrm{D}_q)$ will be called quantum disc algebra. As shown in [2], the C*-closure $C(\mathrm{D}_q)$ of $\mathcal{O}(\mathrm{D}_q)$ is isomorphic to the Toeplitz algebra. Moreover, (5) determines the unique (up to unitary equivalence) faithful irreducible *-representation of $C(\mathrm{D}_q)$, see e. g. [5].

Defining $y \in B(\ell_2(\mathbb{N}))$ by

$$y e_n := q^n \, e_n\,, \quad n \in \mathbb{N}, \tag{7}$$

we have

$$y = \sqrt{1 - zz^*}, \qquad yz = qzy, \qquad z^* y = q y z^*. \tag{8}$$

In particular, $y = \sqrt{1 - zz^*} \in C(\mathrm{D}_q)$. Finally, with u, z and y given in (4), (5) and (7), the operators c and d in (3) can be written

$$c = y \otimes u = \sqrt{1 - zz^*} \otimes u, \qquad d = z \otimes 1. \tag{9}$$

The representation (9) corresponds to the classical parametrization

$$\psi : \bar{\mathrm{D}} \times [-\pi, \pi] \longrightarrow \mathbb{S}^3, \qquad \psi(z, t) := (z, \sqrt{1 - z\bar{z}}\mathrm{e}^{\mathrm{i}t}),$$

The restriction of ψ to the open set $\mathrm{D} \times (-\pi, \pi)$ defines a dense coordinate chart for $\mathbb{S}^3 \cong \mathrm{SU}(2)$ which is compatible with the standard differential structure on $\mathbb{S}^3$. Moreover, the generators c and d correspond in the classical limit $q \to 1$ to the coordinate functions

$$c(z, \sqrt{1 - z\bar{z}}\mathrm{e}^{\mathrm{i}t}) = \sqrt{1 - z\bar{z}}\mathrm{e}^{\mathrm{i}t}, \qquad d(z, \sqrt{1 - z\bar{z}}\mathrm{e}^{\mathrm{i}t}) = z.$$

We call them disc coordinates because the discs $\psi_t(\bar{\mathrm{D}}) := \{\psi(z, t) : z \in \bar{\mathrm{D}}\}$ for fixed $t \in (-\pi, \pi]$ correspond to the 2-dimensional symplectic leaves associated to a known Poisson group structure on SU(2) [1].

3. Differential calculus

One of the differential operators occurring in [4] is the classical partial derivative $-\mathrm{i}\frac{\partial}{\partial t}$ acting on $C^{(1)}(\mathbb{S}^1)$. On the basis vectors b_k from (4), one obviously has

$$-\mathrm{i}\tfrac{\partial}{\partial t} b_k = k\, b_k\,, \qquad k \in \mathbb{Z}. \tag{10}$$

Our next aim is to describe noncommutative partial derivatives $\frac{\partial}{\partial z}$ and $\frac{\partial}{\partial \bar z}$ on the quantum disc algebra. As in [7], consider the first order differential *-calculus $\mathrm{d} : \mathcal{O}(\mathrm{D}_q) \longrightarrow \Omega(\mathrm{D}_q)$ given by $\Omega(\mathrm{D}_q) = \mathrm{d}z\,\mathcal{O}(\mathrm{D}_q) + \mathrm{d}z^*\mathcal{O}(\mathrm{D}_q)$ with $\mathcal{O}(\mathrm{D}_q)$-bimodule structure

$$z\,\mathrm{d}z = q^{-2}\mathrm{d}z\,z, \quad z^*\,\mathrm{d}z = q^2\mathrm{d}z\,z^*, \quad z\,\mathrm{d}z^* = q^{-2}\mathrm{d}z^*\,z, \quad z^*\mathrm{d}z^* = q^2\mathrm{d}z^*\,z^*,$$

satisfying the Leibniz rule $\mathrm{d}(fg) = \mathrm{d}(f)g + f\mathrm{d}(g)$ for all $f, g \in \mathcal{O}(\mathrm{D}_q)$. We define the partial derivatives $\frac{\partial}{\partial z}$ and $\frac{\partial}{\partial \bar z}$ by

$$d(f) = \mathrm{d}z\,\tfrac{\partial}{\partial z}(f) + \mathrm{d}z^*\,\tfrac{\partial}{\partial \bar z}(f), \qquad f \in \mathcal{O}(\mathrm{D}_q).$$

On monomials, one gets $\mathrm{d}(z^n z^{*k}) = \sum_{j=0}^{n-1} q^{-2j}\mathrm{d}z\,z^{n-1}z^{*k} + \sum_{l=0}^{k-1} q^{-2n+2l}\mathrm{d}z\,z^n z^{*k-1}$, therefore

$$\tfrac{\partial}{\partial z}(z^n z^{*k}) = q^{-2(n-1)}\tfrac{1-q^{2n}}{1-q^2}\,z^{n-1}z^{*k}, \qquad \tfrac{\partial}{\partial \bar z}(z^n z^{*k}) = q^{-2n}\tfrac{1-q^{2k}}{1-q^2}\,z^n z^{*k-1}.$$

From (6), it follows that the monomials $z^n z^{*k}$ span the linear space $\mathcal{O}(\mathrm{D}_q)$. Using the facts that commutators satisfy the Leibniz rule, $[z, z] = [z^*, z^*] = 0$, $\frac{1}{1-q^2}[z^*, z] = y^2$, and y^2 commutes with z and z^* in the same way as $\mathrm{d}z$ and $\mathrm{d}z^*$ do, one readily verifies that

$$\tfrac{\partial}{\partial z}(f) = \tfrac{1}{1-q^2}\,y^{-2}\,[z^*, f], \qquad \tfrac{\partial}{\partial \bar z}(f) = \tfrac{-1}{1-q^2}\,y^{-2}\,[z, f], \qquad f \in \mathcal{O}(\mathrm{D}_q). \tag{11}$$

In order extend the partial derivatives to an algebra of differentiable functions, we first observe that, by (6) and (8), each $f \in \mathcal{O}(\mathrm{D}_q)$ can be written

$$f = \sum_{n=0}^{N} z^n p_n(y^2) + \sum_{n=1}^{M} p_{-n}(y^2)\,z^{*n}, \qquad N, M \in \mathbb{N}, \tag{12}$$

with polynomials p_n in one variable. Furthermore, for all functions $f \in C(\mathrm{spec}\{y^2\})$, the formulas in (11) together with (8) give

$$\tfrac{\partial}{\partial z} f(y^2) = \tfrac{1}{1-q^2}\,y^{-2}\,(z^* f(y^2) - f(y^2)z^*) = -\frac{f(y^2) - f(q^2y^2)}{y^2 - q^2y^2}\,z^* = -\nabla_{q^2} f(y^2)\,z^*,$$
$$\tfrac{\partial}{\partial \bar z} f(y^2) = \tfrac{-1}{1-q^2}\,y^{-2}\,(z f(y^2) - f(y^2)z) = -z\,\frac{f(y^2) - f(q^2y^2)}{y^2 - q^2y^2} = -z\,\nabla_{q^2} f(y^2),$$

where

$$\nabla_{q^2} f(x) := \frac{f(x) - f(q^2x)}{x - q^2x}$$

denotes the q^2-difference operator and the operators $f(y^2)$ are defined by the spectral calculus of the self-adjoint operator y^2. The reason why we prefer to consider functions $f = f(y^2)$ instead of $f = f(y)$ is because $z \mapsto 1 - z\bar z$ is a differentiable function on $\bar{\mathrm{D}}$ but $z \mapsto \sqrt{1 - z\bar z}$ is not. Note that

$$\mathrm{spec}(y^2) = \{q^{2n} : n \in \mathbb{N}\} \cup \{0\}$$

and $\lim_{k\to\infty}\frac{f(q^{2k})-f(q^{2k+2})}{q^{2k}-q^{2k+2}} = \lim_{k\to\infty}\big(\frac{1}{1-q^2}\frac{f(q^{2k})-f(0)}{q^{2k}} - \frac{q^2}{1-q^2}\frac{f(q^{2k+2})-f(0)}{q^{2k+2}}\big) = f'(0)$ if the derivative of f in 0 exists. In that case, $\nabla_{q^2}\psi(y^2)$ defines a bounded operator on $\ell_2(\mathbb{N})$. This together with (12) motivates the following definition of an algebra of differentiable functions on the quantum disc:

$$\mathcal{F}^{(1)}(\mathrm{D}_q) := \left\{ \sum_{n=0}^{N} z^n f_n(y^2) + \sum_{n=1}^{M} f_{-n}(y^2)\, z^{*n} \;:\; M,N\in\mathbb{N},\ f_j \in C^{(1)}(\mathrm{spec}(y^2)) \right\}. \tag{13}$$

Observe that differentiability of $f \in C^{(1)}(\mathrm{spec}(y^2))$ amounts to the differentiability of f in 0 which is the only accumulation point of $\mathrm{spec}(y^2)$. Using that $f(y) \in \mathcal{F}^{(1)}(\mathrm{D}_q)$ if and only if $f(q^k y) \in \mathcal{F}^{(1)}(\mathrm{D}_q)$ for any $k \in \mathbb{Z}$, it is easily seen that $\mathcal{F}^{(1)}(\mathrm{D}_q)$ is a *-algebra of bounded operators on $\ell_2(\mathbb{N})$ containing $\mathcal{O}(\mathrm{D}_q)$. It follows from the (twisted) Leibniz rules

$$y^{-2}[\zeta, z^n f(y)] = y^{-2}[\zeta, z^n] f(y) + q^{-2n} z^n y^{-2}[\zeta, f(y)],$$
$$y^{-2}[\zeta, f(y) z^{*n}] = y^{-2}[\zeta, f(y)] z^{*n} + f(y) y^{-2}[\zeta, z^{*n}], \quad f \in C^{(1)}(\mathrm{spec}(y^2)),\ \zeta \in \{z, z^*\},\ n\in\mathbb{N},$$

that the actions of $\frac{\partial}{\partial z}$ and $\frac{\partial}{\partial \bar z}$ given by the expressions in (11) are well defined on $\mathcal{F}^{(1)}(\mathrm{D}_q)$.

4. Noncommutative integration

Let s denote the unilateral shift operator acting on $\ell_2(\mathbb{N})$ by $s e_e = e_{n+1}$. Using the relations

$$z = s\sqrt{1-q^2y^2}, \quad z^* = s^*\sqrt{1-y^2}, \quad f(y)s = s f(qy), \quad s^* f(y) = f(qy) s^*, \tag{14}$$

where $f \in L_\infty(\mathrm{spec}(y)) := \{g : \mathrm{spec}(y) \to \mathbb{C} : \text{bounded}\}$, one sees that each $\psi \in \mathcal{F}^{(1)}(\mathrm{D}_q)$ can be written $\psi = \sum_{n=0}^{N} s^n g_n(y^2) + \sum_{n=1}^{M} g_{-n}(y^2) s^{*n}$ with continuous (but not necessarily differentiable) functions $g_j \in C(\mathrm{spec}(y^2))$. For a noncommutative integration theory on the quantum disc, it happens to be more convenient to work with the *-algebra of bounded operators

$$\mathcal{F}(\mathrm{D}_q) := \left\{ \sum_{n=0}^{N} s^n g_n(y) + \sum_{n=1}^{M} g_{-n}(y)\, s^{*n} \;:\; M,N\in\mathbb{N},\ g_j \in L_\infty(\mathrm{spec}(y)) \right\}. \tag{15}$$

According to the above remark, $\mathcal{F}^{(1)}(\mathrm{D}_q) \subset \mathcal{F}(\mathrm{D}_q)$.

We follow [4, 5, 7] and define for $\alpha > 0$

$$\int_{\mathrm{D}_q}^{\alpha} : \mathcal{F}(\mathrm{D}_q) \longrightarrow \mathbb{C}, \qquad \int_{\mathrm{D}_q}^{\alpha} \psi := (1-q)\,\mathrm{Tr}_{\ell_2(\mathbb{N})}(\psi y^\alpha). \tag{16}$$

Since y^α is a positive compact operator for all $\alpha > 0$, the weighted trace $\int_{\mathrm{D}_q}^{\alpha}$ is a well-defined positive functional on $\mathcal{F}(\mathrm{D}_q)$. As the trace over weighted shift operators vanishes, we have

$$\int_{\mathrm{D}_q}^{\alpha} \psi = \int_{\mathrm{D}_q}\left(\sum_{n=0}^{N} s^n g_n(y) + \sum_{n=1}^{M} g_{-n}(y)\, s^{*n}\right) = (1-q)\sum_{n\in\mathbb{N}} g_0(q^n) q^{\alpha n}. \tag{17}$$

In terms of the Jackson integral $\int_0^1 f(y)\mathrm{d}_q y = (1-q)\sum_{n\in\mathbb{N}} f(q^n)q^n$, the functional $\int_{\mathrm{D}_q}^{\alpha}$ may be written

$$\int_{\mathrm{D}_q}^{\alpha}\left(\sum_{n=0}^{N} s^n g_n(y) + \sum_{n=1}^{M} g_{-n}(y) s^{*n}\right) = \int_0^1\int_{-\pi}^{\pi}\left(\sum_{n=0}^{N} \mathrm{e}^{\mathrm{i}n\theta} g_n(y) + \sum_{n=1}^{M} g_{-n}(y)\mathrm{e}^{-\mathrm{i}n\theta}\right) \mathrm{d}\theta\, y^{\alpha-1}\mathrm{d}_q y.$$

Journal of Physics: Conference Series **670** (2016) 012053 doi:10.1088/1742-6596/670/1/012053

By (14), the commutation relation between y^α and functions from $\mathcal{F}(\mathrm{D}_q)$ can be expressed by the automorphism $\sigma^\alpha : \mathcal{F}(\mathrm{D}_q) \longrightarrow \mathcal{F}(\mathrm{D}_q)$,

$$\sigma^\alpha\Big(\sum_{n=0}^{N} s^n g_n(y) + \sum_{n=1}^{M} g_{-n}(y) s^{*n}\Big) = \sum_{n=0}^{N} (q^{-\alpha}s)^n g_n(y) + \sum_{n=1}^{M} g_{-n}(y)\,(q^\alpha s)^{*n}. \tag{18}$$

Then $hy^\alpha = y^\alpha \sigma^\alpha(h)$ and hence, by the trace property,

$$\int_{\mathrm{D}_q}^{\alpha} gh = (1-q)\,\mathrm{Tr}_{\ell_2(\mathbb{N})}(ghy^\alpha) = (1-q)\,\mathrm{Tr}_{\ell_2(\mathbb{N})}(\sigma^\alpha(h)gy^\alpha) = \int_{\mathrm{D}_q}^{\alpha} \sigma^\alpha(h)g \tag{19}$$

for all $f, g \in \mathcal{F}(\mathrm{D}_q)$. Note also that

$$(\sigma^\alpha(h))^* = \sigma^{-\alpha}(h^*), \qquad h \in \mathcal{F}(\mathrm{D}_q),$$

where $\sigma^{-\alpha}$ denotes the inverse of σ^α.

Using $\mathrm{Tr}_{\ell_2(\mathbb{N})}(s^n s^{*k} f(y)y) = 0$ if $k \neq n$, one easily verifies that $\int_{\mathrm{D}_q}$ is faithful. Therefore

$$\langle f, g\rangle := \int_{\mathrm{D}_q}^{\alpha} f^* g, \qquad f, g \in \mathcal{F}(\mathrm{D}_q), \tag{20}$$

defines an inner product on $\mathcal{F}(\mathrm{D}_q)$. Its Hilbert space closure of will be denoted by $L_2(\mathrm{D}_q)$.

5. Representations on the quantum disc

The left multiplication of $\mathcal{O}(\mathrm{D}_q)$ on $\mathcal{F}(\mathrm{D}_q)$ defines a bounded *-representation of $\mathcal{O}(\mathrm{D}_q)$ on $L_2(\mathrm{D}_q)$ since

$$\langle xf, g\rangle = \int_{\mathrm{D}_q}^{\alpha} f^* x^* g = \langle f, x^* g\rangle, \qquad f, g \in \mathcal{F}(\mathrm{D}_q), \quad x \in \mathcal{O}(\mathrm{D}_q).$$

In [4], representations of the opposite algebra $\mathcal{O}(\mathrm{D}_q)^{\mathrm{op}}$ play also a crucial role in the definition of the Dirac operator. Recall that the opposite algebra $\mathcal{A}^{\mathrm{op}}$ of an algebra $\mathcal{A}$ is the vector space $\mathcal{A}$ with the opposite multiplication $a^{\mathrm{op}} b^{\mathrm{op}} = (ba)^{\mathrm{op}}$. $\mathcal{A}^{\mathrm{op}}$ remains to be a *-algebra if $\mathcal{A}$ is one. The right multiplication of $\mathcal{O}(\mathrm{D}_q)$ on $\mathcal{F}(\mathrm{D}_q)$ defines a representation of $\mathcal{O}(\mathrm{D}_q)^{\mathrm{op}}$ which is not a *-representation because

$$\langle x^{\mathrm{op}} f, g\rangle = \langle fx, g\rangle = \int_{\mathrm{D}_q}^{\alpha} x^* f^* g = \int_{\mathrm{D}_q}^{\alpha} f^* g \sigma^{-\alpha}(x^*) = \langle f, \sigma^{-\alpha}(x^*)^{\mathrm{op}} g\rangle \tag{21}$$

by (19).

Our next aim is to construct an orthonormal basis for $L_2(\mathrm{D}_q)$. To this end, we introduce the following notation: As customary, the symbol δ_{nm} denotes the Kronecker delta. For $k \in \mathbb{Z}$, let $\delta_{q^k} : \mathbb{R} \longrightarrow \{0, 1\}$ be defined by

$$\delta_{q^k}(t) := \begin{cases} 1, & t = q^k, \\ 0, & t \neq q^k. \end{cases}$$

If a is a densely defined operator, we set

$$a^{\#k} := \begin{cases} a^k, & k \geq 0, \\ a^{*k}, & k < 0. \end{cases} \tag{22}$$

Note that δ_{q^n} is continuous on spec(y) and $\delta_{q^n}(y)$ is the orthogonal projection onto the one-dimensional subspace span$\{e_n\} \subset \ell_2(\mathbb{N})$. For any (measurable) function $f : \mathrm{spec}(y) \longrightarrow \mathbb{C}$, we have

$$f(y)\,\delta_{q^n}(y) = \delta_{q^n}(y)\,f(y) = f(q^n)\,\delta_{q^n}(y). \tag{23}$$

In particular, $\delta_{q^m}(y)\delta_{q^n}(y) = \delta_{mn}\,\delta_{q^n}(y)$. Furthermore, it follows from (14) that

$$s\,\delta_{q^n}(y) = \delta_{q^n}(q^{-1}y)\,s = \delta_{q^{n+1}}(y)\,s, \quad s^*\,\delta_{q^n}(y) = \delta_{q^n}(qy)\,s^* = \delta_{q^{n-1}}(y)\,s^*. \tag{24}$$

As a consequence, for all $k, n \in \mathbb{N}$ such that $n < k$,

$$s^{*k}\,\delta_{q^n}(y) = \delta_{q^{n-k}}(y)\,s^{*k} = 0, \quad \delta_{q^n}(y)\,s^k = s^k\,\delta_{q^{n-k}}(y) = 0, \tag{25}$$

since $\delta_{q^{n-k}}(t) = 0$ on spec(y). From $ss^* = 1 - \delta_{q^0}(y)$, we get

$$ss^{*k}\,\delta_{q^n}(y) = (1-\delta_{q^0}(y))s^{*k-1}\,\delta_{q^n}(y) = s^{*k-1}(1-\delta_{q^{k-1}}(y))\,\delta_{q^n}(y) = s^{*k-1}\,\delta_{q^n}(y) \tag{26}$$

as we may assume that $n > k-1$ by (25). Note that $s^{\#k}s^{\#l} = s^{\#k+l}$ does not hold in general, for instance $s^{\#1}s^{\#-1} = 1 - \delta_{q^0}(y) \neq s^{\#0}$. However, from (26) and $s^*s = 1$, we conclude that

$$s^{\#l}s^{\#k}\,\delta_{q^n}(y) = s^{\#l+k}\,\delta_{q^n}(y) \quad \text{for all} \quad k, l \in \mathbb{Z}, \quad n \in \mathbb{N}. \tag{27}$$

Using (20), (27), (19), (23) and (17), we compute

$$\langle s^{\#k}\,\delta_{q^n}(y), s^{\#l}\,\delta_{q^m}(y)\rangle = \int_{\mathrm{D}_q}^{\alpha} \delta_{q^n}(y)\,s^{\#-k}\,s^{\#l}\delta_{q^m}(y) = \int_{\mathrm{D}_q}^{\alpha} s^{\#l-k}\,\delta_{q^m}(y)\,\delta_{q^n}(y) = (1-q)q^{\alpha n}\delta_{nm}\delta_{kl}. \tag{28}$$

The last equation is the key to constructing an orthonormal basis for $L_2(\mathrm{D}_q)$.

Proposition 1. *For $n \in \mathbb{N}$ and $k \in \mathbb{Z}$ such that $k \geq -n$, define*

$$\eta_{nk} := \tfrac{q^{-\alpha n/2}}{\sqrt{1-q}}\,s^{\#k}\,\delta_{q^n}(y). \tag{29}$$

Then $\{\eta_{nk} : n \in \mathbb{N},\ k \in \mathbb{Z},\ k \geq -n\}$ is an orthonormal basis for $L_2(\mathrm{D}_q)$.

Proof. It follows from (28) that the set defined in the proposition is orthonormal. Therefore it only remains to show that it is complete. As $L_2(\mathrm{D}_q)$ is the closure of $\mathcal{F}(\mathrm{D}_q)$, it suffices to show that the elements $s^k f(y)$ and $f(y)\,s^{*k}$ can be expanded in this basis, where $f \in L_\infty(\mathrm{spec}(y))$ and $k \in \mathbb{N}$. Since $\delta_{q^n}(y)$ is the orthogonal projection onto the eigenspace corresponding to the eigenvalue q^n of the self-adjoint operator y, the spectral theorem gives $f(y) = \sum_{n\in\mathbb{N}} f(q^n)\,\delta_{q^n}(y)$. Thus $s^k f(y) = \sqrt{1-q}\,\sum_{n\in\mathbb{N}} q^{\alpha n/2} f(q^n)\,\eta_{nk}$ and the series converges in $L_2(\mathrm{D}_q)$ since f is bounded and $\{q^{\alpha n/2}\}_{n\in\mathbb{N}} \in \ell_2(\mathbb{N})$. Similarly, by (25),

$$f(y)\,s^{*k} = \sum_{n\in\mathbb{N}} f(q^n)\,\delta_{q^n}(y)\,s^{*k} = \sum_{n\in\mathbb{N}} f(q^n)\,s^{*k}\,\delta_{q^{n+k}}(y) = \sqrt{1-q}\,q^{\alpha k/2}\sum_{n\in\mathbb{N}} q^{\alpha n/2} f(q^n)\,\eta_{n+k,k}\,,$$

where the sequence of coefficients $\{q^{\alpha n/2} f(q^n)\}_{n\in\mathbb{N}}$ belongs to $\in \ell_2(\mathbb{N})$. □

We will now compute the actions of our noncommutative multiplication and partial differential operators on this basis. To begin,

$$y\,\eta_{nk} = \tfrac{q^{-\alpha n/2}}{\sqrt{1-q}}\,y\,s^{\#k}\,\delta_{q^n}(y) = q^k\,\tfrac{q^{-\alpha n/2}}{\sqrt{1-q}}\,s^{\#k}\,y\,\delta_{q^n}(y) = q^{n+k}\,\tfrac{q^{-\alpha n/2}}{\sqrt{1-q}}\,s^{\#k}\,\delta_{q^n}(y) = q^{n+k}\,\eta_{nk}\,, \tag{30}$$

where we used (14) in the second equality and (23) in the third. Similarly,

$$y^{\mathrm{op}}\,\eta_{nk} = \eta_{nk}\,y = \tfrac{q^{-\alpha n/2}}{\sqrt{1-q}}\,s^{\#k}\,\delta_{q^n}(y)\,y = q^n\,\tfrac{q^{-\alpha n/2}}{\sqrt{1-q}}\,s^{\#k}\,\delta_{q^n}(y) = q^n\,\eta_{nk}\,. \tag{31}$$

Writing $z = s\sqrt{1-q^2y^2}$, it follows from (30) and (27) that

$$\begin{aligned} z\,\eta_{nk} &= \tfrac{q^{-\alpha n/2}}{\sqrt{1-q}}\,s\sqrt{1-q^2y^2}\,s^{\#k}\,\delta_{q^n}(y) = \sqrt{1-q^{2(n+k+1)}}\,\tfrac{q^{-\alpha n/2}}{\sqrt{1-q}}\,s^{\#k+1}\,\delta_{q^n}(y)\\ &= \sqrt{1-q^{2(n+k+1)}}\,\eta_{n,k+1}\,, \end{aligned} \tag{32}$$

and analogously, for $z^* = s^*\sqrt{1-y^2}$,

$$\begin{aligned} z^*\,\eta_{nk} &= \tfrac{q^{-\alpha n/2}}{\sqrt{1-q}}\,s^*\sqrt{1-y^2}\,s^{\#k}\,\delta_{q^n}(y) = \sqrt{1-q^{2(n+k)}}\,\tfrac{q^{-\alpha n/2}}{\sqrt{1-q}}\,s^{\#k-1}\,\delta_{q^n}(y)\\ &= \sqrt{1-q^{2(n+k)}}\,\eta_{n,k-1}\,, \end{aligned} \tag{33}$$

The actions of z^{op} and $z^{*\mathrm{op}}$ are calculated by applying (24), (23) and (29):

$$\begin{aligned} z^{\mathrm{op}}\,\eta_{nk} = \eta_{nk}\,z &= \tfrac{q^{-\alpha n/2}}{\sqrt{1-q}}\,s^{\#k}\,\delta_{q^n}(y)\,s\sqrt{1-q^2y^2} = \tfrac{q^{-\alpha n/2}}{\sqrt{1-q}}\,s^{\#k+1}\,\delta_{q^{n-1}}(y)\sqrt{1-q^2y^2}\\ &= q^{-\alpha/2}\sqrt{1-q^{2n}}\,\tfrac{q^{-\alpha(n-1)/2}}{\sqrt{1-q}}\,s^{\#k+1}\,\delta_{q^{n-1}}(y)\\ &= q^{-\alpha/2}\sqrt{1-q^{2n}}\,\eta_{n-1,k+1}\,, \end{aligned} \tag{34}$$

$$\begin{aligned} z^{*\mathrm{op}}\,\eta_{nk} = \eta_{nk}\,z^* &= \tfrac{q^{-\alpha n/2}}{\sqrt{1-q}}\,s^{\#k}\,\delta_{q^n}(y)\,s^*\sqrt{1-y^2} = \tfrac{q^{-\alpha n/2}}{\sqrt{1-q}}\,s^{\#k-1}\,\delta_{q^{n+1}}(y)\sqrt{1-y^2}\\ &= q^{\alpha/2}\sqrt{1-q^{2(n+1)}}\,\tfrac{q^{-\alpha(n+1)/2}}{\sqrt{1-q}}\,s^{\#k+1}\,\delta_{q^{n+1}}(y)\\ &= q^{\alpha/2}\sqrt{1-q^{2(n+1)}}\,\eta_{n+1,k-1}\,. \end{aligned} \tag{35}$$

Note that $(z^{\mathrm{op}})^* = q^{-\alpha}z^{*\mathrm{op}} = (\sigma^{-\alpha}(z^*))^{\mathrm{op}}$ as observed in (21).

By Equation (13) and the remark following it, the basis vectors η_{nk} belong to $\mathcal{F}^{(1)}(\mathrm{D}_q)$. As eigenvectors of the self-adjoint operator y, they also belong to the domain of the unbounded operator y^{-2}. We will use the formulas in (11) to compute the actions of $\frac{\partial}{\partial z}$ and $\frac{\partial}{\partial \bar z}$ on η_{nk}. From (30), (33) and (35), it follows that

$$\begin{aligned} \tfrac{\partial}{\partial z}\,\eta_{nk} &= \tfrac{1}{1-q^2}\,y^{-2}(z^*\,\eta_{nk} - \eta_{nk}\,z^*) = \tfrac{1}{1-q^2}\,y^{-2}(z^*\,\eta_{nk} - z^{*\mathrm{op}}\,\eta_{nk})\\ &= \tfrac{q^{-2(n+k)}}{1-q^2}\big(q^2\sqrt{1-q^{2(n+k)}}\,\eta_{n,k-1} - q^{\alpha/2}\sqrt{1-q^{2(n+1)}}\,\eta_{n+1,k-1}\big). \end{aligned} \tag{36}$$

Further, by (30), (32) and (34),

$$\begin{aligned} \tfrac{\partial}{\partial \bar z}\,\eta_{nk} &= \tfrac{-1}{1-q^2}\,y^{-2}(z\,\eta_{nk} - \eta_{nk}\,z) = \tfrac{-1}{1-q^2}\,y^{-2}(z\,\eta_{nk} - z^{\mathrm{op}}\,\eta_{nk})\\ &= -\tfrac{q^{-2(n+k)}}{1-q^2}\big(q^{-2}\sqrt{1-q^{2(n+k+1)}}\,\eta_{n,k+1} - q^{-\alpha/2}\sqrt{1-q^{2n}}\,\eta_{n-1,k+1}\big). \end{aligned} \tag{37}$$

To compare the representations of z and z^* with the irreducible ones from (5), it is convenient to change the basis. For $n, k \in \mathbb{N}$, set $e_{nk} := \eta_{n,k-n}$. As an immediate consequence of Proposition 1, $\{e_{nk} : n, k \in \mathbb{N}\}$ is an orthonormal basis for $L_2(\mathrm{D}_q)$ and the change of basis is given by the unitary operator

$$U : L_2(\mathrm{D}_q) \longrightarrow L_2(\mathrm{D}_q), \quad U\,e_{nk} = \eta_{n,k-n}, \quad U^*\,\eta_{nk} = e_{n,k+n}\,.$$

In the new basis, we have

$$y\,e_{nk} = U^* y\,\eta_{n,k-n} = q^k U^* \eta_{n,k-n} = q^k e_{nk}\,,$$

$$z\,e_{nk} = U^* z\,\eta_{n,k-n} = \sqrt{1-q^{2(k+1)}}\,U^* \eta_{n,k-n+1} = \sqrt{1-q^{2(k+1)}}\,e_{n,k+1}\,, \tag{38}$$

$$z^*\,e_{nk} = U^* z^*\,\eta_{n,k-n} = \sqrt{1-q^{2k}}\,U^* \eta_{n,k-n-1} = \sqrt{1-q^{2k}}\,e_{n,k-1}\,. \tag{39}$$

In particular, on each Hilbert space

$$\mathcal{H}_n := \operatorname{span}\{e_{nk} : k \in \mathbb{N}\} \cong \ell_2(\mathbb{N}), \tag{40}$$

we recover the unique irreducible *-representations (5) of the quantum disc $\mathcal{O}(\mathrm{D}_q)$. Further, for the opposite operators, one gets

$$y^{\mathrm{op}}\,e_{nk} = U^* y^{\mathrm{op}}\,\eta_{n,k-n} = q^n U^* \eta_{n,k-n} = q^n e_{nk}\,, \tag{41}$$

$$z^{\mathrm{op}}\,e_{nk} = U^* z^{\mathrm{op}}\,\eta_{n,k-n} = q^{-\alpha/2}\sqrt{1-q^{2n}}\,U^* \eta_{n-1,k-n+1} = q^{-\alpha/2}\sqrt{1-q^{2n}}\,e_{n-1,k}\,, \tag{42}$$

$$z^{*\mathrm{op}}\,e_{nk} = U^* z^{*\mathrm{op}}\,\eta_{n,k-n} = q^{\alpha/2}\sqrt{1-q^{2(n+1)}}\,U^* \eta_{n+1,k-n-1} = q^{\alpha/2}\sqrt{1-q^{2(n+1)}}\,e_{n+1,k}\,. \tag{43}$$

Now, setting $\zeta^{\mathrm{op}} := q^{\alpha/2} z^{\mathrm{op}}$, $\zeta^{*\mathrm{op}} := q^{-\alpha/2} z^{*\mathrm{op}}$ and

$$\mathcal{H}_k^{\mathrm{op}} := \operatorname{span}\{e_{nk} : n \in \mathbb{N}\} \cong \ell_2(\mathbb{N}), \tag{44}$$

we obtain an irreducible *-representation of $\mathcal{O}(\mathrm{D}_q)^{\mathrm{op}}$ on $\mathcal{H}_k^{\mathrm{op}}$, that is, $(\zeta^{\mathrm{op}})^* = \zeta^{*\mathrm{op}}$ and

$$\zeta^{\mathrm{op}}\zeta^{*\mathrm{op}} - q^2\zeta^{*\mathrm{op}}\zeta^{\mathrm{op}} = 1 - q^2. \tag{45}$$

Note that we have of course

$$[f, g^{\mathrm{op}}] = f\,g^{\mathrm{op}} - g^{\mathrm{op}} f = 0 \quad \text{for all } \; f \in \mathcal{O}(\mathrm{D}_q), \;\; g^{\mathrm{op}} \in \mathcal{O}(\mathrm{D}_q)^{\mathrm{op}}, \tag{46}$$

since the elements of $\mathcal{O}(\mathrm{D}_q)$ act only on the second index of e_{nk} and the elements of $\mathcal{O}(\mathrm{D}_q)^{\mathrm{op}}$ act only on the first.

We compute the actions of $\frac{\partial}{\partial z}$ and $\frac{\partial}{\partial \bar z}$ on e_{nk} by using the commutator representation (11) again. Thus

$$\tfrac{\partial}{\partial z}\,e_{nk} = \tfrac{1}{1-q^2}\,y^{-2}(z^* e_{nk} - z^{*\mathrm{op}} e_{nk}) = \tfrac{q^{-2k}}{1-q^2}\big(q^2\sqrt{1-q^{2k}}\,e_{n,k-1} - q^{\alpha/2}\sqrt{1-q^{2(n+1)}}\,e_{n+1,k}\big),$$

$$\tfrac{\partial}{\partial \bar z}\,e_{nk} = \tfrac{-1}{1-q^2}\,y^{-2}(z\,e_{nk} - z^{\mathrm{op}} e_{nk}) = -\tfrac{q^{-2k}}{1-q^2}\big(q^{-2}\sqrt{1-q^{2(k+1)}}\,e_{n,k+1} - q^{-\alpha/2}\sqrt{1-q^{2n}}\,e_{n-1,k}\big).$$

Finally recall that (3) defines a faithful *-representation of $C(\mathrm{SU}_q(2))$, and consider the Hilbert space $L_2(\mathrm{D}_q)\,\bar\otimes\,L_2(\mathbb{S}^1)$ with orthonormal basis $\{e_{nkl} : n, k \in \mathbb{N},\ l \in \mathbb{Z}\}$, where

$$e_{nkl} := e_{nk} \otimes b_l = \tfrac{q^{-\alpha n/2}}{\sqrt{1-q}}\, s^{\#k-n}\,\delta_{q^n}(y) \otimes \tfrac{1}{\sqrt{2\pi}}\mathrm{e}^{\mathrm{i}lt}. \tag{47}$$

As in (9), we define a *-representation of $\mathcal{O}(\mathrm{SU}_q(2))$ on $L_2(\mathrm{D}_q)\,\bar\otimes\,L_2(\mathbb{S}^1)$ by setting

$$c := y \otimes u = \sqrt{1 - zz^*} \otimes u, \qquad d := z \otimes 1. \tag{48}$$

It follows from (38)–(40) that $L_2(\mathrm{D}_q)\,\bar{\otimes}\,L_2(\mathbb{S}^1)$ decomposes into the direct sum

$$L_2(\mathrm{D}_q)\,\bar{\otimes}\,L_2(\mathbb{S}^1) = \bigoplus_{n\in\mathbb{N}} \mathcal{H}_n\,\bar{\otimes}\,L_2(\mathbb{S}^1) \cong \bigoplus_{n\in\mathbb{N}} \ell_2(\mathbb{N})\,\bar{\otimes}\,\ell_2(\mathbb{Z}),$$

and on each copy of $\ell_2(\mathbb{N})\,\bar{\otimes}\,\ell_2(\mathbb{Z})$, we recover the representation from (3). Hence the representation (48) decomposes into the infinite orthogonal sum of *-representations which are all unitarily equivalent to (3). In particular, (48) defines a faithful *-representation of $\mathcal{O}(\mathrm{SU}_q(2))$ and the C*-closure of the *-algebra generated by c and d from (48) is isomorphic to $C(\mathrm{SU}_q(2))$.

To obtain *-representations of the opposite algebras $\mathcal{O}(\mathrm{SU}_q(2))^{\mathrm{op}}$ and $C(\mathrm{SU}_q(2))^{\mathrm{op}}$, the generators c^{op} and d^{op} have to satisfy the opposite relations of (1) and (2). Analogous to $d = z\otimes 1$ and $c = \sqrt{1-zz^*}\otimes u$ with z satisfying (6) and u being a unitary generator of $C(\mathbb{S}^1)$, a *-representation of $\mathcal{O}(\mathrm{SU}_q(2))^{\mathrm{op}}$ is given by replacing z by ζ^{op} satisfying (45) (the opposite relation of (6)) and u by a unitary operator u^{op} with the same properties as u. For u^{op} we may take u^* which generates the same C*-algebra $C(\mathbb{S}^1)$ as u does. The adjoint of u is chosen for consistency because then c and d act as forward shifts, and c^{op} and d^{op} act as backward shifts. For $\zeta^{\mathrm{op}} = q^{\alpha/2}z^{\mathrm{op}}$ and $\zeta^{*\mathrm{op}} = q^{-\alpha/2}z^{*\mathrm{op}}$, as defined in the paragraph after Equation (43), we get $\sqrt{1-\zeta^{\mathrm{op}}\zeta^{*\mathrm{op}}} = \sqrt{1-z^{\mathrm{op}}z^{*\mathrm{op}}} = y^{\mathrm{op}}$. Thus, setting

$$c^{\mathrm{op}} := y^{\mathrm{op}}\otimes u^* = \sqrt{1-\zeta^{\mathrm{op}}\zeta^{*\mathrm{op}}}\otimes u^*, \qquad d^{\mathrm{op}} := \zeta^{\mathrm{op}}\otimes 1, \tag{49}$$

yields a *-representation of $\mathcal{O}(\mathrm{SU}_q(2))^{\mathrm{op}}$ on $L_2(\mathrm{D}_q)\,\bar{\otimes}\,L_2(\mathbb{S}^1)$. Moreover, with $\mathcal{H}_k^{\mathrm{op}}$ defined in (44), we have the Hilbert space decomposition

$$L_2(\mathrm{D}_q)\,\bar{\otimes}\,L_2(\mathbb{S}^1) = \bigoplus_{k\in\mathbb{N}} \mathcal{H}_k^{\mathrm{op}}\,\bar{\otimes}\,L_2(\mathbb{S}^1) \cong \bigoplus_{k\in\mathbb{N}} \ell_2(\mathbb{N})\,\bar{\otimes}\,\ell_2(\mathbb{Z}),$$

and, by (41)–(43) and (4), c^{op} and d^{op} act on each $\mathcal{H}_{k_0}^{\mathrm{op}}\,\bar{\otimes}\,L_2(\mathbb{S}^1) = \mathrm{span}\{e_{nk_0}\otimes b_l : n\in\mathbb{N},\ l\in\mathbb{Z}\}$ by

$$c^{\mathrm{op}}(e_{nk_0}\otimes b_l) = q^n\, e_{nk_0}\otimes b_{l-1}, \qquad d^{\mathrm{op}}(e_{nk_0}\otimes b_l) = \sqrt{1-q^{2(n-1)}}\, e_{n-1,k_0}\otimes b_l\,. \tag{50}$$

Note that (50) are the adjoint relations of (3), and that taking the opposite relations of (1) and (2) amounts to interchanging c and d with their adjoints c^* and d^*. Therefore, as much as (3) defines a faithful *-representation of $\mathcal{O}(\mathrm{SU}_q(2))$ and $C(\mathrm{SU}_q(2))$, (50) yields a faithful *-representation of $\mathcal{O}(\mathrm{SU}_q(2))^{\mathrm{op}}$ and $C(\mathrm{SU}_q(2))^{\mathrm{op}}$ and so does their direct sum representation on $L_2(\mathrm{D}_q)\,\bar{\otimes}\,L_2(\mathbb{S}^1)$. Finally we remark that the operators c and d from (48) commute with the operators c^{op} and d^{op} from (49) since u and u^* belong to the commutative C*-algebra $C(\mathbb{S}^1)$, and the operators in the left factor of the tensor products commute by (46). As a consequence, the representations of $C(\mathrm{SU}_q(2))$ and $C(\mathrm{SU}_q(2))^{\mathrm{op}}$ on $L_2(\mathrm{D}_q)\,\bar{\otimes}\,L_2(\mathbb{S}^1)$ commute.

For the convenience of the reader, we finish the paper by collecting the most important formulas in a final theorem.

Theorem 2. *Let $z, z^*\in B(\ell_2(\mathbb{N}))$ denote the generators of the quantum disc $\mathcal{O}(\mathrm{D}_q)$ given in (5) and let $y = \sqrt{1-z\,z^*}$. With $\mathcal{F}(\mathrm{D}_q)$ defined in (15), the *-algebra $\mathcal{F}(\mathrm{D}_q)\otimes L_\infty(\mathbb{S}^1)$ is considered an algebra of bounded functions on a local chart for quantum SU(2) in the following sense: Define an inner product on $\mathcal{F}(\mathrm{D}_q)\otimes L_\infty(\mathbb{S}^1)$ by*

$$\langle f\otimes\phi, g\otimes\psi\rangle := \int_{\mathrm{D}_q}^{\alpha} f^*g \int_{\mathbb{S}^1}\bar{\phi}\psi\,\mathrm{d}\lambda = (1-q)\,\mathrm{Tr}_{\ell_2(\mathbb{N})}(f^*gy^\alpha)\int_{\mathbb{S}^1}\bar{\phi}\psi\,\mathrm{d}\lambda, \quad \alpha>0,$$

*where λ stands for the Lebesgue measure on $\mathbb{S}^1$. The Hilbert space completion of $\mathcal{F}(\mathrm{D}_q)\otimes L_\infty(\mathbb{S}^1)$ will be denoted by $L_2(\mathrm{D}_q)\,\bar{\otimes}\,L_2(\mathbb{S}^1)$. Furthermore, with $\mathcal{F}^{(1)}(\mathrm{D}_q)$ described in (13), the *-sub-algebra $\mathcal{F}^{(1)}(\mathrm{D}_q)\otimes C^{(1)}(\mathbb{S}^1)$ of $\mathcal{F}(\mathrm{D}_q)\otimes L_\infty(\mathbb{S}^1)$ is viewed as an algebra of differentiable functions* *on the local chart. The Hilbert space* $L_2(\mathrm{D}_q)\,\bar{\otimes}\,L_2(\mathbb{S}^1)$ *has an orthonormal basis of differentiable functions* $\{e_{nkl} : n, k \in \mathbb{N},\ l \in \mathbb{Z}\} \subset \mathcal{F}^{(1)}(\mathrm{D}_q) \otimes C^{(1)}(\mathbb{S}^1)$*, where*

$$e_{nkl} = \tfrac{q^{-\alpha n/2}}{\sqrt{1-q}}\, s^{\#k-n}\, \delta_{q^n}(y) \otimes \tfrac{1}{\sqrt{2\pi}} \mathrm{e}^{ilt}.$$

The left multiplication $x\,(f\otimes\psi) := xf\otimes\psi$, $x\in\mathcal{O}(\mathrm{D}_q)$, $f\otimes\psi\in\mathcal{F}(\mathrm{D}_q)\otimes L_\infty(\mathbb{S}^1)$*, defines a bounded *-representation of* $\mathcal{O}(\mathrm{D}_q)$ *on* $L_2(\mathrm{D}_q)\,\bar{\otimes}\,L_2(\mathbb{S}^1)$*. On the above orthonormal basis, the operators* z, z^* *and* y *act by*

$$z\,e_{nkl} = \sqrt{1-q^{2(k+1)}}\, e_{n,k+1,l}\,, \quad z^*\, e_{nkl} = \sqrt{1-q^{2k}}\, e_{n,k-1,l}\,, \quad y\,e_{nkl} = q^k\, e_{nkl}\,.$$

The right multiplication $x^{\mathrm{op}}\,(f\otimes\psi) := fx\otimes\psi$, $x\in\mathcal{O}(\mathrm{D}_q)^{\mathrm{op}}$, $f\otimes\psi\in\mathcal{F}(\mathrm{D}_q)\otimes L_\infty(\mathbb{S}^1)$*, defines a bounded representation of* $\mathcal{O}(\mathrm{D}_q)^{\mathrm{op}}$ *on* $L_2(\mathrm{D}_q)\,\bar{\otimes}\,L_2(\mathbb{S}^1)$ *which is not a *-representation. The actions of* z^{op}, $z^{*\mathrm{op}} = q^{\alpha}(z^{\mathrm{op}})^*$ *and* $y^{\mathrm{op}} = (y^{\mathrm{op}})^*$ *are given by*

$$z^{\mathrm{op}}\, e_{nkl} = q^{-\alpha/2}\sqrt{1-q^{2n}}\; e_{n-1,kl}\,, \quad z^{*\mathrm{op}}\, e_{nkl} = q^{\alpha/2}\sqrt{1-q^{2(n+1)}}\; e_{n+1,kl}\,, \quad y^{\mathrm{op}}\, e_{nkl} = q^n\, e_{nkl}\,.$$

Let $u\in C(\mathbb{S}^1)$, $u(\mathrm{e}^{\mathrm{i}t}) = \mathrm{e}^{\mathrm{i}t}$*, denote the unitary generator of* $C(\mathbb{S}^1)$*. Setting*

$$c := y\otimes u = \sqrt{1-zz^*}\otimes u, \qquad d := z\otimes 1,$$
$$c^{\mathrm{op}} := y^{\mathrm{op}}\otimes u^* = \sqrt{1-z^{\mathrm{op}}z^{*\mathrm{op}}}\otimes u^*, \qquad d^{\mathrm{op}} := q^{\alpha/2} z^{\mathrm{op}}\otimes 1,$$

*yields commuting *-representations of* $C(\mathrm{SU}_q(2))$ *and* $C(\mathrm{SU}_q(2))^{\mathrm{op}}$ *on* $L_2(\mathrm{D}_q)\,\bar{\otimes}\,L_2(\mathbb{S}^1)$ *given by left and right multiplication, respectively. On basis vectors, these representations read*

$$c\,e_{nkl} = q^k\, e_{nk,l+1}\,, \qquad d\,e_{nkl} = \sqrt{1-q^{2(k+1)}}\; e_{n,k+1,l}\,,$$
$$c^{\mathrm{op}}\, e_{nkl} = q^n\, e_{nk,l-1}\,, \qquad d^{\mathrm{op}}\, e_{nkl} = \sqrt{1-q^{2n}}\; e_{n-1,kl}\,.$$

The partial derivatives $\frac{\partial}{\partial z}$, $\frac{\partial}{\partial \bar z}$ *and* $\frac{\partial}{\partial t}$ *act on differentiable functions* $g\otimes\phi\in\mathcal{F}^{(1)}(\mathrm{D}_q)\otimes C^{(1)}(\mathbb{S}^1)$ *by* $\frac{\partial}{\partial z}(g\otimes\phi) = \frac{\partial}{\partial z}g\otimes\phi$, $\frac{\partial}{\partial \bar z}(g\otimes\phi) = \frac{\partial}{\partial \bar z}g\otimes\phi$ *and* $\frac{\partial}{\partial t}(g\otimes\phi) = g\otimes\frac{\partial}{\partial t}\phi$*. On basis elements, one obtains*

$$\tfrac{\partial}{\partial z}\, e_{nkl} = \tfrac{q^{-2k}}{1-q^2}\big(q^2\sqrt{1-q^{2k}}\,\eta_{n,k-1,l} - q^{\alpha/2}\sqrt{1-q^{2(n+1)}}\; e_{n+1,kl}\big),$$
$$\tfrac{\partial}{\partial \bar z}\, e_{nkl} = -\tfrac{q^{-2k}}{1-q^2}\big(q^{-2}\sqrt{1-q^{2(k+1)}}\,\eta_{n,k+1,l} - q^{-\alpha/2}\sqrt{1-q^{2n}}\; e_{n-1,kl}\big),$$
$$\tfrac{\partial}{\partial t}\, e_{nkl} = \mathrm{i}l\, e_{nkl}\,.$$

Note that the representations of $C(\mathrm{SU}_q(2))$ and $C(\mathrm{SU}_q(2))^{\mathrm{op}}$ do not depend on α but the actions of $\frac{\partial}{\partial z}$ and $\frac{\partial}{\partial \bar z}$ do so.

Acknowledgments

The author thanks Ulrich Krähmer and Andrzej Sitarz for interesting discussions on the subject, and acknowledges financial support from CIC-UMSNH and the Polish Government Grant 2012/06/M/ST1/00169.

References

[1] Chari V and Pressley A 1994 *Quantum Groups* (Cambridge: Cambridge Univ. Press)
[2] Klimec S and Lesniewski A 1993 J. Funct. Anal. **155** 1
[3] Klimyk A U and Schmüdgen K 1998 *Quantum Groups and their Representations* (New York: Springer)
[4] Krähmer K and Wagner E 2015 *Twisted Dirac operator on quantum SU(2) in disc coordinates* (in preparation)
[5] Kürsten K-D and Wagner E 2007 Publ. RIMS Kyoto Univ. **43** 1
[6] Masuda T, Nakagami Y and Watanabe J 1990 *K*-Theory **4** 157
[7] Shklyarov D L, Sinel'shchikov S D, Vaksman L L 1997 Mat. Fiz. Anal. **4** 286
[8] Woronowicz S L 1987 Publ. RIMS, Kyoto Univ. **23** 117

Crossover temperature of the spin-1/2 XXZ chain with an impurity

Ryoko Yahagi[1], **Jun Sato**[2] **and Tetsuo Deguchi**[3]

[1] Department of Physics, Graduate School of Humanities and Sciences, Ochanomizu University, 2-1-1 Ohtsuka, Bunkyo-ku, Tokyo 112-8610, Japan
[2] Research Center for Advanced Science and Technology, University of Tokyo, 4-6-1 Komaba, Meguro-ku, Tokyo 153-8904, Japan
[3] Department of Physics, Faculty of Core Research, Ochanomizu University, 2-1-1 Ohtsuka, Bunkyo-ku, Tokyo 112-8610, Japan

E-mail: [1]`yahagi@hep.phys.ocha.ac.jp`, [2]`jsato@jamology.rcast.u-tokyo.ac.jp`, [3]`deguchi@phys.ocha.ac.jp`

Abstract. We study exactly the effect of an impurity in the interacting quantum spin chain at low temperature by solving the integrable spin-1 2 Z periodic chain with an impurity through the algebraic and thermal Bethe ansatz methods. In particular, we investigate how the crossover temperature for the impurity specific heat depends on the impurity parameter, i.e. the coupling of the impurity to other spins, and show that it is consistent with the analytic expression that is obtained by setting the impurity susceptibility to be proportional to the inverse of the crossover temperature. In the model, two types of crossover behavior appear: one from the high-temperature regime to the low-temperature Kondo regime and another from the N-site homogeneous chain to the $(N-1)$-site chain with a decoupled free impurity spin, with respect to the temperature and the impurity parameter, respectively.

1. Introduction

The Kondo effect was discovered experimentally in the 1930s [1] and was theoretically explained first in the 1960s [2]. However, the Kondo problem and related subjects in many-body problem still evoke considerable interest in various theoretical and experimental researches [3, 4, 5]. First, quantum impurity systems show universal critical behavior at low temperature, which is characterized by the ratio of the impurity susceptibility $\chi_{\rm imp}$ to the impurity specific heat $c_{\rm imp}$ divided by temperature T: $r = (\pi^2/3)\chi_{\rm imp}/(c_{\rm imp}/T)$ [6]. We call it the Wilson ratio. It was exactly shown by the Bethe ansatz that it is given by 2 for the Kondo model [7, 8]. Furthermore, the effect of an impurity embedded in a one-dimensional interacting quantum system or in a Tomonaga-Luttinger liquid [9] has been one of the topics evoking considerable interest during the 1990s and during the last decade. It has been investigated by different methods such as renormalization group techniques [10, 11, 12], conformal field theories (CFT) [13, 14], numerical techniques with CFT [15, 16], and the Bethe-ansatz method [17, 18, 19, 20, 21, 22, 23, 24, 25, 26]. Recently, it is studied by functional renormalization group [27] and also experimentally in a quasi-one-dimensional conductor [28]. However, it is still rare that the finite-temperature thermodynamic behavior is explicitly and exactly shown by a theoretical method for a large but finite lattice system without making any approximation or assumption.

Journal of Physics: Conference Series **670** (2016) 012054 doi:10.1088/1742-6596/670/1/012054

In this paper we study the finite-temperature behavior of an integrable model of the spin-1/2 anti-ferromagnetic XXZ periodic chain with a spin-1/2 impurity and make an additional report to Ref. [29]. In particular, we investigate how the crossover temperature T_c for the impurity specific heat $c_{\rm imp}$ depends on the impurity parameter x and show that the x-dependence of crossover temperature T_c is consistent with the analytic expression derived by setting the impurity susceptibility $\chi_{\rm imp}$ to be proportional to the inverse of the crossover temperature: $1/T_c \propto \chi_{\rm imp}(x)$. Here the impurity parameter x is related to the coupling strength between the impurity spin and other interacting spins.

In the XXZ impurity model we can exactly study the effect of an impurity embedded in the interacting quantum spins in one dimension, while in the Kondo model the effect of an impurity coupled with itinerant electrons in three dimensions. The XXZ impurity model and its variants have been investigated by many authors. Schlottmann studied the integrable spin-S XXZ spin chain with one spin-S' impurity [23] in association with the multi-channel Kondo effect [18, 19, 25]. The spin-1/2 impurity in the open spin-1/2 XXX chain was studied by Frahm and Zvyagin [21] and it was shown that the Kondo-like temperature exists, while the periodic XXZ chain with an impurity was studied by Eckle et al. [20]. The thermodynamic behavior of the spin-1/2 impurities in the spin-1/2 XXZ chain was shown in Ref. [24] for several distributions of the impurity parameters. However, the analytic expressions of the impurity susceptibility and the impurity specific heat at low temperature in Ref. [23] are not accurate enough to evaluate the Wilson ratio correctly, in particular, when the impurity parameter is small.

The integrable model of the spin-1/2 XXZ periodic chain with a spin-1/2 impurity studied in Ref. [29] corresponds to the case of $S = S' = \frac{1}{2}$ in the XXZ impurity model of Ref. [23]. The finite-temperature behavior of the integrable XXZ impurity model is investigated by numerically solving the truncated integral equations of the thermal Bethe ansatz [30, 31, 32, 33] and evaluate the specific heat and the entropy numerically [29]. By plotting graphs of the impurity specific heat $c_{\rm imp}$ versus temperature T and those of the impurity entropy versus temperature T, it is shown that the impurity spin gradually becomes a free spin through pseudo-decoupling from other spins, while in low temperature it couples strongly to them such as the Kondo effect [29]. Here, we call it *p eud -dec uplin* , if the exchange coupling between the impurity spin and other spins is very small but nonzero. It occurs in the present model when the absolute value of the impurity parameter is very large.

It is shown that the Wilson ratio at low temperature is given by $r = 2\pi/(\pi - \zeta)$ for the spin-1/2 anti-ferromagnetic XXZ periodic chain with a spin-1/2 impurity, where the XXZ anisotropy parameter is given by $= \cos\zeta$ ($0 \le \zeta < \pi$) [29]. It is independent of the impurity parameter x if its absolute value is small enough with respect to the given temperature. Thus, the Hamiltonians of the XXZ impurity model with the same XXZ anisotropy parameter but different values of impurity parameter x are classified in the same universality class [29].

The contents of the paper consist of the following. In section 2 we derive the expression of the XXZ impurity Hamiltonian in terms of the local spin operators S_j^a. We show that the ground-state energy evaluated by diagonalizing the XXZ impurity Hamiltonian expressed in terms of the local spin operators is consistent with that obtained by solving the Bethe ansatz equations numerically for several different values of the impurity parameter. In section 3 we review the analytic expression of the impurity susceptibility at zero temperature derived by the Wiener-Hopf method in Ref. [29]. In section 4, after we briefly review the method for evaluating the specific heat via the thermal Bethe ansatz in Ref. [29], we plot the graphs of the impurity specific heat versus impurity parameter x for five different values of temperature T. We then show numerically that the crossover temperature T_c as a function of impurity parameter x is derived by setting the inverse of T_c to be equal to impurity susceptibility $\chi_{\rm imp}(x)$: $A/T_c = \chi_{\rm imp}(x)$ with some constant A. Finally, in section 5 we give concluding remarks.

2. Integrable model with an im urity

e XXZ Ha il nian wi an i puri

We now formulate the integrable spin-1/2 anti-ferromagnetic XXZ chain with a spin-1/2 impurity through the algebraic Bethe ansatz. Let us denote by $V_0, V_1, \ldots, V_N$ the two-dimensional vector spaces over **C**. For the N-site chain we consider the quantum space $V_1 \otimes V_2 \otimes \cdots \otimes V_N$, where the spin operators on the jth site act on V_j for $j = 1, 2, \ldots, N$. We call V_0 the auxiliary space. We denote the transfer matrix of the XXZ spin chain by $\tau_{1\cdots N}(\lambda|\xi_1, \cdots, \xi_N)$ acting on the quantum space with the spectral parameter λ and inhomogeneity parameters $\xi_1, \xi_2, \ldots, \xi_N$. We shall define it shortly. The XXZ Hamiltonian with an impurity is given by the logarithmic derivative of an inhomogeneous transfer matrix

$$H_{\rm XXZ}(x) = \frac{\sinh(\mathrm{i}\zeta)}{2}\frac{d}{d\lambda}\log \tau_{1\cdots N}\left(\lambda \Big| \frac{\mathrm{i}\zeta}{2} - x, \frac{\mathrm{i}\zeta}{2}, \cdots, \frac{\mathrm{i}\zeta}{2}\right)\Bigg|_{\lambda\to\frac{\mathrm{i}\zeta}{2}}, \tag{2.1}$$

where we call the parameter x the impurity parameter and we assume that it is real. The impurity Hamiltonian $H_{\rm XXZ}(x)$ in (2.1) becomes the standard XXZ Hamiltonian when $x = 0$. We define the inhomogeneous transfer matrix $\tau_{1\cdots N}$ by the following product of the R-matrices

$$\tau_{1\cdots N}(\lambda|\xi_1, \cdots, \xi_N) = \mathrm{tr}_0 R_{0N}(\lambda - \xi_N)\cdots R_{01}(\lambda - \xi_1) \tag{2.2}$$

where inhomogeneity parameters $\xi_1, \xi_2, \ldots, \xi_N$ are arbitrary and the R-matrices of the XXZ model acting on the tensor product $V_0 \otimes V_n$ are given by

$$R_{0n}(\lambda) = \begin{pmatrix} 1 & & & \\ & b(\lambda) & c(\lambda) & \\ & c(\lambda) & b(\lambda) & \\ & & & 1 \end{pmatrix}_{[0,n]}. \tag{2.3}$$

Here, suffix $[0, n]$ means that the four-by-four matrix acts on the tensor product space $V_0 \otimes V_n$ and functions $b(\lambda)$ and $c(\lambda)$ are expressed as

$$b(\lambda) = \frac{\sinh(\lambda)}{\sinh(\lambda + \mathrm{i}\zeta)}, \quad c(\lambda) = \frac{\sinh(\mathrm{i}\zeta)}{\sinh(\lambda + \mathrm{i}\zeta)}, \tag{2.4}$$

and parameter ζ is related to the XXZ anisotropy parameter by $= \cos\zeta$ with $0 \le \zeta < \pi$.

Through straightforward calculation we obtain the following expression of the impurity Hamiltonian $H_{\rm XXZ}(x)$ in terms of local spin operators $S_j^{\pm}$ and S_j^z on the jth sites [29].

$$\begin{aligned} H_{\rm XXZ}(x) &= \sum_{n=2}^{N-1}\left[\frac{1}{2}(S_n^+S_{n+1}^- + S_n^-S_{n+1}^+) + \ \ \left(S_n^zS_{n+1}^z - \frac{1}{4}\right)\right] \\ &+ c^+c^-\left[\frac{\cosh x}{2}(S_N^+S_1^- + S_N^-S_1^+) + \ \ \left(S_N^zS_1^z - \frac{1}{4}\right)\right] \\ &+ c^+c^-\left[\frac{\cosh x}{2}(S_1^+S_2^- + S_1^-S_2^+) + \ \ S_1^zS_2^z\right] \\ &+ b^+b^-\left[\frac{\ }{2}(S_N^+S_2^- + S_N^-S_2^+) + \ \ S_N^zS_2^z\right] - \frac{\ }{4} \\ &+ b^+c^-\Big[\cosh x(S_N^+S_2^- - S_N^-S_2^+)S_1^z \\ &\qquad - (S_N^+S_1^- - S_N^-S_1^+)S_2^z \\ &\qquad - (S_1^+S_2^- - S_1^-S_2^+)S_N^z\Big], \end{aligned} \tag{2.5}$$

where symbols $b^{\pm}$ and $c^{\pm}$ are given by

$$b^{\pm} = b(\pm x) = \frac{\sinh x}{\sinh(x \pm i\zeta)}, \qquad c^{\pm} = c(\pm x) = \pm \frac{\sinh(i\zeta)}{\sinh(x \pm i\zeta)}, \tag{2.6}$$

and the spin operators $S_j^{\pm}$ and S_j^z are expressed in terms of the Pauli matrices as

$$S_j^{\pm} = \frac{1}{2}\sigma_j^x \pm i\frac{1}{2}\sigma_j^y, \quad S_j^z = \frac{1}{2}\sigma_j^z, \quad \text{for} \quad j = 1, 2, \ldots, N. \tag{2.7}$$

We confirm in the explicit expression (2.5) that the impurity Hamiltonian $H_{\rm XXZ}(x)$ gives the spin-1/2 anti-ferromagnetic XXZ chain if impurity parameter x vanishes. It also follows from (2.5) that the impurity Hamiltonian $H_{\rm XXZ}(x)$ is Hermitian for any real value of impurity parameter x.

In the limit of sending the impurity parameter to infinity: $x \to \infty$, the impurity Hamiltonian (2.5) reduces to the following:

$$\begin{aligned} H_{\rm XXZ}(\infty) &= \sum_{n=2}^{N-1} \left[\frac{1}{2}(S_n^+ S_{n+1}^- + S_n^- S_{n+1}^+) + \quad \left(S_n^z S_{n+1}^z - \frac{1}{4}\right)\right] \\ & \quad + \frac{}{2}\left(S_N^+ S_2^- + S_N^- S_2^+\right) + \quad \left(S_N^z S_2^z - \frac{1}{4}\right) \\ & \quad + 2\sqrt{1 - \ ^2}\left(S_N^x S_2^y - S_N^y S_2^x\right) S_1^z. \end{aligned} \tag{2.8}$$

Furthermore, in the XXX limit: $\ \to 1$ it becomes

$$H_{\rm XXX}(\infty) = \sum_{n=2}^{N-1} \left[S_n^x S_{n+1}^x + S_n^y S_{n+1}^y + S_n^z S_{n+1}^z - \frac{1}{4}\right] + \left(S_N^x S_2^x + S_N^y S_2^y + S_N^z S_2^z - \frac{1}{4}\right). \tag{2.9}$$

Thus, the impurity spin operators S_1^a $(a = x, y, z)$ are decoupled from those of other sites.

e Be e an a equa i n wi an i puri

Let us define functions $\theta_n(z)$ for positive integers $n = 1, 2, \ldots,$ by

$$\theta_n(z) = \mathrm{i} \log \left[-\frac{\sinh(z + \frac{in\zeta}{2})}{\sinh(z - \frac{in\zeta}{2})} \right]. \tag{2.10}$$

The Bethe ansatz equations (BAE) for the impurity Hamiltonian (2.1) in the M down-spin sector are given by

$$(N-1)\theta_1(z_l) + \theta_1(z_l + x) = 2\pi I_l + \sum_{j=1; j \neq l}^{M} \theta_2(z_l - z_j), \quad \text{for } l = 1, \ldots, M. \tag{2.11}$$

Here I_l are called the Bethe quantum numbers, and they are given by integers or half-integers according to the following rule:

$$I_l \equiv \frac{N - M + 1}{2} \pmod 1, \quad \text{for } l = 1, 2, \ldots, M. \tag{2.12}$$

We note that functions $\theta_n(z)$ are expressed in terms of the arctangent function as follows.

$$\theta_n(z) = 2\tan^{-1}\left(\cot\left(\frac{n\zeta}{2}\right)\tanh z\right). \tag{2.13}$$

By solving BAE (2.11) numerically we can evaluate the energy eigenvalues of the Bethe ansatz eigenvectors with M down-spins. We call a solution of BAE $\{z_\ell | \ell = 1, 2, \ldots, M\}$ the Bethe roots. The energy E of a Bethe ansatz eigenstate with the Bethe roots $\{z_\ell\}$ is expressed as

$$E = -\sum_{l=1}^{M} \frac{\sin^2 \zeta}{\cosh 2z_l - \cos \zeta} \,. \tag{2.14}$$

We note that the energy depends on impurity parameter x, since the solution $\{z_j\}$ depends on the parameter x through BAE in (2.11).

Gr und a e f e XXZ pin c ain wi an i puri
It is not clear how the Bethe quantum numbers $\{I_j\}$ should depend on the impurity parameter x for a given Bethe ansatz eigenstate of the XXZ impurity Hamiltonian. Here we recall that the Bethe quantum numbers appear in BAE (2.11). It is therefore not trivial whether the Bethe quantum numbers for the ground state with $x = 0$ also gives the Bethe quantum numbers for the ground state with $x \neq 0$.

We have a conjecture that the Bethe quantum numbers of the ground state do not change if impurity parameter x is real even when $x \neq 0$. By taking advantage of the explicit expression (2.5) of the impurity Hamiltonian we can evaluate its eigenvalues through exact numerical diagonalization, i.e. we can numerically diagonalize the impurity Hamiltonian through the expression (2.5). We thus confirm that the set of the Bethe quantum numbers of the ground state for $x = 0$ gives those in the case of $x \neq 0$.

x	0	1	2	3	4	5
exact diagonalization	-4.392695167	-3.961714971	-3.743119726	-3.680990863	-3.660728144	-3.653612539
Bethe ansatz	-4.392695167	-3.961714971	-3.743119726	-3.680990863	-3.660728144	-3.653612539

Table 2.1. Ground-state energy of the XXZ impurity Hamiltonian (2.5) with $N = 8$ and $= 0.6$ for six values of impurity parameter ($x = 0, 1, \cdots, 5$) evaluated by exact numerical diagonalization and by the Bethe ansatz. Here we assume that the Bethe quantum numbers of the ground state do not depend on impurity parameter x and are given by $\{\pm 1/2, \pm 3/2\}$.

We have thus confirmed that the lowest energy level of (2.5) is completely consistent with the eigenvalue evaluated from BAE (2.11) in the cases of $N \leqq 10$ with respect to numerical errors. For an illustration, some results for $N = 8$ are shown up to 10 decimal degits in Table 2.1. They are obtained by assuming the same set of the Bethe quantum numbers. We have another conjecture that this correspondence is valid for any larger number of sites.

3. Im urity susce tibility through the Wiener-Ho method

We now derive the analytic expression of the impurity magnetic susceptibility at zero temperature. We add a small magnetic field h to the impurity Hamiltonian (2.5):

$$H' = H_{\mathrm{XXZ}}(x) - 2h \sum_{n=1}^{N} S_n^z. \tag{3.15}$$

By taking the thermodynamic limit of BAE (2.11) through the Euler-Maclaurin formula, we derive the following integral equation:

$$\rho(z, x) + \int_{-B}^{B} a_2(z - z') \rho(z', x) dz' = \frac{1}{N} \{(N - 1) a_1(z) + a_1(z + x)\} \,. \tag{3.16}$$

Here parameter B denotes the Fermi point determined by the magnetic field h, and $\rho(z,x)$ the density of the Bethe roots in the ground-state solution of BAE (2.11), where functions $a_n(z)$ for $n = 1, 2, \ldots$, denote the derivatives of functions $\theta_n(z)$: $a_n(z) = \theta_n'(z)/2\pi$ and they given by

$$a_n(z) = -\frac{\mathrm{i}}{2\pi} \frac{\sinh(\mathrm{i}n\zeta)}{\sinh(z + \frac{\mathrm{i}n\zeta}{2}) \sinh(z - \frac{\mathrm{i}n\zeta}{2})}. \tag{3.17}$$

We note that we have the infinite Fermi point, $B = \infty$, for $h = 0$.

The magnetization per site $s^z(x)$ and the energy per site $e(x)$ are written in terms of the integrals of the root density $\rho(z,x)$ as follows.

$$\begin{aligned} s^z(x) &= \frac{1}{2} - \int_{|z|<B} \rho(z,x)dz \\ &= \frac{\pi}{2\pi - 2\zeta} \int_{|z|>B} \rho(z,x)dz, \end{aligned} \tag{3.18}$$

$$e(x) = \pi \sin\zeta \int_{|z|>B} \sigma_0(z)\rho(z,x)dz, \tag{3.19}$$

where the bulk root density $\sigma_0(z)$ is given by

$$\sigma_0(z) = \frac{1}{2\zeta}\mathrm{sech}\frac{\pi z}{\zeta}. \tag{3.20}$$

Through the Wiener-Hopf method [29] the magnetization per site $s^z(x)$ is given by

$$\begin{aligned} s^z(x) &\simeq \frac{\mathrm{i}\pi}{N\zeta(\pi - \zeta)} G_+(0) \frac{G_-\left(-\frac{\mathrm{i}\pi}{2}\right)}{\frac{\mathrm{i}\pi}{\zeta}} \frac{G_+(0)h\zeta}{(\pi - \zeta)\sin\zeta G_+\left(\frac{\mathrm{i}\pi}{2}\right)} \left(N - 1 + \cosh\frac{\pi x}{\zeta}\right) \\ &= \frac{2\zeta h}{N\pi(\pi - \zeta)\sin\zeta} \left(N - 1 + \cosh\frac{\pi x}{\zeta}\right). \end{aligned} \tag{3.21}$$

Here, function $G_+(k)$ $(k = \mathrm{i}\pi z)$ is expressed in terms of the gamma function $\Gamma(z)$ by

$$G_+(\mathrm{i}\pi z) = \sqrt{2\pi\left(1 - \frac{1}{\gamma}\right)} \left(\frac{(\gamma - 1)^{\gamma - 1}}{\gamma^\gamma}\right)^z \frac{\Gamma(\gamma z + 1)}{\Gamma(\frac{1}{2} + z)\Gamma((\gamma - 1)z + 1)}, \tag{3.22}$$

and γ is given by $\gamma = \pi/\zeta$, and function $G_-(k)$ is determined by the following relation:

$$G_-(k) = G_+(-k). \tag{3.23}$$

In equation (3.21) we have assumed that magnetic field h is very small, so that we consider only linear terms of h. The Fermi point B is determined as a function of h by the following condition:

$$\frac{\partial}{\partial B}\left(e(x) - 2hs^z(x)\right) = 0. \tag{3.24}$$

We therefore have

$$\exp\left(-\frac{\pi B}{\zeta}\right) = \frac{G_+(0)\zeta h}{(\pi - \zeta)\sin\zeta G_+\left(\frac{\mathrm{i}\pi}{2}\right)}. \tag{3.25}$$

Let us define the impurity magnetization $s^z_{\mathrm{imp}}(x)$ at zero temperature by

$$Ns^z(x) = (N - 1)s^z(0) + s^z_{\mathrm{imp}}(x). \tag{3.26}$$

Here, $Ns^z(x)$ gives the total magnetization of the impurity XXZ spin chain with impurity parameter x. We also express it by $S^z_{\rm tot}$. We then define the impurity susceptibility $\chi_{\rm imp}(x)$ by

$$\chi_{\rm imp} = \left. \frac{\partial\, 2s^z_{\rm imp}(x)}{\partial h} \right|_{h=0} . \tag{3.27}$$

The impurity susceptibility at zero temperature is therefore given by

$$\chi_{\rm imp}(x) = \frac{4\zeta}{\pi(\pi-\zeta)\sin\zeta} \cosh\frac{\pi x}{\zeta} . \tag{3.28}$$

4. Im urity s ecific heat

C upled in e ral equa i n f e er al Be e an a

We now evaluate the crossover temperature by numerically evaluating the specific heat per site $c(x,T)$ of the impurity XXZ spin chain at temperature T. We first define the impurity specific heat $c_{\rm imp}(x,T)$ at temperature T with impurity parameter x by [29]

$$Nc(x,T) = (N-1)c(0,T) + c_{\rm imp}(x,T) . \tag{4.29}$$

Here, $Nc(x,T)$ gives the total specific heat of the impurity XXZ spin chain with impurity parameter x at temperature T.

Let us consider the string solutions for the BAE at $\zeta = \pi/3$. According to the string hypothesis [30, 33], there are only three types of solutions: 1-string solutions with even parity (i.e., real solutions) $\{z_j^{(1+)}\}$, 2-string solutions $\{z_j^{(2)} \pm i\zeta/2\}$, and real solutions of 1-string with odd parity $z_j^{(1-)} + i\pi/2$. Here we assume that all the three types of string centers, $z_j^{(1+)}, z_j^{(2)}$ and $z^{(1-)}$ are given by real numbers.

For the density functions of the three types of string solutions appearing at $\zeta = \pi/3$ [30] a set of coupled integral equations truncated at $\zeta = \pi/3$ are derived through the thermal Bethe ansatz, which will be formulated shortly. We express the free energy in terms of the solution to the thermal Bethe ansatz equations. By taking the second derivative of the free energy with respect to temperature we derive the integral expression for the specific heat of the model [32]. Thus, by solving the thermal Bethe ansatz equations numerically, we obtain the specific heat of the impurity XXZ spin chain denoted by $c(x,T)$ as follows.

$$c(x,T) = \frac{1}{N} \int_{-\infty}^{\infty} \{\alpha_1 v_1 + \alpha_2 (2v_2 + b_1 * v_1 + 2b_2 * v_2)\}\, dz, \tag{4.30}$$

where symbol $*$ denotes the convolution and functions $\alpha_j(z)$ are given by

$$\alpha_j(z) = (N-1)a_j(z) + a_j(z+x) \quad \text{for } j=1,2. \tag{4.31}$$

Here functions $a_j(z)$ and $b_j(z)$ are given by

$$a_j(z) = \frac{1}{2\pi} \frac{2\sin\zeta q_j}{\cosh 2z + \cos\zeta q_j}, \qquad q_1 = 2, \quad q_2 = 1, \tag{4.32}$$

$$b_1(z) = \frac{3\cosh(3z/2)}{\sqrt{2}\pi\cosh 3z}, \qquad b_2(z) = \frac{3}{4\pi\cosh(3z/2)} . \tag{4.33}$$

With $\kappa = -\pi \sin\zeta$, functions v_j in (4.30) are given by the solutions of the following equations.

$$\begin{cases} v_1 = \frac{1}{T^2}\frac{u_1{}^2}{e^{h_1}-1} + \left(1-e^{-h_1}\right)\left[-2b_2 * v_1 - 2b_1 * v_2\right] \\ v_2 = \frac{1}{T^2}\frac{u_2{}^2}{e^{h_2}-1} + \left(1-e^{-h_2}\right)\left[-b_1 * v_1 - 2b_2 * v_2\right] \end{cases}, \tag{4.34}$$

$$\begin{cases} u_1 = \left(1-e^{-h_1}\right)\left[-\kappa b_1 - 2b_2 * u_1 - 2b_1 * u_2\right] \\ u_2 = \left(1-e^{-h_2}\right)\left[-\kappa b_2 - b_1 * u_1 - 2b_2 * u_2\right] \end{cases}, \tag{4.35}$$

$$\begin{cases} h_1 = \log\left[1+\exp\left[-\frac{\kappa}{T}b_1 - 2b_2 * h_1 - 2b_1 * h_2\right]\right] \\ h_2 = \log\left[1+\exp\left[-\frac{\kappa}{T}b_2 - b_1 * h_1 - 2b_2 * h_2\right]\right] \end{cases}. \tag{4.36}$$

We can solve the thermal Bethe ansatz equations (4.34), (4.35) and (4.36) as follows. Firstly, we solve the coupled integral equations (4.36) for h_1 and h_2. Secondly, we solve equations (4.35) for u_1 and u_2 from the result of h_1 and h_2. Thirdly, we solve equations (4.34) for v_1 and v_2, from which we have the estimates of specific heat $c(x,T)$ and then $c_{\text{imp}}(x,T)$.

Cr ver e pera ure f r i puri pecific ea ver u i puri para e er

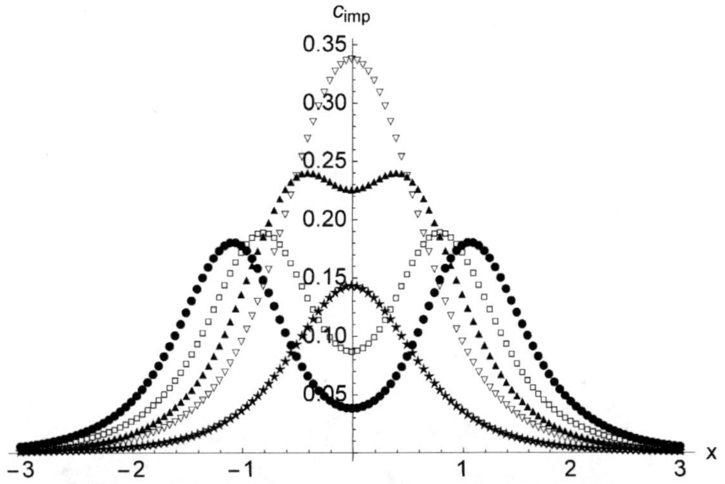

Figure 1. Impurity specific heat c_{imp} versus impurity parameter x at five different temperatures T. The data points for $T = 10^{-4/3}, 10^{-1}, 10^{-2/3}, 10^{-1/3}$ and 10^0 are depicted by filled circles, open squares, filled upper triangles, open lower triangles and filled stars, respectively.

In Figure 1 the estimates of impurity specific heat $c_{\text{imp}}(x,T)$ are plotted against impurity parameter x for five fixed values of temperature T: $T = 10^{-4/3}, 10^{-1}, 10^{-2/3}, 10^{-1/3}$ and 10^0.

Two peaks are clearly seen in the graph of c_{imp} versus x at the lowest temperature $T = 10^{-4/3}$ (depicted by filled circles) in Figure 1. The positions of the two peaks are symmetric with respect to the origin of the x-axis ($x = 0$): $x = x_{\text{peak}}(T)$ and $x = -x_{\text{peak}}(T)$. They are approximately given by $\pm x_{\text{peak}}(T) = \pm 1.1$. It is clear in Figure 1 that as temperature T increases the two peak positions $\pm x_{\text{peak}}(T)$ become closer to the origin: $x = 0$, while the peak height becomes taller. The peak height increases until the two peak positions $\pm x_{\text{peak}}(T)$ merge to the origin ($x = 0$). After the peak positions $\pm x_{\text{peak}}(T)$ reach at the origin ($x = 0$), the peak height decreases rapidly. The peak height of c_{imp} at $x = 0$ decreases from 0.34 to 0.15 when temperature T increases from $10^{-1/3}$ to 10^0, as shown in Figure 1.

Over the most of the region between the two peaks at $x = \pm x_{\text{peak}}$ in the graph of impurity specific heat c_{imp} versus impurity parameter x at temperature T, as shown in Figure 1, we assume that the estimates of $c_{\rm imp}$ as a function of x are consistent with the following expression

$$c_{\rm imp}(x,T) = \frac{2\zeta T}{3\sin\zeta}\cosh\frac{\pi x}{\zeta}\,. \tag{4.37}$$

Here we recall that the consistency has already been shown in Figure 6 of Ref. [29]. Thus, at a given temperature T impurity specific heat $c_{\rm imp}(x,T)$ as a function of impurity parameter x is consistent with that of the low-temperature regime if the absolute value of x is smaller than the peak position $x_{\rm peak}(T)$ of the graph of $c_{\rm imp}$ versus x at the given temperature T, i.e. if $|x| < x_{\rm peak}(T)$. If the absolute value of impurity parameter x is larger than the peak position at temperature T, i.e. if $|x| > x_{\rm peak}(T)$, we expect that the coupling between the impurity spin and other spins is too weak with respect to thermal fluctuations at temperature T, so that impurity specific heat $c_{\rm imp}$ should almost vanish. In Figure 1 impurity specific heat $c_{\rm imp}$ indeed becomes small exponentially with respect to x in the region where we have $|x| > x_{\rm peak}(T)$.

Based on the above observations, we now define the crossover temperature T_c (or $T_c(x)$) as a function of impurity parameter x for impurity specific heat $c_{\rm imp}$. For a given value of impurity parameter x we define the crossover temperature $T_c(x)$ by a temperature T such that the peak position at the temperature T is equal to the given value of the impurity parameter: $x_{\rm peak}(T) = x$. It thus follows that we have $x_{\rm peak}(T_c(x)) = x$.

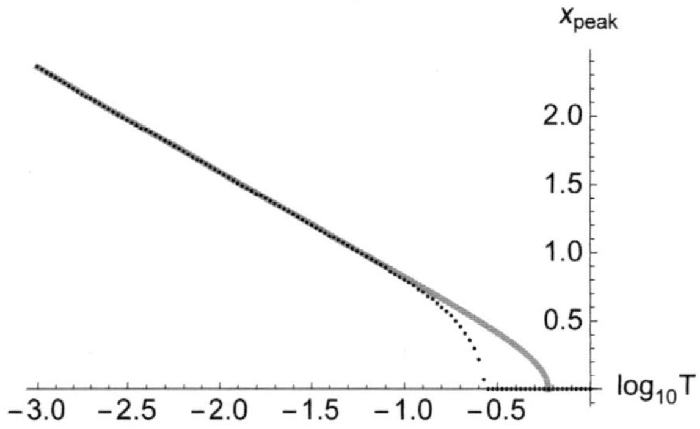

Figure 2. Peak position $x_{\rm peak}(T)$ versus $\log_{10} T$ for temperature T.
The thick gray line is given by eq. (4.39) with $\zeta = \pi/3$ and $A = 0.435$.

In Figure 2 the estimates of peak position $x_{\rm peak}(T)$ in the graph of $c_{\rm imp}$ versus x at temperature T are plotted against $\log_{10} T$. The linear behavior of the data plots in Fig. 2 with lower temperatures is consistent with the inverse of the analytic expression (3.28) of impurity susceptibility $\chi_{\rm imp}(x)$. By setting $x = x_{\rm peak}(T_c)$ and $\chi_{\rm imp}(x) = A/T_c$ with some constant A we have

$$\frac{A}{T_c} = \frac{4\zeta}{\pi(\pi-\zeta)\sin\zeta}\cosh\frac{\pi x_{\rm peak}}{\zeta}. \tag{4.38}$$

We express $x_{\rm peak}$ as a function of T_c as

$$x_{\rm peak} = \frac{\zeta}{\pi}\cosh^{-1}\left(A\frac{\pi(\pi-\zeta)\sin\zeta}{4\zeta T_c}\right)\,. \tag{4.39}$$

Here A denotes a constant to shift the graph of $x_{\rm peak}$ versus $\log T$. As shown in Figure 2, the graph of $x_{\rm peak}$ versus $\log_{10} T$ is consistent with the graph of (4.39) in lower temperatures such as for $\log_{10} T < -1.0$.

Journal of Physics: Conference Series **670** (2016) 012054 doi:10.1088/1742-6596/670/1/012054

When $x \gg 1$ we can approximate $\cosh(\pi x/\zeta)$ by $\exp(\pi x/\zeta)/2$. By expressing $x_{\rm peak}$ as a function of T_c we have

$$x_{\rm peak} \approx -\frac{\zeta}{\pi}\ln T_c + \frac{\zeta}{\pi}\ln\left(A\frac{\pi(\pi-\zeta)\sin\zeta}{2\zeta}\right). \tag{4.40}$$

It is consistent with the linear part of the graph of $x_{\rm peak}$ versus $\log_{10} T$ for lower temperatures such as $\log_{10} T < -1.0$, as shown in Figure 2.

5. Concluding remarks

We have investigated how the crossover temperature T_c for the impurity specific heat $c_{\rm imp}(x,T)$ depends on the impurity parameter x. We have determined T_c through the two peak positions $\pm x_{\rm peak}(T)$ in the graph of $c_{\rm imp}(x,T)$ versus x at temperature T: For a given value of x we define T_c by the condition: $x_{\rm peak}(T_c) = x$. We have shown that the x-dependence of crossover temperature T_c is consistent with the analytic expression derived by setting the inverse of T_c to be proportional to impurity susceptibility $\chi_{\rm imp}(x)$: $1/T_c \propto \chi_{\rm imp}(x)$.

Acknowledgments

The present study is partially supported by Grant-in-Aid for Scientific Research No. 15K05204.

[1] W.J. de Haas, J. de Boer, G.J. van dën Berg, Physica **1**, 1115 (1934).
[2] J. Kondo, Prog. Theor. Phys. **32**, 37 (1964).
[3] P. Nozières, J. Low Temp. Phys. **17**, 31 (1974).
[4] A.C. Hewson, *The Kondo Problem to Heavy Fermions* (Cambridge University Press, Cambridge, 1993).
[5] J. Kondo, *The Physics of Dilute Magnetic Alloys* (Cambridge University Press, Cambridge, 2012).
[6] K.G. Wilson, Rev. Mod. Phys. **47**, 773 (1975).
[7] N. Andrei, K. Furuya, and J. Lowenstein, Rev. Mod. Phys. **55**, 331 (1983).
[8] A.M. Tsvelick and P.B. Wiegmann, Adv. Phys. **32**, 453 (1983).
[9] S. Tomonaga, Prog. Theor. Phys. **5**, 544 (1950); J. M. Luttinger, J. Math. Phys. **4**, 1154 (1963).
[10] C.L. Kane and M.P.A. Fisher, Phys. Rev. B **46**, 15233 (1992); Phys. Rev. Lett. **68**, 1220 (1992).
[11] A. Furusaki and N. Nagaosa, Phys. Rev. B **47**, 4631 (1993).
[12] K.A. Natveev, D. Yue, and L.I. Glazman, Phys. Rev. Lett. **71**, 3351 (1993); D. Yue, L.I. Glazman, and K.A. Natveev, Phys. Rev. B **49**, 1966 (1994).
[13] I. Affleck and A.W.W. Ludwig, Nucl. Phys. **B360**, 641 (1991); Phys. Rev. Lett. **67**, 3160 (1991).
[14] P. Fröjdh and H. Johannesson, Phys. Rev. B **53**, 3211 (1996).
[15] S. Eggert and I. Affleck, Phys. Rev. B **46**, 10866 (1992).
[16] S. Qin, M. Fabrizio, L. Yu, M. Oshikawa and I. Affleck, Phys. Rev. B **56**, 9766 (1997).
[17] N. Andrei and H. Johannesson, Phys. Lett. **100A**, 108 (1984).
[18] P. Schlottmann, J. Phys.: Condens. Matter **3**, 6617 (1991).
[19] P. Schlottmann and P.D. Sacramento, Adv. Phys. **42**, 641 (1993).
[20] H.-P. Eckle, A. Punnoose and R. A. Römer, Europhys. Lett. **39**, 293 (1997).
[21] H. Frahm and A.A. Zvyagin, J. Phys.: Cond. Matter **9**, 9939 (1997).
[22] A. Klümper and A.A. Zvyagin, Phys. Rev. Lett. **81**, 4975 (1998).
[23] P. Schlottmann, Nucl. Phys. B **552**, 727 (1999).
[24] A.A. Zvyagin and A. Klümper, J. Phys.: Cond. Matter **12**, 8705 (2000).
[25] M. Bortz and A. Klümper, Eur. Phys. J. B **40**, 25 (2004).
[26] A.A. Zvyagin, *Finite Size Effects in Correlated Electron Models: Exact Results* (Imperial College Press, London, 2005).
[27] S. Andergassen, T. Enss, V. Meden, Phys. Rev. B **73**, 153308 (2006); S. Andergassen, T. Enss, V. Meden, W. Metzner, U. Schollwöck and K. Schönhammer, Phys. Rev. B **73**, 045125 (2006).
[28] D.S. Shapiro and S. N. Artemenko, J. Commun. Tech. Elect. **6**, 693 (2007).
[29] R. Yahagi, J. Sato and T. Deguchi, J. Stat. Mech., P11020 (2014).
[30] M. Takahashi and M. Suzuki, Prog. Theor. Phys. **48**, 2187 (1972).
[31] M. Takahashi, Prog. Theor. Phys. **50**, 1519 (1973).
[32] M. Fowler and . Zotos, Phys Rev. B **25**, 5806 (1982).
[33] M. Takahashi, *Thermodynamics of One-Dimensional Solvable Models*, Cambridge University Press (1999).

On existence of an x-integral for a semi-discrete chain of hyperbolic type

K Zheltukhin[1] and N Zheltukhina[2]

[1] Department of Mathematics, Middle East Technical University, Ankara, Turkey
[2] Department of Mathematics, Bilkent University, Ankara, Turkey

E-mail: `zheltukh@metu.edu.tr`

Abstract. A class of semi-discrete chains of the form $t_{1x} = f(x, t, t_1, t_x)$ is considered. For the given chains easily verifiable conditions for existence of x-integral of minimal order 4 are obtained.

1. Introduction

In the present paper we consider the integrable differential-difference chains of hyperbolic type

$$t_{1x} = f(x, t, t_1, t_x), \tag{1}$$

where the function $t(n, x)$ depends on discrete variable n and continuous variable x. We use the following notations $t_x = \frac{\partial}{\partial x}t$ and $t_1 = t(n+1, x)$. It is also convenient to denote $t_{[k]} = \frac{\partial^k}{\partial x^k}t$, $k \in \mathbb{N}$ and $t_m = t(n+m, x)$, $m \in \mathbb{Z}$.

The integrability of the chain (1) is understood as Darboux integrability that is existence of so called x- and n-integrals [1, 4]. Let us give the necessary definitions.

Definition 1 *Func i n $F(x, t, t_1, \ldots, t_k)$ i called an x-in e ral f e equa i n ( ) if*

$$D_x F(x, t, t_1, \ldots, t_k) = 0$$

f r all lu i n f ( ) e pera r D_x i e al deriva ive wi re pec x

Definition 2 *Func i n $G(x, t, t_x, \ldots, t_{[m]})$ i called an n-in e ral f e equa i n ( ) if*

$$DG(x, t, t_x, \ldots, t_{[m]}) = G(x, t, t_x, \ldots, t_{[m]})$$

f r all lu i n f ( ) e pera r D i a if pera r

To show the existence of x- and n-integrals we can use the notion of characteristic ring. The notion of characteristic ring was introduced by Shabat to study hyperbolic systems of exponential type (see [11]). This approach turns out to be very convenient to study and classify the integrable equations of hyperbolic type (see [12] and references there in).

For difference and differential-difference chains the notion of characteristic ring was developed by Habibullin (see [3]-[8]). In particular, in [4] the following theorem was proved

Theorem 3 *(ee [ ]) A c ain ( ) ad i a n n- rivial x-in e ral if and nl if i c arac eri ic x-rin i f fini e di en i n*
A c ain ( ) ad i a n n- rivial n-in e ral if and nl if i c arac eri ic n-rin i f fini e di en i n

For known examples of integrable chains the dimension of the characteristic ring is small. The differential-difference chains with three dimensional characteristic x-ring were considered in [6]. We consider chains with four dimensional characteristic x-ring, such chains admit x-integral of minimal order four. That is we obtain necessary and sufficient conditions for a chain to have a four dimensional characteristic x-ring. This conditions can be easily checked by direct calculations.

Note that if a chain (1) admits a nontrivial x-integral $F(x,t,t_1,\dots t_k)$ and a non trivial n-integral $G(x,t,t_x,\dots,t_{[m]})$ its solutions satisfy two ordinary equations

$$F(x,t,t_1,\dots,t_k)=a(n),$$

$$G(x,t,t_x,\dots,t_{[m]})=b(x)$$

for some functions $a(n)$ and $b(x)$. This allows to solve (1) (see [9]).

The paper is organized as follows. In Section 2 we derive necessary and sufficient conditions on function $f(x,t,t_1,t_x)$ so that the chain (1) has four dimensional characteristic ring and in Section 3 we consider some applications of the derived conditions.

2. Chains admitting our dimensional x-algebra.

Suppose F is an x-integral of the chain (1) then its positive shifts and negative shifts $D^k F$, $k\in\mathbb{Z}$, are also x-integrals. So, looking for an x-integral it is convenient to assume that it depends on positive and negative shits of t.

To express x derivatives of negative shifts we can apply D^{-1} to the chain (1) and obtain

$$t_x=f(x,t_{-1},t,t_x).$$

Solving the above equation for t_{-1x} we get

$$t_{-1x}=g(x,t_{-1},t,t_x).$$

Let $F(x,t,t_1,t_{-1},\dots)$ be an x-integral of the chain (1). Then on solutions of (1) we have

$$D_xF=\frac{\partial F}{\partial x}+t_x\frac{\partial F}{\partial t}+t_{1x}\frac{\partial F}{\partial t_1}+t_{-1x}\frac{\partial F}{\partial t_{-1}}+t_{2x}\frac{\partial F}{\partial t_2}+t_{-2x}\frac{\partial F}{\partial t_{-2}}+\dots=0$$

or

$$D_xF=\frac{\partial F}{\partial x}+t_x\frac{\partial F}{\partial t}+f\frac{\partial F}{\partial t_1}+g\frac{\partial F}{\partial t_{-1}}+Df\frac{\partial F}{\partial t_2}+D^{-1}g\frac{\partial F}{\partial t_{-2}}+\dots=0.$$

Define a vector field

$$K=\frac{\partial}{\partial x}+t_x\frac{\partial}{\partial t}+f\frac{\partial}{\partial t_1}+g\frac{\partial}{\partial t_{-1}}+Df\frac{\partial}{\partial t_2}+D^{-1}g\frac{\partial}{\partial t_{-2}}+\dots,\tag{2}$$

then

$$D_xF=K\,F.$$

Note that F does not depend on t_x but the coefficients of K do depend on t_x. So we introduce a vector field

$$X=\frac{\partial}{\partial t_x}\tag{3}$$

The vector fields K and X generate the characteristic x-ring L_x.

Let us introduce some other vector fields from L_x.

$$C_1 = [X, K] \quad \text{and} \quad C_n = [X, C_{n-1}] \quad n = 2, 3, \dots \tag{4}$$

and

$$Z_1 = [K, C_1] \quad \text{and} \quad Z_n = [K, Z_{n-1}] \quad n = 2, 3, \dots \tag{5}$$

Thus

$$C_1 = \frac{\partial}{\partial t} + f_{t_x}\frac{\partial}{\partial t_1} + g_{t_x}\frac{\partial}{\partial t_{-1}} + \dots$$

$$C_2 = f_{t_x t_x}\frac{\partial}{\partial t_1} + g_{t_x t_x}\frac{\partial}{\partial t_{-1}} + \dots$$

$$Z_1 = (f_{t_{xx}} + t_x f_{t_x t} + f f_{t_x t_1} - f_t - f_{t_x} f_{t_1})\frac{\partial}{\partial t_1} + (g_{t_{xx}} + t_x g_{t_x t} + g g_{t_x t_1} - g_t - g_{t_x} g_{t_1})\frac{\partial}{\partial t_{-1}} + \dots$$

and so on.

It is easy to see that if $f_{t_x t_x} \neq 0$ then the vector fields X, K, C_1 and C_2 are linearly independent and must form a basis of L_x provided $dim L_x = 4$. By Lemma 3.6 in [6], if $f_{t_x t_x} = 0$ and $(f_{t_{xx}} + t_x f_{t_x t} + f f_{t_x t_1} - f_t - f_{t_x} f_{t_1}) = 0$ then $dim L_x = 3$. So in the case $f_{t_x t_x} = 0$ we may assume $(f_{t_{xx}} + t_x f_{t_x t} + f f_{t_x t_1} - f_t - f_{t_x} f_{t_1}) \neq 0$. Then the vector fields X, K, C_1 and Z_1 are linearly independent and must form a basis of L_x provided $dim L_x = 4$. We consider this two cases separately.

In the rest of the paper we assume that the characteristic ring L_x is four dimensional.

Remark 4 *I i c nvenien c eck equali ie be ween vec r field u in e au rp i $D(\ )D^{-1}$ Direc calcula i n w a*

$$DXD^{-1} = \frac{1}{f_x}X,$$

$$DKD^{-1} = K - \frac{f_x + t_x f_t + f f_{t_1}}{f_{t_x}}X.$$

e i a e f er vec r field under i au rp i can be b ained b c u in DXD^{-1} and DKD^{-1}

$f(x, t, t_1, t_x)$ i n n linear wi re pec t_x

Let $f(x, t, t_1, t_x)$ be non linear with respect to t_x, $f_{t_x t_x} \neq 0$. Then the vector fields X, K, C_1 and C_2 form a basis of L_x. For the algebra L_x to be spanned by X, K, C_1 and C_2 it is enough that C_3 and Z_1 are linear combinations of X, K, C_1 and C_2. From the form of the vector fields it follows that we must have

$$C_3 = \lambda C_2 \quad \text{and} \quad Z_1 = \mu C_2$$

for some functions μ and λ. The conditions for the above equalities to hold are given by the following theorem.

Theorem 5 *e c ain ( ) wi $f_{t_x t_x} \neq 0$ a c arac eri ic rin L_x f di en i n f ur if and nl if e f ll win c ndi i n ld*

$$D\left(\frac{f_{t_x t_x t_x}}{f_{t_x t_x}}\right) = \frac{f_{t_x t_x t_x} f_{t_x} - 3f_{t_x t_x}^2}{f_{t_x t_x} f_{t_x}^2}. \tag{6}$$

$$D\left(\frac{f_{xt_x}+t_xf_{t_xt}+ff_{t_xt_1}-f_t-f_{t_x}f_{t_1}}{f_{t_xt_x}}\right)=$$
$$\frac{f_{xt_x}+t_xf_{t_xt}+ff_{t_xt_1}-f_t-f_{t_x}f_{t_1}}{f_{t_xt_x}}f_{t_x}-(f_x+t_xf_t+f_{t_1}). \tag{7}$$

e c arac eri ic rin i enera ed b e vec r field X, K, C_1, C_2

Proo . By Remark 4 we have

$$DC_2D^{-1}=\frac{1}{f_{t_x}^2}C_2-\frac{f_{t_xt_x}}{f_{t_x}^3}C_1+\frac{f_{t_xt_x}f_t}{f_{t_x}^4}X$$

$$DC_3D^{-1}=\frac{1}{f_{t_x}^3}C_2-\frac{3f_{t_xt_x}}{f_{t_x}^4}C_2-\frac{f_{t_xt_xt_x}f_{t_x}-3f_{t_xt_x}^2}{f_{t_x}^5}C_1+f_t\frac{f_{t_xt_xt_x}f_{t_x}-3f_{t_xt_x}^2}{f_{t_x}^6}X$$

$$DZ_1D^{-1}=\frac{1}{f_{t_x}}Z_1-\left(\frac{mf_{t_x}+p}{f_{t_x}^2}\right)\left(C_1-\frac{f_t}{f_{t_x}}X\right),$$

where $p=\dfrac{f_x+t_xf_t+ff_{t_1}}{f_{t_x}}$ and $m=\dfrac{-(f_{xt_x}+t_xf_{t_xt}+ff_{t_xt_1})+f_t+f_{t_x}f_{t_1}}{f_{t_x}}$.
The equality $C_3=\lambda C_2$ implies that

$$DC_3D^{-1}=(D\lambda)\,DC_2D^{-1}. \tag{8}$$

Substituting expressions for DC_2D^{-1} and DC_3D^{-1} into (8) and comparing coefficients of C_1, C_2 and X we obtain that λ satisfies

$$\lambda=f_{t_x}(D\lambda)+\frac{3f_{t_xt_x}}{f_{t_x}}$$

$$(D\lambda)=\frac{f_{t_xt_xt_x}f_{t_x}-3f_{t_xt_x}^2}{f_{t_xt_x}f_{t_x}^2}.$$

We can find λ and $D\lambda$ independently and condition that $D\lambda$ is a shift of λ leads to (6).
The equality $Z_1=\mu C_2$ implies that

$$DZ_1D^{-1}=(D\mu)\,DC_2D^{-1}. \tag{9}$$

Substituting expressions for DC_2D^{-1} and DC_3D^{-1} into (9) and comparing coefficients of C_1, C_2 and X we obtain that μ satisfies

$$\mu-\frac{f_x+t_xf_t+ff_{t_1}}{f_{t_x}}=\frac{(D\mu)}{f_{t_x}}$$

and

$$-(f_{xt_x}+t_xf_{t_xt}+ff_{t_xt_1}-f_t-f_{t_x}f_{t_1})+\frac{f_x+t_xf_t+ff_{t_1}}{f_{t_x}}f_{t_xt_x}=-\frac{f_{t_xt_x}(D\mu)}{f_{t_x}}$$

We can find μ and $D\mu$ independently and condition that $D\mu$ is a shift of μ leads to (7). □

Remark 6 *Le* $\dim L_x=4$ *and* $f_{t_{xx}}\neq 0$ *en e c arac eri ic rin* L_x *ave e f ll win ul iplica i n able*

	X	K	C_1	C_2
X	0	C_1	C_2	μC_2
K	$-C_1$	0	λC_2	ρC_2
C_11	$-C_2$	$-\lambda C_2$	0	ηC_2
C_2	$-\mu C_2$	$-\rho C_2$	$-\eta C_2$	0

w ere $\rho = \lambda\mu + X(\lambda)$ *and* $\eta = X(\rho) - K(\mu)$

Exam le 7 *C n ider e f ll win c ain*

$$t_{1x} = \frac{tt_x - \sqrt{t_x^2 - M^2}(t_1 + t)}{t_1}$$

in r duced b Habibullin and Z el uk ina [0] We can ea il c eck a e func i n

$$f(t, t_1, t_x) = \frac{tt_x - \sqrt{t_x^2 - M^2}(t_1 + t)}{t_1}$$

a i fie e c ndi i n f e re 5 Hence e c rre p ndin x-al ebra i f ur di en i nal e c ain a e f ll win x-in e ral

$$F = \frac{(t_1^2 - t^2)(t_1^2 - t_2^2)}{t_1^2}.$$

$f(x, t, t_1, t_x)$ *i linear wi re pec* t_x

Let $f(x, t, t_1, t_x)$ be linear with respect to t_x, $f_{t_x t_x} = 0$. Then vector fields X, K, C_1 and Z_1 form a basis of L_x. The condition $f_{t_x t_x} = 0$ also implies that the vector field $C_2 = 0$, see [6]. For the algebra L_x to be spanned by X, K, C_1 and Z it is enough that Z_2 is a linear combination of X, K, C_1 and Z_1. From the form of the vector fields it follows that we must have

$$Z_2 = \alpha Z_1$$

for some function α. The conditions for the above equality to hold given by the following theorem.

Theorem 8 *e c ain ( ) wi* $f_{t_x t_x} = 0$ *a e c arac eri ic rin* L_x *f di en i n f ur if and nl if e f ll win c ndi i n ld*

$$D\left(\frac{K(m)}{m} - m + \frac{f_t}{f_{t_x}}\right) = \frac{K(m)}{m} + m - f_{t_1}. \qquad (10)$$

w ere $m = \dfrac{-(f_{xt_x} + t_x f_{t_x t} + f f_{t_x t_1}) + f_t + f_{t_x} f_{t_1}}{f_{t_x}}$ *e c arac eri ic rin i enera ed b e vec r field* X, K, C_1, Z_1

Proo . By Remark 4 we have

$$DZ_1D^{-1} = \frac{1}{f_{t_x}} Z_1 - \left(\frac{mf_{t_x} + p}{f_{t_x}^2}\right)\left(C_1 - \frac{f_t}{f_{t_x}} X\right),$$

and

$$DZ_2D^{-1} = \left(K\left(\frac{1}{f_{t_x}}\right) + \frac{\alpha + m}{f_{t_x}}\right) Z_1 + \left(K\left(\frac{m}{f_{t_x}}\right) + \frac{mf_t}{f_{t_x}^2} - pX\left(\frac{m}{f_{t_x}}\right)\right)\left(C_1 - \frac{f_t}{f_{t_x}} X\right)$$

Journal of Physics: Conference Series **670** (2016) 012055 doi:10.1088/1742-6596/670/1/012055

The equality $Z_2 = \alpha Z_1$ implies that

$$DZ_2D^{-1} = (D\alpha)\, DZ_1D^{-1}. \tag{11}$$

Substituting expressions for DZ_1D^{-1} and DZ_2D^{-1} into (11) and comparing coefficients of C_1, Z_1 and X we obtain that α and $D(\alpha)$ satisfy

$$K\left(\frac{1}{f_{t_x}}\right) + \frac{m}{f_{t_x}} + \frac{\alpha}{f_{t_x}} = \frac{D(\alpha)}{f_{t_x}}$$

$$K\left(\frac{m}{f_{t_x}}\right) + \frac{mf_t}{f_{t_x}^2} = \frac{mD(\alpha)}{f_{t_x}}$$

We can find α and $D(\alpha)$ independently and condition that $D(\alpha)$ is a shift of α leads to (10). □

Remark 9 *Le* $\dim L_x = 4$ *and* $f_{t_{xx}} = 0$ *en e c arac eri ic rin* L_x *ave e f ll win ul iplica i n able*

	X	K	C_1	Z_1
X	0	C_1	0	0
K	$-C_1$	0	Z_1	αZ_1
$C1$	0	$-Z_1$	0	$X(\alpha)Z_1$
Z_1	0	$-\alpha Z_1$	$-X(\alpha)Z_1$	0

Exam le 10 *C n ider e f ll win c ain*

$$t_{1x} = t_x + e^{\frac{t+t_1}{2}}$$

in r duced b D dd and Bull u [ ] We can ea il c eck a e func i n

$$f(t, t_1, t_x) = t_x + e^{\frac{t+t_1}{2}}$$

a i fie e c ndi i n f e re 8 Hence e c rre p ndin x-al ebra i f ur di en i nal e c ain a e f ll win x-in e ral

$$F = e^{\frac{t_1-t}{2}} + e^{\frac{t_1-t_2}{2}}$$

3. A lications

The conditions derived in the previous section can be used to determine some restrictions on the form of the function $f(x, t, t_1, t_x)$ in (1).

Lemma 11 *Le e c ain ( ) ave f ur di en i nal c arac eri ic x-rin en*

$$f = M(x, t, t_x)A(x, t, t_1) + t_xB(x, t, t_1) + C(x, t, t_1), \tag{12}$$

w ere M, A, B and C are e func i n

Proo . Let $f_{t_xt_x} \neq 0$ (if $f_{t_xt_x} = 0$ then f obviously has the above form). Since characteristic x-ring has dimension four the condition (6) holds. It is easy to see that (6) implies that $\frac{f_{t_xt_xt_x}}{f_{t_xt_x}}$ does not depend on t_1. Hence

$$X(\ln|f_{t_xt_x}|) = M_1(x, t, t_x) \quad \text{and} \quad \ln|f_{t_xt_x}| = M_2(x, t, t_x) + A_1(x, t, t_1).$$

The last equality implies (12). □

We can also put some restrictions on the shifts of the function $f(x, t, t_1, t_x)$ in (1).

Journal of Physics: Conference Series **670** (2016) 012055 doi:10.1088/1742-6596/670/1/012055

Lemma 12 *Le e c ain ( ) ave f ur di en i nal c arac eri ic x-rin and* $f_{t_x t_x} \neq 0$. *en*

$$Df = -H_1(x,t,t_1,t_2)t_x + H_2(x,t,t_1,t_2)f + H_3(x,t,t_1,t_2), \tag{13}$$

w ere H_1, H_2 *and* H_3 *are e func i n*

Proo . Note that the shift operator D and the vector field X satisfy

$$DX = \frac{1}{f_{t_x}} XD. \tag{14}$$

The condition (6) can be written as

$$DX(\ln|f_{t_x t_x}|) = \frac{1}{f_{t_x}} X(\ln|f_{t_x t_x}| - \ln|f_{t_x}|^3)$$

Using (14) we get

$$\frac{1}{f_{t_x}} XD(\ln|f_{t_x t_x}|) = \frac{1}{f_{t_x}} X(\ln|f_{t_x t_x}| - \ln|f_{t_x}|^3)$$

which implies that

$$X\left(\ln\left|f_{t_x}^3 \frac{Df_{t_x t_x}}{f_{t_x t_x}}\right|\right) = 0 \quad \text{or} \quad X\left(f_{t_x}^3 \frac{Df_{t_x t_x}}{f_{t_x t_x}}\right) = 0.$$

Thus $Df_{t_x t_x} = H_1(x,t,t_1,t_2)\frac{f_{t_x t_x}}{f_{t_x}^3}$. Since $Df_{t_x t_x} = DX(f_{t_x})$ and $\frac{f_{t_x t_x}}{f_{t_x}^3} = -\frac{1}{f_{t_x}} X(\frac{1}{f_{t_x}})$ we can rewrite previous equality using (14) as

$$X\left(Df_{t_x} + H_1(x,t,t_1,t_2)\frac{1}{f_{t_x}}\right) = 0$$

which implies

$$Df_{t_x} = -H_1(x,t,t_1,t_2)\frac{1}{f_{t_x}} + H_2(x,t,t_1,t_2).$$

Writing

$$DX(f) = -H_1(x,t,t_1,t_2)\frac{1}{f_{t_x}} + H_2(x,t,t_1,t_2)\frac{f_{t_x}}{f_{t_x}}$$

and applying (14) as before we get

$$X(Df + H_1(x,t,t_1,t_2)t_x - H_2(x,t,t_1,t_2)f) = 0.$$

The last equality gives (13). □

Note that the equality (13) can be written as

$$t_{2x} = H_2(x,t,t_1,t_2)t_{1x} - H_1(x,t,t_1,t_2)t_x + H_3(x,t,t_1,t_2).$$

Journal of Physics: Conference Series **670** (2016) 012055 doi:10.1088/1742-6596/670/1/012055

Re erences

[1] Darboux G 1915 *Leçons sur la théoriegénérale des surfaces et les applications geometriques du calcul infinitesimal* **2** (Paris: Gautier Villas)

[2] R K Dodd and R K Bullough 1976 *Proc. R. Soc. London* Ser. A **351** 499

[3] Habibullin I T 2005 *SIGMA Symmetry Integrability Geom.: Methods Appl.* **1**

[4] Habibullin I T and Pekcan A 2007 *Theoret. and Math. Phys.* **151** 781790

[5] Habibullin I 2007 *Characteristic algebras of discrete equations. Difference equations, special functions and orthogonal polynomials* (Hackensack NJ World Sci. Publ.) 249-257

[6] Habibullin I Zheltukhina N and Pekcan A 2008 *Turkish J. Math.* **32** 277292

[7] Habibullin I Zheltukhina N and Pekcan A 2008 *J. Math. Phys.***49** 102702

[8] Habibullin I Zheltukhina N and Pekcan A 2009 *J. Math. Phys.* **50** 102710

[9] Habibullin I Zheltukhina N and Sakieva A 2010 *J. Phys.* A **43** 434017

[10] Habibullin I and Zheltukhina N 2014 Discretization of Liouville type nonautonomous equations *Preprint* nlin.SI:1402.3692vl

[11] Shabat A B and Yamilov R I 1981 Exponential systems of type I and Cartan matrices *Preprint* BBAS USSR Ufa

[12] Zhiber A B, Murtazina R D, Habibullin I T and Shabat A B 2012 *Ufa Math. J.* **4** 17-85

A Microscopic Theory of the Neutron

J.X. Zheng-Johansson
Institute of Fundamental Physics Research, 611 93 Nyköping, Sweden

Abstract.
A microscopic theory of the neutron, which consists in a neutron model constructed using key relevant experimental observations as input information and the first principles solutions for the basic properties of the model neutron, is proposed within a framework consistent with the Standard Model. The neutron is composed of an electron e and a proton p that are separated at a distance r_1 of the order 10^{-18} m, and are in relative orbital angular motion and Thomas precession highly relativistically, with their reduced mass moving along a quantised circular orbit $l = 1, j = \frac{1}{2}$ of radius vector $\mathbf{r}_{1\frac{1}{2}} = r_1 \hat{r}_{1\frac{1}{2}}$ about their mass centre. The associated rotational energy flux has a spin $\frac{1}{2}$ and resembles a confined antineutrino. The particles e, p are attracted with one another predominantly by a central magnetic force produced as result of the particles' relative precessional-orbital and intrinsic angular motions. The interaction force (resembling the weak force), potential (resembling the Higgs' field), and a corresponding excitation Hamiltonian (H_I), among others, are derived based directly on first principles laws of electromagnetism, quantum mechanics and relativistic mechanics within a unified framework. In particular, the equation for $\frac{4}{3}\pi r_1^3 H_I$, which is directly comparable with the Fermi constant G_F, is predicted as $G_F = \frac{4}{3}\pi r_1^3 H_I = A_o C_{0\frac{1}{2}}/\gamma_e\gamma_p$, where $A_o = e^2\hbar^2/12\pi\epsilon_0 m_e^0 m_p^0 c^2$, m_e^0, m_p^0 are the e, p rest masses, $C_{0\frac{1}{2}}$ is a geo-magnetic factor, and γ_e, γ_p are the Lorentz factors. Quantitative solution for a stationary meta-stable neutron is found to exist at the extremal point $r_{1m} = 2.537 \times 10^{-18}$ m, at which the G_F is a minimum (whence the neutron lifetime is a maximum) and is equal to the experimental value. Solutions for the magnetic moment, effective spin ($\frac{1}{2}$), fine structure constant, and intermediate vector boson masses of the neutron are also given in this paper.

1. Introduction

The neutron is a building particle of matter, as the proton and electron are. The neutron distinguishes yet from the proton and electron prominently in its undergoing weak decay with a notable non-conservative parity. In inverse proportion to the weak interaction strength represented by the Fermi constant G_F, the lifetime of a free neutron is of a finite 12 minutes only. The Fermi constant G_F combined with the Heisenberg relation indicates moreover a weak interaction distance of an order $10^{-18}m$. Weak decay is a common property of all of the other several hundred elementary matter particles observed in the laboratory except for the proton and electron, by virtue of which process these particles are unstable, short lived. The basic properties of the weak processes, foremost the neutron β decay, have been experimentally studied extensively over the past eight decades or so, and summarised under the Standard Model for elementary particles [1]. Theoretically, the weak decay of neutron and other particles has been accounted satisfactorily for, most notably in quantitative prediction of the branching ratio, by the unified renormalisable theories of weak interaction. The Glashow-Weinberg-Salam (GWS) electroweak theory [2a-c] based on group $SU(2) \times U(1)$ is one of these. This theory in particular also predicts the charged and neutral intermediate vector bosons W^-,W^+ and Z^0

which were confirmed by the experiments at CERN; its renormalisability was proven by t'Hooft in 1971 [2d]. All of the current field theories of the neutron are essentially focused with the neutron β decay, and are rested on the original hypothesis of Fermi[2e]. Namely that, in a neutron (n) β decay reaction $n \to p + e + \bar{\nu}_e$, the matter particles proton p and electron e, and the antineutrino $\bar{\nu}_e$ do not exist until the neutron n decays. And upon neutron decay, these particles are envisaged as simply emitted by the neutron (as a point entity) in an analogous way to an accelerated point charge emitting electromagnetic radiation. The current theory of the neutron remains as a phenomenological one. There remain certain outstanding questions yet to be resolved. In particular, the nature and origin of the weak interaction force are not yet well understood, an equation of the weak force accordingly is yet to be derived, and the Fermi constant G_F is yet to be derived based on the interaction force. At a similar significant level, the nature and the origins of the (anti)neutrino, the intermediate vector bosons, the Weinberg weak mixing angle, and the Higgs mass are not yet fully well understood. One common feature suggestive of the nature of the weak phenomena however is readily recognisable directly from observations, namely that the weak phenomena present with (precede) only the electrons and protons emitted from the baryon (n, Λ, etc) and meson (π, K, etc.) disintegrations or conversely (succeed) ones upon the productions of the n, Λ, π, K, etc., but not with the same electrons and protons in free-particle or bound atomic processes. Weak phenomenon has thus to do with the internal structure of the weak emitting particles. For a more comprehensive understanding of the nature of the weak phenomena, a microscopic theory would be indispensable. The purpose of this paper is to develop a microscopic theory of the neutron, serving as a prototype of the weak interaction (meta-)stabilised systems, based firstly on a realistic real-space model construction of the neutron, such that the fundamental weak force and the variety of weak-interaction related properties and phenomena can be predicted based on first principles solutions within a unified framework of electromagnetism, quantum mechanics, and relativistic mechanics.

Using several key relevant experimental facts, in particular the neutron beta decay reaction equation $n \to p + e + \bar{\nu}_e$, the neutron spin ($\frac{1}{2}$), the order of magnitude of the Fermi constant G_F and the so implied weak interaction distance $\sim 1 \cdot 10^{-18}$ m as direct input information, we propose at the outset of the theory development a real-space (e, p-) neutron model as follows: The *neutron* is composed of an electron e and a proton p separated at a distance $r(= r_1)$ of an order 10^{-18} m; see Fig 1a. The e, p are in relative orbital angular motion and in addition a Thomas precession at a velocity approaching the velocity of light c, under a central force of an electromagnetic origin. The central force is in effect predominantly an attractive magnetic force produced by the magnetic fields ($\mathbf{B}_p, \mathbf{B}_e$) of p, e at e, p as result of their intrinsic spin and relative motions. The z-components (S_{ez}, S_{pz}) of the e, p spin angular momenta are aligned parallel to each other and antiparallel to that of their relative motion ($J_{z\text{-}\frac{1}{2}}$, Figs 1b), so that the magnetic interaction force is maximally attractive. The e, p relative motion is in such a way that their reduced mass ($\mathscr{M}$) moves at a velocity ($\boldsymbol{v}_{\frac{1}{2}}$) accordingly approaching c along a (quantised $l = 1, j = \frac{1}{2}$) circular orbit of radius $r(= r_1)$ about their (the e, p) common centre of mass (CM), with a normal ($\mathbf{n}$) at a precession-modified quantised angle ($\pi - \theta_{\frac{1}{2}}$ for j spin down state) to the z axis; see Fig 1b. The relative precessional–orbital angular momentum projected in z direction ($J_{z\text{-}\frac{1}{2}}$) will show to be a negative half-integer quantum $J_{z\text{-}\frac{1}{2}} = -\frac{1}{2}\hbar$. The corresponding neutral rotational energy flux, or vortex, along the $l = 1, j = \frac{1}{2}$ circular orbit, of accordingly a z-component angular momentum $J_{z\text{-}\frac{1}{2}}$, resembles a "confined antineutrino" ($\bar{\nu}_e$).

It is commented that, the proposed e, p-neutron model suggests also a scheme for the strong force similarly on a unified basis with electromagnetism: A proton p would be attracted with a neutron $n(e, p)$ (mainly) through an electrostatic attraction with the electron e of the neutron at short range; in the same order of the short-range electrostatic interaction, two protons will repel, but never attract with one another. Such characteristics are in accordance with the observational fact that no nucleus exists which is made of more than one protons and protons

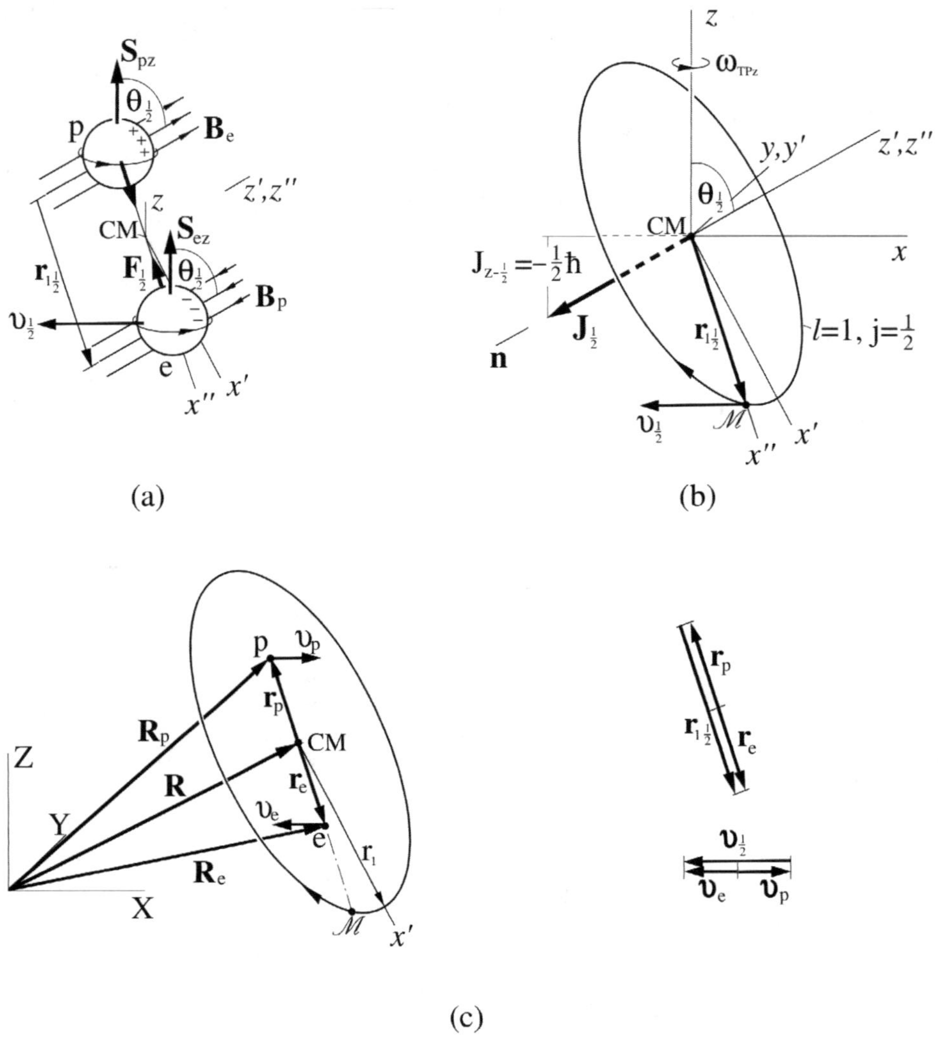

Figure 1. Schematic of the model neutron composed of an electron e and a proton p. (a) The e, p are separated by a distance $\mathbf{r}_{1\frac{1}{2}} = r_1 \hat{r}_{1\frac{1}{2}}$ and are in relative angular motion and a Thomas precession at velocity $\boldsymbol{v}_{\frac{1}{2}}$ under a magnetic interaction force $\mathbf{F}_{\frac{1}{2}}$ in the magnetic fields $\mathbf{B}_p, \mathbf{B}_e$ of p, e at e, p; their spins $\mathbf{S}_{ez}, \mathbf{S}_{pz}$ (in units $\hbar$) are aligned parallel, in the $+z$ direction for the $m_j = -j$ magnetic state shown, and antiparallel to $J_{z\text{-}\frac{1}{2}}$ of graph (b). (b) The reduced mass $\mathscr{M}$ of e, p moves at velocity $\boldsymbol{v}_{\frac{1}{2}}$ about the CM along a $l = 1, j = \frac{1}{2}$ circular orbit of radius vector $\mathbf{r}_{1\frac{1}{2}}$ and normal $\mathbf{n}$ at angle $\pi - \theta_{\frac{1}{2}}$ to the z axis; it has a z-component angular momentum $J_{z\text{-}\frac{1}{2}}$. (c) Left: The e, p are located at positions $\mathbf{r}_e, \mathbf{r}_p$, moving at velocities $\boldsymbol{v}_e, \boldsymbol{v}_p$, relative to the CM in the CM frame (coordinates x, y, z in graph b), and at $\mathbf{R}_e, \mathbf{R}_p$ in the lab frame (coordinates X, Y, Z). Right: vector relations between $\mathbf{r}_e, \mathbf{r}_p$ and $\mathbf{r}_{1\frac{1}{2}}$, and $\boldsymbol{v}_e, \boldsymbol{v}_p$ and $\boldsymbol{v}_{\frac{1}{2}}$. The drawings are made for $m_e \simeq m_p$.

only without neutrons. The author's more recent research (internal work) has further shown that a microscopic representation of the muon and the "muon-emitting" composite elementary particles may be achieved within a consistent scheme with the neutron model. The system of the so-represented elementary particles furthermore is in conformity with the quark model, in a manner that the (internal) spin states of the (composite) elementary particles are in one-to-one correspondences with the configurations of the (observationally-never-isolatable) quarks. The internal spin states of the model neutron under reversed signs (which represents the neutron effective spin, see Sec 3), $-\frac{1}{2}, -\frac{1}{2}, \frac{1}{2}$ (i.e. "down, down, up"), for example, directly correspond with the ddu quarks. The (free) proton, as another example, is a non-composite particle with only one spin state, assigned as $+\frac{1}{2}$ (spin up) by convention. But this may be translated into a systematic three-spin states representation as $\frac{1}{2}, \frac{1}{2}, -\frac{1}{2}$ (i.e. "up, up, down"), by adding two dummy spins $\frac{1}{2}, -\frac{1}{2}$ without changing the original spin $\frac{1}{2}$; the three spin states correspond directly to the uud quarks.

The remainder of this paper gives a first-principles mathematical representation of the model neutron, mainly in respect to the internal relativistic kinematics, dynamics (Secs 2), magnetic structure (Sec 3), and a derivation of the internal e, p interaction force (Sec 4) of the neutron in stationary state, the dynamics upon the neutron β decay (Secs 5), and a quantitative evaluation of the dynamical variables (Sec 6). The (quantitative) predictions of the basic properties of the neutron are subsequently subjected to comparisons with, or constraints by, the available experimental data where in question, so that critical checks and controls of the viability of the neutron model are made as far as possible. Other basic aspects, including the parity associated with the β decay, a direct derivation of the intermediate vector boson masses and Weinberg mixing angle of the neutron, and a corresponding dynamic scheme for the other (composite) elementary particles participating weak interaction, will be elucidated in separate papers.

2. Equations of motion. Coordinate transformations. Solutions

2.1. Transformed Newtonian equations of motion of the mean and instantaneous positions of e, p. Representation in $\mathbf{r}, \mathbf{R}$ coordinates Consider that an electron e and a proton p comprising a neutron are at time t located with the probability densities $|\psi_\alpha(\mathbf{R}_\alpha, t)|^2$ ($\alpha = e, p$) at positions $\mathbf{R}_e, \mathbf{R}_p$ relative to coordinates X, Y, Z fixed in the laboratory (lab) frame; see Fig 1c. (The usual statistical point-particle picture suffices and is referred to here.) The e, p are in relative motion at a velocity to prove high compared to c (Sec 6) under a mutual interaction force $\mathbf{F}$ and gravity $\mathbf{g}$; no applied force presents. Their mean positions, $\bar{\mathbf{R}}_\alpha = \int \mathbf{R}_\alpha |\psi_\alpha|^2 d^3R_\alpha$, evolve according to the transformed Newtonian equations of motion, $\frac{d(m_\alpha(d\bar{\mathbf{R}}_\alpha/dt))}{dt} = \int (m_\alpha \mathbf{g} \pm \mathbf{F}) |\psi_\alpha|^2 d^3R_\alpha$ (the correspondence principle), where m_e, m_p are the e, p masses. The e, p are assumed to form a bound stationary system until Sec 5 and hence feasibly move circularly at constant (tangential) velocities ($\mathbf{u}_\alpha = d\mathbf{R}_\alpha/dt$). The equations of motion thus reduce to

$$m_e \frac{d^2\mathbf{R}_e}{dt^2} = m_e\mathbf{g} + \mathbf{F}, \quad m_p \frac{d^2\mathbf{R}_p}{dt^2} = m_p\mathbf{g} - \mathbf{F}. \quad \text{Or } M\frac{d^2\mathbf{R}}{dt^2} = M\mathbf{g}, \quad \mathscr{M}\frac{d^2\mathbf{r}}{dt^2} = \mathbf{F}, \tag{1}$$

where

$$\mathbf{R} = \frac{m_e\mathbf{R}_e + m_p\mathbf{R}_p}{M}, \quad M = m_e + m_p, \quad \mathbf{r} = \mathbf{R}_e - \mathbf{R}_p = \mathbf{r}_e - \mathbf{r}_p, \quad \mathscr{M} = \left(\frac{1}{m_e} + \frac{1}{m_p}\right)^{-1} = \frac{m_e m_p}{M};$$

$$\mathbf{R}_e = \mathbf{R} + \frac{m_p}{M}\mathbf{r}, \quad \mathbf{R}_p = \mathbf{R} - \frac{m_e}{M}\mathbf{r}; \quad \mathbf{r}_e = \mathbf{R}_e - \mathbf{R} = \frac{m_p}{M}\mathbf{r}, \quad \mathbf{r}_p = \mathbf{R}_p - \mathbf{R} = -\frac{m_e}{M}\mathbf{r}. \tag{2}$$

$\mathbf{R}$ is the position of the centre of mass, CM; M is the total mass located at $\mathbf{R}$; $\mathbf{r}$ is the relative position; $\mathscr{M}$ is the reduced mass (of a fictitious particle) located at $\mathbf{r}$; and $\mathbf{r}_e, \mathbf{r}_p$ are the e, p positions relative to $\mathbf{R}$. Eqs (1c,d) are given for the masses M and $\mathscr{M}$ travelling accordingly circularly at constant velocities ($\mathbf{u}_{\rm cm} = d\mathbf{R}/dt$ relative to the lab frame and $\boldsymbol{v}$ about the CM). A common time t measured by a clock fixed at the CM has been used in order to facilitate the direct transformation of Eqs (1a,b) to (1c,d). Corresponding directly with the dynamic effect of $d^2\mathbf{R}_\alpha/dt^2$ on the left of Eqs (1a,b), this t enters as an independent variable of $\mathbf{g}, \mathbf{F}$: $\mathbf{g} = \mathbf{g}(t)$, $\mathbf{F} = \mathbf{F}(t)$; the relativistic masses m_e, m_p may remain as (implicit) functions of the local times

Journal of Physics: Conference Series **670** (2016) 012056 doi:10.1088/1742-6596/670/1/012056

(t_e, t_p) at $\mathbf{r}_e, \mathbf{r}_p$ (Sec 2.2) in so far as the same masses are used through the equations. For the e, p relative motions internal of a neutron are of major concern in this paper, unless specified otherwise we shall work in the CM frame, i.e. immediately in terms of the relative positions $\mathbf{r}_e, \mathbf{r}_p, \mathbf{r}$ measured with respect to a set of relative coordinate axes x, y, z parallel with the X, Y, Z axes, and with an origin fixed at the CM (cf Fig 1b). This in more general terms means that we shall work with the unsuperscripted variables, including $\mathbf{R}, M, \mathbf{R}_e, m_e, t$ etc, which we hereafter reserve to explicitly refer to ones measured in the CM frame. We shall refer to their counterparts for example measured in the lab frame by $\mathbf{R}^{\rm lab}$, $M^{\rm lab}$, $\mathbf{R}_e^{\rm lab}$, $m_e^{\rm lab}$, $t^{\rm lab}$, etc. where in question.

The partial–relative and relative velocities of the e, p, and the corresponding rotational angular momenta in the CM frame, in terms of the time t, follow as

$$\boldsymbol{v}_e = \frac{d\mathbf{r}_e}{dt} = \frac{m_p}{M}\boldsymbol{v}, \quad \boldsymbol{v}_p = \frac{d\mathbf{r}_p}{dt} = -\frac{m_e}{M}\boldsymbol{v}, \quad \boldsymbol{v} = \frac{d\mathbf{r}}{dt} = \boldsymbol{v}_e - \boldsymbol{v}_p; \tag{3}$$

$$\mathbf{J}_e = \mathbf{r}_e \times (m_e\boldsymbol{v}_e) = \frac{m_p}{M}\mathbf{J}, \quad \mathbf{J}_p = \mathbf{r}_p \times (m_p\boldsymbol{v}_p) = \frac{m_e}{M}\mathbf{J}, \quad \mathbf{J} = \mathbf{J}_e + \mathbf{J}_p = \mathbf{r} \times (\mathscr{M}\boldsymbol{v}) \tag{4}$$

From the relations (2g,h) between the distances $\mathbf{r}_e, \mathbf{r}_p$ of e, p to the CM, and $\mathbf{r}$ of e to p it follows that, by virtue how time in essence is defined, the local times t_e, t_p and the time t for light to traverse the distances $\mathbf{r}_e, \mathbf{r}_p, \mathbf{r}$ at a constant velocity c are related as $t_e = (m_p/M)t$, $t_p = (m_e/M)t$. The partial-relative velocities in terms of t_e, t_p are

$$\boldsymbol{v}'_e = d\mathbf{r}_e/dt_e = \boldsymbol{v}, \quad \boldsymbol{v}'_p = d\mathbf{r}_p/dt_p = -\boldsymbol{v}. \tag{5}$$

Denote $f_t(e) = \frac{t}{t_e} = \frac{|d\mathbf{r}_e/dt_e|}{|d\mathbf{r}_e/dt|} = \frac{v'_e}{v_e}$. Substituting $t = f_t(e)t_e$ in (1a), setting $\mathbf{R} = 0$, we have $m_e\frac{d^2\mathbf{r}_e}{d(f_t^2(e)t_e^2)} = m_e\mathbf{g}(t) + \mathbf{F}(t)$, or $m_e\frac{d^2\mathbf{r}_e}{dt_e^2} = m_e f_t^2(e)\mathbf{g}(t) + f_t^2(e)\mathbf{F}(t) = m_e\mathbf{g}(t_e) + \mathbf{F}(t_e)$, recovering the original form of (1a) expressed by its local time t_e provided $\mathbf{F}(t_e) = f_t^2(e)\mathbf{F}(t)$, $\mathbf{g}(t_e) = f_t^2(e)\mathbf{g}(t)$. Similarly a factor $f_t(p) = \frac{t}{t_p} = \frac{v'_p}{v_p}$ will project (1b) to its original form expressed in t_p. The same projection factors, in the form of geometric mean $f_t^2 = (f_t^2(e)f_t^2(p))^{1/2}$, will be obtained through direct derivation of the magnetic force in Sec 4.

2.2. Lorentz-Einstein transformations The instantaneous rest frame fixed to each rotating particle, e, p, $\mathscr{M}$ or M, may be regarded as an inertial frame for each differential rotation which is essentially linear. (For a complete macroscopic rotation, non-inertial frame effects present and will be included separately, see Eqs (11) vs (12) below and in turn Sec 2.4). Subsequently, the differentials of the space and time coordinates $\mathbf{r}_e$, t_e; $\mathbf{r}_p$, t_p; $\mathbf{r}$, t; $\mathbf{R}$, $\bar{t}_{ep}$ in the CM frame, and their counterparts $\mathbf{r}_e^0$, t_e^0; $\mathbf{r}_p^0$, t_p^0; $\mathbf{r}^0$, t^0; $\mathbf{R}^0$, $\bar{t}_{ep}^0$ in the respective (instantaneous) rest frames are related by the Lorentz-Einstein transformations,

$$\gamma_e(d\mathbf{r}_e - \boldsymbol{v}'_e dt_e) = d\mathbf{r}_e^0, \; \gamma_e(dt_e - \frac{\boldsymbol{v}'_e \cdot d\mathbf{r}_e}{c'^2}) = dt_e^0; \; \gamma_p(d\mathbf{r}_p - \boldsymbol{v}'_p dt_p) = d\mathbf{r}_p^0, \; \gamma_p(dt_p - \frac{\boldsymbol{v}'_p \cdot d\mathbf{r}_p}{c'^2}) = dt_p^0;$$

$$\gamma(d\mathbf{r} - \boldsymbol{v}dt) = d\mathbf{r}^0, \; \gamma(dt - \frac{\boldsymbol{v} \cdot d\mathbf{r}}{c^2}) = dt^0; \; \gamma_{\rm M}(d\mathbf{R} - \boldsymbol{v}_{\rm M}d\bar{t}_{ep}) = d\mathbf{R}^0, \; \gamma_{\rm M}(d\bar{t}_{ep} - \frac{\boldsymbol{v}_{\rm M} \cdot d\mathbf{R}}{c'^2}) = d\bar{t}_{ep}^0 \tag{6}$$

where $\gamma_\alpha = (1 - {v'_\alpha}^2/c'^2)^{-1/2}$ $(\alpha = e, p)$, $\gamma = (1 - v^2/c^2)^{-1/2}$, $\gamma_{\rm M} = (1 - v_{\rm M}^2/c^2)^{-1/2}$; $c' = d\mathbf{r}_{\alpha_{pht}}/dt_\alpha = c = d\mathbf{r}_{pht}/dt$ is the light speed measured in the CM frame; $\gamma_{\rm M}, \gamma$ are the (effective) Lorentz factors of the fictitious particles of masses $M, \mathscr{M}$ moving effectively at the velocities $\boldsymbol{v}_{\rm M}, \boldsymbol{v}$, such that their dynamical consequence is the same as that due to the motions of m_e, m_p relative to the CM. In particular, $\boldsymbol{v}_{\rm M}$ needs be thought of as the speed of the CM relative to the e, p, i.e. $\boldsymbol{v}_{\rm M} = \frac{d\mathbf{R}}{d\bar{t}_{ep}}$ given in terms of a mean local time $\bar{t}_{ep}$ of e, p; the CM is not moving relative to itself.

Transformations from the scalar distances r_e, r_p, R, r to r_e^0, r_p^0, R^0, r^0 at fixed t (hence $t_e, t_p, \bar{t}_{ep}$), from the time t to t^0 at fixed $\mathbf{r}$, and from the CM-frame masses $m_e, m_p, M, \mathscr{M}$ to their respective rest-frame counterparts $m_e^0, m_p^0, M^0 (= m_e^0 + m_p^0), \mathscr{M}^0 (= m_e^0 m_p^0/M^0)$ follow as

$$\gamma_e r_e = r_e^0, \quad \gamma_p r_p = r_p^0, \quad \gamma_{\rm M} R = R^0, \quad \gamma r = r^0; \quad \gamma t = t^0; \tag{7.1}$$
$$m_e = \gamma_e m_e^0, \quad m_p = \gamma_p m_p^0, \quad M = m_e + m_p = \gamma_{\rm M} M^0, \quad \mathscr{M} = \gamma \mathscr{M}^0. \tag{7.2}$$

Using Eqs (7.2) for $m_e, m_p, M, \mathscr{M}$ in (2b),(d) gives (8), and solving gives (9) below:

$$\gamma_{\rm M} M^0 = \gamma_e m_e^0 + \gamma_p m_p^0, \quad \gamma_{\rm M}\gamma = \gamma_e \gamma_p; \quad \text{or} \quad M^0 = m_e^\dagger + m_p^\dagger, \quad \text{where} \tag{8.1}$$
$$m_e^\dagger = \frac{m_e}{\gamma_{\rm M}} = \gamma_e^\dagger m_e^0, \quad m_p^\dagger = \frac{m_p}{\gamma_{\rm M}} = \gamma_p^\dagger m_p^0, \quad \gamma_e^\dagger = \frac{\gamma_e}{\gamma_{\rm M}}, \quad \gamma_p^\dagger = \frac{\gamma_p}{\gamma_{\rm M}}; \quad \gamma_e^\dagger \gamma_p^\dagger = \frac{\gamma_e \gamma_p}{\gamma_{\rm M}^2} = \frac{\gamma}{\gamma_{\rm M}} = \gamma^\dagger \tag{8.2}$$

$$\gamma_e = \frac{\gamma_{\rm M}(M^0 \pm K)}{2m_e^0}, \quad \gamma_p = \frac{\gamma_{\rm M}(M^0 \mp K)}{2m_p^0}, \quad K = \sqrt{(M^0)^2 - 4m_e^0 m_p^0 \gamma^\dagger}. \tag{9}$$

For (9) to have real solutions requires $(M^0)^2 - 4m_e^0 m_p^0 \gamma^\dagger \geq 0$, or $\gamma^\dagger \leq (\gamma^\dagger)_{\rm max} = (M^0)^2/4m_e^0 m_p^0 = 459.556$, where $\gamma^\dagger = (\gamma^\dagger)_{\rm max}$ if $K = 0$, in which case $\gamma_e = \gamma_{\rm M} M^0/2m_e^0$, $\gamma_p = \gamma_{\rm M} M^0/2m_p^0 \simeq \frac{1}{2}\gamma_{\rm M}$, $m_e = m_p$. In general m_e and m_p may not be equal. Let $m_e = km_p$; this combined with (9a) gives $(m_e =)km_p = \gamma_e m_e^0 = \frac{1}{2}\gamma_{\rm M}(M^0 + K)$. Dividing it by (9b) times m_p^0, i.e. $(m_p =)\gamma_p m_p^0 = \frac{1}{2}\gamma_{\rm M}(M^0 - K)$ gives (10a,b), and re-arranging (9c) gives (10c) below,

$$k = \frac{M^0 + K}{M^0 - K}, \quad \text{or} \quad K = \frac{(k-1)M^0}{k+1}; \quad \gamma^\dagger = \frac{(M^0)^2 - K^2}{4m_e^0 m_p^0} = \frac{k(M^0)^2}{(k+1)^2 m_e^0 m_p^0} \tag{10}$$

Substituting in these $k = m_e/m_p = 1.3165$ from the solution for neutron magnetic moment (Sec 3) gives $K = \frac{(1.3165-1)}{1.3165+1} 938.78(3) = 128.26(5)$ GeV, and $\gamma^\dagger = 450.96(0)$. Eqs (2g),(h) and (3a),(b) for this case become $\mathbf{r}_e = \frac{\mathbf{r}}{k+1} = 0.43\mathbf{r}$, $\mathbf{r}_p = -\frac{k\mathbf{r}}{k+1} = -0.57\mathbf{r}$; $\boldsymbol{v}_e = 0.43\boldsymbol{v}$, $\boldsymbol{v}_p = -0.57\boldsymbol{v}$ (cf Fig. 1c, right graph). $k \gtrsim 1$ implies $\gamma_e, \gamma_p >> 1$.

Multiplying $\frac{\gamma_e m_e}{\gamma_e + 1}$ to the quadratics of Eq (3a), and $\frac{\gamma_p m_p}{\gamma_p + 1}$ to that of (3b), adding, we obtain on the left side the total kinetic energy $T_e + T_p$ of e,p measured in the CM frame and in time t,

$$(T_e + T_p \equiv) \frac{\gamma_e m_e v_e^2}{\gamma_e + 1} + \frac{\gamma_p m_p v_p^2}{\gamma_p + 1} = \left[\frac{\gamma_e m_p}{(\gamma_e + 1)M} + \frac{\gamma_p m_e}{(\gamma_p + 1)M}\right] \mathscr{M} v^2 (\equiv T) \tag{11.1}$$
$$\text{for } \gamma_e, \gamma_p >> 1: \quad m_e v_e^2 + m_p v_p^2 = \mathscr{M} v^2 \tag{11.2}$$

The right side of (11.1) or (11.2) expresses the kinetic energy T of the reduced mass relative to the CM. Eq (11.1) or (11.2) expresses invariance of kinetic energy under the $\mathbf{r}_e, \mathbf{r}_p$ to $\mathbf{r}, \mathbf{R}$ coordinate transformation as described in the CM frame and in time t. Performing similar operations to Eqs (5a,b) instead we obtain on the left side the total kinetic energy $T_e' + T_p'$ of e,p measured in the CM frame but in their local times t_e, t_p,

$$(T_e' + T_p' \equiv) \frac{\gamma_e m_e}{\gamma_e + 1} v_e'^2 + \frac{\gamma_p m_p}{\gamma_p + 1} v_p'^2 = \left[\frac{\gamma_e m_e}{(\gamma_e + 1)M} + \frac{\gamma_p m_p}{(\gamma_p + 1)M}\right] M v^2 (\equiv T') \tag{12.1}$$
$$\text{for } \gamma_e, \gamma_p >> 1: \quad m_e v_e'^2 + m_p v_p'^2 = M v^2 \tag{12.2}$$

The right side of (12.1) or (12.2) represents in effect the kinetic energy T' of the total mass at the CM relative to the e, p local space and time coordinates. Since $M > \mathscr{M}$, so $T_e' + T_p' > T_e + T_p$. The difference $(T_e' + T_p') - (T_e + T_p)$ apparently represents a kinetic energy contribution from the non-inertial frame motion at $\mathbf{r}_e, \mathbf{r}_p$ relative to the CM.

Unless specified otherwise we shall hereafter suppose for simplicity the e, p system as a whole, i.e. its CM, to be at rest in the lab frame. Provided further setting the coordinate origins of the CM and lab frames the same, hence $\mathbf{R} = 0$, the relativistic effects in the two frames are the same.

2.3. Total mass as measured in the lab frame. Neutron mass For the centre of mass CM of the e, p system assumed at rest in the lab frame, naturally an observer in the lab frame will measure a rest total mass $M^{\text{lab}} = M^0 = m_e^0 + m_p^0$ of the model neutron. In more elaborate terms, a measurement of the neutron mass in the laboratory is typically made in a specified say X direction over a macroscopic time interval Δt which is $>> 2\pi r/v$, the rotation period of $\mathscr{M}$ (Secs 2.4, 6). During Δt, $\boldsymbol{v}_e, v_p$ explore all directions each with a zero average projection in the X direction. Hence the relativistic augments in the masses of e, p as measured along the instantaneous directions $\boldsymbol{v}_e, v_p$ in the CM frame do not enter the mass $M^{\text{lab}}(= m_e^{\text{lab}} + m_p^{\text{lab}})$ measured in the lab frame. (This mass augment however evidently enters the interaction force or potential of Sec 4, which has a constant magnitude so as to manifestly effectuate a bound e, p in stationary state irrespective of the direction of the e, p separation.)

A dually relevant example here is electron scattering by a target neutron. In respect to the internal dynamics of a target neutron, an incident electron e travelling in a fixed direction is as a (moving) observer in the lab frame. The incident e thus will see the rest (as contrasted to relativistic) masses of the e, p of the neutron. Moreover, the e, p of the neutron are fast rotating along circles of similar radii about their CM and thus about equally exposed to the incident e. So in terms of exposure frequency, the e, p would equally probably scatter with the incident e, through electro and magnetic potentials and naturally at their contracted radii a_e, a_p. The scattering potential from the proton p of the neutron, on the other hand, would dominate because of its much heavier rest mass, which for the electrostatic part at least is attractive. Incidentally, the experimentally measured electron-neutron scattering length is negative and suggests an attractive scattering potential.

The (very large) e, p interaction potential fields within the neutron, on the other hand, are liable to (considerably) modify the vacuum potential surrounding the e, p charges; the effect would be particularly large given the e, p separation distance 10^{-18} m (Sec 6) here is comparable with the inter-vacuuon distance based on the "vacuuonic vacuum structure"[a]. This would consequently further modify the e, p particles' rest masses, in terms of the IED model[b], produced as their generating charges move through this modified vacuum. (a, b: see the author's earlier published work). The above gives a qualitative account for the (order of MeV) larger neutron rest mass over the sum of the e, p rest masses; this difference is relatively small and is ignored where in question throughout this paper.

2.4. Eigenvalue equations. Orbital and precessional angular momenta. Antineutrino In the absence of applied force and omitting the very weak gravity, M is free and hence not directly subject to quantisation condition. We thus need only to establish the relativistic Schrödinger or Klein-Gordon equation (KGE) for the reduced mass $\mathscr{M}$, in terms of the spherical polar coordinates r, ϑ, ϕ transformed directly from x, y, z. The KGE has the usual form $[((E_{tot})_{op} - V)^2 - \mathscr{M}^{0\,2} c^4 - (p^2)_{op} c^2]\psi_{tot} = 0$, where $(E_{tot})_{op} - V = \mathscr{M} c^2$; the associated non-inertial frame effect is not contained in it and will be included separately. Since the mass $\mathscr{M}$ under consideration is moving at velocity exceedingly close to c such that its rest-mass energy is negligibly small compared to its kinetic energy (Sec 6), more relevant here is the square-root (SQR) form of the KGE: $\mathcal{H}_{op}\psi = \mathcal{H}\psi$, where $\mathcal{H}_{op} = ((E_{tot})_{op} - V) - \mathscr{M}^0 c^2 + V = \mathcal{T}_{op} + V$, and $\mathcal{T}_{op} = \mathscr{M} c^2 - \mathscr{M}^0 c^2 = (\gamma - 1)\mathscr{M}^0 c^2 = \frac{(\gamma-1)\gamma(p^2)_{op}}{\gamma^2(v/c)^2\mathscr{M}} = \frac{\gamma(p^2)_{op}}{(\gamma+1)\mathscr{M}}$ (with $\gamma^2(\frac{v}{c})^2 = \gamma^2 - 1$) are the Hamiltonian and kinetic energy operators associated with the kinetic motion of $\mathscr{M}$; $(p^2)_{op} = (p_r^2)_{op} + \frac{(\mathcal{J}^2)_{op}}{r^2}$; $(p_r^2)_{op}$ and $(\mathcal{J}^2)_{op}$ are the squared radial and orbital angular momentum operators. For the e,p interaction potential V being central (Sec 4), hence $V(\mathbf{r}) = V(r)$, the wave function of $\mathscr{M}$, $\psi(r,\vartheta,\phi)$, may be written as $\psi = \mathcal{R}(r)\mathcal{Y}(\vartheta,\phi)$. And the SQR-KGE separates into two eigenvalue equations,

$$\left[-\frac{\gamma\hbar^2}{(\gamma+1)\mathscr{M}r^2}\frac{\partial}{\partial r}(r^2\frac{\partial}{\partial r}) + \frac{\gamma l(l+1)\hbar^2}{(\gamma+1)\mathscr{M}r^2} + V(r)\right]\mathcal{R}(r) = \mathcal{H}\mathcal{R}(r) \tag{13}$$

$$(\mathcal{J}^2)_{op}\mathcal{Y}(\vartheta,\phi) = \mathcal{J}^2\mathcal{Y}(\vartheta,\phi), \quad (\mathcal{J}^2)_{op} = -\hbar^2\left(\frac{\partial^2}{\partial\vartheta^2} + \cot\vartheta\frac{\partial}{\partial\vartheta} + \frac{1}{\sin^2\vartheta}\frac{\partial^2}{\partial\phi^2}\right). \tag{14}$$

(14) may be solved without $V(r)$ being explicitly known. The eigen functions are the spherical harmonics, $\mathcal{Y}_l^{m_l} = C_l^{m_l}P_l^{m_l}(\cos\vartheta)e^{im_l\phi}$. The square-root eigen values and their z components are

$$\mathcal{J}_l = |\mathbf{r}_l\times(\mathscr{M}\boldsymbol{v}_{tl})| = \sqrt{l(l+1)}\,\hbar, \quad \mathcal{J}_{zm_l} = \mathcal{J}_l\cos\vartheta_{m_l} = m_l\hbar, \quad l = 0,\,1,\,\ldots;\, m_l = 0,\,\ldots,\,\mp l. \tag{15}$$

Based on the semiclassical expression $\mathbf{r}_l\times(\mathscr{M}\boldsymbol{v}_{tl})$, the particle of mass $\mathscr{M}$ in lth state executes an orbital angular motion along a circular orbit l of radius vector $\mathbf{r}_l$ at a tangential velocity $\boldsymbol{v}_{tl} = d\mathbf{r}_l/dt = \boldsymbol{\omega}_o\times\mathbf{r}_l$, $\boldsymbol{\omega}_o = \mathbf{r}_l\times\boldsymbol{v}_{tl}/r_l^2$; $\mathbf{r}_l \equiv \mathbf{r}_n$ for all $l(=0,1,\ldots,n-1)$ values of the same principal quantum number n. The normal of the orbital plane or the axis of rotation $\mathbf{n_o}$ passes through the CM and is at a quantised angle ϑ_{m_l} to the z axis.

Owing to their having a finite acceleration $\mathsf{a}_l = -|d^2\mathbf{r}_l/dt^2|(\mathbf{r}_l/r_l)$ in radial direction here, as a well-known non-inertial frame effect the e,p in addition execute a Thomas precession (TP), with an instantaneous angular velocity denoted by $\boldsymbol{\omega}_{TP}$ and thus angular momentum $\mathbf{J}_{TP} = r_l^2\mathscr{M}\boldsymbol{\omega}_{TP}$. $\boldsymbol{\omega}_{TP} = \frac{\gamma^2}{(\gamma+1)}\frac{\mathsf{a}_l\times\boldsymbol{v}_{tl}}{c^2}$ according to LH Thomas (1927), as may be alternatively derived directly based on (transformed) infinitesimal Newtonian inertial-frame and hence linear motion combined with acceleration in infinitesimal time (internal work). $\boldsymbol{\omega}_{TP}$ is in the instantaneous direction $\mathsf{a}_l\times\boldsymbol{v}_{tl} \propto -\mathbf{r}_l\times\boldsymbol{v}_{tl} = -\boldsymbol{\mathcal{J}}_l/\mathscr{M}$, i.e. opposite to $\boldsymbol{\mathcal{J}}_l$, describing an instantaneous rotation in opposite sense to the orbital angular motion underlining $\boldsymbol{\mathcal{J}}_l$. For a quantum system as the bound e,p here, the z component of $\mathbf{J}_{TP}$, J_{TPz}, will be necessarily constrained such that both the space quantisation conditions (15) above and (16) below are met.

The total, precessional-orbital angular momentum $J_j = |\boldsymbol{\mathcal{J}}_l - \mathbf{J}_{TP}|$ and its z component $J_{zm_j} = \mathcal{J}_{zm_l} - J_{TPz} = J_j\cos\theta_{m_j}$ are given according to the quantum vector addition model as

$$J_j = |\mathbf{r}_{lj}\times(\mathscr{M}\boldsymbol{v}_j)| = r_l\mathscr{M}v_j = \sqrt{j(j+1)}\,\hbar, \quad j = l - l_{TP} = 0,\,\frac{1}{2},\,1,\,\frac{3}{2},\,\ldots \tag{16.1}$$

$$J_{zm_j} = J_j\cos\theta_{m_j} = \mp J_j\cos\theta_j = m_j\hbar, \quad m_j = \mp j, \tag{16.2}$$

where the permitted j values are results of the general solutions of the quantum commutation relation for the angular momentum $\mathbf{J}$ here, $\mathbf{J}\times\mathbf{J} = i\hbar\mathbf{J}$. $\mathbf{r}_{lj} = r_l\hat{r}_{lj}$ is the quantised radius vector of the instantaneous circular orbit l of a normal or axis of rotation $\mathbf{n}$; $\mathbf{n}$ is at a fixed angle $\theta_{m_j} = \arccos(J_{zm_j}/J_j)$ to the z axis and rotates about the z axis at the angular velocity $\boldsymbol{\omega}_{TPz}$ in opposite sense to that of the ω_o-orbital angular motion projected in the xy plane, whence the Thomas precession; see Fig 1b. The magnitude of $\mathbf{r}_{lj}$, $|\mathbf{r}_{lj}| = r_l$ is unchanged subjected to the radial eigenvalue equation (13) but immediately to the quantum equation (15a) here. $\boldsymbol{v}_j = d\mathbf{r}_{lj}/dt = \boldsymbol{\omega}\times\mathbf{r}_{lj}$; $\boldsymbol{\omega} = \mathbf{r}_{lj}\times\boldsymbol{v}_j/r_l^2 = |\boldsymbol{\omega}_o - \boldsymbol{\omega}_{TP}|\mathbf{n}$ is the precessional-orbital angular velocity. For facilitating later discussion we attach as in Fig 1b the axes x', y', z' to the instantaneous rest frame of the precessional orbit l, with their origin coinciding with that of x, y, z, i.e. fixed at the CM. So the x', y', z' axes precess about the z axis at the angular velocity ω_{TPz}, in counterclockwise sense for the $m_j = -j$ state in the figure, in such a way that the z' axis maintains at fixed angle θ_j to the z axis, the x' axis at fixed angle θ_j to (its projection x'_{xy} in) the xy plane and along the $\mathbf{r}_{lj}$ direction, and the y' axis in the xy plane. And we attach the

x'', y'', z'' axes to the instantaneous rest frame of the precessing-orbiting particles e, p with an origin fixed at the CM too, the z'' axis coinciding with z', and the x'' axis lying along the line joining the e, p; see Fig 1a,b. So the x'', y'' axes rotate in the $x'y'$ plane and about the z' (z'') axis, in clockwise sense for the $m_j = -j$ state, at the angular velocity $\omega_o + \omega_{TP}$ relative to the x', y' axes.

From Eq (16.1b), it follows that the permitted l_{TP} values are uniquely specified once l, j are specified according to (15),(16): For $l = 0$, only $l_{TP} = 0$ is permitted; and $j = l = 0$. This gives an e, p system not magnetically bound at a separation of the order 10^{-18} m (see further Sec 4); a bound state would in principle be achievable at much larger separation as a hydrogen only. For any non-zero integer l values, $l_{TP} = 0$ is permitted formally by (16.1) but is however unphysical because of the so implied absence of Thomas precession. $l_{TP} = \frac{1}{2}$ is therefore the smallest finite and hence physical value which is also permitted based on the quantum solutions for l, j. Moreover, $l_{TP} = \frac{1}{2}$ is itself a solution for $\mathbf{J}_{TPz}$ to separately satisfy the quantum commutation relation $\mathbf{J}_{TPz} \times \mathbf{J}_{TPz} = i\hbar \mathbf{J}_{TPz}$; this establishes a condition for the carrier of $\mathbf{J}_{TPz}$, the neutral rotational energy flux (to be identified as the antineutrino) to be created or emitted as a quantum particle. Higher half-integer or integer l_{TP} values satisfying (16.1) and (15) are permitted in theory but are not liable for a neutron candidate and will not be considered. For the permitted l and $l_{TP} = \frac{1}{2}$, Eqs (16) are written as

$$J_j = r_l \mathscr{M} v_j = \sqrt{j(j+1)}\,\hbar = \frac{\sqrt{(4l^2-1)}\,\hbar}{2}, \qquad j = l - \tfrac{1}{2} = \tfrac{1}{2}, \tfrac{3}{2}, \ldots; \tag{16.1$'$}$$

$$J_{zm_j} = J_j \cos\theta_{m_j} = m_j \hbar = (\mp|m_l| \pm \tfrac{1}{2})\hbar, \quad m_j = \mp j = \mp|m_l| \pm \tfrac{1}{2} = \mp\tfrac{1}{2}, \ldots, \mp j, \tag{16.2$'$}$$

where $\cos\theta_{m_j} = \frac{J_{zm_j}}{J_j} = \frac{\mp 2l \pm 1}{\sqrt{4l^2-1}}$. For $j = \frac{1}{2}$, $m_j = \mp\frac{1}{2}$:

$$J_{\frac{1}{2}} = |\mathbf{r}_{1\frac{1}{2}} \times (\mathscr{M}\boldsymbol{v}_{\frac{1}{2}})| = r_1 \mathscr{M} v_{\frac{1}{2}} = \frac{\sqrt{3}\,\hbar}{2}, \quad J_{z\mp\frac{1}{2}} = r_1 \mathscr{M} v_{\frac{1}{2}} \cos\theta_{\mp\frac{1}{2}} = \mp\frac{\hbar}{2}; \tag{17}$$

$\cos\theta_{\frac{1}{2}} = J_{z\frac{1}{2}}/J_{\frac{1}{2}} = 1/\sqrt{3}$; $\theta_{\frac{1}{2}} = \arccos(1/\sqrt{3}) = 54.7^o$. The $j = \frac{1}{2}$ ($l = 1$) states describe a ground-state neutron (Secs 3, 4). Eq (17a) thus gives the e, p relative precessional–orbital angular momentum internal of the neutron, and (17b) the two possible z components associated with a minimum-energy ($m_j = -\frac{1}{2}$) and excitation ($m_j = \frac{1}{2}$) state in an applied magnetic field in the $+z$ direction (see Sec 3). The precessing circular orbit of the mass $\mathscr{M}$ has the quantised radius vector $\mathbf{r}_{1\frac{1}{2}}$ about the CM in the $x'y'$ plane. For the $m_j = -\frac{1}{2}$ state, the normal $\mathbf{n}$ of the rotation plane, and hence $\mathbf{J}_{\frac{1}{2}}$, is at angle $\theta_{-\frac{1}{2}} = \pi - \theta_{\frac{1}{2}}$ to the z axis, the rotation being in clockwise sense, as shown in Fig 1b. And conversely for $m_j = \frac{1}{2}$. For the next orbital, $j = \frac{3}{2}$ ($l = 2$), $J_{z\frac{3}{2}}/J_{\frac{3}{2}} = 3/\sqrt{15}$, $\theta_{\frac{3}{2}} = 39.2^o$.

For the neutron existing (in zero applied field) only in a single non-degenerate state $j = l - \frac{1}{2} = \frac{1}{2}$ and presuming that, in terms of the SQR-KGE here, energies of different l and same n are degenerate, then N (the radial degree of freedom)$= 0$ and $n = N + l + 1 = 0 + l + 1 = 2$. So $\mathcal{T}_{rl}|_{l=n-1=1} = 0$; and the total kinetic energy of $\mathscr{M}$ is, with $J_{\frac{1}{2}}^2 (= (\mathcal{J}_1 - \mathbf{J}_{TP})^2)$ for $\mathcal{J}_1^2$ and T in place of $\mathcal{T}$, $T_{\frac{1}{2}} = T_{t\frac{1}{2}} = \frac{\gamma}{(\gamma+1)} \mathscr{M} v_{\frac{1}{2}}^2 = \frac{\gamma}{(\gamma+1)} (J_{\frac{1}{2}}^2 / \mathscr{M} r_1^2) = \frac{3\gamma\hbar^2 M}{4(\gamma+1) m_e m_p r_1^2}$. Accordingly, the eigen energy (Hamiltonian) $H_j = T_j + V_j$, in place of $\mathcal{H}_{l=n-1} = \mathcal{T}_{l=n-1} + V_{l=n-1}$ (where $\mathcal{H}_{l=n-1} \equiv \mathcal{H}_n$, etc.), may be directly computed from the sum of the T_j given above and the V_j to be derived in Sec 4, without formally solving the radial differential equation (13).

The precessional–orbital motion (propagation) of the e, p particle waves ψ_e, ψ_p relative to one another, or alternatively of the matter wave ψ of $\mathscr{M}$ relative to the CM, is associated with a neutral precessing-orbiting — simply rotational kinetic energy flux, or vortex, on disregarding their charges and also their rest-mass energies. This vortex entity carries a spin angular momentum with a z component equal to one unit of half-integer quantum $J_{z\mp\frac{1}{2}} = \mp\frac{1}{2}\hbar$, has apparently a positive helicity, and therefore resembles directly an antineutrino ($\bar{\nu}_e$) confined within the neutron; see further Sec 5. Accordingly the z-component spin angular momenta of $\bar{\nu}_e$ are

$$S_{\bar{\nu}_e z} = J_{z\mp\frac{1}{2}} = \mp\tfrac{1}{2}\hbar = \mp s_{\bar{\nu}_e}\hbar, \quad s_{\bar{\nu}_e} = \tfrac{1}{2}. \tag{18}$$

3. e, p system magnetic structure. Neutron magnetic moment, (effective) spin

3.1 e, p spins and magnetic moments. Neutron internal spin configurations. Total spin Either particle $\alpha = e$ or p has an intrinsic spin $s_\alpha = \frac{1}{2}$, spin angular momentum $S_\alpha = \sqrt{s_\alpha(s_\alpha+1)}\ \hbar = \frac{\sqrt{3}}{2}\hbar$ about a spin axis $\mathbf{n}_\alpha$ passing through $\mathbf{r}_\alpha$, and z component $S_{\alpha z} = S_\alpha \cos\theta^s_\alpha = \pm s_\alpha\hbar$. For the spin up state as in Fig 1a, $\cos\theta^s_\alpha = S_{\alpha z}/S_\alpha = 1/\sqrt{3}$; the spin axis $\mathbf{n}_\alpha$ is at fixed angle $\theta^s_\alpha = 54.73^o$ to a z^s_α axis parallel with z and passing through $\mathbf{r}_\alpha$. Certain external, random environmental in the case of zero applied, magnetic field would always present and thus define the (instantaneous) z direction here. For computing the magnetic field produced by the spin motion of one particle α at the other particle (α'), $\mathbf{B}^s_\alpha$ (Eq 30b for $\alpha = p$, Sec 4) at a separation r_l comparable to the sizes of their charges (Secs. 4, 6), it is appropriate to treat the e, p charges as extended objects, firstly simply as spheres of radii a_e, a_p with specific mass and charge distributions. We assume that the mass m_α of either particle α is distributed predominantly within its charge (and thus by a negligible amount in its wave field), with a volume density $\rho_\alpha(\boldsymbol{\xi}_\alpha)$ at a distance $\boldsymbol{\xi}_\alpha$ from $\mathbf{r}_\alpha$; $dm_\alpha = m_\alpha\rho_\alpha(\boldsymbol{\xi}_\alpha)d^3\boldsymbol{\xi}_\alpha$ is a mass element at $\boldsymbol{\xi}_\alpha$. So S_α formally is given rise to by the angular motion of the sphere about $\mathbf{n}_\alpha$ at the angular velocity $\omega^s_\alpha = d\phi^s_\alpha/dt_\alpha$, tangential velocity $v^s_\alpha = a_\alpha\omega^s_\alpha$ as,

$$S_\alpha = m_\alpha \textstyle\int |\boldsymbol{\xi}_\alpha \times \boldsymbol{v}^s_\alpha|\rho_\alpha(\boldsymbol{\xi}_\alpha)d^3\boldsymbol{\xi}_\alpha = \frac{1}{g_\alpha}a^2_\alpha\omega^s_\alpha m_\alpha = \frac{\sqrt{3}}{2}\ \hbar; \quad S_{\alpha z} = \frac{1}{g_\alpha}a^2_\alpha m_\alpha\omega^s_\alpha\cos\theta^s_\alpha = \frac{1}{2}\hbar \tag{19}$$

where g_α is the Lande g factor of particle α. And the charge q_α of either particle α is distributed along a circular loop of radius a_α with a normal parallel with $\mathbf{n}_\alpha$. The spin (dipole) magnetic moment μ^s_α of particle α is accordingly produced by the current loop of charge q_α, area πa^2_α, and angular velocity ω^s_α in the $\mp S_\alpha$ direction for $q_\alpha = \mp e$. The z components, written for e, p explicitly, are

$$\mu^s_{ez} = e\frac{\omega^s_e}{2\pi}\pi a^2_e\cos(\pi - \theta^s_e) = -\frac{g_e e S_{ez}}{2m_e} = -\frac{g_e e\hbar}{4m_e}; \quad \mu^s_{pz} = \frac{g_p e S_{pz}}{2m_p} = \frac{g_p e\hbar}{4m_p} \tag{20}$$

For either particle α, besides the z^s_α above we further specify the local axes x^s_α, y^s_α to be parallel with the x, y but with an origin located at $\mathbf{r}_\alpha$. Its spin axis $\mathbf{n}_\alpha$ as projected in this $x^s_\alpha y^s_\alpha$ plane is unspecified in orientation according to the uncertainty principle, and in the external magnetic field in z direction would rotate about the z^s_α axis. Accordingly the direction of $\boldsymbol{v}^s_\alpha$ at any fixed point on the current loop varies over time as the current loop precesses. Since the z component $S_{\alpha z}$ of S_α is a constant (Eq 19b), the projection of $\boldsymbol{v}^s_\alpha$ in the $x^s_\alpha y^s_\alpha$ plane is a constant: $v^s_{\alpha_{xy}} = (a_\alpha\cos\theta^s_\alpha)\omega^s_\alpha = v^s_\alpha\cos\theta^s_\alpha$; this gives also the time average of $\boldsymbol{v}^s_\alpha$, for the projection of $\boldsymbol{v}^s_\alpha$ in z direction averages to zero over time. For deriving an effective algebraic equation for the magnetic field produced by the spin current loop (Sec 4), the distinct symmetry property of the system will be utilised to further reduce the system to a two point half-charge representation (Appendix A).

3.2 e, p system spin configuration. Total spin angular momentum For the e, p to be in a bound, minimum (internal magnetic) energy state (Sec 4), apart from the specific $j = \frac{1}{2}$ and $m_j = \mp\frac{1}{2}$ values (Sec 2.4), S_{ez}, S_{pz} need be parallel mutually and each antiparallel to J_{zm_j} (Figs 1a,b; 2a,b). S_{ez}, S_{pz}, J_{zm_j} may therefore assume two possible configurations

$$(i)\ S_{ez} = \tfrac{1}{2}\hbar,\ \ S_{pz} = \tfrac{1}{2}\hbar,\ \ J_{z\text{-}\frac{1}{2}} = -\tfrac{1}{2}\hbar; \quad (ii)\ S_{ez} = -\tfrac{1}{2}\hbar,\ \ S_{pz} = -\tfrac{1}{2}\hbar,\ \ J_{z\frac{1}{2}} = \tfrac{1}{2}\hbar \tag{21}$$

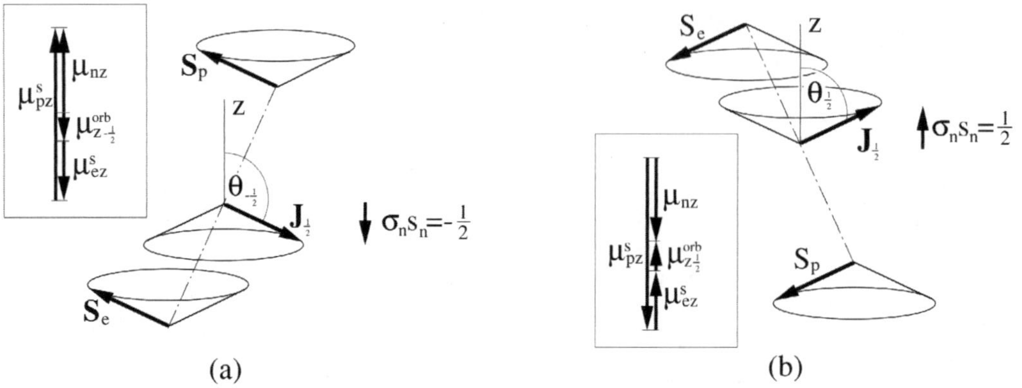

Figure 2. Internal angular momenta S_e, S_p, J_{m_j} of the model neutron with the internal spin configurations (a) $s_e, s_p, m_j = \frac{1}{2}, \frac{1}{2}, -\frac{1}{2}$ in effective spin down state $s_n\sigma_n = m_j = -\frac{1}{2}$, and (b) $s_e, s_p, m_j = -\frac{1}{2}, -\frac{1}{2}, \frac{1}{2}$ in effective up state $\sigma_n s_n = m_j = \frac{1}{2}$. Insets in (a),(b): schematics of the corresponding z-component neutron magnetic moments $\boldsymbol{\mu}_{nz}$ as the respective vector sums of the internal component moments along z direction.

as shown in Figs 2a,b.

Based on the usual vector addition model, the total angular momentum of the e, p system, denoted by $\mathscr{J}_{j_n}$, in the precessional-orbital state $j = \frac{1}{2}$ are

$$\mathscr{J}_{j_n} = \sqrt{j_n(j_n+1)}\hbar = \tfrac{\sqrt{3}}{2}\hbar, \quad j_n = (s_e + s_p) - j = (\frac{1}{2} + \frac{1}{2}) - \frac{1}{2} = \frac{1}{2} \tag{22}$$

Its z components $\mathscr{J}_{zm_{jn}}$ for the $m_{j_n} = \pm j_n$ states are

$$\mathscr{J}_{z\pm\frac{1}{2}} = (S_{pz} + S_{ez})_{\substack{up \\ down}} + J_{z\mp\frac{1}{2}} = [\pm(\tfrac{1}{2} + \tfrac{1}{2}) \mp \tfrac{1}{2}]\hbar = \pm\tfrac{1}{2}\hbar. \tag{23}$$

3.3 Total magnetic moment The differing g factors and potentially also asymmetric relative positions of e, p suggest an asymmetric internal magnetic structure of the e, p system. One thus expects a total magnetic moment which is not simply related to $\mathscr{J}_{zm_{jn}}$ of Eqs (23) as for a simple particle; nor would it be zero as implied by its zero net charge as both given by the direct sum of the e, p charges here and from experimental observation. We shall find the total magnetic moment based directly on vector addition of the individual magnetic moments along the z direction below. For the spin configuration (i) of Eqs (21), the z-component total magnetic moment of the e, p spins is the vector sum

$$\mu_z^s = \mu_{pz}^s + \mu_{ez}^s = \frac{g_p e S_{pz}}{2m_p} - \frac{g_e e S_{ez}}{2m_e} = (g_p - \frac{g_e}{k})\frac{eS_{pz}}{2m_p}, \tag{24}$$

where the last of Eqs (24) is given after substituting $m_e = km_p$ (as in Sec 2.2) and $S_{ez} = S_{pz}$. k may in general differ from 1. So the relative precessional-orbital motion may contribute a finite moment given by the vector sum, for the case (i) or $m_j = -\frac{1}{2}$,

$$\mu_{z\text{-}\frac{1}{2}}^{orb} = \frac{eJ_{pz\text{-}\frac{1}{2}}}{2m_p} - \frac{eJ_{ez\text{-}\frac{1}{2}}}{2m_e} = \frac{e(\frac{m_e}{M})J_{z\text{-}\frac{1}{2}}}{2m_p} - \frac{e(\frac{m_p}{M})J_{z\text{-}\frac{1}{2}}}{2m_e} = \left(\frac{k-1}{k}\right)\frac{eJ_{z\text{-}\frac{1}{2}}}{2m_p} \tag{25}$$

where $J_{\alpha z m_j}$ are the z-projections of $J_{\alpha j}$ given after Eqs (4a,b); a g factor equal to 1 is assumed. The total z-component magnetic moment of the e, p system is the vector sum

$$\mu_{z\text{-}\frac{1}{2}\phi} = \mu_z^s + \mu_{z\text{-}\frac{1}{2}}^{orb} = (g_p - \frac{g_e}{k})\frac{e(-J_{z\text{-}\frac{1}{2}})}{2m_p} + \left(\frac{k-1}{k}\right)\frac{eJ_{z\text{-}\frac{1}{2}}}{2m_p} = -\frac{g_n e J_{z\text{-}\frac{1}{2}}}{2m_p}, \tag{26a}$$

$$g_n = g_p - \frac{g_e}{k} - \frac{k-1}{k} \tag{26b}$$

where in the expression of μ_z^s we substituted $S_{pz} = -J_{z\text{-}\frac{1}{2}}$; J_{zm_j}, instead of $\mathscr{J}_{zm_{jn}}$, is used for reason to be explained below. The subscript ϕ indicates that the total $\mu_{z\text{-}\frac{1}{2}\phi}$, as the μ_{pz}^s, μ_{ez}^s and $\mu_{z\text{-}\frac{1}{2}}^{orb}$, is as directly probed by a detector placed at the CM relative to which the e, p are moving in the $\boldsymbol{v}_e, \boldsymbol{v}_p$ or ϕ directions, in which case the relativistic masses m_e, m_p in Eq (26a) remain physical.

An experimenter (as an external observer) in the laboratory on the other hand commonly probes the neutron magnetic moment by applying a magnetic field to turn the moment along the θ direction, typically at a speed $u_\theta << c$. This is to turn the e, p system as a whole here, or manifestly the e, p precessional-orbital plane about the y' axis along the θ direction in the $x'z'$ plane, i.e. in a direction perpendicular to $\boldsymbol{v}_e, \boldsymbol{v}_p$. So immediately for the proton $\gamma_{p\theta} = (1 - u_{p\theta}^2/c^2)^{-1/2} \simeq 1$, $m_p(u_{p\theta}) = \gamma_{p\theta} m_p^0 \simeq m_p^0$; the proton dominates the turning process because of its much larger rest-mass moment of inertia. The electron can only be turned in the same rigid precessional-orbital plane as the proton in a bound relativistic dynamic process, its mass (hence moment of inertia) must therefore manifestly be weighed by the same factor k as in this relativistic process, i.e. as $m_e = km_p$, not as m_e^0. The only means of correctly carrying the factor k through to the experimenter's result (Eq 27 below) is to convert m_e to km_p before transforming to Eq (27), as has been done in (26a). Substituting m_p^0 in place of m_p in the last of Eqs (26a), and accordingly $\mu_{z\text{-}\frac{1}{2}}$ in place of $\mu_{z\text{-}\frac{1}{2}\phi}$, gives therefore the z-component magnetic moment of the model neutron, μ_{nz}, as probed by the experimenter, for the $m_j = -\frac{1}{2}$ state

$$\mu_{nz}(m_j = -\tfrac{1}{2}) = \mu_{z\text{-}\frac{1}{2}} = -\frac{g_n e J_{z\text{-}\frac{1}{2}}}{2m_p(u_{p\theta})} = -\frac{g_n e J_{z\text{-}\frac{1}{2}}}{2m_p^0} = \frac{g_n e\hbar}{4m_p^0} = \frac{1}{2} g_n \mu_N, \tag{27}$$

where $\mu_N = \frac{e\hbar}{2m_p^0}$ (the nuclear magneton). Similarly for the $m_j = \frac{1}{2}$ state we obtain $\mu_{nz}(m_j = \frac{1}{2}) = \mu_{z\frac{1}{2}} = -g_n e J_{z\frac{1}{2}}/2m_p^0 = -\frac{1}{2} g_n \mu_N$. g_n represents the g factor of the model neutron. Equating g_n of (26b) with the experimental value $g_n^{exp} = 3.8261$, and using the experimental g values of e, p, $g_p = 5.5857$, $g_e = 2$, numerical solution for k is obtained as $k = 1.3165$. According to (27) or (26a), $\mu_{z\text{-}\frac{1}{2}} > 0$, i.e. $\mu_{z\text{-}\frac{1}{2}}$ points in the $+z$ direction, and is in the opposite direction to $J_{z\text{-}\frac{1}{2}}$ (Fig 2a). And $\mu_{z\frac{1}{2}} < 0$ (Fig 2b). Clearly, the magnetic moment of the model neutron is dominated by the proton spin magnetic moment because of the asymmetrically much larger g_p over g_e. From $k > 1$ and hence $m_e > m_p$, $r_p > r_e$ (by a small amount each), it follows that $\mu_{z\text{-}\frac{1}{2}}^{orb}$ points in the $-z$ direction, adding a negative but small term to $\mu_{z\text{-}\frac{1}{2}}$.

3.4 Effective spin of the neutron In an applied magnetic field say $\mathbf{B}_0$ in $+z$ direction, the magnetic-interaction energy of the e, p system with the field is $U_j = -\boldsymbol{\mu}_j \cdot \mathbf{B}_0 = -\mu_{zm_j} B_0$. $U_j < 0$ for the spin configuration (i) of Eqs (21) or the $m_j = -\frac{1}{2}$ state, and $U_j > 0$ for (ii) or the $m_j = \frac{1}{2}$ state. That is, (i) or $m_j = -\frac{1}{2}$ corresponds to the minimum-energy state and (ii) or $m_j = \frac{1}{2}$ the excited state in the applied field. A transition from the minimum-energy to excited state corresponds to a flip of the spin-configuration (i) to (ii) of the bound e, p system as a whole, which is dictated, and thus manifested by the flip of the processional-orbital plane from the $m_j = -\frac{1}{2}$ state to $m_j = +\frac{1}{2}$. In other terms, $\mu_{z_{-\frac{1}{2}}}$ is as if produced by a negative charged current loop in spin up state, or alternatively by a positively charged current loop in spin down state. In so far as the total magnetic moment of the e, p system as a whole, whence the model neutron, is probed, it is therefore physical to assign to it an effective spin s_n and spin vector σ_n, corresponding directly to the j and $m_j = \mp j$ values (rather than the j_n and m_{j_n}). So the effective neutron spin angular momentum S_n, its z components S_{nz}, and accordingly μ_{zm_j}

in relation with S_{nz}, given by substituting S_{nz} for J_{zm_j} in (27), are

$$S_n = \sqrt{s_n(s_n+1)}\hbar \equiv J_j = \frac{\sqrt{3}}{2}\hbar, \quad s_n = j = \frac{1}{2}; \quad S_{nz} = \sigma_n s_n \hbar \equiv J_{z_{m_j}} = m_j\hbar = \mp\frac{1}{2}\hbar,$$
$$\sigma_n = \mp 1 \text{ (for } m_j = \mp\tfrac{1}{2}); \quad \mu_{nz} = -\frac{g_n e S_{nz}}{2m_p^0} = -\frac{g_n \sigma_n s_n e\hbar}{2m_p^0} \equiv \mu_{z_{m_j}} = \pm\frac{g_n e\hbar}{4m_p^0} \tag{28}$$

Notice that for either spin configuration the S_n, S_{nz} are equal to $\mathscr{J}_{j_n}$, $\mathscr{J}_{zm_{jn}}$ in magnitudes but opposite in directions. The magnitude of S_n or $\mathscr{J}_{zm_{jn}}$, being a one-half integer quantum, has the absolute significance that it identifies the neutron as a fermion in accordance with observation. The assignment of the effective spin parameters s_n, σ_n above is in direct conformity with the role of the Standard Model neutron spin with respect to the experimental determination of neutron magnetic monument based on magnetic resonance method [1f].

4. e, p electromagnetic interaction. Weak interaction force

We shall below derive for the electron e and proton p comprising the model neutron in stationary state their interaction force $\mathbf{F}$, the corresponding potential V and Hamiltonian H based on first principles laws of electromagnetism and (the solutions of Sec 2 of) relativistic kinematics and quantum-mechanics. We shall continue to work in the CM frame, i.e. in terms of the unsuperscripted mass and space-time variables which will directly enter the electromagnetic interactions below, and for simplicity the time t instead of t_e, t_p; the local time t_e, t_p effect will be included in the end by a projecting factor (f_t^2). In this section, $\mathbf{r}$ or $\mathbf{r}_{lj}$ refers to the e, p separation distance starting at $\mathbf{r}_p$, ending at $\mathbf{r}_e$, as in Figs 1a, c; its magnitude is equal to that of $\mathbf{r}$ or $\mathbf{r}_{lj}$ of Sec 2.4, Fig 1b.

Consider the e, p system in a precessional-orbital state $j = l - l_{TP}$, $m_j = -j$, with the e, p spins S_{pz}, S_{ez} in the $+z$ direction, i.e. antiparallel with $J_{z\text{-}j}$ (as in Figs 1a,b or Fig 2a for $j = 1 - \frac{1}{2} = \frac{1}{2}$, $m_j = -\frac{1}{2}$). Firstly, the proton of a charge $+e$ produces at the electron at $\mathbf{r}_{lj}$ apart a (transformed) Coulomb field $\mathbf{E}_p(r_l) = (e/4\pi\epsilon_0 r_l^2)\,\hat{\mathbf{r}}_{lj}$ (in SI units here and below); $\hat{r}_{lj} = \mathbf{r}_{lj}/r_l$ is a unit vector pointing from p to e. $|\mathbf{E}_p|$ is amplified from its rest-frame value E_p^0 by a factor $\propto (1/r^2)/(1/(r^o)^2) = \gamma^2 = 1/f_c$ and hence has a narrowed profile at a point $\mathbf{r}$ perpendicular to its motion ϕ direction by an inverse factor, f_c. And so are the magnetic fields below. Furthermore, the proton is in relative precessional–orbital motion in clockwise sense at the tangential velocity $\boldsymbol{v}_p$ about the CM in the $x'y'$ plane, and in spin motion at the tangential velocity $\boldsymbol{v}^s_{p_{xy}}$ in the $x_p^s y_p^s$ plane (Sec 3.1). The latter, after a projection on to the $x'y'$ and hence $x''y''$ plane, may be effectively represented (Eq A.3, Appendix A) as two point half-charges located at $-\bar{a}, \bar{a}$ from $\mathbf{r}_p$ on the x'' axis and moving oppositely at velocities $-\bar{v}_p^{s\prime\prime}, +\bar{v}_p^{s\prime\prime}$ in $-y'', +y''$ directions. So p produces at e magnetic fields $\mathbf{B}_p^{orb}(= -\boldsymbol{v}_p \times \mathbf{E}_p/c^2)$ and $\mathbf{B}_p^s(r_l \pm \bar{a})(= \pm|\bar{v}_p^{s\prime\prime} \times \mathbf{E}_p(r_l \pm \bar{a})|/c^2)$ along the z' direction given as (the transformed Biot-Savart law),

$$\mathbf{B}_p^{orb}(r_l, \theta_j) = \frac{e\boldsymbol{v}_p \times \mathbf{r}_{lj}}{4\pi\epsilon_0 c^2 r_l^3} = -\frac{e\mathbf{r}_{lj} \times (-\frac{m_e m_p}{M})\boldsymbol{v}_j}{4\pi\epsilon_0 m_p c^2 r_l^3} = -\frac{\sqrt{4l^2-1}\, e\hbar\, \hat{z}'}{8\pi\epsilon_0 m_p c^2 r_l^3}; \tag{29}$$

$$\mathbf{B}_p^s(r_l \pm \bar{a}, \theta_j) = \frac{\mp\frac{1}{2} e\bar{\boldsymbol{v}}_p^{s\prime\prime} \times (\mathbf{r}_{lj}/r_l)}{4\pi\epsilon_0 c^2 (r_l \pm \bar{a})^2}, \quad \mathbf{B}_p^s(r_l, \theta_j) = \mathbf{B}_p^s(r_l - \bar{a}, \theta_j) + \mathbf{B}_p^s(r_l + \bar{a}, \theta_j) =$$
$$= \frac{e\bar{a}\bar{\boldsymbol{v}}_p^{s\prime\prime} \times (\mathbf{r}_{lj}/r_l)}{2\pi\epsilon_0 c^2 r_l^3 (1 - \frac{\bar{a}^2}{r_l^2})^2} = \frac{-\eta^2 g_p e\hbar \cos\theta_j \hat{z}'}{4\pi\epsilon_0 m_p c^2 r_l^3 C_{1l}}, \quad C_{1l} = \left(1 - \frac{\bar{a}^2}{r_l^2}\right)^2 \tag{30}$$

The last of Eqs (29) is given after substituting (3b) for $\boldsymbol{v}_p$ and (16.1)$'$ for $\mathbf{r}_{lj} \times (\frac{m_e m_p}{M})\boldsymbol{v}_j$, given for $l_{TP} = \frac{1}{2}$. The last of Eqs (30b) is given after substituting (A.1a), (A.3b) for $\bar{a}$, $\bar{v}_p^{s\prime\prime}$, a for a_p, and (19) for $av_p^s m_p \cos\theta_p^s/g_p(=\frac{1}{2}\hbar)$. The negative signs in the two final results indicate $\mathbf{B}_p^{orb}$, $\mathbf{B}_p^s$ to be in the $-z'$ direction each.

In the $\mathbf{E}_p$ and $\mathbf{B}_p^{orb}+\mathbf{B}_p^s=\mathbf{B}_p$ fields of the proton (cf Fig 1a), the electron at $\mathbf{r}_{lj}$ apart, with an effective charge $q_e=-f_c e$, and in precessional–orbital and spin motions at the tangential velocities $\boldsymbol{v}_e$ about the CM and $\bar{v}_e^{s\prime\prime}$ about $\mathbf{r}_e$ in clockwise and counter-clockwise senses in the $x'y'$ plane, is acted by an electromagnetic force along the $\mathbf{r}_{lj}$ direction according to the Lorentz force law,

$$\mathbf{F}_{pe}(r_l,\theta_j)=-f_c e\mathbf{E}_p(r_l)+f_t^2[\mathbf{F}_{pe,m}^{orb-orb}(r_l,\theta_j,t)+\mathbf{F}_{pe,m}^{s-orb}(r_l,\theta_j,t)+\mathbf{F}_{pe,m}^{s-s}(r_l,\theta_j,t)], \tag{31}$$

$$\text{where}\quad \mathbf{F}_{pe,m}^{orb-orb}=-e\boldsymbol{v}_e\times\mathbf{B}_p^{orb}=-\frac{e|\mathbf{r}_{lj}\times(\frac{m_e m_p}{M}\boldsymbol{v}_j)||\mathbf{B}_p^{orb}|\,\hat{r}_{lj}}{m_e r_l}=-\frac{(4l^2-1)e^2\hbar^2\,\hat{r}_{lj}}{16\pi\epsilon_0 m_e m_p c^2 r_l^4}, \tag{32}$$

$$\mathbf{F}_{pe,m}^{s-orb}=-e\boldsymbol{v}_e\times\mathbf{B}_p^s=-\frac{e|\mathbf{r}_{lj}\times(\frac{m_e m_p}{M}\boldsymbol{v}_j)||\mathbf{B}_p^s|\,\hat{r}_{lj}}{m_e r_l}=-\frac{(2l-1)\eta^2 g_p e^2\hbar^2\,\hat{r}_{lj}}{16\pi\epsilon_0 m_e m_p c^2 r_l^4 C_{1l}}, \tag{33}$$

$$\mathbf{F}_{pe,m}^{s-s}=-\frac{\partial V_{pe,m}^{s-s}}{\partial r_l}\,\hat{r}_{lj}=|\mu_{ez}^s|\frac{\partial|\mathbf{B}_p^s|}{\partial r_l}\cos\theta_j\,\hat{r}_{lj}=-\frac{3\eta^2 g_e g_p e^2\hbar^2\cos^2\theta_j\,\hat{r}_{lj}}{16\pi\epsilon_0 m_e m_p c^2 r_l^4 C_{1l}}. \tag{34}$$

f_c is the fraction of the e-charge sphere momentarily facing the narrowed $\mathbf{E}_p$ profile at $\mathbf{r}_{lj}$. Eq (3a) for $\boldsymbol{v}_e$ and again (16.1)$'$ for $r_l(\frac{m_e m_p}{M})v_j$ are used in (32),(33). In (34), $V_{pe,m}^{s-s}=-\boldsymbol{\mu}_e^s\cdot\mathbf{B}_p^s=-|\mu_{ez}^s||\mathbf{B}_p^s|\cos\theta_j$ is the magnetic potential of the spin-spin interaction; μ_{ez}^s is an intensive quantity at $\mathbf{r}_e$, hence not affected by the B_p profile narrowing, and is given by Eq (20a). $\mathbf{F}_{m0}^{s-s}=-\int_0^{2\pi}e\boldsymbol{v}_e^s\times\mathbf{B}_{pz}^s(r_l)d\phi_e^s=0$; $\frac{\partial|\mathbf{B}_p^s|}{\partial r_l}=-\frac{3B_p^s}{r_l}$. f_t^2 projects the product term $v_e v_p$ contained in each component magnetic force to $v_e'v_p'$ which actually enters the interaction; $\mathbf{F}_{pe}(r_l,\theta_j)\equiv\mathbf{F}_{pe}(r_l,\theta_j,t_e,t_p)$ is implicitly meant. $v_e'v_p'=(v_e M/m_p)(v_p M/m_e)=(M/\mathscr{M})v_e v_p=f_t^2 v_e v_p$ (Sec 2.1), so $f_t^2=M/\mathscr{M}$. A repulsion $\mathbf{F}_{pe}^{rep}=A_{rep}\hat{r}_{lj}/r_l^{N+1}$ at short range, relative to the magnetic interaction strength at the distance $r\sim10^{-18}$ m, may generally also present but is omitted for the intermediate range of interest here. Given the presumed $S_{ez},S_{pz},J_{z\text{-}j}=\frac{1}{2},\frac{1}{2},-j$ configuration in units $\hbar$ here, all the three component magnetic forces (for $l>0$ for $F_{pe,m}^{orb-orb}$, $F_{pe,m}^{s-orb}$) acted by p on e above are optimally in the $-\mathbf{r}_{lj}$ direction and hence attractive. $\mathbf{F}_{pe}$ is therefore in the $-\mathbf{r}_{lj}$ direction and maximally attractive. Any alteration of the relative orientations between $S_{ez},S_{pz},J_{z\text{-}j}$ will render some or all of the component magnetic forces repulsive. An alteration of $S_{ez},S_{pz},J_{z\text{-}j}$ as a whole, i.e. from the configuration (i) to (ii) of Eqs (21) or from Fig 2a to 2b for $j=\frac{1}{2}$, retains all component magnetic forces attractive, and hence a total force $\mathbf{F}_{pe}$ the same as given in (32), or F_j same as given in (35) below.

Similarly, e produces at p at $-\mathbf{r}_{lj}$ apart the electromagnetic fields $\mathbf{E}_e$ and $\mathbf{B}_e$, and forces given as $f_c e\mathbf{E}_e$, $f_t^2\mathbf{F}_{ep,m}^{orb-orb}=-f_t^2\mathbf{F}_{pe,m}^{orb-orb}$, $f_t^2\mathbf{F}_{ep,m}^{s-orb}=-f_t^2\mathbf{F}_{pe,m}^{s-orb}\frac{g_e}{g_p}$, $f_t^2\mathbf{F}_{ep,m}^{s-s}=-f_t^2\mathbf{F}_{pe,m}^{s-s}$. The action and reaction forces for the e,p in equilibrium must be equal in magnitude and opposite in direction (Newton's third law); the magnitude may be here represented by the geometric mean as $F=\sqrt{|\mathbf{F}_{pe}||\mathbf{F}_{ep}|}=\sum_{\lambda,\lambda'}\sqrt{|\mathbf{F}_{pe}^{\lambda}||\mathbf{F}_{ep}^{\lambda'}|}\,\delta_{\lambda\lambda'}$, where λ,λ' indicate the different component forces; $\delta_{\lambda\lambda'}$ is the Kronecker delta. The last equation needs to hold for the action and reaction to maintain detailed balance upon any small variation of the independent variables such as $\mathbf{r}_{lj}$. The final total (attractive) force of p on e in equilibrium in the $j=l-l_{TP},m_j=-j$ state is therefore, suffixing j after $\mathbf{F}$ explicitly, $\mathbf{F}_j(r_l,\theta_j)=-[f_c e\sqrt{|\mathbf{E}_p||\mathbf{E}_e|}+f_t^2\sum_\lambda\sqrt{|\mathbf{F}_{pe,m}^{\lambda}||\mathbf{F}_{ep,m}^{\lambda}|}]\hat{r}=-f_c e\mathbf{E}_p+f_t^2[\mathbf{F}_{pe,m}^{orb-orb}+\mathbf{F}_{pe,m}^{s-orb}\frac{\sqrt{g_e g_e}}{g_p}+\mathbf{F}_{pe,m}^{s-s}]$. Substituting Eqs (32)–(34) into the foregoing we obtain this force in explicit and scalar form (and explicitly for $l_{TP}=\frac{1}{2}$),

$$F_j(r_l,\theta_j)=-\frac{e^2}{4\pi\epsilon_0 r_l^2}(f_c+f_m)\simeq-\frac{e^2 f_m}{4\pi\epsilon_0 r_l^2}=-\frac{f_t^2 e^2\hbar^2\mathcal{C}_{0j}}{16\pi\epsilon_0 m_e m_p c^2 r_l^4}, \tag{35}$$

$$f_m = \frac{f_t^2 \hbar^2 \mathcal{C}_{0j}}{4m_e m_p c^2 r_l^2}, \quad \mathcal{C}_{0j} = (4l^2-1) + \frac{(2l-1)\eta^2\sqrt{g_e g_p}}{C_{1l}} + \frac{3g_e g_p \eta^2 \cos^2\theta_j}{C_{1l}}. \tag{36}$$

The negative sign in Eqs (35) indicates that F_j is attractive. The approximation is given for $f_m >> f_c = 1/\gamma^2$. For $j = \frac{1}{2}$ ($l = 1$), using the solution values from Sec 6 gives $f_m = \hbar^2 c^2 \mathcal{C}_{0\frac{1}{2}}/m_e m_p c^4 r_{1m}^2 \simeq 6.9$ which indeed is $>> f_c = 1/\gamma^2 \simeq 5.7\times 10^{-11}$.

$j = 0$ yields $J_0 = 0$, $B_p^{orb} = 0$, and hence zero orbit-orbit and orbit-spin interactions. The resultant system, even if possible to also form a bound state by the finite spin-spin interaction only, is not a viable candidate of the neutron, at least because it does not contain a confined antineutrino. For $j \geq \frac{1}{2}$, the three component magnetic forces are each finite and attractive. $j = \frac{1}{2}$ therefore is the lowest possible (eigen) state of the e, p bound by an attractive magnetic force at the separation scale $\sim 10^{-18}$ m (Sec 6), has a confined antineutrino, and has the correct spin $\frac{1}{2}$ (Sec 3). The $j = \frac{1}{2}$ state is therefore a liable candidate for (the ground state of) the neutron. For $j = \frac{1}{2}$ ($l = 1$), hence $\cos\theta_{\frac{1}{2}} = 1/\sqrt{3}$, and $m_e = km_p = 1.3165 m_p$ (Sec 3), hence $f_t^2 = \frac{(m_e+m_p)^2}{m_e m_p} = \frac{(k+1)^2}{k} \doteq 4$, Eq (35), the corresponding interaction potential $V_{\frac{1}{2}}$ and Hamiltonian $H_{\frac{1}{2}}$ are written as, with Eqs (7.2a,b) for m_e, m_p, and $T_{\frac{1}{2}}$ given in Sec 2.4,

$$F_{\frac{1}{2}}(r_1,\theta_{\frac{1}{2}}) = -\frac{3A_o\mathcal{C}_{0\frac{1}{2}}}{\gamma_e\gamma_p r_1^4}, \quad A_o = \frac{e^2\hbar^2}{12\pi\epsilon_0 m_e^0 m_p^0 c^2}, \quad \mathcal{C}_{0\frac{1}{2}} = 3 + \frac{\eta^2\sqrt{g_e g_p}}{C_{11}} + \frac{\eta^2 g_e g_p}{C_{11}}; \tag{37}$$

$$V_{\frac{1}{2}}(r_1,\theta_{\frac{1}{2}}) = -\int_\infty^{r_1} F_{\frac{1}{2}}(r,\theta_{\frac{1}{2}})dr = \frac{r_1 F_{\frac{1}{2}}(r_1,\theta_{\frac{1}{2}})}{3} = -\frac{A_o\mathcal{C}_{0\frac{1}{2}}}{\gamma_e\gamma_p r_1^3} = -\frac{e^2\hbar^2\mathcal{C}_{0\frac{1}{2}}}{12\pi\epsilon_0 m_e m_p c^2 r_1^3}; \tag{38}$$

$$T_{\frac{1}{2}} = -C_{k\frac{1}{2}}V_{\frac{1}{2}}, \quad C_{k\frac{1}{2}} = \frac{\gamma 9\pi\epsilon_0 M c^2 r_1}{(\gamma+1)e^2\mathcal{C}_{0\frac{1}{2}}}; \quad H_{\frac{1}{2}}(r_1,\theta_{\frac{1}{2}}) = T_{\frac{1}{2}} + V_{\frac{1}{2}} = V_{\frac{1}{2}}(1 - C_{k\frac{1}{2}}). \tag{39}$$

In terms of the e, p-neutron model, $F_{\frac{1}{2}}$ represents the weak interaction force, $V_{\frac{1}{2}}$ the corresponding interaction potential, and $H_{\frac{1}{2}}$ Hamiltonian of neutron.

5. e, p disintegration. Neutron β decay

Suppose that an afore-described (free) neutron, being initially in stationary state of the Hamiltonian $H_{\frac{1}{2}}$ at a time earlier, is now perturbed by an excitation or external-interaction Hamiltonian $H_I = H_I^0 + H_I^1 = H_I^1$ given in the CM frame; evidently $H_I^0 = 0$. So the bound e, p are in the final (f) state disintegrated into free e, p separated at an effective infinite distance r_∞ such that $V_{\frac{1}{2}f}(r_\infty) = 0$. The removal of the central force, say acted by p on e in the $-\mathbf{r}_{1\frac{1}{2}}$ direction, subjects e to a deceleration along that direction and subsequently deceleration or Bremsstrahlung radiation. Provided no exertion of external torque on the neutron, the angular momentum must be conserved before (being a quantum $S_{\bar{\nu}_e} = -\frac{1}{2}\hbar$ in $-z$ direction) and after the deceleration radiation. The electromagnetic radiation emitted is therefore necessarily in the form of a precessing-orbiting or simply rotational energy flux so as to convey the same angular momentum quantum $-\frac{1}{2}\hbar$ in the z direction, and the same rotational kinetic energy $T_{\bar{\nu}_e} = T_{\frac{1}{2}}$ provided also no exchange of the kinetic energy with the surrounding. The rotational radiation energy flux emitted resembles directly an antineutrino, $\bar{\nu}_e$, which is now free. The equation of the foregoing (disintegration) reaction straightforwardly is

$$n \rightarrow p + e + \bar{\nu}_e$$

i.e. the e, p disintegration resembles a neutron β decay. The final-state total Hamiltonian has the general form $H_{\frac{1}{2}f} = V_{\frac{1}{2}f}(r_\infty) + T_{\frac{1}{2}f} = 0 + T_{\frac{1}{2}f}$. The emitted particles would convey a certain translational kinetic energy T_{tr} as converted from the total mass difference before (assuming the

n being at rest) and after the neutron decay. T_{tr} is of a MeV scale (a scale as is also known from β decay experiment) which is $<< T_{\frac{1}{2}}$ of GeV scale. Omitting this T_{tr}, in the case of $T_{\overline{\nu}_e} = T_{\frac{1}{2}}$, we have $T_{\frac{1}{2}f} = T_{\frac{1}{2}} + T_{tr} \simeq T_{\frac{1}{2}}$, and $H_{\frac{1}{2}f} = 0 + T_{\frac{1}{2}f} \simeq T_{\frac{1}{2}}$.

The energy condition for the neutron β decay to occur is $H_I = H_{\frac{1}{2}f} - H_{\frac{1}{2}}$. Substituting in it the equation for $H_{\frac{1}{2}f}$ above and (39c) for $H_{\frac{1}{2}}$ gives

$$H_I = T_{\frac{1}{2}} - (V_{\frac{1}{2}} + T_{\frac{1}{2}}) = -V_{\frac{1}{2}} = \frac{A_o \mathcal{C}_{0\frac{1}{2}}}{\gamma_e \gamma_p r_1^3} \quad \text{or} \tag{40}$$

$$G_F = H_I \left(\frac{4}{3}\pi r_1^3\right) = \frac{A_o C_{0\frac{1}{2}}}{\gamma_e \gamma_p} = \frac{e^2 \hbar^2 C_{0\frac{1}{2}}}{12\pi\epsilon_0 m_e m_p c^2} = \frac{e^2 \hbar^2 C_{0\frac{1}{2}}}{48\pi\epsilon_0 \mathscr{M}^2 c^2}, \quad C_{0\frac{1}{2}} = \frac{4\pi \mathcal{C}_{0\frac{1}{2}}}{3}, \tag{41}$$

where $\left(\frac{4}{3}\pi r_1^3\right)$ is the volume in which the electron is confined about the proton; the last of Eqs (41a) is given after substituting the relation $m_e m_p = \mathscr{M} M = \mathscr{M}(\frac{(k+1)^2}{k}\mathscr{M}) \doteq 4\mathscr{M}^2$ given for $m_e = k m_p = 1.3165 m_p$. By virtue of its physical significance, the product term $G_F = H_I(\frac{4}{3}\pi r_1^3)$ in (41b) is directly identifiable with the CM-frame counterpart of the Fermi constant $G_F^{\rm lab}$.

$G_F^{\rm lab}$ is experimentally determined (as G_F^{exp}) from the neutron lifetime, denoted by $\tau^{\rm lab}$ here as is usually measured in the lab frame, on the basis of the quantum theoretical relation $G_F^{\rm lab} \propto 1/\sqrt{\tau^{\rm lab}}$. The neutron under consideration may be generally in motion, say at a velocity $u_{\rm cm}^{\rm lab}$ in a fixed X direction. The (model) neutron mass in this direction is (cf Sec 2.3) $m_n^{\rm lab} \doteq M^{\rm lab} = \gamma_{\rm cm}^{\rm lab}\langle M\rangle = \gamma_{\rm cm}^{\rm lab} M^0$, where $\gamma_{\rm cm}^{\rm lab} = (1 - (u_{\rm cm}^{\rm lab})^2/c^2)^{-1/2}$. Its component masses are formally $m_e^{\rm lab} = (\gamma_{\rm cm}^{\rm lab})^\kappa m_e$, $m_p^{\rm lab} = (\gamma_{\rm cm}^{\rm lab})^\kappa m_p$, i.e. each in effect augmented by a factor $(\gamma_{\rm cm}^{\rm lab})^\kappa$, where κ is a certain (positive) exponent resulting from the mapping of $u_{\rm cm}^{\rm lab}$ onto the instantaneous interaction direction of e, p. So Eqs (41a), here re-written from its original form as $G_F = \frac{e^2\hbar^2 C_{0\frac{1}{2}}}{12\pi\epsilon_0 m_e m_p c^2} = \frac{A_1}{m_e m_p}$, $A_1 = \frac{e^2\hbar^2 C_{0\frac{1}{2}}}{12\pi\epsilon_0 c^2}$, transformed to the lab frame is formally

$$G_F^{\rm lab} = \frac{A_1}{m_e^{\rm lab} m_p^{\rm lab}} = \frac{A_1}{(\gamma_{\rm cm}^{\rm lab})^{2\kappa} m_e m_p} = \frac{G_F}{(\gamma_{\rm cm}^{\rm lab})^{2\kappa}} \propto \frac{1}{(\gamma_{\rm cm}^{\rm lab})^{2\kappa}\sqrt{\tau^0}} = \frac{1}{\sqrt{\tau^{\rm lab}}} \tag{42}$$

where τ^0 denotes the lifetime of a neutron at rest. (42) suggests that for a fast moving neutron such that $u_{\rm cm}^{\rm lab 2}/c^2 > 0$ and $\gamma_{\rm cm}^{\rm lab} > 1$ appreciably each, the neutron life time $\tau^{\rm lab} = (\gamma_{\rm cm}^{\rm lab})^{2\kappa}\tau^0$ will appear appreciably "dilated", as the result of a reduced internal (magnetic) interaction strength, or reduced Fermi constant. For a neutron at rest of major concern in this paper, G_F identifies with $G_F^{\rm lab}$ as measured for a rest or slow-moving neutron. We shall continue to speak of G_F.

Multiplying $\frac{137\times 12\mathscr{M}^2 c^2}{\hbar^3 c C_{0\frac{1}{2}}}$ on its both sides, rearranging, the last of Eqs (41a) is written as

$$\frac{G_F(137\times 12\mathscr{M}^2/C_{0\frac{1}{2}})c^2}{\hbar^3 c} = \frac{137e^2}{4\pi\epsilon_0\hbar c}; \quad \text{or} \tag{43}$$

$$\frac{G_F M_{\rm ef}^2 c^2}{\hbar^3 c} = \frac{g_{neu}^2}{\hbar c}, \quad M_{\rm ef} = \left(\frac{137\times 12\mathscr{M}^2}{C_{0\frac{1}{2}}}\right)^{1/2} = \frac{40.546\mathscr{M}}{\sqrt{C_{0\frac{1}{2}}}} \doteq \frac{23.043 m_p}{\sqrt{C_{0\frac{1}{2}}}}, \quad g_{neu}^2 = \frac{137e^2}{4\pi\epsilon_0}; \tag{44}$$

or $G_F = \frac{g_{neu}^2(\hbar c)^2}{M_{\rm ef}^2 c^4}$. $M_{\rm ef}$ is an effective mass; for the last of Eqs (44b) $m_e = 1.3165 m_p$ is used. $\frac{e^2}{4\pi\epsilon_0\hbar c} = \frac{g_H^2}{\hbar c} = \frac{1}{137} = \alpha_H (= \frac{v_{1H}}{c})$ corresponds to the fine structure constant of the hydrogen atom (and v_{1H} the orbiting velocity of electron relative to proton thereof). So the right side of (43) is unity, $\frac{137\times e^2}{4\pi\epsilon_0\hbar c} = 1$. The fine structure constant for the model neutron is accordingly defined by $\alpha_{neu} = v_{\frac{1}{2}}/c$; using $v_{\frac{1}{2}} \doteq c$ from Sec 6 in it gives $\alpha_{neu} \doteq 1$. Based on its unity value, and on the physical significance of the dynamical variables, say $\frac{g_{neu}^2}{\hbar c}$ on its right side compared to $\frac{g_H^2}{\hbar c}$, Eq (44a) is immediately identifiable as the equation for α_{neu},

$$\alpha_{neu} = \frac{\upsilon_{\frac{1}{2}}}{c} \equiv 137\alpha_H(\doteq 1) = \frac{g_{neu}^2}{\hbar c} = \frac{G_F M_{\rm ef}^2 c^2}{\hbar^3 c} \tag{45}$$

In the GWS theory, G_F is given the formula $G_F^{\rm GWS} = \frac{g_{\rm w}^2\sqrt{2}(\hbar c)^2}{M_{\rm w}^2 c^4}$, where $g_{\rm w}^2 = \frac{e^2}{8\epsilon_0 \sin^2\theta_{\rm w}}$. Equating $G_F^{\rm GWS}$ with G_F of (41a) gives a first-principles microscopic expression for $M_{\rm w}$, accordingly M_z,

$$M_{\rm w} = \left(\frac{3\sqrt{2}\,\pi m_e m_p}{2C_{0\frac{1}{2}}\sin^2\theta_{\rm w}}\right)^{1/2} = \left(\frac{3\sqrt{2}\,\pi k}{2C_{0\frac{1}{2}}\sin^2\theta_{\rm w}}\right)^{1/2} m_p; \qquad M_z = M_{\rm w}/\cos\theta_{\rm w} \tag{46}$$

It is well appreciated in the literature that, whilst the G_F value is absolutely determined by the lifetime of the neutron in question, the $M_{\rm w}$ value (or $M_{\rm ef}$ in Eq 45) is dependent on the definition or choice of the coupling constant $g_{\rm w}^2$ (or g_{neu}^2 in Eq 45); g_{neu}^2 in (45) is uniquely specified for $\upsilon_{\frac{1}{2}}$ is separately known. In terms of the e, p-neutron model, $M_{\rm w}$, or $M_{\rm ef}$, represents essentially the (reduced) mass of the e, p particles in the binding and hence manifestly resistive potential field $V_{\frac{1}{2}}$. $V_{\frac{1}{2}}$ resembles the Higgs field. The $M_{\rm w}$, or $M_{\rm ef}$, of a neutron is highly relativistically augmented (Sec 6) over that of a hydrogen, primarily as the result of the relative velocity of e, p within a neutron being so high as to approach c. Moreover, the e, p interaction in a neutron is predominately magnetic, and in a hydrogen electrostatic.

6. Numerical evaluation

Equations (37)–(41) are specified effectively by four independent variables $\bar{a}$, r_1, $\upsilon_{\frac{1}{2}}$, and $\gamma(\upsilon_{\frac{1}{2}}, c)(= \gamma_e\gamma_p/\gamma_{\rm M})$ ($\gamma_{\rm M}$ is given if γ is given), to be determined each. We need four independent constraints for quantitatively determining these and subsequently the remaining dynamical variables. Equation (1d) would ordinarily serve as one basic constraint: it describes a stable state condition under which the central force $\mathbf{F}$ on mass $\mathscr{M}$ counterbalances with the inertial (or here centrifugal) force $\mathscr{M}\frac{d^2\mathbf{r}}{dt^2}$. It may be checked (App Appendix B) that at a r_1 value satisfying Eq (1d), the lifetime of the e, p system however is not an optimum. This suggests that the neutron candidate e, p system, if opted for a maximum lifetime, is not in stable state. We shall choose the maximum lifetime condition here on the basis that a real free neutron indeed is "meta" stable only, with a relatively short lifetime 12 m.

In overall view of the basic solutions from preceding sections, the discussion just made above, and the available key experimental data such as to realistically identify the neutron, we employ (i) the quantisation condition (17a) for $J_{\frac{1}{2}}$[1], (ii) a maximum neutron lifetime, hence a minimum G_F, and (ii) the experimental value of the Fermi constant, G_F^{exp}, as three basic constraints. These are (re-) written as, on dividing (17a) by $r_1\mathscr{M}^0\upsilon_{\frac{1}{2}}$ for (i), and denoting by r_{1m} the extremal value of r_1 at which G_F is a minimum,

$$(i): \qquad \gamma = \gamma_{\rm M}\gamma^\dagger = \frac{\sqrt{3}(\hbar c)c}{2\mathscr{M}^0 c^2 r_1 \upsilon_{\frac{1}{2}}} = \frac{D_o c}{r_1\upsilon_{\frac{1}{2}}}, \quad D_o = \frac{\sqrt{3}\hbar c}{2\mathscr{M}^0 c^2} \tag{47}$$

$$(ii): \qquad G_F(r_{1m}) = G_{F.min} \tag{48}$$

$$(iii): \qquad G_F(r_{1m}) = G_F^{exp} = 1.43585(37)\times 10^{-62}\ \text{Jm}^3 \quad \text{(data from [1e])} \tag{49}$$

[1] The eigen value solution (17a) represents directly a Heisenberg relation for $J_{\frac{1}{2}}$ and the angular interval 2π, or alternatively in theory the Maupertuis-Jacobi's action integral $2T_{\frac{1}{2}} = C_{k\frac{1}{2}}H_{\frac{1}{2}}/(C_{k\frac{1}{2}} - 1)$ and $\Delta t_{\frac{1}{2}}$; $\Delta t_{\frac{1}{2}} = 2\pi r_1/\upsilon_{\frac{1}{2}}$. The excitation Hamiltonian H_I is not necessarily conjugated with the $\Delta t_{\frac{1}{2}}$, but generally with some other time interval subjecting to a Heisenberg relation depending on the excitation scheme.

Since (47a) suggests that $\gamma >> 1$ for any $v_{\frac{1}{2}}$ value not too far below c, and $c = c^{\text{lab}}$ by the standard assumption, so $v_{\frac{1}{2}} = c\sqrt{\gamma^2 - 1}/\gamma \simeq c = c^{\text{lab}}$, which serves as the fourth constraint here. With this $v_{\frac{1}{2}}$ value in (47a), we obtain (50a,b) below; further with (8.1b) for $\gamma_e\gamma_p (= \gamma_{\rm M}\gamma = \gamma_{\rm M}^2\gamma^\dagger)$ and the resultant $\gamma_{\rm M}$ from (50b) in (41a), with $\gamma^\dagger = 450.96$ given in Sec 2.2 (for $m_e = 1.3165 m_p$), we obtain (51) below,

$$\gamma = \frac{D_o}{r_1}, \quad \gamma_{\rm M} = \frac{\gamma}{\gamma^\dagger} = \frac{D_o}{\gamma^\dagger r_1}, \tag{50}$$

$$G_F = \frac{A_o C_{0\frac{1}{2}}}{\gamma_{\rm M}^2\gamma^\dagger} = \frac{\gamma^\dagger A_o C_{0\frac{1}{2}} r_1^2}{D_o^2} = \frac{450.96 A_o C_{0\frac{1}{2}} r_1^2}{D_o^2}. \tag{51}$$

$D_o (= 3.3462 \times 10^{-13}$ m) and $A_o (= 6.2455 \times 10^{-57}$ Jm^3) are constants. For evaluating $C_{0\frac{1}{2}}$ (Eqs 41b, 37c), we shall use the experimental g values of e, p, $g_e = 2$, $g_p = 5.5857$, and $\eta = 1/\sqrt{2}$ (Appendix A). G_F of (51) is then solely dependent on $r_1, \bar{a}$. Characteristically, for a specified

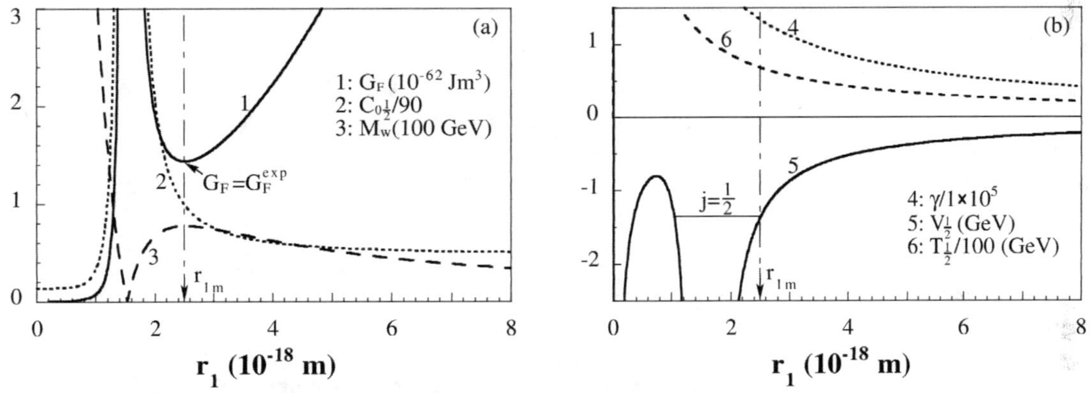

Figure 3. (a) $G_F = H_I r_1^3$, $C_{0\frac{1}{2}}$, $M_{\rm w}$ (curves 1,2,3), and (b) γ, $V_{\frac{1}{2}} = -H_I$, $T_{\frac{1}{2}}$ (curves 4,5,6) as functions of r_1 computed from Eqs (51), (37c), (46), (50a), (38), (39a) for $\bar{a} = 1.5391(8) \times 10^{-18}$ m. At $r_1 = r_{1m} = 2.5369(5) \times 10^{-18}$ m, $G_F = G_F^{exp} = 1.43585(37) \times 10^{-62}$ Jm^3.

$\bar{a}$ value, the $G_F(r_1)$ vs r_1 function presents an extremal point at a (uniquely specified) r_1, r_{1m}, at which $G_F(r_{1m})$ is a minimum satisfying Eq (48), as in Fig 3a, although this is not generally equal to G_F^{exp}. $G_F(r_{1m})$ increases monotonically with $\bar{a}$. Computing $G_F(r_{1m})$ as a function of $\bar{a}$ over a range of $\bar{a}$ values, a unique $\bar{a}$ is found at $\bar{a} = 1.5391(8) \times 10^{-18}$ m at which $G_F(r_{1m}) = G_F^{exp}$ satisfying Eq (49), $r_{1m} = 2.5369(5) \times 10^{-18}$ m, $\gamma = 1.3190 \times 10^5$ (Eq 50a), and $C_{0\frac{1}{2}} = (\frac{4\pi}{3}) \times 3(1 + 1.3952 + 4.6633) = 88.69$ (Eqs 41b, 37c). Note that the $\bar{a}$ value obtained is in accordance with the order of magnitude of the neutron charge radius, $\sim 1.4 \times 10^{-18}$ m, measured by electron-neutron scattering experiment (see also Sec 2.3).

With the $\bar{a}$, r_{1m} (hence $C_{0\frac{1}{2}}$), $v_{\frac{1}{2}}, \gamma$ values obtained, all the remaining dynamical variables and functions may be evaluated. For the fixed $\bar{a} = 1.5392 \times 10^{-18}$ m value, the G_F, $C_{0\frac{1}{2}}$, $M_{\rm w}$ (using the average experimental value $\sin^2\theta_{\rm w} = 0.23$), γ, $V_{\frac{1}{2}} (= -H_{\frac{1}{2}})$, and $T_{\frac{1}{2}}$ vs. r_1 functions, computed from Eqs (51), (41b,37c), (46a), (47a), (38), (39a) are as shown in Figs 3a,b (curves 1–6). $r_1 = r_{1.\text{min}}$ lies as expected in the region where $-\partial V_{\frac{1}{2}}(r)/\partial r = F_{\frac{1}{2}}(r) < 0$, and $V_{\frac{1}{2}}(r) < 0$. At $r_1 = r_{1m}$, $V_{\frac{1}{2}} = -H_I = -1.310$ GeV, $T_{\frac{1}{2}} (\simeq \mathscr{M}c^2) = 67.36$ GeV, $H_{\frac{1}{2}} (\simeq E_{tot.\frac{1}{2}}) = 66.05$ GeV, $M_{\rm w} = 77.23$ GeV. Furthermore specifically, with the γ value in (50b),(9a),(b), we obtain $\gamma_{\rm M} = 292.48(3)$, $\gamma_e = \gamma_{\rm M}\frac{(M^0+K)}{2m_e^0} = 3.0537(6) \times 10^5$, and $\gamma_p = \gamma_{\rm M}\frac{(M^0-K)}{2m_p^0} = 126.33$, which are $>> 1$ each. So the particles e, p within the neutron are travelling at velocities $v_e', v_p' \simeq c$ measured in their local space and time coordinates $\mathbf{r}_e, t_e, \mathbf{r}_p, t_p$ (Eqs 5) in the (non-inertial) CM frame; and so is the total mass M at the CM relative to e, p. The total kinetic energy of e, p in these absolute terms is given by substituting the $v'_e, v'_p, m_e(=km_p), m_p$ values in (12.2) as $T'_e+T'_p \doteq (k+1)m_pc^2 = 2.3165\gamma_p m_p^0 c^2 = 2.3165\times 118.53 = 2\times 137.29$ GeV. Substituting $m_e = km_p$ into Eqs (3a,b) gives the e, p velocities measured in time t, $v_e = \frac{m_p}{M}v = \frac{c}{k+1} = 0.43c$, $v_p = -\frac{m_e}{M}v = -\frac{kc}{k+1} = -0.57c$; and in turn the above values into Eq (11.2) gives the corresponding total kinetic energy $T_e + T_p = km_p(\frac{c}{k+1})^2 + m_p(\frac{kc}{k+1})^2 = \frac{k}{k+1}m_pc^2 = 0.56831\times 118.53 = 67.36$ GeV, equal to the solution value for $T_{\frac{1}{2}} = \mathscr{M}c^2$ earlier. The non-inertial frame motion contributes an amount $(T'_e + T'_p) - (T_e + T_p) = (k+1)m_pc^2 - \frac{k}{k+1}m_pc^2 = \frac{k^2+k+1}{k+1}m_pc^2 = 1.75m_pc^2$. The exceedingly large kinetic energy apparently is mainly consumed to contract the size of the system.

The author expresses thanks to emeritus scientist P-I Johansson for his private financial support of the author's research, to Kissemiss Johansson for his joyful companion when the unification researches were carried out, and to Professor C Burdik for providing the opportunity of presenting this wok at the 23rd International Conference on Integrable Systems and Quantum Symmetries (ISQS23), Tech Univ, Prague, June 2015, during which the author also very much enjoyed interesting discussions with a number of participants.

Appendix A. Mapping of the spin current loop on to reduced geometries

Consider here the circular spin current loop of proton as projected in the $x_p^s y_p^s$ plane, spinning at tangential velocity $v_{p_{xy}}^s$ about the z_p^s axis passing through $\mathbf{r}_p$ in counterclockwise sense, i.e. in spin up state as in Sec 3.1. $dq_p = \rho_{pxy}a_p d\phi_p^s$ is a charge element at $\boldsymbol{\xi}_p(\phi_p^s)$ on it. The magnetic field produced by this spin current loop at a distance $\mathbf{r}$ from $\mathbf{r}_p$ has the general form (Biot-Savart law) $\mathbf{B}_p^s(\mathbf{r}) = \int d\mathbf{B}_{p\theta}^s(\mathbf{r}') = \int \frac{dq_p \boldsymbol{v}_{pxy}^s \times \mathbf{r}'}{4\pi\epsilon_0 c^2 r'^3}$, where $\mathbf{r}' = \mathbf{r} + \boldsymbol{\xi}_p(\phi_p^s)$. The integration in algebraic terms is a complex problem. We below reduce the current loop to simpler geometries to facilitate an effective algebraic expression for the field (Eq 30, Sec 4).

Consider first this field produced at a point at distance $x_p^s = |\mathbf{r}|$ from $\mathbf{r}_p$ on the positive x_p^s axis, $\mathbf{B}_p^s(x_p^s = |\mathbf{r}|)$; x_p^s is on the right side to the z_p^s axis as plotted for the parallel x, y, z axes of x_p^s, y_p^s, z_p^s in Fig 1b. The problem has the obvious symmetry that the left-half current loop produces at x_p^s a magnetic field $\mathbf{B}_{pL}^s(\propto \boldsymbol{v}_{pxy}^s \times \mathbf{x}_p^s) > 0$ in $+z$ direction, and the right-half a field $\mathbf{B}_{pR}^s < 0$ in $-z$ direction. The total field $\mathbf{B}_{pR}^s - \mathbf{B}_{pL}^s = \mathbf{B}_p^s(x_p^s = |\mathbf{r}|)$ is in $-z$ direction. Furthermore, on either half loop the differential $d\mathbf{B}_p^s$ fields produced by all charge elements as associated with the y-component velocities add up, and with the x-component velocities cancel out. So, in so far as the same $\mathbf{B}_{pL}^s, \mathbf{B}_{pR}^s$ are in effect produced, the left- and right- half spin current loops may be further reduced to two point half-charges $+\frac{1}{2}e, +\frac{1}{2}e$ located at effective distances $-\bar{a}, \bar{a}$ from $\mathbf{r}_p$ on the x_p^s axis and moving oppositely at velocities $-\bar{v}_{pxy}^s, \bar{v}_{pxy}^s$ in the $-y_p^s, +y_p^s$ directions, where

$$\bar{a} = \eta a, \quad \bar{v}_{pxy}^s = \bar{v}_p^s \cos\theta_p^s, \quad \bar{v}_p^s = \bar{a}\omega_p^s = \eta a\omega_p^s = \eta v_p^s, \quad v_p^s = a\omega_p^s; \qquad (A.1)$$

η is a coefficient to be determined. We have set $a = a_p = a_e = \frac{1}{2}(a_e + a_p)$ here; so $\bar{a}$ is effective also in its being scaled by η from the average radius a of the e, p charges. In analogy to $r = r^0/\gamma$, a is contracted[2] from its rest value a^0 formally according to $a = a^0/\gamma_a$; γ_a is a factor analogous to γ, being an increasing function with $(v_p/c)^2$.

[2] Contractions in charge radius and in the wavelength of matter wave may be comprehended on a common physical ground as follows. Charges and matter waves are distributed energy entities in space each. Two charges or two matter waves by this nature will inevitably repel with one another when attempting to occupy same space. Higher velocities facilitate two charges or two matter waves to counterbalance such repulsion to a larger extent, manifesting consequently as the contraction in their dimensions.

Journal of Physics: Conference Series **670** (2016) 012056 doi:10.1088/1742-6596/670/1/012056

By virtue of its physical significance, $\bar{a}$ should be such that the moments of inertia of the right and left point half-charges about the z_p^s axis, I_R^{point}, $I_L^{point}(= I_R^{point})$, are equal to those of the half current loops about z_p^s, I_R^{loop}, $I_L^{loop}(= I_R^{loop})$. Let m_p^* be the effective mass uniformly distributed along the full current and in turn on the full charge, and hence a mass $\frac{1}{2}m_p^*$ on each half loop and in turn at each half point charge. The moments of inertia of one point half charge and one half loop are known as, written for the right ones,

$$I_R^{point} = \frac{1}{2}m_p^*\bar{a}^2, \quad I_R^{loop} = \int_0^\pi x^2 dm_p^* = 2\int_0^{\pi/2}(a_p\cos\phi)^2\frac{(m_p^*/2)}{\pi}d\phi = \frac{m_p^* a^2}{4} \qquad (A.2)$$

The equality $I_R^{loop} = I_R^{\rm point}$ gives $\frac{1}{2}\bar{a}^2(=\frac{1}{2}(\eta a)^2) = \frac{1}{4}a^2$, so $\eta = 1/\sqrt{2}$. Accordingly, the right point half-charge is associated with a spin angular momentum $S_{pR} = I_R^{\rm point}\omega_p^s(= I_R^{loop}\omega_p^s) = \frac{1}{2}m_p^*\bar{a}^2\omega_p^s = \frac{1}{2}m_p^*\bar{a}\bar{v}_p^s$, so that the z component is $S_{pRz} = S_{pR}\cos\theta_p^s = \frac{1}{2}m_p^*\bar{a}\bar{v}_{p_{xy}}^s$.

Now with respect to a point at a distance $\mathbf{x}_p^{s\prime\prime}$ from $\mathbf{r}_p$ on the x'' axis (Sec 2.4) lying in a plane whose normal is along the z' or z'' direction at angle θ_j to the z axis, with $|\mathbf{x}_p^{s\prime\prime}| = |\mathbf{x}_p^s| = |\mathbf{r}|$, the component spin angular momentum of the right point half-charge perpendicular to x'' is the projection of S_{pRz} onto the z'' or z' direction,

$$S''_{pRz} = S_{pRz}\cos\theta_j = \frac{1}{2}m_p^*\bar{a}\bar{v}_{p_{xy}}^s\cos\theta_j = \frac{1}{2}m_p^*\bar{a}\bar{v}_p^{s\prime\prime}, \quad \bar{v}_p^{s\prime\prime} = \bar{v}_{p_{xy}}^s\cos\theta_j = \bar{v}_p^s\cos\theta_p^s\cos\theta_j \quad (A.3)$$

Appendix B. Stable-state solution for r_l

Substituting $\frac{d^2\mathbf{r}_{lj}}{dt^2} = -\frac{v_j^2}{r_l}\hat{r}_{lj}$, v_j from (16.1)$'$, and $\mathbf{F}_j$ from (35) into (1d) gives $-\mathscr{M}\frac{(4l^2-1)\hbar^2\hat{r}}{4\mathscr{M}^2 r_l^3} = -\frac{f_t^2 e^2\hbar^2 \mathcal{C}_{0j}\hat{r}}{16\pi\epsilon_0 m_e m_p c^2 r_l^4}$. Cancelling common factors on both sides, sorting, with $f_t^2 = 4$ for $m_e = 1.3165 m_p$ (Sec 4) and $M = \gamma_{\rm M}^{(j)}M^0$ for jth state, we obtain

$$r_l = \frac{e^2\mathcal{C}_{0j}}{(4l^2-1)\pi\epsilon_0\gamma_{\rm M}^{(j)}M^0c^2}. \quad \text{For } j=\frac{1}{2},\ l=1:\ \ r_1 = \frac{e^2\mathcal{C}_{0\frac{1}{2}}}{3\pi\epsilon_0\gamma_{\rm M}^{(\frac{1}{2})}M^0c^2}; \qquad (B.1)$$

for $j = \frac{3}{2}$, $l = 2$, $r_2 = \frac{e^2\mathcal{C}_{0\frac{3}{2}}}{15\pi\epsilon_0\gamma_{\rm M}^{(3/2)}M^0c^2}$, $\mathcal{C}_{0\frac{3}{2}} = 15 + \frac{3\sqrt{g_e g_p}}{2C_{12}} + \frac{9g_e g_p}{10C_{12}}$ given after Eq (36b). For the $j=\frac{1}{2}$ state, with (37b) for $\mathcal{C}_{0\frac{1}{2}}$, (30c) for C_{11}, and $\bar{a} = 1.53918\times10^{-18}$ m, the right side of (B.1b) may be computed as a function of r_1. Two numerical solutions are found at $r_1 = 1.440\times10^{-18}$, $1.661(7)\times10^{-18}$ m, at which the two sides of (B.1b) are equal.

[1] (a) Perkins DH 1982 *Introduction to High Energy Physics*, 2nd ed (Reading: Addison-Wesley); (b) Griffiths D 1987 *Introduction to elementary particles* (New York: Harper & Row); (c) Williams WSC 1992 *Nuclear and Particle Physics* (Oxford: Clarendon); (d) Enge HA 1969 *Introduction to Nuclear Physics* (Massachusetts: Addison-Wesley); (e) Beringer J, et al (Particle Data Group) 2012 *Phys. Rev.* **D86** 010001; (f) Alvarez L and Bloch H 1940 *Phys Rev* **57** 111; Staub HH and Roger EH 1950 *Hel Phys Acta* **23** 63-92.

[2] (a) Higgs P 1964 *Phys. Lett.* **12** 132; 1964 *Phys. Rev. Lett.* **13** 508-9; *ibid* 321; Englert P and Brout R 1964 *Phys. Rev. Lett.* **13** 321; Gutranik GS, Hagen CR and Kibble TWB 1964 *Phys. Rev. Lett.* **13** 585; (b) Weinberg S 1967 *Phys. Rev. Lett.* **19** 1264; Salam A 1968 in *Elementary particle physics: relativistic groups and analicity, Nobel Symp.* **8**, Svartholm N ed (Stockholm: Almquist & Wiksells) p 367; (c) Glashow SL, Lliopoulos L and Maiani I 1970 *Phys. Rev.* **D2** 1285; (d) 'T Hooft G 1971 *Nucl. Phys.* **B33** 173-99; 1971 *Phys. Lett.* **B37** 195; (e) Fermi E 1934 *Zeit. f. Physik* **88**171; tr Wilson FL 1968 *Am. J. Phys.* **36** 1150-60.

Institute of Physics
Dirac House, Temple Back
Bristol BS1 6BE UK

ISSN: 1742-6588
ISBN 978-1-5108-1961-0

CURRAN ASSOCIATES INC.
proceedings.com